International Conference on Engineering, Science and Technology (TALENTA-CEST 2017)

IOP Conference Series: Materials Science and Engineering
Volume 309

Sumatera Utara, Indonesia
7 – 8 September 2017

ISBN: 978-1-5108-5923-4
ISSN: 1757-8981

Printed from e-media with permission by:

Curran Associates, Inc.
57 Morehouse Lane
Red Hook, NY 12571

Some format issues inherent in the e-media version may also appear in this print version.

Printed by Curran Associates, Inc. (2018)

For permission requests, please contact the Institute of Physics
at the address below.

Institute of Physics
Dirac House, Temple Back
Bristol BS1 6BE UK

Phone: 44 1 17 929 7481
Fax: 44 1 17 920 0979

techtracking@iop.org

Additional copies of this publication are available from:

Curran Associates, Inc.
57 Morehouse Lane
Red Hook, NY 12571 USA
Phone: 845-758-0400
Fax: 845-758-2633
Email: curran@proceedings.com
Web: www.proceedings.com

TABLE OF CONTENTS

IOP Conf. Series: Materials Science and Engineering **309** (2018) 011001 doi:10.1088/1757-899X/309/1/011001

INTRODUCTION

TALENTA International Conference on Engineering, Science and Technology (known as TALENTA CEST) is organized by Faculty of Engineering University of Sumatera Utara (USU), Indonesia. As a note, the term of "TALENTA" is known as the competitive advantage of the University of Sumatera. It stands for **T**ropical Science & Medicine, **A**gro-industry, **L**ocal Wisdom, **E**nergy (Sustainable), **N**atural Resources, **T**echnology Appropriate, and **A**rts. This TALENTA is expected will bring USU to a global level of higher education competition. Thus, this conference is the contribution of Faculty of Engineering to support our TALENTA as competitive advantages of USU.

TALENTA CEST 2017 is an International conference that offers opportunity for knowledge sharing, networking and collaboration between engineers, scientists and technologists as well as academician and researcher working in the specific area of engineering. For this year the committee has chosen "Sustainable Energy and Bio materials" as the main theme, with 8 selected tracks including Sustainable Energy Technologies, Sustainable Environmental management, Sustainable Energy Policy, Biomaterial, Green Industrial management and development, Green Infrastructure, Computational for Energy System and Material, and ICT for Energy System and Material. Experts in these areas have joined the conference and selected international keynote speakers are coming from Universiti Sains of Malaysia, University of Sumatera Utara Indonesia, Toyohashi and Charles Darwin University Australia. This conference also fully supported by USU Sustainable Energy and Biomaterial Centre of Excellence (SEBCoE USU).

Finally, the organizing committee express highly gratitude to all distinguished speakers and all respected participants for contributing in this conference. We earnestly hope that this conference and the time you spend here will provide you with a fruitful and rewarding experience.

Thank you for all of you.

TALENTA CEST 2017
Organizing Committee

Conference Activities

TALENTA
CEST
Conference on Engineering, Scien
and Technology (CEST) 2017

IOP Conf. Series: Materials Science and Engineering **309** (2018) 011002 doi:10.1088/1757-899X/309/1/011002

CONFERENCE INFORMATION

Title	: TALENTA - Conference on Engineering, Science and Technology 2017 (TALENTA-CEST 2017)
Date	: 7-8 September 2017
Venue	: Grand Aston Hotel, Jl. Balaikota No 1, Kesawan, Medan 20112, INDONESIA
Organizer	: Faculty of Engineering University of Sumatera Utara
Sponsor	: Sustainable Energy and Biomaterial Centre of Excellent (SEBCoE) USU

Committee

Honorary Board

1. Prof. Dr. Runtung, SH, M.Hum (Rector, University of Sumatera Utara)
2. Ir. Seri Maulina, MSi, PhD (Dean, Engineering Faculty, University of Sumatera Utara)

International Board

1. Prof. Hiromi Homma (Toyohashi University of Technology, Japan)
2. Prof. Hiroyuki Daimon (Toyohashi, Japan)
3. Prof. Masashi Daimaruya (Muroran Institute of Technology, Japan)
4. Prof. Marwan Al-Akaidi (Arab Open University, Kuwait)
5. Prof. Nik Meriam Nik Sulaiman (University of Malaya, Malaysia)

Keynote Speakers

1. Prof. Hanafi Ismail (Universiti Sains Malaysia, Malaysia)
2. Prof. Hiromi Homma (Toyohashi University of Technology, Japan)
3. Dr. Edward Halawa (Charles Darwin University, Australia)
4. Dr. Eng. Ir. Irvan MS. (University of Sumatera Utara)

IOP Conf. Series: Materials Science and Engineering **309** (2018) 011002 doi:10.1088/1757-899X/309/1/011002

Scientific Committee

1. Prof. T. Indra Mahlia (Uniten, Malaysia)
2. Prof. Hiroyuki Daimon (Toyohashi, Japan)
3. Prof. Hideki Kawai (Muroran Institute of Technology, Japan)
4. Assoc. Prof. Edy Harianto Majlan (UKM, Malaysia)
5. Prof. Marwan Al-Akaidi (Arab Open University, Kuwait)
6. Dr. Edward Halawa (Charles Darwin, University, Australia)
7. Dr. Fahmi Ridwan (Universiti Malaysia Perlis, Malaysia)
8. Prof. Bustami Syam (University of Sumatera Utara, Indonesia)
9. Mahyuddin K. Nasution (University of Sumatera Utara,Indonesia)
10. Dwira Aulia, Ph.D (University of Sumatera Utara, Indonesia)
11. Dr .Eng. Irvan (University of Sumatera Utara, Indonesia)
12. Dr. Taufiq Bin Nur (University of Sumatera Utara, Indonesia)
13. Dr. Halimatuddahliana (University of Sumatera Utara, Indonesia)

Organizing Committee

Chairman	: Indra Surya, Ph.D
Co-Chairman	: Dr. Eng. Himsar Ambarita
Secretary	: Suherman, Ph.D
Vice Secretary	: Maya Sarah, Ph.D
Treasury	: Ir. Adina Sari Lubis, MT
Vice Treasury	: Dr. Juliza Hidayati
Member	: 1. Dr. Eng. Rondang Tambun 2. Dr. Eng. Taufiq Bin Nur 3. Dr. Eng. Listiani Nurul Huda 4. Dr. Fahmi 5. Dr. Bode Haryanto 6. Buchari ST, MT

Contact Info

Organizing Committee Engineering Faculty, University of Sumatera Utara

Email: **talentacest@usu.ac.id**

IOP Conf. Series: Materials Science and Engineering **309** (2018) 011003 doi:10.1088/1757-899X/309/1/011003

Peer review statement

All papers published in this volume of *IOP Conference Series: Materials Science and Engineering* have been peer reviewed through processes administered by the proceedings Editors. Reviews were conducted by expert referees to the professional and scientific standards expected of a proceedings journal published by IOP Publishing.

IOP Conf. Series: Materials Science and Engineering **309** (2018) 012001 doi:10.1088/1757-899X/309/1/012001

Layout design in order to improve efficiency in manufacturing

I Siregar[1*], U Tarigan[1] and T H Nasution[2]

[1]Department of Industrial Engineering, Universitas Sumatera Utara Almamater, Jl. Almamater Kampus USU Medan 20155
[2]Department of Electrical Engineering, Universitas Sumatera Utara Almamater, Jl. Almamater Kampus USU Medan 20155

*Email: ikhsan.siregar@usu.ac.id

Abstract. This research was conducted at the company that produces bobbins and ream type cigarette paper. Problems that found on the production process is the back and forth (back tracking) movement. Back and forth (back tracking) movement extending the total distance moved by the material and increase the total moment of transfer materials thus reducing the efficiency of the transfer of materials in the production process. The purpose of this study is to give design for the layout of production facilities in the company, so that the expected production produced by the company can reach the targets set by the management company. The method used in this research is the Graph-Based Construction and Travel Chart Method. The results of the analysis of the proposed layout with Graph-Based Construction was selected with a total value that is equal to the moment of transfer of 780 758 m / year. This result is better than the actual layout in the amount of 1,021,038.12 meters / year and the results of the method Travel Alternative Chart I of 826.236,60 meters/year, Alternative II of 1.004.433,56 meters / year, and Alternative III for 828,467.12 meters/year. The design layout of Graph-Based Construction material increases the transfer efficiency for 23.53%. With this layout proposal, expected production capacity will be increased along with the shortening of the distance of the displacement that must be passed by the material to be processed.

1. Introduction

The ability of a product company in a competitive market supported by the smooth transfer of material from each of the production department. The layout of the facility is an important element which supports the production of a company. The less effective layout of the facility is resulting the products produced by the company less than the maximum. Therefore, it is necessary to design the layout of the facility every appropriate production department to ensure smooth production. Redesigning the layout of the actions needed to smooth the flow of material in order to increase the production capacity of the company.

Previous research mentioned that the design layout can be intended not as an end result, but the artifacts design that gives insight into the solution space for further exploration, so that the optimal layout can be achieved [1]. Other studies mention that redesigning the layout shows the distance traveled in warehouses can be reduced by more than 2,000 km / month by simply reallocating the shop floor to the different [2]. This shows that there is no perfect layout for companies in the industry and can evolve towards the desired optimum point of the company. Another study discusses the layout such as preparing practical proposals for the design method effective layout of the insights gained from the analysis of the simulation. The study contributes to a critical analysis of the conventional layout [3]. Redesigning the layout has a function to search for the optimal layout of existing layout, as mentioned in previous studies that the optimal design of the layout of equipment in multi-station assembly [4]. The design layout is also useful for a new concept in considering simultaneously the economic and safety aspects in the design of the facility layout problem [5]. Concerning the layout design, not only for the above purposes, but also to streamline the flow patterns of materials that exist in the company, this jug can be seen on the research conducted, at which the research shows that inventory levels can be set to decline to 74% and the utilization of rooms was reduced to 18.18% [6]. And research shows that redesigning the layout for track and balancing the use of machinery and equipment manufacturing companies [7].

This research attempts to integrate the two methods that already exist in order to find the optimal solution. This is the difference with studies before.

The problem of this research is the distance moved by the material, the arrangement of the machines and departments that have not regularly result in the production targets, so the companies need to redesign the layout that existed at this current time in order to obtain the layout and material removal is more effective. To overcome these problems, it is necessary to redesign the layout by using Graph-Based Construction and Travel Chart. Method Based Construction and Travel Graph Chart is as a process used to design the layout that is supposed to minimize the distance moved and the production process can run properly.

2. Research method

The data collection phase to collect the information and data required in conducting the research. The method used is the method of Graph-Based Construction and Travel Chart.

The graphical method is a method of designing the layout using the proximity graph (adjacency graph) as a liaison between departments or existing facilities, with the aim of obtaining the greatest weight. Procedure graphical method frequently used in building the graphical method is to create a graph closeness done step by step by putting a couple departments that have the greatest weight proximity.

Travel chart is a conventional technique commonly used in factory planning and material handling in a production process. Travel chart useful while a lot of material movements in an area.

Data collection for this method was done in two ways: in the form of field observations observation and interviews and literature useful as supporting information and materials research literature related to plant facility layout problems.

3. Result and analysist

3.1. The Production Level Department Block Layout Drawing

The layout of each department is made in the form of block layout according to size as on the real situation regardless of the alley between departments. Block Initial production floor layout can be seen in Figure 1.

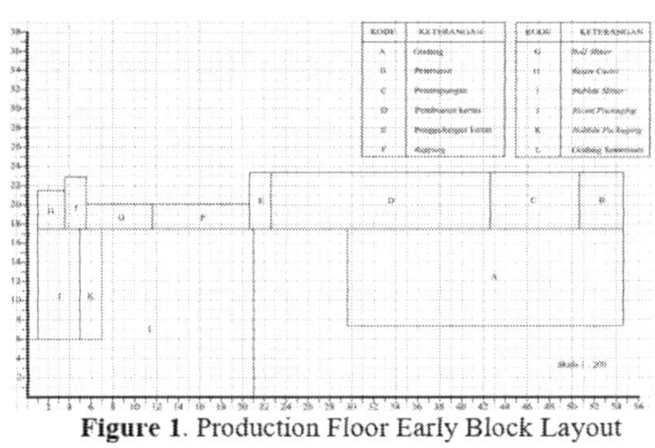

Figure 1. Production Floor Early Block Layout

Based on Figure 1. The depiction block layout on the production company created by the original layout of the production in the company in accordance with the size of each department.

3.2. Determination Department Spacing

Distance between departments is calculated using the formula rectilinear distance, where the distance is measured following a line perpendicular. Distance between departments is calculated based on the

IOP Conf. Series: Materials Science and Engineering **309** (2018) 012001 doi:10.1088/1757-899X/309/1/012001

distance between the center point of the department. The results of the determination of the coordinates of the location for each department can be seen in Table 1.

Table 1. Each Department Coordinate

Department	Coordinate	
	X	Y
A	42	12.5
B	52.5	20.5
C	46.5	20.5
D	32.5	20.5
E	21.5	20.5
F	16	18.75
G	8.5	18.75
H	2.25	19.5
I	3	11.75
J	4.5	20.25
K	6	11.75
L	12.46	7.9

Based on Table 1 shows the coordinates of the location of each department in the company gained from the determination of the midpoint of each department.

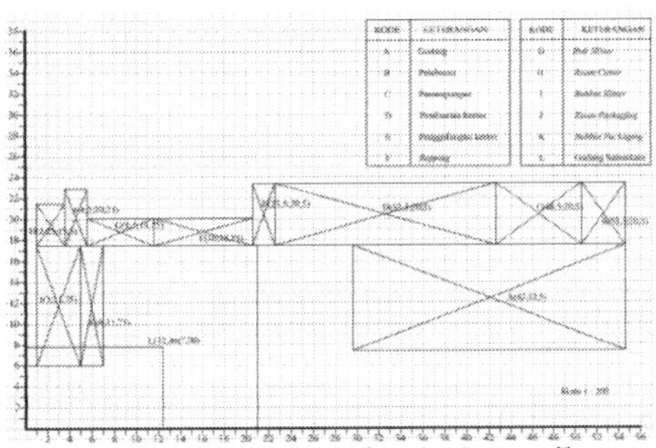

Figure 2. Each Location (x,y) Coordinate

Based on Figure 2 shows the coordinates (x, y) location of each department in the company created by making a diagonal line in each box in order to set the midpoint of each department. Distance between departments is calculated using the formula rectilinear distance. For example, the coordinates A (42 ; 12.5) and B (52.5 ; 20.5), so the distance between A and B is:

$$A\text{-}B = |42-52.5| + |12.5-20.5| = 18.5 \text{ m}$$

The calculations for the distance between the other departments also do like the example above. The result of the calculation of the distance between the departments as a whole to the initial layout can be seen in Table 2. Based on Table 2 shows the distance between the parts department at the company's production gained from the withdrawal line from the midpoint of each department.

Table 2. Each Department Distance (dij) (meter)

	A	B	C	D	E	F	G	H	I	J	K	L
A		18.50	12.5	17.50	28.50	32.25	39.75	46.75	39.75	45.25	36.75	34.14
B	18.5		6.0	20.00	31.00	38.25	45.75	51.25	58.25	48.25	55.25	52.64
C	12.5	6.0		14.00	25.00	32.25	39.75	45.25	52.25	42.25	49.25	46.64
D	17.5	20.00	14.00		11.00	18.25	25.75	31.25	38.25	28.25	35.25	32.64
E	28.5	31.00	25.00	11.00		7.25	14.75	20.25	27.25	17.25	24.25	21.64
F	32.25	38.25	32.25	18.25	7.25		7.50	14.50	20.00	13.00	17.00	14.39
G	39.75	45.75	39.75	25.75	14.75	7.50		7.00	12.50	5.50	9.50	14.81
H	46.75	51.25	45.25	31.25	20.25	14.50	7.00		8.50	3.00	11.50	21.81
I	39.75	58.25	52.25	38.25	27.25	20.00	12.50	8.50		10.00	3.00	13.31
J	45.25	48.25	42.25	28.25	17.25	13.00	5.50	3.00	10.00		10.00	20.31
K	36.75	55.25	49.25	24.25	24.25	17.00	9.50	11.50	3.00	10.00		10.31
L	34.14	52.64	46.54	21.64	21.64	14.39	14.81	21.81	13.31	20.31	10.31	

3.3. *Determination of Interdepartmental Material Transfer Frequency*

Frequency displacement of material needs to be calculated to get a moment of displacement. Frequency displacement of materials on the production floor is obtained through the amount of flow of material displacement occurs. Data transfer frequency data dependent material production volume and capacity transfer equipment used. Volume production of the company in the year 3,098 is a jumbo roll of cigarette paper. The calculation of the frequency of material movements can be seen in Table 3.

Table 3. Material Transfer Frequence

No	Origin Department	Destination Department	Production (unit/year)	Transport Capacity	Frequency of material transfer
1	A	B	3,098	1	3,098
2	B	C	3,098	1	3,098
3	C	D	3,098	1	3,098
4	D	E	3,098	1	3,098
5	E	L	3,098	1	3,098
6	E	F	3,098	1	3,098
7	F	G	3,098	1	3,098
8	L	F	3,098	1	3,098
9	G	H	3,098	1	3,098
10	H	J	3,098	2	6,196
11	G	I	3,098	1	3,098
12	I	K	3,098	2	6,196
13	J	L	3,098	2	6,196
14	K	L	3,098	2	6,196
15	L	A	3,098	4	12,392

Based on Table 3 shows the frequency of material movements on the company by multiplying the volume of production with a payload capacity.

3.4. Calculation of Total Moment Displacement at the Basic Layout

Total moment of displacements on the production floor can be determined by multiplying the frequency of movement of material from one department to another with the distance between related departments. Example calculation of the moment of transfer to the transfer material from department A to department B are as follows:

1. Frequency displacement from A to B = 3,098 times
2. Distance moved from A to B = 18.50 meters
3. Then the moment of transition from A to B ZA-B = 3,098 × 18.50 meters = 57,313 displacement meter / year
4. Complete calculation for any displacement that occurs on the shop floor can be seen in Table 4.

Table 4. Calculation of Beginning Total Moment Displacement

No	Origin Department	Destination Department	Frequency of material transfer (fij) (unit/year)	Department Distance (m)	Moment material transfer (m/year)
1	A	B	3,098	18.50	57,313.00
2	B	C	3,098	6.00	18,588.00
3	C	D	3,098	14.00	43,372.00
4	D	E	3,098	11.00	34,078.00
5	E	L	3,098	21.64	67,040.72
6	E	F	3,098	7.25	22,460.50
7	F	G	3,098	7.50	23,235.00
8	L	F	3,098	14.39	44,580.22
9	G	H	3,098	7.00	21,686.00
10	H	J	6,196	3.00	18,588.00
11	G	I	3,098	12.50	38,725.00
12	I	K	6,196	3.00	18,588.00
13	J	L	6,196	20.31	125,840.76
14	K	L	6,196	10.31	63,880.76
15	L	A	12,392	34.14	423,062.88
		JUMLAH			1,021,038.84

Based on Table 4. The total moment of displacements obtained by multiplying the frequency of movement of material per year that occur with departments within the company.

3.5. Proposed Layout Design

The design layout of the proposal made to each work station first, then after every station has designed the layout of the proposed use of the two methods. Here is an explanation of the process of designing the layout of the proposed using Graph-Based Construction and Travel Chart.

3.5.1 Proposed Layout Design Method Based Graph Construction

The design of the proposed layout using Graph-Based Construction is done by making from to chart that will be designed layout proposal, after it been the greatest value that exist on the map of flights. From to chart can be the value of the degree of proximity of each operation. After selecting one pair of operations that have the greatest weight value then the next step choose the operation that has the greatest weight with two operations that have been previously selected. Here is an example of the work of designing the layout of the proposed using Graph-Based Construction. From moment to chart based transfer between departments can be seen in Table 5.

Table 5. From To Chart Moment Displacement

From \ To	A	B	C	D	E	F	G	H	I	J	K	L	Jumlah
A		57.313										423.062,88	480375,88
B			18.588										18.588
C				43.372									43.372
D					34.078								34.078
E						22.460,5						67.040,72	89.501,22
F							23.235					44.580,22	67815,22
G								21.686	38.725				60.411
H										18.588			18.588
I											18.588		18.588
J												125.840,76	125.840,76
K												63.880,76	63.880,76
L													0
Jumlah	0	57.313	18.588	43.372	34.078	22.460,5	23.235	21.686	38.725	18.588	18.588	724.485,34	[illegible]

From Table 5 above were chosen department of L and A, because it has the greatest weight, namely 423,062.88. Ministry of L and A inserted into the first iteration. Figure iteration 1 Graph Based Construction methods can be seen in Figure 3.

Figure 3. 1st Iteration Graph Based Construction Method

The consideration for selecting the third department can be seen in Table 6.

Table 6. Consideration in Third Department Choosing

Department	L	A	Total	Notes
B	-	57,313	57,313	
C	-	-		
D	-	-		
E	67,040.72	-	67,040.72	
F	44,580.22	-	44,580.22	
G	-	-		
H	-	-		
I	-	-		
J	125,840.76	-	125,840.76	CHOOSEN
K	63,880.76	-	63,880.76	

Based on Table 6. obtained weighting to choose the department into three companies were selected based on the distance the longest distance from the department selected.

A department has the largest weight value so that the department of J into the graph Graph-Based Construction methods like Figure 4.

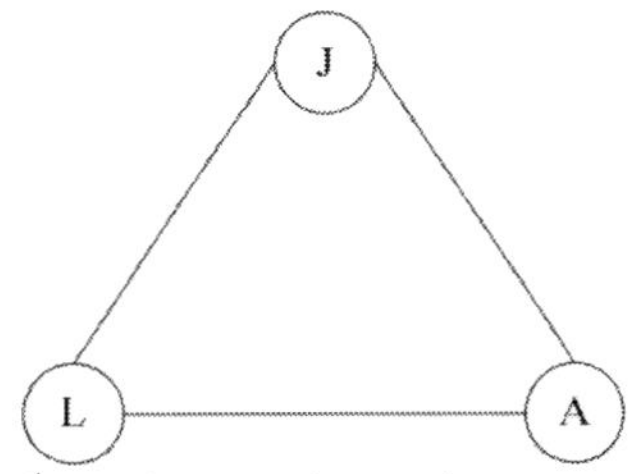

Figure 4. 2nd Iteration Graph Based Construction Method

Based on Figure 4. didapatlah iteration 2 Graph Based Construction methods are selected based on the longest distance the department then created nodes and connected by arc.

The same step conducted on all operations and every department, after doing the design work of the department then it obtained the final results of Graph-Based Construction Methods shown at Figure 5.

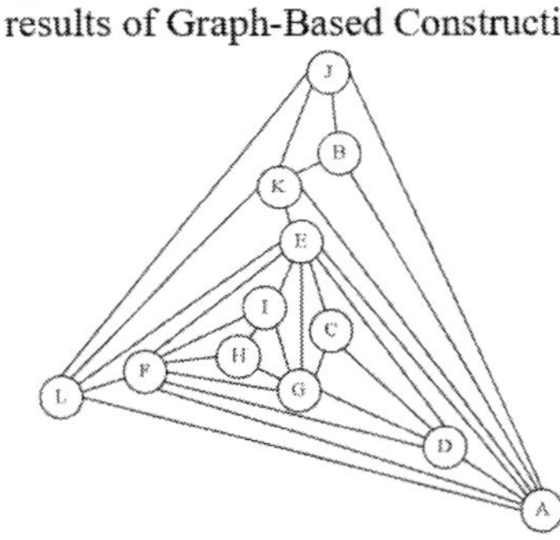

Figure 5. Final Result of Graph Based Construction Method

Based on Figure 5 shows the final results of Graph-Based Construction methods derived from iterations are made continuously until all departments are covered.

Based on the allocation sequence and charts last closeness, it can be made several alternative design layout of new production floor, as described as follows: Block layout design first alternative method Graph-Based Construction can be seen in Figure 6.

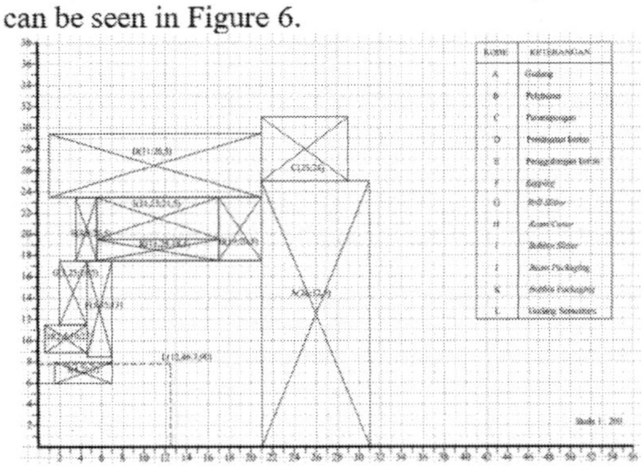

Figure 6. Graph-Based Construction 1st Alternative Block Layout

Based on Figure 6. Above shows the block layout Graph-Based Alternative I consider the moment of transfer of Construction made the shortest. Block layout alternative designs II with Graph-Based Construction methods can be seen in Figure 7.

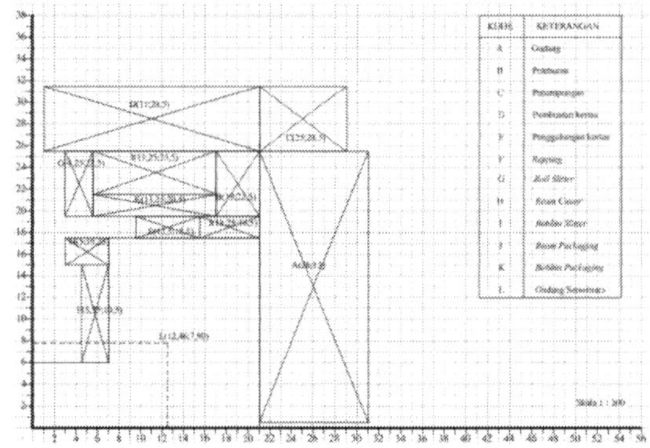

Figure 7. 2nd Alternative Graph-Based Construction Block Layout

Based on Figure 7. Above shows the layout block Alternative II Graph-Based Construction made considering the displacement shortest moment. Block layout alternative design III with Graph-Based Construction methods can be seen in Figure 8.

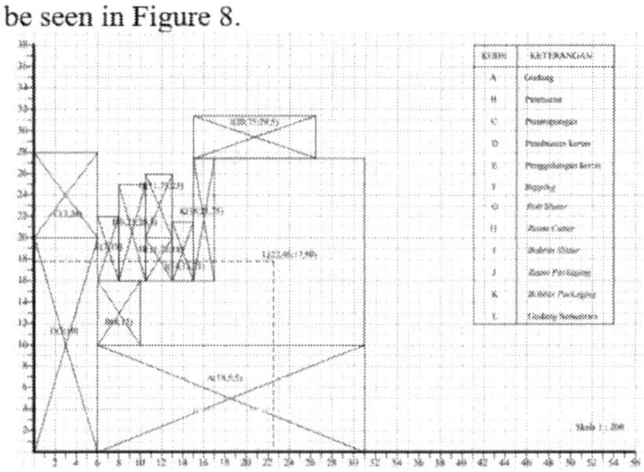

Figure 8. 3rd Alternative Graph-Based Construction Block Layout

Based on Figure 8. The above shows the block layout Graph-Based Alternative Construction III was made considering the displacement shortest moment.

3.5.2. Redesign Proposed Layoud by Using Travel Chart Method

Chart Travel is made using matrix model, where the location department is randomly (for the first time) and thereafter made moments displacement analysis. Based on the method of processing the travel chart, then it can be made several alternative design layout of new production floor, as described as follows: Block layout design first alternative method Chart Traveling can be seen in Figure 9.

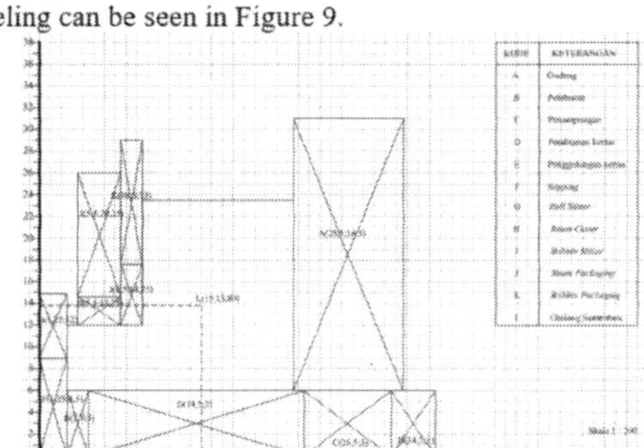

Figure 9. 1st Alternative Travel Chart Block Layout

Based on Figure 9. The above shows the block Layout Alternative I Travel Chart by putting departments have the highest correlation as the center and other departments made arbitrarily by the close relationship between his department. Block layout alternative design Travel Chart II method can be seen in Figure 10.

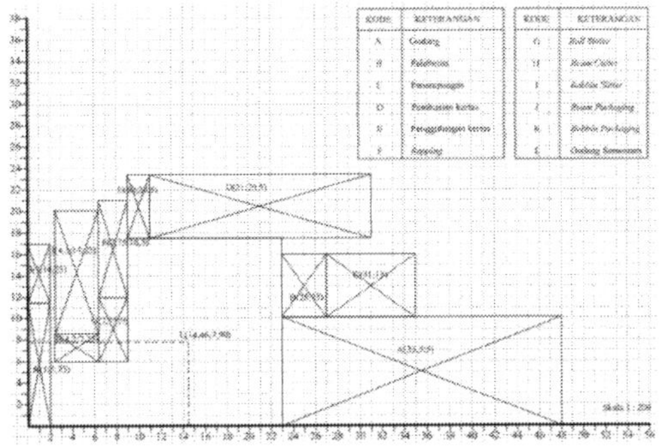

Figure 10. 2nd Alternative Travel Chart Block Layout

Based on the above Figure 10 shows the block Alternative II Travel Chart Layout by putting a department that has the highest correlation as the center and other departments made arbitrarily by the close relationship between his department.Block layout design alternatives Chart Travel III method can be seen in Figure 11.

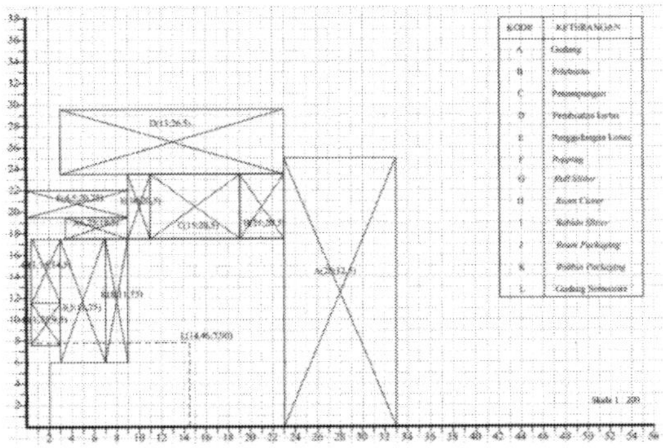

Figure 11. 3rd Alternative Travel Chart Block Layout

Based on Figure 11 above shows the block Layout Alternative Chart Travel III by putting a department that has the highest correlation as the center and other departments made arbitrarily by the close relationship between his department. As for the comparison between each alternative displacement moments of graph-based methods of construction and any alternative method of travel chart can be seen in Table 7.

Table 7. Comparison of Each Alternative Material Moment Transfer

	Graph Based Construction			Travel Chart		
Moment Material Transfer (meter/year)	I	II	III	I	II	III
	780,758.00	895,817.70	807,958.40	826,236.60	1,004,433.56	828,467.20

Based on Table 7. obtained from the calculation of the total transfer between departments within moments of each alternative layout of each method.

IOP Conf. Series: Materials Science and Engineering **309** (2018) 012001 doi:10.1088/1757-899X/309/1/012001

4. Conclusion and discussion

The layout of the selected proposals is having a total moment of material displacement is smaller than the initial layout is the layout of the first alternative method of Graph-Based Construction material movements with a total moment of 780,758 meters/year with an efficiency of 23.53%.

References

[1] Dino, I. G., 2016 *An Evolutionary Approach For 3D Architectural Space Layout Design Exploration, Automation in Construction* **69** 131–150

[2] Horta, M. 2016, *Layout Design Modelling For A Real World Just-In-Time Warehouse, Computers & Industrial Engineering* **101** 1–9

[3] Kim, J. 2016, *Semiconductor FAB Layout Design Analysis With 300-Mm FAB Data: "Is Minimum Distance-Based Layout Design Best For Semiconductor FAB Design?",Computers & Industrial Engineering* **99** 330–346

[4] Tyagi, S. 2016, *Optimal Design Of Fixture Layout In A Multi-Station Assembly Using Highly Optimized Tolerance Inspired Heuristic, Applied Mathematical Modelling* **40** 6134–6147

[5] Neghabi, H. 2016, *A New Concept Of Adjacency For Concurrent Consideration Of Economic And Safety Aspects In Design Of Facility Layout Problem., Journal Of Loss Prevention In The Process Industries* **40** 603-614

[6] Halim, N.H.A. 2015, *Effective Material Handling System for JIT Automotive Production Line, Procedia Manufacturing* **2** 251 – 257

[7] Samya,S.N. 2015, *A granularity model for balancing the structural complexity ofmanufacturing systems equipment and layout, Journal of Manufacturing Systems* **36** 7–19

[8] Apple, M.J. 1990. *Tata letak Pabrik dan Pemindahan Bahan.* Bandung: Penerbit ITB

[9] Purnomo, H. 2004. *Planning and Facility Design* 1st Edition. Yogyakarta: Graha Ilmu

[10] Moore, J. 1959. *Plant Layout and Design* New York: The Macmillan Company

[11] Hadiguna, R. 2008. *Plant Layout.* Yogyakarta : Publisher Andi

[12] Sinulingga, S. 2009. *Production and Planning Control* Yogyakarta: Graha Ilmu

[13] Wignjosoebroto, S. 2003. *Plan Layout and Material* Surabaya: Publisher Guna Widya

[14] Muther, R., 1996. *Practical Plant Layout.* New York McGraw Hill

[15] Sinulingga, S. 2011. *Research Method.* Published I Medan : USU Press

[16] Tompkins, J A and White, J A. 1996. *Facilities Planning* New York: John Wiley & Sons

IOP Conf. Series: Materials Science and Engineering **309** (2018) 012002 doi:10.1088/1757-899X/309/1/012002

Energy consumption model on WiMAX subscriber station

N Mubarakah, Suherman*, M Y Al-Hakim, E Warman

Electrical Engineering Department, Universitas Sumatera Utara, Indonesia

*E-mail: suherman@usu.ac.id

Abstract. Mobile communication technologies move toward miniaturization. Mobile device's energy source relies on its battery endurance. The smaller the mobile device, it is expected the slower the battery drains. Energy consumption reduction in mobile devices has been of interest of researcher. In order to optimize energy consumption, its usage should be predictable. This paper proposes a model of predicted energy amount consumed by the WiMAX subscriber station by using regression analysis of active WiMAX states and their durations. The proposed model was assessed by using NS-2 simulation for more than a hundred thousand of recorded energy consumptions data in every WiMAX states. The assessment show a small average deviation between predicted and measured energy consumptions, about 0.18% for training data and 0.187% and 0.191% for test data.

1. Introduction

Worldwide Interoperability for Microwave Access (WiMAX) is a Broadband Wireless Access (BWA) technology providing high speed connection and wide coverage area standardized by IEEE 802.16 [1]. Its standard varies according to medium access and physical adaptation. WiMAX was initially set for fixed wireless broadband services dominated by point to point configuration on 10-66 Ghz frequency band. In order to enhance its bandwidth, various adaptations were made to work on 2-11 GHz so called 802.16a. Continuous enhancement changes the standard to 802.16d, 802.16e and 802.16m. Handover and roaming procedures were introduced in 802.16e to enable mobile services [2].

WiMAX configuration consists of a single base station (BS) serving many subscriber station (SS) in point-to-multipoint (PMP) operation. WiMAX is also available for mesh/ad hoc configuration [2]. In PMP configuration, BS can be connected to other BSs through TCP/IP networks. Unlike cellular networks, WiMAX is a packet based network. Its technology is compatible to 802.11 with higher data rate and coverage. If 802.11 medium access works based on flexible uplink downlink allocation, WiMAX utilizes a fixed uplink-downlink channel separation which enables WiMAX managing uplink and downlink bandwidth as well as differentiate quality of services. This feature is desired as current subscribers vary in term of QoS requirements [3]. Figure 1 shows how WiMAX allocates slots for both uplink and downlink channels. Uplink and downlink channels are separated by a small time gap. SS is managed by BS, by informing the channel map in downlink channel. SS requests channels through several schemas, from random competition, polling and reservation. All these features make WiMAX be a potential technology replacing the slower Wireless LAN infrastructure and a potential competitor for current cellular infrastructures.

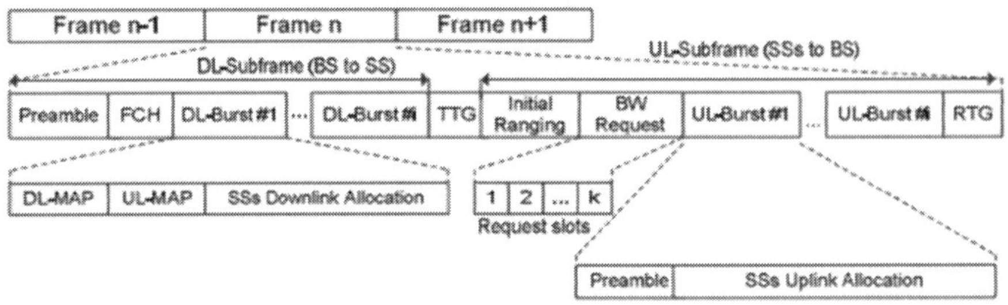

Figure 1. WiMAX Frame Structure [3]

Research on WiMAX varies from bandwidth allocation, physical improvements, to WiMAX applications [4]. As current traffics are dominated by mobile packet switching or TCP/IP services, WiMAX support for mobility based on mobile equipment is open for mobile phone services. However, since WiMAX allows mobile user (SS) to stream high data rates in a longer distance, energy consumption on mobile devices could be a problem. Having a high capacity connection but high energy consumption in not desired by customers. This matter invites researchers to resolve energy consumptions in WiMAX SS. This paper is interested in exploring the possible model of the energy consumption within WiMAX SS. However, there is no much work on it.

This paper found a mathematical model proposed by Bezerra, et al. [5] that normalized energy consumption level in different states such as on downlink (dl) subframe, uplink (ul) subframe, sleep mode, idle mode, turned on, ul burst and dl burst. Each states describe states in the WiMAX transceiver. Table 1 shows the normalized power determined by Bezzera et al [5].

Table 1. WiMAX Normalized Power Consumption [5]

Operation Mode	*NS-2 state*	*Normalized power consumption*
On dl subframe	*While_dl_subframe*	1.00
On ul subframe	*While_ul_subframe*	1.00
On sleep mode	*while_sleep_mode*	0.29
On idle mode	*while_idle_mode*	0.06
Turned on	*while_turned_on*	1.00
Transmitting ul burst	*while_ul_burst ratio*	0.17
Transmitting ul burst	*while_ul_burst_energy ratio*	0.01

Even though the aforementioned model can predict when WiMAX SS consumes power the most, it is not able to predict how much the total power required for sending a certain traffic capacity. In order to do so, this paper proposed the trend based prediction model to determine how much power required by a WiMAX SS, in a certain background network condition for a predetermined traffic rates.

2. Proposed Power Consumption Model

In order to determine the amount of power required to transmit a certain load, this paper proposes a simple trend-based energy consumption model. Initially, states on WiMAX are identified. Bezzera [5] energy consumption model is then applied to each state. A training data on energy consumption is generated by using the same traffic background and network condition as the one that will be predicted. Training data is distributed for each state. The measured duration of each state is plotted against bit rate. Duration of each state may be constant for any bit rate or may follow a trend equation. The amount energy consumed in given bit rate can be calculated by using the Equation 1. E is predicted energy, Di and Pi are duration and normalized power for state i.

$$E = \sum_{i=0}^{i=n} D_i . P_i \quad (1)$$

By using this equation, the predicted energy consumption of any given bit rate can be calculated as long as background traffic and network condition are similar.

3. Research method

In order to evaluate the proposed method, NS-2 simulator is employed. NS-2 is an event-based simulator that is flexible to make changes to some networks and protocols [6]. The simulator is employed for generating training data, forming mathematical model and comparing to test data. Bezzera energy consumption model is applied to Mac802_16SS.cc code. Training data is set for bitrates 40.000 Bps to 680.000 Bps. The model is assessed for bitrates 56.000 Bps to 696.000 Bps and from 64.000 Bps – 704.000 Bps. Both test rates are chosen randomly.

Network condition is set to have 7Mbps bandwidth with 4 (four) subscriber station: one is in a fixed position, (0 m/s); walking speed (1.39 m/s); cycling speed (4.44 m/s) and driving speed (6.67 m/s). WiMAX coverage has 1000 m diameter with 64 QAM and two-ray ground propagation. Simulation runs 20 times. Network configuration is depicted in Figure 2.

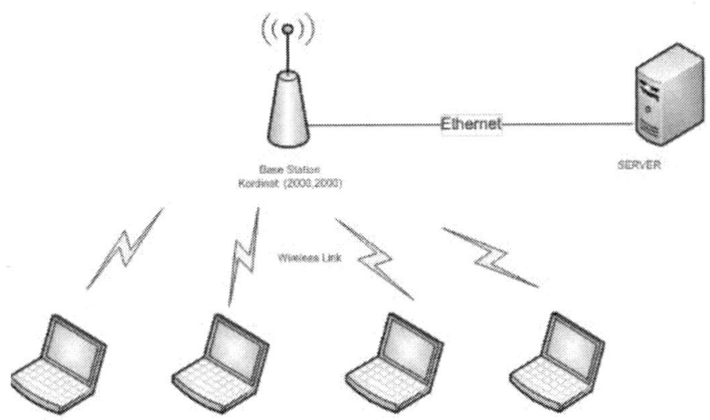

Figure 2. Network configuration

4. Simulation results

4.1. Training Data

Training data generation from 40,000 Bps to 680,000 Bps produces 8-states with constant durations for all bitrates and 7-states with various trend equations. Table 2 and 3 show the state durations.

Table 2. Constant States

No.	Nama *State*	*Equivalent Bezzera Model*	Duration (s)
1	MAC initialization	*TurnOn*	0.001
2	Mac arrangement	*idlePower*	0.000263
3	DL-MAP processing	*DLPower*	1.411993
4	DCD Processing	*idlePower*	0.005981
5	UCD Processing	*idlePower*	0.004005
6	Ranging	*recPower*	0.000026
7	Registration	*idlePower*	0.000005
8	NBR_ADV	*idlePower*	0.003587

Table 3. Duration of variable states

No.	States	Eqivalen Bezzera model	Trend Equation for duration (s)
9	Outgoing Packet Processing	*sendPower*	$y = 58.627\, e^{-0.004\left(\frac{bit\ rate-40000}{40000}\right)}$
10	Sending Packet to Physical layer	*ULPower*	$y = -0.0084\, (\frac{bit\ rate - 40000}{40000})^2 + 1.3799\left(\frac{bit\ rate - 40000}{40000}\right) + 6.4381$
11	Incoming packet processing	*recPower*	$y = 0.0144\, e^{0.0263\left(\frac{bit\ rate-40000}{40000}\right)}$
12	Process packet internally	*DLPower*	$y = 0.0001\, (\frac{bit\ rate - 40000}{40000})^2 + 0.0127\left(\frac{bit\ rate - 40000}{40000}\right) + 3.5168$
13	UL_MAP processing	*ULPower*	$y = 0.0001\, (\frac{bit\ rate - 40000}{40000})^2 + 0.0127\left(\frac{bit\ rate - 40000}{40000}\right) + 2.0896$
14	Time adjustment	*DLPower*	$y = 101.66\, e^{-0.011\left(\frac{bit\ rate-40000}{40000}\right)}$
15	Frame initialization	*ULPower*	$y = -0.0091\left(\frac{bit\ rate - 40000}{40000}\right) + 101.1$

4.2. Energy Consumption Model and Prediction Results

The generated mathematical model follows Equation 1: $= \sum_{i=0}^{i=n} D_i . P_i$. Number of n is 15, *Di* is the fourth columns of Table 2 and 3, *Pi* is the third columns of Table 2 and 3 where the normalized power level for Equivalent Bezzera model is converted based on Table 1.

The comparison of energy calculation between training data and generated mathematical model is plotted in Figure 3. The training data energy consumption is closed to the predicted model with deviation 0.180 %.

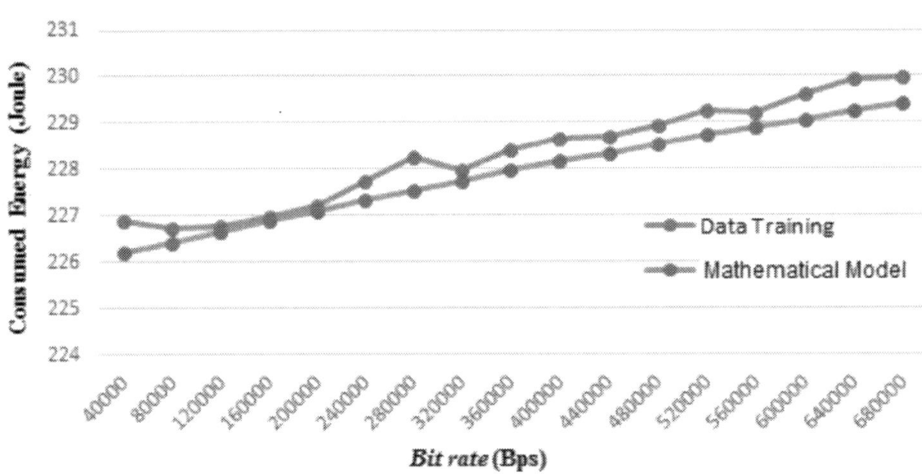

Figure 3. Training data and predicted model

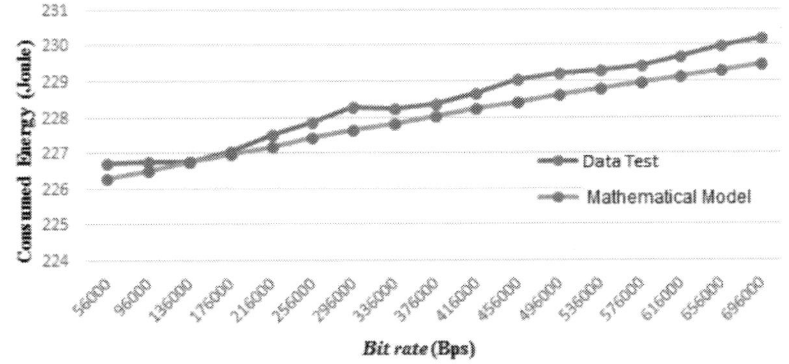

(a) 56000 Bps to 696000 Bps

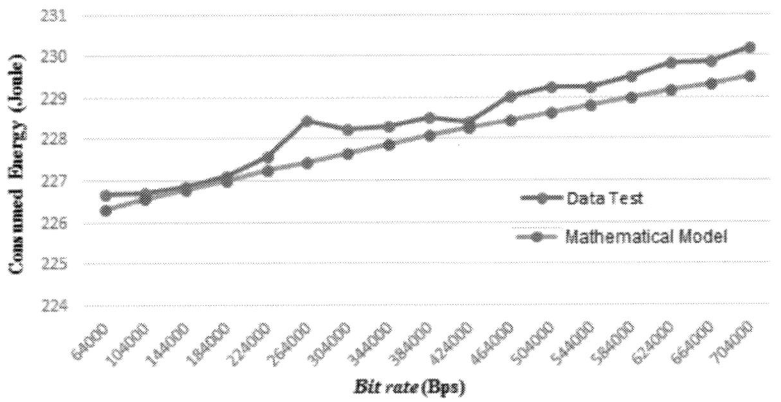

(b) 64000 Bps – 704000 Bps

Figure 4. Test Data and Prediction

4.3. Test Data vs Predicted Energy Consumption

Meanwhile, the test data for bitrates from 56,000 Bps to 696,000 Bps and from 64,000 Bps – 704,000 Bps show close values to prediction results. The comparisons between test data and prediction result average deviation of 0.187% and 0.191% (Figure 4).

5. Conclusion

The assessment show a small average deviation between predicted and measured energy consumptions, about 0.18% for training data and 0.187% and 0.191% for test data.

Based on state calculations, the smallest energy consumption is on state initialisation of MAC 0.001 Joule and the highest consumption is at state Uplink Frame which is 101.017226 Joule.

References

[1]. Vu H L, Chan S, Andrew L 2010 Performance Analysis of Best-Effort Service in Saturated IEEE 802.16 Networks *IEEE Transactions on Vehicular Technology* **59(1)** 460-472

[2] Suherman 2013 *A Novel Approach for Implementing Worldwide Interoperability for Microwave Access for Video Surveillance* Doctoral dissertation De Montfort University

[3] Suherman, Al-Akaidi M and Naemah Mubarakah 2015 A Transport Layer Protocol for uplink WiMAX video streaming *International Journal of Multimedia & Ubiquitous Engineering* **10(1)**
[4] Dhrona P, Ali N and Hassanein H 2008 A performance study of scheduling algorithms in Point-to-Multipoint WiMAX networks *Proc. Local Computer Networks Conference* 843-850
[5] Bezerra N S et al 2010 Modelling Power Consumption in IEEE 802.16e WiMAX Mobile Nodes *The 7th International Telecommunications Symposium*
[6] Issariyakul T and Hossain E 2011 *Introduction to network simulator NS2* Springer Science & Business Media.

IOP Conf. Series: Materials Science and Engineering **309** (2018) 012003 doi:10.1088/1757-899X/309/1/012003

Energy-signal quality trade-offs in a WiMAX mobile station with a booster amplifier

Suherman*, N Mubarakah, O Wiranata, S T Kasim
Electrical Engineering Department, Universitas Sumatera Utara, Indonesia

*E-mail: suherman@usu.ac.id

Abstract. Worldwide Interoperability for Microwave Access (WiMAX) is a broadband wireless access technology that is able to provide high bit rate mobile internet services. Battery endurance remains a problem in current mobile communication. On the other hand, signal quality determines the successful run of the mobile applications. Energy consumption optimization cannot sacrifice the signal level required by the application to run smoothly. On the contrary, the application should consider battery life time. This paper examines the trade-offs between energy and signal quality in WiMAX subscriber station by adjusting signal level using a booster amplifier. Simulation evaluations show that an increment of 0.00000104% energy consumption on using amplifier adaptively produces 16.411% signal to noise ratio (SNR) increment and 10.7% bit error rate (BER) decrement. By keeping the amplifier turned on, energy consumption increases up to 0.00000136%, causing the SNR rises to 17.2638% and BER drops to 11.13%. The evaluated application is video streaming, other application may behave differently.

1. Introduction

Worldwide Interoperability for Microwave Access (WiMAX) is a Broadband Wireless Access (BWA) technology with a wide coverage area standardized by the IEEE 802.16 [1]. WiMAX was initially set for fixed wireless broadband services dominated by point to point configuration on 10-66 Ghz frequency band. In order to enhance its bandwidth, but handover and roaming procedures addition enable WiMAX serving mobile users [2].

WiMAX network components are based on point to point configuration with a single base station (BS) serving many subscriber stations (SSs). BS can be connected to other BSs through TCP/IP networks developing wide cellular-like network. WiMAX works as TCP/IP infrastructure with full support high speed connection and mobility for mobile multimedia application. WiMAX utilizes a fixed uplink-downlink channel separation which enables WiMAX managing uplink and downlink bandwidth as well as differentiate multimedia quality of services. Figure 1 shows WiMAX channel allocations. Uplink and downlink channels are arranged and addressed by a short MAP message. SS requests channels using one of the following schemas: random contention, polling and reservation. These features enable WiMAX to behave differently towards clients.

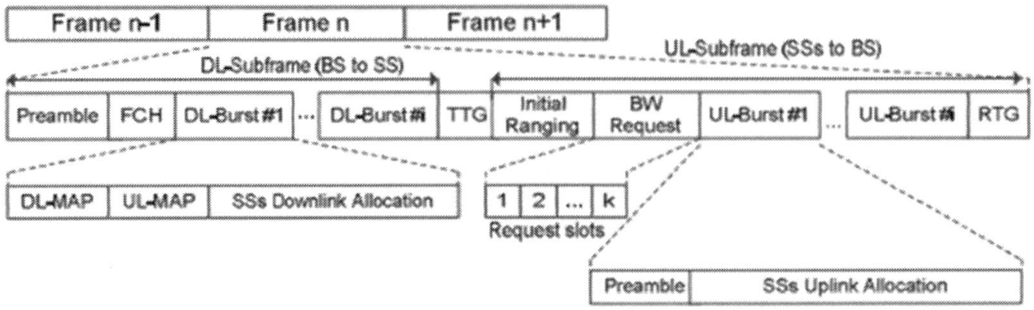

Figure 1. WiMAX Frame Structure [3]

WiMAX supports the multimedia network requirements. Multimedia packet receptions are depending upon the signal strength transmitted by the BS and received by the SS. Moreover if SS is in moving position, propagation media and obstacles influence the received signal very much. Indoor and outdoor changes cause signal variations that may generate packet loss. Amplifying signal at the receiver means increasing signal to noise ratio as well as energy consumption. Increasing signal to noise ratio improves the probability bit detection, improving the performance. However, increasing energy consumed by the amplifier means draining the battery faster. This paper assessed the trade-off between signal amplification and energy consumption on the simulated WiMAX network. Although this simulation does not necessary reflect the real implementation, the idea may be explored further in application environment.

2. Research method

In order to evaluate the trade-off between signal amplification at the receiver and receiver energy consumption, NS-2 simulator is employed. NS-2 is an event-based simulator that is flexible to make changes to some networks and protocols [4]. Network condition is set to have 7Mbps bandwidth with 4 (four) subscriber station one is in a fixed position, (0 m/s); walking speed (1.39 m/s); cycling speed (4.44 m/s) and driving speed (6.67 m/s). WiMAX coverage has 1000 m diameter with 64 QAM and two-ray ground propagation. Network configuration is depicted in Figure 2.

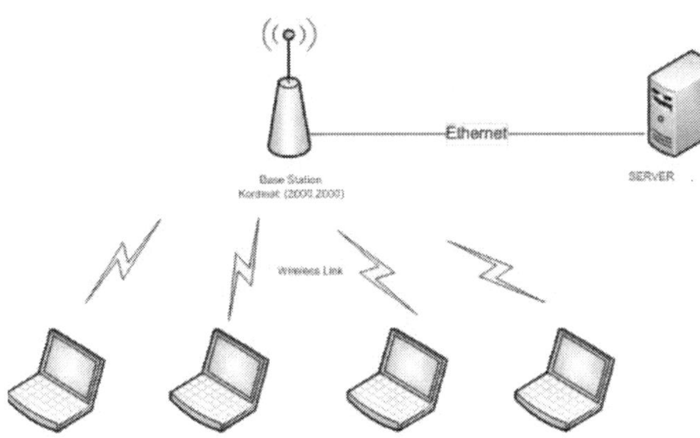

Figure 2. Network configuration

Energy consumption evaluation uses mathematical model proposed by Bezerra, et al. [5] that normalized energy consumption level in some states: downlink (dl) subframe, uplink (ul) subframe, sleep mode, idle mode, turned on, ul burst and dl burst as described in Figure 3. WiMAX scheduler is set to be round robin [7].

Operation Mode	*NS-2 state*	*Normalized power consumption*
On dl subframe	*While_dl_subframe*	1.00
On ul subframe	*While_ul_subframe*	1.00
On sleep mode	*while_sleep_mode*	0.29
On idle mode	*while_idle_mode*	0.06
Turned on	*while_turned_on*	1.00
Transmitting ul burst	*while_ul_burst ratio*	0.17
Transmitting ul burst	*while_ul_burst_energy ratio*	0.01

Figure 3. WiMAX Normalized Power Consumption [5]

The signal amplification was realized by inserting amplifier model proposed by Kumar et al. [6]. The amplification is modelled by using mathematical equation as in Equation 1 [6]. Where x(t) is input signal and y(t) is output signal. Parameter a_1, a_2 and a_3 are designed based on the type of amplifier IC employed.

$$y(t)=a_1 * x(t)+a_2 * x^2(t)+a_3 * x^3(t) \tag{1}$$

Amplifier inserted to SS is designed by using amplifier IC HMC413QS16G/ HMC413QS16GE from Hittit Electronics with specifications [6]:

- Gain, G_db = 23db, Saturated Output Power, P_sat = 29.5dbm
- Output Power at 1db Compression point, P_1db = 27dbm for supply voltage = +2.75V to +5V
- Output power at 3 rd Interception Point, P_IP3 = 40dbm

Kumar et al analyzed that the output depending on a1 and a3 so that Equation 1 is simplified to:

$$y(t)=a_1 * x(t)+a_3 * x^3(t) \tag{2}$$

It can be proved that $a_1 = 14.13$ and $a_3 = -0.4858$. So that Equation 2 be:

$$y(t)=14.13 * x(t)+(-0.4858) * x^3(t) \tag{3}$$

This equation is inserted to WiMAX code in NS-2. Parameter signal to noise ratio (SNR), bit error rate (BER) and energy consumption are evaluated and analyzed.

3. Evaluation results

3.1. SNR

Figure 4 shows that SNR increases as amplifier inserted. For instance, SNR of subscriber 4 increases from 72.31 dBm (without amplifier) to 83.81 dBm (amplifier works continuously). Generally, signal amplification results SNR improvement.

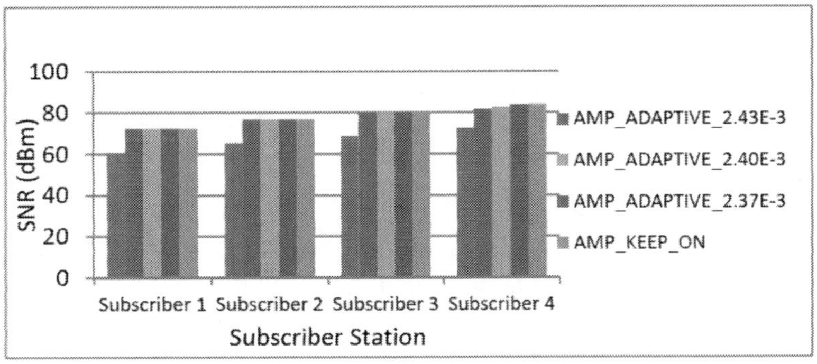

Figure 4. SNR of 4 subscribers

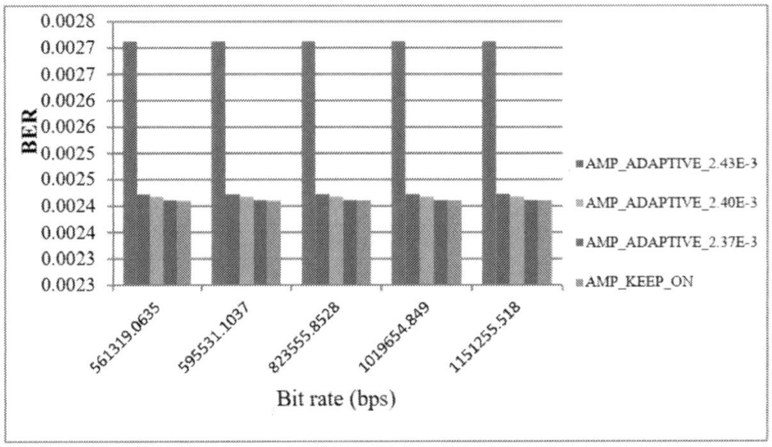

Figure 5. BER of all bitrates

3.2. Bit Error Rate

Bit error rate (BER) deceases when the received signal amplified as shown in Figure 5. BER has the lowest figures when amplifier keeps on. BER remains stable for different traffic rates. BER determines number of error bits. Figure 6 shows that the average error bit are inline to BER values.

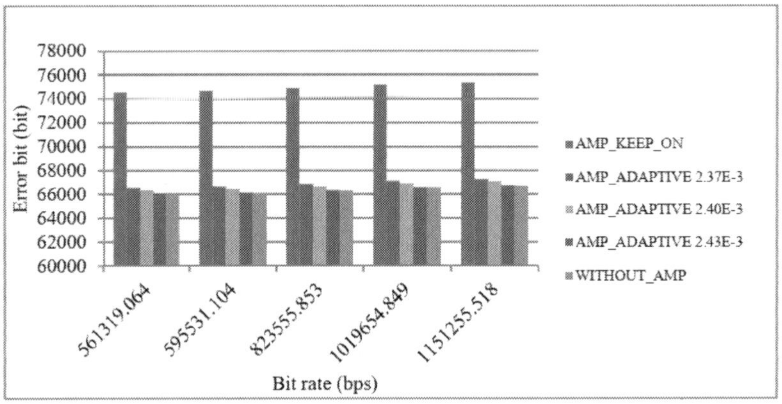

Figure 6. Error bits of all bitrates

3.3. Energy Consumption

Energy consumption increases as amplifier applied. For bit rate 1151255 bps, without amplifier, SS consumes 263.84 Joule. Energy consumption increases to 263.841 Joule and 263.842 Joule when amplifier adaptively applied with BER $2.43e^{-3}$and $2.40e^{-3}$. Keeping the amplifier results the biggest energy consumption (Figure 7).

3.4. Signal Quality vs Energy Consumption

An increase of 16.411% signal to noise ratio (SNR) results 0.00000104% energy consumption increment when using amplifier adaptively. By keeping the amplifier turned on, energy consumption increases up to 0.00000136%, causing the SNR rises to 17.2638% and BER drops to 11.13%.

Based on this figure, the percentage of energy consumption increment is very low compared to SNR increment or BER decrement.

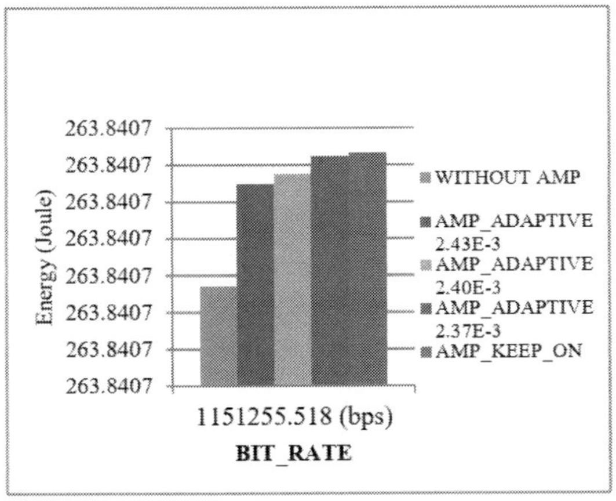

(a)

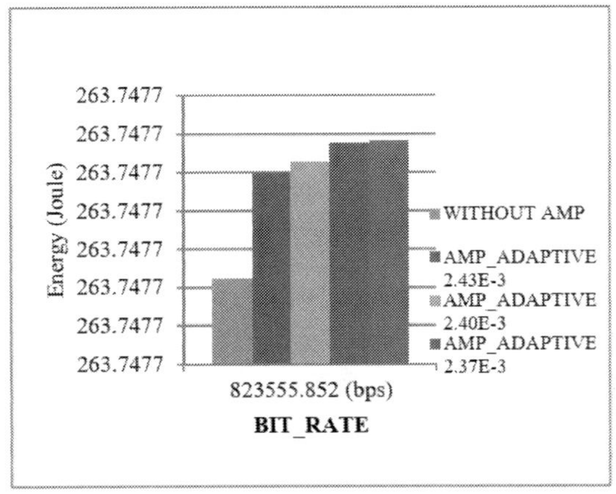

(b)

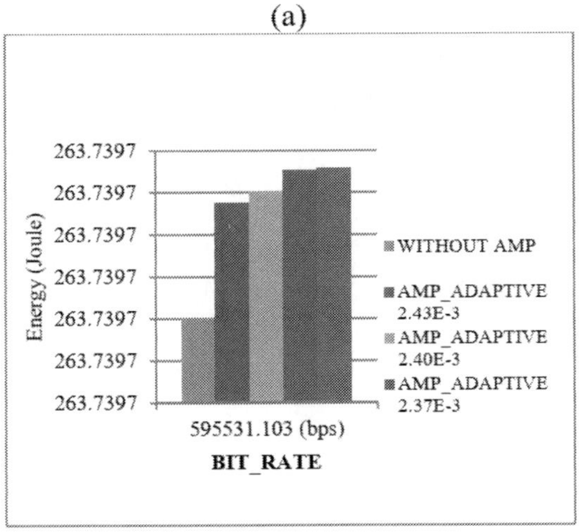

(c)

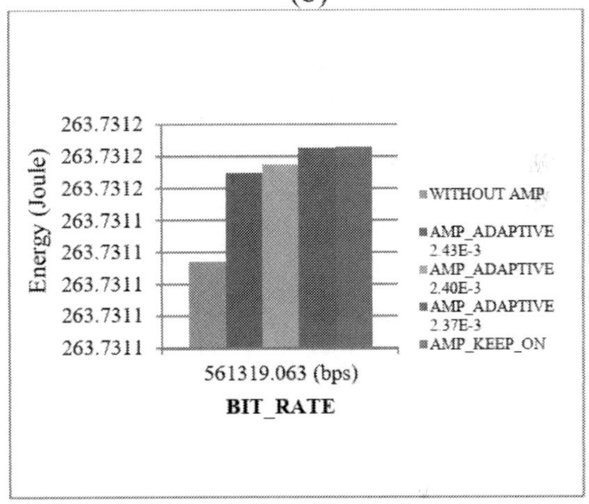

(d)

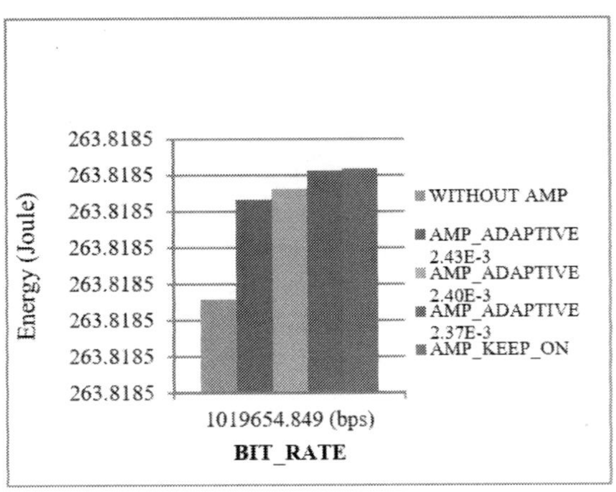

(e)

Figure 7. Energy consumption of all bitrates

IOP Conf. Series: Materials Science and Engineering **309** (2018) 012003 doi:10.1088/1757-899X/309/1/012003

4. Conclusion

Amplifier insertion increases signal quality (SNR) from 66.62 dbm to 78.123 dbm. The consumed energy also increases about 0.0000013589%. Using amplifier adaptively is able to reduce the consumed energy with smaller SNR increment.

References

[1]. Vu H L, Chan S and Andrew L 2010 Performance Analysis of Best-Effort Service in Saturated IEEE 802.16 Networks *IEEE Transactions on Vehicular Technology* **59(1)** 460-472

[2] Suherman 2013 *A Novel Approach for Implementing Worldwide Interoperability for Microwave Access for Video Surveillance* Doctoral dissertation, De Montfort University

[3] Suherman, Al-Akaidi M and Mubarakah N 2015 A Transport Layer Protocol for uplink WiMAX video streaming *International Journal of Multimedia & Ubiquitous Engineering* **10(1)**

[4] Issariyakul T and Hossain E 2011 *Introduction to network simulator NS2* Springer Science & Business Media.

[5] Bezerra N S et al 2010 Modelling Power Consumption in IEEE 802.16e WiMAX Mobile Nodes *The 7th International Telecommunications Symposium*

[6] Kumar A M, Reddy G B and Reddy G KC *Modeling RF Power amplifier to study its non linear effects on rf communication system,with ber as a performance measure* Department of Electronic and Communication Engineering IIIT-Allahabad

[7] Suherman et al 2017 *IOP Conf. Ser.: Mater. Sci. Eng* **237** 012019

IOP Conf. Series: Materials Science and Engineering **309** (2018) 012004 doi:10.1088/1757-899X/309/1/012004

Application work risk of manual material handling operators using different lifting methods

Anizar*, AR Matondang, and JA Sibarani
Department of Industrial Engineering, Faculty of Engineering, Universitas Sumatera Utara, Padang Bulan, Medan 20155, Indonesia

*E-mail: anizar_usu@usu.ac.id

Abstract. This study observed the activity of lifting and moving crates containing bottles at a pallet station in a carbonated drink factory. The activity of moving crates, each weighing 15 kg, is performed by four operators using different methods. An operator manually moves 250 crates daily and often takes rest during working. This study aims to find the most efficient method, that is using the least energy, to move crates. The workload is assessed using the method of Cardiovascular Strain Load (CVL) and energy expenditure. Operators' arterial pulse is measured for 1 minute right before they start working and after they finish working; this is done to obtain work arterial pulse and rest arterial pulse. The way operators lift crates is analyzed using biomechanics. It is found that fatigue is experienced by all operators, with the exception of the 3rd operator who has a %CVL number below 30%. He positions the pallet to be parallel to the conveyor, and then stand in between during the lifting process. He only rotates his body to move the crate from the conveyor onto the pallet, requiring only little energy. This is one of the reasons why the %CVL number is lower than the other operators.

1. Introduction

The production process of soda drink is performed automatically using a machine, starting from washing the bottles until filling the bottles with soda water. However, moving crates containing soda bottles from the conveyor onto the pallet is performed manually by male operators. Each person moves 250 boxes per day; each containing 24 soda bottles drink weighing 15 kg. They often take rest while moving these packing cases, as they feel tired, causing crates to accumulate on the conveyor. Observation conducted on four pallet operators shows that each person uses a slightly different method of heaving and moving crates. Wrong lifting technique risk of causing pain complaint on several body parts.

The workload is of complex construction, and multi-facet, usually defined as part of resource used to develop some activity or work. Another definition is cost spent by an individual while performing specific work[1]. The combination of work demand, the condition where the work is performed, and individual skill, behavior, and perception collectively determine workload [2]. Works requiring simultaneous use of physical and mental power will negatively impact human muscle capacities, such as endurance, fatigue, and recovery. This change depends on the level of physical workload [3]. The high physical workload may lead to poor health, yet it is still not understood completely about time exposure, type of exposure, and the combination of several types of physical workload that can cause poor health [4]. Heart pulse measured and energy consumption calculated are used to classify workload in the category of very light, light, medium, and hard. The balance between sensitivity and specificity for heavy work will contribute to improving workers' safety [5]. Cardiovascular (CVL) workload evaluation consists of an individual factor, work capability, muscle pain, and subjective general health. Distinctive factor that affects CVL of workers is labor type, age, and fitness level [6]. Someone's physical activity, which will affect his energy consumption significantly, can be determined from his/her oxygen consumption and heart pulse [7]. Energy for work is calculated using linear function between heart pulse and energy consumption [8]. Heavy workload possesses ergonomic risk and musculoskeletal problem [9]. Hard work will affect cardiovascular load, hence increasing heart rate.Redesigning work system will reduce pressure and increase productivity [10]. Heavy work will cause excessive muscle contraction, thus increasing the risk of muscle injury. From workers complaints, the sign of hard work side effects can be noticed, therefore work posture is assessed, distance traveled in work, work arterial pulse, and rest arterial pulse are measured. Nordic Body Map questionnaire will show workers' pain complaints on several body parts. Work method assessment using Rapid Entire Body Assessment (REBA) and IRHR can identify problems and risk number that cause workers to suffer from muscle injury [11]. From the high rate of work arterial pulse and short rest of trapezius muscle for the morning shift and afternoon shift workers, a high workload can be observed.For night shift workers, workload and trapezius muscle load are lower [12]. Scheduling is helpful to balance alleviate operators complaints caused by excessive workload and also to increase efficiency [13].

Evaluation of operators work method of lifting and moving the crate is conducted to obtain the biggest energy consumed. The evaluation is also useful to find the best method pallet operators can use. Standardise work method can reduce the amount of energy needed, which will make work less tiring and more efficient.

2. Method

This observation is conducted in a bottling soda factory in Siantar, North Sumatera. Objects of this study are four male pallet operators, whose jobs are to move crate containing 24 bottles weighing 15 kg from the conveyor to pallet. Workers arterial pulses are measured before and after moving these boxes for five days from 08.00 to 16.00 o'clock. Measurement is done manually with indirect assessment method, i.e. measuring operators' arterial pulse for a minute before and after work. The method used to determine workers physical workload is Cardiovascular Strain Load (CVL). CVL is a method used to determine workload classification based on arterial pulse increase compared to the maximum arterial pulse. The measured number is then divided by Rest Arterial Pulse (RAP) and Work Arterial Pulse (WAP). Equation to calculate %CVL is,

$$\%\,\mathrm{CVL} = \frac{100 \times (\text{Work Arterial Pulse} - \text{Rest Arterial Pulse})}{(\text{Max Arterial Pulse} - \text{Rest Arterial Pulse})} \tag{1}$$

Maximum arterial pulse for a man is 220-age and 200-age for a woman. The obtained %CVL number is then compared to the following relation, if %CVL < 30% fatigue happens; 30%<%CVL < 60% improvement needed; 60%<%CVL < 80% work in short time; 80%<%CVL < 100% immediate action needed; %CVL > 100% no activity allowed.

Direct workload assessment means assessing physical workload based on calorie need. In order to determine workload of pallet operator, energy expenditure, and their calorie consumption are calculated. Energy expenditure is determined using interpolation method with the assumption of operators whose heartbeat at 100-125 beats per minute require 5.0 – 7.5 calories per minute. The following equation can be utilized,

$$\frac{100 - 121}{100 - 125} = \frac{5.0 - X}{5.0 - 7.5} \tag{2}$$

By determining energy consumed (Y) in kcal per hour, energy consumption can be calculated. The figure is then used to identify workload received by operators.The formula used to calculate calorie consumption is,

$$Y = 1.80411 - 0.0229038\,X + 4.71711x10^{-4}\,X^2 \qquad (3)$$

where Y is energy consumed (kcal), X is heart pulse rate (beats per minute). Heart pulse data of pallet operators are employed to determine rest time of workers, by using the following equation,

$$R_T = \frac{\left[\frac{K}{S}-1\right]x100 + \frac{T(K-S)}{K-BM}}{2} ......\text{if } S \leq K \leq 2S \qquad (4)$$

where K is total energy consumed, S is standard energy used (kcal/minute), t is total work duration (minute), BM is a basal metabolic number (1.7 kcal/minute).

3. Results and Discussions

The results are to be discussed in 2 subsections; they are work method and workload, energy consumption and rest time.

3.1. Analysis of work method and workload

The 1st operator moves crates from the conveyor on his right side onto the pallet, requiring him to walk. The 2ndperson moves containers from the conveyor on his left side onto the pallet, requiring him to walk forward. The 3rdworker moves packing boxes from the conveyor onto the pallet with his body positioned between the conveyor and the pallet, requiring him only to rotate his body. The 4thlaborer moves bottle chests from the conveyor onto the pallet ±100cm apart, where the conveyor is not parallel to the pallet, also requiring him to walk. The workload of operators is determined based on arterial pulse measured for a minute before and after the work to obtain the average arterial pulse of pallet operators for five days of observation. Workload received by these workers is classified based on %CVL value and energy expenditure. By subtracting age from 220, the maximum arterial pulse number can be obtained. The number of %CVL for each operator is shown in table 1.

Table 1. Pallet operator %CVL number.

No	Name	Age (years)	Operator's arterial pulse (beat/minute)			%CVL	Workload classification
			RAP	WAP	AP_{max}		
1	1st Operator	46	121	77	174	45.36	Improvement needed
2	2nd Operator	45	116	75	175	41.00	Improvement needed
3	3rd Operator	34	106	77	186	26.61	Not experiencing fatigue
4	4th Operator	39	109	74	181	32.71	Improvement needed

The 3rd operator's workload is classified as not experiencing fatigue because his %CVL number is below 30%, while other workers' workloads are classified as immediate improvement needed, as their %CVL number is between 30%-60%. High arterial pulse will affect the %CVL number. The large number of %CVL is also affected by work method.

Energy expenditure is calculated to determine pallet work operator using interpolation method. Pallet operators whose heartbeat at 100-125 beat/minute consume 5.0-7.5 calories/minute as shown in table 2.

Table 2. Energy expenditure calculation of operator.

No	Name	WAP	Energi expenditure (cal/min)	Heart rate during work (beats.min)	Energy consumed (cal/min)	Workload type
1	1st Operator	121	5.0-7.5	100-125	7.1	Medium
2	2nd Operator	116	5.0-7.5	100-125	6.6	Medium
3	3rd Operator	106	5.0-7.5	100-125	5.6	Medium
4	4th Operator	109	5.0-7.5	100-125	5.9	Medium

The result of operators' energy expenditure calculation shows that operators' workload is classified as a medium. It can be clearly seen from the table 2 that the 3rd operator consumes the least energy among the 4 operators, only 5.6 calories per minute. Whereas, the one who consumes the most energy is the 1st operator 7.1 calories/ minute. The difference in %CVL number among operators is affected by internal and external factors. One of them is work method used by each operator in doing their work. Good work method will reduce energy consumption used in work. Fatigue can be caused by either wrong movement or unneeded movement performed repetitively, leading to increased energy consumption.

The difference in work method among operators implies that differences in the technique used affect %CVL number and energy expenditure, which explains why the 3rd operator with the best work method has the lowest %CVL number and does not experience fatigue.Work methods of the 1st, 2nd, and 3rd operator are quite similar and represented by work method of the 1st operator as shown in figure 1.

Figure 1. Work method of the 1st operator

Figure 1 shows work posture and body position of the 1stoperator in five work elements of moving crates containing soda drink from the conveyor onto the pallet. These five work features are operator putting a box onto the pallet, then walking toward conveyor, followed by changing his standing position and lifting the box and operator walking to move a crate from the conveyor onto the pallet, operator putting a crate onto the pallet with his trunk straight. The 1st operator moves a box from the conveyor on his right side onto the pallet, requiring him to walk. He changes his posture and position each time he moves the box. This kind of unorganized movements consumes much energy. The way the other operators perform their work is similar to the 1st person; they do many useless movements, causing their energy consumption to increase. This is a reason why their workloads, based on their %CVL number, are classified as immediate improvement needed and why their energy consumption is higher than the 3rd operator. As more energy is used, oxygen needs and work arterial pulse also increases. Nevertheless, the 3rdoperators has the lowest %CVL number. His work method is better than the other operator. Figure 2 below shows how the 3rd operators do his job.

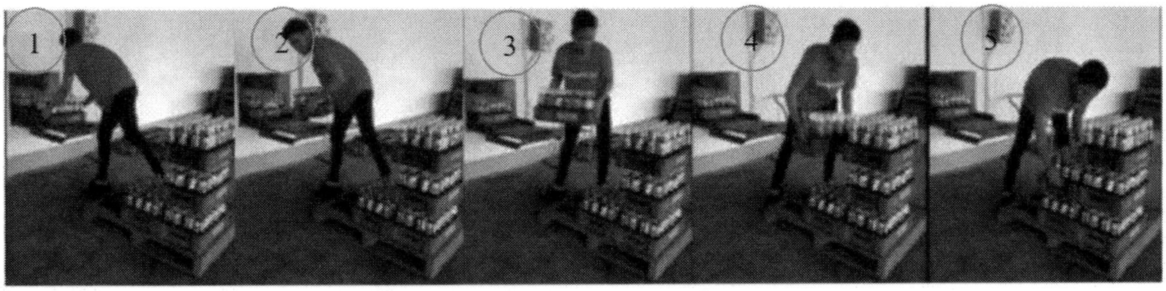

Figure 2. Work method of the 3rd operator

Figure 2 shows work posture and body position of the 3rd operator in 5 work elements of moving crates from the conveyor onto the pallet. They are operator holding a crate, operator lifting a crate in the same standing position as before, operator rotating his body without walking, operator putting a crate onto the pallet. Positions the pallet to be parallel to the conveyor, stands in between, and then lift the containers onto the pallet, requiring him only to rotate his body without moving his legs. Consequently, his energy consumption is low, as no excess energy is consumed. His good procedure is the reason why he has the lowest %CVL number among all four workers. His working method can also be thought to other operators to reduce their workload.

3.2. Analysis of energy consumption and determination of rest time

Based on calorie consumed to move the crates from the conveyor onto the pallet, this work can be categorized as medium work. The 1st operator uses the highest energy to performing this activity, 5.93 kcal/minute or 356.34 kcal/hour, which is categorized as heavy work as the number exceeds 351 kcal/hour. The other operator workloads are categorized as medium work, consuming 201-350 kcal/hour. High energy consumption of the 1st operator is the result of him being older than his colleagues, causing him to utilize more energy. The 3rd operator's workload is categorized as medium work, even though he uses the least calorie, only 280.59 kcal/hour. The reason is he employs different work method that consumes the lowest energy in comparison with the other operators.

Energy consumption calculated is used to determine rest time for each operator. The 1st operator needs 32.65 minutes rest and the 2nd operator requires 18.63 minutes rest for 4 hours of work. Meanwhile, the other two operators do not need any rest because the total energy consumed (K) is less than the standard energy used (S) for a male operator. Operator 1 and 2 require rest time, as both of them are older than the other two operators and doing unnecessary movement during their work. Their work method can be improved to lighten their tasks.

4. Conclusions

This article aims to understand the effect of different work method of pallet operators on physiological workload and energy expenditure. Work procedure of the 3rd operator does not require him to use excess energy because he does not walk or move from his standing position. His efficient method is one factor that causes his %CVL number to be so low in comparison with the other three operators. The more energy used, the more oxygen the body needs, causing an increase in operators heart rate. Efficient work method will reduce the amount of energy used. Tiredness can be caused by either doing the wrong movement or unneeded movement repetitively, leading to an increase in energy expenditure. Work method correction will lighten operators work.

References

[1] Arellano JLH, Martinez JAC, Perez JNS, Alcaraz JLG 2015 *Int J. Phys. Med. Rehabil* **3(6)** 1-6
[2] Wilson MR, Poolton JM, Malhotra N, Ngo K, Bright E, Masters RS 2011 *World Journal of Surgery* **35(9)** 1961-9
[3] Mehta RK, Agnew MJ 2012 *Eur. J. Appl. Physiol* **112(8)** 2891-902

[4] Andersen LL, Fallentin N, Thorsen SV, Holtermann A 2016 *British Medical Journal* **7(4)**
[5] Kolus A, Imbeau D, Dube PA, Dubeau D 2016 *Int.J. Appl. Ergo***54** 158-68
[6] Kristen L, Koch M, Veiersted KB, Moen GH, Waersted M, Knardahl S 2016 *Int. J. Environ. Res. Publ. Health* **13** 356-70
[7] Chen J and Yan L 2017 *Int. J. Recent. Eng. Science (IJRES)* **4(1)** 1-7
[8] Trudeau F, Laurencelle L, Lajoie C 2015 *Int. J. Appl. Ergo* **46** 218-33
[9] Arvidsson I, Balogh I, Hansson GA, Ohlsson K, Akesson I 2012 *J. Appl. Ergo* **43** 1026-32
[10] Ismaila SO 2013 Int. *J. Occup. Safety (JOSE)* **19(4)** 607-11
[11] Widodo L, Sukania IW, Kristiani C 2016 *Proc. Int. Conf. on Industrial Engineering and Management* (Padang) 9th ER1- ER7
[12] Nicoletti C, Spengler CM, Laubli T 2014 *Int. J. Appl. Ergo* **45** 741-6
[13] Dewi DS and Septiana T 2015 Proc. *Int. Conf. on Industrial Engineering and Service Science (IESS)* **4** 445-53

IOP Conf. Series: Materials Science and Engineering **309** (2018) 012005 doi:10.1088/1757-899X/309/1/012005

Numerical study on natural convection between double glasses top cover of a flat plate solar collector

H Ambarita[1]

[1]Sustainable Energy and Biomaterial Centre of Excellent, Universitas Sumatera Utara, Jl. Almamater Kampus USU, Medan 20155, Indonesia

*Email: himsar@usu.ac.id

Abstract. Double glasses cover typically used to decrease top heat loss of a flat plate solar collector. However, reported studies show that the top heat loss of a flat plate solar collector is very significant in comparison with other losses. The present work deals with a numerical study on natural convection between double glasses cover of flat plate solar collector. The main objectives are to investigate the heat transfer characteristics of the covers and to propose heat transfer coefficient correlation. Two-dimensional governing equations and boundary conditions are developed. The governing equations are discretized into linear equation system based on finite volume method. The effects of the aspect ratio (AR) and Rayleigh number are explored and fluid flow in the enclosure are plotted. The results show that AR affects fluid flow and heat transfer characteristics significantly. A correlation proposed by the previous researcher is recommended to calculate top heat loss.

1. Introduction

The world population has increased significantly. It is approximated that the present world population is 7.5 billion people. In the last two decades, the population increased more than 1.5 billion. To support the people activities, the world's energy demand also increasing significantly. Preventing an energy crisis is one of the most casual issues of the 21st century. Solar energy, among other renewable energy sources, is a promising and freely available energy source for managing long-term issues in the energy crisis. The solar industry is developing steadily all over the world because of high demand for energy, on the other hand, major energy source, fossil fuel, is limited and other sources are expensive [1]. Solar energy thermal can be used to power solar desalination [2], adsorption refrigeration [3], solar water heater [4], drying process [5], etc. One of the key components of solar collector system is the collector it'self. Many types of solar collector can be found in literature, and flat-plate type solar collector is the simplest solar collector, and it is very popular [6].

In the flat-plate type solar collector, to decrease the heat loss from the top, a double glass cover is typically employed. Studies on the natural heat transfer mechanism in the enclosure between a cover of flat-plate solar collector have been found in the literature. Varol and Oztop [7] reported a comparative numerical study on natural convection in inclined wavy and flat-plate solar collectors. It was shown that flow and thermal fields are affected by the shape of the enclosure and heat transfer rate increases in the case of wavy enclosure than that of a flat enclosure. The effect of inclination to the wavy and flat-plate solar collectors has also investigated [8]. Kumar [9] investigated natural convective heat transfer in a trapezoidal enclosure of box-type solar cooker. It was shown that the

IOP Conf. Series: Materials Science and Engineering **309** (2018) 012005 doi:10.1088/1757-899X/309/1/012005

values of convective heat transfer coefficient and top loss coefficient for rectangular enclosure are lower by 31-35% and 7 %, respectively.

Recently, Computational Fluid Dynamic (CFD) commercial code has been employed to optimize the design of a flat-plate type solar collector. Martinopoulos et al. [10] investigated the behavior of a polymer solar collector experimentally and numerically. Solar irradiation, as well as convection and heat transfer in the circulating fluid and between the parts of the collectors, is considered in the model. Selmi et al. [11] employed CFD commercial code to simulate heat transfer process in a flat-plate type solar collector. The above-reviewed studies show that heat transfer analysis in the enclosure of double glass cover plays an important role in designing and developing a high-performance flat-plate type solar collector. In fact, only a few of empirical correlations are available in the literature. This paper focuses on numerical analysis in the enclosure between double glass cover of a solar collector. The objective is to explore the heat transfer and flow characteristics in the enclosure. The numerical results will be compared with empirical correlation. The results are expected to support the necessary information in developing high-performance flat-plate type solar collector.

2. Method

Our research group in Sustainable Energy and Biomaterial Centre of Excellent, Faculty of Engineering Universitas Sumatera Utara is developing high-performance flat-plate solar collector for several applications such as drying, adsorption cycle, solar desalination, and solar water heater. These applications use flat-plate type solar collector. Typical of the flat-plate type solar collector is shown in Figure 1. In the figure, photograph of a solar water heater that is also shown. The focus of the present study is the enclosure in the top cover of the solar collector.

2.1 Numerical Method

In this study, only the enclosure between the upper and bottom glasses is taken account into consideration. The computational domain is shown in Figure 1(a).

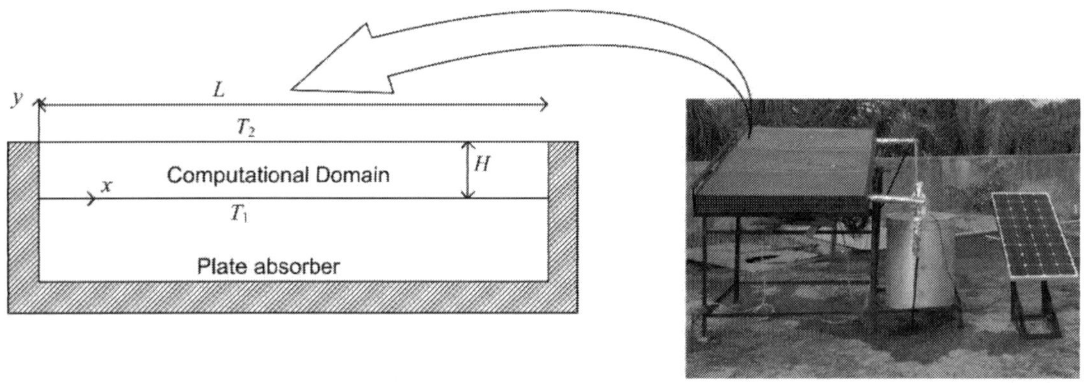

(a) Model for computation (b) Solar Collector System

Figure 1. Computational domain and photograph of a system

In this analysis, the domain is assumed to be the two-dimensional case. The flow in the domain is incompressible laminar flow and steady state. The additional assumptions are, there is no viscous dissipation, the gravity acts in the vertical direction, and fluid properties are constant except in density of the fluid. The Boussinesq approximation is used to model the buoyancy force. By using the above assumptions, the governing equations are as follows.

$$\frac{\partial u}{\partial x}+\frac{\partial v}{\partial y}=0 \tag{1}$$

$$u\frac{\partial u}{\partial x}+v\frac{\partial u}{\partial y}=-\frac{1}{\rho}\frac{\partial P}{\partial x}+\frac{\mu}{\rho}\left(\frac{\partial^2 u}{\partial x^2}+\frac{\partial^2 u}{\partial y^2}\right)+g\beta(T-T_r)\sin\phi \tag{2}$$

$$u\frac{\partial u}{\partial x}+v\frac{\partial u}{\partial y}=-\frac{1}{\rho}\frac{\partial P}{\partial y}+\frac{\mu}{\rho}\left(\frac{\partial^2 v}{\partial x^2}+\frac{\partial^2 v}{\partial y^2}\right)+g\beta(T-T_r)\cos\phi \tag{3}$$

$$u\frac{\partial T}{\partial x}+v\frac{\partial T}{\partial y}=\alpha\left(\frac{\partial^2 T}{\partial x^2}+\frac{\partial^2 T}{\partial y^2}\right) \tag{4}$$

Where u and v are the velocity vector in x- and y-directions, respectively. In the analysis, non-dimensional parameters are stated in the followings.

$$Ra=\frac{g\beta\Delta TH^3}{\nu^2}\Pr \tag{5}$$

Where Pr is Prandtl number and calculated by

$$\Pr=\frac{\nu}{\alpha} \tag{6}$$

Convective heat transfer coefficient (h) is calculated using the non-dimensional Nusselt number. It is given by

$$h=\frac{Nu\times k}{H} \tag{7}$$

Where k [W/mK] is the conductive heat transfer coefficient of the air and H [m] is the distance between the double glasses. The Nusselt number is presented using local and average Nusselt number. The local Nusselt number in the bottom and top surfaces are given by

$$Nu_x=\frac{H}{(T_1-T_2)}\left.\frac{\partial T}{\partial y}\right|_{y=0} \tag{8}$$

and

$$Nu_x=\frac{H}{(T_1-T_2)}\left.\frac{\partial T}{\partial y}\right|_{y=H} \tag{9}$$

, respectively. The average Nusselt number is calculated by

$$\overline{Nu}=\frac{1}{L}\int_{x=0}^{L}Nu_x dx \tag{10}$$

All of those governing equations are converted into linear equations by employing finite volume method. The system of linear equations for all fields is coupled using SIMPLE algorithm. The commercial code of ANSYS FLUENT is used to carry out the simulation.

2.2 Empirical correlation

The results from simulation will be compared with the analytical results. In the analytical one, the below empirical correlation will be used. There several empirical correlations are found in the literature. The first one is given by Jacob [11].

$$\overline{Nu}=0.195Ra_H^{0.25}\ \text{for}\ 10^4<Ra_H<4\times10^5 \tag{11a}$$

$$\overline{Nu}=0.068Ra_H^{1/3}\ \text{for}\ 4\times10^5\le Ra_H<10^7 \tag{11b}$$

In the more recent experimental study using air inside the enclosure, Holland et al. [12] proposed an empirical correlation between the average Nusselt number and Rayleigh number. The equation is given by the following equation.

$$Nu=1+1.44\left[1-\frac{1708(\sin 1.8\phi)^{1.6}}{Ra\times\cos\phi}\right]\left[1-\frac{1708}{Ra\times\cos\phi}\right]^{+}+\left[\left(\frac{Ra\times\cos\phi}{5830}\right)^{1/3}-1\right]^{+} \tag{12}$$

This equation is claimed to be valid for tilt angle from 0° to 75°. In the equation the meaning of the "+" exponent is that only positive values of the terms in the square bracket are to be used and if the value is negative, the value will be converted into zero. The additional condition for using equation (12) is the aspect ration (AR) and the Rayleigh number must be $AR \geq 12$ and $Ra_H < 10^5$.

3. Results and Discussions

3.1. Numerical Validation

The present numerical method is validated by comparing the result with previously reported studies. The selected numerical validation case is a laminar natural convection heat transfer from a square cavity heated and cooled from side walls, respectively and insulated from the top and bottom walls. The results are compared for Rayleigh number 105 and 106. Table 1 shows the numerical validation of the present result. Data of the table reveals that the present results show a very good agreement with the previous results. Thus, the present method can be used to explore the problem.

Table 1. Comparison of averaged Nusselt number

Reference	Parameter	Rayleigh Number	
		$Ra_H = 10^5$	$Ra_H = 10^6$
Present work	$\bar{N}u$	4.518	8.879
De Vahl Davis [13]	$\bar{N}u$	4.519	8.799
	Difference [%]	0.02%	-0.9%
Fusegi et al. [14]	$\bar{N}u$	4.646	9.012
	Difference [%]	2.8%	1.49%
Tiwari et al. [15]	$\bar{N}u$	4.450	8.806
	Difference [%]	-1.5%	-0.8%

3.2. Flow characteristics and heat transfer

Figure 2 shows flow characteristics of the fluid inside the enclosure. The AR of the presented enclosure is 20. In the figure temperature distribution, streamlines, and velocity vector are shown in Figure 2a, 2b, and 2c, respectively. It is shown clearly that inside the enclosure several circulation flows are generated. This circulation flow is known as "Benard cell". The mechanism can be explained as follows. The heated fluid flows upward, and before reaching the top surface, it is divided into two flows. Each flow will generate a circulation flow. This circulation will be repeated. The streamlines (Figure 2b) clearly show the number of the Bernard cell. In this case, there are 14 Benard cells generated in the enclosure. The same pattern also is shown by velocity vector. The velocity of the fluid will follow the Benard cells.

Locally heat transfer coefficient on the bottom surface is calculated and plotted in Figure 3. The horizontal axis is the distance from the left edge. As a note, the dimension of the present enclosure is 1 m$\times$ 5 cm. The figure shows that, as expected, locally heat transfer coefficient is strongly affected by Benard cells. The Benard cells divided the bottom surface into several dead zones, where the velocity almost zero or stagnant. Between two dead zones, the velocity increases and make temperature gradient higher. Heat transfer coefficient in the dead zone will decrease because temperature gradient is very low. On the other hand, the area between dead zones the local heat transfer coefficient increases and reaching a maximum value. This is because temperature gradient is higher. This fact is shown in Figure 3.

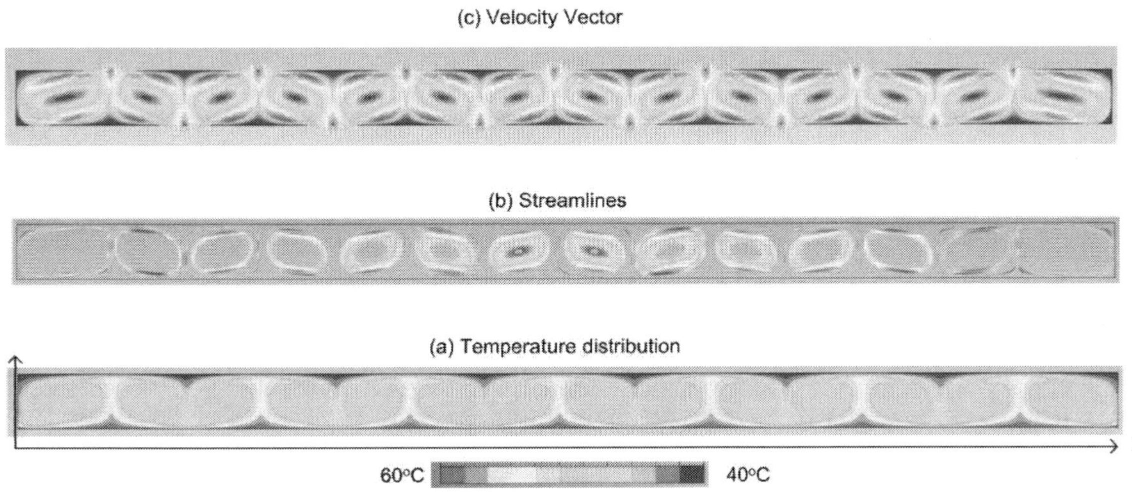

Figure 2. Flow characteristics

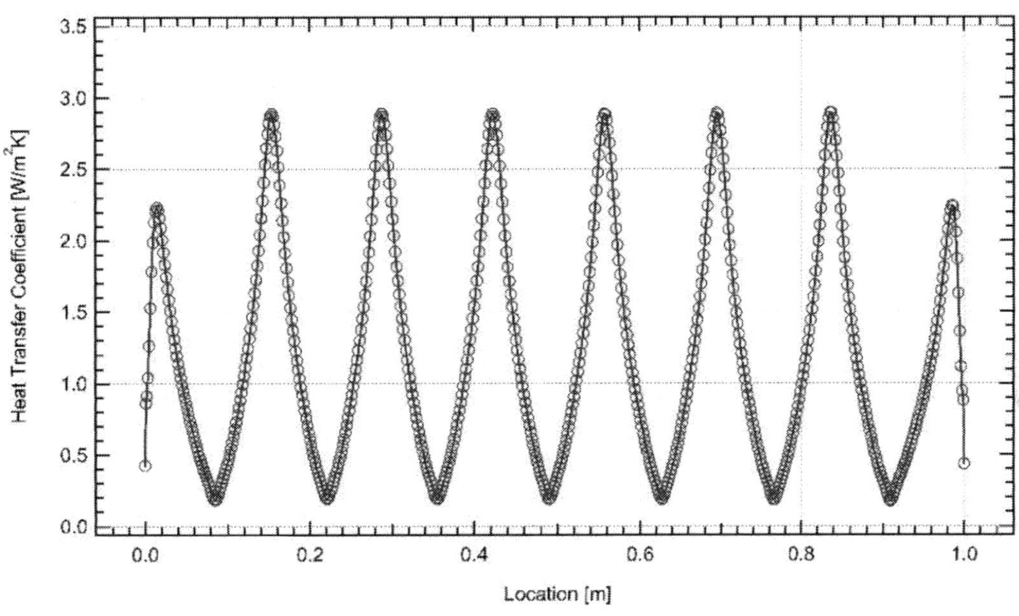

Figure 3. Locally heat transfer Coefficient

3.3. Effect of Rayleigh number

By using equation (7) to (10), the average Nusselt number is calculated. The Nusselt number resulted by empirical correlation is also estimated. The results from the present work and from the empirical estimation are presented in Table 2. The table shows the configuration, Rayleigh number, and averaged heat transfer coefficients, as well as Nusselt number and heat flux. The table shows that there is a clear discrepancy between Nusselt number resulted from the above empirical correlation and with the present results. The difference of the results from Reference [11] and the present results varies from 10.8 – 38.5%, which is lower for higher Rayleigh number. The difference with Reference

[12] varies from 3.5 – 15.2%, this also lower for higher Rayleigh number. This fact suggests that empirical correlation proposed by Jacob [11] shows significant deviation. On the other hand, Holland [12] reveals a better estimation. Thus, it is suggested to use Holland correlation to predict heat transfer rate for a double glass cover of the flat-plate solar collector.

The effect of the Rayleigh number to the performance of a double glass cover can be examined using data shown in Table 2. The Nusselt number increases with increasing Rayleigh number. However, the total heat transfer rate is decreasing.

Table 2. Results comparison

Aspect Ratio	Rayleigh Number	Parameter	Present Results	Equation (11) Ref. [11]	Equation (12) Ref. [12]
$AR = 40$		$\overline{Nu}$	3.545	2.560	3.078
$L = 1m$	$Ra_H = 2.97 \times 10^4$	$\bar{h}$ [W/m²K]	3.432	2.478	2.979
$H = 2.5cm$		Q'' [W/m²]	68.641	49.563	59.587
$AR = 20$		$\overline{Nu}$	5.589	4.305	4.871
$L = 1m$	$Ra_H = 2.38 \times 10^5$	$\bar{h}$ [W/m²K]	2.71	2.084	2.358
$H = 5.0cm$		Q'' [W/m²]	54.10	41.67	47.150
$AR = 10$		$\overline{Nu}$	8.02	7.241	8.321
$L = 1m$	$Ra_H = 1.90 \times 10^6$	$\bar{h}$ [W/m²K]	1.94	1.752	2.014
$H = 10.0cm$		Q'' [W/m²]	38.84	35.046	40.275

3.4. Effect of AR

The data of Table 2 shows the effect of Aspect Ratio (AR) to the performance of a double glass cover. It was shown that increasing AR would increase heat transfer rate. This is because increasing AR will reduce the distance between two glasses and make heat transfer resistance becoming smaller. In contrast, the increasing AR will decrease the Nusselt number. As a note, the Nusselt number is defined by using the distance of the double glasses. Increasing AR will decrease the distance and will decrease the Nusselt number. Since the side wall of the present enclosure is insulated or inactive in transferring heat, it is suggested to define the Nusselt number using the active wall. In this work, as well in the proposed empirical equations [11, 12], the Nusselt number is defined as the distance between double glass cover.

4. Conclusions

A numerical method using commercial code CFD to explore the heat transfer and fluid flow characteristics in the enclosure between double glasses cover of a flat-plate type solar collector has been developed. The method has been validated for a natural convection is a square enclosure heated and cooled from side walls, respectively. The validated method is then used to explore the problem. The results of the developed method are compared with empirical correlations that typically used to analyze the heat transfer in the double glass cover. The conclusions are as follows. In numerical simulation shows that in the enclosure several Benard cells are captured. The Benard cell is a circulation flow and constructs a cell-like flow. The present of the Benard cells strongly affects the local heat transfer coefficient. The average Nusselt number resulted from the present work is higher than resulted from empirical correlations. The discrepancy varies from 10.8 – 38.5% and from 3.5 – 15.2% for empirical correlation proposed by Jacobs [11] and Holland et al. [12], respectively. Based on this fact it is suggested to use the empirical correlation proposed by Holland et al. [12].

IOP Conf. Series: Materials Science and Engineering **309** (2018) 012005 doi:10.1088/1757-899X/309/1/012005

Acknowledgments
The authors gratefully acknowledge the support from my students Eko Y Setiawan, Sintong R Butarbutar Ridho Erlanda S and Agistya Dewi.

References
[1] Kannan N and Vakeesan D 2016 *Renewable and Sustainable Energy Reviews* **62** 1092-1105
[2] Ambarita H 2016 *Case Studies in Thermal Engineering* **8** 346-358
[3] Ambarita H and Kawai H 2016 *Case Studies in Thermal Engineering* **7** 36-46
[4] Jamar A, Majid Z A A, Azmi W H, Norhafana M and Razak A A 2016 *International Communications in Heat and Mass Transfer* **76** 178-187
[5] Dina S F, Ambarita H, Napitupulu F H and Kawai H 2015 *Case Studies in Thermal Engineering* **5** 32-40
[6] Pandey K M and Chaurasiya R 2017 *Renewable and Sustainable Energy Reviews* **67** 641-650
[7] Varol Y and Oztop H F 2008 *Building and Environment* **43** 1535-1544
[8] Varol Y and Oztop H F 2007 *Building and Environment* **42** 2062-2071
[9] Kumar S 2004 *Renewable Energy* **29** 211-222
[10] Martinopoulos G, Missirlis D, Tsilingiridis G, Yakinthos K and Kyriakis N 2010 *Renewable Energy* **35** 1499-1508
[11] Selmi M, Al-Khawaja M J and Marafia A 2008 *Renewable Energy* 33 383-387
[12] Hollands K G T, Unny T E, Raithby G D and Konicek L 1976 *Journal of Heat Transfer* **98**, 189-193
[13] Vahl Davis G De 1983 *International Journal of Numerical Methods Fluids* **3(3)**, 249-264
[14] Fusegi T, Kuwahara K and Farouk B 1991 *International Journal of Heat and Mass Transfer* **34(6)**, 1543-1557
[15] Tiwari R K and Das M K 2007 *International Journal of Heat and Mass Transfer* **50**, 2002-2018

IOP Conf. Series: Materials Science and Engineering **309** (2018) 012006 doi:10.1088/1757-899X/309/1/012006

Effect of engine load and biogas flow rate to the performance of a compression ignition engine run in dual-fuel (diesel-biogas) mode

H Ambarita

Sustainable Energy and Biomaterial Centre of Excellent, Universitas Sumatera Utara, Jl. Almamater Kampus USU, Medan 20155 Indonesia

Email: himsar@usu.ac.id

Abstract. The Government of Indonesia (GoI) has released a target on reduction Green Houses Gases emissions (GHG) by 26% from level business-as-usual by 2020, and the target can be up to 41% by international supports. In the energy sector, this target can be reached effectively by promoting fossil fuel replacement or blending with biofuel. One of the potential solutions is operating compression ignition (CI) engine in dual-fuel (diesel-biogas) mode. In this study effects of engine load and biogas flow rate on the performance and exhaust gas emissions of a compression ignition engine run in dual-fuel mode are investigated. In the present study, the used biogas is refined with methane content 70% of volume. The objectives are to explore the optimum operating condition of the CI engine run in dual-fuel mode. The experiments are performed on a four-strokes CI engine with rated output power of 4.41 kW. The engine is tested at constant speed 1500 rpm. The engine load varied from 600W to 1500W and biogas flow rate varied from 0 L/min to 6 L/min. The results show brake thermal efficiency of the engine run in dual-fuel mode is better than pure diesel mode if the biogas flow rates are 2 L/min and 4 L/min. It is recommended to operate the present engine in a dual-fuel mode with biogas flow rate of 4 L/min. The consumption of diesel fuel can be replaced up to 50%.

1. Introduction

Greenhouse Gases (GHGs) emissions will be a significant problem for the world in supporting life on it. To avoid the world from the catastrophe, many countries have committed to reducing their emissions. The Government of Indonesia (GoI) has released its target on reducing GHG emissions by 26% from level business as usual (BAU) by 2020, and it can be increased up to 41% by international aid [1]. The source of GHG emissions can be divided into four sectors; they are agriculture, energy, land use, land-use change and forestry (LULUCF), and waste sectors. In the energy sector, the emission sources are all activities that burn fossil fuel. In Indonesia, energy sector contributes significant GHG emissions [2]. In specific, the activities that burn fossil fuel using internal combustion (IC) engines to produce power are important in emission reduction actions. Thus, reducing fuel consumption in the IC engines will reduce GHG emissions significantly. Also, in Indonesia, fossil fuel (gasoline, diesel, kerosene) is subsidized, and it is a burden for the GoI budget. These facts suggest that reducing fuel consumption of IC engines will give significant impacts on mitigation GHG emissions and it will help the GoI in reducing fossil fuel subsidy.

Due to the thermodynamic principle of IC engines and the Carnot cycle limit, the emission target can only be achieved with fuel change or blending with biofuel such as biogas. The biogas is produced from anaerobic biodegradation of organic material. Such materials are easily found in the remote area. It consists of methane typically ranges from 40 - 70% and its low heating value are between 15 and 30 MJ/m^3. The IC engines, mostly, can be divided into spark ignition (SI) engines and compression ignition (CI) engines. The CI engines are mainly used in heavy machineries such as a truck, agricultural engines, marine, and power plants. In other word, CI engines are usually used in the remote areas. Considering that biogas is mostly present in the remote areas, thus, fueling CI engines with biogas is a perfect solution for reducing GHG emission. Two methods that can be used to run CI engine with biogas. The first one is by converting the CI engine into SI engine and run on pure biogas. The second one is dual-fuel mode. In this mode, after compression of the charge comprised of biogas and air, a small amount of diesel, called the pilot is injected. This injected pilot fuel gets self-ignited and then becomes the ignition source for the inducted biogas. The main advantage of dual-fuel CI engines is that they can work with a wide variety of gaseous fuels without engine modifications [3].

Study on CI engine run in dual-fuel mode has come under scrutiny in recent year. Ambarita et al. [4] reported a study on the performance and emissions of small CI engine run in dual-fuel mode by using raw biogas with a methane content of 60%. The effect of biogas flow rate is investigated. It was shown that the CI engine without any significant modification can be operated in a dual-fuel mode with raw biogas and diesel oil consumption can be replaced up to 87.5%. In addition, Ambarita et al. [5] reported a study on combustion simulation of a CI engine run in a dual-fuel mode using CFD code. The numerical results and experimental ones show a good agreement. Feroskhan et al. [6] investigate the effects of biogas flow rate and cerium oxide addition on the performance of a dual fuel CI engine. The biogas flow rate and cerium oxide as a diesel additive were investigated. It was found that operating a conventional CI engine on dual fuel mode with biogas flow rate up to 8 L/min improve brake thermal efficiency at high load. The biogas can replace the diesel fuel consumption up to 80%. Cerium oxide affects the brake thermal efficiency, the combination of biogas flow rate 4 L/min and 25 mg/L Cerium oxide give the highest brake thermal efficiency. Rahman and Ramesh [7] studied the effect of reducing methane concentration on the combustion and performance of biogas diesel predominantly premixed charge CI engine. Three biogas compositions with methane proportions of 53 – 58% (raw biogas), 67% (refined), and 22 - 25% (less methane) were tested at constant engine speed of 1800 rpm. Result indicate that extremely low levels of NO and smoke can be reached in the engine at the best efficiency operating condition if the biogas composition is altered based on the output. Verma et al [8] investigated the effect of varying composition of biogas on performance and emission characteristics of compression ignition engine using exergy analysis. Three different composition of biogas are tested, they are 93%, 84%, and 75% of methane by volume. The results show that biogas can substitute 80 – 90% diesel fuel at lower engine loads.

The above literatures show that study on the effect of operational conditions such as biogas flow rate, methane concentration in biogas, and additive in the diesel fuel to the performance of the CI engine have been carried out by several researchers. Those parameters show the strong effect to the performance of the CI engine run in dual-fuel mode. Recently, Ambarita [9] investigated the optimum operational condition of the CI engine run in dual-fuel mode. The results show that brake thermal efficiency of the CI engine run on dual-fuel mode strongly affected by biogas flow rate and methane concentration. And there exist an optimum biogas flow rate for a maximum brake thermal efficiency. It was suggested to operate a CI engine at the optimum flow rate. In this study effect of the engine load and biogas flow rate of the biogas to the performance will be investigate. The main objectives are to explore the optimum operational condition of a CI engine run on dual-fuel mode. The results are expected to supply the necessary information on development of alternative solutions for reducing diesel oil consumption in CI engines.

IOP Conf. Series: Materials Science and Engineering **309** (2018) 012006 doi:10.1088/1757-899X/309/1/012006

2. Method

In order to carry out the study, a CI engine which typically used in a small tractor for agricultural work is used in the experiment. The specifications of the CI engine are presented in Table 1. It is an air cooled, single-cylinder four stroke CI engine fueled by diesel fuel. The maximum output power of the engine is 4.86 W. It is categorized as a small engine with and typically found in small Indonesian farmer.

Table 1. Specification of the CI engine

No	Parameter	Value
1	Commercial name/model	Tiger Diesel Engine R175 AN
2	Number of cylinder/stroke	Single-cylinder/4 strokes and Horizontal
3	Cooling system	Naturally air cooled
4	Bore × Stroke	75 mm × 80 mm
5	Maximum output	4.86 kW
6	Rated output	4.41 kW
7	Rated speed	2600 rpm
8	Engine weight	60 kg

In order to simulate the load, the engine is coupled with a single phase synchronous generator using pulley. The experimental apparatus is shown in Figure 1. To perform experiment, several additional equipments are installed to the experimental apparatus such as a series of lamps, biogas tank, gas mixer, and measurements apparatus. In single fuel mode (pure diesel oil only) the CI engine will be tested without modification. In dual-fuel mode, a gas mixer has been designed and developed in order to mix the fresh air with biogas. The mixture of the fresh air and biogas will be injected into the CI engine.

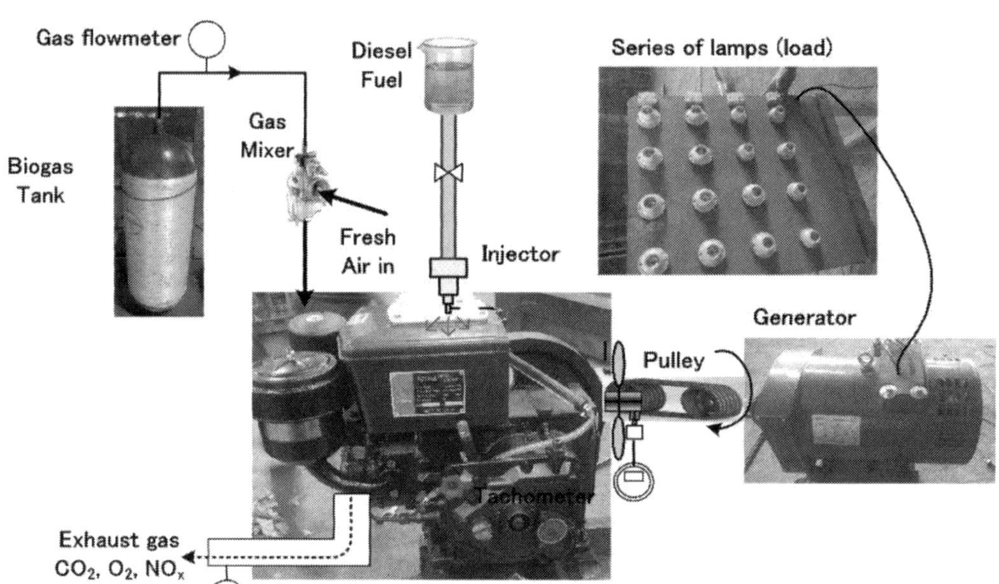

Figure 1. Experimental apparatus

The used generator has the following specification. The maximum power, rating frequency, and rating voltage of the generator are 3 kVA, 50/60 Hz, and 115-230 V, respectively. A series of lamps are used as a load for generator. The load can be operated at load 600W, 900W, 1200W, and 1500W. The biogas is stored in a tank. It is designed and fabricated to store the raw biogas until pressure 20 bar. As a note, Indonesia is known as a biggest produce of Palm Oil in the world [10]. The present biogas is produced from anaerobic biodegradation of Palm oil mill effluent. The effluent is taken from a mill located in Rambutan about 80 km from Medan city and it is operated by PTPN III (state owned plantation III). In this study, the biogas is refined to a higher composition of methane content. The composition of the present biogas consists of CH_4 70% and CO_2 30% of volume. This composition according to Gas Chromatography test. The low heating value (LHV) of the biogas is 20.56 MJ/kg. The diesel oil is purchased from PERTAMINA the government oil retailer in Indonesia. Before used, the LHV of the diesel oil is tested. The LHV is 42.64 MJ/kg.

The characteristics of the exhaust gas are investigated using parameter opacity, HC, and CO number. These parameters are measured using Gas Analyser and engine smoke meter. The model of the engine smoke meter is HD-410. Measuring range, absorption coefficient, oil temperature, and operation temperature of the smoke meter are 0.00-100%, 0.00 - 21.42 m^{-1}, 0 - 800 rpm, 0 - 100°C, and of -10°C - 40°C, respectively. The model of Gas Analyzer is HG-510. The specifications of the Gas Analyser are as follows. Measuring range of CO and HC are 0-9.99% and 0-9999 ppm, respectively. The operating temperature, power, and serial number are -10°C - 40°C, 220V, and 2G9C0101, respectively. Thermometer type of KW 06-278 Krishbow is used to measure the temperature. It's accuracy range is ± 0.5% ± 1°C. The load, electricity resulted by generator, is measured using a Multitaster Meter CD800A. The engine speed is measured using Tachometer.

The experiments are performed in two modes. In the first mode, the CI engine is operated with diesel fuel and named as pure diesel. In this mode, the load is varied from 600 W to 1500 W and engine speed is fixed at 1500 rpm. When the CI engine is stable, the measurement is carried out for 5 minutes for every load. The second mode, the CI engine is operated in the dual-fuel mode, when the CI engine is operated, the biogas from the tank is mixed with the fresh air in the mixer. The pressure of the biogas from the tank is decreased to 1.8 bar by using gas regulator. The flow rate of the biogas is varied from 2 to 6 Litre /minute (L/min). For every biogas flow rate, the load is varied and the speed is adjusted to get constant speed. The same measurements with the pure diesel are performed. Every test is replied for three times and the measurement is averaged.

2.1. Problem formulation

The measured data will be analyzed using several parameters. The performance of the engine will be analyzed using output power, brake thermal efficiency, and specific fuel consumption. The output power P_E (Watt) of the CI engine is calculated by using the following equation

$$P_E = V \times I \tag{1}$$

where V [Volt] and I [Ampere] are voltage and current electricity resulted by the generator, respectively. In the experiment, these parameters are measured using the Multitester Meter. Brake thermal efficiency is defined as electric power resulted by the generator divided by total energy input to the CI engine. For pure diesel mode, brake thermal efficiency is calculated by

$$\eta = \frac{P_E}{\dot{m}_{diesel} \times H_{diesel}} \tag{2}$$

where H_{diesel} is the diesel heating value in kJ/kg. While for dual-fuel mode, the brake thermal efficiency is given by

$$\eta = \frac{P_E}{\dot{m}_{dual} \times H_{diesel} + \dot{m}_{biogas} \times H_{biogas}} \tag{3}$$

Specific fuel consumption sfc [kg/kWh] is a comparison of fuel consumption to the useful energy resulted by the CI engine. Since the useful energy here is the electricity, the specific fuel consumption can be viewed as how many kilograms of fuel is needed to result 1 kWh of electrical energy. When the CI engine run in pure diesel mode, it is given by:

$$sfc = \frac{\dot{m}_{diesel} \times 10^3}{P_E} \tag{4}$$

In the case of dual-fuel mode, the fuel is defined as the total sum of diesel and biogas. The *sfc* will be given by:

$$sfc = \frac{\left(\dot{m}_{dual} + \dot{m}_{biogas}\right) \times 10^3}{P_E} \tag{5}$$

The objective of introducing biogas into the engine is to reduce the diesel fuel consumption. The biogas is used to replace diesel fuel. The percentage of diesel fuel replaced by the biogas is calculated, the replacement ratio [r] is calculated by

$$r = \frac{\dot{m}_{diesel} - \dot{m}_{dual}}{\dot{m}_{diesel}} \times 100\% \tag{6}$$

where $\dot{m}_{diesel}$ (kg/s) is the diesel mass flow rate when the engine is operated in pure diesel mode and $\dot{m}_{dual}$ (kg/s) is the diesel mass flow rate when the engine is operate in dual-fuel mode. Using the above formulated parameters, the performance of the CI engine will be analyzed.

3. Results and Discussions

3.1. Output power

Figure 2 shows the output power of the CI engine as a function of engine load. As a note, the load is varied from 600 W, 900W, 1200W, and 1500W, respectively. For every load, the engine speed is kept constant at 1500 rpm. It can be seen that the output power of the CI engine run in dual-fuel mode is higher than the CI engine run in pure diesel mode. This occurs for all biogas flow rate. There is no significant different of the biogas flow rate to the output power. In other words, the output power shown by the CI engine run in dual-fuel mode with biogas flow rate of 2 L/min is similar with the engine with biogas flow rate 4 L/min and 6 L/min. This is because the input energy to combustion chamber of the CI engine in dual-fuel mode is relatively higher than pure diesel mode. This fact suggest that in term of output power, the CI engine run in dual-fuel mode is better than pure diesel mode.

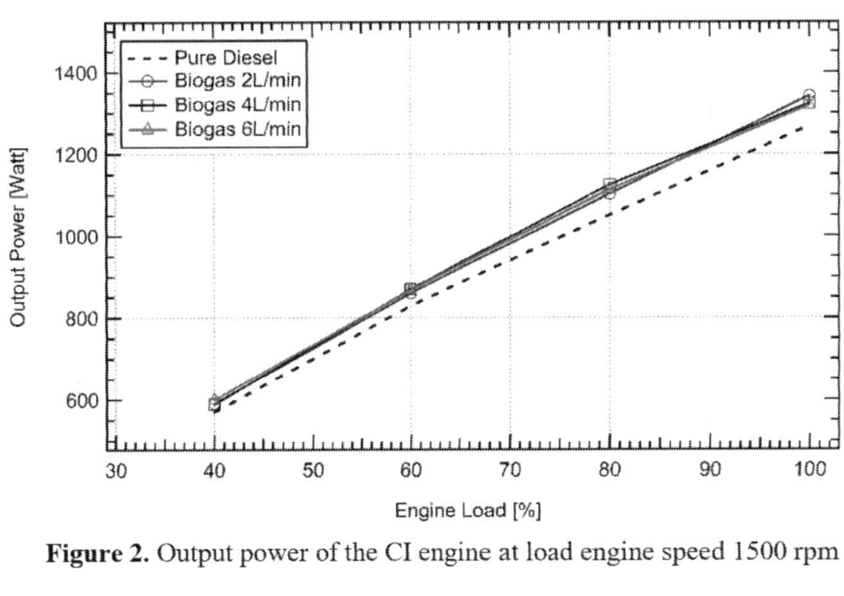

Figure 2. Output power of the CI engine at load engine speed 1500 rpm

3.2. Specific fuel consumption

The specific fuel consumption (*sfc*) as a function of output power is shown in Fig 3. In this case the engine speed is fixed at 1500 rpm. The specific fuel consumption can be viewed as a parameter to show how effective a power generation system to convert an amount of fuel into electrical energy. The lower *sfc* is the better. It can be seen that for all cases, *sfc* decreases as output power increases. This fact suggests that combustion process is better when the CI engine is loaded with high engine load. The effect of biogas flow rate to the *sfc* is clearly shown in the figure. The *sfc* is higher when the CI engine run in dual-fuel mode, in comparison with pure diesel mode. This is because all of the biogas is counted as a fuel. In fact, only 70% of the biogas is methane which content heat energy. This makes the *sfc* of the CI engine run in dual-fuel mode is higher than pure diesel mode. Furthermore, increasing biogas flow rate will increase the *sfc*. Again, this is because CO_2 without energy content is counted as fuel.

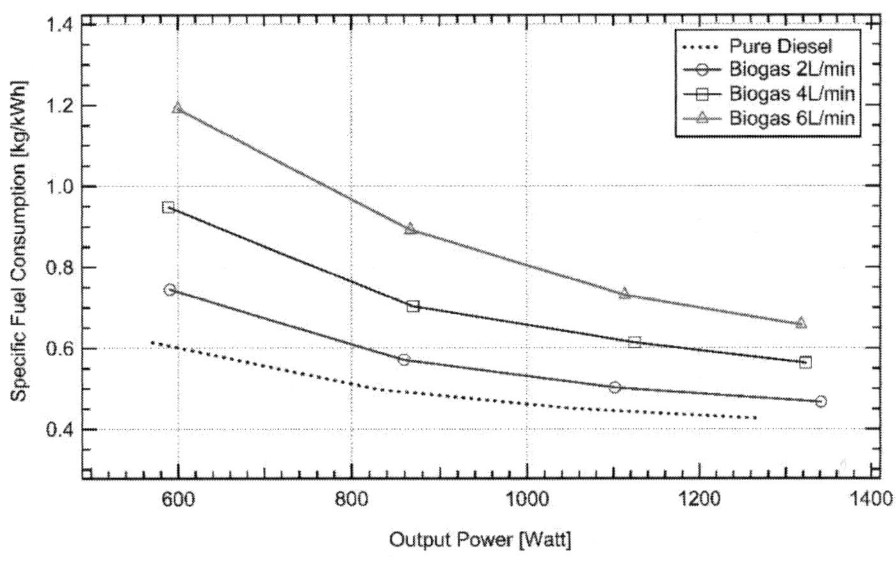

Figure 3. Power vs specific fuel consumption at constant engine speed 1500 rpm

3.3. Brake thermal efficiency

Figure 4 shows brake thermal efficiency of the present CI engine as a function of output power at constant speed of 1500 rpm. It can be seen from the figure that the brake thermal efficiency increases with output power. However, at higher loads the increasing rate of the brake thermal efficiency is decreased. This is because the present CI engine is operated below its maximum rated power. The effect of biogas flow rate to brake thermal efficiency is shown clearly in the figure. Increasing biogas from 2 L/min to 4 L/min will increase the brake thermal efficiency in comparison with CI engine run in pure diesel mode. However, increasing the biogas flow rate to 6 L/min will make the brake thermal efficiency lower than pure diesel mode. It is suggested to operate the present CI engine in dual-fuel mode with flow rate of biogas 2 – 4 L/min.

3.4. Exhaust gas emission

Table 2 shows emissions characteristics of the CI engine run in dual-fuel mode in comparison with pure diesel mode. The opacity numbers show that the CI engine run in dual fuel mode decrease the smoke content in the exhaust gas. The effect of biogas flow rate to the opacity of the exhaust gas can be examined using the table. Increasing biogas flow rate will decrease the opacity number. In the exhaust gas, higher CO related to less effective combustion process in the combustion chamber. In the CI engine run in pure diesel mode, CO number is very low. This is because the engine originally designed for pure diesel mode. Injecting biogas to the engine will decrease fresh air in the combustion chamber. This results in higher CO in the exhaust gas. Increasing biogas flow rate, slightly increases the CO number. The HC number in the exhaust gas related to unburnt fuel that is injected to the combustion chamber. As expected, HC number of the exhaust gas in the CI engine run in dual fuel mode is higher than pure diesel mode. This is because injecting biogas to the CI engine makes more fuel or HC in the combustion chamber. Since not all of the fuel burnt in the combustion chamber, HC number of the exhaust gas will increase. The HC number increases as biogas flow rate increases.

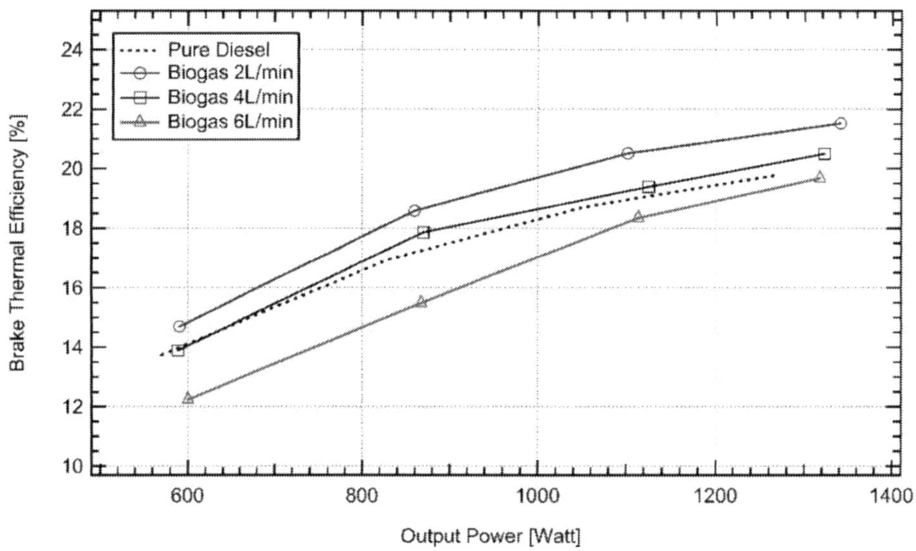

Figure 4. Brake thermal efficiency at constant engine speed 1500 rpm

Table 2. Emissions characteristics of the CI engine

Engine Load [W]	Pure diesel	Biogas 2L/min	Biogas 4L/min	Biogas 6L/min
	OPC Number			
600	13.5	9.0	3.2	1.3
900	26.1	13.9	7.0	6.6
1200	48.8	37.2	10.3	8.9
1500	60.9	64.5	30.2	12.8
	CO number [%]			
600	0.02	0.11	0.16	0.17
900	0.02	0.10	0.13	0.16
1200	0.03	0.08	0.12	0.15
1500	0.03	0.10	0.12	0.17
	HC number [ppm]			
600	6	190	273	325
900	6	176	220	295
1200	6	132	206	251
1500	8	126	173	215

3.5. Diesel replacement ratio

Figure 5 shows Diesel Replacement Ration (DRR) of the CI engine run in dual-fuel mode. When the engine run in single fuel mode, DRR is zero. When the CI engine run in dual-fuel mode the DRR varies from 30% to 20% and it decreases as output power increases. The same trend also shown by the cases with biogas flow rate 4 L/min and flow rate 6 L/min. As expected, increasing biogas flow rate increases DRR. However, the rate of growth is decreased. This suggested that it will be less effective increase the biogas flow rate more than 6 L/min.

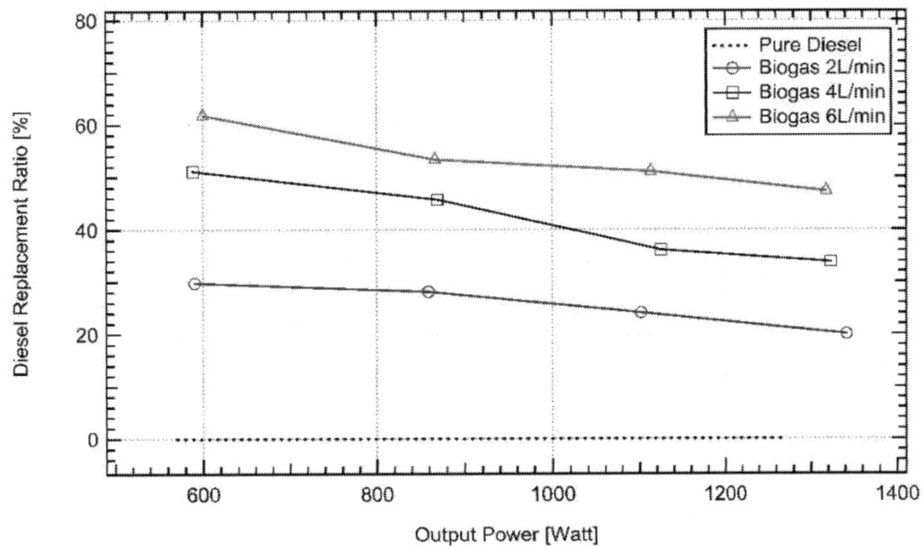

Figure 5. Diesel replacement ratio at 1500 rpm

4. Conclusions

A CI engine with rated power of 4.4 kW has been tested on pure diesel and dual-fuel mode with refined biogas. The methane content of the biogas is 70%. The engine was tested at constant speed 1500 rpm while the load varies from 600 to 1500W. The conclusions are as follows. The output power of the CI engine run in dual-fuel mode is higher than pure diesel mode. Brake thermal efficiency of the CI engine run in dual-fuel mode is better than pure diesel mode if the biogas flow rates are 2 L/min and 4 L/min. The exhaust gas analysis shows biogas reducing opacity of the exhaust gas. On the other hand, the HC number of the engine run in dual-fuel mode is very high in comparison with pure diesel mode. It is recommended to operate the present engine in dual-fuel mode with biogas flow rate of 4 L/min. The consumption of diesel fuel can be replaced up to 50%.

Acknowledgments
The authors gratefully acknowledge that the present research is supported by Ministry of Research and Technology and Higher Education Republic of Indonesia. The support is under the research grant BP-PTN USU of Year 2016. The support from my students Arie Matius P P Bangun and Gunung A. E. Sinambela.

IOP Conf. Series: Materials Science and Engineering **309** (2018) 012006 doi:10.1088/1757-899X/309/1/012006

References

[1] Government of Indonesia Presidential Decree No No 61 Year 2011, National Action Plan For Reducing Greenhouse Gas Emissions
[2] Minister of Energy and Mineral Resources, 2011, Handbook of Energy and Economic Statistics of Indonesia, 2011
[3] Henham A and Makkar M K 1998 *Energy Conversion Management* **39**, 2001-2009
[4] Ambarita A, Sinulingga E P, Nasution M K M and Kawai H 2017 *IOP Conference Series: Material Science and Engineering* **180(1)**, 012025
[5] Ambarita H, Widodo T I and Nasution D M 2017 *Journal of Physics: Conference Series* **801(1)**, 012095
[6] Feroskhan M, Ismail S, Kumar A, Kumar V and Aftab S K 2017 *Biofuels* **8(2)**, 197-205
[7] Abdul Rahman K and Ramesh A 2017 *Fuel* **206**, 117-132
[8] Verma S, Das L M and Kaushik S C 2017 *Energy Conversion and Management* **138**, 346-359
[9] Ambarita H 2017 *Case Studies in Thermal Engineering* **10**, 179-191
[10] Irvan, Trisakti B, Sosanty F, and Tomiuchi Y 2016 *Asian J. Chem.* **28** 377-380

Experimental study on a prototype of heat pipe solar water heater using refrigerant R134a as a transfer fluid

T Sitepu[1], J Sembiring[2] and H Ambarita[1*]

[1]Department of Mechanical Engineering, Faculty of Engineering, Universitas Sumatera Utara, Medan, Indonesia 20155
[2]Department of Mechanical Engineering, Politeknik Santo Thomas, Matahari Raya Street 84A, Medan, Indonesia 20124

* E-mail: himsar@usu.ac.id

Abstract. A prototype of a solar water heater by using refrigerant as a heat transfer fluid is investigated experimentally. The objective is to explore the characteristics and the performance of the prototype. To make heat transfer from the collector to the heated fluid effectively, refrigerant R134a is used as a transfer. In the experiments, the initial pressure inside the heat pipe is varied. The prototype is exposed to solar irradiation in a location in Medan city for three days of the experiment. Solar collector temperatures, solar radiation, water temperature, and ambient temperature are measured. The efficiency of the system is analyzed. The results show that temperature of the hot water increases as the initial pressure of the working fluid increase. However, the increasing is not linear, and there must exist an optimum initial pressure. For the case with the refrigerant pressure of 110 psi, the maximum hot water temperature and maximum thermal efficiency are 45.36oC and 53.23%, respectively. The main conclusion can be drawn here is that solar water heater by using refrigerant R134a should be operated at initial pressure 110 psi.

1. Introduction

Global warming is the increasing average temperature of the atmosphere. This is caused by all activities in the globe that emit Greenhouse Gases (GHGs). To avoid the human being from the catastrophe, many countries have committed in reducing the GHG emissions. According to Intergovernmental Panel on Climate Change (IPCC), the GHG emissions can be divided into four sectors. They are Energy, Industrial Process and Product Use (IPPU), Agriculture, Forestry, and other Land Use, and Waste sectors [1]. In the energy sector emissions resulted by all activities in burning fossil fuel to provide heat and power. One of the activities that use heat is the water heater. The water heater typically powered by fossil fuels and electricity. The subsectors that need water heater includes hotels, hospital, restaurant and residential. Indonesia has released two different targets in reducing its GHG emission [2,3]. To reach the target, one of the potential mitigation actions is using alternative energy to provide solar water heater. The promising alternative energy resource to provide water heater is solar energy. Because it is clean and sustainable. Thus, studies on solar water have received much attention from researchers.

Hossain et al. [4] have reported their review on solar water heater collector and thermal energy performance of circulating pipe. The study found that the solar water heater with siphon system achieves system characteristics efficiency of 18% higher than that of the conventional system by reducing heat loss from the thermos-siphon solar water heater. There are many types of solar water heaters found in literature, such as an evacuated tube, simple box water heater, and heat pipe type solar water heater. In this study, we focus on the heat pipe type solar water heater. Jaisankar et al. [5] reported their study on a comprehensive review on the solar water heater. The conclusions are as follows. More researches may be initiated in thermosyphon solar water heaters to improve the performance. Research work in parallel flow will improve and will give a new insight into thermal performance, variation in flow velocity of the working fluid in the riser tubes can be made uniform using variable headers, and the convective heat loss from glass cover may be reduced using a suitable aero profile design that will prevent the movement of air over glass surface.

The above reviews show that there several types of solar water heater. The solar water heater using two-phase transfer fluid has received more attention due to high performance. Here, a working fluid (or transfer fluid) is employed to transfer the solar heat from solar collector to the hot water. Chen et al. [6] reported an experimental work on the long-term thermal performance of two-phase thermosyphon solar water heater. They proposed a new system in the study. The results show that the proposed system achieved system characteristics efficiency 18% higher than that of the conventional systems is reducing heat loss for the two-phase thermosyphon solar water heater. Chun et al. [7] reported an experimental study on employing heat pipe for solar water heater. The result shows interesting performance data stemming from the difference in working fluids, the presence of wick, and other various design parameters associated with the collection and utilization of solar energy. Arab et al. [8] proposed vibrated heat pipe solar water heater. In the experimental work, it was shown that the best performance of the proposed model resulted by pipe filled with 70% transfer fluid. This model is stable and shows longer durability. The comparison of thermosyphon and vibrated heat pipe solar water heater are made. The average temperatures of hot water resulted by thermosyphon and vibrated heat pipe type is 35.5°C and 34.5°C, respectively. Studies of the performance of different fluid in heat pipe type solar water heater are found in the literature. Esen and Esen [9] studied experimentally the performance of heat pipe solar water heater using different refrigerant as transfer fluid. The refrigerants are R410a, R407C, and R134a. The results showed that higher performance is shown by R410a. Josep and Yakasai [10] reported the performance comparison of R12, R134a, and ethanol as a transfer fluid in the heat pipe type solar water heater. The results revealed that R134a show a better efficiency and higher hot water temperature.

Those reviewed literature showed that transfer fluid plays an important role in the performance of heat pipe solar water heater. It is suggested to employ R134a due to high performance. In this study, we investigate the effect of the internal pressure of the heat pipe when the refrigerant R134a is used as a transfer fluid. The system will be tested by exposing to solar irradiation in Medan city of Indonesia. The objective is to find the optimum operating condition of the heat pipe solar water heater using R134a as a transfer fluid. The results are expected to supply the necessary information on development high-performance solar water heater.

2. Methods

In order to perform the experiments, three identical prototype heat pipe solar water heaters are designed and fabricated. The main reason to fabricate three solar water heaters is to make that all equipment are tested under the same solar irradiation at the same day. As a note, typically the solar irradiation is different every day due to weather condition. Figure 1 shows the schematic diagram of the experimental setup. The solar water heater consists of a flat plate type solar collector, heat pipe, and hot water tank. The collector area of each solar water heater is 1000 mm × 500 mm. The thickness is 200 mm. To reduce heat loss to the surrounding a double-glasses cover and rockwool insulation are used. The insulation of 100 mm thickness is used in the bottom and side walls. The distance of glass cover to plat absorber is 100 mm. The heat pipe is placed on the absorber plate with a distance of 100 mm. Several parts of the heat pipe are passed through a water tank. The water tank with a volume of 5 liters is placed beneath the top part of the solar collector. In order to perform experiment, a data acquisition system is used. The system consists of HOBO micro station data logger and temperature measurement system. There are 20 thermocouples are used to measure temperatures. The sensors are made of *J*-type thermocouples with uncertainty equal to 0.1°C. A datalogger (Agilent 34972) consists of 20 thermocouple channels are used to record temperatures with interval of 1 minute. The solar radiation is measured using HOBO pyranometer smart sensor. The ambient temperature and relative humidity (RH) is measured using HOBO temperature RH smart sensor with an accuracy of 0.2°C and ± 2.5% RH, respectively. The wind speed around the experimental apparatus is measured with HOBO wind speed smart sensor with an accuracy ± 1.1 m/s.

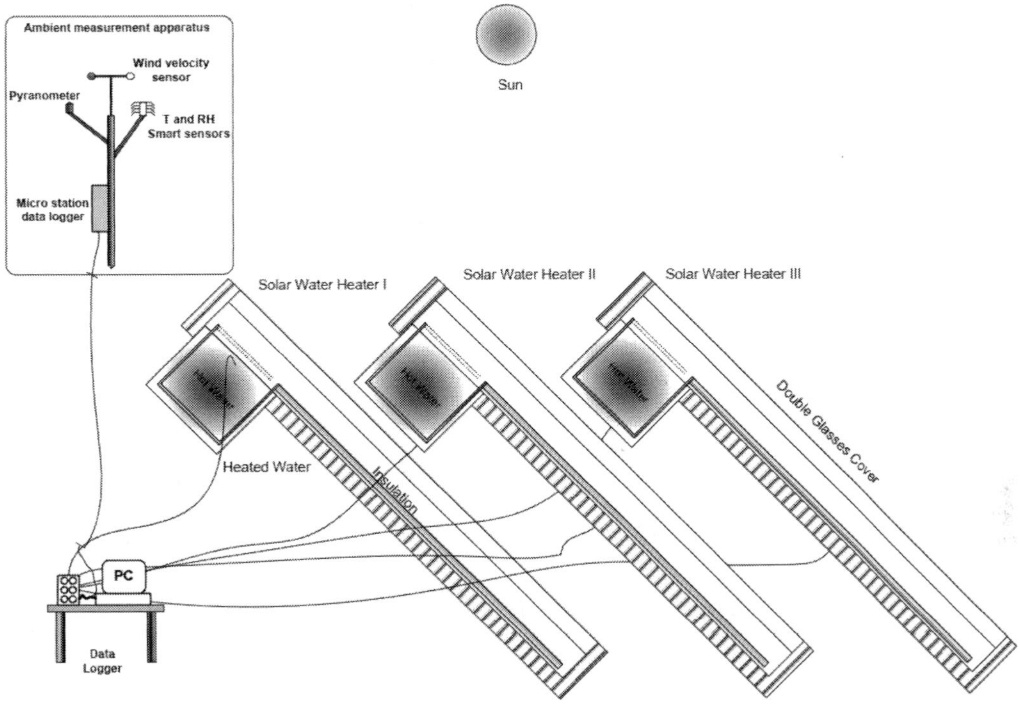

Figure 1. Schematic diagram of solar water heaters and data acquisition system

The working mechanism of a heat pipe solar water heater is explained as follows. The refrigerant as a transfer fluid filled the pipe. Before the system is tested, the pressure of inside the pipe is regulated. Hereafter, the pressure named as initial pressure. In this study, three different initial pressure will be investigated, they are 90 psi, 100 psi, and 110 psi below the atmosphere. When it is exposed to solar irradiation, the received heat will evaporate the refrigerant. The refrigerant vapor flows up to the hot water tank. In the tank, the water cooled the refrigerant vapor and converted it into liquid refrigerant. The liquid refrigerant will flow down and cycle repeats. In the experiments, temperatures and solar irradiation are measured, and the performance is analyzed. The performance of the present water heater will be compared using two parameters; they are maximum temperature and thermal efficiency. The maximum temperature is defined as the highest temperature of the water in the hot water tank. On the other hand, the thermal efficiency is defined as the ratio of useful heat in hot water and solar energy collected by the absorber plate. The thermal efficiency is calculated using the below equation.

$$\eta = \frac{Q_u}{Q_{in}} \tag{1}$$

where Q_u (Joule) and Q_{in} [Joule] are useful energy in hot water and total solar energy incident to the solar absorber plate, respectively. These parameters can be calculated using the following equations.

$$Q_u = m_w c_w \left(T_w - T_i\right) \tag{2}$$

where m_w (kg), c_w (J/kgK), T_w (°C), and T_i (°C) are mass of the hot water, specific heat capacity, temperature of the water, and initial temperature, respectively. While the solar energy incident is given by

$$Q_i = A\int I dt \tag{3}$$

where A (m^2) and I (W/m^2) are the area of the solar collector and solar irradiance, respectively.

3. Results and Discussions

The solar water heaters have been exposed to the solar irradiance in Medan city of Indonesia for three different days. Temperatures and solar irradiance are measured. The results will be discussed in terms of characteristics, maximum temperature comparison, and thermal efficiency.

3.1. Characteristics of the solar water heater

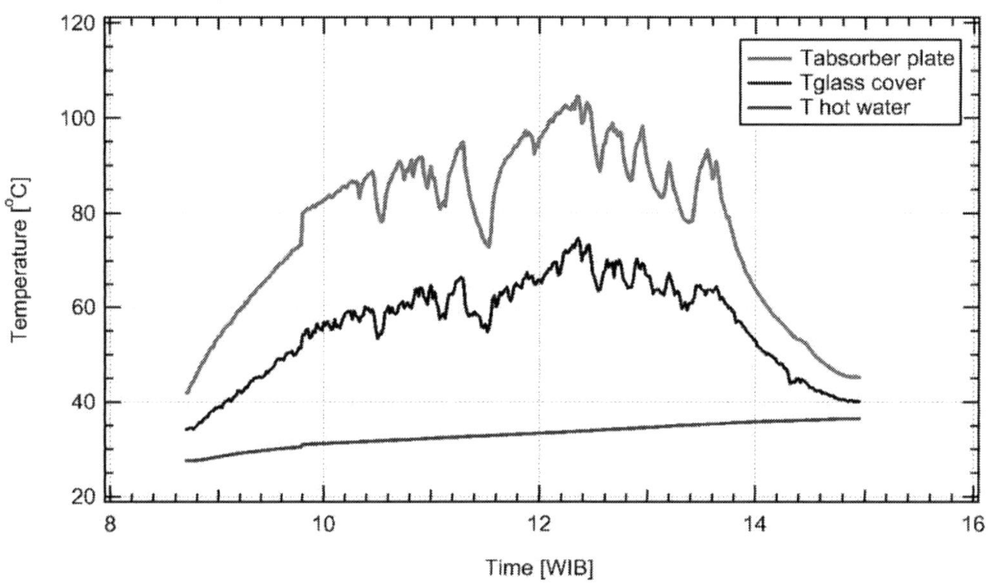

Figure 2. Temperature history of solar water heater during experiment

The characteristic of the solar water heater is presented using temperature history of a solar water heater during the experiment. In the experiment, there are nine different temperature histories have been measured. Figure 2 shows temperature history of the solar water heater for day one at an initial pressure of 90 psi. The measured solar irradiation of day one is started from 8:30 local time. The initial solar irradiation was 161.9 W/m^2, and the maximum solar irradiation was 620 W/m2 captured at 14:00 local time. Total solar energy radiation for the Day 1 of the experiment was 1.204 MJ.

The figure shows that temperature of absorber plate increases with increasing the time until reaching its maximum value of 104°C at 12: 34 local time. After this, the temperature of the absorber plate decreases with increasing time. As expected, the temperature of the plate absorber can reach the temperature above 100°C due to the effectiveness of insulation and glass cover. The temperature of the glass cover shows the similar trend with the temperature of plate absorber but with different maximum value. The maximum temperature of glass cover was 74.6°C at the same time 12:35 local time. During the experiment, the temperature of the glass cover is lower than the temperature of the absorber plate. This is because heat loss from the absorber plate to ambient through the glass cover is still significant. The solar energy collected by solar collector will be used to heat water in the tank, and the temperature of the water will increase. This fact shown in the figure. It can be seen that the temperature of the water increases with increasing time. In this case, it starts from 25.4°C and reaching its maximum value of 36.47°C. Thus, the total energy utilized by the solar collector was 0.186 MJ. By using equation (1) thermal efficiency of the system is 23.25%.

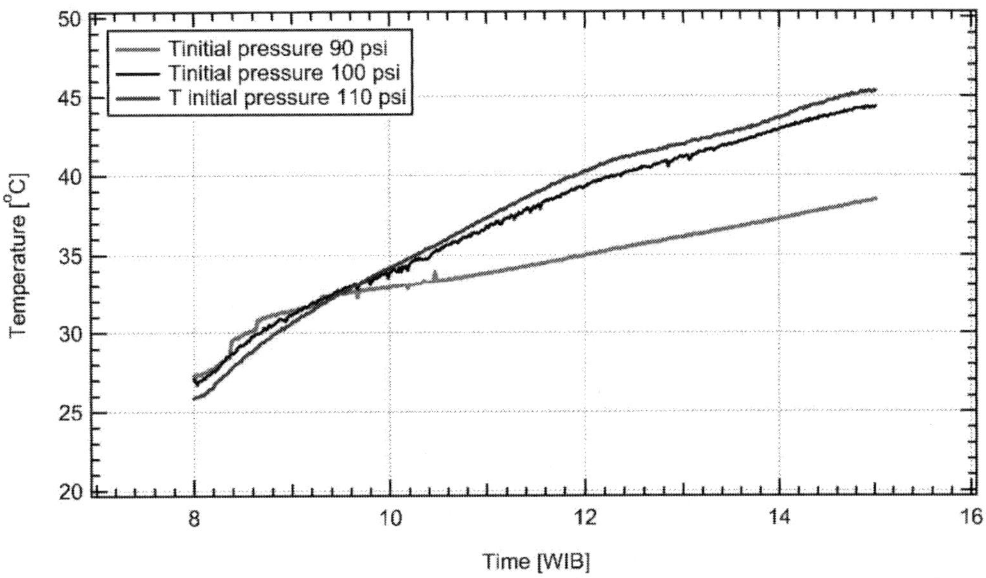

Figure 3. Efect of initial pressure to the maximum hot water temperature

As mentioned in the method, in every experiment three solar water heaters are exposed at the same time. In this experiment, the other solar water heaters are filled with the same refrigerant but different initial pressure. Figure 3 shows temperature history of the hot water heater at initial pressure 90 psi, 100 psi, and 110 psi. It can be seen clearly that, there is a strong effect of initial pressure inside the heat pipe to the performance of the system. The maximum temperatures of the hot water for initial pressure 90 psi, 100 psi, and 110 psi are 36.47°C, 40.55 °C, and 42.87 °C, respectively. This is because when the initial pressure far below the atmosphere, temperature evaporation will be lower. In other word, the amount of refrigerant that evaporated will be higher. It leads to higher solar energy can be transferred to the hot water tank. The increasing of initial pressure is not linier to increasing maximum temperature. Increasing initial pressure 10 psi from 90 to 100 psi will increase maximum temperature 4.08°C. However, increasing the same initial pressure of 10 psi from 100 psi to 110 psi, maximum temperature will increase only 2.32°C. This reveal that there must exist an optimum initial pressure for maximum performance.

3.2. Effects of the initial pressure

In order to make a clear effect of initial pressure to the performance of the solar water heater, additional two similar experiments were performed. In the experiment, all of the condition and initial pressure are the same. The different is only the solar irradiation, which is impossible to provide the similar solar irradiation in the different day. Table 1 shows the performance of the solar water heater during three days of experiment. It can be seen clearly that there a clear effect of initial pressure to the performance of the solar water heater. The table shows that maximum temperature during experiments varies from 36.47°C to 45.36°C. The maximum temperature of the hot water for day 1, day 2, and day 3 of experiments are 42.87°C, 45.36°C, and 44.42°C, respectively. These maximum values are shown by the same initial pressure of 110 psi. The fact that additional initial pressure from 90 psi to 100 psi is more effective than additional initial pressure from 100 psi to 110 psi is also shown in the table.

Table 1 Effects of the initial pressure on the performance of solar water heater filled with R134a

Experiment	Initial Pressure (psi)	Initial Temperature (°C)	Maximum Temperature (°C)	Thermal Efficiency
Day I	90	25.54	36.47	23.25%
	100	25.84	40.55	42.36%
	110	25.01	42.87	53.23%
Day II	90	27.31	38.48	21.86%
	100	27.08	44.35	40.24%
	110	27.87	45.36	45.42%
Day III	90	26.69	37.93	20.13%
	100	26.43	42.28	34.34%
	110	26.16	44.42	43.98%

Thermal efficiency of the solar water heater is also shown in Table 1. As a note, the thermal efficiency is calculated using equation 1. Thus, the thermal efficiency is affected by incident solar energy and maximum temperature of the hot water. In other words, the highest maximum temperature of hot water is not directly the gives the highest thermal efficiency. The present results are comparable with the results reported by Esen and Esen [9]. In their work, by exposing the solar water heater in Elaziq Turkey, for refrigerant R134a the maximum temperature of the hot water varies from 28°C to 50°C while the thermal efficiency varies from 48% to 60%. There are discrepancies of the present result and results reported by Esen and Esen [9]. This is because different of whether condition and different operational method. In their work, the hot water is replaced several times during experiment. In this work, it is without water replacement. As shown in the table the highest maximum temperature of hot water was captured in the Day 2 of experiment it is 45.36 °C. In this case the thermal efficiency is 45.42%. On the other hand, the highest thermal efficiency was captured in Day 1, it is 53.23%. In this case, the maximum hot water temperature is 42.87°C.

4. Conclusions

In this work, a prototype heat pipe type solar water heater filled with R134a as a transfer fluid has been designed and tested by exposing under solar irradiation in Medan city of Indonesia. After exposing for three days of experiments, the conclusions are as follows. The refrigerant R134a as a transfer fluid can deliver the heat from the solar collector plate and transfer it to the hot water. The maximum hot water temperature varies from 36.47°C to 45.36°C and thermal efficiency varies from 20.13% to 53.23%. There is a strong effect of initial pressure inside the heat pipe. Increasing initial pressure will increase maximum temperature of the hot water. It is suggested to operate the present solar water heater with initial pressure of 110 psi.

Acknowledgments
The authors gratefully acknowledge that the present research is supported by Ministry of Research and Technology and Higher Education Republic of Indonesia. The support is under the research grant DRPM of Year 2017.

References
[1] IPCC, 2006, Guidelines for National Greenhouse Gas Inventories.
[2] Gov. of Indonesia PresidentialY Decree No No 61 Year 2011, National Action Plan For ReducingGreenhouse Gas Emissions.
[3] Gov. of Indonesia, 2016, First Nationally Determined Contribution Republic of Indonesia.
[4] Hossain M S, Saidur R, Fayaz H, Rahim N A, Islam M R, Ahamed J U and Rahman M M 2011 *Renewable and Sustainable Energy Reviews* **15**, 3801-2812.
[5] Jaisankar S, Ananth J, Thulasi S, Jayasuthakar S T and Sheeba K N 2011 *Renewable and Sustainable Energy Reviews* **15(6)**, 3045-50.
[6] Chen B R, Chang Y W, Lee W S and Chen S L 2009 *Solar Energy* **83**, 1048-1055
[7] Chun W, Kang Y H, Kwak H Y and Lee Y S 1999 *Applied Thermal Engineering* **19**, 807-17
[8] Arab M, Soltanieh M and Shafii M B, 2012 *Experimental Thermal and Fluid Science* **42**, 6-15
[9] Esen M and Esen H, 2005 *Solar Energy* **79**, 459 - 468
[10] Eneburekhan J and Yakasai U T, 2008 *Desalination* **243**, 208-217

IOP Conf. Series: Materials Science and Engineering **309** (2018) 012008 doi:10.1088/1757-899X/309/1/012008

The study of residential life support environment system to initiate policy on sustainable simple housing

N M Siahaan[1*] , A S Harahap[2], E Nababan[3] , E Siahaan[4]
[1]Department of Architecture, Universitas Sumatera Utara, Medan, Indonesia
[2,3] Department of Mathematics, Universitas Sumatera Utara, Medan, Indonesia
[4] Delft University of Technology, The Netherlands

*E-mail: nelson@usu.ac.id

Abstract. This study aims to initiate sustainable simple housing system based on low CO_2 emissions at Griya Martubung I Housing Medan, Indonesia. Since it was built in 1995, between 2007 until 2016 approximately 89 percent of houses have been doing various home renewal such as restoration, renovation, or reconstruction. Qualitative research conducted in order to obtain insights into the behavior of complex relationship between various components of residential life support environment that relates to CO_2 emissions. Each component is studied by conducting in-depth interviews, observation of the 128 residents. The study used Likert Scale to measure residents' perception about components. The study concludes with a synthesis describing principles for a sustainable simple housing standard that recognizes the whole characteristics of components. This study offers a means for initiating the practice of sustainable simple housing developments and efforts to manage growth and preserve the environment without violating social, economics, and ecology.

1. Introduction

The increase population of major cities in Indonesia demands a huge amount of housing and requires them to be implemented in a sustainable manner that does not harm the environment, social life, economy, and culture [1]. In the meantime, little attention has been given either by government agencies and society to putting sustainable housing policy into practice, particularly for low CO_2 emissions in the middle and lower income residential area. This emission of CO_2 is not merely caused by the process of construction, but also by the whole dynamic of systems in the simple housing areas. [2].

According to Ministry of Housing Strategic Planning Year 2015-2019, Indonesia's upcoming five-year plan is to build 1.367 million units of subsidized housing. This plan brings 2 (two) fundamental questions. First, can the construction of settlements and large-scale housing can be sustained? Second, what is the impact for a sustainable city? These questions raise a number of challenges and difficulties. In particular, cities now grow and develop increasingly complex, and thus self-sustaining housing problem is further complicated [3].

Efforts to minimize the emission of CO_2 in urban residential area is becoming more challenging in terms of planning, designing and building more sustainable housing. The processes of land development for housing, construction, house alterations, and finally house demolitions are also processes that vastly damage the environment, especially related to the CO_2 emission produced [4]. Furthermore, the generation of CO_2 emission is also presented in the implementation of simple housing, due to the manufacturing process of building materials and the transportation of these materials, the use of equipment during the construction process, and household activities when the house is inhabited [5], [2].

This study brings forth issues that are not fully yet adopted by the general design of simple housing, particularly ideas on how components of residential life support environment in simple housing can be regulated in initiating sustainable simple housing. This study develops a standard for introducing sustainable simple housing system based on low CO_2 emissions, using the case of the simple housing at Griya Martubung I Medan, Indonesia in order to obtain insights into the behavior

IOP Conf. Series: Materials Science and Engineering **309** (2018) 012008 doi:10.1088/1757-899X/309/1/012008

of complex relationships between various components of residential life support environment that relates to CO_2 emissions. Through this study, we initiate good environmental planning and design instruments to prevent and mitigate the increase in CO_2 emissions. This study deliberately discusses strategies to initiate low CO_2 emissions of simple housing.

2. Method

This study has been conducted using both quantitative and qualitative techniques. The quantitative technique is applied to study and analyze the residential life support environment system of Griya Martubung I Housing, such that we could determine which components have most to affect the sustainable housing. Descriptive qualitative approaches were then carried out to explain problems which may be encountered by homeowners with respect to the residential life support environment system of Griya Martubung I Housing. Interviews with homeowners have been conducted to get the first hand information on the understanding the need for sustainable simple housing concept and problem in Indonesia. The location of this study was at Griya Martubung I Simple Housing, Medan (see Figure1.)

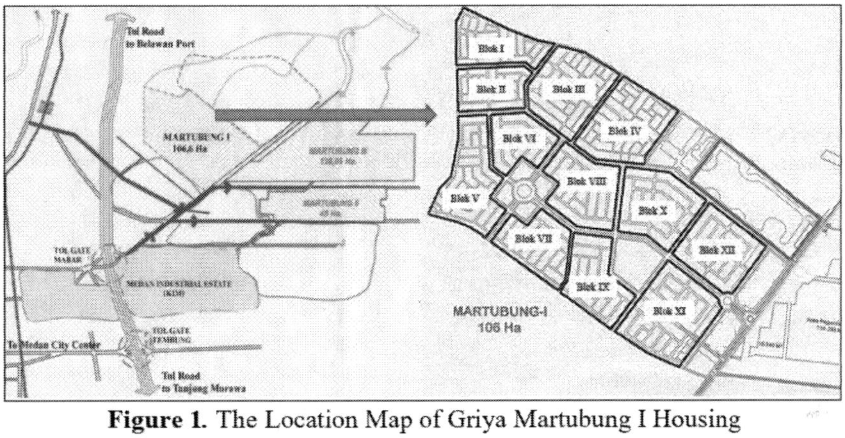

Figure 1. The Location Map of Griya Martubung I Housing

128 questionnaires were distributed to determine residents' level of awareness and understanding on residential life support environment system concepts for sustainable simple housing at Griya Martubung I. The dwellers were selected from 12 blocks of housing at Griya Martubung I as the survey's target. Likert scale was used to measure their perceptions. The variables, indicators and methods used in this study were shown in Table 1.

Table 1. Variables, Indicators & Methods

Dependent Variable	Independent Variable	Methodology
Social and public facilities	• Seating facilities • Lighting facilities • Pedestrian facilities • Accessibility	• Questionnaire • Interview • Field Observation • Photograph
Commercial facilities	• Seating facilities • Lighting facilities • Pedestrian facilities • Accessibility	• Questionnaire • Interview • Field Observation • Photograph
Green open space	• Comfort • Safety • Cleanliness • Interestingness/appeal	• Questionnaire • Interview • Field Observation • Photograph

	• Comfort for pedestrians • Accessibility	
Temporary landfill	• Safety • Cleanliness • Interestingness/appeal • Comfort for pedestrians • Accessibility	• Questionnaire • Interview • Field Observation • Photograph
Man-made pool	• Comfort • Safety • Cleanliness • Interestingness/appeal • Comfort for pedestrians • Accessibility	• Questionnaire • Interview • Field Observation • Photograph

3. Results and Discussions

3.1. Creating Low CO_2 Emissions Of Sustainable Simple Housing

In Indonesia, physical changes of housing in the form of restoration, renovation, or reconstruction are part of housing development dynamics, and they happen in conjunction with the increase of infrastructure demand due to the growth and development of the housing residents' needs. House alterations and the dynamics of residential life in simple housing are a complex interaction system that is associated with the production of CO_2 emission to air [2]. This emission of CO_2 is not merely caused by the process of construction, but also by the whole aspect of utilizing the space in residential areas [6].

The emission of CO_2 produced by a housing is closely related to activities that use non- renewable energy resource. The greater each activity's dependency on energy usage, the greater the emission of CO_2 produced by an urban housing's implementation system [7]. Through utilization and regulation of space, a sustainable simple housing with low CO_2 emission can substantially be built. In relation with Griya Martubung I Housing, the livelihood activities of its residents rely on every resource in and around the housing areas. The links between the various components of the residential life support environment system play a part in producing CO_2 emission, and determining the initiation of sustainable simple housing. The various components arise from space utilization and regulation in the housing implementation and inhabiting, and includes access, green open space, structure plan and hierarchy, pedestrian, utilities, public facilities, social and commercial facilities, temporary landfill, man-made pool.

Observation done from July 2016 until September 2016 on house alterations at Griya Martubung I shows that in the period of 2007 until 2016, 89% of homes (2830 housing units), houses of type 29, type 36 and type 54 went through a variety of physical changes such as; restoration, renovation, and reconstruction. Table 2. shows that the overall changes of these houses are accompanied by the generation of huge amount of CO_2 emissions into the air [8], [9]. It should be noted, however, that the generation of CO_2 emissions is not only caused by the construction of space utilization in the residential area [6], [10].

A study done by Afiaty [11] shows that alterations to bigger types of houses such as type 54 in Griya Martubung I are no longer affected solely by the need of room functionality, but also by the purpose of inner space aesthetics. Alterations in the form of adding rooms or extension of building that differs from the house's original structure will change the existing building structure and material. Table 3 explains some underlying factors that may be controlled to reduce CO_2 emission at Martubung I Housing. The change of building structure and material produces CO_2 emission to air. Therefore, efforts for initiating sustainable housing can be done only if every underlying factors of house alteration can be controlled. The design of home system components can affect the increase of carbon generation in case of repair activities, changes, or additions to area of the home building. In addition, various activities in the house uses air conditioning and light, which also have an mpact on the increase in CO2 emissions.

Tabel 2. Total CO_2 emissions result from Physical Changes At Griya Martubung I Housing from 1995 - August 2016

No	Housing Type	Renovation		Restoration		Reconstruction		Total Unit
		Unit	Kg-C	Unit	Kg-C	Unit	Kg-C	
1.	29	223	121,606	197	259,748	163	690,126	583
2.	36	615	463,366	509	671,113	315	1,336,678	1439
3.	54	348	589,359	255	336,223	205	867,949	808
	Total	1186	**1,174,331**	961	**1,267,084**	683	**2,894,753**	**2830**

Tabel 3. Matrix of Factors of House Alterations

No.	Underlying Factor	Data	Reasoning	Restraining Factors
1.	Number of head of households	Data of the number of head of households at every block and field survey	If the number of head of household in a house increase, then the member of the family and need for room increases too	The awareness that a house should be for one head of household needs to be developed and socialized to members of community.
2.	Average number of residents between 4 and 5 people	Data of the number of head of households at every block and field survey	The bigger the number of house residents, the bigger the number and area of rooms needed	House type does not confirm to the number of family members. The type 29 for 3-4 family members, type 36 for 4-5 family members and type 54 for 5-6 family members
3.	The majority of residents work as civil officers (47%) and private employees (41%)	Field survey August 2016–September 2016	A steady job gives opportunity to residents to do house alternations	Careful access to funding for house construction / repair, even though there is a steady job
4.	55% of Griya Martubung I residents have high school education	Field survey August 2016–September 2016	The higher the resident's education level, the bigger the awareness of need for rooms	High level of education affects the amount of income, and the awareness to save energy
5.	The majority of residents are 40 - 50 years old	Data of number of head of households at every block, and field survey	With the big number of residents in their productive ages, horizontal activities among the residents more intensive, and increase the need for social rooms among residents	Activities for productive ages are mostly outdoor activities. Awareness social-cultural relationships among residents develop

One of the contributing factors of house alteration dynamics at Griya Martubung I is the number of householder, or the increasing number of residents in a house. More residents in a house certainly demands for a larger house area with more complex building scheme. In consequence, house alteration needs construction technology to condition the space, especially to handle ventilation, lighting, and house utility systems that become more complex with the addition of building area. The whole construction technology system affects energy utilization at residential areas, and certainly does not conform to the generation of sustainable simple housing scheme.

Meanwhile, to determine the minimum space requirements for each person that resides in a house required the basic activities of the residents. Table 4 shows a relationship matrix between a house with its residents and the house alterations done by the residents. According to Undang Undang Republik Indonesia No. 1 Tahun 2011, on Housing and Settlement Areas section 22 verse 3, the standard for a house's floor area is at least 36 (thirty six) square meter, with the assumption that the space area requirement per person is 9 (nine) square meter, and the average height of the ceiling is at least 2.80 m.

Since its establishment in 1995, a big number of house alterations have been made at Martubung I Housing. The reason for the alterations is that the residents needed to extend the area of rooms available. The house owner performs alterations to be able to support indoor activity needs of the house residents. Other than the need for additional room, relatively bad building material quality is also one of the underlying factors of house alterations. Table 4 shows some factors that relate with building material changes in correlation with house residents. There are a couple reasons for house residents changing the building material of their house. The reason of material change by residents of Martubung I Housing is mainly because the quality of building material does not yet conform to the residents' preferences, and because of the age of the building material. The whole alteration done needs energy that causes increase of CO2 emission. The residents' accessibility capacity varies and comes from diverse sources. The various types of financing sources for house alterations affects the end result of house alterations. Figure 2 illustrates the relationship scheme between residents' income with building material and CO_2 emission produced from the house alteration.

Tabel 4. Matrix of House Residents and House Alterations Correlation

Underlying Factors	Effect to House	Alteration Done by Residents
Quality of building material generally low and not conforming to standards	Shorter age of building material resulting in recurrent need for changing, and low quality of construction	House residents do repair by changing building material (restoration)
Design of house does not conform to the preference and need of residents	• Low quality of building structure • Insufficient room availability • Increase need of rooms • Decline of comfort inside of house	• Do renovation to change house structure • Redesign by repositioning certain room functions, such as the kitchen, bathrooms. • Adding room sand extending house • Redesign of windows
Implementation of house construction does not conform to building standards	• Bad quality of house • Residents desire for a new house structure	House reconstruction, by building a totally different house from the original (former house structure)

3.2. Griya Martubung I Housing As Residential Life Support Environment System

Since the Griya Martubung I Housing is about 20 kilometers to the city center of Medan, residents always require an easy access in and out of the housing. From observation and interview, it is acknowledged that in general, the dwellers of Griya Martubung I Housing use private transport instead of public transport for their daily activities. Figure. 3 illustrates some of main components of residential life support environmental system of the Griya Martubung I Housing as follows; a main access, temporary land fill or temporary garbage dump site, green open space, streets and drainage.
Qualitative investigations showed that many environmental factors are closely affecting the release of CO_2 to air from the whole dynamics of simple housing system. A unique System Relationship Model (SIM) for Griya Martubung I Housing is then developed with the goal to initiate sustainable simple housing by synthesizing the components of the housing's residential life support environmental system.

Figure 4. visually defines the interactions among the components that sustain the housing and the dwellers' way of life. The conclusion of this study is useable on condition that Griya Martubung I is inhabited. The components of residential life support environmental system are access, structure and hierarchy, distance, gas and water supply, electricity and phone line, streets and drainage, public transport, public facilities, social facilities, commercial facilities, man-made pool or ponds, temporary garbage dump site, and green open space.

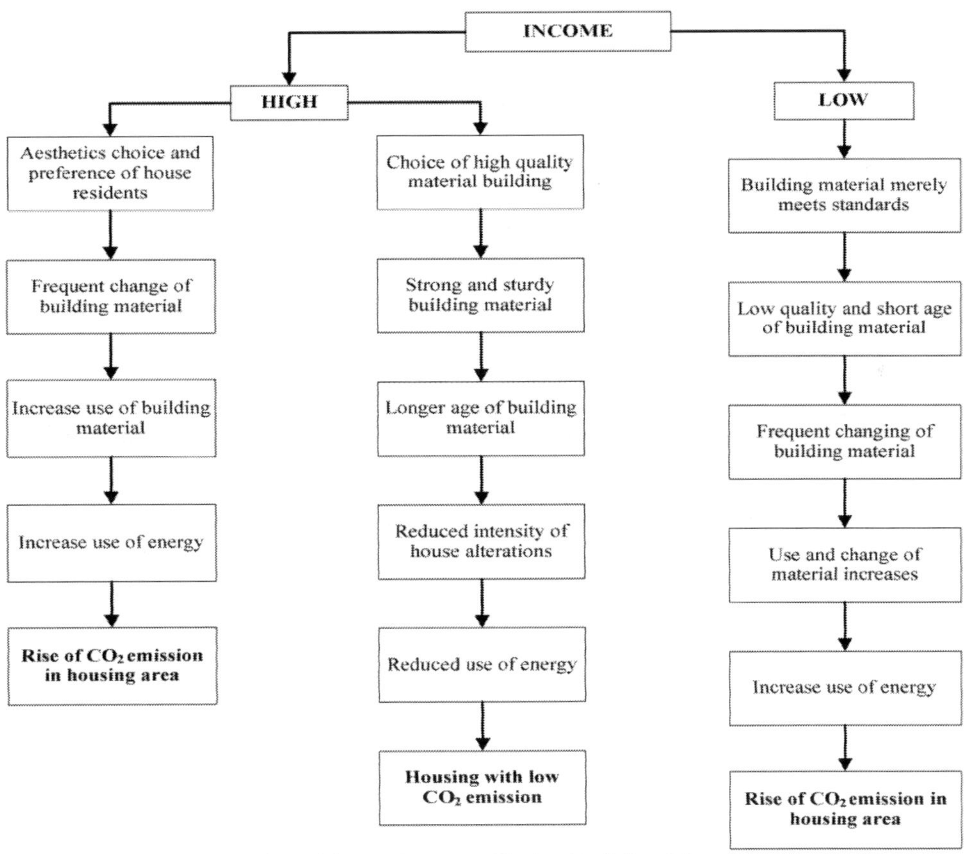

Figure 2. Diagram of Resident's Income, Choice Building Material and CO_2 Emission.

Figure 4 also qualitatively indicates that the system components considerably determine the interrelationship between all areas in and around the housing to initiate sustainable simple housing. The schematic model shows that access is an important component to initiate life support environment in Griya Martubung I housing. Distance, structure and hierarchy will establish a pattern of the dwellers' activities to fulfill their life in and around of housing. Similarly, structure and hierarchy play a significant role in determining the locations of public facilities, social facilities, commercial facilities as well as public transport network to fulfill basic public mobility needs in and around Griya Martubung I housing.

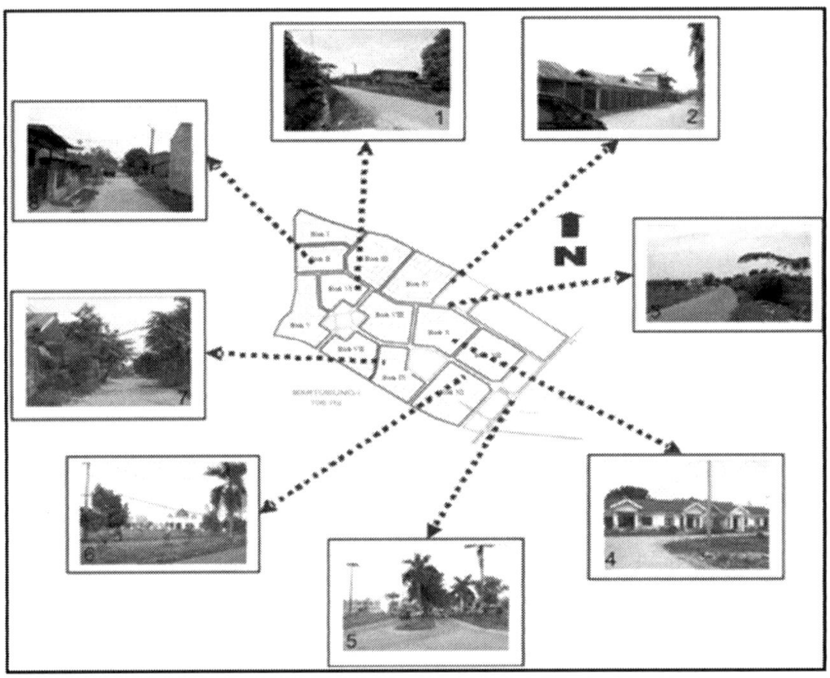

Figure 3. Some of Components of Residential Life Support Environmental System In The Griya Martubung I, Medan

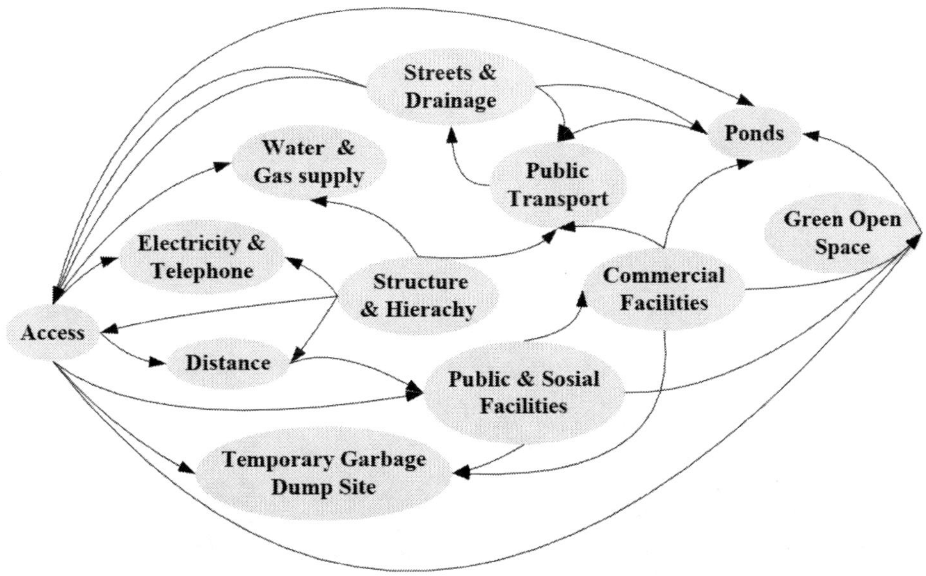

Figure 4. Residential Life Support Environment System of Griya Martubung I

Accessibility to temporary garbage dump site or temporary landfill should be easily reached from each block of the housing. At the moment, the temporary garbage dump is located at the northern part of the housing site, which is only close to block IV and block X. Consequently, dwellers from the other 10 blocks extensively depend on motorcycles to upload their garbage. Having this in mind, it is preferable for the housing management to provide a garbage container around of each block of housing, and later on collect them to dump into temporary landfill.

One remarkable finding from this study relates with the pond that is located in the eastern part of the site (figure 5.). This pond is a man-made pool which exist as a part of cut and fill works for the purpose of Belawan-Tanjung Morawa (Belmera TOL-Road). This pond can be greatly used for social, economic, and environment reasons in developing sustainable simple housing. The pond can functioned not only to collect unnecessary rain water, but also improves the microclimates of the surrounding area of the Griya Martubung I. The pond can effectively create huge impact on living condition in the housing area by filtering out adverse climatic effect.

Figure 5. Man-made ponds in eastern part of Griya Martubung I

3.3. Initiation Strategy to Create Sustainable Simple Housing

This study has been clearly identified that the dynamics of living in Griya Martubung I carry out simply to prepare housing for low income people. What reveals from this study is that there is a need from now on to adopt policy which measured the whole interrelationship of components of residential life support environment system. This measure in need due to the increase requirements for living in more sustainable simple housing in Indonesia.

In general, not many housing planning and design policy exists that directly administers the components of resident life support environmental system to decrease CO_2 emission in the development of simple housings. Some policies that relate with the steps of reducing CO_2 emission in the development of simple housings are: regulations on building density, green open spaces, and regulations on house lot areas. However, there is still need for policies that directly relate to the utilization of components in simple housing, such as regulations on road networks, regulations on centers of activities, control of public and social facilities' locations, control of public transportation, distance between buildings, control of household waste, and control of gas waste caused by the use of household utensils and instruments. To include these in documents of housing development we should refer to policies in the related fields, such as transportation, environment, and energy.

Based on our observation in this study, accessibility and public transport are two important components to initiate sustainable simple housing. From the 128 respondents, it was known that as many as 93% of Griya Martubung I residents use their own transportation mode daily. More specifically, 11% use cars, and 82% use motorcycles as means of transportation to support their daily activities. Residents' perception about characteristics of accessibility in Griya Martubung I were not quite good; the total score for accessibility was 2.52. Based on the correlation of distance, working area, travel time, and transportation mode used, it is shown that the residents of Griya Martubung I Medan work relatively far from their house, and thus do not use the mode of walking to support their activities. Nevertheless, as one determinant factor in initiating sustainable simple housing, it is a misfortune that the housing is not provided with pedestrian facilities for walking activities, especially inside the housing area. Indeed, pedestrian facilities for pedestrians in Griya Martubung I are not yet provided.

Another aspect that has not been acknowledged is the extent to which the green open space policies have been implemented and what are their effects particularly to support the sustainable life in the housing area. The greenness of Martubung I was perceived as quite good by its residents with a total score of 3.10. This perception of greenery stems from the presence of plants in the surrounding area of housing. However, the study shows that green open space in the Martubung I is 3.4% of the total 1,063,099 m^2. This does not yet conform to Undang-Undang no.1 tahun 2011 section 28 and 29, on Housing and Settlements Areas [12], which states that the standard for green open space should be 30% of the total of housing area.

In general, experience of housing development practices only addresses the aspect of correctness or distortion of space utilization, based on land usage that has been established before. Meanwhile, the change of utilization intensity such as usage of roads and man-made pond, that results in the generation of CO_2 emission, is rarely discussed. This observation was made from the usage of existing temporary landfill in Martubung I housing. Residents' perception about the temporary landfill is quite good with a total score of 2.82. Nevertheless, there were still some trash that were not correctly disposed, and the capacity of the trash bins around the housing area were still minimal. For example, there were no the trash bins especially in the playground area.

Attractiveness of man-made pond in Griya Martubung I, Medan was perceived as quite good with a total score of 2.67. This was supported by various events that were frequently held, the existence of many people visiting the pool. However, the availability of playground area next to it is needed as well as the availability of eating and drinking area in the surrounding area of the ponds.

One possible way to initiate sustainable life in simple housing areas is by optimizing the people flows around and inside various land usage for social, public, and commercial reasons. Movements happen when there is constant open space that does not interfere with the speed of passing vehicles. In relation with sustainable housing, the general approach for transportation in land use planning is usually done by distributing centers of activities, public facilities, and social facilities in proximity with housing. This will reduce the operational distance of vehicles and ease residents to walk around the housing area. The road networks generally should go with policies for sustainable housing. For Griya Martubung I, accessible public, social, and commercial facilities around and inside the housing area were perceived as good with the total score of 3.18

This research reveals that 5 (five) main components play a significant role to generate sustainable simple housing such as; accessibility, green open space, temporary landfill, man-made pool or pond, and public, social, and commercial facilities. Figure 6. shows that none of component was perceived as very good. For dwellers, the overall components were perceived as quite good with the total score of 2.86. From the findings, the following things are considered to be important in initiating simple housing developments.

(1) Improvements of housing planning and design policy that directly relate to the utilization of components in simple housing, such as regulations on road networks, regulations on the centers of activities, control of public and social facilities' locations, control of public transportation, distance between buildings, control of household waste, and control of gas waste caused by the use of household utensils and instruments.
(2) Promotion of compact housing development which measured the whole interrelationship of components on the policy making.
(3) Development of control policy to prevent excessive usage of the components.
(4) Promotion of man-made pool as a basic standard of sustainable simple housing.

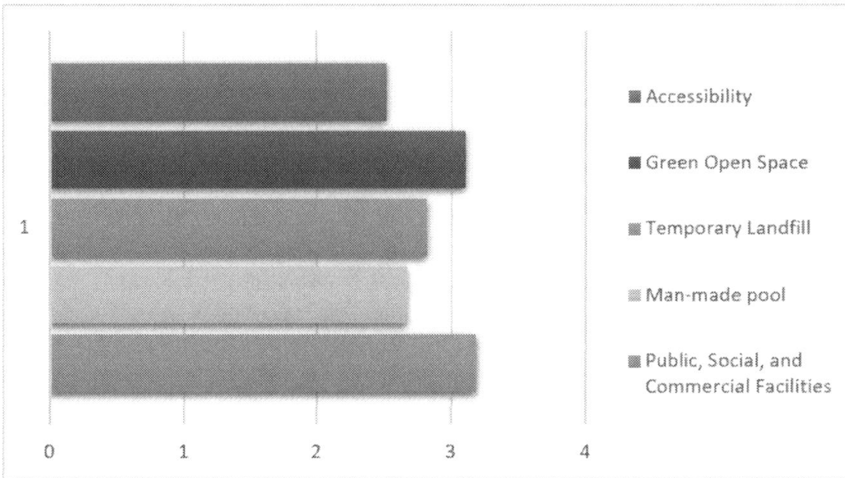

Figure 6. Residents' perceptions about main components of housing life support environment at Griya Martubung I Simple Housing – Medan, Indonesia

Conclusions

This study has shown that 5 main components of residential life support environment system such as accessibility, green open space, temporary landfill, man-made pool, and public, social and commercial facilities play an important roles in proposing sustainable simple housing. Establishment of developing of the components still needs in relation to control residential life support environment. By establishing a tight policy on a construction permit, any changes made would conform to rules, policies and regulations of low CO_2 emissions in this simple housing area.

This study shows that the emission can be kept low if the dynamics of Griya Martubung I Housing is maintained by controlling components of residential life support environment. Based on the components of the system, ideas on policy to reduce CO_2 emission can be realized. These include appropriate concepts on housing alterations; affordable low-emitted construction technology system; establishment of rules and policy to use low-emitted local materials; distribution of centers of activities, public facilities, and social facilities in proximity with housing blocks; enforcement of the role of the management to control excessive usage the overall components required to support activities and the use of public transport for mobility needs and to minimize the use of energy.

Acknowledgments

The author acknowledges the support of the Ministry of Research and Technology and Higher Education Republic of Indonesia. The support was under the BPPTN-Universitas Sumatera Utara 2016 (No.66/UN5.2.3.1/PPM/SP/2016.

References

[1] Bhatti M and Dixon A 2003 Introduction to the Special Focus: Housing, Environment, and Sustainability *Housing Studies* **18** 501-504

[2] Siahaan N 2012 *Model Pengendalian Perumahan Sederhana Dalam Sistem Perumahan Berkelanjutan Perkotaan Berbasis Rendah Emisi CO_2* Disertasi

[3] Tombe D D 2015 *Handling Societal Complexity A Study of the Theory of Methodology of Societal Complexity and the COMPRAM* Methodology Amsterdam Springer.

[4] Priemus H 2005 How to Make Housing Sustainable? The Dutch Experience *Environment and Planning: Planning and Design* **32** 5-19

[5] Priana S, Indira K D and Chendy C O Y 2008 Analysis on Life in Kampong Naga to Deduce Policy on Carbon-Dioxide Emitted from House Construction *International Symposium on*

Climate Change and Human Settlements

[6] Jabareen Y R 2006 Sustainable Urban Forms, Their Typologies, Models, and Concepts *Journal of Planning Education and Research.*

[7] Zubaidah S K 2005 *Lokakarya Faktor-Faktor Penentu Emisi CO_2 Pada Perumahan dan Permukiman Perkotaan* Bandung

[8] Pusat Penelitian dan Pengembangan Pemukiman Departemen Pekerjaan Umum 2007 *Keterkaitan Penyelenggaraan Bangunan dengan Emisi CO2*

[9] Suhedi F 2007 Emisi CO_2 dari Konsumsi Energi Domestik *Seminar Pusat Penelitian dan Pengembangan Pemukiman*

[10] Zhang X, Shen L and Wu Y 2011 Green Strategy for Gaining Competitive Advantage in Housing Development: a China Study *Journal of Cleaner Production* **19** 157-167

[11] Afiaty D F 2003 *Faktor-faktor Penentu Pemilihan Tipe Rumah dan Implikasinya terhadap Tingkat Kepuasan Pemilik serta Jenis dan Frekuensi Perubahan Rumah: Perumnas Sarijadi Bandung* Tesis Program Magister Institut Teknologi Bandung.

[12] Undang-Undang Republik Indonesia No.1 Tahun 2011 Tentang Perumahan dan Kawasan Permukiman.

IOP Conf. Series: Materials Science and Engineering **309** (2018) 012009 doi:10.1088/1757-899X/309/1/012009

The effect of steel slag as a coarse aggregate and Sinabung volcanic ash a filler on high strength concrete

R Karolina, A L A Putra*
Departemen Teknik Sipil, Universitas Sumatera Utara,
Jl. Perpustakaan, Kampus USU, Medan, 20155, INDONESIA

*E-mail: lamra.alif@gmail.com

Abstarct. The Development of concrete technology is continues to grow. The requisite for efficient constructions that are often viewed in terms of concrete mechanical behavior, application on the field , and cost estimation of implementation increasingly require engineers to optimize construction materials, especially for concrete materials. Various types of concrete have now been developed according to their needs, such as high strength concrete. On high strength concrete design, it is necessary to consider several factors that will affect the reach of the quality strength, Those are cement, water cement ratio (w/c), aggregates, and proper admixture. In the use of natural mineral, it is important for an engineer to keep an eye on the natural conditions that have been explored. So the selection of aggregates as possible is a material that is not causing nature destruction. On this experiment the use of steel slag from PT.Growth Sumatra Industry as a substitute of coarse and fine aggregate, and volcanic ash of mount Sinabung as microsilka in concrete mixture substituted to create high strength concrete that is harmless for the environment. The use of mount sinabung volcanic ash as microsilika coupled with the use of Master Glenium Sky 8614 superplasticizer. This experiment intend to compare high strength concrete based slag steel as the main constituent aggregates and high strength concrete with a conventional mixture. The research result for 28 days old concrete shows that conventional concrete compressive strength is 67.567 MPa, slag concrete 75.958 Mpa, conventional tensile strength 5.435 Mpa while slag concrete 5.053 Mpa, conventional concrete bending strength 44064.96 kgcm while concrete slag 51473.94 kgcm and modulus of conventional concrete fracture 124.978 kg / cm^2 while slag concrete 145.956 kg / cm^2. Both concrete slump values shows similar results due to the use of superplasticizer.

1. Introduction

High Quality Concrete is a concrete that has characteristics as a very solid unity of material with a compressive strength ranging from 55.5 to 200 Mpa. This concrete allows the creation of a sleek, lightweight concrete structure that can also save energy and natural materials. High Strength Concrete Density also gives the advantage that it has high resistance to dangerous attack of liquids or gases.

High Quality Concrete is generally used in high-rise buildings and bridges. In high buildings, High Strength Concrete is used to conserve dimensions of columns and beams so that wider space is available between columns to columns, as well as between beams that affect the elevation of each floor. The reduction of the structure component dimension itself will reduce the weight of the structure so that the load on the foundation becomes lighter. On long bridges that generally use pre-cast concrete, High Strength Concrete is required to support larger span loads due to bridge spans and also to overcome the possibility of precast concrete damage that often occurs in the mobilization of precast concrete itself.

The need for these structural components leads to the use of high strength concrete which includes strength, durability, service life and efficiency. High strength concrete is greatly influenced by the constituent material on the concrete. There are several factors that influence the achievement of high quality concrete compressive strength in its design. The preparation of eligible materials is one of the factors that can influence in the manufacture of high quality concrete. Factors affecting the quality of a concrete are cement factors, cement water factor, aggregate factor, microsilic usage and usage of admixture materials.

One material that meets as an alternative material in the achievement of high quality concrete is Steel Slag. Steel Slag is a by-product of steelmaking produced during the separation of molten steel from dirt in a blast furnace (enclosure for metal heating or fireworks). Steel Slag is a substance that remains when the steel has been extracted from its seeds. Steel Slag is also a nonmetallic material in which the main constituent components are calcium, magnesium, and alumanium silicate in several combinations (Burge, T.A, 2004).

Steel slag waste, included in the category of hazardous and toxic waste materials (B3). In 2010, slag production in Indonesia is only about 800 thousand tons per year. Every ton of steel production produces 20 percent of slag waste. PT Krakatau Steel in Cilegon, Banten is one of the steel manufacturer companies in Indonesia that produces at least 150 tons of slag every day. In order not to cause pollution, the steel association asked the government to use steel waste (slag waste). This utilization can be used for infrastructure projects. If not utilized, the waste is included in the category of toxic and hazardous waste (B3). (Puslitbang Jalan and Jembatan, 2011).

The use of steel slag as a coarse aggregate in a concrete mixture is an effort to utilize B3 waste that can cause environmental pollution. Besides, steel slag as a coarse aggregate of concrete mix is expected to be a solution in achieving high quality concrete that is environmentally friendly and economical. This is because the steel slag itself has a rough surface texture and perforated which caused due to gas trapping when the hot slag cooling process, the gas holes are not mutually dependent and not porous, if the slag is split because of the breaking process, then the hardness is not lost even the smallest grains.

Steel slag wastes that normally produced by the Indonesian steel industry are still in the form of chunks. So it still needed other stages before being processed into another product. Slag with a large lump size is through the collision process first to get the size of the aggregate in need. From the crushing process, the slag can be used as a rough aggregate or fine aggregate that is safe to use and have economic value.

Currently in Indonesia the utilization of steel slag has started to be developed and applied in the field of pavement. Specification of pavement design by using slag as its constituent material also being enforced by Pusjatan. Through this research, it is expected the optimization of steel slag can be developed in the field of construction of concrete especially high quality concrete.

Utilizing waste materials for useful things is one of the best ways to tackle environmental problems. Not only reducing the damage to environmental impacts but also become an alternative use of materials that are still commonly used. Various research and experiments in the field of concrete is done as an effort to improve the quality of concrete. Materials technology and implementation techniques obtained from the results of these experiments and the experiments are intended to answer the increasing demands on the use of concrete and overcome the problems that often occur during field work. In the construction of high-rise buildings and other buildings, it needs concrete with high strength, high quality concrete is the most appropriate choice.

The advantages of using slag in a concrete mix are as follows: enhancing the compressive strength of concrete due to the tendency of slow increase in compressivestrength; increasing the ratio between flexibility and compressive strength of concrete; reducing the variation of compressive strength of concrete; increasing the resistance to sulfate in seawater; reducing alkali-silica attack; reducing the heat of hydration and lowering the temperature; fixing the final solution and giving a bright color to the concrete; enhancing the durability due to changes in the volume; reducing porosity and chloride attack.

2. Literature Review

Slag is the waste material from casting iron (piq iron), where the process uses a kitchen (furnance) that the fuel is from the air that blown (blast) (Paul Nugraha and Antoni, 2007). A mixture of calcium, aluminum, silicon and phosphor forms (slag) that acting at a temperature of 1600°C and forming a liquid, when the liquid is cooled it will occur crystals resembling aggregate shapes. Slag waste has porous particle granules on its surface, has a good gradient, with varying particle size variations. In the smelting of steel, ore or scrap metal is liquefied with a combination of limestone, delomite or lime, the manufacture of steel is initiated by removing the impurity ions of steel, including aluminum, silicon and phosphorus. To remove the impurity ions, calcium is required in the limestone. A mixture of calcium, aluminum, silicon and phosphor forms (slag) which reacts at a temperature of 1600 ° C and forms a liquid, when the liquid is cooled there will be crystals, that can be used as mixture of cement and may also be a substitute for aggregate. Factors to determine the cementious properties in slag are chemical composition, alkali concentration and reaction to the system, the glass content in slag, subtlety, and temperature that generated during the hydration process takes place.

Slag is the result of a high blast furnace residue, produced by steel smelting industries that physically resemble rough aggregates. When the liquid is cooled slowly there will be useless crystals as a cement mixture and can be used instead of aggregates. However it forms a highly reactive granulated glass, suitable for the manufacture of slag cement. The slag is then ground to fine, it can be used as a substitute for cement in the manufacture of concrete.

Figure 1. Coarse and fine steel slag of PT. Growth Sumatera Industries

According to Cain (1994: 505) Factors to determine the cementious properties in slag are chemical composition, alkali concentration and reaction to the system, glass content in slag, subtlety and generated temperature during the hydration process. ASTM (1995,494) Slag is a non-metallic product that is a finely shaped to large blocks material, from a cooled combustion product.

Table 1. The chemical element content in steel slag

No.	Parameter	Unit	Result	Method
1	Lead (Pb)	Mg/kg	26.6	AAS
2	Cadmium (Cd)	Mg/kg	<0.003	AAS
3	Copper (Cu)	Mg/kg	97.5	AAS
4	Cromium (Cr)	Mg/kg	5353	AAS
5	Silver (Ag)	Mg/kg	<0.001	AAS
6	Selenium (Se)	Mg/kg	<0.01	AAS
7	Barium (Ba)	Mg/kg	817	AAS
8	Mercury (Hg)	Mg/kg	0.38	AAS
9	Arsen (As)	Mg/kg	0.21	AAS

Source: Test Laboratory of Research and Industrial Standardization Council, Medan 2016

2.1 Stone Ash

Stone ash is an aggregate of filler minerals, obtained from byproducts of cement factories or stone crushing machines. This type of material is much needed as a mix in the process of asphalting and can be used as a substitute for sand. Currently the stone ash is not so good for sale because the usage in the pavement construction industry has very little use of pavement construction with Lapen and has been switched to concrete asphalt layer. On the other hand the use of sand for concrete mixture continues to increase. The similarity of physical properties between stone ash and sand makes the use of sand can be reduced by the use of stone ash.

Here is the usage advantages of a mixture of stone and sand ash in fine aggregates:

- Stone ash generally has excellent grain size, it is very influential to reduce the gap of grain size in the concrete mixing material.
- Stone ash can also reduce air content caused by concrete aggregates. This is caused by the fine size of the stone that fills the cavities between the aggregate granules.
- With the addition of stone ash on the sand used in the concrete mixture, it will reduce the bleeding in the concretemixture. This is in because the better arrangement of gradation of sand that can hold water, so that water to the surface of the concrete can be reduced.

With the advantages, the use of cement can be reduced because the resulting concrete will be more solid and improve its quality.In high quality concrete planning according to ASTM, the fine aggregate used must have a greater modulus of fineness equal to 2.5. Therefore, stone ash can increase the sand modulus of non-compliant sand by substituting sand for a certain amount of sand.

2.2 Sinabung Ash

Volcanic dust is a mineral volcanic rock including glass material that has a size of sand and gravel with a diameter of approximately 2 mm (1/2 inch) which is the result of volcanic eruption. Such very small ash particles can have a cross section smaller than 0.001 mm (1 / 25,000th of an inch).

Volcanic ash has a very hard and insoluble nature in the water so it is often very abrasive and slightly corrosive and capable of conducting electricity when in a wet state.

Table 2 Content of Characteristic Elements of Sinabung Ash

No.	Parameter	Result	Unit	Method
1	Silica as SiO2	79.7	%	Gravimetry
2	Aluminium as AL2O3	3.88	%	Calculation
3	Calcium as CaO	12.10	%	Titrimetry
4	Magnesium as MgO	0.14	%	Gravimetry

Source: Test Laboratory of Research and Industrial Standardization Council, Medan 2016

From the results of examination of chemical content above, it is seen that sinabung ash has a very high content of silica. This percentage of content indicates that the ash of sinabung can be made as a mirosilika in high quality concrete mixing. Thus it is possible to use the volcanic ash of Mount Sinabung as a filler and become a cement substitution so as to reduce the use of cement in concrete mixture.

3. Research Method

Research method and research flow can be seen in Figure 2.

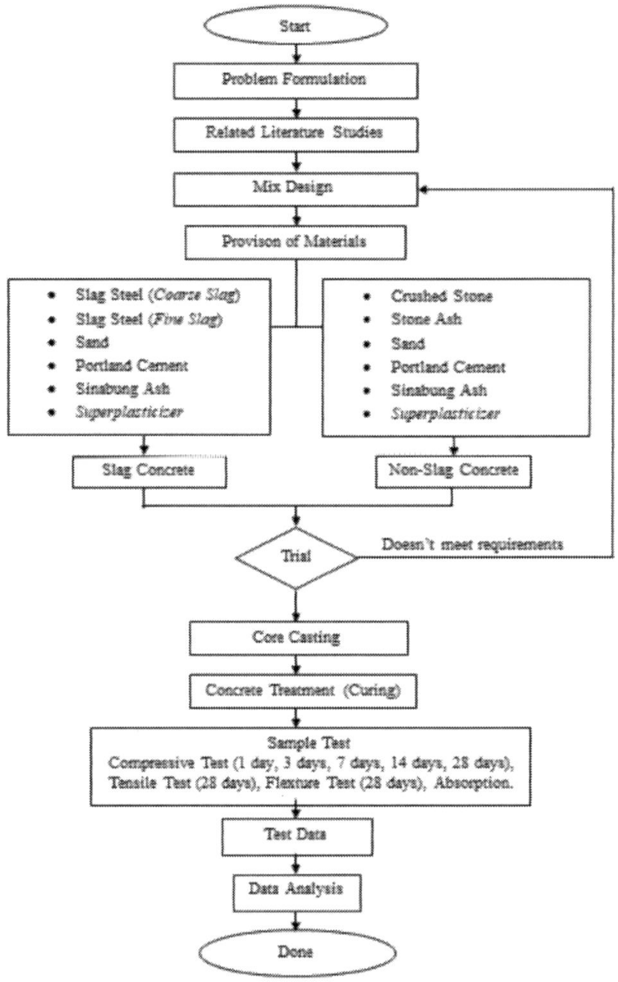

Figure 2. Research flow

4. Result and Discussion

4.1 Slump Test

Test results of slump value for concrete with steel slag and without steel slag can be seen in table.

Table 3. Slump value test results

Mix Variation	Slump Value (cm)
Non-Slag	25
Slag	25

Table 3shows that the slump values of the two concretes do not show any significant differences. High Slump value is caused by the use of Superplasticizer which serves to improve the workability of fresh concrete.

4.2 Compressive Strength of Cylinder Concrete

Tests were performed at 1, 3, 7, 14, and 28 days old. Testing is done based on SNI 1974: 2011, The test method of compressive strength of concrete is with cylindrical test object. The results of the compressive strength test for both concrete variations can be seen in the table below.

Table 4 Compressive Strength Test Results of Slag Steel Concrete

No	Sample	Age of Concrete (Days)	Test Weight (kg)	Actual Compressive Load (kN)	Cross-Sectional Area (cm2)	Actual Compressive Strength (Mpa)	Average Com. Strength (Mpa)
1	Slag 01		4.938	278	78.5	36.831	
2	Slag 02	1	4.886	348	78.5	46.104	43.896
3	Slag 03		4.879	368	78.5	48.754	
4	Slag 04		4.750	490	78.5	64.917	
5	Slag 05	3	4.819	560	78.5	74.191	71.541
6	Slag 06		4.842	570	78.5	75.516	
7	Slag 07		4.849	650	78.5	86.115	
8	Slag 08	7	4.861	550	78.5	72.866	77.282
9	Slag 09		4.783	550	78.5	72.866	
10	Slag 10		4.967	600	78.5	79.490	
11	Slag 11	14	4.929	580	78.5	76.841	70.658
12	Slag 12		4.976	420	78.5	55.643	
13	Slag 13		4.792	600	78.5	79.490	
14	Slag 14	28	5.017	560	78.5	74.191	75.958
15	Slag 15		4.983	560	78.5	74.191	

From the table of test results, the compressive strength above shows that within 1 day the concrete has reached high compressive strength, and continues to increase until the age of 7 days concrete, then the compressive strength growth of the press begin to slow down afterwards.

Table 5 Compressive Strength Test Results of Concrete Without Slag Steel

No	Sample	Age of Concret e (Days)	Test Weight (kg)	Actual Compressive Load (kN)	Cross-Sectional Area (cm2)	Actual Comp. Strength (Mpa)	Average Comp. Strength (Mpa)
1	Non-Slag 01	1	3.923	260	78.5	33.121	33.651
2	Non-Slag 02		3.874	266	78.5	33.885	
3	Non-Slag 03		3.828	236	78.5	30.064	
4	Non-Slag 04	3	3.863	392	78.5	49.936	47.871
5	Non-Slag 05		3.919	314	78.5	40.000	
6	Non-Slag 06		3.923	378	78.5	48.153	
7	Non-Slag 07	7	3.915	440	78.5	56.051	57.321
8	Non-Slag 08		3.966	444	78.5	56.561	
9	Non-Slag 09		4.023	414	78.5	52.739	
10	Non-Slag 10	14	4.061	338	78.5	43.057	53.170
11	Non-Slag 11		3.953	370	78.5	47.134	
12	Non-Slag 12		4.058	496	78.5	63.185	
13	Non-Slag 13	28	4.077	510	78.5	64.968	67.567
14	Non-Slag 14		4.031	520	78.5	66.242	
15	Non-Slag 15		4.097	500	78.5	63.694	

From the table of compressive strength test results above, it is seen that within 1 day the concrete has reached high compressive strength, and continues to increase until the age of 7 days concrete, then go down at day 14 and reach maximum strength at age 28 days. From the results of both compressive strength of steel slag concrete samples and the non-steel slag samples against the age of the test, then both test results of compressive strength can be compared through the following graph

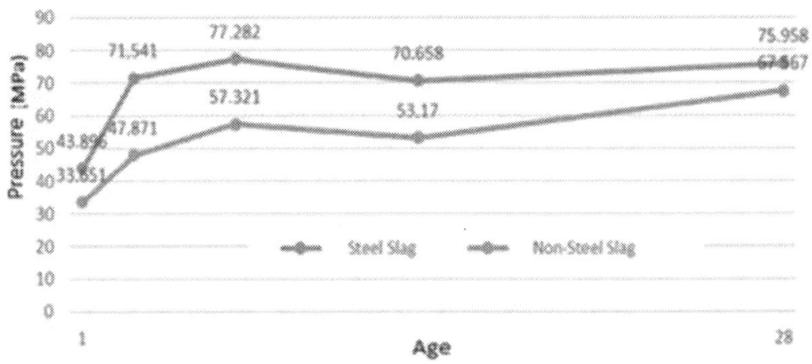

Figure 3. Compressive Strength Comparison of Steel Slag Concrete and Conventional

From the test results of compressive strength of both types of specimens it can be concluded that high quality concrete with main aggregate of slag steel has higher compressive strength than the compressive strength of high quality concrete with main aggregate of crushed stone. With compressive strength of slag concrete at age 28 days is 75.958 Mpa and compressive strength of non slag concrete is 67.567 Mpa. Increasing compressive strength of concrete with the main aggregate of steel slag is between 12 -15% of the concrete with crushed stone aggregate as the main constituent aggregate.

4.3 Tensile Strength of Concrete Cylinders

This test aims to determine the value of compressive strength of high quality concrete with rough aggregate replacement by steel slag compared with high quality concrete without rough aggregate replacement

Table 6 Tensile Strength Test Results of Slag Steel Concrete

No	Sample	Age of Concrete (Days)	Test Weight (kg)	Actual Tensile Load (kN)	Tensile Strength (Mpa)	Average Compressive Strength (Mpa)
1	Slag Concrete 01		4.887	162	5.159	
2	Slag Concrete 02	28	1.911	154	4.904	5.053
3	Slag Concrete 03		4.881	160	5.096	

Table 7 Tensile Strength Test Results of Concrete Without Slag Steel

No	Sample	Age of Concrete (Days)	Test Weight (kg)	Actual Tensile Load (kN)	Tensile Strength (Mpa)	Average Comp.Strength (Mpa)
1	Non-Slag 01	28	3.830	168	5.350	5.435
2	Non-Slag 02		3.931	180	5.732	
3	Non-Slag 03		3.881	164	5.223	

From the comparison of concrete tensile strength it can be seen that the concrete without slag provides greater tensile resistance compared to the concrete with the use of slag steel as its main aggregate.

4.4 Flexture Strength of Concrete Beam (15 x 15 x 60)

The calculation data of flexture capacity based on strain data for reinforced concrete beams can be seen in the table below.

Table 8 Test Results of Fracture Modulus of Steel Slag Concrete

No	Sample	Age of Concrete (Days)	Test Weight (kg)	Actual Tensile Load (kN)	Fracture Modulus (kg/cm^2)	Average Frac. Modulus (kg/cm^2)
1	Slag Concrete 01	28	43.100	81.8	145.422	145.956
2	Slag Con crete 02		42.900	82.4	146.489	

Table 9 Test Results of Fracture Modulus of Concrete Without Steel Slag

No	Sample	Age of Concrete (Days)	Test Weight (kg)	Actual Tensile Load (kN)	Fracture Modulus (kg/cm^2)	Average Frac. Modulus (kg/cm^2)
1	Non-Slag 01	28	34.200	70.2	124.800	124.978
2	Non-Slag 02		32.800	70.4	125.156	

Table 10 Flexture Moment Of Slag Concrete

No.	Sample	Maximum Moment (kgcm)	Average Maximum Moment (kgcm)
1	Slag Flexture Concrete 01	51287.000	51473.938
2	Slag Flexture Concrete 02	51660.875	

Table 11 Flexture Moment Of Non-Slag Concrete

No.	Sample	Maximum Moment (kgcm)	Average Maximum Moment (kgcm)
1	Non-Slag Flexture Concrete 01	44003.250	44064.965
2	Non-Slag Flexture Concrete 02	44126.680	

From the table above it can be concluded that Steel Slag Concrete has a fracture modulus and a higher flexture moment of resistance than the concrete without aggregate replacement.

5. Conclusion

Based on the research that has been done, it can be concluded that:

1. Normal compressive strength of high quality concrete at age 1,3,7,14, and 28 days respectively are 33.651 Mpa. 47.871 Mpa. 57.321 Mpa. 53.170 Mpa and 67.567 Mpa. While compressive strength of concrete with slag substitution at age 1,3,7,14 and 28 days respectively are 43.896 MPa. 71.541 MPa. 77.282 MPa. 70.658Mpa and 75.958. Thus it shows that high quality concrete made of steel slag as the main constituentaggregate has a compressive strength greater than that of conventional high-quality concrete.
2. The average tensile strength of high quality slag concrete with 28 days of concrete is 5.053 MPa. While the average tensile strength of conventional high quality concrete with 28 days of concrete is 5,435 Mpa. It shows that the tensile strength of conventional high quality concrete is stronger than the tensile strength of high quality concrete with slag steel replacement.
3. Fracture Modulus on high quality slag concrete is 145.956 kg / cm2 while the fracture modulus on conventional high quality concrete is 124.978 kg / cm2. So it can be concluded that the fracture modulus of slag concrete is higher than the conventional concrete.
4. The flexture moment on high quality slag concrete is 51473.938 kgcm while the flexture moment on conventional high quality concrete is 44064.965 kgcm. So it can be concluded that the flexture moment of slag concrete is higher than the conventional concrete

References

[1] Gunawan G, Oetojo P D, Kusminingrum N, Rahmawati TandLeksminingsih 2011*Pemanfaatan Slag Baja UntukTeknologiJalan yang Ramah Lingkungan*Bandung :PusatPenelitiandanPengembanganJalandanJembatan

[2] Gustafson D P et al 1990 *Specifications for Structural Concrete for Buildings*USA : American Concrete Institute

[3] Hanif 2012Penggunaan Slag Steel denganVariasi FAS terhadapkuattekanbeton*REINTEK***7(2)**

[4] Lumbangaol TAM 2016*PengaruhPenggunaan Steel Slag Sebagai Agregat Halus Terhadap Kuat Tekan dan Lentur Pada Beton Bertulang Dibandingkan dengan Beton Normal* Medan :Universitas Sumatera Utara

[5] Li Z D 2011 *Advanced Concrete Technology*USA :John Wiley & Sons Inc

[6] Mehta P K and Monteiro P J M 2014 *Concrete : Microstructure, Properties, and Materials, Fourth Edition* USA : McGraw-Hill Education

[7] Mindess S and Young J F 1981 *Concrete*USA : Prentice-Hall Inc

[8] Neville A M 1977 *Properties of Concrete* Great Britain : The Pitman Press

[9] Pandiangan J 2016 *Pengaruh Penggunaan Steel Slag Sebagai Agregat Kasar Terhadap Kuat Tekan dan Lentur Pada Beton Bertulang Dibandingkan dengan Beton Normal* Medan :Universitas Sumatera Utara

[10]Pujianto A 2011 BetonMutuTinggidengan Admixture SuperplasticizerdanAditif Silica Fume*SemestaTeknika***14(2)** 177-185

[11]Pujianto A2010 Beton Mutu Tinggi dengan Admixture Super plasticizer dan Fly Ash*SemestaTeknika***13(2)** 171-180

[12]Rangkuti R 2014PengaruhSifatMekanikBetonpadaPenambahan Master Glenium Sky 8316 Medan :Universitas Sumatera Utara

[13]SNI 1972 2008*Cara Uji Slump Beton* Jakarta : StandarNasional Indonesia

[14]SNI 1974 2011*Cara UjiKuatTekanBetondengan Benda UjiSilinder* Jakarta : StandarNasional Indonesia

IOP Conf. Series: Materials Science and Engineering **309** (2018) 012010 doi:10.1088/1757-899X/309/1/012010

Increased productivity through waste reduction effort in oil and gas company

J Hidayati*, NA Silviana, RA Matondang

Magister in Industrial Engineering, Faculty of Engineering, Universitas Sumatera Utara, Prof. T. Maas Street, Kampus USU

*Email: rivaijuliza@gmail.com

Abstract. National companies engaged in oil and gas activities in the upstream sector. In general, the on going operations include drilling, exploration, and production activities with the result being crude oil channelled for shipment. Production activities produce waste gas (flare) of 0.58 MMSCFD derived from 17.05% of natural gas produced. Gas flares are residual gases that have been burning through flare stacks to avoid toxic gases such as H2S and CO that are harmful to human health and the environment. Therefore, appropriate environmental management is needed; one of them is by doing waste reduction business. Through this approach, it is expected that waste reduction efforts can affect the improvement of environmental conditions while increasing the productivity of the company. In this research begins by identifying the existence of problems on the company related to the amount of waste that is excessive and potentially to be reduced. Alternative improvements are then formulated and selected by their feasibility to be implemented through financial analysis, and the estimation of alternative contributions to the level of productivity. The result of this research is an alternative solution to solve the problem of the company by doing technological based engineering by reusing gas flare into fuel for incinerator machine. This alternative contributes to the increased productivity of material use by 23.32%, humans 83.8%, capital 10.13 %, and waste decreased by 0.11%.

1. Introduction

Flaring are considered the single largest loss in many industrial operations, such as oil-gas production, refinery, chemical plants, natural gas processing plants, the coal industry and landfills. Wastes or losses to the flare include process gases, fuel gas, steam, nitrogen and natural gas [1]. In general, many flare gas comes from the process of exploitation of oil well both onshore and offshore [9]. A flare is normally visible and generates noise and heat. Low quality gas composition that is flared releases many impurities and toxic particles into the atmosphere during the flaring process. The burning of gas flaring produces combustion by-products such as CO_2, CO and NOx that are emitted to the atmosphere [1]. Based on data release by British Petroleum its known that total global gas flare from oil production in 2008 reached 139 Billion cubic meter per year or equal to 13,448.61 mmscfd.[10] Burning flare gas aims to minimize environmental pollution, because if flare gas discharged into the air without being burned first can be a negative impact on the surrounding environment [8].

Waste gases are subjected to such a process either because gases are a waste or it is difficult to store and transport them. Non-waste gases are burnt off to protect the processing equipment when unexpected high pressure develops within them. Gas flaring in oil rigs and wells contribute significantly to greenhouse gases in our atmosphere [2].

The emission of greenhouse gases (GHGs) especially carbon dioxide (CO_2) has been reported to be the major factor behind recently observed changes in climate extremes such as storms, floods, and heat waves as well as increasing global temperatures and rising sea levels. Gas flaring impacts climate change by adding about 400 million tons of CO_2 in annual emissions [3]. CO_2 and CH_4 are GHG that, when released directly into the air, traps heat in the atmosphere. The climate impact is obvious, suggesting a great contribution to global GHS emission [13]. As a result of the environment, gas flaring has raised temperatures and rendered large areas uninhabitable. CO_2 emissions from flaring have high global warming potential and contribute to climate change. CO_2 emissions come from only combustion of fossil fuels for about 75% [14]. CH_4 is actually more harmful than CO_2, it has about 25 times greater global warming potential than CO_2 on mass basis [15]. It is also more prevalent in flares that burn at lower efficiency [13].

Based on the fact that one of the national companies in the oil and gas field proficiency activities in the upstream sector. During the production, produce some waste namely liquid waste, solid and gas. Liquid waste in the form of water, which made the process of treatment back to use again for all stations, while solid waste is in the form of mud to be processed back into use. From observations in the field to the waste gas in the form of gas remains (flare) which is produced from the oil production process and the earth and during this is done burning to free air.

Gas flare produced has not been fully utilized as an environmentally friendly alternative, this is due to the gas flare only thrown into the air by burning so that produce carbon dioxide emissions by 33.5 tons/day. Then, with respect to the composition and the potential for gas that is produced, This object is to increased productivity with utilize flare to be economical and environmentally friendly alternative. [4].

2. Research method

The following are the methodological steps conducted in this study:

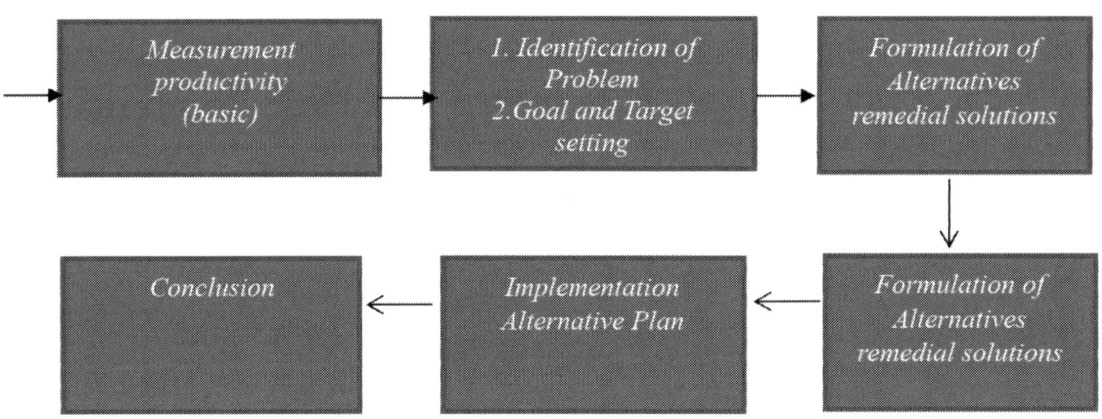

Figure 1. The methodological steps

3. The Result

3.1. Measurement of productivity

The value of productivity can be obtained with the following formula:

$$P = \frac{Output}{Input} \text{ x } 100\ \% \qquad [1]$$

By using the formula, the calculated results of the company's productivity value for period of January 2014 until June 2015 can be seen in Table 1.

Table 1. The Value of Productivity

No	Output	Input	Value of Productivity
1	205,747,717,271	21,000,000,000	9.80
2	169,787,953,986	20,000,000,000	8.49
3	199,054,241,922	21,000,000,000	9.48
4	198,573,141,583	20,000,000,000	9.93
5	196,946,043,423	20,000,000,000	9.85
6	184,674,553,282	21,000,000,000	8.79
7	184,674,553,282	21,000,000,000	8.79
8	216,490,801,574	21,000,000,000	10.31
9	203,497,987,261	20,000,000,000	10.17
10	210,054,974,493	21,000,000,000	10.00
11	190,721,930,159	21,000,000,000	9.08
12	196,146,206,445	20,000,000,000	9.81
13	211,143,576,117	21,000,000,000	10.05
14	195,350,549,141	21,000,000,000	9.30
15	218,407,061,892	21,000,000,000	10.40
16	199,365,193,290	20,000,000,000	9.97
17	198,610,000,817	21,000,000,000	9.46
18	178,311,804,649	21,000,000,000	8.49

After seeing the results of productivity calculations, it is necessary to evaluate productivity that aims to see the productivity changes that have been achieved in order to be further analyzed.

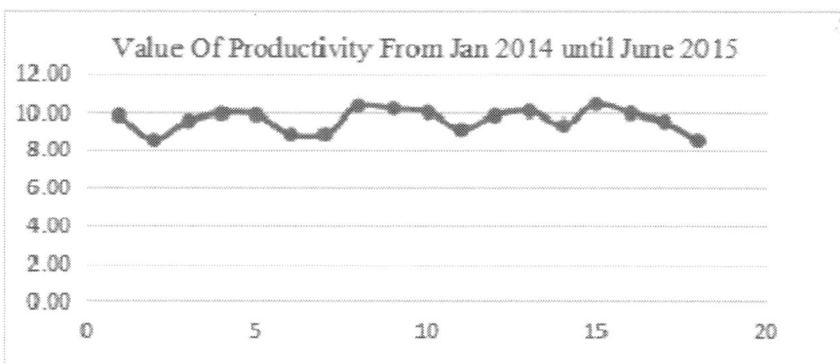

Figure 2. Value of productivity from Jan 2014 until June 2015

3.2. Identification of problem and cause

Identification of this problem is done by using the tool of cause effect diagram. This tool is used to identify the problem in detail so that the root of the problem you want to solve. In searching for the root source of the problem can be used using the Ishikawa Diagram obtained from the interview in the production section. Identification of causes can be seen in the following explanation of cause effect diagram on Figure 3.

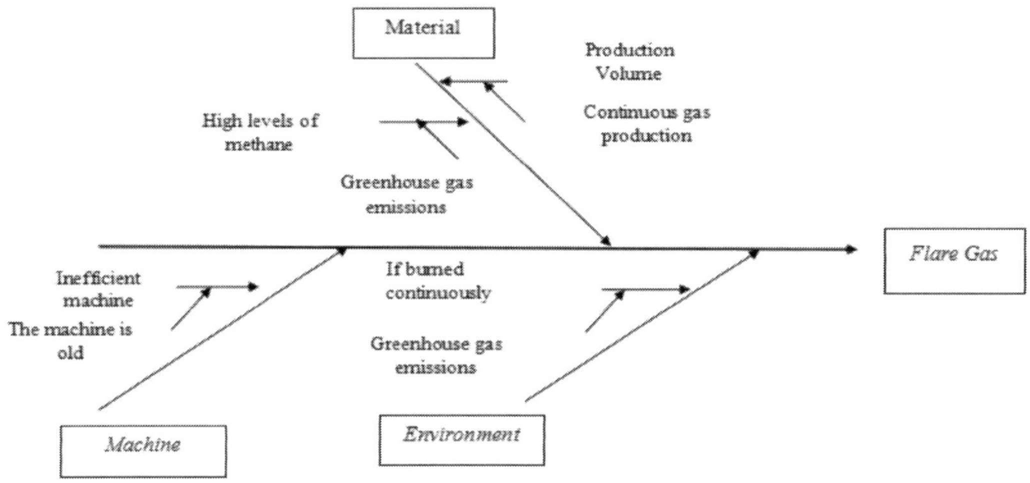

Figure 3. Cause Effect Diagram

3.3. Goal and target setting

Based on the identification of the problem described through the causal diagram it is reducing of flare gas derived from the production process into an economic and environmentally-friendly alternative in order to achieve zero flaring 2030.

3.4. Formulation of alternative remedial solutions

Compilations of proposed alternatives to consider environmental, economic and social aspects are as follows:

a. Purchased gas cylinders to third parties. The purchase of gas cylinders is intended to make the flare gas utilized as LPG. Based on pollution prevention hierarchy with recycling (material reused in the production process). Where waste of residual gas (flare) is compressed into technology in the form of LPG gas cylinders through a series of processes. The application of this technology is based on a study conducted by Sugito in 2011 with a feed gas derived from java field (Jambi) of 2 MMSCFD where gas flares produce LPG of 12.26 ton / day, condensate of 27.24 barrels/day and Lean gas of 1.75 MMSCFD [7].
b. The alternative with recycling technique makes it possible to utilize flare gas as incinerator engine fuel. Where incinerator machine is a solid and liquid waste combustion technology that comes from various industries, buildings and hospitals [11].

3.5. Alternative selection

3.5.1. Alternative Financial Analysis with Annual Worth

a. Utilization as mini LPG [12]

Benefit - cost analysis is estimated five years for utilization as mini LPG

Benefit : Savings (P/A.i%.5)
= Rp 312,890,688,000. (P/A.18%.5)
= Rp 312,890,688,000 (3.12717)
= Rp 978,462,372,793

Cost : Initial investment + Operational cost (P/A.i%.5)
=Rp 2,368,101,600,000 + Rp236,996,160,000 (P/A.18%.5)
= Rp 2,368,101,600,000 + Rp 236,996,160,000 (3.12717)
= Rp 8,146,583,562,139

BCR index for utilization as mini LPG
= **Error! Reference source not found.**
=Rp 978,462,372,793
Rp8,146,583,562,139
Error! Reference source not found.= 0.12

b. Utilization as Incinerator Fuel
Benefit - cost analysis is estimated five years for utilization as incinerator fuel
Benefit : Savings (P/A.i%.5)
= Rp 312,890,688,000. (P/A.18%.5)
= Rp 312,890,688,000 (3.12717)
= Rp 978,462,372,793
Cost : Initial Investment + Operational cost (P/A.i%.5)
= Rp 8,028,165,600+ Rp9,783,509,520(P/A.18%.5)
= Rp 8,028,165,600+ Rp 9,783,509,520 (3.12717)
= Rp 55,700,136,085
BCR index for utilization as incinerator fuel :
= **Error! Reference source not found.**
=**Error! Reference source not found.** Rp 978,462,372,793/Rp 55,700,136,085
= 17.56

3.5.2. Estimate the Contribution of Each Alternative with Level of Productivity
Table 2. shows contribution consideration at each utilization. The selected utilization as incinerator engine fuel with the index of BCR equals to 17.56 can be implemented as repair solution because it has a big contribution from every consideration to company and environmental performances.

Table 2. Contribution for utilization

Contribution	**Utilization as mini LPG (%)**	**Utilization as incinerator fuel (%)**
Productivity		
Material	23.3	83.8
Labor	77.8	23.32
Capital	0.6	10.13
Waste	0.1	0.11
Carbon emission	97.9	97.9
Community income aspect	1.08	1.00
BCR Index	0.12	17.56

3.6. Implementation of the alternative plan
After an alternative solution is available which has the greatest contribution to the improvement of productivity and environmental performance, the next is to drafting the implementation plan of the chosen alternative that is utilization as incinerator engine fuel.

Table 3. Implementation of Alternative Plan [6]

Purpose	Goal	Action	Responsib le Person	Implementer
Reducing of flare gas derived from the production process into an economic and environmentally friendly	Zero flarin g 2030	Purchase of Incinerator Machine	Production	Administration and Financial
		Provide training on incinerator machine operating procedures to workers in charge of the operation machine	Production	Production Workers

4. Conclusion

The chosen alternative to decrease the amount of residual gas (flare) is as fuel in incinerator engine with index benefit cost ratio is 17.56. This alternative contributes to the increased productivity of material used by 23.32%, humans 83.8%, capital 10.13 %, and waste decreased by 0.11%.

References

[1] Emam A 2016 Gas Flaring Reduction Perspective Environmental and economical, Departement of Chemical Eng. And Pet Refinery, *Suez* **2** pp.240-251
[2] Yobo, Moses, Tambari, S L, Paulinus A P 2016 Flare Gas Gathering and Utilization: A Strategic Approach to GreenhouseGas Emission Reduction in Nigeria *International Journal of Environmental Protection and Policy* **4 (1)** pp. 10-15
[3] Oseh, Jeffrey, Onuoma, Alex O I, Olawale O O, Adekanmi A A 2015 *International Journal of Oil, Gas and Coal Engineering* **3(5)** pp. 56-61
[4] Asian Production Organization 2006 Handbook on Green Productivity APO Canada
[5] Sumanth D J 1985 Productivity Engineering and Management Productivity Measurement Evaluation Planning and Improvement in Manufacturing and Service Organizations Mc Graw- Hill Company S
[6] Singgih M L 2011 Waste Reduction with Green Productivity Approach for IncreasingProductivity (Case Study : PT Indopherin Jaya),The 11th Asia Pacific Industrial Engineering and Management Systems Conference, The 14th Asia Pacific Regional Meeting of International Foundation for Production Research, pp.1-6
[7] Sugito 2011 Integrated Feasibility analysis on the solution of the reduction performance of flare gas for power plant Tanjung Jabung UPT University Indonesia Library
[8] Gervet B 2007 Gas Flaring Emission Contribute To Global Warming 2007 *Swedian* : Renewable Energy Reasearch Group
[9] Indriani G 2005 Gas flaring reduction in the Indonesian oil and gas sector, technical and economic potential of clean development mechanism (CDM) project HWWA Hamburg Report Hamburg Institute of International Economics
[10] Courban B 2010 Flare gas and the role of CNG-a challenge and an opportunity, Paper presented atthe3rd Annual CNG/NGV Indonesia Forum, *Jakarta*
[11] Wiranda A 2016 Manufacture of Small scale medical waste incinerator processing apparatus Thesis University Muhammadiyah Yogyakarta
[12] Pramono Y 2011 Case study of Mini LPG Plant development with utilization of flare gas in field X in terms of technical and economical aspects UPT Library of ITB Bandung
[13] Abdulhakem C A 2014 SPE-170211-MS Sep 15-17
[14] Sangrasaki M E, Anajafi K M 2005 International Conference on Chemical, Food, and Environment Engineering (ICCFEE'15) Dubai (UAE), Jan 11-12.
[15] Johnson M R, Coderre A R 2012 International Journal of Greenhouse Gas Control, Canada, p 121-131

IOP Conf. Series: Materials Science and Engineering **309** (2018) 012011 doi:10.1088/1757-899X/309/1/012011

Comparison of compressive strength of paving block with a mixture of Sinabung ash and paving block with a mixture of lime

I P Hastuty*, I S Sembiring and Nursyamsi
Departement of Civil Engineering, Universitas Sumatera Utara, Medan 20155

*Email: ika.hastuty@usu.ac.id

Abstract. Paving block is one of the material used as the top layer of road structure besides asphalt and concrete paving block is usually made of mixed material such as Portland cement or other adhesive material, water, and aggregate. People nowadays prefer paving block compared to other pavement such as concrete or asphalt. Their interest toward the use of paving block increase because paving block is an eco-friendly construction which is very useful in helping soil water conservation, can be done faster, has easier installation and maintenance, has a variety of shades that increase the aesthetic value, also costs cheaper than the other. Preparation of the specimens with a mixture of Sinabung ash and a mixture of Sinabung ash and lime are implemented with a mixture ratio of cement : sand : stone ash is 1: 2 : 3. The mixture is used as a substitute material by reducing the percentage amount of the weight of the cement with the composition ratio variation based on the comparative volume category of the paving block aggregate, i.e. 0%, 5%, 10%, 15%, 20%, and 25%. The result of this research shows that the maximum compressive strength value is 42.27 Mpa, it was obtained from a mixture of 10% lime with curing time 28 days. The maximum compressive strength value which is obtained from the mixture of sinabung ash is 41.60 Mpa, it was obtained from a mixture of 15% sinabung ash. From the use of these two materials, paving blocks produced are classified as paving blocks quality A and B (350 – 400 Mpa) in accordance to specification from SNI 03-0691-1996.

1. Introduction

Sinabung alongside Mount Sibayak nearby are two active volcanoes in North Sumatra and the highest peak in the province. The height of this mountain is 2,460 meters. This mountain has never recorded erupted since 1600, but suddenly active again by erupting in 2010. Until now in 2017 the mountain is still actively releasing lava and hot ash clouds around the site. The eruption of Mount Sinabung has caused damage to facilities and infrastructures such as plantations, houses, roads, and the cessation of normal activities of the community. This will certainly have a negative impact on the economic development of communities around the location of Mount Sinabung. The volcanic ash caused by this eruption covered almost all villages around Mount Sinabung, so there is a need to find ways to overcome the problem of the build-up of volcanic ash.

In addition to increase soil fertility, volcanic ash also contains silica and quartz that makes it can be classified into pozzolan materials. Pozzolan material is defined as non-cement material containing silica and alumina. While the classification of pozzolan material is divided into two parts, natural pozzolan (natural) and artificial (synthetic), natural pozzolan example is: tuff, volcanic ash, diatomaceous soil, and trass is a famous natural pozzolan name in Indonesia. Further examples of

IOP Conf. Series: Materials Science and Engineering **309** (2018) 012011 doi:10.1088/1757-899X/309/1/012011

artificial pozzolan is the result of burning clay, rice husk ash, ash bagasse and the result of burning coal (fly ash)[6].

Paving block is a building material made of a mixture of Portland cement or similar hydraulic binder, water and aggregate with or without other additives (SNI 03 - 0691 - 1996). In order to improve the quality of paving block, we can use materials that contain pozzolan ingredients such as silica, alumina and calcium.

Aggregate materials that are often used in the manufacture of paving blocks are sand which is derived from quarrying sand or river. Based on SNI 03-0691-1996 paving block (concrete blocks) can be classified into four qualities, quality A used for pavement, quality B used for parking areas, quality C used for pedestrian ,and quality D used for parks or other uses. Paving block which is produced manually are usually classified as paving block quality C and D, these types of paving blocks are used for garden and other uses that are not required to withstand the load on it. The quality of the paving block using a press machine can be categorized into concrete quality C to A with a compressive strength above 125 kg/cm^2 depending on the ratio of the mixture of materials used.

A good concrete is a concrete that has a high compressive strength, in other words it can be said that the quality of concrete is reviewed only from the compressive strength alone [3]. Several factors affecting the strength of concrete:

(1) Water Cement Ratio (WCR) and density
The function of the cement water ratio is to allow chemical reactions that cause binding and hardening, as a lubricant of the mixture of gravel, sand and cement to make it easier in concrete molding.
The strength of concrete depends on the ratio of the cement water ratio. The higher the WCR, the lower the strength of the concrete, however, the lower WCR does not alwas mean the higher the strength of the concrete. There is a limit in this case, the low WCR value will cause difficulty in the process, i.e. the difficulty in the implementation of compaction which will eventually cause the quality of concrete declines. Generally, the minimum WCR value is about 0.4 and the maximum WCR value is about 0.65. So it can be concluded that almost for all purposes, concrete that has a minimal WCR and enough to provide the specific workability required for perfect compaction without excessive compaction work, is the best concrete.

(2) Age of concrete
The compressive strength of concrete will increase in accordance with the age of concrete. Comparison of concrete compressive strength can be seen in Indonesia Reinforced Concrete Rules 1971.

(3) Type and quantity of concrete
The type of cement affects the compressive strength of the concrete in accordance with the intended use. The type of cement can be seen in SK SNI M-106-1990-03

(4) Aggregate properties
Surface Hardness: on aggregate with a rough surface there will be a good bond between the cement paste and the aggregate.

2. Method

2.1 General

The method used in this research is experimental research method. While the factor studied is the composition factor of the ash mixture of Sinabung and lime on the paving block of quality A, with the aim to know the effect of Mount Sinabung ash and lime as the additive materials by reducing the amount of cement in the compressive strength. Research flowchart is depicted in Figure 1.

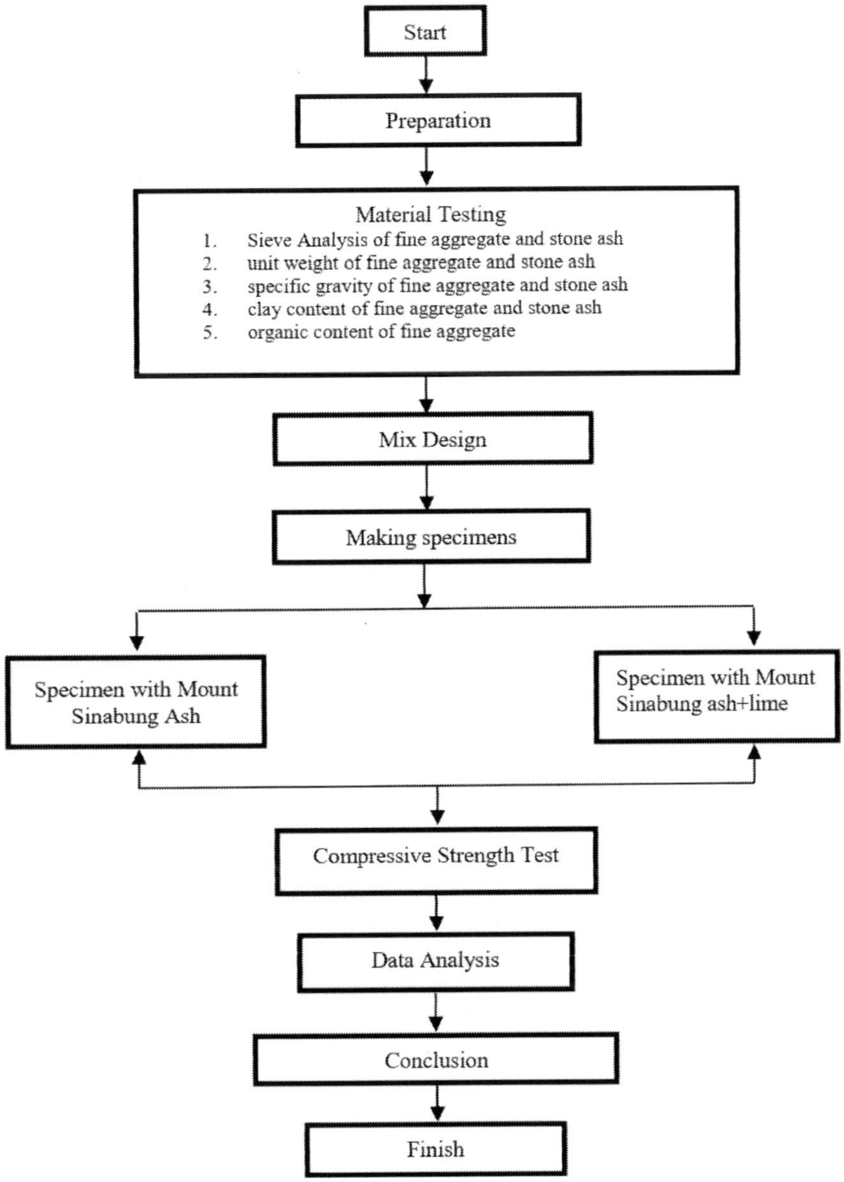

Figure 1. Flowchart Research

2.2 Materials Preparation

(1) Cement, the cement used in this research is Portland Cement type I
(2) Sand, the sand used in this research was taken from Sei Wampu quarry, Binjai.
(3) Water, the water used as the materials mixer was taken from Laboratory of Engineering Material, Departement of Civil Engineering, Faculty of Engineering, University of North Sumatera.
(4) Sinabung Volcanic Ash. The Sinabung volcanic ash was taken directly from Gurukinayan village which is only about 2.2 miles from Mount Sinabung and included as disaster-prone area II [6].

Table 1. Chemical composition of Sinabung Volcanic Ash

No.	Parameter	Unit	Result
1	Silica (SIO_2)	%	74.3
2	Aluminium oxide (AL_2O_3)	%	3.31
3	Calcium (CaO)	%	1.79

Source: Laboratory of Research Center and Industrial Standardization, Medan

(5) Lime

Lime used as a stabilizing material is Tohor lime with a chemical composition as shown in Table 2.

Table 2. Chemical composition of Lime

No.	Parameter	Result
1.	Silica (SiO_2)	3.03 %
2.	Aluminium oxide (Al_2O_3)	1.53 %
2.	Iron (III) Oxide (Fe_2O_3)	0.54 %
3.	Calcium (CaO)	51.8 %
4.	Magnesium (MgO)	0.81 %

Source: Laboratory of Research Center and Industrial Standardization. Medan

3. Results and Discussions

The test results of material can be seen in Table 3.

Table 3. Material Test Data

No.	Test	Result	Unit
1	Sieve Analysis of Fine Aggregate	FM = 2.68	
2	Sieve Analysis of Stone Ash	FM = 4.83	
3	Unit Weight of Fine Aggregate	1.663	g/cm^3
4	Unit Weight of Stone Ash	1.543	g/cm^3
5	Unit Weight of Sinabung Ash	1.342	g/cm^3
6	Specific Gravity of Fine Aggregate	2.52	g/cm^3
7	Specific Gravity of Stone Ash	2.52	g/cm^3
8	Specific Gravity of Sinabung Ash	2.36	g/cm^3
9	Clay Content of Fine Aggregate	2.53	g/cm^3
10	Clay Content of Stone Ash	3.70	%
11	Organic Content of Fine Aggregate	0.80	%

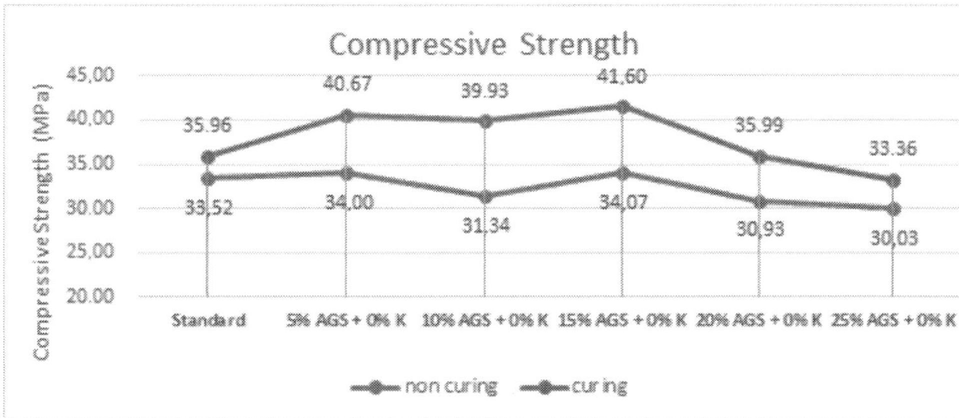

Figure 2. Compressive Strength Test with Sinabung Ash

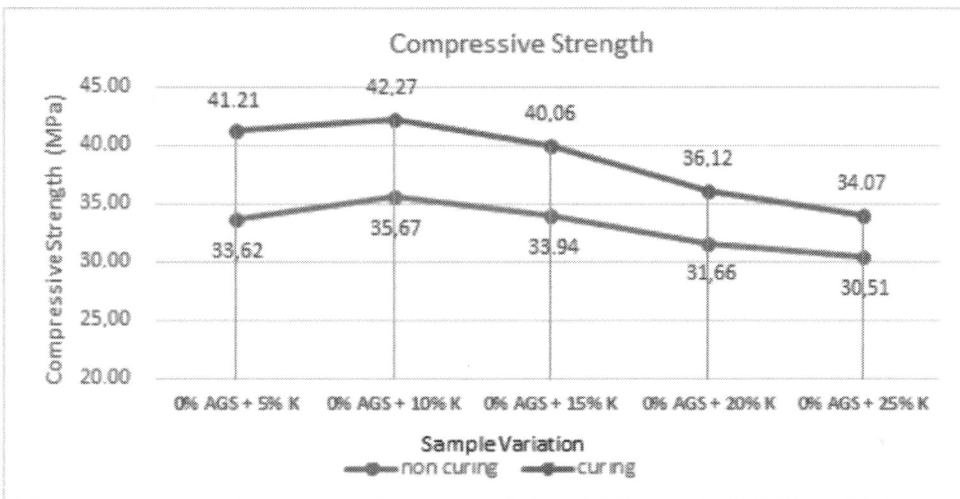

Figure 3. Compressive Strength Test with lime

From Figure 2. the highest compressive strength value is obtained on the use of 15% Sinabung ash with curing and and from Figure 3. it is shown that the compressive strength value of a mixture of 10% lime with curing time 28 days is 42.27 Mpa. Figure 1 shows that normal concrete (with curing) without the addition of sinabung ash and lime has a compressive strength value which is equal to 35.96 Mpa. The increasing of paving block quality is caused by the use of Sinabung and Lime ash. which is known as Sinabung ash has Silica content of 73.4% and Lime has CaO content equal to 51.8%.

Silica gives effect as a filler on the manufacture of paving block where the filler contributes to the paving block density that occurs at the beginning when the chemical reaction is still running slowly. The very fine particles fill the empty cavities containing water and $Ca(OH)_2$ between the aggregate and the binder. With fine particles filling the cavities. the layers of the paving block are saturated so as to increase the compressive strength of the paving block [5]. Lime is also able to improve the quality of paving block. this is due to the pozzolan properties of lime which quickly react to the silica which is released by cement during the process of hydration and forming compounds that are binding at normal temperatures in the presence of water.

IOP Conf. Series: Materials Science and Engineering **309** (2018) 012011 doi:10.1088/1757-899X/309/1/012011

4. Conclusions

Based on the test results it can be concluded as follows:

1. The use of Sinabung ash and lime as a cement substitution material can increase the compressive strength of the paving block.
2. The maximum compressive strength value which is obtained from the mixture of sinabung ash is 41.60 Mpa. it was obtained from a mixture of 15% sinabung ash.
3. The optimum compressive strength value which is obtained from the mixture of lime is 42.27 Mpa. it was obtained from a mixture of 10% lime.

Acknowledgments

The authors gratefully acknowledge that the present research is supported by Ministry of Research and Technology and Higher Education Republic of Indonesia. The support is under the research grant TALENTA USU of The Year 2017 Contract Number: 145/UN5.2.3.1/PPM/KP – TALENTA USU/2017

References

[1] ASTM. ASTM Annual Book of ASTM Standards Section 4 Volume 04.02. ASTM 100 West Conshohocken. PA.

[2] Badan Standarisasi Nasional. SNI 03-0691 1996 Bata Beton (Paving Block) . Bandung

[3] Dipohusodo I 1994 Struktur Beton Bertulang. Gramedia Pustaka Utama. Jakarta

[4] Mulyati 2015 Pengaruh Penggunaan Fly Ash sebagai Pengganti Aggregat Terhadap Kuat Tekan Paving Block. Jurnal Momentum

[5] Ilham A, Media Komunikasi Teknik Sipil. 13 : **75 – 85**

[6] I.S Sembiring I. Hastuty I P 2016 Sinabung Volcanic Ash Utilization As The Additive for Paving Block Quality A and B. IOP Publishing **180**

IOP Conf. Series: Materials Science and Engineering **309** (2018) 012012 doi:10.1088/1757-899X/309/1/012012

Feasibility study on intersection in North Sumatera

I P Hastuty*, I S Sembiring, R Anas, and A S Lubis
Departement Civil Engineering, Universitas Sumatera Utara, Jl. Dr. Mansur Kampus USU Medan 20155

*Email: ika.hastuty@usu.ac.id

Abstract. Traffic congestion is one of the problems faced by big cities, One of them is Binjai city and Medan city in North Sumatera. One of the causes of congestion is intersection of roads with highway and roads with railroads. To ensure the smooth movement of vehicles, technical handling at the intersection is needed. Therefore, it is necessary to pre-study the feasibility of level crossing in Binjai and Medan to be able to assess the investment needs and the level of importance of road development in the region. The development of transportation infrastructure and facilities is based on the thought of improving the transportation network system. The necessity of systemical integrated transportation system handling is needed in creating a transportation efficiency. The purpose of this study is to identify and prioritize the needs of the railway crossings. The objective which we want to be achieved is to obtain a document that contains technical, economic and environmental ability indicators as a reference in the feasibility and planning studies. The methodology used is collecting the primary data, secondary data, introduction of study area as the initial analysis of the problem. From the study it can be concluded that the existence of railway interchange will move the movement through traffic so it will not interfere the movement of traffic within urban areas and it keeps the national road network performance is still good.

1. Introduction

1.1Basic Legal Reviews
The development of transportation infrastructure and facilities is based on the thought of improving the transportation network system. The transportation system evolves to provide a balance between demand and supply. The need for integrated transportation system handling systemically is needed in creating a transportation efficiency [1]. Normative reference in the preparation of this study is the law of each sub-sector of transportation, especially the land Transport sub-sector.

1.2 Level Crossing Concept
Based on the Regulation of the General Director of Land Transportation Number SK.770 / KA.401 / DRJD / 2005 about Technical Guidelines for intersection between Roads and Railways, the concept of intersection and interchange is as follows:

a. The intersection of roads and railways, consisting of:
   1. Intersection with barrier/crossing gate
      a) Automatic;
      b) Not automatic either mechanical or electrical
   2. Intersection without barrier/crossing gate

b. Intersection as referred in a letter a) item 1 if it exceeds the provisions concerning:
   1. The number of trains crossing to the location is at least 25 train/day and at most 50 trains/day;
   2. The average daily traffic volume is 1,000 to 1,500 vehicles on city roads and 300 to 500 vehicles on the road outside the city; or
   3. The multiplication result between the average daily traffic volume and the train frequency is between 12,500 to 35,000. Then it should be upgraded into an interchange.

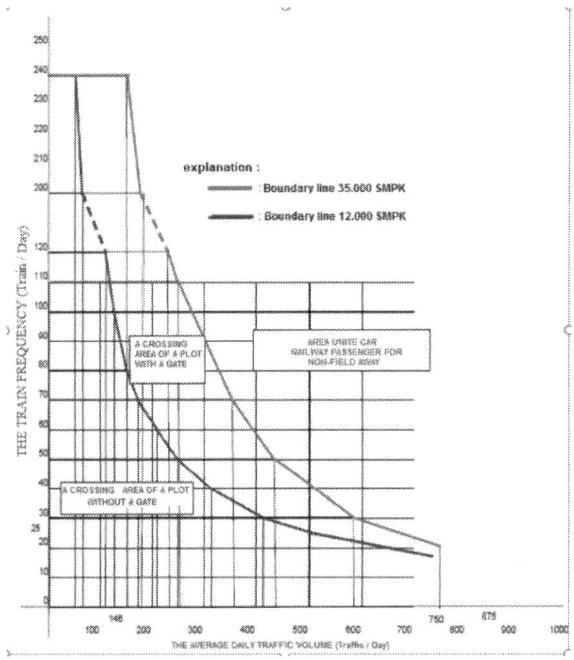

Figure 1. Graph of intersection area according to train traffic per day and average daily traffic volume

1.3 Delay

When the train passes through the interline, all vehicles must stop, this is because the train crossing gate is closed so travel time increases longer than the normal condition (delay occurs).

The delay occurring at the intersection of the highway and the railroad consists of two parts:
A. The delay that occurred when the closing of the crossing gate (see Figure 2.2).
B. The geometric delay as a result of crossing by rail when the crossing gate is opened.
So the delay experienced by any vehicle passing through the railway crossing can be stated as follows:
Delay = (delay when the railway crossing gate is closed) + (geometric delay, delay when the railway crossing gate is closed)

Meanwhile, Robertson (1994), defined the delay as follows:

- Delay is the time lost when driving due to things beyond the ability of the driver.
- Waiting time is a part of the delay process where the vehicle can not run or run slower than planned speed.
- The traffic delay occurs because of the difference between the actual travel time and the travel time based on the vehicle crossing the road segment that is used as the study field with the average speed representing the condition of the non-saturated traffic flow in the study section.

In the case of analytical calculations, the delay on the traffic light is measured on the stop line by assuming deceleration and acceleration. The delay time is identified by the length of time of a vehicle from the time when it stops until it accelerates and reaches its average speed again. This can be explained in Figure 2.2 which shows the curve of the time line trajectory of a vehicle conducting a process of stopping by slowing the vehicle from a constant average travel speed to a halt then accelerating the vehicle until it reaches its average speed again, and maintains zero speed (stop) for a certain time between slowing motion and acceleration (the slope of the time line trajectory represents the speed of the vehicle).

From the above it can be explained that the delay occurred to the vehicle, in this case, the change between the disrupted and undisturbed travel time, can be measured on each stop line with the average value of deceleration and acceleration assumed.

The parameters shown in Figure 2:

Vc	= travel speed (km / h)
L	= mileage
Tc	= 1 / Vc
	= travel time without interference (s)
D	= delay time (s)
T	= tc + d
	= travel time with interference (s)
Ta, tb	= deceleration time and acceleration time (s)
La, lb	= deceleration distance and acceleration distance (meters)
Ds	= stopped delay time(s)

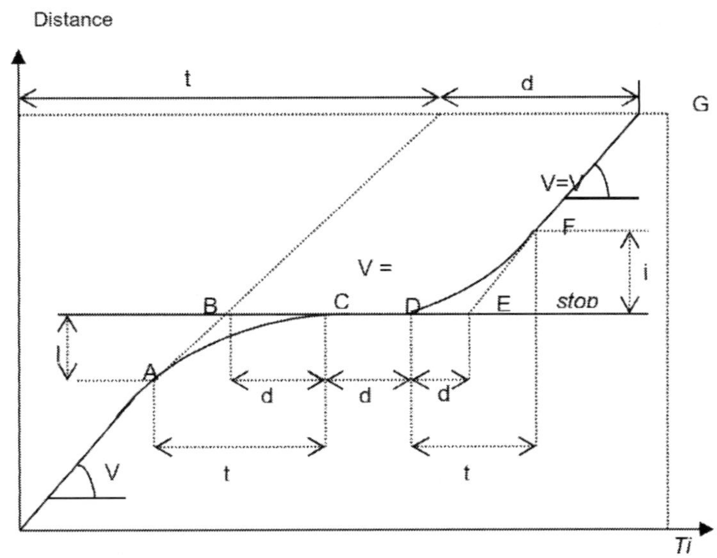

Figure 2. Trajectory of Time Ranges From A Vehicle That Made The Stop Process Due to Delay (AKCELIK 1981)

1.4 Crosses Flow

In calculating the number of currents and the length of the queue on the railway crosses with the road, there are some times that need to be considered, i.e.:

a. When the train begin to be seen by the driver
b. When the vehicle is slowing down
c. When the traffic flows start running the vehicle again
d. When traffic flows back to normal where the influence of the railway crossing gate is gone

By the time a train passes an intersection between railway and road without crossing gate, the driver can only detect by hearing first and then eyesight. When the train starts to be detected, the driver slows down the vehicle speed. As the train begins to pass through the crossings, the vehicle begins to stay motionless. This situation is prevailing during the time of rail traffic. For an intersection with a crossing gate, the train is detected with a automatic signal and then the driver sees the door of the train that began to close. After the train passed the crossing, the vehicle begins to move through the intersection, causing currents to change gradually until a certain time. After passing through the crossing, the speed will gradually rise so that the flow will return to normal conditions. The condition can be shown in Figure 3. The shaded part of Figure 3 is the amount of delay due to the closing of the railway crossing.

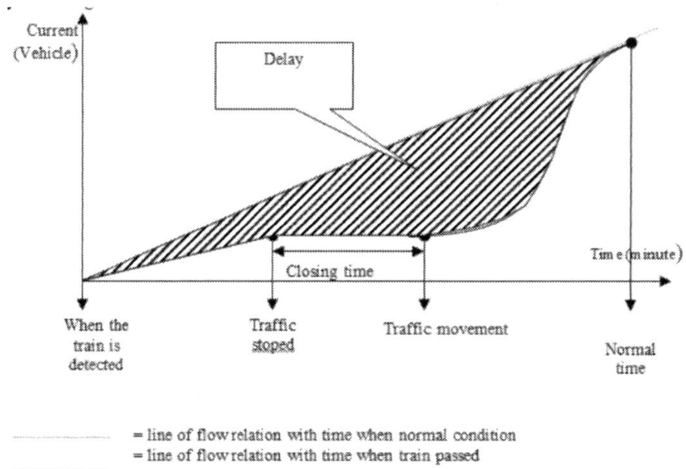

Figure 3. Graph between vehicle flows and crossing time

2. Research Method

2.1 Preparation

a. Consolidating the Methodology, planning detailed implementation phases, collecting relevant agency data, introducing study areas, establishing modelling and analysis, developing technical analysis structures and survey plans.
b. The Literature Study, by reviewing implementation methods, maximizing data usage, and analytical methods.
c. Read the relevant Regulations, by preparing the pre-feasibility study concept, knowing the layout plan and developing some assessment indicators.
d. Identify initial conditions and problems on an intersection with the road network system and the impact on the overall road network system

2.2 Collecting Primary and Secondary Data

Secondary data includes socio-economic, document related to research and road network database. While the primary data are road network survey, condition and soil structure survey, topographic survey, and traffic counting.

2.3 Stages of analysis

Analyzes conducted in this study include:

- Conditions for existing traffic conditions (delay), creating a handling scenario and putting the analytical prediction basis in the planned "time horizon."
- Do Nothing Conditions, predicting traffic performance without changes in network conditions, and then the analysis results will be used as a benchmark for the benefits of various handling scenarios.

- Do something condition, traffic performance prediction performed with changing the network conditions (FO / UP) and the benefits reviewed is a benefit to users and society.

The environmental analysis examines the implementation feasibility of the environment, including estimation of the benefits of air pollution reduction caused by vehicle emissions due to reduced congestion, assessing the hydrological side of the study area and the possible social impacts.

Project Cost Analysis is a calculation of all components of project costs incurred for the project, in this case, carried out on: land acquisition, construction, and maintenance.

The Project Benefit Analysis is a cost calculation that will be a benefit if the upgrading project is built, in this carried out on BOK and time value.

3. Research Result

The survey result of the condition of intersection based on its use of the land can be seen in Table 1.

Table 1. Recapitulation of crossing condition

No		Intersection	District	Land use		Land condition
				Left	Right	
1		JLN. LINGKAR LUAR BINJAI	BINJAI CITY	Industry	Industriy	- - -
2		JLN. ASRAMA (MEDAN)	MEDAN CITY	Commercial	Commercial	-
3		JLN. YOS SUDARSO (MEDAN)	MEDAN CITY	Plantation	Commercial	+
4	a	BTS. KOTA MEDAN – TEMBUNG – LUBUK PAKAM	DELI SERDANG DISTRICT	Commercial	Commercial	-
	b	BTS. KOTA MEDAN – TEMBUNG – LUBUK PAKAM (2)	DELI SERDANG DISTRICT	Vacant land	Vacant land	+ + +
	c	BTS. KOTA MEDAN – TEMBUNG – LUBUK PAKAM (3)	DELI SERDANG DISTRICT	Commercial	Commercial	-

Table 2. Recapitulation of Topographic Survey

No		Intersection	District	Topography conditions
1		JLN. LINGKAR LUAR BINJAI	BINJAI CITY	Flat
2		JLN. ASRAMA (MEDAN)	MEDAN CITY	Flat
3		JLN. YOS SUDARSO (MEDAN)	MEDAN CITY	Flat
4	a	BTS. KOTA MEDAN – TEMBUNG – LUBUK PAKAM	DELI SERDANG DISTRICT	Flat
	b	BTS. KOTA MEDAN – TEMBUNG – LUBUK PAKAM (2)	DELI SERDANG DISTRICT	Flat
	c	BTS. KOTA MEDAN – TEMBUNG – LUBUK PAKAM (3)	DELI SERDANG DISTRICT	Flat

Table 3. Recapitulation of Traffic Counting Survey Results

No		Intersection	District / City	JVP	k	Av. traffic	The frequency of trains	SMPK	delay (m)	Average Delay (second)
1		JLN. LINGKAR LUAR BINJAI	BINJAI CITY	816	9%	9067	24	217 608	99	65
2		JLN. ASRAMA (MEDAN)	MEDAN CITY	1246	9%	13845	24	332 280	135	110
3		JLN. YOS SUDARSO (MEDAN)	MEDAN CITY	971	9%	10789	24	258 936	64	45
4	a	BTS. KOTA MEDAN – TEMBUNG – LUBUK PAKAM	DELI SERDANG DISTRICT	653	12 %	5442	38	206 796	24	40
	b	BTS. KOTA MEDAN – TEMBUNG – LUBUK PAKAM (2)	DELI SERDANG DISTRICT	702	12 %	5850	38	222 300	21	30

c	BTS. KOTA MEDAN – TEMBUNG – LUBUK PAKAM (3)	DELI SERDANG DISTRICT	821	12 %	6842	26	177 892	14	32

The result of transportation modelling obtained from ATTN data in 2011 is calibrated and estimated into 2016 data with Furness method to obtain Ei and Ed value equal to 1 with an accuracy of 3 decimal digits.

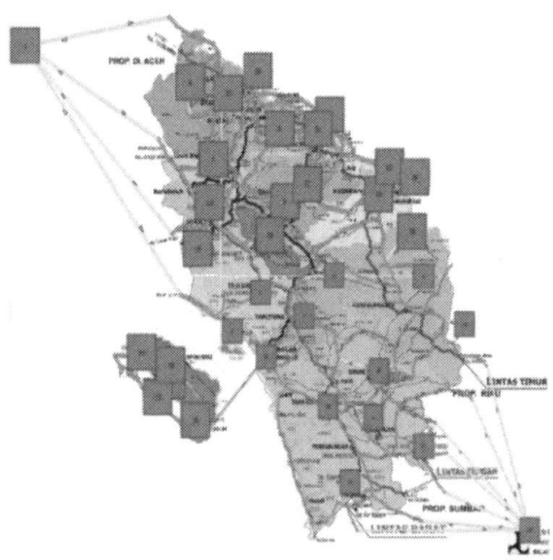

Figure 4. Zoning area in North Sumatera

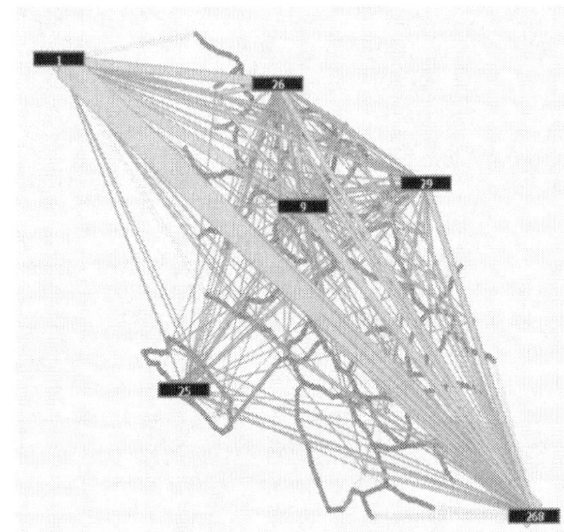

Figure 5. Desire Line in North Sumatera 2016

For modelling, the comparison of the result of the run (estimated) and TC (target) is being done. The result of modelling comparison and survey can be seen in Table 4.

Table 4. Comparison of traffic survey volume and modelling

No		Intersection	TC Volume	Model Volume
1		JLN. LINGKAR LUAR BINJAI	816	1450
			801	1334
2		JLN. ASRAMA (MEDAN)	1246	1450
			1093	1334
3		JLN. YOS SUDARSO (MEDAN)	943	1433
			971	1203
4	a	BTS. KOTA MEDAN – TEMBUNG – LUBUK PAKAM	628	787
			653	957
	b	BTS. KOTA MEDAN – TEMBUNG – LUBUK PAKAM (2)	702	787
			646	957
	c	BTS. KOTA MEDAN – TEMBUNG – LUBUK PAKAM (3)	510	787
			821	957

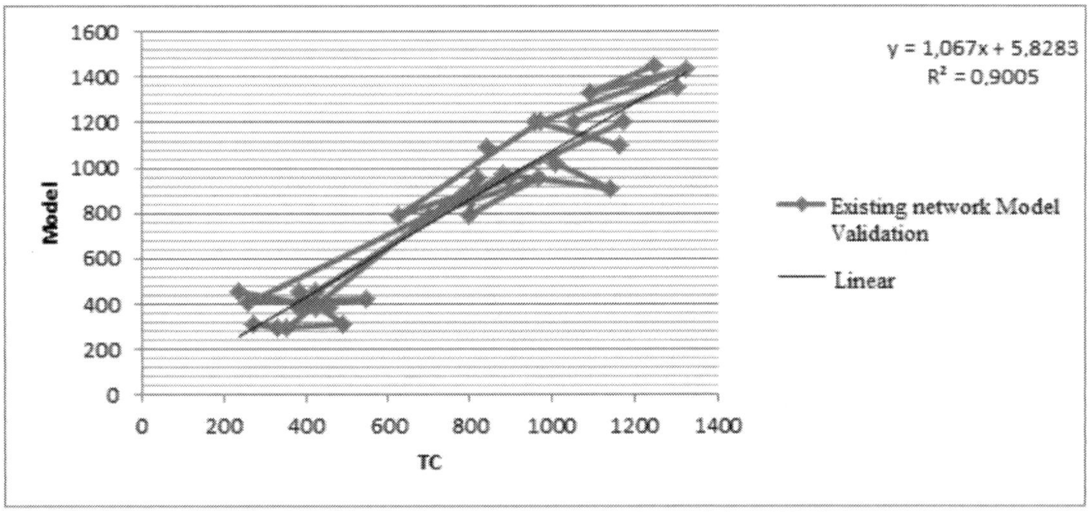

Figure 6. Graph of Existing Network Model Validations

The R2 value is 0.9005; this is indicating that the MAT model estimation results are can be represent more than 90.05% of the traffic flow in the study area. It can be said that this MAT estimation model is good enough to describe the pattern of vehicle travel demand in the study area, so it can be used as a basis for predicting MAT in North Sumatera Province and surrounding areas in the future.

Table 5. Priority-based on Analyze Results

No		Intersection	JVP (amp/hour)	K (%)	Average daily traffic (amp/day)	The frequency of passing trains	SMPK	Delay (m)	Average Delay (second)	Priority
1		JLN. LINGKAR LUAR BINJAI	816	9	9067	24	217608	99	65	2
2		JLN. ASRAMA (MEDAN)	1246	9	13845	24	332280	135	110	1
3		JLN. YOS SUDARSO (MEDAN)	971	9	10789	24	258936	64	45	3
4	a	BTS. KOTA MEDAN – TEMBUNG – LUBUK PAKAM	653	12	5442	38	206796	24	40	4
	b	BTS. KOTA MEDAN – TEMBUNG – LUBUK PAKAM (2)	702	12	5850	38	222300	21	30	5
	c	BTS. KOTA MEDAN – TEMBUNG – LUBUK PAKAM (3)	821	12	6842	26	177892	14	32	6

Based on validation and traffic survey data of the construction of the intersection on Asrama street, Medan, the average delay obtained from the primary data is 110 seconds. The average delay obtained for outer ring of Binjai is 65 seconds, Yos Sudarso street is 45 seconds, and the border of Medan-Tembung is 40 seconds.

4. Conclusion

From the results of the analysis in the previous chapters it can be drawn some conclusions as follows:

- The existence of railway interchange will move the movement through traffic, so it does not interfere with the movement of traffic in urban areas, and this is keeping the national road network performance remains good.

- The construction of railway interchange would benefit the movement of passengers and goods and assist in the development plan of the regions which is passed by those railway interchanges.
- The order of interchange contruction in the study area: Jl. Asrama, Jl. Lingkar Luar Binjai, Jl. Yos Sudarso, Medan – Tembung – Lubuk Pakam (a), Kota Medan – Tembung Lubuk Pakam (b), Medan – Tembung – Lubuk Pakam (c)
- From the analysis of transportation needs, it can be seen that the railway interlined already needed to be built, but it still must be adjusted to the order of priority.

Acknowledgments

The authors gratefully acknowledge that the present research is supported by Ministry of Research and Technology and Higher Education Republic of Indonesia. The support is under the research grant TALENTA USU of The Year 2017 Contract Number : 335/UN5.2.3.1/PPM/KP – TALENTA USU/2017

References

[1] Hobbs F D 1995 Perencanaan dan Teknik Lalu-lintas, Gadjah Mada University Press, Yogyakarta
[2] Khisty C J, Lall B K 2005 Dasar-dasar Rekayasa Transportasi Jilid 1, Erlangga, Jakarta
[3] Miro F 2005 Perencanaan Transportasi, Erlangga, Padang
[4] Morlok E K 1998 Pengantar Teknik dan Perencanaan Transportasi, Terjemahan oleh: J. K. Hainim, Penerbit Erlangga, Jakarta
[5] Nasution M N 2008 *Manajemen Transportasi edisi ketiga*, Penerbit Ghalia Indonesia, Bogor
[6] Tamin O Z 1997 Perencanaan dan Pemodelan Transportasi, ITB, Bandung

IOP Conf. Series: Materials Science and Engineering **309** (2018) 012013 doi:10.1088/1757-899X/309/1/012013

Effect of HDPE plastic waste towards batako properties

N Nursyamsi [1*], **I Indrawan**[1], **V Theresa**[1]
[1]Department of Civil Engineering, Faculty of Engineering, Universitas Sumatera Utara, Padang Bulan, Medan 20155, Indonesia

*E-mail: nursyamsi@usu.ac.id

Abstract. Indonesia is the world's second largest producer of plastic waste to the sea, after China. Most of the plastic waste is polyethylene. Polyethylene is a polymer consisting of long chains of ethylene monomers. Moreover, polyethylene is plastic that has characteristics such as; thermoplastic, elastic, non-translucent, odorless, slightly opaque and transparent, resistant to impact and has a resistance of up to 135 degrees Celsius. The type of HDPE plastic (high-density polyethylene), which has been cleaned and chopped as a substitute of fine aggregate, is used in the brick's making process. HDPE has a stronger, harder, smoother and more resistant to high-temperature properties. In this study, a weight variation of 0%, 10%, and 20% of HDPE plastic wastes was used from the total weight of sand as a substitution. Furthermore, the tensile and compressive strength were tested on day 7. Based on the research, the quality of the specimen achieved was categorized in quality III according to SNI 03-0349-1989.

1. Introduction

Nowadays, the development of construction materials is focused on the use of wastes to produce something that is beneficial to humans. One of them is the plastic waste. In this study, plastics were used as the sand substitute material for the manufacture of concrete brick.

Indonesia is the world's second largest producer of plastic waste to the sea, after the China. Most of the plastic waste is polyethylene. Polyethylene is a polymer consisting of long chains of ethylene monomers. Moreover, polyethylene is plastic that has characteristics such as; thermoplastic, elastic, non-translucent, odorless, slightly opaque and transparent, resistant to impact and has a resistance of up to 135 degrees Celsius [3]. In this study, a type of HDPE plastic (high-density polyethylene) was used. HDPE has stronger material properties, harder, opaque and more resistant to high temperatures[2]. HDPE is commonly used for colored milk bottles, oil container, shampoo bottles, and much more. HDPE is one of the safe plastic materials because of the ability to prevent chemical reactions between HDPE plastic packaging and the packaged food or beverages. However, this type of plastic is recommended only for single use because of the release of antimony trioxide compounds that continue to increase over time.

The use of plastics in daily life is one of the main factors of waste production in Indonesia. Plastics have hard-to-decompose properties where it takes hundreds of years to decompose completely. Therefore, most of the Indonesian people disentangle plastic waste by burning it. In fact, the plastic will cause air pollution during the burning process, and it is quite dangerous to human health, where the burned plastic will release harmful substances into the air such as carbon monoxide, dioxin and furan, volatile and other harmful particles. One way to reduce the plastic waste is reusing the plastic waste itself.

In this study, the waste of HDPE plastic waste will be used in the process of making brick as a substitution of fine aggregate. This research was expected to produce something useful in the manufacture of bricks that have high quality and environmentally friendly.

Figure 1. HDPE plastic chopping

2. Literature Review

Concrete brick is a construction material in the form of alternative replacement of bricks composed of sand, cement, and water. Concrete brick is usually focused on non-structural building wall constructions [7].

According to general requirements of construction materials in Indonesia (1982) article 6, "Concrete brick is a brick made by molding and maintaining in humid conditions." According to SNI 03-0349-1989, "Conblock (concrete block) is a construction component made from a mixture of Portland cement or pozzolan, sand, water and other additives, molded in such way as to qualify and can be used as wall pair materials." While Frick Heinz and Koesmartadi (1999: 96) argued that: "Unburned stones, known as the brick (brick made in solidification of trass, chalk, water)."

Concrete brick must meet the physical requirements by the following table:

Table 1. Physical requirements of concrete bricks

Physical Requirements	Unit	Solid concrete brick quality level				Hollow concrete brick quality level			
		I	II	III	IV	I	II	III	IV
1. Minimum average of gross compressive strength.	Kg/cm^2	100	70	40	25	70	50	35	20
2. Minimum gross compressive strength for each specimens.	Kg/cm^2	90	65	35	21	65	45	30	17
3. Maximum average of water absorption	%	25	35	-	-	25	35	-	-

Polyethylene is a polymer consisting of long chains of ethylene monomers. Polyethylene is plastic that has characteristics such as: thermoplastic, elastic, non-translucent, odorless, slightly opaque and transparent, resistant to impact and has a heat resistance of up to 135 degrees Celsius[3]. High-Density Polyethylene (HDPE) is one type of polyethylene plastic. HDPE has a density exceeding or equal to 0.941 g/cm^3. HDPE has a low degree of branching and has a very high molecular strength and tensile strength. HDPE can be produced with chromium/silica catalysts, Ziegler-Natta catalysts, or metallocene catalysts. HDPE has stronger, harder, opaque and more resistant to high-temperature properties. HDPE is commonly used for colored milk bottles, oil container, shampoo bottles, and many others[2]. HDPE is one of the safe plastic materials to use because of the ability to prevent chemical reactions between HDPE plastic packaging and the packaged foods/beverages. However, this type of plastic is recommended only for single use because of the release of antimony trioxide compounds that continue to increase over time [7].

3. Research Methodology

The method used in this research was experimental research method. In this study, a weight variation of 0%, 10%, and 20% of HDPE plastic wastes was used from the total weight of sand as a substitution with a mixture of bricks was 1: 6 [5]. Furthermore, the tensile and compressive strength was tested on day 7.

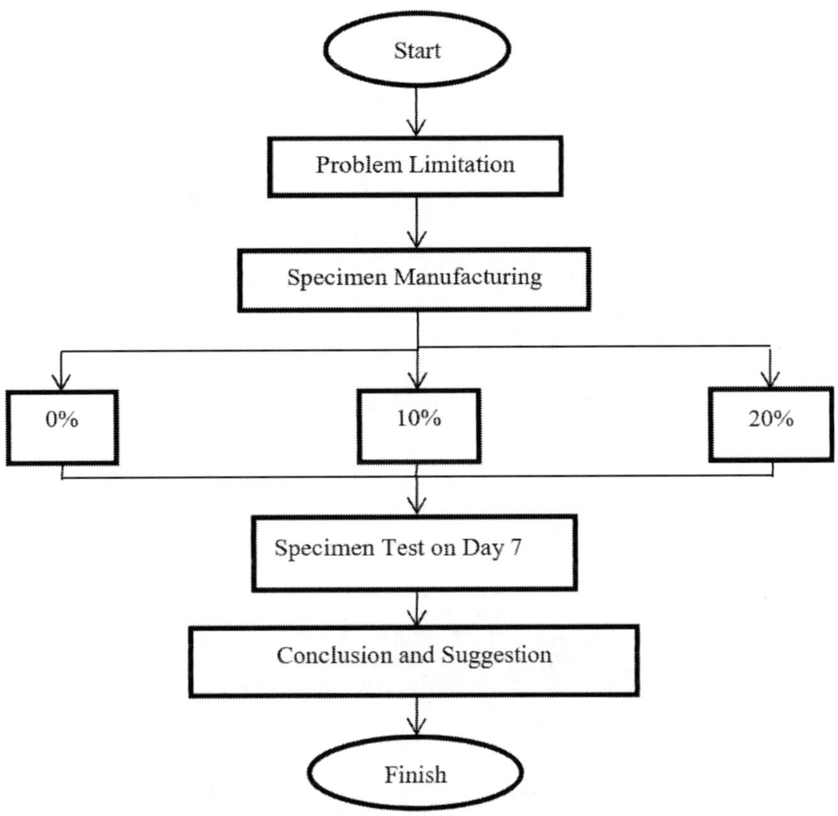

Figure 2. Research Methodology Flowchart

The tensile and compressive strength test of concrete brick were tested at the Concrete Materials and Engineering Laboratory of the University of Sumatra Utara. It was tested on day 7, started from the manufacture of the specimen with three variations. Each of them was amounted to 7 specimens of concrete brick and nine specimens of briquette.

Figure 3. Sample with 20% of HDPE mixture

Figure 4. Sample with 10% of HDPE mixture

Figure 5. Commercial Concrete brick

4. Result and Discussion

The test was done on day 7 in which on the compressive test, it was acquired the maximum load which is when the bricks crushed after given the load. The number of test specimens in the compressive and tensile strength test is seven samples and nine samples for each mixed variation. From the tensile and compressive strength test of concrete brick obtained data as follows:

Table 2. Average compressive strength test result of commercial concrete brick

Weight (g)	P max (kN)	Compressive Area (cm^2)	Compressive Strength (Kg/cm^2)
16446	202	280	73.59
16322	185	280	67.39
16221	196	280	71.40
15971	185	280	67.39
16925	198	280	72.13
16644	190	280	69.21
16265	192	280	69.94
16399.14		**Average**	**70.15**

Table 3. Average compressive strength result of concrete brick with 10% HDPE mixture

Weight (g)	P max (kN)	Compressive Area (cm^2)	Compressive Strength (Kg/cm^2)
14221	130	280	47.36
14458	144	280	52.46
13609	128	280	46.63
13704	132	280	48.09
14461	144	280	52.46
13790	135	280	49.18
13785	134	280	48.81
14004		**Average**	**49.28**

Table 4. Average compressive strength result of concrete brick with 20% HDPE mixture

Weight (g)	P max (kN)	Compressive Area (cm^2)	Compressive Strength (Kg/cm^2)
14034	124	45.17	45.17
14343	130	47.36	47.36
13268	118	42.99	42.99
12783	125	45.54	45.54
12497	113	41.16	41.16
11756	126	45.90	45.90
11754	118	42.99	42.99
12919.29		**Average**	**44.44**

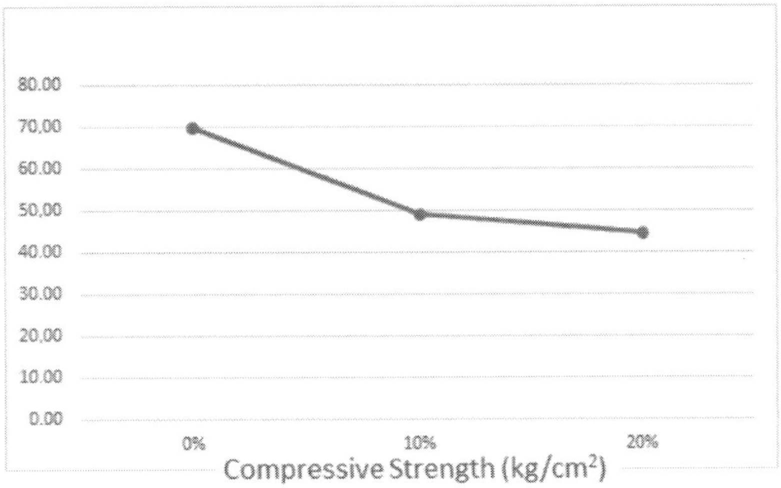

Figure 6. Concrete brick compressive strength graphic

Table 5. Tensile strength test result of the commercial briquette

Weight (g)	Tensile Strength (kN)	Area (cm²)	Tensile Strength (Kg/cm²)
119.000	0.825	6.350	13.252
109.000	0.825	5.240	16.059
99.000	0.700	5.110	13.973
110.000	1.100	5.800	19.345
115.000	1.250	6.860	18.586
106.000	0.800	5.820	14.021
110.000	0.900	6.320	14.525
108.000	1.100	5.200	21.577
104.000	1.200	6.240	19.615
108.889	**Average**		**16.773**

Table 6. Average tensile strength result of the briquette with 10% HDPE mixture

Weight (g)	Tensile Strength (kN)	Area (cm²)	Tensile Strength (Kg/cm²)
114.000	0.550	6.030	9.303
99.000	0.550	5.270	10.645
93.000	0.450	8.025	5.720
103.000	0.500	5.950	8.571
96.000	0.550	5.280	10.625
110.000	0.650	5.250	12.629
91.000	0.550	5.360	10.466
102.000	0.550	5.860	9.573
97.000	0.600	6.025	10.158
100.556	**Average**		**9.743**

Table 7. Average tensile strength result of the briquette with 20% HDPE mixture

Weight (g)	Tensile Strength (kN)	Area (cm²)	Tensile Strength (Kg/cm²)
109.000	0.650	5.650	11.735
110.000	0.625	5.460	11.676
106.000	0.700	5.900	12.102
111.000	0.750	6.525	11.724
112.000	0.625	5.540	11.507
107.000	0.900	5.360	17.127
118.000	1.025	5.420	19.290
113.000	0.650	5.930	11.180
108.000	0.550	6.370	8.807
110.444	**Average**		**12.794**

5. Discussion and Analysis

From the data above, the average compressive strength analysis of concrete brick with 10% and 20% HDPE had a smaller average compressive strength compared to the compressive strength of normal concrete brick. This occurred because the concrete brick with plastic had more cavities than the normal concrete brick. As for the case with the compressive strength of concrete brick, the average tensile strength of the briquette with the plastic waste mixture was also smaller compared to the normal briquette.

6. Conclusion

1. According to SNI 03 - 0349 - 1989, the quality of normal concrete brick in the study was included in the second quality of 70.15 kg/cm^2.
2. According to SNI 03 - 0349 - 1989, the quality of concrete brick with the composition of 10% HDPE plastic mixture in the study was included in the third quality of 49.28 kg/cm^2.
3. According to SNI 03 - 0349 - 1989, the quality of concrete brick with the composition of 20% HDPE plastic mixture in the study was included in the third quality of 44.44 kg/cm^2.
4. The average tensile strength for the normal briquette, briquette with the composition of a plastic mixture of 10% and 20% were 16.773 kg/cm^2, 9.743 kg/cm^2, and 12.794 kg/cm^2 respectively.

References

[1] Ahyat, M. Ridha, Nursyamsi. 2016. *Pemanfaatan Limbah Botol Plastik dan Abu Batu sebagai Bahan Substitusi dalam Pembuatan Concrete brick*. Medan.
[2] Bierley, A.W., R.J. Heat and M.J. Scott. 1988. *Plastic Materials Properties and Aplications*. Chapman and Hall Publishing: New York.
[3] Christopher. H. 1981. *Polymer Materials*. Mac Millan Publishers LTD: London.
[4] Mulyono, Tri. 2005. *Teknologi Beton*. Yogyakarta: Andi.
[5] Ramadhan, Prasetyo. 2016. *Pengaruh Penggunaan Limbah Plastik LDPE sebagai Agregat Halus Pada Concrete brick Beton Ringan*. Medan.
[6] Soebandono, Bagus, As'at Pujianto, Danar Kurniawan. 2013. *Perilaku Kuat Tekan dan Kuat Tarik Beton Campuran Limbah Plastik HDPE*. Yogyakarta: Semesta Teknika. Vol 16,No. 1:**76-82.**
[7] Tjokrodimuljo, Kardiyono. *Teknologi Beton*. Yogyakarta: Biro Penerbit Teknik Sipil Universitas Gajah Mada.

IOP Conf. Series: Materials Science and Engineering **309** (2018) 012014 doi:10.1088/1757-899X/309/1/012014

The use of glass powder in making batako

N Nursyamsi* and **I Indrawan**

Department of Civil Engineering, Faculty of Engineering. Universitas Sumatera Utara, Padang Bulan, Medan 20155, Indonesia

*E-mail: nursyamsi@usu.ac.id

Abstract. Along with the increase in construction materials, innovation is needed to lessen the use of them, and one of them is by using cement [1]. In this research, it is reduced by glass powder; the reason for using it as the substitution of cement is that some chemical elements in cement are similar to those in glass powder such as SiO2, A12o3, Fe2O3, and CaO. The glass powder used was the one who passed sieve no. 100 and was hampered in sieve no. 200. It passed sieve no. 200 with its composition of 0%, 10%, 15%, 20%, 25%, and 30% from the volume of the use of cement. The specimen would treat within 28 days before the testing of compressive strength, water absorption, and tensile strength [2]. The variation which produced optimum result would mix with the foaming agent as the material for reducing the weight of the specimen. After that, the test of compressive strength, water absorption, and tensile strength on the installment of batako walls were done. The data analyzed by using SNI 02-0349-1989[3] reference about concrete brick for wall installment. The variation of 20% of glass powder passing sieve no. 200 gave optimum result. A specimen of the variation on glass powder of 20% which passed sieve no. 200 and the foaming agent was higher than the compressive strength of the specimen which used glass powder substitution of 0% of passing sieve no. 200 and foaming agent. The compressive strength of batako walls which used the batako construction with glass powder substitution of 20% of passing sieve no. 200 and the foaming agent was also higher than the compressive strength of the assaying object which used glass powder substitution of 0% of passing sieve no. 200 and foaming agent.

1. Introduction

Batako is a construction material which made of the mixture of cement, sand, and water which are put into a standard mold [4]. It used as an alternative substitution of brick for making walls. Today, building construction in Indonesia increases rapidly; therefore, the need for construction materials is also increasing. People in Indonesia use bricks as the material for installing walls. As the result of increasing development in construction, the need for bricks also rapidly increases; therefore, alternative materials for installing walls as the substitution of bricks are needed, and one of them is batako. Today, batako is used by many people because it is easily made, without being burned up like bricks. It has bigger than a brick so that installing walls will be faster. The effect is that many people now use batako instead of bricks which eventually there is a rising demand for batako in the market.

Batako constitutes a non-structural material for installing walls although it also has to fulfill the requirement for a standard of strength. It has the standard which has organized for the use and the making of batako which is embody, either directly or indirectly, in the Indonesian National Standard (SNI) 03-0349-1989 [3].

IOP Conf. Series: Materials Science and Engineering **309** (2018) 012014 doi:10.1088/1757-899X/309/1/012014

In this research, glass powder used as the substitution for cement. If it viewed from the similarity of chemical composition between cement and glass powder, it assumed that glass powder could used as the substitute of cement [5]. It has chosen because its chemical elements were similar to those of cement like silica (SiO2), Na2O, and CAO of 60%. It could see in the Table of Portland Cement Elements [4] and the Table of Glass Powder Elements (MIPA Chemistry Laboratory, USU). It expects that the use of glass powder can decrease the use of cement so that it will produce more efficient value.

Table 1. Portland Cement Elements

Elements	Portland Cements
SiO_2	20% - 25%
Al_2O_3	7% - 12%
Fe_2O_3	7% - 12%
CAO	60% -65%

Table 2. Glass Powder Elements

Elements	Glass Powder
SiO_2	97.0080%
Al_2O_3	0.1273%
Fe_2O_3	0.0026%
CAO	0.1084%

The glass is a material which is very difficult to decompose; its decomposition will take million of years. The use of glass powder as the material supplement for batako is aimed either to reduce raw materials of cement for making batako or the impact of environmental damage caused by glass waste. The reuse of glass waste is also one of the methods of reducing it in the environment [6]. It is recycled to become glass powder which will use as the substitution for cement in making batako. The goal of the research was to apply glass powder for using cement and foaming agent as the materials for reducing the weight of batako, to find out compressive strength, absorption, tensile strength, and compressive strength of wall installment.

2. Research Method

This used an experimental method which analyzed the composition of using glass powder in batako. It was aimed to find out the influence of glass powder as the substitution on the amount of using cement in water absorption, compressive strength, and tensile strength [7]. The research design in batako was as follows: testing object would make with the mixture ratio of 1Pc:7Ps in which glass powder as the substitution would put into by reducing some parts of the cement. The variation of the proportion was 0%, 10%, 15%, 20%, 25%, and 30% of glass powder of the weight of cement with two variations of glass grains: glass grains with ϕ passing sieve no. 200 and glass grains with ϕ sieve no. 100 and hampered on the sieve no. 200. The variation of the most optimum result would add with the foaming agent in the mixture of 1:20 (for water weight) to find out compressive strength, absorption, and compressive strength of batako wall installment. The making of the specimen and the procedure of quality testing was by the Indonesian National Standard (NSI) 03-0349-1989.

2.1 Materials

The materials for installing walls were Portland cement, sand, water, and glass powder. Portland cement was Type I Portland Cement [8] with the trade mark, Semen Padang, with the package of 50 kilograms per one sack. The sand used in this research taken from The Wampu River quarry, Binjai. The water which used as the mixing material was taken from the Laboratory of Engineering Materials, Civil Engineering Department, the Faculty of Engineering, University of Sumatera Utara.

2.2 Glass Powder

This research used glass powder which mixed with batako where it obtained from the waste of used drinking glass bottles which had been cleaned up, smashed to pieces, and crushed by pounding [5]. The glass was sifted with sieves no. 100 and no. 200.

3. Research Result

3.1 Batako Compressive Strength Testing

Batako used as the specimen was the one which had 28 day-treatment with two variations of glass powder supplement: glass powder which had passed sieve no. 100, was hampered on sieve no. 200, and sieve no. 200. The specimen was a cube which had been taken during the process of batako die casting and compressed until the maximum load was able to be held back by the batako cube. The result of the testing could see in Picture 1 below:

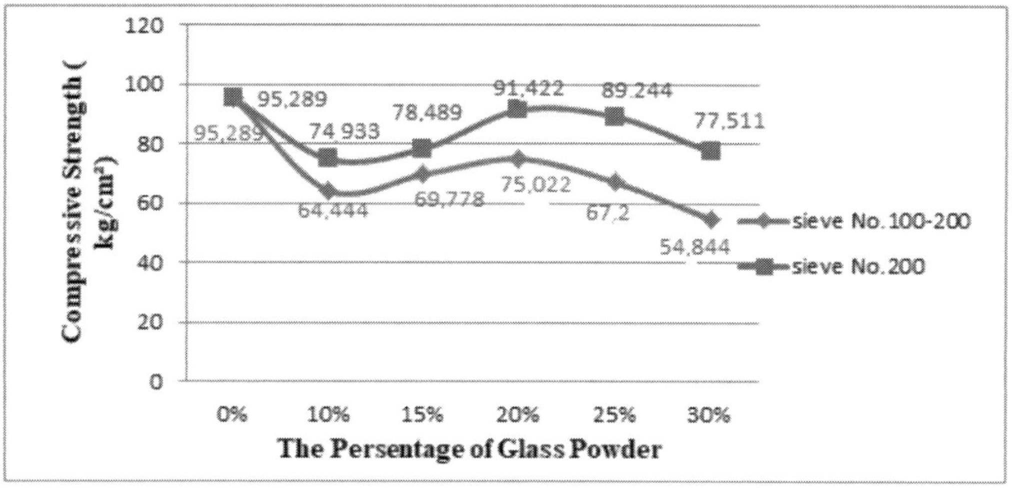

Figure 1. Graph of Compressive Strength

3.2 Tensile Strength Testing

The specimen in the form of briquette used as to test the tensile strength. It taken during the process of die casting which was given maximum load until the specimen broke. A 28 day-old specimen had two variations of glass powder supplement: glass powder which had passed sieve no. 100, was hampered on sieve no. 200, and sieve no. 200 The result of the testing could be seen in Picture 2 below:

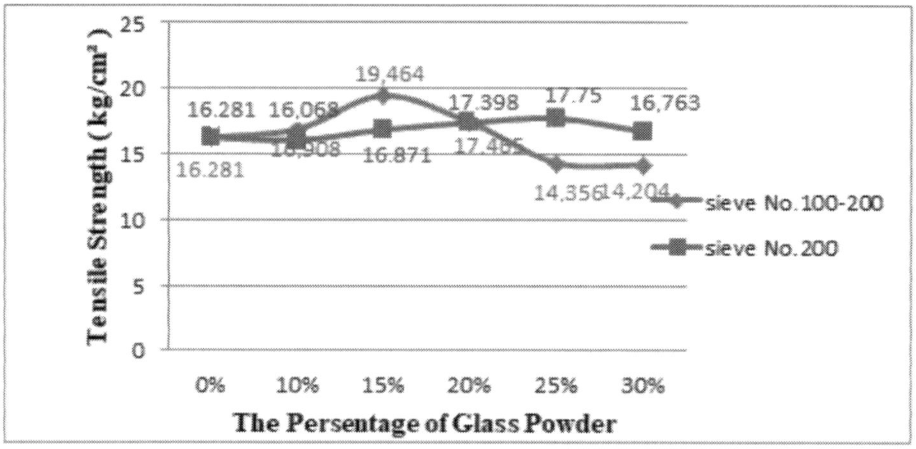

Figure 2. Graph of Tensile Strength

3.3 Absorption Testing

A specimen of 10 x 20 x 40 cm as a size of batako was used for the test. A 28 day-old specimen soaked within 24 hours; then it was weighed to get its weight when it was still wet. After that, it was put into an oven and let it be there for 24 hours to get its weight when it was dry.

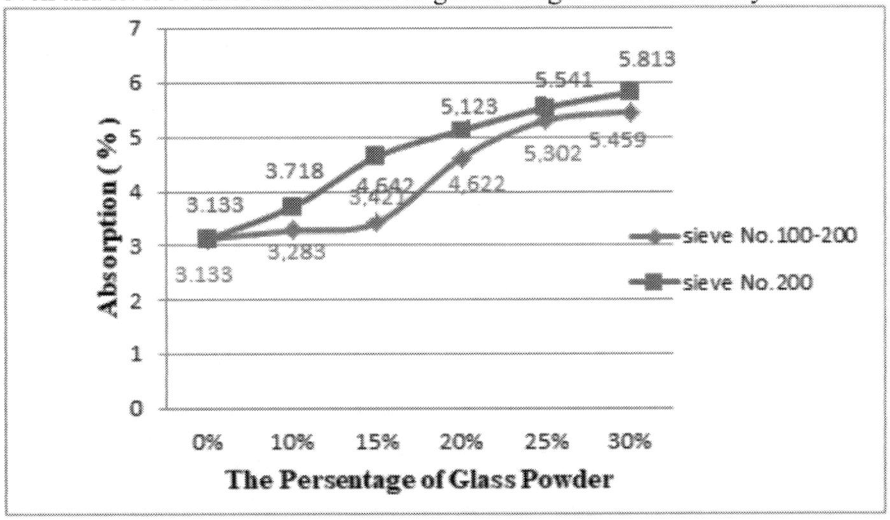

Figure 3. Graph of Absorption

3.4 Optimum Variation Testing (glass powder of 20% which passed sieve no. 200)

The result of assaying absorption, the compressive strength of the sample, cylinder, and wall installment, using glass powder substitution which passed sieve no. 200 with the composition of 0% and 20%, intended to reduce the volume of cement and foaming agent as the reductive substances of the sample load presented in Table 3, Table 4, Table 5 as follows:

Table 3. Comparison of the Mean Absorption of Batako Specimen and Glass Powder Substitution which Passed Sieve no.200 and Foaming Agent with Standard of Quality

No.	Glass Powder Supplement	Mean of Wet Weight (Kg)	Mean of Dry Weight (Kg)	Mean of Water Absorption (%)		Quality
				Assaying Object	SNI 03-0349-1989	
1	0% + Foaming Agent	11.788	10.158	11.715	25	1
2	20% + Foaming Agent	12.354	10.639	16.401	25	1

Table 4. Comparison of Mean Absorption of Cylinder Specimen and Glass Powder Substitution which Passed Sieve no. 200 and Foaming Agent with Standard of Quality

No.	Glass Powder Supplement	Mean of Wet Weight (Kg)	Mean of Dry Weight (Kg)	Mean of Water Absorption (%)		Quality
				Assaying Object	SNI 03-0349-1989	
1	0% + Foaming Agent	2.240	1.738	29.901	35	2
2	20% + Foaming Agent	2.612	2.366	11.404	25	1

Based on Table 3 and Table 4 for the value of water absorption in batako, it was found that there was the difference in water. The smallest mean of water was in the cylinder sample with the mixture of glass powder of 20% and foaming agent of 11.404%, while the value of the largest average water absorption found in the cylinder sample by using the foaming agent of 29.901%.

Table 5. Comparison of Mean Compressive Strength of Batako Specimen and Glass Powder Substitution which Passed Sieve no.200 and Foaming Agent with Standard of Quality

No.	Glass Powder Supplement	Mean Dial Reading (KN)	Comp. Strength Area (cm^2)	Mean Comp. Strength (Kg/cm^2)		Quality
				Assaying Object	SNI	
1	0% + Foaming Agent	69.8	291.4214	23.952	21	4
2	20% + Foaming Agent	87.6	291.4214	30.060	21	4

Table 6. Comparison of Mean Compressive Strength of Cylinder Specimen and Glass Powder Substitution which Passed Sieve no.200 and Foaming Agent with Standard of Quality

No.	Glass Powder Supplement	Mean Dial Reading (KN)	Compressive Strength Area (cm^2)	Mean Compressive Strength (Kg/cm^2)		Quality
				Assaying Object	SNI	
1	0% + Foaming Agent	19.8	78.5	25.223	25	4
2	20% + Foaming Agent	3.7	78.5	40.382	40	3

Based on Table 5 and Table 6 for the value of compressive strength in batako and cylinder specimen, it was found that there was the difference in the value of compressive strength. The smallest mean of compressive strength was found in the batako sample with the foaming agent of 23.952 kg/cm^2, while the greatest value of compressive strength found in the cylinder sample with glass powder of 20% and foaming agent of 40.382 kg/cm^2.

Table 7. Comparison of Mean Compressive Strength in Walls with Glass Powder Substitution which Passed Sieve ø200 and Foaming Agent with Standard of Quality

No.	Glass Powder Supplement	Dial Reading (PSi)	Mean Compressive Strength (Kg/cm^2)		Quality
			Assaying Object	SNI 03-0349-1989	
1	0% + Foaming Agent	360	25.2	25	4
2	20% + Foaming Agent	650	45.5	40	3

Based on Table 7 for the value of compressive strength in batako walls, it was found that there was the difference in the value of compressive strength. The smallest value of compressive strength was found in batako walls with the foaming agent of 25.2 kg/cm^2, while the greatest value of compressive strength found in batako walls with the use of glass powder for cement substitution of 20% and foaming agent of 45.5 kg/cm^2.

4. Conclusion

From the result of the research, it could conclude that

1. Compressive strength (0%) was without glass powder substitution of 95.289 kg/cm^2 and included in the classification of batako level I.
2. Maximum compressive strength with glass powder substitution ø which passed sieve no. 200 found in the percentage of 20% at 75.022 kg/cm^2, included in the classification of Quality III.

3. Maximum compressive strength for glass powder substitution ø which passed sieve no. 200 found in the percentage of 20% at 91.822 kg/cm^2, included in the classification of batako level I, and minimum compressive strength found in the percentage of 10% at 74.933 kg/cm^2, in the quality II.
4. Tensile strength (0%) was without glass powder substitution of 16.268 kg/cm^2.
5. Maximum tensile strength for glass powder substitution ø which passed sieve no. 100 and hampered on sieve no. 200 found in the percentage of 15% at 19.464 kg/cm^2, and minimum tensile strength was found in the percentage of 30% at 14.205 kg/cm^2.
6. Maximum tensile strength for glass powder substitution ø which passed sieve no. 200 was found in the percentage of 25% at 17.75 kg/cm^2, and minimum tensile strength was found in the percentage of 10% at 16.068 kg/cm^2.
7. Water absorption for normal batako (0%) was without glass powder substitution of 3.133%.
8. Maximum water absorption for batako with glass powder substitution ø which passed sieve no. 100 and was hampered on sieve no. 200 found in the percentage of 30% at 5.495 kg/cm^2, and minimum absorption found in the percentage of 10% at 3.28%.
9. Maximum water absorption for batako with glass powder substitution ø which passed sieve no. 200 found in the percentage of 30% at 5.813%, and minimum absorption found in the percentage of 10% at 3.283%.
10. All specimen passed the classification of batako level I, based on absorption smaller than 25%.
11. Absorption of the batako specimen which used glass powder substitution of 20% with the foaming agent of 16.401% and cylinder of 11.404%, was included in the classification quality I.
12. The compressive strength of batako sample which used glass powder substitution of 20% and foaming agent of 30.060 kg/cm^2, was included in the classification of quality IV, while cylinder specimen with the compressive strength of 40.382 kg/cm^2, was in the quality III.
13. The compressive strength of batako walls which used glass powder substitution of 20% and foaming agent found in 650Psi in dial reading or 45.5 kg/cm^2.

References

[1] Neville A M 1987 *Concrete Technology* London: Longman Group Ltd.
[2] Murdock L J and K M Brook 1999 *Bahan dan Praktek Beton* Jakarta: Erlangga.
[3] Departemen P U 1989 SNI 03-0349-1989 *Bata Beton untuk Pasangan Dinding*, Balitbang, Jakarta.
[4] Mulyono T 2004 *Teknologi Beton* Andi: Yogyakarta.
[5] Kasiati E 2011 *Pembuatan Paving Blok dengan Menggunakan Semen Portland dan Semen Pozzolan dengan Bahan Tambahan Serbuk Kaca dan Abu Batu* Seminar Nasional Aplikasi Teknologi Prasarana Wilayah.
[6] Karwur, Yohanes H, Dkk. 2013. *Kuat Tekan Beton Dengan Bahan Tambah Serbuk Kaca Sebagai Substitusi Parsial Semen.* Jurnal Sipil Statik **1**, 276-281.
[7] Latief A 2010 *Kuat Tarik Langsung, Kuat Tarik Lentur, Susut Dan Density Mortar Campuran Semen, Abu Sekam Padi, Dan Precious Slag Ball Dengan Persentase 30%; 30%; 40%,* Skripsi Program Studi Teknik Sipil Depok: UI
[8] Departemen P U 2004 SNI 15-2049-2004 *Semen Portland* Balitbang: Jakarta.

IOP Conf. Series: Materials Science and Engineering **309** (2018) 012015 doi:10.1088/1757-899X/309/1/012015

Hydraulic modeling of flow impact on bridge structures: a case study on Citarum bridge

R I Siregar[1, 2 *]

[1]Civil Engineering Department, Faculty of Engineering, Universitas Sumatera Utara, Jl. Dr. Mansur Kampus USU, Padang Bulan, Medan 20155, Indonesia
[2]Sustainable Energy and Biomaterial Centre of Excellence, Universitas Sumatera Utara, Jl. Almamater Kampus USU, Padang Bulan, Medan 20155, Indonesia

[*]Email: rizawaloed@yahoo.com

Abstract. Flood waves because of the rapid catchment response to high intense rainfall, breaches of flood defenses may induce huge impact forces on structures, causing structural damage or even failures. Overflowing stream that passes over the bridge, it means to discharge flood water level is smaller than the capacity of the river flow. In this study, the researches present the methodological approach of flood modeling on bridge structures. The amount of force that obtained because of the hydrostatic pressure received by the bridge at the time of the flood caused the bridge structure disrupted. This paper presents simulation of flow impact on bridge structures with some event flood conditions. Estimating the hydrostatic pressure developed new model components, to quantify the flow impact on structures. Flow parameters applied the model for analyzing, such as discharge, velocity, and water level or head that effect of bridge structures. The simulation will illustrate the capability of bridge structures with some event flood river and observe the behavior of the flow that occurred during the flood. Hydraulic flood modeling use HEC-RAS for simulation. This modeling will describe the impact on bridge structures. Based on the above modelling resulted, in 2008 has flood effect more than other years on the Citarum Bridge, because its flow overflow on the bridge.

1. Introduction

The flood as result of the rapid catchment response to intense rainfall, breaches of flood defenses, tsunamis or storm surges may induce huge impact forces on structures, causing structural damage or even failures. It may endanger and threaten human safety [1].As part of the effort to return flows to historical levels, several changes to the existing water management infrastructure that implemented or are in the design phase. The percent probability of water levels exceeding baseline conditions was determined for the simulation period. This analysis showed that the length of time of inundation or the hydro period [2].

Current bridge design specifications deal with various extreme hazards independently, which may lead to less economic design and construction practices, and may underestimate failure probabilities [3].Flood simulation present two representations of bridges. They include a novel and simple empirical formulation for total head loss caused by bridges, which covers all flow regimes: free water surface, partially submerged flow, and fully submerged flow [4].The extension of the expected flood affected area and the increase in water surface elevation is highly dependent on the river morphology, bridge geometry, and flow and floodplain characteristics and may originate upstream flooding [5].

This study makes hydraulic modeling of the flow occurring that affects the bridge structure. To obtain such mapping impact, hydraulic modeling is central because it provides water depths and velocities everywhere and along the whole flood period [6].

There are several parameters become indicator of result of hydraulic flood modeling which influence bridge structure. They are flood discharge, velocity, and water level or head. Vehicle loads through the bridge and flood loads during extreme flooding influence Citarum Bridge structural resistance. The objective of this study is to investigate whether and to what extent bridges structures have an influence in flooding.

2. Study Area and Materials

2.1. Study Area

Citarum Bridge is located at upper Citarum watershed. It flows through the southern area of Bandung Regency up to Saguling Reservoir. Close to the Citarum Bridge, there is Dayeuh Kolot that is station off low discharge. In study area, there are seven tributaries in the upper Citarum up to the point of Dayeuh Kolot flow discharge station. Upper Citarum watershed is with a total area of watershed is 1.771 square kilometers.

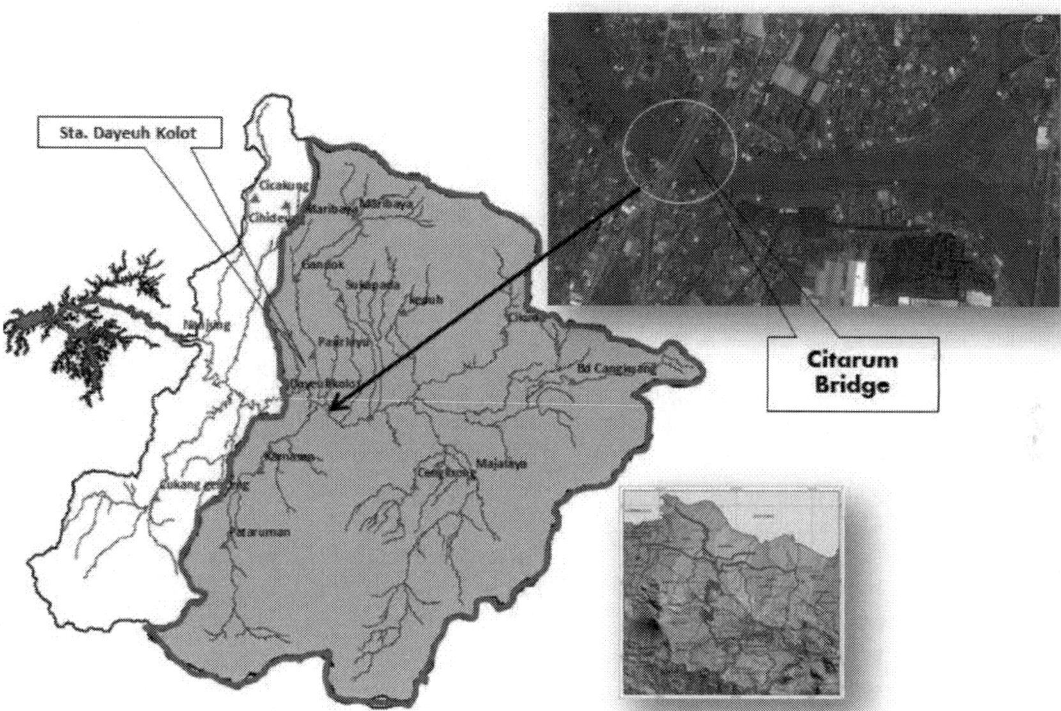

Figure 1. The location map of upper Citarum Watershed in West Java that shows the location of Citarum Bridge as a study area

The Figure 1 shows location map of upper Citarum Watershed that lies in Citarum Watershed in West Java, Indonesia. There is image of the Citarum Bridge obtained from Google Earth.Citarum Bridge called Rancamanyar Bridge because it is located on Rancamanyar Street, Baleendah, Bandung district, Indonesia. The bridge, which has a length about80 meters and a width of 9 meters, is a link between the Rancamanyar Street and Sayuran Street.

2.2. Data Sources

In this study will make simulations for several flood events in 2007, 2008 and 2010. The three flood events based on the measurement data of flow discharge, that there is extreme flow discharge compared to other year. The data required in this model is the upper Citarum watershed data with output at the point of the Citarum Bridge. In analyzing simulation, the flow required the characteristics of upper Citarum watershed.

The data of the debit serves as the calibration and verification reference of the modeling created. The available record data is the measurement data of water level three times a day. Upper Citarum watershed utilizes remote sensing technology to provide information on land coverage by interpretation of objects in the image. Data sources on upper Citarum Watershed based on data from Major River Basin Organization of Citarum, which consists of several types of land cover.

3. Methods

3.1. HEC-RAS

The influence of the minor hydraulic modeling on flood dynamics, analyzed through a combined use of one-dimensional (1D) and two-dimensional (2D) hydraulic models. First, a number of detailed and simplified approaches to represent hydraulic structures in the computational grids analyzed by means of the HEC-RAS 1D model. 2D simulations of the flood eventimplemented and tested in several these approaches [7]. In this study, use HEC-RAS as hydraulic modeling software to help modeling analysis.

HEC-RAS employs 1d flood routing in both steady and unsteady flow conditions by applying an implicit-forward finite difference scheme between successive sections of flexible geometry. Unsteady flow is a physical process in the flow in a channel by adopting the concept of conservation of mass and momentum conservation. This physical process described by mathematical equations, known as the equation St.Venant [8]. The equations consist of continuity equation (mass conservation principle) and the momentum equation(the principle of conservation of momentum), which written in the form of partial differential equations as follows:

Continuity equation:

Momentum equation:

Where: A: area, Q: flow discharge, ql: lateral discharge per unit width of channel, V: velocity, g = acceleration of gravity, x: distance measured in downstream flow direction, z: mean depth, t: time.

3.2. Boundary Conditions

In all the above models, two boundary conditions are required, which are usually set at the upstream end of the channel through an imposed inflow as well as the assumption of uniform water depths at the upstream and downstream end (kinematic wave condition). Although an imposed depth would result in more stable solutions than the condition of uniform flow, we choose the latter since, in practice, it is rare to know the temporal evolution of the water depth at a specific location. The models compute the appropriate time step based on the Courant number stability criteria.

3.3. Calibration and Verification

If compared to the model implemented on the basis of in-situ data only the hydraulic model parameterized on the basis of satellite and in-situ hydrometric data provides a better reproduction of average flow conditions, and it also results in the most accurate representation of the maximum water profile observed during a major flood occurred [9].The Root Mean Squared Error (RMSE) criterion and its related normalization, the Nash–Sutcliffe efficiency (NI) are the two criteria most widely used for calibration and evaluation of hydrological models with observed data.The performance indicators are Nash–Sutcliffe Index (NI) [10].

IOP Conf. Series: Materials Science and Engineering **309** (2018) 012015 doi:10.1088/1757-899X/309/1/012015

4. Results and Discussions

4.1. Flood Modeling

Based on input data in flood modeling use HEC-RAS with 1D (one-dimensional) and 2D(two-dimensional) with hydrograph peak flood discharge: 259.94 m^3/s in 2007; 223.55 m^3/s in 2008, and 298.22 m^3/s in 2010. In those years, the flood modeling on bridge will be analyze

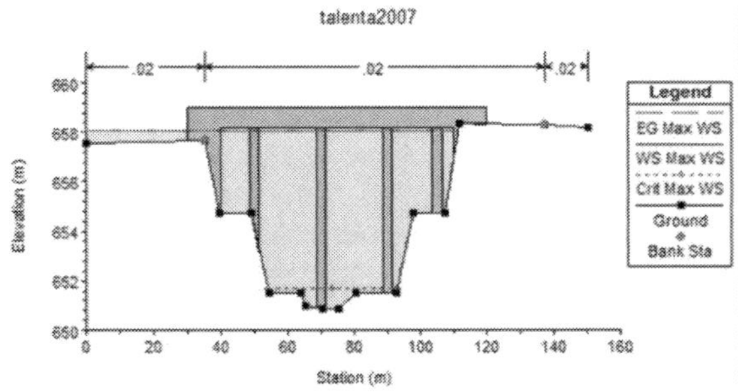

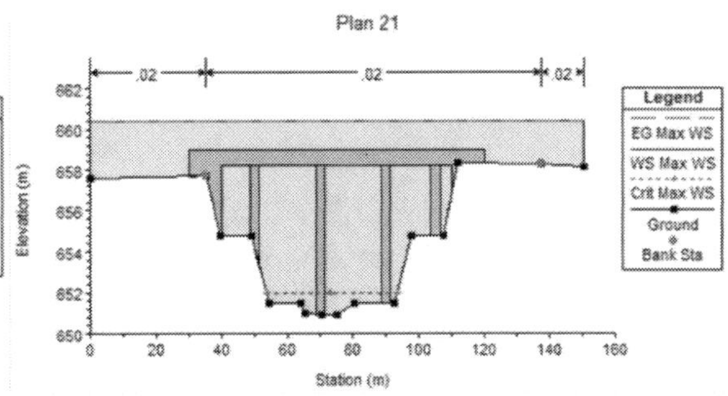

(a) (b)

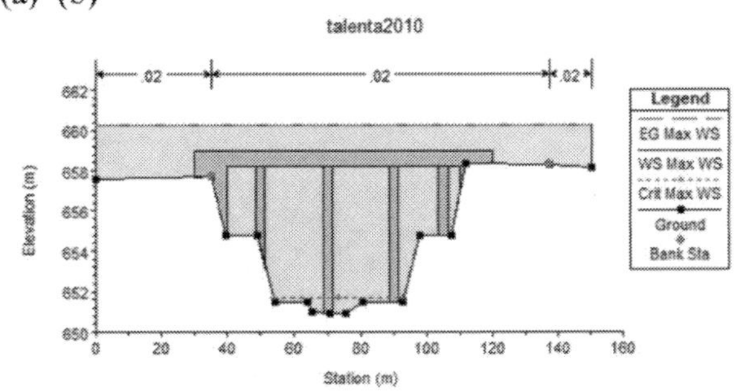

(c)

Figure 2. Citarum Bridge in upper Citarum flood conditions in years (a) 2007; (b) 2008, and (c) 2010

The Figure 2 shows flow parameter that impact on Citarum Bridge. There are three parameters: flow discharge, velocity and water level in years 2007, 2008 and 2010. The figure (a) showed that in 2007, when the flood happened, the flow was not overflow on the Citarum Bridge. The figure (b) showed that in 2008, when the flood happened, the flow was not overflow on the Citarum Bridge. It seems more save than in 2007. In other condition, in 2010, the figure (c) showed that in that year when the flood happened, it was overflowing on Citarum Bridge. It was about a meter above the bridge. It means the area around the bridge will be inundation.

The Figure 3 shows flow parameter that impact on Citarum Bridge. There are three parameters: flow discharge, velocity and water level in years 2007, 2008 and 2010. The figures (a) show the flow discharge in 2010 higher than in 2007 and 2008. The discharge can analyze to other flood condition. The figure (b) shows that velocity in 2010 higher than 2007 and 2008. Based on theory, the speed is directly proportional to the flow discharge for the same cross section and long section. The figure (c) shows that water level in 2010 is higher than in 2007 and 2008. It seems in Figure 2 that the flow are overflowing the bridge.

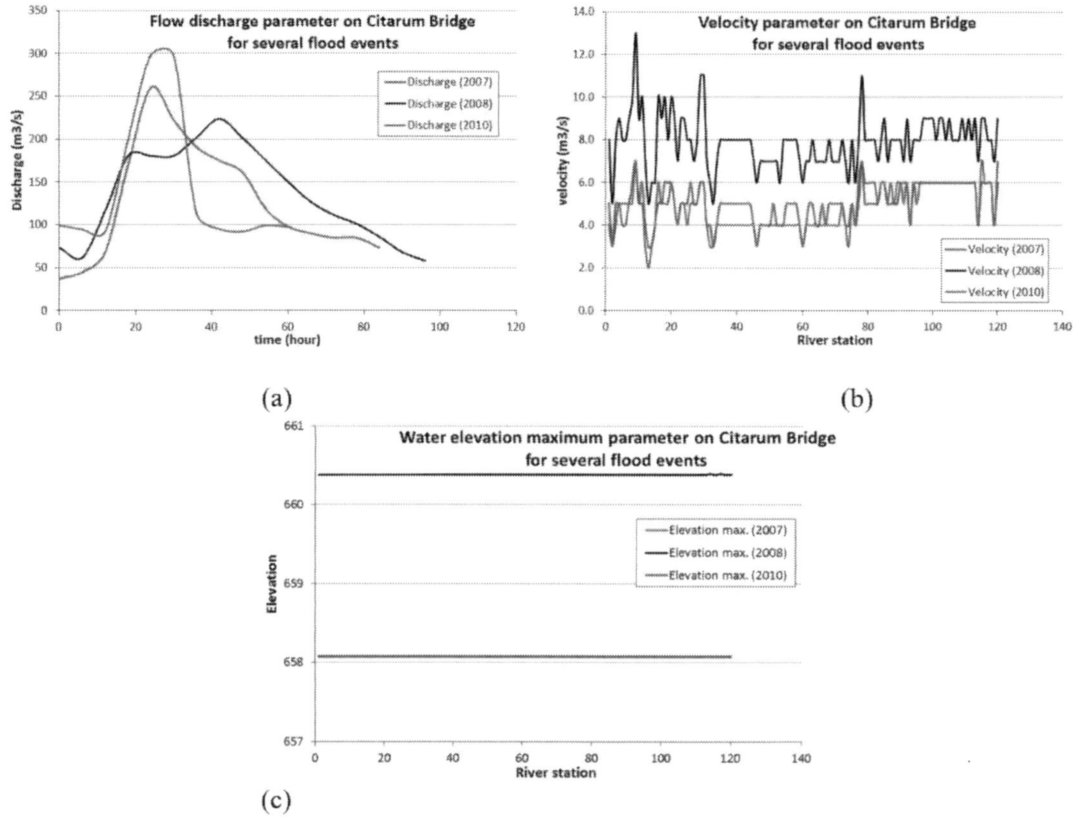

Figure 3. Flow parameter on Citarum Bridge: (a) flow discharge, (b) velocity, and (c) water level in years 2007, 2008 and 2010 for several flood events

4.2. Flow validation

Theflow determined the strength in predicting the modeling by Nash-Sutcliffe model efficiency coefficient (E). Based on the validationof Nash Index method (Index E) obtained the result in years2007 with index 0.76, 2008 with index 0.74 and 2010 with index 0.76. The index values show that the flow modeling, is close to one, that means the modeling is acceptable.

5. Conclusions

Base on this study, it introduced a new methodology for prediction of the design flood discharge. The role of hydraulic modeling in the design of the bridge across the river is on the aspect bridge safety against the flood stream in the river. The bridge over structure should be sufficient High so that pier and deck protected from flood. The base, the pillar, and the foundation of the bridge should be safe against the risk of scouring. This risk increases great if the riverbed decreases (degradation). Hydraulic modeling that analyzed in this study shows the different about three flood conditions in years 2007, 2008 and 2010. Based on the above modeling resulted, in 2008 has flood effect more than other years on the Citarum Bridge, because its flow overflow on the bridge.

Acknowledgments

The author are grateful acknowledge that the research is supported by Universitas Sumatera Utara, Indonesia for the financial support. Thanks to Major River Basin Organization of Citarum, Public Work Ministry, for providing the data and others.

References

[1] Liang Q, Chen K, Hour J, Xing Y, Wang G and Qing J 2016 Hydrodynamic Modeling of Flow Impact on Structures under Extreme Flow Conditions *Elsevier* **28** 267-274

[2] Long S A, Cook A M, Fennema R, Tachiev G I, Villamizar V, Kotun K, Wilhelm F M 2016 Analysis of Bridge Construction as a Restoration Technique for Everglades National Park, FL, USA, Using Hydrological Numerical Modeling *Journal of Hydraulic Engineering ASCE* **142**.

[3] Wang Z, Patgett J E and Osorio L D 2014 Risk-Consistent Calibration of Load Factors for the Design Of Reinforced Concrete Bridges Under The Combined Effects Of Earthquake And Scour Hazards *Elsevier* **79** 86-95

[4] Heilemariam F M, Brandimarte L, Dottori F 2013 Investigating The Influence of Minor Hydraulic Structures on Modeling Flood Events in Lowland Areas *Hydrological Processes* **28** 1742-1755

[5] Shen D, Wang J, Cheng X, Rui Y and Ye S 2015 Integration of 2-D Hydraulic Model and High-Resolution Lidar-Derived DEM for Floodplain Flow *ModelingHydrology and Earth System Sciences* **19** 8

[6] Paquier A, Mignot E, Bazin P H 2015 From Hydraulic Modeling to Urban Flood Risk *Elsevier* **115** 37-44

[7] Brandimarte L, Woldeyes M K 2013 Uncertainty in the Estimation of Backwater Effects at Bridge Crossings *Hydrological Processes* **27** 1292-1300

[8] Dimitriadis P, Tegos A, Oikonomou A, Pagana V, Koukouvinos A, Mamassis N, Koutsoyiannis D and Efstratiadis A 2016 Comparative evaluation of 1D and quasi-2D hydraulic models based on benchmark and real-world applications for uncertainty assessment in flood mapping *Elsevier* **534** 478-492

[9] Domenneghetti A, Tarpanelli A, Brocca L, Barbetta S, Moramarco T, Casstellarin A and Brath A 2014 The Use of Remote Sensing-Derived Water Surface Data for Hydraulic Model Calibration *Elsevier* **149** 130-141

[10] Yucel I, Onen A, Yilmas K K, Gochis D J 2015 Calibration and evaluation of a flood forecasting system: Utility of numerical weather prediction model, data assimilation and satellite-based rainfall *Elsevier* **523** 49-66

IOP Conf. Series: Materials Science and Engineering **309** (2018) 012016 doi:10.1088/1757-899X/309/1/012016

The roles of weather modification technology in mitigation of the peat fires during a period of dry season in Bengkalis, Indonesia

A Sandhyavitri[1*], M A Perdana[2], S Sutikno[3], F H Widodo[4]
[1, 3] Lecture of Engineering Faculty, University of Riau, Indonesia
[2] Student of Engineering Faculty, University of Riau, Indonesia
[4] Agency for the Assessment and Application Technology, BBPT, Indonesia

*Email: ari.sandhyavitri@lecturer.unri.ac.id

Abstract: The annual peat fire disasters have suffered human life, deteriorated peat land ecosystems, and caused severe economic losses in Indonesia, Malaysia, Brunei, and Singapore during a period of 2014-2016. The objective of this study is to investigate to what extent the weather modification technology (WMT) may play its roles in increasing the precipitation rates in order to mitigate the peat fires disaster in Bengkalis, Riau Province, Indonesia. The study obtained the precipitation rates data from the tropical rainfall measuring mission (TRMM) satellite during the period of 2014, 2015 and 2016. The data then statistically analyzed using the Grads software package. While the process of WMT at the designated research location was evaluated together between the Artificial Rain Unit of BPPT (Agency for the Assessment and Application Technology), and the Engineering Faculty, University of Riau, 2016. The research showed that the WMT increased the rainfall rates during the dry season within the study area by 8% (520.7 mm to 557 mm) in the 3 months period (July 14 - October 6, 2016), and reducing significantly hotspots by 88.6% from 6.725 in 2014 to 770 in 2016. Hence, it confirmed that the application of technical WMT procedures may increase precipitation rates and reduce the number of hot spots in Bengkalis.

1. Introduction

The peat fire disasters have been regularly occurred in the Riau Province, Sumatra, Indonesia. In the dry season, the peatlands will become dry and flammable. Peat contains fuel in the form of plant residue. Hence, small fires and smouldering may penetrate below the soil surface, burning peatlands slowly and it was hard to detect, thus causing thick smoke hazards. The smouldering in the peatland areas is also difficult to extinguish so it can be last for weeks [1]. The forest fires have negative impacts on the social and economic environment for Riau province and the regional areas of Indonesia, Malaysia, and Singapore. It was calculated that more than 1.600 of flight schedules were canceled from and to Sumatra during the period of August-October 2015. This affected to economic loss about US$8 million or Rp109 billion for airline industry [2, 3]. The impacts on the public health were even worst as there were rising cases of acute respiratory infection diseases (ISPA) more than 10 folds (normally 6000 cases per month) became 61.017 cases in September 2017 [4].

However, a large number of fire-fighting efforts were conducted in order to extinguish these wild fire trough water spray by the fire brigades from the land surface, and water bombs through the helicopters and airplanes from the sky, these efforts have been considered yield little impacts in reducing the large scale of the forest fire cases. Hence, there is a need to establish a robust precautions effort to mitigate peat forest fires before they occur. Several methods of forest fires prevention have been identified such water spray, water bombing and this study concern in the role of weather modification technology (WMT) [5].

This paper considered the application of weather modification technology is as important approach in mitigating the forest fire disasters. According to T.P. DeFelice, 2005. He stated that the WMT might potentially give positive contribution in alleviating weather disasters such as severe smog hazards caused by the forest fire events and water resource deficiency caused by long dry season periods [6,7].

The WMT is highlighted as a human effort in modifying the existing weather conditions with a specific purpose in order to yield the desired weather conditions for example increasing the intensity of rainfall in a particular location (rain enhancement) otherwise decreasing rainfall intensity (rain reduction) in another one to minimize natural disasters caused by the climate change and weather conditions by utilizing weather parameters [8, 9, 13]

The state-of-the-art weather modification technologies have been applied toward mitigating the effects of peat fires in Indonesia, USA, Australia, and Thailand [5, 9]. The WMT activity is basically based on the effort in manipulating the existing natural conditions so that the concentration of smog hazards in the atmosphere can be reduced. This activity may have a positive impact on the atmospheric dynamics. This effect may include an improvement of precipitation rates consequently extinguish wild fires [6, 10, 11].

The role of WMT in mitigating of the forest fire disaster has been regulated in the Presidential Instruction of the Republic of Indonesia (RI) Number 16 the year 2011 concerning the Improvement of Forest and Land Fire Control in Indonesia. The President of RI since 2011 has instructed the Minister of Research and Technology to mitigate the forest and land fire disasters in Indonesia by conducting artificial rain-making technology (WMT) [12].

2. Literature Review

The terminology of weather Modification Technology (WMT) is defined as a human interventions and efforts in the control of water resources in the atmosphere to increase or decrease rainfall rates in a certain area in order to minimize the natural disasters caused by the climate and weather change [6, 8].

2.1. Cloud Seeding Mechanism.

The air bubbles constantly formed by the sea foam may cause the water particles to rise up to the sky. These particles called aerosols serve as water traps and then form water droplets. Furthermore, this aerosol rises into the atmosphere, and when a large amount of air is lifted to a higher layer, it will cool down and then condense. A collection of water droplets resulting from moisture in the condensing air is called as a cloud [5, 9].

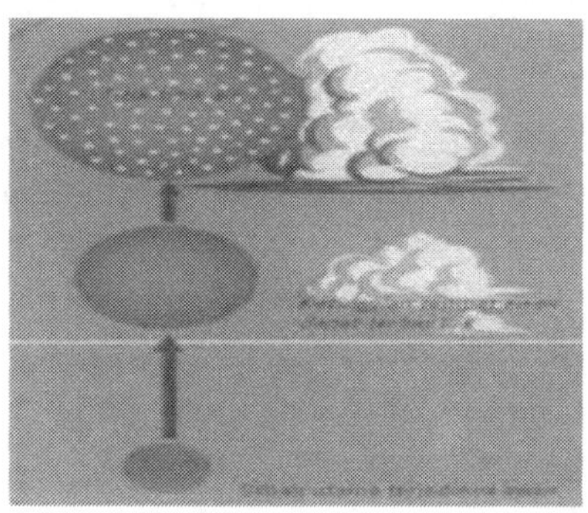

Figure 1. Condensation of clouds

At a certain level, the aerosol particles (0.01-0.01 micron) flying in the air will serve as condensation nuclei that cause the water moisture to condense. The main source of condensation core is salt derived from the seawater around the word. Since they are hygroscopic particles, the particles have become liquid droplets, and the collections of many droplets may form the clouds. The water particles surrounding the salt crystals and dust particles thicken, so the particles become heavier from the air, starting to fall from the clouds in the form of rainfalls. The main objectives of cloud seeding are to accelerate the collision process and the coalescence of the particles in the cloud which is a condition of rain to fall [13, 14].

Typically, clouds used as a seeding object are Cumulus (Cu) clouds which contain abundant of water vapor and has the potential to cause rainfall. The WMT is conducted in the form of cloud seeding by using hygroscopic seeding materials such as kitchen salt or Sodium Chloride (NaCl), $CaCl_2$, and Urea.

a. Jumping Process Mechanism: This method aims to accelerate the process of rainfall production. Initially, stages are by observing the clouds which carry plenty of water vapor from the sea and are considered potentially to cause rainfall will be "intercepted" away from the target area. By using the aircraft, the clouds are seeded with the objective to accelerate rain fall process [8].
b. Competition Mechanism: Applied for ground-based clouding activities with the Ground Based Generator system. For this second method, the goal is to disrupt the physics process in the cloud for growing convective clouds, so that the rain can be shortened in duration and reduced in intensity [13].

2.2. The Roles of WMT in Reducing Forest Fire.

This technology takes advantage of the opportunities that exist in the existing natural weather conditions. For example, with the application of WMT, the existing energy available in nature can be utilized effectively and efficiently to extinguish fires. Such energies may include wind energy, solar radiation energy, and airborne moisture content (potential clouds) which are available in the sky [15].

It is acknowledged that the existing peat fires would not continue smouldering as long as the amount of rain drops sufficiently rewetting the peat land surface. Thus it subsequently extinguished the existing peat fire events. It is necessary that as long as the amount a potential cloud around the fire location sufficient, there would be adequate rain to fall. There are several terms for the presence of potential clouds around the location of forest fires e.g. the amount of moisture content in the air (the conditions of the air column or the degree of air extinction that can support the potential cloud to grow). However, it is no guarantee that the growing cloud will become potential, although the amount of moisture content is in considered high and stable [13]. The accumulation of smoke hazards caused by peat fires prevented the sun's radiation penetrate the soil surface. Hence, the process of ground surface heating would not occur. This affect to stabilize the air column condition which is not possible to develop aerosol, condensation nuclei, and potential cloud. The application of WMT is also may reduce the smoke hazard concentration and improve cloud development processes.

Weather Modification Technology has been applied in Riau during the period of 22 June 2013 on 9 July 2013 and it was continued in the period of 14 June to 6 November 2016. In 2013, the WMT deployed two Hercules C130 TNI AU and Cassa BPPT aircrafts to seed 71.4 tons of NaCl over the Riau area. Based on the BPPT WMT, 2013 report, the WMT might improve rainfall intensity since June 23, 2013, in the Riau region, thus reducing the number of hotspots and reducing smoke [16].

In 2016 the WMT processes were conducted for 2 months period, deploying 1 aircraft CASA 212 from the BPPT to seed hygroscopic particles (NaCl 1-100 micron (μ)) [5].

Figure 2. CASA 212 aircraft for seeding activities 2016

This study investigated the roles of weather modification technology (WMT) may play in mitigating the peat forest disaster in Bengkalis, Riau Province, Indonesia by increasing precipitation intensity.

3. Research Methodology

3.1 Research sites.
This research was conducted in Riau Province, especially in Bengkalis Regency, The location of research is presented in the following figure.

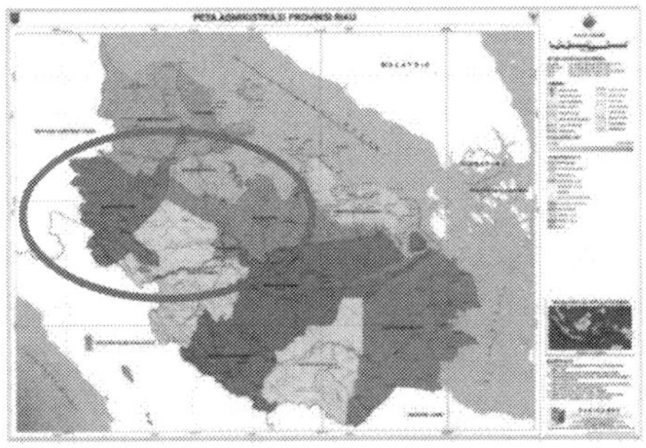

Figure 3. Research location

The Weather Modification Technology (WMT), and track seeding data were obtained from BPPT and Riau University collaboration team 2016. This study also deployed a portable Global Positioning System (GPS) to tracking and recording the aircraft movement during seeding WMT materials.

The TRMM (Tropical Rainfall Measuring Mission) satellite, which was joint operating between NASA and the Japan Aerospace Exploration Agency was dedicated to monitoring and studying the tropical rainfall conditions along the equator were used to obtain Bengkalis precipitation data [17, 18]. The TRMM has been operated with 16 x passing orbit per day (92 minutes per orbit) with a resolution of 5.2 km. The remote sensing images could be obtained from the FileZilla application connected to the TRMM satellite. The precipitation recorded data were collected in this research in the period of July 15, 2016, until October 6, 2016.

3.2 Determining the Grid of Riau Province.
In determining Riau province grid done by using ArcMap version 10.3 software. The grid is used to determine the coordinates of the research area.

3.3 Rainfall Data.
In this study precipitation or rain, datasets were obtained from the satellite (Tropical Rainfall Measuring Mission) TRMM the Japan Aerospace Exploration Agency (JAXA). The TRMM satellite has been joint operated between JAXA and NASA and for monitoring and studying the tropical rainfall conditions along the equator (Figure 4).

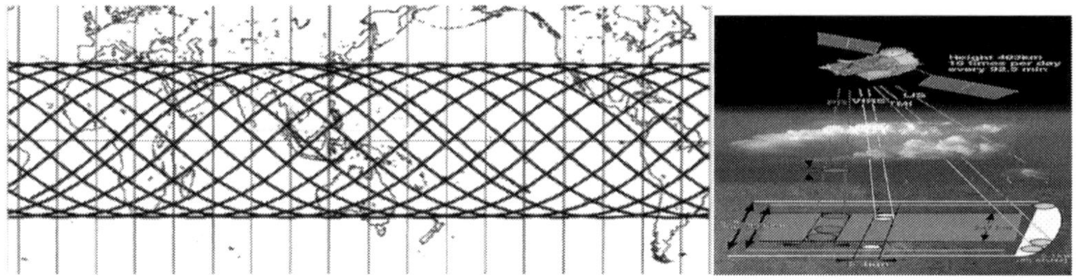

Figure 4. Tropical Rainfall Measuring Mission (TRM) orbit

Rainfall data is downloaded for three years from 2014 until 2016. The TRMM precipitation dataset has been considered as the space standard for measuring rainfall or precipitation rate [17, 18]. The rainfall data obtained in the form of rainfall hourly. FileZilla software was used to download the data. GrAds and Command Prompt software was used to analyze the data.

3.4 Hotspot Data.
This research uses hotspot data for three years from 2014 until 2016 which were obtained from www.fires.forestwatch.com.

4. Results and Discussion

4.1 Weather Modification Technology Activities (WMT) and Its Impact.
The weather modification technology (WMT) was implemented in Riau province from July 15, 2016, to October 5, 2016. Here is an example of the impact distribution of rainfall rates along with the flight route and the seeding location on August 1, 2016 (Figure 5a).

The Figure 5a shows that there was no precipitation around Pekanbaru, Siak and Bengkalis area (white color zone). Then at August 1, 2016, WMT was conducted using CASA 212 aircraft. The flight routes were drawn in red line. On August 2, 2016 there was rainfall within the area Riau encompassing Siak, Bengkalis, Rokan Hulu and some parts of Pelalawan (Figure 5b).

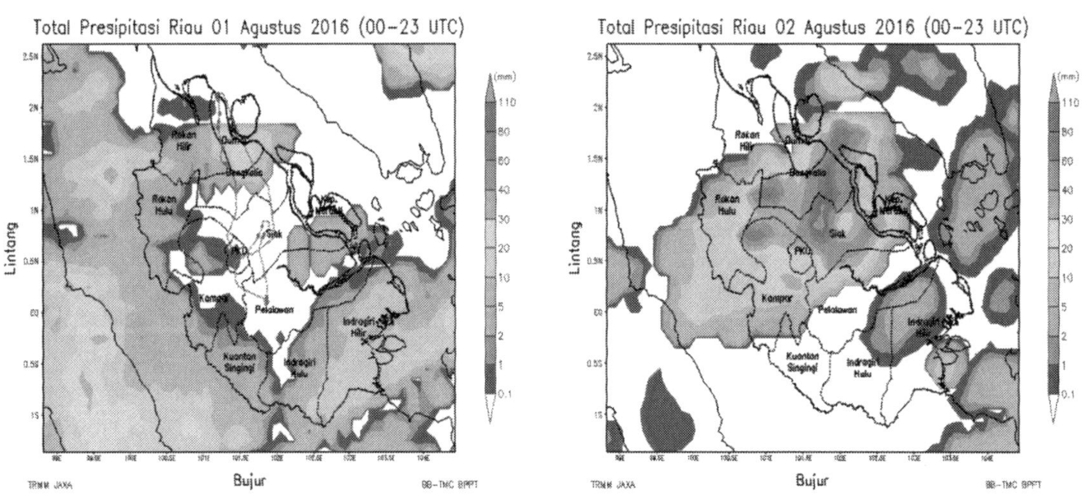

(5a) **(5b)**

Figure 5a, 5b. The impact of rain distribution from the seeding process from 01 to 02 August 2016, in Riau Indonesia

There are 3 cases were evaluated in this study. This study evaluated the effect of weather modification technology (WMT) in an increase of precipitation rates [19, 20]. Based on the historical data an increase in the precipitation rates may reduce the peat fire events in Indonesia [13].

4.2 Case 1 Bengkalis 2014.

The significant number of hotspots were identified on February 23, 2014 with 534 hotspots. There was no rainfall for more than 10 days. During the period of March 5, 2014, to July 24, 2014, the Government conducted weather modification technology (WMT). On March 28, 2014, the number of hotspots decreased along with the rainfall occurrences. The number of hotspots decreased to 45 points with rainfall of 1.492 mm in that date. Until the end of the WMT period of 24 July 2014, the number of hotspots continued to decrease along with an increasing of rainfall rates gradually. Figure 3 shows the fluctuation between the numbers of hotspots with the rainfall intensity in Bengkalis District in 2014 (Figure 6).

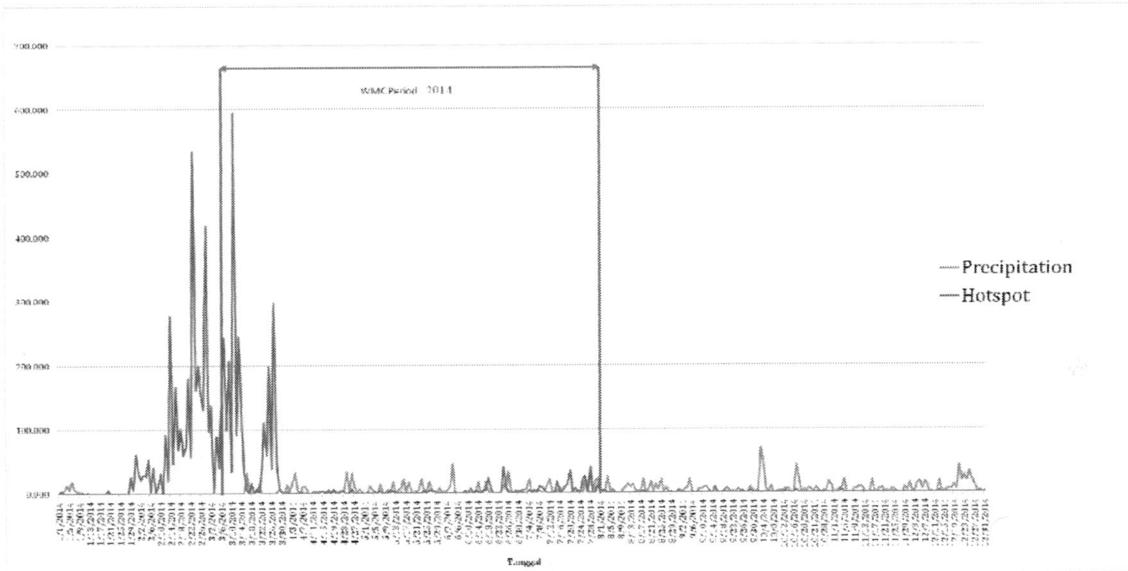

Figure 6. The comparison between the numbers of hotspots with the rainfall intensity in Bengkalis Regency 2014

4.3 Case 2 Bengkalis 2015.

At the beginning of 2015, it was detected as many as 86 hotspots occurred on March 14, 2015. The rainfall rate was recorded 0 mm. Then WMT program (under BPPT) was conducted at the mid of April 2015. As the results on April 23, 2015, the number of hotspots became 0 with the rainfall intensity of 19.59 mm. The hotspots decrease up to April 31, 2015. Then, the WMT was terminated on May 2017 as there was no more hotspots recorded.

Then WMT program was re-continued on July 2017. The number of hotspots on July 11, 2015, were recorded as many as 94 hotspots with the rainfall intensity of 0 mm. On August 2, 2015, the number of hotspots began to decrease significantly to 0 hotspots with the rainfall intensity of 38.071 mm. Hence, the number of hotspots decreased along with an increasing of rainfall intensity until the end of the WMT period of 20 November 2015 (Figure 7).

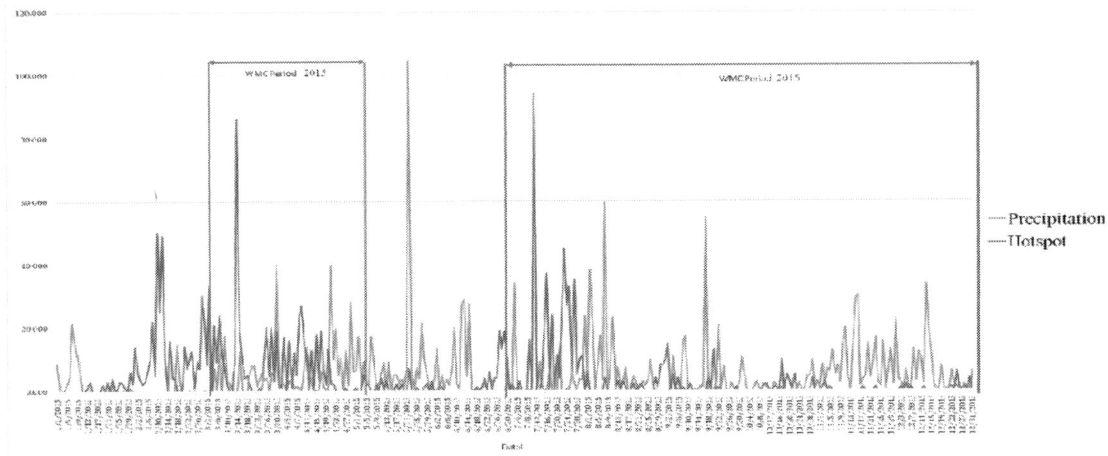

Figure 7. The comparison between the numbers of hotspots with rain intensity in Bengkalis Regency 2015

The figure 7 shows that during 2 periods of WMT (red lines) the number of hotspots tends to decline, and the rain intensity tends to increase.

4.4 Case 3 Bengkalis, 2016.

During the period of February 12 to 25, 2016 there was no rainfall events recorded and it caused the number of hotspots increased to become 17 occurrences. The number of hotspots tends to fluctuating up to March 21, 2016. Then the WMT program was conducted from June 15, 2016, to October 4, 2016. It was reported that on August 6, 2016, the hotspots was not detected due to high rainfall intensity. However, on August 19, 2016 until August 29, 2016 WMT process had to be halted because of there was no an available aircraft to conduct the seeding processes. During this period there was an increase of the number of hotspots up to 25 because there was no rainfall. After August 29, 2016 the WMT program was restarted again and the rain intensity rate increased, then as the consequences the number of hotspots decreased to 0 (Figure 8).

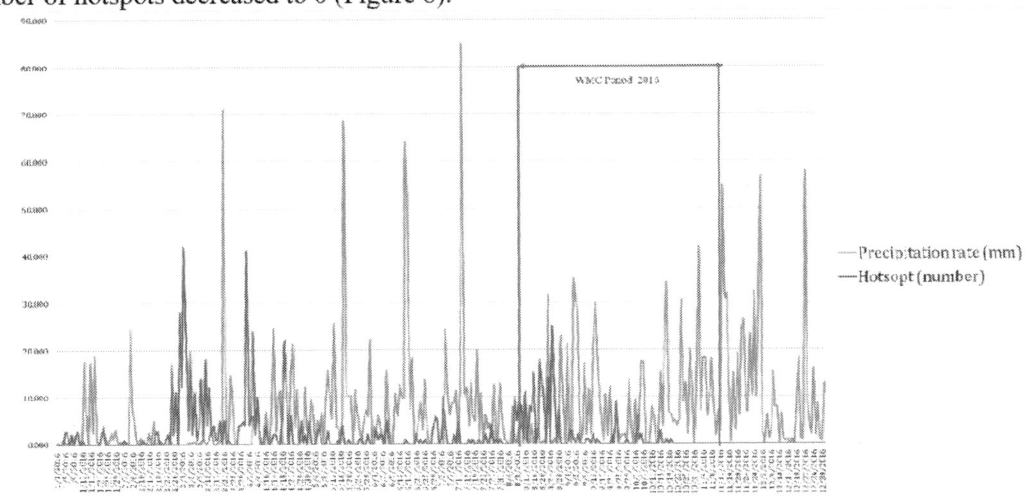

Figure 8. The comparison between the numbers of hotspots with rain intensity in Bengkalis Regency 2016

It was summarized that, the fluctuation of precipitation rates at the periods of 14 July to 6 October 2014, 2015 and 2016 in Bengkalis, Riau Province, has very strong correlation with the number of hotspots (-97.45 %), and the determination coefficient = 94.97% (Figure 9). There was 94.97% of the rainfall in Bengkalis affected to reduce the number hotspots, and 5.03% affected by other variables.

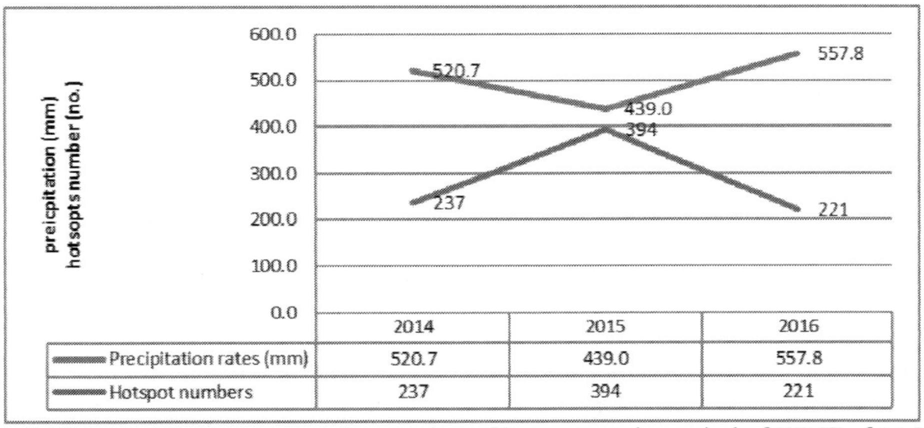

Figure 9. The precipitation rates vs the number of hotspots at the period of WMT of 14 July- 6 October 2016 in Bengkalis, Riau Province.

There was a strong correlation between the precipitation rates with the number of hotspots (adversely) at the peatland of Bengkalis. Hence, it was confirmed that the roles of WMC in improving precipitation rates have proved in reducing peatland fires in Bengkalis 2014-2016.

5. Conclusion

The result shows that weather modification technology (WMT) may enhance precipitation rates, thus this rain wetted the peat soil in Bengkalis, Indonesia thus as consequences the number of hotspots reduced significantly during the period of 2014, 2015 and 2015. It is recommended that this method can be used in mitigating peatland fire disasters in Indonesia.

References

[1] Sandhyavitri A 2015 Flood and Fires Mitigation (Mitigasi Bencana Banjir dan Kebakaran), Unri Press Pekanbaru ISBN:978-979-792-534-5 **1** 151-193
[2] Angkasa Pura II report 2015 Airliner loss affected by forest fires in Sumatra Loose report **1** 1-2
[3] Candradewi R 2014 Kebakaran Hutan dan Kabut Asap di Riau dalam Prespektif Hubungan Internasional *Jurnal Phobia* **1(3)** 12-25
[4] Riau Province Health Office Report 2015 Smog Hazards Effects Pekanbaru **1** 1-2
[5] BPPT report 2016 concerning Implementation of Weather Modification Technology in Riau **1** 1-12.
[6] DeFelice T P 2005 A Plan for the next phase in Weather Modification Science and Technology Development *WMA Annual Meeting, San Diego, Ca. 2005; 16th Weather Modification Conference* **1** 11
[7] Husni M 2001 Kajian Seeding dan Hujan Di DAS Brantas Bagian Penerapan Teknologi Modifikasi Cuaca di Sub Das Kali Brantas *Jurnal Teknologi & Sains Modifikasi Cuaca, Indonesia* **1** 95–100
[8] DeFelice T P 1998 *Introduction to Meteorological Instrumentation and Measurement, Prentice Hall, Saddle Brook New Jersey* ISBN:0-13-243270-6
[9] Weather Modification Association WMA 1993 *Weather Modification capability statement. Weather Modification* **2** 11

[10] American Meteorological Society 1998 Capability statements. *Bulletin American Meteorological Society (BAMS)* **79** 2771
[11] Ahrens D 1994 *Meteorology Today an introduction to weather, climate and the environment. Fifth Edition* - West Publishing Co USA
[12] Presidential Instruction of the Republic of Indonesia (RI) Number 16 the year 2011 concerning the Improvement of Forest and Land Fire Control in Indonesia.
[13] TIM UPT Hujan Buatan BPPT 2014 Peranan Teknologi Modifikasi Cuaca di Indonesia Cetakan 1 ISBN : 978-602-1124-38-3 BPPT PRESS Indonesia
[14] Haryanto U 2000 Teknologi Modifikasi Cuaca yang Efektif dan Efisien *Jurnal Teknologi & Sains Modifikasi Cuaca* **1(1)**
[15] Nugroho, Tikno 2002 Evaluasi Peningkatan Hasil Curah Hujan dan Ketersediaan Air Akibat Kegiatan Modifikasi Cuaca di DAS Citarum *Jurnal Teknologi & Sains Modifikasi Cuaca* **3(1) 71**-78
[16] BNPB 2013 *Ketangguhan Bangsa Dalam Menghadapi Bencana* GEMA BNPB **2(2)**
[17] NASA 2017 National Aeronautic and Space Administration *Goddard Space Flight Center: Satellite Terra&Aqua* http://mirador.gsfc.nasa.gov/ [February 2, 2017]
[18] NASA 2016 National Aeronautic and Space Administration *Goddard Space Flight Center: Tropical Rainfall Measurement Mission* http://trmm.gsfc.nasa.gov/ [October 12, 2016]
[19] Sasmito S D 2016 Rainfall Estimation Based on Geostationary (MTSAT-1R) and Microwave Imager (TRMM) Satellite
[20] Nugroho S P 2002 Ketersediaan Air Akibat Kegiatan Modifikasi Cuaca Di DAS Citarum *Jurnal Teknologi & Sains Modifikasi Cuaca* **2** 15

IOP Conf. Series: Materials Science and Engineering **309** (2018) 012017 doi:10.1088/1757-899X/309/1/012017

Waste management as an effort to improve urban area cleanliness and community income (journal review)

Bag Kinantan [1] **, A Rahim Matondang** [2] **and Juliza Hidayati** [3*]
[1, 2, 3] Industrial Engineering Departement, Universitas Sumatera Utara

*Email: rivaijuliza@gmail.com

Abstract. The problem of urban waste has reached a point of concern. Population and economic growth are thought to be the cause of increasing the waste generation. The major problem related to this condition is the increasing of waste production which is not balance with the increase of its management capacity. Based on the Law Number 18 of 2008 that waste management starts from the source by applying the 3R approach (Reduction, Reuse, Recycle). This regulation provides a way which expect the waste management can be better, so that, the level of waste service can be improved and load on landfills (TPA) can be reduced.The cost of garbage collection and transport are 85% of the total waste management cost, so if this is optimized, it will optimize the system as a whole. Subsequent research focuses on how to optimize the garbage collection and transport sub-systems by finding the shortest route of transportation to the landfill by developing a Vehicle Routing Problem (VRP) model. The development of an urban area leads to the preparation of the best route is no longer an optimal solution. The complexity of the waste problem is not only related to the technical matters, but also the social and economic problems of the community. So, it is necessary to develop a model of waste management which does not only pay attention to the technical aspects, but also the social and economic. Waste is expected to be no longer a burden, but can also be utilized economically to increase community income.

1. Introduction

Waste is the remaining from human activities. The continuity increasing of population growth factor and accompanied by the consumptive lifestyle are thought to be the causes of the high volume of the waste. The consumptive culture of today's society has a huge part in increasing the types and qualities of the waste [1].

The problem in the management of urban waste occurs due to the imbalance between the production and the capability to manage it; the waste volume continues to increase in line with the population growth, changes in the quality of life and the dynamics of community activities. Waste problem relates to other development issues such as population, social, economic and land acquisition. These issues will have a profound impact on health, public order and security, and others. Therefore, the implementation of management of waste should be in an integrated way with other development issues [2].

In addition, the pattern of urban waste management, particularly in Indonesia, still adopts the system of collecting – transporting – disposing [3] or so called end of pipe system, in which some officers collect, transport and dispose the waste to the determined processing location. These activities will only add larger burden to the final landfill (TPA), while the capacity of the landfill has its limitation.

Moreover, TPA with open dumping system also produces various problems in environment and health. Decomposition of organic material can produce methane gas which can cause fires and explosions, and contribute to global warming. The chemical and biological processes in TPA can produce leachates which can contaminate the surface and ground water. Not yet to mention the process of waste burning which produces smoke and contributes to the air pollution [4].

According to [5] that it requires a pioneer to change the mindset of waste management from the end of pipe system, which dispose waste directly into the TPA, to the principle of 3R which stands for Reduce, Reuse, Recycle.

The enactment of Law Number 18 The year of 2008, as the basis of systematic and sustainable of waste management, expects the waste management system will no longer apply the end of pipe system, but the process of waste management starts from the source level by reducing and managing the waste in an integrated way prior to final disposing to the TPA. Also, there is expectation that waste is no longer a burden, but has economic value which can provide additional income for the community.

This paper focuses to examine the development of existing urban waste problem in Indonesia and take the proper approach to overcome the problem. Plus some literature from several relevant international journals. The literature studies conducted within last 15 years show several approaches by researchers to solve the urban waste problem.

The approach used is the operational waste technique and integrated municipal solid waste management.

2. Operational Waste Technique

The operational waste technique starts from the waste placement, collection, removal, transportation, processing and final processing [3].

According to [6], the cost of collecting and transporting of urban waste is 85% of the total system cost. So, if the sub system collection and transportation can be optimized, then indirectly the overall system is optimized.

Based on the Profile of Depok City Waste Management in [7], the quantity of waste transportation system can absorb financial resources higher than others. So that, the improvement of efficiency, effectiveness, and productivity related to waste transportation system is one of the main keys to reduce the total cost of the waste system.

The reduction of these costs will be obtained based on the determination of the shortest travel time (not only the shortest distance) from the waste transport vehicle on its way to Landfill (TPA). It means that the determination of the waste transportation route which will provide the shortest travel time will be the spearhead to execute this [7].

The approach used to prepare the best route is by using Geographic Information System (GIS) and model of Vehicle Routing Problem (VRP).

2.1. The Preparation of the Shortest Route with Geographic Information System (GIS) Approach

The preparation of the optimum transportation route uses the Geographic Information System (GIS) approach. Before the preparation of waste transportation route, firstly the model of domestic waste volume estimation is made, through a quick bird image interpretation process with the help of *ArcGIS* software [8]. While for the optimum route preparation is used Network Analysis method which is the extension of GIS. One of the advantages of GIS is the network analysis application which has been widely used in daily life. One type of network analysis is determining the route [9].

Consider that there are three variables to determine the optimum route in this network application such as: the length of the road, the width of the road, and the direction of the vehicle movement. Each variable has a level of resistance called impedance. The smaller the impedance value of a variable, the more optimal the results will be. The optimum route will be formed from the shortest part segment, the road width is more than 5 meter, and has the direction of movement toward the final disposal.

In the research by [10] using the GIS approach with the help of Route view Pro. Software in the

IOP Conf. Series: Materials Science and Engineering **309** (2018) 012017 doi:10.1088/1757-899X/309/1/012017

preparation of the route, not only paid attention to the shortest route in the waste collection and transportation, but also noted the number of turns whether to the right or left. In this research, turned to the right took more time from left, so that, the optimum route produced a smaller number of turns to the right.

While in the research by [8] the prepared spatial model could estimate the domestic waste volume. While for the transporting route was obtained 52 optimum routes from 52 temporary landfill (TPS) toward the final landfill (TPA). The route preparation by using *ArcGIS 9.3* software with Network Analysis extension facility.

2.2. The Vehicle Routing Problem (VRP) Model Approach

According to [11], waste management is part of supply chain and green logistic management. And according to [12] distribution issues involve several considerations. Including the vehicle routes, vehicle fleets, to the scheduling of vehicles. These problems then become the issues of *Vehicle Routing Problem* (VRP). VRP has several types of goals associated with minimizing distribution costs. The functions of the goals include minimization of the number of vehicles, the travel time, mileage, and other purposes by the characteristic of the issues.

According to [12] stated that the problem of VRP could be defined as the problem of finding the route with the minimum cost from a depot to a customer scattered with the number of different requests. The route is made in such a way that each customer is only visited by one vehicle. The entire routes start and end at the depo, and the number of request should not exceed the vehicle capacity in one–way route.

The purpose of preparing this VRP model in addition to obtain the most optimum route, also can be used to improve urban waste service. Because one of the problems of waste service is the ineffectiveness of transportation (waste collection and transportation) from settlement to landfill which is caused by the maximum function of the temporary landfill (TPS) regarding current location [11].

To design the optimum route model of waste transportation, it needs several stages, namely identification of decision variable, formulation of the objective function, and identification of constraints. The model can be used to find the solution of optimum objective function [13]. The mathematical model formed becomes the basis to construct waste transportation route algorithm [12].

Several types of research used VRP model, such as performed by [12] in Bandung, which developed Vehicle Routing Problem model by adding Intermediate Facility at the end of the route, that is the final landfill (TPA). This model is referred to as *Vehicle Routing Problem With Multi Trips and Intermediate Facility* (VRPWMTIF) and solved by using *sequential insertion* algorithm. In this research was obtained 28 tours for West Bandung, 41 tours for Central Bandung, and 68 tours for East Bandung.

So is the research was done by [14] which used ant colony algorithm to optimize the waste transportation route in Yogyakarta. It was obtained 12 routes optimum for waste transportation in Yogyakarta.

While [11] did not conduct the route preparation based on the existing temporary landfill (TPS) but compiled first the temporary landfill (TPS) then prepared the waste transportation route from the TPS toward TPA by using *Vehicle Routing Problem* (VRP) model. So that, it was obtained a better network of waste logistic.

These two approaches, both GIS and VRP model focus on the preparation of the waste vehicle route, to obtain the optimum route. The goal is to reduce travel time so that the number of release becomes more and reduce the transportation cost.

The study of operational waste technique includes both approaches.With the increasing of city development complexity, also, the unpredictable of the traffic jam, repairing the transportation route is not the optimum solution. Moreover, the capability of TPA is increasingly limited. So it is needed to find other solution which is more comprehensive and sustainable.

3. The Integrated Waste Management

With the issuance of Law Number 18 The year of 2008 concerning waste management, the government has already a legal umbrella to improve the waste management system which was initially used the pattern of the end of pipe to an integrated and sustainable management pattern. This system allows involving all elements to improve the role in solving the urban waste problem.

The integrated waste management is a comprehensive program to prevent the waste generation, recycling, composting and final disposal. An effective system of waste management considers how to prevent, recycle, and manage waste in the most effective way to keep community's health and environment. It involves the evaluation of local needs and condition, then selects management activities that best this condition [4].

In practice, health and safety have become the concern in the waste management. Waste should be managed to reduce the risk to human health. Today, communities want more, not only safe, but also the waste management should be sustainable. Sustainable development is the development which meets the current needs without sacrificing the ability of future generations to meet their own needs [15]. It identifies the synergies among economic development, social equity and environment, therefore, the sustainable waste management should meet three requirements :

1. Affordable economically
2. Accepted socially
3. Effective environmentally

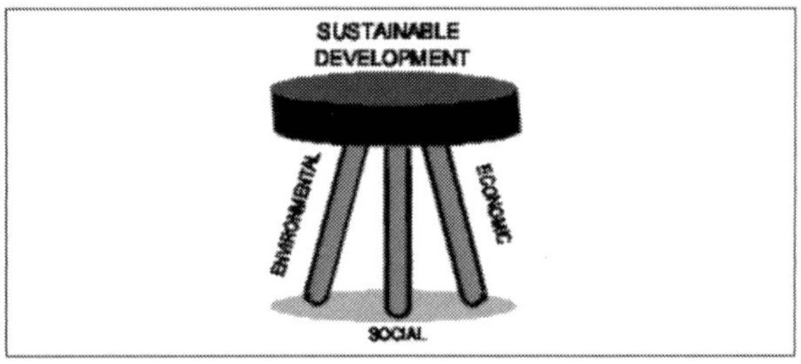

Figure 1. Three pillars of sustainable development

It takes the same attention for each pillar because otherwise there will be an imbalance of the system as a whole.

According to [15], in the past the economic cost of the waste management system is the major controlling factor in the decision-making process, while environmental consideration plays a more role in the current decision-making process. The inclusion of social aspects in the decision-making process in the waste management system, though it is not a new concept, has been the topic of research on how to calculate social awareness that has just begun.

3.1. The Integrated Waste Management Approach

The identification of waste management as integral to sustainable urban development is increasingly recognized by the international aid and development community. The United Nations Conference on Environment and Development stressed that solid waste production should be minimized, reuse and recycling, maximized, environmentally sound waste disposal and treatment promoted and waste service coverage extended. [4].

For the waste management in Indonesia, according to [16] argued that community assessment on the regional government commitment in waste management is relatively less. The related offices in responding waste management always return to the technical problems such as budget, the lack of infrastructure, without speaking the human and cultural aspect. Therefore, the solution made by the regional government does not only prepare the master plan of drainage arrangement, but also the

human and cultural aspects which hold the important role to be noticed, without minimizing technical aspects.
With the issuance of Law Number 18 The year of 2008, the government is not the only party who should be responsible for the waste management. Communities as the source of waste generation should be involved actively. According to [4] in a prosperous country, the ultimate motivation for reducing waste related to high–cost problem and scarcity of area to be the landfill as well as the condition of the environment degradation caused by poisonous material in the collected waste. The same consideration applies to the metropolitan cities in developing countries surrounded by jurisdictions of other residents.

Also, the burden borne by the relevant agencies who responsible for urban waste is very heavy. So that waste management effort should start from household by gradually reducing waste and separating the waste they produce. This separation will support the program of 3 R (Reduction, Reuse, Recycle) as mandated by Law Number 18 The year of 2008.

3.2. The Integrated Waste Management and Local Wisdom
Integrated solid waste management (ISWM) is a comprehensive waste prevention, recycling, composting, and disposal program. An effective ISWM system considers how to prevent, recycle, and manage solid waste in ways that most effectively protect human health and the environment. ISWM involves evaluating local needs and conditions, and then selecting and combining the most appropriate waste management activities for those conditions [4].

In other words, local approaches are needed to make the waste management system work effectively. They are commonly referred to as local wisdom. As an example of the local wisdom of Minangkabau community in Padang. [17] has developed a model of social awareness participation in waste management in Padang City based on the local wisdom *"adat basandi syarak, syarak basandi kitabullah"*. In this research is argued that there are two causes of waste management problems in Padang City, first is the limited facilities and infrastructure and second is the low awareness of social behavior of society in waste management.

While in Purwokerto City, the government reactivated the local wisdom of *"kerigan"* patterns to increase the community participation to manage the household waste. Also, other policy strategies from the government to support the efforts of the realization of a green city that has been done, among others: the formation of green community, waste banks, pilot tree banks, set up some green open spaces [16].

3.3. Management of Household Waste and Increasing Community Income
With the development of cities, population and economy, the waste production also increase. This high waste production has advantages and disadvantages. The advantages of waste can still be utilized and have economic value. While the shortcomings are when not managed properly, waste can cause community health problems and environmental damage.

With the implementation of 3R program: Reduction, Reuse and Recycle, communities together with the government can utilize the waste they generate to be a source of income. There are at least 3 (three) things obtained from the utilization of waste and economic value. First, the making of compost fertilizer from organic waste and the producing of methane gas with anaerobic digestion method. Second, recycling anorganic waste such as plastic, metal, and others. The third, process of burning waste by using incinerator can be utilized to become electrical energy.

As has been done by Purwokerto Municipal Government by running Waste Bank program [16], this program encourages communities to separate self – contained waste in their homes by separating organic and inorganic waste. They sold the recycling waste to the waste bank, and recorded the sales proceeds in their respective savings books, while the organic waste was processed to be the compost.

4. Conclusion
Community waste production will continue to increase in line with the growth of population and urban

economy. Particularly the developing countries like Indonesia, waste management does not meet the expectations, so it can pose a risk to the community such as health problems and environmental damage.

The approach to solve waste problems still applies the old paradigm of focusing on the *end of pipe* or the haul and dump approach. It only provides a temporary solution that is transporting the waste from residential areas. Another problem arises, the capacity of the landfill has its limitation.

An integral approach in waste management is expected to be a long – term solution to solve waste problems. Although theoretically it has been described the integrated waste management steps, in practice, it is adapted to local needs and conditions and selected the most likely combination (feasible) to be applied under these circumstances.

Through an integrated approach also, waste is not the only burden for waste stakeholders, but can also have economic value which can increase community income.

References

[1] Pramitaningrum E 2014 *Pengembangan Model Untuk Memonitor Jumlah Sampah di Kota Yogyakarta*, Jurusan Teknik Mesin dan Industri, Universitas Gajah Mada, Yogyakarta

[2] Purristiyana 2011 *Strategi Peningkatan Pelayanan Dinas Kebersihan dan Pertamanan dalam Pengelolaan Sampah Perumahan (Kajian di Kabupaten Bogor)*, Sekolah Pasca Sarjana, Institut Pertanian Bogor, Bogor

[3] Damanhuri E 1999 *Teknik Pengelolaan Persampahan - Modul A dan Modul B, Disiapkan untuk PT Freeport Indonesia*, Teknik Lingkungan ITB, Bandung

[4] Sridevi V et. al. 2012 *A Review On Integrated Solid Waste Management*, International Journal of Engineering Science and Advanced Technology (IJESAT) Vol **2**, Issue 5, pp 1491 – 1499

[5] Rizani MD, Surjono 2016 *Strategi Pengelolaan Persampahan di Wilayah Perkotaan dalam Mencapai Target Layanan*, The 3rd University Research Coloquium 2016, Universitas Brawijaya, Malang

[6] Apaydin O 2004 *GIS Supported Optimization of Solid Waste Collection in Trabzon*. Journal of Engineering and Natural Sciences, Sigma 2004/4

[7] Suryanto DA, Widjadjakusuma J 2005 *Kajian Sistem Pengangkutan Sampah Kota Depok*, Proceeding Seminar Nasional PESAT 2005, Jakarta

[8] Han V, Raharjo N 2013 *Estimasi Volume Sampah Domestik dan Rekomendasi Rute Pengangkutan Sampah Berdasarkan Analisis Spasial di Kota Surakarta*, Universitas Gadjah Mada, Yogyakarta.

[9] Suprajaka dan Rinaldi, 2014, *Kajian Optimalisasi Model Pengangkutan Sampah di Kota Tangerang Menggunakan Sistem Informasi Geografis (SIG)*, Jurusan Teknik Perencanaan Wilayah dan Kota, Universitas Esa Unggul, Jakarta.

[10] Apaydin, O. and Gonullu, M.T, 2007, *Route Optimization for Solid Waste Collection: Trabzon (Turkey) Case Study*, Global NEST Journal 2007: **9** (1): 6 – 11.

[11] Alfath, Sabda dan Rika Ampuh Hadiguna, 2015, *Usulan Kebijakan Logistik Pengelolaan Sampah di Kota Padang : Lesson Learnt Penentuan Lokasi dan Rute Pengangkutan Sampah*, Jurnal Sains, Teknologi dan Industri Vol **12** No 2, Juni 2015, pp 146 – 154.

[12] Fitria, L. Susanti, S dan Suprayogi, 2009, *Penentuan Rute Truk Pengumpulan dan Pengangkutan Sampah di Bandung*, Jurnal Teknik Industri, Vol **11**, No 1, Juni 2009, pp 51 - 60.

[13] Sutrisno, Dwi, 2014, Optimasi Rute Pengangkutan Sampah Dengan Metode *Vehicle Routing Problem With Time Window* Menggunakan *Binary Integer Programming*, Jurusan Teknik Industri, Universitas Sultan Ageng Tirtayasa, Serang.

[14] Lestari, Himmawan Puji dan Eminugroho Ratna Sari, 2013, *Penerapan Algoritma Koloni Semut untuk Optimisasi Rute Distribusi Pengangkutan Sampah di Kota Yogyakarta*, Jurnal Sains Dasar 2013 Vol **2** (1) hal 13 – 19.

[15] Mc Douglas, Forbes. et. al, 2001, Integrated Solid Waste Management : a Life Cycle

Invetory, Oxford, Blackwell Sciences, 2nd Edition, pp 3.
[16] Suyanto, Edi. et. al, 2015, *Model Kebijakan Pengelolaan Sampah Berbasis Partisipasi "Green Comunity" Mendukung Kota Hijau*, Jurnal Mimbar Vol **31**, No 1, Juni 2015, pp 143 – 152.
[17] Assahary, Salman, 2014, *Model Penyadaran Sosial Masyarakat dalam Pengelolaan Sampah Berbasis Kearifan Budaya Lokal (Adat Basandi Syarak, Syarak Basandi Kitabullah) di Kota Padang*, Prosiding SNSTL I 2014, Padang.

IOP Conf. Series: Materials Science and Engineering **309** (2018) 012018 doi:10.1088/1757-899X/309/1/012018

Deforestation and reforestation analysis from land-use changes in North Sumatran Mangroves, 1990-2015

M Basyuni* **and N Sulistiyono**

Department of Forestry, Faculty of Forestry, Universitas Sumatera Utara, Jl. Tri Dharma Ujung No. 1 Medan, North Sumatra 20155, Indonesia

*Email: m.basyuni@usu.ac.id

Abstract. Mangrove forest plays a critical role in the context of climate change in tropical and subtropical regions. The present study analyzed the deforestation and reforestation from land-use and land-cover changes from 1990, 2000, 2009 and 2015 in North Sumatran mangrove forest, Indonesia. The land-use/land-cover consists of thirteen classes namely, primary mangrove forest, secondary mangrove forest, shrub, swamp shrub, swamp, settlement, paddy field, oil palm plantation, aquaculture, dry land farming, mixed dry land farming, mining, and barren land. Results showed that primary mangrove forests significantly decreased 61.21% from 1990 to 2015, mostly deforestation was derived from 1990 to 2000 to be secondary mangrove forest and swamp shrub. During 25 years observed, no reforestation was noted in the primary mangrove forest. Similarly, secondary mangrove forest had been degraded from 56,128.75 ha in 1990 to only 35,768.48 ha in 2015. Drivers of deforestation found in secondary mangrove forests were aquaculture (43.32%), barren land (32.56%), swamp shrub (10.88%), and oil palm plantation (5.17%). On the other hand, reforested activity was occurred only 701.83 ha from 1990 to 2015, while the nonforest use has been increased. These data are likely to contribute towards coastal management planning, conservation, and rehabilitation of degraded mangrove forests.

1. Introduction

Mangrove forests are widespread in the inter-tidal zone of tropical and sub-tropical climates and play a significant role in the context of climate change in the regions. Given their ecological, economic, and environmental significance, mangroves are currently the most threatened ecosystems due to indirect and direct deforestation and degradation [1]. Aquaculture and oil palm estate have been major drivers of mangrove loss in Indonesia [2-3]. Deforestation negatively impacted on 16% of global mangrove species under risk of extinction [4] and reduced mangrove biodiversity and carbon storage of forest biomass [3,5]. Mangrove conservations as well as sustainable mangrove management, therefore, are needed to maintain the presence of mangrove forest and increased reforestation programs.

In spite of mangrove deforestation globally or regionally was well documented [1,3], deforestation and reforestation as well from land-use and land-cover changes are rarely reported especially at regency level in Indonesia. To get more insight into a better understanding of spatial and temporal variation in the land uses that in displacing mangrove forests, the present study aimed to analyze the drivers of mangrove loss and reforested mangrove from land-use and land-cover changes from 1990, 2000, 2009 and 2015 in North Sumatran mangrove forest, Indonesia.

IOP Conf. Series: Materials Science and Engineering **309** (2018) 012018 doi:10.1088/1757-899X/309/1/012018

2. Materials and Method

2.1. Study area and dataset

The study was carried out in North Sumatran mangrove forest, Indonesia, where primary and secondary mangrove distributed naturally in 13 Regencies of North Sumatra province. These regencies namely Asahan, Batubara, Gunung Sitoli, Medan city, South Nias, North Nias, Central Tapanuli, Serdang Bedagai, Labuhanbatu, Deli Serdang, Langkat, Mandailing Natal, and North Labuhanbatu (Figure 1). Land-use and land-cover changes data period 1990-2015 was obtained from Ministry of Forestry, Government of Indonesia. Landsat 7 Enhanced Thematic Mapper Plus (ETM+) satellite image was acquired from USGS (http://govis.usgs.gov/).

2.2. Analysis of land-use and land-cover changes

Analysis of the Landsat images was carried out by applying supervised classification with maximum likelihood as previously reported [2,6]. Image pre-processing, the process of image interpretation, image classification and change detections were done by ArcGIS 9.3.1 and ArcView 3.3. The ground check was conducted by utilizing Global Positioning System (GPS) to collect information of recent land use/land-cover. The information was used as guidance for image geometric correction and image rectification.

2.3. Analysis of deforestation rate of North Sumatran mangrove

Primary and secondary mangroves existed in 13 regencies in North Sumatra as previously described. Deforestation rate was measured from total mangrove loss (primary and secondary) from the year 2015 to the year 1990. From this measurement, annual deforestation, deforestation proportion, and class of deforestation were determined.

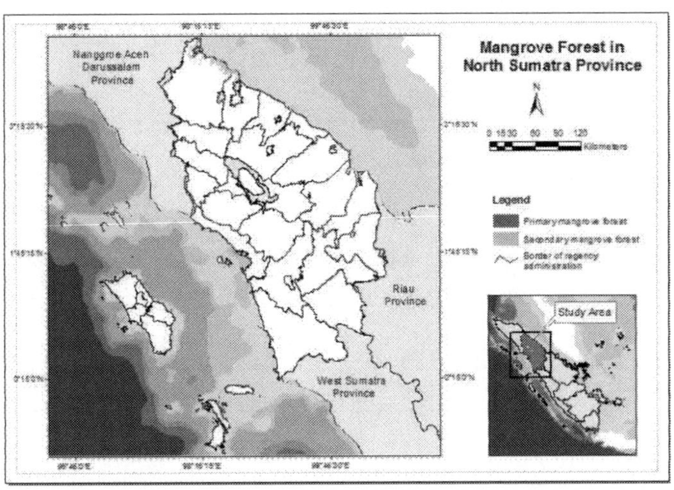

Figure 1. Location of study area showing primary and secondary mangrove forest in North Sumatra

3. Results and Discussion

The results will be discussed in two subsections; they are land-use/land-cover changes between 1990 and 2015 and deforestation and reforestation implication in North Sumatran mangrove.

3.1. Land-use/land-cover changes between 1990 and 2015

Mangrove deforestation remains huge in North Sumatra with more than 22,513.2 ha mangrove lost between 1990 and 2015 (Figure 2). Figure 2 depicts that primary mangrove forests significantly decreased 61.21% from 1990 to 2015, mostly deforestation was derived from 1990 to 2000 to be secondary mangrove forest and swamp shrub. During 25 years observed, no reforestation was noted in the primary mangrove forest. The land-use/land-cover changes in North Sumatran mangrove over period 1990-2015 are shown in Tables 1-2. The land-use/land-cover consists of thirteen classes namely, primary mangrove forest, secondary mangrove forest, shrub, swamp bush, swamp, settlement, paddy field, oil palm plantation, aquaculture, dry land farming, mixed dry land farming, mining, and barren land.

There were four main displacing mangrove land use types identified in the primary forest from the largest area to fewer one: secondary mangrove forest, swamp shrub, barren land, and aquaculture (Table 1). Similarly, secondary mangrove forest has been degraded from 56,128.75 ha in 1990 to only 35,768.48 ha in 2015 (Figure 2). As displayed in Table 2, there are four primary drivers of deforestation found in secondary mangrove forests were aquaculture (43.32%), barren land (32.56%), swamp shrub (10.88%), and oil palm plantation (5.17%).

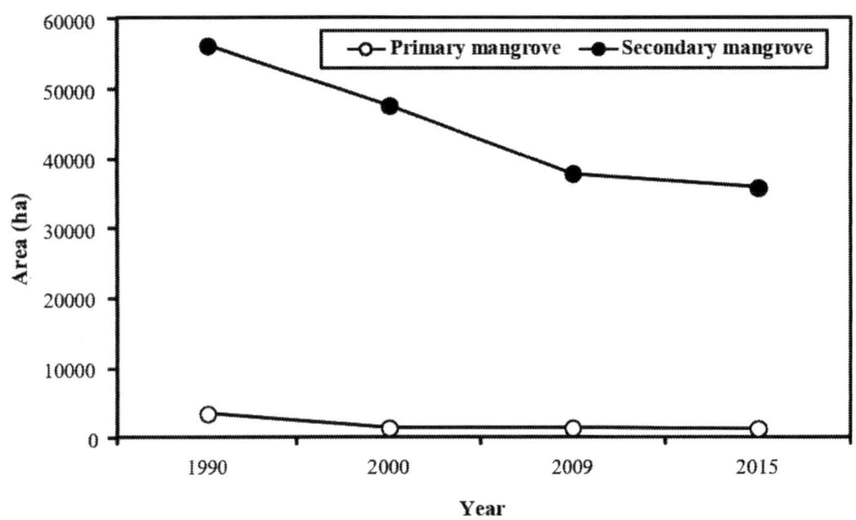

Figure 2. Trend of mangrove deforestation in North Sumatran mangroves over 1990-2015

Table 1. Deforested primary mangrove forest 1990-2015 converted to different land uses

Land-use (1990)	Land-use (2015)	Area (ha)	Proportion (%)
Primary mangrove forest	Shrub	30.9	1.4
	Barren land	366.1	17.0
	Secondary mangrove	1,142.8	53.1
	Swamp shrub	445.9	20.7
	Dryland farming	1.8	0.1
	Aquaculture	165.5	7.7
Total		2,152.9	100.0

Other factors, such as agricultural expansion (dryland and mixed dryland), mining, and settlement, are also responsible for deforestation in the secondary mangrove in North Sumatra. By contrast to this study and in Indonesian mangrove was described [3]. Mangrove conversion to agriculture expansion as major causes of deforestation in some Asia countries such as Thailand (50%), Myanmar (98%), Malaysia (43%), Bangladesh (77%), India (50%), and Sri Lanka (92%) [3].

Table 2. Deforested secondary mangrove from 1990-2015 converted to other land uses

Land-use (1990)	Land-use (2015)	Area (ha)	Proportion (%)
Secondary mangrove forest	Shrub	141.3	0.6
	Oil palm plantation	1,149.0	5.2
	Settlement	37.1	0.2
	Barren land	7,228.9	32.6
	Swamp shrub	2,416.5	10.9
	Dryland farming	803.1	3.6
	Mixed dryland farming	765.0	3.4
	Paddy field	26.8	0.1
	Aquaculture	9,619.6	43.3
	Mining	8.3	0.0
Total	Swamp	9.4	0.0

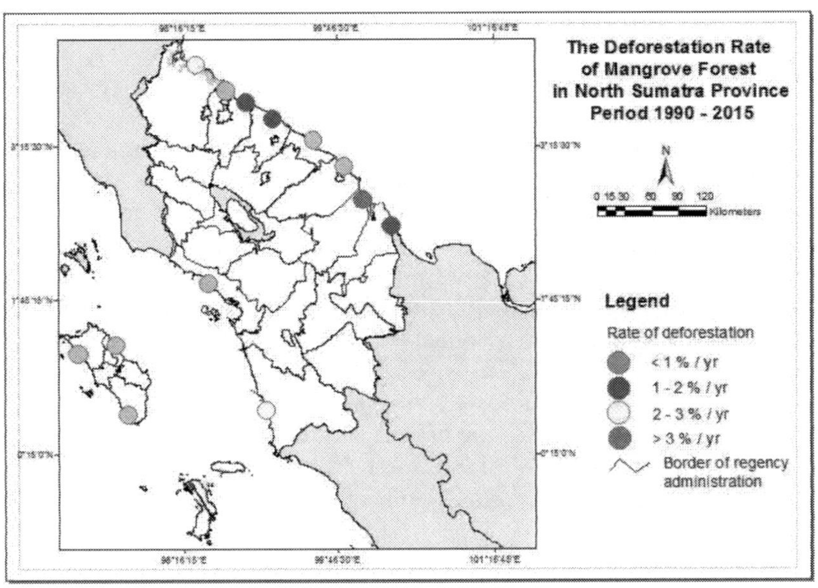

Figure3. Deforestation rate in North Sumatran mangrove over period 1990-2015

3.2. Deforestation and reforestation analysis

Figure 3 shows percentage mangrove deforestation in North Sumatra between 1990 and 2015 was classified into 4 group, the first group with deforestation rate less than 1 % occurred in Asahan, Batubara, Gunung Sitoli, Medan, South Nias, North Nias, and Central Nias. The second group of 1-2 %, where deforestation occurred in Serdang Bedagai, Labuhanbatu, and Deli Serdang. The third group of 2-3% deforestation rate was in Langkat and Mandailing Natal. The last group with more than 3% deforestation rate was only in North Labuhanbatu.

Furthermore, the deforestation occurred in North Sumatra was 900.5 ha/year with percentage rate was 1.2 %. The highest deforestation rate was in Langkat (719.1 ha/year), while the lowest in South Nias (0.11 ha/year). The present result supported the previous report on the drivers of mangrove deforestation in Langkat mostly that mangrove conversion into aquaculture ponds (50 %) and oil palm plantation (28.8%) is one of the primary reasons for such high rates of deforestation [2]. Oil palm plantation also contributed to mangrove deforestation in North Sumatra. The threat of oil palm to mangrove is likely to increase in future [1]. Furthermore, it has been reported that mangrove deforestation between 2000 and 2012 in the east coast of Sumatra around 500-1000 ha/ha [1]. Indonesian mangrove, however, has percentage mangrove loss was 1.7% [1], slightly similar to this study (1.5%). Mangrove conversion to aquaculture also has been reported in Kalimantan and Sulawesi [1]. Our study and other studies suggested that aquacultures are responsible for the significant part of mangrove deforestation.

Table 3. Reforested secondary mangrove from 1990-2015

Land-use (1990)	Land-use (2015)	Area (ha)	Proportion (%)
Barren land	Secondary mangrove	219.2	31.2
Water body	Secondary mangrove	126.6	18.0
Swamp shrub	Secondary mangrove	211.5	30.1
Dryland farming	Secondary mangrove	15.5	2.2
Aquaculture	Secondary mangrove	129.1	18.4
Total		701.8	100.0

Reforestation activity was observed in this study and occurred in secondary mangrove only as 701.83 ha from 1990 to 2015, while the nonforest has been increased (Table 3). The main land uses to be reforested barren land (219. 2 ha), swamp shrub (211.5 ha), aquaculture (129.1), and water body (126.6 ha). Rehabilitation on degraded mangrove was proposed with some considerations such as extent, accessibility, and socioeconomic factors of a local community to get successful mangrove regrowth [3]. Mangrove rehabilitation is significant efforts to restore within the framework of regional development. The rehabilitation program is proposed by the involvement of the local communities dependent on the mangrove ecosystem for sustenance [7]. Recently it has been reported that to reduce global emission from mangrove loss for a cost at less than $ 10 to^{-1} CO_2 [5]. Parallel with this finding, the primary threats to all mangrove species are habitat degradation and mangrove area for conversion to aquaculture, agriculture, urban and coastal development, and overexploitation [4]. Thus, nowadays social induced-rapid developmental changes stipulate a better understanding of the dynamic of mangrove forest [8].

4. Conclusions

Primary drivers of deforestation in North Sumatran mangrove, especially from secondary mangrove forest was derived from land-use of aquaculture. Conservation and reforestation of mangrove forest should be the principal target in North Sumatra. The present data are likely to contribute towards coastal management planning, conservation, and rehabilitation of degraded mangrove forests in North Sumatra.

Acknowledgment

This study was partly supported by an International Research Collaboration and Scientific Publication Grant 2017 (No. 003/SP2H/LT/DRPM/IV/2017 to MB) from the Directorate for Research and Community Service, Ministry of Research, Technology and Higher Education, Republic of Indonesia.

References

[1] Richards DR and Friess DA 2012 *Proc. Natl. Acad. Sci. USA* **113**: 344–3349
[2] Basyuni M, Putri LAP and Murni MB 2015 *J. Man. Hut. Trop.* **21**: 25–35
[3] Giri C, Zhu Z, Tieszen LL, Singh A, Gillette S and Kelmelis JA 2008 *J. Biogeogr.* **35**:519–528
[4] Polidoro BA, Carpenter KE, Collins L, Duke NC, Ellison AM et al. 2010 *Plos ONE* **5 (4)**: e10095
[5] Siikamaki J, Sanchirico N and Jardine SJ. 2011 *Proc. Natl. Acad. Sci. USA* **109**: 14369–14374
[6] Basyuni M, Rouf RA, Saragih M, Asbi AM and Yuriswan W *Adv. Soc. Sci. Educ. Hum. Res.* **81**: 551–555
[7] Field CD 1999 *Marine Poll. Bull.* 37: 383–393
[8] Basyuni M, Oku H, Baba S, Takara K and Iwasaki H 2007 *J. Oceanography* **63**: 601–608

IOP Conf. Series: Materials Science and Engineering **309** (2018) 012019 doi:10.1088/1757-899X/309/1/012019

Impact analysis of leading sub sector on basic sector to regional income in Siak Regency, Riau Province

P Astuti*, I Nugraha, F Abdillah
Urban and Regional Planning Department, Faculty of Engineering, Universitas Islam Riau, Pekanbaru, Indonesia

*E-mail: pujiastutiafrinal@eng.uir.ac.id

Abstract. During this time Siak regency only known as oil producing regency in Riau province, but based on the vision of spatial planning Siak's regency in 2031 there was a shift from petroleum towards to other sectors such as agribusiness, agroindustry and tourism. The purpose of this study was to identify the sector base, the leading subsectors and shift with their characteristics and to identify the leading subsectors development priority. The method used in this research consisted of the method of Location Quotient (LQ, Shift Share, and Overlay method). The research results were used Location Quotient (LQ) to identify sector's base in Siak regency based on the document of PDRB. The sector's refers to the constant prices year of 2000 were mining and quarrying sector (2.25). The sector's base using document of PDRB at constant prices 2000 without oil and gas sector was the agricultural sector with a value of LQ was 2,45. The leading sub sector in the Siak regency with mining and quarrying sector was oil and gas (1.02) and leading sub sector without oil and gas sector was the plantation sector (1.48) and forestry sector (1.73). Overlay analysis results shown that agriculture sector as a sector base and plantation and forestry as a leading sub sector has positive value and categorize as progressive and competitiveness. Because of that, this leading sub sector gets high priority to developing.

1. Introduction

One of the objectives in regional development is the improvement of economic development. The economic development is progress where the local government and the community manage the available resources and form a partnership patterns between local government, the private sector in creating jobs and stimulating economic activities in the region [1]. The increasing of economic development is one of the development's objectives that became an important indication of the success of economic development related to real income and income per capita, take into an account of population increase and fundamental development of the economic structure that can be felt by publics [2].

The important indication to know the economic condition within a certain period based on the amount of Gross Regional Domestic Product or Produk Domestik Regional Bruto (PDRB) that using constant price. [3] explain that PDRB is the indicator to manage the extent of the success of the local government in utilizing available resources and can be used as the main material of planning and decision making. Commonly, PDRB consists of nine business field such as agriculture sector, mining and quarrying sector, manufacturing industry sector, electricity, gas and water supply sector,

IOP Conf. Series: Materials Science and Engineering **309** (2018) 012019 doi:10.1088/1757-899X/309/1/012019

construction sector, trade, hotel and restaurant sector, transport and communication sector, the sector of finance dwelling and business service, and service sector. In other hand, Sukirno in [3], explains that based on business field, the economic sectors in Indonesia's economy divided into three main classes, that is:

Prime sector, consists of the agriculture sector and their subsectors and mining and quarrying sector.Secondary sector, consists of manufacturing industry sector, electricity, gas and water supply sector and construction sector. Tertiary Sector, consists of trade, hotel and restaurant sector, transport and communication sector, the sector of finance dwelling and business service, and others services sectors.

Siak Regency is one of regency that located in Riau province. Siak regency has nine business field that been used as a benchmark for development and income's enhancement. Based on Kabupaten Siak Dalam Angka 2013, Siak regency has a number of populations amount of 491,967 inhabitants. Total revenue of Siak regency in 2013 was 57,515,706.73 (in a million rupiahs). Based on the spatial vision of Siak regency in 2031 that focus on regional development is agribusiness, agroindustry and tourism sector, but it is inversely proportional to the data of Gross Regional Domestic Product or Produk Domestik Regional Bruto (PDRB), that explains the mining and quarrying sector was a sector base in Siak regency. That sector has the biggest amount of income is about 22,935,980.50 (in a million rupiahs).

Refers to nine of business fields in Siak regency, there is no information about the sector base and the leading sub sector. That means yet of the nine business field have known the level of competitiveness, growth values and contributions. With the information of the sector base and the leading sub sector, it can determine the potential sector in the region, development priority and development policy. So the purpose of this research was to identify the sector base, the leading subsectors and shift with their characteristics and to identify the leading subsectors development priority.

2. Methodology

The method in this research was used deductive approach, that means this research objective was to examine the hypothesis with validating the theory or examine application the theory in the certain condition [4]. The data type in this research was secondary data. Secondary data are the sources of research data that was obtained by researcher indirect with intermediaries. Deeper, [4] explained that the secondary data commonly in the form of evidence, notes or historical reports that have been archived and has been published or not. In this research, the secondary data was obtained from Badan Pusat Statistik Kabupaten Siak and Badan Pusat Statistik Provinsi Riau. Analysis techniques were used quantitative analysis with using Location Quotients method, Shift-Share method and Overlay method. For further explanation, the following below explains the research methods, that is :

2.1 Location Quotients (LQ) Method

Location Quotients (LQ) method was used to determine the sector base and the leading sub-sector. This method was a model to help show the potential sector in the region. In this research divided in two parts, first determine LQ using Gross Regional Domestic Product or Produk Domestik Regional Bruto (PDRB) without mining and quarrying sector based on constant price year of 2000. The second is LQ analysis with mining and quarrying sector based on constant price year of 2000. The Location Quotients (LQ) equation is as follows [1]:

$$LQ = \frac{vi/vt}{Vi/Vt} \text{ x } \frac{vi/Vi}{vt/Vt} \tag{1}$$

Remarks :
v_i = the income/revenue from *i* sector in Siak regency
v_t = the total income/revenue in Siak regency
V_i = the income from *i* sector in Riau Province
V_t = the total income of sectors in Riau Province
The Criteria :
LQ = 1, means the domestic products in the regional have been consumed personally.
LQ > 1, means the sector in the regional is sector base that can export the commodity into other.
LQ < 1, means the sector in the regional is a non-base sector and had a trend to import from other.

2.2 Shift Share Method

Shift share analysis is one of analysis technique in regional economic science that has objective to determine the main factors that influence and determine economic growth and sector base in a region. The shift share assumptions are sector growth in the region has an equivalent value with reference region. The Shift Share method was used to determine competitiveness, contribution and economic growth in the leading sub sector in Siak regency. The shift share analysis equation is as follows [7]:

$$SSA = \underbrace{\left(\frac{X_{..(tl)}}{X_{..(t0)}} - 1\right)}_{(a)} + \underbrace{\left(\frac{X_{i(tl)}}{X_{i(t0)}} - \frac{X_{..(tl)}}{X_{..(t0)}}\right)}_{(b)} + \underbrace{\left(\frac{X_{ij\,(tl)}}{X_{ij(t0)}} - \frac{X_{i(tl)}}{X_{i(t0)}}\right)}_{(c)} \quad (2)$$

Rema

a = Economic Growth
b = Proportional Shift
c = Differential Shift
X = Total of Riau Gross Regional Domestic Product in national total.
X_i = Number of Riau Gross Regional Domestic Product in *i* sector in national total.
X_{ij} = Number of Riau Gross Regional Domestic Product in sector *i* in region.
T_1 = The Final Year
t_0 = The Starting Year

Interpretation:
Economic Growth = National Economic Growth
Proportional Shift (b) = if the results have positive value means the sector growth faster than national economic and vise versa.
Differential Shift (c) = If the results have positive value means the region has strong competitiveness and vise versa.

The identification of economic growth in the region also can use the addition between Proportional shift and Differential shift (b+c) that we called net shift (d). If the proportional shift (b) and differential shift (c) have been compared, the characteristic of the shift can be categorized into four types, such as type I (fast growing), type II (Highly potential), type III (Developing) and type IV (Depressed) (table 1).

IOP Conf. Series: Materials Science and Engineering **309** (2018) 012019 doi:10.1088/1757-899X/309/1/012019

Table 1. Relative Position of Leading Sub Sector

Differential Shift (c)	Proportional shift (b)	
	Positive (+)	Negative (-)
Positive (+)	TYPE I (Fast Growing)	TYPE II (Highly Potential)
Negative (-)	TYPE III (Developing)	TYPE IV (Depressed)

2.3 Overlay Method

Overlay method is the analysis that the addition of LQ values with Shift Share values and has the objective to determine the sector will be the choice in determining development priorities. Overlay method divides the economic activities into four categories, namely :

1. **Dominant and Progressive**; if the growth (LQ) and contribution (Shift Share) have positive value, then the sector was dominant and need get high priority in development.
2. **Dominant and Unprogressive**; if the growth (LQ) has a positive value while the contribution (Shift Share) has negative value, that means the sector have been developed so need attention to increase the contribution to Gross Regional Domestic Product.
3. **Not Dominant and Progressive**; if the growth (LQ) has negative value while the contribution (Shift Share) has positive value, then the sector has declined, so the growth needs to be pushed up.
4. **Not Dominant and Nonprogressive**; if the growth (LQ) and contribution (Shift Share) have negative value, means the sector was not potential to be developed.

3. Results and Discussion

3.1. Location Quotients (LQ) Analysis

LQ analysis was used to determine sector base and leading sub sector in Siak regency. Firstly, LQ was used to determine sector base based on Gross Regional Domestic Product Series Data from 2009-2013 with Mining and Quarrying Sector using the constant price year of 2000. The result (Table. 2) shown that the agriculture sector has LQ value > 1 (with the mean value is 2,45) that means the agriculture sector became sector base in Siak regency. The LQ value of the agriculture sector has an increasing trend from 2009-2012 and has been declining in 2013.

Table 2. LQ Value without Mining and Quarrying Sector Using The Constant Price year of 2000

No	Business Field	Location Quotient (LQ) Value					Mean
		2009	2010	2011	2012	2013	
1	Agriculture sector	2.30	2.35	2.40	2.60	2.60	2.45
2	Electricity. gas and water supply sector	0.10	0.11	0.11	0.12	0.13	0.11
3	Construction sector	0.33	0.42	0.48	0.54	0.51	0.46
4	Trade. hotel and restaurant sector	0.32	0.32	0.32	0.30	0.33	0.32
5	Transport and communication sector	0.22	0.22	0.23	0.24	0.25	0.23

No	Business Field	Location Quotient (LQ) Value					Mean
		2009	2010	2011	2012	2013	
6	The sector of finance dwelling and business service	0.33	0.31	0.29	0.27	0.26	0.29
7	Service sector	0.57	0.55	0.54	0.54	0.54	0.55
	Total	4.17	4.28	4.37	4.61	4.62	4.41

Source: Results Analysis. 2016

Based on LQ analysis with mining and quarrying sector based on constant price year of 2000 the mining and quarrying sector has LQ>1 (Table 3). Mining and quarrying sector have LQ with the mean value is 2.25. That means the mining and quarrying sector had become sector base in Siak Regency.

Table 3. LQ Value with Mining and Quarrying Sector Using The Constant Price year of 2000

No	Business Field	Location Quotient (LQ) Value					Mean
		2009	2010	2011	2012	2013	
1	Agriculture sector	0.19	0.26	0.30	0.33	0.38	0.29
2	Mining and Quarrying	2.34	2.20	2.18	2.27	2.24	2.25
3	Manufacturing Industry	1.24	1.62	1.80	2.00	2.22	1.78
4	Electricity. gas and water supply sector	0.01	0.01	0.01	0.02	0.02	0.01
5	Construction sector	0.03	0.05	0.06	0.07	0.07	0.06
6	Trade. hotel and restaurant sector	0.03	0.03	0.04	0.04	0.05	0.04
7	Transport and communication sector	0.02	0.02	0.03	0.03	0.04	0.03
8	The sector of finance dwelling and business service	0.03	0.03	0.04	0.03	0.04	0.03
9	Service sector	0.05	0.06	0.07	0.07	0.08	0.07
	Total	3.94	4.28	4.53	4.86	5.14	4.55

Source : Results Analysis. 2016

After determining the sector base. the next step was determined leading sub sector using LQ method. Based on table 2. the agricultural sector has become sector base. In the agriculture sector. plantation and forestry have become the leading sub sector. While. the mining and quarrying sector. the oil and gas have become the leading sub sector. The LQ analysis in this step shown that plantation. forestry and oil and gas have become the leading sub sector because have LQ value > 1 (Table 4).

Table 4. LQ Value in Sub Sector using The Constant Price year of 2000

No	Business Field	Location Quotient (LQ) Value					Mean
		2009	2010	2011	2012	2013	
1	Plantation	1.48	1.49	1.46	1.50	1.48	1.48
2	Forestry	1.60	1.63	1.78	1.81	1.85	1.73
3	Oil and Gas	1.01	1.01	1.02	1.02	1.02	1.02

Source: Results Analysis. 2016

3.2. Shift Share Analysis

Shift share method was used to determine competitiveness. contribution and economic growth in sub sector. Based on LQ analysis without mining and quarrying sector. produced that plantation (1.48) and forestry (1.73) become the leading sub sector. The plantation sector has included into type I (fast growing) and the forestry sector includes into type III (developing). The results of shift share analysis in the plantation and forestry is as follows :

Table 5. Shift Share Analysis in Plantation and Forestry Sub Sector

No	Plantation	Remarks	Forestry	Remarks
b	+0.193	This sector growth faster than the reference economic	-0.236	This sector has slow growth compare the reference economic growth
c	+0.111	This sector has high competitiveness	+0.149	This sector has high competitiveness
d	+0.35	This sector showed the progressive growth.	+0.087	This sector showed the unprogressive growth.

Source: Results Analysis. 2016

Based on LQ analysis included mining and quarrying sector put oil and gas (1.02) become the leading sub sector. Based on the shift share analysis. the oil and gas sub sector has a negative value on proportional shift (b) and differential shift (c). The results shown that oil and gas included into type IV (Depressed). The results of shift share analysis in the plantation and forestry is as follows :

Table 6. Shift Share Analysis in Oil and Gas Sub Sector

No	Oil and Gas	Remarks
b	-0.043	This sector has slow growth compare the reference economic growth
c	-0.243	This sector has low competitiveness
d	0.286	This sector shown the unprogressive growth.

Source: Results Analysis. 2016

3.3. Overlay Analysis

The overlay analysis objective is to determine the development priority regarding the sector base. leading sub sector and sub sector contribution. Overlay analysis on sub sector of plantation and forestry and oil and gas is as follows:

Table 7. Overlay Analysis

No	Sub Sector	LQ (mean)	a	b	c	SSA
1	Plantation	+1.48	0.146	0.193	0.111	+0.45
2	Forestry	+1.73	0.146	-0.236	0.149	+0.059
3	Oil and Gas	+1.02	0.052	-0.043	-0.243	-0.234

Source: Results Analysis. 2016

Based on table 10. shown that LQ and SSA value of plantation and forestry has a positive value. That means the development of plantation and forestry categorized as dominant and progressive and then need to get high priority to developing from the government of Siak Regency. It is strengthened by the vision of Siak regency in 2031. where the focus of development are agribusiness sector.

agroindustry sector and tourism sector. It is completely different with sub sector of oil and gas. This sub sector has negative value on LQ and shift share analysis. so it categorized as not dominant and not progressive. That means the sub sector of oil and gas is not a potential sub sector and not worth to develop. The reason of oil and gas sub sector has a negative value because oil and gas are not renewable resources with high-cost production and high technology so this sub sector has decreased trend.

4. Conclusion

So the conclusion of this paper is drawn as follows:

1. Based on LQ analysis using Gross Regional Domestic Product Series Data from 2009-2013 without mining and quarrying sector. the agriculture sector has become sector base with plantation and forestry as a leading sub sector.
2. Based on LQ analysis using Gross Regional Domestic Product Series Data from 2009-2013 with mining and quarrying sector. mining and quarrying have become sector base with oil and gas as a leading sub sector.
3. Refers to shift share analysis. agriculture sector with their leading sub sector has a positive value. while the mining and quarrying sector with oil and gas as a leading sub sector has negative value.
4. Based on overlay analysis. Siak regency can change their focus of economic development from mining and quarrying sector into the agriculture sector.

Acknowledgements

This work is partially supported by Urban and Regional Planning Departement. Faculty of Engineering. Universitas Islam Riau.

References

[1] Arsyad L. 2010. *Ekonomi Pembangunan.* Yogyakarta: UPP STIM YKPN. 46
[2] Larasati N D. 2017. *Analisis Sektor Basis dan Sektor Unggulan Pembangunan Daerah dan Strategi Pembangunannya (Studi Kasus : Kabupaten Magelang tahun 2011-2015).*Skripsi. Fakultas Ekonomi dan Bisnis. Universitas Muhammadiyah Yogyakarta
[3] Muta'ali L. 2015. *Teknik Analisis Regional.* Yogyakarta: Fakultas Geografi Universitas Gadjah Mada. 45
[4] Sangadji M E. Sopiah. 2010. *Metode Penelitian.Pendekatan Praktis dalam Penelitian.* Yogyakarta: Andi. 78
[5] Tarigan R 2003 *Perencanaan Pembangunan Wilayah Edisi Revisi.* Jakarta: Bumi Aksara.25
[6] Tarigan R 2005 *Perencanaan Pembangunan Regional.* Jakarta: Bumi Aksara. 56
[7] Rustiadi E. Sunsun S. Dyah R P 2011 *Perencanaan dan Pengembangan Wilayah.* Jakarta: Crestpent Press dan Yayasan Pustaka Obor Indonesia
[7] UU No 33 tahun 2004 tentang penimbangan keuangan antara pemerintah pusat dan daerah
[8] Badan Pusat Statistik Propinsi Riau 2014 PDRB Provinsi Riau Menurut Lapangan Usaha Tahun 2009-2013. Badan Pusat Statistik Provinsi Riau
[9] Badan Pusat Statistik Propinsi Riau 2013 Kabupaten Siak Dalam Angka 2013. Badan Pusat Statistik Provinsi Riau
[10] Badan Pusat Statistik Propinsi Riau 2014 PDRB Kabupaten dan Kota se-Provinsi Riau Menurut Lapangan Usaha Tahun 2009-2013. Badan Pusat Statistik Provinsi Riau

IOP Conf. Series: Materials Science and Engineering **309** (2018) 012020 doi:10.1088/1757-899X/309/1/012020

Analysis of vehicle braking behaviour and distance stopping

M Sabri*, A Fauza
Computational & Experimental System Mechanics Research Group
Mechanical Engineering Department
Universitas Sumatera Utara, Medan Indonesia

*Email: sabrimesin@gmail.com

Abstract. Generally Vehicles show some degree of instability during braking as there are no real vehicles that are completely symmetrical. Especially when there is a component malfunction. There are many reasons for this incident: e.g. Failure of kinematic brake function, asymmetric mass distribution or brake imbalance. One possible approach to eliminating the negative effects of inevitable lifting brakes is to examine the differences in the patterns and characteristics of the vehicle braking process. This study investigates the behaviour of vehicle movement during braking until the vehicle stops. The results show that the force parameters are inversely proportional to the distance and time of braking and the process behaviour has the tendency of various distances and braking times at various power given.

1. Introduction

Braking distances should be understood by the driver as an important basic parameter in order to estimate the visibility of an affordable distance. The braking distance of a vehicle is influenced by a number of factors related to the vehicle, road and driver behaviour. The most important factors are: Vehicle speed, tire-road friction, drivers braking technique, braking system and condition, tire pressure, tire thread depth and road surface capability, also road vertical grade.

These factors affect braking distance to a greater or lesser extent depending on the actual conditions present when decelerating. A general method for determining braking distance at different speeds, which is representative for the composition of cars, drivers and friction, requires knowledge of the significance of the individual factors for overall braking distance.

The purpose of this study is to assess the braking behaviour including braking distance under different physical conditions. The findings will be included in an assessment of the present methods for calculating braking distance.

2. Mathematical Model

There are three forces acting on the tire: (a) The normal or vertical force (F_Z), caused by the weight of the vehicle, and the vertical inertia-oriented force, (b) the longitudinal force (F_x), which is generally due to inertial forces Acceleration or braking and may also be caused by longitudinal components of the vehicle centrifugal force, (c) lateral forces, caused by the centrifugal force of the vehicle.

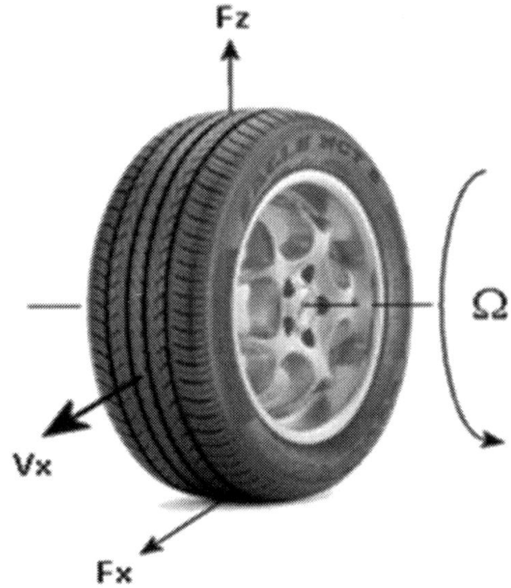

Figure 1. Dynamics of the Moving Wheel

The rolling wheel of Fig. 1 is the stress occurring on the wheels that not in steady state conditions, resulting in a contact slip (κ ') and deformation (u) which is also not constant. In this model, u and κ ' tend to be small. Thus, the relationship between F_x and u and u to κ ' becomes a linear function:

$$F_x = C_{Fx} \cdot u = C_{F\kappa} \cdot \kappa' \text{ , } C_{Fx} = \left(\frac{\partial F_x}{\partial u}\right)_{u=0} \text{ , } C_{F\kappa} = \left(\frac{\partial F_x}{\partial \kappa'}\right)_{\kappa'=0}$$

$$u = \sigma_\kappa \cdot \kappa' \text{ , } \sigma_\kappa = \left(\frac{\partial F_x}{\partial \kappa'}\right)_{\kappa'=0} \Big/ \left(\frac{\partial F_x}{\partial u}\right)_{u=0} = C_{F\kappa}/C_{Fx} \tag{1}$$

Deformation (u) is expressed according to the following equation:

$$\frac{du}{dt} + \left(\frac{1}{\sigma_\kappa}\right) \cdot |V_x| u = -V_{sx} \tag{2}$$

Slip κ' follows σ_κ and u, so the drive wheel has a damping rate of: | V_x | / σ_κ.The braking distance is obtained from the speed, coefficient of friction and the roadway grade by applying the following formula:

$$s = \frac{v^2}{2.\mu.g} \tag{3}$$

s = braking distance (m); v = speed (km/h); g = acceleration due to gravity (9.81 m/s2) and μ = mean coefficient of friction.

Table 1. Coefficient of friction to result measurement of braking distance

No	Speeds (km/h)	Braking Distance (m)	Friction, μ
1	10	0.54	0.73
2	20	2.24	0.70
3	30	5.55	0.64
4	40	10.91	0.58
5	50	14.5	0.68

3. Measurement Programme

To determine the braking distance, a measurement program in which 15 tests performed braking manoeuvres at different velocities (10, 20, 30, 40 and 50 km / h) was performed. Braking maneuvers are performed on different friction paths. Separation of road segments on each test track as in Fig. 3 is used for braking test. Manoeuvre is done by completely stopping the vehicle through comfortable braking. A medium-sized Daihatsu Taft Hiline car is used for braking experiments. The car has ABS brakes and a manual transmission. The same tire is used for all braking experiments. To record the braking behavior, the measuring device is mounted on the car, Figure 2 shows the tap gauge mounted on the brake pedal, then the measurement data is collected. During the braking experiment, mileage data, time and pressure recorded on the brake pedal are recorded and stored. Based on this data, it is possible to calculate the distance and duration of the braking process. The depth of the tread pattern is measured as 7-8 mm and total wear during the entire program is <1 mm. Safe braking maneuvers are carried out on the streets inside the University of North Sumatra (closed for traffic during braking experiments) and a covered parking lot using taxiways is used as a test track.

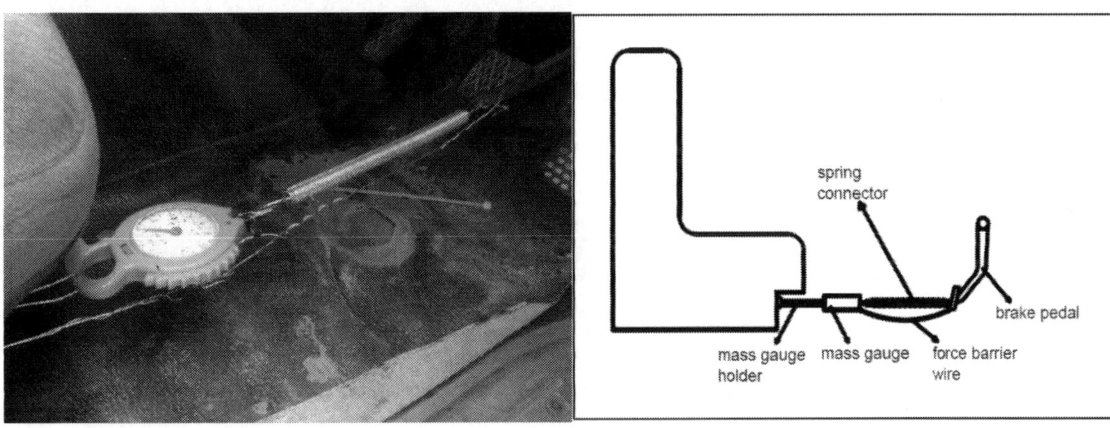

Figure 2. installation and setup dinamometer on the driver seat

IOP Conf. Series: Materials Science and Engineering **309** (2018) 012020 doi:10.1088/1757-899X/309/1/012020

Figure 3. The procesess measurement of braking distance

4. Findings From Braking Trails

Figure 4 shows the measured braking distances for 10 and 20 km / h respectively, on streets with nearly identical stopping distances and times. For speeds 30 and 40 km / h as shown in figure 5 taken with a longer distance. Figure 6 show brake distance achieve during the car speeds of 50 km / h. Thus the cessation distance is almost constant for the friction value in the range 0.58 - 0.73. It shows gradually increase the stopping braking distance will reduce friction. This is happen for all speeds. For example at a distance of 50 km / h, the stop brake distance, increases from 0,54 m to 14,5 m and friction is reduced from 0.73 to 0.58. This corresponds to an increase in brake spacing around 20%. The same percentage increase in brake distance is also seen in other speed cases.

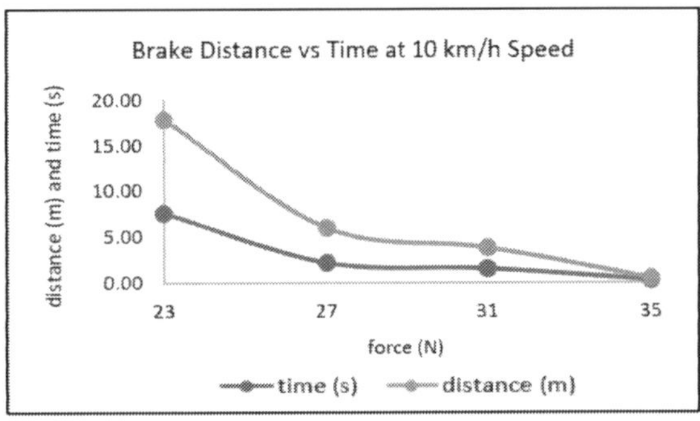

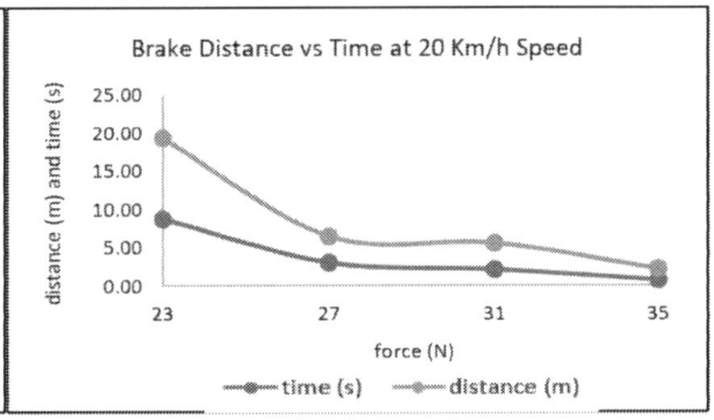

Figure 4. Brake distance vs time at 10 km/h and 20 km/h speed

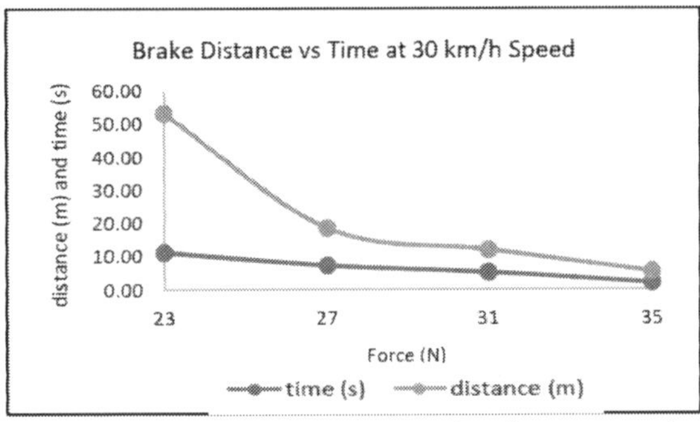

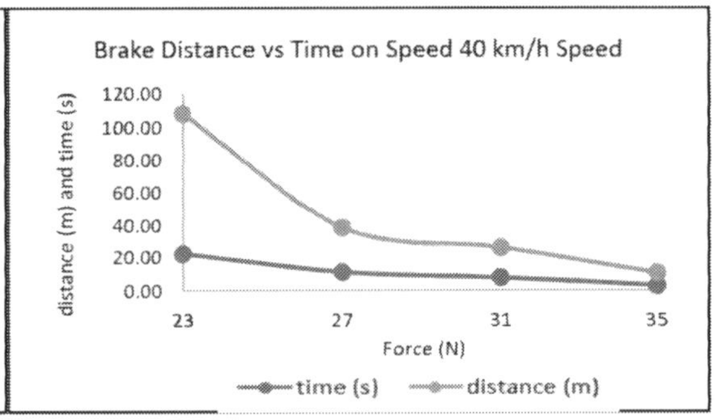

Figure 5. Brake distance vs time at 30 km/h and 40 km/h speed

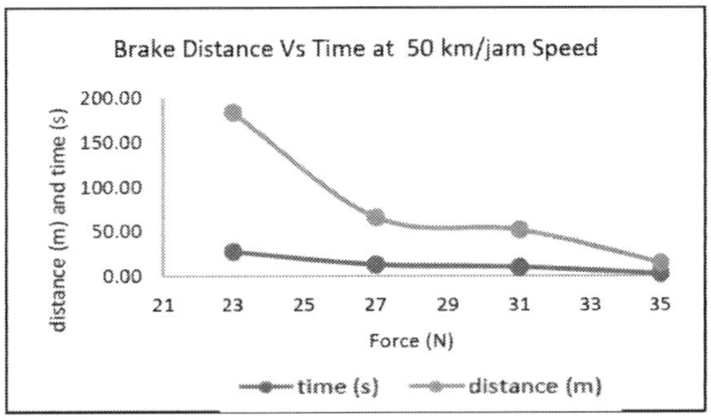

Figure 6. Brake distance vs time at 50 km/h speed

5. Conclusion

Based on the findings of the measurement program (comfort braking maneuver) and the knowledge gained regarding the importance of other parameters for braking distances (tire options, vehicles, etc.), we have attempted to establish recommended braking distances to improve road traffic safety.

Based on the results of research and testing, it is obtained that using pedal pressure forces from 16 N until 17 N to achieve the brake distance around 20 – 200 m.

Braking distances are judged somewhat carefully determined with good safety margins in relation to cars and drivers.

References

[1]. Sabri M 2016 Displacement of a Free Tyre in Natural Frequencies to Determine Noise Structure *Applied Science and Engineering Conference* (AASEC 2016).

[2]. Nuh M, Sabri M 2016 Analisa Interaksi Beban Dinamik Mobil Terhadap Karakter Struktur Permukaan Jalan, prosiding SNTTM XV. No ISBN : 978-602-96269-2-6

[3]. Setiyon R 2015 Analisis Gaya Pengereman Pada Mobil Nasional Mini Truck. Surakarta. Universitas Muhamnadiyah Surakarta

[4]. Layton R, Dixon K 2012 Stoping Sight Distance. Oregon State University

[5]. Evans M 2011 Braking Distance. Australia. Australian mathematic Sciences Institute.

[6]. Sabri M, 2010 Free Vibration of Tire/Road Contact Fracture and Strength of Solids VII, ISBN-13 978-0-87849-210-7 Key Engineering Materials Vol. 462-463, Part 2

[7]. Grelbe P 2007 Braking Distance Friction and Behavior. Denmark. Trafitec Scion – DTU Diplomvej bygning 376 2800 Kgs Lyngby.

IOP Conf. Series: Materials Science and Engineering **309** (2018) 012021 doi:10.1088/1757-899X/309/1/012021

The study of the influence of functionalism and international style on architecture development in Medan City

I F Pane[1,2,*], M N Loebis[1,2], I Azhari[3], N Ginting[1], D D Harisdani[1,2]

[1]Department of Architecture, Faculty of Engineering, Universitas Sumatera Utara, Padang Bulan, Medan 20155 Indonesia.
[2]Laboratory History, Theory, and Criticism of Architecture, Department of Architecture, Universitas Sumatera Utara, Padang Bulan, Medan 20155 Indonesia.
[3]Department of History Education, Universitas Negeri Medan, Jalan Willem Iskandar Pasar V Medan Estate, Indonesia.

*Email: imam.faisal@usu.ac.id

Abstract. As we know before that the development of Modern Architecture especially in Europe was not separated from the influence of *Avant Garde*, this influence brought a remarkable thing to the development of mindset at the time. It was not only in art and its beauty, but also in the development of architectural theories. The development of this new theory produced new forms in the world of architecture. The transition era from Renaissance to Modern was showed with Eclectic Architecture, at that time the shape of the building was still in classical form. The use of classical ornaments was still often used so that the outside appearance of architecture was seen like the previous one. Although skeleton concrete had been found as a new technology at that time, the shape of the building was still looked like in the classical period. This will be different at the time after the Eclectic because the views on the mindset of life have changed. The development of modern architecture in Europe is influenced by the movement of Functionalism and International Style; this movement characterizes a free architecture in the past. This movement also brought to Indonesia and some big cities like Jakarta, Surabaya and Medan. This study aims to see its influence in several colonial buildings built during the reign of the Dutch East Indies in Medan. With a qualitative approach looking at the existing theory and theories used to clarify the effect of the movement and see that the colonial buildings have a significant influence on the development of Functionalism and International Style in Europe.

1. Introduction

The influence of Avant Garde (*an ideology that breaks away from the past*) brings a big change to the development of architecture in the world, especially in Europe and America. In its development, the different views of art and beauty in the past and now bring the changes in the development of architectural theories and their physical manifestations in the field. The beginning modern era is marked by the development of *Eclecticism / Eclectic Architecture.* At this time the physical form of the building still refers to the classical forms and the use of classical ornaments are still often used so that the outer look of the architecture looks like in the classical period. Although reinforced concrete has been found as a new technology at the time, the shape of the building still looks like in the classical period. This will be very different after its time because the different views of life's mindset changed and *Avant Garde's* view became a trend of the moment [1].

The discovery of reinforced concrete makes the architecture develop rapidly. Avant Garde's mindset took an important role in the development of the architecture after that. The discourse of function was surprisingly getting the interest in the hearts of the architects at that time, so the term of *Functionalism* in architecture emerged, it was an early period of Modern architecture, where the mindset of art and beauty and architecture changed [2]. The beauty of Functionalism is showed at the geometrical function and form of the building and the use of limited geometrical and floral ornament.

This form of architecture in modern times is not measured by elements applied to the building, but it has seen the *function* of the building as a factor of its beauty. At this time many ideologies come that refers to the Avant Garde such the ideology of *Art Nouveau* and *Art Deco* that use decorative ornaments, but not same as in the Eclectic period because the ornaments used are usually shaped organic like nature. Modern Functionalism states that the function of building elements such as windows, doors, etc. is a form of beauty and everything is seen as something logical, and beauty appears in it.

The Disconnecting of relation from the past is a concept of Functionalism, there is a pure form that is completely different from the form in the Eclectic period. This change is also called *pure ideology* or *Purism*, because the buildings in this period have been free from the ornaments and shapes used tend to be geometrically shaped like cubes, squares, etc. These shapes are combined so that it produces a building with a new form of arrangement of boxes that are harmonious in one unity of fields that arise from the arrangement of geometrical shapes. No more decorations or ornaments on the cover of the building, so that the beauty arises from the new form without ornament.

In this period, the term Cubism Architecture emerged because the arrangement of geometrical shape was in line with the emergence of the *Cubism* in France in the early decade of the 1900s. This ideology is an ideology of art that incorporates the element of time in its design. Previous space elements added with time elements so that the results can be viewed from various sides, in other words, can be viewed from another perspective.

This influence is also found in architecture, the architect of Functionalism seeks to design the building with the concept of cubism, one of which is *Schroder House* (1924) in Utrecht, Netherlands by Gerrit Rietveld (1888-1964) (Figure 1.). This building looks like boxes which are arranged harmoniously that is completely different from the previous period. The composition of space created from the wide opening on the sides of the building. Harmonization is seen in the side of the building and the use of contrasting colors.

Figure 1. Schroder House, Utrecht, The Netherlands
(Source: en.wikipedia.org)

At the same time in Germany, this ideology also develops which was pioneered by Walter Gropius (1883-1969) who designed the *Bauhaus Building* in Dessau, Germany (1925-1926) (Figure 2.). This building is a school of art and design; the curriculum has included about the modern architecture that

IOP Conf. Series: Materials Science and Engineering **309** (2018) 012021 doi:10.1088/1757-899X/309/1/012021

was developing at the time. The Boxy building looks harmonious with the use of concrete and glass as a building facade. On one of its sides, there is a contrasting emphasis on the other side, indicating the overall harmonization of the shape.

Figure 2. Bauhaus Building, Dessau, Germany
(Source: en.wikipedia.org)

At the same time also in the United States of America, Functionalism develops which has grown in Europe. The style of cubism is used by American architects in planning the building for example, *Lovell House* (1927-1929) located in Los Angeles, USA. The building is designed by Richard Neutra (1892-1970) (Figure 3.). This residential building is built on contoured land, clean without ornament and only uses the composition of typical forms of cubism. Glass and concrete plays show a harmonious relationship between outer space and inner space.

Figure 3. Lovell House, LA,USA
(Source: en.wikipedia.org)

The development of modern architecture at that time in every country in Europe and America is pioneered by various movements that use modern concepts in the mindset. Previously there was *Ecole des Beaux-Arts* in France than in America pioneered by the *Chicago School*, in England there was the *Glasgow School of Fine Arts*, in Germany appeared *Bauhaus* and in the Netherlands developed *De Stijl*.

Movement in each country has almost the same mindset that is rejecting to come back to the old form, in other words, no look back to history. Developing mindsets and rules accentuate the new technology and building constructions that developed at the time. At that time, as the modernization movement grew stronger, a group called MARS (Modern Architectural Research Group) was established in 1932 [3], and then held an International congress for the development of Modern architecture in Athens in 1933 called CIAM (*Congres Internationaux de l'Architecture Modern)*. This

IOP Conf. Series: Materials Science and Engineering **309** (2018) 012021 doi:10.1088/1757-899X/309/1/012021

modernization movement in architecture is finally referred to as the *International Style* movement, with the intention that the developing architecture has the same characteristics even in different places. This movement supported functionalism that developed at the time. This movement did not see art as individual satisfaction anymore, but art can be widely applied to manufacturing and industry.

There are several important figures in this movement such as in Germany there is *Peter Behrens*, in the United States, there are *Louis Henry Sullivan* and *FL. Wright*, in the Netherlands there is *HP. Berlage*, in Austria there is *Adolf Loos* and In England there is *Mackintosh*. This ideology tends to principally architecture as a whole and not separate and considers architecture as volume and beauty arise through the composition of form. This ideology also applies the principle of regality that is the similarity of building forms and patterns seen in the building. In addition, this International Style movement does not use ornaments for decoration so that the overall building looks clean of decorative and decorative elements.

Along with the discussion above, the influence of modernization also hit the Netherlands as one of the countries in Europe. This movement is called *De Stjil* which is a movement pioneered by a group of modernization of art, architecture and interior in the Netherlands. The name of the movement is taken from the name of the first published magazine in 1917 [3], and was founded by Theo Van Doesburg (1883-1931), Piet Mondrian (1872-1944) both painters and several other painters and GerritReitvield (1888- 1964) and JJP.Oud (1890-1963) are both architects along with some other architects and sculptors, poets and photography artists [2].

The movement of De Stjil is important to know because the architects at the time used this movement as a trend in architectural design. This directly affected Indonesia as a colony of the Netherlands. The influence of this movement was also brought to Indonesia by Dutch architects who work in Indonesia. The influence of De Stjil is also inseparable from the influence of Functionalism and Cubism that developed at that time there. With the facts above, the buildings built by the colonial government in Indonesia use the same mindset. Indirectly, this mindset was carried away because the Dutch architects were then accustomed to the powerful movement of modernism and the technology that had existed in the development process.

De Stjil specializes to the visual expression generated by the expression of angles and the play of horizontal and vertical lines and the colors commonly used are red, yellow and blue and added gray and black. From the concept of art that is rolled by De Stjil is the two-dimensional form used in painting so that it creates new forms (*Neo Plasticism*) which are harmonious. The concept of De Stjil's mindset also leads to a pure form so-called Purism, as it developed in early modern times. Purity and clean are elements of beauty, abstract shapes with clear boundaries with the play of visual elements to create harmonization. This ideology is not only limited to the art , but also concerns the development of architecture in the Netherlands, some architects have used this principle on the design.

One of the works of Dutch architect William Marinus Dudok (1884-1974) is Town Hall or Town Hall in Hilversum (1924-1928) this building is patterned Kubism with the spirit of De Stijl (Figure 4.). It is seen the building that implements the function of purity that at that time developed the term Form Follows Function which is a form following function. The design of the building shows a harmonious composition without being decorated by ornaments and carvings on the side of the building. The building looks clean with doors and windows arranged harmoniously. There is a tower as the main point of view to add to the splendor of this building. This tower became one of the hallmarks of De Stjil's design at the time.

In addition to Dudok, there is another architect who uses the spirit of De Stjil in his design. An architect named JJP.Oud (1890-1963) one of the founders of De Stjil designed Cafe De Unie (1925) in the city of Rotterdam (Figure 5.). This building applies the concept of De Stjil art with the use of red, yellow and blue on the facade of the building. It is seen the play of vertical and horizontal fields to get the harmonization from building the shape. The use of glass material is also found on the parts of the facade arranged in harmony to get an attractive appearance. This building also receives the influence of Cubism by using concrete and glass that look abstract, but still shine the beauty by the beauty concept of De Stjil. This proves that the movement of De Stjil has greatly influenced the Dutch architects at the time.

Figure 4. Town Hall, Hilversum, The Netherlands
(Source: www.wikiwand.com)

Figure 5. Café De Unie, Rotterdam, The Netherlands
(Source: www.wikiwand.com)

2. Method

This research uses *the qualitative method* and the facts that exist in the field will try to be elaborated to see the relationship between the influence of modernism in Europe and the existing in the city of Medan. The influence of modernization is devoted to the influence of *Functionalism* (*De Stjil*). For that, the chosen method is to parse the existing facts with a qualitative approach. This method is able to see the object comprehensively and its data analysis using *inductive approach* [4]; [5].

This study uses the existing object in the city of Medan. This city once had become one of the important cities in the Dutch colonial government. The opening of a tobacco plantation in this area in 1863 by *Jacob Nienhuys, Vandervalk, Elliot* made this area (Deli land) reckoned by the colonial government [6]. In parallel, it was established a plantation company in 1866 by the colonial government to run and manage the interests of the Dutch, named Deli Maatschappij by Jannsen, PW.Clemen and Nienhuys. Furthermore, many government and private plantations were built around Medan [7]. With this situation, the city of Medan becomes more popular.

Case studies are used to see the development of architectural modernization whether both the shape of the building and ornament or decoration. Field survey stage is to record the existence of this building and see the relation with the development of modern architecture. In addition to primary data from the field, literature review of modern architectural theories used as a test tool to the object of research. The case study in this study is Medan Post Office which is the main post office in Medan City until now.

Changes that occur to the development of cultural activities in society provide a significant change to life [8]. This change has had a major impact on the overall architecture as a result of the discovery of new materials and technologies. For that, the parameters used in analyzing the research object are:

1. Ornaments of Building

Building with the theory of functionalism in modern times has the intention, not like the building in the past. Both interior and exterior buildings look “cleaner" in other words the use of ornaments is not a trend in this modernization. The building used only simple ornaments in the form of floral and geometric , but does not contain the meaning as in Classic Architecture.

2. Form and Function of Building

The form and function of buildings in modern era are more to geometric form and seem to accumulate each other , but look harmonious with the use of glass windows that tend to be wide. The form of the building is adapted to the function of the building so that space is designed as efficiently as possible by managing the function of space in the building. Function plays an important role in the formation of buildings and building forms will follow the existing function.

3. Result and Discussions

In addition to the eclectic architecture that developed in Indonesia [9], at the same time also developed the modern architecture of functionalism, this architecture developed in the 19th century. In the beginning, it developed the term architecture "*Indisch*" derived from the term "*Nederlansch Indie*" or often called the Dutch East Indies. This term is used by the Dutch for naming the archipelago as its colony. This Indis architectural style is a mixture of local forces with the strong influence of modernism brought by the Dutch to Indonesia. The beginning of the development of this architecture is because the Dutch need shelter and buildings to support their colonization. Then they begin to build by looking at the local potential that exists, especially the tropical climate in Indonesia [10].

Dutch policy in dealing with its colonies, as discussed above that the Dutch colonial government issued a regulation on buildings built in colonies [11]. This rule is used to distinguish the architecture built in the colony, the term used is *Indische Huizen* or *Indo Europeesche Bouwkunst* to mention architecture in the colonies because the architecture is no longer the same as that occurred in Europe because they have been mixed with climate and customs in Indonesia. From this point, the term Indies Architecture began to be widely used , but still within a small scope [12].

This Indis architecture is closely related to the passionate spirit of De Stjil in the nineteenth century and inspired Dutch architects in Indonesia. The influence of modernization they brought along with the emergence of the ideology of Art Deco which uses elements of decoration, but not like the classical form, but a more natural form. The influence of cubism that occurred in Europe also entered into the scope of architectural development in Indonesia.

At that time in Indonesia and Medan City, in particular, this new model of architecture also developed. The style that adopts the spirit of De Stjil is named with the term *Nieuwe Bouwen* which means new building [12]. *Nieuwe Bouwen* style is also paying attention to the climate because it does not merely pursue the form of boxes as contained in Cubism Architecture, but it must consider the local potential, especially the tropical climate in Indonesia.

One of the buildings with the concept of De Stjil, found in Indonesia is Denis Bank Building 1936 (now Bank Jabar) is located at Braga Road Bandung, designed by Albert Frederick Aalbers (1897-1961). The hallmark and form of the building are plaid and use the harmonization of the composition of the horizontal and vertical plane so that this building looks harmonious in its composition. The presence of a tower on the front of the building indicates that the influence of De Stjil is in this building. Art Deco geometric decoration is also contained in the building, so the expression that arises to be typical of modern (Figure 6.).

Figure 6. Bank Jabar Building in Bandung, Indonesia
(Source: en.wikipedia.org)

This style developed after the modernization began to enter into the realm of human life. At that time the influence of modernization is not only belongs to architecture, but it has become a new culture that developed at the time. The breaking of the chain of history colored the development of modern culture because the views of people at that time no longer look back, but think far forward to what is achieved or their future goals. The development of modernization at that time has entered the joints of life because along with the development of science and technology that looked at everything must be effective and efficient.

This modern culture came to Indonesia directly brought by the Dutch architects who got a lot of work designing the building. *Medan Post Office* is a case study to see the influence of *Functionalism* and *International Style* (Figure 7.). It is designed by a Dutch architect named *J. Snuyf*, and its location lies at the crossroads of City Hall and Bukit Barisan Street. The building was founded in 1911 and J.Snuyf was still the Head of BOW throughout Indonesia. It is Influenced Art Deco, L-shaped building and has an atrium on the corner of the building as well as the right and left wing of the building. The walls are not filled with ornaments and use many horizontal and vertical lines on buildings.

Figure 7. Front View of Medan Post Office

3.1. Ornaments of Building

For ornaments of the building do not look as striking as in eclectic buildings, there are some decorations on the front side of the building, but do not dominate the image of the building. Art Deco style is thick in the building. Some decorative wall paintings are on the front facade of the building. On the inside of the building is also not filled with ornaments, there are only pigeon ornaments in the hall of the building. The building is only relying on the play of vertical and horizontal geometric shapes to affirm the overall shape. In general, the building is "clean" of ornaments or decorations that are usually found in the previous architecture (Figure 8.).

Figure 8. Ornaments and vertical and horizontal elements in buildings

3.2. Form and Function of Building

Medan Post Office is *art deco* style and has a linear form of the building following the shape of the road, at the corner of the street there is a different hall height. De Stijl's influence is seen with the height difference in the building element; it is only the tower is replaced by the raised hall. The octagonal hall serves as a catch point at the corner of the building. In this hall, there are postal service counters, and there are several rooms for administrative purposes.

On the roof, especially the octagonal hall, there are additional roofs upstairs, and on some sides of the roof, there is a small roof that protrudes (lucarn) as a natural conditioner buildings. The Strong modern influences affect the building as a whole, a curved opening play for the natural lighting and conditioner of the building. On the side of the building that is just on the corner of the road, there are wide openings with arches and the use of color glass which is so harmonious when it is viewed by eyes (Figure 9.).

Figure 9. Elements of the roof that protrude (lucarn) and openings in buildings

4. Conclusion

The conclusion drawn from this study is that the architectural style of Medan Post Office is very closely related to the development of the ideology of functionalism and International Style in the world. Although it is not entirely using the style of De Stjil, it is seen that the elements of ornaments that are used do not dominate the building and the tower elements manifested to the hall that has a roof that rises upwards and similar to the rotunda in Roman times. So, it can be concluded that the role of functionalism and International Style is in the building. It is advisable to continue researching some of the other colonial buildings to get how much influence the functionalism and international style is to the existing buildings in the city of Medan.

Acknowledgment
On this occasion, I would like to thank the Directorate of General of Higher Education who funded this research and also thank the Department of Architecture students who assist in the process of surveying and searching the data in the field.

References
[1] Heinrich K 1988 Vision Of The Modern *UIA Journal of Architecture Theory and Criticism* **1(1)**
[2] Sumalyo Y 1997 *Arsitektur Modern (Akhir Abad XIX dan Abad XX)* Gadjah Mada University Press Yogyakarta
[3] Frampton K 1992 *Modern Architecture a critical history* Thames and Hudson Germany
[4] Moleong L J 2000 *Metoda Kajian Kualitatif* Remaja Rosdakarya Bandung
[5] Cresswell J W 2010 *Research Design Pendekatan Kualitatif, Kuantitatif, dan Mixed*, Yogyakarta Pustaka Pelajar
[6] Napitupulu S P, Sanusi S S, Lubis, M 1991 S*ejarah Perlawanan Terhadap Kolonialisme dan Imperialisme Di Sumatera Utara* Direktorat Pendidikan dan Kebudayaan Jakarta
[7] Sinar T L 2011 *Sejarah Medan Tempoe Doeloe* Sinar Budaya Group Medan
[8] Ginting N and Rahman N V 2016 Maimoon Palace heritage District in Medan, Indonesia: What we preserve and why we preserve? *Procedia-Social Behavioral Sciences* **222** 332-341
[9] Sumalyo Y 2002 Dutch Colonial Architecture and City Development of Makassar *Jurnal Dimensi Teknik Arsitektur* **30(1)** 46-53
[10] Passchier C 2016 *Bouwen in Indonesië 1600-1960* LM Publishers Volendam
[11] Wignjosoebroto S 2004 *Desentralisasi Dalam Tata Pemerintahan Kolonial Hindia-Belanda* Bayu Media Malang
[12] Suptandar J P 2001 Harian Kompas 14 Oktober 2001 (www.arsitekturindis.com)

IOP Conf. Series: Materials Science and Engineering **309** (2018) 012022 doi:10.1088/1757-899X/309/1/012022

Green design application on campus to enhance student's quality of life

H Tamiami[1]*, F Khaira[1] and A Fachrudin[2]
[1]Department of Architecture, Universitas Sumatera Utara, Medan 20155, Indonesia
[2]Department of Management, Universitas Sumatera Utara, Medan 20155, Indonesia

*E-mail: hilma@usu.ac.id

Abstract. Green design becomes an important thing to applied in the building. Green building will provide comfortability and enhance Quality of Life (QoL) for the users. The purpose of this research is to analyze how green design application on campus to enhance student's QoL. This research conducted in three campuses which located in North Sumatera Province, namely Universitas Sumatera Utara (USU), Universitas Negeri Medan (Unimed) and Universitas Medan Area (UMA) which have a lot of vegetation, open space, and multi-mass buildings. This research compared the green design application to QoL from three universities. Green design in this research that become independent variables focus on the energy efficiency and conservation (EEC), indoor health and comfort (IHC) and building environment management (BEM) with dependent variable is QoL. This research uses quantitative methods with questionnaire survey techniques. The population is students from the three universities with the sample of each University is 50 samples. The analysis uses multiple regression analysis. The results show that green design application may enhance QoL of students. The campus should have a good green design application to enhance QoL of students and give them comfortability.

1. Introduction

Green building becomes important things and by using an integrated approach of design, may reduce negative impacts from building to the environment. Green building also reduces operating costs, increases occupants productivity and create sustainability community [1]. Sustainable, green design, green architecture and eco-friendly architecture becomes the important things to create a better life for humans and contain some universal consent [2]. Green design is a collaboration activity in the business which involves human resource, facility, machine investment and training [3].

The architect may create some features in their building that consider to eco-system [2]. Some planners have tried to adopted green buildings policies in their design to meet environmental and sustainability goals. In green building assessment, there are some indicators that need to be investigated, such as sitting, water consumption that may be reduced, building codes and local situation [4].

Based on Fachrudin *et al* [5], universities should be designed to follow the green concept to provide comfort to the users. The campus which designed as green campus may provide comfort to the users, and they will feel their quality of life is getting better. Universities should design as a green campus to provided comfort environment. Student and staff will comfort and get a better quality of life.

Based on the high importance of applying the green concept to the building, the aim of this research is to analyze the influence of green concept on campus to enhance student's quality of life.

1.1. Green Design
Based on Ali *et al* [1], green buildings concepts are energy efficient, water conservation, durable and non-toxic, high-quality spaces and high-recycled content material. In Jordan, green buildings are needed to improve the environmental and economic performance. The variables that used to assess green building are the site, energy efficiency, water efficiency, material, indoor environment quality, waste and pollution and cost and economic. It found that sustainability strategy and goals should be addressed as a major aim.

Green architecture principles consist of Water features and their management; natural building design; passive solar design; green building materials; living Architecture. Passive solar design refers to the use of the sun's energy for the heating and cooling of living spaces. Then, the green building should use renewable material and material that can reduce maintenance and replacement costs. Green building must provide comfort because have an energy efficient, economically may reduce maintenance costs, aesthetically for passive solar features can make buildings bright and pleasant inside and environmentally responsible [2].

The green buildings rating system is consists of the sustainable site; water efficiency; energy and atmosphere; materials and resources; indoor environmental quality and innovation and design process. Green building rating systems are focusing on high-performance; energy efficient and economic and environment-friendly building. Energy efficiency is major components in designing green building [6].

Based on Setyowati *et al* [7], the architectural approaches on green concept consist of land use efficiency, water savings, energy efficiency, application of the building materials used in the construction process to save the waste that occurs (the concept of recycling materials) and maintenance of healthy indoor air quality. For mechanical and electrical may using an alternative energy such as the solar cell generator system; innovation on natural lighting, energy saving in electricity, and the application of information technology (IT) with the concept of High Technology.

Building assessment system is tools used to rank or assess how buildings or neighborhoods concerns compared to typical practice and to ultimate goals. Then, this will be the basis for green building policies and programs. Building assessment system focus on energy efficiency, water conservation, site selection, building materials, waste management, indoor environmental quality, and education. Some building assessment indicates that water and sitting issues are major problems, but the assessed object will show different results according to its characteristics [4].

Based on [8], green building assessment refers to six parts, namely: appropriate site development; Energy efficiency and conservation; Water conservation; Indoor air health and comfort; Material resources and cycle; and building environment management.

Based on Maditinosa *et al* [9] the quality of life refers to current satisfaction. The university which designed with green design concept will provide a good quality of life. Based on [10] Green University is more comfortable and can significantly enhance the quality of life of its students compared to the Non-Green Campus concept. Students at Green University can experience the quality of life than non-green universities. Some universities that are not designed as green building try to reduce energy consumption. They try to reduce the use of air conditioning, but they do not have ventilation. This situation will make users feel uncomfortable. Students agree that university green spaces can improve their quality of life.

Based on previous studies, this study will focus on three variables from the green design concept that is considered most influential on the quality of life of its users, namely: Energy Efficiency and Conservation (EEC); Indoor Health and Comfort (IHC) and Building Environment Management (BEM).

2. Methods
This research is a quantitative method which uses primary data with questionnaire technique. This research was conducted in three universities, i.e Universitas Sumatera Utara (USU), Universitas Negeri Medan (Unimed) and Universitas Medan Area (UMA). The three universities have similarity i.e., have multi-mass buildings and a lot of vegetation (figure 1).

Figure 1. Research Location (from left to right: USU, Unimed, and UMA)

The population of this study was an undergraduate student with total students Universitas Sumatera Utara is 33,000 students, Universitas Negeri Medan is 16,525 students, and Universitas Medan Area is 6,823 students. The sample is 50 respondents for each university. There are three parts on questionnaire form, namely data of respondents, green design concept and quality of life. Green design concept divided into three variables, namely energy efficiency and conservation (EEC), indoor health and comfort (IHC) and building environment management (BEM). The questionnaire using 1 to 5 points of Likert Scale with 1 for strongly disagreed to 5 for strongly agreed. The analysis was using multiple linear regression with alpha 5% and the confident level is 95%.

3. Results and Discussion

3.1. Green Design

Green design concept in this research as independent variables that consists of three variables and each variable has some indicators. The first variable is Energy Efficiency and Conservation (EEC) as X1 that divided into four indicators namely, lighting in the classroom (EEC1), the campus is free from air pollution (EEC2), the campus is free from noise pollution (EEC3) and campus operates on energy-efficient (EEC4). The second variable is Indoor Health and Comfort (IHC) that consists of clean air (IHC1), healthy air (IHC2), comfort in the classroom (IHC3), comfort air in the corridor (IHC4) and comfortable temperature in the classroom (IHC5). The third variable is Building Environment Management (BEM) that consists of campus has many plants (BEM1), healthy campus environment (BEM2) and separation of waste disposal for organic and an organic waste (BEM3).

Based on Table 1, students from USU, Unimed and UMA have same argumentation about the lighting in the classroom that is bright (EEC1), the campus has clean air (IHC1), and the campus has many plants (BEM1). It means that USU, Unimed, and UMA already apply some concept from green design. This result is in accordance with the research by [8] which states that the three variables are part of the green design concept.

Table 1. Mean Value for Independent Variables

Indicator	N	USU		Unimed		UMA	
		Mean	Std Dev.	Mean	Std Dev.	Mean	Std Dev.
EEC1	50	2.96	0.880	4.06	0.867	3.80	0.833
EEC2	50	2.32	0.999	3.72	1.070	3.74	0.986
EEC3	50	2.36	1.208	3.52	1.147	3.48	1.111
EEC4	50	2.78	1.298	3.38	1.105	3.68	0.957
IHC1	50	2.90	1.015	4.22	0.790	4.28	0.809
IHC2	50	2.84	1.017	4.12	0.824	4.20	0.756
IHC3	50	2.30	1.015	3.64	0.875	3.54	0.994
IHC4	50	2.52	0.931	3.56	0.837	3.82	0.800

Indicator	N	USU		Unimed		UMA	
		Mean	Std Dev.	Mean	Std Dev.	Mean	Std Dev.
IHC5	50	2.18	0.962	3.48	0.909	3.22	1.016
BEM1	50	3.76	1.001	4.66	0.798	4.70	0.763
BEM2	50	2.86	0.969	3.88	0.961	4.08	0.853
BEM3	50	1.92	1.047	3.30	1.147	3.52	1.344

3.2. Quality of Life
Dependent variable in this research is quality of life is consists of ten indicators, namely: convenience on campus (QoL1), good mood (QoL2), plants that may provide comfortability (QoL3), happy feeling (QoL4), friendly campus (QoL5), good interaction (QoL6), interaction with lecturer (QoL7), feel good (QoL8), life quality score (QoL9) and optimistic (QoL10). From Table 2, students from USU, Unimed and UMA have same argumentation about optimistic with mean value respectively 4.50, 4.54 and 4.42 and good interaction with mean value respectively 3.88, 4.32 and 4.30.

Table 2. Mean Value for Independent Variables

Indicator	N	USU		Unimed		UMA	
		Mean	Std Dev.	Mean	Std Dev.	Mean	Std Dev.
QoL1	50	2.70	0.909	3.98	0.892	3.90	0.763
QoL2	50	2.86	0.904	3.72	0.927	4.00	0.833
QoL3	50	2.90	1.015	3.92	0.922	4.02	0.937
QoL4	50	2.94	0.818	3.82	0.896	3.84	0.866
QoL5	50	2.86	0.833	3.42	0.906	3.44	0.884
QoL6	50	3.88	0.895	4.32	0.794	4.30	0.863
QoL7	50	3.74	0.803	4.10	0.735	4.22	0.737
QoL8	50	3.28	0.671	3.98	0.820	4.04	0.755
QoL9	50	3.50	0.814	3.98	0.845	4.06	0.793
QoL10	50	4.50	0.580	4.54	0.646	4.42	0.859

3.3. Influence of Green Design to Quality of Life
Multiple regression analysis conducted to find the influence of green design application to enhance student's quality of life. The independent variables are Energy Efficiency and Conservation (EEC) as X1; Indoor Health and Comfort (IHC) as X2, and Building Environment Management (BEM) as X3, and quality of life as dependent variable (Y).

3.3.1. Universitas Sumatera Utara (USU)
The first analysis was conducted to the data from Universitas Sumatera Utara (USU). R square value is 0.536 and the residual has a normal distribution. All of the independent variables has a VIF value below 10 and tolerance value not less than 0.10 (Table 3 and 4). It shows that in this analysis there is no multicollinearity. Glejser test result shows the highly significant level more than alpha 0.05 that means in regression model there are no heteroskedasciticas in the regression model.

In Table 5, there is one variable that has t value less than 0.05. It means that the variables influence the student's quality of life, namely: Building Environment Management (BEM). While Energy Efficiency and Conservation (EEC) and Indoor Health and Comfort (IHC) have t value more than 0.05, it means that the variables do not influence the student's quality of life.

Table 3. One-Sample Kolmogorov-Smirnov Test

		Unstandardized Residual
N		50
Normal Parameters[a,b]	Mean	.0000000
	Std. Deviation	.36370201
Most Extreme Differences	Absolute	.099
	Positive	.069
	Negative	-.099
Test Statistic		.099
Asymp. Sig. (2-tailed)		.200[c,d]

a. Test distribution is Normal.
b. Calculated from data.

Table 4. Coefficient Determination Test

Model Summary [b]					
Model	R	R Square	Adjusted R Square	Std. Error of the Estimate	Durbin-Watson
1	.732[a]	.536	.506	.37537	1.656

a. Predictors: (Constant), BEM, EEC, IHC
b. Dependent Variable: QoL

Table 5. Coefficient

Model		Unstandardized Coefficients		Standardized Coefficients	t	Sig.	Collinearity Statistics	
		B	Std. Error	Beta			Tolerance	VIF
1	**(Constant)**	1.806	.220		8.191	.000		
	EEC	-.025	.090	-.038	-.282	.779	.545	1.835
	IHC	.142	.101	.208	1.410	.165	.465	2.150
	BEM	.426	.124	.592	3.435	.001	.340	2.943

a. Dependent Variable: QoL

3.3.2. Universitas Negeri Medan (Unimed)

The second analysis is for Unimed and found the residual has a normal distribution with R square value is 0.580. All of the independent variables has a VIF value below 10 and tolerance value not less than 0.10 (Table 6 and 7) that explain in this analysis there is no multicollinearity. Glejser test result shows the highly significant level than alpha 0.05 that means in regression model there are no heteroskedasciticas in the regression model.

From Table 8, there are two variables that have a t value less than 0.05. It means that the variables influence the student's quality of life, namely: Energy Efficiency and Conservation (EEC) and Indoor Health and Comfort (IHC). While Building Environment Management (BEM) have t value more than

0.05, it means that the variables do not influence the student's quality of life. This analysis is opposite with the first analysis which conducted in USU.

Table 6. One-Sample Kolmogorov-Smirnov Test

		Unstandardized Residual
N		50
Normal Parameters[a,b]	Mean	.0000000
	Std. Deviation	.38825023
Most Extreme Differences	Absolute	.074
	Positive	.061
	Negative	-.074
Test Statistic		.074
Asymp. Sig. (2-tailed)		.200[c,d]

a. Test distribution is Normal.
b. Calculated from data.

Table 7. Coefficient Determination Test

Model Summary [b]					
Model	R	R Square	Adjusted R Square	Std. Error of the Estimate	Durbin-Watson
1	.762[a]	.580	.553	.40071	2.289

a. Predictors: (Constant), BEM, IHC, EEC
b. Dependent Variable: QoL

Table 8. Coefficient

Model		Unstandardized Coefficients		Standardized Coefficients	t	Sig.	Collinearity Statistics	
		B	Std. Error	Beta			Tolerance	VIF
1	**(Constant)**	1.156	.365		3.164	.003		
	EEC	.229	.105	.297	2.179	.034	.493	2.029
	IHC	.371	.117	.396	3.177	.003	.587	1.702
	BEM	.145	.102	.189	1.415	.164	.514	1.945

b. Dependent Variable: QoL

3.3.3. Universitas Medan Area (UMA)

The third analysis is for UMA that found the residual has a normal distribution with R square value is 0.407. All of the independent variables has a VIF value below 10 and tolerance value not less than 0.10 (Table 9 and 10) that explain in this analysis there is no multicollinearity. Glejser test result shows the highly significant level than alpha 0.05 that means in regression model there are no heteroskedasciticas in the regression model.

Based on Table 11, there is one variable with t value less than 0.05. It means that the variables influence the student's quality of life, namely: Building Environment Management (BEM). Energy Efficiency and Conservation (EEC) and Indoor Health and Comfort (IHC) t value more than 0.05, it means that the variables do not influence the student's quality of life. This analysis has the same result with the first analysis which conducted in USU.

Table 9. One-Sample Kolmogorov-Smirnov Test

		Unstandardized Residual
N		50
Normal Parameters[a,b]	Mean	.0000000
	Std. Deviation	.44305739
Most Extreme Differences	Absolute	.075
	Positive	.058
	Negative	-.075
Test Statistic		.075
Asymp. Sig. (2-tailed)		.200[c,d]

a. Test distribution is Normal.
b. Calculated from data.

Table 10. Coefficient Determination Test

Model	R	R Square	Adjusted R Square	Std. Error of the Estimate	Durbin-Watson
1	.638[a]	.407	.369	.45728	1.861

a. Predictors: (Constant), BEM, IHC, EEC
b. Dependent Variable: QoL

Table 11. Coefficient

Model		Unstandardized Coefficients		Standardized Coefficients	t	Sig.	Collinearity Statistics	
		B	Std. Error	Beta			Toleran ce	VIF
1	**(Constant)**	1.425	.475		2.999	.004		
	EEC	.162	.119	.199	1.362	.180	.605	1.652
	IHC	.173	.124	.193	1.393	.170	.669	1.496
	BEM	.328	.108	.396	3.045	.004	.763	1.311

a. Dependent Variable: QoL

3.4. Finding

The concept of green design is very appropriate if applied on campus. Students have a good response to buildings and environments in the campus. Although not all green design concepts significantly influence student's quality of life, this research found that the concept of green design that applied on campus may enhance the student's quality of life. This three campus already have a good lighting in the classroom because supported by good ventilation, have clean air and have many plants. Students from Unimed give high appreciation about the application of green design concept in their campus. While UMA students have a high value for quality of Life. It means that they feel that their quality of life is good.

Overall, students have felt their quality of life is good enough, they have good interaction with friends and lecturers and always feel optimistic in life. The findings are appropriate with research by [1], [5], and [10] who stated that green design concept might provide sustainability community and enhance the quality of life of its users.

4. Conclusions

Building assessments that apply green concepts based on several variables. This study only focuses on three variables that are considered important and have a great impact on the building and quality of life of its users. The three variables are energy efficiency and conservation, indoor health and comfort and building environment management. These three variables were found influence the quality of life of students on the campus of USU, Unimed, and UMA. In USU and UMA, building environment management is influence student's quality of life. While in Unimed, energy efficiency and conservation and indoor health and comfort influence student's quality of life. The campus should apply the concept of green design on the building and its environment to enhance the quality of life of its users consisting of students, staff, and lecturers. This study is expected to be a reference for further research using a complete variable.

References

[1] Ali H H and Nsairat S F 2009 Developing a Green Building Assessment Tool for Developing Countries Case of Jordan *Building and Environment* **44** p. 1053-1064

[2] Ragheb A, El-Shimyb H and Raghebb G 2016 Green Architecture: A Concept of Sustainability *Procedia Social and Behavioral Science* **216** p. 778-787

[3] Yazid M, Udin Z, and Abdul G A 2015 The Collaboration of Green Design and Technology Towards Business Sustainability in Malaysian Manufacturing Industry *Procedia Social and Behavioral Science* **211** p. 237-242

[4] Retzlaff R C 2008 Green Building assessment System *Journal of the American Planning Association* vol **74** No 4 p. 506

[5] Fachrudin H T and Fachrudin K A 2016 Influence of Green Campus Application To Quality of Life *Proc. 1st Int. Conference on Liveable Built Environment (Bali, Indonesia)* p. 130-134

[6] Gowri K 2004 Green Building Rating System: An Overview *ASHRAE Journal* **46** 11 p. 56

[7] Setyowati E, Harani A R and Falah Y N 2013 Green Building Design Concepts of Healthcare Facilities On the Orthopedic Hospital in The Tropics *Procedia Social and Behavioral Science* **101** p. 189-199

[8] Green Building Council Indonesia 2015 *Perangkat Penilaian Greenship*

[9] Maditinosa, Papadopoulos and LluisPratsc 2014 The Free Time Allocation and its Relationship with The Perceived Quality of Life (QoL) and Satisfaction with Life (SwL) *Procedia Economics and Finance* Vol. **7** p. 519-532

[10] Tiyarattanachai R and Hollmann N M 2014 Green Campus initiative and its impacts on quality of life of stakeholders in Green and Non-Green Campus Universities *SpringerPlus* Vol. **84** No. 5 p. 1-17

IOP Conf. Series: Materials Science and Engineering **309** (2018) 012023 doi:10.1088/1757-899X/309/1/012023

Study of carbon dioxide emission inventory from transportation sector at Kualanamu International Airport

I Suryati *[1,3], I Indrawan[2], K N Alihta[1]
[1]Environmental Engineering, Faculty of Engineering, University of Sumatera Utara, Medan 20155, Indonesia
[2]Civil Engineering, Faculty of Engineering, University of Sumatera Utara, Medan 20155, Indonesia
[3]Sustainable Energy and Biomaterial Center of Excellence, Universitas Sumatera Utara, Padang Bulan, Medan 20155, Indonesia

*Email: isratl@usu.ac.id

Abstract. Transportation includes sources of greenhouse gas emission contributor in the form of carbon dioxide (CO_2). CO_2 is one of the air pollutant gases that cause climate change. The source of CO_2 emissions at airports comes from road and air transportation. Kualanamu International Airport is one of the public service airports in North Sumatera Province. The purpose of this study is to inventory the emission loads generated by motor vehicles and aircraft and to forecast contributions of CO_2 emissions from motor vehicles and aircraft. The research method used is quantitative and qualitative methods. The quantitative method used is to estimate emission loads of motor vehicles based on vehicle volume and emission factors derived from the literature and using the Tier-2 method to calculate the aircraft emission loads. The results for the maximum CO_2 concentration were 6,206,789.37 $\mu g/m^3$ and the minimal CO_2 concentration was 4,070,674.84 $\mu g/Nm^3$. The highest aircraft CO_2 emission load is 200,164,424.5 kg/hr (1.75 x 109 ton/year) and the lowest is 38,884,064.5 kg/hr (3.40 x 108 ton/year). Meanwhile, the highest CO_2 emission load from motor vehicles was 51,299.25 gr/hr (449,38 ton/year) and the lowest was 38,990.42 gr/hr (341,55 ton/year). CO_2 contribution from a motor vehicle is 65% and 5% from aircraft in Kualanamu International Airport.

1. Introduction

According to [1], the aviation service industry has contributed significantly to the growth of trade, communications, and tourism. In addition to the increasing specialization of air transport services also occurred the addition of the number of vehicles that are very significant. The increase in the number of vehicles will be directly proportional to the emission load, thus increasing air pollution and affecting ambient air quality.

Air pollution at the airport comes from exhaust fumes, aircraft refueling, motor vehicles within the airport area, ground service equipment, heating plants, fuel evaporation spilled on refueling, and emissions from combustion gases in incinerators. The environmental impacts of air traffic are often associated with noise, smoke and emissions of carbon monoxide (CO), unburned hydrocarbons (HC), including methane (CH_4) and nitrogen oxides (NOx), carbon dioxide (CO_2), sulfur dioxides (SO_2) around the airport which can be detrimental [2; 3].

CO_2 is one of the greenhouse gas pollutants. Increased greenhouse gas emissions will cause global warming. Global warming is an international issue that correlates with climate change. Climate change

IOP Conf. Series: Materials Science and Engineering **309** (2018) 012023 doi:10.1088/1757-899X/309/1/012023

is a phenomenon triggered by human activities related to the use of fossil fuels as well as land-use activities.

The Government of Indonesia in the Pittsburg G-20 meeting is committed to reducing greenhouse gas emissions by 26% by self-effort and reaching 41% if received international assistance by 2020.

Following the GHG emission reduction commitment, the government issued Presidential Regulation No. 61 of 2011 on the National Action Plan for Greenhouse Gas Emission Reduction (RAN-GRK) which mandates to provinces to be responsible for the preparation of the Local Green House Gas Emission Reduction Action Plan (RAD-GRK) no later than 12 months after the enactment of the presidential regulation.

The effects of the development of aircraft transport and air quality at airports today are so complex and controversial. This effect occurs not only in the area surrounding the airport but also at the regional and global levels. Emissions from aircraft not only affecting air quality in the airport area but are also driven by other emissions such as vehicles entering and exiting from the airport entrance, power usage, machine use, and other supportive things in the operation and activities of the airport itself [3].

Kualanamu International Airport is the main airport in North Sumatra Province. The airport is expected to become the main entrance, which is an international transit base for the region of Sumatra and its surroundings. Therefore the airport should be supported by the quality of infrastructure, services and in terms of good management and environmental management.

This study aims to calculate the concentration of CO_2 pollutants in the ambient air in Kualanamu International Airport, calculate CO_2 emissions load from aircraft and motor vehicles located in Kualanamu International Airport, and forecast contribution of CO_2 emissions from aircraft and motor vehicles versus ambient air quality in Kualanamu International Airport.

2. Methodology

The study was conducted at Kualanamu International Airport. Administratively, Kualanamu International Airport is in Beringin Village, Beringin District, Deli Serdang Regency. The research location located in Kualanamu International Airport area will be divided into 5 (five) sampling points, namely runway 05, runway 23, apron w, A7 parking lot and the gate of Kualanamu International Airport. Selection of this sampling point using purposive sampling method.

2.1. Data Collection

The data collected in this study consist of primary and secondary data. Primary data is a measurement of CO_2 concentration using monitoring tool CO_2 with analysis method that is NDIR Analyzer, the volume of aircraft traffic and motor vehicle and meteorology condition at the time of research. The duration of the primary data is 1 (one) hour with the measurement frequency 2 (two) times.

Secondary data collected are wind direction and wind speed data to obtain wind rose for consideration in determining sampling location, airplane fuel usage data, plane type data, and landing and take-off (LTO) data.

2.2. Data Analysis

Data analysis used is quantitative analysis. For quantitative analysis used with 4 (four) methods of data processing, namely:

1. Data on CO_2 concentrations measured at 5 (five) sampling points together with Balai Teknik Kesehatan Lingkungan and Pengendalian Penyakit (BTKLPP) Medan City.
2. The calculation of the emission load will use the Tier 2 method. This method is chosen because the Tier in the emission calculation shows the accuracy of the calculation [4]. The Tier-2 methodology is used for estimation of GHGs from aviation-fueled aircraft. In this methodology aircraft operations are divided into landing and take-off (LTO) and flying (cruise). To be able to use Tier-2 data landing and take-off (LTO) must be known. Steps for calculating GHG emissions by the Tier-2 method are as follows:
   a. Estimate aircraft fuel consumption for domestic and international

b. Estimate LTO fuel consumption for domestic and international
c. Calculate current LTO emissions for domestic and international

$$\text{LTO Consumption} = \text{Number of LTO} \times \text{Consumption per LTO} \quad (1)$$

$$\text{LTO Emission} = \text{LTO Consumption} \times \text{Emission Factor of LTC} \quad (2)$$

3. Calculation of vehicle emission load at Kualanamu International Airport. The data of the calculation of the number of vehicles that have been done will be converted into Passenger Car Unit (SMP). The conversion can be done by multiplying the number of vehicles calculated in the study sites with the factor value of SMP in accordance with the Indonesian Road Capacity Standards Manual [5]. After conversion to passenger car units, vehicle activity data is multiplied by the emission factor. The emission factor used refers to Minister of Environment Regulation No.12 Year 2010. The calculation of emission load for a pollutant from a motor vehicle on a road segment using Equation 3:

$$E_j = \sum_{i=1}^{n} E_{ij} = \sum_{i=1}^{n} l.P_i.V.C_{ij} = lV \sum_{i=1}^{n} P_i.C_{ij} \quad (3)$$

Where: E_j = emission load (g / hour); l = the length of the observed road (km); V = total volume of vehicles passing a road (vehicle / hour); P_i = the probability distribution fraction of the type vehicle. (If the number of vehicles per category has been obtained, this probability value is not required); C_{ij} = vehicle emission factor (g / km)
4. Regression test and correlation of emission load on ambient air quality by using SPSS

3. Result and Discussions

3.1. Volume of Aircraft Traffic and Vehicles

Observation of aircraft traffic volume conducted on Wednesday and Thursday with 2 times the measurement frequency that is morning and afternoon. Segments observed on Wednesday are in Apron W and on Thursday are on runway 05 and runway 23. For more details, traffic volume based on observation time of morning and afternoon can be seen in Figure 1.

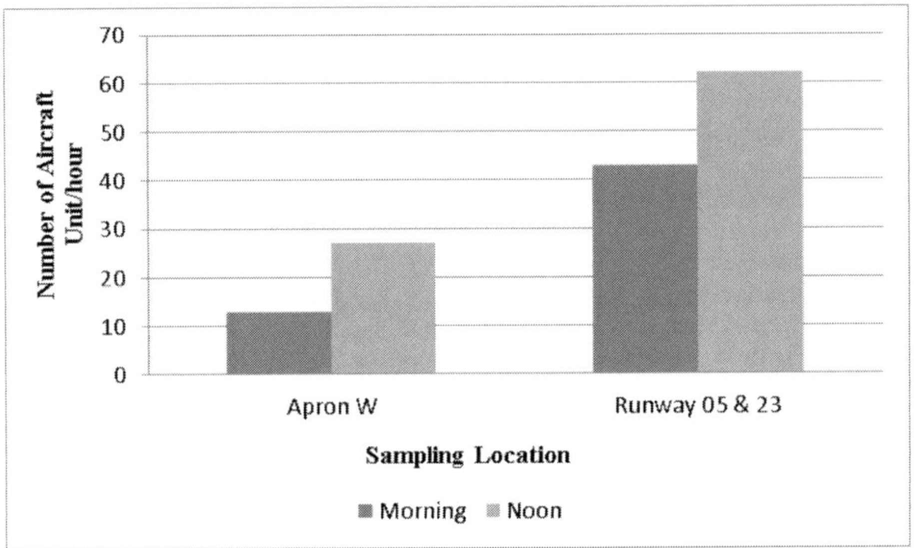

Figure 1. Aircraft Traffic Volume in Apron W and Runway

Based on Figure 1, it is can be seen that the average volume of aircraft is more on daytime measurement than in the morning measurement. This is due to the number of flight planes departing at around 10:00 to 15:00 pm.

Observation of motor vehicle traffic volume was done on Thursday with the frequency of sample measurement 2 times that is morning and afternoon. A segment of roads observed along ± 200 m. The traffic volume during observation can be seen in Figures 2. The location for observing the number of vehicles is at the entrance and exit Kualanamu International Airport.

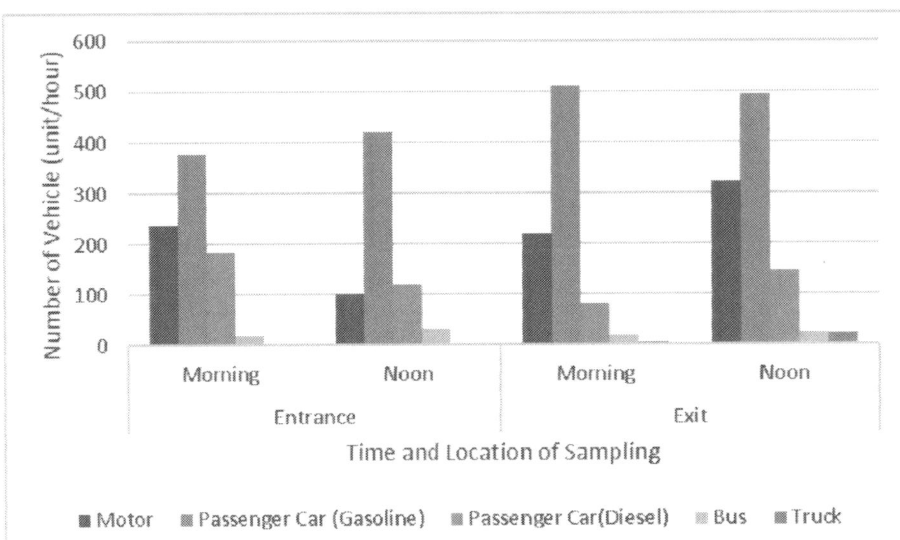

Figure 2. Number of Motor Vehicle Traffic at Entrance and Exit of Kualanamu International Airport

Figure 2 shows the most dominant types of vehicles at the entrance of Kualanamu International Airport in the morning are passenger cars (gasoline) by 45.84%, motorcycles by 28.85%, 22.49% for passenger cars (solar), buses by 2.44% and trucks by 0.37%. Meanwhile, during the day, the number of passenger cars (gasoline) by 62.46%, passenger cars (solar) and motorcycles are less than the morning of 17.80% and 14.84%, the number of buses also increased to 4.45%, and the number of trucks by 0.45%.

Based on Figure 2, the most dominant vehicle types at the exit of Kualanamu International Airport in the morning are passenger cars (gasoline) by 61.20%, motorcycle (26.11%), and diesel cars (9.58%), bus (2.40%) and trucks i.e. 0.72%. While at noon, the number of gasoline cars is less than the morning of 49.21%. In addition, diesel and motorcycle cars more than the morning of 14.48% and 31.94%. While the number of buses by 2.28% and trucks by 2.08%.

3.2. CO_2 Emissions Load From Aircraft and Motor Vehicles

The estimated total CO_2 emission load generated by each measuring time can be seen in Figure 3.

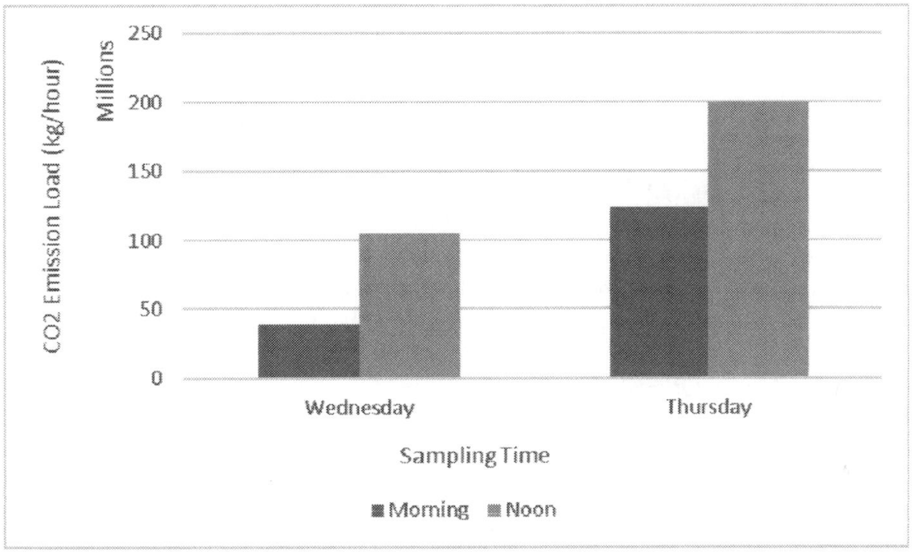

Figure 3. Total CO_2 Emissions Load from Aircraft at Kualanamu International Airport

Figure 3 shows that the highest estimated CO_2 emission load on Thursday at noon is 200.164.424.5 kg/hour, followed by the morning time of 123. 954.502,8 kg / hour. This is also due to the high flow of traffic at the time of measurement on Thursday when compared on Wednesday. The lowest CO_2 emissions burden on Wednesday morning is 38,884,064,5 kg/hour.

Results of research [6] at Soekarno Hatta International Airport show total CO_2 emissions per year is 588.747,3 tons (67.208,6 kg/hour). When compared to the CO_2 emission load results obtained from this research, the CO_2 emissions load at Kualanamu International Airport is higher than the previous study. This difference is caused by meteorological conditions, data retrieval times, and is influenced by machine type and aircraft type [7].

The estimated total CO_2 emission load generated at the entrance and exit of the airport at each measurement time can be seen in Figure 4.

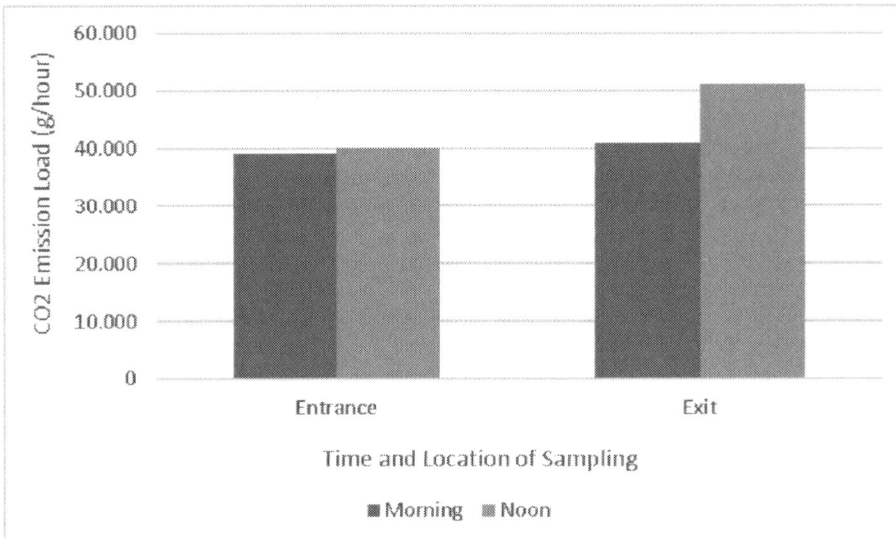

Figure 4. Total Motor Vehicle CO_2 Emissions Load at Kualanamu International Airport

Figure 4 shows that the highest estimated CO_2 emission load from a motor vehicle at the exit of Kualanamu International Airport is 51.299,25 g/hr, followed by the morning time of 41.054,64 g/hr. This is due to the high flow of traffic on both sides of the road. Road segment that produces the lowest CO_2 emission load is at the entrance of Kualanamu International Airport in the morning at 38,990.42 g/hour.

3.3. CO_2 Concentration at Kualanamu International Airport

The result of measurement of CO_2 concentration in ambient air at each measurement location can be seen in Figure 5.

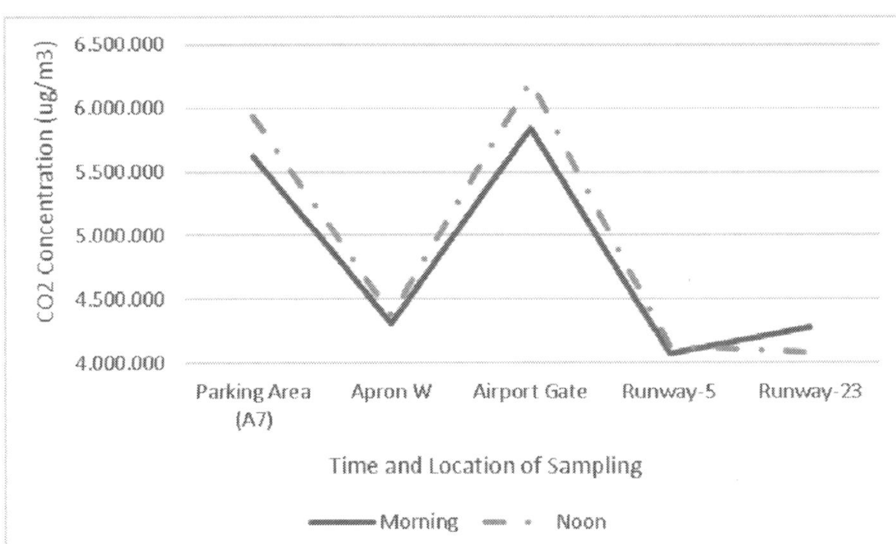

Figure 5. CO_2 Concentration at Kualanamu International Airport

Figure 5 shows the measured CO_2 concentration. The highest CO_2 concentration is at KNO Gate at daytime measurements of 6.206.789,37 µg/m^3 or 3.449 ppm. The lowest concentration of CO_2 is Runway 05 in the morning measurement of 4.070.674,85 µg/m^3 or 2.462 ppm. The CO_2 concentration of the measurement results is still below 5,000 ppm or still within reasonable limits. According to [8], concentrations greater than 5.000 ppm are not good for health while concentrations greater than 50.000 ppm can endanger animal life.

3.4. Effect of CO_2 Emission Load from Aircraft on CO_2 Observation

The result of analysis of CO_2 emission load data from aircraft to measured CO_2 concentration by using statistical analysis got regression equation as follows:

$$y = 4{,}2 + 0{,}001x \tag{4}$$

Where: y = CO_2 concentration; x = CO_2 Emission Load from Aircraft

Based on equation (4), it can be explained that each addition of 1 (one) CO_2 emission load value from the plane there is an increase in the CO_2 concentration of 0.001 µg/m^3. Correlation value (R) and the value of the coefficient of determination (R^2) obtained for 0.224 and 0.050. The R^2 value states that the CO_2 emission load of the aircraft influences the CO_2 concentration of R^2 = 0.05 or 5%, while 95% is influenced by other factors such as meteorological conditions i.e. wind direction and speed, increase of airplane flight frequency, movement of the opposite plane. The amount of power and fuel used will also impact on the emissions load [7; 9; 10].

3.5. Effect of CO_2 Emission Load from Motor Vehicles on CO_2 Observation

The result of data analysis of CO_2 emission load from motor vehicle to measured CO_2 concentration by using statistical analysis got regression equation as follows:

$$y =- 2{,}100 + 942.285x \tag{5}$$

Where: y = CO_2 concentration; x = CO_2 Emission Load from Motor Vehicles

Based on equation (5), it can be explained that each addition of 1 (one) CO_2 emission load value from aircraft then there is an increase of CO_2 concentration as much 942.285,69 µg/m^3. Correlation value (R) and determination coefficient value (R^2) obtained for 0.806 and 0.65. The R^2 value states that the CO_2 emission load of the motor vehicle affects the CO_2 concentration of R^2 = 0.65 or 65%, while 35% is influenced by other factors such as meteorological conditions, CO_2 exchange with both main carbon sources i.e. biosphere, catalytic converter, fuel used and other activities at airports such as electricity usage and waste incineration [7; 11; 12; 13].

CO_2 emissions from motor vehicles and aircraft contribute 65% and 5% to measured CO_2 concentrations. This is due to the technology of motor vehicles that are currently using a catalytic converter. Catalytic converter is a tool that will react the harmful exhaust gases through chemical reactions so that later these gases will turn into a gas that is not harmful to the environment. However, the use of catalytic converter does not guarantee that all exhaust gases issued are not 100% clean of harmful flue gases such as CO that will become CO_2, where the use of catalytic converter reduces CO emissions by up to 94% and HC to 86% but increases CO_2 emissions [14; 15]. This, therefore, causes a greater CO_2 emission load on motor vehicles compared to aircraft emission loads.

4. Conclusion

The results for the maximum CO_2 concentration were 6,206,789.37 µg/m^3 and the minimal CO_2 concentration was 4,070,674.84 µg/m^3. The highest aircraft CO_2 emission load is 200,164,424.5 kg/hour and the lowest is 38,884,064.5 kg/ hour. The highest CO_2 emissions load from motor vehicles amounted to 51,299.25 g/hr and the lowest was 38,990.42 g/hr. The aircraft emission load contributes

IOP Conf. Series: Materials Science and Engineering **309** (2018) 012023 doi:10.1088/1757-899X/309/1/012023

5% to the measured CO_2 concentration. Meanwhile, motor vehicle emissions contribute 65% to the measured CO_2 concentration.

5. References

[1] Singh V, Sharma S K 2015 *Journal. Eur. Transp. Res. Rev.* **7**: 12

[2] Olivier J G, Janssens-Maenhout G, Peters J A H W, Wilson J 2011 *Long-Term Trend in Global CO_2 Emission* (Europian Union: PBL Netherlands Environmental Assesment Agency)

[3] Ashford N J, Mumayiz S A, Wright P 2011 *Aiport Engineering* (USA: John Wiley & Sons Inc)

[4] KLH 2012 *Pedoman Penyelenggaraan Inventarisasi Gas Rumah Kaca Nasional. Buku II-Volume 1 Metodologi Perhitungan Tingkat Emisi Gas Rumah Kaca* (Indonesia: KLH)

[5] Direktorat Bina Marga dan Jalan Kota 1997 Baku Manual Kapasitas Jalan Indonesia (Jakarta: Kementerian Perhubungan)

[6] Adiati R R, Rahardyan B 2011 *Estimasi Kondisi Eksisting sebagai Dasar Rancangan Eco Airport Bandar Udara Soekarno Hatta* (Bandung: FSTL ITB)

[7] Ombasta, Osha 2012 *Pengaruh Siklus Landing Take Off (LTO) Pesawat di Bandar Udara Terhadap Fluktuasi Kadar NOx Pada Udara Ambien Studi Kasus Bandar Udara Internasional Soekarno Hatta di Jakarta* (Depok: Universitas Indonesia)

[8] Sehabudin S 2011 *Penambatan Karbon Dioksida dan Pengaruh Densitas Alga Air Tawar (chlorella sp)* (Jakarta: UIN)

[9] Slamet LS 2012 Potensi dan Dampak Polusi Udara dari Sektor Penerbangan. *Jurnal LAPAN*, 31 -36

[10] Kaleka Y U, Suyasa I W B, Mahendra M S 2014 *Jurnal Ecotrophic* **9 (1)**, 72-79

[11] Samiaji T 2012 *Berita Dirgantara* **12 (2)**, 68 -75

[12] Sanata A 2012 *Jurnal Rotor* **5 (2),** 1-7

[13] Winarno J 2014 *Jurnal Fakultas Teknik Universitas Janabadra*, 1 -7

[14] Ellyanie 2011 *Prosiding Seminar Nasional AVoER UNSRI* **3,** 1 – 9

[15] Sudjada C M, Warju 2014 *Jurnal Teknik Mesin Universitas Negeri Surabaya* **3 (2)**, 104 - 113

IOP Conf. Series: Materials Science and Engineering **309** (2018) 012024 doi:10.1088/1757-899X/309/1/012024

Soil settlement analysis in soft soil by using preloading system and prefabricated vertical draining runway of Kualanamu Airport

Roesyanto*, R Iskandar, S A Silalahi and Fadliansyah
Department of Civil Engineering, Universitas Sumatera Utara, Medan 20155, Indonesia

*Email: roesyanto@usu.ac.id

Abstract. The method of soil improvement, using the combination of prefabricated vertical drain (PVD) and preloading, was used to accelerate the process of consolidation and the consolidation settlement in the runway of Kualanamu International Airport, which was constructed on the soft soil sediment like silty clay. In this research, the investigated area was the runway of Kualanamu International Airport zone I which had 11 meter-thickness of soft soil. Geotechnic instruments surveyed was settlement plate. Monitoring was done toward the behavior of landfill such as basic soil settlement. The result were compared with the analysis of finite element method of full scale in Mohr-Coulomb model by verifying the vertical drain of asymmetric unit cell and equivalent plane strain unit cell condition. The results of the research showed that there were an interesting behavior between the data in field observation and finite element of Mohr-Coulomb model. It was also found that the result of soil settlement of finite element method of Mohr-Coulomb model was closed to the result of settlement plate monitoring.

1. Introduction

Due to the rapid development of urbanization and the population increasealong coastal lines, exploitation of undeveloped low-lying areas has became a necessary strategy for many countries [1]. Development of urban transportation and expansion also occurred in North Sumatra, this occurred at the turn of Polonia Airport with the new Airport, known as Kualanamu Airport, located in Kualanamu area, North Sumatra. The airport construction will include runway, taxiway, exit taxiway, apron and other supporting facilities. Given the problem of soft clay soil that is influential in the success of infrastructure development, so must be done soil repairs so that infrastructure is not damaged before the planned age.

Based on above, it is necessary to improve the soft soil conditions. Preloading of soft clay with vertical drains is one of the most popular methods used to increase the shear strength of soft soil and control its post-construction settlement.Since the permeability of soils is very low, consolidation time to the achieved desired settlement or shear strength may take too long [2]. Using prefabricated vertical drains (PVDs), means that the drainage path is shortened from the thickness of the soil layer to the radius of the drain influence zone, which accelerates consolidation [3]. Introduced prefabricated band shaped drains and cardboard wick drains for ground improvement. Typically, prefabricated band drains consist of a plastic core with a longitudinal channel surrounded by a filter jacket to prevent clogging. Most vertical drains are approximately 100 mm wide and 4 mm thick [4]. The soft soil improvement by prefabricated vertical drains (PVDs) combined with preloading is one of the most successful applications to this site [5].

2. Methodology

2.1. Prefabricated Vertical Drain (PVD)

Barron [6] describes the area of influence (D_e) mounting PVD with a function of vertical drains as shown in Fig.1.

For triangle pattern values [6]:

$$D_e = 1.05S \quad (1)$$

While for rectangular values:

$$D_e = 1.128S \quad (2)$$

Where:

S = Distance spacing drainase vertical

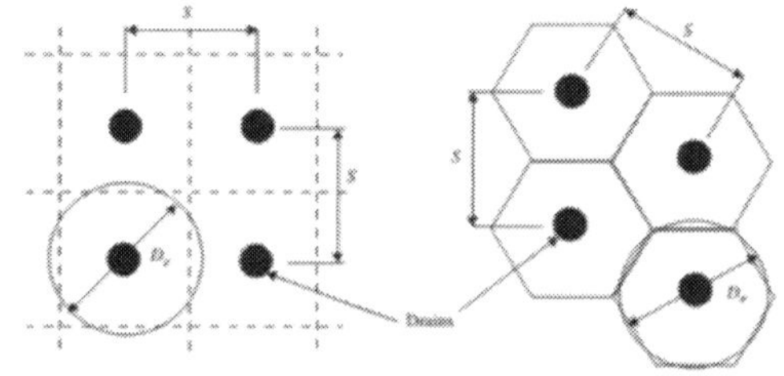

Figure1. Configuring vertical drain; (a) triangle, (b) rectangle

2.2. Equivalent drain diameter of band-shaped vertical drain

The size of the band-drain or PVD is mostly rectangular cross-section. For the purpose of calculating PVD, the cross section of the PVD will be modeled into a circle with the calculation of the equivalent diameter assumed as the circumference of the rectangle as shown in Figure 2.

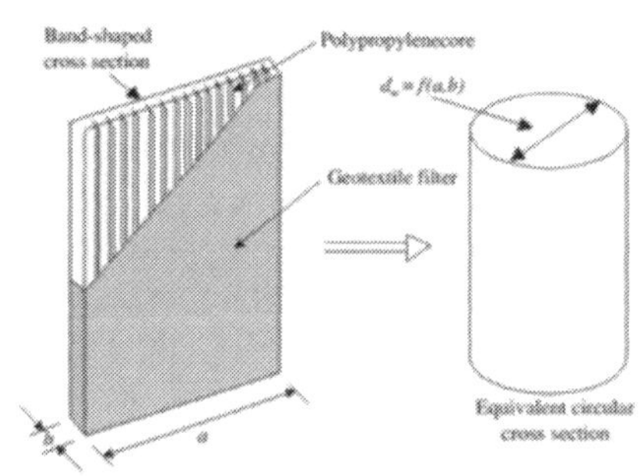

Figure2. Conceptual drawing of a band-shaped PVD and equivalent diameter well.

Assumption is based on Hansbo's formulation as below [7]:

$$d_w = \frac{2(a+b)}{\pi} \quad (3)$$

Where d_w = equivalent diameter vertical drains, a= wide vertical drains, b = thick vertical drains.

2.3. Discharge capacity PVD

Based on the results of laboratory testing Holtz *et al* [2] suggested for large discharge capacities of PVD ranging from 100 to 150 m3/year.

2.4. Smear Zone

PVD is inserted into the ground by using mandrel and at the end of the shoe is given. As a result of this installation, the soil layer stabbed with mandrill will be disrupted. The disturbance that occurs is called smear zone while the result will be a reduction in the coefficient of permeability of soil radial (k_r).Smear zone effect is decrease permeability coefficient value for clay soil near PVD or PVD diameter used is minimized, this is due to process of remolding during installation of PVD.

Assumption is based on Chai and Miura [3] formulation as below:

$$d_s = 3d_m \tag{4}$$

Where d_s= equivalent diametersmear zone, d_m = equivalent diameter mandrel.

2.5. Modeling Verification of PVD

The planeting pattern of PVD installed in the local field, at a certain distance, while in the FEM program the facility of implementing PVD is plane strain as shown in Fig. 3. To be able to implement PVD installed in the field into the program, it must first be verified in the form of plane strain that will produce a new soil permeability coefficient (k), then with the new soil permeability coefficient (k) the new simulation process in the FEM program can do.

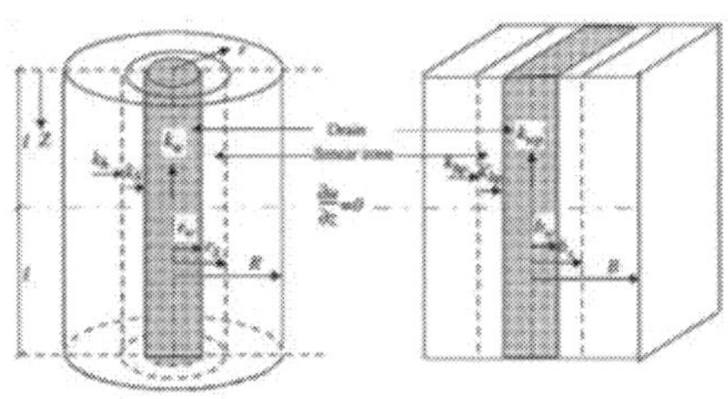

Figure 3. Conversion of an axisymmetric unit cell into plane strain condition.[8, 9].

According to Hird, *et al* [8, 10]. The equivalent process can be done in several ways:

1. The distance between PVD at plane strain condition can be changed (geometric change), with permeability made fixed at axisymmetric and plane strain conditions ($k_{ax} = k_{pl}$).
2. The permeability of plane strain conditions can be altered (permeability change), with the same geometry being created.
3. Combines geometry and permeability changes.

3. **Results and Discussion**

3.1 Field of measurements

Settlements that occured during construction are observed with the help of settlement plate tools. In this study the decrease was reviewed at three points of observation as shown in Figure 4.

The points reviewed in the data monitoring of settlement plate by 3 points, namely point S-19, point S-20, and point S-21. Each point had different settlement at the same consolidation times. Point had S-19 settled 0.578 m, point S-20 settled 0.355 m, and point S-21 settled 0.212 m.

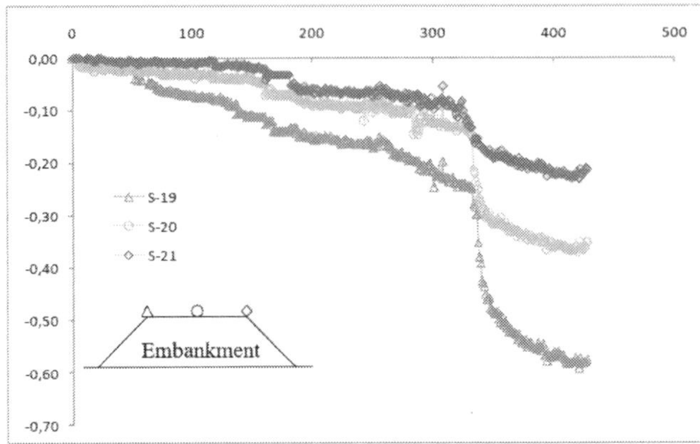

Figure 4. Settlement Comparison of three-point settlement plates

3.2. Finite Element Modeling

Predictions were made using plane strain finite element analyses (FEA). The embankment and soil condition were in model two dimensional.

3.2.1. Unit Cell Analysis

In the field the pattern of PVD planeting is done locally with a certain distance, whereas in the FEM program the implementation of PVD is continuous (plane strain). Using Table 1 and 2, the first stage was verification of the axisymmetric cell unit with the coefficient of permeability of axisymmetric conditions and then equate the same water discharge with the condition of the plane strain in the form of unit cell plane strain (Figure 5).

Table 1. Geometry for Unit Cell Analysis

Depth (m)	w (m)	t (m)	re (m)	rw (m)	dm (m)	rs (m)
11	0,1	0,05	0,6563	0,0334	0,2873	0,1437

Table 2. Coefficient of Permeability for Cell Unit Analysis

Material	Kondisi Axysimetris			Kondisi Plane Strain		
	kh (m/day)	*kv* (m/day)	output	*kh* (m/day)	*kv* (m/day)	output
Silty Sand	2,7376x10-4	1,3688x10-4	*non smear zone*	1,0950x10-4	5,4752x10-5	*non smear zone*
	1,0950x10-4	5,4752x10-5	*smear zone (krs=kr/2,5)*	4,38x10-5	2,19x10-5	*smear zone (krs=kr/2,8)*

IOP Conf. Series: Materials Science and Engineering **309** (2018) 012024 doi:10.1088/1757-899X/309/1/012024

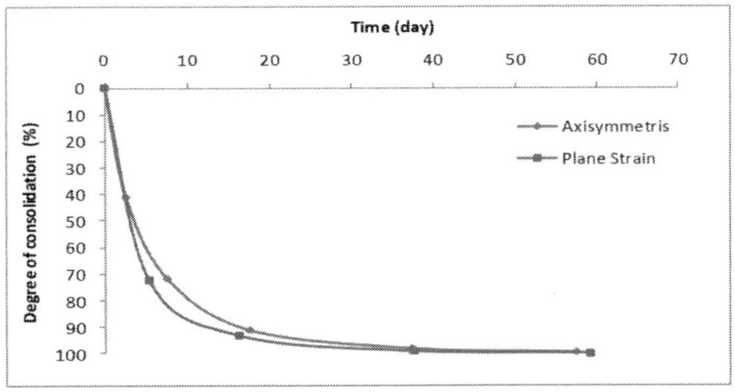

Figure 5. Degree of consolidation versus time of the axisymmetric and plane strain conditions

3.2.2. Full Scale Analysis

Limits modeled on the full-scale element calculations have a length of 150 m in the horizontal direction and 20.26 m in the vertical direction of the lower layer with a heap of 6.02 m. Only half of the heap is modeled because the heap is symmetrical. Horizontal fixities are applied to the finite element mesh on the left and right sides and on vertical movement (Vertical fixities) at the top and bottom of the mesh. In the full-scale analysis of the disturbed area effects considered is k_h / k_s = 2.8 this value is taken from the value of the near-field matching.

Geometry and mesh model of the Sta 0 + 750 runway of landfill is shown in Fig. 6 and Fig. 7. The soil parameters used in the analysis using the finite element method is shown in Table 3.

Table 3. Properties of Land Parameters Mohr-Coulomb Model

		Type of soil				
Properties	Unit	Preloading	Embankment Struktur	Sand Blanket	Silty Sand	Sand
c	kN/m^2	5,000	5,000	10,000	9,600	5,000
φ	derajad	30,00	30,00	30,00	7,750	30,00
ψ	angle	-	-	-	-	-
k_x	m/day	8,640	8,640	8,640	9,78E-05	0,061
k_y	m/day	8,640	8,640	8,640	4,89E-05	0,012
γ_{dry}	kN/m^3	16,00	16,00	16,00	7,7	16
γ_{wet}	kN/m^3	19,00	18,00	18,00	14,6	19
E	kN/m^2	5000	10000	5000	3300	34500
v	-	0,300	0,300	0,300	0,300	0,350

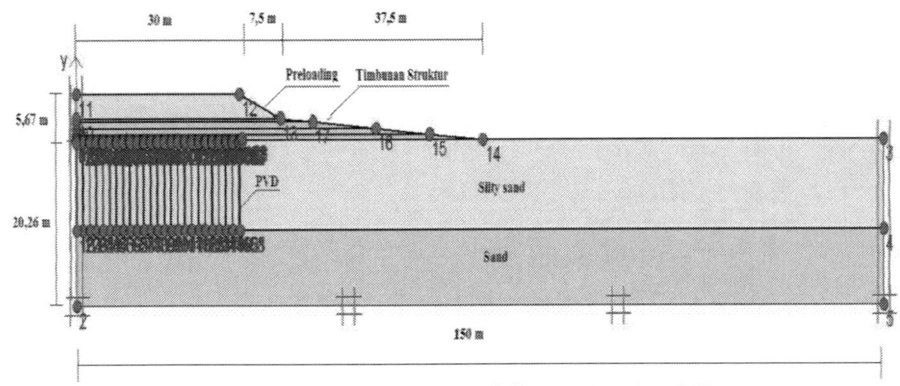

Figure 6. Runway geometry modeling at Sta 0 + 750

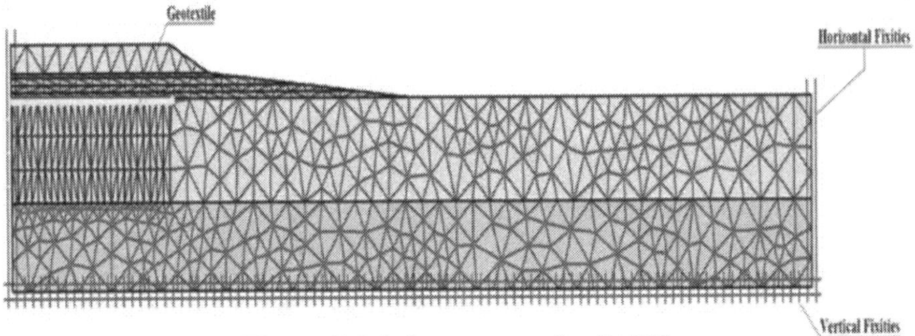

Figure 7. Mesh runway at Sta 0+750

3.3. Deformation pattern

Points reviewed in the calculation of the finite element method are two points, namely point A (0,0) and point B (30,0) indicating the position of the settlement plate location points shown in Fig. 8. Each point has a varying magnitude of decline but the required consolidation time is the same for all points.

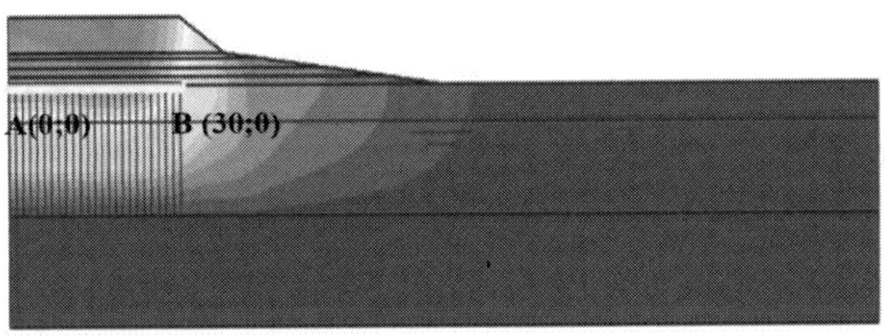

Figure 8. Displacement deformation pattern Mohr-Coloumb Model

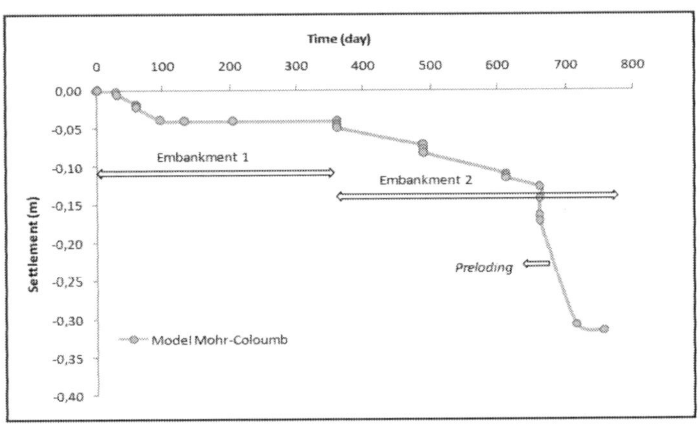

Figure 9. Settlement graph of nodal point A (0;0)

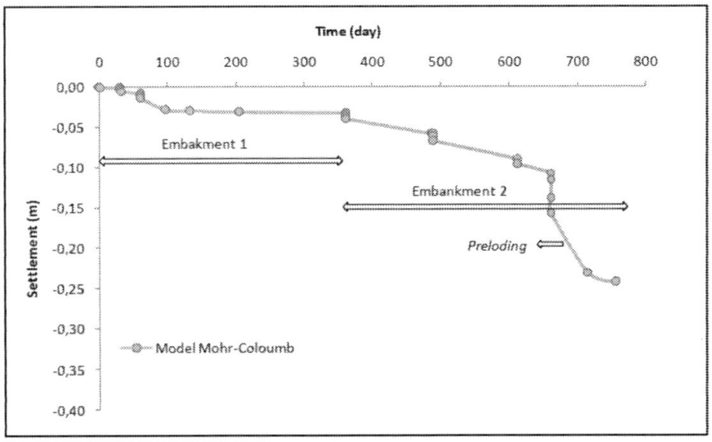

Figure 10. Settlement graph of review point nodal B (30;0)

In Figs. 9 and 10 show the decrease graph that occurs in the silty sand layer with the Mohr-Coloumb model with two points of view i.e. the point node A with a maximum decrease of 0.3162 m, while at the point node B of 0.2427 m with the length of time The same consolidation at both points of review amounted to 758 days.

3.4. Discussion

The settlement comparisons between the field observations and Mohr-Coloumb soil model are shown in Table 4.

Table 4. Settlement Comparison between field observation and Mohr-Coulomb soil Model

Method				
Field Observation (m)			Mohr-Coulomb soil model	
S-19	S-20	S-21	Point A	Point B
0.578	0.355	0.212	0.316	0.243

4. Conclusion

From the results of the research, it could be concluded that:

1. From the embankment analysis using FEM 2D program using Mohr-Coulomb soil model, it was obtained the settlement at both points, point A equal to 0.316 m and point B equal to 0.243 m.
2. From field monitoring results using settlement plate from three points of view, point S-19 settled 0.578 m, point S-20 point settled 0.355 m, and point S-21 settled 0.212 m.
3. The major differences in the settlement at each point could be due to the difference of the soft soil across the runway that was not monitored on bore hole sampling that was performed only on the middle allong the runway.

References

[1] Indraratna, B., Balasubramaniam, A.S., Balachandran, S. 1992 Performance of test embankment constructed to failure on soft marine clay J Geotech Eng ASCE **118 (1)** 12–33.

[2] Holtz, RD 1987 Preloading with prefabricated vertical strip drains, Geotextiles and Geomembranes **6(1–3)** 109–131.

[3] Hansbo S 1981 Consolidation of fine-grained soils by prefabricated drains. In Proceedings of 10thInternational Conference on Soil Mechanics and Foundation Engineering, Stockholm, Balkema, Rotterdam **3** 677-82.
[4] Kjellman W 1948 Accelerating consolidation of fine grain soils by means of cardboard wicks. Proc. 2nd ICSMFE **2** 302-05.
[5] Indraratna, B and Redana, I W 2000 Numerical modeling of vertical drains with smear and well resistance installed in soft clay Canadian Geotechnical Journal **37** 132-45.
[6] Barron R A 1948 Consolidation of fine-grained soils by drain wells Transactions of the American Society of Civil Engineers **113** 718–42.
[7] Hansbo S 1979 Conslidation of clay by band-shaped prefabricated drains Ground Eng **12(5)** 16-25.
[8] Hird C C, Pyrah I C and Ruscell D 1992 Finite element modelling of vertical drains beneath embankments on soft ground. Geotechnique **42(3)** 499-511.
[9] Indraratna B, Balasubramaniam A S and Sivaneswaran N 1997 Analysis of settlement and lateral deformation of soft clay foundation beneath two full-scale embankments International Journal for Numerical and Analytical Methods in Geomechanics **21** 599-618.
[10] Hird C C, Pyrah I C, Russel D, Cinicioglu F 1995 Modelling the Effect of Vertical Drains in Two-Dimensional Finite Element Analyses of Embankment on Soft Ground **32** 795-807 Canada Geotechnical Journal, Canada.

IOP Conf. Series: Materials Science and Engineering **309** (2018) 012025 doi:10.1088/1757-899X/309/1/012025

A study of the effectiveness of the use of gypsum and volcanic ash against the stability of clay soil in terms of UCT and CBR values

Roesyanto*, R Iskandar, IP Hastuty and AIU Lubis
Civil Engineering Department, University of Sumatera Utara, Medan, Indonesia

*Email: roesyanto@usu.ac.id

Abstract. Soil stabilization is an effort to improve engineering properties of soil. The conventional soil stabilization is by adding additives to the soil such as Portland cement, lime, and bitumen. The clay stabilization research was done by adding gypsum and volcanic ash. The research purposes were to find out the value of engineering properties of clay due to the addition of 2% gypsum and 2% - 15% volcanic ash. The soil was classified as Clay – Low Plasticity (CL) based on USCS and was classified as A-7-6 (10) based on AASHTO classification system. The UCT values of original soil and original soil plus 2% gypsum were 1.40 kg/cm^2 and 1.66 kg/cm^2 respectively. The CBR soaked and unsoaked values of original soil were 4.44% and 6.28% correspondingly. Meanwhile, CBR soaked and CBR unsoaked values of original soil plus 2% gypsum were 6.74% and 8.02% respectively. The research results showed that the additives materials of gypsum and volcanic ash improved the engineering properties of clay. The UCT result from the stabilized soil by 2% gypsum and 10% volcanic ash gave value of 2.79 kg/cm^2 (increased 99.28% from original soil). For CBR test, the most effective mixture were in variation of 2% gypsum and 9% volcanic ash which gave value of 9.07% (104.27% increase from original soil) for CBR soaked and 10.29% (63.85% increase from original soil) for CBR unsoaked. The stabilized soil with 2% gypsum and 9% volcanic ash was classified as CL based on USCS and was classified as A-6 (4) based on AASHTO classification system.

1. Introduction

Soil stabilization is a mixture of soil with certain materials, in order to improve the engineering properties of soil [1]. The soil stabilization is by adding chemical materials to the soil. The most common additives are Portland cement, lime, bitumen and tar [2,3]. In this research the stabilization of clay was done by adding gypsum and volcanic ash. Gypsum is mineral with calcium levels that dominate in the minerals. Gypsum as an additive material has better properties than organic additives because it does not cause air pollution, relatively cheap, fire resistant, and resistant to deterioration by biological factors and chemicals [4]. Volcanic ash (VA) used in this research was originated from Mount Sinabung. From the last data, the volume of lava from Mount Sinabung which erupted has reached 2.3 million m^3 [5]. Research about stabilized clay using portland cement and volcanic ash from Mount Sinabung had been proved to enchance the engineering properties of clay [6,7,8]. This research analysed the stabilization of clay which was mixed with gypsum and volcanic ash as stabilizing materials which were expected to improve the engineering properties of the clay sample so that the clay could be used in the construction.

2. Method

The research used experimental method which was conducted in the Laboratory of Soil Mechanics, Civil Engineering Department, University of Sumatera Utara. The research samples were original clay which was taken from PTP II, Patumbak, Deli Serdang, North Sumatera and clay samples which had been mixed with gypsum and volcanic ash erupted from Mount Sinabung.

2.1 Tested Materials

There were 62 soil samples in the research, one of them was the sample of soil without being mixed with gypsum and volcanic ash, while the remaining 14 (fourteen) samples were soil which was mixed with gypsum and volcanic ash for UCT and 48 (forty-eight) soil samples for CBR test. The index properties of clay are shown in Table 1.

2.2 Testing Execution

Proctor Standard test was done to obtain the optimum water content and the maximum dry density. Besides that, there were Unconfined Compression tests and CBR Laboratory test. The curing time of each testing samples was 7 days.

After all data were gathered, they were analyzed. All results would be presented in the form of tables, graphs, and explanation. The flow diagram of the research could be seen in Figure1.

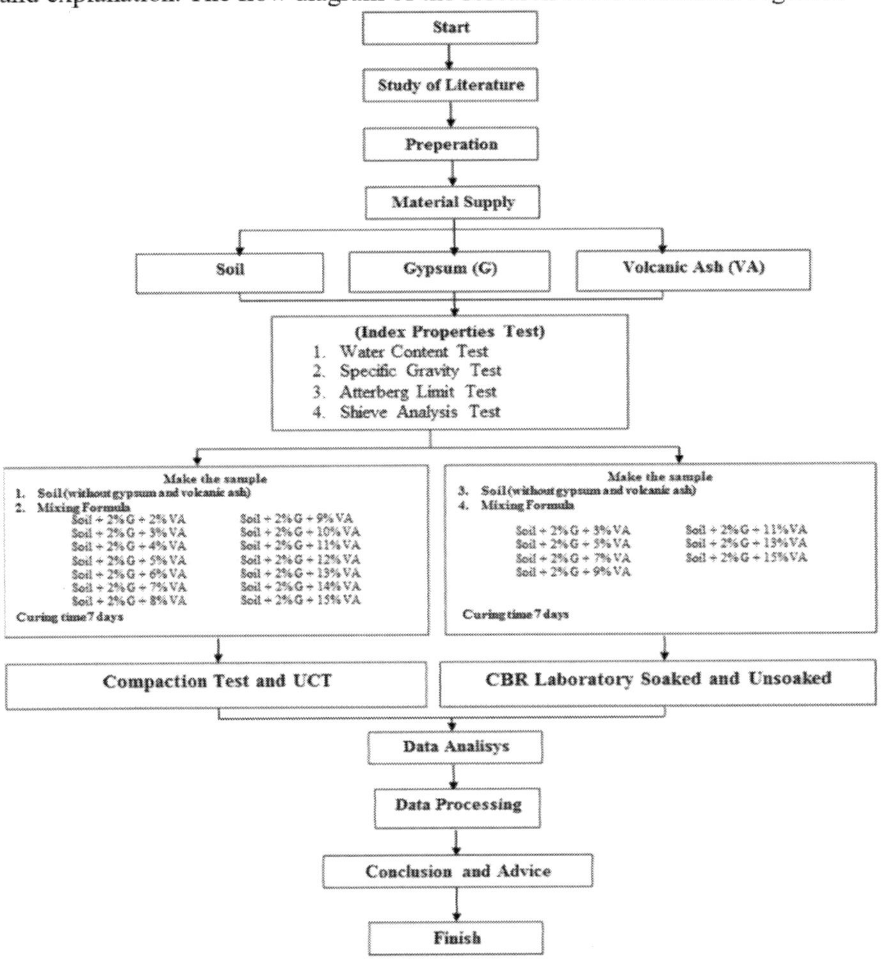

Figure 1. The flow diagram of the research

3. Results and Discussion

3.1 Original Soil Testing
Engineering properties of original soil was shown in Table 1.

Table 1 Engineering properties of original soil

No.	TEST	RESULT
1	Water Content	12.42%
2	Specific Gravity	2.65
3	Liquid Limit	46.82%
4	Plastic Limit	17.42%
5	Plasticity Index	29.40%
6	Sieve Analysis	49.17%
7	Optimum Moisture Content	21.32%
8	Maximum Dry Density	1.31 gr/cm^3
9	CBR soaked	4.44 %
10	CBR unsoaked	6.28%

3.2 Stabilizing Material Testing
Index properties of volcanic ash andgypsum was presented in Table 2.

Table 2 Index properties of volcanic ash and gypsum

No.	Test	Volcanic Ash	Gypsum
1	Specific Gravity	2.62	2.59
2	Liquid Limit	Non Plastis	Non Plastis
3	Plastic Limit		
4	Plasticity Index		
5	Sieve Analysis	22.71%	51.62%

3.3. Engineeringproperties of the stabilized soil
The engineering properties of stabilized soil with gypsum and volcanic ash were presented in Figure 2.

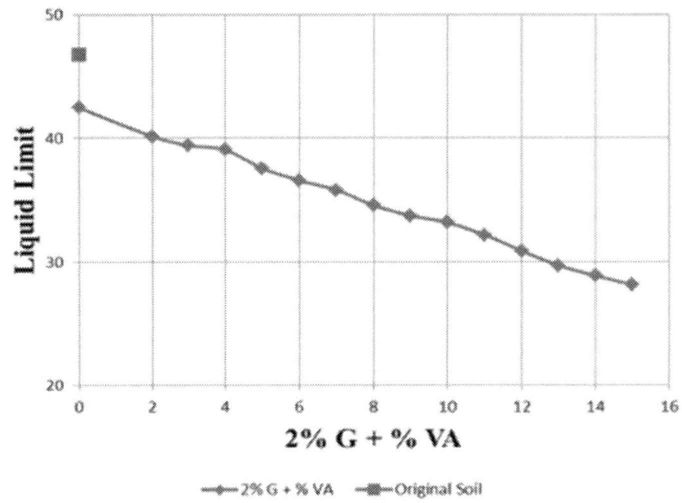

Figure 2. Correlation of Liquid Limit (LL) value with the variation of 2% gypsum in the addition of 0% - 15% of volcanic ash.

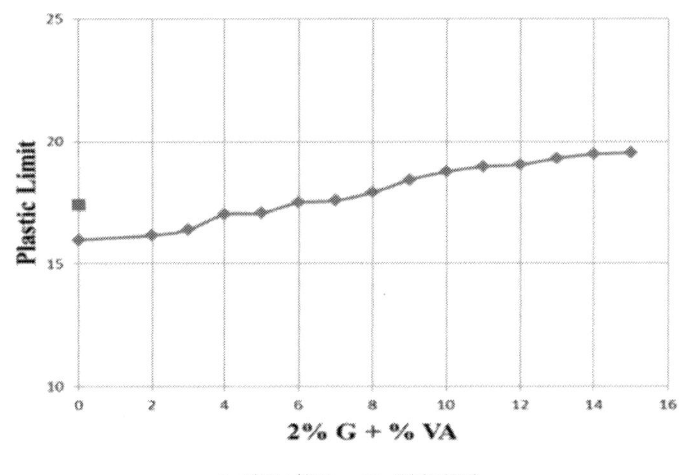

Figure 3. Correlation of Plastic Limit (PL) value with the variation of 2% of gypsum in the addition of 0% - 15% of volcanic ash

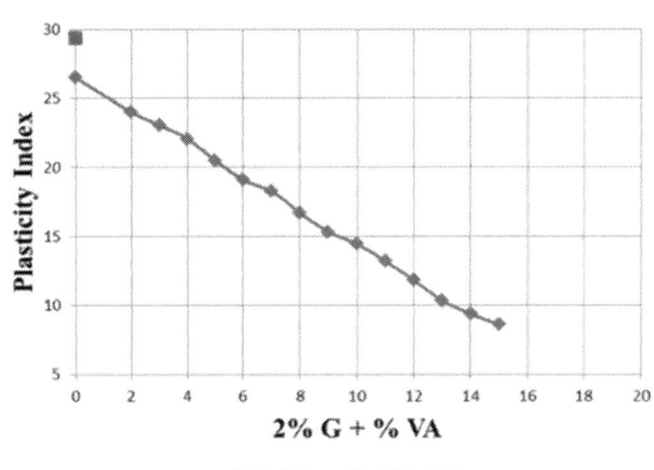

Figure 4. Correlation of Plasticity Index (PI) value with the variation of 2% gypsum in the addition of 0% - 15% of volcanic ash

In Figure 2, it was found that Liquid Limit as the result of the mixture of stabilizing materials, gypsum and volcanic ash decreased. The higher percentage of the volcanic ash was, the lower its liquid limit. In the original soil, the liquid limit reached 46.82%, while the lowest liquid limit was in the variation of the mixture of 2% G + 15% VA with curing time of 7 days at 28.12%. This was because in the short-term reactions the soil underwent flocculation process and was marked bya flocculation of the soil particles, a reduction in plasticity and swelling potential, improvement in workability and its strength. In the longer-term was cementation process in which stabilized soil developed strength as a result of pozzolanic reactions.

Figure3 indicated the increase in the value of Plastic Limit as the result of the addition of gypsum and volcanic ash. For the original soil, plastic limit was 17.42% and was increasing to the highest value of 19.52% in the variation of 2% G + 15% VA.

Figure4 indicated the decrease in the value of Plasticity Index as the result of the addition of gypsum and volcanic ash so that its value decreased. The total decrease in plasticity index was 29.40%. The lowest plasticity index was 8.59% for the mixture of 2% G + 15% VA with the curing time of 7 days.

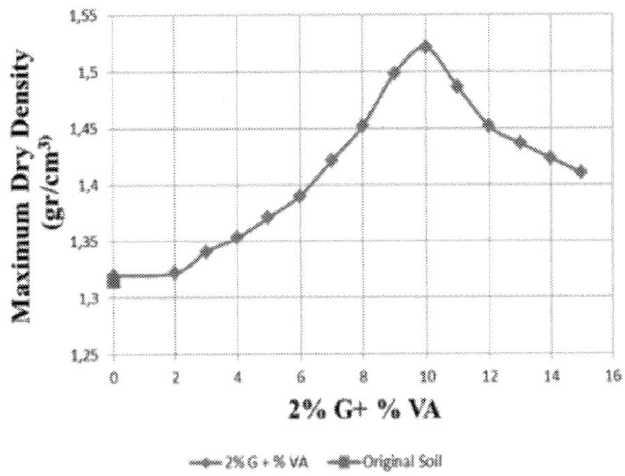

Figure 5 Correlation of γ_{dmax} soil with the variation of 2% gypsum in the addition of 0% - 15% of volcanic ash.

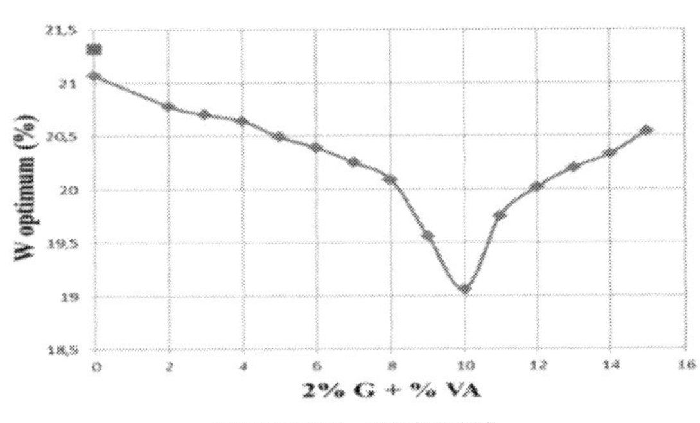

Figure 6. Correlation of w_{opt} with the variation of2% gypsum in the addition of 0% - 15% of volcanic ash.

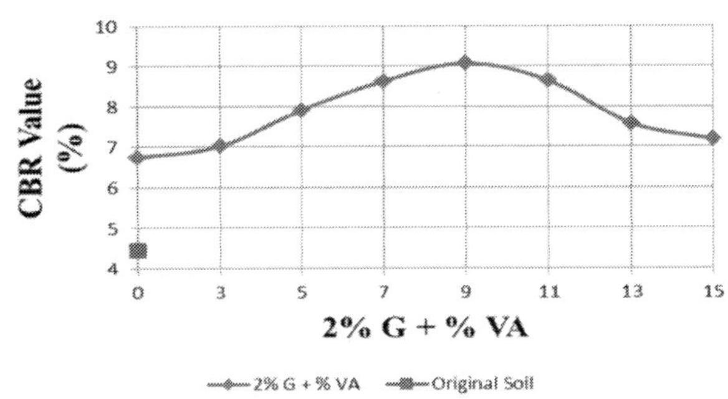

Figure 7. Correlation of soaked CBR value with the variation of2% gypsum in the addition of 0% - 15% of volcanic ash.

The optimum water content of the original soil was 21.32% and decreased until it reached minimum dry density and increased when it reached 10% of volcanic ash.

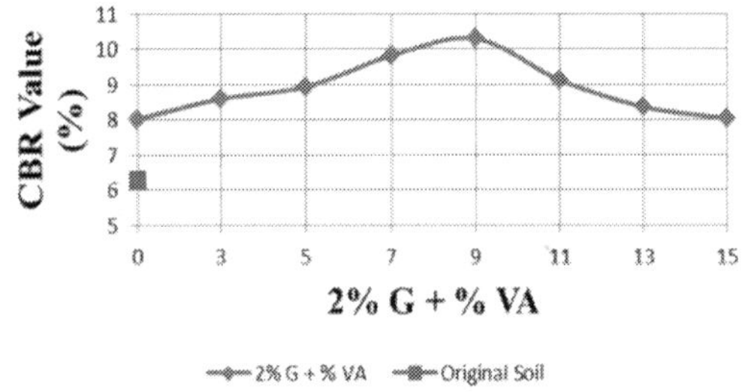

Figure 8. Correlation of unsoaked CBR value with the variation of2% gypsum in the addition of 0% - 15% of volcanic ash.

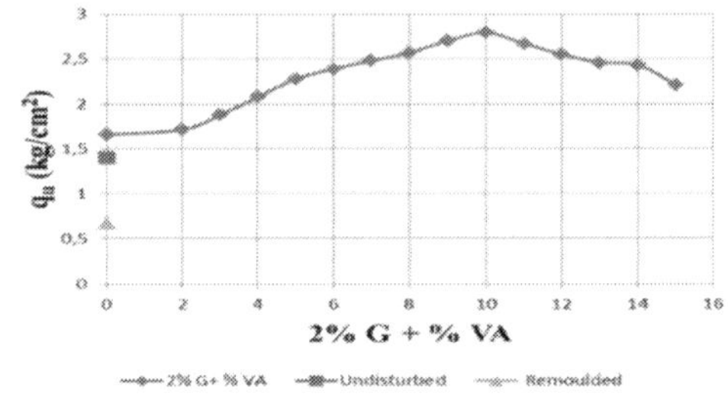

Figure 9 Correlation of UCT value with the variation of2% gypsum in the addition of 0% - 15% of volcanic ash.

Figure 7 and Figure 8 showed that the highest CBR value was in the variation of 2% G + 9% VA at 9.07% for soaked CBR and 10.29% for unsoaked CBR. Addition of volcanic ash more than 9% caused the CBR value decreased because the soil pores were filled with more mixture of volcanic ash and gypsum.

Figure 9 showed that the stabilized clay with variation of 2% G + 10% VA produced highest UCT value of2.79gr/cm^2.

4. Conclusion

From the research results, it could be concluded that:

1. Based on the classification of USCS, the soil sample was classified in the Clay – Low Plasticity (CL).
2. Based on AASHTO classification, the sample of original soil was the A-7-6 (10) type.
3. The specific gravity of original soil was 2.65. The specific gravity of the gypsum was 2.74 and the specific gravity of the volcanic ash was 2.62.
4. The original soil had Liquid Limit (LL) of 46.82% and plasticity index of 29.40%. The mixture of 2% G + 15% VA had the lowest plasticity index of 8.59% and liquid limit (LL) of 28.12%
5. The optimum moisture content of original soil was 21.32% and the maximum dry density was 1.31gr/cm^3. While the stabilized soil mixture of 2% G + 10% VA had optimum moisture content of 19.06% and the maximum dry density of1.52 gr/cm^3.
6. The CBR value of original soil was 4.44% for soaked CBR and 6.28% for unsoaked CBR respectively. The mixture of 2% G + 9% VA produced highest value of CBR of 9.07% for soaked CBR and 10.29% for unsoaked CBR respectively.
7. The value of UCT original soil was 1.40 gr/cm^2. While the stabilized soil mixture with 2% G + 10% VA had resulted the highest UCT value of 2.79 gr/cm^2.

References

[1] Hardiyatmo H C 2010 *Stabilisasi Tanah untuk Pekerasan Jalan* Yogyakarta: Gadjah Mada University Press

[2] Krebs R D and Waker R D 1971 *Highway Material* New York: McGraw-Hill Publishing Company

[3] Hausmann M R 1990 *Engineering Principles of Ground Modification* Singapura: McGraw-Hill Publishing Company

[4] Batubara M H dan Roesyanto Kajian Kuat Tekan Bebas Pada Stabilitas Tanah Lempung Dengan Gipsum dan Serbuk Kaca *Jurnal Online*
[5] Perkembangan Aktivitas Vulkanik G Sinabung 2 Agustus 2017 https://www.esdm.go.id/id/media-center/arsip-berita/perkembangan-aktivitas-vulkanik-g-sinabung-2-agustus-2017
[6] Margaretha I dan Roesyanto Stabilitas Tanah Lempung dengan Meenggunakan Semen Portland Tipe 1 dan abu Vulkanik dengan Pengujian Kuat tekan Bebas *Jurnal Online*
[7] Napitupulu M dan Roesyanto Pengaruh Campuran Semen Portland Tipe I dan Abu Gunung Vulkanik Sebagai Bahan Stabilisasi Pada Tanah Lempung dengan Uji UCT *Jurnal Online*
[8] Sormin, Prince Joy dan Roesyanto Pengujian CBR (California Bearing Ratio) Pada Stabilitas Tanah Lempung dengan Campuran Semen Portland Tipe I dan Abu Gunung Vulkanik *Jurnal Online*

IOP Conf. Series: Materials Science and Engineering **309** (2018) 012026 doi:10.1088/1757-899X/309/1/012026

Clay stabilization by using gypsum and paddy husk ash with reference to UCT and CBR value

Roesyanto*, R Iskandar, I P Hastuty and W O Dianty
Civil Engineering Department, University of Sumatera Utara, Medan, Indonesia

*Email: roesyanto@usu.ac.id

Abstract. Clays that have low shear strength need to be stabilized in order to meet the technical requirements to serve as a subgrade material. One of the usual soil stabilization methods is by adding chemicals such as Portland cement, lime, and bitumen. The clay stabilization research was done by adding gypsum and paddy husk ash. The research goals were to find out the value of engineering properties of clay due to the addition of 2% gypsum and 2% - 15% paddy husk ash. The soil was classified as Clay – Low Plasticity (CL) based on USCS and was classified as A-7-6 (10) based on AASHTO classification system. The UCT value of original soil was 1.41 kg/cm^2. While the CBR soaked and unsoaked values of original soil were 4.41% and 6.23% respectively. The research results showed the addition of paddy husk ash decreased the value of unconfined compressive strength as well as CBR. The stabilized soil by 2% gypsum and 0% paddy husk ash gave maximum UCT value of 1.67 kg/cm^2, while the maximum value of CBR were found 6.71% for CBR soaked and 8.00% for CBR unsoaked. The addition of paddy husk ash did not alter the soil classification according to AASHTO or USCS, even degrade the engineering properties of original soil.

1. Introduction

Soil stabilization is a mixture of soil with chemical materials, in order to upgrade the engineering properties of soil[1]. The usual soil stabilization is by adding chemical materials to the soil. The most common additives are Portland cement, lime, bitumen and tar [2, 3]. The clay stabilization research was done by adding powder gypsum and paddy husk ash. Gypsum is a soft white mineral consisting of hydrated calcium sulfate. The chemical formula is calcium sulfate dehydrate (CaSO4. 2(H2O)). Gypsum has better properties than organic additives because it does not cause air pollution, relatively cheap, fire resistant, and resistant to deterioration by biological factors and chemicals [4]. Paddy husk ash (PHA) used in this research was originated from Secanggang, Langkat, North Sumatera. Paddy husk ash is a by-product of agricultural products, which is considered only waste. However, burned paddy husk has pozzolan properties that have high silicate element, SiO_2 [5]. Percentage of SiO_2 is 86.90 % – 97.30 % and CaO 0.20 % – 1.50 %. This Pozzolan contains cementation properties when mixed with water [6].

2. Research Methods

The research was carried out using experimental method at Soil Mechanics Laboratory, Department of Civil Engineering, University of Sumatera Utara. The research samples were original clay and clay samples which had been mixed with 2% gypsum and paddy husk ash.

IOP Conf. Series: Materials Science and Engineering **309** (2018) 012026 doi:10.1088/1757-899X/309/1/012026

2.1. Preparatory work
Preparations conducted in this research were as follows:
1. First stage was collecting and studying the literature related to this research.
2. Determining the location of clay soil sampling. The clay soil was taken from PTPN II, Patumbak, Deli Serdang.
3. Providing the additives materials (gypsum and paddy husk ash).

2.2. Manufacture of test specimens and testing implementation
The preparation of the specimens was divided into several sorts according to each test with gypsum content and paddy husk ash for all tests.
Testing performed:
1. Test of indigenous soil properties index, ie: water content, soil type, specific gravity, Atterberg limits, and grain size distribution.
2. Test of gypsum properties and raw paddy husk ash index, ie: water content, specific gravity test, Atterberg limits, and grain size distribution.
3. Proctor Standard test to obtain the optimum moisture content and the maximum dry density of indigenous clay.
4. Proctor Standard test for UCT test and soaked and unsoaked CBR laboratory test.
5. Curing time for stabilized clay was 7 days.

After all data were gathered, they were analyzed. All results would be presented in the form of tables, graphs, and explanation.

3. Results and Discussion

3.1 Original Soil Testing
Engineering properties of original soil was shown in Table 1.

Table 1. Soil Physical Properties Test Data

No.	TEST	RESULT
1	Water Content	12.55%
2	Specific Gravity	2.66
3	Liquid Limit	45.76%
4	Plastic Limit	17.72%
5	Plasticity Index	28.05%
6	Sieve Analysis	50.34%
7	Optimum Moisture Content	20.50%
8	Maximum Dry Density	1.31 gr/cm^3
9	CBR soaked	4.41 %
10	CBR unsoaked	6.23%

3.2. Stabilization Material Testing
Index properties of paddy husk ash and gypsum were presented in Table 2 below:

IOP Conf. Series: Materials Science and Engineering **309** (2018) 012026 doi:10.1088/1757-899X/309/1/012026

Table 2. Index properties of paddy husk ash and gypsum

No	TEST	Paddy Husk Ash	Gypsum
1	Specific Gravity	2.55	2.74
2	Liquid Limit		
3	Plastic Limit	Non Plastic	Non Plastic
4	Plasticity Index		
5	Sieve Analysis	8.56%	51.62%

3.3. Engineering properties of the stabilized soil

The engineering properties of stabilized soil with gypsum and paddy husk ash were presented in the following Figures.

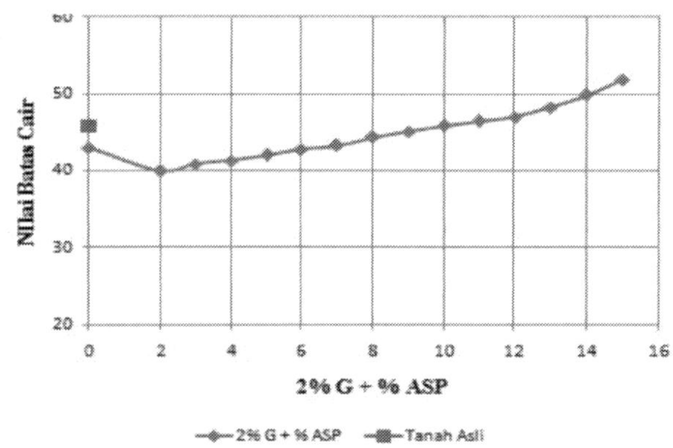

Figure 1. Correlation of Liquid Limit (LL) value with variation of 2% gypsum in the addition 0% - 15% paddy husk ash

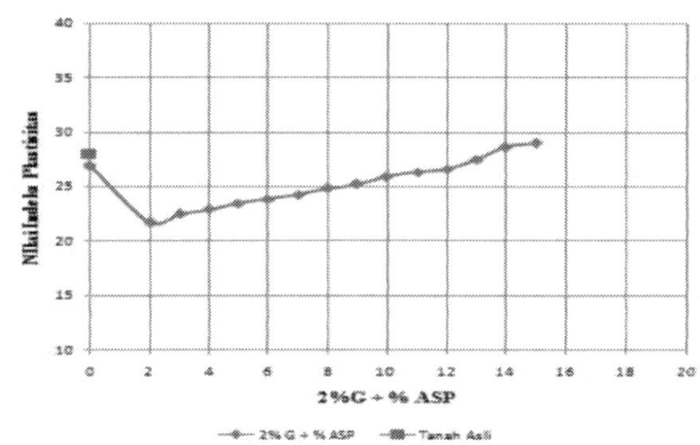

Figure 2. Correlation of Plasticity Index (PI) value with variation of 2% gypsum in the addition 0% - 15% paddy husk ash

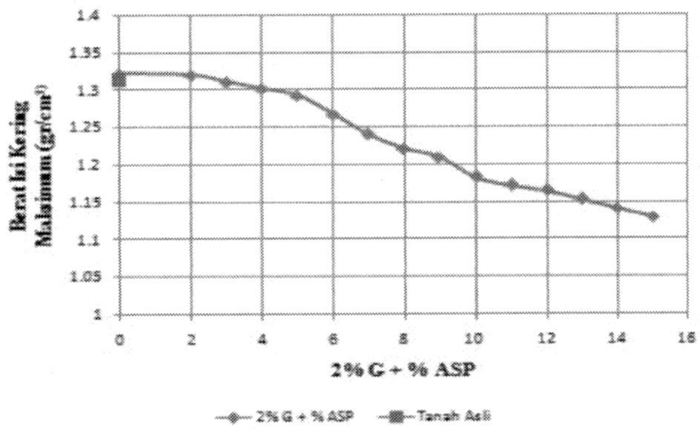

Figure 3. Correlation γd_{max} of soil with the variation of 2% gypsum in the addition 0% - 15% of paddy husk ash

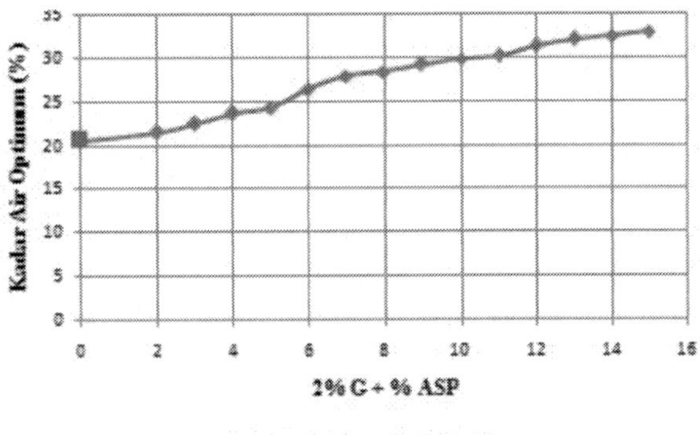

Figure 4. Correlation of w_{opt} with the variation of 2% gypsum in the addition 0% - 15% paddy husk ash

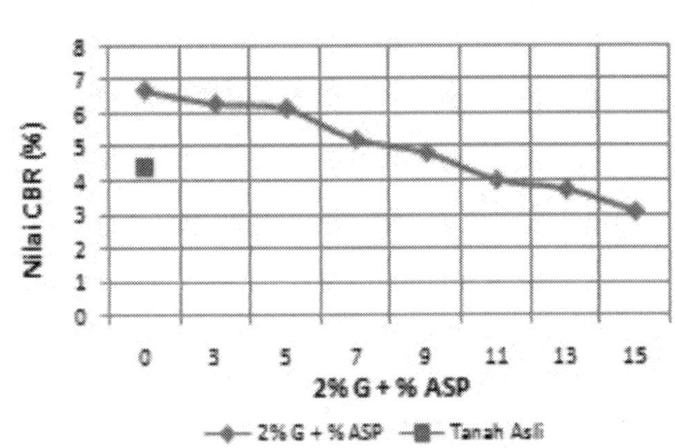

Figure 5. Correlation of soaked CBR value with the variation of 2% gypsum in the addition 0% - 15% paddy husk ash

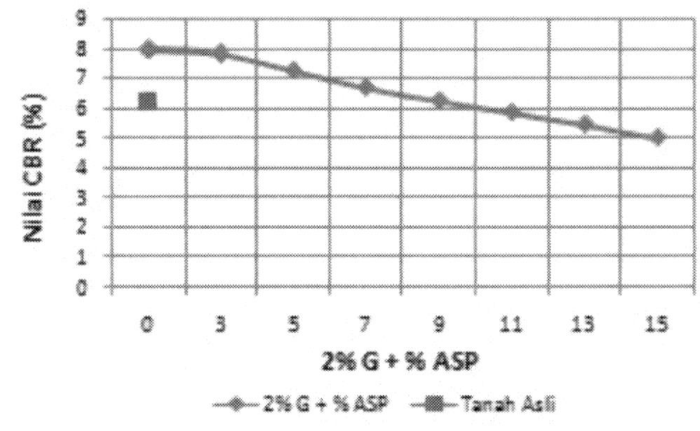

Figure 6. Correlation of unsoaked CBR value with the variation of 2% gypsum in the addition 0% - 15% paddy husk ash

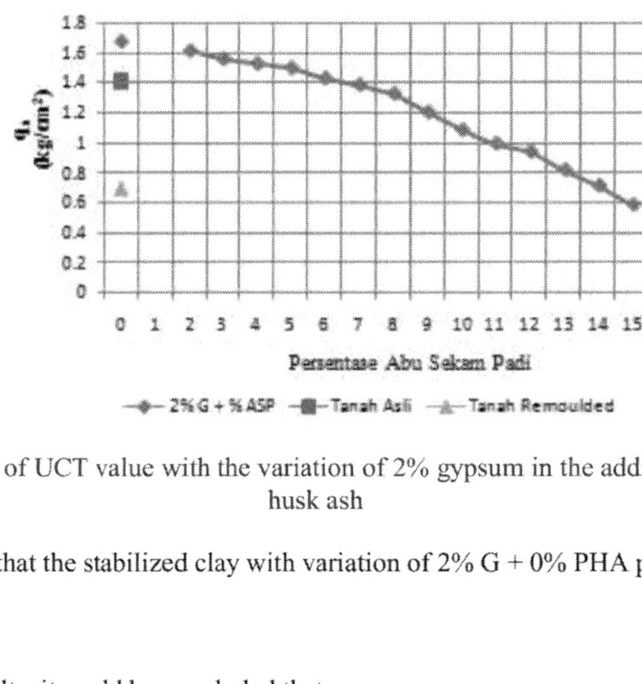

Figure 7. Correlation of UCT value with the variation of 2% gypsum in the addition 0% - 15% paddy husk ash

Figure 7 showed that the stabilized clay with variation of 2% G + 0% PHA produced highest UCT value of 1.67 gr/cm^2.

4. Conclusions

From the research results, it could be concluded that:

1. Based on USCS classification, the soil samples are included in CL (Clay - Low Plasticity).
2. Based on the AASHTO (American Association of State Highway Transportation Official) classification, the original soil samples was A-7-6 (10).
3. From the Proctor standard test, the optimum moisture content of original soil was 20.50% and the maximum dry density was 1.31 gr/cm^3. While the maximum dry density of all mixture was in the variation of 2% gypsum + 0% paddy husk ash which is 1.32 gr / cm^3 and its optimum water content is 20.32%.
4. The specific gravity of original soil was 2.66. The specific gravity of the gypsum was 2.74 and the specific gravity of paddy husk ash was 2.55.

5. The original soil had Liquid Limit (LL) of 45.76% and plasticity index of 28.05% and liquidity index (LI) of -0.18% (LI <0). The mixture of 2% G + 2% PHA had the lowest plasticity index of 22.46%. With a liquid limit value of 40.69%.
6. The CBR value of original soil was 4.41% for soaked CBR and 6.23% for unsoaked CBR respectively. The mixture of 2% G + 0% paddy husk ash produced highest value of CBR of 6.71% for soaked CBR and 8.00% for unsoaked CBR respectively.
7. The value of UCT original soil was 1.41 kg/cm^2. While the stabilized soil mixture with 2% G + 0% PHA had resulted the highest UCT value of 1.67 kg/cm^2.
8. Stabilized clay with 2% gypsum + 0% paddy husk ash, based on USCS classification was classified as CL (Clay - Plasticity) and based on AASHTO was classified as A-7-6 (9). While stabilized clay with 2% gypsum + 2% paddy husk ash was classified as A-7-6 (8). In addition, stabilized clay with 2% gypsum + 3% - 15% paddy husk ash were classified as A-7-6 (11).
9. The stabilized clay with a fixed percentage of gypsum and addition of paddy husk ash caused a decrease in shear strength of soil as shown in reduction of UCT and CBR value.

References

[1] Hardiyatmo H C 2010 *Stabilisasi Tanah untuk Pekerasan Jalan* Yogyakarta: Gadjah Mada University Press
[2] Krebs R D and Waker R D 1971 Highway Material New York: McGraw-Hill Publishing Company
[3] Hausmann M R 1990 Engineering Principles of Ground Modification Singapura: McGraw-Hill Publishing Company
[4] Batubara M H and Roesyanto Kajian Kuat Tekan Bebas Pada Stabilitas Tanah Lempung Dengan Gipsum dan Serbuk Kaca *Jurnal Online*
[5] Fadilla N 2014 *Pengujian Kuat Tekan Bebas (Unconfined Compression Test)Pada Stabilitas Tanah Lempung Dengan Campuran Semen dan Abu Sekam Padi* Program Studi Teknik Sipil Universitas Sumatera Utara Medan
[6] Houston D F 1972 Rice Chemistry and Techonology American Association of Cereal Chemist Inc, Minnesota

IOP Conf. Series: Materials Science and Engineering **309** (2018) 012027 doi:10.1088/1757-899X/309/1/012027

Application of lean six sigma to waste minimization in cigarette paper industry

K Syahputri*, R M Sari, Anizar, I R Tarigan, I Siregar
Department of Industrial Engineering, Universitas Sumatera Utara, Medan 20155, Indonesia

*Email: khalida@usu.co.id

Abstract. The cigarette paper industry is one of the industry that is always experiencing increasing demand from consumers. Consumer expectations for the products produced also increased both in terms of quality and quantity. The company continuously improves the quality of its products by trying to minimize nonconformity, waste and improve the efficiency of the whole production process of the company. In this cigarette industry, there is a disability whose value is above the company's defect tolerance that is 10% of the production amount per month. Another problem also occurs in the production time is too long due to the many activities that are not value added (non value added activities) on the production floor. To overcome this problem, it is necessary to improve the production process of cigarette paper and minimize production time by reducing non value added activities. Repairs done with Lean Six Sigma. Lean Six Sigma is a combination of Lean and Six Sigma concept with DMAIC method (Define, Measure, Analyze, Improve, Control). With this Lean approach, obtained total production time of 1479.13 minutes proposal with cycle efficiency process increased by 12.64%.

1. Introduction

North Sumatra is one of the tobacco producing areas in Indonesia. Tobacco is the main raw material in the cigarette industry. The growth of the cigarette industry is certainly in line with the growth of the cigarette paper industry, as the two are complementary. This research was conducted on one of cigarette paper manufacture industry in North Sumatera. Based on the results of initial observations in the company obtained the number of defective products produced by the company. The number of defective products produced above the company's tolerance of only 5%. Based on the defect product, it can be seen that the level of efficiency is also low and certainly a lot of waste that occurs during the production process of cigarette paper. This defective product will experience rework that will cause waste of energy, energy and time. Based on the *muda* in Japan, there are several categories of waste in the industry, namely: rework, over processing, over production, Overproduction consists of making either excessive quantity, waiting and excessive inventory [1,2] . Based on the seven types of waste above, it appears that there is a lot of waste that occurs in the cigarette paper industry. Therefore, it is necessary to take steps to minimize waste in this industry.

One method that can be used to minimize waste is by lean concept. Lean is an approach that seeks to improve flow in the value stream and eliminate waste. Lean concept has various tools to reduce or eliminate waste, one of them is six sigma. [3] The Lean Management tools focus on speed and efficiency of a process, while those of Six Sigma on its precision and accuracy [4] Lean Six Sigma is dedicated to increase quality, reduce variability and eliminate any wasta from company [5] By eliminating waste then production time and production cost can be reduced so that quality can be improved [6] The lean six sigma concept is a combination of lean concepts into six sigma methodology [7] A key component of lean concept for reduction of waste is flow and smart automation [8]

There are many case studies that have used lean six sigma, both in the service industry [9] and in the manufacturing industry. Case studies in the service industry include: in the field of HR administration [10,11] Other studies in health care delivery services industry. Lean Six Sigma projects [12], with projects executed in the following processes [13]: Meanwhile, lean six sigma has also been used in the manufacturing industry Welding wires [14] and Litchi Juice Production Plant of foods and beverage industries in Bangladesh [15]

Lean Six Sigma

Lean defined as systematic approach to identifying and eliminating non value add (wastes) through continuous im-provement, flowing the product at the pull of the customer in pursuit of perfection. The six sigma define as business process that allows companies to drastically improve their bottom line by designing and monitoring everyday business activities in ways that minimize waste and resources while increasing customer satisfaction by some of its proponents [16]The integration between six sigma and lean manufacturing, there is an ongoing trend of integrating Lean and Six Sigma by adding Six Sigma projects to a Lean initiative. The integration of lean and Six Sigma, The phrase "lean Six Sigma" is used to describe the integration of lean and Six Sig-ma philosophies [17]. Lean and Six Sigma are key business process strategies which may be employed by companies to enhance their manufacturing performance [18]. Integrated Lean Six Sigma (LSS) model for manufacturing industry is illustrated in Figure 1

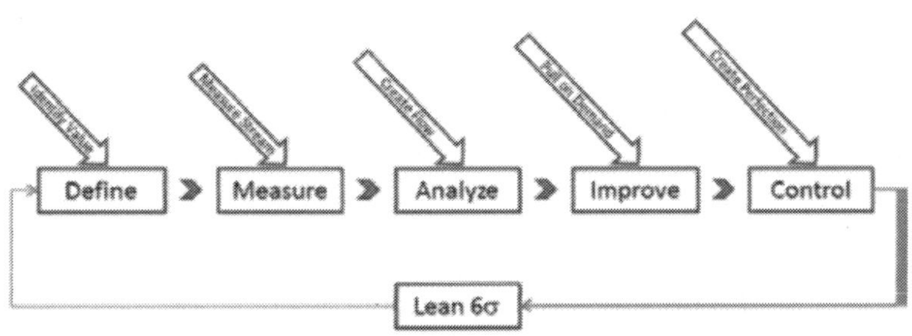

Figure 1. Integrated Lean Six Sigma (LSS) model

The implementation of Six Sigma is always done using DMAIC approach In some of the above mentioned references, the five letters abbreviations are simply explained as follows;D : Define, **M**: Measure, **A**: Analysis, **I**: Improve, and **C**: Control [19,20,21].

Meanwhile, it describes the flow during the production prose for the actual condition and after the improvement is by using value stream mapping. Value stream mapping is a method of presentation of material and information flow. It allows identification of all activities from the moment an order has been made by a customer until it is delivered. It also presents both activities which add value and the ones that do not. Activities which do not add any value and are not indispensable for company to operate on the market must be eliminated [22].

2. Methodology

This study was conducted on the cigarette paper industry in North Sumatra. This study was carried out by following the six sigma methodology (Define, Measure, Analysis, Improvement, and Control). The first step is to define the process of identifying the flow of cigarette paper production. The define stage is done to study the whole process of making cigarette paper. This stage produces a picture of current state value stream mapping. Current state value stream mapping is used to describe the actual

IOP Conf. Series: Materials Science and Engineering **309** (2018) 012027 doi:10.1088/1757-899X/309/1/012027

condition of observation objects based on value added, non value added, and non value added added activities. Classification of company activities provides information on waste / waste that occurs along the flow of production.

Measure stage is done by calculating Manufacturing Lead Time, Process Cycle Efficiency, Process Lead Time, and DPMO, so as to obtain the level of sigma company. The six sigma considers the continuous improvement tool and as the continuous improvement process for reducing variation in process which means the defected products or defected service, which focuses on continuous and breakthrough improvements. Improvement projects are driven in the range of areas and at different levels of complexity, in order to reduce variation. The main purpose of reducing variation on a product or a service is to satisfy customers.

Stages Analysis is done by analyzing the cause of each waste with root cause analysis. This stage is very important to find the process is not appropriate and the cause. In the analysis phase, an evaluation of the current sigma values, production process analysis, information flow, and current problem-causing analysis are presented.

Improvement stage is performed to determine the improvement that will minimize the waste that has been identified. Corrective action is taken to improve the current state. After the selection of alternative improvements, then re-calculation related activities value added, non value added, necessary non value added, Manufacturing Lead Time, Process Cycle Efficiency, Process Lead Time, and DPMO. The goal is to know the positive impact for the company. Furthermore, the Future State Value Stream Mapping.

The last phase is the control phase. This phase is a phase to provide an alternative mechanism of control over the course of alternative improvements. The control phase is the phase to monitor whether the selected repair alternatives are actually applicable.

3. Results and Discussion

In the first phase or phase define, the number of activities that occur during the process of making cigarette paper is sebanyaak 48 working processes with a total processing time of 1645.05 minutes. Out of 48 work processes there are 10 non value added activities and 38 value added activities. At this define stage also obtained the number of defective products as much as 702.13 tons during the study took place.The second stage is the measure phase, as for the steps taken is to do:

a. Manufacturing Lead Time Calculation

   Manufacturing lead time is the time required to perform the production process from start to finish based on standard time. The calculation of manufacturing lead time is done by summing up the entire working process where in making cigarette paper there are 48 working processes with total processing time of 1645.05 minutes.

b. Calculation of Process Cycle Efficiency

   In this calculation, a separation of value added activities or work activities from non-value added activities or work processes takes place. With the separation of these activities, it was found that the time for value added activity was 1119.75 minutes with 20 work processes while non value added activities were 525.30 minutes with 28 work processes. Non value added activities are caused by transportation activities, waiting times, and inspections. Based on the above data, then obtained process cycle efficiency is as follows:

   Process Cycle Efficiency : 68.04%

c. Calculation of Process Lead Time

   Process lead time is a metric used to find out how long it takes to process a number of items from start to finish. Calculation process lead time is as follows:

   Process Lead Time : 26 days

Furthermore, the quality of the product is done by determining the critical to quality and calculating the level of six sigma and DPMO.

1. Critical to quality

IOP Conf. Series: Materials Science and Engineering **309** (2018) 012027 doi:10.1088/1757-899X/309/1/012027

Critical to quality is a quality characteristic that affects the moment in the production process and when used by the customer. Characteristics of quality that affect customers are perforated cigarette paper and cigarette lined paper.

2. Calculation of sigma level
Sigma level calculations are performed to unify the quality measurements occurring during the inspection phase which will improve the outcome of the worst process of the inspection phase. Calculation of sigma level is done by following steps.
Total production amount = 5857.44 tons
Total defective product = 702.13 tons
Disability rate (defect per unit / DPU)

DPU : 0.119869
Defect oppurtunities : 1
Defect per Million Oppurtunities (DPMO)

DPMO : 119869

Calculation of sigma level is done by using Microsoft excel with formula: = 2,676

From the calculation obtained sigma value of 2.676. This shows that the result is still far from the sigma value to be achieved that is 6 sigma.

The next step is analysis, which aims to increase the sigma value obtained. At this stage it is necessary to identify and analyze the cause of the process that generates waste. To find out the root cause of the problem then used 5 why analysis.

The next stage is improvement, at this stage is done elimination activities that are not value-added. The eliminated activities consist of 8 activities, bringing the total to 40 activities. Activities that are eliminated include:

- Transfer of raw materials to the destruction station is eliminated by placing the raw material near the destruction station.
- Activities waiting to be transported are eliminated by adding a forklift to operate.
- Delay activity for wrapping is eliminated by arranging operator placement while working on packing station.

After 8 activities were eliminated, the time for non value added activities was 359.38 minutes with 20 working processes and total processing time of cigarette paper was 1479.13 minutes. Based on these values, the value of process cycle efficiency can be obtained as follows:

Based on the above calculation, it is seen that process cycle efficiency has increased by 12.64% after elimination of some activities. This happens due to non value added activities that are reduced due to elimination so that the total lead time decreases while. Recapitulation of the details of actual lean metric calculations and proposals can be seen in Table 1.

Table 1. Recapitulation of Actual and Proposed Leaf Metrics Calculations

Num	Metric	Actual	Proposed
1	Number of activities in production activities	48	40
2	Total production time	1645.05 minutes	1479.13 minutes
3	*Non value added time*	525.30 minutes	359.38 minutes
4	*Value added time*	1119.75 minutes	1119.75 minutes

The last stage is control. This stage is done by monitoring the result of the proposed improvement by using lean six sigma concept. This stage is needed to assess whether the proposed improvements can actually be applied and can meet the objective of reducing waste that occurs during the production process of cigarette paper.

4. Conclusion

From the current state mapping results obtained 48 activities consisting of 10 non-value added activities and 38 activities of value added. Then after applying the concept of Lean Six Sigma activity can be reduced to 40 activities during the production process. Total production time decreased as much as 165.92 minutes and cycle efficiency process increased by 12.64%.

References

[1] Neha S, Neha S, Singh M G, Simran K, Pramod G 2013 *International Journal of Scientific Research and Reviews* IJSRR **2(1)** 54-63
[2] Melton T 2005 The Benefits of Lean Manufacturing What Lean Thinking has to Offer the Process Industries, Chemical Engineering Research and Design **83(A6)** 662-673
[3] Mousa, Ahmed and Lean 2013 *International Journal of Scientific and Engineering Research* **4(5)**
[4] Laureani A, Antony J 2009 *International Journal of Productivity and Performance Management* **59(8)**
[5] Furterer S, Elshennawy A 2005 *Implementation of TQM and Lean Six Sigma Tools in Local Government: a framework and a case study Total Quality Management* **16(10)**
[6] Apreutesei M, Sucio E, Arvinte I R 2010 *Lean Manufacturing – A Powerfull Tool for Reducing Waste During The Processes* Analele Universita ANUL XVII, NR,2, ISSN 145 3 -7397
[7] Ramaswamy R 2007 *Integrating Lean and Six Sigma Methodologies for Busineess Excellence* ORIEL
[8] Munteanu D, Iteanu C O 2007 Lean Manufacturing – A Success Key Inside of A Industry Company, International Conference of Economic Engineering and Manufacturing Systems *Brasov RECENT* **8**
[9] Laureani A 2012 *Lean Six Sigma in the Service Industry, Advanced Topics in Applied Operations Management* Mr. Yair Holtzman (Ed) ISBN: 978-953-51-0345-5 InTech Available from: http://www.intechopen.com/books/advanced-topics-in-applied-operations-management/lean-six-sigma-in-theservice-industry
[10] Lee Y, Chen L and Chen S 2008 *International Journal of Operations and Quantitative Management* **14(2)** 117-128
[11] Gupta P 2005 *Six Sigma in HR, Quality Digest* QCI International
[12] Antony J, Antony F and Taner T 2006 *The secret of success. Public Service Review: Trade and Industry* **10** 12-14
[13] Taner M T, Sezen B and Antony J 2007 *International Journal of Health Care Quality Assurance* **20(4)** 329-340
[14] Hassan M K 2013 *American Journal of Industrial Engineering* **1(2)** 28-35
[15] Hossain M M, Uddin M K 2015 *International Journal of Scientific & Engineering Research* **6(9)**
[16] Andersson R, Eriksson H, Torstensson H 2006 Similarities and differences between TQM, six sigma and lean *The TQM Magazine* **18(3)** 282 – 296
[17] Pepper M P J, Spedding T A 2010 *International Journal of Scientific & Engineering Research,* **4(5)** May-2013 1153 ISSN 2229-5518 IJSER © 2013 **27(2)** 138 – 155
[18] Thomas A, Barton R and Okafor C C 2009 *Journal of Manufacturing Technology Management* **20(1)** 113-129
[19] Kaushik P, Khanduja D 2008 *Journal of Scientific and Industrial Research* **67(1)** 36-42

[20] Thomas A, Barton R and Byard P 2008 *Journal of Quality in Maintenance Engineering* **14(3)** 262-271
[21] Kumar S and Sosnoski M 2009 *International Journal of Productivity and Performance Management* **58 (3)** 254-273
[22] Maleszka A, Linke M 2016 *Polish Journal Of Natural Sciences* **31(1)** 101–112

IOP Conf. Series: Materials Science and Engineering **309** (2018) 012028 doi:10.1088/1757-899X/309/1/012028

The effect of carbon black (CB) loading on curing characteristics and mechanical properties of virgin acrylonitrile butadiene rubber (Nbrv)/recycled acrylonitrile butadiene rubber (Nbrr) blends

M A Husnan*, H Ismail and R K Shuib
School of Materials and Mineral Resources Engineering, Universiti Sains Malaysia, 14300 Nibong Tebal Pulau Pinang, Malaysia.

*E-mail: anasbhusnan@gmail.com

Abstract. Recently, the interest of polymer industry researchers have grown rapidly on the use of specific techniques which can reduce cost and utilize rubber waste into the processing form. The increasing of cognizance in environmental matters and the desire to sustain the resources had fortified the practice of recycling waste materials. In this work, the effect of carbon black loading on curing characteristics and mechanical properties of virgin acrylonitrile butadiene rubber/recycled acrylonitrile butadiene rubber (NBRv/NBRr) blends were studied. Cure time (t_{90}), scorch time (t_{S2}) and swelling percentage decreased but minimum torque (M_L) and maximum torque (M_H) increased with increasing carbon black (CB) loading in the blends. Increasing CB loading also increased tensile strength, tensile modulus (M100), hardness and compression set but decreased elongation at break (E_b) of NBRv/NBRr blends.

1. Introduction

Currently, Malaysia is among the world biggest supplier and manufacturer of natural rubber and nitrile gloves. The properties of the NBR gloves are vary included an excellent resistance to punctures, tears and chemicals. Due to environment concerns, a lot of research to gives value to the recycled NBR gloves were done [1-3].

The NBR is well known to its excellent oil resistance, superior strength, abrasion, water and heat. Eventhough NBR did not shows any self-reinforcing effect which means no crystallinity but the vulcanizates with excellent mechanical properties can be achieved by using reinforcing filler in the rubber blend [4-6].

The reinforcement of rubber by rigid entities is one of the most important phenomena in material science where carbon black has been found to offer substantial improvement in the mechanical properties of rubber because of it has very good compatibility with rubbers [7, 8]. In this work, carbon black is used as the only reinforcing filler with different loading ranging from 30 up to 70 phr.

One of the most valuable technique to prepare materials with greater properties is blending method, where two or more types of materials will be blended. This method can gives added value of properties to the vulcinates which are lacking in the component of rubber. The best combination of the rubber compounding can be obtained through elastomer blends technique where it is widely used the rubber industry in order to improve physical properties, processability and reduce cost [9-11]. The virgin acrylonitrile butadiene rubber (NBRv) and recycled acrylonitrile butadiene rubber (NBRr) were blended together with an intention to combine the best properties from each rubber component in order to improve the rubber blend properties.

In polymer compound area there are a lot studies had been conducted relate to recycled acrylonitrile butadiene rubber (NBRr) [5, 12]. However, no reports have been made so far on the use of carbon black (CB) filled recycled acrylonitrile butadiene rubber (NBRr) blend with virgin acrylonitrile butadiene matrix (NBRv). This study aimed to investigate the effect of carbon black loading on curing characteristics and mechanical properties of virgin acrylonitrile butadiene rubber/recycled acrylonitrile butadiene rubber (NBRv/NBRr) blends.

2. Materials and Methods

2.1 Materials

The recycled NBR glove was purchased from Juara One Resources (M) Sdn. Bhd., Bukit Mertajam, Penang, Malaysia. Some of ingredients of the NBRr glove obtained from manufacturer are listed as follows: stabilizer 25 % (0.5 phr), colorant 43 % (0.4 phr), accelerator 30 % (0.2 phr), heat sensitive agent 11 % (0.2 phr) and thickener 2 % (0.2 phr). NBR Nipol, DN3335 with ACN content of 33 % is a medium high nitrile rubber with Mooney viscosity [ML (1 + 4) at 100 °C] of 35 Mooney unit was used as a virgin NBR. Carbon black (N330) was supplied by Malayan Carbon (M) Ltd. The compounding ingredients such as stearic acid,, zinc oxide (ZnO), 2,2-methylenebis-(4-methyl-6-tert-butylphenol) (BKF), N-cyclohexyl-2-benzothiazolesulfenamide (CBS), tetramethylthiuram disulfide (TMTD) and sulfur were supplied by Bayer (M) Ltd (Petaling Jaya, Malaysia).

2.2 Preparation of blends

The NBRv/NBRr blends were prepared at fixed recycled NBR content which is 15 phr with different carbon black loading ratio. Table 1 shows rubber blend formulations. In compliance with ASTM method D3184-80, a laboratory-sized two roll mills (160mm x 320 mm) was used to prepare rubber blends with five diverse compositions as shown in table 1. Monsanto Moving Die Rheometer (MDR 2000) was used in order to study the curing characteristics of the rubber blends. Approximately 4g of each compounds was tested where the vulcanization temperature was set at 160 °C and the informations obtained from the test were cure time (t_{90}), scorch time (t_{s2}), maximum torque (M_H) and minimum torque (M_L). At 160°C, the rubber compounds were compressed by using a hot press conferring to each rubber blend cure time, t_{90}.

2.3 Characterization and Testing of blends

Conferring to ASTM D3182, the molded sheets were cut into five samples dumb-bell shaped for each compound. Universal tester tensile machine (Instron 3366) were used to perform tensile test at a cross-head speed of 500 mm/min conferring to ASTM D412-06ae2. Wallace dead load was used to measure the hardness of the samples ranging from 30 to 84 IRHD (International Rubber Hardness Degree) and the hardness test was done according to ASTM D2240-05 (2010).

By conferring to ASTM D471, the swelling test in motor oil was conducted in order to measure the crosslink density of the samples. From the molded sheets, the test pieces with dimensions of 30 mm × 5 mm × 2 mm were prepared. The initial weight of the test samples was recorded before the test. Prior to equilibrium, it takes 48 h at room temperature (25°C) after the test pieces immersed and conditioned in motor oil. The samples were weighed again prior to the conditioned period. The motor oil was removed from the sample surfaces prior to removing the samples from the liquid, and the weight was determined. The swelling percentage (Q) is calculated as follows:

$$Q = \left(\frac{w_s - w_{ds}}{w_{ds}}\right) * 100 \tag{1}$$

where w_s is the weight of the swollen specimen and w_{ds} is the weight of the initial specimen.

Conferring to compression set test ASTM D395, the samples undergo vulcanization by compression molding method where the standard of the test specimens is cylindrical shape of 25 ± 0.1 mm diameter and 12 ± 0.5 mm thickness. The compression device has spacers on each side in order to allow sufficient clearance of bulging when the rubber were compressed and the test samples was placed between the plates of the device. In order for the plates to drawn together uniformly until in contact with the spacers, the bolts were tightened. The percentage of the compression laboring was 25% of the original thickness. For 22 h at 70 °C in oven, the assembled compression device was placed. As for heating period the completed assembly were remain in dry air circulated oven, the device had taken out from the oven and the test specimen removed immediately and allowed to cool for 30 min, using an electronic digital caliber (0.01 mm) accuracy the final thickness was measured prior to rest time. The compression set can be expressed as:

$$C\% = \frac{t_0 - t_1}{t_0 - t_s} * 100 \quad (2)$$

Where t_0 = the original thickness of the sample; t_1 = the thickness of the sample after removed from the clamp; and ts = thickness of the spacer bar used.

Table 1. Formulation of NBRv/NBRr blends with different CB loading

Ingredients	Blends (phr[a])				
NBRv	85	85	85	85	85
NBRr	15	15	15	15	15
Sulfur	1.5	1.5	1.5	1.5	1.5
Zinc Oxide	4.5	4.5	4.5	4.5	4.5
Stearic Acid	2	2	2	2	2
BKF	1	1	1	1	1
CBS	1	1	1	1	1
TMTD	1	1	1	1	1
CB (N330)	30	40	50	60	70

[a] per hundred resin

3. Results and Discussion

3.1 Curing characteristics

The scorch time (t_{s2}), optimum cure time (t_{90}), minimum torque (M_L) and maximum torque (M_H) of NBRv/NBRr blends were studied. Scorch time was known as induction time and it is the time at which crosslink starts to form in the blends. From Figure 1 and 2, the scorch time and cure time decreased with increasing carbon black content. This can be explained by the cross-link and relics additives in the recycled rubber and the presence of curative agent such as sulfur in recycled NBRr that had crosslinked with carbon black.

The effect of the different loading of carbon black that filled NBRv/NBRr blend on minimum torque (M_L) and maximum torque (M_H) is shown in Figure 3 and Figure 4. It has shown that the maximum torque (M_H) and minimum torque (M_L) of NBRv/NBRr blends increase with increasing carbon black loading. This observation was probably due to the cross-linked of NBRv/NBRr rubber blend with carbon black and other additives [13] that caused the flow resistance in the rubber blend increases.

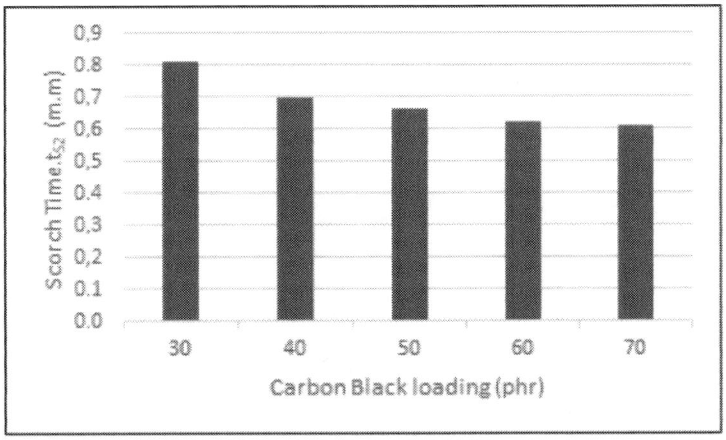

Figure 1. The effect of carbon black loading on the scorch time of carbon black filled NBRv/NBRr blends

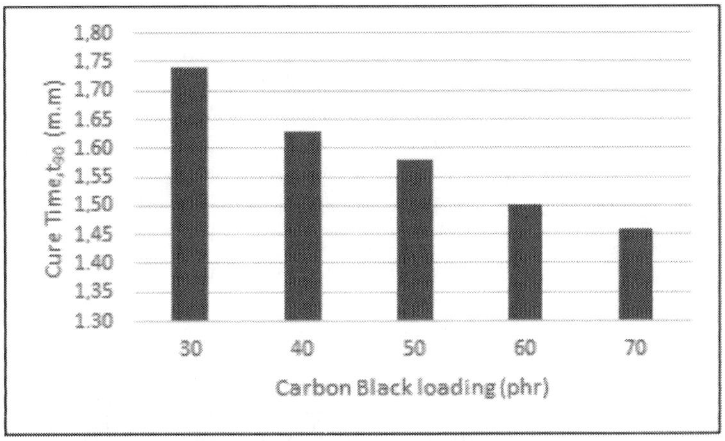

Figure 2. The effect of carbon black loading on the cure time of carbon black filled NBRv/NBRr blends

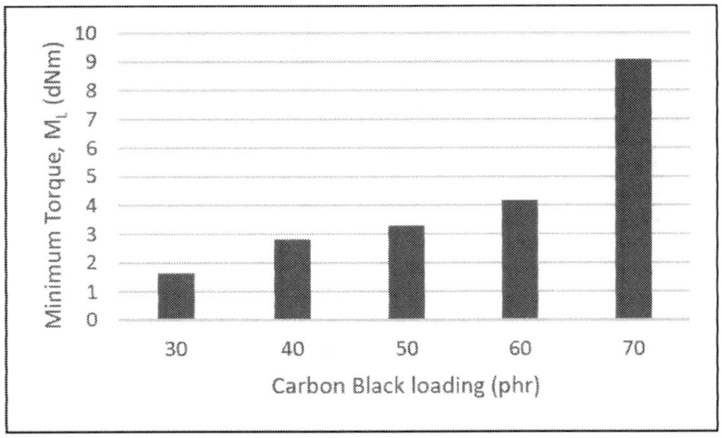

Figure 3. The effect of carbon black loading on the minimum torque of carbon black filled NBRv/NBRr blends

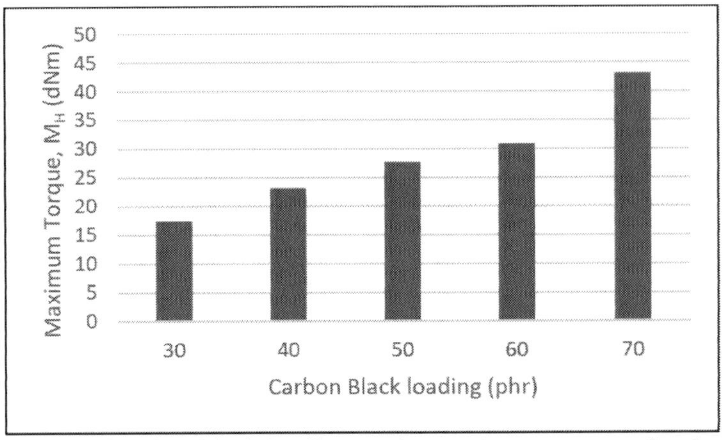

Figure 4. The effect of carbon black loading on the maximum torque of carbon black filled NBRv/NBRr blends

3.2 Mechanical properties

The addition of fillers to polymeric materials leads to improvement in the mechanical properties of the polymer matrix. The reinforcement effect is directly related to the properties of the interphase and depends on the nature of the specific interactions between polymer and reinforcing fillers [14].

The effect of blend ratio on the tensile strength of the carbon black filled NBRv/NBRr blends is shown in Figure 5. It can be seen that with increasing carbon black loading the tensile strength of the rubber blends increases. The reinforcement of carbon black with the NBRv/NBRr blends has caused the increasing of tensile strength. When more carbon black were added into the blends, it increases the crosslinking density of the blends thus increases the tensile strength of the NBRv/NBRr blends [15].

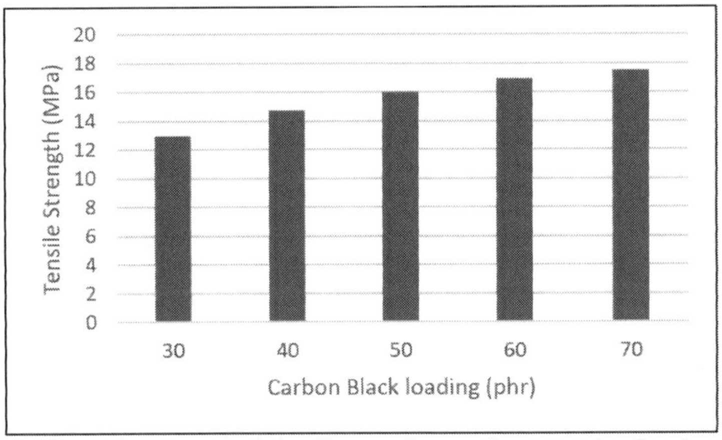

Figure 5. The effect of carbon black loading on the tensile strength of carbon black filled NBRv/NBRr blends

The effects of carbon black filled NBRv/NBRr rubber blends on elongation at break (Eb), M100 (stress at 100 % elongation), hardness and swelling percentage are shown in Figure 6-9. The results indicated that elongation at break decreasing with increasing content of carbon black while M100 and hardness were decreasing. The trend shows that M100 slightly increase with increasing carbon black content. While the increment in crosslink density of the rubber blend caused the increased in modulus and hardness. When more carbon black were incorporated into the NBRv/NBRr blend, the flexibility and elasticity of the rubber chain reduced which caused an increase in hardness and more rigid rubber vulcanizates. The maximum swelling percentage decreases (Figure 6) with increasing of carbon black content. The decreasing in swelling percentage might due to lessen penetration of motor oil into NBRv/NBRr blend. In practice, the density of crosslinking in rubber compound increases drastically with increasing CB content, resulting a rise of network elasticity contributions. These crosslinks restrict extensibility of the rubber chains induced by swelling and make it more difficult for motor oil to diffuse into the gaps between rubber molecules thus decrease the swelling percentage [16-18].

Figure 7 shows the decreasing of elongation at break of the NBRv/NBRr blends with increasing carbon black content. The decreasing trend was due to the presence of the carbon black in the compound which restrict the mobility of NBRv/NBRr blend that cause the sample to fail at lower elongation.

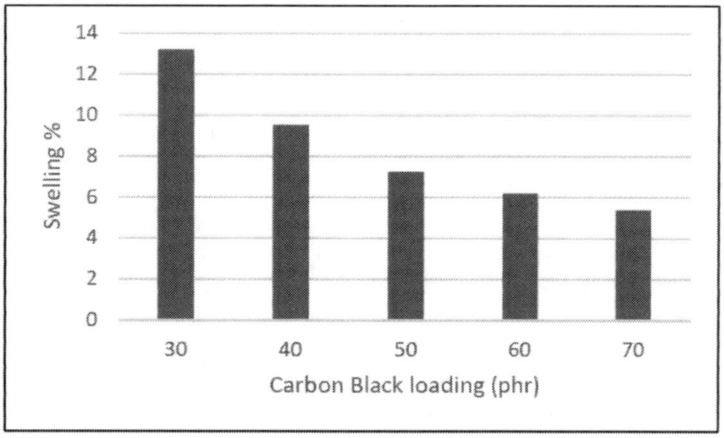

Figure 6. The effect of carbon black loading on the maximum swelling percentage of carbon black filled NBRv/NBRr blends

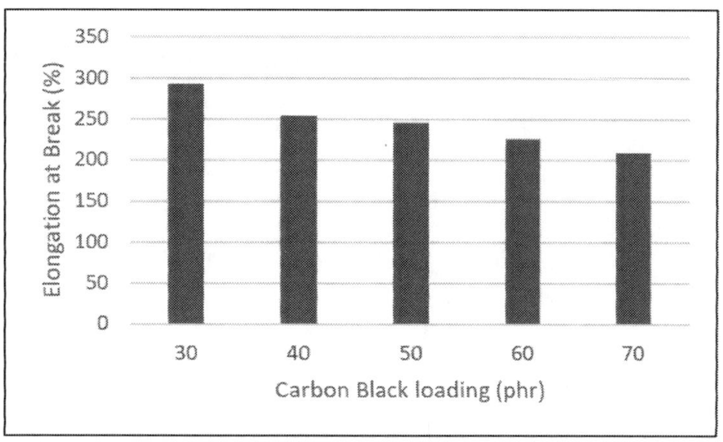

Figure 7. The effect of carbon black loading on the elongation at break of carbon black filled NBRv/NBRr blends

Figure 10 shows the variation of the percentage of compression set (C%) against NBRv/NBRr blend ratio. With increasing the carbon black loading, the compression set of NBRv/NBRr compounds increases. As carbon black loading increases, the mobility of the rubber chains decreased due to the increasing of the crosslink density in NBRv/NBRr blend, subsequently, induce stiffness in the filled compounds [19].

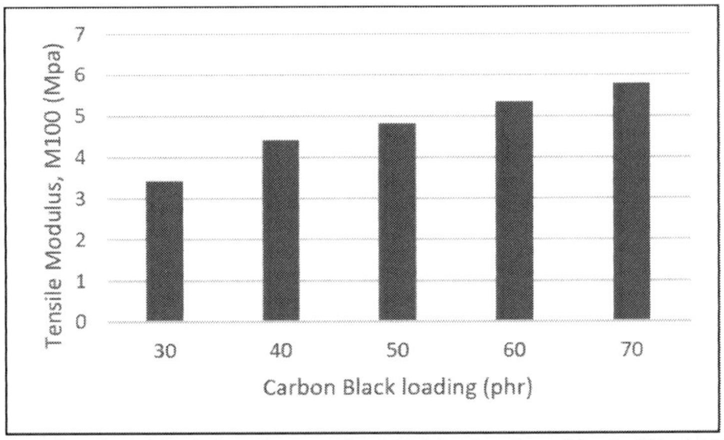

Figure 8. The effect of carbon black loading on the M100 of carbon black filled NBRv/NBRr blends

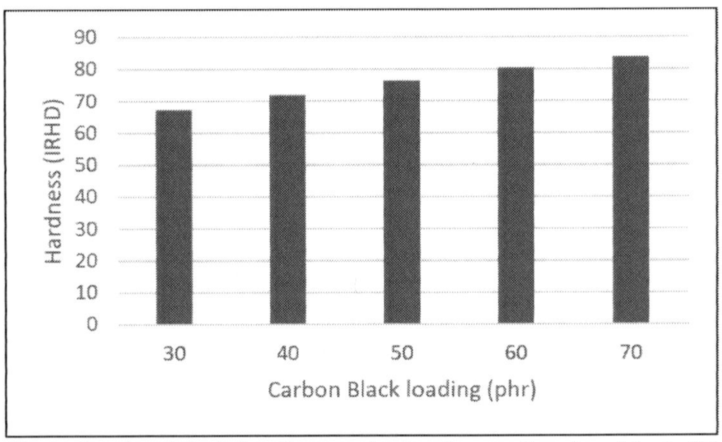

Figure 9: The effect of carbon black loading on the hardness of carbon black filled NBRv/NBRr blends.

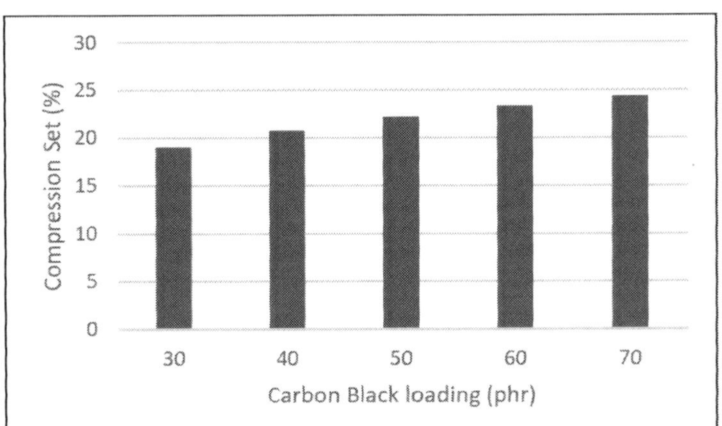

Figure 10. The effect of carbon black loading on the compression set of carbon black filled NBRv/NBRr blends

IOP Conf. Series: Materials Science and Engineering **309** (2018) 012028 doi:10.1088/1757-899X/309/1/012028

4. Conclusions

The following conclusions were derived from the experimental results:

1. The scorch time, cure time and swelling percentage decreased with increasing carbon black loading in the NBRv/NBRr blends.
2. The ML and MH increased with increasing carbon black loading in the NBRv/NBRr blends.
3. The elongation at break decreased meanwhile tensile strength, M100 and hardness increased with increasing carbon black loading content.
4. The compression set of the NBRv/NBRr blends increased as carbon black loading content increasing. The higher compression set value is due to higher crosslinking density of the NBRv/NBRr blends.

References

[1] Noriman N, Ismail H, Rashid A 2010 Characterization of styrene butadiene rubber/recycled acrylonitrile–butadiene rubber (SBR/NBRr) blends: The effects of epoxidized natural rubber (ENR-50) as a compatibilizer *Polymer Testing*, **29**: p. 200-208

[2] Zulkepli N, Ismail H 2012 A study of FTIR, thermal properties and natural weathering test on NBR virgin/recycled with SBR blends *Polymer Plastic Technology Engineering* 2012. **51**: p. 350-357

[3] Zulkepli N, Ismail H, Rashid A 2009 Effects of different particle sizes of recycled acrylonitrile–butadiene rubber and its blend ratios on mechanical and morphological properties and curing characteristics of SBR/NBRr blends *Iran Polym J* **18**: p. 139-148

[4] El-Nemr K 2011 Effect of different curing systems on the mechanical and physico-chemical properties of acrylonitrile butadiene rubber vulcanizates *Mater. Des.* **32**: p. 3361-3369

[5] Ismail H, Galpaya D, Ahmad Z 2009 Comparison of properties of polypropylene (PP)/virgin acrylonitrile butadiene rubber (NBRv) and polypropylene (PP)/recycled acrylonitrile butadiene rubber (NBRr) blends, *Polymer Plastic Technology Engineering* **48**: p. 440-445

[6] Ismail H, Tan S, Poh B 2001 Curing and mechanical properties of nitrile and natural rubber blends *J. Elastomers. Plast.* **33**: p. 251-262

[7] Pandey K N, Setua D K, Mathur G N 2003 Material behaviour *Polymer Testing* **22**(3): p. 353-359

[8] Zhang A, Wang L, Zhou Y 2003 A study on rheological properties of carbon black extended powdered SBR using a torque rheometer *Polymer Testing* **22**(2): p. 133-141

[9] Ismail H, Hairunezam H 2001 The effect of a compatibilizer on curing characteristics, mechanical properties and oil resistance of styrene butadiene rubber/epoxidized natural rubber blends *Europe Polymer Journal* **37**: p. 39-44

[10] Ismail H, Leong H 2001 Curing characteristics and mechanical properties of natural rubber/chloroprene rubber and epoxidized natural rubber/chloroprene rubber blends *Polymer Testing* **20**: p. 509-516

[11] Ismail H, et al. 2003 Effect of filler loading on cure time and swelling behaviour of SMR L/ENR 25 and SMR L/SBR blends *Polymer International* **52**: p. 685-691

[12] Noriman, N., H. Ismail, and A. Rashid, *Curing characteristics, mechanical and morphological properties of styrene butadiene rubber/virgin acrylonitrile–butadiene rubber (SBR/vNBR) and styrene butadiene rubber/recycled acrylonitrile–butadiene rubber (SBR/rNBR) blends.* . Polymer Plastic Technology Engineering, 2008. **47**(1016-1023).

[13] Salleh, S., H. Ismail, and Z. Ahmad, *Properties of natural rubber latex-compatibilized natural rubber/recycled chloroprene rubber blends.* Journal of Elastomers & Plastics, 2016. **48**(7): p. 640-655.

[14] Arrighi, V., et al., *Quasielastic neutron scattering as a probe of molecular motion in polymer-filler systems.* E-MRS spring meeting, 2002. **Strasbourg**(France): p. N-15.

[15] Samaržija-Jovanović, S., et al., *Nanocomposites based on silica-reinforced ethylene–propylene–diene–monomer/acrylonitrile–butadiene rubber blends.* Composites Part B: Engineering, 2011. **42**(5): p. 1244-1250.

[16] Kader, M.A. and A.K. Bhowmick, *Thermal ageing, degradation and swelling of acrylate rubber, fluororubber and their blends containing polyfunctional acrylates.* Polymer Degradation and Stability, 2003. **79**(2): p. 283-295.
[17] Alam, M.M., M.F. Mina, and F. Akhtar, *Swelling and Hydration Properties of Acrylamide Hydrogel in Distilled Water.* Polymer-Plastics Technology and Engineering, 2003. **42**(4): p. 533-542.
[18] Gwaily, S.E., et al., *Influence of thermal aging on crosslinking density of boron carbide/natural rubber composites.* Polymer Testing, 2003. **22**(1): p. 3-7.
[19] Schuur, M. and R. Gaymans, *Influence of chemical crosslinks on the elastic behavior of segmented block copolymers.* Polymer, 2005. **46**: p. 6862-8.

IOP Conf. Series: Materials Science and Engineering **309** (2018) 012029 doi:10.1088/1757-899X/309/1/012029

Synthesized of PEG-6000 coated $MgFe_2O_4$ nanoparticles based on natural iron sand by co-precipitation method

E A Setiadi[1,*], S Simbolon[1,3], A S P Saputra[2], E Marlianto[2], Djuhana[3], C Kurniawan[1], M Yunus[1], and P Sebayang[1,3]

[1]Research Center for Physics, Indonesian Institute of Sciences (LIPI), Puspiptek Office Area, Tangerang Selatan, Banten, 15314-Indonesia

[2]Department of Physics, University of North Sumatera, Jalan Bioteknologi No.1, Medan, Indonesia

[3]Department of Mechanical Engineering, Universitas Pamulang, Jalan Surya Kencana No. 1, Tangerang Selatan, Banten, Indonesia

*E-mail: eko.arief.setiadi@lipi.go.id

Abstract. The polymer coated Magnesium Ferrite nanoparticles ($MgFe_2O_4$) based on natural iron sand, $Mg(CH_3COO)_2.4H_2O$, and PEG-6000 have been successfully prepared by co-precipitation method. The mass variation of PEG-6000 content was from 0 to 12 gram. It was prepared at synthesize temperature of 70°C. The PEG coating reduced the effect of agglomeration, so the coercivity value can be closed to soft magnets. The nanoparticle of synthesized has $MgFe_2O_4$ single phase and cubic spinel structure. The bonding of $MgFe_2O_4$ and PEG-6000 as a coating material was confirmed by FTIR curve. The $MgFe_2O_4$ density decreased with the increasing of PEG 6000 content. On the other hand, the coercivity value was slightly reduced as the addition of PEG-6000, with the lowest value was obtained on 8 gram PEG content. The optimum condition is obtained at addition of 8 gram PEG 6000 to $MgFe_2O_4$, with coercivity, saturation, and remanence are 198.41 Oe, 52.53 emu/g, and 8.51 emu/g, respectively. So that, the sample is widely used as absorbance material of heavy metal.

1. Introduction

The magnetic nanoparticles are widely used in immobilization and separation of proteins or enzymes, drug administration or purification of Deoxyribonucleic Acid (DNA) [1]. It is also used to drugs delivery, hyperthermia treatment, magnetic resonant imaging [2], enhancement agent, manipulating cell membranes, biosensors, bio-labelling [3] and adsorption of metal ions [4]. One of magnetic nanoparticles application which widely developed is nanoparticles based on spinel ferrite. Spinel ferrite is a crystalline structure composed of two substructures such as tetrahedral and octahedral. The chemical formula of spinel ferrite nanoparticles is $M\text{-}Fe_2O_4$ which M represents a divalent of metal ion. The ionic radius about 0.6-1.0 Å. For spinel ferrite nanoparticles, M is a transition metal such as Cu, Zn, Ni, Co, Fe, Mn, Mg, or a combination of them [5]. The behavior of this material has a high permeability and resistance, but low coercivity. The spinel ferrite nanoparticles is soft ferrites having a cubic crystal structure [6]. Some methods developed for synthesizing magnetic nanoparticles is such as hydrothermal synthesis, thermal decomposition, microwave plasma synthesis [7], sol gel, sonochemical, co-precipitation and others [6]. Morever this nanoparticles has superparamagnetic behavior when the particles size is nanometer. Superparamagnetic properties of materials have a high saturation magnetization, and the average coercivity equal to zero when those material are not influenced by an external magnetic field.

Natural iron sand is a mineral with a high iron element content. It can be utilized as precursor in the synthesis of magnetic nanoparticles based on ferrite. Many researchers interest to study Magnesium Ferrite ($MgFe_2O_4$) because of its great potential for many applications. The $MgFe_2O_4$ nanoparticles has high saturation magnetization, Currie temperature and electrical resistivity. The $MgFe_2O_4$ nanoparticles are soft magnetic materials (low coercivity) and one of the most important inverse spinel groups [5]. It is also an n-type semiconductor material that can be applied as adsorption, sensor, and used in magnetic technology [8].

The magnetic nanoparticles based on ferrite are obtained under agglomerated conditions. In the other hand, the air which contacts to the surface of the nanoparticles causes oxidation process easily [9]. To overcome these effects, it necessary surface modification of nanoparticles which is coated with a polymer material or an organic material. Some materials that can be used as surface coatings include polyethyleneglycol (PEG), polyvinyl alcohol (PVA), polyethyleneimine (PEI), polyvinylpyrolidone (PVP) and other ingredients [10]. While reducing agglomeration, these coatings also increase the dispersibility, biocompatibility of chemical stability [11].

Modification of PEG coated magnetite (Fe_3O_4) is widely used in drugs delivery application. In their research, Bazaan et. al synthesized PEG coated Fe_3O_4. The Results revealed that the mean diameter of pure Fe_3O_4 was very high because of the strong interaction between the particles and aggregation phenomena that match well with TEM photograph. By modifying Fe_3O_4 with different types of PEG, due to the presence of hydrophilic polyethylene glycol on the surface of particles, the mean diameter of nanoparticles significantly decreased and significantly improved their dispersibility which were very favorable for biomedical application [12]. Meanwhile, $MgFe_2O_4$ nanoparticles modified with PEG can increased crystallinity by using sol gel method.

In this paper, $MgFe_2O_4$ nanoparticles was synthesized with natural iron sand as the precursor. Natural iron sand has an abundance and has a large iron content. The material was prepared by co-precipitation method which one method of synthesis of inorganic compounds based on precipitation of more than one substance. Another advantage of co-precipitation method, it can be processed at room temperature easily and control the particle size, so that the required time is relatively short [6,9]. The addition polyethyleneglycol 6000 (PEG-6000) as the coating is used for reducing the effect of nanoparticle agglomeration at the synthesizing process.

2. Experiment Method

Main material that was used for synthesizing $MgFe_2O_4$ nanoparticles coated PEG-6000 is natural iron sand from Buaya River, Deli serdang, North Sumatera, Indonesia. The composition of the natural iron sand is Fe > 90 % and < 9% of Ti, Si and others. Mass of PEG-6000 was a varied variable. Iron sand powder of 2.72 gram and 5.91 gram of $Mg(CH_3COO)_2.4H_2O$ was added into 30 mL of 5.1 M HCl solution, followed by stirring at 500 rpm until the mixture of these solution was homogeneous. After that, the mixture of this solution was added dropwise into the mixture of 99 mL of 2 M NH_4OH and variation of PEG-6000 of 0, 4, 8, and 12 gram which marked as Sample A, B, C, and D, respectively. The mixture of all these solutions were stirred by using magnetic stirrer at 70°C and 500 rpm for 120 minutes. The result of the mixture was cleaned with aquades until pH neutral. Then, the precipitate sample was dried in an oven at 100°C for 1 hour. After that, the result of $MgFe_2O_4$ powder was characterized by using pycnometer, X-ray Diffraction (XRD - Rigaku SmartLab, λ = 1.5418 Å), Fourier Transform Infrared Spectrometer (FTIR – Thermo Scientific Nicolet iS10), Field Emission Scanning Electron Microscopy (FE SEM – Jeol JIB 4610F), and Vibrating Sample Magnetometer (VSM – Electromagnetic VSM250).

3. Results and Discussion

Diameter $MgFe_2O_4$ nanoparticles are black and responsive to the magnetic field. The effect of PEG-6000 addition for powder density of $MgFe_2O_4$ nanoparticles is shown in Figure 1. The addition of PEG-6000 decreases powder density value. The powder density value is about 4.41-4.66 g/cm^3. The decreasing of powder density is resulted by PEG-6000 density that is lower than powder density of $MgFe_2O_4$.

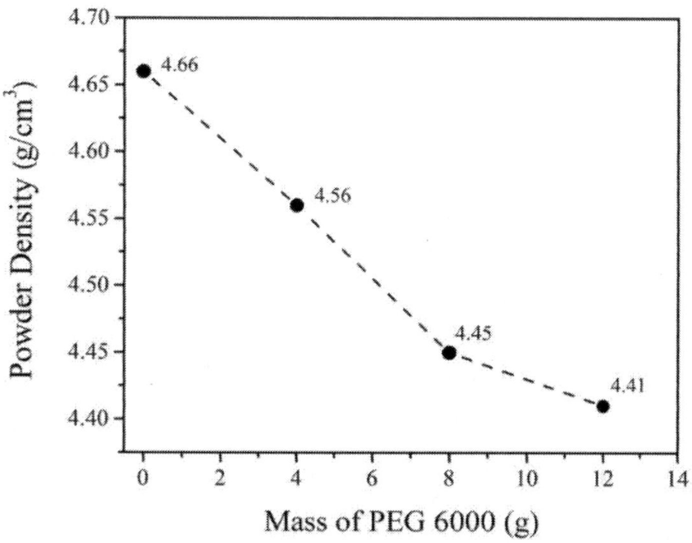

Figure 1. The relation between the powder density of $MgFe_2O_4$ nanoparticles and addition of PEG- 6000

XRD diffraction pattern for $MgFe_2O_4$ nanoparticles is shown in Figure 2. From the result, there is cubic spinel single phase of $MgFe_2O_4$ with crystal planes of (220), (311), (440), (511) and (440).

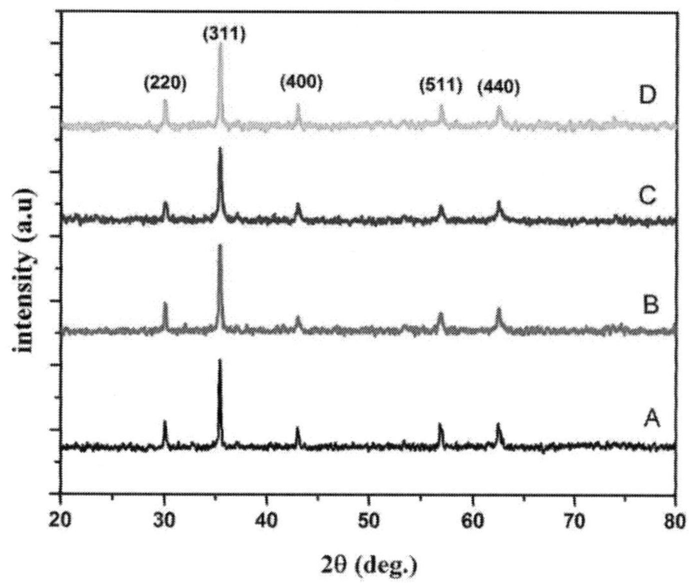

Figure 2. XRD diffraction pattern of $MgFe_2O_4$ nanoparticles coated PEG-6000.

Lattice parameter and crystalline diameter of $MgFe_2O_4$ nanoparticles coated PEG-6000 can show in Table 1.

Table 1. Lattice parameter and crystalline diameter of $MgFe_2O_4$ nanoparticles.

Sample	Composition of PEG -6000 (gram)	Lattice Parameter (Å)	Diameter of Crystalline (nm)
A	0	8.411	40.20
B	4	8.412	32.22
C	8	8.409	28.22
D	12	8.410	34.02

According to Table 2, it shows that lattice parameter value of all samples is close enough to lattice parameter value of $MgFe_2O_4$ standard (a = 8.410 Å). It indicates that the crystal is formed well and the addition of PEG-6000 doesn't affect to the crystal structure of $MgFe_2O_4$. Crystalline diameter is change, and the smallest diameter value is obtained in sample C. This is due to without addition of PEG-6000, nanoparticles will be through agglomeration. Meanwhile, the addition of PEG-6000 reduces effect of agglomeration, so the crystalline size of nanoparticles decreases. However, on the addition of 12 gram PEG-6000, the diameter of crystalline increases because the surface of nanoparticles is getting thicker.

The result of FTIR characterization of sample B (4 gram of PEG-6000) can show in Figure 3. The functional group with wavenumber around 841.94, 962.67, 1111.91, 1242.24, 1280.67, 1342.66, 1466.97, and 2888.83 cm^{-1} is characteristic of PEG [1,13,14]. It indicates that $MgFe_2O_4$ particles have been coated. While wavenumber around 528.55 cm^{-1} and 578.45 cm^{-1} is the functional group of the $MgFe_2O_4$. The wavenumber is the functional group of metal ion and oxygen (M-O). The functional group of metal and oxygen is on site octahedral (528.55 cm^{-1}) and tetrahedral (578.45 cm^{-1}) [13,14]. Wavenumber around 1685.28 and 3447.28 cm^{-1} is the functional group of H-O. From the results, it can conclude that sample C and D also has been coated.

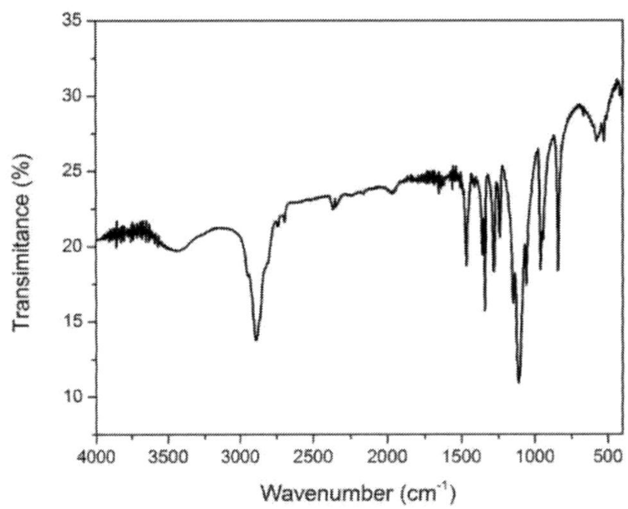

Figure 3. FTIR spectra of PEG-6000 coated $MgFe_2O_4$ (Sample B).

The FE-SEM characterization of sample A shows particle size and elements contained in the sample as shown in Figure 4. The Figure 4a shows that most of the particles undergo agglomeration. Meanwhile, the coating of PEG-6000 prevented agglomeration effect. The particle size can be showed in Figure 4b which is around 65.192 nm. Figure 4c shows composition of O = 76.8, Fe = 22.5 and Mg = 0.7 %At.

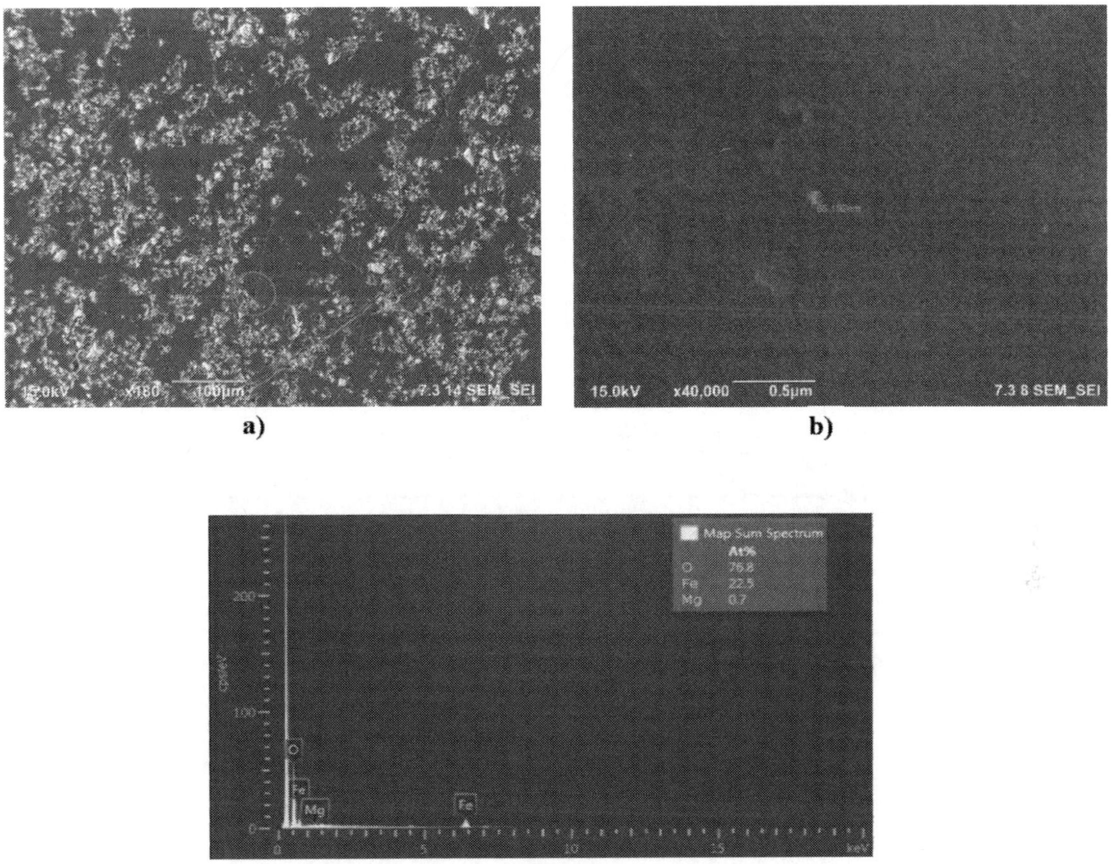

a) **b)**

c)

Figure 4. FE-SEM imaging of sample A a) morphology, b) morphology zoom, and c) EDX.

Hysteresis curve of VSM characterization results is shown in Figure 5. The coercivity and remanence tend to increase at sample A, B, and C. While for sample D, coercivity value increased to 232.26 Oe and remanence value increased to 11 emu/g. Furthermore, the result is shown in Table 2. The increasing of coercivity and remanence on sample D is influenced by crystallite magnification.

In application, the absorbance material of heavy metal need the smaller particle size with the magnetic properties as super paramagnetic for getting the optimum condition. In this research, the optimum condition is obtained on sample C with coercivity value about 198.41 Oe, saturation value about 52.53 emu/g, and remanence value about 8.51 emu/g. Magnetic characterization is influenced by particle size. The increasing of crystalline diameter is in proportion to decreasing of coercivity and remanence value. Meanwhile, the saturation value tends to increase. This is caused by the smaller particle size will be closer to the single domain. Morever, barrier energy is lower [15]. In previous research, the nanoparticle $MgFe_2O_4$ is prepared by sol-gel combustion method with calcination temperature variation of 400, 500, and 600 °C. The $MgFe_2O_4$ material has two phases which are cubic and tetragonal phases. In lower temperature, the $MgFe_2O_4$ material has properties as the smaller diameter of crystallite and the lower coercivity and saturation.

IOP Conf. Series: Materials Science and Engineering **309** (2018) 012029 doi:10.1088/1757-899X/309/1/012029

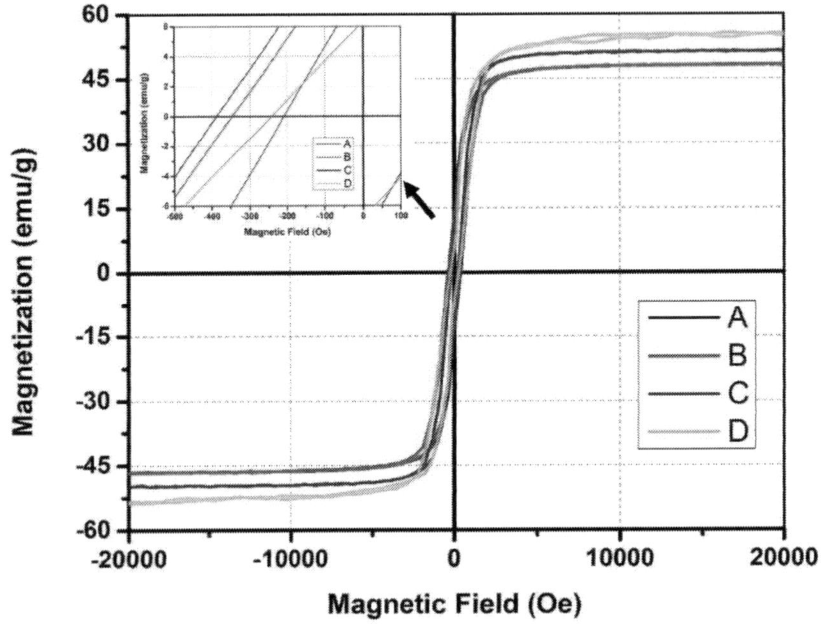

Figure 5. Hysteresis curve of $MgFe_2O_4$ nanoparticles

Table 2. Magnetic properties of $MgFe_2O_4$ nanoparticles

Sample	Coercivity (Oe)	Saturation (emu/g)	Remanence (emu/g)
A	382.31	48.37	13.79
B	335.21	51.69	12.65
C	198.41	52.53	8.51
D	232.26	51.13	11.00

4. Conclusion

The $MgFe_2O_4$ nanoparticles have been created out with PEG-6000 additive. The powder density of $MgFe_2O_4$ nanoparticles is around 4.41-4.66 g/cm^3. The $MgFe_2O_4$ nanoparticles have a single phase with cubic spinel structure. The lattice parameter of nanoparticles is around 8.410 Å. The addition of PEG6000 additive decreases agglomeration effect and crystalline size. The PEG-6000 coating on $MgFe_2O_4$ nanoparticles can be proof with the existing of PEG-6000's functional group. The particles size is about 65.192 nm with consisting of elements such as O = 76.8, Fe = 22.5 and Mg = 0.7 %At. The optimum condition is obtained at $MgFe_2O_4$ with 8 gram PEG 6000 with coercivity, saturation, and remanence are 198.41 Oe, 52.53 emu/g and 8.51 emu/g respectively. So that, the sample is widely used as absorbance material of heavy metal.

5. Acknowledgement

The authors acknowledge to Research Center for Physics, Indonesian Institute of Sciences 2017 for financial support and characterization facility.

6. Reference

[1] Saoud F 2010 *Superparamagnetic Nanoparticles for Synthesis and Purification of Polymers Prepared Via Controlled Living Radical Polymerization (CLRP)* (Stellenbosch : The University of Stellenbosch)

[2] Feng B, Hong R Y, Wang L S, Guo L, Li H Z Ding J Zheng Y and Wei D G 2008 *Colloids Surfaces A Physicochem. Eng. Asp.* **328** 52

[3] Anbarasu M, Anandan M, Chinnasamy E, Gopinath V and Balamurugan K 2015 *Spectrochim. ACTA Part A Mol. Biomol. Spectrosc.* **135** 536

[4] Larraza I, López-gónzalez M, Corrales T and Marcelo G 2012 *J. Colloid Interface Sci.* **38** 24

[5] Chen Q, Rondinone A J, Chakoumakos B C and Zhang Z J 1999 *Journal of Magnetism and Magnetic Materials* **194** 1

[6] Suharyadi E, Setiadi E A, Shabrina N, Kato T and Iwata S 2014 *Adv. Mater. Res.* **896** 126

[7] Liang X, Shi H, Jia X, Yang Y, and Liu X 2011 *Mater. Sci. Appl.* **2** 1644

[8] Bangale S V, Patil D R and Bamane S R 2011 *Archives of Applied Science Research* **3** 506

[9] Setiadi E A, Sebayang P, Ginting M, Sari A Y, Kurniawan C, Saragih C S and Simamora P 2016 *Journal of Physics: Conference Series* **776** 012020

[10] Umut E 2013 Surface Modification of Nanoparticles Used in Biomedical Applications *Modern Surface Engineering Treatments* ed by M Aliofkhazraei (Rijeka: Intech) chapter 8

[11] Shen Y F, Tang J, Nie Z H, Wang Y D, Ren Y and Zuo L 2009 *Bioresource Technology* **100** 4139

[12] Khoee S, Saadatinia A and Bafkary R 2017 *Ultrasonics Sonochemistry* **39** 144-152

[13] Junejo Y and Baykal A 2013 *Cent. Eur. J. Chem.* **11** 1527.

[14] Gao G, Qiu P, Qian Q, Zhou N, Wang K, Song H, Fu H and Cui D 2013 *J. Alloys Compd.* **574** 340

[15] Goya G F, Fonseca F C and Jardim R F 2003 *J. Appl. Phys.* **93** 6531

IOP Conf. Series: Materials Science and Engineering **309** (2018) 012030 doi:10.1088/1757-899X/309/1/012030

Design of mathematical models assessment of working achievements based on spencer competency in PT. Z

K Siregar[1,*] and S F Siregar[2]
[1]Department of Industrial Engineering, Universitas Sumatera Utara, Medan 20155, Indonesia
[2]Department of Chemical Engineering, Universitas Sumatera Utara, Medan 20155, Indonesia

[*]E-mail: khawaritasiregar@yahoo.com

Abstract. This research is design employee performance assessment by considering work result of employee based on competency. Relevant competencies are identified according to Spencer's competence of employees that subsequently processed by Analytical Hierarchy Process (AHP) method. The results of weighting AHP indicate the highest priority order of criteria, there are; concern of customer satisfaction (0.1325), group work (0.1324) and technical expertise (0.0826). The weight of the criteria is used to design the Work Performance Value (WPV) to be used as the basis for calculating the incentive index. The higher incentive index of an employee, the greater amount of incentives was earned. The calculation of incentives is made to four employees of chopsticks production. From employee incentives A, B, C and D, employee D has the highest incentive index and increment of IDR 2,700,675 compared to previous incentive system. The incentive division system based on the Work Performance Values (WPV) of this proposal reflects a real incentive so that the incapacity of incentive can be reduced.

1. Introduction

In both government organizations and private organizations are now increasingly required to maintain the existence of the company. Therefore, it is necessary Human Resources (HR) who have high competence because of the expertise or competence will be able to support the improvement of employee performance. Employee performance appraisal is the way companies do to improve and develop human resources. The results of the performance appraisal, the company is expected to be able to know the advantages and weaknesses of employees as a basis to maintain the advantages and reduce the existing weaknesses in an effort to improve productivity for the sake of corporate progress.

PT. Z is one of the companies engaged in the paper industry (Joss Paper), skewers and chopsticks. In general, employees work because they are encouraged to meet the needs of his life so that employees work harder when the results achieved from his work to get reward or satisfactory remuneration. One form of reward that can be given is incentives. PT. Z provides bonuses or incentives to all employees based on work performance as measured by absenteeism. This has an impact on the lack of enthusiasm and motivation of employees in the work. Therefore, it is necessary to design a mathematical model of performance appraisal by taking into account the factors of employee work result to calculate incentive giving.

The purpose of this research is to identify the competence of production employees at PT. Z as a basis for providing incentives based on performance appraisal using the Analytical Hierarchy Process (AHP) method. The benefits of this research are to be considered in determining the decision making strategy about planning, designing and developing the provision of incentive with optimal employee performance.

The first step taken in this research is:

- Identify the factors that become the assessment of employee performance.
- Determine factor weights by using the Analytical Hierarchy Process (AHP) Method.
- Establish criteria of assessment which is the competence to be calculated.
- Designing models mathematic assessment of employee performance as a basic model of incentives

1.1. Analytical Hierarchy Process (AHP)

According to Wang *et al.* [1], the contractor selection process is a multi-criteria decision-making (MCDM) problem. Numerous MCDM models including utility theory [2], fuzzy theory [3], and performance-based modeling [4] have been developed for contractor qualification or final selection. The AHP [5, 6] is a decision-making method that was developed by Saaty. This technique calculates the qualified priorities of a given set of alternatives on a scale based on the judgment of the decision-maker. The process stresses the importance of the intuitive judgments of a decision-maker and consistency in the comparison of alternatives in the decision-making process. Skibniewski and Chao [7] suggested that the strength of this approach is that it organizes tangible and intangible factors in a systematic way and offers a structured, simple solution to decision-making problems. The AHP-based approach has become quite popular primarily due to its simple and systematic implementation steps [8, 9].

2. Methodology

The research type is descriptive research. In this case, the researchers conducted a survey by distributing questionnaires to the respondents. Questionnaires were filled with or without the help of surveyors depending on their needs. Respondents in this research are employees of PT. Z and the variables contained in this study are competency criteria Lyle M. Spencer and Signe M. Spencer which is the factor of performance appraisal from the results of employee questionnaires and the results of the value of performance appraisal by using competency criteria Lyle M. Spencer and Signe M. Spencer. The procedure of research implementation consists of several stages: At the beginning of the study conducted a preliminary study to identify the problems contained in the company; After identification then the research problem was formulated; Preparation of theoretical basis, using appropriate theories to solve problems for the problem; Data collection consisting of observation and research directly in the field; Data processing by Analytical Hierarchy Process (AHP) method; Discussed the results of data processing; Drawn conclusions and given suggestions.

3. Results and Discussion

Questionnaires understanding a preliminary questionnaire that collects performance assessment criteria are considered relevant in supporting the work of production employees of PT. Z. In addition, the understanding questionnaire aims to disseminate performance appraisal by using Spencer's competence as an effort to demonstrate and understand the advantages of the Spencer method. The criteria chosen based on the questionnaire were then grouped into six groups of Spencer criteria.

Table 1. Spencer Criteria Grouping

No	Criteria Gouping	Criteria
1	Actions and Success	Passion for achievement Attention to tidiness, quality and thoroughness Initiative / Proactive Spooler information, problem definition
2	Helping and Serving Consumers	Concern for customer satisfaction
3	Impact and Influence	Awareness of organization Building relationships
4	Managerial	Group work and cooperation
5	Cognitif	Analytical thinking Conceptual thinking Technical / Professional / Managerial expertise
6	Personal Effectiveness	Self-control Confidence Flexibility / Ability to adapt Commitment to the organization

Based on Table 1, it is shown that there are six groups of spencers consisting of 15 competencies that will be used in performance appraisal. After the criteria are grouped according to the Spencer criteria grouping, hierarchies of employee performance appraisals are made.
Analytical Hierarchy Process (AHP) is a method used to obtain the weight of each competency criterion. Ordering the 15 criteria of competence from the highest score to the lowest value seen in Table 2.

Table 2. Competency Criteria of the Highest Weight to the Lowest

No	Criteria	Weight Competency Criteria
1	Concern for customer satisfaction	0.1325
2	Group work and cooperation	0.1324
3	Technical Expertise	0.0826
4	Confidence	0.0688
5	Self-control	0.0683
6	Building relationships	0.0658
7	Organizational Awareness	0.0639
8	Information Collection	0.0636
9	Conceptual Thinking	0.0589
10	Analytical thinking	0.0545
11	Excitement to Achievement	0.0499
12	Flexibility	0.0444
13	Initiative / Proactive	0.0442
14	ommitment to the Organization	0.0362
15	Attention, quality, thoroughness, tidiness	0.0338

After obtaining the weight of each criterion, then designed an employee performance appraisal format. The performance appraisal format using Spencer criteria can be seen in Table 2 and Table 3.

The calculation of priority weighting for alternatives are done to determine which alternatives are selected from the lowest level and then proceed to the next level. Recapitulation of weights at each level. Having obtained the criteria of general competence criteria, then designed employee performance appraisal system that is able to reduce weaknesses and provide convenience when compared with previous employee performance. This will make it easier for assessors to value employees more objectively.

The proposed scoring scale is based on the Rating Scales rating method, Spatially:
Range = Highest value - Lowest value

Number of interval classes = 5

Based on the formula above then the interval class length is The interval class length = (5 – 1)/5= 0.8.

The grouping of employee performance based on the grade range is shown in Table 3.

Tabel 3. Grouping Employee Performance

Scoring Scale	Information
1.00-1.79	*Unsatisfactory Performance*
1.80-2.59	*Improvement Needed*
2.60-3.39	*Meets Expectation*
3.40-4.19	*Exceeds Expectation*
4.20-5.00	*Outstanding Performance*

After obtaining the weight of each criterion, then designed an employee performance appraisal format. The format of performance appraisal using Spencer criteria can be seen in Table 4.

Table 4. Employee Performance Appraisal Format Using Spencer Criteria

	Grouping Criteria	Weight of Group Criteria	Criteria	Parcial Weight	Weight	Weight of Group Criteria
1	Action and Succcess	0.1916	Passion for achievement	0.3320	0.0636	0.1916
			Attention to tidiness. quality and thoroughness	0.2604	0.0499	
			Initiative/ Proactive	0.1766	0.0338	
			Spooler information. problem definition	0.2309	0.0442	
2	Helpin and Serving Consumers	0.1325	Concern for customer satisfaction	0.1325	0.1325	0.1325
3	Impact and Influence	0.1297	Awareness of organization	0.4924	0.0639	0.1297
			Building relationships	0.5076	0.0658	
4	Managerial	0.1324	Group work and coorporation	0.1324	0.1324	0.1324
5	Cognitive	0.1961	Analytical thinking	0.2781	0.0545	0.1961
			Conceptual thinking	0.3005	0.0589	
			Technical expertise/ Professional/ Managerial	0.4214	0.0826	
6	Personal Effectiveness	0.2177	Self – control	0.3138	0.0683	0.2177
			Confidence	0.3159	0.0688	
			Flexibility/ ability to adapt	0.2039	0.0444	
			Commitment to the organization	0.1664	0.0362	

The mathematical model of performance appraisal can be formulated as follows.

For example: Y = Work Performance Value

X = Compentency criteria

Y = f(X)

$$Y = 0.0636\ X_1 + 0.0499\ X_2 + 0.0338\ X_3 + 0.0442\ X_4 + 0.1325\ X_5 + 0.0639\ X_6 + 0.0658\ X_7 + 0.1324\ X_8 + 0.0545X_9 + 0.0589X_{10} + 0.0826X_{11} + 0.0683X_{12} + 0.0688X_{13} + 0.0444\ X_{14} + 0.0362X_{15}$$

If the Y result or Work Performance Value $\leq$ 3,000 then the value of Y is considered to be 0 because the value is below the standard value of the company's incentives, so that the employee is not entitled to receive the incentive.

From the value of achievement is then being calculated incentives received by employees of production PT Z. Calculation of the incentives of both chopstick production and paper production prayer (joss paper).

The source of funds for incentive division comes from the profits owned by the company. Profit achieved by the company reflects the achievement of the company and the achievement of the company itself can be achieved if the employee performance is good even increased.

$$\sum WPV = WPV_1 + WPV_2 + WPV_3 + ... + WPV_n$$

$$EI_1 = \frac{WPV_1}{\sum WPV} = \text{Employee incentive index 1}$$

Then obtained,

$$\text{Employee incentive index} = \frac{WPV_1}{\sum WPV} \text{ x TI}$$

Or,

$$\text{Employee incentive index} = EI_1 \text{ x TI}$$

Explanation:

TI = Total Incentives given (Rp)

WPV_1 = Work Performance Value 1

The amount of incentive fund distributed by the company is the same, but giving incentives based on the acquisition of Work Performance Values (WPV) has more impact on the development of potential and employee work.

It shows the higher the value of performance achievement, then the employees will get a greater incentive.

4. Conclusion

The weighting results shown that the criteria that have the highest weight of awareness of customer satisfaction has the highest weight (0.1325). Companies require employees to be more sensitive to the needs of consumers so as to support the success of the company in serving the needs of consumers. Then followed by the criteria of group work / cooperation (0.1324). The importance of exchanging information so as to form employees who work in teams at each work station can smoothen the way its production process. Criteria of technical expertise (0.0826). The company requires employees who have practical skills in managing and proficiency in operating the production equipment. The calculation of incentives used by the company has shown a fluctuating result between incentive calculation based on WPV and incentive calculation by the company. From the calculation result, employee D who has the biggest incentive gain increment value incentive based on WPV that is Rp200.675 compared to the original incentive by the company. The amount of incentive fund distributed by the company is IDR 10,000,000 but the incentives based on the acquisition of Work Performance Value (WPV) give more impact to the development of the potential and the work of the employees.

References

[1] Anagnostopoulos K P, Vavatsikos A P *Operational Research. Anais: An International Journal* **6** 333-346

[2] Balubaid M, Alamoudi R 2015 *American Journal of Industrial and Business Management* **5**

581-589
[3] Dobi K, Gugic J, Kancijan D 2010 *The 32nd International Conference on Information Technology Interfaces* (*ITI* 2010)
[4] Li Y, Nie X, Chen S 2007 *Journal of Construction Engineering and Management ASCE* **133** 40-49
[5] Padhi S S, Mohapatra P K J 2009 Contractor Selection in Government Procurement Auctions: A Case Study European *Journal of Industrial Engineering* **3** 170-186
[6] Triantaphyllou E, Man S H *Inter'l Journal of Industrial Engineering: Applications and Practice* **2(1)** 35-44
[7] Verma S D, Pateriya A 2013 *International Journal of Engineering Trends and Technology (IJETT)* **4**
[8] Wang W C, Yu W D, Yang I T, Lin C C, Lee M T, Cheng Y Y 2013 *Journal of Civil Engineering and Management* **19** 24-36
[9] Fong P S, Choi S K 2007 Final Contractor Selection Using the Analytical Hierarchy Process *Construction Management and Economics* **18** 547-557

IOP Conf. Series: Materials Science and Engineering **309** (2018) 012031 doi:10.1088/1757-899X/309/1/012031

Application study of evolutionary operation methods in optimization of process parameters for mosquito coils industry

E Ginting*, M M Tambunan and K Syahputri
Department of Industrial Engineering, Faculty of Engineering, Universitas Sumatera Utara, Medan 20155, Indonesia.

*Email: ir.elisabethginting@gmail.com

Abstract: Evolutionary Operation Methods (EVOP) is a method that is designed used in the process of running or operating routinely in the company to enables high productivity. Quality is one of the critical factors for a company to win the competition. Because of these conditions, the research for products quality has been done by gathering the production data of the company and make a direct observation to the factory floor especially the drying department to identify the problem which is the high water content in the mosquito incense coil. PT.X which is producing mosquito coils attempted to reduce product defects caused by the inaccuracy of operating conditions. One of the parameters of good quality insect repellent that is water content, that if the moisture content is too high then the product easy to mold and broken, and vice versa if it is too low the products are easily broken and burn shorter hours. Three factors that affect the value of the optimal water content, the stirring time, drying temperature and drying time. To obtain the required conditions Evolutionary Operation (EVOP) methods is used. Evolutionary Operation (EVOP) is used as an efficient technique for optimization of two or three variable experimental parameters using two-level factorial designs with center point. Optimal operating conditions in the experiment are stirring time performed for 20 minutes, drying temperature at 65°C, and drying time for 130 minutes. The results of the analysis based on the method of Evolutionary Operation (EVOP) value is the optimum water content of 6.90%, which indicates the value has approached the optimal in a production plant that is 7%.

1. Introduction

The EVOP method [1] was introduced in the late 1950s as a field application technique for improving existing industrial processes. It was to be applied to an existing manufacturing process that was currently producing acceptable product. By exploring small incremental changes in an existing set of process conditions, the process could be improved and moved in the direction of some process optimum [2]. There are other advanced statistical methods, such as Artificial Neural Network. [3] simplex optimization [4, 5] response surface methodology, [6] and advanced factorial design but they are more complex and require a great deal of training for reliable application and intepretation. Most of the methods deal with an intial strategy of experimentation when formulating a set of bench-scale experimental runs.[7]

Evolutionary Operation Methods (EVOP) is designed to use in the process of running or operating routinely in the company to enables high productivity [8]. This method embedded in the activities carried out by employees with minimum assistance from the research or development of the company.

PT.X.is one of the private companies engaged in manufacturing, which produces anti-mosquito drugs in big quantities and has cooperated with many other similar companies to producing anti-mosquito drugs in various brands. In carrying out its production activities, the company faces a problem with a water content of products produced mosquito repellent. With the operating conditions in the drying section that runs the company today is stirring time 15 minutes, drying temperature 65°C, and the drying time for 120 minutes produces water content approximately 7.4% - 7.6%, while the optimum value is 7%, this leads to around 8% (10-20 pieces) broken products, where the mosquito coils produces is 240 pieces in one drying. To overcome this problem researchers has been trying to reduce the value of the water content by testing it in a laboratory to determine the better-operating conditions and to apply it in the production floor. Therefore a way to get information on how to process improvement so that the product is better than the current conditions Improving quality and productivity will be effective if it is an integral part of the product development cycle and production processes. The Design of experiments on line can be used for this purpose, because the experimental design online application used in manufacturing processes to identify the relationship between the variables in a process. One method of design of experiments on line that can be used to obtain the desired operating condition information is Evolutionary Operation Method (EVOP).

This paper describes the application of Evolutionary Operation Method to research the impact of water content towards mosquito coils product quality.

2. Methodology

Since EVOP's introduction in the late 1950's, many books and journal articles have been published discussing the method. This article is not intended as a survey review of EVOP; the reader is invited to consult the original publication or other excellent discussions [9-12].

2.1. *Evolutionary Operation Method and Design of Experiment*

In EVOP method, there are stages of planning which are done, where the definition of planning is the process, how, or plan, organize, and outlines the steps the implementation of activity. Planning stages to begin the implementation of EVOP is:

1. Select two, at most three variables of the process and select a level for each process variable.
2. Select the suitable criteria for testing / treatment combination.
3. Estimate the number of cycles required to obtain a significant change in the process.

EVOP worksheets consist of five parts, each of which requires calculation. The sections are : Average calculation, Effects calculation, Standard deviation calculation, Error limit calculation and Decision Making Procedure.

2.2. *Methods of Evolutionary Operation (EVOP) for Three Variables*

Design of experiments performed or information obtained from the production of the three variables that influence, Va, Vb, and Vc. Then the effects were calculated using an EVOP three variables worksheet. The position 0,1,2,3, and 4 show different combinations of these three variables.

In EVOP three variables program, a cycle is broken down into two blocks, each of which consists of a condition. The condition of zero, one, two, three, and four are inserted into block I, and the condition A, five, six, seven, eight is inserted into the block II. Reference conditions to run only once in each block (twice per cycle), and is expressed by the condition 0 and A [1].

Calculation of effect value that use as comparation to error limit in EVOP, Yates Algoritm in experimental design is used.

2.3. *Yates Algorithm*

Yates Algorithm method used to calculate the contrast and the sum of squares (SS) for each treatment combination in factorial design experiments 2^3.

2.4. The Analysis of Variance

Analysis of variance is a method of decomposing the total variability in a set of observations, as measured by the sum of the squares of these observations from their average, into component sums of squares that are associated with specifically defined sources of variation [13,14].

2.5. Normality Test Data

To check whether or not the population is normally distribution, then tested for normality using Chi Square test (X^2).

3. Results and Discussion

3.1. Analysis of Variance

The calculation of ANOVA test using the ANOVA formula can be seen in Table 1.

Table 1. Analysis of Variance

Source of Variation	**Degree of Freedom**	**Sum of Squares**	**Central Square**	**F**
Treatment	9	3.71	0.412	
Error	30	0.18	0.006	68.66
Total	39	3.89		

F table (9, 30) for the level of significance (α) = 0.05, F table = 2.21, F count > F table meaning that there is a real treatment effect or there is a difference between real treatment.

3.2. Normality Test

Test normality of data on the processing done by using Chi Square (X2). Based on the data obtained, the calculated X^2 value is 1.7014. According to the Chi Square table for α = 0.05 df = k-r-1. Because X^2 calculated = 1.7014 < X^2 table = 5.991, it means the data distributed normally.

3.3. Calculation of Average, Standard Deviation, effects, and Error Limits

In continuing EVOP program that considered is the change in the level of factor does not affect the quality of the product. EVOP calculations performed is a factorial design with three factors and two level factors, namely: stirring time (A) with a level of 10 minutes and 20 minutes; Drying temperature (B) with a level of 55° C and 65° C; and the drying time (C) with a level of 110 minutes and 130 minutes and a currently performed in the factory as a reference, namely the condition of stirring time (A) 15 minutes; drying temperature (B) 60 ° C; and the drying time (C) 120 minutes. Therefore, there are ten combinations of operation conditions. The results of 10 experiment conditions seen in Table 2.

Table 2. Data Observations of Water Content

Operation Condition	Replication			
	1	2	3	4
(1)	7.41	7.44	7.54	7.52
(2)	8.12	8.03	8.02	7.94
(3)	7.18	7.11	7.13	7.13
(4)	7.25	7.25	7.27	7.25
(5)	7.18	7.26	7.03	7.00
(6)	7.53	7.48	7.50	7.42
(7)	7.24	7.31	7.34	7.20
(8)	6.84	6.97	6.90	6.96
(9)	7.61	7.82	7.65	7.82
(10)	7.31	7.24	7.47	7.42

The EVOP calculation begins by calculating the amount of water content for each cycle and operating conditions, then calculate average as follows:

Average cycle n = (total n) / n

The next step is to calculate the difference in the value of the average cycle by cycle water content values for block I and II.

Difference = average cycle (n-1) - the water content of the cycle.

Next step is to calculate the standard deviation and the average standard deviation of Block I and Block II are as follow [15]:

Standard deviation = Range x f_{FK} n = cycle

$F_{K,n}$ = constant depending on number of replication and experiment per-cycle

Range = the difference value - the value of the smallest difference

The following table shows the value of range, standard deviation the experiment

Table 3. The Calculation of Range and Standard Deviation

Block	**Cycle**	**Range**	**Standard Deviation**
	1	-	-
	2	0.17	0.051
I	3	0.305	0.106
	4	0.216	0.08
	1	-	-
	2	0.28	0.084
II	3	0.26	0.091
	4	0.223	0.0825

The calculation of average standard deviation for all cycle can be seen below :

Cycle 2:

Total S $= S_1 + S_2 = 0{,}051 + 0{,}084 = 0.135$

Average S $= \frac{0{,}135}{2n-2} = \frac{0{,}135}{(2x2)-2} = \frac{0{,}135}{2} = 0.0675$

The calculation of average standard deviation for all cycles carried out with the same formula for the entire block.

Cycle 3: 0.083 Cycle 4: 0.0824

The limit calculated by the following formula : Limit $= \left(\frac{1{,}41}{\sqrt{n}}\right) \text{x S}$ (1)

The calculation of the limit of the main effects and interaction effects for each cycle happened by using the value of deviation's standards The calculation of the limit of the following effects:

Cycle 2 Limit = 0,0673, Cycle 3 Limit = 0,0675 and Cycle 4 Limit = 0,058

By using Yates Method effect value for each cycle found compare to the limit as shown in Table 4. From the Table 4 effect value factor A (stirring time), factor B (drying temperature), Factor C (drying time) and interaction among Factor ABC is smaller compared to error limit. This shows that factor stirring time, drying time and drying temperature are significance of confidence level.

Table 4. The Comparison of the Limit and the Factor of the Experiment

Cycle	**Factor**	**Effect**	**Limit**
2	A	-0.208	0.0673
	B	-0.443	
	C	-0.390	
	ABC	-0.130	
3	A	-0.21	0.0675
	B	-0.441	
	C	-0.405	
	ABC	-0.094	
4	A	-0.186	0.058
	B	-0.437	
	C	-0.422	
	ABC	-0.066	

4. Conclusion

The conditions before using EVOP are time 15 minutes, drying temperature 60 °C, and the drying time of 120 minutes, yielding a value of water content averaging 7,51%, far from the desired standard value is 7%. The required operating conditions to obtain the value of the standard water content of the calculation by the EVOP method is 65°C drying temperature, stirring time of 20 minutes and 130 minutes of drying time.

Determination of the conditions using Evolutionary Operation (EVOP) produces product water content of 6.90%. And the effect of factor A, B, C and interaction of ABC are significance to the confidence level.

The Company did not change its production capacity is 240 cartons per production cycle in by what has been done, so that operating conditions obtained from the calculation EVOP can produce a product quality standard value of water content. Companies can apply the method EVOP in the repair process continuing basis to maintain product quality and to be able to improve processes quickly, accurately, and continuous corresponding calculation results EVOP.

Acknowledgements

The author would like to thank University of Sumatera Utara for funding the financial support for publishing this paper.

References

[1] Rutten K De Baedemaeker. 2014. A comparison of Evolutionary Operation and Simplex for process improvement, Chemom. Intel. Lab. Syst. **139** 109-120
[2] Box GEP Hunter W.G. 1978. Statistics for Experiments. An Introduction to Design, Data Analysis and Model Building. John Wiley & Sons, Inc, USA 653
[3] S Vats. 2013. Use of Artificial Neural Network (ANN) for the development of bioprocess using Pinus roxburghii fallen foliages for the release of polyphenols and reducing sugars, Bioresour. Technol **140** 392-398
[4] Montgomery D C Design and Analysis of Experiments 2001 John Wiley & Sons, New INC.427.
[5] Box GEP Hunter W.G. 1978. Statistics for Experiments. An Introduction to Design, Data Analysis and Model Building. John Wiley & Sons, Inc, USA
[6] Smart, Jimmy L 2003 Use of an Applied Statistical Method to Optimize Efficiency of an Air Pollitiion Scrubber Within an Undergraduate Laboratory, ASEE Conf. Proc., Nashville, TN
[7] Hoerner G. M, and W E Shiesser 1965 Simultaneous Optimization and Transient Response Evaluation of Packed-Tower Gas Absorption. Chem. Eng. Prog. Symp. Ser. **61**(55) 115
[8] Juran, J M 1974. *Quality Control Handbook*, 3rd ed. New York: MC Graw Hill Book Company

26 30-26
[9] Klingel A R and R G McIntyre 1962 An Experimental Strategy for Investigating Commercial Processes, Appl. Statisticcs, **11**(2) 79.
[10] Spendley W, G R Hext and F Himsworth 1962 Sequential Application of Simpelx Designs in Optimisation and Evolutionary Operation Technometrics, **4**(4) 441
[11] Carpenter B H, and H C Sweeny 1965 Process Improvement with Simplex Self-Directing Evolutionary Operation Chem. Engr. 117
[12] Scarrah w p 1987 Improve Production Efficiency via Evolutionary Operation, Chem. Engr. 131
[13] Montgomery Douglas C 2009. *Statistical Quality Control*. New York: John Wiley & Sons. 560-1.
[14]Montgomery Douglas C 1985. *Design and Analysis of Experiments*. New York: John Wiley & Sons INC. 231-3.
[15] Banerjee R and Bhattacharyya B C 1992. *Evolutionary Operation (EVOP) to Optimize Three Dimensional Biological Experiments. Bioprocess Engineering* **8**: 151

IOP Conf. Series: Materials Science and Engineering **309** (2018) 012032 doi:10.1088/1757-899X/309/1/012032

Production facility layout by comparing moment displacement using BLOCPLAN and ALDEP Algorithms

M Tambunan*, E Ginting and R M Sari
Lecturer in Industrial Engineering Department, Faculty of Engineering, Universitas Sumatera Utara, Medan, Indonesia.

*Email : araapul_tambunan@yahoo.com

Abstract. Production floor layout settings include the organizing of machinery, materials, and all the equipments used in the production process in the available area. PT. XYZ is a company that manufactures rubber and rubber compounds for retreading tire threaded with hot and cold cooking system. In the production of PT. XYZ is divided into three interrelated parts, namely Masterbatch Department, Department Compound, and Procured Thread Line Department. PT. XYZ has a production process with material flow is irregular and the arrangement of machine is complicated and need to be redesigned. The purpose of this study is comparing movement displacement using BLOCPLAN and ALDEP algorithm in order to redesign existing layout. Redesigning the layout of the production floor is done by applying algorithms of BLOCPLAN and ALDEP. The algorithm used to find the best layout design by comparing the moment displacement and the flow pattern. Moment displacement on the floor layout of the company's production currently amounts to 2,090,578.5 meters per year and material flow pattern is irregular. Based on the calculation, the moment displacement for the BLOCPLAN is 1,551,344.82 meter per year and ALDEP is 1,600,179 meter per year. Flow Material resulted is in the form of straight the line.

1. Introduction

The main thing to consider in designing a layout of the production floor is the material movement system [1]. Effective and efficient material movement systems will affect a company's production capability to provide products on time to consumer [2]. Conversely, a poor material movement system will cause production to be hampered and will give a loss to the company. Good layout is the layout that can handle the material handling system as a whole.

PT. XYZ is a company engaged in the processing of *rubber* products with raw materials into rubber compounds and pre Cured Tread liner for tires with hot and cold cooking system.
Some problems were found in the company such as no spesific raw materials and products warehouse were available and hence, the raw materials and finished products were placed on the empty space of the production floor. Placement of raw materials and products hamper the production process. Also, the flow patterns of material contained in the production floor in the form of irregular flow patterns (odd angel) and production machines not placed in the order of processes. The using of BLOCPLAN algorithm solves the problem of backtracking, irregular flow material, distance the movement, and add one machine. The using of ALDEP algorithm solves the problem of a far material displacement [3].

2. Methodology

The research was conducted at PT. XYZ from September 2016 to January 2017. The research object observed on the production layout. The research describes troubleshooting to a problem that exists now and get the findings of practical as operational decision making. So this study included the collection, presentation and data processing as well as Analysis and interpretation on PT. XYZ

This research has activities in the form of data collection and data processing. The variables are material flow type, production process sequence, material movement, material movement distance, and production capacity [4]. After data collection, data processing calculated by using software BLOCPLAN 90 and ALDEP algorithms. The design results of the two algorithms calculated the distance of material movement and calculated the moment of material displacement per year. The design results with the smallest displacement moment values are the best and with other assessments such as the flow patterns of the materials produced.

3. Results and Discussion

The initial phase of the study is the preliminary to find out the initial conditions for observation and interviews and literature studies on the methods of problem solving used and other supporting theories. The problem of this research is a long distance movement of materials and material flow away, causing big displacements moment can reduce yield. Therefore, it is necessary to design a layout which has a regular flow of material and minimum material movement distance so the moment is minimum. o the next stage is collecting data by observing and measuring directly on the production floor. Primary data obtained from observations, and direct measurements of the object of research in the field are: the data size of the production department, block the company's production floor layout and sequence of the process. Secondary data is in the form of production capacity [5].
The initial layout of PT XYZ can be see in Figure 1.

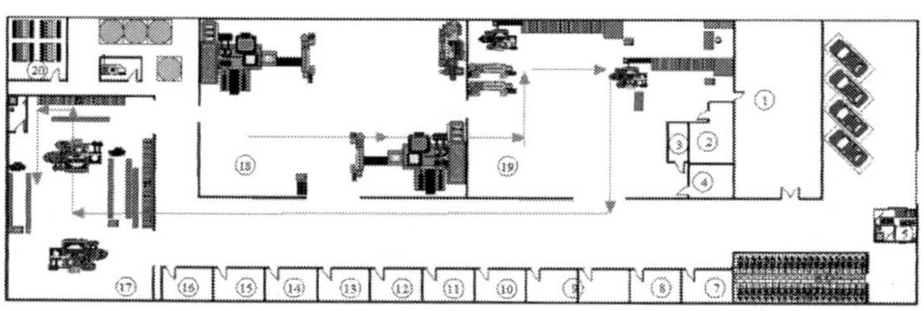

Symbol	Description	Symbol	Description
1	Office	11	Clinics
2	Cold Lateks Room	12	Rest Room
3	Plastic Storage	13	Mushollah
4	Warehouse	14	Wudhu Place
5	Security Room	15	Tankos Tank
6	Motorcycle Area	16	Boiler
7	arbon Balack Storage	17	Material Warehouse
8	Waste Storage	18	Production Floor 1
9	roduct Reject Storage	19	Production Floor 2
10	Chemical Storage	20	Canteen

Figure 1. Initial Layout PT XYZ

The work station size and encoding data can be seen at Table 1.

Table 1. Work Station Size and Encoding Data

Workstation	Size of Area (m)	Area (m^2)	Code
Main ingredient and cuts	11.15 x 7.65	84.9	A
Appolo Machine (External Mixer 22 Inch and Internal Mixer) 1	8.8 x 14.5	127.6	B1
Appolo Machine (External Mixer 22 Inch and Internal Mixer) 2	8.8 x 14.5	127.6	B2
Milling rubber hard	3.3 x 6.9	22.7	C
Milling with External Mixer 18 Inch	5.5 x 3.8	29.9	D

Formation with Extruder 1	16.2 x 3.5	56.7	E1
Formation with Extruder 2	14 x 3.5	49	E2
Pressing and Formation Screw	7.5 x 16.4	123	F
Smoothing with grader machine 1	10 x 1.9	19	G1
Smoothing with grader machine 2	10 x 1.9	19	G2
Quality Control and Packing 1	8 x 3	24	H1
Quality Control and Packing 2	8 x 3	24	H2

After collecting the size data collected other data such as the production capacity of 5.4 tons per day, the sequence data of the process and the depiction of Block Layout from the initial production floor. Next calculation of determining the point of coordinates X and Y on the block layout from the initial production floor with the formula:

- $X = X_0 + \frac{(X_1 - X_0)}{2}$
- $Y = Y_0 + \frac{(Y_1 - Y_0)}{2}$

The degree of proximity relationship or Activity Relationship Chart [5-8] can be seen at Figure 2.

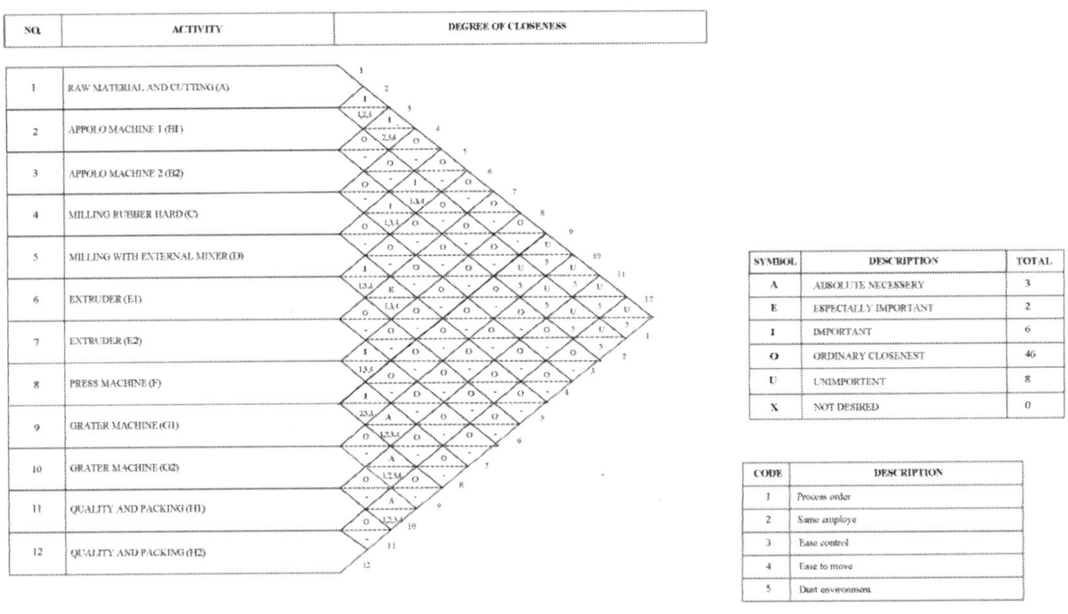

Figure 2. ARC PT. XYZ

Once obtained ARC PT. XYZ, then calculated by using BLOCPLAN algorithm. This calculation is done with BLOCPLAN90 software.

Then followed data with BLOCPLAN90 software to get the best results. Coordinate Point of software result BLOCPLAN90 is used to find the distance between work stations and total moment of movement.

The design result of BLOCPLAN algorithm resulted in total displacement moment of 1.551.344,82 m / year. Layout design with ALDEP algorithm is done by using ALDEP Software. ALDEP software is used to find the most optimal TCR (Total Closeness Rating) and layout values by entering the number of departments and production floor area [9].

From a distance between the work stations, calculated the total moment of material movement of ALDEP algorithm design results. The total movement time of ALDEP algorithm design is 1.600.179 meters per year. Total of moment displacement calculation can be seen at Table 2.

Table 2. Total Moment of Displacement Results of BLOCPLAN Algorithm and ALDEP Algorithm

No	First Station	Destination Station (m)	Distance of Displacement (m)		Moment of Displacement (m/year)	
			BLOCPLAN	ALDEP	BLOCPLAN	ALDEP
1	A	B1	9.75	26.12	43724.9	43724.9
2	A	B2	9.74	20.62	34517.9	34517.9
3	B_1	D	9.64	29.45	49299.3	49299.3
4	B_2	D	29.13	23.95	40092.3	40092.3
5	D	E1	23.02	17.1	204482	204482
		E2	10.92	9.6	114797	114797
6	E_2	F	29.69	6.7	11215.8	11215.8
7	F	G1	8.48	7.9	220410	220410
		G2	11.9	11.4	318060	318060
8	G_1	H1	8.69	10.65	297135	297135
9	G_2	H2	6.71	9.55	266445	266445
		Total			1.551.344,82	1.600.179

The results of the algorithm compared with the company's initial layout. The comparison made by assigning weight to each category. The weighting is:

1. Category moment of movement, weighted values are:
   A. 1.000.000 m / year – 1.500.000 m / year weight value = 3
   B. 1.500.001 m / year – 2.000.000 m / year weight value = 2
   C. 2.000.001 m / year – 2.500.000 m / year weight value = 1
2. Categories of building changes, weighted values are:
   A. The shape of the building does not change; the weight value is = 2
   B. The shape of the building changed; the value of weight that is = 1
3. Category of material flow patterns generated, weighted values are:
   A. The pattern of regular flow; the value of weight that is = 2
   B. Irregular flow pattern; weight value i.e., = 1
4. Space category for material handling, weighted values are:
   A There is a space for material handling; the weight value is = 2
   B. There is no space for material handling; the weight value is = 1

The results of the comparison with the weighting can be seen at Table 3.

Table 3. Best Layout Comparison

	First Layout (Company Layout)	Result From BLOCPLAN Algoritm	Result From ALDEP Algoritm
Moment Displacement (m/year)	2.090.578,5 (1)	1.551.334,82 (2)	1.600.179 (2)
Changing Shape Of Building	Not Changed (2)	Changed (1)	Not Changed (2)
Flow System	Not Directed (1)	Not Directed (1)	Directed (2)

Material Handling Space	Exist (2)	None (1)	Exist (2)
Total Weight	6	5	8

From Table 3, the highest weighting values contained in the draft outcome Algorithm total moment ALDEP value material movements is 1.600.179 meters/year. Form of the building has not changed, the flow pattern which is irregular and there is still space for the material handling. Also, each design result of the ALDEP algorithm provides a warehouse facility for raw materials or a finished product. Layout result for both BLOCPLAN Algorithm and ALDEP Algorithm can be seen at Figures 3 and 4.

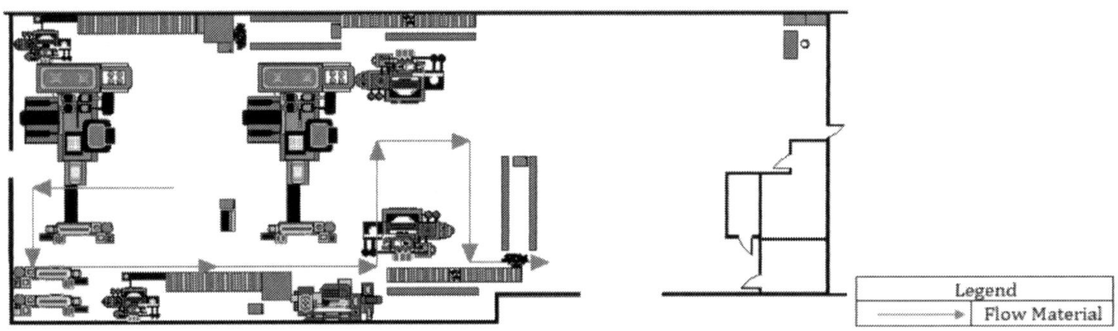

Figure 3. Layout Result BLOCPLAN Algorithm

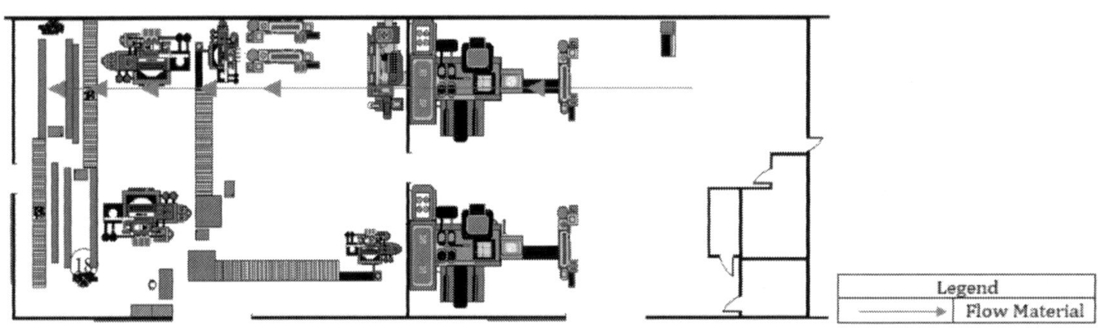

Figure 4. Layout Result ALDEP Algorithm

4. Conclusion

After data collection, data processing, and problem solving as well as alternative search algorithms using BLOCPLAN layout and algorithms ALDEP in redesigning the layout of the production floor to the PT. XYZ, it can take several conclusions, moment displacement of material from the production floor PT. XYZ is currently 2.090.578,5 meters per year. Moment of the existing layout of greater value than the displacement moment with an algorithm designed by BLOCPLAN which has a value of 1.551.344,82 meters per year and ALDEP is 1.600.179 meter per year. Comparison of results with BLOCPLAN algorithm and ALDEP has a smaller value than the initial layout. BLOCPLAN has the smallest displacement but must change the shape of the building and production flow pattern while ALDEP has a value of displacement is smaller than initial layout but bigger than BLOCPLAN, if ALDEP is adopted the shape of the building is not changed and production floor has a flow patern of straight line.

Acknowledgement

Author would like to thank Universitas Sumatera Utara for funding the financial support for publishing this paper.

References

[1] W, Sritomo 2009 *Tata Letak Pabrik dan Pemindahan Bahan*. Surabaya : Technology Institute of 10 November 67.
[2] S, Merry. 2016. *Redesign and Production Floor Layout Simulation Method Using Pairwise Exchange PT. Permata Alam Riau,* Pekanbaru: Journal of Science and Technology Industry | No. 02 | **13** 1-4.
[3] S, Ivama Christine. 2014. *The Proposed Facilities Layout Design of PT. X Using ALDEP Algorithm*. Journal of Ma Chung University| No. 01 | **04** 17-18.
[4] Andryzio. 2015. *Proposed Facility Layout Design with Method Using Automated Layout Design Program (ALDEP) CV. Kawani Tekno Nusantara*. Bandung: Online Journal of National Technology Institute | No.04| **02** 8-9.
[5] M, Fajar. 2015. *Workstation Layout Redesign Injection Molding Process Based on Workload Analysis and Process Simulation Component Manufacture LCDTV*. Malang: JEMIS. Jurnal Teknik Industri Brawijaya | No. 02 | **03** 4-7.
[6] Barathi, G. 2015 *Comparative Study Of Craft And QAP Techniques Applied To Redesigning Of Layout For Cub-Board Manufacturing*. International Journal of Invormative & Futuristic Research | No.11 | **02** 6-9.
[7] F, Angga Daula. 2015. *The Design of The Production Floor Layout Fender Using The Automated Layout Design Program at PT. AGRONESIA RUBBER TECHNIQUES DIVISION*. Bandung: Journal of Industrial Engineering Itenas | No.02 | **03** 8-12.
[8] J, Inaki Maulida. 2015. *Production Facilities Improvement Of Layout For A Secondary Packaging Area Of A Pharmaceutical Company In Indonesia Using The CORELAP Method*. International Journal of Technology 4-8
[9] P, Tika. Et. 2015. *Proposed Facility Layout Design Method Using Automated Layout Design Program (ALDEP) In Ceramic Edem*. Itenas Journal. Jurnal Online Institut Teknologi Nasional |No. 02| **03** 8-9.

IOP Conf. Series: Materials Science and Engineering **309** (2018) 012033 doi:10.1088/1757-899X/309/1/012033

Physical properties of sago starch biocomposite filled with Nanocrystalline Cellulose (NCC) from rattan biomass: the effect of filler loading and co-plasticizer addition

H Nasution*[1], H Harahap[1], M T Al Fath[1] and Y Afandy[1]
[1]Department of Chemical Engineering, Faculty of Engineering, Universitas Sumatera Utara, Padang Bulan, Medan 20155, Indonesia.

*Email: h_dahliana@yahoo.com

Abstract. Rattan biomass is an abundant bioresources from processing industry of rattan which contains 37.6% cellulose. The high cellulose contents of rattan biomass make it a source of nanocrystalline cellulose as a filler in biocomposites. Isolation of alpha cellulose from rattan biomass was being prepared by using three stages: delignification, alkalization, and bleaching. It was delignificated with 3.5% HNO_3 and $NaNO_2$, precipitated with 17.5% NaOH, bleaching process with 10% H_2O_2. Nanocrystal obtained through the hydrolysis of alpha cellulose using 45% H_2SO_4 and followed by mechanical steps of ultrasonication, centrifugation, and filtration with a dialysis membrane. Biocomposite was being prepared by using a solution casting method, which includes 1–4 wt% nanocrystalline cellulose from rattan biomass as fillers, 10-40 wt% acetic acid as co-plasticizer and 30 wt% glycerol as plasticizer. The biocomposite characteristic consists of density, water absorption, and water vapors transmission rate. The results showed the highest density values was 0.266 gram/cm^3 obtained at an additional of 3 wt% nanocrystalline cellulose from rattan biomass and 30 wt% acetic acid. The lowest water absorption was 9.37% at an additional of 3 wt% nanocrystalline cellulose from rattan biomass and 10 wt% acetic acid. It was observed by the addition of nanocrystalline cellulose might also decrease the rate of water vapor transmission that compared to the non-filler biocomposite.

1. Introduction

Cellulose is the main component of plant cell walls and it is regarded as the most abundant natural resource. Cellulose has potential applications in the development of new high-performance materials with low environmental impact. Cellulose in particular size has large surface areas, relatively high strength and low density [1]. As a result, cellulose can be used to make green composite polymers [2], surfactant [3] and membrane [4].

Rattan biomass is the agricultural waste that can be used as a source of cellulose fiber for biocomposite. The high availability of rattan makes this natural resource can be engineered into a national technology product using a nanotechnology system [5]. Rattan consists of 37.6% cellulose, 41% hemicellulose and 22.6% lignin [6]. The cellulose content of rattan can be used as nanocrystalline cellulose. Based on the function of cellulose, it can be a reinforcement filler on biopolymer [7]. Reports have shown that cellulose as a filler has been done by Halimatuddahliana, et al. (2017) that made microcrystalline cellulose (MCC) from empty bunches of oil palm [8] and Harahap, et al. (2015) made nanocrystalline cellulose (NCC) from sugarcane bagasse [9]. NCC and MCC have potential as reinforcement in biopolymers.

Over the years, all people are concerns about environmental problems with conventional plastics. This is causes the production of the petroleum-based plastics from non-renewable resources. Many researchers have been focused on materials that have properties comparable with those of the conventional polymeric materials. From this perspective, biocomposite based on starch are seen as the candidates to be developed as biodegradable materials.

Starch is a natural renewable polymer obtained from the variety of agricultural products like tubers and seeds. Among its advantages, starch is cheap, abundant and biodegrades quickly [10]. Biodegradable packaging can be produced from cassava starch [11], corn [12] and sago starch [13]. Sago starch has unique characteristics: it is easy to gelatinize and molded, availability as a renewable resource, cheap and has a high starch content of 82.94% [13]. Sago starch can be processed into thermoplastic starch or biodegradable plastic under the action of temperature and shearing action [13].

The bioplastic consisting of starch as an ingredient requires a mix of additives that are used to produce mechanical properties of soft, and resilient. The additives as plasticizer, co-plasticizer, and filler are typically used to be less rigid from starch [14]. Plasticizers such as glycerol are more commonly used because when processing there is no glycerol evaporate so that the mechanical process is easy to do. Acetic acid can be used as a plasticizer coating (co-plasticizer). The carboxyl group of acetic acid will make a strong hydrogen bond with the hydroxyl group of starch [13]. Tawakaltu, et al. (2015) reports the effect of acetic acid and citric acid modification on biodegradability of cassava starch bio-nanocomposite films [15].

Based on the explanation above, will research the effect of filler loading and co-plasticizer addition on physical properties of sago starch biocomposite.

2. Method

2.1 Material

Rattan biomass was obtained from Kota Pinang, North Sumatera, Indonesia. Sago starch was being purchased by the local market. Sago starch particles were being screened by through a mesh size of 200. The chemicals, nitric acid (HNO_3), sodium nitrite ($NaNO_2$), sodium hydroxide (NaOH), sodium sulfite (Na_2SO_3), sodium hypochlorite (NaOCl), hydrogen peroxide (H_2O_2), sulfuric acid (H_2SO_4), glycerol and acetic acid were supplied by MERCK and used as received.

2.2 Production of Alpha Cellulose

Alpha cellulose was obtained from rattan biomass with a delignification process by using 3.5% nitric acid (HNO_3) at 90 °C. Followed by washing and filtrating and digested with 2% w/v sodium hydroxide (NaOH) and sodium nitrite ($NaNO_2$). Bleaching was being treated by two steps. The first bleaching step was carried out with 1.75% sodium hypochlorite (NaOCl) at boiling temperature for 0.5h. Alkalization was being treated by 17.5% sodium hydroxide (NaOH) at 80 °C for 0.5 h. The second bleaching step was carried out with 10% hydrogen peroxide (H_2O_2) at 60 °C for 1h.

2.3 Production of NCC

The alpha cellulose derived from cellulose was hydrolyzed using 49% sulfuric acid (H_2SO_4) and stirred at high-speed for 45 min at 45 °C. The white suspension was diluted with distilled water and allowed to settle overnight. The clear top layer was being decanted, and the remain of a cloudy white layer was being washed by distilled water and centrifuged at 10000 rpm for 25 min at several times to remove the solvent and other organics soluble. The white precipitation was diluted with distilled water and dispersed by ultrasonication for 10 min. The suspension was dialyzed using activated dialysis tubing for 96 hours. NCCs were dried in vacuum oven at 25 °C until the water was evaporated.

2.4 Production of Biocomposite

Biocomposite films were being based by a solution casting and evaporation process. The variation amount of NCCs (1, 2, 3, 4 wt% of starch) was dispersed into the 95 ml of distilled water under continuous stirring at 1000 rpm for 15 min. After that designated weights of 10 gr of sago starch were mixed and stirred for the complete dissolution of starch. Glycerol content based on starch was being fixed by 30 wt% and Acetic acid with the variation amount (10, 20, 30, 40 wt% of starch). Starch, glycerol and acetic acid were mixed and heated at 70 °C for 30 min with stirring until the mixture gelatinized. The water was evaporated at room temperature overnight after that solution poured into an acrylic mold. The Film formed uniformly and removed smoothly.

2.5 Characterization

2.5.1. Density. The density of biocomposite has been investigated by the standard of ASTM D792-91 on film with size approximately 5x5 cm. The film's mass was measured using analytical balance. The Density calculated as follows:

$$\text{Density} = \frac{\text{mass (gram)}}{\text{volume (cm}^3\text{)}} \quad (1)$$

2.5.2. Water Uptake. Water uptake was investigated by cutting film with size approximately 2x2 cm and then weighed the mass. The film put into a container filled with distilled water for 24 hours. After immersion in water, a film was removed from the water and weighed to measure the wet weight. Water uptake calculated as follows:

$$\text{Water Uptake} = \frac{\text{Wet Weight-Dry Weight}}{\text{Wet Weight}} \times 100\% \quad (2)$$

2.5.3. Water Vapor Transmission Rate (WVTR). The Film to be tested in the mouth of a porcelain bowl with an inside diameter of 7 cm, an outside diameter of 8 cm and a depth of 2 cm which contained 10 gram of silica gel. The edges of the porcelain bowl and film covered with wax or insulation. The porcelain bowl inserted into the jar that containing 40% sodium chloride (NaCl) (w/v). Water vapor diffuses through the film will be absorbed by silica gel and will add weight to the silica gel. The condition of the water vapor transmission rate of equilibrium is reached by 7-8 hours. The films were weighed periodically every 1 hour. The data obtained made linear regression equation and WVTR values can be determined by the equation:

$$\text{WVTR} = \frac{\text{linear slope for the change in weight of a jar (g/hour)}}{\text{Surface Area (m}^2\text{)}} \quad (3)$$

3. Results and Discussion

3.1 The Effect of NCC Content and Acetic Acid Addition on the Density of Biocomposites

The density of the biocomposite shows in Figure 1. The density increased considerably with the filler and co-plasticizer content, at 3 wt% NCCs and 30 wt% acetic acid addition, the biocomposite was showed an optimum value of 0.26 gram/cm³. The improvement indicated that nanocrystalline cellulose (NCC) is evenly distributed causing and increase the density of the biocomposite structure [16,17]. At 4 wt% NCC loading, a slight decrease in the density was observed, it indicated that nanocrystalline cellulose (NCC) formed aggregates that reducing the effectiveness of biocomposite density [14,18].

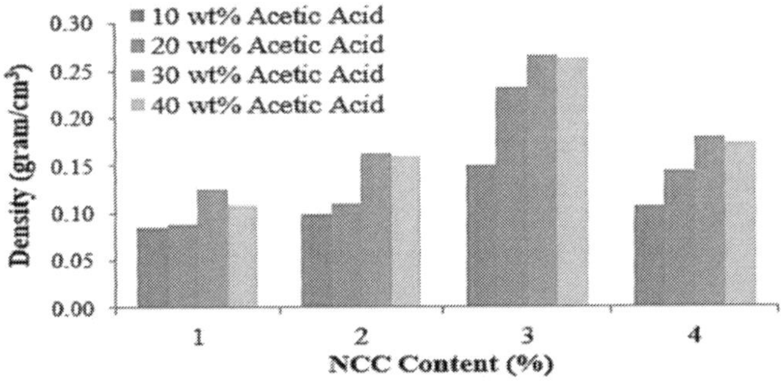

Figure 1. The Effect of NCC Content and Acetic Acid Addition on the Density of Biocomposites

As shown in Figure 1, at 30 wt% acetic acid addition, biocomposite has a high-density value because the acetic acid will increase the physical interaction that allows the formation of hydrogen bonds into more chains of starch which increases the density of the biocomposite. When the acetic acid percentage increased to 40 wt%, acetic acid will loosen the bonds between molecules in biocomposite, reducing the stiffness of the biocomposite structure that can reduce the density of biocomposite [19,20]. The mechanism of acetic acid and cellulose interaction show in the Figure 2 [20]:

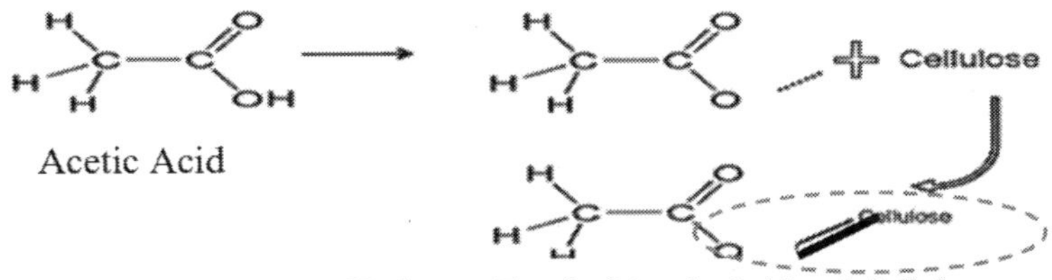

Acetic Acid

Hydrogen Bond of Acetic Acid and Cellulose

Figure 2. Mechanism of Acetic Acid and Cellulose Interaction

The distribution and dispersion were being supported by scanning electron microscope (SEM) of bioplastic and biocomposite that showed the surface of the matrix is more smooth and coherent. It shows that nanocrystalline cellulose (NCC) has been uniformly distributed on the sago starch matrix. As co-plasticizer agent, the acetic acid has good interface bondage with the filler and matrix as shown in Figure 3.

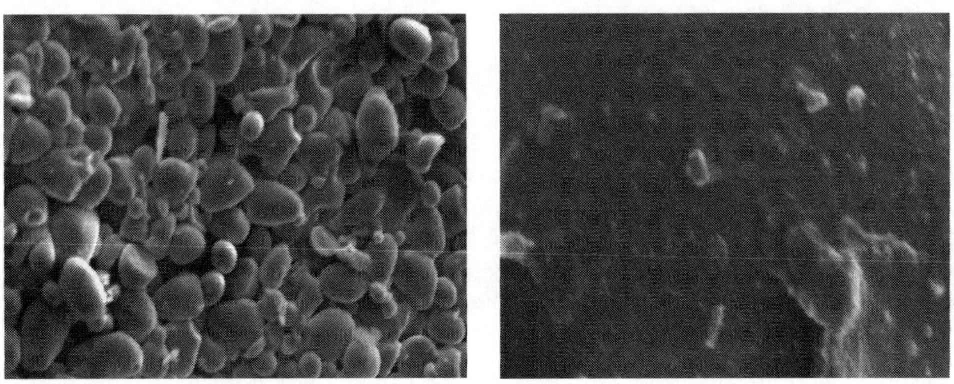

(a) **(b)**

Figure 3. SEM (1000x) of Bioplastic (a) and Biocomposite (b)

3.2 The Effect of NCC Content and Acetic Acid Addition on the Water Uptake of Biocomposites

Figure 4 shows the water uptake of biocomposites with NCCs and acetic acid addition. The addition of NCCs led to a decrease in water uptake, which attributes to the formation of a cellulose nanoparticle network that prevented swelling of the starch and therefore water uptake as well. At 4 wt% nanocrystalline cellulose (NCC) addition, the water uptake biocomposite increased due to the agglomeration of nanocrystalline (NCC) cellulose and the presence of amorphous zones in nanocrystalline cellulose (NCC) making it easier for water to diffuse into the biocomposite that increasing the properties of water uptake [21,22]. At 30 wt% acetic acid addition, the biocomposite water uptake becomes increased. The increased percentage of water uptake is due to an increase in the concentration of co-plasticizer, this is thought to cause the bonding that occurs between the plasticizer, and the polymer becomes brittle when exposed to water [23].

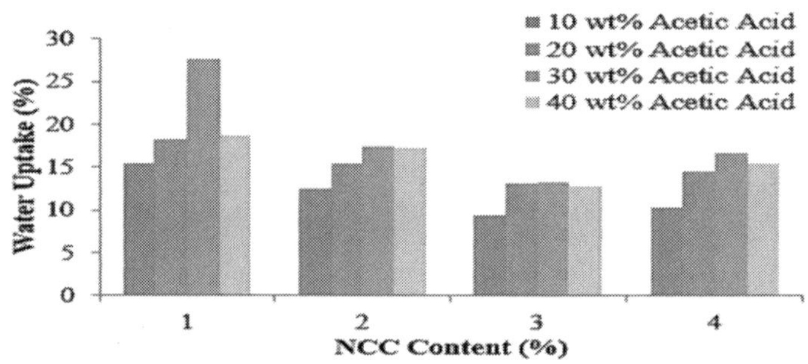

Figure 4. The Effect of NCC Content and Acetic Acid Addition on the Water Uptake of Biocomposites

3.3 Water Vapor Transmission Rate (WVTR)

The water vapor transmission rate is the important parameter for packaging film materials on biocomposite. Table 1 shows the water vapor transmission rate of bioplastic and biocomposite which is calculated by according to the equation (3).

Table 1. Water Vapor Transmission Rate of Bioplastic and Biocomposite with 3 wt% NCC and 10 wt% Acetic Acid Addition

Sample	**WVTR (g/m^2.hour)**
Bioplastic	10.33
Biocomposite	5.61

The diffusivity and solubility of water molecules in the film matrix will affect the water vapor transmission rate [24]. Starch is hydrophilic, and hence it has high WVTR. The high WVTR is decreased by the presence of dispersed phase of NCC into the starch biopolymer matrix [14]. Biocomposite has better dispersion and lower aggregation of NCCs particles contributing to generating a barrier for the passage of water molecules evoking a decrease in WVTR [24]. On the other hand, NCC was well dispersed and covered by the matrix, suggesting that the introduction of NCC into the starch and plasticizer films resulted in improved adhesion between the NCC and the polymer matrix [25]. This result was supported by the morphology image as seen in Figure 3b.

4. Conclusion

The density value increases with NCC and acetic acid additions. Furthermore, the addition of NCC content decreases water uptake of biocomposites. Contrary from NCC content addition, an increases in water uptake value was being reported by the acetic acid addition. It observes that the addition of nanocrystalline cellulose might also decrease the rate of water vapor transmission that compared to the non-filler biocomposite.

Acknowledgement

The authors gratefully acknowledge that the present research supported by the Lembaga Penelitian Universitas Sumatera Utara. The support is under the auspices of the research grant TALENTA USU of the year 2017 contract number 5338/UN5.1.R/PPM/2017 Date 22 Mei 2017.

References

[1] Peng B L, Dhar N, Liu H L and Tam K C 2011 *The Canadian Journal of Chemical Engineering* **89** 1-16

[2] Xiaodong C, Yun C, Peter R C, Mark S and Michel A H 2008 *J. Appl. Poly. Sci.* **109** 3804-3810

[3] Evdokia K O, Nikolay C, Galder C, Annie V, Marc A, Claudie B, Laurent H and Jean F B 2017 *J. Phys.Chem.* **10** 2299-2307
[4] Hongyang M, Christian B, Benjamin S and Benjamin C 2011 *Biomacromolecules* **13** 180-186
[5] Siti N, Sudirman, Lisdar I and Mersi K 2014 *Jurnal Ilmu Pertanian Indonesia (JIPI)* **3** 163-168
[6] Siti N, Setyo P, Akhiruddin M, Tieneke M and Aris P 2010 *Indonesian Journal of Materials Science* **3** 97-102
[7] Chen D, Lawtona D, Thompson M R and Liu Q 2012 *Carbohyd. Polym.* **90** 709– 716
[8] Halimatuddahliana N, Yurnaliza, Veronicha, Irmadani and Sitompul S 2017 *Materials Science and Engineering: IOP proceeding* **180** 1-8
[9] Hamidah H, Marfuah L, Nuim H and Sukardi 2006 *Inter. Chem. Tech Res.* **9** 201-205
[10] Johar N and Ahmad I 2012 *Bioresources* **4** 5469-5477
[11] Zainuddin, Ahmad I and Kargarzadeh H 2014 *Composite Interfaces* **3** 189-199
[12] Xiaofei M, Jiugao Y and John F K 2005 *Carbohyd. Polym.* **62** 19–24
[13] Zuraida A, Hazleen A, Yusliza Y 2011 *Key Engineering Materials* **471-472** 397
[14] Savadekar N R and Mhaske S T 2012 *Carbohyd. Polym.* **89** 146-151
[15] Abdul Rasheed-Adeleke T, Egwin E C, Ochigbo S S and Ossai P C 2015 *Journal of Materials Science and Engineering* **5** 372-379
[16] Zimmermann T, Bordeanu N and Strub E 2010 *Carbohyd. Polym.* **79** 1086-1093
[17] Bilbao S C, Bras J, Williams T, Sénechal T and Orts W 2011 *Carbohyd. Polym.* **86** 1549-1557
[18] Kengkhetkit N and Amornsakchai T 2012 *Ind Crop Prod* **40** 55-61
[19] Babak G, Hadi A and Ali A E 2011 *Ind Crop Prod* **33** 229-235
[20] Martha H R, Aji P M and Kristiina O 2014 *Carbohyd. Polym.* **112** 494–501
[21] Dagang L, Tuhua Z, Peter R, Chang b, Kaifu L and Qinglin W 2010 *Bioresource Technology* **101** 2529–2536
[22] Yongshang L, Lihui W and Xiaodong C 2006 *Carbohyd. Polym.* **63** 198–204
[23] Nur A B, Khaswar S and Akhiruddin Maddu 2014 *Jurnal Teknologi Industri Pertanian* **3** 226-234
[24] Feng Y H, Xiao J W, Yan Y and Yan-Hua L 2015 *Polym.* **7** 1331-1345
[25] Pedro N, Renan M, Giziline C and Suzana M 2016 *Material Research* **1** 167-174

IOP Conf. Series: Materials Science and Engineering **309** (2018) 012034 doi:10.1088/1757-899X/309/1/012034

Effect of pressing temperature on the mechanical properties of waste styrofoam filled sawdust composite

H Nasution*, H Harahap, R Riani and A I Pelawi
Department of Chemical Engineering, Faculty of Engineering, Universitas Sumatera Utara, Padang Bulan, Medan 20155, Indonesia.

*E-mail: h_dahliana@yahoo.com

Abstract. This study has investigated the effect of pressing temperature on mechanical properties of waste styrofoam composite filled with sawdust. The waste styrofoam as the matrix was mixed with sawdust as filler and maleic anhydride (6%wt) as a compatibilizer. The weight fraction ratio between matrix and filler 70:30 (wt) and wood fiber size of 100 mesh were conducted. The pressing temperatures were investigated using a hot press with temperatures varied viz. 120, 130, 150, and 170 °C. Surface modification was applied to sawdust to diminish its polarity so that it could be compatible with the non-polar waste styrofoam matrix. Composites were evaluated using Instron and impact tester machine to investigate the tensile strength and impact strength of the material, respectively. The result indicated that tensile strength has decreased with the increase of pressing temperature where the largest tensile strength is at 130 °C of 33 MPa. The same trend has occurred on impact strength, where the value has reached of 300 J/cm^2 on pressing temperature of 130 °C. From scanning electron microscopy (SEM) analysis it is also confirmed that during impact test, the resistance of the composite which has been pressed at the temperature of 130 °C have given better morphology than the composite at 170 °C.

1. Introduction

The development of composite technology has been starting to experience a shift from the synthetic fibers composite material into natural fiber composite materials. Natural fiber is considered one of the environmentally friendly materials which have good properties compared to synthetic fiber. Some natural fibers as fillers researches have been done, among others: green coconut fiber [1], banana fibers [2], and empty fruit bunch palm oil [3, 4], Sugarcane bagasse [5]. Wood is one of Indonesia's natural resource. Each timber processing into semi-finished materials (e.g., in the form of boards or beams) or finished goods (furniture) always produces byproducts that form sawdust.

The sawdust contains high cellulose content of 69%. Besides, sawdust contains high levels of hemicellulose in general, also contains lignin in an amount of about 15-30% by weight [6]. Therefore, when added to the mixed composite as filler, it will absorb the matrix and provide additional tensile strength between the filler due to adhesion properties and dispersion [7]. Moreover, the matrix will be encapsulated the filler as well as inhibit the diffusion of water in the material due to the nature of matrix hydrophobic.

Aside from being a major product of wood, Indonesia is also among the most widely used plastic in everyday life. One type of plastic that is often used is styrofoam. Styrofoam is made from polystyrene (PS) which is plastic that is brittle.Styrofoam is produced from a mixture of 90-95% polystyrene and

5-10% of gases such as n-butane or n-pentane. Previously, the blowing agent used is CFC (Freon), because this class of compounds can damage the ozone layer it is currently not used anymore, now blowing agent used is more environmentally friendly. Styrofoam is made from styrene monomer by suspension polymerization at a certain temperature and pressure, further heating to soften resins and evaporate the residual blowing agent. Polystyrene is a rigid, transparent, fragile, chemically inert, and is a good insulator. While styrofoam is a plastic material that has special properties with a structure composed of grains with low density, has a light weight, and there is space between granules that contain air and can melt in the temperature around 120 – 180 °C [8].

On the other hand, maleic anhydride (MAH) is a kind of compatibilizer which is often used in the composite. This compatibilizer could modify the adhesion between molecules that are not similar/incompatible. Incompatibility occurs due to the differences in polarity between the materials, in this study, the styrofoam is nonpolar while sawdust containing lignocellulose tends to be polar. This may cause a high surface tension between the matrix-filler interfaces. Therefore, the addition of MAH in this composite may be located between the sawdust and styrofoam to form a bridge thus improve the adhesion between matrix-filler interface [9].

Manufacture of composites from waste in the form of sawdust and plastic waste food wrappers types of polystyrene (styrofoam) is expected to reduce the adverse environmental impacts generated by both these materials. In this study, the effect of pressing temperature on the mechanical properties of waste styrofoam filled sawdust composite was observed.

2. Methods

2.1 Material
Sawdust was obtained from the carpenter. The chemicals such as sodium hydroxide (NaOH), maleic anhydride (MAH), and toluene were supplied by MERCK and used as received.

2.2 Modification Sawdust
Sawdust was soaked with 5% of sodium hydroxide (NaOH) at 30 °C for 4 hours. Then washed with water and dried with oven temperature 100-110 °C. Sawdust particle was screened through a mesh size of 100.

2.3 Composite Manufacturing Process
25 g pieces of styrofoam that cut with around 1,5 x 1,5 cm was diluted with toluene 20% with the addition of 6% maleic anhydride. Diluted styrofoam was mixed with sawdust with a composition ratio of 70:30 (wt). The mixture was poured into a metal mold which has been formed according to the standard tensile strength test and impact strength test. Samples were then pressed by using a hot press with a pressure of 125 psi, preheat for 5 minutes and continue with compression for 8 minutes with varied the temperature120, 130, 150 and 170°C.

2.4 Characterization

2.4.1 Fourier Transform Infra-Red (FTIR).
FTIR analysis was characterized using Shimadzu IR-21.

2.4.2 Tensile Properties.
Tensile strength and elongation at break were investigated by the standard of ASTM D 638 - 10 Type IV using Instron Machine.

2.4.3 Impact Strength.
Impact strength was investigated by the standard of ASTM D 4812 – 11 with *Unnotched Izod* method using Impact Testing Machine.

2.4.4 Scanning Electron Microscopy.
Scanning electron microscopy was employed for microscopic structure of composites fracture from impact test. The instrument used for morphology observation was SEM Evo MA 10 Zeiss.

3. Results and Discussion

3.1 Fourier Transform Infra-Red (FTIR)
FTIR spectra of the different samples are shown in Figure 1.
From the figure it shows the spectra were recorded in the range of 4000-500 cm^{-1}. Styrofoam spectra show the band at 3020, 14355, 1014 and 694 cm^{-1} were attributed to C-H groups from benzene, C-H from alkyl, C-X fluoride, and chloride from HCFC gas. Sawdust spectra show that there are some absorption bands at 3428 and 1570 cm^{-1} were attributed to O-H hydroxyl group and N-H amides band bonding the presence of those peaks have supported that the sample consists cellulose and lignin.

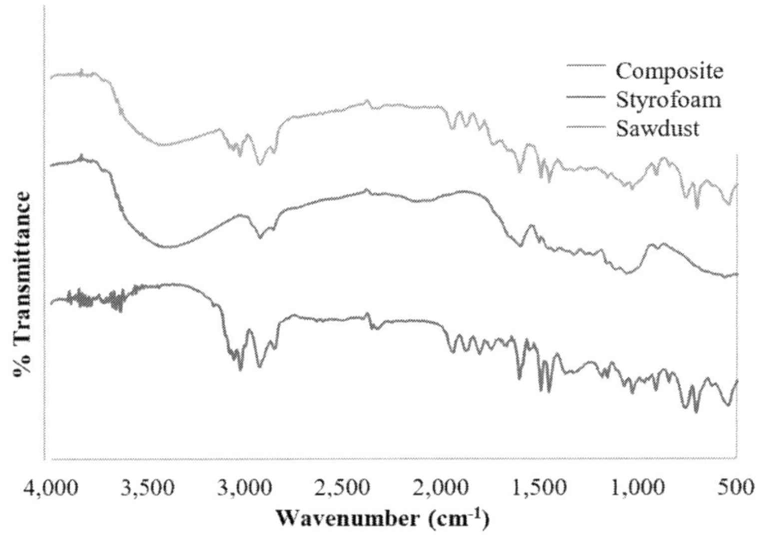

Figure 1. Fourier Transform Infra-Red (FTIR) of Samples

In composite spectra, the presence of MAH in the wavenumber of 1712 cm^{-1} which indicates the existence of ketone groups (C=H). However, the composite FTIR spectra show the combination of all styrofoam, sawdust, and MAH. This spectrum shows no new peaks appeared in the figure, indicated in composite only a physical interaction has occurred.

3.2 Effect of Pressing Temperature on the Tensile Properties of Composite
The tensile strength and elongation at break of the composites are presented in Figure 2. It shows the tensile strength increases from 120 to 130 °C of pressing temperature and decreases at 150 and continued to pressing temperature of 170 °C. Heat transfer is occurred by conduction from the hot platens hot press to the composite surface. The increase of temperature during hot press process vaporizes the moisture within the fillers at the surface. The air, which also contains water vapor that resides in the voids of the composite also increases in temperature and hence vaporizes. On the other hand, the filler (sawdust) exhibits elastic expansion and voids between the filler open, thus increasing composite permeability and gas flow. As the composite is compressed, the gaps are diminished, and the permeability is considerably smaller thus the adhesion between filler and matrix increase [10]. In this case, an optimum condition to produce a composite with high tensile strength was at pressing temperature of 130 °C at 33 MPa. Here, the adhesive bonding between matrix and filler increased and subsequently has enhanced the strength.

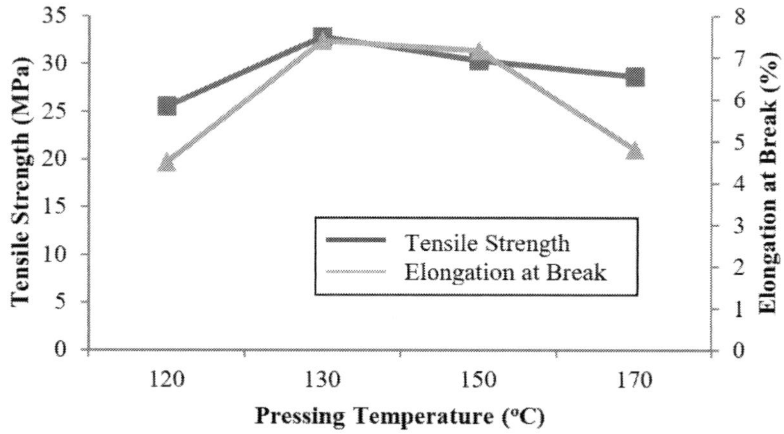

Figure 2. Tensile Strength and Elongation at Break of Composite: Effect of pressing temperature

At lower pressing temperature (120 °C) during hot press, the process has resulted from the low strength that is (25 MPa). Here, at lower pressing temperature the ability of water vapor or gas which have trapped inside the composite has reduced to vaporize. Thus the gaps and voids remain. The consequences were decreasing the mechanical bonding strength, although the pronounced of MAH.

The reduction of tensile strength at pressing temperature of 150°C (30 MPa) and continued until 170 °C (29 MPa) could be due to the presence of voids. The voids have expanded during pressing process in higher temperature. This caused the pull-out filler during the test. In this case, the filler unable to support stress transferred from the matrix and this weaken composite. Moreover, when the very high pressing temperature was used, the degradation of the matrix may cause the brittleness of composite.

As for elongation at break, the curve shows the same pattern as tensile strength. Here, at pressing temperature of 130 °C the elongation at break of the composite is higher as compared to 120 °C. This shows the composite has relatively higher elasticity level. This optimum level of elongation at break has decreased again with the increase in pressing a temperature of 150 and continuing to the temperature of 170 °C. The presence of gaps and void have weakened interfacial regions between the filler and the matrix; then the crack propagates conveniently through the voids. This declined the elongation at break of the composite.

3.3 Effect of Pressing Temperature on The Impact Strength ofThe Composite

Fig 3 shows the effect of pressing temperature on the impact strength of the composite.

It can be seen that the lower value of impact strength composite on the pressing temperature of 120 °C has occurred (230 J/cm^2). As the pressing temperature increase of 130 °C the impact strength has improved significantly at the value of 300 J/cm^2. A slight decrement of impact strength value has evaluated on the pressing temperature reached of 150 °C (293 J/cm^2) and continued at 170 °C 285 J/cm^2).

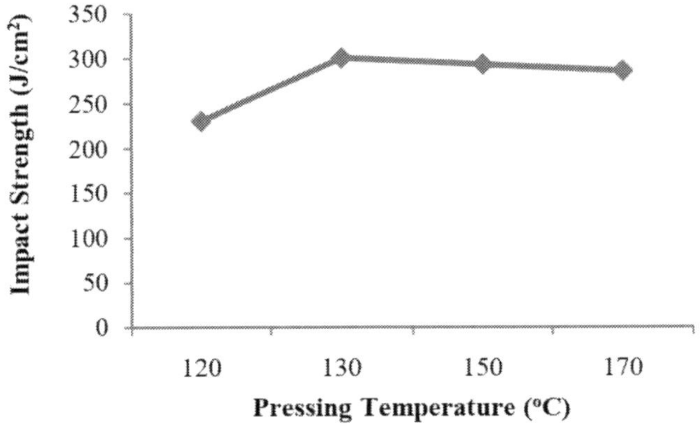

Figure 3. Impact Strength of Composite: the effect of pressing temperature

This trend of the curve has almost similar with has been shown in tensile strength curve as well as elongation at break. This is because the pressing temperature of 130 °C was an optimum temperature where the bonding between matrix and filler increased and subsequently has enhanced the strength. To support the analysis on the impact strength improvement in pressing temperature of 130 °C, the microstructure of the fractured surface of composite specimens tested in impact test is examined using scanning electron microscopy (SEM). Figure 4a and 4b show the morphology of the composite at pressing temperature of 130 °C and 170 °C, respectively.

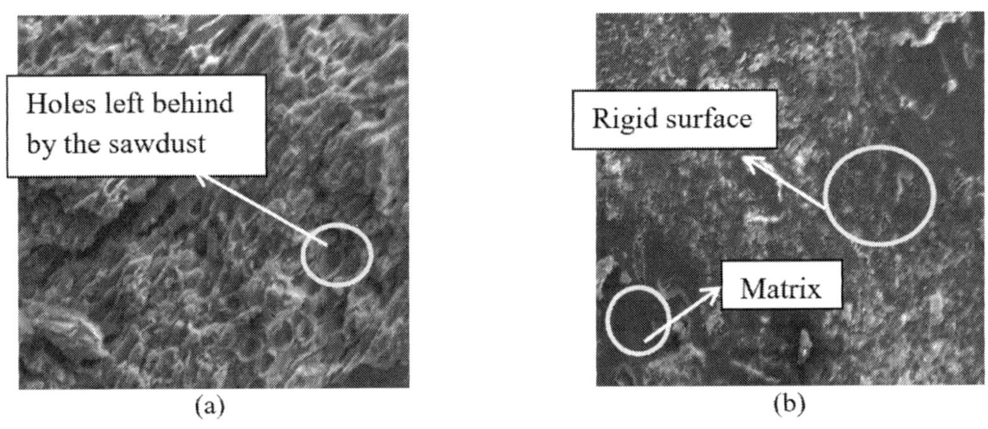

Figure 4. Scanning Microscopy Analysis of Composites at (a). Pressing temperature of 130 °C (b) pressing temperature of 170 °C

It is shown in Figure 4a the uniform distribution of sawdust in styrofoam matrix and also cavities correspond to empty holes left behind by the sawdust during fracture. Moreover, it observes rougher surface that proves the matrix shows more resistance against load given. This surface of the composite confirming its effect on promoting adhesion in the interfacial region, indicating stress transfer from the matrix to the filler.

However, Figure 4b shows smooth fractured surface, and the fillers have not distributed homogeneously. The morphology has also indicated the rigid composite that corresponds to less resistance of composite during fracture. In this case, during higher pressing temperature (170 °C) the presence of voids as well as degradation of the matrix have occurred thus the consequences is the reduction in impact strength. Therefore, Figure 4a and 4b could lead in a difference of energy absorbed in each of composites which composite at pressing temperature of 130 °C would absorb more energy to break it down compared to 170 °C.

4. Conclusion

Results have shown styrofoam filled with sawdust composite only present a physical interaction supporting from FTIR spectra. The highest impact strength, elongation at break and impact strength was achieved in pressing temperature 130 °C. Moreover, SEM micrographs showed the difference of energy absorbed in each composite which composite at pressing temperature of 130 °C would absorb more energy to break it down compared to 170 °C.

Acknowledgement

The authors gratefully acknowledge that the present research is supported by Directorate of Research and Community Service Director General Strengthening Research and Development Ministry of Research and Technology and The Higher Education Republic of Indonesia on grant Contract Number: 003/SP2H/LT/DRPM/IV/2017 Date 20 April 2017.

References

[1] Hussain S A, Pandurangadu V and Palanikuamr K 2011 *International Journal of Engineering Science and Technology* **3** 7942–7952
[2] Singh V K, Gope P C, Sakshi C and Singh B D 2012 *J. Mater. Environ. Sci.* **1** 187-194.
[3] Halimatuddahliana N, Yurnaliza, Veronicha, Irmadani and Sitompul S 2017 *Materials Science and Engineering: IOP Proceeding* **180** 1-8.
[4] Ewulonu C M and Isaac O I 2011 *Inter. J. Eng.Tech* **3** 458-471
[5] Zizumbo A, Angel L C, Eder L M, Edgar G H Domingo M and Roberto Z 2011 *J. Mex. Chem. Soc.* **55** 33-41
[6] Al-Obaidi M F and Al-Ajaj I A 2015 *Int. J. Innov. Res. Sci, Eng. Tech.* **4** 433-442
[7] Idrus M M, Sini H, Muhammad R R and Muhammad S I 2011 *J. Bio. Nanobiotech.* **2** 435-444
[8] Demirkir C, Semra C and Ismail A 2013 *Composites: Part B* **55** 513—517
[9] Halimatuddahliana N, Castiqliana, Silvia, Nuim H 2016 *J. Polym. Mater* **33** 667-676
[10] Winandy, Jerrold E, Kamke and Frederick A 2004 *Proceedings Of A Workshop* Department Of Agriculture, Forest Service, Forest Products Laboratory **118**

IOP Conf. Series: Materials Science and Engineering **309** (2018) 012035 doi:10.1088/1757-899X/309/1/012035

High Voltage Distribution System (HVDS) as a better system compared to Low Voltage Distribution System (LVDS) applied at Medan city power network

R Dinzi, TS Hamonangan, F Fahmi*

Department of Electrical Engineering, Faculty of Engineering, University of Sumatera Utara, Jl. Dr. Mansur 9 Medan 20155, Indonesia

*E-mail: fahmimn@usu.ac.id

Abstract. In the current distribution system, a large-capacity distribution transformer supplies loads to remote locations. The use of 220/380 V network is nowadays less common compared to 20 kV network. This results in losses due to the non-optimal distribution transformer, which neglected the load location, poor consumer profile, and large power losses along the carrier. This paper discusses how high voltage distribution systems (HVDS) can be a better system used in distribution networks than the currently used distribution system (Low Voltage Distribution System, LVDS). The proposed change of the system into the new configuration is done by replacing a large-capacity distribution transformer with some smaller-capacity distribution transformers and installed them in positions that closest to the load. The use of high voltage distribution systems will result in better voltage profiles and fewer power losses. From the non-technical side, the annual savings and payback periods on high voltage distribution systems will also be the advantage.

1. Introduction

The distribution system is the most critical part of electrical power systems performance as it is close to the users. AC Voltage (Alternating Current) with 3 phase 20 kV supplied from the transmission line will be received by the primary distribution system. The power transformer at the distribution substation will provide 20 kV voltage to the industrial loads. The distribution system will then lower the voltage of 20 kV into 220/380 V by using distribution transformers scattered at various points to provide power to household loads.

The main problem with current configuration is too many technical losses in the distribution system especially when it is planning with insufficient knowledge and cost, including low voltage channel 380 V (LT / Low Tension) compared to 20 kV channel which in this case is considered high voltage (HT / High Tension). This system we assumed as LVDS (Low Voltage Distribution System) because the majority of the voltage that is distributed is a voltage of 380 V rather than 20 kV. This results in a loss in the location of a non-optimal distribution transformer that ignores the load location, the development of irregular transmission and distribution networks, and so forth.

A high voltage distribution system or HVDS (High Voltage Distribution System) will use 20 kV voltage lines longer than the 380 V voltage line. In other words, this HVDS system will distribute in the voltage 20 kV; then it will be lowered to 380 V at certain positions. This HVDS system is basically the location change and the capacity of the adjusted distribution transformer. We did a simulation to see the comparison of the old system (LVDS) with the new system (HVDS). The simulation will be

IOP Conf. Series: Materials Science and Engineering **309** (2018) 012035 doi:10.1088/1757-899X/309/1/012035

done using software for a case study of a large capacity distribution transformer of the large load in Medan city.

2. Materials and Methods

2.1 Distribution System

The distribution system is defined as the part of the power system located between the distribution substation and the point of entry of customer service [1].In general, the distribution system can be grouped into two levels, namely primary distribution system or medium voltage network (JTM) and secondary distribution system or low voltage network (JTR). The part of the electrical service system that lies between the distribution substation and the distribution transformer is called the primary distribution system. Meanwhile, the part that lies between the distribution transformer and the service load drop point is called the secondary distribution system.

2.2 Distribution Transformer

The distribution transformer is a step-down transformer that serves to lower the voltage from the primary distribution system (20 kV) into a voltage for the secondary distribution system (220/380 V).

2.3 Loss

The losses to be addressed herein are losses in the distribution transformer, power losses, and non-technical losses. The losses contained in the distribution transformer are divided into copper losses (full load loss) and iron loss (loss without load). Electric theft belongs to the important part as a non-technical loss.

2.4 Cost

To ensure that the HVDS system is better than the LVDS system in non-technical aspect, it is also necessary to determine the advantages of the HVDS system in terms of economic or financial, such as the amount of annual savings, the capital outlay, and the payback period.

2.5 Low Voltage Distribution System (LVDS)

The existing distribution system in Indonesia uses a three-phase mainframe distribution of 20 kV and a three-phase distribution transformer that converts a voltage of 20 kV to 220/380 V. This system is referred to as LVDS (Low Voltage Distribution System) or low voltage distribution system because the majority of the voltage that is distributed is a voltage of 380 V rather than 20 kV. This distribution system includes a ratio of almost 2: 1 at low and high voltage network lengths [2-4].

2.6 High Voltage Distribution System (HVDS)

This High Voltage Distribution System (HVDS) system will use a 20kV voltage line longer than the 220/380 V line. In this system, the 20 kV network extends as close to the load as possible and installs a small capacity distribution transformer which extends the supply of electricity to the consumer via a short low-voltage network [2].

2.7 Research Design

This research was conducted on a network of one distribution transformer installed at Jalan Kartini (Kantor PTPN IV), Medan City and the network of distribution transformer installed at the intersection between Jalan Diponegoro and Jalan Sudirman, Medan City. The research was conducted for 2 (two) months. The variables observed in this study are the voltage profile, power loss, and transformer loss generated on LVDS and HVDS. The flow diagram can be seen in Figure 1.

IOP Conf. Series: Materials Science and Engineering **309** (2018) 012035 doi:10.1088/1757-899X/309/1/012035

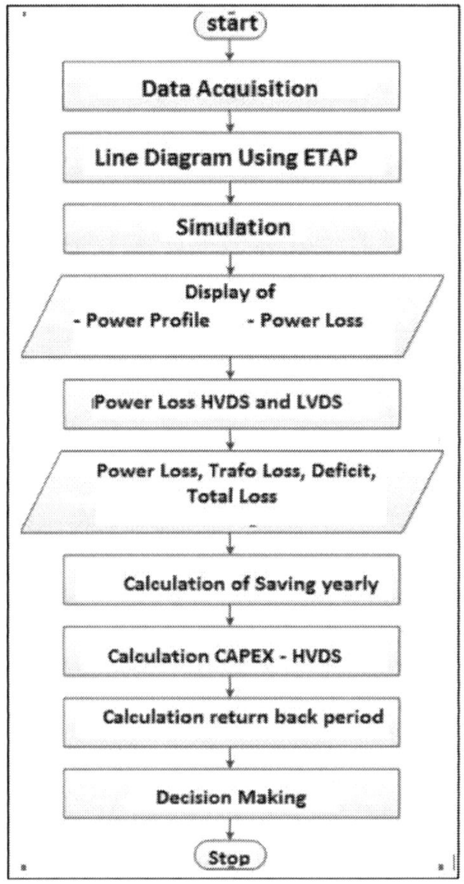

Figure 1. Flow Diagram of Research

2.8 Data Analysis

Data collection of distribution of secondary distribution network is done in PT PLN (PERSERO) AREA MEDAN. ETAP (Electrical Transient and Analysis Program) Power Station is a program that displays the GUI (Graphical User Interface) on power system analysis. The ETAP program was created by the company Operation Technology, Inc. (OTI) from 1995 [5,6]. ETAP calculates the voltage of each bus, the power factor of each branch, the current, and the flow of power throughout the electrical system [7]. A simulated power flow for the old and new system is done using ETAP software on the computer. The result of a voltage profile and simulated losses is used to perform manual side technical calculations on both systems and the new non-technical side of the system

3. Results and Discussions

The results of this study are consumer voltage profiles on both systems, losses on both systems, annual savings on both systems, capital expenditure on HVDS, and payback times on HVDS. There are two distribution network configurations studied in this paper based on the degree of load capacity variation. The first configuration (the level of variation of large load capacity) is on the network of one distribution transformer with code MK053 installed at Jalan Kartini (Kantor PTPN IV), Medan City. While the second configuration (the level of variation in load capacity is small or almost uniform) is on the network of one distribution transformer with code MK098 installed at the intersection between Jalan Diponegoro and Jalan Sudirman, Medan City. Both of these transformers are equally attached to LK 6 repeater, Electric Substation. Both transformers have double pole construction (fitted with two poles) and a capacity of 200 kVA. This 3 phases transformer delivers power to 13 loads (MK053 transformers) and 9 loads (MK098 transformers) with varying usage.

3.1 First Configuration (Variation Rate of Large Load Capacity)

There were loads of very small capacity and loads with very large capacity on this network. The one-line diagram of MK053 transformer network diagram can be seen in Figure 2. The new one-line diagram of the system, after changes in the use of conductors and transformers, can be seen in Figure 3. To reduce technical losses, LVDS (Low Voltage Distribution System) is converted to HVDS (High Voltage Distribution System). This case study was made by considering the length of the primary and secondary distribution networks.

The long-run low-voltage line to the consumer was replaced by a long high-voltage line and then converted into a low voltage conduit at the point closest to the consumer by a small-capacity transformer. For comparison of the length of conductor and transformer used LVDS and HVDS in configuration 1 can be seen in Table 1.

From the overall analysis result in configuration 1, we can get comparison table between LVDS and HVDS like Table 2 and Table 3. The annual savings on HVDS can be determined to multiply the power purchase price of electricity, in this case, the Rp 1,352 / kWh, the difference between the total loss amounted HVDS with LVDS 69334.3488 kWh, in order to obtain the results of Rp 93,740,039. Capital expenditure on HVDS is determined by summing the total cost for a transformer purchase that is assumed to be Rp 300,000,000, and the total other assumed costs of Rp 400,000,000. So obtained capital expenditure of Rp 700,000,000.

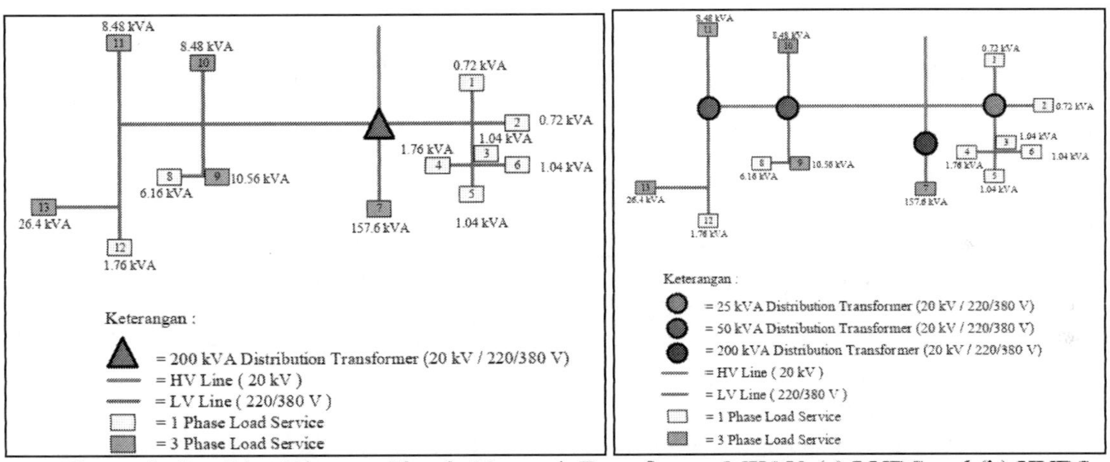

Figure 2. One Line Diagram Simple Network Transformer MK053 (a) LVDS and (b) HVDS

The duration of the payback period can be determined by dividing the capital expenditure by the annual savings so that the result is 7.47 years.

3.2 Second Configuration (Variation Level of Small or Almost Small Load Capacity)

There are no loads with very small or very large capacity on this network. The one-line diagram of MK098 transformer network diagram is simply shown in Figure 4. While the new one-line diagram form of the system, i.e., after changing the use of conductors and transformers, can be seen in Figure 5. For the comparison of the length of the conductor and transformer used LVDS and HVDS on Configuration 2 can be seen in Table 4.From the overall analysis result in configuration 2, we can get comparison table between LVDS and HVDS like Table 5 and Table 6.

The annual savings on HVDS can be determined by multiplying the purchase price of electric power from PLN, in this case, Rp 1,352 / kWh, with the total difference between HVDS and LVDS losses of 56475.3696 kWh, resulting in Rp 76,354,700.Capital expenditures on HVDS are determined by summing the total cost for a transformer purchase that is assumed to be Rp 250,000,000, and the total assumed the cost of Rp 500,000,000. So obtained capital expenditure of Rp 750,000,000. The duration of the payback period can be determined by dividing the capital expenditure by the annual savings so that the result is 9.82 years.

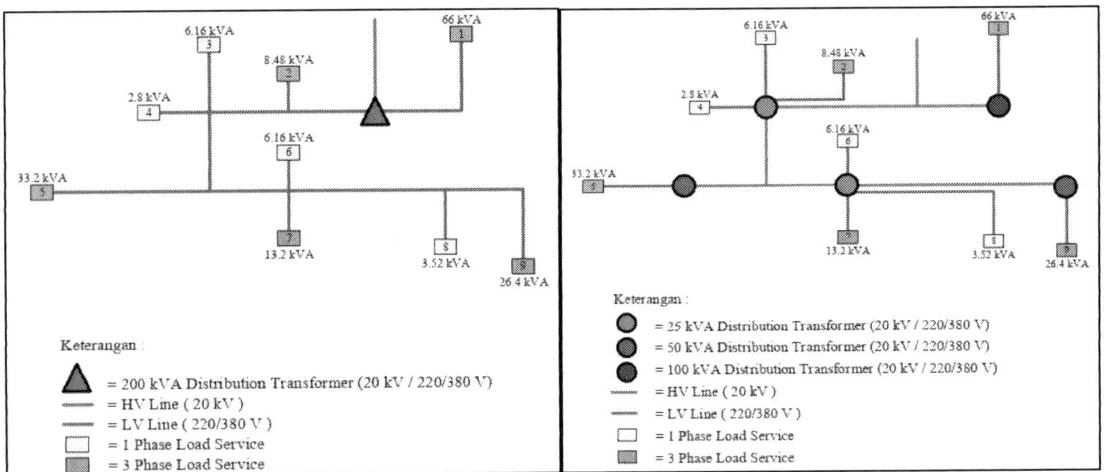

Figure 3. One Line Diagram Simple Network Transformer MK098 (a) LVDS and (b)HVDS

Results of measurements are shown in Table 1, 2, and 3.

Table 1. Comparison of Conductor and Transformer

Parameter	Conf 1		Conf 2	
	LVDS	HVDS	LVDS	HVDS
Conductor Length 20 kV	0.5 Km	0.675 Km	0.5 Km	0.86 Km
Conductor Length 220/380 V	0.52 Km	0.353 Km	0.62 Km	0.38 Km
Total Conductor Length	1.02 Km	1.028 Km	1.12 Km	1.24 Km
Distribution Transformator	200 kVA	25 kVA 50 kVA 50 kVA 200 kVA	200 kVA	25 kVA 25 kVA 50 kVA 50 kVA 200 kVA

Table 2. Comparison of Consumer Voltage

No.	Consumers	VOLTAGE	
		LVDS	HVDS
1	0857121	209.154	217.206
2	0906560	209.154	217.206
3	0809975	209	217.03
4	0827987	208.648	216.678
5	0809987	208.934	216.964
6	0809999	208.736	216.766
7	0139297	359.0696	367.388
8	0915072	202.004	211.86
9	0136634	354.5628	371.879
10	0139563	354.8022	372.134
11	0126602	352.3056	368.722
12	0169169	202.84	212.3
13	0661509	350.36	366.685

Table 3. Comparison of System Loss

No.	Parameter	Conf 1		Conf 2	
		LVDS (kW)	HVDS (kW)	LVDS (kW)	HVDS (kW)
1	Total Power Losses	9.234	4.726	7.562	3.571
2	Total Iron Losses	0.355	0.68	0.355	0.61
3	Total Copper Losses	2.35	4.375	2.35	3.87
4	Total Transformer Losses	2.705	5.055	2.705	4.48
5	Total Theft Losses	5.75688	0	4.23096	0
6	Total Losses	17.69588	9.781	14.49796	8.051

4. Conclusions

The voltage profile of the consumer-generated by the high voltage distribution system (HVDS) was proved to be better than the low-voltage distribution system (LVDS). In configuration 1, the total loss of power at LVDS is 9,234 kW, whereas in HVDS it is 4,726 kW, almost half of current configuration. In configuration 2, the total loss of power in the LVDS is 7,562 kW, while the HVDS is 3.571 kW. In this case, HVDS produces power losses. Then, in configuration 1, the total loss of the transformer in LVDS is 2,705 kW, whereas in HVDS it is 5,055 kW. In configuration 2, the total loss of the transformer in the LVDS is 2,705 kW, while the HVDS is 4.48 kW. HVDS has larger transformer losses due to the use of more transformers.

By using high voltage distribution system, it can be obtained cost savings of Rp 93,740,039 per year in configuration 1 and Rp 76,354,700 per year in configuration 2. While for large capital expenditure on HVDS, the length of time required for payback is 7.47 Years in configurations 1 and 9.82 years in configuration 2.

Although it requires a large cost at the beginning, for long-term high voltage distribution system proved to be better applied to the distribution network than the low-voltage distribution system because it has lower losses and better voltage profile.

IOP Conf. Series: Materials Science and Engineering **309** (2018) 012035 doi:10.1088/1757-899X/309/1/012035

Once converted to HVDS, in configuration 1, the old transformer still cannot be replaced into another smaller capacity transformer because there is a very large load capacity. While in configuration 2, the old pure transformer is replaced by transformers that have smaller capacities due to variations in the small or nearly uniform load capacity for transformer loading.

References

[1]. Gönen and Turan 2008 Electric Power Distribution System Engineering, *Second Edition, Boca Raton, CRC Press*

[2]. Spandana K and Varsha RA 2014 Restructuring of a Low Voltage Distribution System into a High Voltage Distribution System for an Improved Voltage and Power Loss Profile *International Conference and Utility Exhibition 2014 on Green Energy for Sustainable Development (ICUE 2014)*

[3] Salomonsson D and Sannino A 2007 Low-voltage DC distribution system for commercial power systems with sensitive electronic loads. *IEEE Transactions on Power Delivery*, **22**(3), 1620-27

[4] Richardson P, Flynn D and Keane A 2012 Optimal charging of electric vehicles in low-voltage distribution systems. *IEEE Transactions on Power Systems*, **27**(1), 268-9

[5] Jusmedy and Fery 2007 Studi Aliran Daya Sistem 115 KV PT. Chevron Pacific Indonesia. *Teknik Elektro Universitas Sumatera Utara*

[6] Sutherland P E, Waclawiak M and McGranaghan M F 2006 System impacts evaluation of a single-phase traction load on a 115-kV transmission system. *IEEE Transactions on power delivery*, **21**(2), 837-44

[7] ETAP 2015 ETAP Products Overview *Operation Technology, Inc*

IOP Conf. Series: Materials Science and Engineering **309** (2018) 012036 doi:10.1088/1757-899X/309/1/012036

Analysis of temperature changes on three-phase synchronous generator using infrared: comparison between balanced and unbalanced load

S Amien[1], W Yoga[1], F Fahmi*[1,2]

[1]Department of Electrical Engineering, Faculty of Engineering, University of Sumatera Utara, Jl. Dr. Mansur 9 Medan 20155, Indonesia
[2]Sustainable Energy and Biomaterial Centre of Excellent, University of Sumatera Utara, Jl. Almamater Kampus USU, Medan 20155, Indonesia

*E-mail: fahmimn@usu.ac.id

Abstract. Synchronous generators are a major tool in an electrical energy generating systems, the load supplied by the generator is unbalanced. This paper discusses the effect of synchronous generator temperature on the condition of balanced load and unbalanced load, which will then be compared with the measurement result of both states of the generator. Unbalanced loads can be caused by various asymmetric disturbances in the power system and the failure of load forecasting studies so that the load distribution in each phase is not the same and causing the excessive heat of the generator. The method used in data collection was by using an infrared thermometer and resistance calculation method. The temperature comparison result between the resistive, inductive and capacitive loads in the highest temperature balance occured when the generator is loaded with a resistive load, where T = 31.9 ° C and t = 65 minutes. While in a state of unbalanced load the highest temperature occured when the generator is loaded with a capacitive load, where T = 40.1 ° C and t = 60 minutes. By understanding this behavior, we can maintain the generator for longer operation life.

1. Introduction

Synchronous generators are a major tool in electrical energy generation systems. In this case, the load supplied by the generator is generally unbalanced. The unbalanced load problem supplied by the three-phase sync generator is one of the problems in operating a three-phase sync generator. Unbalanced loads can be caused by various asymmetry disturbances in the power system and the failure of the load forecasting study so that the load distribution in each phase is not the same.

The load unbalances supplied by the three-phase sync generator will result in overheating of the generator. This is due to an unbalanced load condition that causes the sequence of components of the channel flow. In a three-phase three-wire system, unbalanced loads will generate positive sequence current and negative sequence current. While on a three-phase three-wire system, unbalanced loads will generate positive, negative and zero sequence currents [1].

The temperature rise of the generator can be measured by using the Infrared Thermometer or by other methods. Therefore it is necessary to do a good study in the form of analysis and research in the laboratory to see how the load imbalance affects the temperature rise in the synchronous generator. And comparing synchronous temperature generator measurement results based on loads that burden the synchronous generator. The load borne by the generator is resistive, inductive and capacitive. Our study aims to provide a thorough understanding of this behavior in order to maintain the generator for longer operation life.

2. Materials and Methods

2.1 Sync Generator

The synchronous generator is an alternating current electric engine that converts mechanical energy into alternating current electrical energy. Mechanical energy derived from prime mover coupled with generator rotor, while electric energy obtained from an electromagnetic induction process involving rotor coils and stator coils. This alternating current electric machine is called synchronous because the rotor rotates synchronously or rotates at speed equal to the speed of the rotating magnetic field [2]. In the synchronous generator, direct current is applied to the rotor coil which then produces the rotor magnetic field. The rotor of the generator will be rotated by a prime mover, producing a rotating magnetic field inside the machine. In the stator generator, there is also a coil. The rotating magnetic field causes the magnetic field surrounding the stator coil to change continuously. This continuous magnetic field change induces the voltage on the stator coil. This induced voltage will be sinusoidal, and the magnitude depends on the strength of the magnetic field and the rotational speed of the rotor. To create a three-phase generator, the stator is placed three separate coils 120o each other, so that the induced voltage will be separated as far as 120 ° to each other.

2.2 Synchronous Three Phase Load Generator Balanced

What is meant by a balanced state is a state where [3,4]:

- The three vectors of current and voltage are equal
- The three vectors form an angle 120^0 with each other.

The vector current diagram in a load balanced state is shown in Figure 1a. When the load current is in the balance it is seen that the sum of the three current vectors (IR, IS, IT) is equal to zero, so that neutral current (I) does not appear. The unbalanced state is a state in which one or both conditions of a balanced state are not fulfilled. The possibility of unbalanced state there are three namely:

- The three vectors are equal but do not form angles 120^0 with each other.
- The three vectors are not as large but form angles 120^0 with each other.
- The three vectors are not as large and do not form angles 120^0 with each other.

Figure 1b shows the vector current diagram when the load is unbalanced.

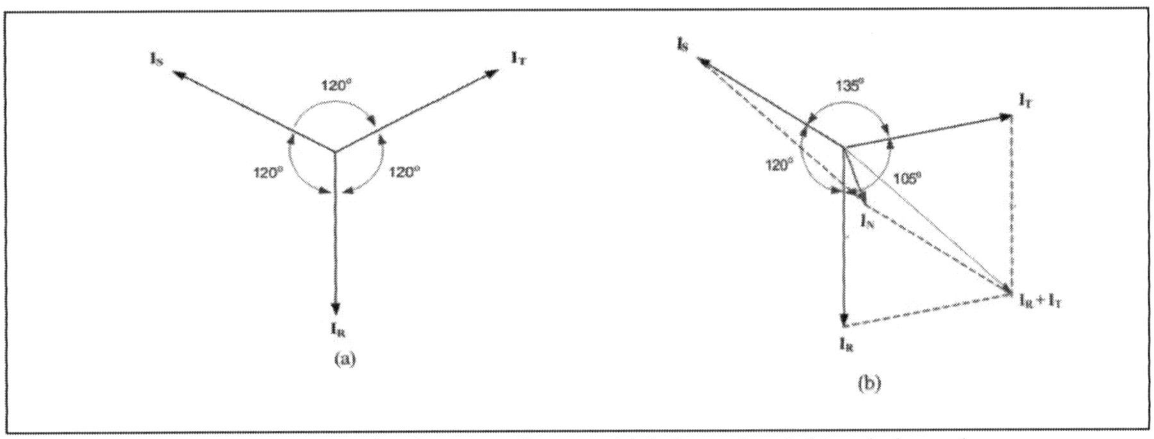

Figure 1. Vector load current diagram (a) balanced and (b) unbalanced

IOP Conf. Series: Materials Science and Engineering **309** (2018) 012036 doi:10.1088/1757-899X/309/1/012036

The N vector unbalanced current diagram shows that the sum of the three current vectors (IR, IS, IT) is not equal to zero so that a neutral current (IN) is generated depending on the magnitude of the imbalance factor.

2.3 Infrared Thermometer

The infrared thermometer measures the surface temperature of the object (target). The optics of the thermometer detect the infrared energy that has been collected and focused on the detector and translated into the form of temperature information that can be read on display (interface). The laser here is used to shoot the target to be measured [5,6]. The infrared thermometer is characterized by specifications including accuracy and angular coverage. Simple instruments may have measurement errors of about ± 2 ° C / ± 4 ° F).

There are the advantages of infrared thermometer:

- Non-contact temperature measurement has no effect on the measured object.
- Quick response and movement of objects can be measured and transient temperatures.
- The measurement accuracy, high resolution is small. Large measurement range
- Small area measurement temperature. And can be the same time point, line, surface temperature.
- Can be measured the absolute temperature, relative humidity can be measured.

Some disadvantage of infrared thermometer includes:

- The effect of temperature on the measured object.
- Not suitable for measuring transient temperature.
- It is not easy to measure moving objects.
- Measuring range is not wide enough, and equipment.
- Not suitable for measuring toxic, high pressure, and dangerous opportunities

2.4 Research Design

Based on the research flowchart, the technique of calculation and data processing can be seen in Figure 2.

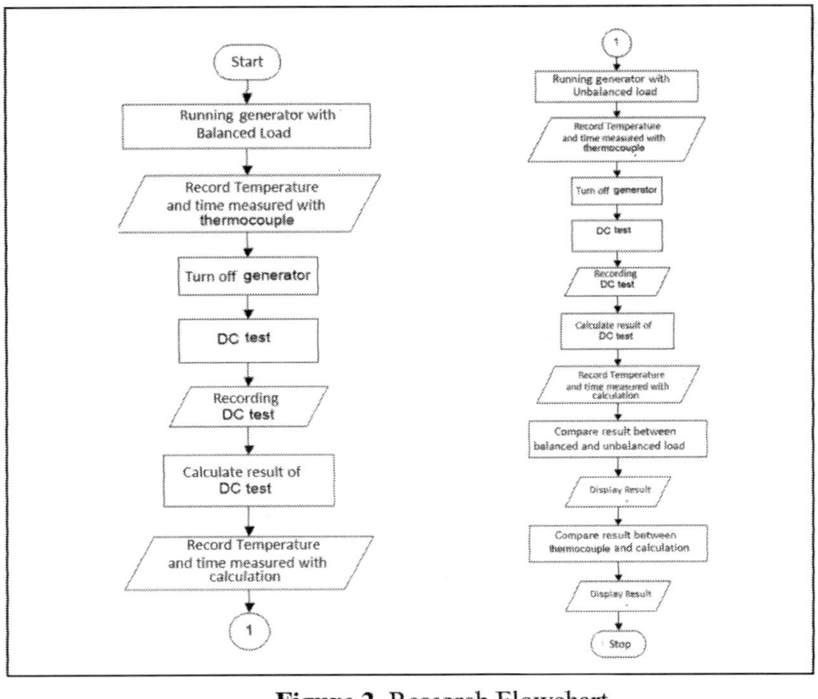

Figure 2. Research Flowchart

IOP Conf. Series: Materials Science and Engineering **309** (2018) 012036 doi:10.1088/1757-899X/309/1/012036

The data was collected at the Laboratory of Electrical Energy Conversion of the Department of Electrical Engineering, Faculty of Engineering, University of Sumatera Utara. In the implementation of the research, the required data is then analyzed and calculated according to the corresponding formula. Then the results obtained are presented in the form of tables and curves.

The variables observed in this study include:

- The type of load that weighs on the generator
- The magnitude of the load imbalance that weighs on the generator
- Length of operation time of generator
- The change in the value of the generator resistance measured by the DC test at the time of balanced load and unbalanced load.
- Changes that are measured by the infrared thermometer when the load is balanced, and the load is unbalanced

Figure 3 shows a series of synchronous generator tests underbalanced or unbalanced load conditions.

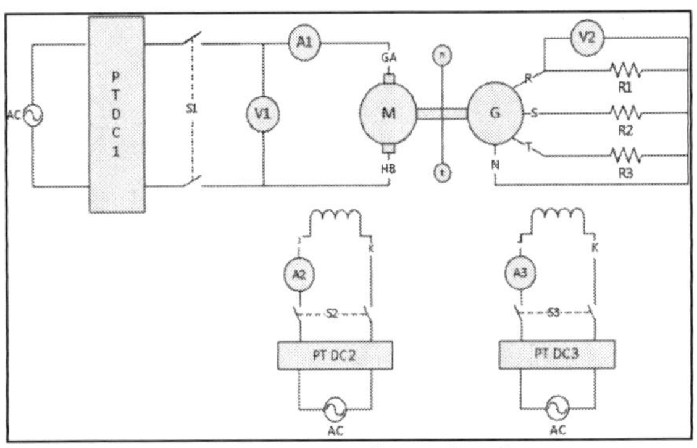

Figure 3. Testing Circuit

The amount of load and the type of load is adjusted based on the measurements to be done first.

3. Results and Discussions

Three different load tests were performed, i.e., synchronous generator test with Resistive load, synchronous generator test with Inductive load and synchronous generator test with a Capacitive load. The results are shown in Table 1-3.

Table 1. Data of temperature generator test results with resistive load

t (minute)	Temp (°C) / *Thermometer Infrared*		Temp (°C) / Resistance	
	Unbalanced	Balanced	Unbalanced	Balanced
0	26	26	26	26
5	28.6	26.8	28.63797	27.00192
10	29.5	27.6	29.95696	27.66987
15	30.2	28.5	30.61646	29.33974
20	30.4	28.8	31.6057	30.67564
25	30.8	29.2	32.26519	31.67756
30	31.5	29.4	33.58418	32.34551
35	32.1	29.8	34.24367	33.34744
40	32.7	30.3	34.90316	34.34936

45	33.1	30.8	35.56266	34.68333
50	33.5	31	35.89241	35.35128
55	33.7	31.6	36.5519	35.68526
60	33.8	31.8	36.88165	36.35321
65	33.8	31.9	36.88165	36.68718

Table 2. Generator temperature test data with inductive load

t (minute)	Temp (°C) / *Thermometer Infrared*		Temp (°C) / Resistance	
	Unbalanced	Balanced	Unbalanced	Balanced
0	26	26	26	26
5	26.6	26.7	26.9892	26.6654
10	27.1	27.1	27.9785	27.3308
15	27.3	27.3	28.638	27.6635
20	27.9	27.7	29.2975	28.3289
25	28.3	27.8	29.6272	28.6616
30	28.5	27.9	30.2867	28.9943
35	28.6	27.9	30.6165	28.9943
40	28.7	27.9	30.9462	28.9943
45	28.7	27.9	30.9462	28.9943

Table 3. Data of temperature generator test result with capacitive load

t (minute)	Temp (°C) / *Thermometer Infrared*		Temp (°C) / Resistance	
	Unbalanced	Balanced	Unbalanced	Balanced
0	26	26	26	26
5	29.1	27.4	29.2975	27.6916
10	30.4	27.7	31.276	28.3682
15	31.7	28	32.9247	29.0448
20	32.8	28.5	35.5627	29.3831
25	33.7	29	37.2114	30.0597
30	34.3	29.2	38.2006	30.3981
35	35.3	29.4	38.8601	31.0747
40	37.4	30	39.1899	31.7513
45	38.2	30.2	40.1791	32.0896
50	38.8	30.3	41.8279	32.7662
55	39.5	30.3	42.8171	32.7662

Under unbalanced load conditions, the generator will experience a greater temperature rise than when the generator is in a state of the balanced load. Under unbalanced load conditions, the generator will experience a greater temperature rise than when the generator is in a load-balanced state even though the temperature rise is not too large. Under unbalanced load conditions, the generator will experience a greater temperature rise than when the generator is in a state of the balanced load.

In order to obtain temperature based on resistance calculation method, equations 1 and 2 are used [7].

$$R_{dc} = \frac{V_{dc}}{2I_{dc}} \quad (1)$$

$$T_t = T_b \left(\frac{R_t - R_b}{R_b}\right)(T_b + k) \qquad (2)$$

Where Tt: Total winding temperature (°C), Tb: Temperature at the time of cold motor (°C), Rt: Resistance at the time of hot motor (ohms), Rb: Prisoner during cold motor (ohms), K: 234.5 (copper material constant) (°C), 225 (aluminum material constant) (°C).

The temperature comparison curves of each load type between the generator temperature under balanced load conditions and the temperature of the generator under unbalanced load conditions. The resistive load comparison curve.

Figure 4 shows the temperature comparison curve between balanced load and unbalanced load where the measurement uses Infrared Thermometer (Left) and using Resistance measurement (Right).

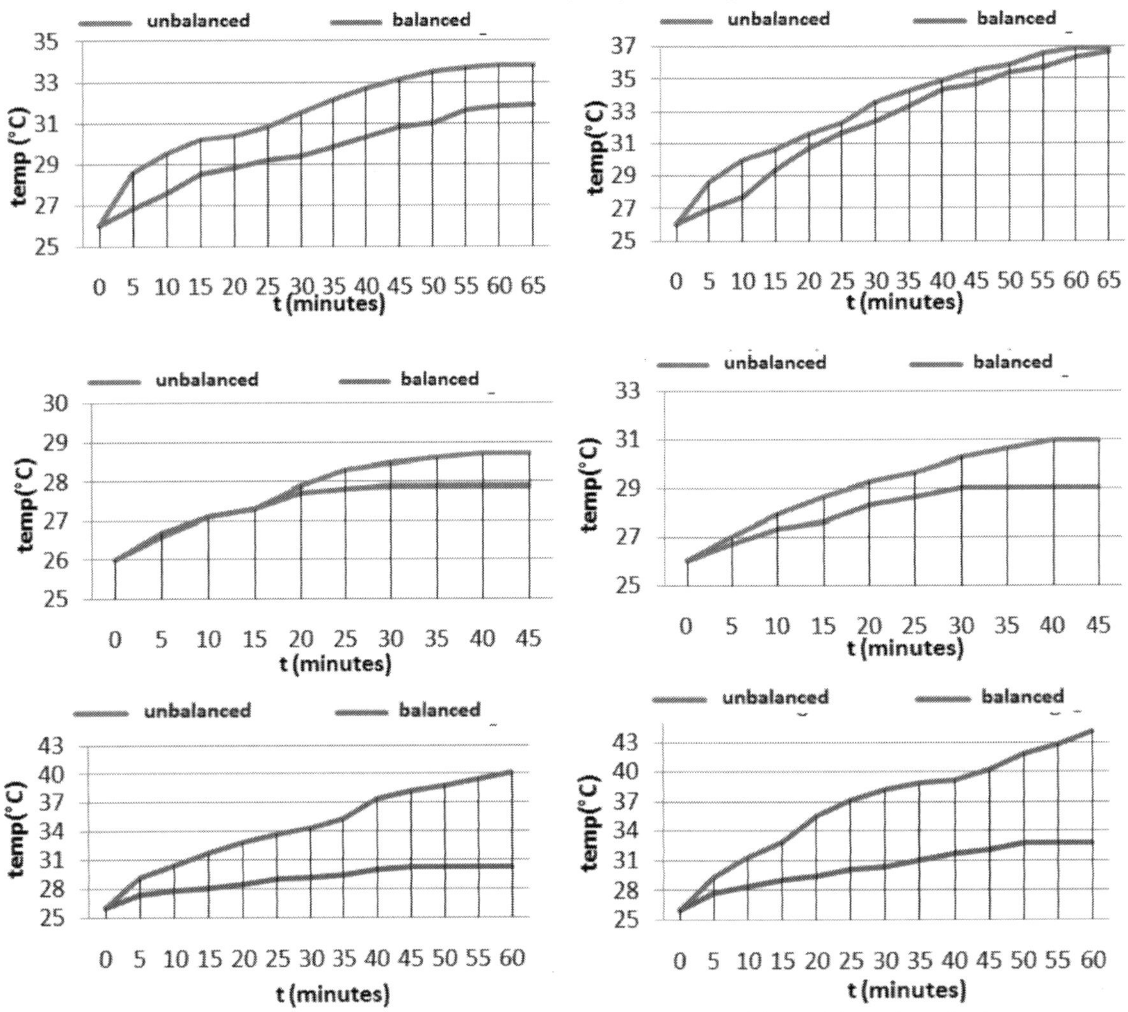

Figure 4. Temperature comparison: Resistive, Inductive and Capacitive with Infrared Thermometer (Left) and using Resistance measurement (Right)

4. Conclusions

In an unbalanced load state, the synchronous generator experiences higher heating compared to when the generator is in balanced loaded. This occurs in the three types of loads, namely, resistive, inductive and capacitive loads. With balanced resistive load measurement 31.9 ^{0}C and unbalance 33.8 ^{0}C. at balanced inductive load 27.9 ^{0}C and unbalance 28.7 ^{0}C. at balanced load capacitive 30.3 ^{0}C and Unbalanced 40.1 ^{0}C.

IOP Conf. Series: Materials Science and Engineering **309** (2018) 012036 doi:10.1088/1757-899X/309/1/012036

Of the three types of loads, at the highest load balances, the highest temperatures occur when the generator is loaded with resistive loads, while at unbalanced loads the highest temperature is present when the generator is loaded with a capacitive load.

Comparison of the temperature of each load between infrared thermometer measurement methods with resistance calculation method, there is a comparison of temperature rise of measurement. This is because the measurement position using an infrared thermometer is incorrect in the stator coil of the synchronous generator, which is blocked by the frame of the generator and because of the tolerance of the accuracy of the infrared thermometer-measuring instrument of ± 2° C.

References

[1] Chapman S J 1999 Electric Machinery Fundamentals, *Third Edition Mc Graw Hill Companies, New York*

[2] Zuhal 1988 Dasar Teknik Tenaga Listrik dan Elektronika Daya, *Penerbit ITB, Bandung*

[3] IEEE Guides: Test Procedures for Synchronous Machines, *IEEE Std 115-1995* (R2002)

[4] De Mello FP and Ribeiro JR 1977 Derivation of synchronous machine parameters from tests. *IEEE Transactions on Power Apparatus and Systems* **96**(4) 1211-18

[5] Stipetic S, Kovacic M, Hanic Z and Vrazic M 2012 Measurement of excitation winding temperature on synchronous generator in rotation using infrared thermography. *IEEE Transactions on Industrial Electronics*, **59**(5), 2288-98

[6] Fraden J 1989 *U.S. Patent No. 4,797,840*. Washington, DC: U.S. Patent and Trademark Office

[7] Theraja B L and Theraja A K 2001 A Text Book of Electrical Technology, *New Delhi, S.Chand and Company Ltd*

IOP Conf. Series: Materials Science and Engineering **309** (2018) 012037 doi:10.1088/1757-899X/309/1/012037

Morphological study of fluorescent carbon Nanoparticles (F-CNPs) from ground coffee waste soot oxidation by diluted acid

S Gea*, S Tjandra, J Joshua, B Wirjosentono
Department of Chemistry, Faculty of Mathematics and Natural Sciences, University of Sumatera Utara, Jl. Bioteknologi No. 1 Padang Bulan Medan 20155, Indonesia

*E-mail: s.gea@usu.ac.id

Abstract. Coffee ground waste utilization for fluorescent carbon nanoparticles (F-CNPs) through soot oxidation with diluted HNO_3 has been conducted. Soot was obtained through three different treatments to coffee ground waste; which was burned in furnaceat 550°C and 650°C and directly burned in a heat-proofcontainer. Then they were analyzed morphologically with Scanning Electron Microscope (SEM) instrument. Soot from direct burning indicated the optimum result where it has denser pores compared to other two soots. Soot obtained from direct burning was refluxed in diluted HNO_3 for 12 hours to perform the oxidation. Yellowish brown supernatant was later observed which lead to green fluorescent under the UV light. F-CNPs characterization was done in Transmission Electron Microscopy, which showed that 7.4-23.4 nm of particle size were distributed.

1. Introduction

Coffee for the last decade has become an important beverage around the globe. Drinking coffee has become a lifestyle for most societies either to kick start the morning or as afternoon snack companion. Since 2014, Indonesia has become one of the top countries for producing and exporting coffee beans. For the last decade,business in coffee world has also been growing in Indonesia which can be indicated from the increasing number of international coffee tenants and local coffee shops. This consequently increases the amount of coffee ground waste produced. Survey conducted to international and local coffee shops, in a day, at least 1 kg of coffee waste can be collected.

Ground coffee waste is the primary by-product derived from espresso extraction process at high temperature. It is considered as a potential waste with negative impacts. The big amount of coffee waste in landfill can produce methane and carbon dioxide, greenhouse gasses that contribute to global warming.[1] Furthermore, it also contributes to huge financial cost for its deposit. Although with the disadvantages it causes, ground coffee waste is a relativelyunique organic waste. Elemental analysis to ground coffee waste shows coffee samples to posses high carbon (>58%), low nitrogen (<2%), and low ash (<1%) contents and low polarity coefficient.[2] With that high organic and low inorganic material, ground coffee waste has plenty of uses.

Via pyrolysis, groundcoffee waste can be made as soot for fluorescent carbon nanoparticles (F-CNPs) through soot oxidation methods. F-CNPs are biocompatible and chemically inert, which gives more advantages than the cadmium based fluorescent nanoparticles (F-NPs).[3] As a result, the emergence of F-CNPs with better properties has increased the interest for F-CNPs synthesis.

F-CNPs derived from soot oxidation with nitric acid will give a green fluorescence and soluble in water. There are two steps of F-CNPs synthesis involving soot oxidation in nitric acid for 12 hours at 100°C by reflux and size separation of carbon particles.[4] However, instead of using soot derived from candle burning, soot used in the synthesis process was obtained from the ground coffee waste collected. There are three different methods used to gain groundcoffee waste soot which were then observed and compared by scanning electron microcopy (SEM) and F-CNPs/coffee soot was analyzed by transmission electron microscopy to observe the particle size distribution and using the UV light to check the fluorescence properties.

2. Materials and methods

2.1. Soot derived from ground coffee waste

There are two approaches to get soot from ground coffeewaste; by furnace burning and direct burning in a heat proof container. Soot was black in colour and came in as powdery material. First, for furnace burning soot,ground coffee waste was dried in an oven at 120°C to lose the water content and putted into two porcelain dishes. Two furnaces were used, one was set to 550°C and the other was set to 650°C and ground coffee waste was putted for 4 hours to obtain a black powdery material. White powder might be formed due to the inorganic content in coffee waste collected. To separate the white powder, it was dilutedinto weak acid, HCl 2N and then was centrifuge at 4000 rpm for 10 minutes. The black soot was collected and dried in oven and characterized with SEM. Direct burning soot was obtained from ground coffeewaste soaked in kerosene inside a heat-proof container.It was burned for approximately 20 minutes or until fire was out. Soot can be collected at the side of the container and finally characterized with SEM.

2.2. Soot oxidation

As many as 25 mg carbon soot derived from ground coffee waste was mixed with 15 ml of 5M nitric acid and refluxed at 100°C for 12 hours with magnetic stirrer. The solution was cooled and centrifuged at 3000 rpm for 10 minutes to separate the unreacted carbon soot. Then, yellowish brown supernatant was collected which gave a green fluorescence under UV exposure. It was mixed with acetone and centrifuged at 14000 rpm for 10 minutes to separate the excess nitric acid. Black precipitation was collected and the aqueous solution was discarded and dissolved in 5-10 ml water.

2.3. Size-based separation of carbon particles

Size-based separation of carbon particles was performed in a solvent mixture with the combination of high speed centrifugation. Carbon particles are soluble in water, ethanol and acetone but insoluble in chloroform. Therefore, solvent mixture of water-ethanol-chloroform was used where water-ethanol helped to solubilize particle whereas chloroform decreased the solubility. It was then followed by a step-by-step separation in different centrifugation speed from 4000-16000 rpm. The F-CNPs in water synthesized was then characterized by TEM to see the particle size distribution.

3. Result and Discussion

3.1. Soot morphology

With three different treatments to ground coffeewaste, giving three different soot which for each were characterized by SEM to observe their morphology. There was a huge significant differentamong them. The coffee soot which was in furnace burning at 550°C has 8 pores per 10 cm length, while at 650°C, it has 18 pores per 10 cm. In contrast, the coffee soot obtained from direct burning has smaller pores per 10 cm length as shown in Figure 1.Based on the morphological study of the soot, it can be concluded that coffee soot from direct burning gave the most suitable carbon source to the synthesis of F-CNPs as the most pores it has, the more carbons can bind the nitrogen and oxygen from nitric acid.

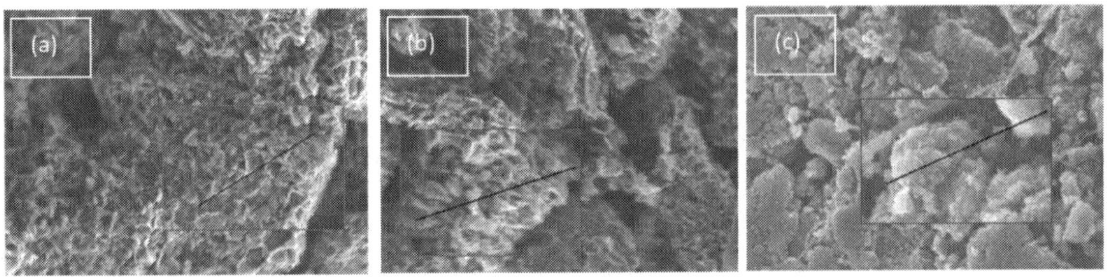

Figure 1. The coffee ground waste soot morphology: (a) Soot derived from furnace burning at 550 °C, (b) Soot derived from furnace burning at 650 °C, (c) Soot derived from direct burning

3.2. F-CNPs from diluted HNO_3 oxidized ground coffee waste soot

After 12 hours refluxing the coffee soot with nitric acid, black solution was formed and was centrifuged at 3000 rpm for 10 minutes. Then, the yellowish brown supernatant was yielded. Under the UV light, the green fluorescent was observed as shown in Figure 2.

Figure 2. Fluorescent properties of F-CNPs obtained from ground coffee waste soot oxidation: (a) Yellowish brown supernatant, (b) Green fluorescent under the UV light

F-CNPs/coffee soot was also characterized by TEM to see the particle size distribution. The result showed that nanoparticles were formed with the size averageof 7.4–23.4 nm. The result is shown in Figure 3.

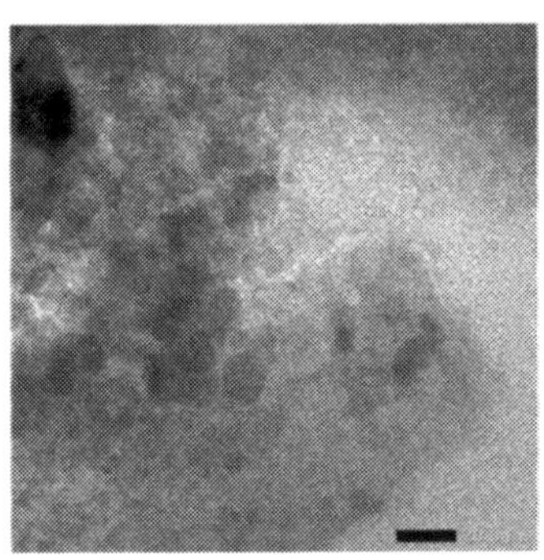

Figure 3. TEM image of F-CNPs/coffee soot

4. Conclusion

Soot derived from ground coffee waste with direct burning method was the best method to the by-product to get soot for F-CNPs. Although F-CNPs/coffee soot formed still had large distribution particle size (7.4–23.4 nm), it still shows that coffee waste is promising as carbon source for F-CNPs. However, further research to get better method and alternative for coffee waste utilization are still needed to be conducted.

5. Acknowledgement

Authors would like to acknowledge Ministry of Research Technology and Higher Education, Indonesia for the research funding support via Student's Creativity Program Funding (PKM).

6. References

[1] Cameron A and O'Malley S 2016 *Coffee Ground Recovery Program Summary Report*, (January) 1–17 Retrieved from http://planetark.org/documents/doc-1397-summary-report-of-feasibility-study-april-2016.pdf
[2] Pujol D, Liu C, Gominho J, Olivella M, Fiol N,Villaescusa I and Pereira H 2013 *Industrial Crops and Products*. **50** 423–9
[3] Lai C W, Hsiao Y H, Peng Y K, Chou P T 2012 *J. Mater. Chem* **22(29)** 14403–9
[4] Ray S C, Saha A, Jana N R, Sarkar R 2009 *J. Phys. Chem* **113(43)**18546–51

IOP Conf. Series: Materials Science and Engineering **309** (2018) 012038 doi:10.1088/1757-899X/309/1/012038

Oriented particleboard made from tali bamboo (*Gigantochloa Apus*): effect of particle length on physical and mechanical properties

A H Iswanto*
Department of Forest Product, Faculty of Forestry, Universitas Sumatera Utara, Padang Bulan, Medan 20155, Indonesia

*Email: apriheri@yahoo.com

Abstract. Strength properties are one of the problems of particleboard. The objective of this research was to analyze the effect of particle length on physical and mechanical properties oriented particleboard (OPB). The variation particle length size in this experiment namely 3, 5, and 7 cm. The width and thickness size of all bamboo particles were 1 and 0.1 cm respectively. 12% mixed resin of UF and MDI (70/30 %w/w) used for binding. Board size fabricated in 25 by 25 cm^2 with thickness and density target of 1 cm and 0.75 gcm^{-3}. The OPB layers for face and back layers aligned perpendicular to the core layer. The weight ratio of the face–to–core–to–back layers were set at 1:2:1. Mat was pressed at 160 °C under 30 $kgcm^{-2}$ as the pressure for 10 minutes. The results showed that 7 cm length particle produced of the best strength and dimensional stability. The increase of particle length resulted in increasing of strength properties. Over all the parameters of physical and mechanical properties fulfill requirements of JIS A 5908 (2003) excepted of thickness swelling and modulus of elasticity.

1. Introduction

Forest degradation in Indonesia resulted in the lack of timber materials for wood industry. Therefore, it is necessary to find alternative materials like non-wood materials to substituting of wood. Bamboo is a non-wood raw material that is often found and used by the community. One type of bamboo that can utilized is bamboo Tali (Gigantochloa apus J.A & J.H. Schult.Kalz).

The type of bamboo that is widely used for building materials is bamboo Tali or bamboo apus (G. apus) because it has better physical and mechanical properties than other types [1]. Bamboo Tali has a diameter 4 - 8 cm, stem length 6 - 13 m, and very elastic. The price of bamboo Tali is quite low compared to wood or with other types of bamboo. According to that, bamboo Tali could be considered as raw material of particleboard.

Maloney [2] stated that the preparation and orientation of particle direction as one of the factors that affected of particleboard quality. Ordinary particleboard using UF resin has a low dimensional stability [3] and strength properties [4]. Through the orientation of perpendicular crossed on oriented particleboard (OPB) was expected to minimize the weakness of dimensional stabilization and particle board strength.

2. Materials and methods

2.1. Materials
The materials used in this research included of Bamboo Tali, Urea Formaldehyde (UF) resin, and Isocyanate.

2.2. Methods

2.2.1. Particle preparation
Bamboo converted to be particle in size of 3, 5, dan 7 cm (length), 1 cm (width), dan 0,1 cm (thickness). Therefore, the particle was dried at 103°C temperature for 24 hours up to 5% moisture content.

2.2.2. Manufacturing of board
OPB was produced in size of 25 by 25 cm^2 with the density and thickness target were 0.75 gcm^{-3} and 1 cm respectively. Therefore particle and UF/Isocyanate (70/30 %w/w) at 12% level were blended using the rotary bending machine. After that, the particle was placed into mat former. The next step is hot pressing process at 160°C temperature and 30 $kgcm^{-2}$ pressure for ten minute. After board conditioning for seven days, OPB evaluated that refers to JIS A 5908 (2003).

3. Result and discussions

3.1. Density

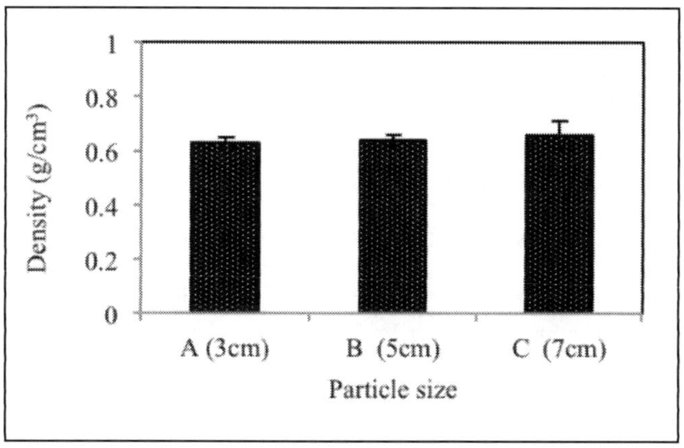

Figure 1. Density of oriented particleboard

The target density has not achieved because of the spring back effect. In the other hand, it was also caused by the loss of particle weight during the blending process. Bufalino et al. [5] stated that board density values lower than the target becaused of the loss of particles during the manufacturing process. Kelley [6] reported that several factors affecting board density include of wood species, pressure in the pressing process, particle quantity, type and quantity of adhesives. Over all, The OPB produced meets the JIS A 5908-2003 standards [7] that requirement of the density values ranges between of 0.40 to 0.90 gcm^{-3}.

3.2. Thickness swelling (TS) and water absorption (WA)
The highest and lowest of TS values were produced by particles of 3 and 7 cm in size respectively (Fig.2). Bektas et al. [8] stated that the TS of the panel depends on several factors such as the amount of adhesive content and its spread, furnish moisture, furnish compatibility and adhesive, the chemical composition of the furnish, etc. Therefore, Garay et al. [9] stated that the particles which are resistant

to moisture content were able to produce of boards with good dimensional stabilization. The TS value in this research had not met the JIS A 5908-2003 standards [7] which require a TS value of less than 12%.

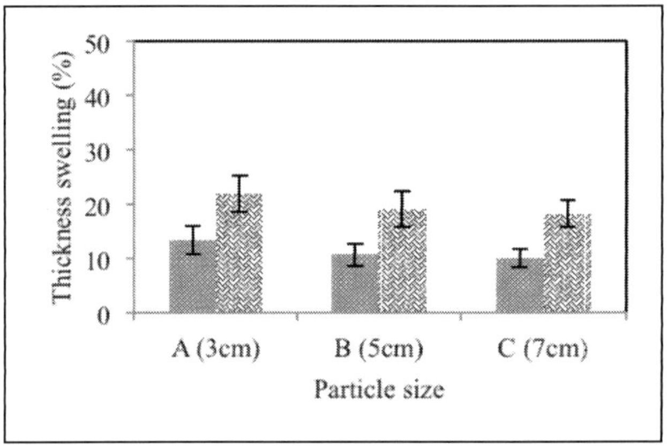

Figure 2. Thickness swelling of oriented particleboard

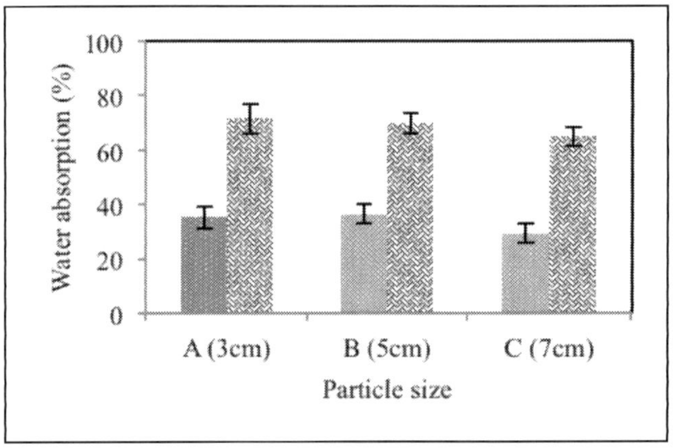

Figure 3. Water absorption of oriented particleboard

The highest and lowest WA values were produced by particles of 3 and 7 cm in size respectively (Fig. 3). WA is not required in JIS standards. OPB with 3 cm particle absorbs more water than particles length 5 and 7 cm. WA produced is still high because in the manufacture of the board did not utilize of water repellent additives. In the other hand according to Rowell [10] that hemicellulose is most responsible for water absorption, however, cellulose, lignin and the surface of crystalline cellulose also affect.

3.3. Internal bond (IB)

The average value of IB presented in Figure 4. The highest and lowest IB values were produced by particles of 3 and 5 cm in size respectively. The high-value IB of OPB was caused the presence of an isocyanate adhesive wherein the isocyanate adhesive has a highly reactive chemical group, that is R-N = C = O [11].

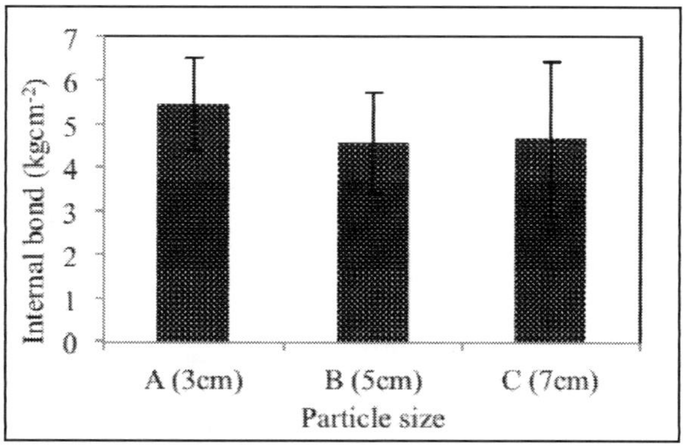

Figure 4. Internal bond of oriented particleboard

IB decreased with increasing of slenderness ratio (SR) particle. In this study, SR values for particle 3, 5, and 7 cm length were 27.60, 49.01, and 62.72, respectively. Lin et al. [12] state that the increasing of SR causes the percentage of cavity area larger than the resin closing area so that it has a negative effect on IB. Likewise, the increasing size of length and width causes the uneven resin distribution on the surface. All of the resulting particle board meets the JIS A 5908-2003 standards [7] was requiring an IB minimum of 1.50 $kgcm^{-2}$.

3.4. Modulus of rupture (MOR)

A 7 cm length produced a higher MOR value than another size. These were supported by the highest value of slenderness ratio and aspect ratio of 7 cm particle in length size. Arabi et al. [13] stated that large particle size increases the mechanical strength of particleboard over small particles. Maloney [14]; Sun and Arima [15] suggest that particles with the bigger surface area have larger resin closures and this leads to an increase in mechanical strength. All of the resulting particleboard meets the JIS A 5908-2003 standards [7] was requiring a minimum MOR value of 82 $kgcm^{-2}$.

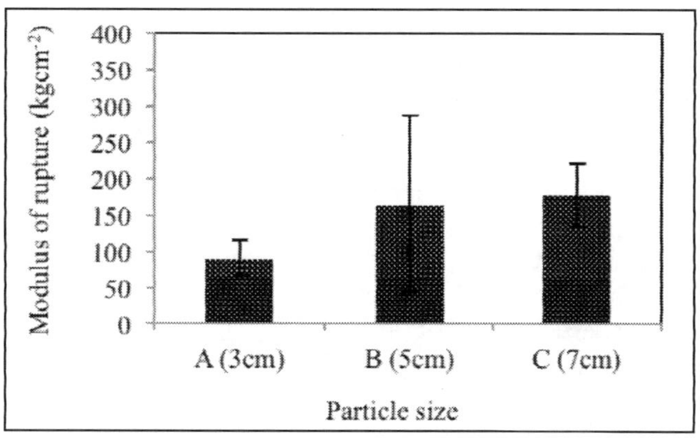

Figure 5. Modulus of rupture of oriented particleboard

3.5. Modulus of elasticity (MOE)

The highest and lowest MOE values were produced by particles of 7 and 3 cm in size respectively.

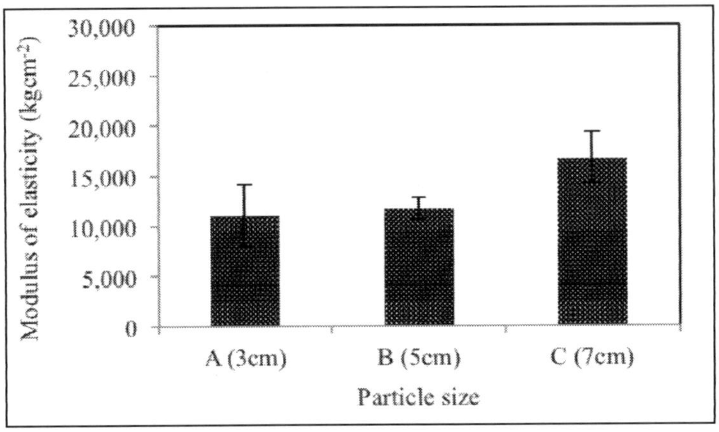

Figure 6. Modulus of elasticity of oriented particleboard

Particles with a length of 7 cm yield the highest MoE value because the particle had the highest SR value, high SR will be easier to orientate so that the strength will increase and it requires a little surface gluing adhesive to bind the particles. According to Maloney [2], SR's ideal value for flake particles is 150. Overall results show that the resulting of MOE value had not met the JIS A 5908-2003 standards [7] was requiring a minimum MOE value of 20,400 kgcm^{-2}.

4. Conclusion

The particle size had a significant effect on MOE parameters. The treatment of particle size produced of density, IB, and MOR parameters had complied with JIS A 5908 (2003) standard. While the parameters TS and MOE had not met the JIS A 5908 (2003). Over all the particle of 7 cm was the best length to produced of OPB.

Acknowledgment

We would like to express my sincerely thanks to Directorate General Higher Education - Ministry of Research, Technology, and Higher Education - Republic of Indonesia for funding support through to the College Excellent Research Grant 2016.

References

[1]. Berlian NVA, and Rahayu E 1995 Jenis dan Prospek Bisnis Bambu (Indonesia: Penebar Swadaya Jakarta)
[2]. Maloney TM 1993 Modern Particleboard and Dry-Process Fiberboard Manufacturing (USA: Miller Freeman Inc. San Fransisco)
[3]. Iswanto AH, Sucipto T, Nadeak SSD, Fatriasari W 2017 *Conf. Ser.: Mater. Sci. Eng.* **180** 012015
[4]. Iswanto AH, Aritonang W, Azhar I, Supriyanto, Fatriasari W 2017 *J Indian Acad Wood Sci* **14** 1–8
[5]. Bufalino L, Albino VCS, VA de Sá, Corrêa AAR, Mendes LM, Almeida NA 2012 *J. Trop. For. Sci.* **24** 162–172
[6]. Kelly MW 1977 Critical Literature Review of Relationship Between Processing Parameter and Physical Properties of particleboard, General Technical Report FPL-10, US Department of Agriculture Forest Service and Forest Products Laboratory University of Wisconsin
[7]. [JSA] Japanese Standard Association 2003 Japanese Industrial Standard Particle Board JIS A 5908, Tokyo
[8]. Bektas I, Guler C, Kalaycioglu H, Mengeloglu F, Nacar M 2005 *J. Comp. Mat.* **39** 467-473
[9]. Garay RM, MacDonald F, Acevedo ML, Calderón B, Araya JE 2009 *J. BioResources* **4** 1396-1408
[10].Rowell RM 2006 *Forest Products Journal* **56** 4-12

[11].Nuryawan A, Massijaya MY, and Hadi YS 2008 *J. Ilmu dan Teknologi Hasil Hutan* **1** 60-66
[12].Lin HC, Huang JC 2004 *Taiwan J. For. Sci.* **19** 109–117
[13].Arabi M, Faezipour M, Layeghi M, Akbar A 2011 *J. For. Research* **22** 461–464
[14].Maloney TM 1977 Modern particleboard and dry process fiberboard manufacturing. (USA: Miller Freeman Inc. San Fransisco)
[15].Sun YG, Arima T 1999 *J. Wood Sci.* **45** 221–226

IOP Conf. Series: Materials Science and Engineering **309** (2018) 012039 doi:10.1088/1757-899X/309/1/012039

Performance evaluation of existing building structure with pushover analysis

MAP Handana*, R Karolina, Steven

Civil Engineering Department, Universitas Sumatera Utara, Jl. Perpustakaan Kampus USU, Medan 20155, Indonesia

*Email: agung22handana@yahoo.com

Abstract. In the management of the infrastructure of the building, during the period of buildings common building damage as a result of several reasons, earthquakes are common. The building is planned to work for a certain service life. But during the certain service life, the building vulnerable to damage due to various things. Any damage to cultivate can be detected as early as possible, because the damage could spread, triggering and exacerbating the latest. The newest concept to earthquake engineering is Performance Based Earthquake Engineering (PBEE). PBEE divided into two, namely Performance Based Seismic Design (PBSD) and Performance Based Seismic Evaluation (PBSE). Evaluation on PBSE one of which is the analysis of nonlinear pushover. Pushover analysis is a static analysis of nonlinear where the influence of the earthquake plan on building structure is considered as burdens static catch at the center of mass of each floor, which it was increased gradually until the loading causing the melting (plastic hinge) first within the building structure, then the load increases further changes the shapes of post-elastic large it reached the condition of elastic. Then followed melting (plastic hinge) in the location of the other structured.

1. Introduction

A building is planned to function during certain service period. However, during the service period, buildings are vulnerable to damage caused by various things. Any damage attempted can be detected as early as possible, because one damage can spread, trigger and aggravate other damage. Thus re-evaluating the seismic performance of the building structure, is an urgent matter as part of concrete steps in handling the impact of earthquakes. Potential collapse of the structure will endanger the safety of the occupants or users of the structure. Therefore, engineers are required to design structures with earthquake-resistant capability. The design of earthquake-proof buildings should pay attention to the criteria and details according to the applicable code.

The latest concept for earthquake engineering is Performance Based Earthquake Engineering (PBEE). PBEE is divided into two, which is Performance Based Seismic Design (PBSD) and Performance Based Seismic Evaluation (PBSE). In this study, the seismic engineering concept used is Performance Based Seismic Evaluation (PBSE). Evaluation on PBSE one of them is with nonlinear pushover analysis. Pushover analysis is a nonlinear static analysis in which the effect of the earthquake plan on the structure of the building is considered as static loads that capture at the center of each floor mass, whose value is gradually increased to exceed the loading causing the first melting (plastic joint) Within the structure of the building, then with further load increases undergoes a large post-elastic shape change until it reaches the elastic state. Then followed by melting (plastic joints) in another location of the structure. Therefore, the authors conducted a study of seismic building performance evaluation with a pushover analysis using the help of SAP 2000 program which then examine and discuss the output produced by the program. In this study, Wari Medan Building became the research object of the writer. Wari Medan building consists of 3-storey building (3) Front view of the old building facing Jln. Brigjend Katamso.

Figure 1. Front View of Wari Medan Building

The assumptions used in modelling and 3D structure analysis include:
1. The structure model under consideration is Wari Medan Building.
2. The structure of the building serves as an office building consists of 3 floors.
3. The structure used is a concrete structure, including:
   a. Structure of reinforced concrete portal
   b. Floor plates 1,2,3 and reinforced concrete roof
   c. Non structural components such as stairs
4. Building loading includes:
   a. Dead load (in the form of own weight structure)
   b. Life expenses (in the form of load due to building function in accordance with SNI 2002)
   c. Lateral load (in the form of earthquake load in accordance with SNI 03-1726-2002 Tata Cara Perencanaan Ketahanan Gempa untuk Gedung, without taking into account wind load).
   d. Rules of imposition based on Peraturan Pembebanan Indonesia untuk Rumah dan Gedung SNI 03-1727-1989.
5. The performance of the building will be determined based on the guidance contained in FEMA-356 and ATC-40.
6. Behavioural structures are analysed using pushover method with the help of SAP 2000 program.
7. If the structure is not safe, the recommended retrofitting method for column structure with concrete jacketing and beam reinforcement method with Fiber Reinforced Polymer (FRP)

2. Purpose and Objectives
This study is intended to evaluate whether the structure system is still safe or insecure against workloads and know the performance of the building structure when the plan earthquake occurs. If the existing structural system is not secure then what retrofitting action should be done to make the structure system safe. Then the performance of this building will be expressed in the form of plan displacement which will be calculated based on the guidance of FEMA-356 and in the form of performance point which will be calculated based on ATC-40 guideline.

3. Structure modeling
In this journal, the model of the building structure is using an open frame system that made of reinforced concrete which is a unified structural system (monolith) consisting of columns, beams, and

plates. It is assumed that the lateral load is borne by the parent column and the parent beam as the unity of the moment retaining portal system (resisting frame moment).

Table 1. General data of Wari Medan Building

No	Parameter	Value
1	Length of the building	15.22 m
2	Width of the building	10.04 m
3	Primary factor of the building, *I*	1.0
4	Type of earthquake retardant system	Reinforced concrete frame of ordinary moments
5	Coefficient of response modification, R	3

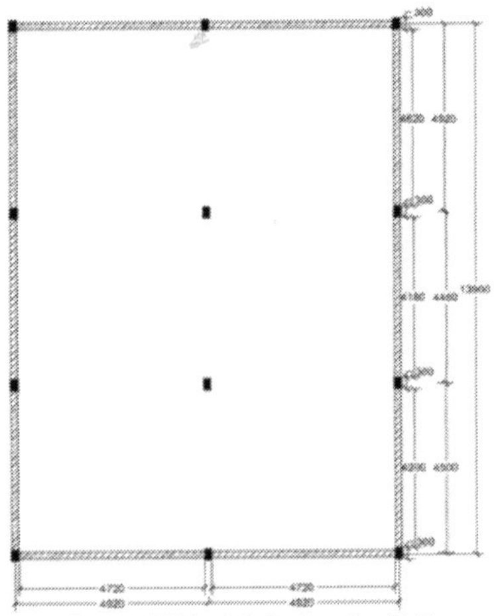

Figure 2. Floor 1 Plan of Wari Medan building

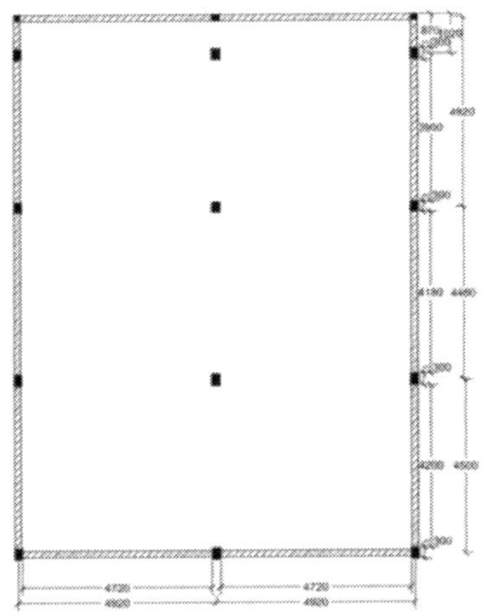

Figure 3. Floor 2&3 Plan of Wari Medan building

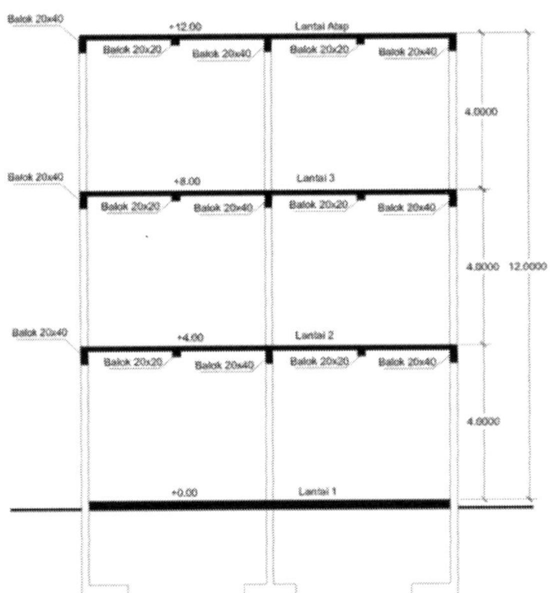

Figure 4. Section of Wari Medan building

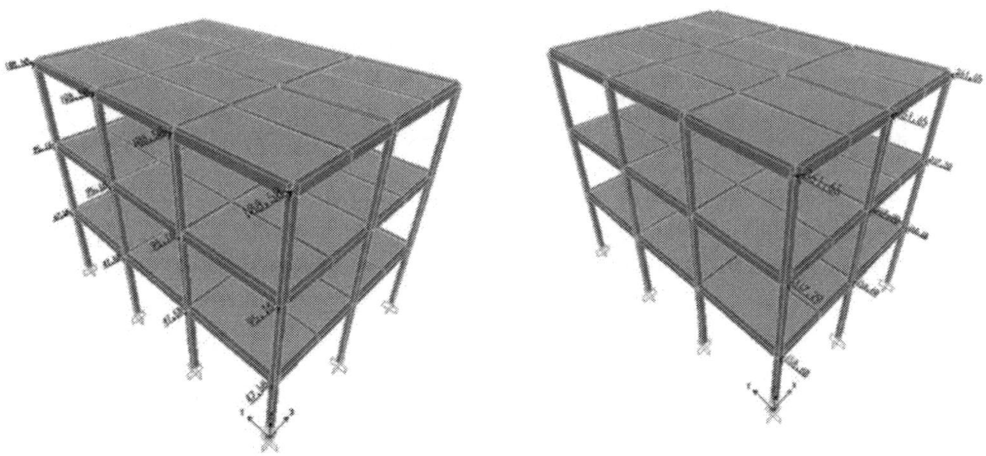

Figure 5. Lateral load modeling of X and Y direction in SAP 2000 program.

The building to be analyzed is a regular building so that the equivalent static analysis method will be used to calculate the effect of earthquake load and the building will be analyzed with pushover analysis.

4. Analysis of thrust load and building performance

Pushover analysis is performed to obtain the capacity curve of the building structure which will then be used to analyze the performance of the building in the event of a certain scale earthquake. Thrust load analysis is a type of static analysis procedure that takes into account the nonlinear state of the structure component. The plastic joint property of each structure component plays an important role in determining the behaviour of the building structure under the influence of the load acting on the structure.

Then the performance of the building will be determined using the help of SAP2000 program . Several types of methods can be used by SAP2000 to determine the performance of the building:

1. ATC-40 Capacity Spectrum Method.
2. FEMA 356 Coefficient Method.
3. FEMA 440 Equivalent Linearization Method.
4. FEMA 440 Displacement Modification Method

According to ATC-40, the criteria of earthquake- resistant structures are as follows:

1. Immediate Occupancy, "IO", When an earthquake occurs, the structure is able to withstand the earthquake, the structure does not suffer structural damage and does not experience non structural damage. So it can be directly used.
2. Level of damage control (Damage Control), "DC". At this stage the structure of the building may be damaged, but not collapsed. The risk of casualties is very low. The damage varies between IO and LS categories. This is useful where the performance targets to be achieved have more strict criteria than the LS level, but habitation is not the main problem. Examples are historic buildings.
3. Level of life safety (Life Safety), "LS". When an earthquake occurs, the structure is able to withstand earthquakes, with minimum structural damage, humans living / residing in the building is safeguarded from earthquakes.
4. Level of structural stability (Collapse Prevention or Structural Stability), "CP". When an earthquake occurs, the structure undergoes severe structural damage, but has not collapsed.

According to ATC-40, the drift ratio limit is as follows:

Table 2. Roof drift ratio limit according to ATC-40.

Parameter	Perfomance Level			
	IO	Damage Control	LS	Structural Stability
Maksimum Total Drift	0.01	0.01 s.d 0.02	0.02	$0.33\frac{Vi}{Pi}$
Maksimum Total Inelastik Drift	0.005	0.005 s.d 0.015	No limit	No limit

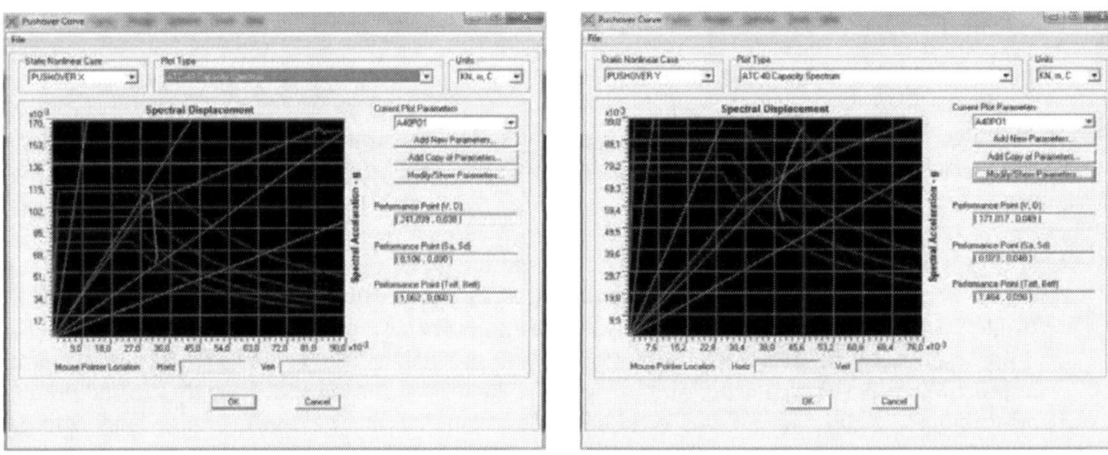

Figure 6. The capacity curves of X and Y Direction

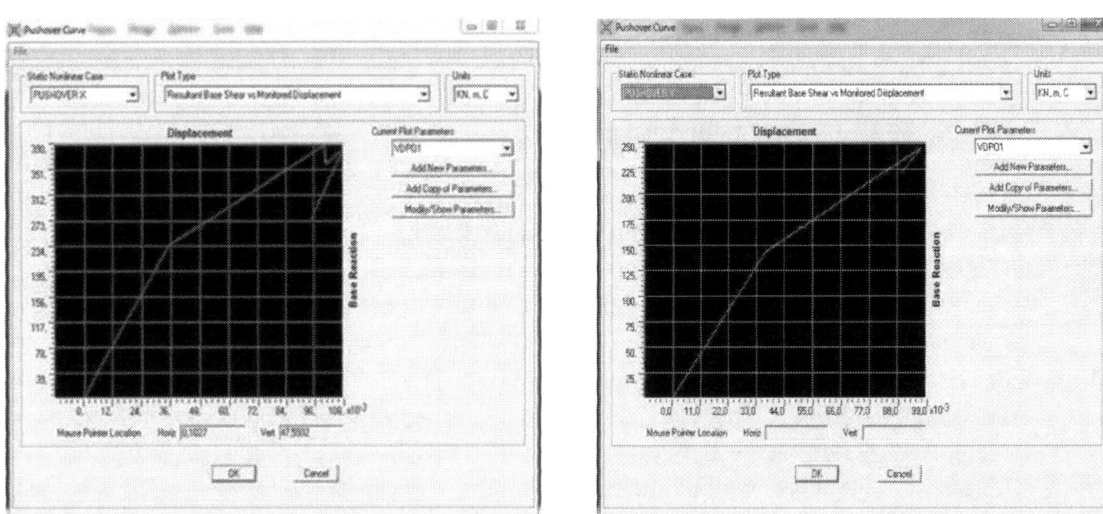

Figure 7. Graph of capacity spectrum of X and Y Direction

IOP Conf. Series: Materials Science and Engineering **309** (2018) 012039 doi:10.1088/1757-899X/309/1/012039

5. Results and Discussion

Based on the results of iteration calculations, the pushover analysis stops at step 17. Figure 1 shows that after a maximum displacement of structures of 0.1006 m, the structure shifts in the opposite direction and the base shear force decreases and undergoes collapse. In the spectrum capacity (ATC-40) method, the pushover curve with certain modifications is transformed into the capacity spectrum as shown in Figure 6. From the plan spectral response curve of SNI 1726-2002 for earthquake 3, soft soil conditions. The result of pushover analysis also shows the output of performance point of building structure with effective bath value (βeff).

Table 3. X direction of Performance evaluation of structure according to ATC-40

Base shear force (ton)	Performance Point			
	V_1 (ton)	D_1 (m)	β_{eff} (%)	T_{eff} (second)
25.1235	241.039	0.038	0.06	1.06

Table 4. X direction of Performance evaluation of structure according to ATC-40

Base shear force (ton)	Performance Point			
	V_1 (ton)	D_1 (m)	β_{eff} (%)	T_{eff} (second)
83.745	171.017	0.049	0.0098	1.484

In Table 3 and 4 it can be seen that the value of base shear force Vtx = 241.039 ton> Vyx = 25.1235 ton and Vty = 171.017 ton> Vyx = 83.745 ton. Thus, based on the capacity spectrum method (ATC-40) the behavior of X and Y direction structures in the plan earthquake has undergone an in-elastic condition caused by melting on the plastic joints. Maximum displacement limit = 0.02 H = 0.02. 12 m = 0.24 m. Target displacement result of pushover analysis for X direction = 0,038 m and Y direction = 0,049 m <0,24 m, so the security requirement is fulfilled.

Based on table 4:13 and 4:14, it can be seen the limits of the roof drift ratio evaluated on the performance point (PP), whose parameters are the maximum drift and maximum inelastic drift. The calculation is as follows:

$$Maksimum\ total\ drift = \frac{D_t}{H_{total}} = \frac{0{,}049}{12} = 0.00408 < 0{,}01\ (IO)$$

$$Maksimum\ inelastik\ drift = \frac{D_t - D_1}{H_{total}} = \frac{0.049 - 0.0095}{12} = 0.004 < 0.005\ (IO)$$

Based on Table 2 the limit of roof drift ratio according to ATC-40, the calculation results above shows that the building reviewed in this study is included in the level of performance of Immediate Occupancy (OI).

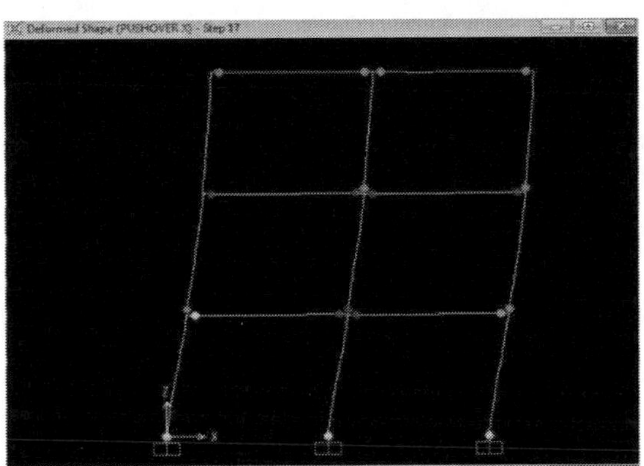

Figure 8. Spreading of Plastic Joints on Buildings Due to Push Load of X Direction

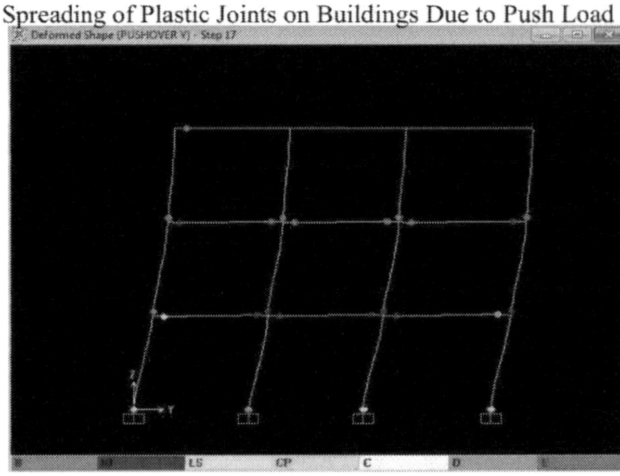

Figure 9. Spreading of Plastic Joints on Buildings Due to Push Load of Y Direction

6. Conclusion

Based on the evaluation of the existing structure of buildings, which is Wari Medan Building in Brigjen Katamso road, it is obtained the following conclusions.

Based on the performance point, the base shear force was 241,039 ton, displacement at step 3> 0.038 m (Dt), from the plastic joint distribution of structural performance did not pass LS (Life Safety). Maximum total value drift = 0.00408 and maximum inelastic drift = 0.004. This indicates that the building under review is included in the performance of Immediate Occupancy (IO), then in case of earthquake, the building does not suffer structural and non structural damage, so the building is safe and can be directly used.

If Wari Medan Building is not safe, it is necessary to repair the existing structure of Wari Medan Building. The improvements made in this research are using performance improvement methods of existing structural elements in the form of concrete jacketing techniques on columns and composite fiber coating (FRP) on the beams that lack the nominal strength in bending and shear to withstand the seismic loads caused by the earthquake plan.

References

[1] Anonim 1996 *Seismic Evaluation and Retrofit of Concrete Buildings*, Applied Technology Council (ATC-40), Volume 1, Report No. SSC 96-01, California

[2] Anonim 2 2000 Prestandard and Commentary for The Seismic Rehabilitation of Buildings, FEMA-356, Reston.Virginia

[3] Anonim 3 2005 Improvement of Nonlinear Static Seismic Analysis Procedure, FEMA-440, California.

[4] FEMA-273 1997 *NEHRP Guidelines For the Seismic Rehabilitation of Buildings*. Virginia. American Society of Civil Engineers

[5] Badan Standarisasi Nasional 2002 Tata cara perencanaan ketahanan gempa untuk bangunan gedung , SNI-1756-2002, Jakarta

[6] Pranata YA Evaluasi Kinerja Gedung Beton Bertulang Tahan Gempa dengan Pushover Analysis (Sesuai ATC-40, FEMA 356, dan FEMA 440)

[7] Dewobroto W 2000 Evaluasi Kinerja Bangunan Baja Tahan Gempa dengan SAP 2000

[8] Nugroho F 2008 *Evaluasi Kinerja Seismik dengan Analisis Statik Nonlinear Pushover Pada Gedung Berdenah,* Skripsi Jurusan Teknik Sipil, Universitas Sebelas Maret, Surakarta

IOP Conf. Series: Materials Science and Engineering **309** (2018) 012040 doi:10.1088/1757-899X/309/1/012040

Tensile properties and water absorption assessment of linear low-Density Polyethylene/Poly (Vinyl Alcohol)/Kenaf composites: effect of eco-friendly coupling agent

A L Pang[*1], **H Ismail**[1,2] **and A Abu Bakar**[1]

[1] School of Materials and Mineral Resources Engineering, Universiti Sains Malaysia (USM), Engineering Campus, 14300 Nibong Tebal, Penang, Malaysia
[2] Cluster for Polymer Composites (CPC), Science and Engineering Research Centre (SERC), Universiti Sains Malaysia (USM), Engineering Campus, 14300 Nibong Tebal, Penang, Malaysia

*E-mail: pangailing87@yahoo.com

Abstract. Linear low-density polyethylene (LLDPE)/poly (vinyl alcohol) (PVOH) filled with untreated kenaf (UT-KNF) and eco-friendly coupling agent (ECA)-treated kenaf (ECAT-KNF) were prepared using ThermoHaake internal mixer, respectively. Filler loadings of UT-KNF and ECAT-KNF used in this study are 10 and 40 parts per hundred parts of resin (phr). The effect of ECA on tensile properties and water absorption of LLDPE/PVOH/KNF composites were investigated. Field emission scanning electron microscopy (FESEM) analysis was applied to visualize filler-matrix adhesion. The results indicate LLDPE/PVOH/ECAT-KNF composites possess higher tensile strength and tensile modulus, but lower elongation at break compared to LLDPE/PVOH/UT-KNF composites. The morphological studies of tensile fractured surfaces using FESEM support the increment in tensile properties of LLDPE/PVOH/ECAT-KNF composites. Nevertheless, LLDPE/PVOH/UT-KNF composites reveal higher water absorption compared to LLDPE/PVOH/ECAT-KNF composites.

1. Introduction

It is a regular trend at present among researchers, industries and governments focusing on developing new products based on natural resources that minimize the impact to environment [1]. The use of natural resources such as kenaf (KNF) as reinforcement or filler in polymers is being actively investigated [2-4]. Kenaf offers a wide range of merits such as light in weight, environmental friendly, available in abundance and so on [5-7].

Due to the fact that KNF is naturally hydrophilic, it tends to absorb moisture and exhibit poor filler-matrix adhesion in polymer composites [6,8]. Subsequently, the mechanical properties of KNF-polymer composites are affected and to tackle this weakness, chemical treatment of KNF and/or modification of polymers can be applied to enhance filler-matrix adhesion [3]. There are vast studies on chemical treatment of KNF composites conducted in the interest of enhancing adhesion between KNF and matrix materials [9-13].

In this work, surface modification of KNF with eco-friendly coupling agent (ECA) was performed, before using it as filler into LLDPE/PVOH blend. The effect of ECA on tensile properties, morphological properties, and water absorption of LLDPE/PVOH/KNF composites were examined.

2. Experimental

2.1. Materials

LLDPE was supplied by PT. Lotte Chemical Titan Nusantara, Indonesia with melt flow rate of 1 g/10 min at 190 °C and a density of 0.92 g/cm^3. PVOH was purchased from Sigma-Aldrich (M) Sdn. Bhd., with molecular weight of 89,000 to 98,000 g/mol and density of 1.269 g/cm^3. Epichlorohydrin and benzyl-trimethylammonium chloride were products from Sigma-Aldrich (M) Sdn. Bhd., with molecular weight of 92.52 and 185.69 g/mol, respectively. Ethanol was supplied by Merck (M) Sdn. Bhd. with molar mass of 46.07 g/mol. KNF was obtained from National Kenaf and Tobacco Board (LKTN), Kelantan, Malaysia.

2.2. Characterizations

Universal Testing Machine (Model: Instron 3366) was used to determine tensile properties (strength, modulus and elongation at break) of composites, according American Society for Testing and Materials (ASTM D638-Type IV). The tensile fractured surfaces morphology of composites was examined under a FESEM (Zeiss Supra-35VP). Water absorption was carried out in accordance to ASTM D570, and the percentage of water absorption was calculated as Equation 1:

$$W_t(\%) = \frac{W_2 - W_1}{W_1} \times 100 \tag{1}$$

where W_t is the total water absorption by composites, W_1 and W_2 are the weights of composites before and after immersion in distilled water, respectively.

3. Result and discussion

3.1. *Tensile properties*

Table 1 displayed the tensile properties of LLDPE/PVOH/UT-KNF and LLDPE/PVOH/ECAT-KNF composites. The tensile strength and elongation at break was found to decrease, while the tensile modulus increases with respect to KNF loading (from 10 phr to 40 phr). This is because of the inability to transfer the load effectively resulting from weak interfacial adhesion between filler and matrix, as well as poor wettability of filler by the matrix [2,4]. Furthermore, it is visible that composites with ECAT-KNF exhibit higher tensile strength and tensile modulus, but lower elongation at break when compared to composites with UT-KNF. The increase in tensile strength and tensile modulus implied that interfacial adhesion between KNF and LLDPE/PVOH was improved in the presence of ECA, and hence permitting efficient stress transfer to occur. Similar findings were observed by Nurfatimah et al. [9] and Majid et al. [12]. Study by Majid et al. [9] shows that benzoyl treatment of kenaf core powder (KCP) improved the adhesion of KCP and polyvinyl chloride/epoxidized natural rubber (PVC/ENR), thereby increase the composites' tensile properties. Nevertheless, the drop in elongation at break is expected, mainly due to the decrement in the deformability of composites after ECA treatment. The possible reactions between ECA, KNF and LLDPE/PVOH matrices are as follow: (i) The interaction of hydroxyl groups of KNF with the oxirane group of ECA [14] (ii) The hydrophobic chain in ECAT-KNF is likely to interact with the hydrocarbon chains of LLDPE/PVOH matrices by intermolecular attraction.

Table 1. Tensile properties of LLDPE/PVOH/UT-KNF and LLDPE/PVOH/ECAT-KNF composites.

KNF loadings (phr)	LLDPE/PVOH/UT-KNF composites			LLDPE/PVOH/ECAT-KNF composites		
	Tensile strength (MPa)	Tensile modulus (MPa)	Elongation at break (%)	Tensile strength (MPa)	Tensile modulus (MPa)	Elongation at break (%)
10	8.6 ± 0.17	796.5 ± 23.71	6.5 ± 0.56	9.2 ± 0.06	921.3 ± 9.25	5.5 ± 0.13
40	7.1 ± 0.15	1169.0 ± 25.54	3.0 ± 0.13	7.8 ± 0.27	1397.0 ± 25.17	2.2 ± 0.34

3.2. *Morphological study*

Figure 1 illustrates tensile fracture surfaces morphology of LLDPE/PVOH/UT-KNF and LLDPE/PVOH/ECAT-KNF composites at 10 phr KNF loading. As seen in Figure 1a, there are many voids presence on composites fractured surfaces owing to the detachment of UT-KNF from LLDPE/PVOH matrices after stress application. This observation confirms the poor compatibility between UT-KNF and LLDPE/PVOH caused by their different polarity. Nevertheless, it was found that interfacial adhesion between ECAT-KNF and LLDPE/PVOH matrices was improved, as shown by less voids presence on fractured surfaces. ECAT-KNF was seen well dispersed and embedded firmly within LLDPE/PVOH matrices after application of tensile stress.

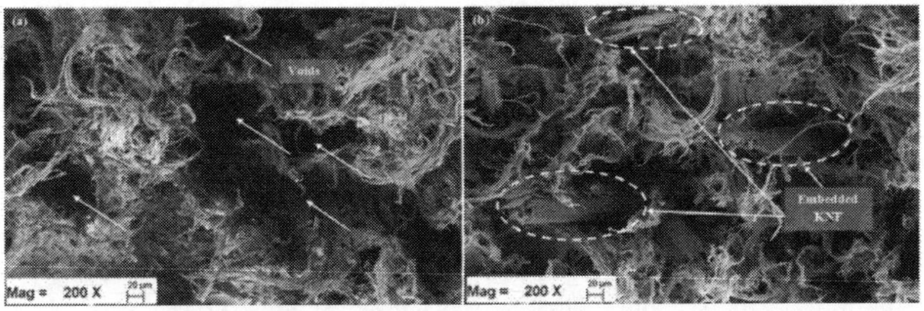

Figure 1. FESEM micrographs on tensile fractured surfaces of LLDPE/PVOH/10 phr KNF composites with (a) UT-KNF (b) ECAT-KNF, respectively.

3.3. *Water absorption*

Figure 2 shows the equilibrium water absorption of LLDPE/PVOH/KNF composites under the effect of ECA, respectively. It shows ECAT-KNF composites exhibit lower equilibrium water absorption than that of UT-KNF composites, at both 10 phr and 40 phr KNF loading, respectively. This indicates that ECA treatment has enhanced the adhesion between KNF and LLDPE/PVOH matrix, thus decreasing the formation of micro-voids between them. Subsequently, it limits water from diffuse into ECAT-KNF composites. Based on Figure 2 also, equilibrium water absorption of UT-KNF and ECAT-KNF composites was increased with increasing KNF loading. This is attributed to more hydroxyl groups from KNF to form hydrogen bonding with water, and also presence of micro-voids between KNF and LLDPE/PVOH matrices at higher filler loading.

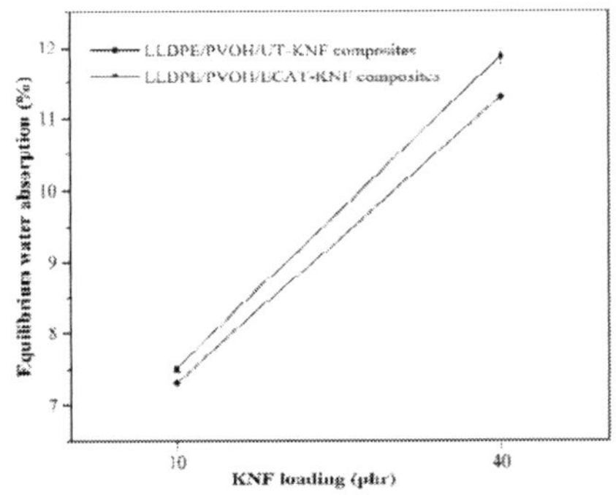

Figure 2. Effect of ECA on equilibrium water absorption of LLDPE/PVOH/KNF composites.

4. Conclusion

The effect of ECA on the tensile properties and water absorption of LLDPE/PVOH/KNF composites were studied, and it is apparent that LLDPE/PVOH/ECAT-KNF composites have more optimal results than that of LLDPE/PVOH/UT-KNF composites. The tensile strength and tensile modulus show increment with the ECA treatment. ECA was found effective in enhancing KNF-LLDPE/PVOH interfacial adhesion, as proven in the FESEM micrographs. Furthermore, LLDPE/PVOH/ECAT-KNF composites exhibit lower equilibrium water absorption, mainly attributable to the good adhesion between KNF and LLDPE/PVOH.

Acknowledgement

The authors are grateful to Malaysia Higher Education, RUC research grant (No: 1001/PKT/8640014) from Universiti Sains Malaysia and National Kenaf and Tobacco Board (LKTN), Kelantan, Malaysia for sponsoring and supporting this research.

References

[1] Mariotti N, Wang X M, Rodrigue D and Stevanovic T 2014 J. *Mater. Sci. Res.* **3** 8
[2] Sarifuddin N, Ismail H and Ahmad Z 2013 *J. Phys. Sci.* **24** 97
[3] Saba N, Paridah M T and Jawaid M 2015 *Const. Build. Mater.* **76** 87
[4] Pang A L, Ismail H and Abu Bakar A 2015 *Biores.* **10** 7302
[5] Akil H M, Omar M F, Mazuki A A M, Safiee S, Ishak Z A M and Abu Bakar A 2011 *Mater. Des.* **32** 4107
[6] Pang A L and Ismail H 2013 *J. Thermoplast. Compos. Mater.* **27** 1607
[7] Kamal I, Thirmizir M Z, Beyer G, Saad M J, Abdul Rashid N A and Abdul Kadir Y 2014 *J. Sci. Technol.* **6** 41
[8] Tan B K, Ching Y C, Gan S N, Ramesh S and Rahman M R 2014 *Mater. Res. Inno.* **18** 144
[9] Nurfatimah B, Chee C Y, Abdullah L C, Ratnam C T and Azowa N 2014 *Compos. Part A. Appl. Sci. Manuf.* **63** 45
[10] Tawakkal I S M A, Marlene J C and Stephen W B 2014 *Ind. Crops. Prod.* **61** 74
[11] Datta J and Kopczynska P 2015 *Ind. Crop. Prod.* **74** 566
[12] Majid R A, Ismail H and Taib R M 2016 *Procedia. Chem.* **19** 803
[13] Pang A L, Ismail H and Abu Bakar A 2016 *Biores.* **11** 5889
[14] Muresan E I, Piroi C, Creanga D, Stelea L, Oprica L and Sandu I 2016 *Rev. Chim.* **67** 871

IOP Conf. Series: Materials Science and Engineering **309** (2018) 012041 doi:10.1088/1757-899X/309/1/012041

Alternative of raw material's suppliers using TOPSIS method in chicken slaughterhouse industry

R M Sari*[1], I Rizkya[2], K Syahputri[3], Anizar[4], I Siregar[5]
Universitas Sumatera Utara
Jalan Almamater Kampus USU, Padang Bulan Medan 20155, Indonesia

*E-mail: rahmi.m.sari@usu.ac.id

Abstract. Chicken slaughterhouse industry is one of the fastest growing industries that depends on the freshness of raw materials. The raw materials quality arrive at the company depends heavily on the suppliers. Fresh chicken and frozen chicken meat are the main raw materials for this industry. Problems occurred by the suppliers are catering the amount of raw material needs that are not appropriate and also delay during delivery process. This condition causes disruption of the production process in the company. Therefore, it is necessary to determine the best suppliers to supply the main raw materials of fresh and frozen chicken meat on the slaughterhouse chicken industry. This study analyze the supplier's capability by using TOPSIS method. This method use to find out the best supplier. The TOPSIS method is performed using the principle that chosen alternative must have the shortest distance from the positive solution and furthest from the ideal solution of the geometric point by using the Euclidean distance to determine the relative proximity of the optimum solution alternative. TOPSIS method found the rank of best supplier's order is supplier A followed by supplier D, supplier B, supplier C, supplier E, supplier F, and supplier G. Based on the rank order obtained from each company, it will assist the company in prioritizing the order to the supplier with the best rank. Total supply from All suppliers are 885,994 kg per month. Based on the results of research, the top five suppliers have been sufficient to meet the needs of the company.

1. Introduction

In modern supply chain management, an appropriate evaluation of alternative suppliers' choices is the most common type of problem. The right suppliers will greatly assist the industry directly or indirectly. Supplier's choice is the first step in production activities from raw material purchases to delivery of finished products [1]. In big industries, the role of suppliers is not only limited to raw materials, but also a place for warehousing locations [2]. Therefore, suppliers are very important in most production activities. Most production activities depend on it. If the supplier is late to fulfill the demand, then the production activity will be disrupted. The selection of suppliers evaluated is a critical factor for the industry to win the competition in today's industry competition [3]. The study of many alternative selections have discussed in various studies. Starting from the selection of a car purchased [4], the most accurate selection of management factors [6], including alternative options in the energy field [7, 8, 9, 10, 11]. Study of suppliers' alternative selections in chicken slaughterhouse industry discusses by using TOPSIS method. In using TOPSIS, there are four major differences in the use of normalization [12, 13, 14]. The end result of TOPSIS may differ depending on the normalization method so it will impact on the alternatives order [12].

This result will select the alternative suppliers that supply raw materials to the chicken slaughterhouse industry.

2. Literature Review

One of the most popular in Multi-criteria Decision Making (MCDM) is TOPSIS (Technique for Order Preference by Similarity to Ideal Solution). Based on the best chosen alternative concept not only has the shortest distance from the ideal solution, but also has the longest distance from the ideal solution. This method was first developed by Hwang and Yon [13] and has been widely applied in various research studies [12]. This concept is widely used on Multi-criteria Decision Making (MCDM) models to solve practical decision problems. The concept of TOPSIS method is simple and easy to understand; computing is efficient and has the ability to measure the relative performance of decision alternatives in simple mathematical form.

TOPSIS uses the principle that the chosen alternative must have the closest and furthest distance from a geometric point of view by using the Euclidean distance to determine the relative proximity of an alternative with the optimal solution. In general, TOPSIS procedure follows the following steps:

- Make a normalized decision matrix

$$r_{ij} = \frac{Xij}{\sqrt{\sum_{i=1}^{m} Xij^2}}, \; i = 1, \ldots, m; \; j = 1, \ldots, n \qquad (1)$$

- Create a weighted normalized decision matrix

$$y_{ij} = wij, \; i= 1, \ldots, m \; ; j=1, \ldots, n \qquad (2)$$

- Determine the matrix of positive ideal solutions and the ideal negative solution matrix

$$A^+ = (y_1^+, y_2^+, \ldots, y_n^+) \qquad (3)$$

$$A^- = (y_1^-, y_2^-, \ldots, y_n^-) \qquad (4)$$

- Determine the distance between the score for each alternative and the matrix of positive ideal solutions and the ideal negative solution matrix

$$D_i^+ = \sqrt{\sum_{j=1}^{n} (y_i^+ - y_{ij})^2}, \; i = 1, \ldots, m \qquad (5)$$

$$D_i^- = \sqrt{\sum_{j=1}^{n} (y_{ij} - y_i^-)^2}, \; i = 1, \ldots, m \qquad (6)$$

- Determine the preference value for each alternative

$$V_i = \frac{D_i^-}{D_i^- + D_i^+} \qquad , i = 1, \ldots, m \qquad (7)$$

3. Methodology

This study was conducted in one of chicken slaughterhouse industry in North Sumatra. This study will produce an alternative supplier that will supply raw materials to the chicken slaughterhouse industry. To solve this problem used TOPSIS method. In general the procedure in TOPSIS method consists of five steps. At the beginning, it is necessary building a decision matrix that refers to the number of alternatives to be evaluated based on the number of criteria. The first step is making a normalized decision matrix, followed by the second step of weighting a normalized decision matrix. The third step is the determination of positive ideal solution matrix and ideal negative solution. The fourth step is the determination of the distance between the values of each alternative with the matrix of positive ideal solutions and the ideal negative solution matrix. The final step is the determination of the preference value for each alternative. Having obtained the value of preference for each alternative, then got the suplier alternative needed.

4. Result and Discussion

Based on the ranking order obtained from each company then this sequence will help the company in giving priority to the supplier who has the best ranking. Reservations can be made in accordance with the capacity that can be given the best supplier 1 for the next shortage is supplied from the best supplier

2 and so on until the capacity of the company can be met. Capacity of raw material capacity of each supplier can be seen in Table 1.

Table 1. Supplier Capacity

Supplier	Delivery Capacity (Times/ Month)	Capacity Each Delivery (Kg)	Capacity (Kg/Month)
Supplier A	23	10.838	281.783
Supplier D	20	8.128	186.942
Supplier B	20	5.419	119.216
Supplier C	19	5.419	113.797
Supplier E	17	5.419	102.959
Supplier F	17	2.710	51.488
Supplier G	9	2.710	29.809
Total			885.994

Based on Table 1 it can be concluded that the seven suppliers are able to supply raw materials of 885.994 kg / month. Production capacity of chicken slaughterhouse industry is 711.947 kg/month, based on the above data it can be concluded that the raw material capacity of the supplier is greater than the company's production capacity. Based on the results of the research has been obtained the best supplier order, therefore it is recommended that the company make an order to the supplier based on the order. To meet the company's production capacity of 711,947 kg/month, the selected supplier can be seen in Table 2.

Table 2. Optimum Raw Material Supplier Alternative

Supplier	Capacity (Kg/Months)
Supplier A	281.783
Supplier D	186.942
Supplier B	119.216
Supplier C	113.797
Supplier E	102.959
Total	711.947

Based on Table 2 it can be concluded that the capacity of the five suppliers ie 711.947 kg / month has been able to meet the needs of the company. The best supplier alternative distribution network for chicken slaughterhouse industry can be seen in Figure 1.

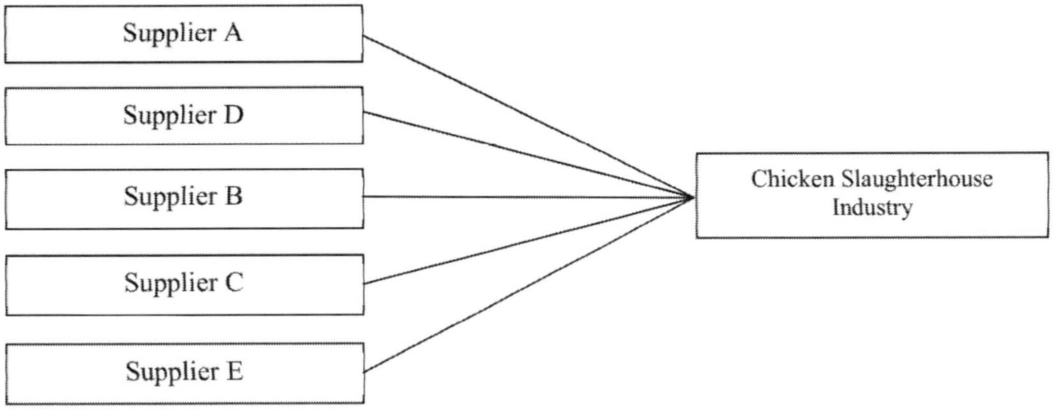

Figure 1. The Best Alternative Supplier Distribution Network for Chicken Slaughterhouse Industry

5. Conclusion

The conclusion in the research using TOPSIS Method on chicken slaughterhouse industry is the chosen supplier alternative for the fulfilment of fresh chicken raw materials needed i.e. Supplier A, Supplier D, Supplier B, Supplier C, Supplier E. While two more suppliers are not needed because the five suppliers There already meet the needs of industrial raw materials chicken slaughterhouse.

Acknowledgement

Author would like to thank to Universitas Sumatera Utara for funding the financial support in publishing this paper. We also thanks to Miss Ria for collecting and helping authors to finish this research. We really appreciate the efforts from all of the team that have cooperated in conducting this study.

References

[1] Lavanpriya C, Muralidaran V M and Lakshmanpriya C 2013 *The SIJ Transactions on IFBM.* **1(1)**
[2] Ashrafzadeh M, Rafiei F M, Isfahani N M, Zare Z 2012 *Interdis. Jourl of Contemporary Research in Business* **3(9)**
[3] Christian A V, Zhang Y and Salifou C K 2016 *Modern Economy* **7** 470-6
[4] Kwangsun Y 1987 *The Journal of Operation Research* **38** 277-86.
[5] Zadeh S A 2013 *Journal of Industrial Engineering and Management* **6** 860-75.
[6] Li W H 2012 *International Journal of Advancements in Computing Technology* **4** 304-10
[7] Chamodrakas I and Martakos D 2011 *Applied Soft Computing* **11(4)** 3734-43
[8] Cavallaro F 2010 *Applied Energy* **87(2)** 496-503
[9] Behzadian M, Otaghsara S K, Yazdani M, Ignatius J 2012 *A State-of-The-Art Survey of TOPSIS Applications* **39(17)** 13051-69
[10] Kaya T, Kahraman C 2011 *Expert Systems with Applications* **38(6)** 6577-85
[11] Yang L, Deuse J, Jiang P 2013 *International Journal of Advanced Manufacturing Technology* **66(5-8)** 795-807
[12] Wojciech S 2013 *J. of Theoretical and Applied Computer Science* **7(3)** 40-50
[13] Milani A S, Shanian A, Madoliat R, Nemes J A 2005 *Structural and Multidisciplinary Optimization* **29(4)** 312–18
[14] Shih H S, Shyur H J, Lee E S 2007 *Mathematical and Computer Modelling* **45(7-8)** 801–13
[15] Hwang C L, Yoon K P 1981 Multiple attribute decision making: Methods and applications. New York: Springer-Verlag

IOP Conf. Series: Materials Science and Engineering **309** (2018) 012042 doi:10.1088/1757-899X/309/1/012042

Foam capacity and stability of Sodium Dodecyl Sulfate (SDS) on the presence of contaminant coffee and Cd ions in solution

B Haryanto[1*], C H Chang [2], A T Kuo[2], M Z Siswarni [1] and T M A Sinaga[1]

[1]Department of Chemical Engineering, Universitas Sumatera Utara, Padang Bulan, Medan 20155, Indonesia
[2]Department of Chemical Engineering, National Cheng Kung University, Tainan 701, Taiwan

*E-mail: bode.haryanto@usu.ac.id

Abstract. In this study, the effect of the coffee colloidal particle and Cd ion contaminant on the foam capacity and stability of sodium dodecyl sulfate (SDS) solution was investigated. The foam was generated by using a foam generator. The foam capacity of SDS was first evaluated at different concentrations. After the foam capacity reaching a constant value, the foam stability was then measured by flowing to a column. The results showed that the presence the coffee colloidal particles or Cd ions in the solution would decrease the foam capacity and stability of SDS. In addition, the decreased foam capacity and stability was more pronounced in the presence of coffee colloidal particles than Cd ions. The colloidal particles may have stronger interaction with SDS and thus reduce the formation of the foam.

1. Introduction

The foam may be described as a dispersed system in which the bubbles of gas are encircled and stabilized by surfactant molecules absorbed in the liquid-air interface in a continuous liquid medium [1,2,3,4]. The foam is a dispersed fluid of a dispersed fluid containing small air bubbles with a large surface area that can be stabilized by surfactant molecules [5].

Surfactants are heterogeneous molecules that have long chains in which the head has a water-like and the tail has a water-dislike character. In aqueous phase, when the concentration of surfactant over than a certain critical value, the monomer molecule forms an organized group of large numbers of molecules called micelles. This particular concentration is called critical micelle concentration (CMC). Physical properties such as: surface tension, interface tension, adsorption, and detergency have changed to concentrations if below CMC, but have no change in these properties if above CMC[6,7]. The surfactant solution exhibits significant changes on others physical properties such as density, equivalent conductivity and organic solubility at below and above the CMC when compared to concentration [2].

Foam capacity is one of the important interfacial properties possessed by surfactants. The amount of foam capacity will affect the surfactant's ability to spread and press down to the pores of contaminated material [8]. Foaming ability can be seen from the increase in volume, once the gas is fed into the solution. Foam stability is associated with a decrease in the height of the foam with time [9].

Foam stability is the ability to retain gas for a certain time. Foaming ability can be seen from the increase in volume, once the gas is fed into the solution. Foam stability is associated with a decrease in the volume of the foam with time [2]. The efficiency of the surface active agent to form and stabilize the foam depends mainly on the molecular structure and intrinsic properties of the surfactant [2].

The capacity and stability of the surfactant is the most important factor in applying it to the remediation process using foam. The capacity of the foam to be associated with the surfactant's ability to produce foam and foam stability is the ability of the fixed surfactant in the form of foam and its ability not to break.

Surfactant foam applications in recovery of wasted oil to the environment are reported by Urum, et al. 2004 [5]. Surfactants are used to remediate mediacontaminated by metal-ions with surfactant foam applications [8,10,11,12]. Surfactant applications in the remediation process for transporting nanoparticle material to porous media [13]. Ability of surfactant on interactions with mercury in liquids thencan be separated from the fractional fraction [14].

A number of studies and theories in evaluating foam capability and stability [15]. There are several methods to measure the nature of this foam capacity, one of which is the dynamic foam capacity determined by dividing the constant volume of the foam (mL) with the gas flow rate N2 (mL / min)[4]. Razafindralambo et al. [2] measures the foam capacity by the formula:

The purpose of this research is to investigate the foam capacity and stability of SDS as anionic sutfactanton the effect of the presence a contaminants in the liquid phase. The contaminants are Cd metal ions and coffee powder.

2. Material and Method

Sodium dodecyl sulfate (SDS) was purchased from Sigma–Aldrich. Cd (II) was used as the model contaminant. Cadmium acetate (Cd $(CH_3COO)_2$. $2H_2O$) purchased from Merck KgaA, Darmstadt, Germany and used as a source of cadmium (Cd^{2+}). The sodium hydroxide (NaOH) purchased from Merck KgaA, Darmstadt, Germany and the hydrochloric acid (HCl) purchased from Mallinckrodt Baker, Inc, Paris. Both chemicals were used to control the pH variation of the deionized water and used to dissolve SDS and both contaminants. The black coffee (Kopi KapalApiSpesial) purchased from PT Santos Jaya Abadi Indonesia.

Foam-generator applied to measure the foam dynamic capacity (FDC) of SDS. It was designed and made by laboratory of surfactant and application, Chemical Engineering Department University of Sumatera Utara (USU). The size dimension of foam-generator is (L= 20 cm, OD=3.5 cm), equipped with circular porous ceramic, which enabled the foam generation dynamically in the presence of surfactant solution and N_2gas as shown in the Figure 1. Peristaltic pump was used to delivers the surfactant solution to foam generator (Figure 1A). The instrument prepared with progressed cylinder to measure foam stability (Figure 1 B).

In this study, the activity measurements for each run on the variables which have been determined for the dynamics of the formation of foam from the beginning until to achieve constant foam capacity for the time needed (T) as shown in Figure 1A. For the foam stability as shown in Figure 1B, the measurement after the initial foam height 7 cm in cylinder column then was beginning to measure the foam height for the certain time. Measure the height of foam in column generator to get the dynamics foam capacity of surfactant. The foam capacity (FC) is the constant foam volume in the column foam generator. The analysis tools used are scale measuring the height of dynamic foam and Atomic Absorption Spectrometer (AAS) for metal ion concentration. Measuring the time of foam dynamic to reach constant capacity was used timer [16].

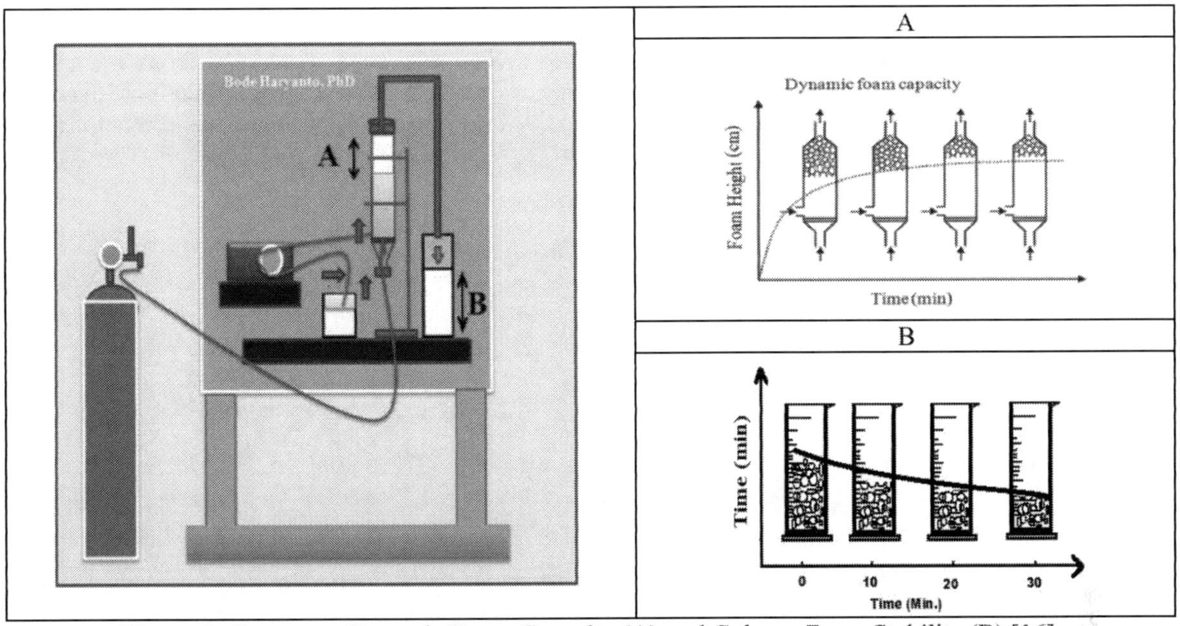

Figure 1. Generator Dynamic Foam Capacity (A) and Column Foam Stability (B) [16].

In this study, the concentration variations used for Cd^{2+} were 10 ppm, 20 ppm, 30 ppm, 40 ppm and 50 ppm. Variations of coffee concentration used were 10 ppm, 20 ppm, 30 ppm, 40 ppm and 50 ppm. SDS concentration variations were 1x cmc, 2x cmc and 3x cmc at pH 7.

3. Result and Discussion

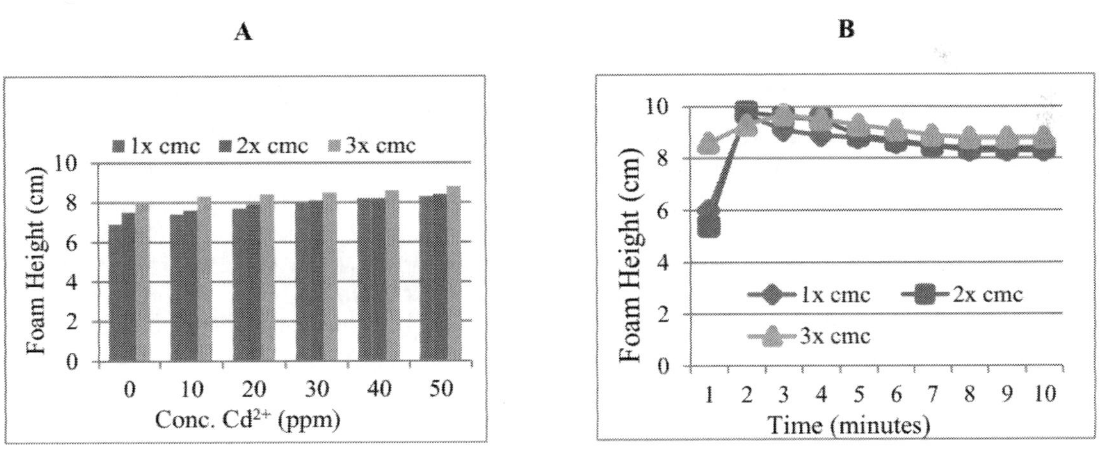

Figure 2. Foam Capacity with Variation Concentration of Cd (A) and Foam Dynamic Capacity with Concentration of Cd ions 50 ppm (B)

The result of the measurement of the height of the SDS foam capacity on the effect of variation concentration of Cd^{2+} ions in the presence on the solvent is shown in Figure 4.A. The greater of SDS concentration that will cause increasing the foam height obtained. The greater metal concentration Cd^{2+}

was tending to produce higher foam capacity. The presence of Cd^{2+} ions in a liquid phase with its positive charge and the SDS surfactant having a negative charge will increase the interaction on the formation of micelles. So when the gas bubbles flow into the column, the micelle's ability to absorb the N_2 gas in producing the foam is tends to more stable and quicker to reach a constant foam capacity as shown in Figure 2.B. [16,17,18]. Figure 2.B shows graphic kinetics of foam capacity by adding the time against variation of concentration SDS in the presence of contaminant Cd^{2+} 50 ppm. The higher concentration has higher foam capacity produced [19].

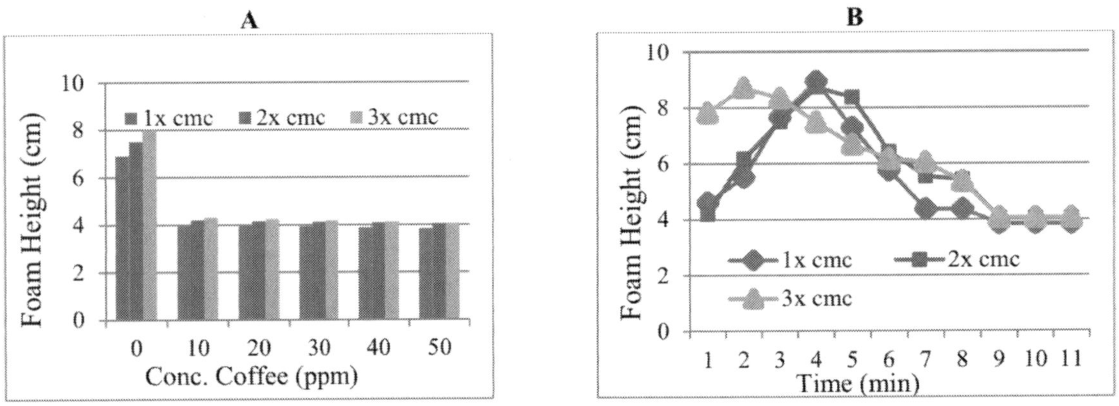

Figure 3. Foam Capacity with Variation Concentration of Cd (A) and Concentration of Coffee 50 ppm (B)

Figure 3A shows the result of variation concentration of coffee on the effect of the foam capacity of SDS. The result obtained that coffee colloidalparticle has ability to decrease foam ability of SDS. The presence of the contaminant was decreased the foam capacity of SDS significantly. The foam capacity of SDS was decreased by increasing the concentration of coffee.

Black coffee as reported has pH 4.5 to 5 [19,20] with negative charge on its surface particle [21]. Thekinetics of dynamic foam capacity of the SDS in the presence the coffee particle50 ppm shows in Figure 3B.Initially, when the N_2gas starts to flow into the foam generator to produce bubbles, the foam capacity was tend to normal ability. The condition was possible to produce higher foam capacity becausethe interaction of the contaminant black coffee with the micelles is not significant in impacting the foam dynamic capacity [16]. After a few minutes, the presence of coffee colloidalparticle in the solution may create complexity on the SDS micelles in the solution. The coffee particles with negative charge interact with micelles, the foam capacity start to reduce then after 9 minutestend to reach constant height.Theoretically, the bubble N_2 gas flowing in promising for wet foam lamellae from the micelle where the both layers as fairly thick water slab become more complex by presence of coffee colloidalparticle. Then it impacted the dynamic foam capacity of variation SDS concentration even though with the higher SDS concentration has higher foam capacity [19]. One can see the results of this study that the coffee contaminant has ability to decrease the foam capacity in comparing with Cd^{2+} ions.

Measurement the stability of the foam is carried out by flowing the foam generated from the foam generator at a constant foam height to the foam stability measurement column. Measurement of foam stability on the measuring cup column began to be done to the height of the foam at 7 cm. Figure 4 A and B show the foam stability measurement. It has done to find the time that the foam needed to return to the liquid phase. In surfactants, the bubble interface stabilization mechanism is concerned with the ability of micelles to withstand surface tension gradient.Resistance to changes in surface area or surface concentrations are described by interface resistance [22].From the results of foam stability show that coffee contaminant has similarstability in comparing with Cd^{2+}ions.On the other hand still more stable than SDS without contaminants.The presence of coffeecolloidalparticles with negatively charged and has a low pH that may cause lower stability as slightly lower at low SDS concentration (Figure 4B).

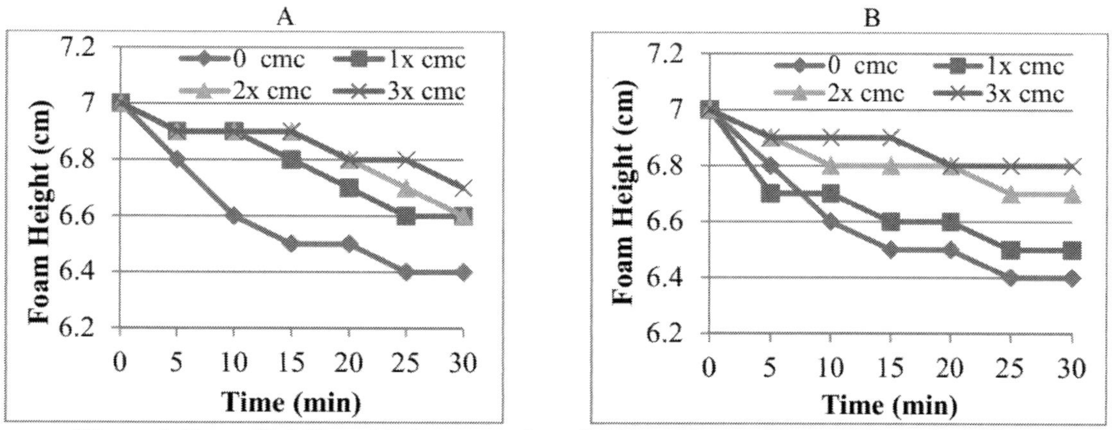

Figure 4. Foam Stability with Concentration of Cd ion 50 ppm Figure 7. Foam Stability with Concentration of Coffee 50 ppm

4. Conclusion

The contaminants Cd^{2+} and coffee is an important element that affects the capacity of the foam. The foam capacity of SDS at 3x cmc with presence of concentration Cd^{2+} 50 ppm was obtained the highest foam height 8.8 cm. Coffee has ability to decrease the foam capacity with the high of foam about 4 cm to all variation concentration of coffee contaminant and SDS. The foam dynamics capacity of SDS in the presence of Cd^{2+} reaches a constant height is about 8 minutes. The height of foam dynamic capacity with Cd^{2+} is more stable and quicker to reach a constant foam capacity. The fluctuation of the foam height of SDS was occurred at the beginning of measurements the kinetics of foam dynamic before reach the constant height in the presence of coffee. Presence of coffee particle in the solution may create complexity on the SDS micelles in the solution. The foam stability shows that coffee and Cd^{2+} ions have similar stability.The presence of coffee colloidalparticles may cause lower stability at lower SDS concentration.

References

[1] Halling PJ 1981 Protein-stabilized foams and emulsions *Critical Reviews in Food Science and Nutrition* **15** 155
[2] Razafindralambo H, Popineau Y, Deleu M, Hbid C, Jacques P, Thonart P and Paquot M 1998 Foaming Properties of Lipopeptides Produced by Bacillus subtilis: Effect of Lipid and Peptide Structural Attributes *Journal of Agricultural and Food Chemistry* **46** 911-916
[3] Durand M 2010 Statistical mechanics of two-dimensional foams *Europhys Lett* **90 (6)** 60002 1-6
[4] Haryanto B 2014 Exploring potential application of surfactin and rhamnolipid on adsorbed metal ion removal from sand surface with continuous flushing column PhD Thesis, NCKU, Tainan, Taiwan
[5] Urum K and Pekdemir T 2004 Evaluation of biosurfactants for crude oil contaminated soil washing *Chemosphere* **57** 1139-1150
[6] Mulligana C N, Yong R N and Gibbs B F 2001 Heavy metal removal from sediments by biosurfactants. *Journal of Hazardous Materials* **85** 111-125
[7] Mulligan C N 2005 Environmental applications for biosurfactants *Environmental Pollution* **133** 183-198
[8] Mulligan C N and Gibbs B F 2004 Types, production and applications of biosurfactants *Proceedings-Indian National Science Academy*

[9] Zhang Y, Miller R M 1992 Enhanced octadecane dispersion and biodegradation by a Pseudomonas rhamnolipid surfactant (biosurfactant) *Appl Environ Microbiol* **58** 3276–82
[10]Haryanto B, Chang J S, Chang C H 2014 Application of biosurfactant surfactin on copper ion removal from sand surfaces with continuous flushing technique *Tenside Surfactants Detergents* **51(5)** 407-414
[11]Haryanto B, Chang C H 2014 Foam-enhanced removal of adsorbed metal ions from packed sands with biosurfactant solution flushing *Journal of the Taiwan Institute of Chemical Engineers* **45(5)** 2170-2175
[12] Haryanto B, Chang C H 2015 Removing adsorbed heavy metal ions from sand surfaces via applying interfacial properties of rhamnolipid *Journal of oleo science* **64(2)** 161-168
[13] Raymundo A, Empis J and Sousa I 1998 Method to Evaluate Foaming Performance *Jounml of Food Engineering* **36** 445-452
[14] Paria S 2008 Surfactant-enhanced remediation of organic contaminated soil and water *Advances in Colloid and Interface Science* **138** 24-58
[15] Ross J and Miles G D 1941 An Apparatus for Comparison of Foaming Properties of Soaps and Detergents Colgate Palmolive Peet Company, Jersey City, New Jersey
[16] Haryanto B, Siswarni M Z, Sianipar Y C H, Sinaga T M A, Bestari I 2017 Investigation the foam dynamics capacity of SDS in foam generator by affecting the presence of organic and inorganic contaminant *AIP Conference Proceedings* **1840(1)** 110009
[17] Pandey S, Bagwe R P and Shah D O 2003 Effect of counterions on surface and foaming properties of dodecyl sulfate *Journal of Colloid and Interface Science* **267** 160-166
[18] Wang L and Yoon R H 2004 *Hydrophobic Forces in the Foam Films Stabilized by Sodium Dodecyl Sulfate: Effect of Electrolyte* Center for Advanced Separation Technologies Virginia Polytechnic Institute and State University, Blacksburg, Virginia 24061
[19] Petkova R, Tcholakova S and Denkov N D 2013 Role of polymer–surfactant interactions in foams: effects of pH and surfactant head group for cationic polyvinylamine and anionic surfactants *Colloids and Surfaces A: Physicochem. Eng. Aspects* **438** 174–185
[20] Aditya I W 2015 *Kajian Kandungan Kafein Kopi Bubuk, NilaiPh Dan Karakteristik Aroma Dan Rasa Seduhan Kopi Jantan (Pea Berry Coffee) Dan Betina (Flat Beans Coffee) Jenis Arabika Dan Robusta* Jurusan Ilmu Dan Teknologi Pangan Fakultas Teknologi Pertanian Universitas Udayana Bukit Jimbaran
[21] Julia Hinde 2000 Melbourne www.newscientist.com
[22] Fameau, A L and Salonen A 2014 *Effect of Particles and Aggregated Structures on the Foam Stability Aging* Biopolymeres Interactions Assemblages, INRA, rue de la Geraudiere, 44316 Nantes, France.

IOP Conf. Series: Materials Science and Engineering **309** (2018) 012043 doi:10.1088/1757-899X/309/1/012043

Study on Energy Productivity Ratio (EPR) at palm kernel oil processing factory: case study on PT-X at Sumatera Utara Plantation

B Haryanto[1*], R Br Bukit[2], E M Situmeang[1], Christina E P[1], F Pandiangan[1]
[1] Department of Chemical Engineering, Universitas Sumatera Utara, Padang Bulan, Medan 20155, Indonesia
[2] Department of Accounting, Universitas Sumatera Utara, Padang Bulan, Medan 20155, Indonesia

*E-mail: bode.haryanto@usu.ac.id

Abstract. The purpose of this study was to determine the performance, productivity and feasibility of the operation of palm kernel processing plant based on Energy Productivity Ratio (EPR). EPR is expressed as the ratio of output to input energy and by-product. Palm Kernel plan is process in palm kernel to become palm kernel oil. The procedure started from collecting data needed as energy input such as: palm kernel prices, energy demand and depreciation of the factory. The energy output and its by-product comprise the whole production price such as: palm kernel oil price and the remaining products such as shells and pulp price. Calculation the equality of energy of palm kernel oil is to analyze the value of Energy Productivity Ratio (EPR) bases on processing capacity per year. The investigation has been done in Kernel Oil Processing Plant PT-X at Sumatera Utara plantation. The value of EPR was 1.54 (EPR > 1), which indicated that the processing of palm kernel into palm kernel oil is feasible to be operated based on the energy productivity.

1. Introduction

Palm kernel oil (PKO) is one of the oil produced by oil palm crop beside CPO products. The palm kernel oil is derived from the seeds inside the palm fruit called palm kernel. Usually, PKO is more widely used for oleo chemical industry [1].

In Sumatera Utara plantations there are a number of companies produce palm kernel oil (PKO) and crude palm oil (CPO). One of the companies that produce palm kernel oil was taken as sample to get the data for this study. Palm kernel is processed at palm kernel processing plant to produce palm kernel oil as main product and palm oil cake as by product [2]. The raw material of palm kernel is obtained as by product of palm oil processing plants of one state factory (PT-X) and some private companies in Sumatera Utara area.

Productivity is defined as the relationship between input and output of a production system. This relationship is often more commonly expressed as the ratio of what is produced (output) to the total resources used (input) or simply the ratio of output divided by input [3,4]. Productivity is one factor that is important in influencing the performance of a company. The EPR formulation is as shows in equation 1.

$$\text{EPR} = \frac{\text{Output Energy}}{\text{Input Energy} - \text{Output Energy of side products}} \quad (1)$$

The PT-X Company is used as a sample to get data to calculate the value of EPR. The company processes palm kernel into palm kernel oil. The target data required is all the energy used in doing the processing activities to produce palm kernel oil and its by-products. The results of this EPR calculation can be used as one of the indicators to analyze the performance and productivity of the plant. Certainly this analysis can be used to improve for example in the use of electrical energy as an input during the process and take advantage of by-products in the output. This will certainly increase the company's profit.

The purpose of this study is as information to assist in analyzing the performance, productivity and feasibility of the operation of a factory based on the assessment of the value of Energy Productivity Ratio (EPR) at PT X Plantation Processing Factory in Sumatera Utara plantation area.

2. Methodology

The methodology is first to collect the required data. The data is then calculated the equal energy for each input and energy output in the processing of palm kernel into palm kernel oil. Then calculated Energy Productivity Ratio (EPR) on a ton / year basis. The stages are as shown in Figure 1

The EPR of a product can be defined as the ratio of primary product energy plus its by-product energy in comparing to the total energy of raw materials, electricity and others energy used and the tools process depreciation. The calculation is based on the product capacity per year.

Calculation result of the energy input: energy output and if the energy ratio of 1 > 1 then the product production can be continued. Whereas if the energy ratio of 1 < 1 then the production of renewable fuel product become a loss and can be declared as not a feasible source of renewable energy in an energy review [5]. Calculation of equal energy, for value of unit of rupiah price used base of 1 liter diesel with price Rp. 5650 [7] with energy equal to 42.96 MJ/kg [8].

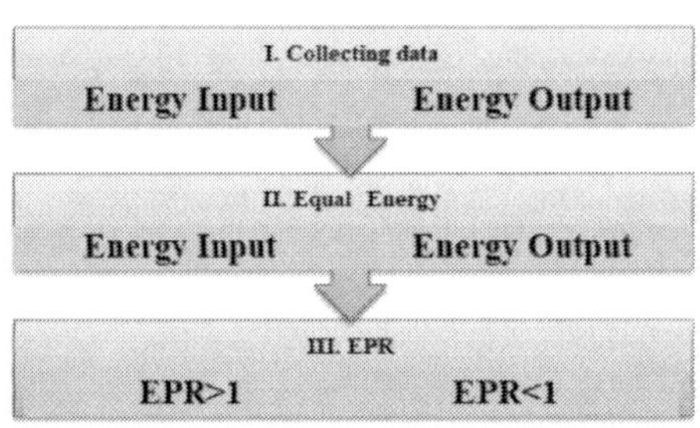

Figure 1. Research steps: collecting data, equal energy quantity and EPR calculation

3. Results and Discussion

Energy input which consists of equal energy from palm kernel prices, electrical energy requirements in the processing and equal energy from the equipment depreciation. For energy output is extracted from equal energy selling price of palm kernel oil and the rest is shell and pulp core [9].

The following is the calculation of equal energy input and output at the palm kernel processing plant: The following data is used to find equal energy input with working time per year 300 days.

Table 1. The information on energy input in producing Palm Kernel Oil

No	The input resource	The Price or Energy Used
1	Price of palm kernel 120.000 tons	Rp, 487,920,000,000
2	Annual electricity usage	14,400,000 kWh
3	Depreciation of equipment	Rp, 765,400,348

The raw material of palm kernel that enter annually 120,000 tons, and the investment cost of factory is Rp. 64,958,402,381. The data from Table 1 to obtain the price of each of the input then was calculated for equal energy by dividing the price of each input with the energy base of 1 liter diesel (42.96 MJ/kg) [8].

- Equal energy from palm kernel is for total 120.000 ton per year:

$$Energy = \frac{\text{Rp. }487.920.000.000}{\text{Rp. }5650} x\ 42.96\ MJ/kg$$
$$Energy = 3{,}709{,}919{,}150{,}442\ MJ/kg$$
$$Energy = 3{,}709{,}919{,}150{,}442\ MJ/ton$$

- Electrical energy used per year:

14,400,000 kWh x 3.6 MJ = 51,840,000 MJ

- Equal energy from depreciation:

$$Energy = \frac{\text{Rp. }765400348}{\text{Rp. }5650} x\ 42.96\ MJ/kg$$
$$Energy = 5{,}819{,}752{,}026.5628\ MJ/ton$$

Table 2. The equal energy input in producing Palm Kernel Oil

No	The input resource	The Equal Energy (MJ/Year)
1	Energy of palm kernel	3,709,919,150,442
2	Annual electricity usage	51,840,000
3	Depreciation of equipment	5,819,752,026
	Total	3,715,790,742,468

From the above calculation the total input energy equal to 3,715,790,742,468.9 MJ/ year. The following data is used to find equal energy output. The raw material of palm kernel seeds per year is 120000 tons with the percent of palm kernel oil yield 43% is 51600 tons. The palm kernel shell is 68400 tons. The kernel pulp is used as fertilizer in self-plantation.

Table 3. The information on energy output in producing Palm Kernel Oil

No	The output resource	The Price
1	Palm kernel oil (51600 tons)	Rp 604,088,424,000
2	Palm kernel shell (68400 tons)	Rp95,760,000,000

The output energy data obtained is continued by calculating equal energy. The basis for determining equal energy is from the energy price of 1 liter of diesel oil.

The energy produced by palm kernel oil is:

$$Energy = \frac{604088424000}{5650} x\ 42{,}96\ MJ/kg$$
$$Energy = 4{,}593{,}210{,}388.5026\ MJ/kg$$
$$Energy = 4{,}593{,}210{,}388.502{,}6\ \ MJ/ton$$

The energy generated by palm kernel shell is:

$$Energy = \frac{95{,}760{,}000{,}000}{5650} x\ 42.96\ MJ/kg$$
$$Energy = 728{,}114{,}973{,}45132\ MJ/kg$$
$$Energy = 728{,}114{,}973{,}451.32\ MJ/ton$$

Table 4, The equal energy output in producing Palm Kernel Oil

No	The output resource	The Equal Energy (MJ/Year)
1	Energy of palm kernel	4,593,210,388,502.6
2	Energy of palm kernel shell	728,114,973,451.32
	Total	5,321,325,361,953.9

From the above calculation results obtained the total output energy equal to 5,321,325,361,953.9 MJ / ton,

Once in can equalize the input and output energy, the next will be calculating the EPR (Energy Productivity Ratio),

$$\text{EPR}=\frac{Output\ product}{(Input - by\ product)}$$

$$\text{EPR}=\frac{4{,}593{,}210{,}388{,}502.6}{(3{,}715{,}790{,}742{,}468.9 - 728{,}114{,}973{,}451.32)}$$

$$\text{EPR}=\frac{4{,}593{,}210{,}388{,}502.6}{2{,}987{,}675{,}769{,}017.5}$$

$$\text{EPR}=1.54$$

Energy value productivity ratio obtained from the calculation > 1, This indicates that the output energy requirement is greater than the input in the PT-X plant, In accordance with theory then EPR indicates that this factory provides benefits to the company, From the results of calculations performed, it obtained the EPR value of 1,54, The results of this EPR value included in the low EPR category is due to the use of electricity input energy during the process of inefficient processing, which amounted to 51,84 million MJ / ton per year, However, improved EPR can be performed even better by increasing the efficiency of the use of electricity needs in the processing, In addition, by calculating the pulp from the palm kernel as a fertilizer with a certain sale value as output energy of side product was ignored in this calculation, because it is used as fertilizer by the company itself, This study can be concluded that the processing of palm kernel PT-X into palm kernel oil meets the feasibility of production and can provide benefits to the company,

4, Conclusions

The value of EPR obtained is 1.54, These results indicate that the processing of palm kernel PT-X into palm kernel oil is still feasible to operate, This indicates that the value of EPR is higher than 1 that can still provide benefits to the company, In the processing of palm kernel into palm kernel oil, electrical energy consumption is still less efficient, The value of the pulp from the rest of the palm kernel is still negligible in the value of its output energy equivalent,

Acknowledgment

To PT-X as the company with kernel processing factory that contributed the data on this case study,

References

[1] Widyastuti S 2009 Analisis Pengendalian Persediaan Inti Sawit, *Fakultas Pertanian, Institut Pertanian Bogor, Indonesia,*

[2] Ketaren S 1986 Pengantar Teknologi Minyak dan Lemak Pangan *Cetakan Pertama, Jakarta: UI-Press,*

[3] Kusmindari D and Aprianto 2009 Produktivitas dan Pengukuran Kerja Proses Produksi Medium Dencity Fibreboard (MDF), *Jurnal Ilmiah Tekno*, **6**: 2, 85-96

[4] Muhammad H E 2009 Analisis Pengukuran Produktivitas Perusahaan dengan Menggunakan Metode Marvin E, Mundel di PTPN IV PKS PABATU, Tebing Tinggi, *Medan : Fakultas Teknik, Universitas Sumatera utara*

[5] Batchelor, Sheila E, Elaine J, Booth, Kerr and Walker 1995 Energy Analysis of Rape Methyl Ester (RME) Production from Winter Oilseed Rape, *Industrial Crops and Products an International Journal*, UK :**4**:2, 193-202, Elsevier

[6] Bode H 2000 Studi Neraca Energi Pembuatan Biodiesel dari Minyak Sawit, *Thesis Magister, ITB*

[7] Pertamina, 2016, *Harga Bahan Bakar Minyak*, http://www,pertamina,com/news-room/info-pertamina/pengumuman/,

[8] Felten D, Fröba N, Fries J and Emmerling C 2013 Energy Balances and Greenhouse Gas-Mitigation Potentials of Bioenergy Cropping Systems (Miscanthus, Rapeseed, and maize) Based on Farming Conditions in Western Germany, *Renewable Energy an international Journal*, Germany: **55**, 160 - 167, Elsevier

[9] PT, X,, 2013, Tebing Tinggi,

[10] Hulya Uysal and Gamze Saner 2016 Energy balance and cost analysis for raisin production in Aegean Region in Turkey, *39th World Congress of Vine and Wine*, 7, 03020 DOI: 10,1051/bioconf/20160703020

IOP Conf. Series: Materials Science and Engineering **309** (2018) 012044 doi:10.1088/1757-899X/309/1/012044

The cost analysis of material handling in Chinese traditional praying paper production plant

H Nasution*[1], I Budiman[1] , A Salim[2]
[1]Industrial Engineering Department, Faculty of Engineering, Universitas Sumatera Utara, Jl. Almamater Kampus USU Medan 20155, Indonesia
[2]Industrial Engineering Department, Faculty of Technology and Computer Sciences, University of Prima Indonesia, Jl. Sekip sp. Sikambing Medan, Indonesia

*E-mail: harmein_nasution@yahoo.com

Abstract. Chinese traditional praying paper industry is an industry which produced Chinese traditional religion praying paper. This kind of industry is rarely examined since it was only in Small and Medium Enterprise (SME's- form). This industry produced various kinds of Chinese traditional paper products. The purpose of this research is to increase the amount of production, reduce waiting time and moving time, and reduce material handling cost. The research was conducted at prime production activities, consists of: calculate the capacity of the material handler, the frequency of movement, cost of material handling, and total cost of material handling. This displacement condition leads to an ineffective and inefficient production process. The alternative was developed using production judgment and aisle standard. Based on the observation results, it is possible to reduce displacement in the production. Using alternative which by-passed displacement from a rolled paper in the temporary warehouse to cutting and printing workstation, it can reduce material handling cost from 2.26 million rupiahs to 2.00 million rupiahs only for each batch of production. This result leads to increasing of production quantity, reducing waiting and moving time about 10% from the current condition.

1. Introduction

From a worldwide perspective, it has been recognized that small and medium enterprises (SMEs) play a vital role in economic development, as they have been the primary source of job/employment creation and output growth, not only in developing but also in developed countries [1]. SME also play a vital role in the rural income because they scattered widely and very labor-intensive in the rural area [2]. Tambunan identifies several key issues for improving SME competitiveness in Indonesia, they are human resource, working capital, management and technological skills. These key factors are important to improve SME's business performance [3].

Chinese traditional praying paper industry is an industry which produced Chinese traditional religion praying paper. This kind of industry is rarely examined since it was only in Small and Medium Enterprise form. This industry produced various kinds of Chinese traditional paper products. In producing products, manufacturer has to follow several steps, which is, establishing paper from pulp, tin printing, cutting, screen printing and packing process. Paper is cut and printed as per product specification.

Material Handling cost is an expense arising from the activities of material from one machine to another or from one department to another which are determined to arrive at a particular [4, 5]. Units used was rupiah per meter of movement. Expected result in planning material handling is increasing capacity, improving a working condition, customer service, room space, and reduce expenses [5, 6, 7, 8, 9].

In this industry examined, it has 4.5 tons of daily capacity. To produce daily, it needs 30 to 75 times of repeated material handling and 130.03 meters for a day of material handling in the main production line. Therefore, before doing re-layout for the existing main production line, the researcher needs to analyze the existing material handling condition.

2. Method and equipment

The method used in this paper is motion and time study (descriptive research) [10]. Steps in collecting primary data are field survey (preliminary survey), problem identification, collecting primary data and secondary data such as production process, material handling movement, the frequency of material handling, allowance, the capacity of handling, and time needed for material handling.

In this research, there were several movement in the department such as: material handling from temporary warehouse to printing and cutting, material handling from printing and cutting to re-cutting machine, material handling from re-cutting to screen printing, material handling from screen printing to packing workstation, and material handling from packing to final product warehouse.

According to Alreck & Settle, generally, the number of samples is 10% of the population is recommended within 30 to 500 parent population. In this research, the number of samples is 30 for each movement of material handling [11].

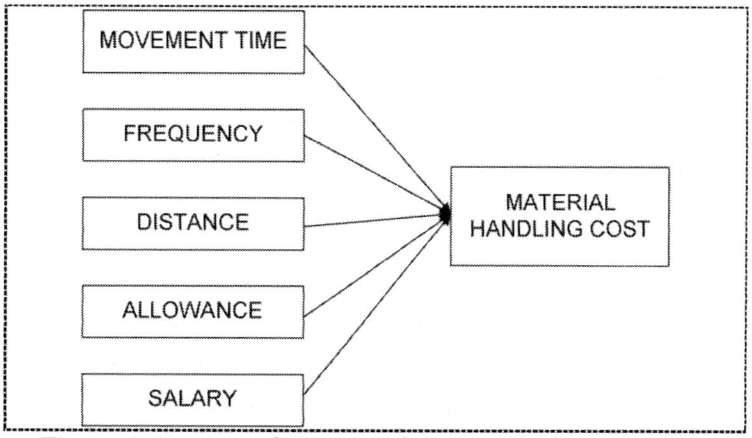

Figure 1. Conceptual Framework in Material Handling Cost

Whether movement time is time required to do movement of material from one workstation to another workstation, frequency is how many times it is needed to move material between workstations in a day at normal condition, distance is the length between workstations, allowance is given to the operator to release fatigue, and salary is used as the cost needed for the activities.

Material handling cost needed to calculate as follow: [12]

1. Firstly, it is needed to calculate the capacity of the material handler/tool (*C*) by dividing the volume of tool (v_{tool}) and material can be moved ($v_{material}$) as follow:

$$C = \frac{v_{tool}}{v_{material}} \quad (1)$$

2. Then, the frequency of movement (*F*) can be calculated by dividing material need to be moved with the capacity of material handler as follow:

$$F = \frac{n_{mat}}{C} \quad (2)$$

3. Cost of material handling (*OMH/m*) can be calculated by dividing cost with distance (*d*) as follow:

$$OMH\ (m) = \frac{Cost}{d} \tag{3}$$

4. The total cost of material handling (*OMH*) can be calculated by multiplying distance (r), frequency (*F*) and cost of material handling (*OMH/m*) as follow:

$$OMH\ = r \times F \times OMH(m) \tag{4}$$

Since the flow pattern in the production is odd angle, then the researcher developed an alternative route by cutting off the material handling from temporary warehouse to printing and cutting workstation by creating a door directly to the workstation as can be seen in Figure 2.

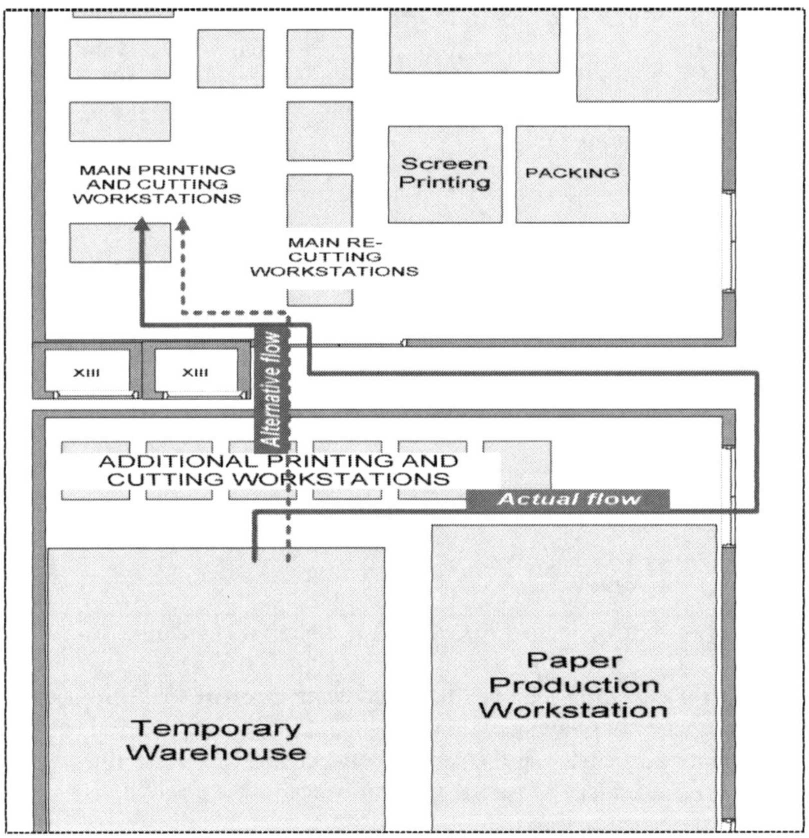

Figure 2. Creating an Alternative Route

Figure 2 shows an alternative used to calculate the cost of material handling so it can clearly educate the industry that even with a single improvement, they can reduce material handling cost.

3. Results and Discussions

Cost of material handling was calculated through the steps as stated in the equation (1) to (4) for each displacement from one to another workstation through several steps of process and result as follow:

IOP Conf. Series: Materials Science and Engineering **309** (2018) 012044 doi:10.1088/1757-899X/309/1/012044

Table 1. Recapitulation for Total Cost of Material Handling

Step of Process	EXISTING		ALTERNATIVE	
	Route	OMH (Rp/ month)	Route	OMH (Rp/ month)
1	A	655,824	F	478,582
2	B	238,368	B	238,368
3	C	884,416	C	884,416
4	D	277,542	D	277,542
5	E	204,630	E	204,630
TOTAL		2,260,780		2,083,538

Where A refers to displacement from temporary warehouse to printing and cutting workstation, B refers to displacement from printing and cutting to re-cutting workstation, C refers to displacement from re-cutting to screen printing workstation, D refers to displacement from screen printing to packing workstation, E refers to displacement from packing workstation to final product warehouse, and F is alternative route for displacement from temporary warehouse to printing and cutting warehouse. Based on Table 1, we can see that Route A and Route C is the most priority route to be reduced with value of OMH as high as 68% of the total cost. However, route C is not possible to be re-route because there's no other way to place the re-cutting to the nearer place with screen printing workstation.

Table 2. Recapitulation of Displacement in Process

Step of Process	EXISTING		ALTERNATIVE	
	Route	Distance (m)	Route	Distance (m)
1	A	87.63	F	66.00
2	B	8.19	B	8.19
3	C	11.91	C	11.91
4	D	4.30	D	4.30
5	E	18.00	E	18.00
TOTAL		130.03		108.40

As can be seen in Table 1, Alternative gives a better material handling cost to the production. Changing the route for one step give about 10% of cost reducing in this case. The reducing of material handling cost occurs because there are also reducing in the displacement in route A. This will also giving impact to increase in the production time so that with this improvement, production will increase to about 5 to 8% from the experience from the production manager.

For a better result in the material handling, re-layout is highly recommended for further research so that the industry can operate in an effective and efficient way, not only in the production as in this research.

4. Conclusions

The conclusion from this paper is reducing displacement in material handling will reduce total cost of material handling about 10%, from IDR 2,260,780 to IDR 2,083,538. this result will give direct impact to lower production time and giving increase for a number of product about 5 to 8%. Route A and C is Further research is needed for a better result in the total cost of material handling.

Acknowledgments

In this research, the authors would like to thank profusely to the industry which has been willing to be a place of research for the writing of this paper and highly support researcher for a better result

References

[1] Tambunan T, Liu X F 2007 SME Development in Indonesia and China, *ERIA Related Joint Research of SME Project, IDE-JETRO*, pp. 1-29

[2] Tambunan T 2006 Development of Small and Medium-Scale Industry Clusters in Indonesia, *Kadin Indonesia-JETRO*

[3] Setyawan A, Isa M, Wajdi MF, Syamsudin and Nugroho 2015 An Assessment of SME Competitiveness in Indonesia, *Journal of Competitiveness* Vol. **7**, pp. 60-74

[4] Putra OS Analisis dan Rancangan Ulang Sistem Perpindahan Material Di PT. Dwi Indah Menggunakan Material Handling General Analysis Procedure, *Bandung: University of Telkom*

[5] Wignjosoebroto S 2009 Tata Letak Pabrik dan Pemindahan Bahan, Plant Layout and Material Handling, 3rd Ed., *Surabaya: Widya Guna*

[6] Ali M 2012 Impact of routing and pallet flexibility on flexible manufacturing system", *Global Journal of Flexible Systems Management,* **13**(3), 141–9

[7] Ali M and Ahmad Z 2014 Simulation study of FMS under routing and part mix flexibility, *Global Journal of Flexible Systems Management,* **15**(4), 277–94.

[8] He N D, Zhang Z and Li Q 2014 Agent-based hierarchical production planning and scheduling in make-to-order manufacturing system, *International Journal of Production Economics,* **149**, 117–130.

[9] Fazlollahtabar H and Olya MH 2013 A cross-entropy heuristic statistical modeling for determining total stochastic material handling time, *Int. J. Adv. Technol, 2013*, Vol. **67**, pp. 1631-41

[10] Sinulingga S 2011 Metode Penelitian, Research Methodology, 1st Ed., *Indonesia: USU Press*

[11] Hill R 1998 What Sample Size is Enough in Survey, *Interpersonal Computing and Technology: An Electronic Journal for the 21st Century*, ISSN 1064-4326, Vol. **6** No. 3-4, 1998.

[12] Apple JM Material Handling Systems Design, *USA: The Ronald Press Company*.

IOP Conf. Series: Materials Science and Engineering **309** (2018) 012045 doi:10.1088/1757-899X/309/1/012045

Enhancing efficiency of production cost on seafood process with activity based management method

U Tarigan*[1] and U P P Tarigan[1]

[1]Industrial Engineering Department, University of Sumatera Utara, Jalan Almamater USU Campus Medan 20155, Indonesia

*E-mail: ukurta.tarigan@yahoo.com

Abstract. The efficiency of production costs has an important impact maintaining company presence in the business world, as well as in the face of increasingly sharp global competition. It was done by identifying and reducing non-value-added activities to decrease production costs and increase profits. The study was conducted at a company engaged in the production of squid (seafood). It has a higher product price than the market as Rp 50,000 per kg while the market price of squid is only Rp 35,000 per kg. The price of the product to be more expensive compared with market price, and thereby a lot more consumers choose the lower market price. Based on the discussions conducted, the implementation of Activity Based Management was seen in the reduction of activities that are not added value in the production process. Since each activities consumers cost, the reduction of nonvalue-added activities has effects on the decline of production cost. The production's decline costs mainly occur in the reduction of material transfer costs. The results showed that there was an increase after the improvement of 2.60%. Increased production cost efficiency causes decreased production costs and increased profits.

1. Introduction

Business competition in the industrial world that becoming very intense, forces any company to have a high competitiveness to maintain its existence. Therefore, needed business strategies to win the competition among companies. A company can minimize its production costs through efficiency in non-value- added activities. Activities of an organization or unit can be said efficient if in the execution of it consumed fewer resources or lower costs to produce a greater amount of output, by reducing non-added-value activities. By reducing non-value-added activities, the costs of the production process will reduce and in turn, the reduction of production costs will increase the profit of the company [1] Packing industry of seafood products is one industry that has to prepare itself to implement that strategy. The XYZ company is an industry processing seafood products such as squids, crabs, prawns, tuna and other seafood products and each seafood product has its own processing plant. The price of processed squids of this company is Rp 50,000 per kg while the price in the market is only Rp 35,000 per kg. The company is facing an increase in production cost each year and consequently reduces the company's profit. The price of the product to be more expensive compared with market price, and thereby a lot more consumers choose the lower market price.

Activity Based Management (ABM) is an integrated approach in the whole system, which focused the management's attention on the various activities, with the goal to increase value for the customers and in turn increase profit as its result [2]. Activity based management is putting efforts to identify and in the end to abolish all non-added-activities and at the same time focused the management's attention on the needed activities to increase value for the customers (cost value) and in turn increase profit.

The success of ABM depends on the business unit being used. ABM method research is used at the organizational Life Cycle Stage (OLC). Variables used in this study: business unit maturity, birth, growth and decline. The result shows that ABM business maturity level is higher than other variables in OLC [3]

In this study discussed the use of Activity Based Management in a fictitious company Business-to Customer (B2C). Initially the company must determine the purpose of the research. Variables used in this study are Objectives, requirements and cost objects. The analysis used is by using matrixes in order to trace overhead, resulting in a proposed implementation procedure of this study [4]

The main purpose of this study is to determine the scientifically designed the financial costs of each model that has been established. In this study combined the financial model of activity-based cost system with financial model of the standard cost system. It also calculates the integration of the advantages of the two models. This research covers the entire production and operation process from start to finish. In this study used the R & D expenses as the accounting starting point. And after-sale service expenses as the terminal point. This research is conducted in China because the development of production in China is at the stage of mechanical and semi-automation [5-6]

In this study compared traditional systems with Cost of Quality (COQ) system oriented Activity Based Costing. COQ is a tool used to measure company performance. This research attempts to analyze the shortcomings of traditional methods by using the advantages of the activity based costing method. Activity based costing is an alternative to overcome the weaknesses of traditional systems so that the results obtained the detailed cost of quality according to the case studies taken. The weakness of this research is limited to high priority process that is functional opinion management, so it has a bad risk and data quality [7-8]

2. Methodology/Experimental

The conducted research belongs to the applied research. Applied research is basic research in solving a real existing problem [9] In this research an analysis of factual situation and variables of a certain object was conducted, compare them with a condition, situation or variables that been set for the object of research, and then find the solution by reducing the non-added-value activities in the company, so that the production cost will be efficient.

2.1. Data Collection Methods

The technic of data collection in this research, started with conducting observation by direct observing the actual activities in processing squids and counting the time of production process directly by using Stopwatch. The next step is conducting interviews with the management and the operators that work in the company concerning all things connected with the research object and to complete data collected by observation. Alongside, documentation was conducted by collecting secondary data within the company, that is data related to the research being conducted.

2.2. Data Processing Methods

The technic of data processing in this research, begins with the identification of all activities that happen during the production process. Identification of activities was conducted on the activities in packing the marine products, particular squids by setting up the flow process chart. Classification of activities was done by dividing the company activities in two that is those that have value-added and those that do not have value added. The classification of activities is conducted by using process value analyses that consist of value-added activities and non-value added activities. The allocation of production costs to each activity is being done to find out the cost of each activity allocated by the company during the process. The allocated cost was defined by measuring cycle time, needed space, and other needs of the plant in advance. Activities analyses are the key to reach the goal efficient costs. Activities analysis is a process of elimination, reducing non-value-added activities to reduce production costs of the company.

3. Results and Discussion

3.1. Business Process

The aim of business process analysis is to map out various activities conducted by XYZ company to produce output and at the same time provide information related to resource driver, activity resource, value and non-value added activities. Aside of that, the business process analysis are also aimed at finding out various opportunities for improvement of the process practiced by the company in producing value for the consumers.

Business process analysis consist of 4 stages as follows:

a. Business Process Identification.

The identification process is conducted to identify the whole processes that exist in the company business. The activities in a company are so complex, that it will be necessary to decide the business process that being the priorities of the management of this company. The analysis tool that was used in the diagram value chain as illustrated in Figure 1.

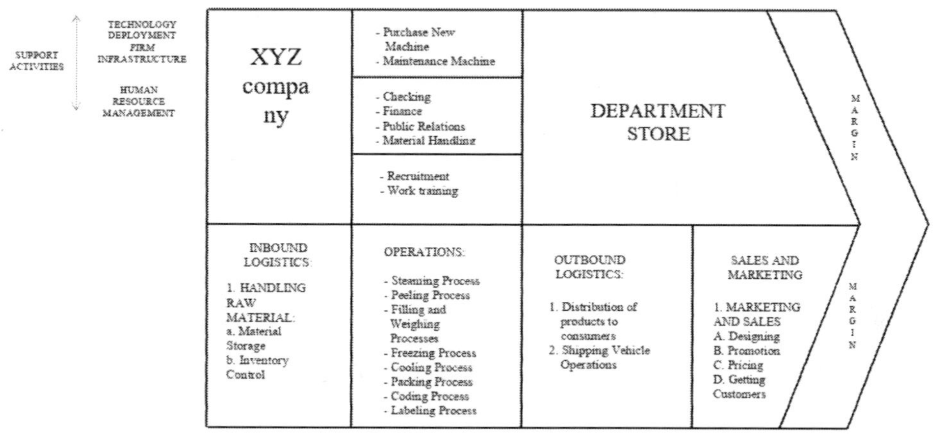

Figure 1. Diagram Value Chain for Business Process Identification

b Sub-Process Identification

From the above-described steps, it was found that most activities are in the production process, and the goal of the company is to produce squids with low production cost. Subsequently, analysis will be centered on the production section, and then continue to identify the sub process in the production process that can be seen in Table 1 below:

Table 1. Sub Processes in Production Section

Business Process	Sub Process
Production	Steaming Process
	Peeling Process
	Filling and Weighing Processes
	Freezing Process
	Cooling Process
	Packing Process
	Coding Process
	Labeling Process

c. Activities Identification
Activities identification started with the observation the diagram of the flow of production of the squids. It is followed by interviews to find out the activities in supporting sections such as storage of production materials and finished products.

d. Identification of Information related to the activities
The following steps are identifying some information related to the activities namely resources being consumed covering the total employee resources (human resources) and time needed to produce the squids.

3.2. Cost Driver Analysis

Cost Driver Analysis is used to find out the driver of costs consumption. Cost driver analysis scan be seen in Table 2 below.

Table 2. Activities, Activities Level and Cost Driver

Activites	Activities Level	Cost Driver
Material Cost	Batch	Total Amount Consumed
Direct Labor Cost	Batch	Total Direct Labor
Supporting Material Cost	Batch	Total Amount Consumed
Electricity Cost	Batch	Total Kilowatt Hours
Water Cost	Batch	Water Usage
Handling Cost	Batch	Amount of Goods being Handled
Costs of Machine and Tools	Unit	
Maintenance		Direct Allocation
Labor Insurance	Unit	Direct Allocation
Plant Building Insurance	Unit	Direct Allocation

3.3. Defining Allocation for Standard Wages

The improvement that can be made is by defining a transparent and universal wages standard for labor. This can be done by using Merit Pay System, that is wages are based on performance and length of time in performing one task that can be seen in Table 3.

Table 3. Allocation of Merit Pay Wages

Activities	Starting Wages (Rp)	UC (Rp)	UM (Rp)	UC (Rp)
Steaming	87	1.13	73.3	83
Peeling	75	1.12	73.3	82
Filling	85	1.09	73.3	80
Weighing	60	0.81	73.3	59
Freezing	59	0.77	73.3	56
Cooling	26	0.41	73.3	30
Packing	25	0.20	73.3	15
Coding	17	0.35	73.3	26
Label	57	0.94	73.3	69
Total	500	6.82	73.3	500

3.4. Value Process Analysis

At this stage, various activities that formed the sub process are defined according to the value considered by the customers. Each activity in each sub process will be divided into three categories: value-added activity and non-value-added activity. According to Hansen and Mowen, there are many criteria that can be used to define weather one activity adds value or not. The result of the above-mentioned categorized activities shows that the total production activities in this company are 20 activities consist of 14 value-added activities and six non-added value activities. From all those categories of activities is found out that the handling of the material is non-value-added activities.

3,5. Improvement Process

At this stage, efforts are taken to improve each activity of each sub process in producing squids. Added value and non-added value activities are analyzed to find out which of these two categories indicate wastage that being practiced by the company. After knowing the aim of the company in conducting those activities, the next step is to find the solution for the existing problems.

3.6. Material Handling Analysis

Actions proposed to minimize the cost of material handling is to shorten the distance between work-centers. Minimizing the distance is done by calculating the distance between activities in the work-center, so that related activities are close to each other. By taking this actions, handling distance is shorter than the actual condition. The difference of handling distance in actual condition and the propose condition can be seen in Table 4 below.

Table 4. Shortening Distance

Condition	Distance (m)	Shortened Distance (m)
Actual	291.25	
Proposed	113.72	87.53

3.7. Production Costs Efficiency

Costs reduction that occurred by shortening the handling distances has an impact on the total cost spent by the company. Handling distances needed by the worker are shortened, and thereby costs of direct labor for hauling activities are also lower. Comparison between the cost of hauling activities at actual and proposed condition can be seen in Table 5 below.

Table 5. Cost of Squids Transportation

Actual Condition (Rp)	Proposed Condition (Rp)
Rp 242.155.000,00	Rp 39,496,653.00

From Table 5 it is seen that total costs spent by the company for handling activities at actual condition are Rp 242,155,000.00. After improvement of hauling distances, costs spent by the company are Rp 39,466,653.00 less than then actual condition costs. Saving occurred in direct labor costs for hauling activities, and consequently reducing the total costs of production. Comparison between production costs at the actual condition and production costs after improvement can be seen in Table 6 below:

Table 6. Recapitulation of Financial and Non-Financial Performances

Situation Indicator	Costs
Actual	Rp 36,848,218,000.00
Proposed	Rp 35,889,280,653.00

From Table 6 it can be seen that direct labor costs are reduced. This also means a reduction of production costs. The costs to produce squids at actual condition are Rp 36,848,218,000.00; with costs saving amounted to Rp 958,937,347.00, thereby the actual costs spent in producing squids are reduced amounted to Rp 35,889,280,000.00.

The calculation of costs efficiency as result of improved activities is as follows :

Original Production Costs	Rp. 36,848,218,000.00
Proposed ABM Production Costs	Rp. 35,889,280,653.00
Reduction of Costs	Rp. 958,937,347.00

4. Conclusion.

From all data processing and analysis and taking into consideration the aim of this research, it can be concluded that material handling activities are activities that do not have any added value. Therefore costs saving that can be achieved by the management of this company is by minimizing the hauling distances, so that the reduction of production costs amounted to Rp 958,937,347.00. Consequently, an efficiency increase of 2.60% is achieved. With the increase of production costs efficiency, the production costs will be reduced, and in turn, the profit of the company will increase. Aside of that, the other benefit gained by the company is the increase of product value in the eyes of the consumers. By that, the company will be able to win all aspects of competition. The proposed improvement for material handling is regarding the capacities of transportation equipment and shortening hauling distances. Alongside with minimizing material hauling distances, the allocation of wages should also be improved based on performance and there by justice will be achieved.

References

[1] Edward B 2000 *Cost Management: A Strategic Emphasis* **1**
[2] Hansen, Don R and Mowen M 2006 *Manajerial Accounting* **1**
[3] Tan N, Kevin B and Bill B 2013 *The Use and Success of Activity-Based Management Practices at Different Organisational Life Cycle Stages, International Journal of Production Research*
[4] Narcyz R 2010 *Activity-Based Management for Electronic Commerce: A Structured Implementation Procedure Journal of Theoretical and Applied Electronic Commerce Research* **5** 718-25
[5] Zhang Q 2015 *A New Activity-Based Financial Cost Management Method Physics Procedia* **33** 1906-12
[6] Jana P 2015 *The Expansion of Activity Based Costing/Management in The Czech Republic, Central European Journal of Management* **2**

[7] Sailaja, Basak and Visnawandhan 2013 *Activity Based Cost Management: An Effective Tool for Quality Performance Measurement in Manufacturing Industries International Journal of Scientific & Engineering Research* **4** 2229-35
[8] Angappa G, R. Mcneil, D. Singh 2000 *Activity Based Management in A Small Company: Study Case, Production Planning, and Control* **11** 391-9
[9] Sukaria 2011 *Metodologi Penelitian USU Press of Indonesia*

A low profile rectangular patch microstrip antenna for dual-band operation of wireless communication system

A H Rambe[1,*] and K Abdillah[1]
[1]Department of Electrical Engineering University of Sumatera Utara, Indonesia.

*Email : ali3@usu.ac.id

Abstract. This paper discussed a low profile rectangular patch microstrip antenna design working on dual-band 1.8 GHz and 2.4 GHz. Dual-band characteristic is achieved by using inset-feed point and slot size adjustment. The designed antenna was printed on a FR4 substrate with relative permittivity of 4.4 and a thickness of 1.6 mm with patch size 40 x 29 mm. The measurement results show that the realized antenna successfully working on dual-band, achieving bandwidth of 45 MHz and 95 MHz, gain of 4.08 dBi and 5.79 dBi for 1.8 GHz and 2.4 GHz subsequently.

1. Introduction

Microstrip antenna (MSA) is widely used in various wireless communication systems. MSA has advantages such as low volume, light weight, thin profile configuration and can be easy to make of dual-band [1]. Dual-band antenna is used to streamline devices of two different wireless communication services. The design and development of RMSA for dual-band are shown in [2-4]. Many techniques can be used to obtain dual-band of rectangular microstrip patch antennas such as multi-patch, orthogonal-mode and reactively load [5]. This paper discussed a low profile rectangular patch microstrip antenna design working on familiar dual-band of wireless communication system, 1.8 GHz (for mobile communication) and 2.4 GHz (for wireless local area network).

2. Method

Antenna Design

Figure 1 shows the rectangular patch microstrip antenna (RPMSA) configuration that consists of a radiating patch, dielectric substrate, a feed line and a ground plane.

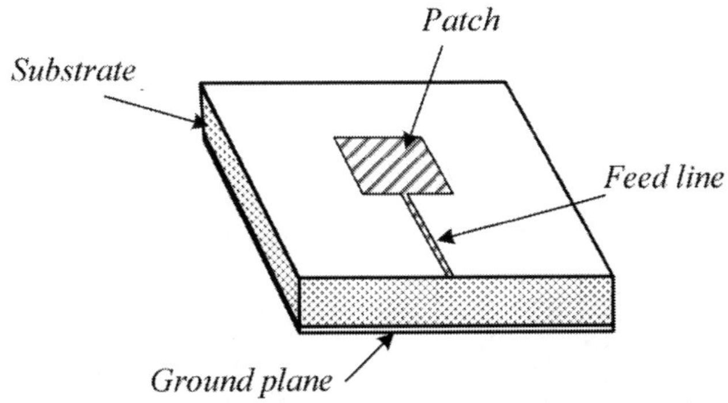

Figure 1. The RPMSA configuration

The width (W) and length (L) properties of the rectangular patch antenna are given by [1][6]:

$$W = \frac{c}{2f_r\sqrt{\frac{(\varepsilon_r+1)}{2}}} \tag{1}$$

$$L = L_{eff} - 2\Delta L \tag{2}$$

where

$$L_{eff} = \frac{c}{2f_r\sqrt{\varepsilon_{reff}}} \tag{3}$$

$$\Delta L = 0.412h\frac{(\varepsilon_{reff}+0.3)\left(\frac{W}{h}+0.264\right)}{(\varepsilon_{reff}-0.258)\left(\frac{W}{h}+0.8\right)} \tag{4}$$

$$\varepsilon_{reff} = \frac{\varepsilon_r+1}{2}+\frac{\varepsilon_r-1}{2}\left(\frac{1}{\sqrt{1+12h/W}}\right) \tag{5}$$

Constant c is velocity of light ($3x10^8$ m/s), ε_r is the dielectric constant of the substrate, f_r is resonant frequency, h is thickness of the substrate, L_{eff} is the effective patch length, ΔL is the length extension and ε_{reff} is the effective dielectric constant of the substrate. FR4 substrate of dimensions 56 x 56 mm^2 with a height (h) of 1.6 mm and permittivity (ε_r) of 4.4 is chosen as the dielectric material for the proposed design. From (1) and (2) arectangular patch is designed to resonant (f_r) at 2.44 GHz (center frequency of bands 2.4 GHz – 2.48 GHz). To obtain dual band operation in the same antenna, the position of feeding line and slot size (inset-feed) is arranged that is discovered by iterating simulation. A RPMSA for 2.4 GHz was designed in [7].The geometry and fabricated antenna is shown in Figure 2.

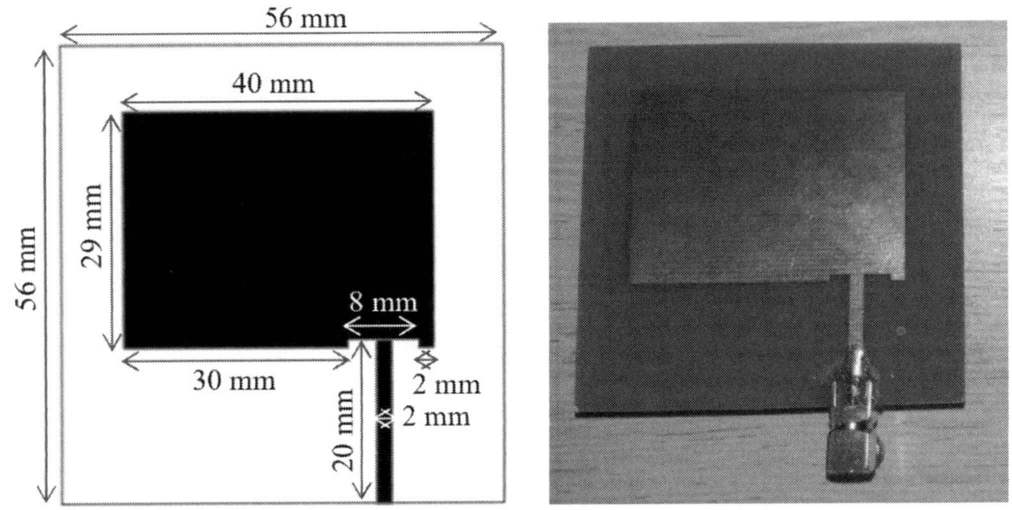

(a) (b)

Figure 2. Antenna designed: (a). Geometry of antenna, (b). Fabricated antenna

3. Results and Discussions

The RPMSA design with slot dimension 3x1 mm^2has been fabricated and measured using VNA (vector network analyzer), Anritsu MT8222A. Figure 3 shows comparison between measured and simulated return loss. Return loss of both frequency band is under –10dB at both simulated and measured results. The measured VSWR is shown in Figure 4.

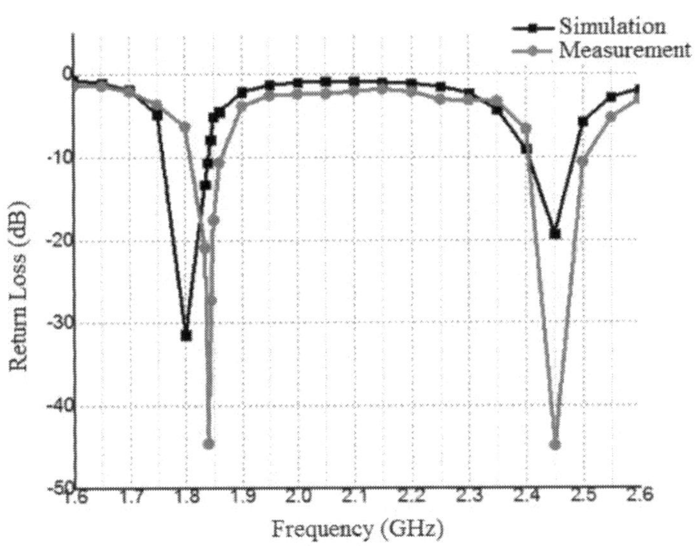

Figure 3. Comparison between measured and simulated return loss

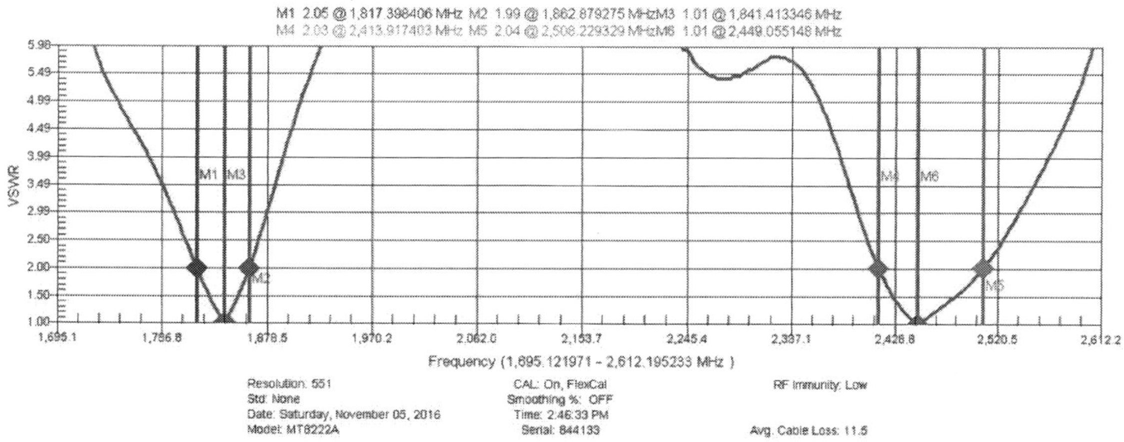

Figure 4. Measured VSWR

VSWR of both frequency band is under VSWR = 2. The RPMSA has operating frekuency at 1.8 GHz and 2.4 GHz. Achieving bandwidth in 1.8 GHz band for VSWR ≤ 2 is shown from M2 and M1 substraction, which are:

$$\text{Bandwidth} = 1{,}862 \text{ MHz} - 1{,}817 \text{ MHz} = 45 \text{ MHz}.$$

and achieving bandwidth in 2.4 GHz band for VSWR ≤ 2 is shown from M5 and M4 substraction, which are:

$$\text{Bandwidth} = 2{,}508 \text{ MHz} - 2{,}413 \text{ MHz} = 95 \text{ MHz}.$$

Measured normalized radiation pattern for 1.8 GHz and 2.4 GHz band is plotted in Fig. 5. Gain of 1.8 GHz is 4.08 dBi and 2.4 GHz is 5.79 dBi at θ=0°.

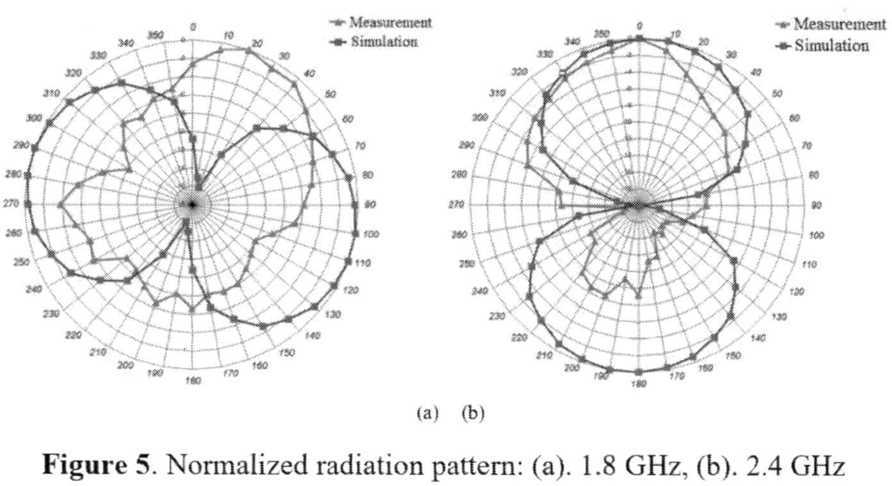

(a) (b)

Figure 5. Normalized radiation pattern: (a). 1.8 GHz, (b). 2.4 GHz

4. Conclusion

A low profile rectangular patch microstrip antenna is designed for dual-band operation of wireless communication system, which operates at 1.8 GHz and 2.4 GHz. The designed can be improved by array and defected structure ground techniques to widen the bandwidth and increase the gain.

5. Acknowledgements

This work was supported by Lembaga Penelitian University of Sumatera Utara, under the Talenta Research project 2017.

6. References

[1] Garg R *et al* 2001 *Microstrip Design Handbook* Norwood: Artech House. Inc
[2] Prakash V and Khanna R 2011 Dual Band Aperture Coupled Rectangular Patch Antenna For PCS and WLAN Applications *International Journal of Communication Engineering Applications IJCEA*
[3] Kavya A *et al* 2015 Design of a Miniaturized Dual Band Patch Antenna for WLAN Applications *ICACS*
[4] Wu Z *et al* 2016 Dual-Band Antenna Integrating with Rectangular Mushroom-Like Superstrate for WLAN Applications *IEEE Antennas And Wireless Propagation Letters* **15**
[5] Maci S and Gentili G B 1997 Dual-Frequency Patch Antennas *IEEE Antennas and Propagation Magazine* **39(6)**
[6] Balanis C A 2005 *Antenna Theory : Analysis and Design* Canada: Jhon Wiley & Sons
[7] Rambe A H *et al* 2013 Optimizing Rectangular Patch Antenna with Microstrip Line Feed Using Single Stub *International Journal of Engineering Research and Technology IJERT*

IOP Conf. Series: Materials Science and Engineering **309** (2018) 012047 doi:10.1088/1757-899X/309/1/012047

Design and analysis microstrip dipole using fractal Koch for 433 MHz applications

M Zulfin, A H Rambe* and B Budi
Department of Electrical Engineering, Universitas Sumatera Utara, Indonesia

*E-mail: ali3@usu.ac.id

Abstract. This paper discussed the dipole microstrip antenna design using fractal Koch for working on frequency of 433 MHz. The fractal Koch was used to reduce the size of the microstrip antenna. The smaller the antenna size, the lighter the equipment. AWR simulator was employed to evaluate antenna parameters such as return loss, gain and radiation pattern. The antenna was designed on a FR4 substrate with relative permittivity of 4.4 and thickness 1.6 mm. The result shows that the fractal Koch reduce antenna size about 41.2% and decrease return loss about 30%.

1. Introduction

Fractal means broken or irregular fragment. A fractal can be described as a rough fragmented geometric shape that can be separated into parts which are an approximation to the whole geometry but in reduced size. There are many fractal geometries that have been found to be useful developing new and innovative design for antenna. The Koch curve was firstly introduced by the Swedish Mathematician Helge von Koch [1][2]. Fractal Koch has been particularly useful in reducing the size of the antenna, while other designs aim at incorporating wide-band and circularly polarized [2-5]. This paper discussed of microstrip dipole using fractal Koch for 433 MHz Applications on FR4 substrate with a height (h) of 1.6 mm and permittivity (ε_r) of 4.4. The fractal Koch has been chosen due to its capabilities to reduce the size of the element by applying different number of iteration.

2. Antenna Design

The geometrical construction of Koch curve is shown in figure 1 [4]. The Koch curves generated using an iterative function system. L_0 is called the initiator, which is equal to the half-wave length of dipole antenna. L_1 is the first iteration of Koch curve, which is achieved by partitioning the L_0 into three parts of same length at certain angle. L_2 and L_3 is the second iteration and the third iteration of Koch curve. The total length (L_k) of Koch curve is shown as equation (1) [4].

$$L_k = L_0 \left(\frac{4}{3}\right)^n \tag{1}$$

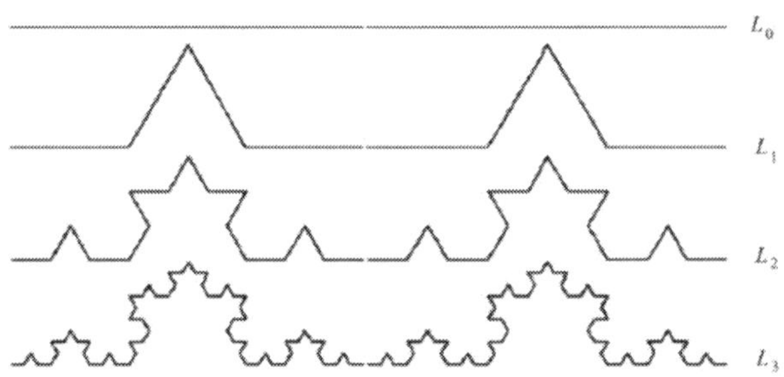

Figure 1. Fractal Koch construction

In this works, the width of microstrip dipole is 2 mm and the initiator length (L_0) is 148 mm. Based on the calculation using equation (1) obtained $L_1 \approx 197$ mm, $L_2 \approx 263$ mm and $L_3 \approx 351$ mm. Using AWR simulator, the optimum total dipole length and the size of antenna design are shown in table 1. The optimum design is achieving after tuned the length of dipole each section. The geometry of the antenna design is shown in figure 2.

Table 1. The optimum antenna design

Microstrip dipole	Total length	Size of the antenna (mm)
first iteration of Koch curve	$L_1 = 144$ mm	92
second iteration of Koch curve	$L_2 = 192$ mm	91
third iteration of Koch curve	$L_3 = 256$ mm	87

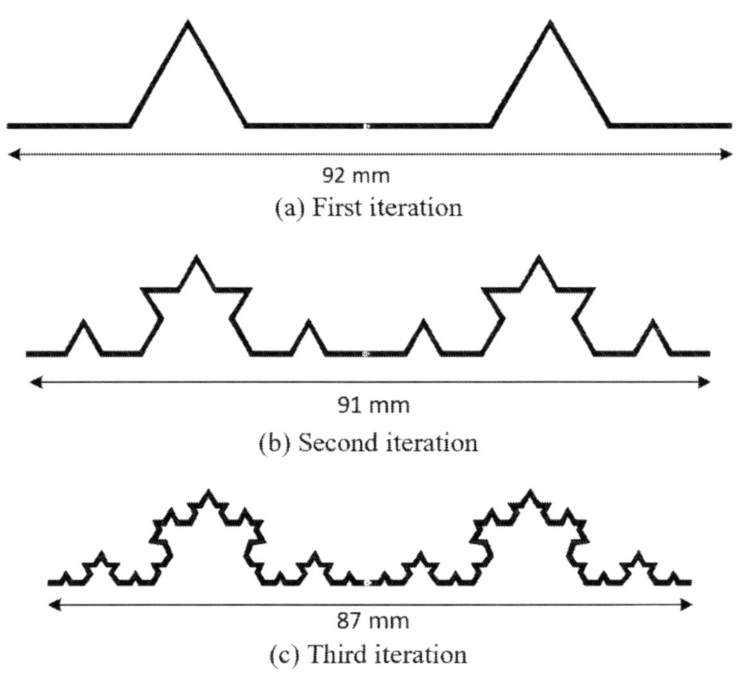

Figure 2. Size of the antenna design using fractal Koch

3. Discussions and Analysis

AWR simulator is used to simulate and design the microstrip dipole antenna using fractal Koch. Figure 3 is shown return loss of the optimum antenna design. Length of the section fractal is tuned to get optimum design.

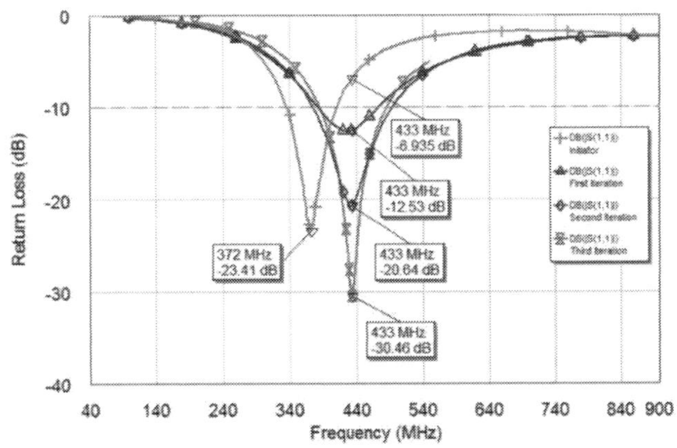

Figure 3. Return loss of the optimum antenna design using first iteration, second and third iteration of fractal Koch

Fractals Koch are workable to reduce the size of microstrip dipole antenna as shown in table 1. Size reduction of the antenna between straight dipole and fractal dipole can be calculate as:

$$Size\ reduction = \frac{S_{sd}-S_{fd}}{S_{sd}} \times 100\% \qquad (2)$$

Where S_{sd} = size of straight dipole antenna
S_{fd} = size of fractal antenna

The size of straight dipole antenna is 148 mm or the initiator (L_0). For the first iteration, size reduction is 37.8%. The size reduction for second and third iteration are 38.5% and 41.2%. However, the fractal Koch can be decrease return loss as shown in figure 3. The minimum return loss is achieved -30.46 dB for third iteration of fractal Koch. This decrease in value reaches 30% compare to the minimum return loss of the initiator. Moreover, gain of each antenna design is about 4.7 dB for single element. To achieve gain exceeding about 10 dB, antenna preparation of about 21 elements array is required as shown in figure 4.

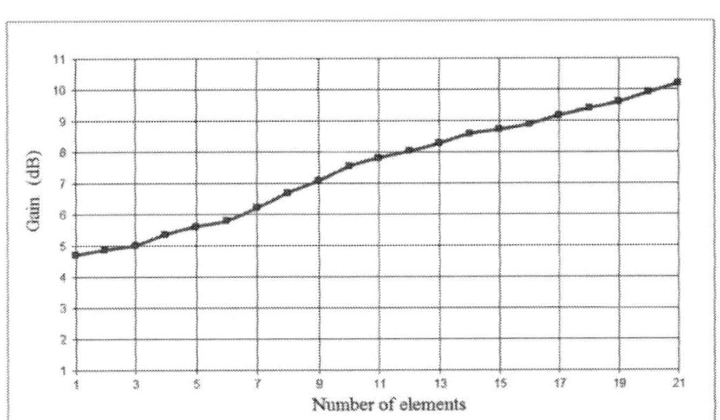

Figure 4. Gain of the microstrip dipole antenna

IOP Conf. Series: Materials Science and Engineering **309** (2018) 012047 doi:10.1088/1757-899X/309/1/012047

4. Conclusion

The antenna microstrip dipole using fractal Koch is designed and analyzed for 433MHz applications such as radiosonde. The designed can be improved by fractal Koch to reduced antenna size. The 10 dB of the gain antenna can be achieved by over 21 elements array.

Acknowledgements

This work was supported by Lembaga Penelitian University of Sumatera Utara, under the Talenta Research project 2017 (No. 148/UN5.2.3.1/PPM/KP-TALENTA USU/2017).

References

[1] F M Lopes 2009 *Fractal Antenna for Wireless Communication Systems* (Portugal: Universidade do Porto)

[2] O H Niang 2013 *Fractal Koch Antenna for Indoor TV and FM Reception* (Malaysia: Universiti Tun Hussein Onn Malaysia)

[3] M N A Karim, M K A Rahim, H A Majid, O Ayop, M Abu and F Zubir 2010 *Progress In Electromagnetics Research* **100** p 201-218

[4] C Thaijiam, K Subprasert and K Tieamkru 2016 *2nd Asia Pacific Conf. on Multimedia and Broadcasting* (Indonesia: Bali/IEEE)

[5] A Farswan, A K Gautam, B K Kanaujia and K Rambabu 2016 *IEEE Trans. on Antennas and Propagation* **64** (2) 771-775

Application of Quality Function Deployment (QFD) method and kano model to redesign fresh fruit bunches sorting tool

Anizar[*]**, I Siregar, I Yahya, and N Yesika**
[1]Department of Industrial Engineering, Faculty of Engineering, Universitas Sumatera Utara, Padang Bulan, Medan 20155, Indonesia

[*]E-mail: anizar_usu@usu.ac.id

Abstract. The activity of lowering fresh fruit bunches (FFB) from truck to sorting floor is performed manually by workers using a sorting tool. Previously, the sorting tool used is a pointed iron bar with a T-shaped handle. Changes made to the sorting tool causes several complaints on worker and affect the time to lower the fruit. The purpose of this article is to obtain the design of an FFB sorting tool that suits the needs of these workers by applying the *Quality Function Deployment* (QFD) and Kano Model methods. Both of the two methods will be integrated to find the design that matches workers' image and psychological feeling. The main parameters are to obtain the customer requirements of the palm fruit loading workers, to find the most important technical characteristics and critical part affecting the quality of the FFB sorting tool. The customer requirements of the palm loading workers are the following : the color of the coating paint is gray, the bar material is made of stainless pipe, the main grip coating material is made of grip, the tip material is made of the spring steel, the additional grip is made of rubber and the handle is of triangular shape.

1. Introduction

Application of QFD in the development process can provide various information about the needs and desires of consumers, the needs of producers, as well as product needs. It can also help the company to evaluate the competition from both technical standpoint and consumer's point of view so that long-term orientation can be applied, communications processes can be formalized, and institutionalization of continuous improvement can be established.

Consumer products depend on the engineer as a determinant of product function and industrial designer role to increase the aesthetics on the design results. The design for the manufacturing process is an integrated practice involved in product development [1]. Application of QFD on development stage can give information regarding the needs and wants of consumers, producers needs, and product needs [2]. It can also help the company to evaluate competition from technical and consumer's point of view, so that formalize long-term orientation, formalize communication process, and Institutionalization of continuous improvement can be applied. QFD method can translate consumers wishes into product design [3,4]. The development of Kano-QFD models was successfully developed and successfully applied to product design and development [5]. The development of a conceptual model that proposes the quality of initiatives as alternatives using the QFD-Kano method [6]. This study develops the integration model by considering cost limitation and the different effects each attribute has through Kano concept in QFD framework. Understanding the level of importance of the technical characteristics of a product can reduce the costs incurred to produce products in accordance with the wishes of consumers [7-9]. The Kano and QFD models could be applied to identify customer needs and can be calculated in customer satisfaction [10]. Integration of the Kano model with HOQ can meet the wants and needs of consumers [11]. The Kano model also can be used for the development of product design based on user needs by evaluating product effectiveness [12]. Increased ability of the company to be able to quickly respond to customer demand by taking into account the relationship between the company's matrix and service quality. Increases sales, products become healthy, comfortable and customer satisfaction increases customer loyalty [13].

The author has done a preliminary study of applying her design of FFB sorting tools on palm fruit workers. Based on the study, the faults of the previous design are the tip of the sorting tool is too spherical and slippery, a diameter of the bar is a bit too big which makes it heavy, the border between the bar and the tip is not firm enough, and extra lining on the handle moves easily. Based on these faults, this study will analyze and develop new sorting tools based on the characteristics needed by workers by applying QFD method and Kano model. Kano model is used to clarify element design according to the needs and wishes of workers, and identify a technical characteristic of sorting tool to improve workers performance using QFD. Palm fruit workers need towards a convenient FFB sorting tool to lift palm fruit is correlated with its technical characteristic making.

2. Method

This study is conducted in a palm factory in Tebing Tinggi, Sumatera Utara. This is a descriptive study, describing an object characteristic systematically and factually. The object observed in this study is the previous sorting tool design used by palm fruit workers. Variables used in this study are technical characteristics, product attributes, difficulty level, cost estimation, and importance degree. The sampling method used is functional and dysfunctional Kano questionnaire with a total sampling of 6 workers.

Palm fruit lifting activity is observed using Canon camera for four work days from 12.00 to 17.00 o'clock. The preliminary questionnaire is used to gather complaints from workers regarding the usage of sorting tool. Kano questionnaire is used to categorize sorting tool attributes based on the satisfaction level of palm fruit workers. Kano model has three categories, i.e. dissatisfaction, satisfied, and delighted. Palm fruit workers' attributes will be categorized into attractive (A) must be (M), one-dimensional (O), indifferent (I), reverse (R), and questionable (Q) characteristics [14]. This questionnaire contains 6 questions regarding attributes, i.e. coating paint color, a bar material, tip material, main handle coating material, extra coating material, and handle shape. Attributes quality scoring of sorting tool uses a Likert scale, i.e. like, wish, neutral, tolerate, and does not like. Total sampling is the sampling method used, so questionnaire is distributed to all workers. Modus in Kano questionnaire is turned into question in a closed questionnaire with 5 Likert scale categories, i.e. very important, important, quite important, less important, and not important. Technical characteristics questionnaire is filled by the production manager. The critical part questionnaire is given to factory engineering division who understand attributes affecting sorting tool quality, so it is worth choosing to determine the relationship between variables. The sampling method used is judgment sampling [15]. Pearson method is used for validation test of sorting tool performance data and expectation, whereas Alpha Cronbach with a significant level of 5% is used for reliability test of Kano model data. House of Quality (HOQ) matrix is made based on technical characteristics questionnaire scores 6 points. HOQ is made based on palm fruit workers needs obtained from the semantic differential questionnaire. Determination of customer importance from technical characteristics is based on modus score from each question item in a closed questionnaire. Determination of sales point and absolute value from workers need are input to make the 2nd phase of QFD. The relation between 5 critical part can strongly positive, moderate positive, strong negative, moderate negative, or not related. HOQ is made by calculating working performance consists of three aspects, i.e. difficulty level, importance level, and cost estimation.

3. Results and Discussions

The results will be discussed in 3 subsections, they are Kano questionnaire, QFD matrix, QFD method analysis of 1st and 2nd phase.

3.1. Kano questionnaire

Kano questionnaire categories attributes of sorting tool according to its ability to satisfy the need of palm fruit workers. Based on validity test conducted on functional and dysfunctional Kano data of sorting tool, r counted> r table = 0.706, which means it is valid. From reliability test for Kano model data, r counted = 0.935 > r table = 0.706, which means the questionnaire is reliable. Data from Kano model questionnaire is processed to categorize each attribute based on Kano model. Kano category values on each attribute to all respondents who have been obtained will be determined its Kano categories by using Blauth's formula as shown in Table 1.

Table 1. Kano category mapping of each attribute.

No	Attribute	Kano Category
1	Coating paint color: gray	R
2	Sorting toolbar material: stainless pipe	A
3	Material tip sorting tool FFB: steel spring	R
4	Main handle coating material: grip	R
5	Extra handle coating material: rubber	I
6	Handle shape: triangle	R

Assessment using Blauth's formula shows that workers do not satisfy with the attributes of coating paint, tip material, main handle coating material, and material shape of the sorting tool, which is then categorized as a reverse. Extra handle coating material does not affect workers satisfaction, therefore categorized as indifferent. Bar material greatly affects workers satisfaction, hence categorized as attractive. Meanwhile, calculation using the method of customer satisfaction (CS) and customer dissatisfaction (CD) obtain attribute's score as shown in Table 2.

Table 2. Score point of CS and CD for each attribute.

CR	Kano Kategory	CS	DS	CS Point (1, CS_i)	DS Point (0,DS_i)
1	R	0.0000	0.0000	(1, 0.0000)	(0, 0.0000)
2	A	0.7500	-1.5000	(1, 0.7500)	(0,-1.5000)
3	R	0.0000	0.0000	(1, 0.0000)	(0, 0.0000)
4	R	0.6667	0.0000	(1, 0.6667)	(0, 0.0000)
5	I	0.5000	0.0000	(1, 0.5000)	(0, 0.0000)
6	R	0.0000	-1.0000	(1, 0.0000)	(0, 1.0000)

Kano category mapping is made using Minitab software and the output obtained is shown in Figure 1.

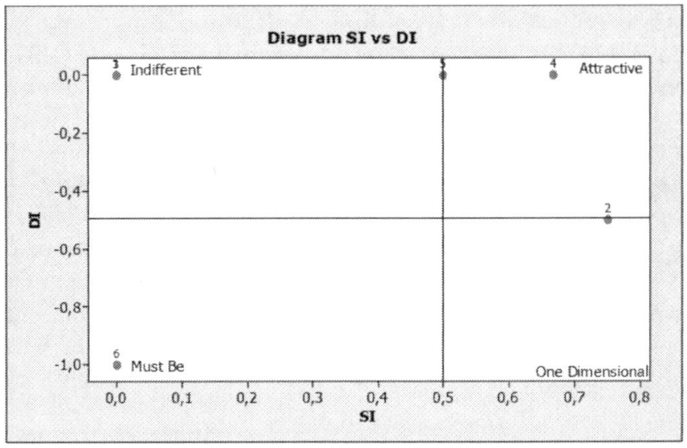

Figure 1. Kano category mapping

The 1st quadrants are attractive category area containing factors that increase selling points, even though does not decrease satisfaction level if the product is not there. Attributes of main handle coating material made of fabric and extra handle coating material made of rubber are categorized as attractive. This means that if the sorting tool has both attributes, workers satisfaction will increase and if not, workers satisfaction will not decrease. The 2nd quadrant is in different category area containing factors that do not affect consumer satisfaction. The attribute of tip material made of spring steel and coating paint color is categorized as indifferent, which does not affect workers satisfaction. The 3rd quadrant is must be category area containing attributes that product must meet and is an absolute requirement for a good assessment of the product. Triangle handle shape attribute is categorized as must-be, thus this is an absolute requirement of sorting tool. If this attribute does not exist, then consumers will lose interest altogether. The 4th quadrant is a one-dimensional area showing the level of performance of linear attributes with customer satisfaction. The attribute of bar material made of a stainless pipe is categorized as one-dimensional, which means The performance level of the attribute is linear with workers satisfaction.A good attribute performance level will increase the satisfaction level of the palm oil loading worker and vice versa.

3.2. Quality Function Deployment (QFD) Matrix

The needs and wishes of palm fruit workers toward the sorting tool are determined using Customer Requirement (CR). Workers wishes in form of attribute will be adjusted to technical characteristics of sorting tool using HOQ. Importance level of workers need will be determined based on modus score from each question in the closed questionnaire. Workers satisfaction level for each attribute is determined by multiplying the scale by frequency and then divided by the number of workers. The number of customer requirement, customer importance, and workers satisfaction level are shown in Table 3.

Table 3. Sorting tool designing process.

No	Customer Requirement	Customer Importance (CI)	Level of Satisfaction
1	Coating paint color: gray	2	1.500
2	Sorting toolbar material: stainless pipe	4	1.833
3	Material tip sorting tool FFB: spring steel	1	2.056
4	Main handle coating material: grip	2	1.500
5	Extra handle coating material: rubber	4	2.167
6	Handle shape: triangle	2	2.571

Attributes importance level of sorting tool is indicated by scores of net sales, importance level, relative weight. Net sales score is obtained from an interview with the manufacturer of Fresh fruit bunches sorting tools, while importance weight and relative weight score are obtained from a closed questionnaire distributed to 6 workers. The questionnaire contains 6 questions regarding importance level of sorting tool attributes. Net sales, importance weight, and relative weight scores are shown in Table 4.

Table 4. A score of net sales, importance weight, and relative weight.

Customer Requirement	Importance Weight	Relative Weight	Net Sales Point
Coating paint color: gray	1.333	3.200	1.2
Sorting toolbar material: stainless pipe	2.182	13.091	1.5
Material tip sorting tool FFB: spring steel	0.486	0.370	1.5
Main handle coating material: grip	1.333	3.200	1.2
Extra handle coating material: rubber	0.385	1.538	1.0
Handle shape: triangle	0.429	1.286	1.5

A variable with the highest score of relative weight, importance level, and net sales is bar material that uses a stainless pipe. This is the most important variable to improve company performance.

3.3. QFD method analysis

QFD 1st phase is used to determine the difficulty level and cost estimation of technical characteristics. The level of relationship between the technical characteristics is determined based on the interview with the manufacture of FFB sorting tools. The most important technical characteristics are the heating process and the printing process with each difficulty level 5, a degree of interest 21% and 32% as well as the estimated cost of each 25%. Technical characteristics obtained from the QFD 1st phase used as inputs to carry out the processing on QFD 2nd phase. QFD 2nd phase is used to determine the critical part affecting the quality of FFB sorting tools. The critical part is the most important part or component characteristic of FFB sorting design obtained from an interview with the workshop that makes FFB sorting tool. Calculation of performance measures of QFD 2nd phase characteristics is aimed at obtaining a value of difficulty level, a degree of importance, and cost estimation. The degree of difficulty, a degree of importance and highest cost estimate is the melting point of material with difficulty level 5, a degree of interest 25% and estimated cost 28%.

4. Conclusions

Based on KANO model mapping, the design attributes of FFB sorting tools categorized as attractive, indifferent and must be. The attractive category is the attributes of the main grip coating material made of fabric and the additional grip coating material made of foam. Indifferent category is the attributes of the material of FFB sorting tool made of As iron and black coating paint. The category must be is the attribute of the triangle-shaped handle, whereas the one-dimensional category is the attribute of FFB material made of a hollow iron bar. Based on difficulty level, a degree of importance, cost estimation, the highest technical characteristics of the QFD phase I are heating and casting process. Meanwhile, based on difficulty level, a degree of importance, and cost estimation, the highest technical characteristics of QFD phase II is the melting point of the material.

In the next study, a sampling of palm fruit workers from several palm oil companies can be done to obtain a more accurate ergonomic FFB soring tool. More studies need to be conducted using the method of QFD Phase III to evaluate the "manufacturing product" and phase IV to review the "production product".

IOP Conf. Series: Materials Science and Engineering **309** (2018) 012048 doi:10.1088/1757-899X/309/1/012048

References

[1] Ulrich KT and Eppinger SD 2008 Product Design and Development (Irwin Mc Graw Hill)
[2] Paryani K 2007 Product Development Decisions Support System Customers Based (Southfield)
[3] Mital A et al 2008 *Product Development A Structured Approach to Consumer, Product Development, Design, and Manufacture* (Elsevier)
[4] ES Jaiswal, A Case study on Quality Function Deployment (QFD). IOSR Journal of Mechanical and Civil Engineering 3(6) (2012)pp.27-35
[5] Nordin, N., & Razak, R. C. (2014). The Concept of Kano-QFD Integration for Non-linear Customer Needs in Product and Service Design. *Adv. Review on Scientific Research*, *4*(1), 1-6.
[6] Suef M *et al* 2013 *Proceeding of Industrial and Service Science* G12-1-G12-6
[7] Kecek G and Akinci OC 2016 *International Journal of Academic Research in Business and Social Science6*(4) 112-135
[8] Israr M and Gangele A 2014 A Quality Function Deployment Methodology for Product Development, *Int. Journal of Emerging Trends in Engineering Research2(*11) 52-63
[9] Singgih ML *et al* 2013 *Proceeding International Conference on Engineering and Technology Development* 22-38
[10] Gupta P and Srivastava RK 2011, Customer satisfaction for designing attractive qualities of healthcare service in India using Kano model and quality function deployment, *International Journal of Mechanical Engineering1*(2) 101-107
[11] Saadon MSI *et al* 2013, Quality Management Program Development Using Servqual and Kano Model, *Journal of Human Capital Development* 1-14
[12] Mohd Yusof IS *et al* 2014 *Iranian Journal of Public Health*43(3):72-80
[13] Baran Z and Yildiz MS 2015 *International Journal of Business and Social Science6*(9) 122-131.
[14] Cohen L 1995 Quality Function Deployment: How to Make QFD Work For You (Addison-Wesley)
[15] Day RG1993 Quality Function Deployment: Linking a Company With its Customers (ASQC. QP)

Quality improvement of laminated board made from oil palm trunk at various outer layer using phenol formaldehyde adhesive

R Hartono* and T Sucipto
Faculty of Forestry, Universitas Sumatera Utara Jl. dr. Mansur Kampus USU Medan, Indonesia 20155

*E-mail: rudihartono@usu.ac.id; titomedan@yahoo.com

Abstract. Characteristic of laminated board from oil palm trunk (OPT) is very low in quality. The effort to improved it's quality done by using the outer layer from high density wood. The purpose of this experiment was to analyzed the effects of the outer layer on physical and mechanical properties of OPT and to obtain optimum treatment to fulfills JAS 234:2003. All of laminated board was made of 3 layers, and for the middle layer was made by densified-OPT. Then for the outer layer was made of sengon and meranti wood. The sample size was 5 cm (width) x 3 cm (thick) x 45 cm (length). The various outer layer of laminated board were A (OPT/densified OPT/OPT); B (Sengon/densified OPT/OPT); C (Sengon/densified OPT/ sengon); D (Meranti/densified OPT/OPT) and E (Meranti/densified OPT/meranti). The results showed that the moisture content, density, thickness swelling, delamination, MOR and MOE were 6.10-8.48%; 0.40-0.63 g/cm^3; 6.43-13.20%; 0%; 168.79-438.29 kg/cm^2 and 30115-100454 kg/cm^2, respectively. The moisture content and delamination fulfills JAS 234:2003, while density and thickness swelling did not fulfill standard. Whereas for MOR and MOE value, only type D and E that fulfill standard. There are strength relationship between density and mechanical properties, such as MOR and MOE value. The optimum treatment in this reseach to made laminated board made from OPT was type D that using the meranti as outer layer.

1. Introduction

North Sumatera Province is the second largest province that has oil palm plantation in Indonesia. In 2015, the area has reached 1,06 million ha [1]. When replanting, waste is generated in the form of trunk. The oil palm trunk (OPT) waste are very potential as raw material for wood product. The other hand, OPT has many weaknesses, such as low the density, high shrinkage-swelling, high moisture content and low strength. Erwinsyah [2] reported that 25 years old OPT density was 0.14-0.60 g/cm^3, while 40 years old OPT was 0.23-0.74 g/cm^3 [3]. The other weaknesses of OPT was very high moisture content, low strength and low durability class [4]. The efforts to improve OPT quality was done by compressed using close system compression [5] and by impregnated of phenol formaldehyde [6]. Through compression and impregnation of phenol formaldehyde into OPT can improve the physical and mechanical properties.

The other efforts to improved OPT quality were made a laminated board. According to Serrano [7], laminated board is a product that was made by making layers and fasten layers and bond together with parallel direction. Many researchers study about laminated boards made from wood, such as laminated

IOP Conf. Series: Materials Science and Engineering **309** (2018) 012049 doi:10.1088/1757-899X/309/1/012049

board made from mangium [8], laminated blocks made from oily keruing wood [9], lamina blocks made from Bengkirai wood [10], laminated blocks made from Balsa wood [11], Laminated block made of Malaysian Hardwood of Resak and Kruing [12]. Aside that, laminated board can also made from non wood material such as Petung bamboo [13], *Gigantochloa apus* dan *Gigantochloa robusta* [14], bamboo laminated board made from Andong and Mayan bamboo [15].

Research of laminated boards of OPT was also conducted. Previous research about laminated board made from OPT shows that all of the physical and mechanical properties, such as MOR and MOE did not fulfill standard [16], [17]. This was caused by low quality of OPT, thus laminated board produced using it was also low of quality.

One way to improve laminated boards quality of OPT can be done using outer layer (face/back) from solid wood and densified OPT for core (middle) layer of laminated board. Outer layer using commercial wood such as Sengon and Meranti to improve its mechanical properties. Also the outer layer used to improve decorative values of laminated board.

Adhesives used is phenol formaldehyde (PF) adhesive. The PF is an exterior adhesive that durable against humidity [18]. Delamination test of laminated board made of OPT using PF was 0%. This means that PF is very suitable to be used as adhesive in laminated board production.

Based on that reasons, this research entitled Quality improvement of laminated board at variouse layer using PF adhesive. It was expected that variouse layer can improve quality of laminated board made from OPT to fulfills JAS 234:2003.

2. Material and Methods

Materials used in this research was boards of OPT, Meranti, Sengon and PF adhesives. OPT logs were obtained from Medan Tuntungan district, Medan city, while meranti and sengon wood was obtained from sawmill around Medan. Adhesives were obtained from PT Pamolite Adhesive Industry. Initial density of inner part of OPT, outer part of OPT, Sengon and Meranti wood were 0.28 g/cm^3; 0.39-0.43 g/cm^3; 0.41 g/cm^3 and 0.59-0 0.64 g/cm^3, respectively.

OPT was cut into boards and separating the inner part (soft density) and the outer part (hard density). The the inner part of OPT were dried and cut into 5 cm (width) x 2 cm (thickness) x 45 cm (length) and densified with compression level of 50%, thus resizing it to 5 cm x 1 cm x 45 cm. The inner part of OPT can be compressed until compression level of 50% [19]. While the outer part of OPT, meranti and sengon wood was cut into 5 cm x 1 cm x 45 cm. After compression, the density of densified OPT became 0.45-0.48 g/cm^3.

Laminated boards made of three layers with final size of 5 cm x 3 cm x 45 cm. Middle layer of laminated board was made of densified OPT. Spread weight of PF was 240 g/m^2. PF spread using kape with double spreading technique. PF adhesive maturing was done at 150°C for 15 minutes. Variouse layer for laminated boards was made into five type as seen in Figure 1.

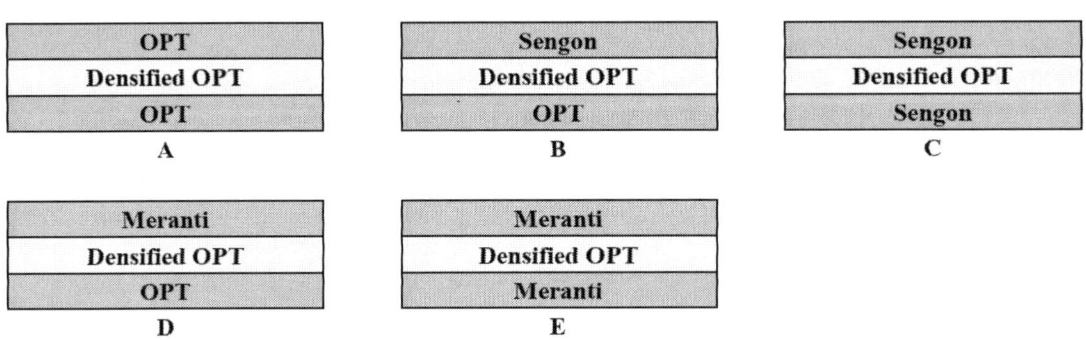

Figure 1. Variouse Outer Layer of OPT laminated boards

Physical and mechanical properties tested were moisture content, density, thickness swelling, delamination, MOR and MOE. Test of laminated board quality was referring to JAS 234:2003, Standard of Glued Laminated Timber [20]

3. Results and Discussion

The physical and mechanical properties in this reseach were moisture content, density, delamination, thickness swelling, MOR, and MOE. The result of physical and mechanical properties were as shown in Table 1.

Table 1. Physical and Mechanical properties of lamina board

Treatment	MC (%)	Density (g/cm^3)	Delamination (%)	TS (%)	MOR (kg/cm^2)	MOE (kg/cm^2)
A	8.48	0.42	0	13.20	168.79	30115
	(0.21)	(0.01)	(0)	(2.79)	(29.28)	(5166)
B	6.10	0.40	0	12.00	186.52	33375
	(0.80)	(0.03)	(0)	(3.02)	(25.47)	(7256)
C	8.06	0.44	0	10.60	251.55	51808
	(0.35)	(0.06)	(0)	(1.16)	(25.89)	(7795)
D	6.29	0.60	0	10.79	327.24	82947
	(1.35)	(0.04)	(0)	(2.03)	(54.69)	(9041)
E	6.85	0.63	0	6.43	438.29	100454
	(0.37)	(0.02)	(0)	(1.17)	(24.10)	(11118)
JAS 234: 2003	< 15	-	< 10	-	> 300	> 75000

Every value were mean value of three samples. Values inside brackets reflect one standard deviation, MC = moisture content, TS = thickness swelling, MOR = modulus of rupture and MOE = modulus of elasticity,

Moisture content of five type of laminated boards ranged from 6.10% to 8.48%. All laminated board moisture content fulfills JAS 234:2003 which set forth maximum of 15%. The Laminated boards using the outer layer of OPT was tend to have highest moisture content when compared to laminated boards using sengon and meranti. The moisture content is one factor of physical properties which affect the mechanical properties. When moisture content is reduced, the mechanical properties increased. Moody and Liu [21] reported that maximum range in moisture content between layer is 5% to minimize thickness swelling changes dimention.

Laminated board density ranged from 0.33 g/cm^3 to 0.44 g/cm^3. The highest density value was type E with face and back layer from Meranti wood which is having a density of 0.63 g/cm^3. The lowest density value was type B with face layer from Sengon with density of 0.40 g/cm^3. Type E with face and back layer from meranti was higher than sengon and OPT layer. This caused by difference of density of materials used in each lamina. Meranti density is the higher than OPT and sengon density. In the contrast, Sengon wood density was lower than OPT and Meranti density.

Meranti wood density was categorized high with the value of 0.63 g/cm^3. This wood will have cellular structure with thicker cell wall and smaller pore. According to Mandang and Pandit [22], Meranti wood was categorized to strength class of II-III, and durability class of III-IV and was suitable to be used as door and window, ship material, furniture, musical instrument, structural building.

Thickness swelling of laminated boards was around 6.43–13.2%. The highest value of thickness swelling was laminated board of OPT with type A using face and back outer layer from OPT, which is 13.20%. The lowest thickness swelling was laminated boards of OPT with type E using face and back outer layer from Meranti wood, which is 6.43%.

Thickness swelling was affected by the type of laminated board composition. Woods with low density will absorb more water when compared to those of higher density. The laminated boards composed from OPT layers, there were highest thickness swelling, such as type A has high water absorbance and moisture content. Bakar [4] stated that OPT will be able to sustain moisture content of 250 to 350%.

If compared the outer layer of laminated board between Sengon and Meranti wood, the thickness swellling of laminated boards using Meranti wood as outer layer was lower than Sengon wood. This was caused by the difference of density of each laminated board material. Meranti wood density was highest when compared to OPT and sengon wood. In contrast, Sengon wood has lowest density when compared to OPT and meranti wood.

Delamination reflects the quality of adhesive and bonding strength of laminated boards. Results showed that delamination value of all treatment was 0% and fulfills JAS 234:2003 which set forth maximum of 10%. Its means there is no bonding that was damaged or breached during immersion in water for 6 hours. This was caused by the use of PF adhesive was higher qualities compared to UF or natural adhesives. Ruhendi et al. [18] stated that PF adhesives has resistance from water, high humidity and temperature. Then, double spread technique was also enabling adhesives to prefectly spread on both fastened surfaces. This technique have good adhesion between adhesive and boards, it is preferable to use double spreading techinque on both side of the surface. Variance of the outer layer and amount of adhesives does not cause different value. The OPT, Sengon and Meranti was able to be fastened with a high bonding strength using PF adhesives.

The mechanical properties in this reseach were modulus of rupture (MOR) and modulus of elasticity (MOE). Table 1 shown that MOR values of laminated board produced was around 168.79–438.29 kg/cm^2. MOR value for laminated board type D and E with meranti as the outer layer fulfills JAS 234:2003. This type exceed the minimum requirement of 300 kg/cm^2 set forth by JAS 234:2003, with MOR value were 327.24 kg/cm^2 (type D) and 438,29 kg/cm^2 (type E). This was caused that meranti has higer density than OPT and sengon density. Wood density value contributed and had relatioship with MOR value.

Herawati et al. [8] stated that MOR of woods was affected by some factors, namely density or spesific gravity, wood knot and oblique fibres. A study by Karlinasari [23] shown that african wood had many knots and MOR value affected by design of laminations arrangement.

MOE values of laminated boards was 30115–100454 kg/cm^2. MOE value for laminated board type D and E exceed the minimum requirement of 75000 kg/cm^2 set forth by JAS 234:2003, with MOE value was 82947 kg/cm^2 and 100545 kg/cm^2, respectively. It caused that outer layer of meranti which high density. The initial density of meranti in this study was 0.61 g/cm^3. Meranti layer contributed to MOE value, both to type D (only top layer) and also type E (top and bottom layer).

Previous research by Herawati et al. [8] showed that laminated board of African wood was having MOE of 7.3 x 10^4 kg.cm^{-2} and those of mangium wood was having values ranged 8.41 x 10^4 to 3.67 x 10^4 kg/cm^2. They stated that MOE was affected by density, same as MOR value. Based on MOR and MOE values it is can be concluded that laminated boards of OPT with Meranti wood as the outer layer was higher than the outer layers made from OPT and Sengon wood. Wood density was proportional with MOR and MOE. The use of high density wood will produce high mechanical properties, whereas low-density wood will produce low mechanical value. Bodig and Jayne [24] stated that factor that affected wood strength was density. The higher density, the higher strength will be resulted.

Mechanical properties were attributable to density. The higher density resulted in the higher MOR and MOE. Figure 2 shows the relationship between density with mechanical properties. The regression analysis showed that the correlation between density and MOR, also density and MOR could be described by a second-order polynomial.

Figure 2 showed that density increased the MOR and MOE, which resulted in a strong relationship, as shown by high values of the coefficient of correlation (R2) of 0.99 and 0.97 for the MOR and MOE, respectively. Similar statement was uttered by Bowyer et al. [25] that mechanical properties will increase with the increase of wood density.

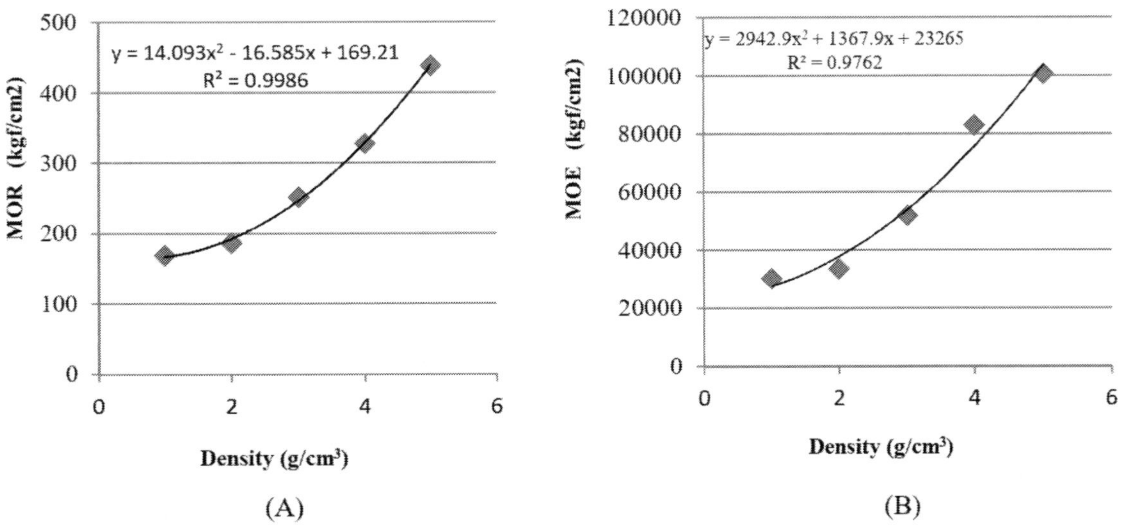

Figure 2. Relationship between density and MOR (A) and density and MOE (B)

Based on the mechanical properties, to improve the quality of laminated board made from OPT must used wood with high density for the outer layer, such as meranti wood. From this result showed that optimum of treatment was type E with used meranti as the outer layer for laminated board.

4. Conclusion

The physical properties such as moisture content and delamination of laminated boards of all variaous outer layer fulfilled JAS 234:2003. The other hand, density and thickness swelling has not fulfilled standard. Whereas mechanical properties (MOR and MOE values), only type D and E that fulfilled JAS 234:2003. There are strength relationship between density and mechanical properties (MOR and MOE). The optimum treatment in this reseach to made laminated board from OPT was type D that using the meranti as outer layer.

References

[1] Republic of Indonesian Ministry of Agriculture 2015 Center for Agricultural Data and Information (Indonesia, Jakarta)
[2] Erwinsyah 2008 Improvment of Oil Palm Wood Properties Using Bioresin, Dissertation of graduated school of Universität Dresden.
[3] Hartono R., Wahyudi I, Febrianto F, Dwianto W 2011 *J. Ilmu dan Teknologi Kayu Tropis* **9**(1) 73-83
[4] Bakar ES 2003 *Forum Komunikasi Teknologi dan Industri Kayu* **2**(1) 2003
[5] Hartono R, Wahyudi I, Febrianto F, Dwianto W, Hidayat W, Jang JH, Lee SH, Park SH, Kim NH 2016 *J. Korean Wood Sci. Technol* **44**(2): 172-183.
[6] Hartono R, Hidayat W, Wahyudi I, Febrianto F, Dwianto W, Jang JH, Kim NH 2016 *J. Korean Wood Sci. Technol* **44**(6): 842-851
[7] Serrano E 2003 Mechanical Performance and Modeling of Glulam in Timber Engineering Thelandesson S, Larsen HJ (USA: Jhon Willy & Sons, New York)
[8] Herawati E, Massijaya MJ, and Nugroho N 2010 *J. Biological Sciences* **10**(1): 37-42
[9] Malik J, Santoso A 2011 *Jurnal Penelitian Hasil Hutan* **29**(3): 271-277.
[10] Setiawan DB 2011 Jurnal TEKNIS **6**(2): 61-65
[11] Gusamo BK, Semeli M, Ozarksa B 2013 *Academic Journal* **8**(29): 3888-3892.

[12] Mohamad WHW, Razlan MA, Ahmad Z 2011 *International Journal of Civil & Environmental Engineering* **11**(04): 7-12.
[13] Oka GM 2008 *Jurnal SMARTek* **6**(2): 94-103.
[14] Sulastiningsih IM, Nurwati 2009 *J. Tropical Forest Science* **21**(3): 246–251.
[15] Sulastiningsih IM, Santoso A 2012 *J. Penelitian Hasil Hutan* 30(3): 199-207.
[16] Hartono R, Sucipto T, Dwianto D, Darmawan T 2013 *J. Penelitian Rekayasa* **6**(2): 54-60
[17] Sucipto T, Hartono R, Dwianto D, Darmawan T 2014 Karakteristik Papan Laminasi Batang Kelapa Sawit dengan Variasi Pelapis Luar dan Berat Labur Perekat, In: Prosiding Seminar Nasional Mapeki XVII, Medan, pp 96-104
[18] Ruhendi S, Koroh DN, Syamani FA, Yanti H, Nurhaida, Saad S, Sucipto T 2007 Analisis Perekatan Kayu (Indonesia: Fakultas Kehutanan IPB, Bogor)
[19] Hartono R, Dwianto W, Wahyudi I, Febrianto F, Morooka T 2017 *IOP Conf. Ser.: Mater. Sci. Eng* **180** 012016
[20] Japan Plywood Inspection Corporation 2003 Japanese Agricultural standard for glued laminated timber (Japan: Tokyo)
[21] Moody RC and Liu JY 1999 Glued Structural Members in Wood Handbook: Wood as an Engineering Materials, pp 11.1-11.24 (USA: Forest Products Laboratory, Madison).
[22] Mandang YI, Pandit IKN 2002 Pedoman Identifikasi Kayu di Lapangan (Indonesia: Yayasan Prosea dan Pusat Diklat Pegawai dan SDM Kehutanan, Bogor)
[23] Karlinasari, L 2007 Wood Stiffness and Bending Strength Analysis Based on Non Destructive Testing, Thesis of Graduated School of Bogor Agricultural University.
[24] Bodig J and Jayne BA 1982 Mechanics of Wood and Wood Composites (USA: Van Nostrand Reinhold Company, New York)
[25] Bowyer JL, Shmulsky R, Haygreen JG 2003 Forest Products and Wood Science: An Introduction Fourth edition (USA: Lowa State University Pr.)

IOP Conf. Series: Materials Science and Engineering **309** (2018) 012050 doi:10.1088/1757-899X/309/1/012050

Examining of solid waste generation and community awareness between city center and suburban area in Medan City, Indonesia

H Khair[1, 2,*], C N Putri[1], R A Dalimunthe[1], T Matsumoto[2]

Department of Environmental Engineering, Faculty of Engineering, University of Sumatera Utara, Medan, Indonesia
Graduate School of Environmental Engineering, the University of Kitakyushu, Kitakyushu, Japan

*E-mail: hafizhul@usu.ac.id

Abstract. Municipal solid waste (MSW) management is still an issue in many cities in Indonesia including Medan. Understanding the waste generation, its characteristic and communities involvement could provide effective solid waste management. This research compares waste generation from people who live in the city center and suburban area. The research also examines the willingness and participation of community about environmental aspect, especially solid waste management. The method of waste generation used Indonesian Nasional Standard 19-3964-1994. The city center generates 0.295 kg/person/day of solid waste and 0.180 kg/person/day for suburbs. The result showed that there are the common amount of waste compositions between the city center and suburban area. The majority waste composition was an organic fraction. Questionnaires were distributed to examine the community awareness. The descriptive statistic used to analyze the data. The result showed that people living in the city center are slightly higher in community awareness than in the suburb. This paper highlights that area of living could give some effect to solid waste generation, waste composition and rate of awareness.

1. Introduction

Increasing generation of municipal solid waste (MSW) is a logical consequence of increasing city population. However, to treat the municipal solid waste is still an issue. This condition will lead to bad impact to human and environment. In the other hand, treating properly municipal solid waste will improve the quality of life. MSW management is an effort to eliminate waste from its sources and disposed to environmental securely.

Present day, solid waste in Medan are collected in mixed scale, put all solid waste in one plastic bag and municipality disposed it into landfill directly without any treatment. This action will reduce the capacity and the age of landfill. Municipalities should upgrade these condition into integrated sustainable solid waste management (ISSWM).

The concept of the integrated solid waste management system has gained acceptance. Integrated solid waste management has been defined as the selection and application of suitable techniques, technologies, and management approach to achieve specific objectives and goals. MSW management is not just a technological system facilitating the handling and disposal of MSW. MSW management deals with many other factors such as socioeconomic conditions, operating environment and actions of the municipal government [1].

In developed countries, citizens are supposed to drop their waste at specific locations by the municipalities on a particular day of the week at a given moment and waste categories. Public participation, awareness, and corporation are the main keys to success in solid waste management. In ISSWM, public participation is a necessary part to make the people aware of active participation in the system. Without public participation, it may be difficult to make city cleaner and may also become less effective in using of cost resource.

Research conducted in several cities in Java and Sumatera. It showed that city with high population density, a high total of population, and high of economic growth characteristics have high waste generation. Meanwhile, the city with low population density and economic growth, its waste generation is also low [2]. Environment problems associated with the generation of waste are part of societal changes where households play an important role. These societal changes influence the characteristics of given households, including family size, income, education, occupation and residential location [3].

The main objective of this paper is to examine the waste generation, characteristic and community involvement in Medan city to reduce solid waste generation. This research compares waste generation from people who live in the city center and suburban area. The research also examines the level of awareness of the community about environmental aspect, especially solid waste management.

2. Methods

Sample collection and measurement of solid waste generation and composition were referred to the Indonesian National Standard SNI 19-3964-1994, Method of Collecting and Measurement of Samples and Composition of Urban Wastes. Based on this method, waste generation and composition were measured in eight days period[4]. Solid waste generation was calculated through load count analysis method, which calculates the total of waste collection every day. Waste samples are weighted and sorted based on pre-defined components. Each element is also weighted afterward.

Two sub-district of Medan City is selected as areas of study. Medan Polonia represents as the city center and Medan Helvetia as a suburban area. Based on SNI 19-3964-1994, Number of the sample in Medan Polonia is 29 households and 72 households in Medan Helvetia. To measure the level of public awareness, questioners distribute into the area of study. Each selected households are also as respondent. The analysis was based on the descriptive statistic.

3. Results and discussions

This study found that the city center generates waste 0.295 kg/person/day. Then, suburban generates 0.180 kg/person/day of solid waste. The study showed that there is a different amount of solid waste generation between two areas. Medan Helvetia represents as suburban area and Polonia as the city center. The amount of solid waste generates in the city center is higher than suburban, nearly 64%. This difference could be due to income, education, food consumption, and residential location.

Results further in the waste composition in two area described that dominantly household waste comes from food waste, where approximately 80% domestic type waste is organic. The other types of waste are plastic, paper, glass, metal, LWTR and other material. The results of the solid waste composition are shown in table 1.

Table 1. Solid waste composition.

Composition	The city center (%)	Suburban (%)
Paper	4.14	1.75
Plastic	5.43	4.91
Organic	79.16	86.29
LWTR	1.56	1.61
Glass	3.70	1.60
Metal	0.67	1.12
Aluminum	1.04	0.53
Inert material	0.19	0.06
Other	4.12	2.13

In this research, respondents were also asked to answer several questions related public awareness in solid waste management. Respondents were asked about:

1. Level of acceptance in paying retribution fee monthly
2. Willingness to separate waste at their home
3. Their habit, to bring own plastic bag while shopping
4. Willingness to get involved in solid waste management community in their area
5. Their habit dumping the waste in proper manner, not littering it

The result of their answers can be shown from figure 1 to figure 5. In figure 1 describes that respondents were asked about the fees of the waste retribution charged to the community monthly. 74-84% of respondents in two areas claimed to accept the retribution fees and said 'it does not matter.' However, when asked to separate the waste at home (figure 2), the respondents in the two study areas were divided, the respondents who live in the suburbs seem reluctant, while the residents in the city center had a willingness to do it.

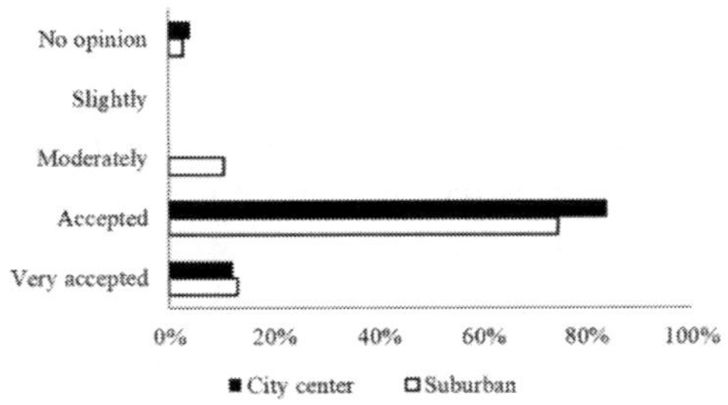

Figure 1. Responses of respondent 'acceptance of paying retribution fee.'

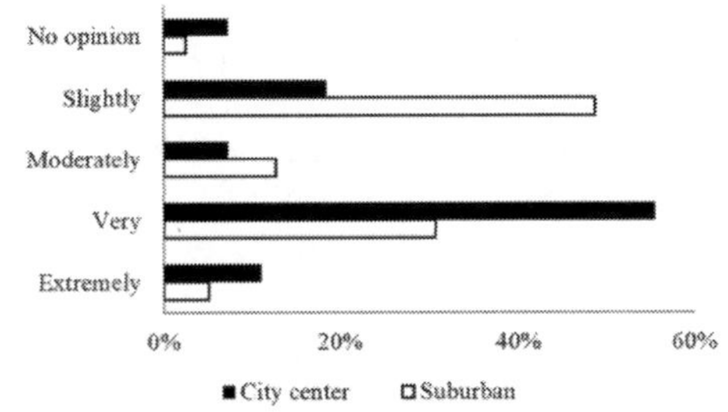

Figure 2. Responses of respondent 'willingness to separate their waste at home.'

Respondents were asked their responses about the habit of carrying their plastic bag while shopping (Figure 3), both areas of the study showed the same responses that more than 50% did not carry it. Asking about their participation in waste management activities (Figure 4), the majority of respondents answered pleased and very pleased. 52% of respondents in city center answered pleased and 19% very pleased. While in the suburbs only 46% who answered pleased, 10% very pleased and the rest hesitate and without opinion.

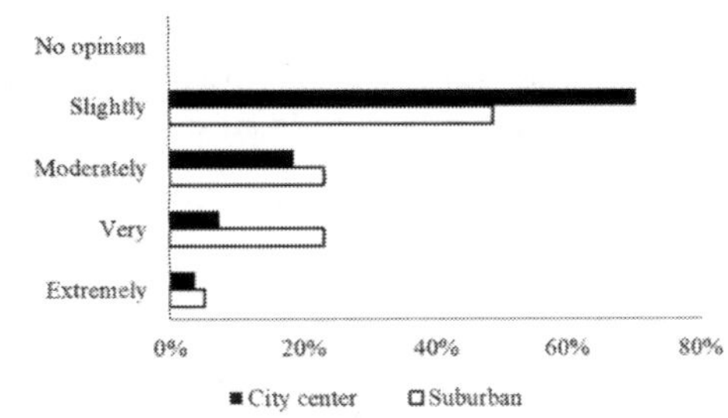

Figure 3. Responses of respondent 'to bring their plastic bag while shopping.'

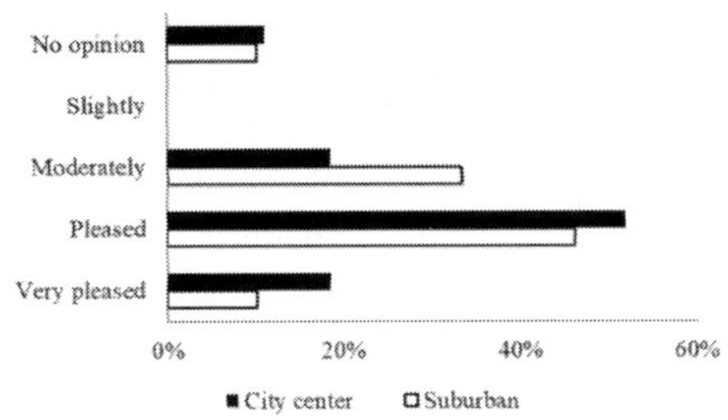

Figure 4. Responses of respondent 'willingness to get involved in solid waste management community.'

People who live in the city center and suburban, both have a willingness to maintain cleanliness. There is not any different answer in this case. It can be seen in figure 5. In the city center, 74% answered agree, and 19% answered very agree. In the suburbs, 67% answered agree, and 26% answered very agree.

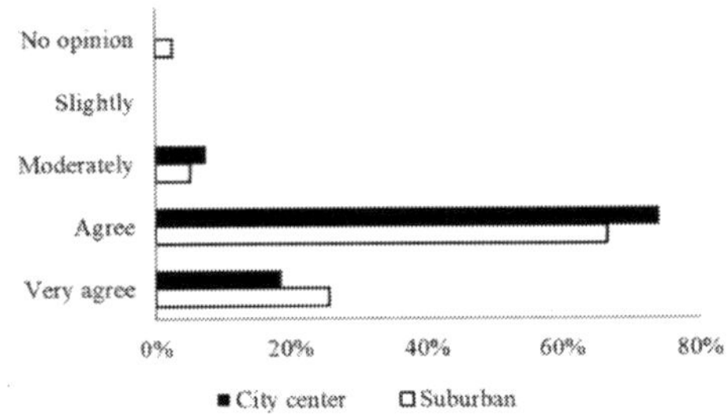

Figure 5. Responses of respondent 'willingness to put the solid waste into the proper manner, not littering it.

All responses data were analyzed using weighting analysis. The score was divided into 5 level: poor < = 20%, fair < = 40%, good <= 60%, very good <= 80% and excellent <= 100%. The results of weighting analysis how those people living in the city center were slightly higher in community awareness than those in the suburbs. The total score for city center was 70% (very good) while in the suburbs was 69% (very good). The T-test was conducted to assess the difference of response from people living in city center and suburbs. The results of the analysis showed that level of awareness of residents was not different between two areas ($\alpha = 0.05$, P-value = 0.258). The Indonesian government has implemented 3R policy through Waste Law 2008. One of the factors influencing the success of the policy is public participation. Emphasis towards greater public awareness would enhance the outcome. This study shows that in general there is no location influence on the level of public awareness. It should be noted; the final score is the result of a combination of several problem points. That is, to solve the problem, specific treatment is required for each area.

4. Conclusions

The majority composition of waste is a biodegradable waste of 79.16% to 86.3%. Thus, the municipality should treat this waste as soon as possible. It also indicates a real prospect of organic waste recycling through composting. The municipality could distribute some compost center points around the city to treat organic waste. This effort could reduce up to 80% waste to landfill. The composition of solid waste also could give information the trends of waste that may be reused/recycling. The fraction of economic/reusable/recyclable items, e.g., paper, plastic, metal, aluminum, glasses.

Statistically, the rate of public awareness in the city center and a suburban are same in 'good level.' The environmental concern is a potential variable in influencing the successful integrated solid waste management. Thus, the municipality should allocate its budget to non-technical purposes, educate the people and approach the people based on their characteristic and improve the awareness by educational promotion programs. The municipality needs to build partnerships in term of community-based management. Through this study, the stakeholders could formulate better policies, plans, and strategies in solid waste management.

References

[1] Shekdar A V 2009 Sustainable solid waste management: An integrated approach for Asian countries *Waste Manag.* **29** 1438–48
[2] Benno Rahardyan, Gita Prajati T P 2015 The Influence of Economic and Domographic Factors to Waste Generation in Capital City of Java and Sumatera *The Third Joint Seminar of Japan and Indonesia Envi. Sustainability and Disaster Prevention (3rd ESDP-2015)* pp 69–80
[3] Monavari S M, Omrani G A, Karbassi A and Raof F F 2012 The effects of socioeconomic parameters on household solid-waste generation and composition in developing countries (a case study: Ahvaz, Iran) *Environ. Monit. Assess.* **184** 1841–6
[4] Badan Standarisasi Nasional 1995 *SNI 19-3964-1995 Metode pengambilan dan pengukuran contoh timbulan dan komposisi sampah perkotaan*

IOP Conf. Series: Materials Science and Engineering **309** (2018) 012051 doi:10.1088/1757-899X/309/1/012051

Evaluation of work posture and quantification of fatigue by Rapid Entire Body Assessment (REBA)

I Rizkya*, K Syahputri, R M Sari, Anizar, I Siregar
Universitas Sumatera Utara
Almater Street Kampus USU, Padang Bulan Medan 20155, Indonesia

*E-mail: indahrizkya@usu.ac.id

Abstract. Work related musculoskeletal disorders (MSDs), poor body postures, and low back injuries are the most common problems occurring in many industries including small-medium industries. This study presents assessment and evaluation of ergonomic postures of material handling worker. That evaluation was carried out using REBA (Rapid Entire Body Assessment). REBA is a technique to quantize the fatigue experienced by the worker while manually lifting loads. Fatigue due to abnormal work posture leads to complaints of labor-perceived pain. REBA methods were used to an assessment of working postures for the existing process by a procedural analysis of body postures involved. This study shows that parts of the body have a high risk of work are the back, neck, and upper arms with REBA score 9, so action should be taken as soon as possible. Controlling actions were implemented to those process with high risk then substantial risk reduction was achieved.

1. Introduction

Musculoskeletal disorders (MSDs) are the most common types of health problems in the industry especially the small-medium scale. In the small-medium industries still rely heavily on human labor to run most of the production activities. Most of the work is done manually. So musculoskeletal disorder often occur due to work in particular position, in a long time and in incorrect working posture. Posture is defined as the physical position of the body, such as standing, sitting or lying down. Correct posture maintain a natural curvature of the spine, minimizing tension and preserve the balance of muscle and skeletal body [1]. Correct posture does not make the body leaning forward, backward, left, or right [2]. Workers in performing their work are in an uncomfortable condition. Eventually, cause various musculoskeletal disorders. Abnormalities generally arise due to the recurrent lifting activity, the elevation, the weight of removed material, and the ambient conditions, etc. [3]. Musculoskeletal disorders can be minimized through improved techniques, working methods, and facilities [4]. Postural analysis can be a powerful technique for assessing work activity. The risk of musculoskeletal injury associated with posture recorded in the context of assessment in the workplace.

Many studies have been conducted regarding the assessment of work posture and the condition of musculoskeletal abnormalities performed in the automobile assembly industry in India [5], pressing stations with spiral press machines used for capacitor smoothing [6], porcelain industry operators [7], and operators of Lathe Machine. [8]. The evaluation of work postures in small-scale industries concludes that there are opportunities for musculoskeletal disorders at moderate to high levels. This study recommends the application of ergonomic interventions with the right knowledge among workers and health education regarding posture, implementation, and monitoring of legislation to reduce pain due to musculoskeletal disorders [9]. A study conducted by Agrawal found musculoskeletal disorder occurring in the welding process whereby the worker did the job with a kneeling position that required a change of posture in work [10]. Another study on posture assessment is done by Lakhwinder in the foundry industry. Rigid posture, removal, forcible removal, and other manual work at high speed and frequencies often cause musculoskeletal disorders [11].

The frequency of work and the level of worker discomfort leads to the risk of fatigue and abnormalities in some parts of the body. The application of a combination of static and dynamic work systems, varying leg position while standing, slightly helps to reduce the discomfort of working in a standing position [12]. Many studies have also found that poor work posture is cause of back pain, waist, and stress at work. As a result, the time lost, the deterioration of employee health and the decreased morale, ultimately lowering the productivity of the company. Implementation of ergonomic principles will help improve quality and productivity, but more importantly, help workers to feel comfortable and secure while doing the job. Implementation of ergonomics associated with the design of methods and processes may reduce job-related risk [13,14,15]. However, until now there are not many studies that analyze the posture of material handling workers especially in small-medium industries. This study was conducted to evaluate the postures material handling workers with minimal facilities in food industry.

2. Rapid Entire Body Assessment (REBA)

From an ergonomic standpoint, manual handling is a high-risk activity that can cause back pain. From a physiological perspective, manual handling of materials requires high energy and muscle strength. Generally, manual handling of materials is done inaccurately and can cause inflammation of the nerves and muscles. The main source of this problem is the static load carried, frequencies, and long duration, causing tension or disturbance to the joints and ligaments.

The Rapid Entire Body Assessment was first introduced by Lynn Mc Atamney and Sue Hignett (1995), a relatively simple way to assess the risk of whole-body abnormalities due to work [16]. REBA method is used as a tool to measure fatigue that occurs due to work. Many studies have applied the REBA method to assess and analyze posture. REBA can reduce 85% employment risk to a low risk of musculoskeletal abnormalities and is negligible [9]. REBA is a sensitive tool for assessing abnormalities in all parts of the body (wrists, upper arms, forearms, neck, trunk, and legs)) [16]. REBA is easy to use and useful for task risk assessment.

REBA techniques classify the body in two parts, namely A and B. Group A consists of the neck, legs, and torso. While group B consists of the forearm, upper arm, and wrist. The REBA technique uses a body assessment sheet to analyze work risk. REBA provides a score to describe the posture and the level of risk. REBA scoring sheets can be seen in Figure 1. High-risk jobs rated with higher scores and those with low or low-risk involvement score close to zero. Immediate improvement actions and necessary changes are recommended for higher score activities to avoid working risks.

3. Methodology

This study was conducted in one of the small- medium industries that produce food in North Sumatra. Fourteen material handling operators are sampled. Observations are made on the printing station to the drying station. The first step is to explore information about musculoskeletal disorders and symptoms felt by workers using Standard Nordic Questionaire (SNQ). Data collection with SNQ is widely used because it is fast, easy, and not costly. SNQ is used to identify parts of the body that have complaints due to ergonomic working conditions and awkward work postures [17].

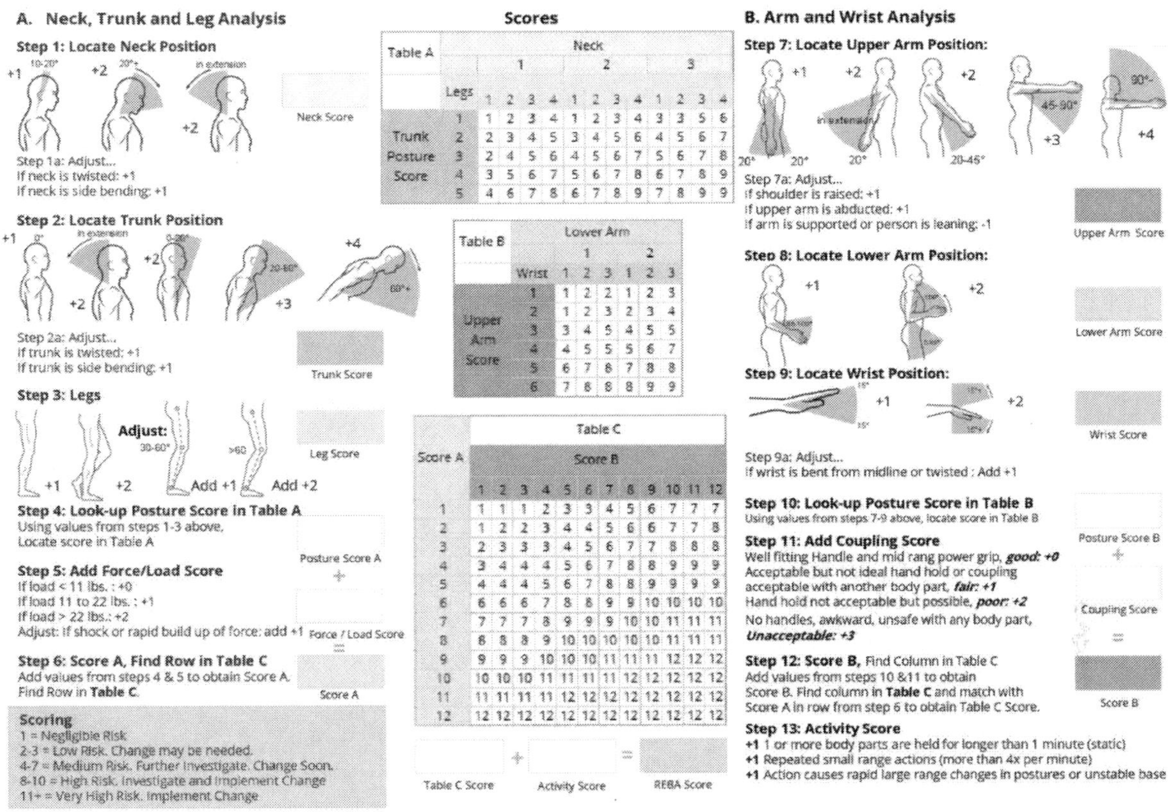

Figure 1. REBA Assessment Sheet [16]

The second step is to record with video all worker activity to get their posture. The video is then cut within the framework of each job, then assessed using the REBA technique. The REBA assessment sheet [16] is filled with the corresponding values obtained based on body position, movement of movement, body movement (gesture) for each body part based on the relative table. Was first performed in the charge sheet REBA assessment is to provide an assessment on the body part belonging to the group A (Trunk, Neck, Legs) and group B (Upper arm, Lower arm, Wrists right, and left). The sum of the values in group A is added by a force/load score. Furthermore, the sum of the values in group B is summed with the coupling/grip score. Total final value of group A and B is a score C. In static posture and/or repetitive actions (more than four times per minute), or the change in posture faster than the Scots in Table C summed with activity scores to obtain the final value REBA. Repair activities are performed on activities/tasks that have substantial risk levels. The REBA scores can be seen on REBA's rating sheet.

4. Result and Discussion

In this study, 12 workers with a mean the age of 34 ± 5.91 years, the height of 158.83 ± 4:15 cm, and the job experience of 4:17 ± 2.89 years were studied. Material handling operators from the printing station to the stations are observed. The observed operators 58.33% were male and 41.67% female. The main task of this operator is to bring the printed work in process and has been arranged on a square plastic sheet. The products that have been prepared on the plastic sheet are brought by trolley to the drying station. The first activity observed in this study was to take plastic sheets from the trolley. The

IOP Conf. Series: Materials Science and Engineering **309** (2018) 012051 doi:10.1088/1757-899X/309/1/012051

position of the body of the operator when doing this activity that is standing with both hands are higher than the head. The second activity is bringing the sheet to the drying area. The position of the body of the operator while performing this activity is to stand with the elbow at an angle of 900. The position is done while walking. The third activity is to put the product in the drying area. The position of the body when the operator performs this activity is the body bent by an angle of 750 when putting it on the ground for drying. The results of the Standard Nordic Questionaire (SNQ) distributed to workers to extract information on body parts that the workers are sickly sick with can be seen in Table 1.

Table 1. Result of Standard Nordic Questionaire (SNQ)

No	Complaint Type	Level of Complaint (%)			
		A	B	C	D
0	Pain/stiff in the upper neck	66.67	33.33		
1	Pain/stiff in the lower neck	58.33	41.67		
2	Pain in the left shoulder	16.67	33.33	50.00	
3	Pain in the right shoulder	16.67	33.33	50.00	
4	Pain in left upper arm		66.66	16.67	16.67
5	Pain in the back			66.67	33.33
6	Pain on the right upper arm		33.33	50.00	16.67
7	Pain at the waist			56.33	41.67
8	Pain on the buttocks	16.67	16.67	50.00	16.67
9	Pain on the butt	83.33	16.67		
10	Pain on the left elbow	41.67	58.33		
11	Pain on the right elbow	33.33	33.33	33.33	
12	Pain in the left forearm		50.00	33.33	16.67
13	Pain in the right forearm	16.67	33.33	33.33	16.67
14	Pain on the left wrist	16.67	41.67	41.67	
15	Pain on right wrist		58.33	41.67	
16	Pain on the left hand		50.00	50.00	
17	Pain on the right hand		41.67	58.33	
18	Pain in the left thigh	16.67	25.00	58.33	
19	Pain in the right thigh	25.00	16.67	58.33	
20	Pain in the left knee		41.67	58.33	
21	Pain in the right knee	33.33	33.33	33.33	
22	Pain in the left calf	33.33	33.33	33.33	
23	Pain in the right calf	16.67	50.00	33.33	
24	Pain in the left ankle	33.33	50.00	16.67	
25	Pain in the right ankle	33.33	50.00	16.67	
26	Pain on the left leg	33.33	33.33	33.33	
27	Pain in the right leg	33,33	33,33	33,33	

Of the 28 questions about pain complaints felt by workers on the body, known five parts of the body that complained workers "quite painful." Upper arm body part, back, upper right arm, waist, left forearm, and right lower arm. Based on the results of SNQ, complaints of the pain in the body of the operator is influenced by factors of sex, age, and duration of work. Women operators complain more about pain than men. Over 35 years of workers who complain of pain than under 35 years. Workers who have worked for more than four years complained of pain in parts of their bodies. Assessment of worker's posture is done by filling in the REBA assessment sheet for each activity. Table 2 illustrates the results of posture assessment sheet with REBA.

Table 2. The Result of Posture Assessment Sheet of Body with REBA

Parameters	Activity 1	Activity 2	Activity 3
Neck	2	1	2
Trunk	2	1	4
Legs	1	1	4
Table A Posture Score	3	1	8
Force/Load Score	0	0	0
SCORE A	3	1	9
Upper Arm Position	4	2	2
Lower Arm Position	2	1	2
Wrist Position	1	1	2
Table B Posture Score	5	1	3
Coupling Score	3	3	3
SCORE B	8	4	6
Activity Score	1	1	1
Table C	8	3	8
REBA SCORE	9	4	10
RISK LEVEL	High	Medium	High

In the first activity, the neck is bent at the top of the corner 300, so given the score 2. The upper arm is raised higher than the head and form an angle of 110^0 so given a score 4. The neck on the 3rd activity bends down about 25^0 so given a score 2. At the activity of three bodies bending With an angle of 75^0 so given a score 4. REBA scores on the first activity and the third is 9 and 10 means that the posture of the worker's work has high risk to the musculoskeletal disorder. The first and third activities need to be investigated and implemented changes/improvements. While the second activity gets a score of four means having a medium risk. This activity needs further investigation, and changes can be done immediately.

As the results of the present study indicate, most cases of discomfort of body parts due to problems such as pain, discomfort, stiff and numbness were related to the waist. This result finding was in agreement with that of a study by Ahmadi [7]. Their research assessed the risk of MSDs with the Nordic Musculoskeletal Questionnaire (NMQ), Job Content Questionnaire (JCQ), and REBA. Due to inappropriate posture, workers feel pain in their body parts. To eliminate it, this study need the new work facilities to assist workers' activities. Besides, it is necessary to improve the working method to minimize their complaints.

5. Conclusion

The manual handling of material where operators work in a standing position and bend. Activities undertaken by operators have a high risk of disruption, pain, stiffness, and discomfort. Musculoskeletal disorders are found in these activities by complaints of perceived pain workers. It shows that there is a need to apply ergonomic knowledge as well as repair work facilities to prevent workers from Musculoskeletal Disorders.

Acknowledgement

Author would like to thank to Universitas Sumatera Utara for funding the financial support to publish this paper. The speech also to Mrs. Widya and small- medium scale industry of food processing who are willing to be the object research. We appreciate the efforts of all those who have cooperated in conducting this study.

References

[1] Moon HH 2007 J Sports Leis Stud. 31 1033-1041.
[2] Chen KM, Chen MH, Hong SM 2008 J Clin Nurs. **17** 2634-2646.
[3] Aman S et al 2011 International Journal of Ergonomics. **1** 2.
[4] Isa H and Abdul R O 2011 International Journal of Research and Review in Applied Science **8** 1.
[5] Srikanth P, Chakravarthya1, Subbaiah K M, and Shekar G L 2015 International Journal of Research Science & Management **2** 38-42.
[6] Bhandarl A, Bahirat P, Nagarkar V, and Bewoor A 2013 International Journal of Mechanical and Production Engineering. **1** 3.
[7] Ahmadi A, Mirzaei R, and Ansari H 2015 JOHE. **4** 146-153.
[8] Ingale P A and Salunke P V 2016 International Journal of Mechanical Engineering and Information Technology. **4** 1636-1641.
[9] Ansari N A and Sheikh M J 2014 IOSR Journal of Mechanical and Civil Engineering (IOSR-JMCE). **11** 18-23.
[10] Agarwal A, D N et al 2011 International Journal of Engineering Science and Technology. **3** 10.
[11] Lakhwinder P S 2012 Asian Journal of Managerial Science. **1** 1.
[12] Zahari Taha et al 2008 The 9th Asia Pacific Industrial Engineering & Management Systems Conference. APIEMS.
[13] Maldonado A and Macias 2009 XV Congreso Internacional De Ergonomia Semac.
[14] Javier A, Holguín L, et al 2008 *Ergonomics and its relationship to the design of production processes* (Ponencia: Conergo)
[15] Asim Z et al 2012 International Posture Journal of Science & Technology. **2** 1.
[16] Hignett S and McAtamney L 2000 *REBA: A Survey Method for the Investigation of Work-Related Upper Limb Disorders Applied Ergonomics.*
[17] Kazemi S E, Savas S, and Aydos L 2016 International Journal of Science Culture and Sports. **4** 260-266.

IOP Conf. Series: Materials Science and Engineering **309** (2018) 012052 doi:10.1088/1757-899X/309/1/012052

Sustainable brooder for supporting local chicken in North Sumatera Province, Indonesia

N Ginting[1 2,*], R Hidayat[1 2]

[1]Faculty of Agriculture, University of Sumatera Utara, Indonesia
[2]Animal Production Laboratory, Faculty of Agriculture, University of Sumatera Utara, Indonesia

*Email:nurzainahginting@gmail.com

Abstract. Demand for local chicken has lasted for ages in Sumatera Utara Province, Indonesia. Demand is higher than supply because local chicken has low productivity which caused by several factors, for example heater or brooder for day old chicken (DOC). People who live in villages where mostly local chicken are raised have difficulty in serving brooder as electricity is expensive. In Biological Laboratory, Animal Production Programme Study, University of Sumatera Utara, a study has been conducted from August until November 2016. The objective of this study is to investigate sustainable brooder for local chicken. A study of 300 local DOC with designs: T1) local DOC + electric lamp brooder, T2) local DOC + biogas lamp brooder, T3) local DOC + propane lamp brooder has conducted. The key parameters measured were temperature stability, DOC spread, body weight gain, and mortality. Designs T1, T2 and T3 were almost equal in temperature stability, DOC spread, body weight gain and mortality while T2 was superior as it was environmentally friendly. A recommendation for local people to use biogas brooder instead of electric or LPG brooder.

1. Introduction

Local chickens are native chickens of Indonesia that have long been nurtured and developed by the community, especially rural communities who maintain them as a source of family food for eggs, meat, and as savings that at any time can help finance Suprijatno and Atmomarsono [1]. Brooding time was an important moment in nurturing local chicken so handling intensively should be taken at this period. Failure in this period will minimize the achievement. In the brooding period, the temperature needs to be well noticed, especially the heating process.

Local chicken breeders who have not developed well, still use oil lamps, LPG gas and in small amounts using an incandescent lamp as a source of energy for the brooder. Although the breeders complained about brooder operational cost, they kept using them like no other option. Electricity consumption was growing along with the increase of population while raw materials to produce electricity such as fossil fuel was limited and LPG gas was expensive.

On the other hand, fossil fuel emits green house gases, and Indonesia committed to reducing its greenhouse gas (GHG) by 26% in 2020 one way by promoting clean energy use for example biogas. Ginting [2] found biogas gave more benefits in the agricultural areas such as energy which was environmentally friendly energy, slurry which was organic fertilizer and clean environment as agricultural waste which usually discharged into the environment become biogas input. Ginting [3] mentioned that biogas support farmers in the rural area when farmers use biogas instead of firewood in the process of coffee bean roasting. Biogas was capable of producing good energy; the flames are blue, odorless and smokeless. Therefore it was expected that biogas could be used during brooding as source of heat. Balde [4[found that biogas produced renewable electricity and heat. According to Rajakovic [5] in comparison, 1 m3 biogas was equivalent to 0.48 kg LPG, 0.42 l solar, 0.8 l gasoline, 0.62 l kerosene, 0.6 crude oil, 1.4 kg coal, 4.7 kWh electricity and 3.5 kg firewood.

2. Method

Research material was chicks cage, 40% formalin for cage fumigation, 300 (DOC), feed, vaccine. The tool was used brooder for chicken, i.e., incandescent lamp 60 Watt, LPG lamp, biogas lamp, thermometer, digital scales for weighing chickens. The design used in this study is Completely Randomized Design (RAL) with three treatments, namely: P1 = electric brooder, P2 = biogas brooder, P3 = LPG brooder.

Parameter of Research was as follows: Temperature Stability, DOC spread, body weight gain

3. Results and Discussions

The results will be discussed in 4 subsections, they are temperature stability, spreading of DOC, body weight gain and mortality.

3.1 Temperature Stability

The temperature stability was to evaluate the temperature change of each brooder. The results showed that the average temperature stability on the starter phase of brooder was 30.42 °C for 14 days while the average stability of chicken brooder temperature should be about 30-32 °C. The highest temperature stability on the LPG brooder was 33.45 °C, followed by biogas brooder which was 29.97 °C and the lowest on electric brooder which was 27.84 °C. Ideal temperature caused activity of DOC grow normally and they spread evenly throughout the cage. According to Hajar Fawaz et al. [6] using the solar cell as a source of energy for brooder caused stability of DOC brooder.

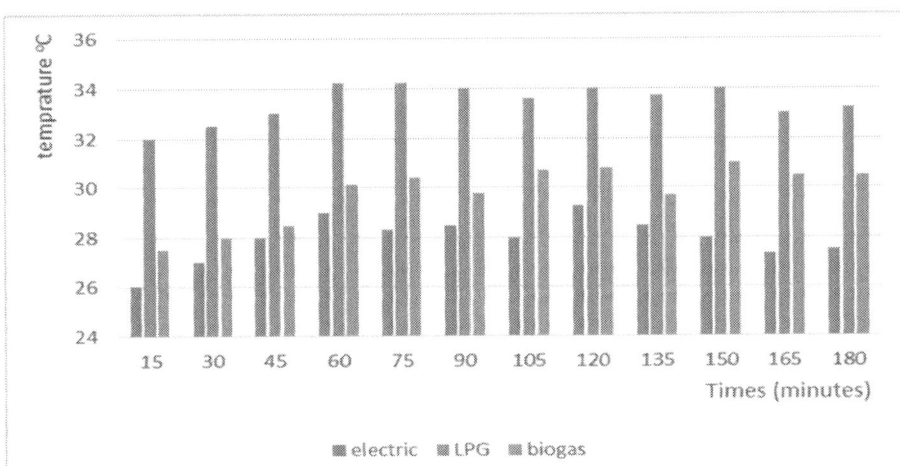

Figure 1. Temperature Stability of Electric, Biogas and LPG Brooder

3.2 Spreading of DOC

Spreading of DOC were depend on brooder temperature. LPG brooder caused DOC spread out, they left the brooder because it was too hot which triggered DOC consumed the feed less and they more drank water. Electric brooder brought about DOC cuddled together for seeking heat. Only biogas brooder effect normal activity of DOC. R. Denny [7] mentioned that feed efficiency was influenced by DOC activity; DOC should consume more feed than drink. Normal activity caused better DOC growth.

According to Zhenghou Jiang et al. [8] biogas brooder on young chicks significantly reduce LPG cost. The use of biogas for brooding young chicks on a poultry farm can significantly reduce the natural gas or propane fuel cost. Jiang also mentioned that biogas brooder was an alternative for electric brooder. However, people in the rural area were not familiar with biogas technology.

3.3 Body Weight gain

A calculation on body weight gain was held on a weekly basis based on the difference between the final body weight weighing and the weight of the initial weight divided by the length of the study in g/head/day. From Table 1 it can be seen that the analysis of the diversity of gain of chicken body weight during the study gave no significant effect (P> 0.05). The absence of this difference was due to almost stable cage temperature and ration was given which was equal in nutrition content. This was by the statement of Jull [9] which stated that the percentage increase in body weight gain from week to week following period was not the same and in this research, body weight gain tends to increase on the sixth week and by P2 which was biogas brooder treatment.

Table 1. DOC Body Weight Gain (g).

Treatment	Replication						Total	Mean±sd
	1	2	3	4	5	6		
P1	59.14	57.19	52.43	50.95	59.57	60.81	340.10	56.68±4.07
P2	59.91	65.81	57.52	61.14	57.14	65.05	366.57	61.10±3.68
P3	57.24	63.62	62.29	57.76	54.33	68.00	363.23	60.54±5.01
Total	176.29	186.62	172.24	169.86	171.05	193.86	1069.91	
Mean	58.76	62.21	57.41	56.62	57.02	64.62		59.44

3.4 Mortality

From table 2 it can see that the highest mortality was in P1(electric brooder) that were 13,3%, whereas in P2 (biogas brooder) and P3 (LPG gas brooder) there was no mortality. Mortality was occurred due to the temperature of P1 which was not hot enough for DOC while the temperature of P2 and P3 were stable. Usually, people in a rural area who raise local chicken use 60 Watt lamp for DOC brooder. This was the reason why in this research at P1 used 60 Watt lamp. This experiment proved what the caused of DOC mortality in a rural area which was electric brooder not hot enough for DOC. The people in the rural area did not want to use 100 Watt lamp, because it was too expensive.

Overall mortality rate should be less than 5%. The first week of death rate during the gain period should be no more than 1%, deaths in the following week should be relatively low until the end of the week and continue in a constant state until the end of the gain period.

Table 2. Percentage of DOC Mortality

Treatment	**Replication**						**Total**	**Mean±sd**
	1	2	3	4	5	6		
P1	3.3	0	10	0	0	0	13.3	2.21 ± 0
P2	0	0	0	0	0	0	0	0±0
P3	0	0	0	0	0	0	0	0±0
Total	3.3	0	10	0	0	0	13.3	
Mean	1.1	0	3.3	0	0	0		2.21

Factors that affect the percentage of death include brooder temperature, body weight, strains, types of chickens, climate, environmental hygiene and diseases Wiedosari, E and Wahyuwardan, S [10]. Also they mention the death of DOC were largely determined by the density of DOC in the cage, the air circulation in the cage which caused the incoming O2 was less while CO2 was produced more from DOC respiration.

4. Conclusions

Based on the research, biogas and LPG brooder were better than electric brooder as they had better on temperature stability, chicken spreading, body weight gain, and mortality. Biogas brooder was better than LPG as it was environmentally friendly.

References

[1] Suprijatno and Atmomarsono 2005 *Ilmu Dasar Ternak Unggas* Jakarta: Penebar Swadaya
[2] Ginting N 2017 *IOP Conf. Ser.: Earth and Environmental Sci.* **65** 012007
[3] Ginting N 2017 *IOP Conf. Ser.: Mater. Sci. Eng.* **180** 012112
[4] Balde, Hambaliou, Andrew C, VanderZaag, Burtt S D, Wagner-Riddle C, Crolla A, Desjardins R L and MacDonald D J 2016 Methane emissions from digestate at an agricultural biogas plant *Bioresource Technology* **216** 914-22
[5] Rajakovic N and Milomir K 2006 Biogas-energi instead of waste *6th Int. Symposium Nikola Tesla*
[6] Fawaz H, Abiad M G, Ghaddar N and Ghali K 2014 Solar-assisted localized ventilation system for poultry brooding *Energy and Buildings* **(71)** 142-54
[7] Purnama R D 2002 Respon pertumbuhan ayam broiler periode starter yang dipelihara dalam ruang dengan suhu terkontrol *Temu Teknis Fungsional Non Peneliti* Bogor: Balai Penelitian Ternak
[8] Jull M A 1982 *Poultry Husbandry* New Delhi: Tata Mc Graw-Hill
[9] Jiang Z, Steinsberger S C and Shih J C H 1987 *In situ* utilization of biogas on a poultry farm: Heating, drying and animal brooding *Biomass* **14(4)** 269-81
[10] Wiedosari E and Wahyuwardan S 2015 Studi kasus penyakit ayam pedaging di Kabupaten Sukabumi dan Bogor *Jurnal Kedokteran Hewan* **9(1)** 9-113

IOP Conf. Series: Materials Science and Engineering **309** (2018) 012053 doi:10.1088/1757-899X/309/1/012053

Substitution of pig feed preparation using firewood with biogas in Samosir Island: transforming an environmental catastrophe into environmental, social, economic benefits

N Ginting[1*]

[1]Faculty of Agriculture, Universitas Sumatera Utara, Kampus USU, Medan 20155, Indonesia

*E-mail: nurzainahginting@gmail.com

Abstract. In Samosir Island, Indonesia pigs care was not environmentally friendly as people were used firewood in pig feed preparation. A series of research has been conducted from March until September 2017 which was preceded by survey. It was found that people cut tree for firewood. As Samosir Island was under Toba Go Green Project which was a tree planting project so feed pig preparation was in contrast to the project. More over, Indonesia has been committed to reduce its green house gases (GHG) by 26% in 2020, any mitigation on GHG was strongly recommended. One way of mitigation in Samosir was by installing biogas for pig feed preparation. 5 biodigesters 500 liters capacity each were installed in Parbaba Village, Samosir Island and biodigester input were pig manure, water hyacinth. Research design was Randomized Completely Design. Parameters were gas production, pH, temperature and C/N ratio. Biogas than used to cook feed pig. It was known that to cook for 5 finisher pigs, 3 kg firewood could be substituted by 250 liters of biogas.

1. Introduction

Samosir Island is one of the most beautiful islands in Indonesia. It is a large volcanic island in Lake Toba, located in the north of the island of Sumatera in Indonesia. The lake and island were formed after the eruption of a supervolcano some 75,000 years ago. At 640 square kilometres (247 sq mi), Samosir and Lake Toba are a popular tourist destination due to exotic history and the vistas they offers.

The majority of Samosir's inhabitants were Christians and therefore pigs were one of the most well-kept animals in Samosir. Most of the pigs were extensively kept, mostly penned and partly free. In Samosir, livestock manure were left in the environment and naturally degraded. For livestock which were penned next to the lake, usually manure were discharged into the lake. According to Li Fei et.al [1] livestock manure caused surface water eutrophication and groundwater nitrate enrichment. In the management of feeding on pigs, people used fish from Lake Toba as a source of protein. Currently silver fish commonly used for pig feed because the fish was not consumed by humans. Before it was given to pigs, silver fish was blended or finely ground, then cooked. Usually the community cook pig feed in the yard, near the pigsty. Duration of cooking pig feed between 30 - 60 minutes depending on the amount of the pigs that were kept and they used firewood for cooking fuel for pig feed. Ginting [2[mentioned that using firewood harmed the environment as wood burning produces greenhouse gas, i.e CO2 and recommended for local coffee farmer in Karo District, Indonesia that instead of using firewood in roasting coffee bean they should used biogas. Since 2011, Toba Go Green program has been implemented which was a reforestation program to improve the Lake Toba Water Catchment Area due 61.23% or 235,217 ha in critical condition. Cutting tree will bring a great influence on the condition of lake Toba. It will also had a significant impact on the aesthetic side of Samosir and surrounding islands that could have an impact on tourist arrivals Ginting [3]. Also worrisome was the invasion of water hyacinth that had spread in the waters of Lake Toba. Water hyacinth caused siltation of the waters and depresses fish populations. The purpose of this study was to find alternative fuel to cook pig feed by utilizing biogas technology. The input used was pig manure and water hyacinth. Thus would be obtained several other benefits that include the benefits of environmental, social and economic.

2. Materials and Method

The study lasted from March until May 2017. The tools used in the research were five-unit installation of biogas 500 litres capacities, five units of gas volume measuring device with a 1 bar pressure. Organic waste which were pig manure, water hyacinth and buffalo manure for starter were used as an input. The research method was used Completely Randomized Design (RAL). HRT (Hydraulic Retention Time) 45 days. Treatments were as follows: T1 (100% water hyacinth), T2 (25% pig manure + 75% water hyacinth), T3 (50% pig manure + 50% water hyacinth), T4 (75% pig manure + 25% water hyacinth) and T5 (100% pig manure). Parameter research were gas production, pH temperature and C/N ratio. Pig manure and choppered water hyacinth with size 2 cm diluted with water in ratio of 1: 2. Methane production was measured by using a 3 inch PVC pipe with 1 atm pressure. pH was measured by using pH meter on slurry in outlet. Temperature was measured by inserting thermometer from inlet onto bottom of digester. The research of biogas done by filling the bio gas digester according to treatment and left for 30 days of HRT *(Hydraulic Retention Time)*. Refilling raw ingredients per day was necessary. Data collection was performed by calculating the volume of biogas, pH, the temperature and C/N ratio.

3. Results and Discussion

3.1. Gas production

Gas production from this study was presented in table 1. There were differences in gas production between T1 and other treatments. Data analysis was then followed by Duncan Test where T1 and T2 showed gas production were significantly different (5%) from other treatments. Jagadish H. Patil et al [4] mentioned that water hyacinth produced a good biogas as it contained nitrogen, essential nutrients and has a high content of fermentable matter. Water hyacinth produced biogas better than pig manure. Biogas production depended on nutrition content in biogas input. Nutrition content in manure already absorbed in livestock stomach and this was the reason why biogas production was bigger by water hyacinth than pig manure.

Table 1. Gas production (CH_4) and pH: input pig manure and water hyacinth.

Treatments	Replication					Total	Mean	pH
	1	2	3	4	5			
T1	376.5	360.3	382.2	390.5	383.8	1893.3	378.66a	7.8
T2	215.2	216.7	210.6	215.7	218.1	1076.3	215.26b	7.6
T3	155.52	155.4	153.6	153.7	152.9	771.12	154.224c	6.4
T4	106.2	105.4	104.6	106.1	106.0	528.3	105.66d	4.7
T5	72.1	72.4	73.1	72.8	72.6	363	72.6e	4.3
						4632.02		

3.2. pH

pH in this research was took from samples in the outlet of biodigesters/slurry. pH were range from 4 to7. pH is influenced by combination of substrates materials. Lower pH such 4 may affect microorganisms metabolism thus affect gas production and pH of 6 to 7 are preferable. According to Ware [5] it was important to measure pH in order to ensure that there was no disturbance on gas production thus gas production could ceased due to acidification or alkaline condition. Ginting [6] found that addition rotten orange decrease pH thus decrease gas production. pH of biogas slurry is a critical point when slurry will be used as fertilizer which will be conducted after this research on multi purposes plants. According to Ginting and Rauf [7] pH of soil was 5.5 – 7.0. and application of slurry should be on pH 6-7 in order to support a normal metabolism in soil. In this research, pH from T1, T2 and T3 were recommended to be used as liquid fertilizer. More over Yun Cao et.al. [8] found that application of bio gas slurry as soil amendment significantly could help in maintaining fertility and healthy soil.

3.3. Temperature

In this research, data of temperature were taken as temperature as part of climate effected gas production. Temperature in this research was about 19°C which was same as temperature of environment. Location of the research in Samosir Island where ambient temperature was 22 - 27°C in July and August thus temperature of underground where biodigester was set up was below 20°C and automatically effect temperature inside biodigester. According to Cantrell et al. [9] there were three temperature ranges in the anaerobic digestion process which were 1) < 20°C; 2) 20 - 45°C; 3) 45 - 60°C and optimal temperature was 35°C. High gas production were occurred in July, August and September where temperature were 19 - 21°C. Extreme temperature affected methane bacterium thus cause gas production stopped.

3.4. C/N ratio

C/N indicates nutrient content and how far degradation process has already taken placed on materials which in this research pig manure and water hyacinth. C/N in this research in T1 was 27 while T2 until until T5 were around 24 until 15. C/N in T1 was high as C value of water hyacinth was high, i.e. 21.25 and N value was 0.29 While C/N of T5 was low and this was related with C and N value of pig manure which was 18.17 and 1.25 respectively. According to Ginting and Mustamu [10], slurry with C/N was high, i.e. more than 20 was suggested to get maturity by keeping the slurry in a container for at least one month. In this time, N value was getting better as population of microorganism still in form. Microorganism itself contains of amino acid.

3.5. Environmental benefits

As mentioned before, in Samosir Island, people used firewood to cook pig feed. However, this practice harm the environment as wood burning produces greenhouse gas, i. e CO_2. This research was try to provide energy alternative for pig feed cooking. A simple research was conducted by cooking feed for 5 finisher pigs with fire wood. Livestockers cooked about 4 kg pig feed which a formula of silver fish and sweet potato or cassava waste. Cooking took about 40 minutes and were used 3 kg firewood. The same process was repeated but with biogas and it was found that for cooking 4 kg pig feed, it was needed 250 liters of biogas. Gas container was a plastic tube and gas could be calculated by formula $V = \pi r^2 . t$.

Combustion of CH_4 produces much less CO_2 than combustion of firewood thus it was suggested to use biogas in order to reduce greenhouse gases. Moreover, as biogas is a renewable energy, it serves energy consumption widely. IPCC (*Intergovernmental Panel on Climate Change*) predicted that in 2050 global energy consumption from biomass become 180 EJ (Eksa Joule) or 30% from total energy consumption and in 2100 become 325 EJ (Eksa Joule) or 45% from total energy consumption. It means that energy from biomass were significantly supported energy consumption. Li Fei et.al [1] in China found that energy from manure alone was equivalent to 20% of natural gas used. As mentioned before livestock manure were left on the environment or discharged into Lake Toba. Manure trigger eutrophication and as consequences water hyacinth spread over the lake especially on the edge of the lake. This research proved that a combination of 1.875 kg of water hyacinth together with 0.625 kg of pig manure diluted with water 1:2 produced the highest biogas as much as 214 litres CH_4. From Statistic of Samosir in 2015 it was mentioned that there were 42,352 pigs. As 1 pig produced as much as 2.5 kg manure per day and 42,352 pigs produced 105,880 kg manure so totally 7,623,360 CH_4 were captured instead they emit into the atmosphere. This was some kind of contribution on Indonesia willingness to mitigate GHG. Indonesia committed to reduce its greenhouse gas (GHG) by 26% in 2020. At the UNFCCC (Conference of the United Nation Framework Convention on Climate Change) held in Paris in December 2015 Indonesia committed to reduce GHG; one way by promoting clean energy use for example biogas. If all pig manure combined with water hyacinth were processed in biogas, a totally 36,253,312 litres CH_4 will be produced which were more than enough to cook pig feed all over Samosir Island without the need to use firewood anymore. Even more, biogas could be used as source of energy in rural area, for example for cooking and pumping water from Lake Toba in dry season for watering horticulture. Instead of produced renewable energy, biogas technology also produced slurry. Slurry was an organic fertilizer and associated with type of soil in Samosir Island which was originally volcanic soil and thin of organic matter, the best treatment was by applying good organic matter such as slurry.

3.6. Social benefits

Either livestock manure or agro by-product which are utilized for example through biogas technology is ecologically promoting as cleanliness of the environment is one of biogas technology production. Samosir Island was planned by Indonesian government as one of five outstanding tourist destination in Indonesia. In 2019, it was planned that there will be about 1 million tourists will come to Samosir Island and 250,000 will be overseas tourist. To be a tourist destination, cleanness of the environment is a prerequisite. In Samosir, the highest pig population were in Simanindo sub-district. Simanindo was located next to the lake. So biogas was highly recommended to be implemented in Simanindo in order to control pig manure discharging to the lake and eutrophication. Abdeshahian P [11] mentioned in Malaysia, biogas technology give benefits such as tremendous renewable energy, fertilizer and could enhancing public health whereas anaerobic digestion reduces the impacts of waste on the environment. Associates with public health, biogas in Samosir could minimize skin irritation cases.

3.7. Economic benefits

Biogas cause economic benefits because people do not need to buy fire wood. In this research, it was found that biogas could cover energy for cooking pig feed in Samosir which were 2,117,600 litres CH_4 for 42,352 pigs. There were still 34,135,712 litres CH_4 which were enough for 27,308 household for cooking as for 1 household 1250 litres CH_4 were enough for cooking per day. Usually 1 household need 1 litres kerosene for cooking for 1 day which was cost Rp. 7500,- It could be concluded that there were Rp. 204,810,000 saving if people used biogas instead of kerosene. Further more, people also got benefit from slurry. According to Ginting and Mustamu [10], application of biogas slurry as 250 ml could be an alternative of 2.5 g NPK as early fertilizer on the growth of Spinach Plant (*Amaranthus tricolor*). For every household, generally it need 1250 litres of biogas. This was come from the used 5x 1.875 kg of water hyacinth together with 0.625 kg of pig manure diluted with water 1:2 so there were 30 litres slurry or Rp. 4,000,- everyday as every kg NPK was cost Rp. 12.000,- Every household get benefits as much as Rp. 11,500,- everyday if they used biogas which came from saving from not buying firewood or kerosene and saving from fertilizer.

4. Conclusion

Water hyacinth was combined with pig manure in Samosir Island produced biogas more than enough on supporting sustainable pig feed preparation. People did not have to use fire wood anymore, instead they still had renewable energy for other purposes. In addition, people had slurry which was a good organic fertilizer.

Acknowledgements

The author gratefully acknowledges that financial support from Universitas Sumatera Utara.

IOP Conf. Series: Materials Science and Engineering **309** (2018) 012053 doi:10.1088/1757-899X/309/1/012053

References

[1] Fei L, Cheng S, Yu H and Yang D 2016 Waste from livestock and poultry breeding and its potential assessment of biogas energy in rural china *J. of Cleaner Production* **126** 451–60.
[2] Ginting N, Rahmayadi E, Sirait N N, Tobing C and Purba E 2015 *Proc. Toba Go Green* Indonesia: Universitas Sumatera Utara Kodam I/Bukit Barisan
[3] Ginting N 2017 Biogas technology on supporting "sustainable" coffee farmers in North Sumatera Province Indonesia *IOP Conf. Ser.: Mat. Sci. Eng.*
[4] Patil J H, Antonyraj M A L, Shankar B B, Shetty M K and Kumar B P P 2014 Anaerobic co-digestion of water hyacinth and sheep waste *Energy Procedia* **52** 572-8
[5] Ware A and Niamh P 2016 Biogas from cattle slaughterhouse waste: Energy recovery towards energy self-sufficient industry in Ireland *Renewable Energy* **97** 541-9
[6] Ginting N 2017 *IOP Conf. Ser.: Earth and Environmental Sci.* **65** 012007
[7] Ginting N and Rauf A 2017 Benefits of using biogas technology in agricultural area in supporting organic fertilizer for local farmers in Karo District, North Sumatra Province, Indonesia. *Int. J. Env. Eng.* **4(1)**
[8] Cao Y, Wang J, Wu H, Yan S, Guo D, Wang G and Ma Y 2016 Soil chemical and microbial responses to biogas slurry amendment and its effect on fusarium wilt suppression *Appl. Soil Ecol.* **107** 116-23
[9] Cantrell, Keri B, Thomas D, Kyoung S R and Patrick G H 2008 Livestock waste-to-bioenergy generation opportunities *Bioresource Technology* **99** 7941-53
[10] Ginting N and Mustamu N 2012 The application of biogas sludge as organic fertilizer on the growth of spinach plant (*Amaranthus tricolor*) *Proc. Annual Int. Conf. Syiah Kuala University in 2012 and The 8th IMT-GT Uninet Biosciences Conf.* 154-6
[11] Abdeshahian P, Jeng S L, Wai S H, Haslenda H and Chew T L 2016 Potential of biogas production from farm animal waste in Malaysia *Renewable and Sustainable Energy Reviews* 714-23

IOP Conf. Series: Materials Science and Engineering **309** (2018) 012054 doi:10.1088/1757-899X/309/1/012054

Implementation of bayesian model averaging on the weather data forecasting applications utilizing open weather map

R F Rahmat*, F R Nasution, Seniman, M F Syahputra and O S Sitompul
Department of Information Technology, Faculty of Computer Science and Information Technology, Universitas Sumatera Utara, Medan, Indonesia

*E-mail: romi.fadillah@usu.ac.id

Abstract. Weather is condition of air in a certain region at a relatively short period of time, measured with various parameters such as; temperature, air preasure, wind velocity, humidity and another phenomenons in the atmosphere. In fact, extreme weather due to global warming would lead to drought, flood, hurricane and other forms of weather occasion, which directly affects social andeconomic activities. Hence, a forecasting technique is to predict weather with distinctive output, particullary mapping process based on GIS with information about current weather status in certain cordinates of each region with capability to forecast for seven days afterward. Data used in this research are retrieved in real time from the server openweathermap and BMKG. In order to obtain a low error rate and high accuracy of forecasting, the authors use Bayesian Model Averaging (BMA) method. The result shows that the BMA method has good accuracy. Forecasting error value is calculated by mean square error shows (MSE). The error value emerges at minumum temperature rated at 0.28 and maximum temperature rated at 0.15. Meanwhile, the error value of minimum humidity rates at 0.38 and the error value of maximum humidity rates at 0.04. Afterall, the forecasting error rate of wind speed is at 0.076. The lower the forecasting error rate, the more optimized the accuracy is.

1. Introduction

Global warming is the uprising of greenhouse gas in the atmosphere factored by human activities leading mainly to weather inconsistency. It also allows extreme weather that will bring more severe storm, drought, floods, hurricanes and more weather phenomenon which not only impact the social and economic life directly, but also bring difficulties for mankind on agriculture, water supply and forestry [1]. The weather is a condition of air in a relatively short time at a specific area [9], which is depicted by the rates of various parameters such as; temperature, air pressure, wind speed, humidity, and other atmospheric phenomena. The selected method to properly determine the weather conditions is with the current activities frequently performed by atmospheric or weather researchers [2]. The current activities frequently performed by previous researchers are critical since there are greater demands from various stakeholders for a more rapid, accurate, and detailed information about the atmospheric conditions. Even some other parties demand the availability of forecasting over the atmospheric conditions with a fairly short time span within daily, hourly, or even in minutes. Given this need, researchers are encouraged to continuously conduct atmospheric studies related to atmospheric weather and develop even more sophisticated methods to determine the weather conditions for a better and more accurate results [3].

There are several methods to do forecasting, for example Evolving Connectionist System [7] or Exponential Smoothing Methods [8], and Bayesian Model Averaging to predict cases of pneumonia in infants [4]. Since the data pre-processing in this study is a long process, we prefer to use BMA as our proposed method. In this study, Bayesian Model Averaging(BMA) method is performed to forecast the weather data from three to seven days onwards. BMA is a method that is capable to predict the best model based on the weighted average of all models. BMA evenly averages the posterior distribution of all the possibly formed models. The purpose of the BMA is to combine uncertain models in order to get the best model. The estimation result includes all models that may have formed so they can get a better estimation [5].

Impacts of global warming on the climate and weather conditions that occur at these days are unstable and fluctuate. This causes many environmental problems and social issues which directly impact livings. Therefore, forecasting techniques to predict weather with diverging outputs are necessary; such method which output is mapped into each coordinate of a region presenting detailed information on the status of the weather. Weather forecast within three to seven days ahead in a particular area would also be provided, so that the environmental and social problems could be anticipated.

Previous research forecasted the monthly rainfall in Medan, Northern Sumatera, using box-Jenkins method. This study was conducted to analyse, implement and apply the Box-Jenkins method in order to predict the rate of monthly rainfall in Medan. The data used is the historical data of relative humidity and rainfall in Medan from 2005 to 2009. The forecasting model used for this data is according to the ARIMA model identification by using the order (111; 111). In the evaluation process the estimated value of the constant is tested against the value zero by using statistical t-test. [6].

Following research about the weather forecast was based on Time Series Data using Adaptive Neuro Fuzzy Inference System (ANFIS). The data used in this research sourced from secondary observational data of Badan Meteorologi, Klimatologi dan Geofisika (BMKG) from Karangploso sub district, Malang regency, from January 2011 to May 2012 and was obtained with the accuracy ranging from 48.39% to 76.67 % [3]. Bayesian Model Averaging method has also been employed for other researches. The method is used to predict cases of pneumonia in infants. This method is used to predict the cases of pneumonia in children under the age of five in Situbondo [4].

2. Methodology

Weather forecasting system using Bayesian Model Averaging is a system that serves to generate forecasting from three to seven days. The employed data are weather data obtained from combination of BMKG OpenWeatherMap API and API. The collected data are passed as preprocessing data. Preprocessing consists of three stages; namely data cleaning, data integration, and data selection. Then subsequently these data are next entered into the forecasting process consisting of several stages of training data normalization, determining activation function process, calculating on each weight differences and calculating on net weight and output. The testing process is then performed. Forecasting results on testing process are analyzed using the forecasting error rate Mean Square Error (MSE) and the analysis of forecasting results. Furthermore, the forecasting results are sent to the web server.

2.1. Data acquisitions

In early stage prior to data processing, collection of data from the openweathermap server and BMKG are implemented. In data collection process, historical data and classification data are required.

2.1.1. Historical data. Historical data acquired is the weather data at some geographical points of the city's districts in North Sumatra. The collection comes from raw data list retrieved from the site in openweathermap and data.bmkg.go.id. The next data sources are sorted and stored into a database application. List of historical raw data used in this study is shown in figure 1 below.

"{\"coord\":{\"lon\":98.49,\"lat\":3.6},\"weather\":[{\"id\":500,\"main\":\"Rain\",\"description\":\"light rain\",\"ico
{\"temp\":295.53,\"pressure\":1003.45,\"humidity\":97,\"temp_min\":295.53,\"temp_max\":295.53,\"sea_level\":1
{\"3h\":2.505},\"clouds\":{\"all\":88},\"dt\":1475621002,\"sys\":{\"message\":0.5095,\"country\":\"ID\",\"sunrise\

Figure 1. List of raw data from Openweathermap

Data sources from bmkg.go.id are updated daily that triggered by values of weather parameters which is later processed in this research. The sample of the data are shown in figure 3 below. Then the web server sends data to the web client forecasting view. Eventually, the output is informed in PDF-based reports and graphs. The general architecture of this research can be seen in figure 2.

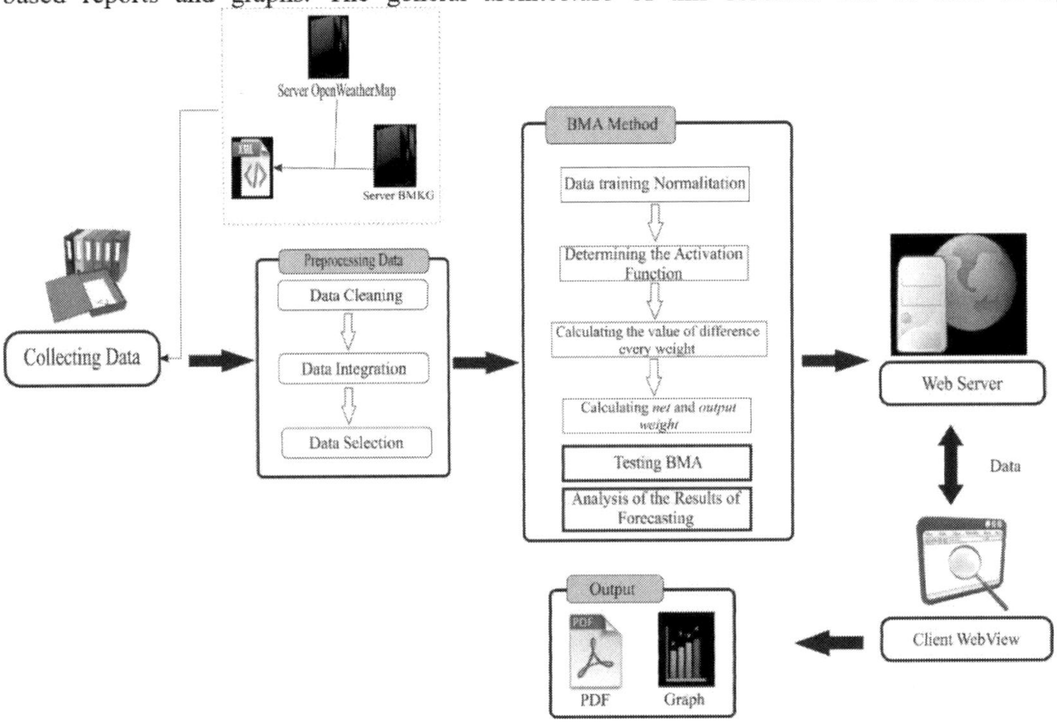

Figure 2. General architecture

data.bmkg.go.id - /

11/14/2014	6:55 AM	1354	alamatbalai.xml
4/1/2011	3:31 PM	1696	alamatbalai2.xml
11/14/2014	6:55 AM	64206	alamatstasiun.xml
11/14/2014	6:56 AM	75882	alamatstasiun_NEW.xml
5/23/2014	11:11 AM	4048	artikel.xml
9/24/2016	6:35 PM	670	autogempa.xml
5/7/2015	11:53 AM	901240	blueMarble_20130120_0747.png
9/24/2016	6:35 PM	752	bmgdiy.xml
9/24/2016	6:35 PM	160	csveqslastest.txt
9/24/2016	6:35 PM	6531	csvlast60event.txt
9/29/2016	9:37 PM	3454	cuaca_dunia_1.xml
9/29/2016	9:37 PM	3469	cuaca_dunia_2.xml
8/18/2015	6:04 AM	1191	cuaca_harian_en.xml
3/22/2015	1:48 PM	1220	cuaca_harian_id.xml
9/29/2016	9:37 PM	6756	cuaca_indo_1.xml
9/29/2016	9:37 PM	6752	cuaca_indo_2.xml
9/29/2016	9:37 PM	1312	cuaca_jabodetabek_1.xml
9/29/2016	9:37 PM	1308	cuaca_jabodetabek_2.xml
5/7/2013	6:01 PM	1456	cuaca_mingguan.xml
2/24/2013	9:01 AM	4781	cuaca_sulselbar.xml
8/23/2015	7:04 AM	3209	cuaca_tiga_harian.xml
4/1/2013	7:57 PM	941	daerah_gelombang_tinggi.xml

Figure 3. List of raw data from bmkg.go.id

2.1.2. Data classification. In this stage, the data is identified. Such identification can be accomplished for a group or several groups of data so that later becomes characteristic of the data concerned. In this study, there are several categories or types of data from data sources, such as satellite imagery, world-weather data, list of locations or areas with high-potential waves, and so forth. For this study, the data used is particularly catagorized on weather in North Sumatra.

2.2. Pre-processing
On this stage, the process has been focusing to pre-processing steps in the application. The collected data are processed to enhance information from these data using Bayesian Model Averaging. The techniques are staged as follows:

2.2.1. Data cleaning. This stage is to eliminate data noise or unintelligible data which lead to inconsistency on the ultimate goal of forecasting. Data from the data source is composed using numerous attributes. Some unnecessary attributes in the forecasting process are removed. Several attributes are taken, such as; weather status, minimum temperature, maximum temperature, id city, latitude and longitude, minimum humidity and maximum humidity, and wind speed. Raw data obtained from the data source is formatted in XML, as shown in figure 4 below.

2.2.2. Data integration. This stage is to merge the collected data from the openweathermap server and BMKG to avoid dataduplication.

2.2.3. Data selection. This stage is to select relevant data, such as; air temperature, humidity, wind speed, wind direction, and weather status which are essential for weather forecasting necessity. At this stage, the data collected from data source will be reformatted with defined attributes and converted into XML format as shown in figure 4 below.

2.3. Bayesian model averaging forecasting
Once the data has undergone a series of pre-processing, the data is then stored as initial data to be used in the forecasting process. There is a chronology of the forecasting process using Bayesian Model Averaging (BMA).Dataset used as training data on this flow is the weather data obtained from BMKG and OpenWeatherMap consistingly; minimum temperature, maximum temperature, minumum humidity, maximum humidity, wind speed and weather status. The weather data is normalized to facilitate the calculation process. In the process, max_forecast_day variable, also specified as the maximum value, is set to determine the results of forecasting. In the calculation of forecasting the weather data, the value of the max_forecast_day variable is one week since the results of forecasting the weather data are short-term.

2.3.1. Data normalization. Normalization allows to obtain values with a certain range tailored to the output of the activation function. The aim of data normalization is to reduce data complexity, eliminate duplicated data, and ease in modifying data. In this study, normalization is set with value ranging 0 to 1. To obtain the value of the normalization used the following equation:

Where: X_n = Value of data normalization

X_0 = The Actual Data

X_{min}= The minimum value of the overall actual data

X_{max}= The maximum value of the overall actual data

```
▼<markers>
  <marker kota="Pandan" lat="1.86878" lng="98.60940" cuaca="Hujan Ringan" suhuMin="23" suhu
  tanggalMulai="29 September 2016" tanggalSampai="30 September 2016" waktuMulai="07.00" wak
  <marker kota="Tarutung" lat="2.01956" lng="99.07629" cuaca="Hujan Ringan" suhuMin="18" su
  tanggalMulai="29 September 2016" tanggalSampai="30 September 2016" waktuMulai="07.00" wak
  <marker kota="Sipirok" lat="1.43298" lng="99.55917" cuaca="Hujan Ringan" suhuMin="23" suh
  tanggalMulai="29 September 2016" tanggalSampai="30 September 2016" waktuMulai="07.00" wak
  <marker kota="Gunung Sitoli" lat="1.17692" lng="97.49009" cuaca="Hujan Sedang" suhuMin="2
  kecepatanAngin="20" tanggalMulai="29 September 2016" tanggalSampai="30 September 2016" wa
  <marker kota="Stabat" lat="3.76600" lng="98.23067" cuaca="Hujan Sedang" suhuMin="23" suhu
  tanggalMulai="29 September 2016" tanggalSampai="30 September 2016" waktuMulai="07.00" wak
  <marker kota="Kabanjahe" lat="3.09635" lng="98.25144" cuaca="Hujan Ringan" suhuMin="17" s
  tanggalMulai="29 September 2016" tanggalSampai="30 September 2016" waktuMulai="07.00" wak
  <marker kota="Lubuk Pakam" lat="3.48347" lng="98.70535" cuaca="Hujan Sedang" suhuMin="23"
  tanggalMulai="29 September 2016" tanggalSampai="30 September 2016" waktuMulai="07.00" wak
  <marker kota="Pematang Raya" lat="2.95304" lng="98.78540" cuaca="Hujan Sedang" suhuMin="1
  kecepatanAngin="20" tanggalMulai="29 September 2016" tanggalSampai="30 September 2016" wa
  <marker kota="Kisaran" lat="2.98305" lng="99.53104" cuaca="Hujan Sedang" suhuMin="23" suh
  tanggalMulai="29 September 2016" tanggalSampai="30 September 2016" waktuMulai="07.00" wak
  <marker kota="Rantau Prapat" lat="2.14783" lng="99.92183" cuaca="Hujan Sedang" suhuMin="2
  kecepatanAngin="20" tanggalMulai="29 September 2016" tanggalSampai="30 September 2016" wa
  <marker kota="Sidikalang" lat="2.82712" lng="98.27832" cuaca="Hujan Ringan" suhuMin="18"
  tanggalMulai="29 September 2016" tanggalSampai="30 September 2016" waktuMulai="07.00" wak
  <marker kota="Balige" lat="2.38032" lng="99.24647" cuaca="Hujan Ringan" suhuMin="19" suhu
  tanggalMulai="29 September 2016" tanggalSampai="30 September 2016" waktuMulai="07.00" wak
  <marker kota="Panyabungan" lat="0.71628" lng="99.41830" cuaca="Hujan Ringan" suhuMin="23"
  tanggalMulai="29 September 2016" tanggalSampai="30 September 2016" waktuMulai="07.00" wak
  <marker kota="Teluk Dalam" lat="0.78668" lng="97.71663" cuaca="Hujan Ringan" suhuMin="23"
  tanggalMulai="29 September 2016" tanggalSampai="30 September 2016" waktuMulai="07.00" wak
```


Figure 4. Raw data after data cleaning process

2.3.2. Determination of activation function. The type of activation function for forecasting process is bipolar threshold activation function. The modification of the threshold function has three output values, rangely -1, 0 or 1. Calculating the value of each node activation function is by setting with random value that outputs the value of -1,0 or 1.

2.3.3. Determination of probability values, weights, and difference. The next process is probability calculation of each node stored in an array or a list. In addition, the outcome error value by each node and t the averaging data value in entire nodes are also determined using the MSE. Both of these values are to update weights in order to obtain forecasting value for the next few days in a row. Weight results saved into a list are able to be stored into a database and to generate forecasting process report.

In determining the difference of each node, actual data value of the previous day istrimmed withcurrent data value. This determination occurs in each category list of parameters being calculated.

The process of calculating the difference is then accumulated and is then divided by the number of nodes initiated (in this study, 26 nodes are intiated with 26 locations,also known as id based on the city, in the province of North Sumatra). Thus, averaging the value is essential to be an addition to or deduction factor with the intial weight value. Each valuedetermines whether the final weight will be added or subtracted depending on the appearing of activation value f(a). If the activation value f(a) is greater than 0 ($f(a) > 0$), the value will be added by a factor enhancer.Otherwise, if the activation value f(a) is smaller than 0, it will be reduced by adder value. So the final weights are obtained with the following formulas.

2.4. Web service, web server, and client

The Results of statistical information and charts formerly drawn from the forecast are then compressed into JSON (Java Script Object Notation) and XML files. JSON and XML files are proceedingly sent to web service. Next, the web Service receives JSON and XML files from the Open Weather Map and sends it to a web server.

Web server receives JSON and XML files which are to be delivered by the Web Service. The data are then processed and made into a web-based application Dashboard which is accessible for clients online Graphs and charts are displayed according to client request. Clients select the forecast from a dropdown menu that contains the forecast of previous or next few days. After the client chooses request, the Web Service sends the requested data to the web server and displays it. Graphs and charts are steadily updated as the client chooses new option.

Clients are allowed to access web pages on a web server dashboard. This page specifically contains graphs and charts of forecasting results in real time. When the clients select their desired forecasting data, the graphs and charts will be updated and the data obtained is available to be reported as a printable PDF file format.

3. Result and discussion

In the testing process, data used for testing in system is the weather data retrieved from the OpenWeatherMap server and BMKG which has been stored in the database earlier. Forecasting the weather data begins by selecting forecasting the weather menu on the main page. Then on the same page, system requires users to choose amount of forwarding days that forecasting will be performed. Results of forecasting the weather data are shown in table 1..

Table 1. Weather forecasting result.

Day	City ID	Min. Temperature	Max. Temperature	Min. Humidity	Max. Humidity	Wind Speed
Day 1	24	23.02	32.15	50.38	92.04	20.35
Day 1	25	23	32.3	50.76	92.08	20
Day 1	26	23.02	32.45	51.14	92.04	19.65
Day 1	27	23.04	32.6	50.76	92.08	20
Day 1	28	23.06	32.45	50.38	92.04	19.65
Day 1	29	23.04	32.3	50.76	92.08	20
Day 1	30	23.02	32.45	50.38	92.12	19.65
Day 1	31	23	32.3	50.76	92.16	19.3
Day 1	32	23.02	32.45	50.38	92.12	19.65
Day 1	33	23.04	32.6	50	92.16	19.3
.....	......	....	....	.....	....	......
Day 3	24	22.86	33.35	48.86	92.12	21.75
Day 3	25	22.88	33.5	49.24	92.16	22.1
Day 3	26	22.86	33.65	48.86	92.2	22.45
Day 3	27	22.84	33.8	49.24	92.16	22.8
Day 3	28	22.86	33.65	49.62	92.12	22.45
Day 3	29	22.84	33.8	50	92.16	22.8
Day 3	30	22.86	33.65	49.62	92.2	22.45
Day 3	31	22.84	33.8	50	92.16	22.1

In table 1, It is shown that the BMA method performs forecasting the weather data which has been recorded into database for three days. Forecasting error is calculated by MSE divergingly at 0.2% of minimum temperature parameter, 0.15% of maximum temperature parameter, 0.38% of minimum humidity parameter, 0.04% of maximum humidity parameter, 0.076% of wind speed parameter. MSE is another method for measuring forecasting error. In this method, each error or residual is squared, then summed or divided by the number of observations. Thus, MSE calculations also show averaged absolute forecast error against its actual data. Error calculation results using the MSE is are provided in table 2.

Table 2. Error results using MSE.

Node	Min Temp	Max Temp	Min Humidity	Max Humidity	Wind Speed
1	0	0	0	0	-1
2	-5	-2	0	3	-2
3	5	0	15	-5	-2
4	0	1	-10	2	4
5	0	2	-5	0	0
6	-6	-4	20	4	0
7	6	4	-20	-6	0
8	-4	-2	18	4	-5
9	4	2	-13	-4	5
10	0	1	0	2	0
11	-5	-4	15	4	0
12	1	0	-5	-1	-6
13	4	2	-10	-2	6
14	0	0	0	-2	-4
15	-4	-2	10	4	4
16	0	0	5	0	0
17	-1	0	0	0	0
18	5	3	-15	-4	1
19	0	1	0	0	1
20	0	0	0	1	-1
21	-3	-3	5	1	-2
22	3	1	-5	-1	3
23	0	2	-5	-2	-2
24	0	-1	0	3	2
25	0	0	5	-2	0
26	0	-1	5	2	1
MSE *Error*	**0.28**	**0.15**	**0.38**	**0.04**	**0.076**

In the real time-forecasting chart, parameters of minimum and maximum temperature are shown in figure 5 and figure 6.In actual-forecasting chart of minimum and maximum, humidity parameters can be seen in figure 7 and figure 8. Graphic for actual- forecasting of wind speed is shown in figure 9

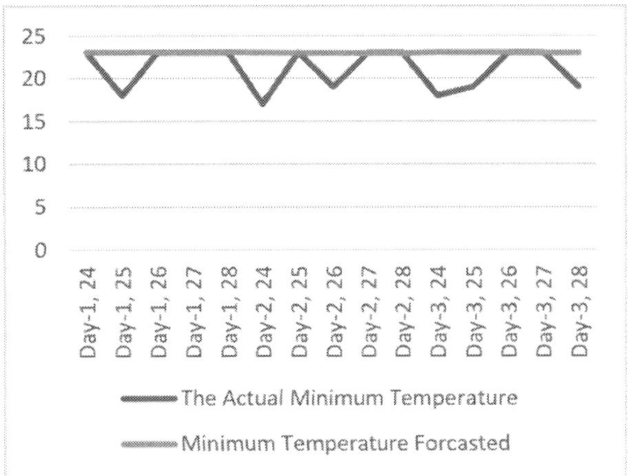

Figure 5. Real time-forecasting chart for minimum temperature

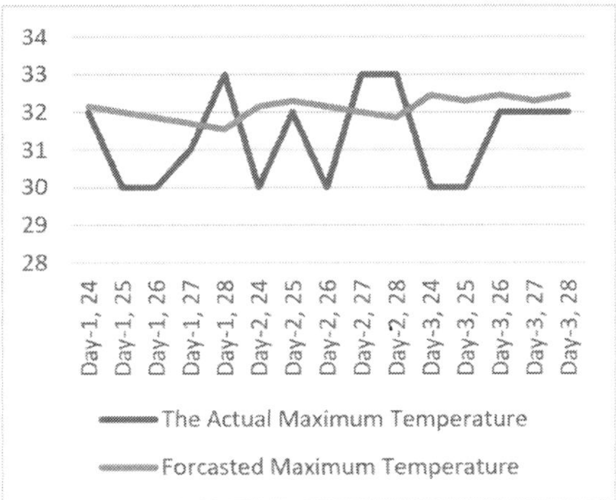

Figure 6. Real time-forecasting chart for maximum temperature

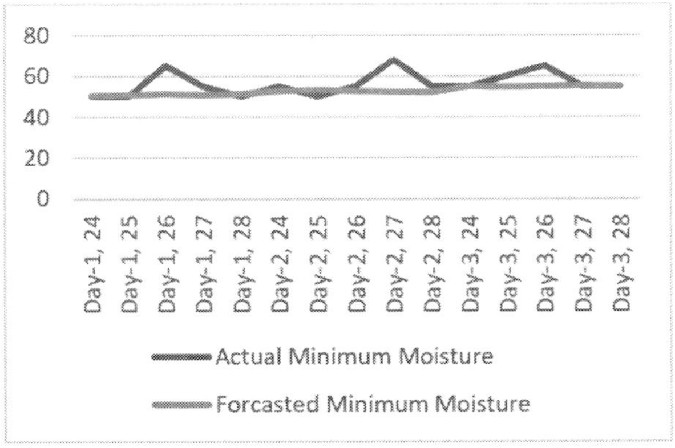

Figure 7. Real time-forecasting chart for minimum humidity

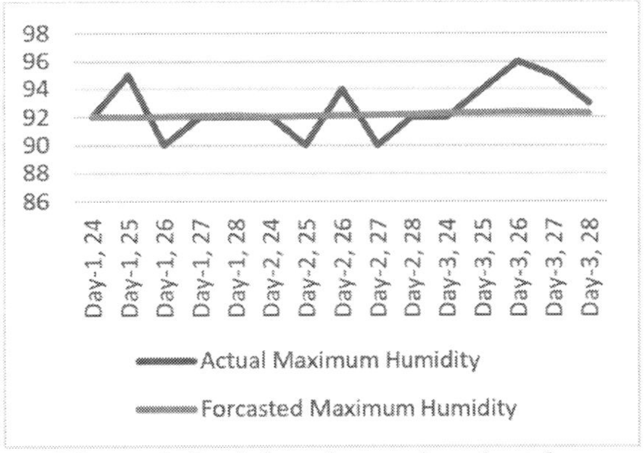

Figure 8. Real time-forecasting chart for maximum humidity

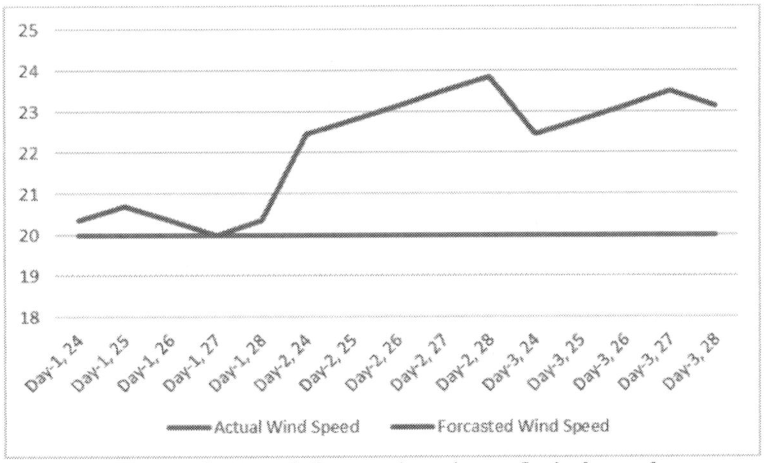

Figure 9. Actual-forecasting chart of wind speed

4. Conclusion

Based on weather-data forecasting system tested using the BMA, forecasting error values generated by MSE are 0.28 for minimum temperature, maximum temperature 0:15 to 0:38, and 0:04 for minimum humidity in maximum moisture and 0076 for wind speed. The lower the value of MSE shows, the more optimal the accuracy in forecasting is. Forecasting process with BMA has met the objective function of the problem which is to help the process of forecasting in period of 3 to 7 days.

In the future study, we would like to do forecasting using our classification engine such as Distributed Adaptive Neural Network [10-12] and improved by using fast learning algorithm [13].

5. References

[1] Cahyono W E 2010 Pengaruh pemanasan global terhadap lingkungan bumi *Jurnal Sains Dirgantara* **8**(2) 28-31

[2] Kresnawan A 2008 Penerapan model jaringan saraf tiruan untuk memprediksi gangguan cuaca maritim *Tugas Akhir Jurusan Teknik Fisika* – FTI - ITS. Surabaya

[3] Dewi C, Kartikasari D P and Mursityo Y T 2014 Prediksi cuaca pada data time series menggunakan adaptive neuro fuzzy inference system (ANFIS) *Jurnal Teknologi Informasi dan Ilmu Komputer* **1**(1): 18-24

[4] Sofia D, Kuntoro and Soenarnatalina 2014 Pemodelan bayesian model averaging (BMA) pada kasus pneumonia pada belita *Jurnal biometrika dan kependudukan* **3**(1): 34-42

[5] Montgomery J and Nyhan B 2010 Bayesian model averaging: theoretical developments and practical applications *Political Analysis* **18**(2): 245-270

[6] Sembiring J H 2010 Peramalan curah hujan bulanan di kota medan dengan metode box-jenkins. Skripsi. Universitas Sumatera Utara. R. Nicole, "Title of paper with only first word capitalized," J. Name Stand. Abbrev., in press

[7] Rahmat R F, Rizki A, Alharthi A F and Budiarto R 2016 *4th Saudi Int. Conf. on Information Technology* (Big Data Analysis) 1–5

[8] Siregar B, Butar-Butar I A, Rahmat R F, Andayani U and Fahmi F 2017 Comparison of exponential smoothing methods in forecasting palm oil real production *J. Phys.: Conf. ser.* **801**(1) 12004

[9] Rahmat R F, Syahputra M F and Lydia M S 2016 *International Conf. on Informatics and Computing* 383–38

[10] Rahmat R, Pasha M, Syukur M and Budiarto R 2015 A gene-regulated nested neural network *Int. Arab J. of Information Technology* **12**(6) 532–539

[11] Aaron O S, Sitompul and Rahmat R F 2014 *Int. Conf. on Cyber and IT Service Management* 63–68

[12] Pasha M F, Rahmat R F, Budiarto R and Syukur M 2010 A distributed autonomous neuro-gen learning engine and its application to the lattice analysis of cubic structure identification problem *Int. J. of Innovative Computing, Information and Control* **6**(3) 1005–1022

[13] Alwajeeh T, Alharthi A F, Rahmat R F and Budiarto R 2017 *IOP Conf. Ser.: Mater. Sci. Eng.* **190** 12015

IOP Conf. Series: Materials Science and Engineering **309** (2018) 012055 doi:10.1088/1757-899X/309/1/012055

Design and process integration of organic Rankine cycle utilizing biomass for power generation

S Ependi[1], T B Nur [1,2*]

[1]Department of Mechanical Engineering, Faculty of Engineering, Universitas Sumatera Utara, Padang Bulan, Medan 20155, Indonesia.
[2]Sustainable Energy and Biomaterial Center of Excellence, Universitas Sumatera Utara, Padang Bulan, Medan 20155, Indonesia.

*E-mail: taufiq.bin_nur@usu.ac.id

Abstract. Indonesia has high potential biomass energy sources from palm oil mill industry activities. The growing interest on Organic Rankine Cycle (ORC) application to produce electricity by utilizing biomass energy sources are increasingly due to its successfully used for generating electricity from rejected waste heat to the environment in industrial processes. In this study, the potential of the palm oil empty fruit bunch, and wood chip have been used as fuel for biomass to generate electricity based ORC with combustion processes. The heat from combustion burner was transfer by thermal oil heater to evaporate ORC working fluid in the evaporator unit. The Syltherm-XLT thermal oil was used as the heat carrier from combustion burner, while R245fa was used as the working fluid for ORC unit. Appropriate designs integration from biomass combustion unit to ORC unit have been analyzed and proposed to generate expander shaft-work. Moreover, the effect of recuperator on the total system efficiency has also been investigated. It was observed that the fuel consumption was increased when the ORC unit equipped recuperator operated until certain pressure and decreased when operated at high pressure.

1. Introduction

Renewable energy sources, such as geothermal energy, solar energy, wind energy and biomass energy are the promising energy sources to replace fossil fuel energy. The implementation of renewable energy will not only reduce the usage of fossil energy but also opportunities in reducing greenhouse gases (GHG) emissions. Referring to Indonesia condition as one of the biggest countries as palm biomass producer, research on utilizing of biomass energy to supply power generation is great significance.

One of promising technology to extract biomass energy is by using biomass as fuel for organic Rankine cycle. The ORC uses an organic working fluid which is able to condense at a lower pressure and evaporate at a higher pressure. The working fluid as the organic substance can be better adapted than water for lower heat source temperatures. The heat releases during combustion process can be used to heat up system working fluid prior entering expander to produce electrical energy. Generally, the selection of working fluids to be used for ORC systems are based on their appropriate thermochemical properties, safety, cost, and environmental impact [1-4]. Numerous works in literature studied the performance evaluation of the ORC systems with different configurations. A review on the ORC architectures is given by Lecompte et al. [5].

Palm oil has been one of the success stories of the North Sumatera Province of Indonesian agricultural sector. The palm oil mills (POM) have depended on their own biomass, such as shells, fibers, empty fruit bunches (EFBs) as fuel to generate power. The power plant is usually located in the same area of POM itself. However, the EFBs have not been optimally used due to wet and bulky which are unfavorable during their handling in the power plant. However, the EFBs should be utilized effectively to overcome its disposal problem in the POM. A preliminary study to use the EFBs in the combustion processes have been analyzed [4]. In this study, we extend the analysis to more details integration processes of the ORC plant by utilizing EFBs and wood chip as biomass fuels.

2. System Description

In this section, two different cycle architectures of ORC systems as shown in figure 1 are discussed bearing in mind the performance evaluation criteria. Each system consists of two unit: (i) thermal oil heater (TOH) unit, (ii) organic Rankine cycle (ORC) unit. The processes start from burning biomass fuel in TOH to generate heat which is then transferred to an ORC working fluid by the thermal oil that goes in the evaporator section. The compressed working fluid will evaporate in evaporator prior entering to expander producing work by rotating shaft which then converted to electricity by generator.

Several researchers [4,6-8] suggest a recuperator to reuse the heat after the expander to preheat the working fluid going to the evaporator by utilizing heat energy contained by the working fluid before entering the condenser. A recuperator essentially increases the thermal efficiency thus a high-power output can be maintained for a decreased heat input to the ORC unit [4,5].

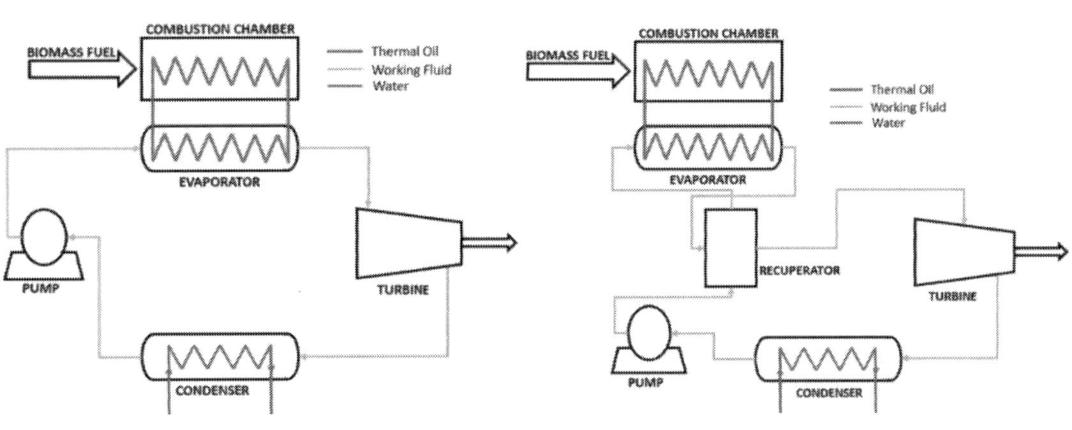

(a) (b)

Figure 1. Schematic diagram of the ORC system, (a) Basic ORC, (b) ORC with recuperator.

3. Methodology

3.1. System Design, Integration, and Efficiency

The Aspen Plus flowsheets of the ORC configurations with biomass combustion system are depicted in figures 2-3. The systems are based on the following main assumptions: steady state operation, pressure drop neglected, working fluid is R-245fa, thermal oil is Syltherm XLT, and the biomass used is EFB and wood chip.

The biomass combustion will be simulated at near atmospheric condition. The Aspen Plus block units used to simulate combustion process are RYield, RGibbs, SSplit, Compr, Heater [4]. The biomass (FEED) was sent to RYield, labeled with DECOMPOSER, to predict the decomposition of the biomass into the reference components such as C, H_2, S, O_2, N_2, etc. DECOMPOSER is a dummy tool and not used in the actual condition. The RGibbs, labelled with COMBUSTOR, used to react biomass with the oxygen, by Gibbs free energy minimization and it does not require the stoichiometry reaction. The unit blocks of Pump, HeatX, Compr, and Valve were used to simulate and analyse the TOH and ORC units within the Aspen Plus. The oxygen was taken from the ambient air which supplied by a blower or compressor.

The biomass combustion produced heat that can be used to heat up the thermal oil (Syltherm XLT) until 188°C. A high temperature thermal oil will enter the evaporator (EVP) and distribute the heat to R245fa rising its temperature and become a vapor. The low temperature thermal oil will be pumped to the burning heat source while the R245fa will enter and rotate the shaft on the expander. The rotation of this shaft will be converted to electrical energy by the generator. Low-pressure and temperature of R245fa will enter the condenser before being pumped back into the evaporator. To exploit the energy potential still contained in R245fa leaving the expander, a recuperator is installed. The use of a recuperator will increase the thermal efficiency of the system thus a high power output can be maintained for a decrease heat input to the system [4-6].

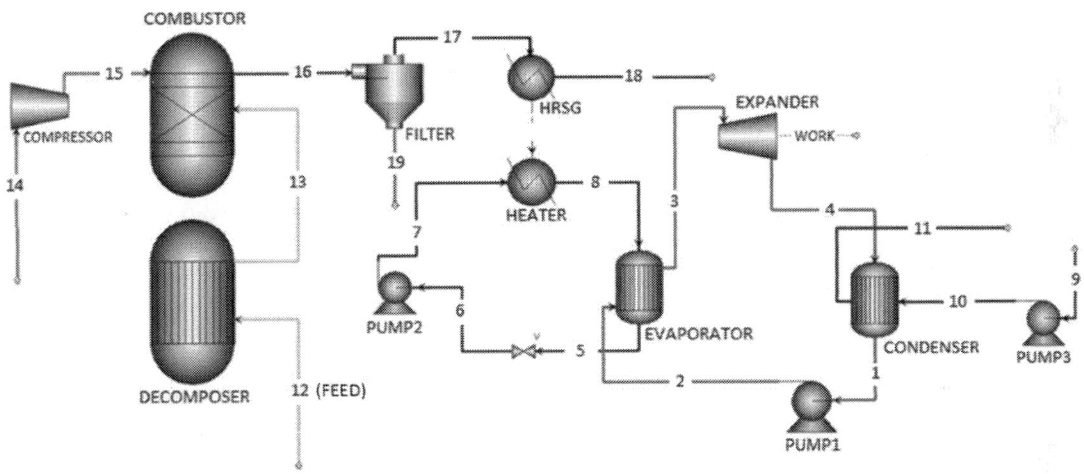

Figure 2. Aspen Plus simulation configuration for ORC plant without recuperator unit.

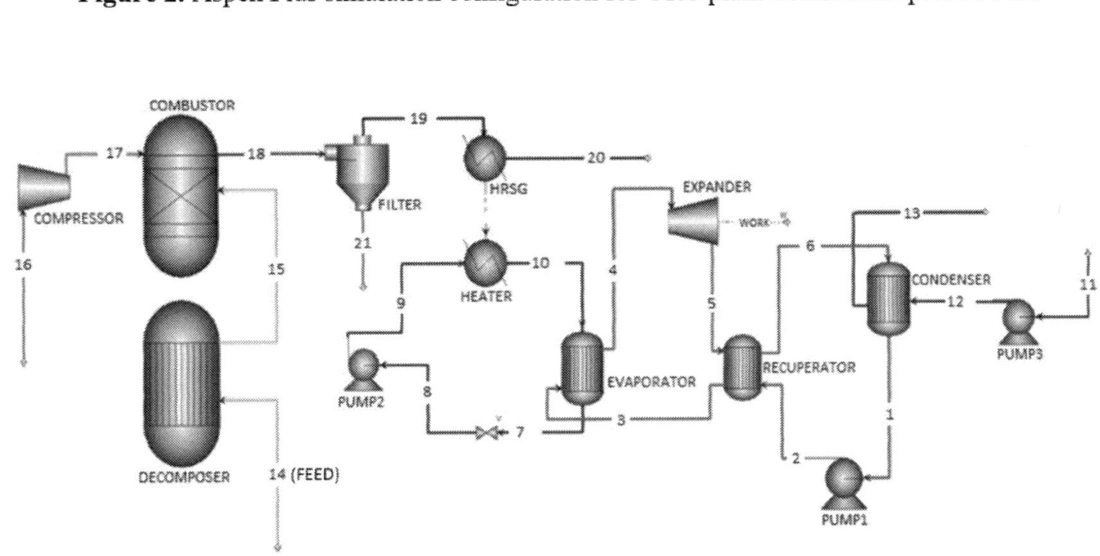

IOP Conf. Series: Materials Science and Engineering **309** (2018) 012055 doi:10.1088/1757-899X/309/1/012055

Figure 3. Aspen Plus simulation configuration for ORC plant with a regenerator unit.

The system efficiency will be analyzed based on results from Aspen Plus. The total efficiency of the system can be calculated by using:

$$\eta_{system} = (W_{expander} - E_{consumed})/Q_{in} \quad (1)$$

where $E_{consumed}$ is total energy consumed during the process.

3.2. *Input data*

EFB and wood chip were selected as the main fuel in this study because of its abundant availability in Indonesia, especially the province of North Sumatera. The feed composition of the EFB and wood chip are specified as described in table 1, while the main assumptions for analysis are shown in table 2.

Table 1. Composition of EFF and wood chip [10,11].

	EFB	Wood Chip
Proximate analysis (wt.%)		
Fixed carbon	18.84	17.94
Volatile matter	80	76.17
Moisture content	20	20
Ash content	5.89	1.16
Ultimate analysis (wt.%)		
Ash content	5.89	1.16
Carbon	48.63	51.19
Hydrogen	5.81	6.08
Sulphur	0.09	0.02
Nitrogen	0.63	0.2
Chlorine	0	0.05
Oxygen	38.95	41.3
Heating values (MJ/kg)		
Lower heating value (LHV)	18.5	19.09

The properties of Syltherm XLT as working fluid for TOH are given by Aspen Plus. Each biomass fuel will be burned with 20% excess air in a furnace operating at near ambient pressure.

Table 2. Main operational conditions and assumptions for plant calculation.

Flue gas temperature (°C)	60
Combustion temperature (°C)	400
Mass flow rate of thermal oil Syltherm XLT (kg/hr)	40.000
Outlet temperature of oil heater from evaporator (°C)	88
Mass flow rate of working fluid R-245fa (kg/hr)	2500
Outlet pressure of expander (bar)	2

4. Results and discussion

4.1. *Design characteristic*

Configuration 2, described in figure 2, was simulated with the inlet pressure of working fluid (R245fa) at the expander varied from 9 to 13 bar. According to the need of air required during combustion process, there are differences in air needs for each fuel. The above conditions are caused by differences in the composition of each fuel, especially carbon and hydrogen. The required air during combustion for EFB and wood chip are 7.28 kg air/kg of fuel and 7.54 kg air/kg of fuel, respectively. This indicates that the air supply in the use of fuel from wood chip leads to the use of more blower energy compared to the case of EFB as fuel.

In the case of EFB as fuel, results shown that 188.39 kW energy was used by the ORC system without recuperator configuration to reheat the working fluid up to 130°C at a pressure of 9 bar. Increasing the pressure of working fluid at the inlet of expander up to 13 bar, decreases the energy required to 184.02 kW. However, by introducing recuperator to the system (as shown in figure 3), the energy required can be reduced to 164.79 kW when operated at 9 bar and 164.77 kW when operated at 13 bar. It can be said that the recuperator provides an advantage in the integration of ORC system thus reducing energy losses during the process. In addition, the Temperature-Entropy diagram (T-S diagram) for both system configurations and selected properties for ORC system with recuperator are shown in figure 4 and table 3, respectively.

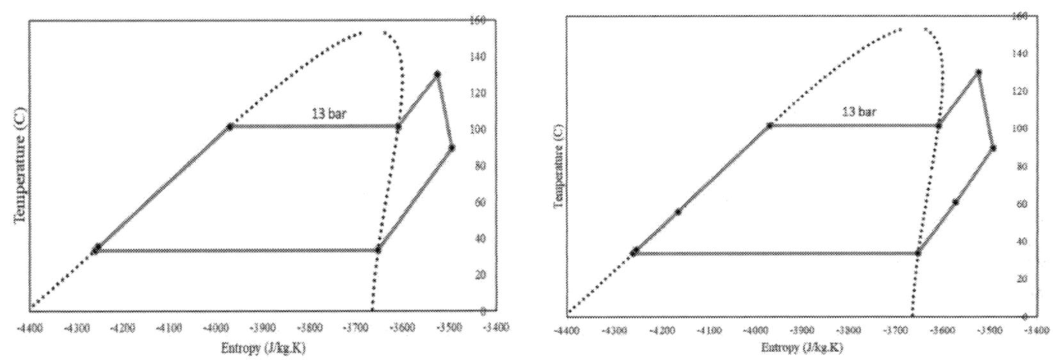

(a) ORC system without recuperation (b) ORC system with recuperation

Figure 4. Temperature-Entropy (T-S) Diagram for both system configurations.

Table 3. Selected properties of the ORC system with recuperator unit.

CODE	FLUID	T (°C)	P (BAR)	ENTHALPY	QUALITY
1	R245fa	33.4	2	-8940.3	0
2	R245fa	35.4	13	-8937.5	0
3	R245fa	55.9	13	-8909.7	0
4	R245fa	130	13	-8672.5	1
5	R245fa	89.5	2	-8700.9	1
6	R245fa	60.9	2	-8728.6	1
10	Syltherm- XLT	188.04	1.5	-101.1	0
7	Syltherm- XLT	88	1.2	-115.9	0
12	Air	27	1.2	-15963	0
13	Air	30.9	1.2	-15945	0

4.2. System Efficiency

The total net power produced by ORC system without recuperator is 11.71 kW, and 12.25 kW for ORC system with recuperator in the case of EFB as fuel. The systems efficiency without recuperator and with recuperator are 6.84% and 7.91%, on the basis of LHV respectively. It can be seen that adding recuperator into system contribute to increasing system efficiency of 1.07 point. As shown in figure 5, it is also found that increasing the inlet expander pressure will increase the output power from the expander, the system efficiency, as well as decreasing fuel consumption rate during processes especially in the case of ORC system without recuperator unit. The recuperator plays an important role in lowering the fuel consumption of the system during the process by utilizing the latent heat of R245fa contained as it exits the expander before entering the condenser. The effects of adding recuperator unit to the system on reducing fuel consumption are shown in figure 6.

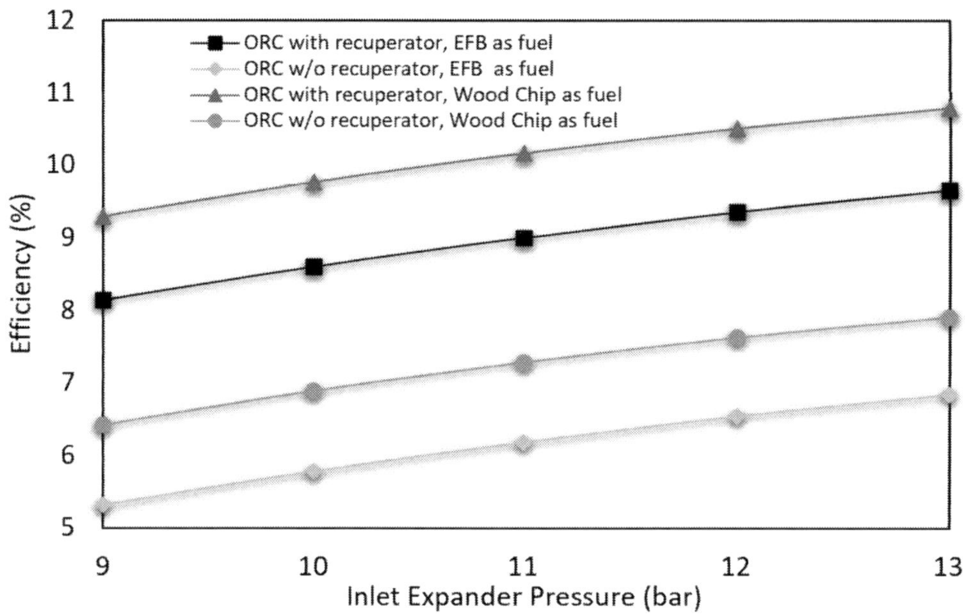

Figure 5. Effect of various inlet expander pressure on system efficiency.

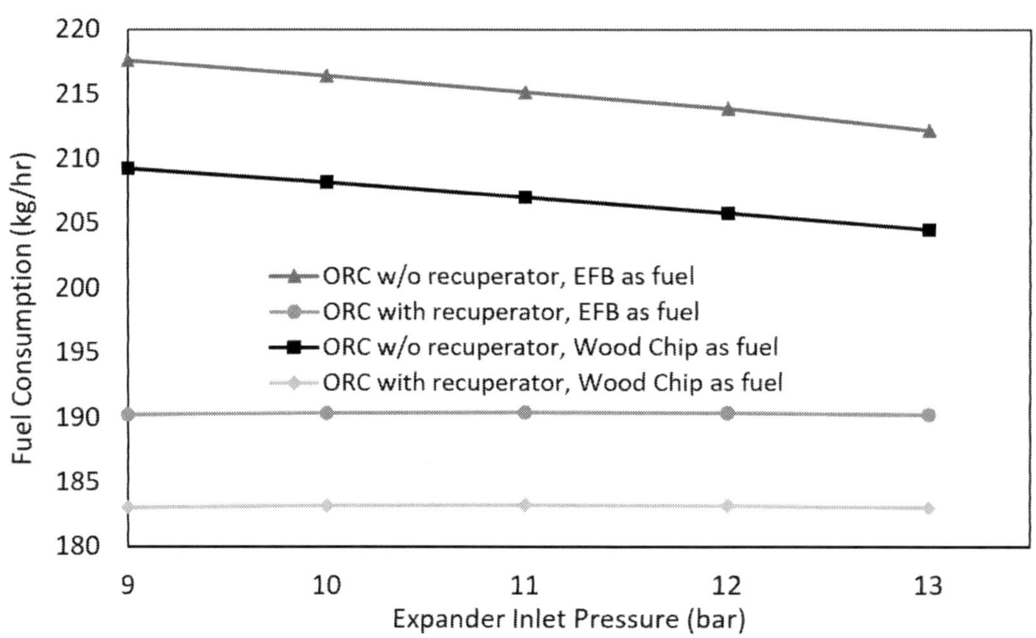

Figure 6. Effect of expander inlet pressure on fuel consumption.

5. Conclusions

The biomass combustion integrated to ORC to generated the electricity was analyzed in this study. Aspen plus was used to calculate the thermodynamic characteristics of the proposed design and integration models. It was found that the system efficiency will increase with increasing the expander inlet pressure. Adding a recuperator unit between expander and condenser will increase the system efficiency, and reducing fuel consumption during the process. However, the using of EFB as biomass fuel is good options, remembering this much available in Indonesia. Utilizing wood chip as biomass fuel for ORC system is better than biomass from EFB due to wood chip has highest LHV. However, due to the greater availability of EFB at the site, the use of biomass from EFB is more preferred.

Acknowledgements
This work has been fully supported by BP-PTN USU (Universitas Sumatera Utara), No: 6049/UN5.1.R/PPM/2016, July 19th, 2016.

References
[1] Li L, Ge Y T, Luo X, Tassou S A 2017 *App. Therm. Eng.* **115** pp 815-824
[2] Li G 2016 *Renew. Sust. Energ. Rev.* **53** pp 477-499
[3] Tanczuk M, Ulbrich R 2013 *Energy* **62** pp 132-141
[4] Nur T B, Pane Z, Amin M N 2017 *IOP Conf. Series: Materials Science and Engineering* **180** pp 1-7
[5] Lecompte S, Huisseune H, van den Broek M, Vanslambrouck B, Paepe M D 2015 *Renew. Sust. Energ. Rev.* **47** pp 448-461
[6] Saleh B, Koglbauer G, Wendland M, Fischer J 2007 *Energy* **32** pp 1210-1221
[7] Chen H, Goswami D Y, Stefanakos E K 2010 *Renew. Sust. Energ. Rev.* **14** pp 3059-3067
[8] Quoilin S, van den Broek M, Declay S, Dewallef P, Lemort V 2013 *Renew. Sust. Energ. Rev.* **22** pp 168-186
[9] Pezzuolo A, Benato A, Stoppato A, Mirandola A 2016 *Energy* **102** pp 605-620
[10] Wijono A 2014 *Simposium Nasional [National Symposium] RAPI XIII-2014 FT UMS ISSN* 1412-9612 pp 111-118
[11] Doherty W, Reynolds A, Kennedy D 2013 *Book Chapter:Material and Processes for Energy: communicating current research and technological development* ed A M Vilas (Formatex Research Centre) pp 212-220

IOP Conf. Series: Materials Science and Engineering **309** (2018) 012056 doi:10.1088/1757-899X/309/1/012056

Effect of water addition in a microwave assisted thermal cracking of biomass tar gasification

A Warsita*[1] and I Surya[2,3]
[1]Department of Mechanical Engineering, Sekolah Tinggi Teknologi Nasional Yogyakarta, Depok, Sleman, Yogyakarta, 55281, Indonesia
[2]Department of Chemical Engineering, Faculty of Engineering, Universitas Sumatera Utara, Medan, Sumatera Utara, Indonesia
[3]Sustainable Energy and Biomaterial Center of Excellence, Faculty of Engineering, Universitas Sumatera Utara, Medan, 20155, Indonesia

*Email: aris-warsita1@yahoo.com

Abstract. Producer gas from biomass gasification is plagued by the presence of tar which causes pipe blockages. Thermal and catalytic treatments in a microwave reactor have been shown to be effective methods for removing tar from producer gas. A question arises as to the possibility of enhancing the removal mechanism by adding water into the reactor. Thermal treatment with a various amount of water was added at temperatures in the range of 800–1200°C. The tar removal efficiency obtained 96.32% at the optimum temperature of 1200°C at the water to tar ratio (W/T) of 0.3. This study shows that the removal of tar by microwave irradiation with water addition is a significant and effective method in tar cracking.

1. Introduction

Biomass gasification is an efficient and environmentally friendly method to produce gaseous fuel (producer gas) for heat and power generation through internal combustion engine, gas turbine, and fuel cell as well as for chemical synthesis applications such as methanol, methane and Fischer–Tropsch liquids. However, producer gas is always accompanied by undesirable products such as tar and particulates. Until now, tar presents the main obstacle in biomass gasification for not only causing serious operational problems in a downstream pipeline and end-user application but also affecting the energy efficiency of the overall process. Therefore, removal of tar from producer gas is indispensable.

Since three decades ago, various producer gas cleaning methods have been developed and reported in some reports with the aim to produce high-quality producer gas for end-user application. In general, these methods are classified into two categories, namely: primary methods that consist of gasifier design and optimization of operating conditions and secondary methods that consist of mechanical and thermocatalytic treatments [1,2]. From the technical point of view, a combination of the two methods guarantees a more satisfactory quality of producer gas [3]. Moreover, appropriate implementation of thermocatalytic treatment is more favorable due to converting tar into useful gases such as hydrogen, carbon monoxide or hydrocarbon gases which can improve the producer gas energy content.

Until now, it is undeniable that thermal treatment process improves the composition of the producer gas, however, little attention has been paid to its development since it requires high additional energy to achieve the desired operating temperatures. Reports show that most of the heavy tar can be cracked at a temperature of 900 °C [4]. Even so, to achieve sufficiently high tar conversion efficiency, temperatures of more than 1100 °C are needed [5]. In another approach, catalytic treatment processes have gained more attention. In particular, there have been ongoing efforts for developing more economical catalysts for tar conversion.For this reason, the utilization of natural catalysts such as dolomite and zeolite or their impregnation on metal catalysts would be a wise choice.

Based on the preceding concerns, implementation of radio frequency (RF) energy for thermocatalytic treatment of tar would be a more realistic option. In this method which commonly utilizes microwave, the transfer of energy into the material occurs instantaneously through molecular interaction with the electromagnetic field [6]. The unique feature of volumetric heating of this technique results the more rapid heating process of the reactor in the presence of susceptor material [1] resulting in significant energy saving, reduce process time, increase process yield and environmental compatibility [7]. For this reason, a simple and rapid test technique has been developed by using RF energy which is not only effective but has also low energy consumption. The performances of this technique have been demonstrated and reported in our previous paper for thermocatalytic treatment of tar using toluene and naphthalene as tar model compounds [1].

This work was conducted to examine the capability of RF thermocatalytic treatment reactor in improving the quality and production of producer gas. The producer gas containing tar and particulates was continuously produced from rubber woodblocks gasification in a suction fixed bed downdraft gasifier. The effects of temperatures and catalysts (calcined dolomite and Y-zeolite) on the yield of products including tar, particles, and gas composition as well as high heating value (HHV) of producer gas were studied.

2. Materials and Methods

2.1. Materials

In this study, pellets wood as biomass material for the production of producer gas in a suction throat-less downdraft fixed bed gasifier. The feed material was prepared into small pieces with a size 5 mm and 3 cm length. Proximate analysis showed that the feed material was comprisedof11.4 wt.% of fixed carbon,78.3 wt.% of volatile matter, 0.2 wt.% of a shand 10.29 wt.% of moisture. Whilst elemental analysis revealed that it was composed of 44.80 wt.% of carbon,12.19 wt.% of hydrogen, 0.45 wt.% of nitrogen, 0.88wt.% of sulfur and 41.68 wt.% of oxygen. The HHV of the feed material determined using an automatic bomb calorimeter was 20.6 MJ kg^{-1}.

2.2. Experimental apparatus

The schematic of overall experimental setup is given in supplementary material (Fig. 1). It consists of three main units: gasifier, producer gas treatment, and gasification product collection systems. The system includes a suction throat-less downdraft fixed bed gasifier, cyclone separator, condenser, blower, and a flare port. The gasifier with an inner diameter of 0.15m and height of 1.05 m has a thermal power output of 10 kW corresponding to 6 kgh^{-1} biomass feeding rate. An orifice meter with a differential pressure transducer was used to measure the pressure drop for flow rate quantification of the producer gas.

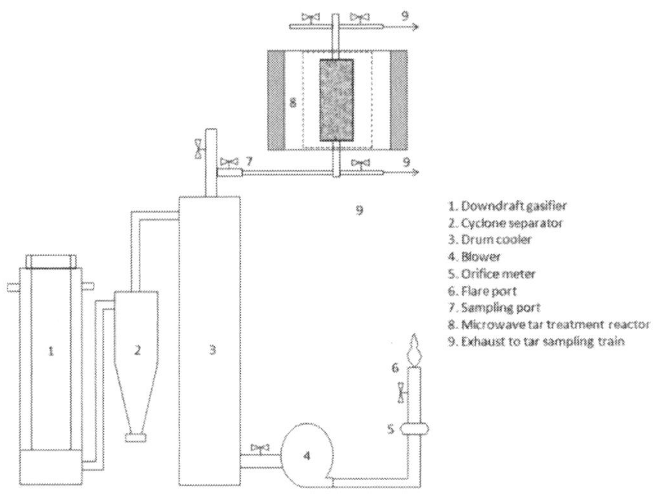

Figure 1. A schematic of the experimental apparatus for tar conversion

Producer gas treatment system includes a modified RF oven and a reactor containing silicon carbide (SiC) as susceptor material. The modified RF oven (Panasonic, NN-SM330 M) has a frequency of 2.45 GHz corresponding to wave-length of 12.23cm. The maximum power consumption and maximum output power of the RF oven were 1125 and 700 W, respectively. The alumina reactor (25.4mm i.d. and 160 mm length) was installed vertically in the RF chamber and designed as afixed bed reactor. To absorb and convert RF energy into heat, SiC with a particle size of 2.085 mm and bulk density of 1.48 g cm^{-3} was employed. This system is also called RF tar treatment system and has been fully described in a previous study [1]. For maintaining the temperature in side the reactor, a temperature controller was also attached to the system.

Gasification product collection system consists of a tar sampling train, flow meter, vacuum pump and a gas sampling bag. The tar sampling train has been modified along with the guideline for sampling and analysis of tar and particles in producer gas [8]. The producer gas passes through has series of six impinger bottles. The first two bottles were placed in an atmospheric environment and the second four bottles were immersed in a mixture of ice and salt bath with a temperature of about-22 °C. In addition, the first five bottles were filled with 50mL isopropanol whilst the last was empty. The consideration of the modified tar sampling train is based on the behavior of tar produced by downdraft gasifier that is mainly composed of class 3 and 4 tar [9]. The vacuum pump was used to extract the producer gas into the RF tar treatment system.A flow meter measures the flow rate of dry clean producer gas through the RF reactor and gasification product collection system.

2.3. Experimental procedure

2.3.1. Thermal treatment
For each experiment, a portion of the raw producer gas generated from pellets wood gasification was taken after the condenser (see Fig. 1) and then extracted into the RF tar treatment fixed bed reactor at a particular flow rate to give the desired residence time. The rest of the raw producer gas was burned on the flare port. In thermal treatment, the temperature was varied from 900 to 1200 °C whereas the residence time was in the range of 0.12–0.13s. In catalytic treatment, the catalyst with SiC bed is sandwiched between SiC beds, forming a total bed height of 120 mm within the reactor.

2.3.2. Sampling and analysis of the product
Before and after leaving the RF tar treatment reactor, the gas stream was passed into a tar sampling train containing organic solvent of isopropanol to condense and absorb the tar as well as particulates.

Once the samples have been collected, they were mixed together and filtered to separate the particles through the preweighed qualitative filter paper (Whatman, 90 mm diameter) into a flask. The filter paper containing particles was dried in an oven and then weighed to obtain final weight. The difference of initial and final weight of the filter paper was considered as a yield of particles. The filtered solvent was evaporated by a standard rotary evaporator equipped with solvent library software. Gravimetric tar yield was obtained by the weight of the dry residue normalized by the collected gas volume. Tar samples were analyzed using gas chromatography - - mass spectrometry (GC–MS) analyzer combined with NIST MS 2.0 software. The dry clean gas product was collected using a gas sampling bag and then analyzed in a gas chromatography–thermal conductivity detector (GC–TCD) to quantify producer gas composition using He as a carrier gas. Three samples were taken to obtain the average.

2.3.3. Calculation of equivalence ratio and gas heating value

In this work, the biomass gasification was run under a fixed condition for each experimental study. In order to reduce the number of parameters affecting the performance of the biomass gasifier, an equivalence ratio (ER) of 0.26 was maintained constant. ER reflects the combined effect of airflow rate, the rate of wood supply and duration of the run that can be calculated based on the following equation [1]:

$$ER=\frac{Air\ flow\ rate}{Biomass\ consumption\ rate} / \frac{air\ flow\ rate}{biomass\ consumption\ rate}|Stoichiometric \quad (1)$$

The stoichiometric ratio of air flow rate to biomass consumption rate is 5.22 m^3 air/kg of wood [1].

Evaluations of the gasification performance such as gas yield, carbon conversion efficiency, and cold gas efficiency were excluded because only a fraction of the producer gas can be processed in the RF reactor. Thus, in this study, the producer gas quality was characterized by the tar and particles contents, producer gas compositions and high heating values (HHV) of the producer gas. HHV (MJ Nm^{-3}) of the producer gas is dependent on the percentage volume fraction (X) of H_2, CO, and CH_4 and can be calculated by the following equation:

$$HHV = 12.766X_{H2} + 12.6441X_{CO} + 39.847X_{CH4} \quad (2)$$

3. Results and Discussion

Since the raw producer gas is taken after the condenser and directly introduced into the RF reactor without gas pre-filter, it contains not only producer gas (H_2, O_2, N_2, CO, CH_4, and CO_2) and tar but also particulates and water. The result showed that the raw producer gas consisted of 94.25 wt.% of gases, 0.19 wt.% of tar, 0.04 wt.% of particles and 5.51 wt.% of water. Meanwhile, the producer gas was mainly composed of 12.96 vol.% of H2, 16.67 vol.% of CO, 1.77 vol.% of CH_4, 12.89 vol.% of CO_2, 3.16 vol.%of O_2 and 52.55 vol.%of N_2. Consequently, several reactions take place simultaneously during thermocatalytic treatment of producer gas tar in the RF reactor.

Thermal treatment of tar

Gas compositions and HHV of producer gas from thermal treatment process at various temperatures are shown in Figure 2. H_2 content seems to decrease from the initial condition after thermal treatment at 900^{0}C. Hydrocracking of tar and hydrogasification of solid particles take a more prominent role at this condition that consumes more H_2, resulting in increasing CH_4 production.

The increase of CO formation is mainly formed by partial oxidation of solid particles. Fig. 2 displays the yields of tar and particle as a function of temperature. More than 85% of the particles were consumed and converted into gases whilst tar conversion reached only about 50% at 900^{0}C showing less contribution of tar cracking and reforming reactions for H_2 production. Compared to other similar studies, this system resulted in lower tar conversion efficiency due to the presence of light aromatic and light poly-aromatic hydrocarbon which is relatively stable.

The gas composition shows different behavior at higher temperatures where the formation of combustible gases become more intense. Thermodynamically, when the reaction temperature increases, CO, and H_2 formation increases whilst CO_2 and CH_4 formation decreases. The presences of O_2, H_2O, and particles support the production of combustible gases by means of water gas reaction, Boudouard reaction, and partial oxidation.

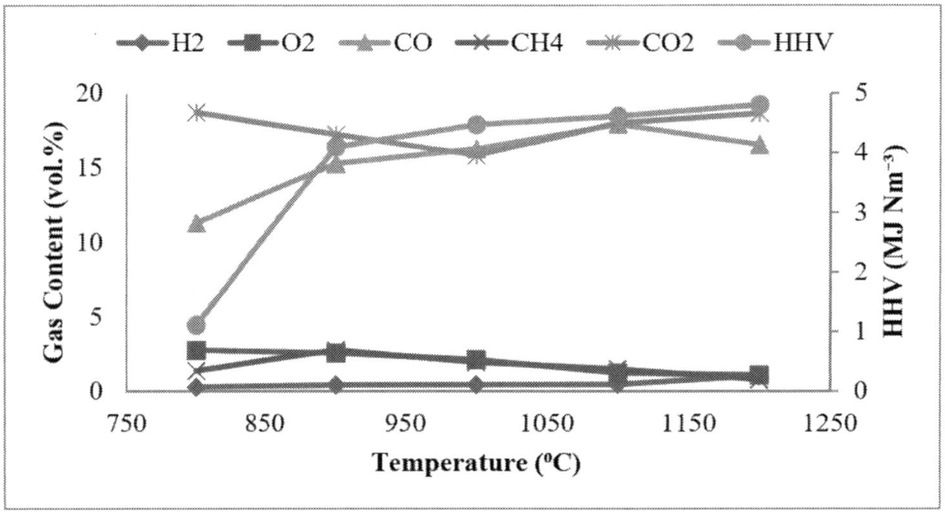

Figure 2. Tar and particle concentration with thermal treatment of producer gas

According to Le Chatelier's principle, the high reaction temperatures favor the first two endothermic reactions. Moreover, the conditions also occur due to the high activity of tar cracking and reforming reactions. In the meantime, the CH_4 content shows a slight improvement from 900 to 1000 ^{0}C and then decreases as the reaction temperature increases. The reduction of CH_4 content at higher temperature is mainly caused by steam and dry methane reforming reactions. These reactions are also endothermic, meaning the forward reaction is preferred at higher temperatures. As a result, the gas heating value is enhanced by about 18% at the highest reaction temperature of 1200 ^{0}C. More than 90% and 98% tar and particle conversion efficiencies were obtained, respectively. These results correspond to the tar and particle concentration of 147 and 4 mg Nm^{-3}, respectively. This achievement is in line with other studies where complete conversion of tar and particle in the producer gas requires extremely high temperatures above 1200 °C. Tar compounds identified by GC–MS from raw producer gas and after thermal treatment processes of producer gas tar are shown in Table 1. Tar species contained in raw producer gas generated by downdraft gasification of rubber woodblocks include heterocyclic compounds (e.g., phenol and benzofuran), aromatic compounds (e.g., benzene, toluene, xylene, ethylbenzene, styrene, indene and methylidene), light poly-aromatic hydrocarbons (PAHs) (e.g., naphthalene, methylnaphthalene, fluorene and anthracene) and heavy PAHs (e.g., pyrene, fluoranthene, chrysene and acepyrene). It can be noted that these compounds are commonly indentified in the case of biomass downdraft gasification [8,9].

The identified tar compounds changed considerably when the producer gas tar was allowed to pass within the high temperature of RF reactor. Some of the tar species especially aromatic compounds (e.g. xylene, ethylbenzene, and styrene), light PAHs (e.g. methylnaphthalene, fluorene and phenanthrene) and heavy PAHs (e.g. benzo|b|fluorine, chrysene, and acepyrene) were not detected during thermal treatment at 900 °C. Increasing the reaction temperature up to 1200^{0}C, it can be seen that only benzene, toluene, indene, naphthalene, biphenyl, and anthracene can be found in tar species. In addition, more than 70% of tar species was occupied by benzene and naphthalene. This result confirms the high stability of these compounds during thermal treatment process [1].

Table 1. Relative area percentage of major tar compounds resulting from thermal treatment of producer gas tar

Compound name	Molecular weight	Boiling point (^{0}C)	% of compound in producer gas tar				
			Untreated	Thermal reaction temperature (^{0}C)			
				900	1000	1100	1200
Benzene	78	80	11.01	23.97	25.06	31.64	35.52
Toluene	92	110.6	13.10	11.62	15.20	14.18	4.40
Styrene	104	145	0.40	-	-	-	-
Ethylbenzene	106	145	0.65	-	-	-	-
p-Xylene	106	138.3	1.26	-	-	-	-
o-Xylene	106	144.5	0.72	-	-	-	-
Benzene, 1,2,3-trimethyl	120	98	0.27	8.81	6.68	-	-
Benzene, (1-methylethyl)-	120	172.8	2.65	-	-	-	-
Benzene, 1-ethenyl-2-methyl-	118	172.8	1.47	-	-	-	-
Benzene, 1-ethenyl-3-methyl-	118	172.8	3.31	9.53	6.03	-	-
Benzofuran	118	174	0.26	-	-	-	-
Phenol	94	181.8	4.82	2.02	-	-	-
Indene	116	182	6.31	3.76	7.16	7.77	10.23
Methylindene	130	199	3.88	7.16	-	-	-
Naphthalene	128	217.9	4.46	13.80	23.98	35.64	34.80
2-Methylnaphthalene	142	241.1	2.19	4.57	4.33	-	-
1-Methylnaphthalene	142	244.7	2.47	3.43	4.41	-	-
4-Butyl-1,1'-biphenyl	210	318	1.46	3.74	4.03	6.23	8.48
Anthracene	178	339.9	1.32	3.14	3.10	4.54	6.57
Diphenylethyne	178	256.1	2.13	-	-	-	-
9H-Fluorene-9-methylene-	178	295	0.39	-	-	-	-
Phenanthrene	178	340	0.49	-	-	-	-
Methylenephenanthrene	190	353	2.54	-	-	-	-
Anthracene, 9-methyl-	192	196	1.00	-	-	-	-
4-Methylenephenanthrene	192	197	0.66	-	-	-	-
Fluoranthene	202	384	2.68	2.01	-	-	-
Pyrene	202	404	11.07	2.45	-	-	-
Pyrene, 4,5-dihydro-	204	404	2.92	-	-	-	-
11H-Benzo[b]fluorene	216	405	3.81	-	-	-	-
Pyrene, 1-methyl-	216	405	3.50	-	-	-	-
Chrysene	228	448	1.15	-	-	-	-
Acepyrene	226	448	5.31	-	-	-	-

4. Conclusions

The performance of RF tar thermocatalytic treatment system was investigated. Temperature plays a crucial role for tar and particle conversions under the investigated conditions. The thermal treatment

IOP Conf. Series: Materials Science and Engineering **309** (2018) 012056 doi:10.1088/1757-899X/309/1/012056

produces a higher heating value of 5.76 MJ Nm^{-3} at 1200°C. Both dolomite and Y-zeolite offer better tar and particles conversion efficiencies of around 97%. Even at a higher temperature of thermocatalytic cracking benzene, naphthalene and toluene were still found. The presence of reforming products, water, and particulates in the producer gas has a major impact on upgrading producer gas quality.

References

[1] Anis S and Zainal ZA 2011 *Renew Sust Energy Rev* **15** 2355– 77.
[2] Devi L *et al* 2003 *Biomass Bioenergy* **24** 125– 140.
[3] Gil J *et al* 1999 *Chem Res* **38** 4226–35.
[4] Qin Y *et al* 2007 *Biomass Bioenergy* **31** 243– 49.
[5] Jess A 1996 *Fuel* **75** 1441–48.
[6] Thostenson ET and Chou TW 1999 *Appl Sci Manuf* **30** 1055–71.
[7] Yin C 2012 *Bioresour Technol* **120** 273–84.
[8] Paasen SV BV *et al Guideline for Sampling and Analysis of Tars and Particulates in Biomass Producer Gases ECN-C-02-090 ECN Petten* The Netherlands 2002.
[9] Milne TA *et al Biomass Gasifier "tars": Their Nature, Formation, and Conversion* NREL Golden CO USA (Report No. NREL/TP-57025357) 1998.

IOP Conf. Series: Materials Science and Engineering **309** (2018) 012057 doi:10.1088/1757-899X/309/1/012057

Exergy analysis of biomass organic Rankine cycle for power generation

T B Nur [1,2,*], Sunoto[1]

[1]Department of Mechanical Engineering, Faculty of Engineering, Universitas Sumatera Utara, Padang Bulan, Medan 20155, Indonesia.
[2]Sustainable Energy and Biomaterial Center of Excellence, Universitas Sumatera Utara, Padang Bulan, Medan 20155, Indonesia.

*E-mail: taufiq.bin_nur@usu.ac.id

Abstract. The study examines proposed small biomass-fed Organic Rankine Cycle (ORC) power plant through exergy analysis. The system consists of combustion burner unit to utilize biomass as fuel, and organic Rankine cycle unit to produce power from the expander. The heat from combustion burner was transfered by thermal oil heater to evaporate ORC working fluid in the evaporator part. The effects of adding recuperator into exergy destruction were investigated. Furthermore, the results of the variations of system configurations with different operating parameters, such as the evaporating pressures, ambient temperatures, and expander pressures were analyzed. It was found that the largest exergy destruction occurs during processes are at combustion part, followed by evaporator, condenser, expander, and pump. The ORC system equipped with a recuperator unit exhibited good operational characteristics under wide range conditions compared to the one without recuperator.

1. Introduction

The energy crisis that has occurred in Indonesia over the years has hampered national development, upgrading and developing technology. The rapid growth of national population leads to increasing the national energy demand. It has been predicted that the global energy consumption is likely to grow faster than the population growth [1-3]. The fuel consumption was growing from 777.925 million barrels of oil equivalent (MBOE) in 2000 to 1040.677 MBOE in 2015. Around 74% of Indonesia's energy supplies come from non-renewable energy. The primary energy supply by source for 2015 is 364.62 MBOE from coal, 444.81 MBOE from crude oil and products, 279.63 from natural gas and product, 34.597 MBOE from hydro power, 16.34 MBOE from geothermal, 309.73 MBOE from biomass, and 19.11 MBOE from biofuel [4]. The power plant installed capacity in Indonesia during 2000 is 37,287.19 MW, and have increased to 55,531.20 MW in 2015, where the power plant production reaches 176,472 GWh in 2015 [4]. About 88% of Indonesia's population has access to electricity compared to less than 68% in 2010 [5]. The most of Indonesian power plants are using non-renewable sources fuel such as natural gas, fuel oil, coal and diesel to generate electricity. Therefore, an important priority is to increase the national power generation capacity to complete the electrification of the country and meet increasing electricity consumption.

Due to the environmental issues from burning fossil fuels like greenhouse gas emission [6] and the fact that fossil fuel will be depleted one day, the dependency of Indonesian energy scenario to fossil fuels will have an adverse effect on the economy of Indonesia [7]. Therefore, action toward using renewable energy source should be put as the priority by the government.

Renewable energy sources in Indonesia, such as geothermal energy, solar energy, wind energy and biomass energy are the promising energy sources to replace fossil fuel energy. Indonesian's biomass energy policy basically follows government regulation No. 79/2014 on National Energy Policy. It set the targets for an optional energy mix in 2025, where renewables contribute at least around 23% of the total energy mix and should increase to at least 31% in 2050.

Referring to Indonesia condition as one of the biggest countries as palm biomass producer, research on utilizing of biomass energy to supply power generation is great significance. One of promising technology to extract biomass energy is by using biomass as fuel for organic Rankine cycle (ORC). The ORC uses an organic working fluid which is able to condense at a lower pressure and evaporate at a higher pressure. The working fluid as the organic substance can be better adapted than water for lower heat source temperatures. In the previous work (8,9), the biomass from empty fruit bunch (EFB) of palm oil has been used for ORC power plant to generate electricity. To study the quality of energy used during processes on the system to improve the performance of the ORC system is by using the exergy analysis method. Therefore, the objective of the present work is to perform exergy analysis on the design of ORC fueled by palm oil EFB. The configurations analyzed are similar to the proposed design in the preceding study [9].

2. Technology Description

Two ORC configurations were simulated in this analysis, (i) ORC without recuperator unit, and (2) ORC with recuperator unit, as shown in figure 1. The EFB of palm oil used as fuel in the thermal oil heater (TOH) combustion unit. The processes start from burning biomass fuel in TOH to generate heat which is then transferred to ORC working fluid by the thermal oil that goes in the evaporator section. The compressed ORC working fluid, R245fa, will evaporate in the evaporator prior entering to expander producing work by rotating shaft which then converted to electricity by a generator. A recuperator essentially increases the thermal efficiency thus a high-power output can be maintained for a decreased heat input to the ORC unit [8-10].

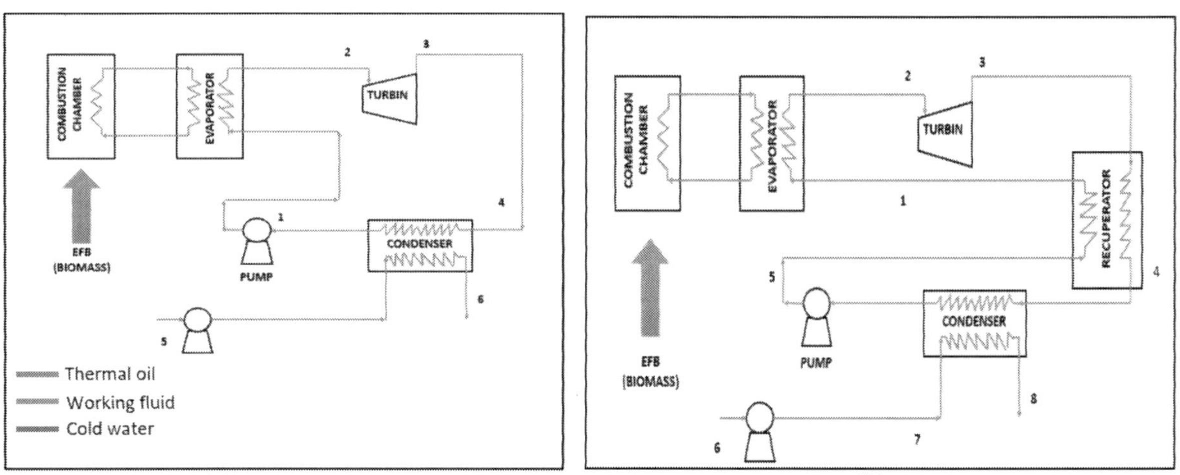

(a) (b)

Figure 1. Schematic diagram of the ORC system, (a) ORC without recuperator unit, (b) ORC with recuperator unit.

3. Methodology

3.1. Aspen Plus flowsheet

The Aspen Plus flowsheets of the ORC with biomass combustion system are depicted in figures 2-3. The systems are based on the following main assumptions: steady state operation, pressure drop neglected, working fluid is R-245fa, thermal oil is Syltherm XLT. More detailed about the configuration and firs law thermodynamic analysis can be found in the preceding study [8,9]. Here, in order to facilitate the exergy analysis, we display the schemes that have been used previously.

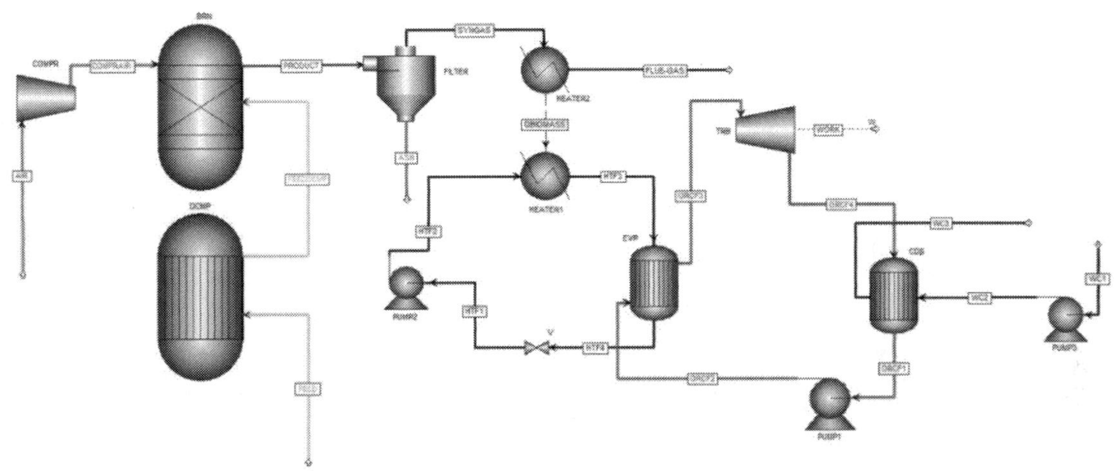

Figure 2. Aspen Plus simulation model for ORC plant without recuperator unit.

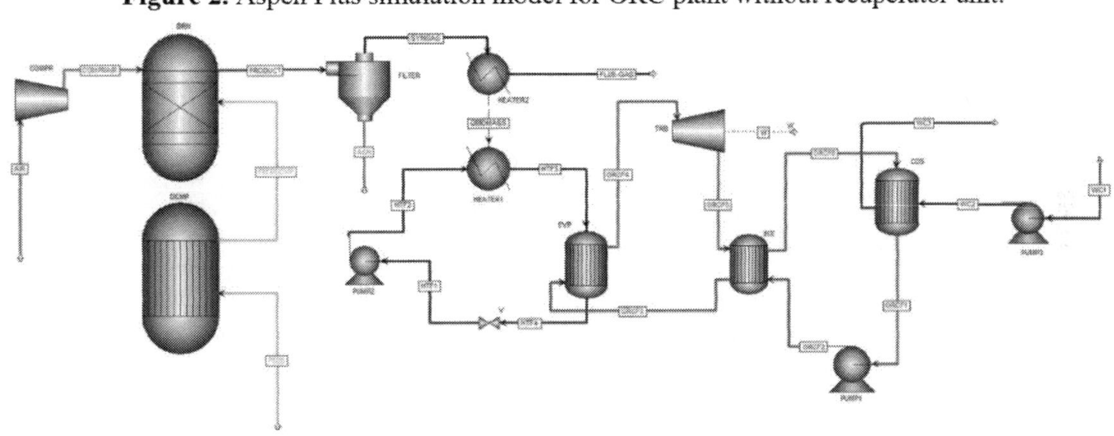

Figure 3. Aspen Plus simulation model for ORC plant with a regenerator unit.

3.2. System and Exergy Analysis

Exergy can exist in the forms of thermo-mechanical exergy and chemical exergy; there are also corresponding forms of exergy for kinetic, potential, and electrical energy. More details on exergy analysis method can be found in available references [11-14]. The exergy evaluation of the biomass fuel can be calculated by using the following [15]:

$$e = 1812.5 + 295.606C + 587.354H + 17.506O + 17.735N + 95.615S - 31.8A \quad (1)$$

Exergy destruction of each individual unit can be calculated by finding the difference between the exergy of input and output streams of this unit [12]. The total exergy destruction of the ORC system can be calculated as follow:

IOP Conf. Series: Materials Science and Engineering **309** (2018) 012057 doi:10.1088/1757-899X/309/1/012057

$$I_{total} = I_{pump} + I_{evaporatore} + I_{turbine} + I_{condenser} + I_{recuperator} + I_{combustion} \quad (2)$$

where I is the exergy destruction by each component. The main operational conditions for analysis are shown in table 1.

Table 1. Main operational conditions and assumptions for plant calculation.

EFB feed to boiler (kg hr^{-1})	100
Air feed to boiler (kg hr^{-1})	604
Outlet temperature of oil heater from evaporator (°C)	115
Combustion temperature (°C)	300
Type of working fluid for ORC unit	R245fa
Mass flow rate of working fluid (kg hr^{-1})	750
Temperature of ORC working fluid at the inlet evaporator (°C)	70
Outlet pressure of expander (bar)	2
Pressure losses during process (bar)	0
Electric generator efficiency (%)	98.7
Dead state temperature (°C)	27-31

4. Results and discussion

The exergy destruction during processes on the ORC system without recuperator unit and system with recuperator unit are 176.311 kW and 156.51 kW, respectively. Figure 4-5 show the schematic diagram of both configurations with their temperature (°C), mass flow rate (kg/h), and exergy flow (kW). The exergy destruction on the combustion unit decreased due to reducing fuel consumption rate at the same system power output level by adding a recuperator unit. Combustion unit contributes to highest exergy destruction on the systems during processes. More results on effect of adding recuperator to fuel consumption rate can be found in the preceding study [9].

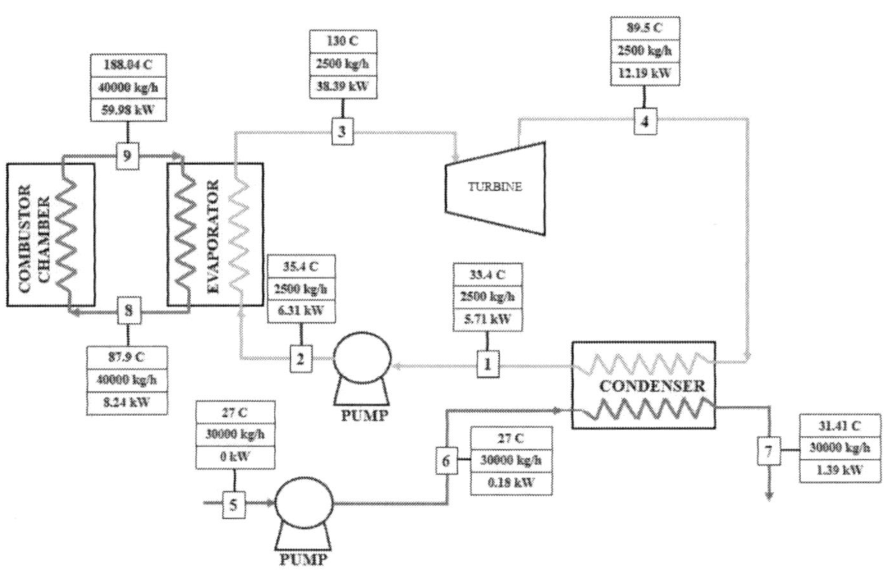

Figure 4. ORC system without recuperator with selective properties during processes whereas the expander inlet pressure of 13 bar.

IOP Conf. Series: Materials Science and Engineering **309** (2018) 012057 doi:10.1088/1757-899X/309/1/012057

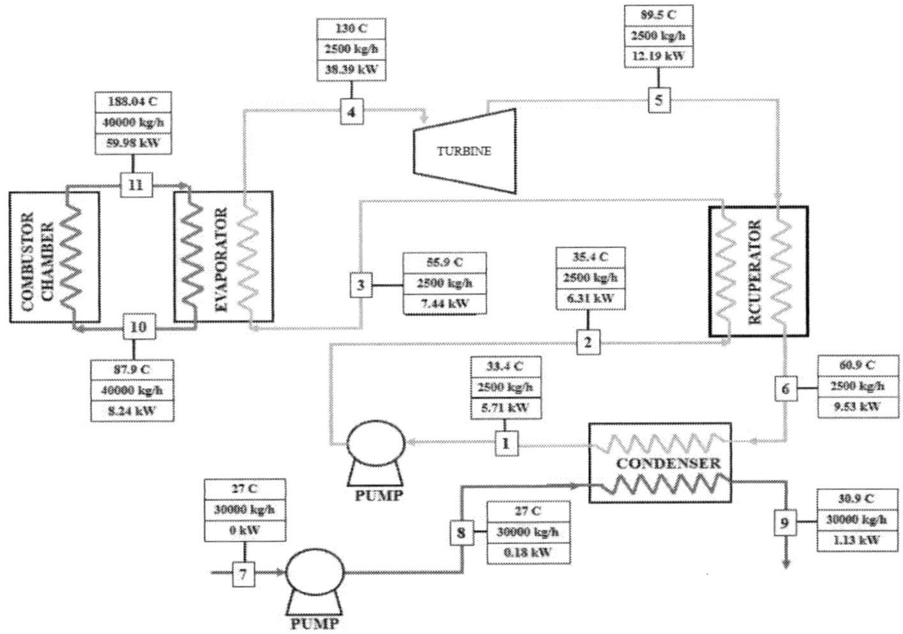

Figure 5. ORC system with recuperator with selective properties during processes whereas the expander inlet pressure of 13 bar.

4.1. Effect of various expander inlet pressure on the exergy destruction

The total exergy destruction on ORC system without recuperator unit decreased when increasing the expander inlet pressure, as shown in figure 6. Total exergy destruction decreased from 184.773 kW when expander inlet pressure at 9 bar to 176.311 if expander inlet pressure operated at 13 bar. Meanwhile, the total exergy destruction on ORC system with recuperator also decreased from 157.885 kW to 156.515 kW at the same conditions.

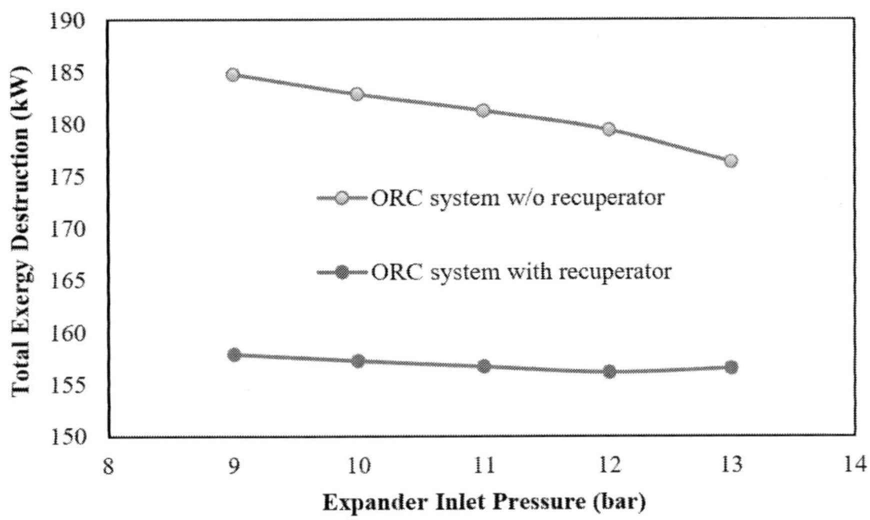

Figure 6. Effect of various expander inlet pressure on system total exergy destruction.

4.2. Effect of ambient temperature on exergy destruction

The effect of various ambient temperatures on the total system exergy destruction can be found in figure 7. It can be seen that when the ambient temperature increased, the total system exergy destruction for ORC system without recuperator decreased as from 176.311 kW at 27 °C to 171.499 kW with an ambient temperature of 31 °C. Meanwhile, for the ORC system with recuperator, the total system exergy destruction decreased from 156.515 kW to 152.833 kW at the same conditions.

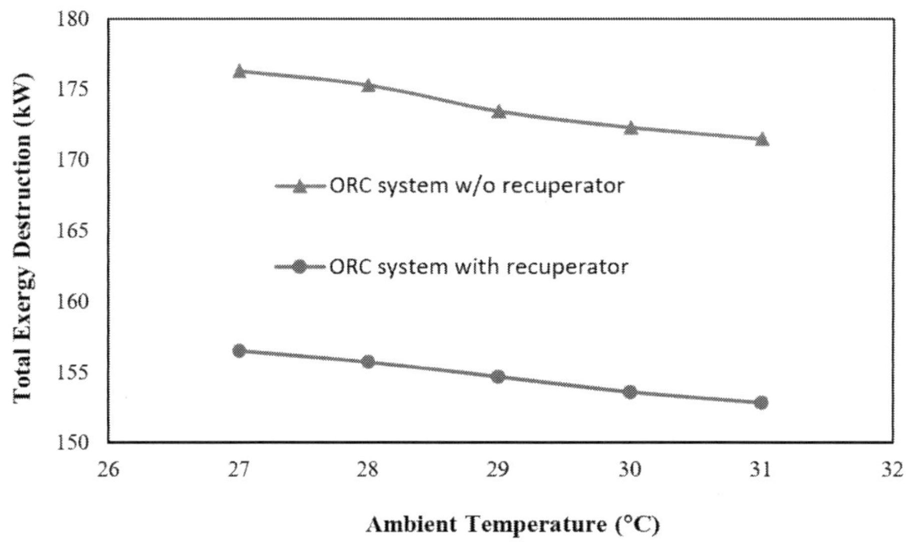

Figure 7. Total exergy destruction and environmental temperatur.

5. Conclusions

We performed exergy analyzed on the proposed ORC system fueled EFB. It was found that the highest losses by the destruction of exergy occurred in the combustion chamber unit, followed by the evaporator, condenser, expander, and pump. Adding a recuperator contributes to lower fuel consumption rate at the same level system power output hence decreasing exergy destruction on the combustion chamber. As the ambient temperature increased, the exergy destruction in the combustion chamber and condenser decreased, reducing the total exergy destruction of the system during the process. Although the exergy destruction in other units such as expander, pump, recuperator, and evaporator have increased, the total increase is lower than the rate of decrease of destruction exergy in the combustion chamber and condenser. A further analysis of the systems needs to be done, especially to see the effect of different working fluid to the system performance.

Acknowledgements

This work has been fully supported by BP-PTN USU (Universitas Sumatera Utara), No: 6049/UN5.1.R/PPM/2016, July 19th, 2016.

References

[1] Hassan M H, Mahlia T M I, Nur H 2012 *Renew. Sust. Energ. Rev.* **16** pp 2316-2328
[2] Asif M, Muneer T 2007 *Renew. Sust. Energ. Rev.* **11** pp 1388-413
[3] Omer A M 2008 *Renew. Sust. Energ. Rev.* **12** pp 2265-300
[4] Handbook of Energy & Economic Statistics of Indonesia, Jakarta, Indonesia: Ministry of

Energy and Mineral Resources; 2016
[5] Indonesia's electricity demand and the coal sector: Export or meet domestic demand?, Oxford Institute for Energy Studies; 2017
[6] Dincer I 2000 *Renew. Sust. Energ. Rev.* **4** pp 157-75
[7] Hasan M H, Muzammil W K, Mahlia T M I, Jannifar A, Hasanuddin I 2012 *Renew. Sust. Energ. Rev.* **16** pp 3206-3219
[8] Nur T B, Pane Z, Amin M N 2017 *IOP Conf. Series: Materials Science and Engineering* **180** pp 1-7
[9] Ependi S, Nur T B 2017 Design and process integration of organic Rankine cycle utilizing biomass for power generation (*submitted to IOP Conf. Series: Materials Science and Engineering*)
[10] Lecompte S, Huisseune H, van den Broek M, Vanslambrouck B, Paepe M D 2015 *Renew. Sust. Energ. Rev.* **47** pp 448-461
[11] Saidur R, Boroumandjazi G, Mekhilef S, Mohammed H A 2012 *Renew. Sust. Energ. Rev.* **16** pp 1217-1222
[12] Mukherjee S, Kumar P, Yang A, Fennell P 2015 *J. Environm. Chem. Eng.* **3** pp 2104-2114
[13] Querol E, Gonzalez-Regueral B, Ramos A, Perez-Benedito J L 2011 *Energy* **36** pp 964-974
[14] Li G 2016 *Renew. Sust. Energ. Rev.* **53** pp 477-499
[15] Song G, Shen L, Xiao J 2011 *Int. Eng. Chem. Res.* **50** pp 9758-9766

IOP Conf. Series: Materials Science and Engineering **309** (2018) 012058 doi:10.1088/1757-899X/309/1/012058

Re-designing illumination level in printing working area

D Wahyuni*, M Tambunan, N Panjaitan, and I Budiman

Industrial Engineering Department, Faculty of Engineering, Universitas Sumatera Utara, Jl. Almamater Kampus USU Medan 20155, Indonesia

*Email: diniwahyuni2015@gmail.com

Abstract. This research was conducted in four printing business in Medan city. The illumination level on the research object is very low around 30 Lux far below the required value of government regulation at 200 Lux. Poor lighting has an impact on the number of defective products that pass the inspection, so we need to improve the workspace lighting to improve the quality of work. The method of determining the measuring point follows SNI 16-7062-2004, and the measuring instrument used is 4 in 1 Environment Meter. The results show that almost all workspaces under study require improved lighting, because the light bulbs used are not able to meet the lighting needs. This research recommends improving the workspace lighting using LED (Light Emitting Diode) lights because it has high energy efficiency and relatively more lifetime compared to the existing lamp.

1. Introduction

The results of good working need a conducive environment conditions. Operators who work in bad environment, then the quality of their work can decrease. Research conducted previously shows that lighting, humidity, and temperature of the work environment affect the productivity of workers [1-6].

Lighting greatly affects the human ability to see objects clearly, quickly, and without error. The need for good lighting will be more necessary when humans do work that requires precision of vision. The lighting is too grim to cause the eyes of the workers get tired quickly because the eyes will accommodate maximum to be able to see clearly. Tired eyes can lead to mental fatigue, further the situation can cause damage to the eyes. The ability of the eyes to see objects is clearly determined by the size of the object, the degree of contrast between the object and its surroundings, the luminance, and the duration of view [5-11].

Measurements of illumination at several measuring points in the work area scattered throughout production floor of the printing business under study showed a number with a range of 22 lux - 34 lux, while the lighting level standards for rough and continuous work specified in Health Minister decree amounted to 200 lux [12]. Illumination that doesn't meet existing lighting level standards can be said to be bad lighting. Bad lighting will impair vision so as to decrease the concentration of workers and result in poor work results. Furthermore, the result of the disturbance has an impact on the finding of defective products that pass the inspection [13]. This is in line with the opinion of operators who complain about the disruption of vision when doing inspection of the product result.

Recommendation of the requirements and procedures for the implementation of industrial working environment health according to decision of Health Minister No.45 year 2002 for work with machines and rough assembly requires 200 lux illumination. CIE (Commission International de l'Eclairage) and

IES (Illuminating Engineers Society) set the general lighting for the work space is 200 lux, as well as the SNI 03-6197-2000 [10].

2. Method and equipment

Illumination is many light currents that come in a unit of field and have a unit of lux (lumen/m2) or in other words illumination is the coming of light to an object. Luminance is the intensity of light emitted, reflected, or transmitted by an illuminated field unit or luminance is the passage of light from an object. The IES Lighting Handbook states that each object reflects a portion of the light on it [10]. Depending on the geometric arrangement, the exact size can be either total light reflectance, regular light reflectance, diffuse light reflectance, light reflectance or luminance. The light reflectance scale is between 0 and 100% from black to white [14].

Illumination for the field of work is measured horizontally as far as 75 cm above the surface of the floor, while for a certain area the illumination is obtained by taking the average value from several points of measurement (SNI 03-6575-2001) [15].

The determination of the point of illumination measurement is regulated in SNI 16-7062-2004 concerning Measurement of Intensity of Information in the workplace. Determination of measurement point on the lighting is divided into 2 categories:

1. Local lighting: work objects, in the form of desks and equipment; measurements can be made on an existing table. Measurement of illuminance and luminance can be done using luxmeter or 4 in 1 environment meter. Illumination measurement steps are as follows:
   a. Turn on the calibrated luxmeter by opening the sensor cover.
   b. Bring tool to a predefined point of measurement, either measurement for the intensity of local or general illumination.
   c. Read the result of measurements on the monitor screen after waiting for a while to get a stable value.
   d. Record the measurement results on the recording sheet for the local lighting intensity.
   e. Turn off the luxmeter after completion of the measurement of the intensity of lighting.
2. General illumination: the point of horizontal line length and width of the room at any given distance of one meter from the floor. Certain distances are distinguished by the wide of room. There are 3 categories of determination of the measuring point based on the area of the room that is:
   a. The space is less than 10 square meters: horizontal cutting point length and width of the room is at a distance of every 1 (one) meter.
   b. The space between 10 square meters to 100 square meters: horizontal cutting point length and width of the room is at a distance of every 3 (three) meters.
   c. Room area of more than 100 square meters: horizontal cutting point length and width of the room is at a distance of 6 meters.

The luminance for the work plane is measured using a lux meter. Measurement of luminance is done by placing the light sensor facing the surface of the object to be measured luminance level at a distance of 2 to 4 inches until the reading rate on the screen table is stable. The position of the sensor should be arranged in such a way as to avoid falling shadow of the tool or operator on the area to be measured.

The reflectance measurement method is divided into two ways, namely the sample comparison method is known and the light-coming light-reflection method. The sample comparison method is known to use a reflectance measurement card and is used to measure reflectance on surfaces that reflect light diffusely (spread). The light-coming light-reflection method is used to determine the reflectance (in percent) on a reflective or non-glossy surface. This method consists of three steps, that is:

1. Measuring the intensity of light that falls to the surface of the object.
2. Measuring the intensity of reflected light from the surface of the object.

3. Measuring the reflectance of the object's surface by dividing the intensity number of reflected light with the received light intensity.

The calculation of illumination at a point is influenced by the total light current in the relevant plane and the wide area. In addition, it is also influenced by the light intensity distribution of the luminaire, efficiency, shape, size of space, surface reflections, and the height of the lamps of the field of work.

During lamp usage, the light intensity will be reduced by the dust heap and the lumen values will shrink. Therefore, we need to consider the value of Light Lost Factor (LLF).

The value of CU (Coefficient of Utilization) is highly dependent on the surface reflectance. The higher the reflected surface of the roof, space, and floor, the higher the CU value will be.

The formula for calculating the required illumination level, F (flux luminuous) is as follows:

$$F = \frac{(E) \times (A)}{(CU) \times (LLF)} \quad (1)$$

Where:
E = Recommended illumination level on surface (*lux*)
F = Required *flux luminous* (illumination level) (lumen)
CU = *coefficient of utilization*
LLF = *light loss factor*
A = Area surface (m^2)

3. Results and Discussions

This research was conducted in four printing business in Medan City. Printing becomes the object of observation to produce printed products such as forms, letterhead, invitation cards, books, newspapers, banners, or garment products. Sample printing 1 has several work stations including cutting work stations, printing work stations, numbering stations, and binding stations. Research in each work station is done to illumination level, reflectance number, luminal count calculation and lamp type selection.

For the first area under study is the area of the cutting station. The cutting station has an area of 11,2 m^2, the number of lamps 1 piece. The operator's work area coverage is only 1 m, then the four-point retrieval is taken along 0.5 m from the right and left front and rear operator shoulders and a point directly in front of the operator's head. So that obtained five measurement points approaching cutting operators as in Figure 1.

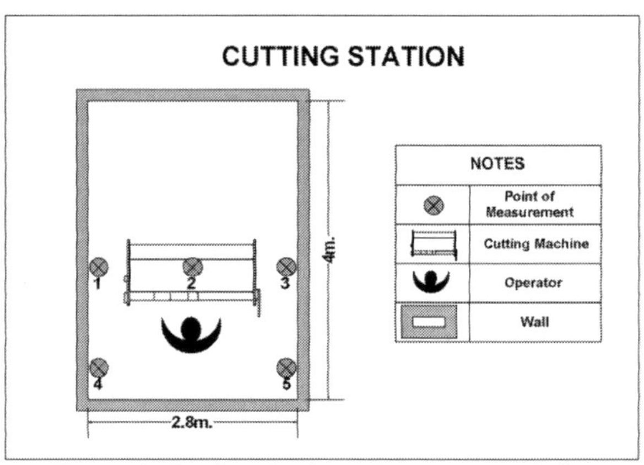

Figure1. Cutting Station Measurement Point

Illumination values at five points of measurement at cutting stations can be seen in Table 1.

Table1. Ilumination of Cutting Station

No	Ilumination of Cutting Station (Lux)						
	A	**B**	**C**	**D**	**E**	**F**	**G**
1	70.0	75.0	72.0	65.0	67.1	73.0	66.0
2	49.2	47.9	45.7	50.3	60.1	49.2	41.5
3	42.5	42.1	46.1	41.9	40.0	43.4	46.2
4	40.0	41.5	42.7	38.6	37.3	42.1	43.4
5	76.0	77.9	75.7	80.3	80.1	79.2	81.5

Note:
A = Time 09:00 – 10:00 WIB
B = Time 10:00 – 11:00 WIB
C = Time 11:00 – 12:00 WIB
D = Time 13:00 – 14:00 WIB
E = Time 14:00 – 15:00 WIB
F = Time 15:00 – 16:00 WIB
G = Time 16:00 – 17:00 WIB

The level of luminance and the level of illumination is also measured on the object material located at cutting station. Appropriate data measurement procedures for each object material at cutting stations, numbering and binding.

The calculation of the total reflectance number of the cutting station walls is as follows:

$$\text{Reflectance number } (\rho) = \frac{\sum_{k=1}^{n} \rho k \text{ x Lk}}{\sum_{k=1}^{n} \text{Lk}} \text{ x } 100\%$$

$$\text{Reflectance number } (\rho) = \frac{(50.5\text{x}12.0) + (47.7\text{x}12.0)}{(12.0 + 12.0)} \text{ x } 100\% = 49\%$$

The total reflectance number of cutting station can be seen in Table 2.

Table 2. The Total Reflectance Number Of Cutting Station

Dept.	Measurement Field	Total Reflection (ρ total)
Cutting Station	Wall	0,49
	Floor	0,39
	Roof	0,5

From the data collection of the characteristics of the room and the room layout then it can be determined the number of light bulbs to illuminate each room for lights can be bright in accordance with the standard of Kepmenkes No 1405/MENKES/SK/2002, using the formula:

$$F = \frac{(E) \text{ x } (A)}{(CU) \text{x } (LLF)}$$

$$F = \frac{(200) \text{ x } (11.2)}{(0.72) \text{x } (0.76)}$$

$$F = 4093.56$$

The type of lamp used earlier in the factory is Philips Essential lamp 23 W with *nominal luminous flux* = @1900 lumen. Then the number of light bulbs used should be

$$= \frac{F}{F1} = \frac{4093.56}{1900} = 2 \text{ lamps}$$

The research was also conducted on the printing work stations, numbering, and binding so as to obtain the recapitulation for all four stations as in Table 3.

Table 3. Recapitulation Illumination Level for Each Work Station

Time	K	L	M	N
A	55.6	209.5	90	97
B	56.9	218.5	90.4	98.4
C	56.5	207.9	89	99.4
D	55.3	215.2	85	98
E	57	216.6	87.1	96.9
F	57.4	221.1	87.3	95.7
G	55.8	213.8	86	94.8
Average	**56.4**	**214.7**	**87.9**	**97.2**

Note:
A = Time 09:00 – 10:00 WIB
B = Time 10:00 – 11:00 WIB
C = Time 11:00 – 12:00 WIB
D = Time 13:00 – 14:00 WIB
E = Time 14:00 – 15:00 WIB
F = Time 15:00 – 16:00 WIB
G = Time 16:00 – 17:00 WIB
K = Cutting Station
L = Printing Station
M = Numbering Station
N = Binding Station

The Illumination level graphs for each are shown in Figure 2. Based on the graph it is seen that only printing stations whose work stations meet the lighting standards set by Kepmenkes No.1405.

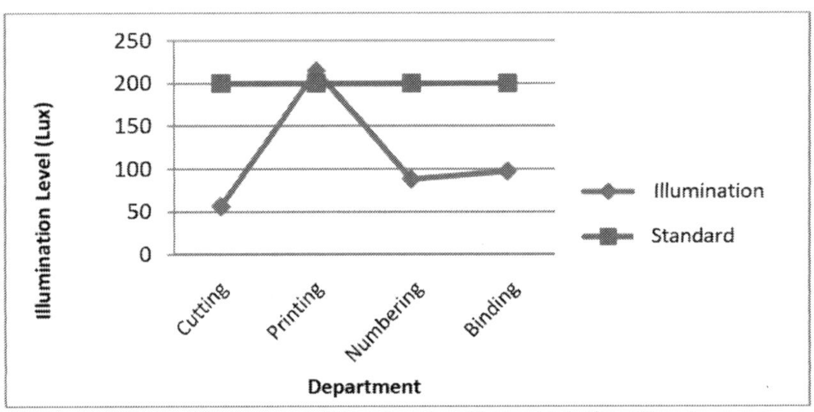

Figure 2. Average Illumination Recapitulation (Lux)

Therefore, design of work area lighting can be performed on cutting station, numbering, and binding.

In the same way as the calculations performed at the cutting station, then obtained the recapitulation of the number of lamp lumen and the number of bulbs for all samples reviewed is shown in Table 4.

Table4. Recapitulation of The Number of Lamp Lumen Requirements and Number of Lamp at The Four Sample Locations

Sample	Station	Number of Lumen	Number of Lamp (unit)	
		Needs	Needs	Actual
1	K	4.093,56	2	1
	M	11.685,90	6	1
	N	23.106,54	6	4
2		48.309,17	19	18
3		92.380,00	19	15
4		40.436,45	16	8

Note:
K = Cutting Station
M = Numbering Station
N = Binding Station

Sample 1 has a separate work space between workstations, while atSample 2 to 4 all workstations are in one work space.

Some possible causes of low levels of illumination on the floor of production, that is:

1. Lack of illumination level (E) towards the area of room (A), this is influenced by the number of luminaires and the number of initial lumen of the luminaire.
2. The low value of CU (Coefficient of Utilization), this is influenced by luminaire's height from the work field, the proportion of the room, and the reflection of material objects.
3. The low value of LLF (Lost Light Factor), this is influenced by nonrecoverable factor and recoverable factor

4. Conclusions

The conclusions from this research is that almost all work areas are reviewed require improvement of lighting, because light bulbs used aren't able to meet the needs of lighting.

This research recommends improving the workspace lighting using LED (Light Emitting Diode) lights because it has high energy efficiency and relatively more lifetime compared to the existing lamp.

Acknowledgments

The authors gratefully acknowledge that the present research is supported by Ministry of Research and Technology and Higher Education Republic of Indonesia under the research grant of Year 2016.

References

[1] Ismail A R 2014 The Impact Of Workers Productivity under Simulated Enviromental Factor by Taguchi Analysis
[2] Egan M D 1983 Concepts in Architectural Lighting. New York: Mc. Graw Hill School Education Group
[3] Flory I L 2008 High-Intensity Discharge Lighting Design Strategies for the Minimization of Energy Usage and Life-Cycle Cost Dissertation. Doctor of Philosophy in Electrical Engineering Virginia Polytechnic Institute and State University
[4] Hawes B K 2012 Effect of Four Workplace Lighting Technologies on Perception, Congnition and Effective State
[5] Light Research Center 2000 Illumination fundamentals. Rensselaer Polytechnic Institute

[6] Macorol E 2015 Design and Implementation of A PID-based Luminance Control and Light Sensing Using Smoothing Averaging Technique
[7] Nugrahani E F 2015 Design of Solar Transmission System using Fiber Optic for Indoor Lighting.
[8] Stanton N A 2004 Handbook of Human Factors and Ergonomics Methods. Cetakan I. New York: CRC Press
[9] Tuan T B 2015 Energy Analysis for Lighting and Air-Conditioning System of An Academic Building
[10] Zumtobel S 2004 The lighting handbook, 1st edition
[11] Luciana K 2004 Penelitian Terhadap Kuat Penerangan dan Hubungannya dengan Angka Reflektansi Warna Dinding: Studi Kasus Ruang Kelas Unika Widya Mandala Surabaya
[12] Keputusan Menteri Kesehatan Republik Indonesia. (2002). No. 1405/MENKES/SK/XI/2002 tentang Persyaratan Kesehatan Lingkungan Kerja Perkantoran dan Industri
[13] Wahyuni D 2002 Pengaruh Pencahayaan terhadap Kualitas Kerja Operator
[14] Sinulingga S 2012 Metode Penelitian. Edisi II. Medan: USU Press
[15] Standar Nasional Indonesia. (2004). SNI 16-7062-2004: Pengukuran Intensitas Penerangan di Tempat Kerja. Jakarta

IOP Conf. Series: Materials Science and Engineering **309** (2018) 012059 doi:10.1088/1757-899X/309/1/012059

Study of activity based costing implementation for palm oil production using value-added and non-value-added activity consideration in PT XYZ palm oil mill

M T Sembiring[1,2,*], D Wahyuni[1], T S Sinaga[1], A Silaban[1]
[1]Department of Industrial Engineering, Faculty of Engineering, University of North Sumatera, Jl. Prof. T. Maas Kampus USU, North Sumatera Province, Indonesia
[2]Sustainable Energy and Biomaterial Center of Excellence, Universitas Sumatera Utara, Padang Bulan, Medan 20155, Indonesia.

*E-mail: meilita_tryana@yahoo.co.id

Abstract. Cost allocation at manufacturing industry particularly in Palm Oil Mill still widely practiced based on estimation. It leads to cost distortion. Besides, processing time determined by company is not in accordance with actual processing time in work station. Hence, the purpose of this study is to eliminates non-value-added activities therefore processing time could be shortened and production cost could be reduced. Activity Based Costing Method is used in this research to calculate production cost with Value Added and Non-Value-Added Activities consideration. The result of this study is processing time decreasing for 35.75% at Weighting Bridge Station, 29.77% at Sorting Station, 5.05% at Loading Ramp Station, and 0.79% at Sterilizer Station. Cost of Manufactured for Crude Palm Oil are IDR 5.236,81/kg calculated by Traditional Method, IDR 4.583,37/kg calculated by Activity Based Costing Method before implementation of Activity Improvement and IDR 4.581,71/kg after implementation of Activity Improvement Meanwhile Cost of Manufactured for Palm Kernel are IDR 2.159,50/kg calculated by Traditional Method, IDR 4.584,63/kg calculated by Activity Based Costing Method before implementation of Activity Improvement and IDR 4.582,97/kg after implementation of Activity Improvement.

1. Introduction

Rapid development of technology in the global era resulted in an intensified usage of production machines then it replaced the use of labor. With the increasing use of the machine then the composition of the company's cost of goods manufactured be changed i.e. decreasing in the direct labor costs and increasing factory overhead costs. PT XYZ Palm Oil Mill is one of the manufacturing companies that are engaged in the processing of Palm Oil Fresh Fruit Bunches (FFB) into Crude Palm Oil (CPO) and Palm Kernel [1]. The production process at the PT XYZ Palm Oil Mill using various kinds of production machines so it absorbs high costs. Therefore, the accuracy in calculating and charging according to the amount consumed by the activities of product manufacturing is absolutely required.

Currently, calculation of cost of goods manufactured in PT XYZ Palm Oil Mill still using the traditional method. Traditional method tends to charge higher overhead costs against its production volume of products more than any other products that produced less. In addition, the results of the preliminary observations are conducted, there is a waste of processing time in PT XYZ Palm Oil Mill's work stations. From the results of the preliminary observations, the actual processing time company as measured using Stopwatch Time Study method shows the result of a longer processing time than based on company data. To solve both problems that a calculation of product pricing based on its activities need to be done.

Research about cost based on activity has been conducted by [2] who estimate cost based on activity of pyro processing. From the results of the calculations there are 4 pieces of activity that consumes the costs i.e. the Pretreatment (19%), Electrochemical Reduction (29%), Electro fining (22%) and Electro wiring (30%). [3] used Activity Based Costing (ABC) method to measure the cost of products in real-time and [4] used Activity Based Costing (ABC) method able to save the cost of a transplant stem cell. The application of ABC Method in Service Examination MR. Enterography who conducted by [5] able to reduce Non-Value-Added stages and charged 13%, employee work time amounting to 16%, and the processing time of the patient by 17%. While research on Value-added and Non-Value-added Activities was done by [6] on iron ore Industry resulted there are 33% of the Non-Value-Added Activities and 14.20% Necessary Non-Value-Added Activities on production process that result in production efficiency reached only 52%.

2. Methods

Research conducted on Palm Oil Mill in North Sumatra in 2016. This research includes descriptive (descriptive research) because it can solve the problem that exists in a systematic and factual. The object of the research is the observed production activities at the Weighting Bridge work station, Sorting work station, Loading Ramp work station and Sterilizer work station.

The data used are the Production Activity, Production Processing Time, Production Cost and Production Volume. Data processing method is described as follow:

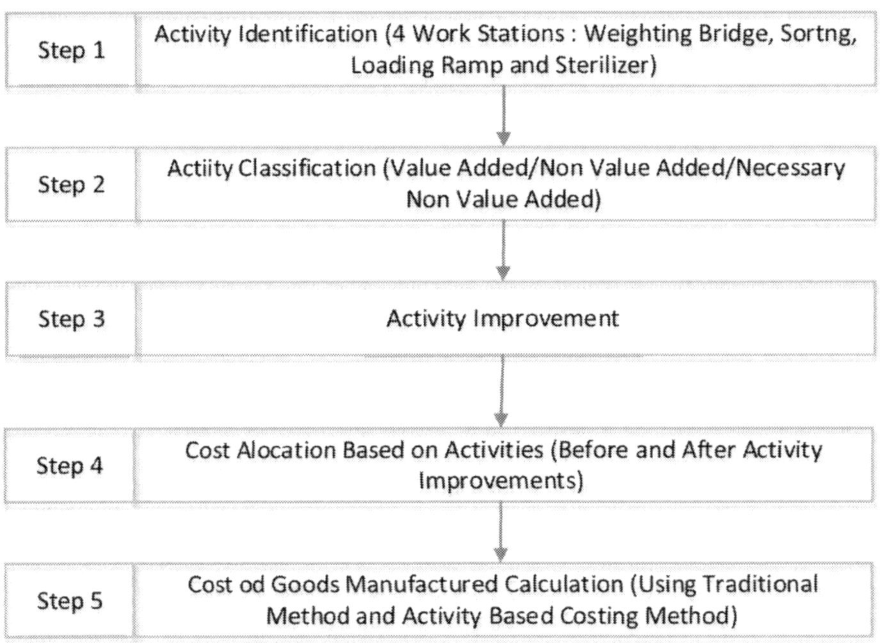

Figure 1. The Stages of Data Processing

Calculation of cost of goods production is conducted using traditional method and Activity Based Costing method

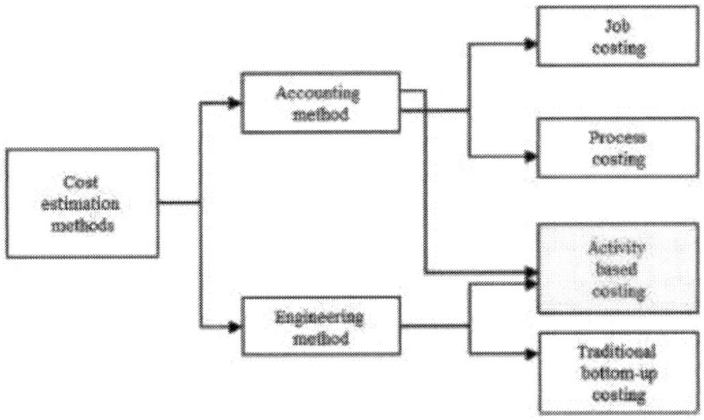

Figure 2. Cost Estimation Methods

Activity-Based Costing (ABC) is an accounting information system that identifies the various activities that are carried out within an organization and collect fees on the basis and nature of existing and expansion of its activities. ABC focuses on the costs inherent in the activity based products to produce, distribute or support the products concerned

3. Results and Discussions

3.1. Process Activity Mapping Arrangement

Process Activity Mapping (PAM) was arranged to identify Value Added and Non-Value-Added Activities. The results obtained can be seen in Table 1. Then the Activity Improvement need to be implemented on each work station

Table 1. Process Activity Mapping

No	Work Station	Amount of Activity	Details	Time Average (minutes)
1	Weighting Bridge	7	3 VA, 2 NVA, 2 NNVA	52.53
2	Sorting	4	1 VA, 3 NNVA	50.21
3	Loading Ramp	6	3 NVA, 3 NNVA	40.59
4	Sterilizer	7	1 VA, 3 NVA, 3 NNVA	115.35

3.2. Activity Improvement

Activity Improvement performed in two ways:

- Doubled the capacity of Fresh Fruit Bunch (FFB) Truck Loader so that the frequency Fresh Fruit Bunch transportation and the average time of activity can be reduced by half. This improvement can only be applied to Weighting Bridge Work Station and Sorting Work Station.
- Improvements to the facilities, equipment and working to smoothen production activity. This improvement leads to smallest production cycle time as the best result. This improvement can only be applied to Loading Ramp Work Station and Sterilizer Work Station.

The results of the activity improvements can be seen in Table 2. The decrease in production activity time will be directly proportional to the decrease in the cost elements involved, i.e. the direct labor cost, as can be seen in Table 3.

Table 2. Results of Activity Improvements

No	Work Station	Average Times		Percentage
		Before	After	
1	Weighting Bridge	52.53	33.75	35.75%
2	Sorting	50.21	35.26	29.77%
3	Loading Ramp	40.59	38.54	5.05%
4	Sterilizer	115.35	114.43	0.79%

Table 3. Decreasing of Direct Labor Cost

No	Work Station	Direct Labor Cost (IDR)	
		Before	After
1	Weighting Bridge	10,043,300	6,452,720
2	Sorting	11,383,190	7,993,851
3	Loading Ramp	6,662,000	6,325,535
4	Sterilizer	17,590,700	17,450,401

3.3. Calculation of Cost of Goods Manufactured Using Traditional Methods

The imposition of costs to calculate the cost of goods production uses a percentage of 90% for Crude Palm Oil products and 10% for palm kernel. The total cost is the cost that is used before activity improvement, as can be seen in Table 4

Table 4. Cost of Goods Manufactured Calculation using Traditional Method

Cost Sub Element	Product	
	Crude Palm Oil (90%)	Palm Kernel (10%)
Raw Material Cost	15,840,531.03	1,760,059.00
Direct Labor Cost	122,498,000.10	13,610,888.90
Indirect Cost	2,574,267.40	286,029,711.30
Total Allocation	18,537,296.43	2,059,699.60
Production Volume	3,539.81	953.78
COGM (IDR/kg)	5,236.81	2,159.50

3.4. Calculation of Cost of Goods Manufactured Using Activity Based Costing Methods.

First overhead percentage allocation for each product was calculated by dividing Production Volume with Cycle Time

$$CPO\ production\ volume = \frac{\text{CPO}\ production\ volume}{Cycle\ time} = 67{,}386.53 \quad (1)$$

$$Palm\ Kernel\ Allocation = \frac{\text{Palm Kernel}\ production\ volume}{Cycle\ time} = 18{,}156.88 \quad (2)$$

So, the overhead percentage for each product can be calculated as follows:

$$\% \, CPO \, Allocation = \frac{CPO \, Allocation}{CPO \text{ Allocation+Palm Kernel Allocation}} x100\% = 78.77\% \quad (3)$$

$$\% \, Palm \, Kernel \, Allocation = \frac{Palm \, Kernel \, Allocation}{CPO \text{ Allocation+Palm Kernel Allocation}} x100\% = 21.23\% \quad (4)$$

Results the percentage of allocation can be seen in Figure 2.

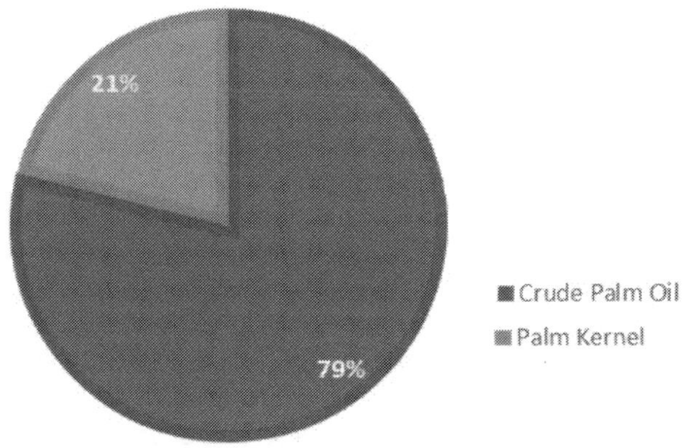

Figure 3. Allocation Percentage

3.4.1. Calculation of Cost of Goods Manufactured Using Activity Based Costing Method Before Activity Improvement. Calculation cost of goods manufactured with the approach of the activity before the Activity Improvement was performed using sub cost element before the activity improvement was implemented and using imposition of percentage that has been calculated above, as can be seen in Table 5.

Table 5. Calculation of Cost of Goods Production with Activity Based Costing Before Improvement Activity

Cost Sub Element	Product	
	Crude Palm Oil (90%)	**Palm Kernel (10%)**
Raw Material Cost	13,863,984.77	1,760,059.00
Direct Labor Cost	107,212,971,90	13,610,888.90
Indirect Cost	2,253,056.04	286,029,711.30
Total Allocation	16,224,253,778,90	2,059,699.60
Production Volume	3,539.81	953.78
COGM (IDR/kg)	4,583,37	2,159.50

3.4.2. Calculation of Cost Of Goods Manufactured Using Activity Based Costing Method After Activity Improvement. Calculation cost of goods manufactured with the approach of the activity after the Activity Improvement was performed using sub cost element after the activity improvement was implemented and using imposition of percentage that has been calculated above, as can be seen in Table 6.

Table 6. Calculation of Cost of Goods Manufactured with Activity Based Costing After Improvement Activity

Cost Sub Element	Product	
	Crude Palm Oil (78.77%)	**Palm Kernel (21.23%)**
Raw Material Cost	13,863,984.77	3,736,605.26
Direct Labor Cost	101,339,342.70	27,312,863.33
Indirect Cost	2,253,056.04	607,241,077.10
Total Allocation	16,218,359,966.55	4,371,149,709.57
Production Volume	3,539.81	953.78
COGM (IDR/kg)	4,581.71	4,582.97

4. Conclusion

Based on the results and discussion so it can be taken a few conclusions are as follows:

- There are 24 activities observed at the Weighting Bridge Works Station, Sorting Works Station, Loading Ramp Works Station, and Sterilizer Works Station with details as follow: 5 Value Added Activities, 8 Non-Value-Added Activities and 11 Necessary Non-Value-Added Activities.
- Improvement activities by producing a decrease in the activity of production time 35.75% in Weighting Bridge Works Station, 29.77% at the Sorting Works Station, 5.05% in Loading Ramp Works Station and 0.79% of the Sterilizer Works Station.
- Cost of goods the manufactured of which is calculated using Activity-Based Costing method after activity improvement is IDR. 4,581 .71/kg for Crude Palm Oil and IDR 4.582,97/kg for palm kernel.
- Cost of goods manufactured of Crude Palm Oil calculated using Activity-Based Costing method shows cheaper result compared with that calculated using the traditional method, meanwhile Cost of goods manufactured for palm kernel shows vice versa result.

References

[1]. Irvan T B, Sosanty F, and Tomiuchi Y 2016 *Asian J. Chem.* **28** 377-0
[2]. Kim S 2015 *Nuc. Eng. And Tech.* **47.7** 849-8
[3]. Wouters W, Stecher J 2016 *Int J. of Prod. Econ.* **183** 235-4
[4]. Massimo M 2017 *Clin. Lymp. Myel. and Leuk.* **1** 1
[5]. Tibor L C 2017 *J of the Ame Col of Rad.* **14(3)** 353-8
[6]. Indrawati S, Ridwansyah M 2015 *Proc. Manuf.* **4** 528-4
[7] Tunggal A W 1993 *Akuntansi Manajemen Kontemporer.* Jakarta, Indonesia: PT Rineka Cipta

IOP Conf. Series: Materials Science and Engineering **309** (2018) 012060 doi:10.1088/1757-899X/309/1/012060

Studies on cure index, swelling behaviour, tensile and thermo-oxidative properties of natural rubber compounds in the presence of alkanolamide

I Surya[1,2*], L Sukeksi[1] and N Hayeemasae[3]
[1]Department of Chemical Engineering, Universitas Sumatera Utara, Medan, 20155, Indonesia
[2]Sustainable Energy and Biomaterial Center of Excellence, Faculty of Engineering, Universitas Sumatera Utara, Medan, 20155, Indonesia
[3]Department of Rubber Technology and Polymer Science, Faculty of Science and Technology, Prince of Songkla University, Pattani Campus, 94000, Pattani, Thailand

*E-mail: indradanas@yahoo.com

Abstract. The studies on cure index, swelling behaviour, tensile and thermo-oxidative properties of unfilled natural rubber (NR) compounds in the presence of alkanolamide (ALK) were carried out. As a rubber additive, the ALK was added into the NR compounds at various loadings (0.2, 0.4, 0.6, 0.8, and 1.0 phr). It was found that the cure indexes of NR compounds with ALK were higher than that of NR compound without ALK (control compound). It was also found that ALK exhibited higher tensile modulus and tensile strength up to 0.6 phr of loading and then decreased with further increases in the loading. The swelling test revealed that 0.6 phr of ALK in the unfilled NR compound exhibited the highest degree of crosslink density which correlated to the highest tensile modulus and tensile strength of the NR vulcanisates. The ALK increased the heat aging resistance of the NR vulcanisates.

1. Introduction

Rubber compounding is an activity that performs the art and science of rubber technology. The rubber compounds, as the output of rubber compounding, are composed by a group of base rubber and rubber additives with their suitable quantities. Each of them has a particular function either in compounding, vulcanisation or end use of the rubber articles. The base rubber will give some rubbery characteristics; rubber additives cause in a wide range of physical, chemical and mechanical properties to the rubber articles.

Rubber additives are curative and non-curative additives. Curative additives affect the cure characteristics of rubber compounds. They are sulphur or other curing agents such as zinc oxide, stearic acid, and accelerators.

Alkanolamide (ALK) is considered as a new rubber additive, relatively. It can be prepared from Refined Bleached Deodorized Palm Stearin (RBDPS), a by-product of cooking oil production, and diethanolamine. It improved the cure characteristics and tensile/mechanical properties of the unfilled NR compounds [1], silica [2] and carbon black filled NR compounds [3]. It also can be utilised as a coupling agent in silica reinforcement of NR [4] and as a curing agent of CB filled chloroprene rubber compounds [5].

It is important to explore more about the function and application of ALK in NR compounds. Therefore, this study reports the cure index, swelling behaviour, tensile and thermo-oxidative properties of unfilled NR compounds in the presence of ALK.

2. Materials and Methods

2.1. Materials

NR grade SMR L was used and obtained from Guthrie (M) Sdn. Bhd., Seremban, Malaysia. Other compounding ingredients such as sulphur, zinc oxide, stearic acid, N-isopropyl-N'-phenyl-p-phenylenediamine (IPPD), and Benzothiazolil disulfide (MBTS) were supplied by Bayer (M) Ltd., Sdn. Bhd., Petaling Jaya, Selangor, Malaysia. All the materials were freshly supplied and utilised as supplied. The ALK was prepared in laboratory utilising Refined Bleached Deodorized Palm stearin (RBDPS) and diethanolamine. The reaction procedures and molecular characterisations of the ALK were given in our previous report [2]. The molecular structure of ALK is presented in Fig. 1.

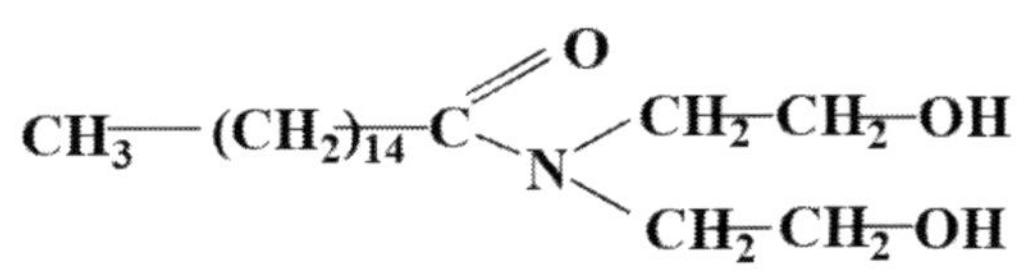

Figure 1. The molecular structure of alkanolamide.

2.2. Compounding

A semi-efficient vulcanization system was used for the compounding. The recipe for the preparation of the NR compound is given in Table 1. The compounding procedure was done in accordance with the American Society for Testing and Material (ASTM) – Designation D 3184 – 80. The compounding was done on a two-roll mill. Table 1 also shows the designation and composition of the NR-based recipes used in this study.

Table 1. The composition and designation of the NR compounds

Ingredients	Content (phr)*	Designation					
		A/0.0 (Control)	B/0.2	C/0.4	D/0.6	E/0.8	F/1.0
SMR L	100.0	100.0	100.0	100.0	100.0	100.0	100.0
Zinc oxide	5.0	5.0	5.0	5.0	5.0	5.0	5.0
Stearic acid	2.0	2.0	2.0	2.0	2.0	2.0	2.0
IPPD	2.0	2.0	2.0	2.0	2.0	2.0	2.0
MBTS	1.5	1.5	1.5	1.5	1.5	1.5	1.5
Sulphur	1.5	1.5	1.5	1.5	1.5	1.5	1.5
ALK	0.0 – 1.0	0.0	0.2	0.4	0.6	0.8	1.0

*parts per hundred parts of rubber

2.3. Cure characteristics

The cure characteristics of the NR compounds were obtained using a Monsanto Moving Die Rheometer (MDR 2000) which was employed to determine the scorch time (ts_2) and cure time (t_{90}) according to ISO 3417. Samples of the respective compounds were tested at 150 °C. The NR compounds were subsequently compression-moulded using a stainless steel mould at 150 °C, with a pressure of 10 MPa and applying a laboratory hot-press based on respective curing times. The cure index or cure rate index (CRI) of the compounds was calculated as follows:

$$CRI = \frac{100}{(t_{90} - ts_2)} \tag{1}$$

2.4. Swelling behaviour

The swelling test was performed in toluene according to ISO 1817. Cured test pieces of the compounds of dimension (30 mm x 5 mm x 2 mm) were weighed using an electrical balance and swollen in toluene until equilibrium, which took 72 h at room temperature. The samples were removed from the liquid, the toluene was removed from the samples surfaces, and the weight was determined. Calculation of the change in mass is as follows;

$$\text{Swelling (\%)} = \frac{W_2 - W_1}{W_1} \times 100\% \tag{2}$$

where W_1 is the initial mass of specimen (gram) and W_2 is the mass of specimen (gram) after immersion.

2.5. Tensile properties

Dumbbell-shaped samples were cut from the moulded sheets. The tensile test was performed at a cross-head speed of 500 mm/min using an Instron 3366 universal tensile machine, according to ISO 37. The tensile strength (TS), stress at 100% elongation (M100), stress at 300% elongation (M300), and elongation at break (EB) were determined.

2.6 Thermo-oxidative Ageing.

The thermo-oxidative ageing tests were done according to ASTM D573-04. Test samples were placed in an air circulating oven at 100 °C for 48 hours. At the end of the period, test samples were removed from the oven, cooled for 30 minutes at room temperature and allowed to rest from 16 to 96 hours before determination of the tensile properties. Five specimens were used and the average was determined and reported. Retention in tensile properties was calculated as follows:

$$\text{Retention (\%)} = \frac{\text{Value After Ageing}}{\text{Value Before Ageing}} \times 100 \tag{3}$$

3. Results and Discussion

3.1. The cure index and swelling behaviours

The cure index and swelling behaviour of the NR compound with and without ALK are shown in Table 2. The cure rate index (CRI) is a measurement of the cure rate based on the differences between the optimum cure time (t_{90}) and incipient scorch time (ts_2) [6]. The CRI of the NR compound with ALK was higher than that of the NR compound without ALK. It indicated that the rate of curing increased slightly with increasing the ALK loading. It was due to the function of ALK as an additional accelerator [2, 7] which increased in the number of reactive sites on the rubber molecules used for the cross-linking reactions.

Table 2. The cure index and swelling behaviour of NR compounds with and without ALK.

CRI and Swelling Properties	NR Compounds					
	A/0.0 (Control)	A/0.2	A/0.4	A/0.6	A/0.8	A/1.0
CRI, min.	39.4	39.5	41.5	42.2	44.2	47.6
Swelling, %	396.2	395.9	395.2	375.5	384.8	399.2

As shown in Table 2, ALK decreased the swelling percentage of NR compounds up to 0.6 phr. Further increases in the ALK loading (0.8 phr) increased the swelling percentage. It is believed that the swelling percentage is directly related to the crosslink density of a rubber vulcanisate [6]. The less is the percentage; the higher is the crosslink density. The ALK affected the crosslink density of the NR vulcanisate. It was attributed to the action of the ALK as an additional accelerator during the curing of the NR compounds [2-4, 7]. The amine constituent of the ALK activated not only the rubber but also the elemental sulphur during curing. The amine and others curatives formed intermediate complexes which attached the available elemental sulphur to rubber chains more efficient [8], causing a higher degree of crosslink density.

The decrease in crosslink density beyond a 0.6 phr loading of ALK was most probably due to the excessive amount of ALK which dissolved a part of the elemental sulphur and as a consequent, a less sulphur was attached to the NR chains.

3.2. The tensile properties

The tensile properties of the NR vulcanisates with and without ALK are presented in Table 3. As shown, the EB of B/0.2-vulcanisate was higher than that of A/0.0-vulcanisate. Increasing the ALK loading caused further increases in the EB of the NR vulcanisate.

Table 3. The tensile and thermo-oxidative aging properties of NR compounds with and without ALK

Compounds	Tensile properties				Retention (%)		
	M100 (MPa)	TS (MPa)	EB (%)	EB Increment (%)	M100 (MPa)	TS (MPa)	EB (%)
A/0.0	0.538	17.7	1070.0	0.00	129.4	92.9	93.1
B/0.2	0.544	19.3	1095.7	2.40	133.1	93.9	92.4
C/0.4	0.561	19.5	1116.5	4.35	136.7	94.7	88.0
D/0.6	0.586	20.1	1141.5	6.68	137.5	97.5	78.5
E/0.8	0.545	18.4	1154.0	7.85	139.9	99.0	75.5
F/1.0	0.516	16.5	1172.0	9.53	147.5	99.1	74.2

It was attributed to the function of ALK as an internal plasticiser to the unfilled-NR compound [2-3, 7]. A plasticiser is a rubber additive which improves rubber compound processing and modifies hardness, flexibility or distensibility of a rubber vulcanisate [8]. The ALK plasticised and provided some layers in NR compound which provided free volume, thus allowing more mobility/flexibility for the NR chains [8-9]. Increasing the ALK amount has the same effect as an increase in free volume that enhanced the flexibility or the extensibility of the NR vulcanisates. Therefore, a higher ALK loading caused in a higher percentage of EB increment.

Table 3 shows that the ALK increased the M100 and M300 slightly up to a maximum level at 0.6 phr and then decreased with further increases in the loading. The TS also exhibit a similar trend. M100 and M300 are typical measures stiffness or hardness of a rubber vulcanisate [10-12]. Modulus or stiffness/hardness and tensile properties are dependent mainly on the degree of crosslink [9, 11]. The improvements in M100, M300, and TS up to 0.6 phr of ALK were due to a higher degree of crosslink density. The decreases in those properties, beyond 0.6 phr, were attributed to a lower degree of crosslink density and a more pronounced the plasticizing effect of the ALK.

3.3. The thermo-oxidative properties

The thermo-oxidative ageing properties of the NR vulcanisates with and without ALK were also studied. During the ageing process, NR vulcanisates were subjected to the hot air and this leads to the diffusion of oxygen into the outer part of NR vulcanisates. Some parts of the NR vulcanisates were immediately consumed by oxidation reaction to produce a further crosslinking, scission of rubber chains and/or crosslinking and combination of the reactions [13].

IOP Conf. Series: Materials Science and Engineering **309** (2018) 012060 doi:10.1088/1757-899X/309/1/012060

The ageing process is also like post-curing, the phenomenon clearly indicates that some microstructures such as the formation of new intermolecular bonds have taken place after ageing and resulted in higher crosslink density of the rubber vulcanisates.

The ageing properties of a rubber vulcanisate might be presented in terms of retention values (the property ratio of the aged to unaged specimens). The retention values of M100, TS, and EB of the NR vulcanisates with and without ALK are shown in Table 3, respectively. As shown in Table 3, the M100 of the aged vulcanisates increased due to the formation of some new crosslinks and enhancement the degree of crosslink density [13]. Hence, the increased in M100 was simply attributed to post-curing effect which enhanced the crosslink density of the aged unfilled NR vulcanisates.

It was also observed that the retention values of M100 of the aged vulcanisates with ALK were greater than that of the aged vulcanisate without ALK. The higher the ALK loading, the greater was the value of M100 retention. It was clearly due to the function of ALK as an additional accelerator which greatly affected the post-curing process of the unfilled NR compounds. Increasing the amount of accelerator of a Semi-EV curing system will have the same with changes the curing system into an EV system. The increase in the accelerator to sulphur ratio, as in EV system, is believed could reduce the main chain modifications, the formation of cyclic structures and rank of sulphur crosslinks and improve the vulcanisation efficiency and crosslink density [8]. These changes in microstructures have been correlated with improved of mechanical and ageing properties. According to Rattanasom et al. [14], a higher content of monosulphidic crosslinks of EV system would provide higher thermal stability. This contributes to increasing the retention of M100 with higher ALK loading for the unfilled NR vulcanisates with ALK.

Table 3 shows that the TS of the NR vulcanisates with and without ALK decreased after the ageing process. This was attributed to the degradation process of NR vulcanisates were dominant by the main chain scission which leading to the main chains modification which reduced the ability of NR vulcanisates to undergo strain-induced crystallization. This phenomenon decreased the EB of the aged NR vulcanisates.

It was also observed with the addition of ALK provided better TS retention to unfilled NR vulcanisates. The higher the ALK loading increased the TS retention. Again, this was due to the function of ALK as an accelerator which modified the microstructures of the aged unfilled NR vulcanisates during the post-curing process. The nitrogen atoms of ALK will react with sulphur atoms of disulphidic crosslinks and immediately formed more monosulphidic crosslinks in the aged vulcanisates which gives better ageing resistance compared to disulphidic crosslinks.

From Table 3, the EB of the NR vulcanisates with and without ALK decreased significantly after 3 days of thermo-oxidative ageing at 100 °C. The higher the ALK loading, the lower the EB retention was observed. It simply was attributed to the increased crosslink density which mobilized the rubber chains of the aged vulcanisates less freely. Similar with M100, EB also depends mainly on the degree of crosslink density [13,15].

4. Conclusions

From the studies, the following conclusions were drawn:

1. Alkanolamide enhanced the cure index of unfilled natural rubber compounds. The cure indexes of natural rubber compounds with alkanolamide were higher than that of without alkanolamide. The higher the alkanolamide loading, the higher was the cure index.
2. Alkanolamide decreased the swelling percentage and increased the crosslink density of unfilled natural rubber compound up to 0.6 phr loading. The cure indexes of natural rubber compounds with alkanolamide were higher than that of without alkanolamide.
3. Alkanolamide enhanced the tensile properties of unfilled natural rubber vulcanisate up to 0.6 phr loading.
4. Alkanolamide enhanced the thermo-oxidative ageing resistances of unfilled natural rubber compounds.

Acknowledgements
The authors would like to thank Universiti Sains Malaysia, for providing the research facilities for carrying out the experiment and for making this research work possible.

References

[1] Surya I *et al* 2017 *IOP Conf. Ser., Mater. Sci. Eng* **223** 012012.
[2] Surya I *et al* 2013 *Polym Test* **32** 1313-21.
[3] Surya I *et al* 2015 *Polym Test* **42** 208-14.
[4] Surya I *et al* 2014 *Polym Test* **40** 24-32.
[5] Surya I and Ismail H 2016 *Plast Rubber and Compos* **45** 287-93.
[6] Nabil H *et al* 2013 *Polym Test* **32** 385-93.
[7] Surya I and Ismail H 2016 *Polym Test* **50** 276-82.
[8] Rodgers B 2004 *Rubber Compounding: Chemistry and Applications* CRC: 2004.
[9] Ismail H and Chia H 1998 *Polym Test* 7 199-210.
[10] Mousa A and Karger-Kocsis J 2001 *Nanocompos Macromol Mater Eng* **286** 260-66
[11] Sloan J.M. *et al* 1986 *Rubber Chem Technol* **59** 800-08
[12] Teh P *et al* 2004 *Eur Polym J* **40** 2513-21
[13] Azura A.R *et al* 2006 *Polym Plast Technol Eng* **45.7** 893-96
[14] Rattanasom N *et al* 2007 *Polym Test* **26** 369-77
[15] Surya I *et al* 2017 *IOP Conf. Ser., Mater. Sci. Eng* **223** 012012

IOP Conf. Series: Materials Science and Engineering **309** (2018) 012061 doi:10.1088/1757-899X/309/1/012061

How manageable a person feels the place to be: self-efficacy in supporting tourism

N Ginting[1*], N V Rahman[2], A D Nasution[2] and A T Widya[2]
[1,2] Department of Architecture, Faculty of Engineering, University of Sumatera Utara, Jl. Dr. T. Mansyur, Medan 20155, Indonesia

[*]E-mail: nurlisa@usu.ac.id

Abstract. How manageable a person feels the place exceedingly influences people's satisfaction toward supporting tourism development with the respect of self-efficacy. Individuals with self-efficacious can perform or complete a task as a measure of personal agency. This research aims to investigate the influences of self-efficacy principle in defining place identity on tourism environment in Karo Regency, North Sumatera, Indonesia. A mixed-method was employed to obtain local people' and tourists' perception of comfort, safety, accessibility, and confidence. The result shows that confidence contributes in maintaining self-efficacy in Karo Regency. Meanwhile, the absence of public facilities constitutes the feeling of self-efficacy and threatens the place identity.

1. Introduction

As the leading sector of economic, many countries have strived to promote economic development by developing tourism. Significant tourism growth makes it as one of the largest foreign exchange sources for a country that increase in community income, provide for employment, maintain the authenticity and culture of a region, and preserve the environment thus many parties are encouraged to develop it.

The development of tourism is also a foremost priority in achieving Indonesia's economic goals where various unique and distinct natural environments provide tourism resource with noticeable attraction. Its development is potential to accelerate national economic growth which aims to improve people's productivity and competitiveness in international markets. It is also contained in a national development agenda called Nawa Cita, President Joko Widodo's Nine Priorities.

1.1. Issue and Objective

By the Regional Spatial Plan (Rencana Tata Ruang Wilayah) of Karo Regency in 2010-2030, where the current research took place, the tourism sector is one of priority development plannings after the agricultural and industrial sectors. The Karo Regency government website lists 20 tourist destinations in the area, including Sipiso-Piso Waterfall, Mount Sibayak, Mount Sinabung, Tongging, Gundaling Hill, Lake Lau Kawar, Lau Debuk-Debuk Hot Spring, Bukit Barisan Forest Park, Mount Sipiso-Piso, Desa Lingga (Desa Lingga), Pasar Buah Berastagi (Berastagi Fruit Market), and many more [1].

[1] Corresponding author. Tel.: +62811608102; fax: +62618219525

Moreover, Karo regency is also rich in cultural heritage such as traditional ceremonies and dances that are still ongoing until today. This makes it worth mentioning as a natural and cultural tourism destination [2]. Karo Regency also has incredible natural resources. However, this potential has not been well-developed. Many tourism objects are not well-organized and well-maintained, whereas tourism destination that has great natural scenery can affect tourism development. Tourism is an attraction for visitors who come to see with all interesting and valuable things [3][4] which are influenced by tourists' perception toward peculiarity of a tourism environment. It is a mutual relation between a person and his or her environment, within a framework that conceptualizes the environment as a psychosocial construct rather an objective fact [5]. Positive perception indicates satisfaction of local people and tourist alike. Regarding with the topic of sustainable tourism, people satisfaction in tourist destination lies as the main movement toward more socially sustainable development [6].

Personal perception of an area is affected by both of direct and indirect experience, where the character of a tourism environment impacts on embedding positive perception. Tourists will differently value experience depending on their expectation, situation, context, and resources offered. Thus, understanding people's subjective experience is crucial in tourism. The perceived experience indicating the character of a place is important in tourism because it would build meaning of an environment and attachment to the region [7]. Place identity affects the visitor's attitude towards the impact of tourism resulting in a positive and negative evaluation [6][8]. It will support the quality of life and bring the happiness to the residents and tourists alike [8].

The current research aims to examine self-efficacy principle, as one of place identity principles [9], within Karo Regency to analyze associations between self-efficacy and its support for tourism. Previously, researchers examined self-efficacy in a historic corridor in an urban area [8]. The current research took place in tourism environment that has diverse natural resources. Different city population density will affect different perception and concept of self-efficacy in taking a role in a community. The findings will complete prior research (Ginting et al., 2017) that examines distinctiveness principle within Karo Regency [10] toward supporting tourism based on place identity. It will accomplish assessment and reference for the government of Karo Regency in planning tourism development.

2. Literature Review

2.1. Self-efficacy and Manageable Environment

Self-efficacy refers to personal judgments of someone's capabilities in organizing and executing courses of action to attain specific task in a successful way [11][12]. It is a belief about whether a person can carry out the capabilities and abilities to perform a specific goal. Lalli (1992) used "perception of familiarity" term in defining self-efficacy [13]. This perception is an impression of life experience in a place as a result of one's actions in an urban environment which is an expression of the success of one's cognitive orientation. Self-efficacy affects how people feel, think, and perform [12]. Moreover, it significantly influences someone's behavior through cognitive, motivational, affective, and decision processes [14]. A higher feeling of self-efficacy will result in greater effort, performance, and satisfaction [15]. It is closely related to goal orientation, performance evaluation, and job satisfaction [16] while a low sense of self-efficacy is associated with depression, anxiety, and helplessness [12].

Self-efficacy is used to measure personal agency [17][18] which indicates someone's capability to originate and direct actions for given goals. The feeling of self-efficacy must be generated to control the surrounding environment. It is maintained when an environment facilitates people daily life or at least not hinder them from achieving a positive identification [19]. People can perform a task or control an environment in a place so they can carry out their chosen activities. Winkel (1981) stated it as "a manageable environment" where comfort, safety, accessibility, and confidence contribute in controlling an environment [19]. To have a sense of self-efficacy requires direct experience in overcoming obstacles. In the current research, comfort, safety, accessibility, and confidence indicates as the guiding variables in measuring self-efficacy in tourism environment (Table 1).

Table 1. Self- Efficacy Variables

Self- Efficacy Variables	**Parameter**
Comfort	– The feeling of comfort and relax in a place
	– The facilities relates to comfort (dustbins, sitting areas, public toilets)
Safety	– The feeling of security
	– The facilities relates to safety (information center, and night-time lighting)
Accessibility	– The facilities relates to accessibility (pedestrian path, public transportation, and parking spaces)
Confidence	– Ability to complete and perform tasks
	– Positive information obtained

3. Methodology

3.1. Research Area

The research was conducted in 2016. Five tourism destinations in Karo Regency, North Sumatera Province, Indonesia were chosen as research area include Bukit Gundaling (Gundaling Hill), Pasar Buah (Fruit Market), Desa Lingga (Lingga Village), Sipiso-Piso Waterfall, and Bukit Kubu (Kubu Hill) (figure 1). The places represent the major destination that has the highest number of visitors among nine samples of tourism destinations in Karo Regency based on the pilot research conducted by Architecture students in the University of Sumatera Utara.

Bukit Gundaling (Gundaling Hill) is one of the popular destinations due to the natural panorama that can be enjoyed from the height of 1,575 meter above sea level. Before arriving Bukit Gundaling, there is a row of souvenir shops and traditional market that sell local commodities in Pasar Buah. Meanwhile, Desa Lingga, cultural heritage area, is a historical village that has diverse historical and cultural heritage. Moreover, Sipiso-Piso in Merek Sub-district is a plunge waterfall and a nearby vantage point offering great vistas of the waterfall and Lake Toba. It is located at the northernmost tip of the Lake Toba caldera. The last location is Bukit Kubu (Kubu Hill), a green hill with open space.

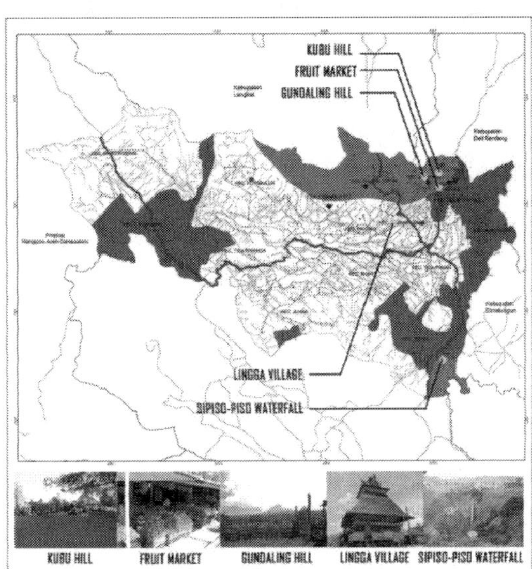

Figure 1. Research Area

3.2. The Method

This research adopts a mixed method -qualitative and quantitative method approach- to obtain respondents' perception of self-efficacy. The qualitative data were conducted through depth interview and field observation at five areas. The depth interview involves eight key respondents consists of tourism stakeholders, local figures, governments, academics, and people who directly or indirectly involved in Karo tourism. Moreover, field observation was done by collecting the physical characteristics, activities of people, the provision of public facilities, the intensity of pedestrians and all things required.

Meanwhile, the quantitative method through questionnaire survey was measured on a five-point scale which ranges from strongly disagree (1) to strongly agree (5) to assess the self-efficacy variables. The measurements include comfort, safety, accessibility, and confident [18][19]. The 360 questionnaires were equally distributed to local people and tourists alike at the five tourism area. The results of questionnaire survey were processed using the Statistical Package for the Social Sciences (SPSS 20) to get an overall sense of the mean ratings of the perceptual items.

Furthermore, the results of qualitative and quantitative data are triangulated to find out the convergence. The result of qualitative data will be coded and separated into the same variables. Meanwhile, the quantitative data used percentage and average value to analyze the level of respondents' perception.

4. Result and Discussion

4.1. Respondents

Each area represents 72 respondents: (36) local people; (22) domestic tourists; and (14) foreign tourists. As a result, of the 360 respondents, 47% were women (n=192), and 53% were men (n=168), the largest age group was 25-49 years old (46%). The other age groups were 18-24 (36%), 50-64 (15%), and 65 years old or above (2%). Tourists were dominated by Malaysian (54%), followed by Singapore (17%). When asked how long they have already visited Karo, 63% of respondents reported one day and 37% two days or more. Regarding travel, 54% of respondents stated that having sightseeing is the main motivation visiting Karo regency where it is their first visit to Karo Regency (53%). Among the respondents, 53% traveled with friends, followed by 32% with groups.

4.2. Self-efficacy Principle

The mean scores of 19 items are displayed in table 2. The item with the highest mean scores is: "I would recommend this place to everyone (table 2). This item indicates respondents look forward to Karo regency in developing its tourism. In contrary, the provision of public facilities gets the lowest score (table 2).

Table 2. Mean Ratings of Self-efficacy Principle (n=360)

	Statements	Pasar Buah		Gundaling		Sipiso-piso		Desa Lingga		Bukit Kubu	
		Local People	Tourist	Local People	Tourist	Local People	Tourist	Local People	Tourist	Local People	Tourist
		Mean	Mean	Mean	Mean	Mean	Mean	Mean	Mean	Mean	Mean
Comfort	I feel comfortable here	3.39	3.42	3.36	3.39	3.36	3.36	3.36	3.58	3.47	3.83
	I feel relax here	3.33	3.44	3.30	3.50	3.33	3.33	3.44	3.28	3.50	3.83
	This place makes me feel enthusiastic	3.22	3.31	3.19	3.28	3.42	3.36	3.38	3.25	3.58	3.58
	I feel at home here	3.11	3.03	3.00	2.92	3.22	3.11	3.50	2.42	3.50	3.50
	There are adequate number of dustbins here	2.69	2.97	2.91	2.83	3.06	2.97	3.00	2.94	3.44	3.44
	There are cozy and relaxing sitting areas here	2.47	2.89	2.91	3.08	2.97	3.14	2.97	2.78	3.39	3.39
	There are sufficient number of good public toilet	2.56	2.36	2.52	2.36	2.86	2.75	2.52	2.42	3.36	3.36

	Statements	Pasar Buah		Gundaling		Sipiso-piso		Desa Lingga		Bukit Kubu	
		Local People	Tourist	Local People	Tourist	Local People	Tourist	Local People	Tourist	Local People	Tourist
		Mean	Mean	Mean	Mean	Mean	Mean	Mean	Mean	Mean	Mean
Safety	I feel safe while visiting this place	3.03	3.31	3.02	3.17	3.14	3.31	2.94	3.56	3.50	3.53
	There is an information center here	2.92	2.72	2.72	2.78	2.69	2.89	2.55	3.39	3.42	3.39
	There is night-time lighting	2.53	2.69	2.63	3.00	2.36	2.89	2.86	3.22	3.17	3.28
Accessibility	There is comfortable pedestrian path	2.81	2.67	2.8	2.97	2.94	3.03	2.69	3.11	3.31	3.17
	There is good public transportation system	2.89	2.83	2.97	2.94	2.69	3.00	3.11	3.22	3.39	3.17
	There are sufficient number of good parking spaces here	2.86	2.78	2.58	2.86	2.86	3.42	3.08	2.89	3.39	3.72
Confidence	I can perform my activities well here	3.11	3.42	3.22	3.11	3.22	3.19	3.30	3.39	3.39	3.42
	I feel satisfied while traveling here	3.06	3.50	3.19	3.39	3.25	3.36	3.16	3.67	3.44	3.61
	This place is the best tourist destination to do the things I want	2.83	3.14	2.97	3.17	3.33	2.97	3.08	2.72	3.25	3.31
	This place gives experience and new insights	3.00	3.47	3.25	3.42	3.22	3.39	3.27	3.81	3.31	3.53
	Positive information obtained reveals the condition of the tourism destination	3.03	3.44	3.08	3.22	3.11	3.22	3.19	3.22	3.33	3.50
	I would recommend this place to everyone	3.25	3.58	3.19	3.44	3.47	3.36	3.44	3.75	3.64	3.81
Value Format: 1 = strongly disagree, 5 = strongly agree		2.95	3.10	2.99	3.10	3.08	3.16	3.10	3.19	3.41	3.49

4.2.1. Comfort

Comfort is the pleasant feeling when interacting in an environment [20]. It will be created when an individual feels satisfied with his or her need that s/he can perform an act rationally and efficiently [8][18]. The comfort level within a place can be determined from the length of time s/he would spend [20]. With the provision of adequate facilities, tourists will be comfortable and satisfied to stay at the tourist attraction because it can influence visitors' satisfaction and well-being [4][12][21].

Comfort contributes to the positive and negative feelings about the place. Awareness of the comfort will affect someone performs rationally [8][18]. The results indicate that most respondents feel comfortable when they are on their destination (table 2). Comfort will generate satisfaction towards an area [7][8]. They also feel relaxed when they are there and feel enthusiastic when being there (table 2). However, they do not feel like at home (table 2). In this case, facilities and environmental factors affect respondents' perception. Desa Linga and Bukit Gundaling received the lowest response due to fewer facilities in Desa Lingga and sexual immorality by adolescents in Bukit Gundaling. This makes people bothered and feels uncomfortable. This is also supported by depth interviews with the key respondent:

> "Sexual immorality by adolescents in Bukit Gundaling makes people uncomfortable here" (Key Respondent: Cultural Figure/Academics)

An uncomfortable environment is closely associated with sitting area, dustbins, and public toilets. According to most respondents, provision of dustbins in five tourism destinations is still inadequate (table 2). Of the five tourism environments, only Bukit Kubu gets good ratings because there is a sufficient number of a well-maintained dustbin. There are cleaners moving it at any time so that no trash can be found. The dirty environment will affect one's psychology and physical [22] hence the provision of dustbins in the tourist attraction is indispensable whereas general cleanliness and its maintenance of an area will encourage longer place engagement and feeling of safety and security [7]. Its absence will generate uncomfortable environment. A clean environment can make a place more attractive [10][20]. Moreover, respondents also complain to horses dung in Pasar Buah and Bukit

IOP Conf. Series: Materials Science and Engineering **309** (2018) 012061 doi:10.1088/1757-899X/309/1/012061

Gundaling which causes a stink. This creates inconvenience feeling both physically and visually as well. This is supported by the key respondents:

> "Environment surrounding Bukit Gundaling is dirty and not well-maintained thus required appropriate maintenance and support from various parties, especially government" (Key Respondent: Cultural Figure/Academics)

> "There are a lot of rubbish previously in Sipiso-Piso due to unavailable dustbins. However, in last three years, the provision of dustbins is available though still any rubbish at the area today" (Key Respondent: Cultural Figure/Academics)

People's comfort is also essentially generated by the provision of sitting area and public toilets. The findings show that sitting areas in five tourism environment are still lacking (table 2). Respondents argue that they are difficult to find a place to sit. Inconvenience environment is also caused by the inadequate and poor condition of the public toilets (table 2). Moreover, the location of the public facilities is less strategic. It is also in line with interviews with the key respondent:

> "Toilets in Desa Lingga are still inadequate; we have to go down the stairs to get there" (Key Respondent: Cultural Figure/Academics)

Only Bukit Kubu gets good ratings because people can easily find a seat and public toilets. Sitting area will build a sense of personalization which encourages human use, social behavior, and social activities [23]. In this case, public toilets contribute to the comfort of an environment. Moreover, the place should also provide public toilets to facilitate people with different abilities. The physical, psychological and environmental comfort contributes to people 'satisfaction and convenience being in place. The degree of comfort will develop an attachment to that place [7].

4.2.2. Safety

Safety is essential to achieve the comfort and satisfaction of visitors [24]. Lack of security will damage place identity and affect the desire of tourists to come. It is also stated that the environment with a high level of crime makes one uncomfortable to stay longer [8][18]. Safe environments make people able to freely move because they feel safe and protected from harm.

The existence of security will make one aware of the efficacy in performing a task [8]. When traveling on five tourism environment, most tourists feel safe (table 2). However, different perception comes from local people in Desa Lingga. The disagreement is supported by interviews with the key respondent:

> "We should keep an eye on local people to avoid crime, so tourists are not disappointed (Key Respondent: Cultural Figure/Academics)

The tourism industry is fragile to the cases such as crime, terrorism, or road accidents. The success of tourism depends on being able to provide safe and secure on tourism destination [25]. The increasing of criminal cases will contribute the feeling of anxiety and scare among tourists [25][26]. In this case, the existence of information center and night-time lighting also plays an important role in the safety of visitors of Karo Regency.

Information center should ensure the current and reliable information on important safety for tourists. However, most respondents argue that information centers are difficult to find in research areas (table 2). Its location is not strategic and has fewer facilities. They also argue that night-time lightings in the research are still inadequate (table 2). Appropriate street lighting in tourism environment is important to create a safe environment [23]. It can eliminate crime and violence and decrease the likelihood of pedestrian, bike, and vehicle accidents. The findings show many tourists leave when the day goes to the night because of inadequate night-time lighting. Night-time lighting is an important public facility that affects human activities and promotes transportation and pedestrian safety [27]. This is also supported by the interview:

> "To develop tourism in Karo, the first which is crucial to be improved is accessibility. When we access the tourism destination on the night, surely crime is more potential happened. Thus, required good accessibility where the provision of signages and night-time lighting in tourism destination are exceedingly important" (Key Respondent: Head of Tourism Planning Program)

Bukit Kubu is the only tourist attraction of the five tourist attractions that receives a positive response due to strategic location and cleanliness of public toilet and adequate street lights. Its existence will help them to understand about the place and feel confident while being in place [8]. People can easily access the existing facilities in the area.

4.2.3. Accessibility
Accessibility is one of the important elements in a tourism area [28] and affects one's interests to visit somewhere [8]. The easier access is then, the higher a person's interest to visit. On the efficacy, the existence of good accessibility also helps people to fulfill their work. It is defined as the availability in reaching the destination such connectivity and continuity of pedestrian path, public transportation, parking facilities, and traffic flow. It should be supported by transportation facilities and road infrastructure [4]. The results show that pedestrian path is not good (table 2). The absence of pedestrian path makes tourist uncomfortable whereas it is an essential element of walkability [29]. Width, roofs & shades, and surface condition of path, drainage & lighting systems, grade separation, landscaping & amenities appropriately addressed to provide comfort and safety for the pedestrians [29].

The lengthy travel times, high costs, and entry barriers can impact on the willingness of tourists visit the place. The results find that public transportation is still poor (table 2). Sipiso-Piso and Desa Lingga particularly need additional transport modes. The provision of parking space also needs appropriate management due to the off-street parking. The parking space can help someone to perform his or her activities. In all five tourism destinations, it also lacked positive response from respondents (table 2) due to the less space. This is also supported by depth interview:

> "The less spacious parking space in Brastagi becomes a crucial problem. No space to pick-up and set-down, so it often disturbs tourists. To overcome the crowded traffic flow, commonly local police will conduct traffic simulation. Hence, car is not able to pass and park any longer in Brastagi" (Key Respondent: Hotel Manager)

Moreover, all off-street parking areas need more reviewed to achieve and maintain secure car park status though most respondents have not fully considered it. Hence, improving security means reducing criminal activity, the fear of crime, and the perception of crime. Only Bukit Kubu gets positive responses because of sufficient capacity of the parking space.

4.2.4. Confidence
Self-confidence is one's belief that s/he is capable of behaving to obtain the desired result [11]. Everybody has a different level of confidence, depending on how they assess the environment [18]. Confidence will influence behavior in a social environment [30] where human behavior contributes in embedding a concept of a place s/he lives in [21]. It comes from the awareness when one decides to do something. The results show that most respondents can perform their activities well (table 2). They were also satisfied when traveling (table 2). Nevertheless, they do not consider that the place was the best place to do what they wanted due to the lack public facilities (table 2). However, Bukit Kubu gets the highest rating. People can perform effectively which is expressed by key respondent:

> "Bukit Kubu has a beautiful natural landscape. People can perform effectively. Every kid also can play in full of happiness there" (Key Respondent: Cultural Figure/Academics)

Though getting negative assessment among tourists, the five tourism environment areas provide new experiences and insights (table 2). The positive information obtained reveals the condition of the tourism destination. The necessity to achieve confidence and comfortable place is a psychological motive for every human being to increase the QoL and preserve culture [12][21]. Hence, respondents would also recommend the five tourist attractions (table 2). Bandura (1986) used term "vicarious experience" to reveal it [31]. In the context of tourism, others' experience or can influence individuals' observation or visualization of a place and raise perceptions of self-efficacy in travel planning. People may believe that they can perform and master similar activities successfully as others' positive experience influences self-confident in travel planning [15].

5. Discussion and Conclusion

This research examined self-efficacy principle through respondents' perception toward supporting tourism. The perception of being comfortable, safe, provision of accessibility, and confidence observed are influential in indicating satisfaction of local people and tourist alike. Also, self-efficacy constructs personal agency in performing specific goal. Perceived experience in a place affects self-efficacious. Both direct and indirect experience contributed in embedding self-efficacy during traveling creates the image of an area which defines place identity.

Among four variables of self-efficacy principles, only one variable, "confidence" showed an impact on respondents' efficacy. Meanwhile, the public facilities and service of tourism related to comfort, safety, and accessibility got negative responses. Due to the increasing number of travelers and demands from several tourism' parties, a tourism environment cannot just merely offer the natural attraction [10]. The lack of the public facilities and service of tourism will influence the travel planning and someone's performance in completing certain goals. The public facilities and potential tourist attractions of Karo Regency need to be explored continually to empower the improvement of people's welfare.

The current findings will complete prior research (Ginting et al., 2017) in accomplishing reference for the government of Karo Regency in spatial planning based on place identity. More studies are recommended to outline and conceptualize the influential pattern exerted by self-efficacy principle. Believed that different population density, the size of a city, and different status and stage of tourism development may impact on self-efficacy principle in embedding place identity [6]. It is suggested that the interpretation of further research be more comprehensive by including moderating variable that account of this limitation. Moreover, the generalizability of findings may be limited in another place due to a different level of knowledge, attitude, behavior, and culture [10]. Social-cultural factor becomes a border that indicates the distinct and unique character of a settlement [21].

This research has one implication for detail of implementation. Most of the local people do not recognize the provision of public facilities toward supporting tourism development. The results show that the absence of public facilities is not a crucial problem for most local people and tourists alike. Thus other practical implication of this research for tourism planners specifically is that they should identify which cluster of local people are more self-efficacy concerned to further refine tourism development.

Acknowledgements

This paper presents part of the main findings from research recently funded by Ministry of Education and Culture of Indonesia under Superior Research of Higher Education in 2016 with the contract number 221/UN5.2.3/PPM/SP/2016. The researchers would like to express their gratitude to the University of Sumatera Utara and all people contributed to this research.

Appendix A. Demographic Profile of the Sample

Demographics	Respondents (n=360)	%	Demographics	Respondents (n=360)	%
Gender			Nationality		
Male	192	53%	Asia		
Female	168	47%	Malaysia	38	54%
Age			Singapore	12	17%
18-24 years	130	36%	Japan	2	3%
25-49 years	167	46%	Others	1	1%
50-64 years	55	15%	Europe		
Above 65 years	8	2%	Netherland	5	7%
Status			Germany	6	9%
Single	185	51%	Others	4	6%
Married	175	49%	America	1	1%
Education			Others	1	1%
High School	194	54%	Occupation		
Graduate	143	40%	Alone	7	4%
Post Graduate	23	6%	With family	37	21%
Job			With friends	79	44%
Government	27	8%	With group	57	32%
Private company	119	33%	Goals		
Self-employement	148	41%	Leisure/sightseeing	107	59%
Unemployement	66	18%	Shopping	13	7%
Residential status			Seeing heritage building	34	19%
Karo Regency	160	44%	Visiting friends/relative	8	4%
Outside Karo	130	36%	Business	5	3%
Foregin Tourists	70	19%	Other	13	7%
Intention			Duration		
First Visit	96	53%	One day or less	113	63%
Second Visit or more	84	47%	2 days or more	67	37%

References

[1] Anon 2017 20 Tempat Wisata di Kabupaten Karo *Pariwisatasumut.net*
[2] YG S 2017 Website Kabupaten Karo - Potensi Wisata Karo Menjanjikan *Karokab.go.id*
[3] Pendit S 1994 *Ilmu Pariwisata: Sebuah Pengantar Perdana* (Jakarta: PT. Pradnya Paramita)
[4] Suwantoro G 2004 *Dasar-dasar Pariwisata* (Penerbit Andi: Yogyakarta)
[5] Rampullo A, Licciardello O and Castiglione C 2015 Intrapersonal Factors Effects on Professional Orientation and Environmental Representations *Procedia-Social and Behavioral Sciences* ***205*** 422-428
[6] Wang S, and Xu H 2015 Influence of Place-based Senses of Distinctiveness, Continuity, Self-esteem and Self-efficacy on Residents' Attitudes toward Tourism *Tourism Management* **47** 241-250
[7] Ujang N 2012 Place Attachment and Continuity of Urban Place Identity *Procedia-Social and Behavioural Sciences* **49** 156-167
[8] Ginting N 2016 How Self-Efficacy Enhance Heritage Tourism in Medan Historical Corridor, Indonesia *Procedia - Social and Behavioral Sciences,* **234** 193 – 200
[9] Breakwell G 1986 *Coping with Threatened Identities (*London: Methuen)

[10] Ginting N, Nasution A , Rahman N 2017 More Attractive More Identified: Distinctiveness in Embedding Place Identity *Procedia Environmental Sciences* **37** 408 – 419
[11] Bandura A 1977 Self-efficacy: Toward A Unifying Theory of Behavioral Change. *Psychological Review* **84** 191-215.
[12] Bandura A 1997 *Self-efficacy: The exercise of control.* In Lunenburg, F. C. 2011 Self-efficacy in the Workplace: Implications for Motivation and Performance. *International Journal of Management, Business, and Administration* **14** 1-6
[13] Lalli M 1992 Urban-related Identity: Theory, Measurement, and Ampirical Findings *Journal of Environmental Psychology* **12** 285-303
[14] Bandura A and Locke E 2003 Negative Self-Efficacy and Goal Effects Revisited *Journal of Applied Psychology* **88** 87-99
[15] Jin X, Wu L, Becken S and Ding P 2016 How do Worry, Self-efficacy, and Coping Interact? Examining Chinese tourists to Australia *Journal of China Tourism Research* **12** 374-393
[16] Klassen R and Chiu M 2010 Effects on teachers' self-efficacy and job satisfaction: Teacher gender, years of experience, and job stress *Journal of Educational Psychology* **102** 741–756
[17] Smith G, Khon S, Savage-Stevens S, Finch J, Ingate R and Lim Y 2000 The Effects of Interpersonal and Personal Agency on Perceived Control and Psychological Well-being in Adulthood *The Gerontologist* **40** 458-468
[18] Twigger-Ross C and Uzzell D 1996 Place and Identity Processes. *Journal of environmental psychology* **16** 205-220
[19] Winkel G 1981 *The Perception of Neigbourhood Change.* In J Harvey Ed. Cognition and Social Behaviour & the Environment New York: Enbaum
[20] Zakaria J and Ujang N 2015 Comfort of Walking in the City Center of Kuala Lumpur *Procedia - Social and Behavioral Sciences* **170** 642 – 652
[21] Marpaung B 2017 *IOP Conf. Ser.: Mater. Sci. Eng.* **180** 012083
[22] Alfonzo M 2005 To Walk or Not To Walk? The Hierarchy of Walking Needs *Environment and Behaviour* **37** 808-836
[23] Ja'afar N, Sulaima A and Shamsuddin S 2012 The Contribution of Landscape Features on Traditional Streets in Malaysia. *Procedia - Social and Behavioural Sciences* **50** 643-656
[24] Firouzmakan S and Daneshpour S 2015 Promotion Quality of Life by Increasing Place Attachment in Public Places *Procedia-Social and Behavioral Sciences* **201** 418-428.
[25] Amir A, Ismail M, and See T 2015 Sustainable Tourist Environment: Perception of International Women Travelers on Safety and Security in Kuala Lumpur *Procedia-Social and Behavioral Sciences* ***168*** 123-133
[26] Anuar N, Bookhari S, and Aziz N 2012 The Effectiveness of Safe City Programme as Safety Basic in Tourism Industry: Case Study in Putrajaya *Procedia - Social and Behavioral Sciences* **42** 477 – 485
[27] Murray A and Feng X 2015 Public Street Lighting Standart Assesment and Achievement *Socio-Economic Planning Sciences* **53** 14-22
[28] Ashworth G and Page S 2011 Urban tourism research: Recent Progress and Current Paradoxes. *Tourism Management* **32** 1-15
[29] Bhattacharyya D and Mitra S 2013 Making Siliguri A Walkable City *Procedia- Social and Behavioral Sciences* **96** 2737-2744
[30] Sawitri D, Hadiyanto H and Hadi S 2015 Pro-Environmental Behavior from a Social Cognitive Theory Perspective. *Procedia Environmental Science* **23** 27 – 33
[31] Bandura A 1986 *Social foundations of thought and action: A social cognitive approach* (Englewood Cliffs, NJ: Prentice Hall)

IOP Conf. Series: Materials Science and Engineering **309** (2018) 012062 doi:10.1088/1757-899X/309/1/012062

Impact of pozzolanic binder addition on stabilization of polluted dredged sediments on its potential reuse as a new material resource for road construction in Basse Normandie, France

E Silitonga*
Lecturer, Universitas Negeri Medan, Jl Willem Iskandar Pasar V Medan Estate 20221 North Sumatera, Indonesia

*E-mail: ernestosilitonga@unimed.ac.id

Abstract. Due to the increase of the amount of marine dredged sediments (MDS) and the cost of managing the disposal site is very high, a reutilization of dredged sediment is urgently needed in France. The primary goal of this research is to find a domain for reuse of MDS materials as a new material regarding the environmental issues, and other requirement needed. In this study, the MDS was used as replacement material in road construction. Hence, various tests need to be realized to identify if MDS could achieve the requirement needed in this domain. The secondary objective is to enhance the geotechnical characteristic and to reduce the pollution content of the MDS, by incorporating binders and sediments, and revealed the geotechnical characteristic. The geotechnical test result shows that the stabilization by hydraulics binders improve the geotechnical characteristic and could fulfil the requirement needed. The present of Fly Ash and Silica Fume in this study is aimed to reduce the pollution level, especially the heavy metal content. After the geotechnical study in laboratory results shows as expected then the study to identify the chemical characteristic was realized. In order to evaluate the environmental impacts, leaching test was performed. The Leaching test result shows that with 7% of Silica Fume and 10% of Fly Ash capable to fulfil the criteria needed, hence, the use of MDS is consider safe in term of geotechnical and environmental impact, as a new material in road construction

1. Introduction

In 2003 the European Council established a regulation stated that operation of rejecting the contaminated dredged sediment into the sea is not allowed. This decision is taken regarding to the environmental issues. Available or attainable, there are many constraints on policy implementation need to be designed. To decide if the dredged material is acceptable for a beneficial use, it is recommended to realize a study to evaluate the capability of marine dredged sediment in its each utilization domain. This goal of this study is to determine the type and composition of binder to stabilize marine dredged sediments to enhance the geotechnical characteristic and reduce the pollutants content in the MDS. Previous studies with different alternative treatment method, to find the domain application for the reuse of marine dredged sediments have been investigated in Europe. Colin [1] treated Marine dredged sediment with hydraulic binders and the result show an improvement of Physical characteristic. Silitonga [2] stated that the ideal amount of hydraulic binder

IOP Conf. Series: Materials Science and Engineering **309** (2018) 012062 doi:10.1088/1757-899X/309/1/012062

needed to stabilized dredged sediment was between 7%-9%, this percentage was able to enhance the Geotechnical characteristic but could not reduce the amount of pollutant content in the MDS. A study by D Zhibo [3] show that to enhance the Geotechnical characteristic of MDS of Port Cherbourg, the amount of hydraulic binder utilized was 10%. E Silitonga [4] [5] utilized Fly Ash to enhance the geotechnical characteristic of MDS of Port de Cherbourg. The percentage of Fly Ash need to stabilize MDS is around 11%. In this study, A serial test were realized to identify the potential reuse of MDS of Port en Bessin as replacement material in road construction. To meet the requirements needed in road construction (geotechnical and environmental) pozzolanic binders such as Silica Fume and Fly Ash were utilized in this study.

2. Material and Methods

2.1. Dredged Sediment and Silica Fume

The MDS used in this study was collected from Port en Bessin. The harbour Port-en-Bessin (PEB) is located in the West of France. The sediments dredged from four different, Bassin no.1 (A, see figure 1), Bassin no.2. (B), Avant Port 1 (C) and Avant Port 2 (D).

Figure 1. Location of material sampling (Port en Bessin)

The hydraulic binder utilized in this study is Cement and Lime that normally used in road construction. The Silica Fumes utilized are not commercialized yet. In fact, the Silica Fumes used in this study are a prototype of low-cost Silica Fume (SF). The SF utilized in this study has a coarser particle size and less Silica content than normal SF commercialized. Three different types of SF were utilized in this study. Two types of Fly Ash utilized in the mixture (Sodeline and Soproline). Sodeline and Soproline are Fly Ash that produced especially for stabilization work.

2.1.1. Particle size distribution

Due to its particle size distribution, a laser diffractometer Coulter LS200 was used to identify the particle size of marine sediments utilized. The results as shown in Table 1, confirm that the particle distribution of marine dredged sediments taken from four different locations (PeB-1, PeB-2, PeB-3 and PeB-4) is almost similar, this result shows the homogeneity of the marine sediment of Port en Bessin. The result shows that MDS constituted of a large amount particle with the size between 66,7-78,4 μm. The Silica Fume type 2 (SF2) has a finer diameter of particle compared to Silica Fume type 2 (SF2) and Silica Fume type 1 (SF1). The particle size distribution of Sodeline and Soproline does not show an important difference. The two types of Fly Ash and Silica Fume constituted a large amount of particle size between 2-63 μm.

Table 1. Particle size distribution of MDS, Fly Ash and Silica Fume.

	A	B	C	D	Sodeline	Soproline	SF1	SF2
<2 μm (%)	10.7	10.6	13.8	9	3.7	5.4	2.3	13.8
2 - 63 μm (%)	77.2	78.8	74.8	78.8	61.3	58.7	78.8	68
> 63 μm (%)	12.6	10.6	11.5	12.2	35	35.8	18.9	18.2

2.1.2. Chemical Characteristics (dredged sediments and Silica Fume)

The level of pollution was identified using Leaching Test. The results of chemical characteristics are shown in Table 2. The samples were taken from 4 different locations (A, B, C and D) in Port en Bessin area. In order to identify the level of pollution of the marine sediment, the researcher used the reference established by European Council 2003/33/EC (Table 2).

Table 2. Leaching test of dredged MDS of Port en Bessin

Pollutants	A	B	C	D	European Council Reference (2003//33/EC)		
					Inert Waste	Non-Hazardous Waste	Hazardous Waste
As (mg/kg)	14	8.9	8.9	4.9	0.5	2	25
Cd (mg/kg)	1.33	1.59	1.6	1.18	0.04	1	5
Cr (mg/kg)	77	58	63	70.4	2	50	100
Cu (mg/kg)	0.63	1.41	0.94	0.7	0.01	0.2	2
Hg (mg/kg)	0.63	0.73	0.3	0.58	0.01	0.2	2
Pb (mg/kg)	20.2	17.9	22	17.3	0.5	10	50
Ni (mg/kg)	29.4	30.2	26.9	19.8	0.4	10	40
Zn (mg/kg)	137	148	109	77.5	4	50	200

Table 3. Chemical characteristic of Silica Fume utilized

Parameters	SiO_2 (%)	Fe_2O_3 (%)	Al_2O_3 (%)	CaO (%)	MgO (%)	Na_2O (%)	C (%)	free Si (%)	Free CaO (%)	SO_3 (%)
Silica Fume 1 (SF1)	90-92	1.5-2	1	0.5-1	1-1.5	0.5-1	0.5-1	< 0.2	< 1	< 1
Silica Fume 2 (SF2)	90-95	1.5-2	1-1.5	0.5-1	1	0.8-1	1-1.5	<0.4	<1	<1
Sodeline	40-45	10-12	20-22	8-10	2	2	5-5.5	>1	>1	<1
Soproline	23-25	23-25	23-25	5-7	5	3	7-9	>1	>1	<1

This reference values relate to the elements contained in the leachate and not in the raw material. According to the European Reference, all MDS considered as a waste material thus it must be disposed in a special area to measure the pollution level, if the MDS categorized as Inert Waste, the MDS can directly be reused. On the other hand, if the pollution level of MDS categorized as Non-Hazardous or Hazardous Waste, it needs to be stabilized until a number of pollutants reach Inert Waste level. From the leaching test, we can observe that the amount of micro pollutants from four different locations (A, B, C and D) categorized as non-hazardous waste. The chemical characteristic of Pozzolanic binders utilized in this study is shown in Table 3.

2.2. Preparation of samples

After being dredged, physical characteristics of marine sediment were measured. The measured initial water content is about 169%. The dewatering process was realized, the marine sediment, oven-dried

IOP Conf. Series: Materials Science and Engineering **309** (2018) 012062 doi:10.1088/1757-899X/309/1/012062

for 5 days at 60°C was pulverized to 2mm sieve size, it was initially mixed with determined quantities of Silica Fume, lime and cement as a binder, in a dry state and subsequently mixed with water by a mechanical mix. The sample was realized by using cylindrical specimens (ø = 40mm, h = 80mm). Samples with SF1, SF2, Sodeline and Soproline addition were set in order to investigate the influence of the Silica Fume on the strength gained, the mixtures were realized with various percentage of Silica Fume content. The mixture without any Silica Fume and Fly Ash content (LIM-CEM) was realized as a reference to identify the effect of Pozzolanic binder (Fly Ash and Silica Fume) addition.

3. Results and Analysis

3.1. Tensile Strength

In order to identify the capability of MDS as replacement material in road construction work, the Tensile Strength test needs to be realized. The results of the tensile test are presented in figure 2.

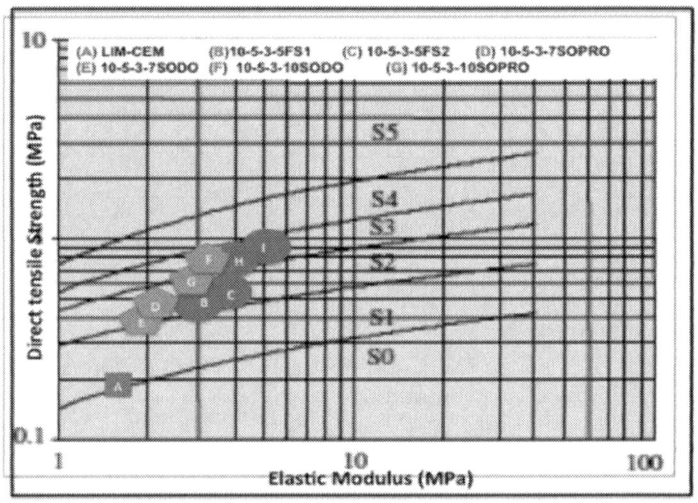

Figure 2. Tensile test result for treated Marine Dredged Sediment (MDS)

As shown in the Figure 1, only the samples with 10% Fly Ash (Sodeline and Soproline) and with 7% of Silica Fume located in class S3 (F, G, H and I). According to the Scheme of road section and recommended engineering characteristics, the samples with 10% of Fly Ash (Sodeline) is consider safe to be utilized as material as material on sub base and Foundation. The result showed that the sample with Silica Fume addition, produces a higher Elastic Modulus and higher Tensile strength value than sample without Silica Fume (LIM-CEM). The sample without Silica fume LIM-CEM (A) located within the limit of S0 and S1 classes, it means the reuse for sub base need to be considered. On the other hand the samples with 5% Silica fume (B and C) and 7% of Fly Ash (D and E) located in the same class (S2), yet the Elastic Modulus values of sample with 5% of Silica Fume (B and C) is higher than samples with 5% Silica Fume. The presents of Silica Fume with its filler effect, due to its fineness can fit into space and fill the empty space between sediment grains and produces more solid bonds between sediment grain, this is the reason the enhancement of Elastic Modulus because a stiffer material will have higher Elastic Modulus. The results show that the samples with 5% of Silica fume (B and C) and 7% of Fly Ash (D and E) could be used as material only in sub base. On the other hand, the samples with 7% of Silica fume (H and G) and 10% of Fly Ash (G and F) could be reused in sub base and foundation.

3.2. Leaching Test

The main objective of using Pozzolanic binder (such as Silica Fume and Fly Ash) is to fulfil the road regulation criteria and is consider harmless to the environment. However, due to drying and oxidation process of polluted marine dredged sediment may affect mobility and bioavailability of the land utilized.

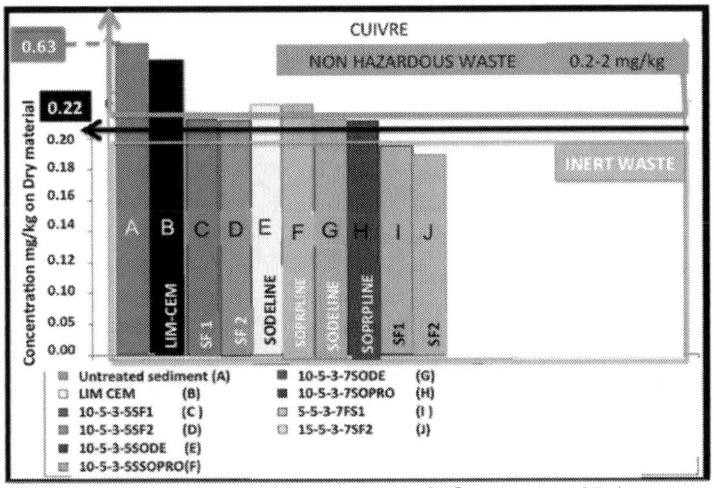

Figure 3. Leaching Test result for Copper (Cu)

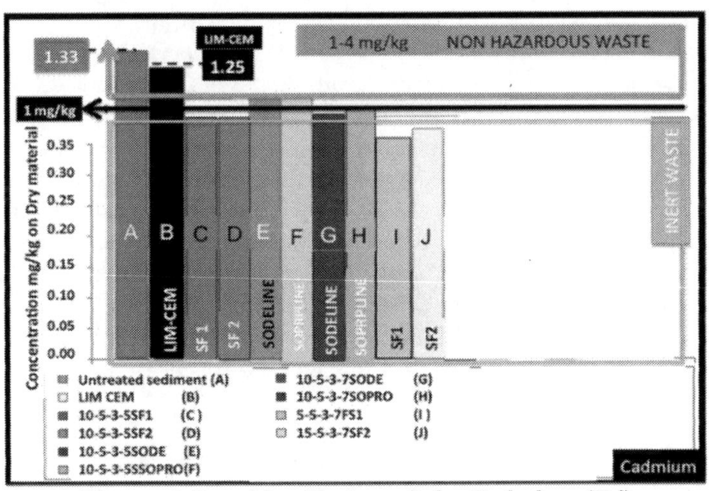

Figure 4. Leaching Test result for Cadmium (Cd)

Lixiviation Test was realized to identify the strength or reaction of sample to aggressive environment, such as the extreme change of pH in environment. The Leaching Test refers to the solution containing the solubilized elements during the test, which are performed on the analytical characterization. According to the result, the amount of Cadmium (Figure 4) in MDS of Port en Bessin is 1.33 mg/kg, and for Copper is 0.633 mg/kg. In order to help to classify the pollution level, the reference values established by European Council No. 2003/33 / EC are utilized. According to European reference; the quantity of Cadmium and Copper are categorized as non-hazardous waste. According to European Council, the non-hazardous waste should be stabilized and after categorized in to inert waste, than the material could be reused as new material. The result of leaching test (figure 2) shows that, in the case of Cadmium (Cd), sample with or without Silica fume show a reduction of Cadmium content. The sample without Silica fume; LIM_CEM (B in Figure 4), show a slight reduction of amount of Cadmium (from 1.33 to 1.25) but it is still in the same level of pollution as the original amount of pollutants (non-hazardous level). All the samples mixed with Silica fume established a major reduction on amount of Cadmium; all the samples are situated in Inert Waste. The samples with 5% percentage of Silica fume (C and D) show minimal reduction of Cadmium content compared to samples with 5% (E and F) and 7% Silica Fume (I and J). The samples with 7% of Silica fume show a very remarkable reduction of Cadmium content and established a group with lowest amount of Cadmium. One of the advantage of using silica fume is the filler effect, because of its fineness, Silica Fume can fit into space between sediment grains in the same way that sand fills the space between particles of coarse aggregates and cement grains, fill the space between sand grains. After the pozzolan reaction, Silica Fume is capable to trap the pollutants, so that the pollutants could not escape from the matrix and pollute the environment. This is the main reason why the Silica Fume is being used in this study. The samples with Fly Ash show a reduction of Cadmium as well, although after stabilization with 7% of Fly Ash the Cadmium content are still located in level Non-Hazardous Waste. On the other hand samples with 10% Fly Ash established to reduce the Copper content (Figure 3) and located in border between (Inert waste and Non Hazardous Waste). From this study we can observe that, the addition of Silica Fume (5% to 7%) has greater impact on reduction the pollutant content than Fly Ash. This result due to the filler effect of Silica Fume and the content of Silica thus produces more dense material, that capable to trap the pollutant in the matrix. According to European reference value for dredged sediment, from the environment point of view, after stabilization process with 7% of Silica fume and 10 % of Fly Ash, the MDS of Port en Bessin categorized as inert waste, hence harmless to environment and could be reused as a material in road construction

4. Conclusion

The objective of this research is to fulfil the criteria (geotechnical and environment impact point of view) for the reutilization of polluted dredged sediment as material in Road construction. The granulometric test shows that dredged sediment from Port en Bessin constituted a large amount particle with silt fraction (2 - 63 µm). The leaching test result state that dredged sediment of Port en Bessin content amount of pollutants is categorized as non-hazardous waste (according to European Reference). Test on tensile strength confirmed that with the sample with addition 7% Silica fume and 10% of Fly Ash could be used as material in sub base and foundation. In the other hand, sample with 5% Silica fume and 7% of Fly Ash only can be used as material in sub base. Leaching test determine that the present of 7% Silica Fume and 10% of Fly Ash in the specimens, remarkably capable to reduce the amount of pollutant (Cadmium and Copper) and passed from level non-hazardous waste to categorize as inert waste. Hence the MDS stabilized with 7% Silica fume and 10%of Fly Ash can be used in road construction and is consider harmless to the environment.

References

[1] Colin D 2003. *Thèse de doctorat, Université de Caen France,* 181p
[2] Silitonga E 2010. *Thèse de doctorat, Université de Caen France*, 160p.
[3] Zhibo D 2008 *Thèse de doctorat, Université de Caen*, 145p
[4] Silitonga E, Levacher D, Mezazigh S 2009 *Environmental Technology* **30 (8)** 799-807
[5] Silitonga E, Levacher D, Mezazigh S 2010 *Revue Européenne de Génie Civil et Environment* **14 (2)** 253-265.
[6] Silitonga E, Levacher D, Mezazigh S 2009 *Environmental technology* **30 (8)** 799-807.

IOP Conf. Series: Materials Science and Engineering **309** (2018) 012063 doi:10.1088/1757-899X/309/1/012063

Determination of service standard time for liquid waste parameter in certification institution

M T Sembiring* and D Kusumawaty
Department of Industrial Engineering, Faculty of Engineering, Universitas Sumatera Utara
Jl. Prof. T. Maas Kampus USU

*Email : meilita_tryana@yahoo.co.id

Abstract. Baristand Industry Medan is a technical implementation unit under the Industrial and Research and Development Agency, the Ministry of Industry. One of the services often used in Baristand Industry Medan is liquid waste testing service. The company set the standard of service 9 working days for testing services. At 2015, 89.66% on testing services liquid waste does not meet the specified standard of services company. The purpose of this research is to specify the standard time of each parameter in testing services liquid waste. The method used is the stopwatch time study. There are 45 test parameters in liquid waste laboratory. The measurement of the time done 4 samples per test parameters using the stopwatch. From the measurement results obtained standard time that the standard Minimum Service test of liquid waste is 13 working days if there is testing E. coli.

1. Introduction

Baristand Industry Medan is a technical implementation unit under the Industrial and Research and Development Agency, the Ministry of Industry. One of the services often used in Baristand Industry Medan is liquid waste testing service.

The testing is the determination of one or more of the characteristics of an object the assessment according to the procedure. The procedure itself is a way to implement an activity or process. The testing is usually applied to the ingredients, products or process according to the procedure. Meanwhile, laboratory is the place of scientific research, experiments or measurements scientific training done. Service activities and laboratory testing in general is to meet the standardization, where all types of ingredients, products and the process is in accordance with certain standards that are assigned to maintain the quality a product/service in the community/consumers. Liquid Waste is a joint or a mixture of water and pollutants carried by the good in the situation dissolved and dispersed suspended domestic (offices, housing and trade), the source of industry and on a specific time mixed with the ground water surface water or rain water.

Baristand Industry Medan specify the standard services for at least 9 working days for testing services. Working days in Baristand Medan is Monday to Friday. 2015, 89.66% on testing services liquid waste does not meet the specified standard of company. The cause of not achieving the minimum standard of services is the old testing time. Testing time is determined by how many test parameters that will be tested in accordance with the customer request. There are 45 effluent testing parameters that can be tested in Baristand Medan Industry that consists of 8 test parameters the nature of physics, 22 test parameters chemical nature, 12 test parameters of metal and 3 parameters a microbiology test. To know how long time testing each parameter, will be done measurement standard time. The measurement of the time is done 4 samples per test parameters, because the capacity of the laboratory only can test 4 sample daily.

2. Research Method

a. The measurement of the time.

The measurement of the working time (Time Study) was an attempt to determine the length of time the work that is required by an operator to complete a work. The measurement of the time consists of 2 types, namely the measurement of the time directly and the measurement of time indirectly 1]. The measurement of the time that is done in this research is the measurement of the time directly using the stopwatch.

b. Testing the uniformity of the data

A data said uniform if all of the data between the two border controls, namely the limit of control over and the under-control limit. the formulation of the limit of control over and under control limit is as follows [1]

c. Testing the adequacy of the data

The adequacy of the data is done to get whether the amount of data observation result enough to do research. To count the number of measurement is required for the level of carefulness 5 percent and the level of confidence 95% is as follows [5]:
When $N' \leq N$, then the amount of data is enough
When $N' > N$, then the amount of data is not enough

d. Rating Factor and Allowance

Rating factor is a technique to equate the result of observation of the operator in completing a work with the time required by the normal operator in completing the job. According to [2] rating factor (P) has three limitations, namely:
$p > 1$ when the operator work too fast
$p < 1$ when the operator working under normal (too slowly)
$p = 1$ when the operator to work normal
There are many methods used to determine rating factors. The following is some method in determining the rating factor that generally used [1]:

1. The Method Skill and Effort Rating
2. The Method Westinghouse
3. The Method Synthetic Rating
4. Performance Rating or Speed Rating
5. The Method Objective

The giving of the allowance is intended to provide the opportunity for the operator to do the things that must be done so that the raw material is obtained can be said the complete working time data and represent the working system that is observed. The concession is given among others:

1. Allowance for personal needs
2. Allowance to eliminate fatigue (fatigue)
3. Allowance to the things that could not be avoided

e. The cycle times.

The cycle times is the time required to create one product unit on a single workstation [3]. The time required to implement the elements of the typical work element- will differ slightly from the cycle to other cycle, even if the operator is working at normal speed or uniform, each of the elements in a different cycle does not always will be completed in the same time.

f. The Normal time.

The normal time for a working operating element is shows that a good qualified operator will work completed the work on the tempo work normal [1].

g. The Standard Time.

The determination of the time standard for each parameter test is done with the way the measurement directly using the stopwatch. Standard Time is obtained by multiplying the normal time with allowance[4].

3. The results

Testing the liquid waste consists of 5 operators. Each operator to test different parameters, division test parameters based on the operator is divided into:

1. The Operator A can test for smell parameters, BOD, conductivity, turbidity, dissolved oxygen (O2), slurries dissolved, pH, taste, temperature, Total Solid Suspension (TSS), colors and organic substances.
2. The Operator B can test for fluoride parameters (F), oil/fat, NH3-N, nitrate, nitric, nitrogen total, phosphate, cyanide and sulfide.
3. The Operator C can test alkalinity parameters, COD, calcium, hardness, chloride, magnesium, salinity, silica, mercury, arsenic, iron, cadmium, chrome, cobalt, potassium, manganese, nickel, zinc, copper, and lead.
4. The Operator D can test for total coliform parameters, E. coli, and numbers total fins.
5. The Operator E can test for boron parameters.

Testing cycle time consisted of several activities namely:

1. Preparing the equipment and the chemicals for the testing process.
2. Testing process. On this testing process for each different test parameters. The activities on this testing process consists of weigh the chemical, make the solution, add some ml solution, do titration, do distillation, heat sample, create calibration curve and read the examples on spectrophotometer test.
3. Perform the calculation of the test results and record the test results.

For example, activities test parameters smell.

Table 1. The Cycle Times to Smell Test Parameters

Activities	The Measurement of The Time (Minutes)									
	1	2	3	4	5	6	7	8	9	10
Took the beaker glass.	0.30	0.30	0.40	0.35	0.30	0.40	0.30	0.30	0.30	0.30
Pour test example into the beaker glass.	2.50	2.40	2.38	2.20	2.25	2.57	2.60	2.50	2.10	2.50
Scent.	0.50	0.55	0.60	0.45	0.63	0.50	0.50	0.40	0.43	0.50
Record results.	0.57	0.50	0.60	0.50	0.57	0.57	0.57	0.60	0.42	0.51
Total Time (minutes)	3.87	3.75	3.98	3.50	3.75	4.04	3.97	3.90	3.25	3.81

Test the uniformity of data for smell parameter and the average time measurements smell is:

$$\bar{x} = \frac{\sum x_i}{N} \quad (1)$$

$$\bar{x} = \frac{3{,}87+3{,}75+3{,}98+3{,}50+3{,}75+4{,}04+3{,}97+3{,}90+3{,}25+3{,}81}{10} = 3{,}782 \quad (2)$$

The Standard deviation of time measurement smell parameter is :

$$\sigma = \sqrt{\frac{\sum(x_i - \bar{x})^2}{N-1}} \quad (3)$$

$$\sigma = \sqrt{\frac{0{,}527}{10-}} = 0{,}242 \quad (4)$$

The calculation of the value of control over (BKA), the limit under control (BKB) is:

$$\text{BKA} = \bar{x} + 2\sigma = 3{,}782 + 2(0{,}242) = 4{,}266 \quad (5)$$

$$\text{BKB} = \bar{x} - 2\sigma = 3{,}782 - 2(0{,}242) = 3{,}298 \quad (6)$$

Map control test uniformity odor data can be seen in Figure 1.

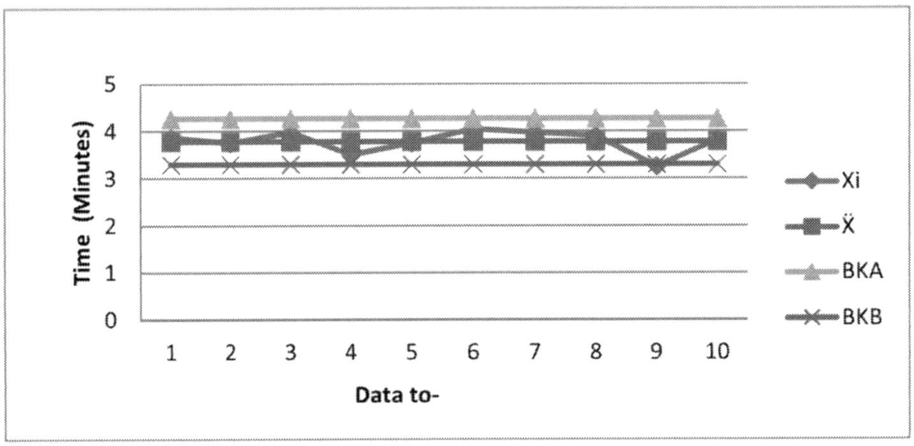

Figure 1. Control Map Test Data Uniformity Smell

Test the adequacy data is done to see whether the data taken from the observation was sufficient for use. In this research, the level of confidence that is used is 95% then k is 2, and high precision used is 5%, then the value of s is 0.05. Calculation formula N' is as follows:

$$N^I = \left[\frac{k/s\sqrt{N\sum xi^2 - (\sum xi)^2}}{\sum xi}\right] \quad (7)$$

If the value of N'< N then enough data, and if N' > N then the data is not enough. For example, test the adequacy data smell parameter:

$$N^I = \left[\frac{k/s\sqrt{N\sum xi^2 - (\sum xi)^2}}{\sum xi}\right]$$

$$N^I = \left[\frac{2/0{,}05\sqrt{(10x143{,}563) - 37{,}820^2)}}{37{,}820}\right] = 5{,}908$$

The value of N' = 5,908 < N worth 10, then observations are enough to smell test parameters. The recapitulation of the uniformity test data and test the adequacy of the data for the measurement of all test parameters can be seen in the Table 2.

IOP Conf. Series: Materials Science and Engineering **309** (2018) 012063 doi:10.1088/1757-899X/309/1/012063

Table 2. Recapitulation Of Uniformity Tests And Test The Adequacy Of Data

The Operator	Test Parameters	Test the uniformity of data.				Test adequacy of data		
		Average	BKA	BKB	Description	N'	N	Description
A	Smell	3.782	4.266	3.298	Uniform	5.908	10	Enough
	BOD	5371.650	5487.383	5255.917	Uniform	0.149	5	Enough
	Conductivity	10.924	11.724	10.124	Uniform	1.930	10	Enough
	Turbidity	40.367	42.359	38.375	Uniform	0.876	10	Enough
	Dissolved Oxygen	31.706	34.152	29.260	Uniform	2.041	7	Enough
	Slurries Dissolves	22.184	24.807	19.561	Uniform	4.474	5	Enough
	pH	36.313	40.313	32.313	Uniform	4.367	10	Enough
	Taste	3.344	3.846	2.842	Uniform	8.119	10	Enough
	Temperature	9.573	10.294	8.852	Uniform	1.891	6	Enough
	Total Solid Suspension (TSS)	850.360	929.255	771.465	Uniform	3.099	10	Enough
	Color	33.873	36.761	30.984	Uniform	2.181	4	Enough
	Organic Substances	36.948	40.488	33.407	Uniform	3.213	8	Enough
B	Fluoride (F)	104.890	109.837	99.943	Uniform	0.712	5	Enough
	Oil and Fat	1622.963	1645.762	1600.164	Uniform	0.068	7	Enough
	NH3-N	273.453	292.928	253.977	Uniform	1.522	4	Enough
	Nitrate (NO3)	228.563	231.422	225.704	Uniform	0.042	3	Enough
	Nitric (NO2)	76.054	78.766	73.342	Uniform	0.407	5	Enough
	Nitrogen Total	859.238	885.643	832.834	Uniform	0.315	6	Enough
	Phosphate (PO4)	114.963	121.306	108.620	Uniform	0.812	3	Enough
	Cyanide (CN)	197.808	209.394	186.223	Uniform	1.143	6	Enough
	Sulfate (SO4)	101.205	107.182	95.228	Uniform	1.046	4	Enough
C	Alkalinity	40.778	44.659	36.896	Uniform	2.718	4	Enough
	COD	178.137	199.334	156.940	Uniform	5.097	10	Enough
	Calcium (Ca)	91.240	98.881	83.599	Uniform	2.244	5	Enough
	Hardness	27.786	30.016	25.556	Uniform	2.061	5	Enough
	Chloride (Cl)	30.240	32.584	27.896	Uniform	1.922	5	Enough
	Magnesium (Mg)	89.410	92.079	86.741	Uniform	0.238	3	Enough
	Salinity	42.747	45.926	39.567	Uniform	1.475	3	Enough
	Silica (SiO2)	430.347	432.281	428.413	Uniform	0.005	3	Enough
	Mercury (Hg)	142.427	145.972	138.881	Uniform	0.165	3	Enough
	Arsan (As)	147.354	154.263	140.445	Uniform	0.704	5	Enough
	Iron (Fe)	135.893	140.699	131.088	Uniform	0.333	3	Enough
	Cadmium (Cd)	138.106	144.889	131.323	Uniform	0.844	8	Enough
	Chrome l	126.300	132.220	120.380	Uniform	0.659	4	Enough
	Cobalt (Co)	143.237	150.502	135.972	Uniform	0.686	3	Enough
	Potassium (K)	132.778	142.304	123.252	Uniform	1.647	5	Enough
	Mangan (Mn)	136.608	142.745	130.472	Uniform	0.673	6	Enough
	Nickel (Ni)	127.836	137.564	118.108	Uniform	1.853	5	Enough
	Zinc(Zn)	128.837	135.088	122.586	Uniform	0.807	7	Enough
	Copper (Cu)	134.909	141.592	128.225	Uniform	0.841	7	Enough
	Lead (Pb)	134.745	142.016	127.474	Uniform	1.019	8	Enough
D	Total Coliform	4558.853	4720.732	4396.973	Uniform	0.378	4	Enough
	E.coli	17064.550	17193.260	16935.840	Uniform	0.015	3	Enough
	Number Total Fins	2275.870	2315.589	2236.151	Uniform	0.081	3	Enough
E	Boron	88.137	89.418	86.855	Uniform	0.056	3	Enough

The formula to find the normal time:

$$Wn = Ws \ x \ p \tag{8}$$

Where:
Wn = the normal Time
Ws = cycle times
p = adjustment factor, if p = 1 working fair, p<1 working too slowly and p>1 working too fast.
The formula to find standard time:

$$Wb = Wn\,x\,(1 + All) \tag{9}$$

Where:
Wb = Time raw material/standard
Wn = the normal Time
All = Allowance
Recapitulation Result of standard time for all test parameters can be seen in Table 3.

Table 3. Recapitulation of Standard Time

The Operator	Test Parameter	Cycle Time (minutes)	p	Normal Time (minutes)	(1+All)	Standard Time (minutes)
A	Smell	3.782	1	3.78	1.11	4.20
	BOD	5371.650	1	5371.65	1.11	5962.53
	Conductivity	10.924	1	10.92	1.11	12.13
	Turbidity	40.367	1	40.37	1.11	44.81
	Dissolved Oxygen	31.706	1	31.71	1.11	35.19
	Slurries Dissolves	22.184	1	22.18	1.11	24.62
	pH	36.313	1	36.31	1.11	40.31
	Taste	3.344	1	3.34	1.11	3.71
	Temperature	9.573	1	9.57	1.11	10.63
	Total Solid Dud pension (TSS)	850.360	1	850.36	1.11	943.90
	Color	33.873	1	33.87	1.11	37.60
	Organic Substances	36.948	1	36.95	1.11	41.01
B	Fluoride (F)	104.890	1	104.89	1.13	118.53
	Oil and Fat	1622.963	1	1622.96	1.13	1833.95
	NH3-N	273.453	1	273.45	1.13	309.00
	Nitrate (NO3)	228.563	1	228.56	1.13	258.28
	Nitric (NO2)	76.054	1	76.05	1.13	85.94
	Nitrogen Total	859.238	1	859.24	1.13	970.94
	Phosphate (PO4)	114.963	1	114.96	1.13	129.91
	Cyanide (CN)	197.808	1	197.81	1.13	223.52
	Sulfate (SO4)	101.205	1	101.21	1.13	114.36
C	Alkalinity	40.778	1	40.78	1.31	53.42
	COD	178.137	1	178.14	1.31	233.36
	Calcium (Ca)	91.240	1	91.24	1.31	119.52
	Hardness	27.786	1	27.79	1.31	36.40
	Chloride (Cl)	30.240	1	30.24	1.31	39.61
	Magnesium (Mg)	89.410	1	89.41	1.31	117.13
	Salinity	42.747	1	42.75	1.31	56.00
	Silica (SiO2)	430.347	1	430.35	1.31	563.75
	Mercury (Hg)	142.427	1	142.43	1.31	186.58
	Arsan (As)	147.354	1	147.35	1.31	193.03
	Iron (Fe)	135.893	1	135.89	1.31	178.02
	Cadmium (Cd)	138.106	1	138.11	1.31	180.92
	Chrome I	126.300	1	126.30	1.31	165.45
	Cobalt (Co)	143.237	1	143.24	1.31	187.64

	Potassium (K)	132.778	1	132.78	1.31	173.94
	Mangan (Mn)	136.608	1	136.61	1.31	178.96
	Nickel (Ni)	127.836	1	127.84	1.31	167.47
	Zinc(Zn)	128.837	1	128.84	1.31	168.78
	Copper (Cu)	134.909	1	134.91	1.31	176.73
	Lead (Pb)	134.745	1	134.75	1.31	176.52
D	Total Coliform	4558.853	1	4558.85	1.11	5060.33
	E.coli	17064.550	1	17064.55	1.11	18941.65
	Number Total Fins	2275.870	1	2275.87	1.11	2526.22
E	Boron	88.137	1	88.14	1.11	97.83

4. Conclusion

Test parameters that have the largest time is E. coli namely 18941,65 minutes = 315,69 hours = 13 days, so if the customer wants to test E. Coli then the minimum standard of services testing of liquid waste is 13 days, if there is no testing of e. coli, minimum standard of services remain 9 days.

References

[1] Wignjosoebroto S 2000 *Ergonomics Study Of Movement And Time: Analysis Techniques To Increase The Productivity Of Labor* Surabaya: Guna Widya

[2] Sutalaksana I Z, Anggawisastra R and Tjakraatmadja J H 2006. *Teknik Perancangan Sistem Kerja*. Bandung, Indonesia: ITB.

[3] Purnomo H D and Wee H M 2014 *Comp. & Ind. Eng*, **76**, 222-0.

[4] Jamal M 2004 *Str. and Hea.* **20(3)**, 113-9.

[5] Denison E F 1961. *Measurement of labor input: some questions of definition and the adequacy of data. In Output, Input, and Productivity Measurement.* New Jersey, USA: Princeton University Press.

[6] Parihar S, Bhar C, and Srivastava N 2016 *A J. of Man. Eth. and Spir.* **7(2)**

IOP Conf. Series: Materials Science and Engineering **309** (2018) 012064 doi:10.1088/1757-899X/309/1/012064

Activated carbon production from bagasse and banana stem at various times of carbonization

E Misran[1*], S Maulina[1], S F Dina[2], A Nazar[1], S A Harahap[1]

[1]Department of Chemical Engineering, Universitas Sumatera Utara,
Jl. Almamater Kampus, Padang Bulan, Medan 20155, Indonesia
[2]Center for Research and Standardization Industry of Medan, Agency for R & D of Industry, Ministry of Industry
Jl. Sisingamangaraja No. 24, Medan 20217, Indonesia

[*] E-mail: erni_misran@yahoo.com

Abstract. The utilization of bagasse and banana stem as activated carbon precursors has been conducted. In this study, the dried samples were impregnated using phosphoric acid (H_3PO_4) solution as the activator at a ratio of sample to the activator (w/w) was 1:1. The impregnation was conducted at room temperature for 24 hours. The samples then carbonized at 400 °C for 30, 45 and 60 minutes and finally washed and dried to obtain the activated carbon. The research aimed to investigate the effects of time of carbonization on the characteristics of activated carbon produced from bagasse and banana stem. The result showed that yield of activated carbon was in the range of 40.03 – 46.73 % with a high content of carbon as high 90.33 %. The result of BET analysis showed that the highest surface area reached1130.465 m^2/g.

1. Introduction

Activated carbons are materials containing the large surface area, well-developed porosity and rich surface group [1]. The beneficial properties of activated carbon have been recognized since ancient times. Today, activated carbons are very useful in wide variety of applications such as decolorization, decontamination, respirators for protection against toxic gases, as desiccants and or medicinal agents in detoxification [2].

The factors such as the nature of raw materials, the activation method, the activating agent and the condition of process affect the textural and chemical characteristic of the activated carbons. In recent years, activated carbon is being produced from a large number of easily available and low-cost materials such as agricultural products like coconut shells, reedy grass leaves, fire wood, and marigold straw [3].

Bagasse and banana stem are very potential to be used as raw material to produce activated carbon due to its components. The content of bagasse is 42.16 % cellulose, 36.0 % hemicellulose, and 19.30 % lignin [4], whereas banana stem contains 43.3% cellulose, 20.6% hemicellulose and 27.8% lignin [5].

In chemical activation, the precursor is mixed with a certain amount of active agents such as KOH, NaOH, $ZnCl_2$, and H_3PO_4, which lead to the development of porous structures in the material. Among the various dehydrating agents used for chemical activation, phosphoric acid is preferred recently due to environmental and economic concerns. Moreover, the carbon obtained by using zinc chloride as an activating agent cannot be used in pharmaceutical and food industries [3]. In this study, we investigatedthe effects of carbonization time on the characteristics of activated carbon produced from bagasse and banana stem using H_3PO_4 as the activator.

2. Experimental Procedure

2.1. Activated Carbon Preparation
The preparation of activated carbon from precursors of bagasse and banana stem was conducted with the same procedure as published before [6]. The raw material that has been washed was then dried under sun light and continued by oven drying at 105°C. The samples were then sieved using sieve size of 32 mesh. Particles that pass the sieve of 32 mesh was mixed with H_3PO_4solution at an impregnation ratio of 1:1 and impregnation time of 24 h. The filtered sample was then carbonized at 400 °C for 30, 45 and 60 minutes under nitrogen (N_2) flow. The product was washed several times with warm distilled water until neutral solution was achieved. Finally, the sample was dried until the mass was constant.

2.2. Activated Carbon Characterizations
The yield of activated carbon was calculated by dividing the mass of the resultant activated carbon by the initial mass of the raw material used. Proximate analysis was conducted to determine the moisture content, ash content and volatile matter content. The SEM-EDX analysis was used to get the surface morphology image and the composition of elements of sample. The Brunauer-Emmett-Teller (BET) surface area of sample was determined by applying the standard BET method to the nitrogen adsorption.

3. Result and Discussion

3.1. Yield of Activated Carbon
Yield of activated carbon was in the range of 40.03-46.73% as shown in Figure 1. The yield of activated carbon from banana stem was slight higher than activated carbon from bagasse. The longer the carbonization time, the lower the yield of activated carbon produced. This was due to more water and volatile matters removed from the sample thus less amount of the obtained product. The best result was achieved when banana stem was carbonized for 30 minutes.

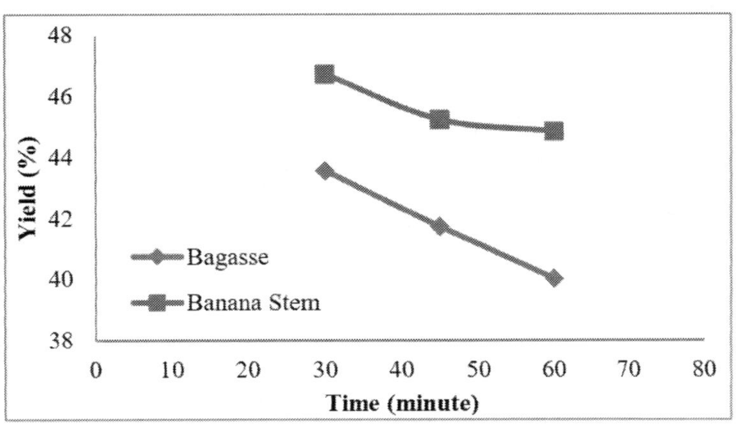

Figure 1. Yield of activated carbon produced from bagasse and banana stem at various time of carbonization

3.2. Proximate and Ultimate Analysis

The proximate and ultimate analysis of the activated carbon produced by carbonization and activation of the material (bagasse and banana stem) is presented in Table 1. The moisture, ash and volatile matter content were quite higher for all products. However, the longer time of carbonization caused the decrease of moisture, ash and volatile matter content; thus increasing the carbon content of the activated carbon.

The carbon content of both activated carbons was in the range of 84.21-90.33%. The large amount of carbon in this study was suitable for activated carbon requirement as stated in [7] that carbon content should be in the range of 80-95%. This result indicated that both materials are potential as activated carbon precursor.

Table 1. Proximate and Ultimate Analysis for Produced Activated Carbon

Activated Carbon	Proximate Analysis (% w/w)			Ultimate Analysis (% w/w)						
	Moisture	Ash	Volatile Matter	C	O	Si	P	F	Ca	Na
Bagasse 30 min	5.13	9.61	68.58	84.21	13.88	0.56	0.57	0.87	-	-
Bagasse 45 min	4.08	8.76	57.46	84.96	9.37	-	4.14	0.89	-	0.64
Bagasse 60 min	7.31	8.28	55.35	87.70	9.75	1.04	0.48	0.89	0.20	-
Banana Stem 30 min	8.02	12.80	73.12	84.55	13.20	0.24	0.83	1.02	-	0.07
Banana Stem 45 min	6.15	9.00	72.69	85.38	12.02	1.06	0.41	1.13	-	-
Banana Stem 60 min	5.32	8.49	63.14	90.33	7.71	0.74	1.20	1.67	0.14	-

3.3. Surface Morphology

The surface morphology of the dried bagasse and banana stem and activated carbon was studied by SEM image. Fig. 2a and 2b show the SEM micrographs of dried bagasse and banana stem. The surface of dried bagasse and banana stem appeared to be covered with thick foreign embodiment. Upon H_3PO_4 activation, more porous structures begin to appear which can be attributed to the dehydration effect of H_3PO_4 and the oxidation of organic compounds in the carbonization step (Fig. 3a and 3b). Concentration of H_3PO_4 used in activation was expected to increase volume due its dehydrating effect, thus providing larger surface area.

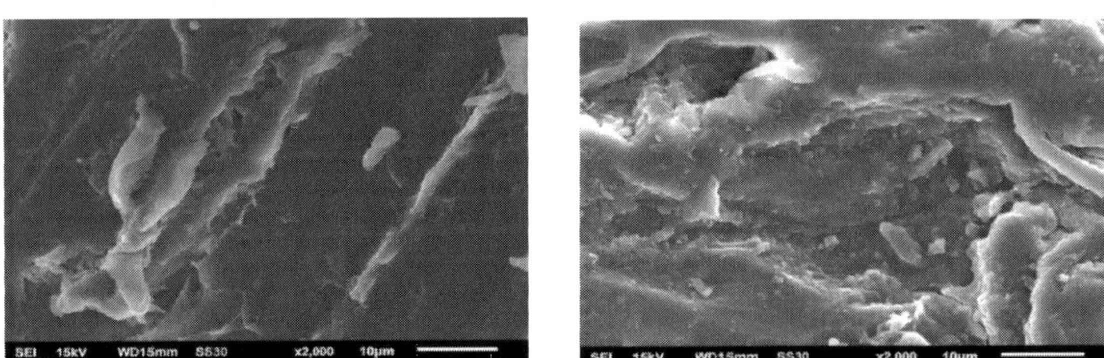

(a) (b)

Figure 2. Scanning electron micrographs of dried (a) bagasse and (b) banana stem

IOP Conf. Series: Materials Science and Engineering **309** (2018) 012064 doi:10.1088/1757-899X/309/1/012064

(a) (b)

Figure 3. Scanning electron micrographs of (a) bagasse and (b) banana stem activated carbon produced at 400 °C and carbonization time of 30 minutes

3.4. Porous Structure

Analysis results of surface area, pore volume, and average pore size using BET are presented in Table 2. It demonstrated that activated carbons have much larger surface area than the raw materials. Principally, based on its pore diameter, activated carbons are classified into three groups by IUPAC: micropore (diameter < 2 nm), mesopore (2-50 nm) and macropore (> 50 nm) [8]. All activated carbons exhibited the presence of micropore but not more than 32.85%. The highest portion of micropore was obtained at carbonization of banana stem for 30 minutes. The longer carbonization time increased the surface area of activated carbon.

Table 2. Porous Structure of Raw Materials and Produced Activated Carbons

Sample	Pore Volume (cc/g)	Pore radius (nm)	BET Surface Area (m^2/g)	Micropore Area (m^2/g)
Bagasse	0.008	2.6917	0	0
Bagasse 30 min	0.186	1.8193	952.611	164.913
Bagasse 45 min	0.188	1.7785	1062.206	154.426
Bagasse 60 min	0.476	1.7925	1130.465	328.588
Banana Stem	0.021	2.7365	22.421	6.622
Banana Stem 30 min	0.486	1.8112	846.531	278.085
Banana Stem 45 min	0.462	1.8183	849.387	273.187
Banana Stem 60 min	0.450	1.7889	876.590	268.619

4. Conclusion

The longer the carbonization time then the higher the yield, carbon content and total surface area. The achieved yield and carbon content of activated carbon from banana stem was slightly higher than that of activated carbon from bagasse. However, activated carbon from bagasse had higher total surface area. The highest yield was 46.73%, carbon content was 90.33% and surface area was 1130.465 m^2/g. All activated carbons exhibited the presence of micropore but not more than 32.85%. Both raw materials of bagasse and banana stem are potential to be utilized as activated carbon precursor.

Acknowledgement
The authors gratefully acknowledge The Ministry of Research Technology and Higher Education of The Republic of Indonesia for the financial support for this research under Fundamental Research Grant 2016.

References

[1] Xu J Z, Chen L Z, Qu H Q, Jiao Y H, Xie J X, Xing G 2014 Preparation and characterization of activated carbon from reedy grass leaves by chemical activation with H_3PO_4 *Applied Surface Science* 320 674-680
[2] Shalna T, Yogamoorthi A 2015 Preparation and characterization of activated carbon from used tea dust in comparison with commercial activated carbon *International Journal of Recent Scientific Reasearch* 6 2750-2755
[3] Kumar A, Jena H M 2016 Preparation and characterization of high surface area activated carbon from fox nut (euryle ferox) shell by chemical activation with H_3PO_4 *Results in Physics* 6 651-658
[4] Chen C X, Huang B, Li T, Wu G F 2012 Preparation of phosphoric acid activated carbon from sugar cane bagasse by mechanochemical processing *BioResources* 7 5109–5116
[5] Noeline B F, Manohar D M, Anirudhan T S 2005 Kinetic and equilibrium modelling of lead(ii) sorption from water and wastewater by polymerized banana stem in a batch reactor *Separation and Purification Technology* 45 131–140
[6] Misran E, Maulina S, Dina S F, Harahap S A A, Nazar A 2017 Characterizations of activated carbon produced from bagasse and banana stem using H_3PO_4 as activating agent *Journal of Engineering and Applied Sciences 12(15) 3839-3842*
[7] Suhas, Gupta V K, Carrott P J M, Singh R, Chaudhary M, Kushwaha S 2016 Cellulose: A review as natural, modified and activated carbon adsorbent *Bioresource Technology* 216 1066–1076
[8] Kaouah F, Boumaza S, Berrama T, Trari M, Bendjama Z 2013 Preparation and characterization of activated carbon from wild olive core (*Oleaster*) by H_3PO_4 for the removal of basic red 46 *Journal of Cleaner Production* 54 296-306

IOP Conf. Series: Materials Science and Engineering **309** (2018) 012065 doi:10.1088/1757-899X/309/1/012065

Productivity improvement with green approach to palm oil factory productivity

N Matondang*

Department of Industrial Engineering, Universitas Sumatera Utara, Jl. Almamater Kampus USU, Medan 20155 Indonesia

*Email: nazaruddin_matondang@yahoo.com

Abstract. The palm oil factory (POF) processes fresh fruit bunches into *crude palm oil* (CPO) and *palm kernel oil* (PKO) by products in the form of liquid and solid waste. One of the solid wastes produced in POF Tanjung Kasau is empty fruit bunches of palm oil (FBPO) which have been burned completely on incinerator tubes so that potentially produces pollutants that pollute the environment. If FBPO waste is managed properly, it will improve the productivity of the company. Therefore, it is necessary to conduct a study to find out how far the increased productivity of the company can reduce their impact on the environment, if FBPO is used as raw material of liquid smoke. The productivity improvement approach is done by Green Productivity concept, by looking at three aspects: environmental, social and economical. Green Productivity aims to protect the environment simultaneously by increasing the productivity of the company. One way is to turn FBPO waste into liquid smoke product is by pyrolysis process. The results showed that turning FBPO solid waste into liquid smoke will increase productivity by 18.18%. Implementation of Green Productivity can improve productivity through the improvement of FBPO waste treatment process which has been done by perfect combustion by pyrolysis process so that waste can be minimized to create environment industry POF clean and friendly environment.

1. Introduction

The palm oil industry plays an important role in the development of the national economy in general and in particular the Province of North Sumatra. Despite the obvious economic benefits from the development of palm oil-based industries, on the other hand, the industry also contributes significantly to environmental degradation, both on inputs and on the outputs of industrial activities. Palm Oil Factory (POF) processes raw materials in the form of fresh fruit bunches (FFB) and produces crude palm oil (CPO) and palm kernel oil (PKO). In addition to the main product (main product), also generated waste in the form of liquid and solid waste in the form of shells, fibers and empty bunches of palm oil (FBPO) [1]. POF Tanjung Kasau located in North Sumatra Batubara District, is one of the POF sites that still processes waste FBPO by burning perfectly through the innecerator. The remaining burning of FBPO ash is dumped into oil palm plantation area as land application. In the combustion process smoke is thrown into free air. The process of handling the FBPO has the potential to pollute the environment. On the other hand FBPO still have economic value when processed further, one alternative is to make it as liquid smoke. Liquid Smoke Rendemen from FBPO pyrolysis process reached 29.45 with selling price in the market around Rp. 30,000, - per liter [7].

Managing FBPO well in addition to contributing to revenue generation can also minimize environmental impacts to create a clean MCC industry. To see how far the improvement of

IOP Conf. Series: Materials Science and Engineering **309** (2018) 012065 doi:10.1088/1757-899X/309/1/012065

productivity and environmental improvement by processing FBPO into liquid smoke, it is necessary to conduct a study using Green Productivity (GP) approach. GP is a strategy to increase productivity and environmental performance for overall socio-economic development. It is an appropriate productivity application and environmental management of policies, tools, techniques and technologies in order to reduce the environmental impact of organizational activities that produce goods or services.

To measure GP, Hur, et al. (2004) [6] developed an economic and environmental performance measurement tool in a single index called the Green Productivity Index (GPI), as the system's productivity ratio for its environmental impact. With this index, companies can estimate economic and environmental performance at once. Adopting this index, Gandhi, et al. (2006) [5], used a weighted environmental impact. However, research related to GP measurement or GP keywords is still quite rare, the same thing as eco-efficiency and environmental productivity are more popular for use in some studies [4]. GP index is actually developed from the concept of eco-efficiency. Initially, eco-efficiency is the ratio of product or service value added to its environmental influences [3,11].

As an indicator of economic performance, the added value of a product or service is modified into productivity as a broader sense of the efficient use of resources with improved quality. To see the extent to which improvement of company productivity and environmental improvement, the study of GPI processing FBPO POF Tanjung Kasau factory as a pilot project, so it can be used as a recommendation for the management of POF industry.

2. Method

The research was conducted at Palm Oil Factory (POF) Tanjung Kasau Sub-Province of Batubara Sumatera Utara. The method used by Green Productivity by measuring and comparing Green Productivity Ratio (GPR) and Green Productivity Index (GPI) before and after processing FBPO into liquid smoke.

Three important steps in conducting analysis using Green Productivity concept include:

1. Getting Started

   The beginnings of the Green Productivity process are walk through surveys and information gathering. Walk through surveys are conducted to identify the sequence of production processes. At this stage must have determined the block diagram of the process and material balance so that known operations that produce waste including estimates or estimates of the amount of waste generated. Early productivity calculations are performed as a basis for consideration in estimating the solution to be obtained to determine whether productivity is improving or otherwise.

2. Planning

   In this planning stage, it is divided into two steps, namely:

   A. Identify the problem and its causes

      Information obtained from walk through surveys will be used to identify problems and causes.

   B. Define goals and targets

      After knowing the problems and causes, it is necessary to set goals and targets as a pointer to select solutions that can eliminate the problem.

3. Generation and Evaluation of GP Options

   This step includes developing solutions to meet the objectives and targets that have been formulated in the previous step. This includes a viewpoint on pollution prevention and planned control procedures. Options are generated and prioritized based on the Green Productivity Indicator. All of that is then synthesized into the implementation plan.

Green Productivity indicator is a strategy, where an indicator is needed that can measure the performance of the strategy quantitatively. Measurements can be made using Green Productivity Index (GPI) and Green Productivity Ratio (GPR) as indicators [6].

$$GP_{index} = \frac{productivity}{the\ impact\ is\ determined} \quad (1)$$

The calculation of Green Productivity Ratio for labor, energy, material, and maintenance is done using the following formula.

$$GPP_{waste} = \frac{kg\ Waste}{kg\ input\ material} \quad (2)$$

4. Results and Discussions

4.1.Existing Condition (Getting Started)

Existing condition is used to make comparisons of productivity calculations before and after processing of FBPO into liquid smoke as a solution to solve FBPO waste problem with green productivity approach. The average input and output of the company during Year 2015 with the processing of FBPO burned perfectly can be seen in Table 1.

Table 1. Average Input and Output of POF Tanjung Kasau in 2015

Output	**Value (Rp)**	**Volume (ton)**
Oil and Core Sales Sawit	28,961,217,685	
Input		
Material	363,794,103	11,293.19
Labor	579,352,468	
Energy	107,080,985	
Maintenance	215,146,432	
Waste		2.37

Based on Table 1 data, GPR calculations for each of the input and effluent are presented in Table 2.

Table 2. GPR POF Tanjung Kasau, 2015

Masukan *(input)*	***Green Productivity Ratio***
Material	79.61
Labor	49.99
Energy	270.46
Maintenance	134.61
Waste	0.21

4.2. Liquid Smoke Processing (Planning)

Based on the pilot project by developing equipment that is able to process FBPO with the calculation of 40 kg of FBPO per day, the following results are obtained:

Liquid smoke produced during one year (350 days) is = 350 days x 11,.82 l / day x Rp. 30,000, - / l = Rp. 124,110,000, - In addition, a by-product of FBPO ash can still be sold for around Rp. 500, - per kilogram. For every 20 kg of FBPO produced 3.36 kg or a day is generated 6.72 kg.

4.3. Generation and Evaluation of GP Options

Using the pilot project basis and potential FBPO waste if all FBPO is treated as liquid smoke, the productivity estimate of POF Tanjung Kasau for one year of production is 27.04. The calculation of productivity by processing all FBPO into liquid smoke is presented in Table 3. While the calculation of GPR and GPI for the use of all FBPO can be seen in Table 4.

Table 3. Productivity of FBPO Processing into Liquid Smoke

Factor	Existing Conditions	Application of Liquid Smoke
Sales	28,961,217,685	63,413,217,685
Input Materials	363,794,103	363,794,103
Input Labor	579,352,468	1,239,352,468
Input Energy	107,080,985	227,080,985
Input maintance	215,146,432	515,146,432
Total productivity	**22.88**	**27.04**

Table 4. Existing GPR Calculations and GPI

Green Productivity Indicators	*Green Proructivity Ratio* **Existing**	*Green Productivity Ratio* **with liquid smoke**	*Green Produtivity Index*
Materials	79.61	170.25	2.13
Labor	49.99	50.24	1.00
Energy	270.46	274.19	1.01
Maintance	134.61	136.81	1.02
Waste	0.21	0	0

The palm oil industry as one of the mainstay and strategic commodities in creating food and energy security is often positioned as an environmentally unfriendly industry. This is because the resulting waste has not been properly managed. Processing of FBPO which is one of solid waste into liquid smoke contribute to the improvement of company performance and also performance to environment. POF Tanjung Kasau if processing FBPO waste overall will increase the productivity of 18.18% ie from 22.88 to 27.04 and GPI to zero which means the impact of FBPO waste on the environment can be minimized. Liquid smoke produced can be used as a substitute for chemical food preservatives that have been used by many people, such as formalin. So that by doing FBPO processing into liquid smoke, impact on the improvement of company performance, create palm oil industry clean and society can use preservative which is safe for health.

5. Conclusions

Based on the result of the research through the pilot project of liquid smoke making from FBPO waste, it can be concluded that:

1. Handling of FBPO solid waste into liquid smoke will increase the productivity of the company from the previous by 22.88% to 27.04% or an increase in productivity level of 18.18%.
2. Implementation of Green Productivity can improve productivity through process improvement. With the improvement of FBPO waste treatment process which has been done perfect combustion with pyrolysis process, the waste can be minimized so that it creates a clean and environmentally friendly industrial environment POF.
3. The resulting liquid smoke can be used as a food preservative that is safe for humans, thus reducing the use of chemical preservatives.

IOP Conf. Series: Materials Science and Engineering **309** (2018) 012065 doi:10.1088/1757-899X/309/1/012065

References

[1] Trisakti B, Lubis J, Husaini T and Irvan 2017 Effect of Turning Frequency on Composting of Empty Fruit Bunches Mixed with Activated Liquid Organic Fertilizer *IOP Conf. Ser. Mater. Sci. Eng.* **180** 8

[2] Himmelblau D M 1999 *Prinsip Dasar dan Kalkulasi dalam Teknik Kimia Jilid 1* Trans Ita Ananta Jakarta : PT. Prenhalido Jakarta

[3] DeSimone L D and Popoff F 1997 *Eco-efficiency: The business link to sustainable development* London: The MIT press

[4] Findiastuti W, Anityasari, Singgih M and Moses L 2011 Green Productivity Index: Do Different Terms Measure The Same Things *Proceeding of Industrial Engineering and Service Science?*

[5] Gandhi N M D, Selladurai V and Santhi P 2006 Green productivity indexing: A practical step toward integrating environmental protection into corporate performance *International Journal of Productivity and Performance Management* **55(7)** 594-606

[6] Hur T, Kim I and Yamamoto R 2004 Measurement of green productivity and its improvement *Journal of Cleaner Production* **12(7)** 673–683

[7] Nazaruddin *et al* 2015 Industrialisasi Asap Cair Berbasis Limbah Padat Kelapa Sawit Medan Lembaga Penelitian USU

[8] Pineda-Henson R and Culaba A B 2004 A diagnostic model for green productivity assessment of manufacturing processes *The International Journal of Life Cycle Assessment* **9(6)** 379-386

[9] Santoso N 2015 *Penerapan Green Productivity Untuk Peningkatan Produktivtas dan Kinerja Lingkungan di Pabrik Gula Sragi* Semarang: UNDIP

[10] Sinulingga S 2010 *Rekayasa Produktivitas* Medan: USUPress

[11] Vervaillie H A and Bidwell R 2000 *Measuring eco-efficiency: A guide to reporting company performance*

IOP Conf. Series: Materials Science and Engineering **309** (2018) 012066 doi:10.1088/1757-899X/309/1/012066

Composting of empty fruit bunches in the tower composter – effect of air intake holes

Irvan[1,2], T Husaini[1], B Trisakti[1,2], F Batubara[1], and H Daimon[3]

[1]Chemical Engineering Department, Universitas Sumatera Utara, Indonesia
[2]Sustainable Energy and Biomaterial Center of Excellence, Universitas Sumatera Utara, Medan, 20155, Indonesia
[3]Department of Environmental and Life Sciences, Toyohashi University of Technology, Toyohashi, 441-8580, Japan

*Email: i_v_a_n_mz@usu.ac.id; irvan@usu.ac.id

Abstract. The process of composting empty fruit bunches (EFB) by mixing with activated liquid organic fertilizer (ALOF) is an alternative utilization of solid waste generated from palm oil mill. This study aims to find composting techniques of EFB and to obtain degradation data of composting EFB by varying the air intake holes to produce good quality compost. Composting process was carried out by tearing the EFB into four shreds, then put into the tower composter while adding ALOF until it reached the optimum moisture content of 55 - 65%. During the composting process, we maintained moisture content at optimum conditions by adding ALOF. Variations of air intake holes area to the outer surface area of the composter are 0/44.314; 72.39/44.314 and 144.78/44.314 (cm^2/cm^2). Composting is carried out for forty days, however, based on the result, compost began to mature on the 10th day. The results revealed that there was an influence of air intake holes to the composting process. The best degradation of EFB was obtained on the variation of air intake holes 72.39/44.314 (cm^2/cm^2), pH 8.1, moisture content 79.14%, water holding capacity 60%, electrical conductivity 4.725 dS/m and C/N ratio 20.97.

1. Introduction

During its operation, besides producing crude palm oil (CPO) and palm kernel, palm oil mill (POM) also produces waste, i.e., solid wastes such as empty fruit bunches (EFB) and palm oil mill effluent (POME). According to Kerdsuwan and Krongkaew, about 20 - 22 tons of EFB are produced from 100 tons of fresh fruit bunches [1]. Indonesia is estimated to produce 90 million metric tons of fresh fruit bunches, resulting in EFB waste about 20.7 million metric tons per year [2]. EFB contains high cellulose, which slightly easy decomposed with a combination of physical, chemical and biological processes [3]. Several studies have been conducted for the processing of EFB into compost by adding additional materials such as animal wastes [4], and by mixing with POME [5]. In this research, the addition was carried out with active organic fertilizer (AOF) that serves as an activator and also maintains the moisture content (MC).

Active organic fertilizer (AOF) is a further processing of POME fermentation into biogas [6]. Effluent used as raw material of active organic fertilizer has passed anaerobic degradation process in the biogas processing in digester tank. The effluent from the biogas digester still contains methane degrading bacteria and carbon dioxide degradation, which is one of the nitrogen-fixing bacteria [7].

IOP Conf. Series: Materials Science and Engineering **309** (2018) 012066 doi:10.1088/1757-899X/309/1/012066

Methanobacterium and *Methanobacillus* contained in the effluent are known to form N_2 and to add other macro elements such as phosphate needed by the processing bacteria *Bacillus sp*, which is achieved by the addition of effective microorganism 4 (EM-4). The purpose of this research is to obtain degradation data of EFB composting using active organic fertilizer inside drum tower composter.

2. Method

2.1. Empty fruit bunches

The raw material of this research was EFB obtained from Palm Oil Mill Sei Mangkei belongs to PTPN III. EFB were torn into 4 (four) shreds. Characteristic of EFB are presented in table 1.

Table 1. Characteristic of EFB and AOF

Parameter	Unit	EFB	AOF
Moisture	(%)	43.83	-
pH		9.10	8.09
C	(%)	33.15	0.58
N	(%)	0.99	0.10
C/N	-	33.48	5.8
Water Holding Capacity	(%)	62	-
P_2O_5	(%)	-	0.016
K_2O	(%)	-	0.167
COD	(mg/l)	-	1,580

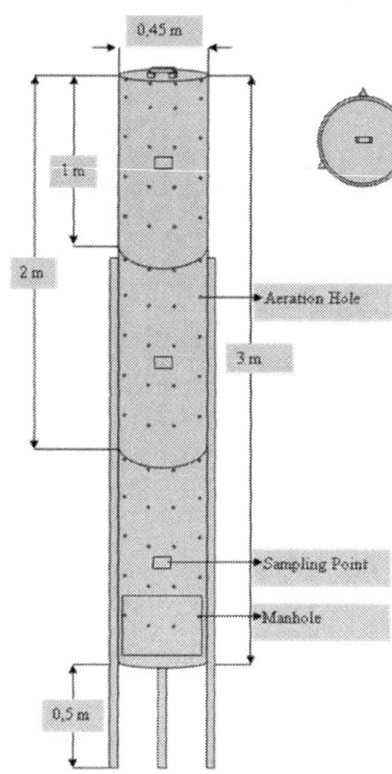

Figure 1. Scheme and dimension of tower composter

2.2. Active organic fertilizer

Active organic fertilizer, as a source of nutrition and maintenance of moisture content, was obtained from POME processing at Biogas Power Plant, Universitas Sumatera Utara, Medan. Characteristic of AOF used can be seen in table 1.

2.3. Tower composter

Main equipment used was a drum tower composter with a height of 3.00 m and diameter of 0.45 m. The composter was equipped with a sampling hole and an air intake hole. The air intake hole was 1.25 cm in diameter. Scheme and dimension of tower composter can be seen in figure 1.

2.4. Procedure

This research was conducted at Biogas Pilot Plant, Universitas Sumatera Utara, Medan. Composting process was carried out by measuring the total weight of EFB initially, after that inserted into the tower composter, then AOF was added until the moisture content reached 55-65% and maintained at that range during the composting process.

Compost sampling was conducted by taking the compost from sampling holes which were available at each composter height. Temperature measurement was recorded by inserting the thermometer into the hole and performed two times a day, in the morning and afternoon. While, moisture content was measured daily using five grams of compost sample, then pH measurements were performed daily using potentiometric methods. Analysis of C and N were performed every ten days by using Walkley & Black and Kjeldahl method, respectively. Electrical conductivity analysis was carried out every ten days using conductivity meter. Bacterial count analysis was performed every ten days, compost quality analysis was done at the beginning and the end of the composting process.

3. Results and Discussions

3.1. Analysis of temperature and moisture content profiles at any height of compost pile

The most essential parameter to be measured in the composting process is temperature [8]. Figure 2 shows the effect of air intake hole on average temperature during the composting process. The average temperature of composter I, II and III at each tower height of 1, 2 and 3 m were 45.48; 44.92; 45.66 °C, 45.46; 44.61; 45.25 °C and 43.01; 42.93; 42.71°C, respectively.

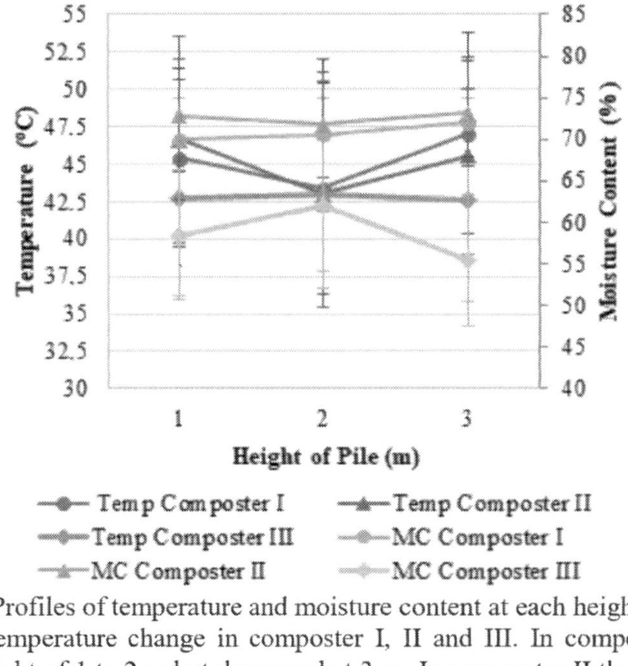

Figure 2. Profiles of temperature and moisture content at each height composter

Figure 2 shows the temperature change in composter I, II and III. In composter I the temperature increased from the height of 1 to 2 m but decreased at 3 m. In composter II the temperature decreased from the height of 1 to 2 m but increased at 3 m, while in composter III the temperature tends to decrease from the height of 1 to 3 m. It can be seen in figure 2, the deeper the compost pile, the higher the temperature. Temperature between composter I and II looks not much different.

Increasing of temperatures in the deeper compost pile was caused by the pile porosity, this was in accordance with those mentioned by Van Ginkel et al. 1999, the mass and heat transfer in the compost pile is significantly influenced by the geometry and the pore size distribution and also composition of the compost material [9]. With the deeper the pile, the porosity in a pile will be smaller, causing the amount of heat generated during the decomposition process was trapped inside the pile.

Moisture content is one of the important factors that must be maintained in the composting process. Baharuddin et al. 2010 stated that the best condition of MC for composting was 55 - 65% [10]. The average MC of composter I, II and III at each height of 1, 2 and 3 m were 68.35, 68.74, 71.99 %; 71.19, 69.84, 73.10 % and 58.09, 61.24, 55.47 %.

As shown in figure 2, the average of moisture content in the composter I and II is almost the same in each height of composters, compared to the moisture content in the composter III. This was due to the liquid that dropped from composter I and II was recycled every day during the composting process, and less air intake hole was used, compared to the composter III, so that moisture content looks much higher. While fluid in composter III was recycled from the bottom of composter at the time of moisture content < 55 % and a number of air intake holes used were more than the number of holes in composter I and II. It caused loss of a large amount of water at vapor state then it can be seen moisture content in composter III was in the range of 55 – 65 % in accordance as reported by Baharuddin et al. 2009. The decrease of average MC value in a pile was also affected by the water evaporation caused by the heat from the pile as reported by Tiquia et al. (2001) where high temperatures in composting process can cause water loss in the vapor state [11].

3.2. Analysis of pH and profiles at any height of compost pile

Figure 3 shows the average pH of composter I, II and III, where the average pH of composter I, II and III at each 1, 2, and 3 m elevations were 7.91, 7.96, 8.02; 8.00, 7.89, 7.92 and 8.83, 8.77, 8.31. The consequence of pH changes during the composting process is due to microbial activity [12, 13].

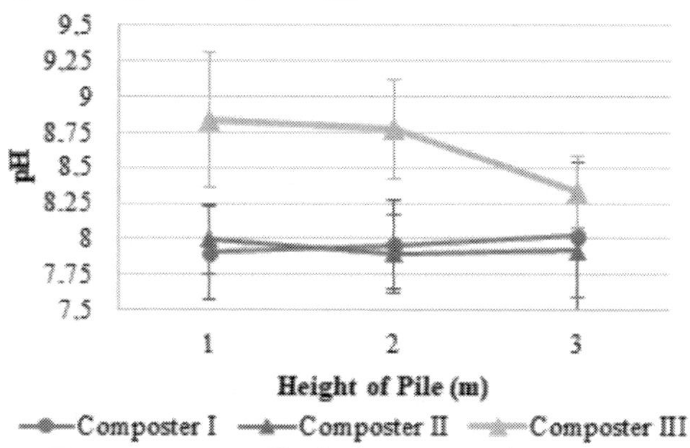

Figure 3. pH profile during composting process

The average pH between composter I and II looks not much different than composter III, but the highest value among them is pH 8.83 in composter III at the height of 1 m. The increase in pH becomes base condition is good for the composting process, because alkaline conditions can inhibit the growth of pathogens such as fungi that can live in acidic conditions. The addition of aeration holes to the composter helps to reduce the anaerobic phase and helps remove CO_2 trapped in the free space between the compost particles, thus preventing the occurrence of acidic conditions in the pile or significant decrease in pH [14, 15].

3.3. Analysis of compost based on bacterial count and temperature

Microbial activity, quantity, and biomass are parameters that can also be used to explain the composting process [16]. Changes in the number of microbial colonies during the composting process can be seen in Figure 4.

Based on the temperature profile in Fig. 4, two phases are observed during the composting process, namely thermophilic phase (from day 1 to day 20) and mesophilic (day 16 to day 40). The initial value of bacterial count is 26 x 10^7. At thermophilic phase, the value of bacterial count relatively increased in all composters. This is due to the availability of nutrients, the mesophilic bacteria that survive in the thermophilic phase were growing. On the 20th day until the 40th, after being in the mesophilic phase, the number of bacteria slowly begins to decrease as the availability of nutrients decreased. There are currently no reports of acceptable amounts of microbes on mature compost [17] but Atkinson et al. (1996) emphasized that the number of microbes should be low and contain no significant number of pathogens [18].

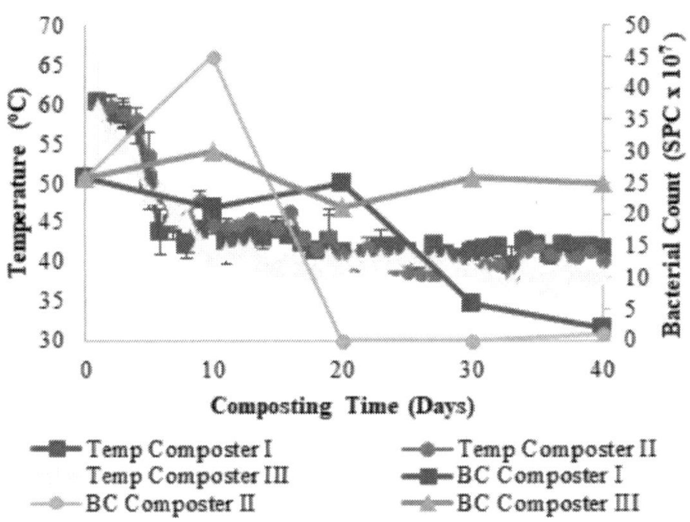

Figure 4. Bacterial count and average temperature during the composting process

3.4. Analysis of compost based on change of C/N

The C/N ratio is an essential indicator of compost maturity [19]. It was necessary to measure the C/N ratio performed five times during this study as shown in Figure 5.

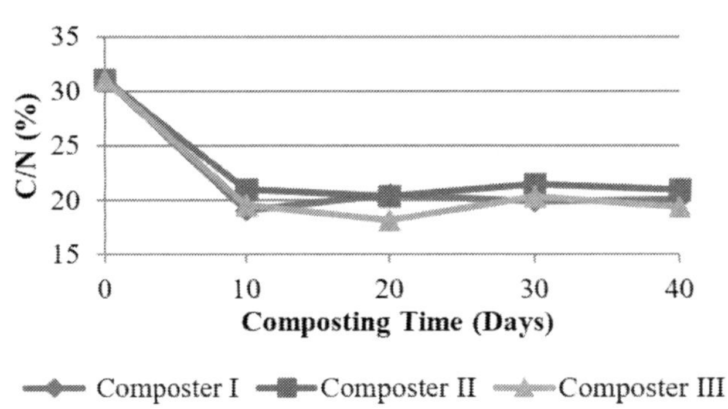

Figure 5. C/N ratio changes during the composting process

The C/N was analysed five times during the composting process from day 0 to day 40. The initial C/N value of EFB was 31.06. On day ten there was a drastic decrease of C/N on composter I, II and III namely 19.08; 20.99 and 19.57. On the 40th day, C/N ratio of each composter was constant in the range 18.13 - 21.47. The decrease in the C/N ratio is due to the decrease in C content during the composting process. This occurs because of the decomposition process of organic matter from the results of microbial activity [20].

3.5. Analysis of compost based on change of electrical conductivity
Electrical conductivity (EC) reflects the level of salinity in a compost product, which suggests the possibility of phytotoxic or phyto-inhibitory effects [21]. The electrical conductivity value changes during composting are shown in Figure 6. The initial value of electrical conductivity compost on composter I, II and III are 4.42, 3.43 and 1.74 $dS.m^{-1}$, respectively. On day ten until day 20, the value of compost EC has increased in composter I, II, and III, however, on day 30, the value of EC decreased and then constant at day 40 in the range 4.57 – 4.72 $dS.m^{-1}$. The decline in EC values during the composting process is a direct result of increased concentrations of nutrients such as nitrates and nitrites [22] whereas the increase in EC values can be caused by the release of mineral salts such as phosphate and ammonia ions through the decomposition of organic substances [21]. The final EC value obtained after the composting process was 3.73 $dS.m^{-1}$. Characteristic of final compost is shown in table 2.

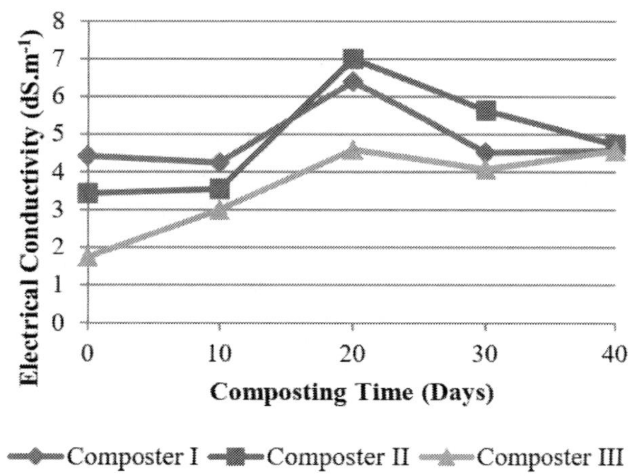

Figure 6. Electrical conductivity changes during the composting process

Table 2. Characteristic of final compost

Parameter / Composition	Unit	Initial EFB	Final		
			Composter I	Composter II	Composter III
Moisture Content	%	43.83	73.35	79.14	60.74
pH		9.1	8.1	8.1	8.8
Water Holding Capacity	%	62	60	60	58
C	%	37.89	22.10	25.16	26.10
N	%	1.22	1.10	1.20	1.35
C/N		31.06	20.09	20.97	19.35
Bacterial Count	SPC x 10^7	26	2	1	25
Electrical Conductivity	$dS.m^{-1}$	6.719	4.568	4.725	4.568
Na	%	0.102	0.059	0.059	0.059

CaO	%	1.759	2.043	2.043	2.043
MgO	%	0.537	0.437	0.250	0.403
P_2O_5	%	0.201	0.091	0.091	0.091
Cd	ppm	0.894	0.437	0.437	0.437
Cu	ppm	0.673	0.078	0.078	0.078
Fe	ppm	1.073	0.051	0.051	0.051
Pb	ppm	1.246	0.080	0.080	0.080
Zn	ppm	2.370	0.052	0.052	0.052
Fat	%	2.150	1.430	1.430	1.430

4. Conclusion

The conclusions obtained from research are as follows:

1. EFB with the mixture of AOF as a microbial source, the nutrient source, and moisture content buffer were proven to produce compost about ten days.
2. The height of the compost pile affected the composting process as shown on some parameters that have been analyzed
3. The composting process using drum tower composter successfully reached thermophilic temperature.
4. The quality of compost on the 40th day was not so much different from compost produced on the 10th day, as indicated by the change in C/N value from 21.47 to 20.97 on composter II.

Acknowledgment

This research was supported by Hibah Penelitian Talenta Universitas Sumatera Utara No: 5338/UN5.1.R/PPM/2017, May 22nd, 2017.

References

[1] Kerdsuwan S, Krongkaew L 2011 Renewable Energy – Trends and Applications, Intech
[2] Ishola M M, Isroi, Taherzadeh M J 2014 *Bioresour. Technol.* **165** 9-12
[3] Alkarimiah R, Rahman RA 2014 *Int. J. of Eng. Sci. Innov. Tech.* **3** 132-145
[4] Ghazali M R, Mutalib S A, Abdullah A 2016 *AIP Conference Proceedings,* **1784** 030042
[5] Baharuddin A S, Wakisaka M, Shirai Y, Abd-Aziz S, Abdul Rahman A N, Hassan, M A 2009 *Int. J. Agri. Res.* **4** 69-78
[6] Irvan, Trisakti B, Sosanty F and Tomiuchi Y 2016, *Asian J. Chem.* 28 377-380
[7] Kyuma K, 2005 Paddy Soil Science, Kyoto University Press and Trans Pacific Press
[8] Saiidi S, Hasani M, Hashemi J, Amini-rad M 2012 The 1th International and the 4th National Congress on Recycling of Organic Waste in Agrilcultural, Isfahan-Iran
[9] Van Ginkel J T, Raats P A C, Van Haneghem I A 1999 *Netherlands J. Agri. Sci.* **47** 105-121
[10] Baharuddin A S, Lim S H, Md Yusof M Z, Rahman N A A, Md Shah U K, Hassan M A, Wakisaka M, Sakai K, Shirai Y 2010 *African J. Biotech.* **9** 2427-2436
[11] Tiquia S M, Wan J H C, Judy H.C. Wan, Tam N F Y 2001 *Proc. Biochem.* **37** 1057-1065
[12] Kananam W, Suksaroj T T, Suksaroj C 2011 *Sci. Asia* **37** 17-23
[13] Sundberg C, Smars S, Jonsson H 2004 *Bioresour Tech.* **95** 145-150
[14] Liu H, Cai L 2014 *BioResources* **9** 4862-4872
[15] Sekman E, Top S, Varank G, Bilgili M S 2011 *Freseniu Env. Bulletin* **20** 1841-1852
[16] Tiquia S M, Sonia M 2005 *J. Appl. Microb.* **99** 816-828
[17] Ogefere H O, Ogbimi A O, Omoregie R 2010 *Malaysian J. Microb.* **6** 25-29
[18] Atkinson C F, Jones D D, Gauthier J J 1996 *Poultry Sci.* **75** 608-617
[19] Nutongkaew T, Duangsuwan W, Prasertsan S, Prasertsan P 2011 *Tiche International Conference,* fb106

IOP Conf. Series: Materials Science and Engineering **309** (2018) 012066 doi:10.1088/1757-899X/309/1/012066

[20] Hock L S, Baharuddin A S, Ahmad M N, Md Shah U K, Rahman N A A, Abd-Aziz S, Hassan M A, Shirai Y 2009 *Australian J. Basic Appl. Sci.* **3** 2809-2816
[21] Gao M, Liang F, Yu A, Li B, Yang L 2009 *Chemosphere* **78** 614-619
[22] Pathak A K, Singh M M, Kumara V, Arya S, Trivedi A K 2012 *Recent Res. Sci. Tech.* **4** 10-14

IOP Conf. Series: Materials Science and Engineering **309** (2018) 012067 doi:10.1088/1757-899X/309/1/012067

Influence of the weighing bar size to determine optimal time of biodiesel-glycerol separation by using the buoyancy weighing-bar method

R Tambun*, Y Sibagariang, J Manurung
Department of Chemical Engineering, Universitas Sumatera Utara, Padang Bulan, Medan 20155, Indonesia

*E-mail: rondang@usu.ac.id

Abstract. The buoyancy weighing-bar method is a novel method in the particle size distribution measurement. This method can measure particle size distributions of the settling particles and floating particles. In this study, the buoyancy weighing-bar method is applied to determine optimal time of biodiesel-glycerol separation. The buoyancy weighing-bar method can be applied to determine the separation time because biodiesel and glycerol have the different densities. The influences of diameter of weighing-bar by using the buoyancy weighing-bar method would be experimentally investigated. The diameters of weighing-bar in this experiment are 8 mm, 10 mm, 15 mm and 20 mm, while the graduated cylinder (diameter : 65 mm) is used as vessel. The samples used in this experiment are the mixture of 95 % of biodiesel and 5 % of glycerol. The data obtained by the buoyancy weighing-bar method are analized by using the gas chromatography to determine the purity of biodiesel. Based on the data obtained, the buoyancy weighing–bar method can be used to detect the separation time of biodiesel-glycerol by using the weighing-bar diameter of 8 mm, 10 mm, 15 mm and 20 mm, but the most accuracy in determination the biodiesel-glycerol separation time is obtained by using the weighing-bar diameter of 20 mm. The biodiesel purity of 97.97 % could be detected at 64 minutes by using the buoyancy weighing-bar method when the weighing-bar diameter of 20 mm is used.

1. Introduction

The determination of biodiesel purity is one of the important stages in the biodiesel industry. During this time, generally the determination the purity of biodiesel is analyzed by using the gas chromatography [1, 2]. The gas chromatography can give the accuracy result in determination the purity of biodiesel, but the price of the equipment is expensive. To overcome this problem, the buoyancy weighing-bar method (BWM) will be used to predict the separation time of biodiesel and glycerol mixtures. The BWM is a new sedimentation method which is developed from the sedimentation balance, and this method has proven accuracy in determination particle size distribution. This method is simple in operation and low-cost of equipment. The principle of this method that measurement the density change in a suspension due to liquid migration is measured by weighing buoyancy against a weighing bar hung in the suspension. This apparatus consists of an analytical balance with a hook for under-floor weighing and a weighing bar, which is used to detect the density change of suspension [3-8]. The BWM can also be used to predict the fine particles by combinating with the Rosin-Rammler equation [9]. In the present study, we are going to develop the BWM for liquid-liquid systems with the different density to determine the separation time of biodiesel and glycerol mixtures by investigating the influence of the weighing bar size. The principle of this experiment is analogous to distributions of the settling particles and floating particles. In this experiment, the initial buoyant mass of the weighing bar depends on the mixtures of liquid between the top and bottom of the weighing bar in a suspension [3-8].

2. Material and Methods

The experiment is illustrated at figure 1. The weighing–tools is weighing–bar which is composed of aluminum (density: 2700 kg/m^3). The diameters of weighing-bar in this experiment are 8 mm, 10 mm, 15 mm and 20 mm, and the length is 210.0 mm and submerged length is 200.0 mm. The samples consist of 95 % of biodiesel (methyl ester, $\rho = 0.8791$ g/cm^3) and 5 % of glycerol ($\rho = 1.26$ g/cm^3) are placed in a 1000 ml graduated cylinder (diameter: 65.0 mm). The analytical balance (minimum readout mass 0.1 mg) has a below–balance–weighing hook for hanging measurement. Using a hanging wire, which does not extend due to the weight of the weighing bar, the weighing bar is hung from the analytical balance. The room temperature is approximately 298 K. After thoroughly stirring the suspension using an agitator, the weighing bar is set with the balance. The measuring data, which consist of time and the corresponding mass of the bar, are recorded. The measuring time is 2 hours and the data are collected in 60–second intervals. As comparison method, the gas chromatography is used to analyze the purity of biodiesel.

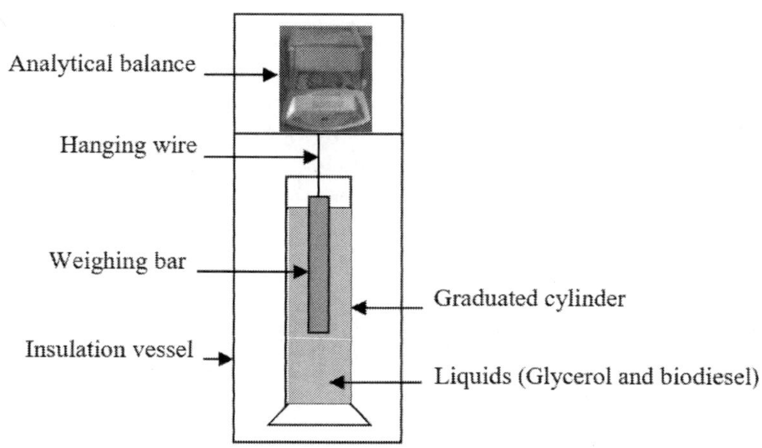

Figure 1. Illustration of experiment apparatus

3. Results and Discussion

Figure 2 shows the change with time in the apparent mass of weighing-bar when the weighing-bar diameter of 8 mm is used. The figure shows that the apparent mass of the weighing-bar increased until all the glycerol settled below the lower end of the weighing-bar, and then the apparent mass of the weighing-bar become constant. The change in the apparent mass is due to the change in the buoyant mass against the weighing-bar as well as glycerol settling. Figure 2 shows us that the BWM can detect the separation time of biodiesel-glycerol after 18 minutes of measurement, where at this time the mass of weighing-bar diameter of 8 mm starts to become constant and the purity of biodiesel is 96.92%. In this experiment, the initial of biodiesel purity is 99.85%, so the measurement error of BWM using weighing-bar diameter of 8 mm to detect the separation time of biodiesel-glycerol is 2.93%. Figure 3 shows the change with time in the apparent mass of weighing-bar when the diameter of weighing-bar of 10 mm is used. The figure shows that the apparent mass of the weighing-bar increased until all the glycerol settled below the lower end of the weighing-bar, and then the apparent mass of the weighing-bar become constant. The change in the apparent mass is due to the change in the buoyant mass against the weighing-bar as well as glycerol settling. Figure 3 shows us that the BWM can detect the separation time of biodiesel-glycerol after 14 minutes of measurement, where at this time the mass of weighing-bar diameter of 10 mm starts to become constant and the purity of biodiesel is 96.81%. In this experiment, the measurement error of BWM using weighing-bar diameter of 10 mm to detect the separation time of biodiesel-glycerol is 3.04 %.

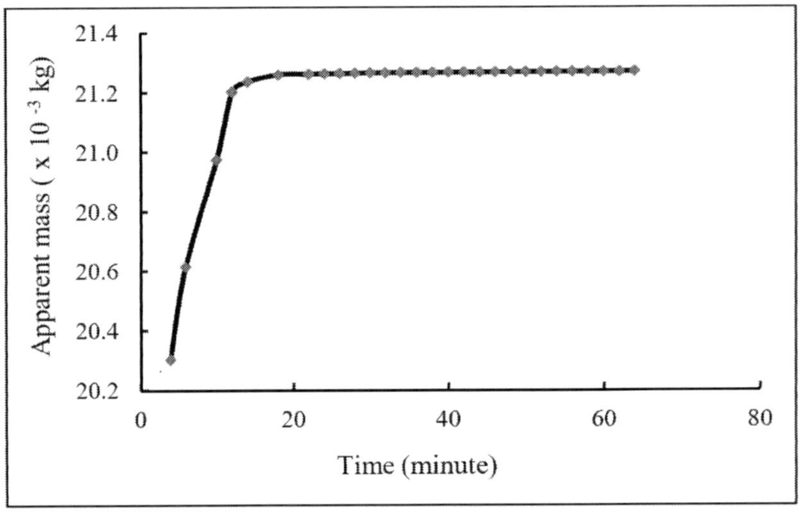

Figure 2. Apparent mass of the weighing-bar as a function of time when using the weighing-bar diameter of 8 mm

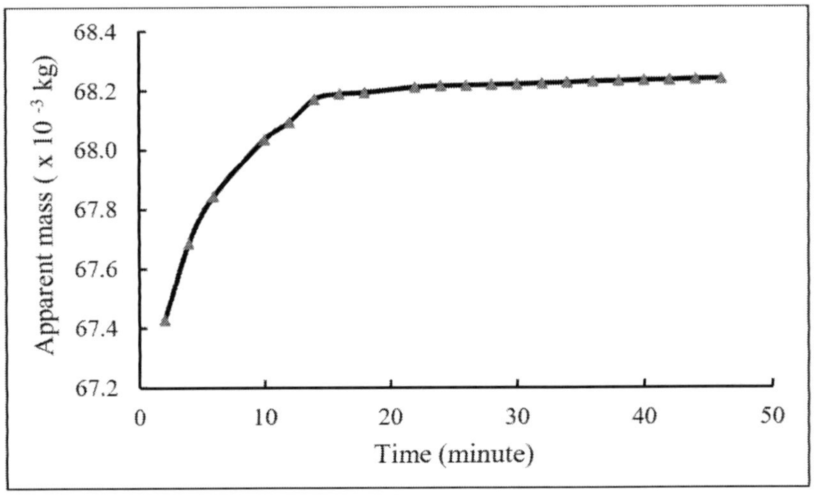

Figure 3. Apparent mass of the weighing-bar as a function of time when using the weighing-bar diameter of 10 mm

Figure 4 shows the change with time in the apparent mass of weighing-bar when the weighing-bar diameter of 15 mm is used. The figure shows that the apparent mass of the weighing-bar increased until all the glycerol settled below the lower end of the weighing-bar, and then the apparent mass of the weighing-bar become constant. The change in the apparent mass is due to the change in the buoyant mass against the weighing-bar as well as glycerol settling. Figure 4 shows us that the BWM can detect the separation time of biodiesel-glycerol after 36 minutes of measurement, where at this time the mass of weighing-bar diameter of 15 mm starts to become constant and the purity of biodiesel is 97.15%. In this experiment, the measurement error of BWM using weighing-bar diameter of 15 mm to detect the separation time of biodiesel-glycerol is 2.70 %.

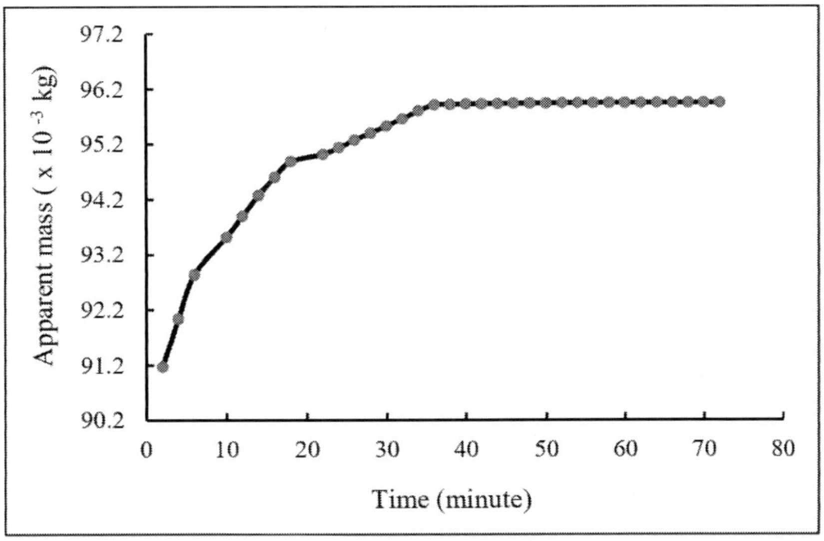

Figure 4. Apparent mass of the weighing-bar as a function of time when using the weighing-bar diameter of 15 mm

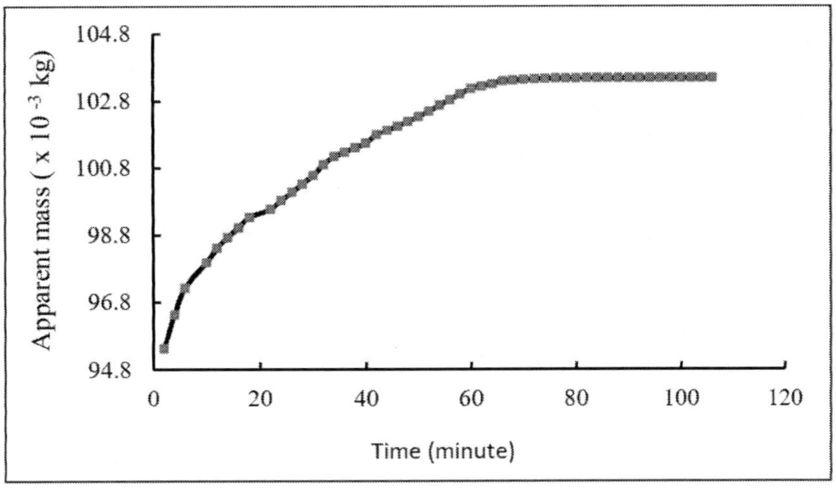

Figure 5. Apparent mass of the weighing-bar as a function of time when using the weighing-bar diameter of 20 mm

Figure 5 shows the change with time in the apparent mass of weighing-bar when the weighing-bar diameter of 20 mm is used. The figure shows that the apparent mass of the weighing-bar increased until all the glycerol settled below the lower end of the weighing-bar, and then the apparent mass of the weighing-bar become constant. The change in the apparent mass is due to the change in the buoyant mass against the weighing-bar as well as glycerol settling. Figure 4 shows us that the BWM can detect the separation time of biodiesel-glycerol after 64 minutes of measurement, where at this time the mass of weighing-bar diameter of 20 mm starts to become constant and the purity of biodiesel is 97.97%. In this experiment, the measurement error of BWM using weighing-bar diameter of 20 mm to detect the separation time of biodiesel-glycerol is 1.88 %. Thus, we confirm that the separation time of biodiesel-glycerol can be detect by using BWM when the weighing-bar diameters of 8 mm, 10 mm, 15 mm and 20 mm and ratio 95% : 5% of biodiesel and glycerol are used, but the most accuracy result is obtained by using the weighing-bar diameter of 20 mm.

4. Conclusions

This study has experimentally investigated the application of BWM to detect the time separation of biodiesel-glycerol mixtures, the following results are obtained:

1. The buoyancy weighing-bar method could determine the separation time of biodiesel-glycerol when ratio 95% : 5% of biodiesel and glycerol is used.
2. The weighing-bar diameter of 20 mm give the more accuracy result than the weighing-bar diameter of 8 mm, 10 mm, and 15 mm in detecting the separation time of biodiesel-glycerol when ratio 95% : 5% of biodiesel and glycerol is used.
3. The mass of weighing-bar diameter of 20 mm starts to become constant when the biodiesel purity is 97.97%, while the weighing-bar diameter of 8 mm, 10 mm, and 15 mm start to become constant when the biodiesel purity are 96.92 %, 96.81 %, 97.15 %, respectively.

References

[1]. M Holčapek, P Jandera, J Fischer, B Prokeš 1999 Analytical monitoring of the production of biodiesel by high-performance liquid chromatography with various detection methods *Journal of Chromatography A* **858 (1)** pp 13-31

[2]. S N Fedosov, J Brask, X Xu 2011 Analysis of biodiesel conversion using thin layer chromatography and nonlinear calibration curves *Journal of Chromatography A* **1218 (19)** pp 2785-2792

[3]. Motoi T, Ohira Y, Obata E 2010 Measurement of the floating particle size distribution by buoyancy weighing–bar method *Powder Technology* **201** pp 283–288

[4]. Obata E, Ohira Y, Ohta M 2009 New measurement of particle size distribution by buoyancy weighing–bar method *Powder Technology* **196** pp 163–168

[5]. Ohira Y, Furukawa K, Tambun R, Shimadzu M, Obata E 2010 Buoyancy weighing-bar method: a particle size distribution measurement using new settling method *Journal of the Sedimentological Society of Japan* **69** pp 17-26

[6]. Tambun R, Motoi T, Shimadzu M, Ohira Y, Obata E 2011 Size distribution measurement of floating particles in the allen region by a buoyancy weighing–bar method *Advanced Powder Technology* **22** pp 548–552.

[7]. Tambun R, Nakano K, Shimadzu M, Ohira Y, Obata E 2012 Sizes influences of weighing bar and vessel in the buoyancy weighing-bar method on floating particle size distribution measurements, *Advanced Powder Technology* **23** pp 855-860

[8]. Tambun R, Shimadzu M, Ohira Y, Obata E 2012 Definition of the new mean particle size based on the settling velocity in liquid *Journal of Chemical Engineering of Japan* **45 (4)** pp 279-284

[9]. Tambun R, Furukawa K, Hirayama M, Shimadzu M, Yamanaka S, Ohira Y 2016 Measurement and estimation of the particle size distribution by the buoyancy weighing-bar method and the Rosin-Rammler equation *Journal of Chemical Engineering of Japan* **49 (2)** pp 229-233

Utilization of mango seed starch in manufacture of bioplastic reinforced with microparticle clay using glycerol as plasticizer

Maulida*[1], T Kartika[1], M B Harahap[2] and M H S Ginting[1]

[1]Department of Chemical Engineering, University of North Sumatera, Medan, Indonesia
[2]Department of Physics, State University of Medan, Medan, Indonesia

*E-mail: maulida@usu.ac.id

Abstract. Bioplastics are plastics that can be used just like conventional plastics but will disintegrate by the activity of microorganisms into water and carbon dioxide. Starch is a natural polymer material that can used for bioplastic production. The addition of reinforcing particles has been shown to improve the mechanical properties of bioplastics. The aim of this research is to know the potency of mango seed starch and microparticle clay as filler and glycerol concentration as plasticizer on tensile strength and elongation at break, functional group (FTIR) and surface morphology (SEM). In this study used mango seed starch size of 5 grams, with the variation of clay filler mass of 0; 3; 6 and nine wt%, while the mass of glycerol with a variation of 0; 20; 25; 30; And 35% wt. The heating temperature of the bioplastics solution used was 80.53 ^{0}C. The resulting bioplastics was analyzed for their physical and chemical properties, including FTIR, SEM, tensile strength, elongation at break. The FTIR analysis shows that no new functional groups was formed. From the analysis of mango starch content obtained 62.82%, 44.0% amylopectin content, amylose content 14.82%, and water content 12.65%. In this study obtained bioplastics with the best conditions on the use of 6% clay and 25% glycerol, with a tensile strength of 5.657MPa, percent elongation at breakup 43.431%.

1. Introduction

Packaging plays a significant part in recent years as over 67 millions of tonnes of packaging waste resulting in environmental concerns. That packaging had much different processing, for example, the using of additives such as fillers, colorants and plasticizers to produce polymers as packaging applications [1]. Environmental, economic and safety challenges have caused many scientists to invent biodegradable polymers as a substitute for petrochemical based polymers. This biodegradable polymers or bioplastics indicate more eco-friendly environment than the conventional plastics which lead to environmental issues [2]. But there is still lack in the development of bioplastics as its high cost of production leads to certain limitations. An alternative was proposed for low cost and the renewable substrate so that would be used in agriculture waste (AW) [3]. In addition to performance and abundant sources, starch from agriculture waste is the solution for an alternative. Starch as biodegradable polymer are proper to use as material for production of bioplastics because of its low cost [4]. Therefore, production of bioplastics is the breakthrough innovation to solve the environmental issues by using renewable and degradable natural resources and to provide more cost effective bioplastics. Raw materials for bioplastics originate from natural constituents such as polysaccharides (e.g. starch, cellulose, chitin, and lignin), proteins (e.g. gelatine, casein and wheat gluten) and lipids (e.g. plant oils and animals fats) [1]. In Indonesia, Mango production in Indonesia is fluctuating but tends to be high every year. There are still many mango seeds that are just left alone, so it only becomes a waste that pollutes the environment. The seed contains macromolecule reserves in large quantities and distinctive and has a starch content that can be utilized and is not a substitute for staple food. Trash of mango seeds that have not been utilized to the maximum this will result in an economic value. Also, mango seeds have a high enough starch content that potentially as an alternative substitute material in the manufacture of bioplastics. Amylose levels in mango seeds are expected to provide optimal mechanical properties, and amylopectin levels provide optimum stickiness [5]. In this study, the fillers used were organic clay with the size of microparticles which were then combined with mango seed starch as the matrix for bioplastic formation was biodegradable and had better mechanical properties. The addition of plasticizers or plasticizers into biodegradable film blends is necessary to overcome the fragility of films caused by high intermolecular forces [6].

Utilization of organic waste such as mango seeds for production of starch based bioplastic can help reduce the environmental damages that are caused by conventional plastics. Higher value bioplastics can be obtained by improving their properties with the most abundant and biodegradable reinforcing filler like cellulose. The goal of this work is to study the properties of bioplastics from agriculture waste. Effect of reinforcement fillers and plasticizers on bioplastics are also examined.

2. Materials and Methods

2.1. Materials

Starch derived from mango seeds was obtained from juice traders in Medan, clay with particle size 10.813 μm, and glycerol 99% was obtained from Rudang Jaya Medan.

2.2. Preparation Starch of Mango Seed

Mango seeds (100 gram) obtained from the waste container at local juice traders was washed with clean water before shredded to small pieces. The shredded mango seeds were later peeled and placed in mixing blender and soaked in water for about 100 ml. After mixing process, starch slurry was filtered and later placed in the tank for settling that took at least 30 minutes. Starch sediment was separated from the slurry and then washed again with distilled water. After the second settling, starch sediment was dried using an oven with temperature ±60 °C for removal of free water. Starch was sieved with strainer 100 mesh/inch for better homogeneous size.

2.3. Bioplastic Preparation

A solution containing 0, 3, 6 and 9% wt/wt of fillers to starch was prepared by dispersing 100 ml distilled water and glycerol with concentration varied from 0, 20, 25, 30, and 35% wt/v of plasticizer to starch. The solution was placed into ultrasonic homogenizer KUDOS tank and processed for about 50 minutes. After ultrasonication, the solution was removed from the tank. Starch (5 gram) was added to the solution and heated using hot plate while stirred until it gelatinized (80.53 °C). After mixing, the solution was cast onto flat and dried with temperature 60 °C for 24 hours. Once set, bioplastic was cooled to ambient temperature before peeled off the flat.

2.4. Bioplastic Characterizations

2.4.1. Tensile Strength and Elongation at Break.

Tensile strength was measured with GoTech Universal Testing Machine using the standard of ASTM D882-91. Tensile strength value was obtained from the observed data. Elongation at break is an indication of bioplastics flexibility and is expressed as a percentage.

IOP Conf. Series: Materials Science and Engineering **309** (2018) 012068 doi:10.1088/1757-899X/309/1/012068

2.4.2. Density.
The density of bioplastic was investigated by the standard of ASTM D792-91 on film with size approximately 5x5 cm. The film's mass was measured using analytical balance. Density was calculated as follows :

$$\text{Density} = \frac{\text{mass (gram)}}{\text{volume} (\text{cm}^3)} \tag{1}$$

2.4.3. Water Uptake.
Water uptake was investigated by cutting film with size approximately 2x2 cm and then weighed the mass. The film was put into a container filled with distilled water for 24 hours. After immersion in water, the film was removed from the water and weighed to measure the wet weight. Water uptake was calculated as follows :

$$\text{Water uptake} = \frac{\text{wet weight - dry weight}}{\text{wet weight}} x100 \tag{2}$$

2.4.4. Fourier Transform Infrared Spectroscopy (FT-IR).
Functional groups of bioplastics were analyzed by using IR Prestige-21 Shimadzu. The analysis using FT-IR represented spectrum data in graphic and wave numbers of each data that provided functional groups of bioplastics.

3. Results

3.1 The Result Starch Extraction from Mango Seed
In this study, the raw materials for bioplastics are starch extracted from mango seeds. Mango seeds obtained from juice traders located at the Pembangunan and Padang Bulan, Medan. The resulting starch grayish powder form with a particle size ± 100 mesh. The results starch extraction from the mango seeds yield of starch obtained by 43.2 %, which of 100 g dried starch mango seeds gained as much as 43.2 g and then further analysis of starch obtained.

(a) (b)

Figure 1. (a) the results extraction (b) starch the mango seeds of starch from mango seeds

3.2 Fourier Transform Infrared Spectroscopy (FT-IR)

Based on Fig. 7, the bioplastic starch of mango seed with glycerol plasticizer generated several peaks of wave numbers in each range of region. There is a peak of the wave number 3527.02 cm^{-1} to 2677.20 cm-[1]. The peak corresponds to the absorption caused by the C-H bond (type alkane compound), and OH (type of phenol compound, hydrogen bonding alcohol). At peak with wave numbers 2140.99 cm^{-1} and 2059.98 cm^{-1}, corresponds to the absorption caused by the triple bonds of C≡C (alkaline type compounds). At the peak with the wave numbers 1948.10 cm^{-1} and 1550.77 cm^{-1}, corresponding to the absorption caused by the bonding C = O (a type of aldehyde, ketone, carboxylic acid, esters) and C = C (a type of alkene compound). At peak with wave numbers of 1477.47 cm^{-1} to 459.06 cm^{-1}, corresponds to absorption caused by the C-H bond (alkane type compound), and C-O (a type of alcohol, carboxylic acid, ester).

Bioplastic mango starch starch has a functional group which is a combination of specific functional groups contained in its constituent components such as C-H, O-H, C-O, C≡C, C = O, and C = C. The wavelength of the functional group OH in clay differs from the bioplastic starch of mango seed with the addition of glycerol plasticizer wherein the addition of glycerol causes the functional group OH to shift with wave length 3622.32 cm^{-1} to 3522.02 cm^{-1}. It identifies that there is a process of forming new bonds between matrix mango starch and glycerol to weaken the OH bond on the matrix of mango seed starch. In bioplastics with the addition of glycerol, the C-O and OH groups form a random carbon chain, thus causing the sample to become more elastic [12]. Chemical interactions are reflected by changes in the characteristics of the absorption peak spectrum after mixing of two or more matrix intermediates with added chemicals such as clay and glycerol [15].

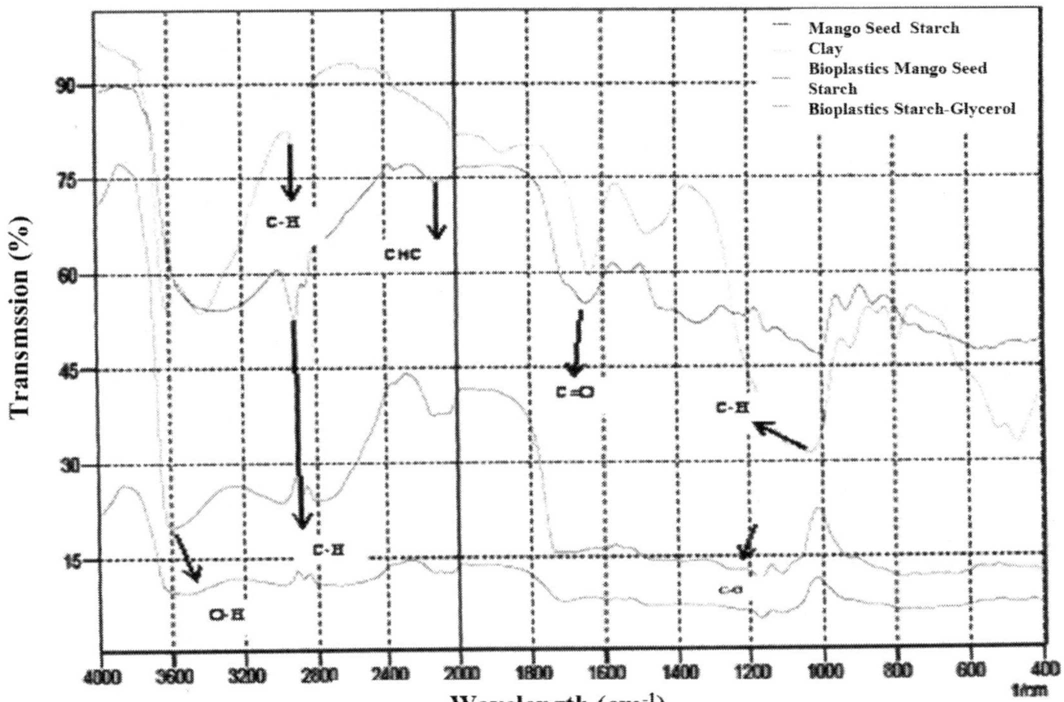

Figure 2. FTIR specs of mango seeds starch, clay, pure starch bioplastic, starch/clay/glycerol bioplastic and starch/glycerol bioplastic

3.3 Tensile Strength and Elongation at Break

In Fig. 3 and four can be seen the effect of adding clay fillers and glycerol plasticizers to tensile strength and elongation at breaks from bioplastics. The highest bioplastic tensile strength value was obtained also of 3% clay filler and 30% glycerol plasticizer with 5.657 MPa value. For the elongation value at the highest breakup bioplastic was obtained by the addition of 6% clay filler with the value of 47.553%. While also of glycerol elongation value at breakup bioplastic at concentration 35% glycerol with value 49.139%.

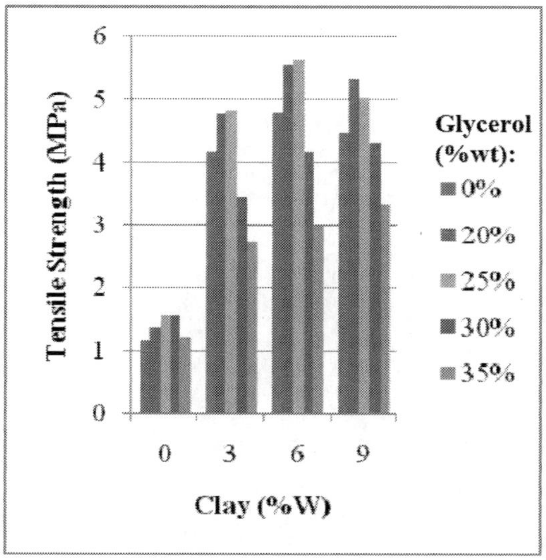

Figure 3. Tensile strength of starch based bioplastic reinforced with clay sing glycerol as the plasticizer.

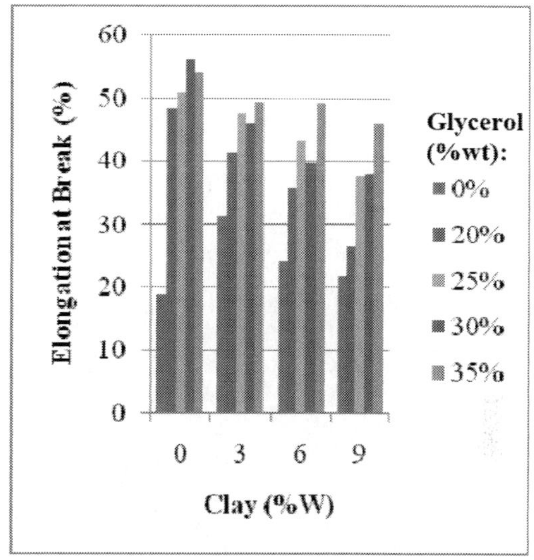

Figure 4. Elongation at break of starch based bioplastic reinforced clay using clay as the plasticizer.

The influence of the addition of organic clay causes more hydrogen bonds in the biocomposite so that the chemical bonds are stronger and harder to break because they require a large amount of energy to break the bond [10]. While the percentage elongation is inversely proportional to the addition of natural clay. Where the natural addition of clay, causing percentage elongation decreased. This is due to the decreasing intermolecular bonding distance [11]. Clay has a micron-size surface area large enough to be able to interact effectively with the polymer matrix at low concentrations (5-8%). As a result, polymer clay showed an increase in modulus, thermal stability, and barrier properties without an increase in density and loss of optical properties [7].

Also, the high mass of clay filler can contribute to a slowdown in the interaction between molecules of starch bioplastics. This leads to the development of bioplastic structures becoming heterogeneous and the results being discontinuities [10]. With the addition of glycerol to the plasticizer, plasticizer molecules in bioplastics located between biopolymer bonding chain and be able to interact by forming hydrogen bonds in the bond between the polymer chains, causing interactions between the molecules of biopolymers have become less and may cause a decrease in biocomposite material stiffness. This is due to the increased speed of viscoelastic response and molecular mobility of the polymer chain [6]. The addition of plasticizers serves as a conduit on the elastic properties of bioplastics, so more plasticizer provided will increase the value of the plastic extension [12]. The plasticizer can reduce internal hydrogen bonds of molecules and lead to the weakening of the intermolecular attractive force adjacent polymer chains, thereby reducing the tensile breaking. The plasticizer can reduce internal hydrogen bonds of molecules and lead to the weakening of the intermolecular attractive force adjacent polymer chains, thereby reducing the tensile breaking [13]. Based on this, it can be seen that elongation percentage is inversely proportional to tensile strength.

3.4 Results Analysis Morphological Characteristics Fault Bioplastics Starch Extraction Mango Seeds with Clay as Filler and Plasticizers Glycerol

Characteristics of fracture surface morphology are shown by analysis Scanning Electron Microscopy (SEM). Presented below are the characteristics of the results fracture morphology analysis bioplastics with the composition of the starch content of skin 5 grams of mango seeds, clay 6 % and 25 % glyserol at 1000x magnification. Characteristics SEM morphology analysis carried out in the laboratory of SEM, University of Diponegoro.

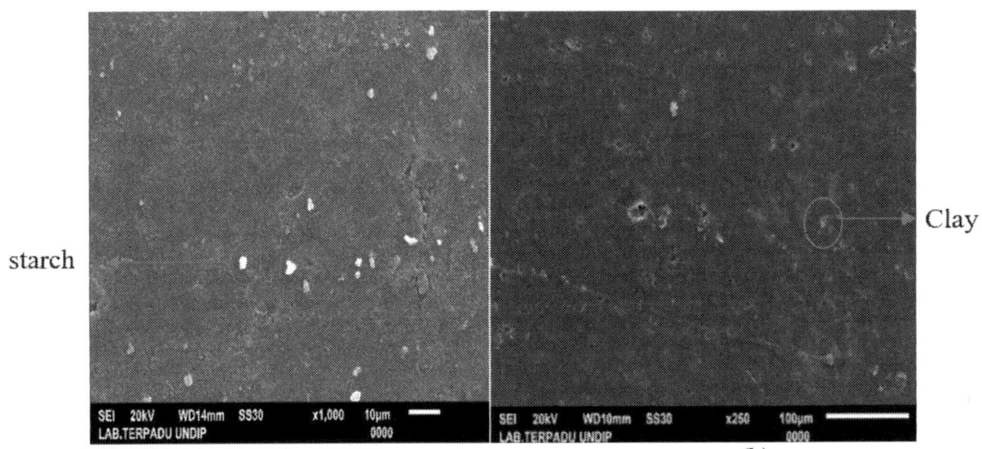

(a) (b)

Figure 5. Morphology Analysis Report Fault (a) Bioplastics Mango Seeds Starch and (b) Bioplastics from Mango Seeds Starch With Microparticles Clay as Filler and Glycerol Plasticizer in Magnification 1000x

From Figure 5 it can be seen analysis fracture morphology, strength testing. The attraction of bioplastics. On the results of SEM analysis of mango seeds, starch bioplastics without filler microparticle clay and glycerol (Figure 5 (a)) shows are clumps of insoluble mango seed starch during plastic film making. It identifies that the starch in the matrix is poorly dissolved but is spread fairly evenly. The lack of sufficiently strong power in the heating and stirring process between the filler and the starch matrix is likely to cause the starch blob not fully dissolved properly. If there is a sufficiently strong force such as good agitation in the blending process at gelatinization temperature and glass transition will easily incorporate insoluble starch particles resulting in better distribution [7].

While at bioplastics mango seeds starch with microparticle clay and glycerol (Figure 5 (b)) SEM analysis results show white spots that are clay microparticle fillers and describe the clay particle size distribution. The above results show a fairly uniform dispersion of clay fillers into the matrix. The uniform dispersion shown from the SEM results shows that there is good adhesion between the micro-sized filler and the matrix. By the research Azizi et al. (2014) which states that clay films in matrices are uniformly dispersed, and the uniform distribution of micro-filler in matrices plays an important role in improving mechanical properties. The ultrasonic process of clay microparticles is intended to produce materials with unusual properties, which lead to the formation of larger particle surfaces, smaller particle size [8]. On the surface of the bioplastic, morphology does not show any empty fraction. This is due to the influence of microparticle-sized clay fillers so that the structure is more compact. [9]. From the above SEM analysis it can be concluded that the addition of clay microparticles dispersed well enough in the starch matrix is indicated by the optimum result of bioplastic mechanical characteristics with tensile strength value 5.657 MPa, percent extension at break off 43.431%, density 1.315 gr / cm^3 and percent Water absorption 32.282%.

4. Conclusion

The mango seeds starch/Clay bioplastics increased in tensile strength from 1,567 MPa for pure starch bioplastic to 5.797 MPa. Contrary to tensile strength, a decrease in elongation at break was reported. The improvement in bioplastics with reinforcing clay could be attributed to the strong hydrogen bond between hydroxyl groups of the interface of both clay fillers and starch matrix. This formation is also influenced by the percolation mechanism. The density of bioplastics increased as clay content increased. Furthermore, an addition of clay content decreased water uptake of bioplastics. The incorporation of clay particles into starch matrix resulted in agglomerates that caused deflections in bioplastics. FT-IR spectra provided information about hydrogen bond through its characteristic peak. Bioplastic with the highest tensile strength assumed to have better characteristics than the other bioplastics was found at clay content 6% and 25% glycerol.

Acknowledgments
The authors gratefully acknowledge that the present research is supported by the University of Sumatera Utara through Non-PNBP funds by the talent research contract 2017.

Referrences

[1] Song J H, Murphy R J, Narayan R, Davies G B H 2009 *Phil. Trans. R. Soc. B* **364** 2127–39
[2] Reddy L, Reddy S, Gupta A 2013 *IJETAE* **3** 76–81
[3] Jain R, Tiwari A 2015 *J Environ Health Sci Eng* **13**
[4] Mose B R, Maranga S M 2011 *J Mat Sci Eng* **1** 239-245
[5] Septiosari A, Latifah E K 2014 *Indonesian Journal Of Chemical Science*
[6] Maulida, Siagian M, Tarigan P 2016 *Journal of Physics: Conference Series* **710** 012012
[7] Nugroho A F 2012 *Sintesis Bioplastik dari Pati Ubi Jalar Menggunakan Penguat Logam ZnO dan Penguat Alami Clay*, Universitas Indonesia, Depok
[8] Widiyana K 2011 *Penumbuhan Nanopartikel Seng Oksida (Zno)Yang Disintesis Dengan Metode Sonokimia Dan Pemanfaatannya Sebagai Tinta*, Pengaman, Jurusan Kimia, Fakultas Matematika Dan Ilmu Pengetahuan Alam, Universitas Negeri Semarang, Semarang
[9] Kittiphoom S 2012 *International Food Research Journal* **4** 1325.1335.
[10] Dani M, Parikin A H, Ismoyo A, Fisli S, Sugiyantoro S 2003 *Karakterisasi Sifat Termal, Struktur, Kristal, Dan Warna Produk Bakar Clay-Sindanglaut.* Poslitbang IPTEK Bahan Batan. Serpong
[11] Krisna D D A 2011 *Pengaruh Regelatinasi Dan Modifikasi Hidrotermal Terhadap Sifat Fisik Pada Pembuatan Edible Film Dari Pati Kacang Merah (Vigna Angularis sp.).* Program Studi Magister Teknik Kimia UNDIP, Semarang
[12] Sulthoni A F, Mawarani L J, dan Budiono A 2013 *Jurnal Teknik Pomits* 2 (**1**): 163-168
[13] Ardiansyah R 2011 *Pemanfaatan Pati Umbi Garut untuk Pembuatan Plastik Biodegradable.* Fakultas Teknik. Departemen Teknik Kimia. Universitas Sumatera Utara
[14] Vega Y P 2016. *Optimasi Formula Film Nanokomposit Berbasis Polivinil Alkohol (Pva) Dengan Penambahan Nanopartikel Seng Oksida (ZnO) Dan Asam Lemak Stearat.* Sekolah Pascasarjana Institut Pertanian Bogor. Bogor
[15] Ata, A W 2016 *Pembuatan Pelapis Bionanokomposit Dari Tapioka, Nanopartikel ZnO Dan Asam Stearat Serta Aplikasinya Pada Mangga Terolah Minimal.* Sekolah Pascasarjana Institut Pertanian Bogor
[16] Janah L 2014 *Penambahan Nanopartikel Seng Oksida (ZnO) Dalam film Bioplastik.* Institut Pertanian Bogor
[17] Herawati H 2011 *Jurnal Litbang Pertanian* **30**: 31-39

IOP Conf. Series: Materials Science and Engineering **309** (2018) 012069 doi:10.1088/1757-899X/309/1/012069

De-embedding technique for accurate modeling of compact 3D MMIC CPW transmission lines

U H Pohan[1], P B KKyabaggu[2] and E P Sinulingga*[1]
[1] Electrical Engineering, Universitas Sumatera Utara, Medan, Indonesia
[2] Bukoola General Enterprises, Kampala, Uganda

*E-mail: emerson.sinulingga@usu.ac.id

Abstract. Requirement for high-density and high-functionality microwave and millimeter-wave circuits have led to the innovative circuit architectures such as three-dimensional multilayer MMICs. The major advantage of the multilayer techniques is that one can employ passive and active components based on CPW technology. In this work, MMIC Coplanar Waveguide(CPW)components such as Transmission Line (TL) are modeled in their 3D layouts. Main characteristics of CPWTL suffered from the probe pads' parasitic and resonant frequency effects have been studied. By understanding the parasitic effects, then the novel de-embedding technique are developed accurately in order to predict high frequency characteristics of the designed MMICs. The novel de-embedding technique has shown to be critical in reducing the probe pad parasitic significantly from the model. As results, high frequency characteristics of the designed MMICs have been presented with minimumparasitic effects of the probe pads. The de-embedding process optimises the determination of main characteristics of Compact 3D MMIC CPW transmission lines.

1. Introduction

Improved characteristics of Monolithic Microwave Integrated Circuit (MMIC) components are highly desirable at high frequency. The solution would be to utilize a three-dimensional (3D) Multilayer Coplanar Waveguide (CPW) technique. By employing 3D CPW technique that has features such as design flexibility and highly integration between passive and active components in MMIC technology, optimization of the current and novel components with much improved performance can be realized. [1–7].As precise and reliable models are of significant importance in RF/microwave circuit design, with the increasing of operation frequency, the influence of probing pads and interconnect parasitic on MMIC components performance has become more and more important [8-11].This research concerns with the understanding of the parasitic effects i.e. probe pads parasitic occur in the characteristics of MMIC components. The parasitic effect of pads had been gradually visible as frequency increases.

Novel de-embedding technique for high frequency characterization of MMIC components is developed and has been proposed in this work to support accurate on-wafer measurements. The effect of frequency resonance and its relation with the pads parasitic effect in CPW transmission line are analyzed. Then the pads effects are reduced by applying the new simplified de-embedding technique to produce more evidence of relation between the pads parasitic with the frequency resonance phenomena in the transmission line. Two types of transmission lines, the conventional planar and the planar with polyimide (3D) CPW transmission line are modeled and tested. These two common CPW transmission lines are chosen to give the fully understanding on the de-embedding technique and the phenomena of resonance frequency for particular transmission lines in microwave circuits.

2. COST De-embedding Technique

In this work, de-embedding technique has been developed from the Cascade Open Short Thru (COST) Method [9] with simplified pad parasitic matrices for Microwave On-Wafer Measurement. This technique utilizes the open, short and thru dummy structures to de-embed the probe pad parasitic of a Device Under Test (DUT). The suggested parasitic model of DUT is shown in figure 1.

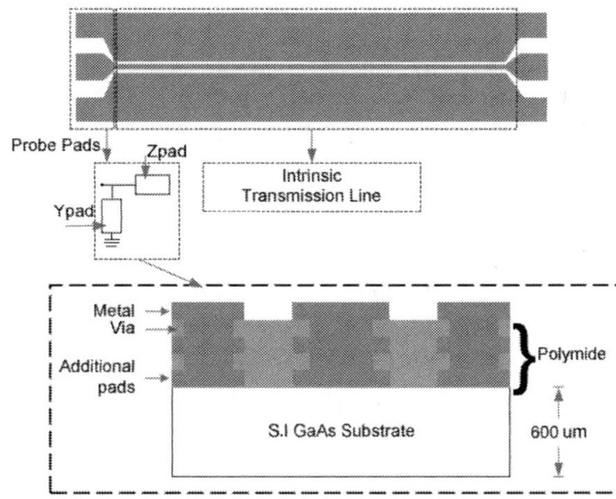

Figure 1. (a) Top view of M3 PI 1,2 CPW 3D Transmission Line and
(b) The cross-sectional view of the left probe pad of M3 PI 1,2 CPW Transmission Line.

In the COST Method, the resistive and inductive effects which originated from the probe pads are taken into account [9]. The ABCD-matrix representations of the probe pads' parasitic in Port-1 and Port-2 are simplified and defined as

It should be noted that the Y and Z-parameters of open and short dummies can be converted from S-parameters measurements and simulations. In this work, the tested dummy structures to acquire the pads parasitic matrices values for open, short and thru conditions are shown in Figure 2.

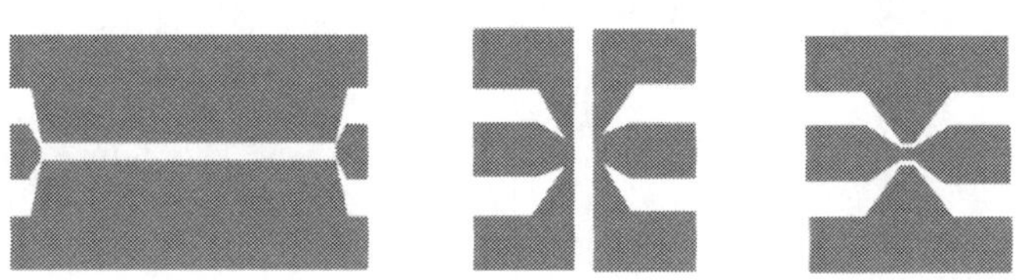

(a) (b) (c)

Figure 2. Dummies test for the new de-embedding technique
(a) 0.5 mm Open, (b) Short and (c) Thru structures

As the operation frequency becomes higher, the influence of resonance frequency, probe pads and interconnect parasitic on transmission line performance become more important for the transmission line characterization in the circuits. By understanding the parasitic effects, then the de-embedding ttechnique can be developed accurately in order to predict high frequency characteristics of the designed MMICs.

3. Results and analysis

In this work, planar CPW transmission line with length 2.4 mm is designed and tested first before the de-embedding technique applied to the planar on Polyimide CPW transmission line. Problem will occur when the length of the chosen transmission line is greater than λ/2 as the structure suffered a resonant effect and the electrical behavior will change after the resonance frequency [8]. This is shown in the simulation in figure 3 and 4 before applied de-embedding. To understand more of the relation, one should analyze the comparison of characteristic impedance and relative effective permittivity data of the planar CPW transmission line before and after applying the simplified COST de-embedding process as shown in figures 3 and 4.

One can see that the half-wavelength resonances' phenomena (λ/2 effect) prevent or limit the determination of characteristic impedance of the transmission line starting at resonant frequency of 24 GHz as shown in figure 3. It becomes more apparent in terms of its relative effective permittivity shown in figure 4. This is due to the length dependencies of the propagation constants. The measured and simulated data after the de-embedding process for planar CPW transmission lines are compared and shown in figures 3 and 4.

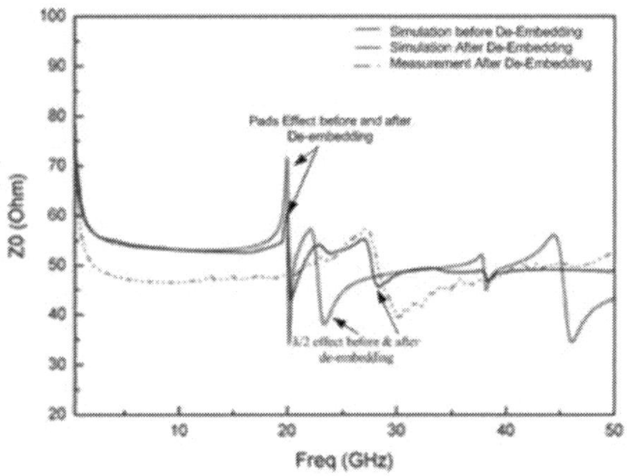

Figure 3. Comparison of characteristic impedance before and after de-embedding process fora planar CPW transmission line.

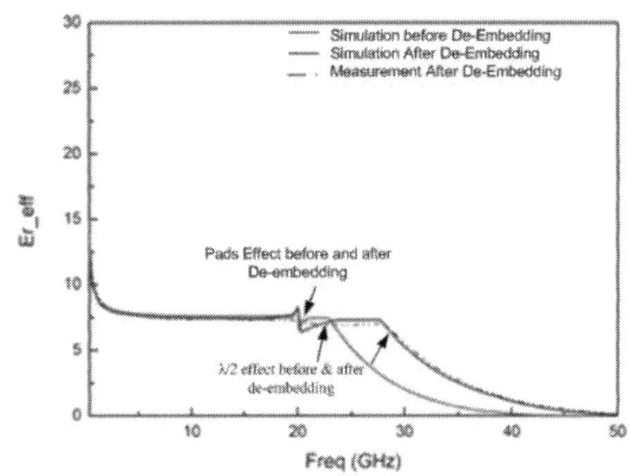

Figure 4. Comparison of relative effective permittivity before and after de-embedding process for a planar CPW transmission line.

As the purpose of the de-embedding process is to maximize the determination of characteristic impedance and relative effective permittivity of the transmission line, the proposed method has de-embedded the parasitic effect. By applying the new technique one can understands better and observes the influence of the parasitic pads in the measurement and take it into consideration for further improvement in designing the microwave circuits.

IOP Conf. Series: Materials Science and Engineering **309** (2018) 012069 doi:10.1088/1757-899X/309/1/012069

De-embedding process shows that the influence of pads parasitic has been reduced. As a result, resonance frequency effect on transmission line has been reduced as well. Thus one can conclude that the parasitic pads increase the resonance effect on transmission line. Another important finding from this work that the resonance frequency has been shifted from 23.5 GHz to 28.5 GHz and these findings are consistence for both measurement and simulation cases. One can states that length and size of the pads in the designed transmission line need to be considered carefully in order to reduce the effect of resonance frequency that can disturb the determination of line's characteristics.

Results in figure 5 and 6 show the characteristic impedance and relative effective permittivity of a planar on Polyimide (M3 PI 1,2) CPW (3D) transmission line respectively. Due to the de-embedding technique, resonance frequency in the de-embedded simulation and measurement has shifted and only occurred at around 38 GHz. Thus one can say that the new technique has yielded the expected improvement characteristics. Again, the electrical behavior will change for transmission line with length greater than $\lambda/2$ [8] as confirmed by data in figure 5. For planar with polyimide, this electrical behavior takes place after 38 GHz.

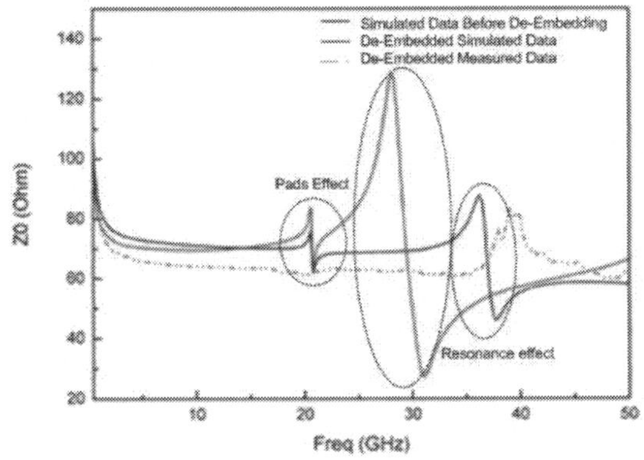

Figure 5. Comparison of characteristic impedance before and after de-embedding process for simulated and measured planar with polyimide (3D) CPW transmission line.

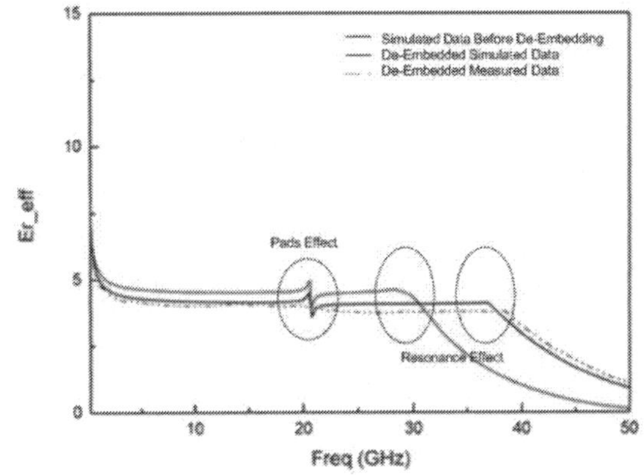

Figure 6. Comparison of relative effective permittivity before and after de-embedding process for simulated and measured planar with polyimide (3D) CPW transmission line.

The novel de-embedding technique has produced the expected improvement by reducing the pads effect significantly. As results, the resonance effect also reduced and this finding confirms the aforementioned statement about the relation of parasitic pads and the resonance effect. The measured data after applying the de-embedding technique have been produced and consistence with the simulated one.

4. Conclusions

It has been confirmed from this work that if one intends to insert transmission line between two parts of a circuit without changing the electrical behavior at particular frequency then one should choose transmission line with length equal to $\lambda/2$. The de-embedding technique is an analytical method therefore the purpose of the de-embedding process is to reduce the parasitic effect and to maximize the determination of characteristic impedance and relative effective permittivity of the Compact 3D CPW MMIC components. The novel de-embedding technique has shown to be critical in reducing the probe pad parasitic significantly from the model. As results, high frequency characteristics of the designed MMICs have been presented with minimum effects of probe pads' parasitic.

5. References

[1] Vo T V, Krishnamurthy L, Sun Q and Rezazadeh A A 2006 3-D low loss coplanar waveguide transmission lines in multilayer MMICs *Microwave Theory and Techniques IEEE Transactions On* **54** 2864-71

[2] Marsh S 2006 *Practical MMIC Guide* Norwood: Artech House, Inc

[3] Ezzedine A K 2007 Advances in microwave and millimeter-wave integrated circuits *Radio Sci. Conf* 1-8

[4] Mukherjee M 2010 Advanced microwave and millimeter wave technologies semiconductor devices circuits and systems *InTech* 1-14

[5] Diebold S, Weber R, Seelmann-Eggebert M, Massler H, Tessmann A, Leuther A, Kallfass I 2011 A fully-scalable coplanar waveguide passive library for millimeter-wave monolithic integrated circuit design *The 41st European Microwave Conf* 293-6

[6] Hong W et al 2012 Research advances in microwave and millimeter wave circuits and systems in the SKLMMW *19th Int. Conf. on Microwave Radar and Wireless Communications* 807-9

[7] Sinulingga E P, Rezazadeh A A, Kyabaggu P B K, Alim M A and Zhang Y 2015 3D momentum modeling technique and measurements of CPW compact gas multilayer MMICS *Asia-Pacific Microwave Conf* 1-3

[8] Heymann P, Prinzler H and Schnieder F 1994 De-embedding of MMIC transmission-line measurements *IEEE MTT-S Digest* 1045-8

[9] Cho M-H and Huang G-W 2006 A cascade open-short-thru (cost) de-embedding method for microwave on-wafer characterization and automatic measurement *IEEE Trans.* **E88-C(5)** 845-9

[10] Issaoun A and Xiong Y Z 2007 On the deembedding issue of cmos multigigahertz measurements *IEEE Transactions on Microwave and Techniques* **55(9)**

[11] Xi-Sung L 2013 THRU-based cascade de-embedding technique for on-wafer characterization of RF CMOS devices *IEEE Transactions on Electron Devices* **60(9)**

IOP Conf. Series: Materials Science and Engineering **309** (2018) 012070 doi:10.1088/1757-899X/309/1/012070

Characterization of low impedance 3d coplanar waveguide interconnects for compact MMICs

E P Sinulingga[1*], P B K Kyabaggu[2], C D Sedayu[1], and A A Rezazadeh[3]
[1] Electrical Engineering Department, Universitas Sumatera Utara, Indonesia
[2] Bukoola General Enterprises, Kampala, Uganda
[3] The University of Manchester, Manchester, UK

*E-mail: emerson.sinulingga@usu.ac.id

Abstract. Low characteristic impedance coplanar waveguide (CPW) interconnects are desirable for microwave applications such as power FET's and photodiodes. In conventional planar MMICs, typical characteristic impedances are in the range of 40-100Ω, such that in order to realize 20Ω interconnect, one has to fabricate a very narrow slot width with a wider signal conductor width. As the objective is to achieve low characteristic impedance interconnects (<20Ω), the proposed techniques are by employing an additional metal layerto increasethe intrinsic distributed capacitance within interconnect. In this work, realization of low impedance 3D CPWinterconnect has been done by employing the overlapped metal techniques. The microwave characteristics are studied with different signal conductor widths and it has shown that the characteristic impedance as low as 10Ω can be achieved for MMIC interconnects with slot width of 15μm. This illustrates the great flexibility of the multilayer technology allowing various parameters to be achieved with small modifications in the design structure.

1. Introduction

Since the requirement of having more compact MMICs, by increasing the packing density and functionality of chips become stronger, the conventional planar CPW interconnects are not suitable to realize the desired low characteristic impedance interconnects. The solution would be to employ the 3D multilayer interconnects technique [2]. Gillick and Robertson [1] demonstrated that CPW structure with Z_0= 9Ω and loss of 0.2dB/mm can be achieved at 10 GHz. Vo et al [2] also demonstrated using another design approach to yield low loss low impedance transmission line. However there is still much information which can be extracted from the comparison based on the CPWs' various design parameters [1,2] and various metal overlapped dimension (between signal conductor tracks in different layer positions with the ground planes). Detailed investigation such as verifications of the extracted parameters using S-parameter measurements in any published data is also necessary to be carried-out. These are presented in this paper. We demonstrate that the losses can be reduced significantly by utilizing additional metal layer with different gap sizes. These have been analyzed by extracting various multilayer interconnects' characteristics with respect to the overlapped metals width.

IOP Conf. Series: Materials Science and Engineering **309** (2018) 012070 doi:10.1088/1757-899X/309/1/012070

2. Interconnect Characterization

An interconnect is a distributed physical system which conveys a communication signal and transmits bulk power over a particular distance. Due to its important characteristics, many researches have been devoted to develop various types of CPW interconnect for microwave and millimetre-wave integrated circuits [5-9]. Numerous researches have been dedicated to understand the electrical behavior of this structure, especially at microwave frequencies [7,9,10]. However, most of these studies have been performed using analytical approximation and full-wave analysis [10,11]. Therefore, it is essential to obtain results using S-parameters characterization technique especially for 3D CPW MMIC interconnects. Interconnect's signal transmission is defined using the classical Telegrapher's transmission line equation [12-15]. The S-parameter responses from a lossy unmatched transmission line can be derived from the ABCD parameters of the particular transmission line where terminal conditions are implemented at the two ends of the line [13,14].

$$\begin{bmatrix} A & B \\ C & D \end{bmatrix} = \begin{bmatrix} \cosh(\gamma l) & Z_0 \sinh(\gamma l) \\ \dfrac{\sinh(\gamma l)}{Z_0} & \cosh(\gamma l) \end{bmatrix} \tag{1}$$

The matrix above is defined as transfer matrix of the transmission line with length ι and γ is the complex propagation constant of the signal travelling in transmission line. The ABCD parameters relate the voltage and current of one port namely load to the voltage and current of the other port namely source. Then, characteristic impedance of transmission line ι can be expressed in S-parameters by converting the parameters above. Since $B=Z_0\sinh(\gamma\iota)$ and $C=\sinh(\gamma\iota)/Z_0$, one can convert these parameters into S-parameters as follow:

$$Z_0 \sinh(\gamma l) = Z_{sys} \frac{(1+S_{11})(1+S_{22}) - S_{12}S_{21}}{2S_{21}} \tag{2}$$

and

$$\frac{\sinh(\gamma l)}{Z_0} = \frac{1}{Z_{sys}} \frac{(1-S_{11})(1-S_{22}) - S_{12}S_{21}}{2S_{21}} \tag{3}$$

where S11=S22 and S12=S21 due to symmetrical interconnect. Simplify and combine equation (2) → (3) one can yield

$$Z_0^{\ 2} = Z_{sys}^{\ 2} \frac{(1+S_{11})^2 - S_{21}^{\ 2}}{(1-S_{11})^2 - S_{21}^{\ 2}} \tag{4}$$

In order to express relative effective permittivity formula of particular transmission line in terms of S-parameters, one has to derive the complex propagation constant (γ) equation from the simulated or measured of S-parameters. By converting A parameter to S-parameters and combining B to C parameters equations with terminal conditions of transmission line [14] we can have

$$A = \cosh(\gamma l) = \frac{(1+S_{11} - S_{22} - \Delta s)}{2S_{21}} \tag{5}$$

where

$$\begin{aligned} S_{11} &= S_{22}, S_{12} = S_{21} \\ \Delta s &= S_{11}S_{22} - S_{21}S_{12} \end{aligned} \tag{6}$$

Then

$$\cosh(\gamma l) = \frac{(1 - S_{11}^{\ 2} + S_{21}^{\ 2})}{2S_{21}}. \tag{7}$$

Also by combining equation (2) → (3) where

$$\sinh(\gamma l) = \frac{Z_{sys}}{Z_0} \frac{(1 + S_{11})(1 + S_{22}) - S_{12}S_{21}}{2S_{21}} \tag{8}$$

and

$$\frac{Z_{sys}}{Z_0} = \frac{1}{\sinh(\gamma l)} \frac{(1 - S_{11})(1 - S_{22}) - S_{12}S_{21}}{2S_{21}} \tag{9}$$

Thus, one can yield

$$\sinh^2(\gamma l) = \frac{((1 - S_{11})(1 - S_{22}) - S_{12}S_{21})((1 + S_{11})(1 + S_{22}) - S_{12}S_{21})}{(2S_{21})^2}.$$

then,

$$\sinh(\gamma l) = \left\{ \frac{\left(S_{11}^{\ 2} - S_{21}^{\ 2} + 1\right)^2 - (2S_{11})^2}{(2S_{21})^2} \right\}^{\frac{1}{2}} \tag{10}$$

Therefore, the equation of propagation constant of transmission line with length l in term of S-parameters can be expressed by adding or substracting both equations (7) and (10) as follow

$$\cosh(\gamma l) \pm \sinh(\gamma l) = e^{(-\gamma l)} = \left\{ \frac{(1 - S_{11}^{\ 2} + S_{21}^{\ 2})}{2S_{21}} \pm K \right\} \tag{11}$$

where

$$K = \sinh(\gamma l) = \left\{ \frac{\left(S_{11}^{\ 2} - S_{21}^{\ 2} + 1\right)^2 - (2S_{11})^2}{(2S_{21})^2} \right\}^{\frac{1}{2}}$$

After acquiring the value of β=imag(γ) from S-parameters of measured or simulated structure then we can express the effective dielectric constant based on its S-parameter using equation

$$\varepsilon_{r,effective} = \left(\frac{\beta c}{2\pi f} \right)^2 \tag{12}$$

The common approach to plot the variation of interconnects' dissipation loss is to consider the two-port network principle where the output power and the transmitted power can be approximated as $|S_{21}|^2$ and $1 - |S_{11}|^2$respectively. Then the dissipation loss can be expressed by [2]

$$Dissipation\_Loss = 10log\left(\frac{1-|S_{11}|^2}{|S_{21}|^2}\right) \tag{13}$$

We have used this procedure to determine various parameters of the transmission lines studied.

IOP Conf. Series: Materials Science and Engineering **309** (2018) 012070 doi:10.1088/1757-899X/309/1/012070

3. Experimental Details And Results

Several 3D multilayer interconnects have been designed, fabricated and tested. Semi-insulating GaAs substrate with a 600µm thickness has been used to fabricate the structures using photolithographic technique. The conducting electrodes were Ti(0.05µm) and Au(0.8µm). The thickness of isolating polyimide layers were 2.5µm with dielectric constant approximately 3.7. Agilent's ADS Momentum was used for electromagnetic modeling and the fabricated interconnects were characterized at microwave frequencies using a Cascade Microtech probe station and HP8510C vector network analyzer.

The structure that has been considered to verify the characterization of the two different approaches is shown in figure 1. As for realizing the low impedance interconnects, structures in figure 2 were modeled, fabricated and tested with various signal conductors (W), gap (G) and different overlap (O) widths. The results in figure 3 show that a good agreement has been achieved between the simulated and modeled 3D interconnection term of effective dielectric constant and the characteristic impedance. For dispersion characteristics and dissipation loss analysis, we modeled and measured structure as shown in figure 2.

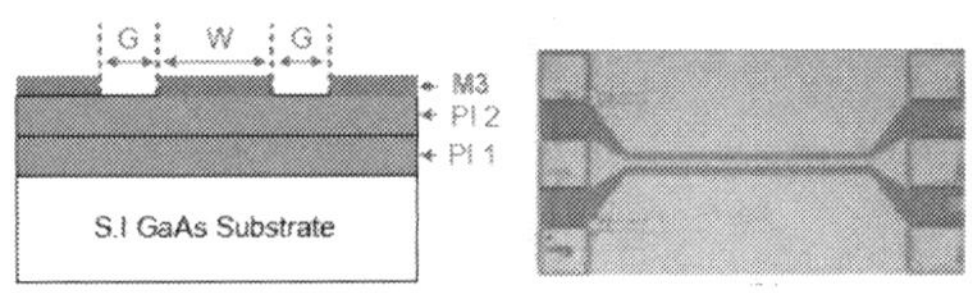

(a) (b)

Figure 1. (a) Cross-sectional view and (b) micrograph of fabricated multilayer interconnect with polyimide thickness 3µm, length of 0.5 mm and W= 20µm, G= 15µm

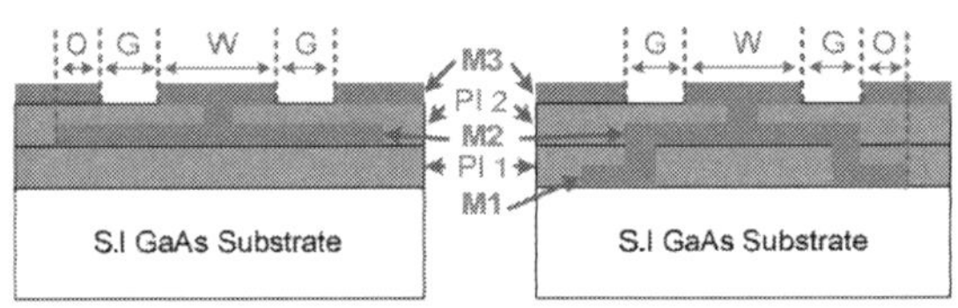

(a) (b)

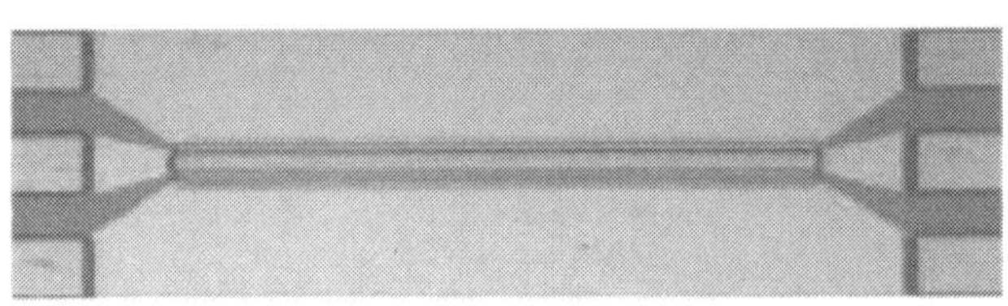

(c)

Figure 2. Cross-sectional view of (a) overlap M23 (b) overlap M123 interconnects and (c) the micrograph of fabricated overlap metal interconnect with polyimide thickness of 5-µm, length of 2.398mm

The interconnects with two overlapped metal layers are examined with different signal conductor and overlap widths. The lengths for all the structures are kept constant at 2.398mm. The dispersion characteristics and the dissipation loss are shown in figures 4a and 4b respectively. The dispersion showed that as the overlap width between metal 2 and 3 increases by 10 times then the characteristic impedance decreases on average by about 37%. As the overlap increases further by 20 times the characteristic impedance decreases on average by about 52%. The reduction of the Z_0 is mainly due to the increased capacitance to the ground of the structure. This argument can be validated by observing the trends of effective dielectric constant in figure 4a.

The effective dielectric constant is the ratio of total capacitance per unit length of the coplanar line with the capacitance of the same line while all the dielectrics replaced by air [11]. Since the capacitance with air as the dielectric does not change but at the same time the total capacitance of the line increases thus the effective dielectric constant increases. Utilizing the common approach used in designing conventional planar interconnect i.e. by increasing the signal conductor width does not decrease the characteristic impedance significantly since the capacitance only increases slightly. It also does not give benefit in decreasing the dissipation loss.

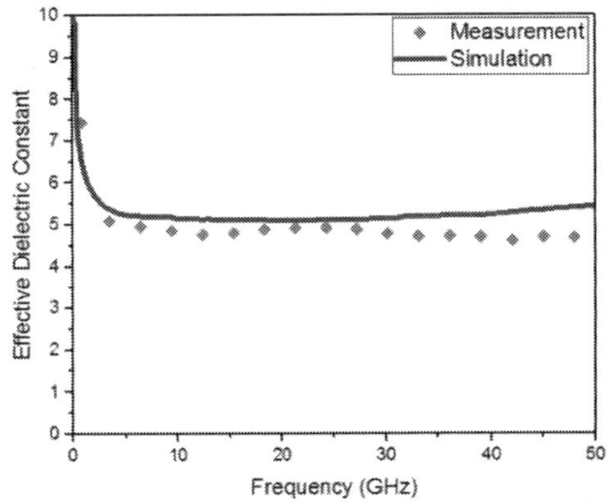

(a)

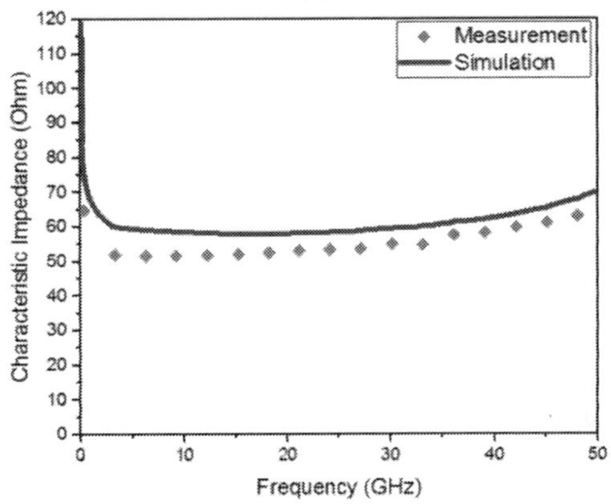

(b)

Figure 3. Comparison data of (a) effective dielectric constant and (b) characteristic impedance of the multilayer M3 PI12 transmission line with length 0.5mm

Wider signal conductor width only reduces the dissipation loss by 0.05dB/mm. Significant reduced characteristic impedance can be achieved by increasing the overlap width i.e. increasing the width of metal 2 in figure 2a. However it also increases the dissipation loss of the structure by 0.1–0.6dB/mm.Since the objective is to achieve low loss low characteristic impedance interconnect then reducing the dissipation loss can be performed easily in the multilayer structure by employing an additional metal layer and using a wider gap size i.e. reducing ground width as shown in figure 2b. These results in lowering the dielectric losses due to reduced electric field at the overlapping area and also more dispersed current in the conductor leads to reduce the current crowding effect. Using this design approach, the dissipation loss decreases by 0.6dB/mm for structure with the same overlap width. The experimental data showed that the two overlapped metal layers technique is more sensitive to losses while the three overlapped metal layers with different gap size has lower dissipation loss by 0.26dB/mm for the same characteristic impedance value. This illustrates the great flexibility of the

IOP Conf. Series: Materials Science and Engineering **309** (2018) 012070 doi:10.1088/1757-899X/309/1/012070

multilayer technology allowing various parameters to be achieved with only small modifications to the design structure.

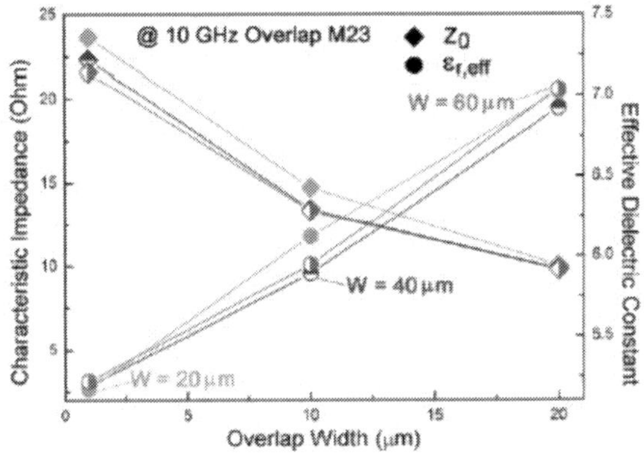

(a)

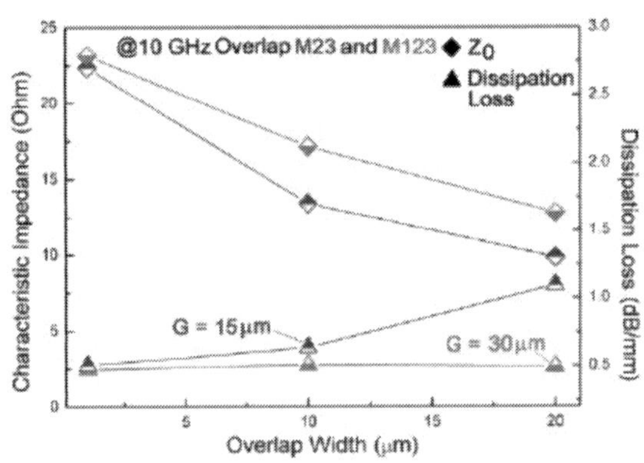

(b)

Figure 4. (a) Characteristic impedance and effective dielectric constant (b) dissipation loss and total power loss of overlap M23 transmission lineG=15μm with various width W, (c) characteristic impedance and dissipation loss of overlap M23, W=40μm, G=15μm and overlap M123, W=40μm, and G=30μm

4. Conclusions

Good agreement between the measured S-parameter data and those obtained from the analytical results validates the characterization of 3D CPW interconnects based on S-parameters extraction technique. By utilizing the 3D multilayer CPW technique, low impedance interconnects with improved dissipation loss can easily be realized without increasing the size of the components in compact MMICs. The trend of dispersion characteristics show that this technique gives microwave engineers more flexibility in designing passive components with improved performances.

References

[1] Gillick M and I D Robertsons 1993 Ultra low impedance CPW transmission lines for multilayer MMICs *IEEE MTT-S Int Microwave Symp Dig* 145–148

[2] Vo T V, Krishnamurthy L, Sun Q, Rezazadeh A A 2006 3-D Low Loss Coplanar Waveguide Transmission Lines in Multilayer MMICs *IEEE Trans. Microw. Theory and Tech* **54(6)** 2864–2871

[3] Sul W S, Kang T S and Dan A 2002 Low Characteristic Impedance Transmission Line Fabricated Using Benzo-cyclo-butene Thin Film *Optoelectronic and Microelectronic Materials and Devices Conference* 353–356

[4] Tanabe M, Nishitsuji M, Anda Y and Ota Y A Low-Impedance Coplanar Waveguide Using SrTiO3 Thin Film for GaAs Power MMIC's *IEEE Trans. Microw. Theory and Tech* **48(5)**

[5] Wen C P 1969 Coplanar waveguide, a surface strip transmission line suitable for nonreciprocal gyromagnetic device applications in *IEEE G-MTT Symp. Dig., Dallas, TX* 110–115

[6] Veyres C and Hanna V F 1980 Extension of the application of conformal mapping techniques to coplanar lines with finite dimensions *Int. J. Electron* **48** 47–56

[7] Hasnain G, Dienes A and Whinnery J R 1986 Dispersion of picosecond pulses in coplanar transmission lines *IEEE Trans. Microw. Theory Tech* **34(6)** 738–741

[8] Ghione G and Naldi C U 1987 Coplanar waveguides for MMIC applications: Effect of upper shielding, conductor backing, finite-extent ground planes, and line-to-line coupling *IEEE Transactions on Microwave Theory and Techniques* **35** 260–267

[9] Frankel M Y, Gupta S, Valdmanis J A and Mourou G A 1991 Terahertz attenuation and dispersion characteristics of coplanar transmission lines *IEEE Trans. Microw. Theory Tech* **39(6)** 910–916

[10] Bedair S S and Wolff I 1992 Fast, accurate and simple approximate analytic formulas for calculating the parameters of supported coplanar waveguides for (M)MIC's *IEEE Trans.Microw. Theory Tech* **40(1)** 41–48

[11] Gupta K C, Garg R, Bahl I and Bhartia P 1996 *Microstrip Lines and Slotlines* Norwood, MA: Artech House 375–397.

[12] Pozar D M 2005 *Microwave Engineering* Norwood, MA: John Wiley and Sons 77-83

[13] Eisenstadt W R and Eo Y 1992 S-parameter-based IC interconnect transmission line characterisation *IEEE Transactions on Components, Hybrids, and Manufacturing Technology* **15(4)** 483–490

[14] Gupta K C, Grag R and Chada R 1981 *Computer Aided Design of Microwave Circuits* Dedham, MA: Artech House 25–43

[15] Simons R N 2001 *Coplanar Waveguide Circuits, Components and Systems* John Wiley and Sons 11–26

Effect of feeders in 3D modeling of low impedance multilayer CPW transmission line

R I Zaini[1], P B K Kyabaggu[2], and E P Sinulingga[1*]
[1] Electrical Engineering Department, Universitas Sumatera Utara, Indonesia
[2] Bukoola General Enterprises, Kampala, Uganda

*Email: emerson.sinulingga@usu.ac.id

Abstract. Improved characteristics with low dissipation loss MMICs are highly desirable for wireless communications. However, the current industrial MMIC design is mainly based on microstrip concept which suffered from parasitic and unwanted phenomenon especially at higher frequency (>20 GHz). On the other hand, for future wireless technology, higher frequency operation is required and on-wafer microwave characterizations as well as precise modeling of 3D Multilayer CPW components are vital. This project concerns with understanding of the microwave characteristics behavior of Multilayer CPW components in MMIC applications. Feeder effect as unwanted parts in the characteristics has been investigated to determine its relation with the half wavelength resonance of the Multilayer CPW Low Impedance Transmission Line.

1. Introduction

Considerations to embed probe pads and launch transitions (feeders) in MMICs are due to requirement to have an interface between the probes and Device under Test (DUT) for testing purpose. Probing onto via-hole in multilayer structure would give more accurate results but with some consequences. The probe tip contacts may puncture and damage the gold pads on top of the via-holes [1-3]. This leads to damage the probe tips or destroy the MMICs. In addition, layers of metal in the probe pads are implemented for several reasons such as fabrication flexibility, better grip of the metal contacts on polyimide dielectric layers and maintaining 50 Ω of characteristic impedance for measurement matching purpose [1-3].

In some fabricated multilayer transmission line structures [4], most of feeders are designed with matched taper and only few have been designed with straight launcher. Furthermore, if the 50 Ω characteristic impedance is not maintained throughout the transitions even though the launcher is very short, significant parasitic capacitance or inductance can occur [2,5,6]. In order to obtain the targeted overall performance of MMICs, components make up the circuit must be accurately characterized including their probe pads and launchers.

In transmission lines, mainly there are three main characteristics i.e. characteristic impedance, effective dielectric constant and dissipation loss that required analyses in order to obtain the targeted specifications. The MMIC multilayer CPW transmission lines with a wide range of characteristic impedances have been fabricated [4]. However, the reported analysis only for frequency up to 20 GHz, while some phenomena characteristics occurs above 20 GHz have not been reported yet. Therefore, this paper will provide some analyses on the effect of feeders for characteristics of multilayer CPW transmission lines up to 50 GHz. Feeder effect as unwanted parts in the characteristics has also been investigated to determine its relation with half wavelength resonance.

2. Conventional Planar MMIC Transmission Lines

In order to obtain the overall accurate performance of MMICs, fabricated components and circuits must be accurately characterised. It is essential, for example, to check that components are within the target specifications and to observe any variations of the characteristics arising from the fabrication process. Some researches [7-10] have suggested that probe pads (bonding pads and feeders) which provide space to accommodate probes in on-wafer measurements may exert a significant parasitic effect. To address this concern, one can observe the effect of probe pads from the compared on-wafer measurements of fabricated non-multilayer conventional planar transmission lines as shown in figures 1 and 2. The probe pads utilised in this research are designed to provide 50 ohms impedance. This is to minimise reflection arising from any mismatch with the measurement system impedance. Dimension of the fabricated CPW transmission lines both with and without probe pads are shown in table 1. Measured characteristic impedance and effective dielectric constant are shown in figures 3 and 4 respectively.

As can be observed in figures 3 and 4, the parasitic effect of the probe pads on the characteristics of the intrinsic transmission lines is not significant. This is clearly demonstrated by the compared characteristic impedance and effective dielectric constant of both structures. However, the fabricated transmission lines experience $\lambda/2$ resonance effects at different frequencies due to their different lengths. Results in figures 3 and 4 indicate that the bonding pads and feeders increase the influence of $\lambda/2$ resonance to the characteristics of the line. This is also demonstrated in their dissipation loss characteristics, as shown in figure 5. The transmission line with probe pads experiences slightly higher losses at $\lambda/2$ resonance. As mentioned earlier, in order to obtain the desired performance of MMIC, components making up the circuit must be accurately simulated and modelled. An accurate layout simulation technique for passive MMIC components is therefore very much needed and this is provided in the following discussion.

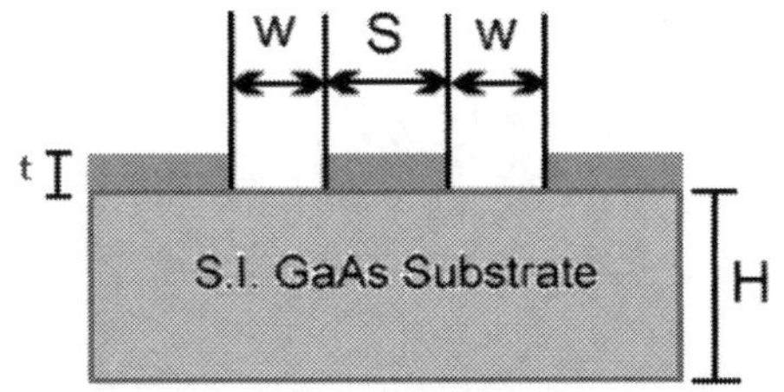

Figure 1. Cross-sectional view of a non-multilayer planar CPW transmission line.

(a)

(b)

Figure 2. Top views of the non-multilayer planar CPW transmission lines (a) without probe pads and (b) with probe pads.

Table 1. Dimension of the fabricated non-multilayer planar CPW transmission line.

Parameters	Dimension (μm)	Remarks
S	60	signal conductor width
W	15	gap
L	2000 (no pads) 2398 (with pads)	length of the line
H	600	substrate thickness
T	0.8	Ti/Au conductor thickness

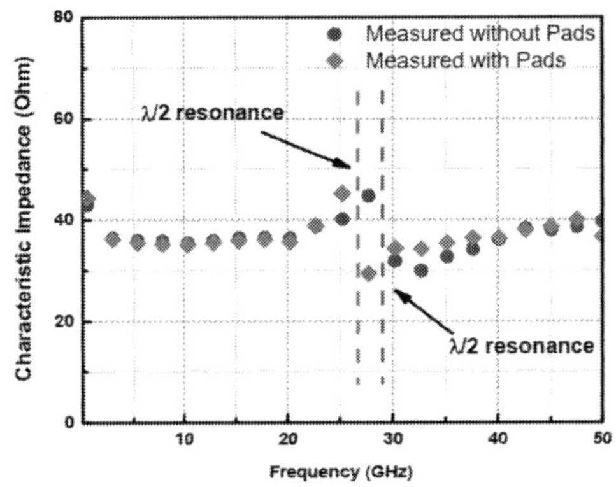

Figure 3. Comparison of measured characteristic impedance of the fabricated non-multilayer planar CPW transmission lines with and without probe pads.

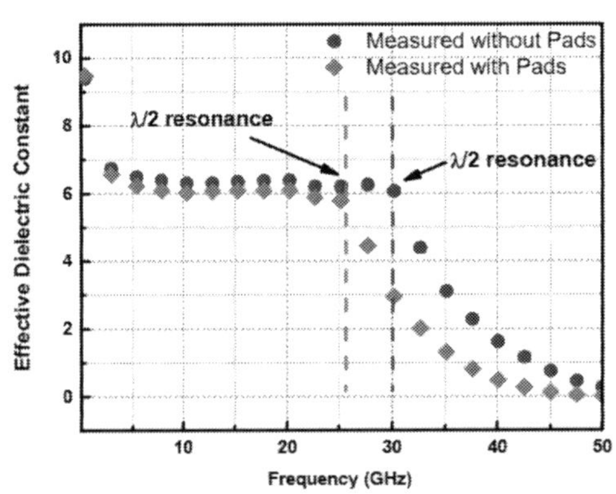

Figure 4. Comparison of measured effective dielectric constant of the fabricated non-multilayer planar CPW transmission lines with and without probe pads.

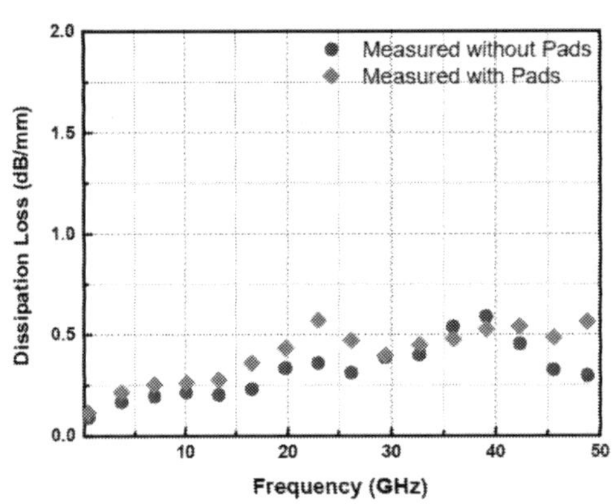

Figure 5. Comparison of measured dissipation loss of the fabricated non-multilayer planar CPW transmission lines with and without probe pads.

3. Multilayer MMIC Transmission Lines

After implementing particular size of feeders i.e. 200μm for the purpose of measurement, it is important to observe the deviation (if any) that may occur due to impedance mismatch of this feeder at the input and output port. It is also important to analyze other effects at higher frequency such as the resonance effect. By investigating this problem, one can assure that the characterization results presented in this report are accurate. As shown in figurés 6 and 7, multilayer transmission lines with feeders experiences higher resonance effect.

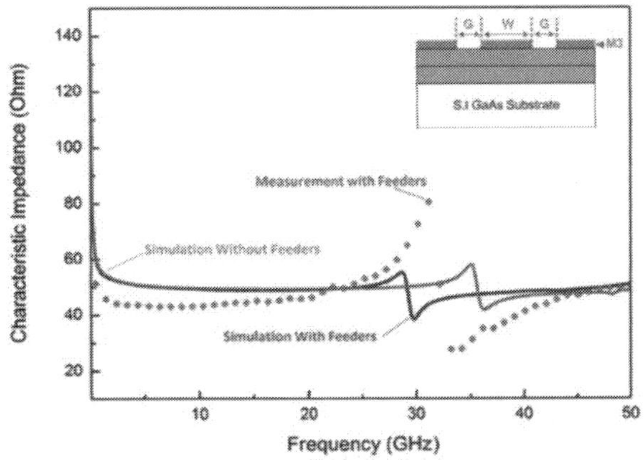

Figure 6. Comparison of Characteristic Impedance of Multilayer Planar CPW transmission line with and without feeders

In the results as shown in figure 6, structure with feeder with length of 2398μm suffers earlier resonance effect compare to the simulation with feeder with length of 2000μm. This is due to resonance effect depends very much on the length of the structure (λ/2). In term of magnitude, both simulations have almost similar resonance effect. But in the measured data, the resonance magnitude is larger with those in the simulation. This means the feeders contributes in increasing the resonance effect. This is due to tapered shape of the feeders.

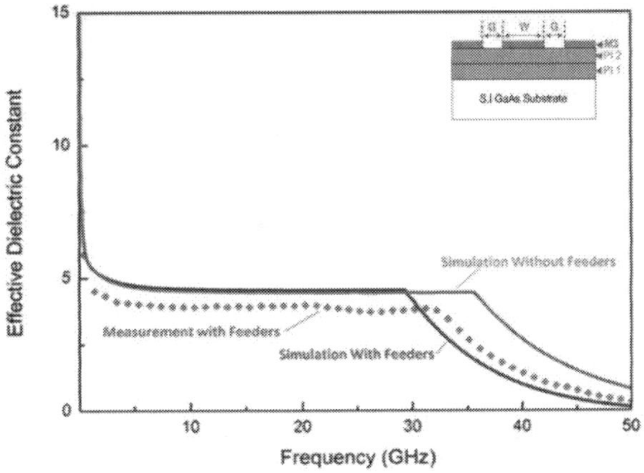

Figure 7. Comparison of Effective Dielectric Constant of Multilayer Planar CPW transmission line with and without feeders

In order to provide additional accurate arguments for the effect of feeders in the multilayer transmission line, one needs to investigate the characteristics of more complex structure such as the multilayer overlapped transmission line. The results are shown in figures 8 and 9. Once again, the feeders show significant effect in the characteristics deviation values at lower frequency and it also affects the half wavelength resonance of the transmission line. This can be verified by comparing the simulation with and without feeders in figure 8. In this figure, feeders disrupt the characteristic impedance of the line by enlarging the resonance effect. At frequency of 25 GHz for simulation without feeders, one can observe that Z_0= 32Ω. But for simulation with feeders and verified by the measured data too, Z_0= 10Ω. One can conclude that below frequency of resonance, structure with feeders do not have significant different characteristics values compare with structure without feeders. But to avoid inaccuracy characterization, especially for the overlapped structure with length of 2398μm, the accurate reading of characteristics values should not beyond frequency of 15 GHz. This means that feeders limit the operating frequency of the line.

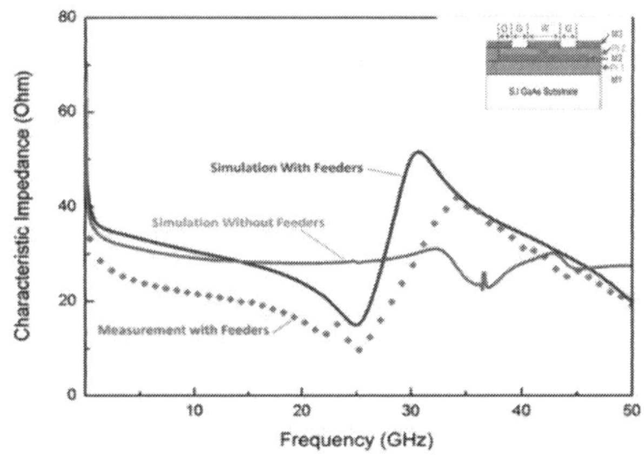

Figure 8. Comparison of Characteristic Impedance of Multilayer Overlapped CPW transmission line with and without feeders

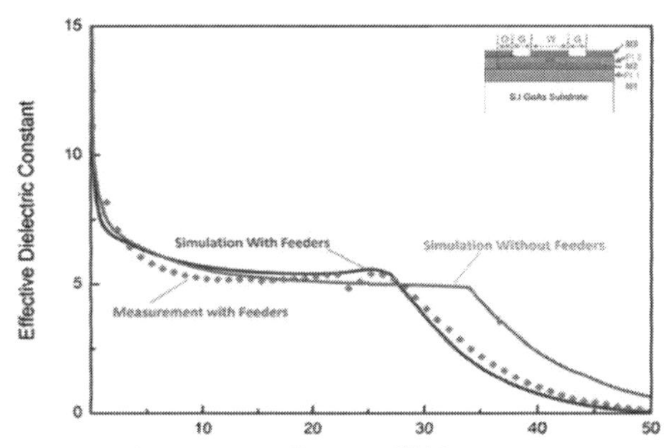

Figure 9. Comparison of Effective Dielectric Constant of Multilayer Overlapped CPW transmission line with and without feeders

Other knowledge that one can gain from this characterization is by observing the trend of effective dielectric constant as shown in figure 9. One can observe that the effective dielectric constant decreases as frequency increases. This happens at the frequency where resonance effect occurs as shown in the characteristic impedance. One can conclude that transmission line will have electrical behavioral change after the resonance effect (λ/2).

4. Conclusion

In the multilayer transmission line with feeders, the resonance frequency increases due to the effect of multilayer feeders. Furthermore, as the structure become more capacitive (overlapped metals structure), the resonance effect even significantly increases due to the presence of feeders. This means feeders contribute to the resonance effect especially for more capacitive structure. This proves that one need to treat the effect of feeders in the structure more carefully.

5. References

[1] Jackson RW 1986 Considerations in the use of Coplanar Waveguide for Millimeter-Wave Integrated Circuits, *Microwave Theory and Techniques, IEEE Transactions On*, vol. 34, pp. 1450 - 1456

[2] Robertson ID and Lucyszyn S 2001 *RFIC and MMIC Design and Technology*. The Institution of Electrical Engineers

[3] Marsh S 2006 *Practical MMIC Guide*. Artech House Inc.

[4] Krishnamurthy TVL, Sun Q, Rezazadeh AA 2006 3-D Low Loss Coplanar Waveguide Transmission Lines in Multilayer MMICs, IEEE Trans. Microw. Theory and Tech., vol 54, no. 6, pp. 2864–2871

[5] Gillick M and Robertsons ID 1993 Ultra low impedance CPW transmission lines for multilayer MMICs, IEEE MTT-S Int. Microwave Symp. Dig. pp.145–148.

[6] Sinulingga EP, Rezazadeh AA, and Kyabaggu PB 2015 3D momentum modeling technique and measurements of CPW compact GaAs multilayer MMICs, Microwave Conference (APMC), 2015 Asia-Pacific

[7] Sul WS, Kang TS, and Dan A 2002 Low Characteristic Impedance Transmission Line Fabricated Using Benzo-cyclo-butene Thin Film, Optoelectronic and Microelectronic Materials and Devices, Conference, pp. 353–356.M.

[8] Tanabe, Nishitsuji M, Anda Y, Ota Y 1997 A Low-Impedance Coplanar Waveguide Using SrTiO3 Thin Film for GaAs Power MMIC's, IEEE Trans. Microw. Theory and Tech., vol., 48 no. 5

[9] Simons RN 2001 Coplanar Waveguide Circuits, Components and Systems, John Wiley and Sons, pp. 11–26

[10] Heymann P, Prinzler H, Schnieder F 1994 De-embedding of MMIC transmission-line measurements IEEE MTT-S Digest, pp. 1045-1048

IOP Conf. Series: Materials Science and Engineering **309** (2018) 012072 doi:10.1088/1757-899X/309/1/012072

Characteristics of activated carbon resulted from pyrolysis of the oil palm fronds powder

S Maulina[*1] **, M Iriansyah**[2]
[1]Department of Chemical Engineering, Faculty of Engineering, Universitas Sumatera Utara
[2]Sustainable Energy and Biomaterial Center of Excellence, Universitas Sumatera Utara

*E-mail: maulina_harahap@yahoo.com

Abstract. Activated carbon is the product of a charcoal impregnation process that has a higher absorption capacity and has more benefits than regular char. Therefore, this study aims to cultivate the powder of oil palm fronds into activated carbon that meets the requirements of Standard National Indonesia 06-3730-1995. To do so, the carbonization process of the powder of oil palm fronds was carried out using a pyrolysis reactor for 30 minutes at a temperature of 150 °C, 200 °C, and 250 °C in order to produce activated char. Then, the char was impregnated using Phosphoric Acid activator (H_3PO_4) for 24 hours. Characteristics of activated carbon indicate that the treatment of char by chemical activation of oil palm fronds powder has an effect on the properties of activated carbon. The activated carbons that has the highest absorption properties to Iodine (822.91 mg/g) were obtained from the impregnation process with 15% concentration of Phosphoric Acid (H_3PO_4) at pyrolysis temperature of 200 °C. Furthermore, the activation process resulted in activated carbon with water content of 8%, ash content of 4%, volatile matter 39%, and fixed carbon 75%, Iodine number 822.91 mg/g.

1. Introduction

Oil palm is a type of plant that produces various kinds of byproducts, such as empty fruit bunches, shells or oil palm fronds [1]. Unfortunately, these products have not been utilized into products that have high economic value. As a result, these products only become useless garden waste [2]. One product with huge potential is oil palm fronds, as it is the largest biomass resources are available in oil palm plantations. The average weight of palm fronds that had entered the cutting period is equal to 4 kg per dry frond, with total production of palm fronds around 5500 kg per hectare per year [3].

Recognizing the enormous potential of palm oil stem, this study aims to utilize palm oil fronds into a more valuable products, such as activated carbon. Furthermore, this study also tries to produce activated carbon in accordance with Indonesian National Standard SNI (06-3730 - 1995), where there is a minimum value of quality that must be possessed in order to meet the criteria of high quality activated carbon.

2. Theory

Palm oil fronds is one of the solid waste resulted from oil palm plantations that can be obtained throughout the year during harvest time. It consists of three main elements: cellulose (31.5%), hemicellulose (19.2%) and lignin (14%) [4]. The potential of palm oil fronds in Indonesia is very promising, with the total production in Indonesia reaching 10.14 million tons in 2016 [5].

There are two stages of the manufacture of activated carbon, which is the stage of carbonization and activation [6-7]. Carbonization is defined as a process of destruction of organic matter by heating to charcoal in the absence of oxygen [8]. Meanwhile, the activation process is required to change the results of carbonization into the adsorbent which has a high porosity and a large surface area. The quality of activated carbon depends on the type of raw materials and processing technology used.

The quality standards for activated carbon has been set in the Indonesian National Standard, SNI 06-3730-1995 [9] as described in Table 1.

Table 1. Quality standards of activated carbon according to SNI (06 - 3730 - 1995)

No	Description	Unit	Requirements
1	Water content	%	Max. 15
2	Ash content	%	Max. 10
3	Volatile matter	%	Max. 25
4	Fixed carbon	%	Min. 60
5	Iodine absorption	mg/g	Min. 750

3. Method

The production of activated carbon from palm oil fronds powder was carried out through four stages: (1) preparation of palm oil fronds; (2) charcoal production (carbonization) through pyrolysis process; (3) immersion (impregnation); and (4) analysis of products (activated carbon). At the preparatory stage, the palm oil fronds was cut into small pieces and dried in an oven at a temperature of 105°C. The resulting product was then mashed using a ball mill and sieved using a 50 mesh filter.

Thereafter, the chemical activation was performed using a Phosphate Acid activator (H_3PO_4) with concentrations of 10%, 15% and 20% in the oven at a temperature of 80°C for 2 hours. Following this, the impregnation process was performed for 24 hours at a room temperature. Then, the activated carbon was filtered and washed using distilled water until it reached a pH of 7, to be dried. Finally, two types of analyses were conducted on the activated carbon: (1) proximate analysis in the form of water content, ash content, volatile matter, and fixed carbon; and (2) qualitative analysis in the form of FTIR and Scanning Electron Microscope (SEM) analysis.

4. Result

The results of the study based on the parameters in SNI (06 - 3730 - 1995) can be seen on Table 2.

Table 2. Results of the research on the activated carbon from oil palm fronds

Time (Minutes)	Temperature (°C)	Concentration of Activator (%)	Water Content (%)	Ash Content (%)	Volatile Matter (%)	Fixed Carbon (%)	Iodine Number (mg/g)
30	150	10	1.48	8.03	13.59	76.90	597.52
		15	1.72	8.90	16.00	73.38	566.02
		20	3.36	7.55	16.04	73.04	603.99
	200	10	5.00	6.00	12.00	77.00	685.70
		15	8.00	4.00	30.39	75.00	822.91
		20	7.00	6.00	13.46	73.54	741.47
	250	10	5.34	8.73	29.41	56.52	429.02
		15	6.70	7.13	17.00	55.78	402.46
		20	3.27	7.90	39.00	49.83	404.67

4.1 Water content

Water content increases with higher activator concentration. This is due to the dissolving process of residual pyrolysis and organic minerals on the surface of activated carbon which will get better as the activator concentration increases. Increased concentration of activator will increase the amount and volume of pores and surface area, which cause an increase in the performance of activated carbon to absorb water from air [10].

The increase in pyrolysis temperature will cause the C and H bonds in charcoal to be released completely. Thus, a shift of the crystallite carbon plate forms a new pore and develops a pore that has been formed [11]. These results indicate that activated carbon has met the requirements of the National Industrial Standard 06-3730-1995.

4.2 Ash content

Ash content has an influential role in affecting the quality of activated carbon produced. The presence of excessive ash can lead to clogging of the pores on the activated carbon so that the surface area of activated carbon becomes reduced [12]. Ash content increases with increasing concentration of activator so that organic compound of activated carbon decreases, however the content of inorganic compound is relatively fixed.

The interaction in the reactor during the pyrolysis process causes the inorganic compound to trigger the formation of the metal oxide. The increasing number of metal oxide causes the ash content in the activated carbon to be increased, especially with the rise in temperature. There is an increase in ash content at 250°C, indicating that there is an increase of metal oxide formation, which resulted in the increase of ash content as well. The concentration of activators that are too high can also cause damage to the structure of activated carbon, which resulted in the increase of ash content [7]. The increase in ash content can also be caused by mineral salt formation during the pyrolysis process. In addition, the ash content is influenced by the silica content of the raw material. This would mean that the higher the silica content, ash content is also higher [7]. Thus, in general the ash content of activated carbon in all variables has met the SNI standards of 06-3730-1995.

4.3 Volatile Matter

The levels of volatile substances decrease as activator concentration increases. This is because the addition of activators causes changes in the structure and properties of activated carbon absorption. H_3PO_4 activators cause degradation of organic material that weakens the surface structure of activated carbon. In addition, this activator also releases volatile substances and develops an active carbon micro pore structure [13].

At higher temperatures, volatile materials will be released into more and cause less carbon to be formed [13]. This is due to higher pyrolysis temperatures, so that the decomposition of components contained in raw materials such as water, tar, and volatile materials is also higher [15].

The impregnation process using the activator is capable of reducing the non-carbon compound attached to the surface of the activated carbon and entering the bottom surface of the charcoal through the pores of the charcoal. Thus, the presence of the activator will cleanse and enlarge the pore surface. Activators will free non-carbon elements, especially hydrogen, oxygen and nitrogen in liquid form known as tar and gases [16].

4.4 Fixed Carbon

The value of the resulting fixed carbon content fluctuates as the pyrolysis temperature increases. Fixed carbon content is also affected by cellulose and lignin content that can be converted to carbon atoms. At a temperature of 250 ° C there is a decrease in fixed carbon caused by the breakdown of activated carbon structures with high temperatures or burning of carbon [16]. Decrease in carbon content can also be caused by the reaction between the carbon and the activator at high concentrations, which can damage the micropore on the carbon surface [16].

The findings show that fixed carbon which meets SNI 06-3730-1995 is at operating condition 150°C and 200°C for all activator concentration.

IOP Conf. Series: Materials Science and Engineering **309** (2018) 012072 doi:10.1088/1757-899X/309/1/012072

4.5 Iodine Number

The highest iodine number of the activated carbon is at a temperature of 200°C and drops to 250°C for all concentrations of activator. This shows that there is damage to the structure of the activated carbon in the form of the collapse of the pore wall so that the iodine number decreases at a temperature of 250°C. Besides pore structure damage, there is also a burning carbon, as well as a micro pore expansion to the macro pore, all of which cause a decrease in the iodine number [18]. All this is due to high activator concentration and / or too high carbonization temperature.

The findings show that the iodine number fulfilling SNI 06-3730-1995 is under operating conditions 200°C for all activator concentrations.

4.6 Scanning Electron Microscope (SEM) Analysis

SEM analysis is used to visualize the surface morphology of the product which has been carbonized and activated. Figure 1 shows the surface characteristics of the activated carbon of the study that has a pore and cavity distribution on its surface.

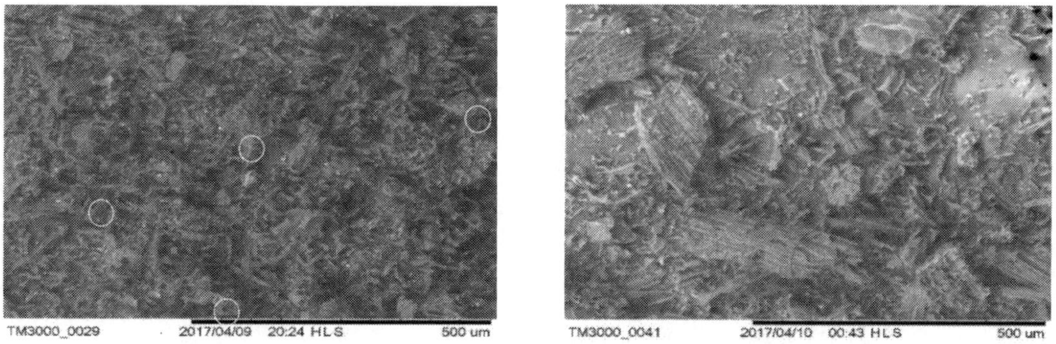

(b) Charcoal (a) Activated Carbon

Figure 1. Scanning Electron Microscope (200x) of Activated Carbon

The structure of the pores formed can result from evaporation and breakdown of the non-carbon compounds contained in the feedstock. Activated carbon also has a rougher surface and a wider irregularity. The presence of the activator can enlarge the pores of activated carbon and expand the surface.

IOP Conf. Series: Materials Science and Engineering **309** (2018) 012072 doi:10.1088/1757-899X/309/1/012072

4.7 Fourier Transform Infra-Red (FTIR) Analysis

FTIR analysis is used to characterize the functional group on the surface of activated carbon. Figure 2 shows the characteristics between Palm Oil (a) and Active Carbon (b).

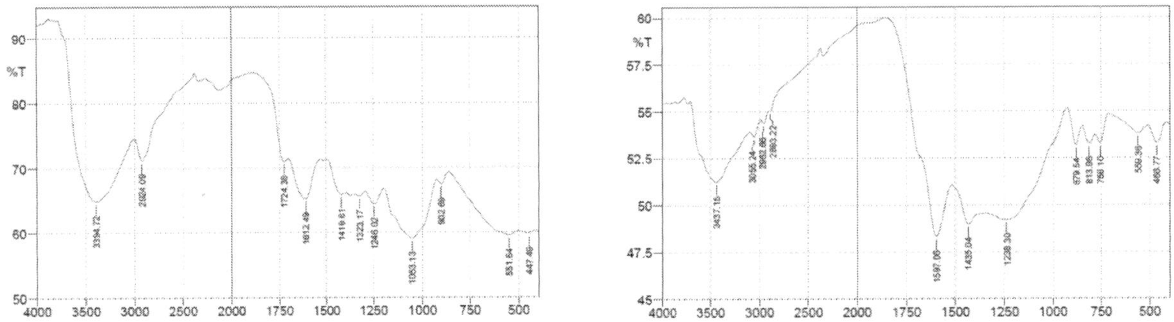

(a) Oil Palm Fronds (b) Activated Carbon

Figure 2. FTIR spectrum of palm oil fronds and activated carbon

As can be seen in Table 3, the wave numbers between palm fronds and activated carbon has a peak with no significant difference. The presence of OH bonds and C-O indicated that activated carbon tends to be more polar. Therefore, the FTIR results show that the activated carbon produced can be said as activated carbon.

Table 3. Comparison between infrared fraction of oil palm fronds and activated carbon

No	Functional Group	Oil Palm Frond	Activated Carbon
1.	O-H	3394.72	3437.15
2.	C-H	2924.09	2962.66
3.	C-O	1323.17	1238.30
4.	C-X	447.49	559.36

5. Conclusion

This study obtained several highest test value for indicators mentioned in the National Industrial Standards (SNI) 06-3730-1995. In particular, the following test value were discovered: water content of 8.00%, ash content of 8.90%, volatile matter of 39.00%, fixed carbon of 77.00%, and iodine number of 822.91 mg/g. Hence, it can be concluded that the activated carbon produced from palm oil fronds has met the requirements of the National Industrial Standard 06-3730-1995.

6. Acknowledgments

Authors would like to thank the Chemical Engineering Department, Faculty of Engineering, Universitas Sumatera Utara, Medan, Indonesia for the support during the process of research and writing this paper.

IOP Conf. Series: Materials Science and Engineering **309** (2018) 012072 doi:10.1088/1757-899X/309/1/012072

7. Referrences

[1] Trisakti B, Lubis J, Husaini T and Irvan 2017 Effect of Turning Frequency on Composting of Empty Fruit Bunches Mixed with Activated Liquid Organic Fertilizer *IOP Conf. Ser. Mater. Sci. Eng.* **180** 8

[2] Salman J M 2014 Optimization of preparation conditions for activated carbon from palm oil fronds using response surface methodology on removal of pesticides from aqueous solution *Arabian Journal of Chemistry* **7** 101-108

[3] Dahlan 2000 Oil Palm Frond, a Feed for Herbivores *Asian-Australian Journal Animal* 300-303

[4] Maulina S, Pandang I and Ambarita Y P 2015 Comparative Study Of Utilization Of Oil Palm Frond to Produce Oxalic Acid by Using Alkali Fusion and Oxidation Method *Proceeding ICIC.*

[5] Winanti W S 2014 *Pengembangan dan Alih Teknologi untuk Mitigasi dan Adaptasi Perubahan Iklim* Dewan Nasional Perubahan Iklim

[6] Kvech S and Tull E 1998 *Activated Carbon in Water Treatment Primer* Environmental Information Management Civil Engineering Dept, Virginia Tech

[7] Mopoung S, Phansiri M, Wanwimon P, Sataporn K 2015 Characterization and Properties of Activated Carbon Prepared from Tamarind Seeds by KOH Activation for Fe(III) Adsorption from Aqueous Solution *The Scientific World Journal*

[8] Menendez R and Alvarez R 2003 *Coal Carbonization : Current and Future Applications* Instituto Nacional Del Carbon y sus Derivados

[9] [BSN] Badan Standarisasi Nasional 1995 *Arang Aktif Teknis* Jakarta: BSN (*SNI 06-3730-95*)

[10] Pambayun G S, Yulianto R Y E, Rachimoellah M and Putri E M M 2013 Pembuatan Karbon Aktif dari Arang Tempurung Kelapa dengan Aktivator $ZnCl_2$ dan Na_2CO_3 sebagai Adsorben untuk Mengurangi Kadar Fenol dalam Air Limbah *Jurnal Teknik POMITS* **2(1)** ISSN: 2337-3539

[11] Noer A A, Awitdrus and Usman Malik 2014 Pembuatan Karbon Aktif dari Pelepah Kelapa Sawit Menggunakan Aktivator H_2O sebagai Adsorben *Jurnal Online Mahasiswa Fakultas Matematika dan Ilmu Pengetahuan Alam Universitas Riau* **1(2)**

[12] Scroder E *et al* 2006 *Experiment on the Generation of activated carbon from Biomass* Institute for Nuclear ang energy Technologies Forschungs Karlsruhe Germany 106-111

[13] Yakout S M and Deen G S E 2016 Characterization of activated carbon prepared by phosphoric acid activation of olive stones *Arabian Journal of Chemistry* **9**

[14] Kumar M, Gupta R C and Sharma T 1992 Effects of Carbonisation Conditions on the Yield and Chemical Composition of Acacia and Eucalyptus Wood Chars *Biomass and Bioenergy* **3(6)** 411 - 417

[15] Latifan R and Susanti D 2012 Aplikasi Karbon Aktif dari Tempurung Kluwak (Pangium Edule) dengan Variasi Temperatur Karbonisasi dan Aktifasi Fisika Sebagai Electric Double Layer Capasitor (EDLC) *Jurnal Teknik Material Dan Metalurgi ITS* **1(1)**1-6

[16] Jagtoyen M, Derbyshire F Activated carbons from yellow poplar and white oak by H_3PO_4 activation *Carbon, Science Direct* **36** 7-8

[17] Saka C 2012 BET, TG-DTG, FT-IR, SEM, iodine number analysis and preparation of activated carbon from acorn shell by chemical activation with ZnCl2 *Journal of Analytical and Applied Pyrolysis* **95** 21–24

[18] Mitome T, Uchida Y, Egashira Y, Hayashi K, Nishiura A and Nishiyama N 2013 Adsorption of indole on KOH-activated mesoporous carbon *Colloids and Surfaces A: Physicochemical and Engineering Aspects* **424** 89–95

IOP Conf. Series: Materials Science and Engineering **309** (2018) 012073 doi:10.1088/1757-899X/309/1/012073

Liquid smoke characteristics from the pyrolysis of oil palm fronds

S Maulina[*,1,2], **F Silia**[1]

[1]Department of Chemical Engineering, Faculty of Engineering, Universitas Sumatera Utara

[2]Sustainable Energy and Biomaterial Center of Excellence, Universitas Sumatera Utara

*E-mail : serimaulina@usu.ac.id

Abstract. This study was conducted as means to characterize the pyrolysis of oil palm fronds into more economical products. In particular, this study was focused on pyrolysis of oil palm fronds, which could generate products such as liquid smoke, tar and char. Four characteristics of liquid smoke were examined in this study, namely the yield of liquid smoke, phenolic content, total acid content and pH. These characteristics were examined in a temperature of 150 °C, 200 °C and 250 °C with processing time of 60 minutes, 90 minutes and 120 minutes. This study revealed that the highest yield of liquid smoke was equal to 43.47% at a temperature of 150 °C for approximately 2 hours, while the highest level of phenolic was obtained at a temperature of 250 °C for approximately 1 hour. Moreover, the highest total acid content obtained was 11.23% at a temperature of 150 °C with a time of 1 hour. In addition, all operating conditions has produced liquid smoke with an average pH value of 3.

1. Introduction

Oil palm plantations in Indonesia have begun to grow rapidly since the early 80s. Approximately, 23.5 million tonnes of crude palm oil was produced in 2012 [1]. Currently, oil palm has become one of the plantation commodities that play a very important role in stimulating the economy in Indonesia, such as increasing foreign exchange, employment, and improving the living standards of people in the region. In addition, producing Crude Palm Oil (CPO) as the main commodity, the palm oil plantation also produces several types of potential wasted, namely Oil Palm Frond (OPF). The plantations will produce 22 stems per tree every year, and each stem has a weight about 2.2 kg. So that in 1 ha of oil palm plantations will generate 6.3 tonnes oil palm fronds each year [2]. Oil Palm Fronds (OPF) is one of the organic waste produced from oil palm plantations that are processed by burning or just throwing away [4]. The moisture content reaching 75 %, OPF is considered as a category of wet by-product. The large amount of waste generated in these oil palm plantations makes palm oil has a great potential to be processed into useful product. It is also driven by the content present in palm oil, which consists of hemicellulose (34.89 %), cellulose (27.14%), and lignin (19.87 %) [3]. In order to produce an economical product, one of the methods that can be used is through pyrolysis of raw materials containing hemicellulose, cellulose, and lignin that would be condensed to become liquid smoke.

Pyrolysis is defined as the thermal degradation process of solids in the absence of oxygen, allowing the occurrence of several thermochemical conversion pathways so the solid becomes permanent gasses, pyrolysis liquid and char [5]. Pyrolysis can be classified as slow pyrolysis and fast pyrolysis [6]. Slow pyrolysis takes place at low temperature with longer time. In general, aimed to produce char. Fast pyrolysis requires the material cutting into small size with special equipment to remove moisture content quickly.

Liquid smoke is a compound that vaporizes simultaneously from a heat reactor through pyrolysis technique (decomposition by heat) and condenses on a cooling system [7]. Liquid smoke has a major component of acidic compounds, phenol derivatives, and carbonyls that act as a flavoring, coloring, antibacterial, and antioxidant [8]. The chemical composition of liquid smoke depends primarily on the type of wood and water content of wood, the influence of both pyrolysis temperature and the duration of smoke generation. The disadvantage of the full strength of liquid smoke is the high content of the active compounds of flavor and color, which limits their applications when used for specific purposes such as antimicrobial agents [9]. Therefore, the purpose of this study was to determine the effect of temperature, pyrolysis time, and moisture content on the yield and pH of liquid smoke produced.

2. Methodology

The pyrolysis process takes place at the pyrolysis reactor under predefined operating conditions. Oil palm fronds is first dried, then weighed. The pulverized oil palm fronds is introduced into the reactor and pyrolized according to predetermined time and temperature, then condensed. The generated smoke will be condensed by using condensor and accommodated in a storage tank. Liquid smoke is allowed for 2 x 24 hours then filtered with filter paper. The set of equipment in the pyrolysis process is shown in Fig. 1.

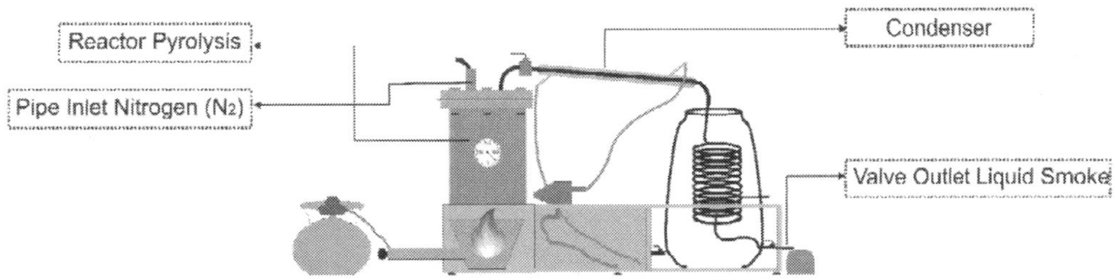

Figure 1. Pyrolysis Reactor

Analyzed parameters to obtain the quality of liquid smoke are total acid, phenol content, pH and yield of liquid smoke.

3. Result

The pyrolysis temperature is allow the breaking of chemical bonds on cellulose, hemicelluloce and lignin compound. Higher pyrolysis temperature leads to more yield of liquid smoke [10]. This is because the oil palm fronds get a sufficient amount of heat to breaking of chemical bonds so that the compound in the palm oil bark increasingly decomposes and condensed into liquid smoke [11]. High temperatures and long periods of time on pyrolysis is led more complete decomposition of raw materials to obtain higher yields of liquid smoke [12]. The highest yield was obtained at pyrolysis temperature of 150 °C with pyrolysis time for 120 minutes, i.e. 43.47%.

Table 1. Yield of Smoke Liquid

Experiment	Water content (%)	Pyrolysis temperature (^{0}C)	Pyrolysis time (minutes)	Yield of Liquid (%)
1	29.64		60	35.187
2	27.25	150	90	41.646
3	27.96		120	43.470
4	28.34		60	28.133
5	26.64	200	90	30.588
6	23.62		120	38.142
7	21.66		60	29.203
8	21.53	250	90	31.970
9	21.28		120	34.001

Cellulosic pyrolysis will produce acetic acid compounds and carbonyl compounds. The acid compounds formed from the pyrolysis process are organic acid compounds [13] among which are large amounts of acetic acid, propionic, butyric and valeric [14]. Other than that, there are carbonyl compounds present in the liquid smoke including vanillin and syringaldehyde [14]. Hemicellulose pyrolysis will produce furfural, furan, acetic acid and its derivatives [13], thus the acid obtained in liquid smoke is obtained from the decomposition of hemicellulose and cellulose. The highest total acid was 11.23%, which was obtained at pyrolysis temperature of 150 °C with a pyrolysis time of 60 minutes, as shown in Figure 2.

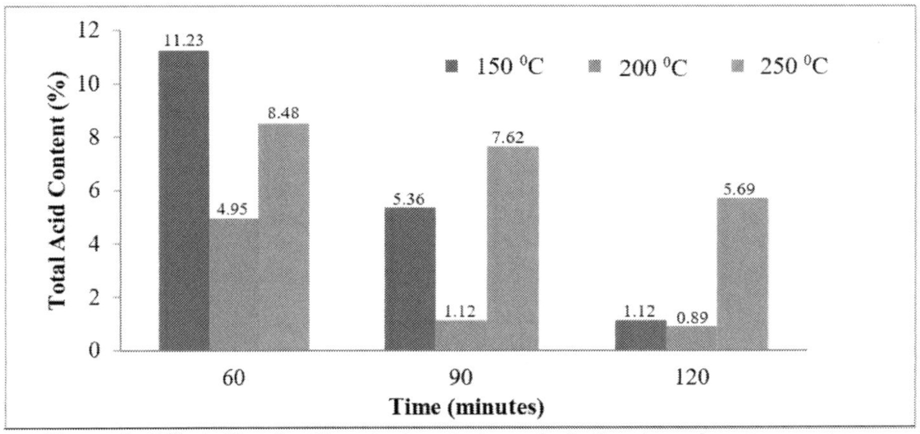

Figure 2. Effect of total acid content against pyrolysis time in various temperatures

Phenol is the result of degradation of a wood component called lignin. The more lignin content in the wood, the greater the phenol content obtained in liquid smoke [15]. The phenol compounds present in wood smoke are generally aromatic hydrocarbons composed of benzene rings with a number of bound hydroxyl groups [14]. The amount and quality of phenol in the liquid smoke depends on the temperature of the pyrolysis and the lignin content of the feedstock [16]. The highest phenol content obtained was 12.28% at pyrolysis temperature of 250 °C and pyrolysis time of 60 minutes.

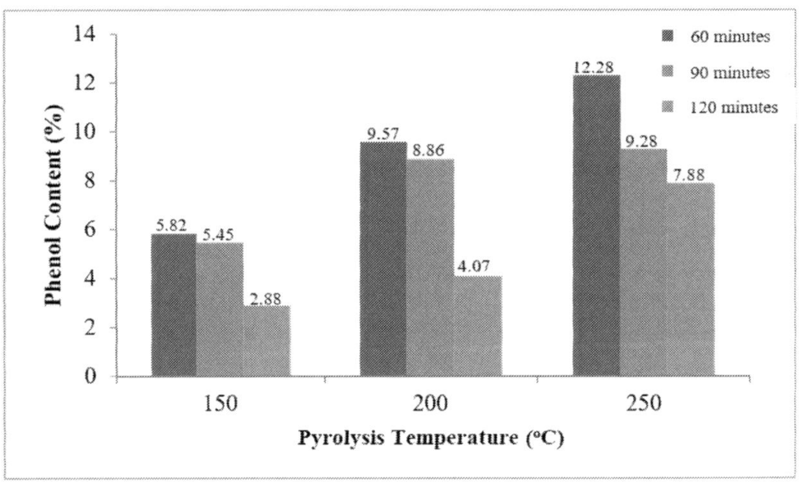

Figure 3. Effect of phenol content against pyrolysis temperature in various times

The quality of the resulting liquid smoke can be determined by measuring the degree of acidity (pH). The pH value indicates the rate of decomposition process of the wood chemical component occured can produce acid and phenol in liquid smoke [17]. The result of measurement of pH of liquid smoke ranged between 3.1 and 3.8. This shows that liquid smoke products are acidic.

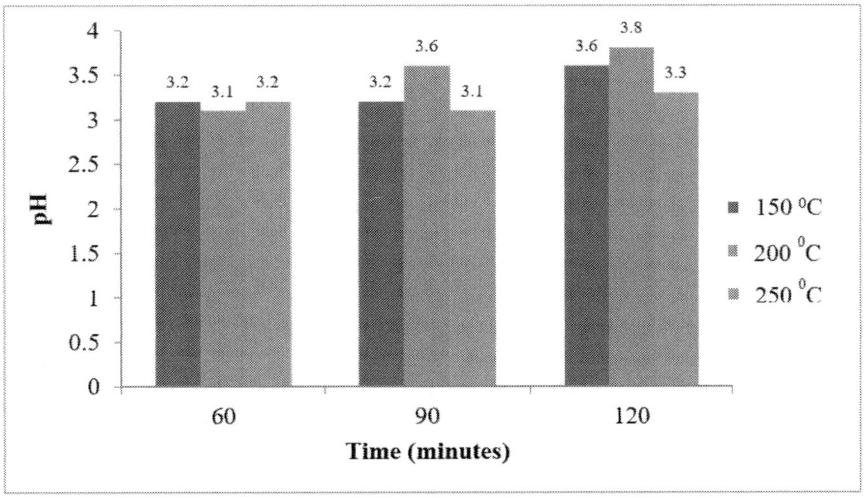

Figure 4. Effect of pH against pyrolysis time in various temperatures

Conclusion
Based on the purpose of this research, it was found that the highest yield of liquid smoke was equal to 43.47% at a temperature 150 °C in 2 hours, while the highest total acid content obtained was 11.23% at a temperature of 150 °C in 1 hour. Moreover, the highest phenolics were obtained at a temperature 250 °C in 1 hour and pH is around 3.1 to 3.8.

References
[1] Irvan, Trisakti B, Tomiuchi Y, Harahap U and Daimon H 2017 Effect of Recycle Sludge on Anaerobic Digestion of Palm Oil Mill Effluent in A Thermophilic Continuous Digester *IOP Conf. Series: Materials Science and Engineering* **206** 012094

[2] Litbang Deptan. 2010. *Pengolahan Pelepah Kelapa Sawit Menjadi Pakan.* Loka Penelitian Kambing Potong

[3] Maulina S, Pandang I and Ambarita Y P 2015 Comparative Study Of Utilization Of Oil Palm Frond to Produce Oxalic Acid by Using Alkali Fusion and Oxidation Method *Proceeding ICICS*

[4] Vakili M, Haque A A M and Gholami Z 2012 Effect of Manual Turning Frequency on Physico-Chemical Parameters During the Oil Palm Frond and Cow Dung Composting *Caspian Journal of Applied Science Research* **1(12)** 49-59 ISSN: 2251-9114

[5] Blasi C D 2008 Modeling Chemical and Physical Processes of Wood and Biomass Pyrolisis *Progress in Energy and Combution Science* **34** 47-99

[6] Martinez J D, Puy N, Murillo R, Garcia T, Navarro M V and Mastral A M 2013 Waste Tyre Pyrolisis – A review *Renewable and Sustainable Energy Reviews* **23** 179-213

[7] Simon R, Caile B, Palme S, Meler D and Anklam E 2005 Compotition and Analysis of Liquid Smoke Flavoring Primary Product *J. Food Sci.* **28** 871-882

[8] Ayudiarti L D and Sari R N 2010 *Asap Cair Dan Aplikasinya Pada Produk Perikanan* Squalen **5(3)**

[9] Montazeri N, Oliveira A C M, Himelbloom B H, leigh M B and Crapo C A 2012 Chemical Characterization of Comercial Liquid Smoke Product *Food Science and Nutrition* **1(1)** 102-115

[10] Ramadhan A P and Ali M 2012 Pengolahan Sampah Plastik Menjadi Minyak Menggunakan Proses Pirolisis *Jurnal Ilmiah Teknik Lingkungan* **4(1)**

[11] Akbar A, Paindoman R and Coniwanti P 2013 Pengaruh Variabel Waktu dan Temepratur terhadap Pembuatan Asap Cair dari Limbah Kayu Pelawan (Cyanometra Cauliflora) *Jurnal Teknik Kimia* **19(1)**

[12] Fadillah H and Alfiarty A 2015 The Influence of Pyrolysis Temperature and Time To The Yield and Quality of Rubber Fruit (Hevea brasiliensis) Shell Liquid Smoke *Prosiding Seminar Nasional Teknik Kimia "Kejuangan" Pengembangan Teknologi Kimia Untuk Pengolahan Sumber Daya Alam Indonesia* ISSN: 1693-493

[13] Wijaya M, Noor E, Irawadi T T and Pari G 2008 Karakteristik Komponen Kimia Asap Cair dan Pemanfaatannya sebagai Biopestisida *Bionature* **9(1)** 34-40 ISSN:1411-4720

[14] Sunarsih S, Pratiwi Y and Sunarto Y 2012 Pengaruh Suhu, Waktu, dan Kadar Air Pada Pembuatan Asap Cair dari Limbah Padat Pati Aren *Prosiding Seminar Nasional Aplikasi Sains dan Teknologi (SNAST) Periode III* ISSN:1979-911X

[15] Budaraga I K, Arnim, Marlida Y and Bulanin U 2016 Liquid Smoke Production Quality from Raw Materials Variation And Different Pyroliysis Temperature *International Journal on Advanced Science Engineering Information Technology* **6(3)** ISSN: 2088-5334

[16] Girard J P 1992 *Smoking in Technology of Meat Product* New York: Clermont Ferrand, Erllis Horwood

[17] Wijaya, M, Noor E, Irwadi T T and Pari G 2008 Perubahan Suhu Pirolisis Terhadap Struktur Kimia Asap Cair dari Serbuk Gergaji Kayu Pinus *Jurnal Ilmu dan Teknologi Hasil Hutan* **1(2)** 73-77

IOP Conf. Series: Materials Science and Engineering **309** (2018) 012074 doi:10.1088/1757-899X/309/1/012074

Optimization of palm fruit sterilization by microwave irradiation using response surface methodology

M Sarah[1,2*], I Madinah[1] and S Salamah[1]

[1]Departement of Chemical Engineering, Universitas Sumatera Utara, Padang Bulan Medan, Indonesia
[2]Sustainable Energy and Biomaterial Centre of Excellent, Universitas Sumatera Utara

*E-mail: Maya3@usu.ac.id

Abstract. This study reported optimization of palm fruit sterilization process by microwave irradiation. The results of fractional factorial experiments showed no significant external factors affecting temperature of microwave sterilization (MS). Response surface methodology (RSM) was employed and model equation of MS of palm fruit was built. Response surface plots and their corresponding contour plots were analyzed as well as solving model equation. The optimum process parameters for lipase reduction were obtained from MS of 1 kg palm fruit at microwave power of 486 Watt and heating time of 14 minutes. The experimental results showed reduction of lipase activity in the present work under MS treatment. The adequacy of the model equation for predicting the optimum response value was verified by validation data (P>0.15).

1. Introduction

Processing of oil palm fruits involves operations of fruit sterilization, threshing, digestion, oil expression, and oil clarification. Palm oil mill sterilizes fresh palm fruit bunches (FFB) within 24-48 hours prior extraction to avoid palm oil deterioration. The aim of sterilization is to protect palm oil quality especially content of free fatty acid (FFA) less than 5-6% [1]. According to Malaysian Palm Oil Board (MPOB), the maximum FFA grade in CPO is 5% [2]. Sterilization of palm fruit carries out in batches with high presurized steam (15 – 45 psi) for 90 minutes. Digestion and pressing extracts oil from mesocarp of sterilized palm fruits. Overall, processing of oil palm fruits generates wastewater with high organic concentration [3]. Current sterilization technology has weakeness because it requires long operation time, needs high energy to produce pressurized steam and discharge palm oil mill effluent (POME). To improve the sterilization process, this study proposes microwave sterilization to reduce sterilization time and avoids POME generation. Several authors reported heating or sterilization of oil palm fruit by microwave irradiation [4, 5, 6, 7, 8, 9, 10].

Sterilization objective in this study is to inactivate lipase. Lipase may hydrolize oil into free fatty acid in the presence of water. Parameters importance to sterilization process are time and temperature. Sterilization time should considers D-value of lipase. The D-value of lipase defines as time to inactivate 90% of lipase activity at certain temperature. Sarah and Taib (2013) reported D-value of lipase from microwave sterilization of oil palm fruit ranges between 8.3 minutes (T= 68°C) to 16.9 min (T=82°C) [8]. Sarah et al. also reported performance of steam batch process and microwave irradiation during sterilization of palm fruits. Microwave sterilization carried out for 14 to 17 minutes at temperature of 70 to 76°C, while laboratory scale steam batch sterilization proceded for 2 hours at temperature of 105°C. The previous published experimental results for microwave sterilization of oil palm fruit showed that the optimum process parameters are time, mass and power [9]. Based on the results, this study investigates effect of main constituent like sterilization time, mass and power on heat generation during sterilization and quality of palm oil. For this study, we therefore evaluated the effects of sterilization time, mass and power on microwave sterilization process, and develop a response surface model using a Central Composite Design (CCD) for lipase inactivation. The development of response surface prediction to describe lipase inactivation in this study should be very beneficial to construct application of microwave sterilization.

2. Materials and Methods

2.1. Materials

This study used oil palm fruit bunch *Dura* variety taken around campus of University Sumatera Utara-Medan as raw material, and chemicals such as ethanol, sodium hydroxide etc for analysis of FFA content. Experiments carried out in microwave oven (R-249 IN (S)/(W)) with frequency of 2450 MHz. We measured temperature using thermocouple type K (Krupp and Closs size diameter of 3 x 300 mm (Mineral Insulated) C/w Cable 2 m) connected with thermo controller (Shimaden).

2.2. Sterilization of oil palm fruit by microwave irradiation

Prior sterilization process, we placed oil palm fruitlets inside microwave cavity and adjusted microwave power and heating time. Temperature measurement conducted during sterilization period. Temperature of oil palm fruitlet increased due to polar water molecules of oil palm fruitlet were rotating and vibrating to align with the changing polarity around them when expose to microwaves. The movement of the molecule creates heat [11].

2.3 Design of experiment

This study applied response surface methodology (RSM) to optimize microwave sterilization process using software Minitab 2017 trial version. The most common experimental design in RSM is CCD. CCD is optimized design for fitting quadratic models because it has equal predictability in all direction from the centre [12]. In order to find out the best microwave sterilization condition, a series of experiments conducted by three independent variables to obtain predictive model allowed for the design of a minimal number of experimental runs. We choose three different parameters like sterilization time, mass and power as main variables and CCD arrangement for this study is shown in Table 1. In this study, significant level was adjust to 5% error ($\alpha = 0.05$) with confidence level of 95%.

3. Results and Discussions

3.1 Predictive models to evaluate effect of sterilization time, mass and microwave power to temperature

The values of temperature obtained under the different experimental conditions (sterilization time (X_1), mass (X_2) and microwave power(X_3)) is related by the following second order polynomial equation:

$$Y = 62.17 - 4.01X_1 + 6.27X_2 + 4.40X_3 - 0.06X_1^2 - 3.11X_2^2 - 0.14X_3^2 - 1.29X_1X_2 + 1.14X_1X_3 - 0.19X_2X_3 \quad (1)$$

A summary of the analysis of variance (ANOVA) for model (1) is shown in Table 2. The value of adjusted determination coefficient (R^2_{Adj}) (0.16) and determination coefficient (R^2) (0.55) indicated low degree of correlation between the observed and predicted value. The ANOVA test of regression model demonstrated insignificant correlation between sterilization time, mass and to temperature (F-value = 1.39 and probability > 0.306).

Figure 1(a) show relationship of residual from ANOVA test plotted against fitted value. Residual data points for temperature are not uniformly or disperse randomly and do not form a specific pattern. Figure 1(a) indicates residuals model are equally dispersed and show clear correlation with the fitted value. Figure 1(b) show distribution of residuals data versus sequences (order) tends to be random which indicates independent variables are related or correlated.

Table 1. Experimental design

Mass (g)	Power Level	Time (min)
750	Medium	12
1000	Medium Low	8
750	Medium	12
330	Medium	12
750	Medium	12
1000	Medium High	16
500	Medium High	8
500	Medium High	16
750	Low	12
750	Medium	18.728
750	High	12
1000	Medium Low	16
500	Medium Low	8
750	Medium	12
750	Medium	12
500	Medium Low	16
1171	Medium	12
750	Medium	12
1000	Medium High	8
750	Medium	5.272

Figure 1(c) shows normalized curve of microwave power and temperature relationship. The linear regression line is a predictor of temperature data set during observation. The residual point is corresponding to straight line which indicates the residual is normally distributed. Thus supported by the results of Kolmogorov-Smirnov (KS). Figure 1(c) shows value of KS is 0.097 while KS table is 0.294. This concludes $KS_{calculation} < KS_{table}$ which indicates no deviation with the model.

Table 2. Results of ANOVA test

Source	Sum of squares	Degree of freedom	Mean squares	*F*-value	Prob > *F*
Model	1186.26	9	131.806	1.39	0.306
Lack of fit	519.24	5	103.849	1.22	0.418
Pure error	427.07	5	85.414		
Corrected total	2132.57	19			

3.2 Effect of sterilization time, mass and microwave power on temperature

The graphical representations of regression Eq.(1) as response surface and contour plots were obtained using Minitab software. As maximum initial and final sterilization time, mass and microwave power list in Table 1, optimum parameters for microwave sterilization of oil palm fruit were optimized. Three dimensional response surface plots generated by Minitab were analyzed to evaluate the impact of two dependent variable while holding the third variable constant as shown in Figure 2.

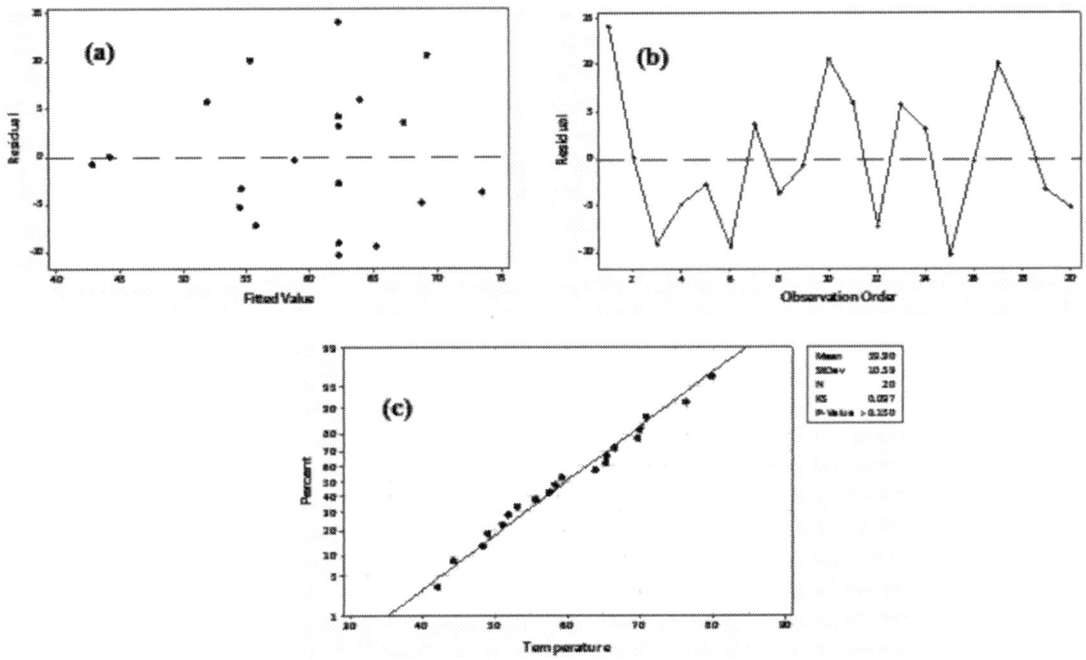

Figure 1. Linear regression modelling of temperature data

Figure 2 shows relationship sterilization time, mass and microwave power on temperature of sterilization process. Figure 2(a) shows microwave power and sterilization time influenced temperature of sterilization at constant sample mass. Microwave power and sterilization time influenced temperature in both linear and quadratic positions. The increase of temperature corresponded to the rise of microwave power level and sterilization time. Increment of microwave power resulted high intensity of microwave energy applied into the oil palm fruit material. The microwave energy penetrates into palm fuits core and water molecules of palm fruits re-oriented themselves as response to microwave energy, creating internal kinetic energy of inter moleculars friction. The kinetic energy instantly converted into thermal energy which indicates by the rise of fruit interior temperature, and then exterior temperature. Thus created instant heating and sterilization occurred very fast. Sarah, M et al.(2014) reported similar phenomena while investigated effectivity of steam batch process and microwave irradiation to sterilize oil palm fruit. Microwave irradiation required only 14 to 17 minutes to inactivate lipase activity as compared to conventional steam batch process that carried out for 1.5 to 2 hours. They also reported temperature for irradiation process (70 – 76.5°C) lower than conventional process (105°C). Steam batch sterilization process disposed waste water, while microwave irradiation none [12].

Figure 2(c) shows mass and microwave power had less correlation with temperature at constant sterilization time. This study observed increment of sample mass did not influenced the rise of temperature at linear position (please see Figure 2(c)), or during sterilization at constant time and microwave power. Figure 2(c) also show sample mass and microwave power influenced the rise of temperature at quadratic position while sterilization time was constant. Increment mass of sample resulted lower intensity of microwave energy applied into the oil palm fruit material and power density was low, thus lower temperature of sterilization process obtained. On the contrary, increment of microwave power increase microwave intensity resulted greater power density and high temperature obtained for sterilization process. Power density express the number of microwave energy utilized by each palm fruit samples.

IOP Conf. Series: Materials Science and Engineering **309** (2018) 012074 doi:10.1088/1757-899X/309/1/012074

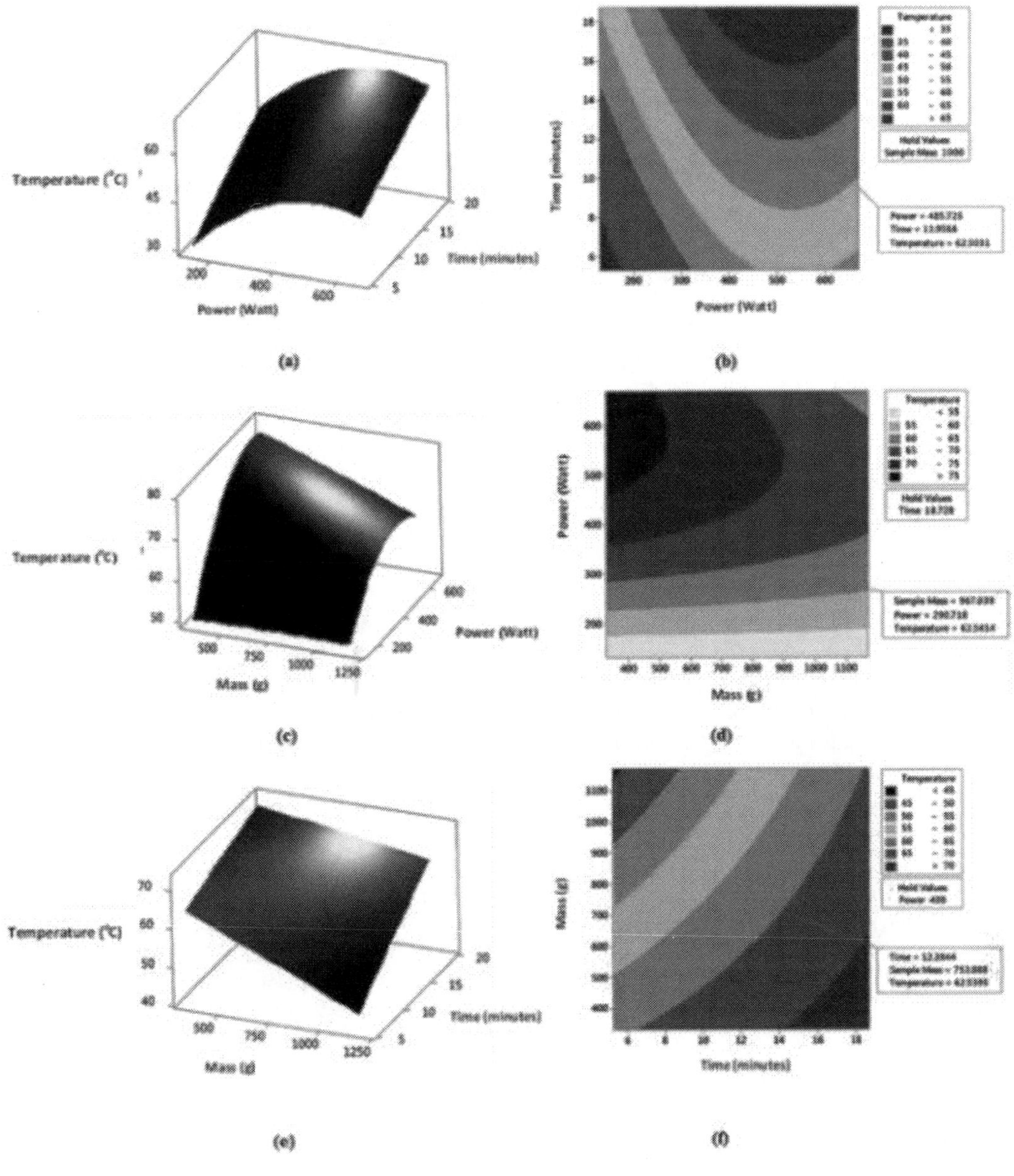

Figure 2. Response surface plots and its contour plots of effect: sterilization time and microwave power (a and b), mass and microwave power (c dan d) and mass and sterilization time (e and f) on temperature

Figure 2(e) shows mass and sterilization time influenced temperature of sterilization at constant microwave power. Mass and sterilization time influenced temperature in both linear and quadratic positions. The increase of temperature corresponded to the reduction of mass and increment of sterilization time. Time play importance role in the sterilization process. Prolonged heating of low power density treatment might obtained high temperature, on the contrary it might caused overheating process for small sample during treatment with high power density that destructed the sample. Other study reported overheating as thermal runway which defines as uncontrolled rise of temperature during microwave irradiation [13]

3.3. *Optimum conditions*

Figure 2(b, d and f) show contour plots of correlations and relationships between sterilization time and microwave power (Figure 2(b)), mass and microwave power (Figure 2(d)) and mass and sterilization time (Figure 2(f)) to temperature of sterilization. This study observed temperature for sterilization of oil palm fruits or lipase inactivation process should proceed between temperature of 44.2°C to 79.8°C with average temperature of 60.8°C.

The temperature of sterilization resulted from this study was lower as compared to commercial sterilization of FFB which is generally carried out at temperature of 131°C and pressure of 40 psi. The temperature of 60.8°C is used as a benchmark to determine the range of contour plots in which the optimal temperature is located as shown in Figure 2(b, d and f). The optimal conditions of palm fruits sterilization are tabulated in Table 3.

Table 3. The optimum conditions of palm fruit sterilization

Independent variables	Temperature (°C)	Power (Watt)	Time (minutes)	Mass (g)
Time and Power	62.50	485.72	13.96	1000
Mass and Power	62.54	290.72	18.73	967.04
Mass and Time	62.54	400	12.39	753.89

Based of RSM study, the best operating conditions to obtain optimum temperature of 62.5°C for oil palm sterilization in this study were sterilization using 1000 g oil palm fruitlets heated by 485.7 Watt microwave power for 13.96 minutes. Meanwhile, during data verification in laboratory, this study observed the best combination of mass, power and time to obtain optimum temperature from experiments oil palm sterilization were 1000 g, 240 Watt and 8 minutes respectively. Temperature obtained from this treatment was 44.2°C. By using those process combination, the FFA content in oil palm fruit in this study were observed below standard.

4. Conclussion

Overall, this study concluded microwave power and sterilization time strongly influenced temperature required to sterilize oil palm fruitlets by microwave irradiation. While mass less influenced temperature sterilization. The optimum temperature for microwave sterilization of palm fruit obtained from RSM were 1000 g, 485.7 Watt, and 13.96 minutes with temperature of 62.5°C. From energy saving point of view, the optimum temperature obtained was 44.2°C from experiment with combination of mass, power and time were 967.04 g, 290.72 Watt, was 18.73 minutes respectively.

References

[1] Odunfa S A 1989 *International Biodeterioration* **25** 393-405
[2] Bahadi, Awadh M, Japir A W, Salih N and Salimon J 2016 *Malaysian Journal of Analytical Sciences* **20(5)** 1042-1051
[3] Irvan, Trisakti B, Tomiuchi Y, Harahap U, Daimon H 2017 Effect of Recycle Sludge on Anaerobic Digestion of Palm Oil Mill Effluent in A Thermophilic Continuous Digester, *IOP Conf. Series: Materials Science and Engineering* **206** 012094
[4] Chow M C and Ma A N 2001 *Proc. PIPOC International Palm Oil Congress (Chemistry and Technology)*
[4] Chow M C and Ma A N 2007 *Journal of Microwave Power & Electromagnetic Energy* **40(3)** 165-173
[5] Sukaribin N and Khalid K 2009 *Industrial Crops and Products* **30(2)** 179-183
[6] Cheng S F, Nor L M and Chuah C H 2011 *Industrial Crops and Products* **34(1)** 967-971
[7] Sarah M and Taib M R 2013 *International Journal of Chemical Engineering and Applications* **4(3)** 129-133
[8] Sarah M and Taib M R 2014 *Journal of Medical and Bioengineering* **2(3)** 153-156

[9] Carrick I T 2016 *Microwave Heating as an Experimental Model for Spontaneous Combustion in Organic Materials* (New York: SUNY Cobleskill)
[10] Umudee I, Chongcheawchamnan M, Kiatweerasakul M and Tongurai C 2013 *International Journal of Chemical Engineering and Applications* **4(3)** 111-113
[11] Gao Y L and Jiang H H 2006 *Biochemical Engineering Journal* **24** 43-48
[12] Sarah M, Taib M R and Adamu A 2014 *Jurnal Teknologi* **69(2)** 55-60
[13] Clemens J. and Saltiel C 1996 *International Journal Heat Mass Transfer* **39(8)** 11
[14] Noerhidajat, Yunus R, Zurina Z A, Syafiie S, Ramanaidu V and Rashid U 2016 *International Food Research Journal* **23(1)** 129-134

IOP Conf. Series: Materials Science and Engineering **309** (2018) 012075 doi:10.1088/1757-899X/309/1/012075

The use of the Kelor Seeds (*Moringa oleifera*) as alternative coagulant in waste delivery process of textile industrial waste

AM Rambe*, **S Pandia, MHS Ginting, R Tambun, B Haryanto**
Chemical Engineering Department, Universitas Sumatera Utara, Padang Bulan, Medan 20155, Indonesia

*E-mail: ahmadmulia76@yahoo.com

Abstract. This research is to know the influence of moringa seed as coagulant, pH of liquid waste textile industry (jeans wash), size of moringa seed particles to decrease of turbidity percentage. Measurements were made to Total Suspended Solid, Color Rate and Chemical Oxygen Demand for wastewater textile industry by coagulation - flocculation method. Variables of this study were conducted on dosage of moringa, with particle size 212 mesh. The results showed that moringa seeds as coagulant dose optimum is 1250 mg/L for the textile industry wastewater at pH 7.8. Moringa seed powder is about 212 mesh with a dose of 1250 mg/L can lower the turbidity of 77.77%, Total Suspended Solid amounted to 83.69% and Chemical Oxygen Demand amounted to 75.86%.

1. Introduction

Coagulation is a liquid water treatment process by stabilizing colloid particles to facilitate the growth of particles during flocculation, while flocculation is a process of water treatment by making contact between colloidal particles that have been destabilized so that the size of the particles grows into larger particles [1].

According to Davis and Cornwell (1991), there are two important factors in the addition of coagulant those are pH and dose. To regulate the pH of the wastewater into the optimal range of coagulation, a coagulant aid is required in the form of an acid or an alkali. The most commonly used acid to decrease the pH is sulfuric acid and to raise the pH is usually used lime [$Ca(OH)_2$], soda ash (Na_2CO_3) or NaOH [2].

The dye is a compound which can be used to form a solution or dispersion to another material so that it is colored. Colors in water can be caused by the presence of natural metal ions, namely iron (Fe) and manganese (Mn). The color that is usually measured is the real color that is the color after the turbidity is removed. While the color appears is a color that is not only caused by the solute in water but also suspended substances. The major polluter of the jeans washing industry is waste water. Wastewater from each process contains residual chemicals used and materials used and materials removed from fibers such as starches and dyestuffs.

Kelor (*Moringa oleifera*) belongs to the Moringaceace family, is a single genus of bush tree families cultivated throughout the tropics and is used for various purposes [3]. Figure 1 is an example of a moringa seed image. The use of moringa seeds as coagualant is done by first making a moringa seed solution, then the mixture is used as coagulant as is commonly used other coagulant. In this case can be used dry beans with skin and dry seeds without skin. However, dry seeds with skin are more effective than skinless seeds.

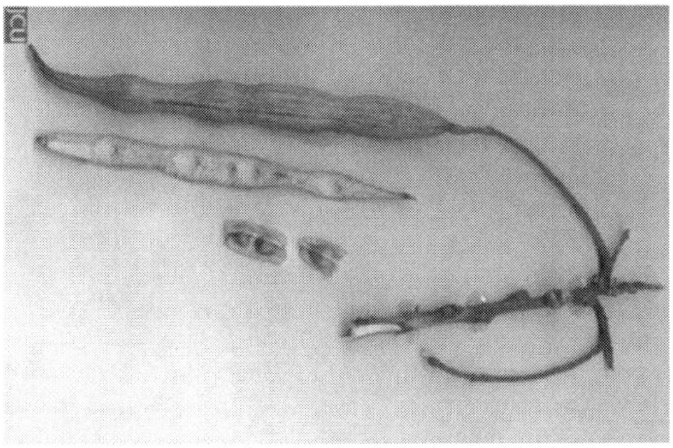

Figure 1. Kelor Seeds (*Moringa Oleifera*)

The purpose of this research is to know the ability of kelor seed as coagulant to percentage of turbidity decrease, total suspended solid (TSS), color content of textile industry liquid waste with pH of textile industry liquid waste and dosage of moringa seeds used.

2. Methodology

Materials processed in this experiment are textile wastewater of textile industry (laundering of Jeans) Liquid waste taken from one of the textile industries in Medan and brought to the laboratory, Turbidity, TSS, color and initial pH of liquid waste used first measured. The coagulant material used in this experiment was the Kelor seed.

The coagulant material used in this experiment was the Kelor seed. To make moringa seed coagulant, ripe mature fruit (brown) and dry naturally in the tree is taken then the seeds are removed from the fruit. Seeds with clean shell and then blend into powder and sifted with starch sieves and then dried in a hot oven at 105°C for 30 minutes to homogenize and lower the water content until constant. Moringa seed powder is then ready for use as a coagulant.

3. Result and Discussion

The effect of coagulation dose on the removal of turbidity wastewater (%) in all experiments can be seen in Figure 2. In moringa seeds 1250 mg / L turbidity decreased to 77.77%. In Moringa seeds also seen a noticeable change in Figure 2 shows that the decrease in turbidity on the working moringa seeds decreases turbidity by absorbing and collecting. So formed clumps followed by a quick stirring of 100 rpm which resulted in the formation of small grains and followed by stirring slowly 40 rpm potentially the occurrence of contact between the granules with each other so as to form large clumps that will settle with the force of gravity [4].

Turbidity is said to decrease when crude materials in waste have been reduced because turbidity is a measure that states how much light can penetrate water, where light penetrates water will be reflected by suspended materials and colloidal materials. Tubidity is caused by abrasive materials that are dispersed [5]. In general, if the value of turbidity increases then water filtration is more difficult and its effectiveness decreases [6].

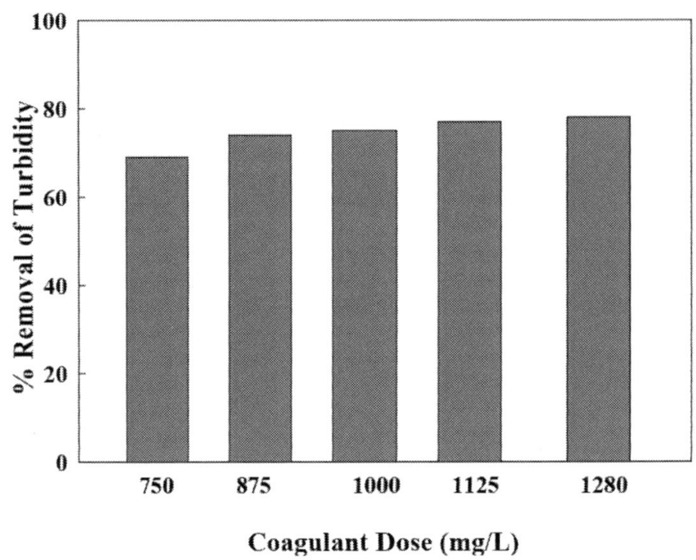

Figure 2. Effect of moringa coagulant dose on turbidity removal (%)

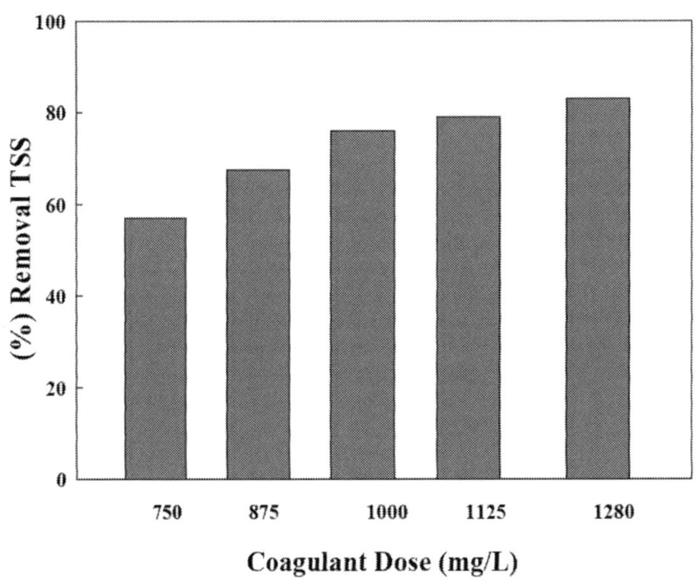

Figure 3. Effect of moringa coagulant dose on TSS removal (%).

The effect of coagulant on the reduction of TSS on the wastewater of the jeans washing industry in the coagulation / flocculation process. After observing the reduction of turbidity level to the coagulant dose used, it is further observed the effect of coagulant dosage used on TSS on industrial wastewater of jeans leaching, the complete graph can be seen in Figure 3.

By using the dosage of moringa seed of 1250 mg / L, TSS was excluded (83.69%). TSS is a solid contained in water and not a solution, this material is distinguished from dissolved solids by laboratory test. TSS usually contains organic and inorganic substances. The effect of coagulant on the decreasing of color content of industrial liquid leaching of jeans on coagulation / flocculation process. Figures 4, 5 and 6 are the relationship between coagulants to decrease the color content of industrial wastewater of jeans leach in coagulation/flocculation process.

From the research, it can be seen that the decrease of color content in waste after coagulation process with the addition of coagulant is very real. By using the color measured lovibond tool is the actual color or real color, i.e. the color after the turbidity is removed while the color appears to be a color not only caused by the solute in water but the suspended substances [7].
From Figures 4to Figure 6 it is seen that the degradation of color content is very real by using Moringa seed coagulation. On the moringa seed dropped to: red = 0.05, blue = 0.20 and yellow = 0.37 with a dose of 1250 mg / L. The process of coagulation and flocculation is a very good process for decreasing color levels to stop at a certain concentration (not working) [8].

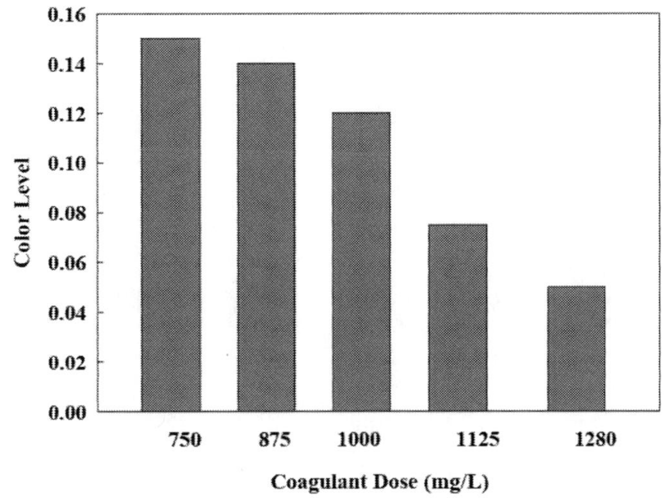

Figure 4. Effect of coagulant dose of moringa seed on the red color.

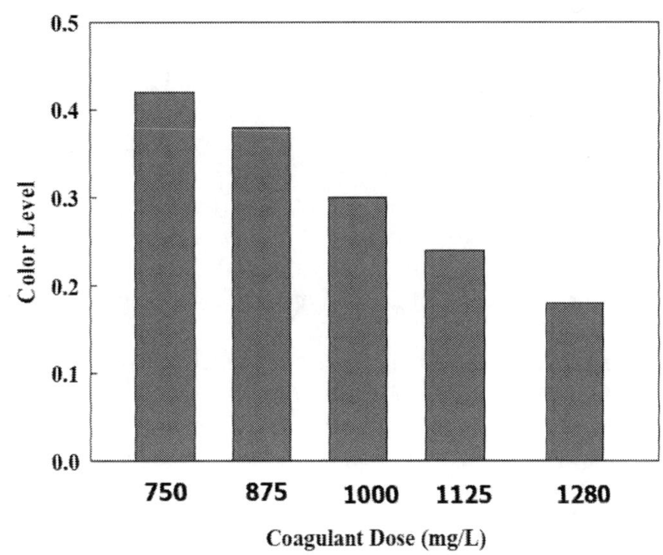

Figure 5. Effect of coagulant dose of moringa seed on the blue color.

The optimum result obtained from the research of dosage of coagulant dosage of moringa seeds to the removal of turbidity of liquid waste of jeans washing industry at pH 7.8 is dose 1250 mg / L turbidity 77,77%. According to K Amdani (2004), the use of moringa seeds, pH and coagulant dose is very real to the turbidity marginalized wastewater washing jeans [9]. The optimum pH of the coagulation 3 (three) turbidities was set aside 83.03%. The optimum dose is 120 mg / 250 mL or 480 mg / L with the excluded turbidity is 92.21%.

For comparison, according to B Enrico (2008), the use of java seeds as a coagulant to the removal of tofu industry liquid waste liquidity at pH 4 doses of 3000 mg / L capable of removing the turbidity of 83.61%, while for alum pH 6 dose 1000 mg / L can eliminate turbidity of 95.73% [10].

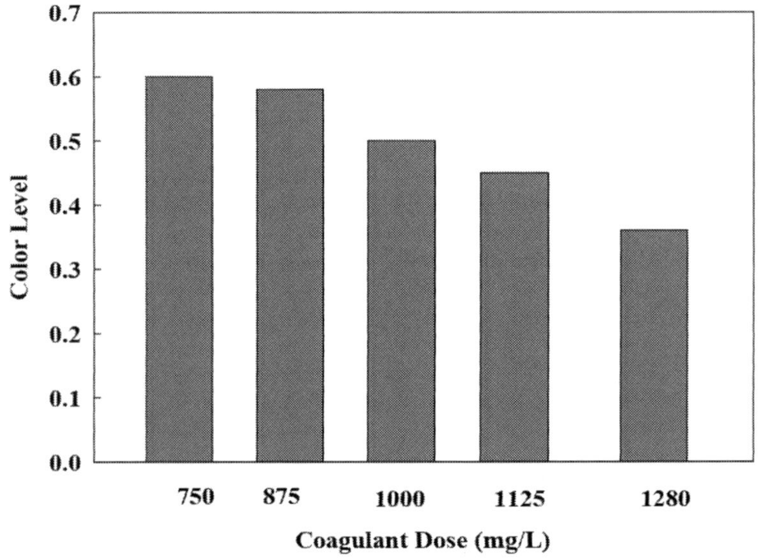

Figure 6. Effect of coagulant dose of moringa seed on the yellow color.

4. Conclusion

From the research of the utilization of moringa seeds as coagulant on coagulation / flocculation of liquid waste of jeans leaching industry can be drawn conclusion as follows The optimum dose of moringa seed coagulant was 1250 mg / L in pH of liquid waste of jeans washing industry, capable of removing turbidity 77,77%, TSS 83,69%, and red color 0,05, blue 0,20, yellow 0, 37 on the lovibond device at pH 7.8. The use of kelor/moringa seeds as coagulant is quite effective in terms of turbidity removal, TSS, and color content with a pH of 7.8.

References

[1] Kiely G 1998 *Environmental Engineering*. Irwin McGraw-Hill. Boston
[2] Davis M L and D A Cornwell 1991 *Introduction to Environmental Engineering*. 2nd ed. McGraw-Hill, Inc. New York
[3] Jahn SAA 1988 Using Moringa Seeds of Coagulants in Developing Countries *Journal of The Water Works Association*
[4] Farooq S and SG Velioglu 1989 *Physico-Chemical Treatment of Domestic Wastewater*. Dalam P.N. Cheremisinoff (Editor). Encyclopedia of Environmental Control Technology **3** Wastewater Treatment Technology. Gulf Publishing Company Book Division. Houston
[5] Prasetya R 2003 *Pemanfaatan Kitosan Kulit Udang Untuk Koagulan Dalam Proses Penjenihan Air*
[6] Koohestanian M H and Abbasian Z 2008 The Separation Method for Removing of Colloidal Particles from Raw Water, *American-Eurasian J. Agric. & Environ. Sci.*, **4 (2)** 266-273
[7] Alaerts, Srisumesti S 1987 *Metode Penelitian Air Usaha Nasional*, Surabaya
[8] Muyasaroh I 2007 *Penurunan Kekeruhan Warna Limbah Industri Tekstil dengan Proses Koagulasi dan Flokulasi*
[9] Amdani K 2004 Pemanfaatan Biji Kelor (Moringa oleifera) sebagai Koagulan pada Proses Koagulasi/Flokulasi dan Sedimentasi Limbah Cair Industri Pencucian Jeans. Magister thesis, Universitas Sumatera Utara
[10] Enrico B 2008 Pemanfaatan Biji Asam Jawa (Tamarindus indica) sebagai Koagulan Alternatif dalam Proses Penjernihan Limbah Cair Industri Tahu (Master's thesis)

IOP Conf. Series: Materials Science and Engineering **309** (2018) 012076 doi:10.1088/1757-899X/309/1/012076

California Bearing Ratio (CBR) test on stabilization of clay with lime addition

I P Hastuty*[1], Roesyanto[1], M N Limbong[1] and J Oberlyn S[2]
[1]Department of Civil Engineering, Universitas Sumatera Utara, Medan, Indonesia
[2]Department of Civil Engineering, Universitas Nommensen, Medan, Indonesia

*ika.hastuty@usu.ac.id

Abstract. Clay is a type of soil with particles of certain minerals giving plastic properties when mixed with water. Soil has an important role in a construction, besides as a building material in a wide variety of civil engineering works, soil is also used as supporting foundation of the building. Basic properties of clay are rock-solid in dry and plastic with medium water content. In high water content, clay becomes sticky like (cohesive) and soften. Therefore, clay stabilization is necessary to repair soil's mechanical properties. In this research, lime is use as a stabilizer that contains the Ca^+ element to bond bigger particles. Lime used is slaked lime $Ca(OH)_2$. Clay used has liquid limitation (LL) value of 47.33%, plasticity index of 29.88% and CBR value 6.29. The results explain about 10% lime mixture variation gives the optimum stabilized clay with CBR value of 8.75%.

1. Introduction

Based on particle size, the soil is classified into four classifications that are gravel, sand, silt, and clay. In this final project report use is clay. Das (1994) explained that microscopic and sub microscopic (too small to be seen in an ordinary microscope) particles consist the most in clay. These particles are the plaque and constitute mica, clay and other fine minerals. Clay is one type of soils which often stabilized due to its plastic and cohesive properties when wet. Consequently, there is a change in volume influenced by water content and results in the soil to swelling and shrinkage relatively fast. Stabilization is necessary to repair those properties to increase bearing capacity. Soil stabilization is by mixing soil with lime and water in working site to alter soil properties to become a better material for construction as what is allowed in the planning. Lime reacts with soil water and changes its properties, reduces cohesiveness and softness in the soil. When lime and clay mineral reacts, it forms hard gel known as calcium silicate which binds with soil fine-grains or particles. Silica gel reacts by coating, and binding clay particles then covered the pores in the soil and reduces plasticity index in the soil. Plasticity index reduction is due to the increase of plastic limit and decrease of liquid limit.

Clay has plastic properties which are the ability to adjust shape changes in the soil after it is mixed with water. The soil can be liquid, plastic, semi solid or solid depends on the water content in it. Plasticity index is the difference between liquid and plastic limit in the soil:

$$PI = LL - PL$$

Compaction is a process where the air is removed mechanically from the pores, therefore, particles become solid. If the soil is solidified with varied water content, then in certain water value it will reach maximum density (γ_d max). Water value that gives maximum density is called optimum water content (w_{opt}).

where : γ_d = dry unit weight in field (kg/cm^3)
ω = water content (%)
V = mould volume (cm^3)
W = total soil weight in mould (kg)

Sub grade strength depends on the water content in which the higher it is, the smaller the CBR value. Nevertheless, this does not mean that sub grade is solidified better with low water content because of its instability. If resistance is given to the soil, water is absorbed into the soil and decreases CBR value until a constant water level is reached which is known as balanced water content.

2. Material and Methods

- Clay used in the research is obtained from PTPN II, Patumbak Deli Serdang. The soil here is often used as piled soil in road construction, Binjai - Tebing Tinggi highway for instance.
- Lime used as the stabilizer is slaked lime with the composition shown in Table 2.1. Slaked lime, CaO, is a white amorphous material. During storing, CaO transforms slowly to a white powder which is Ca $(OH)_2$ because it reacts with water vapour in the air.

Table 1. Lime chemical composition

No.	Parameter	Result
1.	Silicon (SiO_2)	3.03 %
2.	Aluminium (Al_2O_3)	1.53 %
2.	Iron (Fe_2O_3)	0.54 %
3.	Calcium (CaO)	51.8 %
4.	Magnesium (MgO)	0.81 %

Source: Laboratory of Research and Industrial Standard Center North Sumatera

The research method used is the experimental method in Soil Mechanics Laboratory, Civil Engineering Department University of Sumatera Utara. Observation is done by using two samples; soil and soil with lime addition.

There are 21 samples in total, where one is soil and 20 are the varied mixture of soil with 1 until 20% lime addition.

Testing conducted includes:

1. Soil index properties, includes water content, density, Atterberg value (consistency limit) and grain size distribution or sieve analysis.
2. Slaked lime index properties, include chemical composition, density, Atterberg value (consistency limit) and grain size distribution or sieve analysis.
3. Proctor Standard test to determine the optimal water content and maximum dry density.
4. A laboratory test for Proctor Standard and CBR analysis.
5. Curing time for seven days to each sample to get a homogenized mixture.

Total sample used in the research can be seen in **Table 2.**

Table 2. Total Sample

No	Testing	Soil	Lime Addition	Total Samples
1	Index Properties	1	20	21
2	Proctor Standard	1	20	21
3	CBR	1	20	21

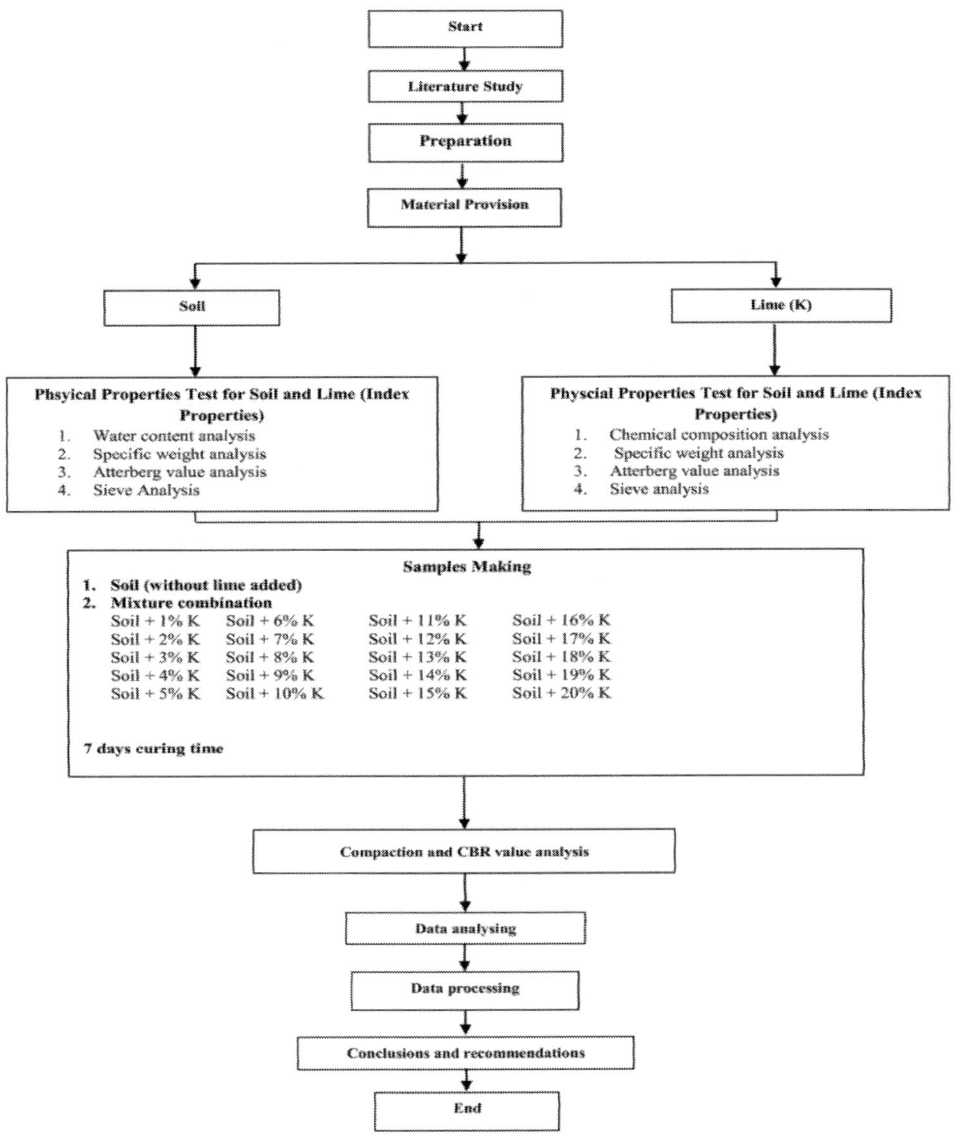

Figure 1. Flowchart Research

IOP Conf. Series: Materials Science and Engineering **309** (2018) 012076 doi:10.1088/1757-899X/309/1/012076

3. Results and Discussions

3.1 Soil Analysis

Physical properties tests for soil are shown in **Table 3.**

Table 3. Soil Physical Tests Data

No.	TEST	RESULT
1	Water Content	34.33%
2	Specific Gravity	2.65
3	Liquid Limit	47.33%
4	Plastic Limit	17.45%
5	Plasticity Index	29.88%
6	Sieve Analysis	48.81%
7	Optimum Moisture Content	21.12%
8	Maximum Dry Density	1.340 gr/cm^3

3.2 Stabilizer Analysis

Physical properties tests for lime are shown in **Table 4.**

Table 4. Lime Index Properties Data

No.	TEST	RESULT
1	Specific Gravity	2.59
2	Liquid Limit	
3	Plastic Limit	Non Plastic
4	Plasticity Index	
5	Sieve Analysis	30.05%

3.3 Physical Properties and Mechanic Tests for Soil with Stabilizer

Test results for physical properties and mechanic test for soil combined with lime as stabilizer are shown in the graphs below.

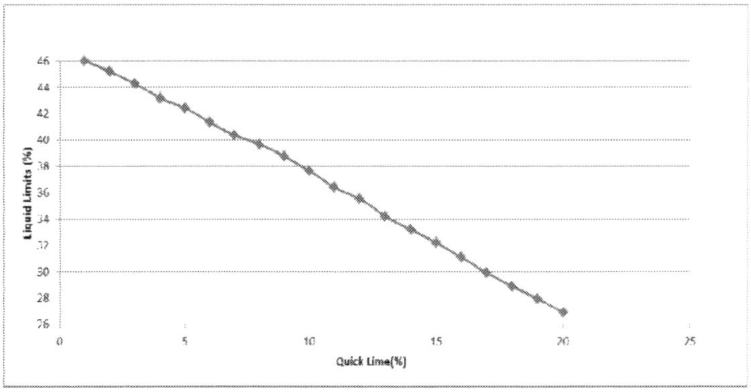

Figure 2. Liquid Limitaion (LL) versus Lime Variation Graph

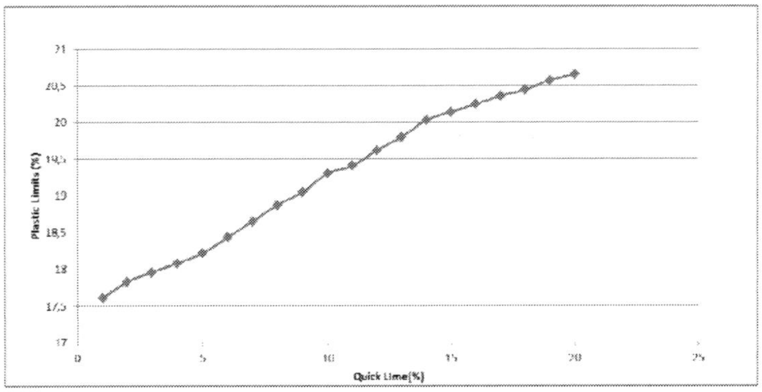

Figure 3. Plastic Limit versus Lime Variation Graph

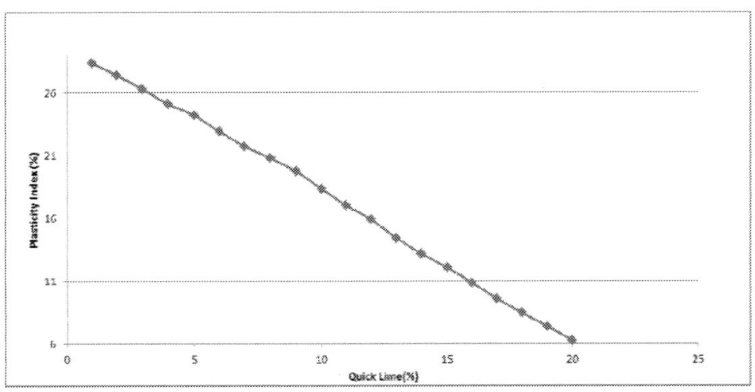

Figure 4. Plasticity Index versus Lime Variation of 1 – 20% in 7 Days Curing Time Graph

In **Figure 2** it is shown that due to lime stabilizer addition, liquid limitation decreases. The bigger the percentage of lime, the lower liquid limitation becomes. In soil, liquid limitation reaches 47.33% whereas the lowest value is 20% K mixture variation with seven days curing time, which is 26.87% and this is because soil underwent lime cementation process where soil becomes bigger grain that decreases attraction force between particles. **Figure 3** shows increases in plastic limitation due to lime addition. For soil, the plastic limitation is 17.45% and keeps on increasing until the highest value is shown with 20% K addition at 20.65%.

Figure 4 shows the decreasing plasticity index due to lime addition. The decrease can reduce soil swelling and shrinking potential. This is due to hydration process in lime added to the soil. These processes strengthen bonds between particles in the soil, which forms bigger and more stable grains. With pores filled, it reduces the risk of seepage in soil-lime mixture that affects the swelling and shrinking potential. Silicate and alumina in lime combined with water to form the paste that binds clay particles and covered the pores. Pores cavity surrounded by water-penetration-proof cementation material makes the mixture last longer towards water absorption and therefore reduces its plastic properties. With the addition of lime, plasticity index keeps decreasing from 29.88% to its lowest point, 6.22% at 20% K mixture at seven days curing time.

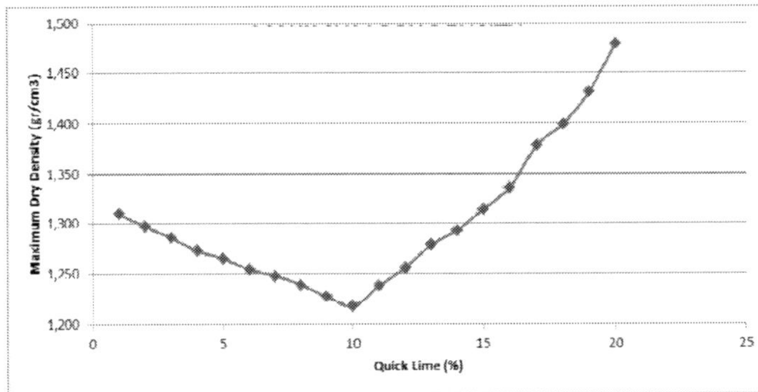

Figure 5. Soil γ_{dmaks} versus 1 – 20% Lime Variation at seven days curing time

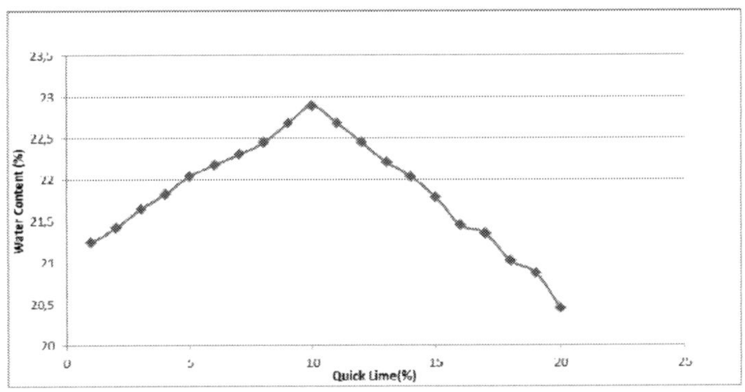

Figure 6. w_{opt} versus 1 – 20% Lime variation at seven days curing time

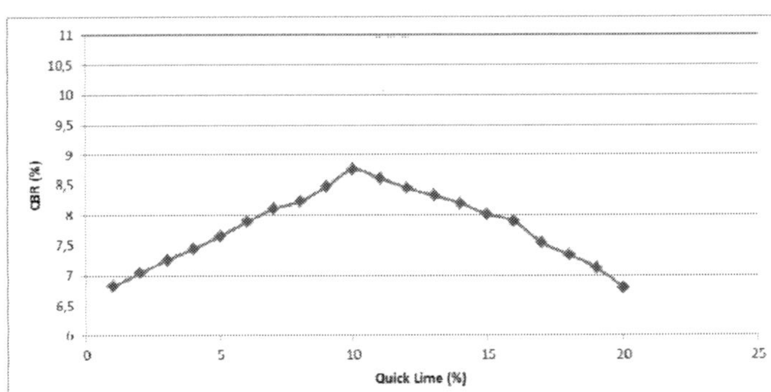

Figure 7. CBR Value versus 1 – 20% lime variation at seven days curing time

In **Figure 5** it can be seen that dry unit weight decreases with lime added, where 10% lime it gives the lowest value 1.218 g/cm^3. Then the value increases afterward with more lime added. The optimum water content in research done in soil is 34.33%, and it increases after lime is added. **Figure 6** shows optimum water content when reaching the maximum dry unit weight and decreases when passing maximum dry unit weight. The graph in figure 3.6 above shows the highest CBR value obtained was at 10% K variation, which is 8.75%. But with further addition, CBR value decreases and this is because soil weight volume reduced because pores are filled with lime mixture. Consequently, penetration reduces in CBR test.

IOP Conf. Series: Materials Science and Engineering **309** (2018) 012076 doi:10.1088/1757-899X/309/1/012076

4. Conclusions

The result shows:

1. Based on USCS classification, the soil sample is considered as CL (Clay-Low Plasticity) which is inorganic clay with low to medium plasticity.
2. Based on AASHTO classification, the soil is considered as a type of A-7-6.
3. Water content analysis gives the water content in soil, which is 34.33%.
4. Specific gravity test shows soil specific weight is 2.65% and that of lime is 2.59.
5. Atterberg value analysis for soil gives liquid limitation (LL) and plasticity index at 47.33% and 29.88% respectively. It is then known:
   - With 20% K added gives the lowest plasticity index (PI), 6.22% with liquid limitation value is 26.87%.
6. Proctor standard test gives soil optimum content at 21.12% and maximum dry unit weight at 1.340 g/cm³, whereas maximum dry unit weight for all mixture is:
   - At 10% variation, dry unit weight is 1.218 g/cm³ and optimum water content is 22.89%.
7. Laboratory CBR value test done gives the highest value for soil is at 6.29%. From the research done the highest CBR value is:
   - At 10% K variation, which gives 8.75
8. From the result at no. 7 above, it can be concluded that 10% K variation gives the most stabilized clay.

References

[1] Hardiyatmo H C 2010 *Stabilisasi Tanah untuk Pekerasan Jalan* Yogyakarta: Gajah Mada University Press

[2] Hastuty I P, Sembiring I S, Abidin M I 2017 The Utilization of Volcanic Ash and High Rusk Ash as Material Stabilization in Clay by Unconfined Compression Test (UCT) and California Bearing Ratio (CBR) *IOP Conf Series* **180**

[3] Hastuty I P, Roesyanto, Jeriko 2017 A Study of the Effectiveness of the Use of Cement and Bottom Ash Towards The Stability of Clay in Term of UCT Value *Science Direct, Procedia Engineering* **171** 484 – 491

[4] Shankar, S 2016 Mount Sinabung Volcano Eruption: Over 1,200 Villagers In Indonesia Forced To Evacuate *Jurnal Online* http://www.ibtimes.com/mount-sinabung-volcano-eruption-over-1200-villagers-indonesia-forced-evacuate-1968566

[5] Margaretha, Irene, dan Roesyanto *Stabilitas Tanah Lempung dengan Menggunakan Semen Portland Tipe 1 dan abu Vulkanik dengan Pengujian Kuat tekan Bebas* Medan: Universitas Sumatra Utara Jurnal Online http://repository.usu.ac.id/handle/123456789/57136

[6] Jatmoko H D *Tinjauan Sifat Plastisitas Tanah Lempung Yang Distabilisasi Dengan Kapur.* Program Studi Teknik Sipil, Fakultas Teknik, Universitas Muhammadiyah Purworejo. Jurnal Online http://download.portalgaruda.org/article.php?article=9559&val=61

IOP Conf. Series: Materials Science and Engineering **309** (2018) 012077 doi:10.1088/1757-899X/309/1/012077

The effect of shear wall location in resisting earthquake

J Tarigan*[1], J Manggala[1], T Sitorus[1]

[1]Department of Civil Engineering, Faculty of Engineering, Universitas Sumatera Utara

*Email: johannes.tarigan@usu.ac.id

Abstract. Shear wall is one of lateral resisting structure which is used commonly. Shear wall gives high stiffness to the structure so as the structure will be stable. Applying shear wall can effectively reduce the displacement and story-drift of the structure. This will reduce the destruction comes from lateral loads such as an earthquake. Earlier studies showed that shear wall gives different performance based on its position in structures. In this paper, seismic analysis has been performed using response spectrum method for different Model of structures; they are the open frame, the shear wall at core symmetrically, the shear wall at periphery symmetrically, and the shear wall at periphery asymmetrically. The results are observed by comparing the displacement and story-drift. Based on the analysis, the placement of shear wall at the core of structure symmetrically gives the best performance to reduce the displacement and story-drift. It can reduce the displacement up to 61.16% (X-dir) and 70.60% (Y-dir). The placement of shear wall at periphery symmetrically will reduce the displacement up to 53.85% (X-dir) and 47.87% (Y-dir) while the placement of shear wall at periphery asymmetrically reducing the displacement up to 59.42% (X-dir) and 66.99% (Y-dir).

1. Introduction

RC multi-story buildings are designed for resisting both the vertical and horizontal load. A taller structure will undergo the greater lateral load. Every structural Engineer is met with the problem of giving sufficient strength and stability of these tall buildings against lateral load thus the effect of lateral loads like wind load, earthquake load and blast forces are attaining escalating importance. One of the solution to give better stability for the structure is the utilization of shear wall [1].

Shear wall is one of the most commonly used lateral load resisting systems in buildings. Shear wall has high plane stiffness and strength which can be simultaneously resist large horizontal loads and support gravity loads, which significantly reduces the lateral sway of the building and thereby reduces damage to structures and its contents. When the shear wall is strong enough, it will transfer the horizontal load to the next element in the load path below them such as floors, other shear walls, slabs or footings. Shear wall also provides lateral stiffness to prevent the roof or floor above from large lateral sway. When shear wall is stiff enough, they will prevent floor and roof from moving off their supports. Also, buildings that are sufficiently stiff will usually suffer less non structural damage.

Since shear wall carry large horizontal earthquake forces, the overturning effects on it is large. Shear wall in buildings must be symmetrically located in plan to reduce ill-effects of twist in buildings. When shear wall are placed in advantageous positions in the building, they can form an efficient lateral force resisting system by reducing lateral displacements under earthquake loads. Therefore it is very necessary to determine effective, efficient and ideal location of shear wall.

Tuppad and Fernandes [2] presented a study of the effect of shear wall placement in the 10-story RC structure at 6 alternative placement. The structure located in India at earthquake zone V. Analysis of earthquake using static equivalent method by ETABS 2015. It is found that the story displacement of the structure without shear wall shows maximum displacement compared to the other Model having shear wall. Placing shear wall near the core of the building was the best location to reduce the story displacement. It also found that by providing shear wall to the high rise building, structural behavior will be affected to a great extent and also the stiffness and the strength of the building will be increased.

Agrawal and Charkha [3] considered 5 different Models of shear wall in 25-story building in India at seismic zone V. Building consists of 7 bays of 7.5M in X- direction and 5 bays of 6.5 M in Y-direction. Earthquake loading calculated using ETABS and applied to the mass centre of the structure. By the results, it is clear that by shifting the shear wall location gave the significant effect on deflection. It is observed that placing shear wall away from centre of gravity resulted in increase in most of the member forces. With the increase in eccentricity, the building shows non-uniform movement of right and left edges of roof due to torsion and induces excessive moment and forces in member.

Based on the studies above, this paper discussed the effect of shear wall location in 4-story building in Pekanbaru. It will analyze 4 Models of structure by shifting the location of shear wall. The objectives of this study is to analyze the effect of shear wall to the structure and find the optimum location of shear wall to reduce the lateral displacement of structure. The result will be displayed in term of natural period, displacement, and story-drift.

2. Modelling and method

For the analysis purpose, the Model of RC structure (G+4) story and 79.85 m x19.85 m plan area has selected which is located in Pekanbaru. The ground to the (G+3) story height is 4 m and the roof height is 9 m. The grade of the concrete used is M20 and for the structural steel is Fe 400. Building description is shown in Table 1.

Table 1. Building description

Particular	RC Structure
Number of Story	G+4
Height	3 m (Ground to G+3) 9 m (G+4)
Grade of Concrete	M20
Grade of Steel	Fe 400
Shear Wall Thickness	0.3 m
Seismic Design Category	D (as per SNI 1726:2012)
Type of Soil	Medium Soil
Importance Factor, I_e	1.25

Structure Model can be seen in Figure 1. Model 1 is RC structure without shear wall. Shear wall the added to the structure in various location as shown in Figure 2. In Model 2, shear wall located near the core of the structure symmetrically, for Model 3 shear wall located at the peripheral of the structure symmetrically and for Model 4 shear wall located at outer location of structure asymmetrically.

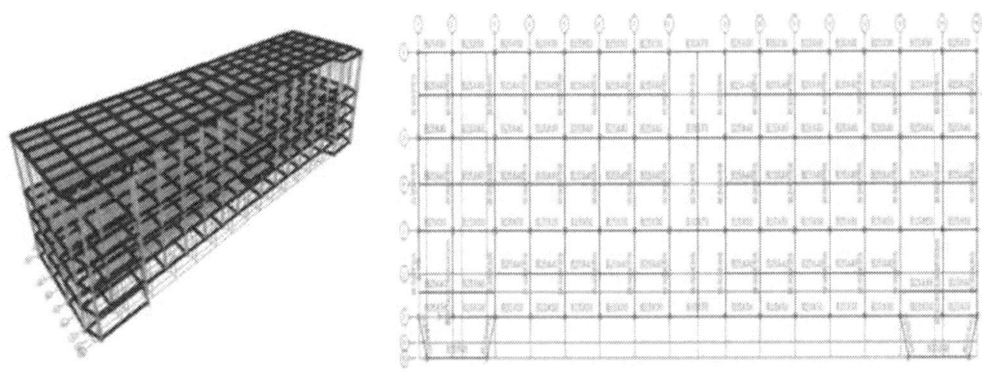

Figure 1. Structure Model 1 (without shear wall)

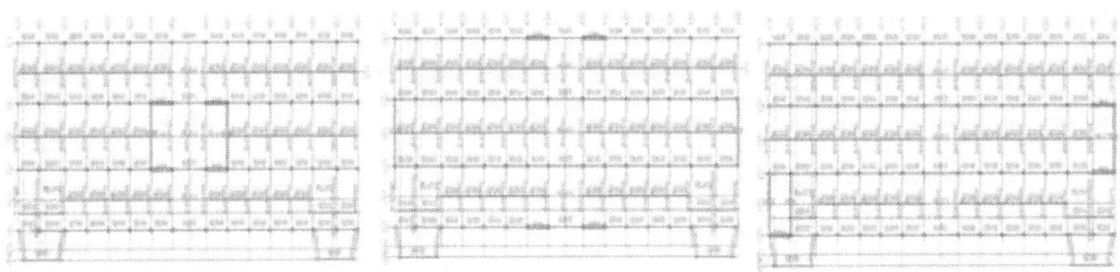

a) Model 2 b) Model 3 c) Model 4

Figure 2. Structures with shear wall; a) Model 2 structure (shear wall at core symmetrically), b) Model 3 structure (shear wall at peripheral symmetrically), and c) Model 4 structure (Shear wall at peripheral asymmetrically)

2.1 Lateral loading

Lateral loading consist of earthquake loading which has been calculated by program and using the response spectrum analysis. Earthquake analysis completed by using the rules in code SNI 1726: 2012 (based of IBC 2009) [4, 5, 6]. By inputted the spectral response curve, program will calculate the earthquake load. For Pekanbaru, the value of acceleration response, $S_s = 0.435$ and $S_1 = 0.273$, then the spectral response curve can be seen in Figure 3.

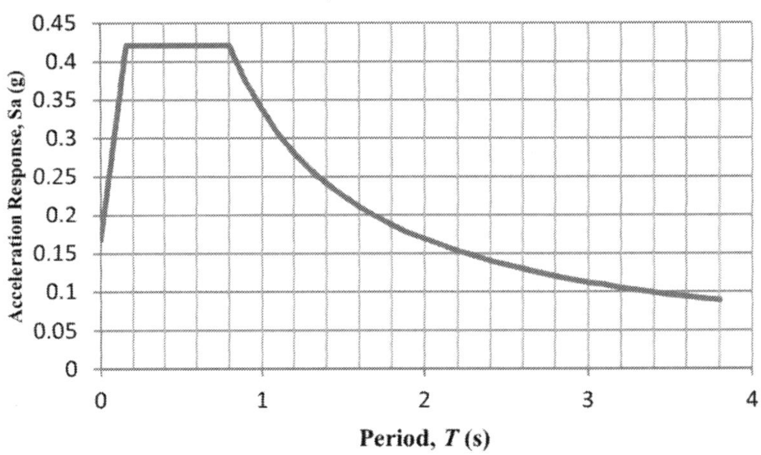

Figure 3. Spectral response design for Pekanbaru

3. Results and discussion

Result obtained from the analysis are recorded in tabular and graphic form for the 4 Model of structures for comparison of maximum natural period, displacement, and story-drift.

3.1 Maximum natural period

The natural period of all Model structures as shown in Table 2.

Table 2. Natural period of structures

Model	Tmax	Reduction Percentage (%)
1	1.690	-
2	0.669	60.41
3	0.751	55.55
4	0.694	58.90

It has been found that the maximum natural period is in Model 1 structure without shear wall. Shear wall gives high stiffness to the structure that can reduce the period of structure. It has been found that the Model 2, Model 3, and Model 4 shows the shorter natural period of structures compared to the Model 1. Model 2 shows the shortest period among the other Models.

3.2 Displacement

Lateral displacement of all Model structures as shown in Table 3 and Figure 4.

Table 3. Lateral displacement of structures

Story	X-Direction				Y-Direction			
	Model 1	Model 2	Model 3	Model 4	Model 1	Model 2	Model 3	Model 4
	mm	mm	mm	mm	mm	mm	mm	mm
Roof	229.68	89.2	106	93.2	221.76	65.2	115.6	73.2
Story 3	113.96	44.8	49.6	46	103.84	32.8	54.4	42
Story 2	79.64	26	27.2	26.4	72.6	19.6	30.8	26.8
Story 1	35.2	9.6	9.2	9.6	32.56	7.6	11.2	11.2

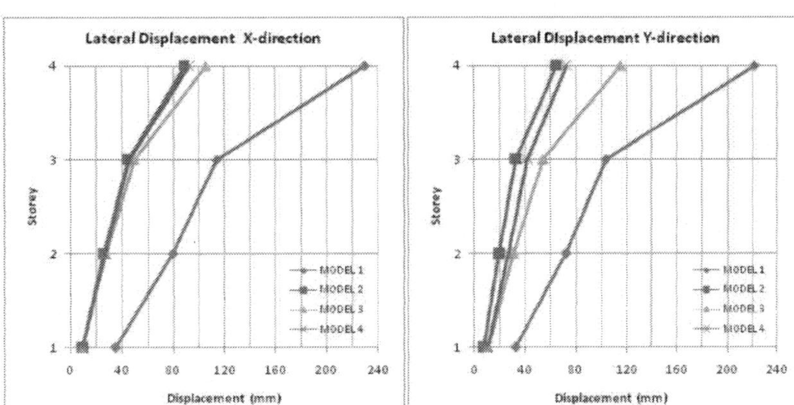

Figure 4. Comparison of displacement structures

From the above results it can be observed that the maximum displacement occurred in Model 1 structures. Shear wall can reduces lateral displacement of structures as it can see in Model that uses shear wall. The maximum reduction of displacement is obtained for Model 2 structure.

3.3 Story-drift

Story-drift of all Model structures as shown in Table 4.

Table 4. Story-drift of structures

Story	X-Direction				Y-Direction				Allowable Drift
	Model 1	Model 2	Model 3	Model 4	Model 1	Model 2	Model 3	Model 4	
	mm	mm	mm	mm	mm	mm	mm	mm	mm
Roof	115.72	44.4	56.4	47.2	117.92	32.4	61.2	31.2	103.85
Story 3	34.32	18.8	22.4	19.6	31.24	13.2	23.6	15.2	46.15
Story 2	44.44	16.4	18	16.8	40.04	12	19.6	15.6	46.15
Story 1	35.2	9.6	9.2	9.6	32.56	7.6	11.2	11.2	46.15

From the above result it can be found that the maximum story-drift occurred in Model 1 structure. It is found that the drift in the roof even higher than the allowable drift which is permitted by SNI 1726:2012 (Figure 5). Utilization of shear wall reduces story-drift below the allowable drift. It is found that Model 2 structure gives the optimum performance to reduce the story-drift.

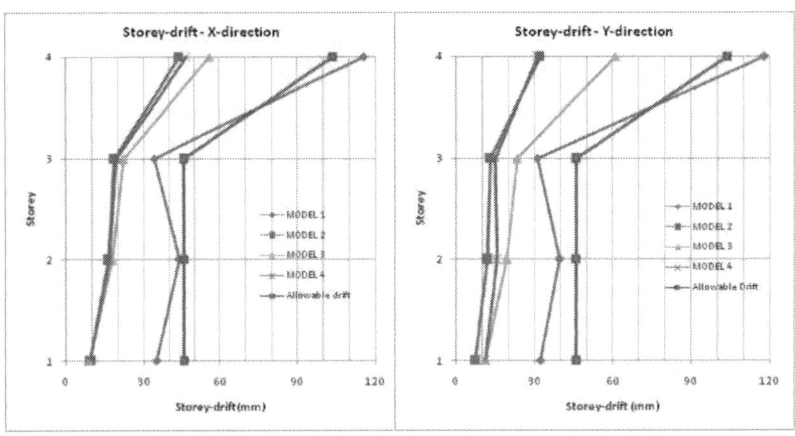

Figure 5. Comparison of story-drift structures

4. Conclusions

Based on this study, it has been observed that the utilization of shear wall can contribute in increasing stiffness of structure. It reduces the natural period of structure, lateral displacement and story-drift significantly. Position of shear wall need to be considered carefully because it gives difference performance to resisting earthquake load. In this investigation, it is found that the optimum location for the structure is Model 2 structure (shear wall at the core symmetrically).

References

[1] Schodek D 1999 *Struktur Edisi kedua Jakarta Erlangga*

[2] Fernandes R and Tuppad S 2015 Optimum location of shear wall in a multi-storey building subjected to seismic behavior using genetic algorithm *International Research Journal of Engineering and Technology* **02** 236-40

[3] Agrawal A and Charkha S 2012 Study of optimizing configuration of multi-storey building subjected to lateral load by changing shear wall location *Proceeding for International Conference on Advances in Architecture and Civil Engineering* **1** 287-289

[4] Badan Standardisasi Nasional 2012 *Tata Cara Perencanaan Ketahanan Gempa untuk Struktur Bangunan dan Non Gedung SNI 03-1726-2012*

[5] Badan Standardisasi Nasional 1989 *Pedoman Perencanaan Pembebanan untuk Rumah dan Gedung SNI 03-1727-1989*
[6] Badan Standardisasi Nasional 2002 *Tata Cara Perhitungan Struktur beton Untuk Bangunan Gedung SNI 03-2847-2002*

IOP Conf. Series: Materials Science and Engineering **309** (2018) 012078 doi:10.1088/1757-899X/309/1/012078

Supercapacitor performance evaluation in replacing battery based on charging and discharging current characteristics

A Sani*, S Siahaan, N Mubarakah, Suherman

Electrical Engineering Department, Universitas Sumatera Utara, Indonesia

*E-mail: armansani.usu@gmail.com

Abstract. Supercapacitor is a new device of energy storage, which has much difference between ordinary capacitors and batteries. Supercapacitor have higher capacitance and energy density than regular capacitors. The supercapacitor also has a fast charging time, as well as a long life. To be used as a battery replacement please note the internal parameters of the battery to be replaced. In this paper conducted a simulation study to utilize supercapacitor as a replacement battery. The internal parameters of the battery and the supercapacitor are obtained based on the characteristics of charging and discharging current using a predefined equivalent circuit model. The battery to be replaced is a 12-volt lead-acid type, 6.5 Ah which is used on motorcycles with 6A charging and discharging currents. Super capacitor replacement capacitor is a capacity of 1600F, 2.7V which is connected in series as many as 6 pieces with 16.2 volt terminal voltage and charging current 12A. To obtain the same supercapacitor characteristic as the battery characteristic to be replaced, modification of its internal parameters is made. The results show that the super-capacitor can replace the battery function for 1000 seconds.

1. Introduction

Various energy storage technologies have been developed for various application, including supercapacitor. Basically, supercapacitor is modeled and analyzed as regular capacitor. However, its physical size and capacity are larger. This makes supercapacitor has higher energy density than the regular one and higher charge density than battery. Furthermore, its charging and discharging time are faster with longer lifetime [1]. As results, supercapacitor can be used as energy storage for various applications.

In order to use supercapacitor as a battery replacement, internal parameters should match with the equivalent battery. This parameters are approximated based on a predetermined equivalent circuit. If necessary, these parameters are modified accordingly [2]. This paper employs a simulation approach to determine the energy supply duration based on charging and discharging current characteristics.

2. Internal Parameters Modelling

In order to obtain the internal parameters of a supercapacitor, an equivalent circuit model is employed [3] as depicted in Figure 1.

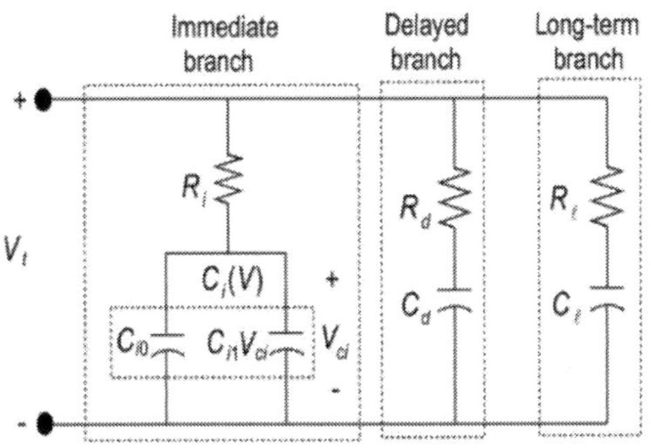

Figure 1. Equivalent Circuit for Battery and Supercapacitor

The model consists of three branches of RC circuits arranged in parallel manner [4]. Each branch has different time rate. Immediate branch has component of Ri, C_{i0} and C_{i1} with a very short charging response. Delayed branch has components of Rd and Cd with response time several minutes. Finally, long-term branch with R_l and C_l determines the characteristics of battery and supercapacitor with response time longer than 10 minutes. In order to obtain equivalent parameters, each branch should be analysed one by one by consider that voltage is time-dependant [4].

2.1. Immediate Branch

Immediate branch parameters are identified by performing charging on both devices by using a constant current. The parameter is then calculated from current and voltage relation on the branch by using Equation 1.

$$Ri = \frac{V1}{Ich} \quad (1)$$

where V_1 is voltage at t1 after charging current is supplied. Variable Ich is the constant charging current. V_1 is the reference point to determine C_{i0} and C_{i1}. C_{i0} is calculated by using voltage difference, ΔV so that $V_2 = V_1 + \Delta V$. The charging time (t_2) is calculated to make the voltage across the battery or supercapacitor achieve V_2, then the time difference is $\Delta t = t_2 - t_1$. C_{i0} is calculated by using Equation 2.

$$Cio = Ich\frac{\Delta t}{\Delta V} \quad (2)$$

Charging time (t_3) is the time required to achieve fully charged voltage (V_3), so that C_{i1} be calculated using Equation 3.

$$C_{i1} = \frac{2}{V4}\left(\frac{Qtot}{V4} - Cio\right) \quad (3)$$

where $Q_{tot} = Ich \times (t_4 - t_1)$ and $t_4 = t_3 + t_1$. V_4 is the remaining voltage on battery or supercapacitor after discharging process for duration of t_1.

2.2. Delayed Branch

Rd and Cd are calculated for delayed branch determination. In order to obtain Rd, discharging time for both battery and capacitor should be calculated which is t_5 for given voltage V_5, where $V_5 = V_4 - \Delta V$. Rd is calculated by using:

$$Rd = \frac{(V4 - \frac{\Delta V}{2}) * \Delta t}{Cdiff * \Delta V} \quad (4)$$

where $C_{diff} = C_{i0} + (C_{i1} \times V_3)$. Afterward, t_6 is calculated by using $t_6 = t_5 + 3\,(Rd \times C_{il})$, Cd is obtained by using Equation 5.

$$Cd = \frac{Qtot}{V6} - \left(Ci0 + \left(\frac{Ci1}{2} * V6\right)\right) \quad (5)$$

2.3. Long-term Branch

For *long-term branch* with $V_7 = V_6 - \Delta V$, and t_7 is capacitor or battery discharging time where the voltage decreases to V_7, then, Rl is calculated by using Equation 6.

$$Rl = \frac{\left(V6 - \frac{\Delta V}{2}\right) * \Delta t}{Cdiff * \Delta V} \quad (6)$$

where $\Delta t = t_7 - t_6$.

At the end of measurement stage, C_l is calculated by charging the capacitor for 30 minutes. C_l is calculated by using Equation 7.

$$Cl = \frac{Qtot}{V8} - \left(Ci0 + \frac{Ci1}{2} * V8 \right) \quad (7)$$

where V_8 is the capacitor voltage after being recharged for 30 minutes.

3. Method

In order to evaluate the supercapacitor performance, this paper employs Matlab by developing the replacement circuit to replace charging discharging branch circuits. The battery intended to be replaces is a motor cycle battery: lead-acid 6.5Ah, 12 V with equivalent circuit depicted in Figure 2a and internal parameters in Figure 2b.

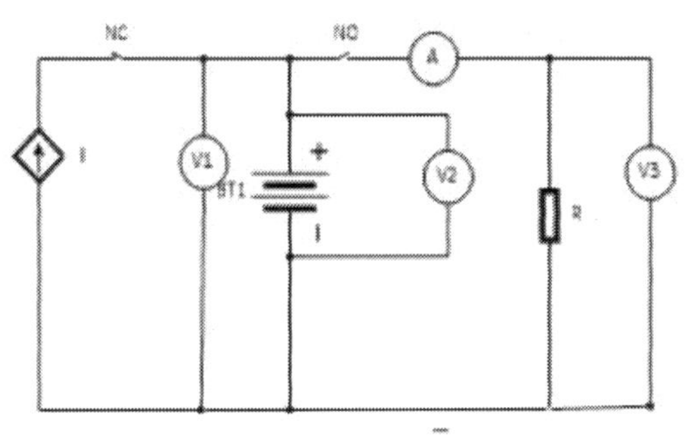

(a) Battery circuit

Parameter	Battery, *lead-acid* 6.5Ah, 12 V
R_i	3,609 m?
C_{i1}	405.612 F
Rd	2,55?
Cd	52.86 F
R_l	0,84?
C_l	24.24 F

(b) Battery parameters

Figure 2. The battery circuit and parameters

Discharging circuit uses 4 rectifiers to ensure that charging current is DC. There are 6 series supercapacitors to achieve similar or larger voltage across the terminal of supercapacitor as the replaced battery. The parameters are calculated using Equation 8 to 11 [4].

IOP Conf. Series: Materials Science and Engineering **309** (2018) 012078 doi:10.1088/1757-899X/309/1/012078

$$R_{\text{series}} = R_x(I) \times S^{n_r} \tag{8}$$

$$C_{\text{series}} = C_x(I) \times S^{n_c}. \tag{9}$$

$$R_{\text{parallel}} = R_x(I) \times P^{n_r} \tag{10}$$

$$C_{\text{parallel}} = C_x(I) \times P^{n_c} \tag{11}$$

where P is number of paralleled supercapacitor, S is number of serried supercapacitor, Nr and Nc are resistor and capacitor correction factors.

4. Result and Discussions

The modelled analysed circuit is on Figure 3a. After carefully consider the internal parameters of the battery, NessCap 1600 F capacitor is chosen. The internal parameters of asingle capacitor is depicted in Figure 3b, and the 6 series supercapacitor is depicted in Figure 3c.

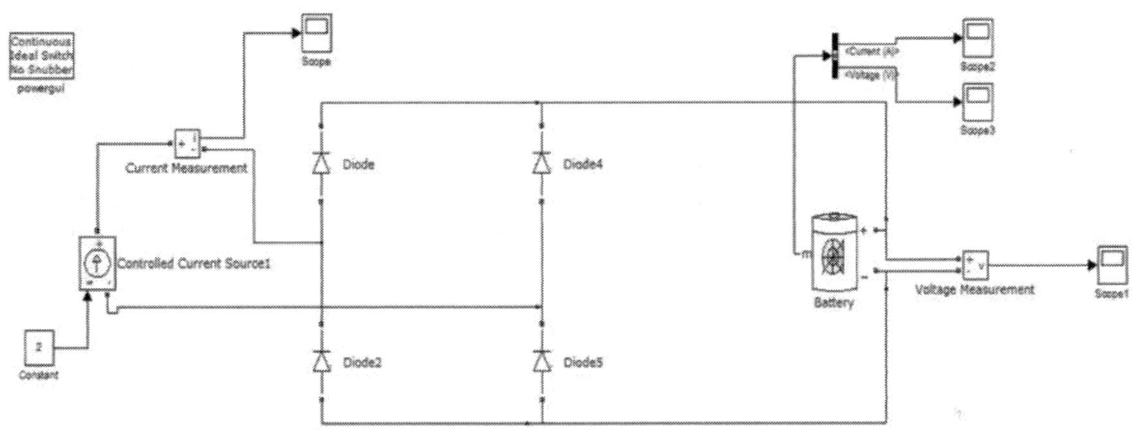

(a) Simulated charging circuit

Parameter	Baterai (6.5 Ah, 12 V)
Ri	4,11 m?
Ci1	101,53F
Rd	3,3 ?
Cd	393,65 F
Rl	1,289 ?
Cl	16,469 F

(b) single supercapacitor

Parameter	Super kapasitor (NessCap1600F, 2.7V)
Ri	0,94 m?
Ci1	4049 F
Rd	0,37 ?
Cd	254,8 F
Rl	0,09 ?
Cl	182,7 F

(c) series supercapacitor

Figure 3. The evaluation model and supercapacitor internal parameters

Increasing series resistor within capacitors changes discharging characteristic where supercapacitor discharging time getting slower. As result, the capacitance of series capacitor is larger.

In order to obtain supercapacitor performance in replacing battery, the discharging simulations are performed by using a model depicted in Figure 4.

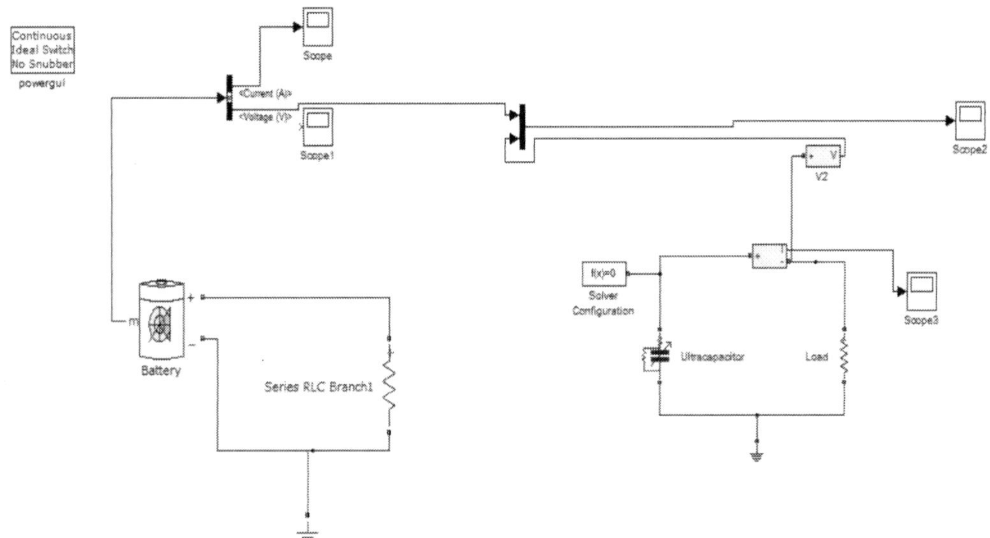

Figure 4. Simulated charging circuit

The discharging voltages are shown in Figure 5. Battery sustains 1800 second with 6 A constant current when connected to a motorcycle starter circuit. On the other hand, supercapacitor experiences rapid discharging and sustain only about 1000 s to supply at least 12 Volt.

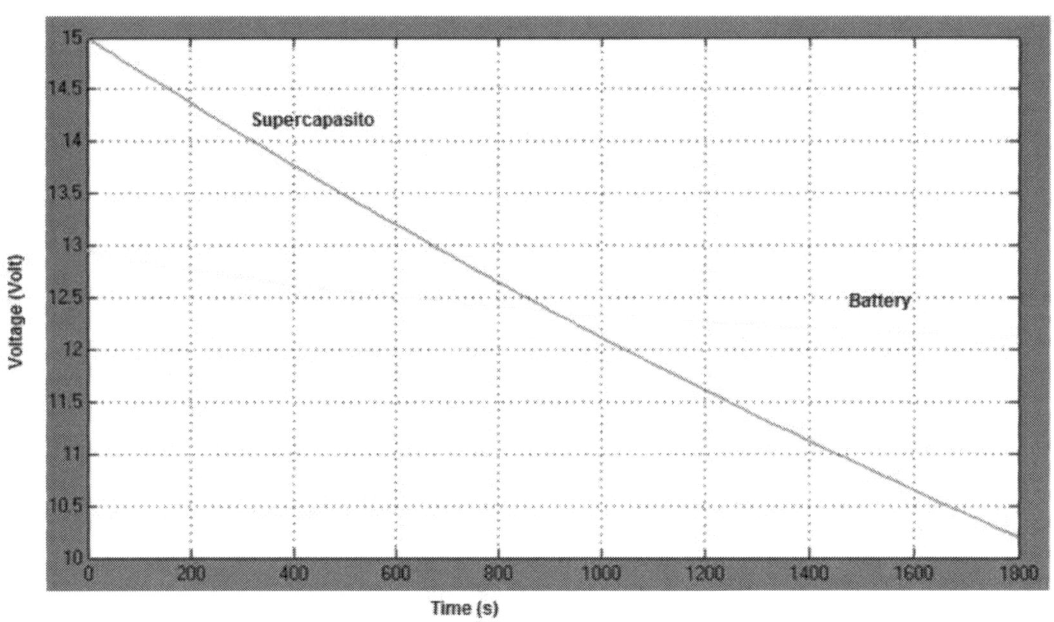

Figure 5. Discharging graph

5. Conclusion

This paper has shown that a *lead-acid* 12 volt, 6.5Ah battery with R_{i1} 4,11 mΩ, Rd 3,3 Ω, and R_l 1.289 Ω C_{i1} =Cd = 393,65 F, and C_l 16,469 F can be replaced by using 6 series capacitor NessCap 1600F. The 6 series 1600F, 2.7V capacitor sustains 1000 s supplying a 6 A, 12 Volt motor cycle starter. The battery itself supplies the same current and voltage for about 1800 s. Supercapacitor discharging time is more rapidly than the battery discharging time. Future works may deal with solutions to enlarge discharging time of supercapacitor.

References

[1] Patel K R and Rushi R D 2012 Calculation of Internal Parameters of Super Capacitor to Replace Battery by Using Charging and Discharging Characteristics *International Journal of Engineering and Innovative Technology* (IJEIT) **2** 1-5

[2] Zubieta L and Bonert R 2000 Characterization of double-layer capacitors for power electronics applications *IEEE Trans. Industry Applications* **36** 199-205

[3] Woollard B 1993 *Practical Electronics* McGraw-Hill Book Company Limited UK England **2**

[4] Hsing L C and Hsing H S 2013 Prediction of Equivalent-Circuit Parameters for Double-Layer Capacitors Module *IEEE Trans Industry On Energy Conversion* **7** 913-920.

[5] Yang H, Zhang Y. Characterization of supercapacitor models for analyzing supercapacitors connected to constant power elements. Journal of Power Sources. 2016 Apr 30;**312**:165-71.

[6] Piórkowski P. Zastosowanie superkondensatorów do rozruchu silników spalinowych w trudnych warunkach. Logistyka. 2015(**3**, CD 1):3918-27.

IOP Conf. Series: Materials Science and Engineering **309** (2018) 012079 doi:10.1088/1757-899X/309/1/012079

Influence of the weighing bar position in vessel on measurement of cement's particle size distribution by using the buoyancy weighing-bar method

R Tambun*, R O Sihombing, A Simanjuntak, F Hanum
Department of Chemical Engineering, Universitas Sumatera Utara, Padang Bulan, Medan 20155, Indonesia

*E-mail: rondang@usu.ac.id

Abstract. The buoyancy weighing-bar method is a new simple and cost-effective method to determine the particle size distribution both settling and floating particle. In this method, the density change in a suspension due to particle migration is measured by weighing buoyancy against a weighing–bar hung in the suspension, and then the particle size distribution is calculated using the length of the bar and the time–course change in the mass of the bar. The apparatus of this method consists of a weighing–bar and an analytical balance with a hook for under-floor weighing. The weighing bar is used to detect the density change in suspension. In this study we investigate the influences of position of weighing bar in vessel on settling particle size distribution measurements of cement by using the buoyancy weighing-bar method. The vessel used in this experiment is graduated cylinder with the diameter of 65 mm and the position of weighing bar is in center and off center of vessel. The diameter of weighing bar in this experiment is 10 mm, and the kerosene is used as a dispersion liquids. The results obtained show that the positions of weighing bar in vessel have no significant effect on determination the cement's particle size distribution by using buoyancy weighing-bar method, and the results obtained are comparable to those measured by using settling balance method.

1. Introduction

The buoyancy weighing-bar method (BWM) is a new method in determination the particle size distribution both settling particle and floating particle [1–2]. The BWM has been proven capable to measure size distribution of some particles and the results are proportional to those measured by sedimentation balance method, centifugal sedimentation method, laser diffraction method, Coulter counter method and microscopy method [3–7]. In addition, to estimate the size distribution measurement of fine particles, the BWM could be combinated with the Rosin-Rammler equation [8]. Besides, the particle size distribution also can be measured in Allen region by using BWM [9]. In this research, the BWM is used to determine particle size distribution of cement by investigating the influences of weighing bar position in vessel. The principal of this method that the density change in a suspension due to particle migration is measured by weighing buoyancy against a weighing–bar hung in the suspension, and the particle size distribution is calculated using the length of the bar and the time–course change in the mass of the bar [1–2, 10–11]. In this study, the initial buoyant mass of the weighing-bar depends on the particles between the top and bottom of the weighing-bar in kerosene as suspension.

IOP Conf. Series: Materials Science and Engineering **309** (2018) 012079 doi:10.1088/1757-899X/309/1/012079

2. Material and Methods

Figure 1 schematically illustrates the experiment. The weighing–tools is weighing–bar (diameter: 10 mm, length: 210.0 mm, submerged length: 200.0 mm) which is composed of aluminum (density: 2700 kg/m^3). The analytical balance (minimum readout mass 0.1 mg) have a below–balance–weighing hook for hanging measurement. The sample material is cement (density: 2500 kg/m^3). The particle suspension is placed in a vessel. Vessel used in this experiment is graduated cylinders with diameter of 65 mm and volume of 1000 ml. The position of weighing bar are in center and off center of vessel. In this experiment, kerosene is used as the dispersion liquids. The suspensions have a solid concentration of 10 kg/m^3 (ca. 1 wt.%) [12]. To prepare a suspension, a 1000 ml liquid and the particles to be tested are mixed in a glass cylinder. Using a hanging wire, which does not extend due to the weight of the weighing-bar, the weighing-bar is hung from the analytical balance. The room temperature is approximately 298 K. After thoroughly stirring the suspension using an agitator, the weighing-bar is set with the balance. The measuring data, which consist of time and the corresponding mass of the bar $_B$, are recorded. The measuring time is less than two hours and the data are collected every 60–second intervals. After the measurement, the particle size distribution is calculated based on the theory [1-2, 10-11]. As comparison method, the particle size distributions are also measured by using the settling balance method.

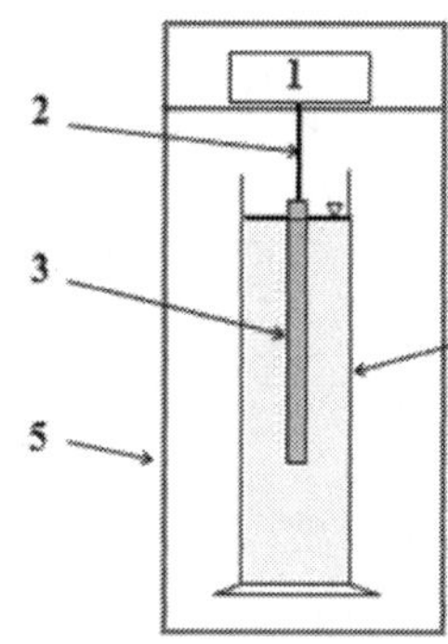

1. Analytical balance 4. Measuring glass cylinder
2. Hanging wire 5. Thermal insulation vessel
3. Weighing-bar

Figure 1. Illustration of the experimental apparatus

3. Results and Discussion

Figure 2 shows the change with time in the apparent mass of weighing-bar the when the position of weighing bar in the center of vessel.

The figure 2 shows that the apparent mass of the weighing-bar increase until all the cement particles settle below the lower end of the weighing-bar, and then the apparent mass of the weighing-bar become constant. The change in the apparent mass is due to the change in the buoyant mass against the weighing-bar as well as particle settling.

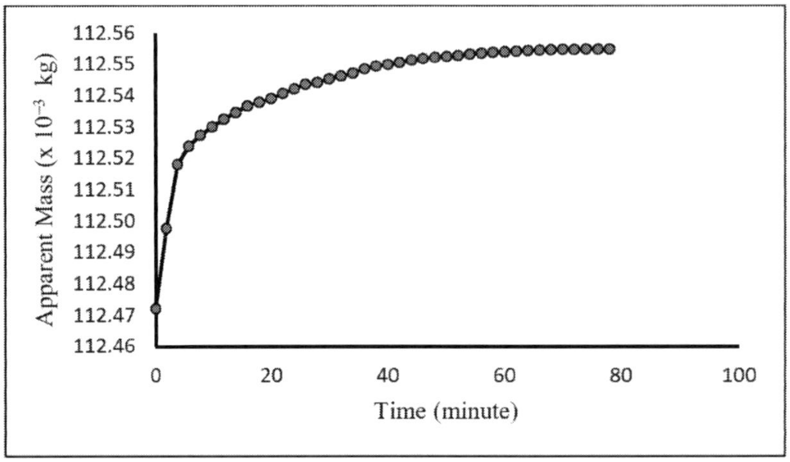

Figure 2. Apparent mass of the weighing-bar as a function of time when the position of weighing bar in the center of vessel

Figure 3 shows the change with time in the apparent mass of weighing-bar the when the position of weighing bar in the off center of vessel. Similar to figure 2, the figure 3 shows that the apparent mass of the weighing-bar increase until all the cement particles settle below the lower end of the weighing-bar, and then the apparent mass of the weighing-bar become constant. The change in the apparent mass is due to the change in the buoyant mass against the weighing-bar as well as particle settling.

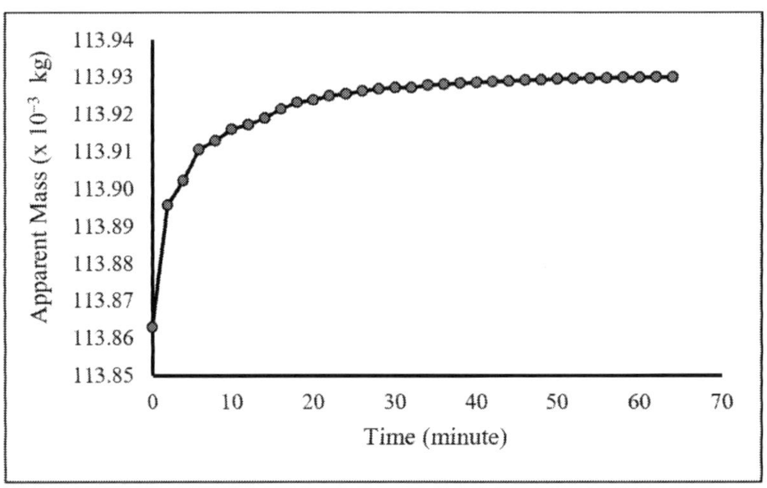

Figure 3. Apparent mass of the weighing-bar as a function of time when the position of weighing bar in the off center of vessel

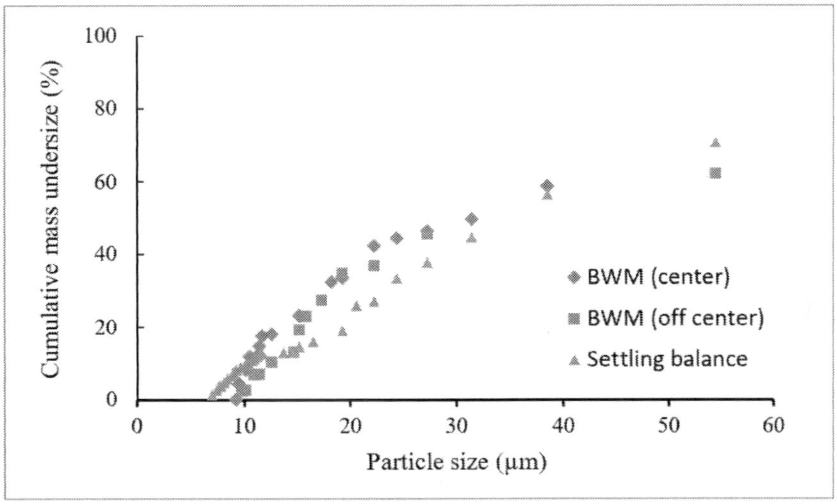

Figure 4. Particle size distributions measurement of cement using kerosene as liquid

Figure 4 shows the influence of the weighing bar position on settling particle size distribution measurements of cement by using BWM. In this experiment, the liquid used is kerosene and the samples are sieved by using sieve tray 80 mesh, but does not pass through sieve tray 140 mesh. The results show that the particle size distribution measured using the weighing bar in the center of vessel is simililar to that the particle size distribution measured using the weighing bar in the off center of vessel, and the results are comparable to those measured by settling balance method. Hence, the location of the weighing bar does not influence the particle size distributions of cement by using kerosene as liquid.

4. Conclusions

The study investigates the influences of position of weighing bar in vessel on settling particle size distribution measurements of cement by using the BWM. The conclusions of this study are:

1. The particle size distribution of cement could be measured by BWM using kerosene as liquid, and the particle size distribution obtained are proportional to those measured by settling balance method.
2. The particle size distribution measured by BWM using the weighing bar in the center of vessel is close to that the particle size distribution measured using the weighing bar in the off center of vessel.

Acknowledgement

The authors would like to thank Ministry of Research, Technology, and Higher Education of Republic Indonesia for Applied Research Grant 2017.

References

[1]. Motoi T, Ohira Y, Obata E 2010 Measurement of the floating particle size distribution by buoyancy weighing–bar method *Powder Technology* **201** pp 283–288

[2]. Obata E, Ohira Y, Ohta M 2009 New measurement of particle size distribution by buoyancy weighing–bar method *Powder Technology* **196** pp 163–168

[3]. Arakawa M, Shimomura G, Imamura A, Yazawa N, Yokoyama T, Kaya N 1984 A new apparatus for measuring particle size distribution based on centrifugal sedimentation, *Journal of The Society of Materials Science of Japan* **33** pp 1141–1145

[4]. Fukui K, Yoshida H, Shiba M, Tokunaga Y 2000 Investigation about data reduction and sedimentation distance of sedimentation balance method *Journal of Chemical Engineering of Japan* **33** pp 393–399

[5]. Kuriyama M, Tokanai H, Harada E 2000 Maximum stable drop size of pseudoplastic dispersed–phase in agitation dispersion *Kagaku Kogaku Ronbunshu* **26** pp 745–748

[6]. Minoshima M, Matsushima K, Shinohara K 2005 Experimental Study on Size Distribution of Granules Prepared by Spray Drying: The case of a dispersed slurry containing binder *Kagaku Kogaku Ronbunshu* **31** pp 102–107

[7]. Ohira Y, Takahashi H, Takahashi M, Ando K 2004 Wall heat transfer in a double-tube coal-slurry bubble column *Kagaku Kogaku Ronbunshu* **30** pp 360–367

[8]. Tambun R, Furukawa K, Hirayama M, Shimadzu M, Yamanaka S, And Ohira Y 2016 Measurement and estimation of the particle size distribution by the buoyancy weighing-bar method and the Rosin-Rammler equation *Journal of Chemical Engineering of Japan* **49 (2)** pp 229-233

[9]. Tambun R, Motoi T, Shimadzu M, Ohira Y, Obata E 2011 Size distribution measurement of floating particles in the allen region by a buoyancy weighing–bar method *Advanced Powder Technology* **22** pp 548–552.

[10]. Tambun R, Nakano K, Shimadzu M, Ohira Y, Obata E 2012 Sizes influences of weighing bar and vessel in the buoyancy weighing-bar method on floating particle size distribution measurements, *Advanced Powder Technology* **23** pp 855-860

[11]. Tambun R, Shimadzu M, Ohira Y, Obata E 2012 Definition of the new mean particle size based on the settling velocity in liquid *Journal of Chemical Engineering of Japan* **45 (4)** pp 279-284

[12]. Ohira Y, Furukawa K, Tambun R, Shimadzu M, Obata E 2010 Buoyancy weighing-bar method: a particle size distribution measurement using new settling method *Journal of the Sedimentological Society of Japan* **69** pp 17-26

IOP Conf. Series: Materials Science and Engineering **309** (2018) 012080 doi:10.1088/1757-899X/309/1/012080

Humidity detection using chitosan film based sensor

T I Nasution[1,3,4*], I Nainggolan[1,3], D Dalimunthe[2], M Balyan[1,4], R Cuana[2], S Khanifah[2]

[1]Centre of Excellence for Sensor Systems and Technology, Universitas Sumatera Utara, Medan 20155, Sumatera Utara, Indonesia
[2]Physics Department, Faculty of Mathematics and Natural Sciences, Universitas Sumatera Utara, Medan 2055, Sumatera Utara, Indonesia
[3]Chemistry Department, Faculty of Mathematics and Natural Sciences, Universitas Sumatera Utara, Medan 2055, Sumatera Utara, Indonesia
[4]Integrated Laboratory, Universitas Sumatera Utara, Medan 20155, Sumatera Utara, Indonesia

*E-mail: ikhsan_05@yahoo.com

Abstract. A humidity sensor made of the natural polymer chitosan has been successfully fabricated in the film form by a solution casting method. Humidity testing was performed by placing a chitosan film sensor in a cooling machine room, model KT-2000 Ahu. The testing results showed that the output voltage values of chitosan film sensor increased with the increase in humidity percentage. For the increase in humidity percentage from 30 to 90% showed that the output voltage of chitosan film sensor increased from 32.19 to 138.75 mV. It was also found that the sensor evidenced good repeatability and stability during the testing. Therefore, chitosan has a great potential to be used as new sensing material for the humidity detection of which was cheaper and environmentally friendly.

1. Introduction

The humidity sensor has a very important role in many aspects in our daily life such as meteorology, agriculture, medicine, biology, and others. Because of the various characteristics of humidity, humidity sensors are usually applied in industrial processes, environment monitoring, and equipment maintenance, all of which require accurate humidity control. Due to many fields involve humidity sensors, so it is necessary to make continuous improvements to the sensor performance, especially on the selection of sensor-making materials [1,2]. MOFs are rarely applied as a material for humidity sensor fabrication because MOFs framework structure can collapse in high humidity. Even if some MOFs are used, it still produces a low sensitivity because its limited specific surface [5,6]. In this study, humidity sensor is based on chitosan. When compared to MOFs then chitosan has several advantages as a sensor material. Chitosan is one of the natural biopolymers derived from chitin, as well as an abundant natural polymer on earth after cellulose [7]. Chitosan is also non-toxic, renewable, environmentally friendly, and doesn't have a bad effect on humans and the environment [8,9]. Chitosan has very high sensitivity to humidity because it has hydrogen bonds that is formed when the process of absorption of

IOP Conf. Series: Materials Science and Engineering **309** (2018) 012080 doi:10.1088/1757-899X/309/1/012080

water molecules and hydrophilic groups in chitosan [10]. The hydrogen bond serves as the transport of protons in chitosan [11], thus making chitosan become a good conductor [12]. Chitosan which is a natural polymer material has amino groups acting as active sites that can interact with water vapor [13]. Besides that, chitosan has unique characteristics that can be formed into a film [14]. Therefore, these characteristics makes chitosan can be used as a humidity sensor designed to work with some advantages such as cheaper, accurate and practice.

2. Methodology

Chitosan powder with medium molecular weight (85% distillation rate) was purchased from Sigma-Aldrich Chemical. Chitosan powder was dissolved using a magnetic stirrer with the rotation speed of 300 rpm for 6 hours at 30oC. The chitosan solution was then tested first using a particle size analyzer (PSA) to determine its particle size and homogeneity. In this work, the chitosan film was formed by a solution casting method in which the chitosan solution was put in a glass petri dish and followed by drying it at 65oC. Furthermore, the film was characterized by using scanning electron microscopy (SEM) to observe its surface morphology and chemical cluster analysis of chitosan film using FTIR. To test the electrical properties of chitosan film as humidity sensor, the film was placed in a cooling machine room model KT-2000Ahu equipped with two electrodes (positive and negative). The humidity level in the room was set from 30 to 90%. Both electrodes function to connect chitosan film sensor to an Arduino-based electronic readout system that was interfaced to the computer via USB TTL and PLX-DAQ software.

3. Results and Discussion

Figure 1 shows that the chitosan particles are evenly distributed. This is indicated by the same diameter value of 562.49 nm. This value also indicates good homogeneity of chitosan solution. In addition, the particle size obtained at 562.49 nm is a fairly small particle size in which a small particle size determines the stability of the chitosan solution and the smoothness of the chitosan film surface produced from the drying process.

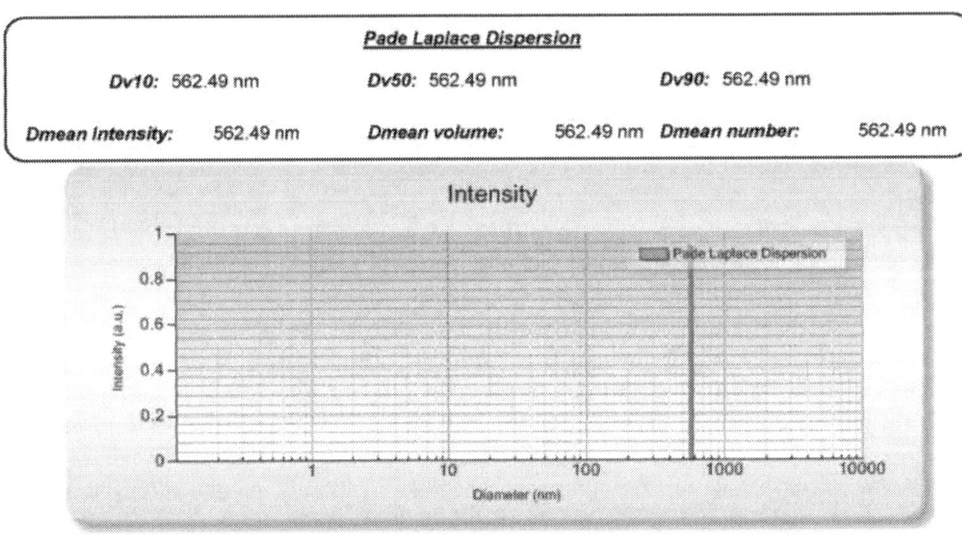

Figure 1. Particle size distribution of chitosan solution.

The surface morphology of the chitosan film was tested using a Scanning Electron Microscopy (SEM). Figure 2 shows the surface morphology of the chitosan film. The smooth surface morphology of the chitosan film shows the distance between the chitosan particles are tightly. The smooth surface of the chitosan film allows the chitosan film to be strongly bonded to the electrode surface during the humidity detection process. This strong bond between the layers of chitosan and the electrode is important to avoid the swelling of the chitosan film layer due to the absorption of water molecules after exposure to humidity.

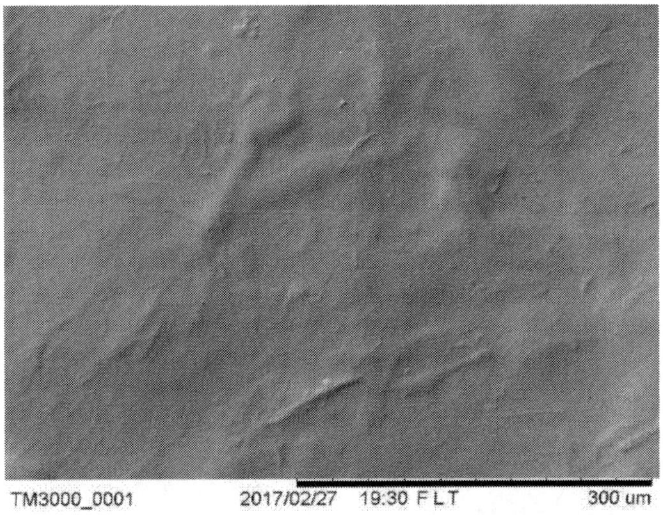

Figure 2. Analysis of chitosan film surface using Scanning electron microscope

The FT-IR spectrum (Figure 3) shows the OH groups at absorption band of 3614 cm^{-1} and NH_2 groups at the absorption band of 3001 cm^{-1}. The absorption peak at 1681 cm^{-1} indicates conjugated ketone C=O. Meanwhile, the peaks at 1593 cm^{-1} and 1438 cm^{-1} show the existence of N-H and –OH. This indicates both peaks are functional groups of chitosan which is having an important role as active sites to react with water vapor to generate electrical energy. It is also found that the C-N, C-O-C and C-O absorption groups appear at peaks of 1172, 1126 and 1045 cm^{-1}, respectively.

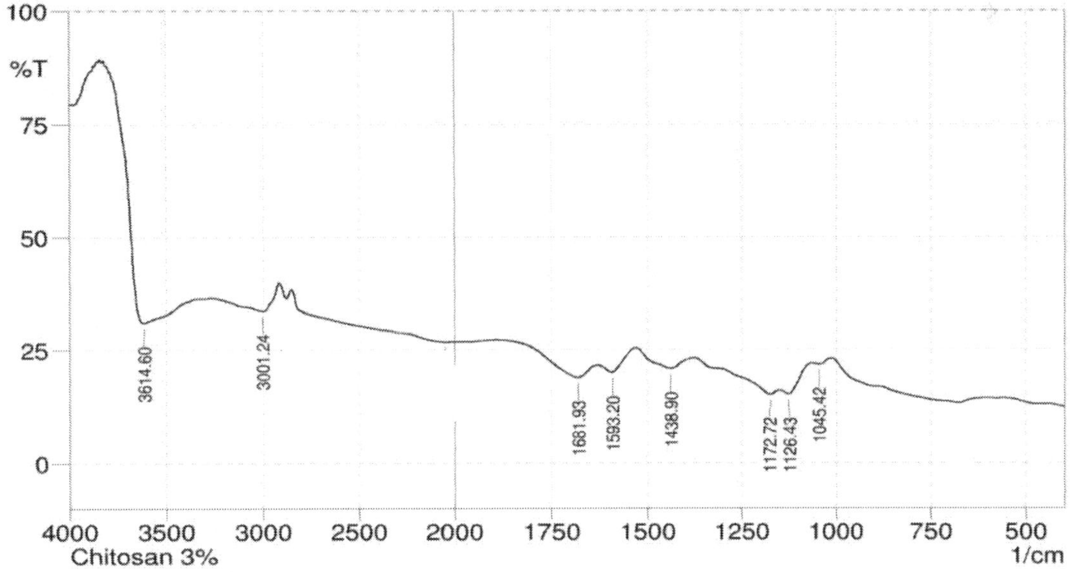

Figure 3. Chemical cluster analysis of chitosan film using FT-I

IOP Conf. Series: Materials Science and Engineering **309** (2018) 012080 doi:10.1088/1757-899X/309/1/012080

The ability of chitosan film sensors to detect humidity is shown in Fig. 3. It is obtained that there is an increase in the output voltage of chitosan film along with the increase of humidity in the cooling machine. This linear value can be explained because there is an interaction between the chitosan film sensor with water vapor that results in the output voltage of chitosan film.

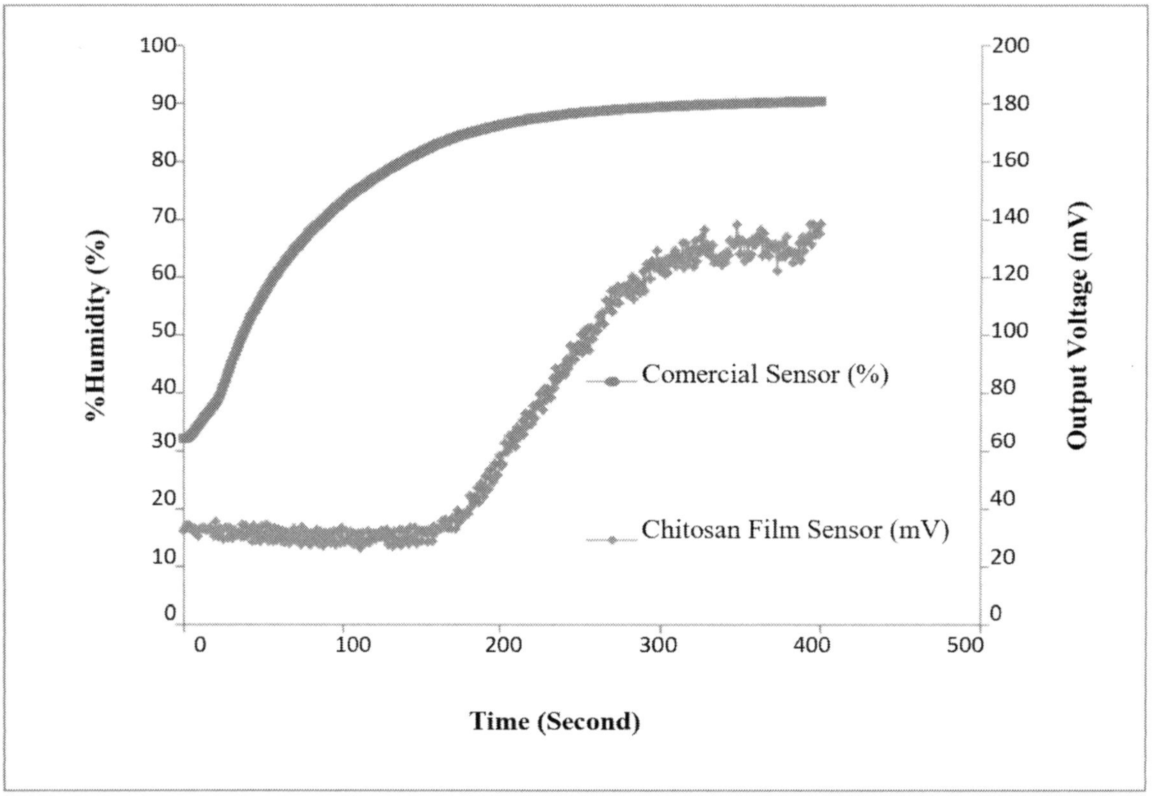

Figure 4. Comparation graph of output voltage from chitosan film and output value of commercial sensor

According to Ikhsan et al., chitosan film in the open air will absorb oxygen (O2). The absorbed oxygen trap electrons contained in chitosan and then form O- and O2- oxygen species as shown in the following reaction:

O_2 (gas) ↔ O_2 (absorbed) (1)

O_2 (absorbed) + e- ↔ O_2^- (2)

O_2 + e- ↔ 2O^- (3)

And when water molecules (H2O) stick on the surface chitosan film, the water molecules react with O- and produce O2 gases. This causes the release of electrons as described by the equation:

Chit-NH_2 + H_2O ↔ 2O^- + Chit-NH_2 + O_2+ H_2O + 2e^- (4)

This reaction raises the surface tension of chitosan film that helps the movement of electrons to escape the trap of oxygen species. The free electrons move randomly in the conduction band and increase the electrical conductivity of the chitosan film [14].

The graphs in Fig. 3 describe that the chitosan film sensor has good capability to detect different humidity level in the cooling machine. Besides that, it is also evidenced that sensors have as good characteristics which include short response time and good sensitivity. Even in terms of the range output voltage values, chitosan film sensor has a wide range of reading resolution, so the chitosan film sensor is excellent in terms of reading resolution. With a wide range of reading resolution, then the accuracy of reading humidity using chitosan film sensor is better than commercial sensor.

IOP Conf. Series: Materials Science and Engineering **309** (2018) 012080 doi:10.1088/1757-899X/309/1/012080

4. Conclusion

The chitosan film sensor shows the ability to detect air humidity. Humidity detection with chitosan film using this method has balanced the ability of general commercial sensor which has an expensive price. Then, the chitosan film sensor has potentially capability to detect humidity due to it is more accurate, cheaper, more durable and friendly environment.

Reference

[1] Wang R, Liu X W, He Y, Yuan Q, Li X L, Lu G Y, Zhang T 2010 *Sens. Actuators B* **145** 386e393
[2] Zhang Y, Chen Y, Zhang Y, Cheng X, Feng C, Chen L, Zhou J, Ruan S 2012 *Sens Actuator B* **174** 485e489
[3] Dan-Hardi M, Serre C, Frot T, Rozes L, Maurin G, Sanchez C, Ferey G, Am J 2009 *Chem Soc* **131** 10857e10859
[4] Zhang Z, Zhao Y, Gong Q, Li J 2013 *Chem. Commun* **49** 653e661
[5] Fu Y, Sun D, Chen Y, Huang R, Ding Z, Fu X, Li Z, Angew 2013 *Chem.Commun* **124** 3420e3423
[6] Segakweng T, Musyoka N M, Ren J, Crouse P R 2016 *Chem Intermed* **42** 4951e4961
[7] Somashekar D and Joseph R 1996 *Technol* **55 (1)** 35–45
[8] Kumar M N V R 2000 *React Funct Polym* **46** 1-27
[9] Martin H J, Schulz K H, Bumgardner J D, Walters K B 2008 *Surf Sci* **254** 4599-4605
[10] Murray C A and Dutcher J R 2006 *Biomacromolecules* **7 (12)** 3460-3465
[11] Hemmatian Z et al 2015 *J Mater Chem C* **3 (25)** 6407-6412
[12] Zhong C, Deng Y, Roudsari A R, Kapetanovic A, Anantram M P, and Rolandi M 2011 *Nat. Commun* **2** p 476
[13] Hassan S, Suzuki M and El-Moneim A A 2014 *J Power Sources* **246** 68-73
[14] Nasution T I, Nainggolan I, Hutagalung S D, Ahmad K R and Ahmad Z A 2013 *Sens Actuators B* **177** 522–528

IOP Conf. Series: Materials Science and Engineering **309** (2018) 012081 doi:10.1088/1757-899X/309/1/012081

Study on chitosan film properties as a green dielectric

I Nainggolan[2,3,4*], T I Nasution[1,2,5], S R E Putri[1], D Azdena[1], M Balyan[2,5] and H Agusnar[3,4]

[1]Physics Department, Faculty of Mathematics and Natural Science, Universitas Sumatera Utara, Medan 20155, Sumatera Utara, Indonesia
[2]Centre of Excellence for Sensor Systems and Technology, Universitas Sumatera Utara, Medan 20155, Sumatera Utara, Indonesia
[3]Centre of Excellence for Green Chitin and Chitosan, Universitas Sumatera Utara, Medan 20155, Sumatera Utara, Indonesia
[4]Chemical Department, Faculty of Mathematics and Natural Science, Universitas Sumatera Utara, Medan 20155, Sumatera Utara, Indonesia
[5]Integrated Laboratory, Universitas Sumatera Utara, Medan 20155, Sumatera Utara, Indonesia

*Email: ikhsan_05@yahoo.com

Abstract. Chitosan film dielectrics to produce an electrostatic capacitor were prepared by the solution cast technique. The charging and discharging of the capacitor were done using RC series circuit with DC voltage supply because chitosan has bipolar properties. First testing was by varying supply voltage of 1, 3, 5, 10 and 15 V, respectively, and could be determined that the most effective voltage for chitosan film can be well polarised is 5 V. The results of second testing for the use of 5 V supply showed that the capacitance of a chitosan film capacitor decreased with the increase in load value. For loads of 100, 1K, 10K, 100K and 1M Ω, the capacitance values of the chitosan film capacitor were 3.1725, 0.4136, 0.05379, 0.007917 and 0.001522 F, respectively. It was also found that the increase in voltage of the capacitor at charging process was faster for the lower load. Therefore, the research result has corresponded to the general formula that used to calculate the capacitance value and thus, the biopolymer chitosan has potential as a sustainable green dielectric.

1. Introduction

Natural polymers are becoming the new belle in the development of electronic devices because they come from renewable sources that are reproducible. While synthetic polymers begin to anxiety about the availability of the source of its raw materials, coupled with its poor degradability [1]. Among the most popular electronic components based natural polymer today are capacitors or supercapacitors, these passive storage components are beginning to experience renewal in their constituent parts. Various natural polymers such as cellulose [2], chitosan and starch [3] with the addition of dopant and plasticizer have been developed in the fabrication of solid electrolyte materials. The combination of two polymers was done by many researchers because they are capable to produce solid electrolyte with better mechanical and chemical properties [4].

In this study, chitosan was chosen as a dielectric material for electrostatic capacitor applications because have many interesting properties among others are biodegradable, biocompatible, biofunctional and low cost [5-7]. Moreover, chitosan is ranked second largest as the most abundant polymer in nature [8]. Chitosan has a bipolar charge, which is a negative charge in the carboxylate group and a positive charge in the NH group [9]. The existence of these two groups makes chitosan film can be placed between two metal plates to produce a bipolar-type capacitor. This is due to the chitosan film is capable to act as a dielectric material that can experience dipole induction when an electric field is applied.

So far we know, the research on the potential of pure chitosan as a dielectric material for electrostatic capacitors has not been found. However, the chitosan-ι-Carrageenan-H_3PO_4-PEG [10] and Poly (Vinyl Alcohol) -Chitosan-Ammonium Nitrate (NH_4NO_3) - Ethylene Carbonate [11] as solid electrolyte materials for supercapacitor applications have been done by A.K.Arof and his team. In this study, the ability of chitosan film to function as a dielectric in electrostatic capacitor applications is explained from charging and discharging test results by varying the supply voltage to obtain the optimal operational voltage and loads to compare the dielectric capability of chitosan film in saving electricity.

2. Experimental

The sample was prepared using the solution cast technique by stirring 1.5 grammes of chitosan powder in 50 mL 1% (v/v) acetic acid solution. The chitosan powder was purchased from Sigma-Aldrich with the type of medium molecular weight and deacetylation rate of 75% - 85%. While the acetic acid was purchased from EMD Millipore. The stirring process was done using a Scilogex magnetic stirrer with the rotation speed of 300 rpm for 6 hours at room temperature (23 ± 2 °C). After the stirring process, the homogeneous solution was poured into a Normax Petri dish and then, dried in the Digital Oven (Memmert UN 55) at 60 °C for 16 hours. The dried chitosan thick film was cut into a smaller thick film with the dimension of 1.8 x 1.8 cm^2.

The experiment setup was carried out using charging and discharging test. The Chitosan film with a surface area of 3.24 cm^2 and a thickness of 200 μm was placed between two copper electrodes in parallel to produce an electrostatic capacitor setup. Charging-discharging test was performed by using series RC circuit where the supply voltage was varied from 1, 3, 5, 10 and 15 V. The measurement data was recorded using a digital multimeter Hydra Series III fluke 87 V. Hereafter, the charging-discharging test was done at voltage value of 5 V by varying the load value of 100, 1K, 10K, 100K and 1M Ω. The dielectric chitosan film was characterised using Differential Scanning Calorimetry (SDT Q100) to determine the temperature of the transition glass, the chitosan film was heated from room temperature to 500 °C with a heating rate of 10 °C/minute. While, the Functional group analysis was conducted using a Fourier Transform Infrared Spectroscopy (SHIMADZU model IRPrestige 21).

3. Results And Discussion

Figure 1 shows the charging and discharging processes of dielectric film chitosan in electrostatic capacitor setup that are done with series RC circuit. Variation of supply voltage aims to see how the value of the voltage that can be accepted dielectric without reducing the ability dielectric in storing the charge.

IOP Conf. Series: Materials Science and Engineering **309** (2018) 012081 doi:10.1088/1757-899X/309/1/012081

a)

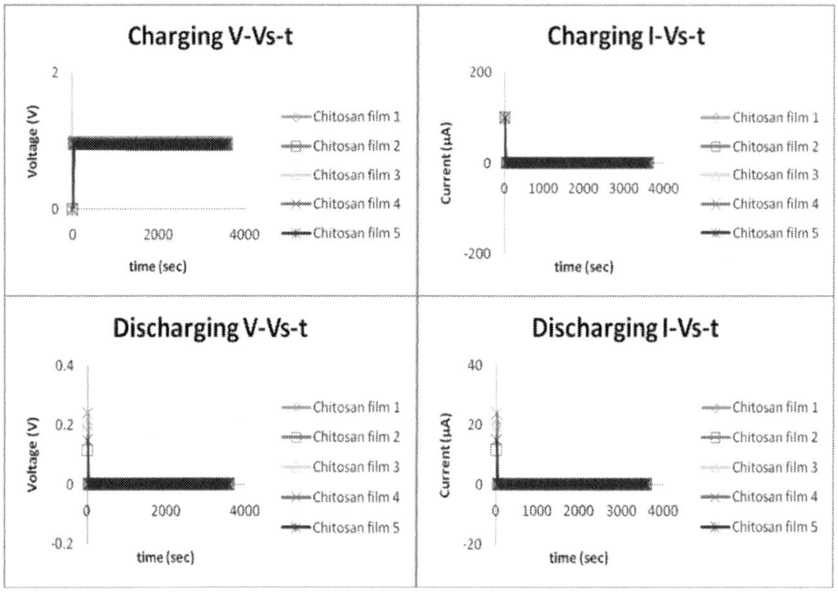

b)

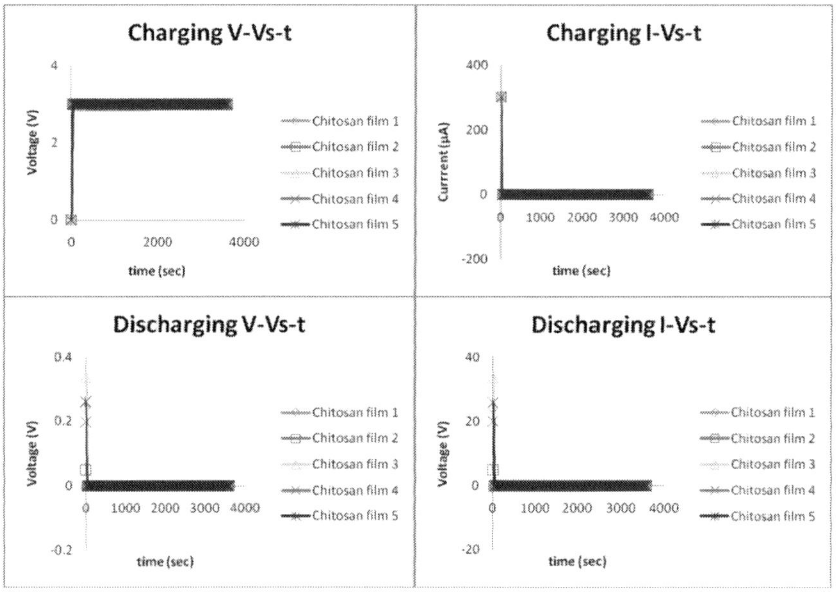

IOP Conf. Series: Materials Science and Engineering **309** (2018) 012081 doi:10.1088/1757-899X/309/1/012081

c)

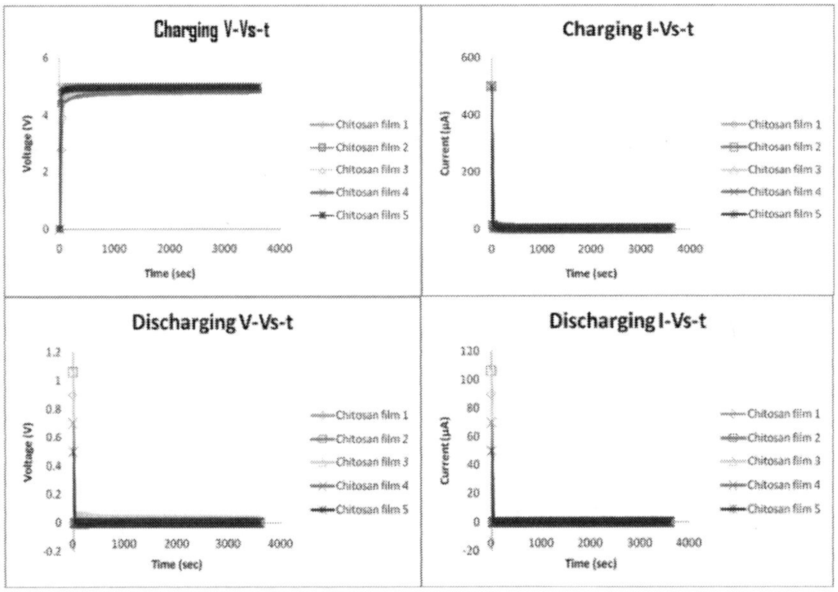

d)

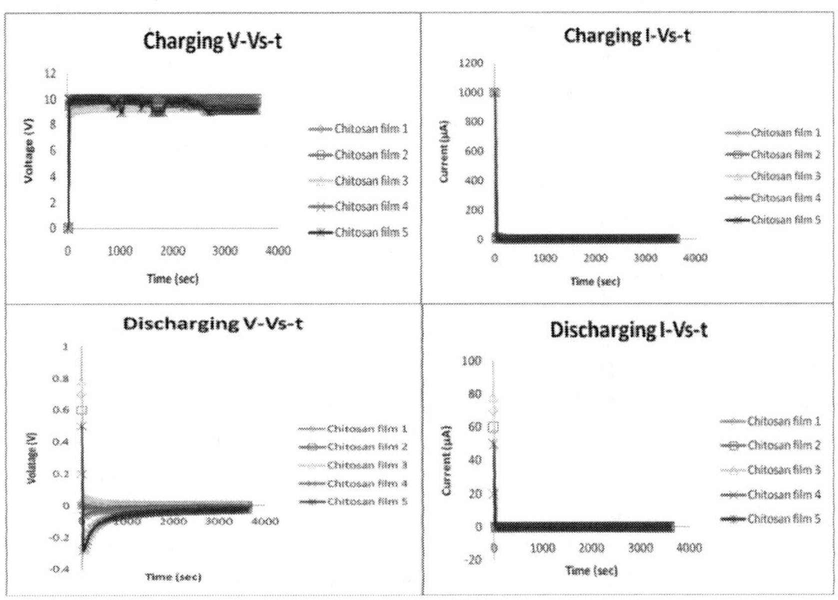

e)

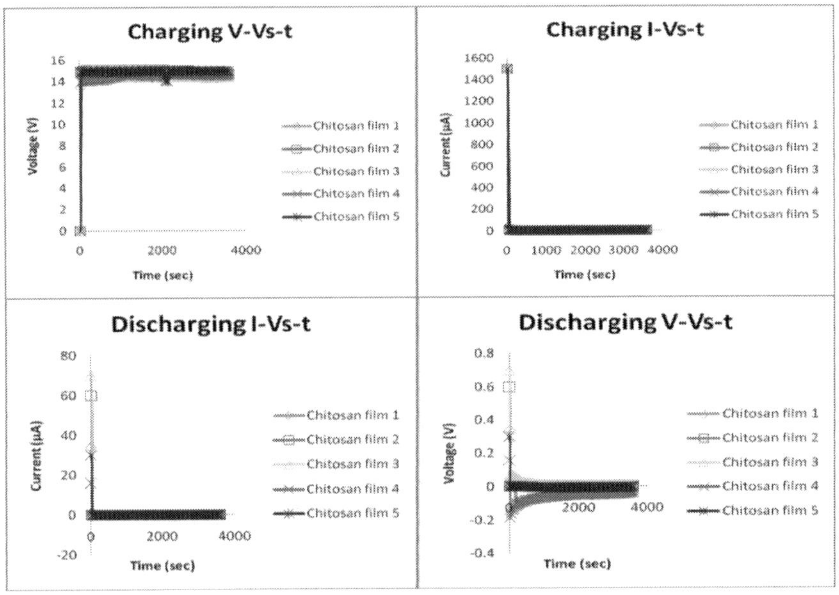

Figure 1. The charge and discharge testing of dielectric chitosan films using a series RC circuit at supply voltages of (a) 1 V, (b) 3 V, (c) 5V, (d) 10 V and (e) 15 V.

In every supply voltage variation, the chitosan dielectric film in the electrostatic capacitor setup shows the increase and decrease in the current and voltage of the charging process called the transient period, this is the state when an object undergoes a sudden change under the influence of current or voltage. In the dielectric, these changes sometimes can lead to fatal damage that is the breaking of the chemical bond of dielectric materials that under special conditions (such as C = O group of primary amide on the chitosan film) Can cause the free electrons are formed which can lead to dielectric failure. Next, in a few seconds the dielectric chitosan film in the electrostatic capacitor setup experiencing steady state, this condition generally indicates that the capacitor is fully charged. A stable voltage and current value from the beginning of the capacitor entering steady state until the specified charging time are over indicate that the transient state creates no damage to the bonding of the chitosan molecule. Then in the discharging process, the transfer of charge on load is so fast this is seen from the sharply decreasing graph. This shows that the chitosan film molecule has a short response time. Some anomalies occur in samples 4 and 5 on variations in supply voltages of 10 and 15 volts, indicating that there is a breaking of the functional group of chitosan molecules when the voltage enters the range of 10-15 volts. Therefore it can be concluded that the 5 volt voltage is the most appropriate voltage for dielectric to work with the optimum, this is proven by calculating the value of the charge that can be stored by the capacitor after the charging process (Figure 2)

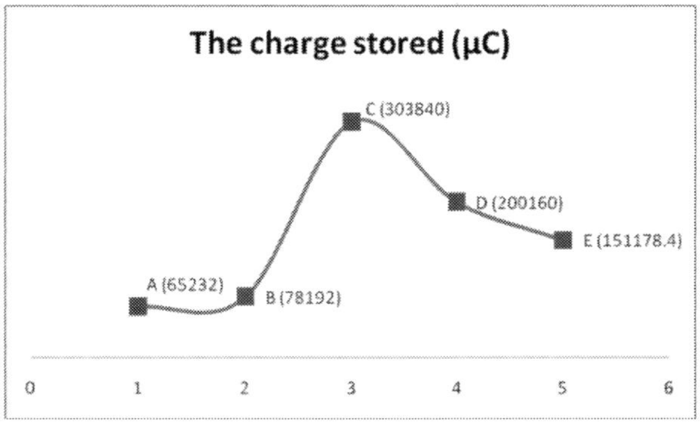

Figure 2. The charge stored after the charging process

The DSC analysis result (Figure 3) shows the endothermic peak value of T_g is 103.681°C and the exothermic peak value T_m is 297.737°C, indicating that the dielectric chitosan film has good thermal resistance. At the charging process, the dielectric chitosan film will experience an increase in temperature due to the given voltage. The temperature effect can cause vibrations in the lattice causing the electrons in the atoms to get extra energy so that the electrons escape from the lattice. This process is referred as the ionization event and then, these free electrons collide and get trapped in the skin of the atom with a higher electrical potential energy than the energy it has. This state causes a chemical change in the dielectric material resulting in deterioration of the dielectric electrical properties that called macroscopic electrical damage, in which dielectrics made from natural polymers such as chitosan is susceptible to this condition. Under allowable voltage conditions, repetition effects may also cause macroscopic electrical damage. In a situation where dielectric can not long withstand the applied voltage due to too many free electrons in the material, it will eventually make the dielectric lose its insulating properties which make it be a dielectric failure.

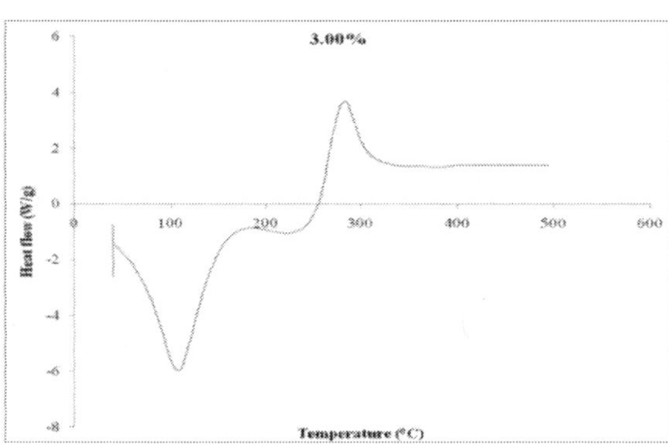

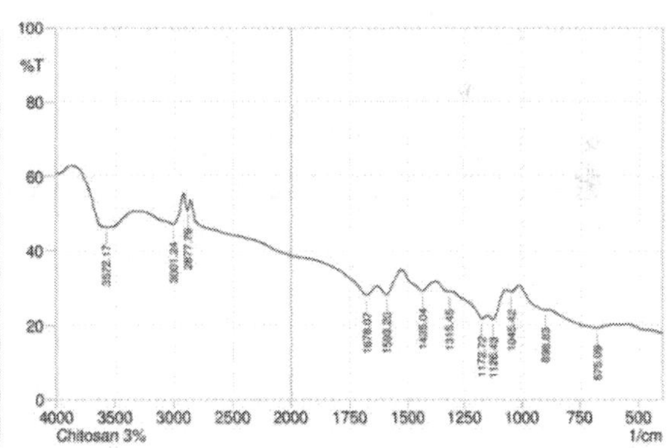

Figure 3. DSC analysis of chitosan film

Figure 4. FTIR analysis of chitosan film

The absorption band at wavenumber 3572 cm^{-1} is the vibration of the overlapping -OH group range with -NH of the primary amine, whereas the absorption band at wavenumber 1593 cm^{-1} is –NH vibration of the secondary amine (Figure 3). Both functional groups are the main actors in the polarization of dielectric film chitosan. During the charging process, the negative charge of an electric current on one of the copper plates attract the positive charge in the NH group in the chitosan film molecule. This causes the negative charge in the carboxylate group to shift towards the positive charge of the electric current on the other plate.

Figure 4 shows the effect of load value variation on the charging and discharging processes of dielectric film chitosan in electrostatic capacitor setup with series RC circuit. It is seen that the capacitance value is decreased when load value is increased (Table 1). This result has corresponded with the general formula of capacitance in the RC circuit,

$$C = -\frac{t}{R\ln\left(\frac{IR}{\varepsilon}\right)} \quad (1)$$

where it looks that the addition of the load value is inversely proportional to the value of the capacitance.

Table 1. Change the value of the capacitance at each load value

R (Ω)	C (F)
100	3.1725
1000	0.4136
10000	0.05379
100000	0.007917
1000000	0.001522

4. Conclusion

Chitosan film that is prepared by the solution cast technique can be used as a green dielectric to produce an electrostatic capacitor. The optimal operational voltage of dielectric chitosan film for storing the most amount of electrical charge is 5 V DC. Therefore, the chitosan film can be applied as an environmental friendly dielectric material with low supplied voltage consumption.

5. Acknowledgements

We would like to acknowledge the funding and support from the Centre of Excellence for Green Chitin and Chitosan Universitas Sumatera Utara.

6. References

[1] Shukur M F, Ithnin R and Kadir M F Z 2014 *Electrochim Acta* **136** 204–216
[2] Shuhaimi N E A, Teo L P, Woo H J, Majid S R and Arof A K 2012 *Polym. Bull* **69** 807–826
[3] Teoh K H, Lim C S, Liew C W, Ramesh S and Ramesh 2015 *Ionics* **21** **(7)** 2061–2068
[4] Mahendran O and S Rajendran 2003 *Ionics* **9** 282–288
[5] Dutta P K, Dutta J D and Tripathi V S 2004 *J Sci Ind Res* **63** 20 – 31
[6] Aldana A A, Toselli R, Strumia M C and Martinelli M 2012 *J. Mater. Chem.* **22** 22670- 22677
[7] Brown M A, Daya M R and Worley J A 2009 *J. Emerg. Med.* **37(1)** 1–7
[8] Kumar M N V R 2000 *React. Funct. Polym.* **46** 1–27
[9] Bansal V, Sharma P K, Sharma N, Pal O P and Malviya R 2011 *Adv. Biol. Res.* **5(1)** 28–37
[10] Arof A K, Shuhaimi N E A, Alias N A, Kufian M Z and Majid S R 2010 *J. Solid State Electrochem.* **14** 2145–2152
[11] Kadir M F Z and Arof A K 2011 *Mater. Res. Innov.* **15** sup2, s217-s220

IOP Conf. Series: Materials Science and Engineering 309 (2018) 012082 doi:10.1088/1757-899X/309/1/012082

Automatic control model of water filling system with Allen Bradley Micrologix 1400 PLC

R Harahap[1], AF Adyatma[1] and F Fahmi*[1,2]

[1]Department of Electrical Engineering, Faculty of Engineering, University of Sumatera Utara, Jl. Dr. Mansur 9 Medan 20155, Indonesia
[2]Sustainable Energy and Biomaterial Centre of Excellent, University of Sumatera Utara, Jl. Almamater Kampus USU, Medan 20155, Indonesia

*Email: fahmimn@usu.ac.id

Abstract. Programmable Logic Controller or PLC today plays an important role in most industrial control systems. PLC usage can be encountered in almost all fields of industry, not only in the manufacturing world but also on many other things such as elevators in office buildings, hotels hospitals, and others. PLC is an electronic control tool that operates in logic that its programming can be modified with relative ease. As with any controller in general, the PLC processes input signals to further discharge output according to the desired program. PLC usage is very broad because of its high reliability, can be reprogrammed or modified with relative ease, and very helpful in the tracking troubleshooting. One type of existing PLC is Allen Bradley PLC. Allen Bradley PLC program is commonly used in various industries. PLC Allen Bradley (AB) has several types, and one of them is the type of Micrologic 1400. In this study we design a system as a comparison with the conventional system. For that to explore the use of a PLC program which will be supported by a simulator tool, including a program to RSLogic 500, how to programming, monitoring via RSView32, and modification. It is expected to understand the application aspect of operation and programming of this specific PLC and its potential. The purpose of this research is to design water filling automation system by using Allen Bradley Micrologic 1400 type 1766-L32BXB PLC, empowering the use of Allen Bradley Micrologic 1400 PLC and to regulate the desired process to obtain efficiency and effectiveness compared with conventional system arrangement using Relay.

1. Introduction

Factors that affect the efficiency and productivity of the industry itself, such as the human error factor and the level of excellence offered by the control system. One of the most widely used control systems is Programmable Logic Controller (PLC). Its applications cover a wide range of industries ranging from cigarette, automotive, petrochemical, paper, and even to the mining industry, for example in the control of gas turbines and advanced industrial units of mining products. The ease of transitioning from previous control systems (e.g., from mechanical relay-based control systems) and the ease of troubleshooting in system configurations are the two main factors driving the popularity of this PLC. PLC is a system that can manipulate, execute, and or monitor the state of the process at a very fast rate, on the basis of programmable data in an integral microprocessor-based system. PLC receives inputs and produces an output electrical signals to control a system. Thus the controlled quantities of physics and chemistry, before being processed by the PLC, will be converted into electrical signals both analog and digital, which is the basic data.

The formulation of the problem is to know how to enter and run ladder logic PLC program Allen Bradley and understand lender logic system Allen Bradley PLC with computer software. Besides, it will explain how to connect PLC with an induction motor.

The purpose of this research is to design water filling automation system by using Allen Bradley Micrologic 1400 type 1766-L32BXB PLC, empowering the use of Allen Bradley Micrologic 1400 PLC to know how PLC works in order to regulate the desired process so as to obtain efficiency and effectiveness compared with conventional system arrangement using Relay.

2. Materials and Methods

2.1. Allen Bradley PLC

There are two forms of Allen Bradley PLC namely compact and modular form. For a compact form, its PLC uses a rack system (CPU and I/O unit) with limited memory capacity. An example of a compact form is AB MicroLogic 1400 . As for the modular form PLC consists of CPU module and I / O module (an integral part). Example PLC AB Modular type is SLC 5/05 (Figure 1).

a. b.

Figure 1. (a) PLC Allen Bradley MicroLogic 1400 and (b) PLC modular series SLC 5/05

Table 1. Type of PLC Micrologix 1400

No	Model	Power Supply	Input	Output
1	1766-L32BWA	120/240V AC	12 Fast 24V DC, 8 Normal 24V DC	12 Relay
2	1766-L32AWA	120/240V AC	20 120V AC	12 Relay
3	1766-L32BXB	24V DC	12 Fast 24V DC, 8 Normal 24V DC	6 Relay, 3 Fast DC, 3 Normal DC
4	1766-L32BWAA	120/240V AC	12 Fast 24V DC, 8 Normal 24V DC, 4 Analog V input	12 Relay, 2 Analog V output
5	1766-L32AWAA	120/240V AC	20 120V AC Input, 4 Analog V input	12 Relay, 2 Analog V output
6	1766-L32BXBA	24V DC	12 Fast 24V DC, 8 Normal 24V DC, 4 Analog V input	6 Relay, 3 Fast DC, 3 Normal DC 2 Analog V output

There are several variants of Micrologix 1400 as shown in table 1. RSLinx is software used for setting communication between PLC and PC.RSView32 is one of the HMI software that allows monitoring of existing data in the field, in this case for example data from sensors and transducers in the field controlled by PLC. RSLogic is software used to create programs in PLC.

2.2 SCADA on Clean Water Treatment System
Commonly in a water treatment, there are three buildings or construction, namely:

a. Intake building is the first building or construction for the entry of water from a water source.
b. Water Treatment Plant (WTP), is a system or facility that works to treat water from contaminated water quality (influent) to get the desired water quality treatment as per quality standard or ready for consumption.
c. Reservoir, as a temporary shelter for clean water before being distributed.

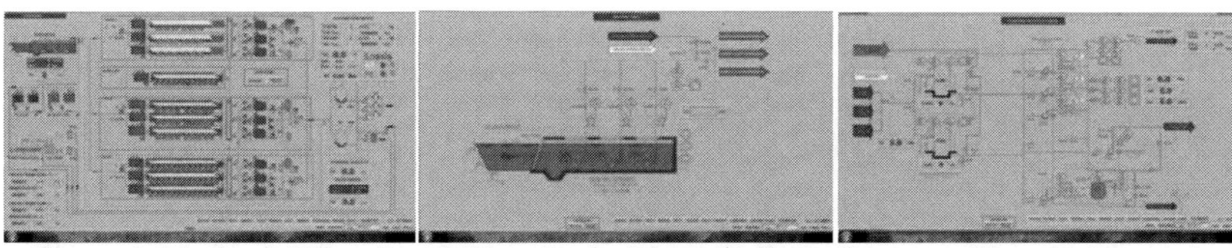

Figure 2. Display RSView for (a) Water Intake River, (b) Water Treatment Plant, (c) Reservoir

2.3. Designing
Simulation model of automation control of water filling system using Allen Bradley MicroLogix 1400 PLC is adjusted with conventional water charging system. The design includes the example of conventional system design, automation system design, creation of automation system program based on PLC Allen Bradley type Micrologic 1400 with Rockwell Software. In automatic water filling system with the conventional system, the 1st and 2nd buoy models of type RADAR are employed. In general, the tools and materials for this conventional design model are tank/tub/water container wells, RADAR buoys, motors / Omron water relay pumps, for the semi-auto system, control panel, and rail cable.

Here is how the conventional water filling systems work. In normal conditions (the tank is empty), the water pump stops working (Off). Water enters through the inlet hole. Underwater conditions are still at low level, the water pump is not working yet. When the water level reaches high level, the water pump starts to work. The water pump will continue working until the water level is at low level. At the water position at low level, the pump stops working, and water enters through the inlet to fill the storage tank. When the water level is in high level position, the water pump starts to work. In the control panel, use the ON-OFF button to turn on the system, selector switch is used to make system Auto-Manual. Auto, where the system works automatically using the RADAR buoy as a switch. Manually, where the water pump can be run using the pushbutton, and the starting motor water pump works with the Direct on Line (DOL) system.

2.4. Automation system design
The simulation will use a board to embed the simulation tools. The substitute of the float water level setting is a light switch. The H1 mark light as a marker for the inlet water pump works. The H2 mark light as a marker for the distribution water pump works. H3 mark lights as markers for open inlet valve. H4 mark lights as markers for open valve outlets. H5 sign light as a marker for maximum water level. The H6 Sign lights as a marker for half the water level of the container. H7 mark lights as markers for minimum water level. To know the water level in the container, it will be shown on the RSView32 Work 100 display. The equipment is used to start/stop the induction motor 1 phase of the

water pump. The simulation is not equipped with the speed setting of 1 phase induction motor. Operation system used to fuse and mcb 1 phase. The materials used are as shown in table 2.

Table 2. Material Used

No.	Component	Specification	Amount
1	Simulator Board	50cm x 30cm	1 pc
2	pilot lamp	Green-red	8 pcs
3	Fuse	2 ampere	2 pcs
4	MCB 1 phasa	2 ampere	10 pcs
5	Relay	24 VDC	2 pcs
6	Contractor	Schneider	2 pcs
7	Relay	-	4 pcs
8	Ethernet	-	10 meter
9	Cable	NYAF 0,5mm	1 roll
10	power supply	Output 24 VDC	1 pc
11	rel rank	1 meter	1 pc

Cables are installed according to the system work function. When the first level switch sends a 24 VDC signal to the PLC, the indicator light will illuminate in accordance with its replacement function, indicating the water level is at 25%. The switch will then be activated sequentially up to 100% level (full water, then sends the input signal to PLC to read and processed to next system) HMI display will show the position of water level in the shelter.For the wiring diagram of Input / Output PLC, and detail of relay/indicator lamp, are shown in figure 4.

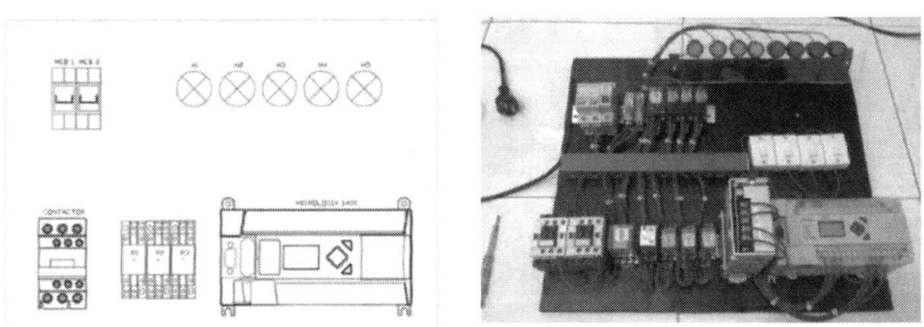

Figure 3. (a) Design of Simulation and (b) the output.

2.5. Rockwell Software

Description of the work of this system is a specification of the design of the desired control as follows: The overall display of the system is shown in RSView32. To simulate the water filling, the valve inlet on the diagram in the open position (ON), the intake pump will work. At the water filling position, the valve outlet is closed, and the distribution pump works. Shortly after the water level of 25%, it will be displayed on the screen HMI monitor, distribution pump position and valve outlet not working (OFF). The water level will be shown with the switch as a simulation. Once the water is at its maximum level, the valve outlet opens, and the distribution pump will light up. When water level is sent from the distribution pump, the intake pump position remains working. When the water level decreases to a height of half the shelter (50%), the system condition is still running normally. If the water level is at minimum height (0%), then the distribution pump will stop working, and the valve outlet will be closed. When the distribution pump turns off, the system will work from the start again. If one or both pumps are experiencing a technical malfunction, then the system will shut down the program automatically.

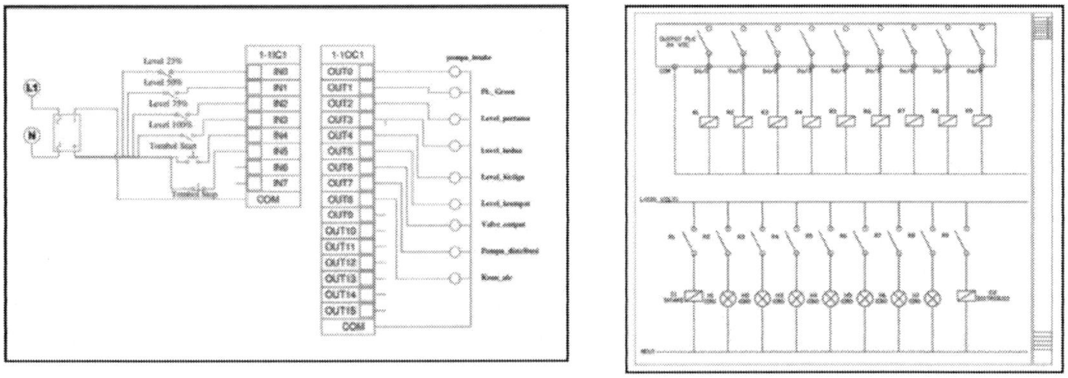

Figure 4. Wiring Diagram for (a) I/O system and (b) Relay and Indicator Lamp.

Here we will use 7 inputs and 5 outputs. All input outputs are digital input/output with 24VDC discrete voltage. The name and address of the input-output are shown in the following table.

Table 3. List Address Input / Output

No.	Input		No.	Output	
	Input Address	Name Tag		Output Address	Name Tag
1	I:O/0	Level 25 %	1	O:O/0	Pump Intake
2	I:O/1	Level 50 %	2	O:O/1	PL Green
3	I:O/2	Level 75 %	3	O:O/2	Level 25%
4	I:O/3	Level 100%	4	O:O/3	Level 50 %
5	I:O/17	Start Button	5	O:O/4	Level 75 %
6	I:O/18	Stop Button	6	O:O/5	Level 100%
			7	O:O/6	Valve Output
			8	O:O/7	Distribution Pump
			9	O:O/8	Water Tap

3. Results and Discussions

Testing was done to the check program that has been made. The control program works based on a conventional water fill system that uses a buoy to insert corrections. From here, we can see the performance of the simulation tool whether it works properly and correctly.

3.1. Normal Operation Test

In normal testing, the system will work when the intake pump and start button has been activated. Then in sequence, the switch is turned on as a water level limit marker. At the time the water is still within 25%, the distribution pump is not working yet. Up to a height of 100%, then gradually open valve outlets. The distribution pump works within 3 seconds after the open valve outlet. Then lapse 3-second water tap in open position. It appears in Figure that when the intake pump works, it will change color. At this moment, it is simulated that the new water goes into the storage tank. In the figure above, the water position that has entered into the shelter reaches 50% (half full). The HMI display is indicated by the color change in the storage tank. In the picture above, it shows the position of the already full water in the storage tank. The figure also shows the color change of the valve outlet, the distribution pump, and the water indicating the working position. Furthermore, when the water intake pump stops working, then in sequence the position of the water level will decrease. When the water level has run out, then the distribution pump, valve outlet, and water faucet will be off automatically.

3.2. Testing software operations

In software testing, the three software components RSLogix500, RSView32, and RSLinx will be communication with 1 piece of computer/laptop. At RSLogix500, the way of running has been discussed earlier. The experimental in the form of input signal and output (I / O 24 VDC) was done. Here we use 6 inputs and 9 outputs. When the start button is pressed, then the rung will read signals enter. It is characterized by a green discoloration on the XIO rung. At the time of the stop, the button is pressed also happened that way. In the input table, when the button is pressed will read signal enter 1 for digital.

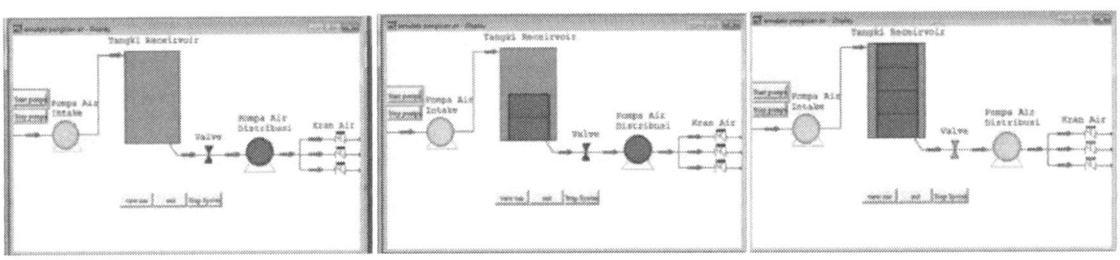

Figure 5. Pump position at intake

When the water level is full, it will activate the timer on delay (TON). TON works to activate the distribution pump motor after the valve outlet works. The following table shows the Timer On Delay (TON) operand.

Table 4. Ladder Table Diagram TON

Operand	Type	Format	Description
Timer	TIMER	Tag	Timer format
Preset	DINT	Soon	How long delay
Accum	DINT	Soon	Millisecond recorded with timer

4. Conclusions

Systems designed by using switches instead of entering digital data for watertight monitoring can work and send signals to the PLC. Rockwell Software as Allen Bradley's PLC Alliance can work as desired and make it easier to apply PLC. PLC Allen Bradley Micrologic 1400 type 1766-L32BXB used to run and process the data has a lack of unavailability Input / Output analog. As for the CPU unit using 42 VDC. It is required to add a power supply of 220 VAC input and 24 VDC output. To make the view on RSView32 better, it is recommended to use one unit of type RADAR sensor that can send signal enter analog to PLC.

IOP Conf. Series: Materials Science and Engineering **309** (2018) 012082 doi:10.1088/1757-899X/309/1/012082

For the development of the system in the next application is expected to use Micrologic 1400 series 1766-L32AWAA or 1766-L32BWAA which has analog I / O socket and enter the 220 VAC power supply.

References

[1] Ogata and Katsuhito 1997 *Teknik Kendali Automatik* vol **1** (Jakarta: Erlangga)
[2] PT. Tirta Lyonnaise Medan 2014 *Maintenance And Troubleshooting for Allen-Bradley SLC 500 Programmable Control* (Medan)
[3] Rashid MH 2001 *Power Electronics Handbook* (Canada: Academic Press)
[4] Theraja AK and Theraja BL 2005 *A Textbook Of Electrical Technology* (New Delhi: S. Chand & Company Ltd)
[5] Stevenson WD 2000 *Analisis sistem tenaga* (Jakarta: Erlangga)
[6] Allen-Bradley AB 2016 *Micrologix 1400 Programmable Controllers Manual User* (Istambul: Rockwell Automation)
[7] Allen-Bradley AB 2016 *Logix500 Controllers General Instructions Reference Manual* (Istambul: Rockwell Automation)
[8] Allen-Bradley AB 2011 *RSView User's Guide* (USA: Rockwell Automation)

IOP Conf. Series: Materials Science and Engineering **309** (2018) 012083 doi:10.1088/1757-899X/309/1/012083

Sodium tripolyphosphate cross-linked chitosan based sensor for enhacing sensing properties towards acetone

T I Nasution[1,2,5*], R Asrosa[2], I Nainggolan[3,4], M Balyan[1,5], R Indah[2] and A Wahyudi[2]

[1]Centre of Excellence for Sensor Systems and Technology, Universitas Sumatera Utara, Medan 20155, Sumatera Utara, Indonesia
[2]Physics Department, Faculty of Mathematics and Natural Science, Universitas Sumatera Utara, Medan 20155, Sumatera Utara, Indonesia
[3]Chemistry Department, Faculty of Mathematics and Natural Science, Universitas Sumatera Utara, Medan 20155, Sumatera Utara, Indonesia
[4]Centre of Green Chitin and Chitosan, Universitas Sumatera Utara, Medan 20155, Sumatera Utara, Indonesia
[5]Integrated Laboratory of Physics, Universitas Sumatera Utara, Medan 20155, Sumatera Utara, Indonesia

*Email: ikhsan_05@yahoo.com

Abstract. In this report, sensing properties of sodium tripolyphosphate (TPP) cross-linked chitosan based sensor has been successfully enhanced towards acetone. Chitosan solutions were cross-linked with sodium TPP in variation of 0.1%, 0.5%, 1% and 1.5% w/v, respectively. The sensors were fabricated in film form using an electrochemical deposition method. The sensing properties of the sensors were observed by exposing the pure chitosan and sodium TPP cross-linked chitosan sensors towards acetone concentrations of 5, 10, 50, 100 and 200 ppm. The measurement results revealed that the maximum response in output voltage value of pure chitosan sensor was 0.35 V while sodium TPP crosslinked chitosan sensors were above 0.35 V towards 5 ppm acetone concentration. When the sensors were exposed towards acetone concentration of 200 ppm, the maximum response of pure chitosan was 0.45 V while sodium TPP crosslinked chitosan sensors were above 0.45 V. Amongst the variation of sodium TPP, the maximum response of 1% sodium TPP was the highest since the maximum response was 0.4 V and 0.6 V towards 5 ppm and 200 ppm acetone concentration, respectively. While the maximum responses of other sodium TPP concentrations were under 0.4 V and 0.6 V towards 5 ppm and 200 ppm acetone concentration. Moreover, 1% sodium TPP cross-linked chitosan based sensor showed good reproducibility and outstanding lifetime. Therefore, 1% sodium TPP cross-linked chitosan based sensor has exhibited remarkable sensing properties as a novel acetone sensor.

1. Introduction

In recent years, the detection of volatile organic compounds (VOCs) has become a serious task all over the world owing to their manifold hazardous effects on environmental safety and human health.

Acetone is one of the amongst VOC which is extensively used to purify paraffin, dissolve plastic and dehydrate tissues in pharmaceutics. Although acetone is regarded as one having relatively low toxic effects, but some studies reveal that it has anesthetic effects on the central nervous system and may cause damage to the liver, kidney and pancreas in the human beings. Futhermore, it is known acetone is a product obtained from fatty acid metabolism, which is an important biomarker for diabetes diagnosis. Hence, a new effective method with low cost, portable and real-time detection of acetone concentration is extremely expected [1–3].

Chitosan is an abundant cationic natural biopolymer [4] derived simply by alkaline N-deacetylation from chitin [5]. Chitosan has been used widely in sensor application [6] due to its good film-forming properties [7–9], low cost and biodegradability [10]. However, limited electrical conductivity [11], low mechanical strength [12], high swelling [13] and poor stability [14] have limited its application in sensor application. Chitosan can be modified by chemical modifications through a cross-linking to overcome these obstacles [15,16]. Chitosan contains abundant free $-NH_2$ and $-OH$ groups [8,17] that provides to enhance solubility and reactivity [16]. Chitosan is soluble in acidic condition due to the protonation of its amino groups ($-NH_3^+$) with pK_a value ~6.3 so it behaves as a polycation and may favor electrostatic interactions [18]. In cross-linking, the available $-NH_2$ groups act as active sites and are capable of forming a number of linkages in three dimensional network structure [19]. However, cross-linking agents that are commonly used such as glutaraldehyde, epichlorohydrine and ethylene glycol diglycidyl ether are not preferred because of their toxicity [20]. Thus, the preference is given to non-toxic polyanion crosslinkers such as sodium tripolyphosphate (TPP) [21]. Sodium tripolyphosphate (TPP) is a non-toxic polyanion which can interact with positively charged amino groups of chitosan via electrostatic forces to form ionic cross-linked networks due to its high net negative charges [22,23]. Hence, the aim of this study is focus on interaction of sodium TPP cross-linked chitosan as a novel acetone sensor which leads to the improvement sensing properties such as conductivity, swelling, repeatability, lifetime, reproducibility and stability since the sensing mechanism toward acetone has been discussed in our previous work [24].

2. Methodology

2.1. Materials and Solution Preparation

Chitosan powder (medium molecular grade) purchased from Sigma Aldrich was used as a sensor material. Chitosan solution was prepared (1.75% w/v) by dissolving chitosan powder in 2% acetic acied (Glacial) obtained from Merck Millipore. The solution was stirred using a magnetic stirrer with rotation speed of 300 rpm for 24h at room temperature. While sodium TPP (technical grade) purchased from Merck was used as a cross-linker in variation of 0.1%, 0.5%, 1% and 1.5% w/v, respectively. The pH of pure chitosan and 0.1%, 0.5%, 1% and 1.5% w/v sodium TPP cross-linked chitosan are 3.4, 7.3, 8.3, 8.6 and 9.0, respectively.

2.2. Sensor Fabrication

A gold layer was coated onto an epoxy substrate surface and patterned by a printing method. Then, pure chitosan and sodium TPP cross-linked chitosan solutions were coated onto the patterned gold layer using an electrochemical deposition method [25,26] as shown in figure 1. The deposited wet film was annealed in a vacuum oven Memert at 205°C for 20 minutes.

2.3. Characterization

The surface morphology of pure chitosan and sodium TPP cross-linked chitosan films were observed using Scanning Electron Microscopy (SEM: Hitachi TM3000) while Fourier Transform Infrared Spectroscopy (FTIR: SHIMADZU IRPrestige21) was used to investigate the functional group of pure chitosan and sodium TPP cross-linked chitosan films.

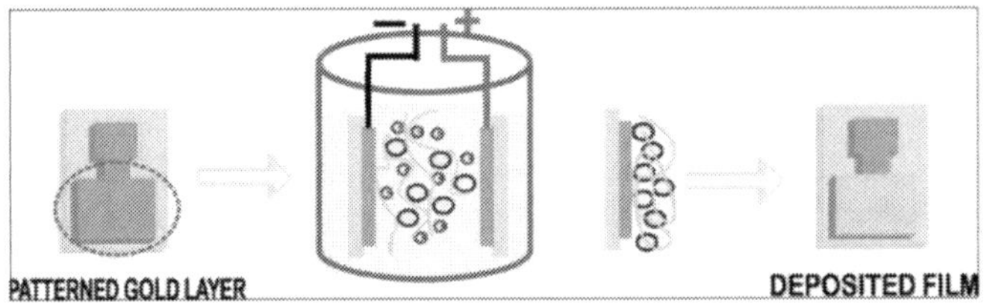

Figure 1. Illustration of electrochemical deposition method.

2.4. Electrical Testing

The pure chitosan and sodium TPP cross-linked chitosan sensors was inserted into a testing chamber and equipped with positive and negative electrode. Both electrode points were connected to a high resolution multimeter. Then, the sensors were exposed to acetone gas with concentration of 5, 10, 50, 100, 200 ppm respectively to observe the sensing properties such as response, recovery, conductivity, repeatability, reproducibility, stability and lifetime.

3. Results and Discussion

Surface morphology of pure chitosan and sodium TPP cross-linked chitosan films using SEM are shown in figure 2. It clearly shows that surface of pure chitosan (1.75% w/v) film seems to be rough and porous due to the grain growth which was largely distributed over the surface. Meanwile, when chitosan was cross-linked by OH^- and $P_3O_{10}^{5-}$ ions in sodium TPP solution, its grain size reduced. As a result, the sodium TPP cross-linked chitosan films have a low surface roughness and form a solid surface. It can be seen that the higher concentration of sodium TPP, the more even and solid surface of films occur. When the sensors were exposed towards acetone, this solid surface may enhance electrons to be transferred faster from one particle to another and leads to produce higher electrical conductivity. However, the surface of 1% sodium TPP cross-linked chitosan film appears the most even and solid since the surface of 1.5% sodium TPP cross-linked chitosan film becomes uneven or wavy. This wavy surface occured due to weak bond between sodium TPP cross-linked chitosan film and gold layer electrode which may cause the film is lower in conductivity. Therefore, it attests that 1% sodium TPP cross-linked chitosan film is applicable for a novel acetone sensor.

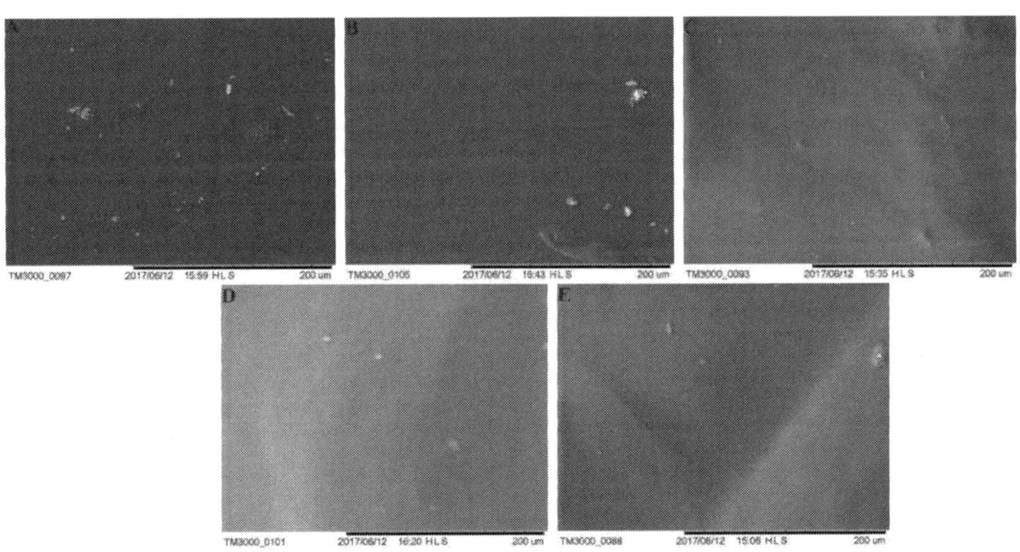

Figure 2. Surface Morphology of (A) pure chitosan, (B) 0.1 %, (C) 0.5%, (D) 1% and (E) 1.5% sodium TPP cross-linked chitosan films using SEM.

IOP Conf. Series: Materials Science and Engineering **309** (2018) 012083 doi:10.1088/1757-899X/309/1/012083

Figure 3 shows the FTIR spectrum of pure chitosan and sodium TPP cross-linked chitosan films. The FTIR spectrum of pure chitosan shows the absorption band at 3275cm^{-1} which indicates the combined broad non-symmetric band of –OH and –NH groups [27] while the peak of absorption band at 2877 cm^{-1} shows the –CH stretching vibration. C=O streching in amide I vibration group appears at the absorption band of 1651 cm^{-1} and N-H deformation in amide II group vibration (NH_2) as the active site of the film occurs at 1566 cm^{-1}. The peak at 1315 cm^{-1} identifies vibration range of the C-N amine and abrorption band at 1150 cm^{-1} indicates C-O asymmetric. Meanwhile, the FTIR spectra of sodium TPP cross-linked chitosan films becomes wider at the band of 3421 cm^{-1} which indicates hydrogen bonding is enhanced [31]. The peaks at 1651 cm^{-1} and 1566 cm^{-1} dissapear and appear new bands at 1662 cm^{-1} and 1589 cm^{-1}. This could be attributed to the linkage between the OH^- and phosporic ions with ammonium ions of chitosan. Sodium TPP cross-linked chitosan films also shows a peak for P=O at 1157 cm^{-1}.

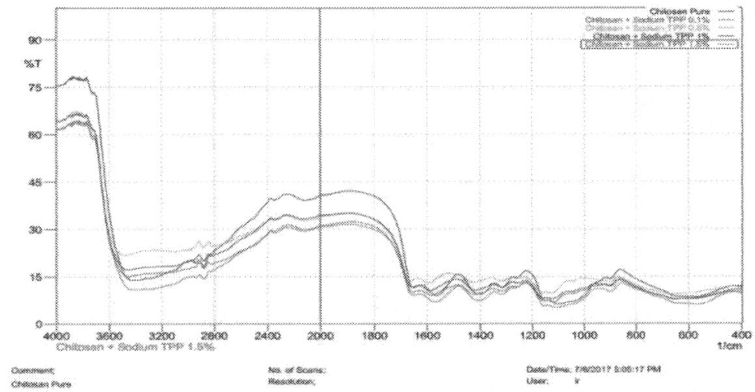

Figure 3. FTIR spectra of (A) pure chitosan, (B) 0.1 %, (C) 0.5%, (D) 1% and (E) 1.5% sodium TPP cross-linked chitosan films

Electrical characteristics of pure chitosan and sodium TPP cross-linked chitosan sensors were observed by exposing the sensors towards acetone concentrations of 5, 10, 50, 100, and 200 ppm, respectively. As clearly seen in figure 4, both pure chitosan and sodium cross-linked chitosan sensors showed fast response, complete recovery and excellent stability. It is also found that sodium TPP cross-linked chitosan sensors increased in conductivity which is clearly seen from the maximum responses in output voltage value of sodium TPP cross-linked chitosan sensors are much higher than pure chitosan. Amongst the variation of sodium TPP, the maximum response of 1% sodium TPP was the highest. This can be explained by the interaction between chitosan and sodium TPP. Since the pKa of chitosan is ~6.3, the hydrogen bonds was broken in acid and electrostatic repulsion would arise between protonated amine groups (NH_3^-) [28]. When sodium TPP was added, the NH_3^- would interact with OH^- and $P_3O_{10}^{5-}$ ions [29]. The OH^- ions are mostly linked to NH_3^- by deprotonation produce –NH_2 while $P_3O_{10}^{5-}$ ions linked by ionic crosslinked produce H^+ ions. This interactions lead to increase the electrical conductivity of the films. Whereas, for 1.5% sodium TPP cross-linked chitosan, OH^- and $P_3O_{10}^{5-}$ ions has got saturated to interact with NH_3^- and resulting in lower conductivity. Futhermore, 1% sodium TPP cross-linked chitosan sensor exhibits the best repeatability since the maximum responses were relative similar for five times continuously measurement within one day. This is worth noting that 1% sodium TPP cross-linked chitosan sensor offers the best electrical characteristics to be used as a novel acetone sensor.

IOP Conf. Series: Materials Science and Engineering **309** (2018) 012083 doi:10.1088/1757-899X/309/1/012083

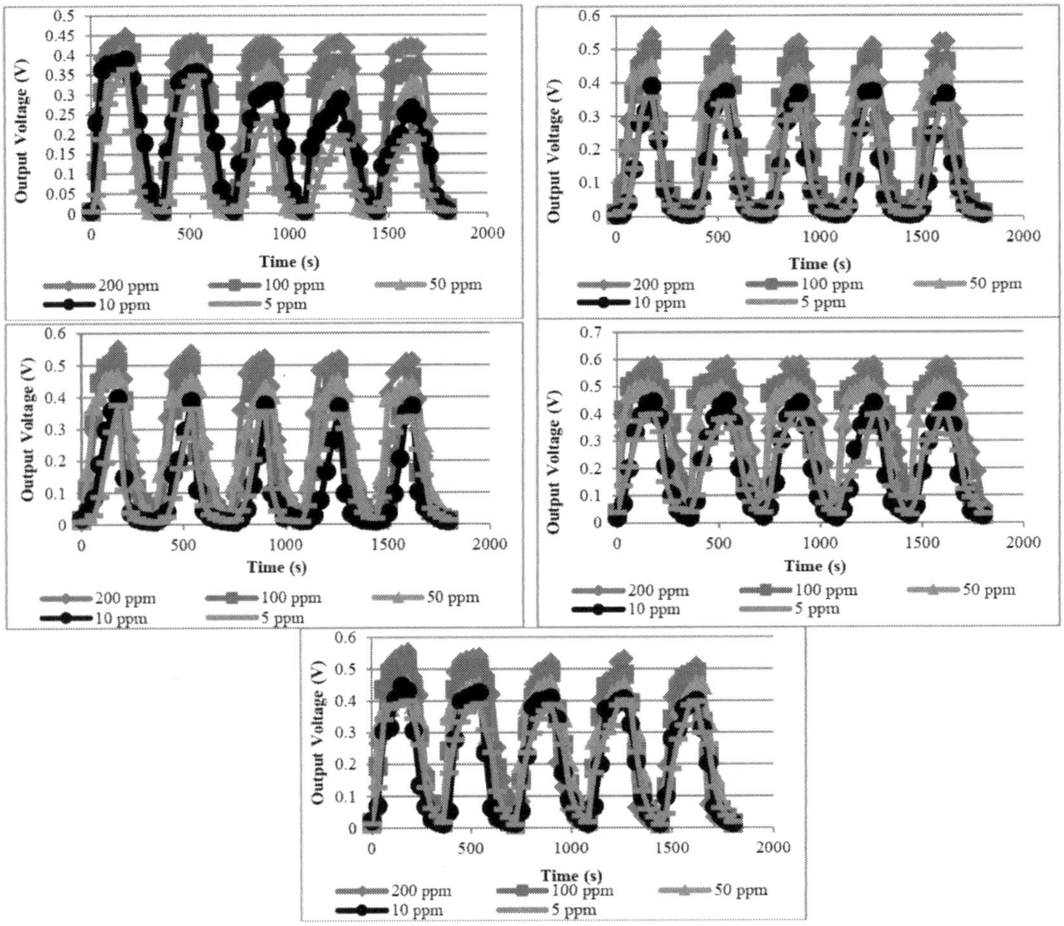

Figure 4. Electrical characteristics of (A) pure chitosan, (B) 0.1 %, (C) 0.5%, (D) 1% and (E) 1.5% sodium TPP cross-linked chitosan film sensors.

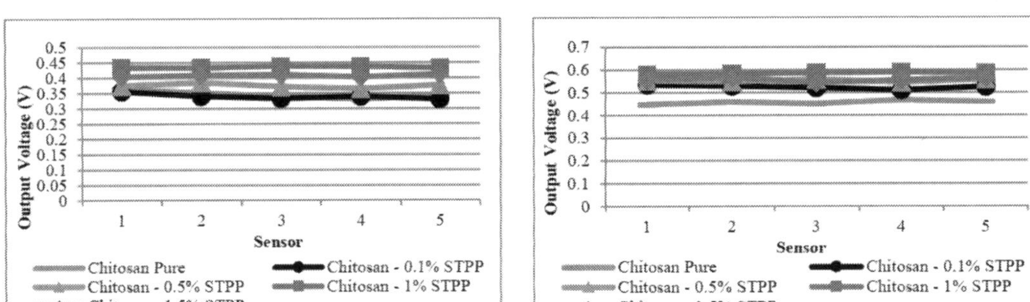

Figure 5. Reproducibility of pure chitosan and sodium TPP cross-linked chitosan film sensors when were exposed towards (A) 5 ppm and (B) 200 ppm acetone concentration.

The reproducibility of pure chitosan and sodium TPP cross-linked chitosan sensors were observed by exposing five different sensors towards 5 ppm and 200 ppm acetone concentration. It is found that sodium TPP cross-linked chitosan sensors are more reproducible than pure chitosan. Nonetheless, 1% sodium TPP cross-linked chitosan demonstrates the most excellent reproducibility which can be seen in figure 5. This indicates that 1% sodium TPP was homogeneously cross-linked with chitosan solution and produce the most reproducible sensors.

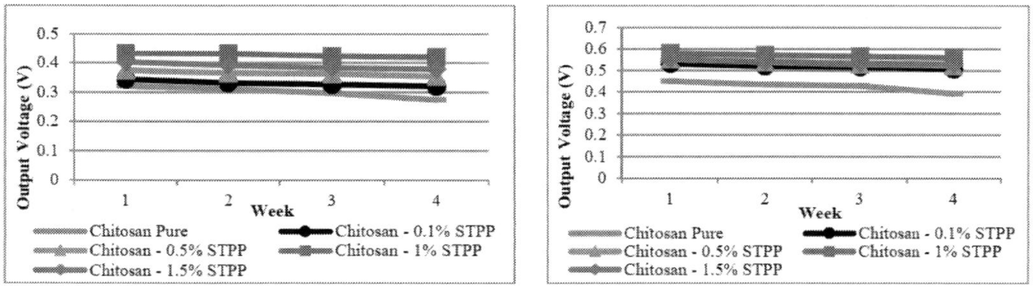

Figure 6. Lifetime of of pure chitosan and sodium TPP cross-linked chitosan film sensors when were exposed towards (A) 5 ppm and (B) 200 ppm acetone concentration.

While figure 6 shows the lifetime of pure chitosan and sodium TPP cross-linked chitosan sensors. Sodium TPP cross-linked chitosan sensors operate more well for 4 weeks compare to pure chitosan since the maximum responses in output voltage value tend to be the same in every week. It is worth noting that crosslinking degree increased as the concentration of sodium TPP increased when it was crosslinked in acidic. The increases crosslinking degree is capable to decrease the swelling of the films. Therefore, the sodium TPP cross-linked chitosan sensors exhibit outstanding lifetime due to the low swelling of the films.

4. Conclusion

In this report, sodium TPP has been successfully cross-linked to chitosan solution in order to enhance the sensing properties as a novel acetone sensor. Based on measurement results, 1% w/v sodium TPP cross-linked chitosan film exhibits the best sensing properties such as fast response, complete recovery, high conductivity, good repeatability, excellent reproducibility and outstanding lifetime. Therefore, the crosslinking of sodium TPP has enhanced the performances of the sensor towards acetone.

Acknowledgement

This study was supported by Centre of Excellence for Sensor Systems and Technology, Universitas Sumatera Utara.

References

[1] Xu X, Chen Y, Zhang G, Ma S, Lu Y, Bian H and Chen Q 2017 *J. Alloys Compd.* **703** 572–9
[2] Lian X, Li Y, Tong X, Zou Y, Liu X, An D and Wang Q 2017 *Appl. Surf. Sci.* **407** 447–55
[3] Shaikh F I, Chikhale L P, Mulla I S and Suryavanshi S S 2017 *Ceram. Int.* **43** 10307–15
[4] Gogoi P, Thakur A J, Devi R R, Das B and Maji T K 2016 *J. Environ. Chem. Eng.* **4** 4248–57
[5] Shi T, Yang D, Yang H, Ye J and Cheng Q 2017 *Appl. Clay Sci.* **142** 100–8
[6] Ni K, Chan C C, Chen L, Dong X, Huang R and Ma Q 2017 *Opt. Fiber Technol.* **33** 56–9
[7] Rui L, Xie M, Hu B, Zhou L, Yin D and Zeng X 2017 *Carbohydr. Polym.* **173** 473–81
[8] Zhang K, Hu R, Fan G and Li G 2017 *Sensors Actuators B Chem.* **243** 721–30
[9] Leceta I, Arana P, Guerrero P and de la Caba K 2014 *Mater. Lett.* **128** 125–7
[10] Fen Y W, Yunus W M M, Yusof N A, Ishak N S, Omar N A S and Zainudin A A 2015 *Opt. - Int. J. Light Electron Opt.* **126** 4688–92
[11] Janaki V, Oh B-T, Shanthi K, Lee K-J, Ramasamy A K and Kamala-Kannan S 2012 *Synth. Met.* **162** 974–80

IOP Conf. Series: Materials Science and Engineering **309** (2018) 012083 doi:10.1088/1757-899X/309/1/012083

[12] Zhang C, Wen H, Huang Y and Shi W 2017 *Int. J. Biol. Macromol.* **97** 635–41
[13] Jawad A H, Azharul Islam M and Hameed B H 2017 *Int. J. Biol. Macromol.* **95** 743–9
[14] Kim U-J, Lee Y R, Kang T H, Choi J W, Kimura S and Wada M 2017 *Carbohydr. Polym.* **163** 34–42
[15] Alshahrani A A, Al-Zoubi H, Nghiem L D and in het Panhuis M 2017 *Desalination* **418** 60–70
[16] Mokhena T C and Luyt A S 2017 *J. Clean. Prod.* **156** 470–9
[17] Ayad M M, Salahuddin N A, Minisy I M and Amer W A 2014 *Sensors Actuators B Chem.* **202** 144–53
[18] Sacco P, Paoletti S, Cok M, Asaro F, Abrami M, Grassi M and Donati I 2016 *Int. J. Biol. Macromol.* **92** 476–83
[19] Delavar Z and Shojaei A 2017 *Carbohydr. Polym.* **167** 219–28
[20] Ma S, Chen Z, Qiao F, Sun Y, Yang X, Deng X, Cen L, Cai Q, Wu M, Zhang X and Gao P 2014 *J. Dent.* **42** 1603–12
[21] Sarkar S D, Farrugia B L, Dargaville T R and Dhara S 2013 *Mater. Sci. Eng. C* **33** 1446–54
[22] Hejjaji E M A, Smith A M and Morris G A 2017 *Int. J. Biol. Macromol.* **95** 564–73
[23] Xiao B, Ma P, Ma L, Chen Q, Si X, Walter L and Merlin D 2017 *J. Colloid Interface Sci.* **490** 520–8
[24] Nasution T I, Nainggolan I, Hutagalung S D, Ahmad K R and Ahmad Z A 2013 *Sensors Actuators B Chem.* **177** 522–8
[25] Liu H, Lu D, Li P, Chen Y, Zhou Y and Lu T 2014 *J. Electroanal. Chem.* **717–718** 125–30
[26] Wang Z, Zhang X, Gu J, Yang H, Nie J and Ma G 2014 *Carbohydr. Polym.* **103** 38–45
[27] Rumansyah T I N and R A and Y M and I N and M B and R 2017 *IOP Conf. Ser. Mater. Sci. Eng.* **180** 12018
[28] Emanet M, Kazanç E, Çobandede Z and Çulha M 2016 *Carbohydr. Polym.* **151** 313–20
[29] Lee S-T, Mi F-L, Shen Y-J and Shyu S-S 2001 *Polymer (Guildf).* **42** 1879–92

IOP Conf. Series: Materials Science and Engineering **309** (2018) 012084 doi:10.1088/1757-899X/309/1/012084

The utilization of stone ash on cellular lightweight concrete

R Karolina*,Y G C Sianipar
Civil Engineering, Universitas Sumatera Utara

*E-mail: rachmie_caroline@yahoo.co.id

Abstract: Lightweight concrete brick is a brick which made of cement, sand, water, and foam as the basic composition. This brick are divided into 2, based on the foam used such as AAC (Autoclave Aerated Concrete) that use aluminium paste and CLC(Cellular Lightweight Concrete) that use foaming agent from BASF as its foaming material. In this trial, the lightweight brick that are ging to be use are the CLC with foaming agent as its foaming material with the mixture of stone ash that are produced by the Stone Crusher with spesific gravity 2666 kg/m^3 as their partly sand substitution . In this research, the stone ash variant that are used are 10%, 15%, and 20% from the amount of sand that planned before. After casting, the result of the 10% will receive a reduction of compressive strength while an increasing in absorption as 25.07% and 39.005% and the 15% variant will recieve a reduction of compressive strength as much as 65.8% and a reduction of absorption as much as 17.441% and the 20% variant will recieve a reduction of compressive strength as much as 67.4% while an increasing of absorption as much as 17.956%.

1. Introduction

The use of bricks as a building material (nonstructural) has long been used throughout the world. The use of bricks in Indonesia has been widely used as a building wall. Almost all buildings in the city and surrounding areas have been using bricks as a building wall. The bricks within the field of construction are a non strutural element, in which the bricks in the building do not accept / withstand the load. The bricks in the building are categorized as dead loads, because if the bricks receive the load there will be a crack.

Broadly speaking, the bricks are divided into 2 types, conventional bricks and concrete bricks. Conventional brick is a red brick made of clay that is dried or burned as a drying process, while concrete bricks are bricks that use cement, sand, water, foaming agent as the material in this research. Lightweight concrete bricks have some advantages over conventional bricks. On the mounting side, concrete bricks are quicker to install because the size of lightweight concrete bricks has larger dimensions than conventional bricks, which can save installation time. Concrete bricks also have fire resistant properties, water resistant, and lighter than conventional bricks.

By the making process, concrete bricks are divided into two, namely: AAC (Autoclaved aereted concrete)/ALC(Aereted Light-weight concrete) and CLC (Cellular Light-weight concrete)' In this research we use stone ash in the manufacturing of lightweight concrete bricks CLC type which has a high compressive strength but has a economical price and environmentally friendly.

Stone ash is a waste of stone crushing from stone crusher machine. The use of stone ashes is usually used as a filler in flexible pavement, rockfill type dam, etc. In this study, I used stone ash as a substitute material for a mixture of lightweight concrete bricks because the stone ash has very fine grains and is easily bonded by cement and for aggregate gradation variations in lightweight concrete bricks. There are several stages on making stone ash on stone crusher machine. In the first stage, the crushed stone is inserted into the primary crusher, then insert it into the secondary crusher and at the final stage insert it into the tertiary crusher. Where ash stones are obtained after passing through tertiary crusher.

IOP Conf. Series: Materials Science and Engineering **309** (2018) 012084 doi:10.1088/1757-899X/309/1/012084

2. Literature Review

Concrete is very commonly used as a building material. Concrete is generally composed of several constituents of fine aggregate (sand), gravel (coarse aggregate), portland cement, and water with a combination of comparisons of each material. Where the mixed cement and water will become a paste material that serves as a binder while fine aggregates and coarse aggregates serve as fillers. When all the mixtures are mixed in accordance with the planned composition will harden like a rock, where the process occurs due to the hydration process (chemical reaction between water and cement).

The quality of the concrete can be influenced by several factors, like the quality of the material from the constituent material, the composition ratio, the way of stirring, the way of treatment, and the way of treatment during the hardening process.

Strength, durability and other concrete properties depend on the nature of the base material, the value of the material composition, the way of stirring and the way of processing during the concrete mixing, the method of compaction, and the way of treatment during the hardening process.

According to (Tri Mulyono, 2005) as a construction material, concrete has advantages and disadvantages, the advantages of concrete are :

1. The plan quality can be adjusted according to the plan.
2. The age of concrete is higher (if not experiencing a greater load of the plan quality.
3. Easily formed according to the needs of construction.
4. Maintenance / treatment costs are small.

The disadvantages of concrete are :

1. Concrete has a low tensile strength, making it easy to crack. Therefore, it is necessary to be given reinforced steel, or gauze reinforcement (meshes).
2. Concrete is difficult to be perfectly waterproof, so it can always inserted by the water, and water carrying salt content can damage the concrete.
3. The shape that has been made is difficult to change.

2.1 Fresh Concrete

A good fresh concrete is the one that can be stirred, transported, poured, compacted, no tendency for segregation (separation of gravel from mortar) and bleeding (separation of water and cement from mortar). This is because segregation and bleeding will make the obtained concrete bad.

2.2 Definition of Lightweight Concrete Brick

The development of construction especially in Indonesia has been growing rapidly. One of the developments is a lightweight concrete brick. Lightweight concrete brick is a brick that has density of conventional brick (red brick).

2.3 Stone Ash

Stone ash is an aggregate that acts as a filler. Stone ash used in this experiment is taken from AMP factory located in Galang area (North Sumatera) and already through a crushing process from stone crusher machine. Usually this stone ash is often used for asphalt mixtures, but nowadays the road construction using asphalt is decreasing so that the use of stone ash is very little because the role of stone ash in the construction is taken over by sand. The collection of sand in the manufacture of construction is using a lot of sand that can damage the environment in the sand dune area. This stone ash is formed from the stone crusher machine which is produced on tertiary crusher.

Stone ash used in this research is produced from AMP factory which has gone through crushing process on stone crusher machine, where the stone ash used is passes filter No.200 and stuck in pan. This stone ash has a density of 2666 kg / m3. The reason for using stone ash which passes filter No.200 is to be able to fill or become a filler in lightweight concrete brick which aim to improve the quality of concrete.

3. Research Method

Research steps were employed in this research depicted in Figure 1. Provision and inspection of materials were performed at first place, following mixing decision and 14 days test. If the result is satisfying, the sample is manufactured, followed by 28 days treatment. Test results were analysed and conclusion was take.

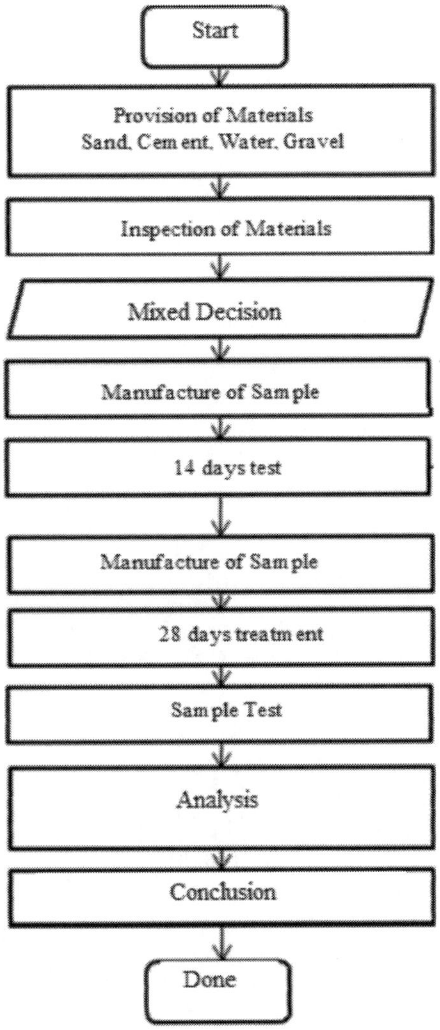

Figure 1. Research Steps

4. Results

4.1 Testing of compressive strength and specific gravity

Testing of compressive strength of lightweight concrete brick was done at age 3, 7, 14, 21, 28 days with the intention to get a picture of the development of the compressive strength of lightweight concrete brick using stone ash as additional material and compare it with the compressive strength of lightweight concrete brick without using stone ash. Table 1 shows that the use of stone ash in a mixture of lightweight concrete bricks showed a significant result in terms of compressive strength. The data is plotted at Figure 1.

IOP Conf. Series: Materials Science and Engineering **309** (2018) 012084 doi:10.1088/1757-899X/309/1/012084

Table 1. Compressive (Pressure) strength and specificgravity (weight)

Sample 1		Weight			Pressure			Average Weight	Average Pressure	Kpa	Average Specific Weight
		BU 1	BU 2	BU 3	BU 1	BU 2	BU 3				
3		8,450	8,083	8,363	118,300	98,400	88,300	8,299	101,667	5,756	1,566
7	0%	7,636	8,156	8,051	138,000	139,000	128,500	7,948	135,167	7,653	1,500
14		7,562	7,449	7,682	159,800	158,000	154,300	7,564	157,367	8,910	1,428
21		7,553	7,410	7,610	162,300	165,400	160,000	7,524	162,567	9,204	1,420
28		7,507	7,398	7,590	163,400	165,800	161,200	7,498	163,467	9,255	1,415

Sample 2		Weight			Pressure			Average Weight	Average Pressure	Kpa	Average Specific Weight
		BU 1	BU 2	BU 3	BU 1	BU 2	BU 3				
3		8,460	8,670	8,520	69,200	69,500	68,300	8,550	69,000	3,907	1,614
7	10%	8,187	8,263	8,297	118,500	112,700	110,600	8,249	113,933	6,451	1,557
14		8,155	8,166	8,200	120,000	122,200	118,700	8,174	120,300	6,811	1,543
21		7,889	8,052	7,900	120,400	122,300	119,800	7,947	120,833	6,841	1,500
28		7,661	7,919	7,887	121,600	123,800	120,400	7,822	121,933	6,904	1,476

Sample 3		Weight			Pressure			Average Weight	Average Pressure	Kpa	Average Specific Weight
		BU 1	BU 2	BU 3	BU 1	BU 2	BU 3				
3		7,353	7,460	7,822	46,600	43,300	39,800	7,545	43,233	2,448	1,424
7	15%	7,340	7,327	7,497	46,900	48,700	41,100	7,388	45,567	2,580	1,394
14		7,280	7,316	7,390	57,200	56,200	50,900	7,329	54,767	3,101	1,383
21		7,220	7,310	7,380	57,400	56,700	51,300	7,303	55,133	3,121	1,378
28		7,308	7,294	7,393	58,100	57,300	52,300	7,332	55,[illegible]	3,165	1,384

Sample 4		Berat			Pressure			Average Weight	Average Pressure	Kpa	Average Specific Weight
		BU 1	BU 2	BU 3	BU 1	BU 2	BU 3				
3		8,225	8,263	8,228	35,400	37,300	34,300	8,239	35,667	2,019	1,555
7	20%	8,110	8,010	7,990	48,200	49,200	38,900	8,037	45,433	2,572	1,517
14		7,890	7,980	7,890	52,460	52,100	52,090	7,920	52,217	2,956	1,495
21		7,640	7,540	7,700	53,500	52,990	52,700	7,627	53,063	3,004	1,439
28		7,294	7,519	7,364	53,600	53,200	52,900	7,392	53,233	3,014	1,395

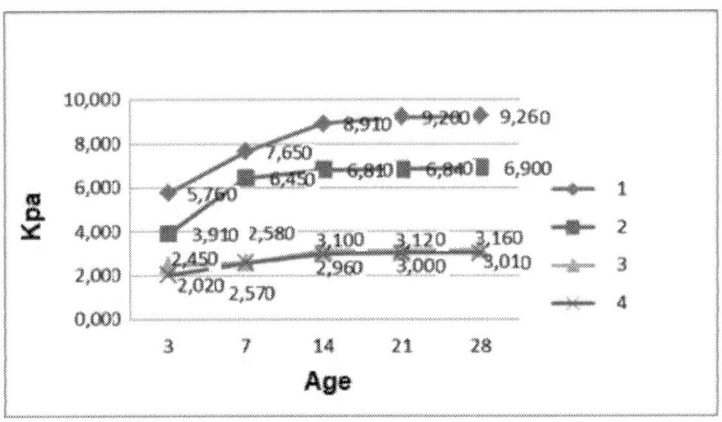

Figure 1. Compressive strength and specific gravity

The absorbtion Test result is enlisted in Table 2 and plotted in Figure 2.

Table 2. The absorption results

Variation 1		Dry Weight	Wet Weight	Absorption Value
14	0%	7,682	8,0930	5,3502
21		7,610	8,2150	7,9501
28		7,590	8,3100	9,4862

Variation 2		Dry Weight	Wet Weight	Absorption Value
14	10%	8,200	8,5940	4,8049
21		7,900	8,7060	10,2025
28		7,887	8,9270	13,1863

Variation 3		Dry Weight	Wet Weight	Absorption Value
14	15%	7,390	7,8230	5,8593
21		7,380	7,9560	7,8049
28		7,393	7,9720	7,8317

Variation 4		Dry Weight	Wet Weight	Absorption Value
14	20%	7,890	8,1390	3,1559
21		7,700	8,2090	6,6104
28		7,364	8,1880	11,1896

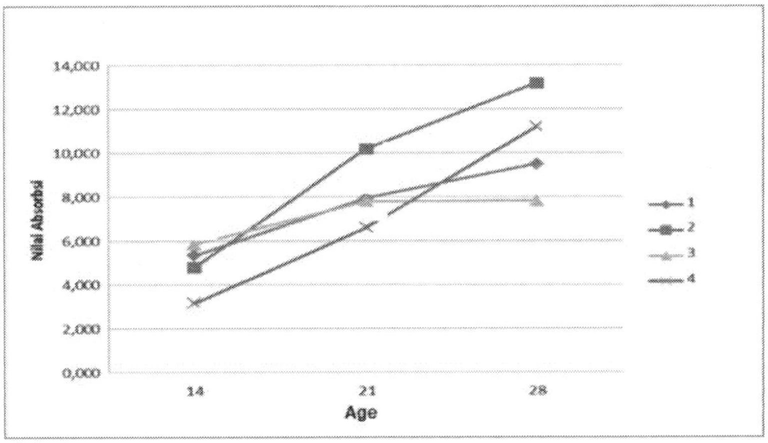

Figure 2. Absorbtion plots

5. Conclusion

Based on the research that has been done, it can be concluded that:

1. The use of stone ash as sand substitution in lightweight concrete brick can decrease the quality of lightweight concrete brick.
2. The use of stone ash up to 15% of the amount of sand is still within the safe limits of use.
3. Curing process should be done when lightweight concrete brick is 7 days old.
4. The quality of lightweight concrete brick depends on the foaming agent used.
5. Samples that have been removed from the mold should be stored in a room that is protected from sunlight.
6. The cost of production of Cuban Light Concrete Type CLC is the most expensive between Red Brick and Lightweight Concrete Brick Type AAC Grand Elephant.

References

[1] Hadi T 2012 *Jurnal ORBITH* **8(1)**
[2] Nugraha P and Antoni 2007 *Teknologi Beton* Yogyakarta:Penerbit Andi
[3] Sutarno 2007 *Jurnal Wahana TEKNIK SIPIL* **12(3)**
[4] Mulyono T 2005 *Teknologi Beton* Yogyakarta:Penerbit Andi.
[5] Zongjin D L 2011 *Advanced Concrete Technology* USA : John Wiley & Sons Inc
[6] Kishore K 2007 *Foamed Cellular Light Weight Concrete*
[7] Santhakumar A R 2007 *ConcreteTechnology* India Oxford University Press
[8] SNI 03-0349-1989 Bata Beton Untuk Pasangan Dinding
[9] BASF *Aplikasi Light Weight Concrete Untuk Bata Ringan (Aplikasi Blok LWC)*

Rubber Fruit Shell (*Hevea brasiliensis*) as bio sorbent to remove FFA (Free Fatty Acid) content in CPO (Crude Palm Oil)

S Pandia*, M S Sinaga, Z Masyithah, A Husin, S Nurfadilla, Fitriani and B K S Sipahutar

Department of Chemical Engineering, Faculty of Engineering, Universitas Sumatera Utara, Padang Bulan, Medan 20155, Indonesia.

*Email: setiaty_pandia@yahoo.com

Abstract. This study aimed to discover the effectiveness of the shell of rubber fruit as bio sorbent for removing FFA (Free Fatty Acid) content in CPO (Crude Palm Oil). Methods used in this study were pretreatment, activation (carbonating and chemically) and adsorption process at room temperature. In the beginning, the shell of rubber fruit was cleaned and dried under the sun. Then the shell was cut for about 0.5 cm of length and carbonated in a furnace for 1h at 600°C. After that, they were crushed to pass through 140 meshes and activated using three variations of chemical such as 6 of HNO_3, 6N of KOH and 6N of H_3PO_4 at certain ratio as 1:3, 1:4, and 1:5 (b/v). The adsorption process was carried out using bio sorbent with the highest iodine number in varying bio sorbent dosage and contact time. The highest iodine number was 913.680 mg/g and obtained at the ratio of bio sorbent to 6N of KOH as 1:5. The best removal of FFA content was 91.94% and at 1% bio sorbent dose and 30 min of contact time.

1. Introduction

Rubber plantation is one of a plantation which is well developing in agriculture in Indonesia. Physically the shell of rubber fruit (*Hevea brasiliensis*) is known as a plant with lignin contained. According are Zakaria *et al* (2015) [1], the shell of rubber fruit contains active compound which is lignin. The use of the shell of rubber fruit is not optimal yet although its content of lignin is relatively (35% - 54%) [1] Therefore it is potential to become an activated carbon product, namely bio sorbent. Several studies have been conduct that used part of a rubber tree, such as the powder of rubber wood sewed to diminish chromium (VI) content in a solution and the leaf of rubber plant to minimize cup rum ion using sodium hydroxide in solution [1].

The initial stage of CPO (Crude Palm Oil) rafi-nation is the removal of gum (degumming) followed by bleaching, deodorization and fractionation [2]. Conventionally, degumming is the flock's formation of substance that is colloidal in crude oil. Crude Palm Oil contains FFA (Free Fatty Acid) which has to be diminishing in a purification process. Free Fatty Acid is an acid which is liberate in fat hydrolysis by an enzyme. The disadvantage of the FFA presence in CPO is it will affect the oil stability so then the further process in CPO treatment. For example to produce biodiesel, if the FFA content more than 1% it will become soap instead of methyl ester [3]. Based on the preceding, the bio sorbent from the shell of rubber fruit must be attempted to be applied as an alternative bio sorbent to diminish FFA content in CPO.

2. Material and Method

2.1 Material

The main materials were the shell of rubber fruit and crude palm oil that were obtain from the neighborhood around the University of Sumatera Utara. Potassium hydroxide, phosphate acid, and nitric acid were use in bio sorbent activation. Phenolphthalein, ethanol, and potassium hydroxide were used for analysis purpose.

2.2 Bio Sorbent Activation

The shell of rubber fruit was washed with water and dried in the sunlight. They were cutting in 0.5 cm long in average and then they were carbonated in a furnace for 1 hour at 600°C. After that, they were crushed to powder form and passed through 140 meshes. Bio sorbent was activateing with a solution of 6 N H_3PO_4. Other solutions for activating use were 6 N KOH and 6 N HNO_3. It was follow by washing several times with distilled water to remove the activating solution. The procedure was repeate for the ratio of bio sorbent: solution of 1:3, 1:4 and 1:5 (m/v).

2.3 Analysis of Iodine Number

Iodine number shows the unsaturated fatty acid composed of oil and fat. The unsaturated fatty acid can bind iodine and creates a saturated compound. Iodine number is stated as the number of iodin compound which is bound by 100 grams of oil or fat.

3. Results and Discussion

3.1 Effect of Type of Activating Solution and Ratio of Bio Sorbent to Iodine Number

Before and after activation, an iodine number of the bio sorbent was analyzed. The iodine number before activation was 545.799 mg/g, while the iodine numbers after activation were given in table 1 below:

Table 1. Iodine number of bio sorbent for any condition of activation at 600°C.

Kind of Solution	Ratio Bio sorbent: Solution (m/v)	Iodine Number (mg/g)
6 N KOH	1: 3	744.887
	1: 4	736.134
	1: 5	913.680
6 N H_3PO_4	1: 3	825.045
	1: 4	672.729
	1: 5	725.928
6 N HNO_3	1: 3	431.562
	1: 4	431.562
	1: 5	482.334

From table 1, that irregularity occurred. It is maybe due to a water content of the after-water bio sorbent cannot be controlled. In general, it is shown that KOH as activating solution is more effective compared with H_3PO_4 and HNO_3. KOH is a strong base solution, therefore it is more affecting the active pore information. Furthermore the increase of the ratio of bio sorbent to activating solution fond increases the iodine number. A very *low* concentration of activator may cause the in complete formation of the active sites where as a very high ratio of activator may cause damage to the structure of the bio sorbent [4]. By comparing theory and result obtained from the research, it can be concluded that the most favorable conditions to produce the bio sorbent from the shell of rubber fruit with the highest iodine number are the ratio of bio sorbent: potassium hydroxide of the 1:5 at 600°C.

According to SNI-06-3730-1995, the resulting bio sorbent is already qualified as activated carbon. It is seen from iodine number 913.68 mg / g, vapor content 2.64%, ash content 9.88% and water content 14.77%. Where the requirements given by SNI in the table below:

Table 2. Standard of activated carbon from SNI.

Requirements	Content
Iodine Number	> 750 mg/g
Vapor Content	< 25%
Ash Content	<10%
Water Content	<15%

Bio sorbent characterization using FTIR Spectrophotometry the characterization was conducted fix bio sorbent before and after the activation, and after used in the adsorption process. Functional groups on bio sorbent can be inferred by comparing result on the graph with literature or IR correlation [6]. The graphs are in figure 1, 2 and 3.

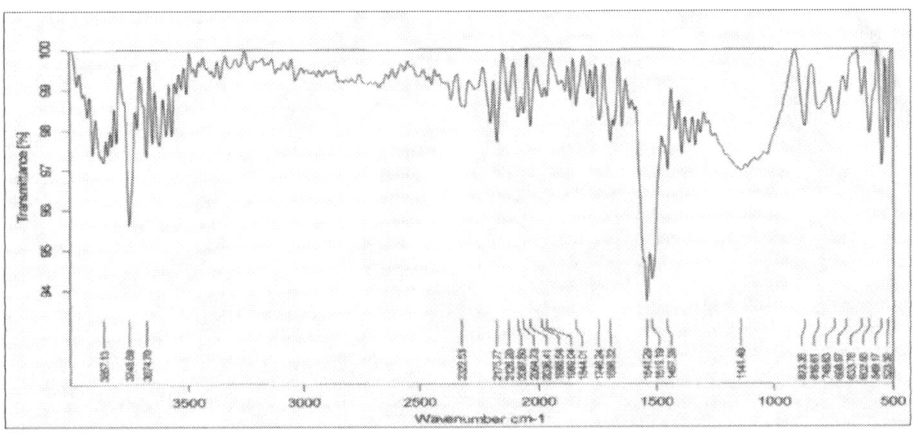

Figure 1. FTIR spectrophotometry result for bio sorbent before activation.

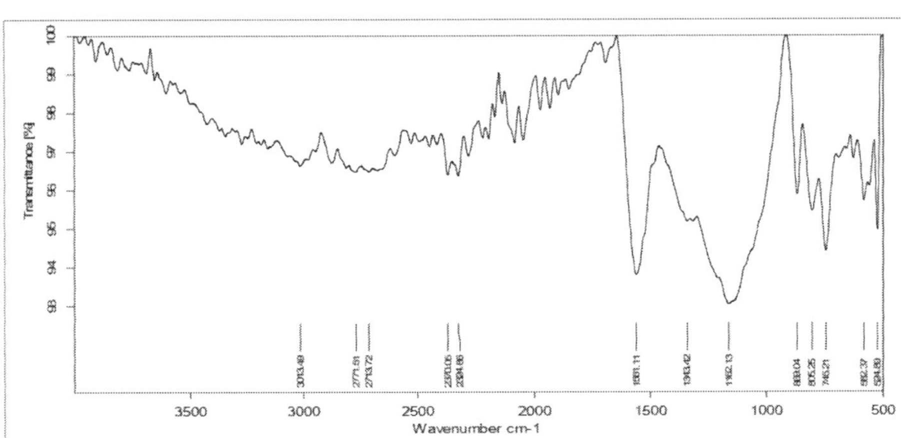

Figure 2. FTIR spectrophotometry result for bio sorbent after activation.

IOP Conf. Series: Materials Science and Engineering **309** (2018) 012085 doi:10.1088/1757-899X/309/1/012085

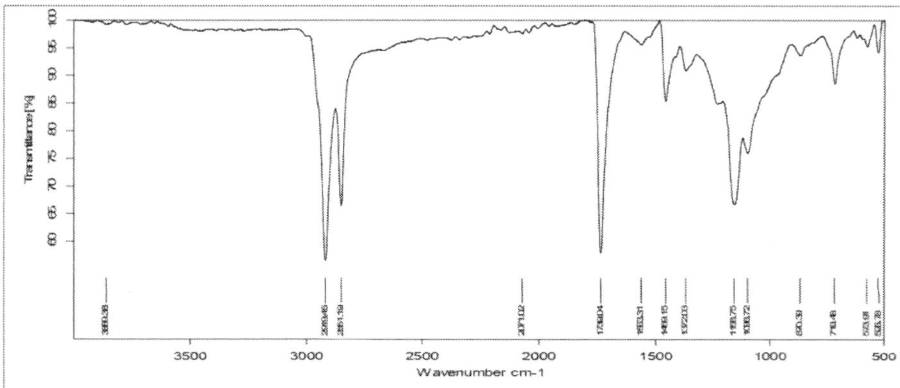

Figure 3. FTIR spectrophotometry result for used bio sorbent.

3.2 Effect of Bio Sorbent Dose and Contact Time on FFA Content in CPO

The initial FFA content in CPO used was 2.48%, while the FFA content in CPO after being contacted with bio sorbent is given in table 3 and can also in figure 4.

Table 3. FFA content in CPO after adsorption (%).

Dose of Bio sorbent (%)	Time (minutes)		
	30	40	50
0.5	1.037	0.259	0.332
1.0	0.200	0.360	0.460
1.5	0.590	0.920	0.860

3.3 Characteristics of Functional Group on Bio Sorbent from the Shell of a Rubber Fruit with FTIR Spectrophotometry

Bio sorbent before and after activation and bio sorbent that has been used for adsorption of impurity contents on CPO subsequently characterized its functional groups by FTIR spectrophotometry. The groups present in the bio sorbent can be summarizes by comparing the wave peaks formed with the literature, which is from the IR correlation table [5]. From figures 1 and two can be compared to the functional groups present in the bio sorbent, before and after activation. The bio sorbent before to activation contains an alkene group, an aromatic carbon ring compound, a carbon dioxide, an -OH group of alcohols, a nitro compound, an alkyne and an alkane. After activation, there is no more alkyne in the bio sorbent. Activating agent (KOH) will oxidize carbon and damage the inner surface of the carbon. Then there may be the possibility of loosening of the carbon double bond in the bio sorbent after it is activated. The content of the hydroxyl -OH group in the rubber shell tends to interact with adsorbat [6].

Alkanol or alcohol groups affect the bio sorbent, free fatty acids, peroxides, and polar organic substances. KOH will react with carbon so that it will form new pores and produce carbon dioxide that diffuses to the carbon surface, where the activating agent (KOH) will oxidize the carbon and damage the inner surface of the carbon so that pore will be formed and increase adsorption [7]. In bio sorbents that have been contact with CPO, there are several functional groups bound by bio sorbents after contacting. At the wave number 2919.45 cm^{-1}, there is a C-O group of carboxylic acid which is an indication of free fatty acid compound. In the wave number 3859.38 cm^{-1} there is -OH group which is an indication of the attached polar peroxide compound HOOH. Aldehydes are compounds that cause turbidity and rancid odor or rancid in oil. In the wave number 1739.04 cm^{-1}, there is an ester group. The amine and nitro compounds present in the bio sorbent, both before and after activation, and after contact with CPO, are most likely to be obtained from rubber shells from soil and crop fertilizers.

The major fat or oil damage is due to oxidation and hydrolytic events, both enzymatic and no enzymatic. Damage to palm oil causes rancid or rancid odors mainly caused by aldehydes and ketones [3]. From the results of IR spectrophotometry, it can be concluded that the bio sorbent from the shell of the rubber fruit can adsorb the carboxylic acid content in the form of free fatty acids, peroxide compounds, and aldehydes from palm oil.

3.4 Adsorption of Free Fatty Acid Content (FFA) in CPO

From this study can be seen the decrease in free fatty acid (FFA) in the optimum CPO of various variations of contact time and a dose of bio sorbent. At 40 minutes with the addition of 1%, bio sorbent dose resulted in the greatest decrease in FFA levels. The results in the graph in the following figure.

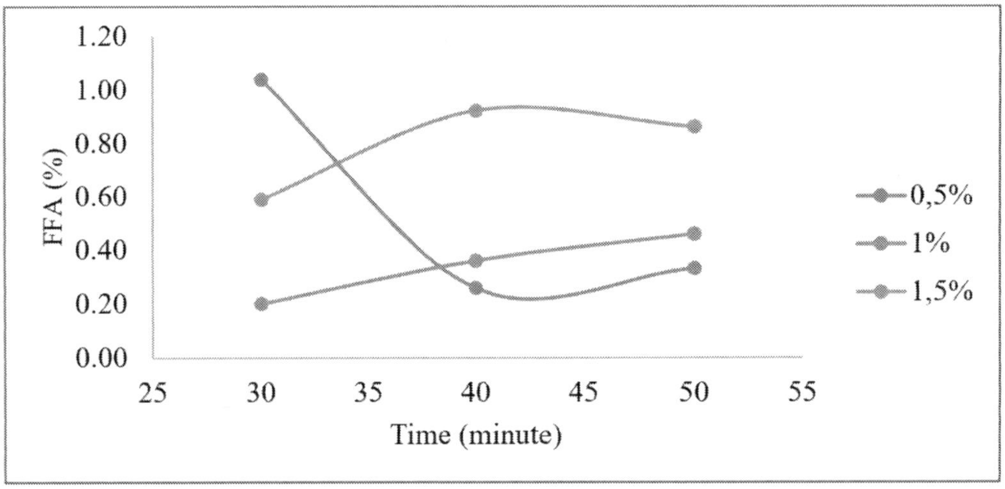

Figure 4. Graph of decreasing fatty acid content (FFA) of oil palm at a specific time.

The initial free fatty acid content in palm oil was 2.48%. For the addition of a dose of 0.5 mg of rubber shell bio sorbent in the oil at 30 minutes, the fatty acid content in the oil decreased to 1.037%. At 40 minutes the free fatty acid content in the oil drops to 0.259%. And at 50 minutes the levels of free fatty acids in the oil increased to 0.332%. Furthermore, for the addition of the dosage of the bio sorbent of the rubber shell of 1.0% in the oil at 30 minutes, the free fatty acid content in the oil decreased to 0.20%. At 40 minutes the free fatty acids content in the oil increased to 0.36%. And at 50 minutes the levels of free fatty acids in the oil increased to 0.46%. Then for the addition of dosage of the bio sorbent of the shell of the rubber fruits by 1.5% in the oil at 30 minutes, the fatty acid content in the oil decreased to 0.59%. At 40 minutes the free fatty acid content in the oil increased to 0.92%. And at 50 minutes the levels of free fatty acids in the oil decreased again to 0.86%. According to Puah *et al.* [3], the best-activated dose of the bio sorbent to remove the undesirable content of CPO is > 0.5%. The standard fatty acid content of CPO adsorption using degumming agent and bleaching earth is 3 - 5% [8]. In the palm oil refining industry using conventional degumming agent (85% phosphoric acid) with a dose of 1% by weight of oil, free fatty acid content increased from 5% to 5.05%. The FFA concentration at the 0.5% bio sorbent dose decreased by 40 minutes and increased at 50. Also, at 1% and 1.5% of the free fatty acid doses increased by 30 to 50 minutes.

This may be due to the longer contact time of palm oil with bio sorbents resulting in longer oil interacting with the heat that triggers the formation of free fatty acids. From the results of the research, the utilization of the activated rubber shell activator with potassium hydroxide with a ratio of (b: v) 1: 5 with a dose of addition of bio sorbent to oil of 1.0% of the weight of the oil with contact time for 30 minutes yielded decrease in free fatty acid content is greatest from 2.48% to 0.20. Therefore, given the economic consideration of the amount of bio sorbent used, a dosage of 1.0% was selected as the best dose for the adsorption of free fatty acid content in palm oil and has met the standard free fatty acid content in palm oil after adsorption. The percentage decrease of FFA content is 91.93%.

4. Conclusion

The iodine number of the rubber shell of 913.680 mg/ g is obtained from activation at 600°C using 6N KOH with bio sorbent ratio: KOH is 1: 5. The minimum Free Fatty Acid content obtained was 0.20% with a decreasing percentage of 91.94%, at a dosage of 1% bio sorbent and a contact time of 30 minutes.

References

[1] Zakaria, Akmar Z, Surtmaw M, Mohammed N and Achmad W A 2008 Chromium (VI) Removal from Aqueous Solution by Untreated Rubber Wood Sawdust. Elsevier Science Direct
[2] Copeland D and Maurice B W 2005 Vegetable Oil Refining *US Patent* 68444 58
[3] Puah C W, Choo Y M, Ma A N and Chuah C 2004 Degumming and Bleaching: Effect on Selected Constituents of Palm Oil *Journal of Oil Palm Research* **16** 2
[4] Fereidoon S 2005 *Bayley's Industrial Oil and Fat Products* ed 6 Jhon Wiley and Sons. Inc
[5] Skoog D A, Holler F J and Niesman T A 1998 *Principles of Instrumental Analysis* ed 5 Orlando Harcourt Brace
[6] Nilaujana R 2005 Use of Plant Material as Natural Coagulants for Treatment of Wastewater Access 1st of March 2014 from Vision review Point http:www.visionreviewpoint.com/article.asp
[7] Thomas B M, Maby M M and Shaleesha 2013 Comparative Analysis of Coast Effective Bio sorbents in the Removal of Chromium and Iron from Aqueous Solution *Advanced Bio Tech Article* **12**
[8] K L Er 1985 Quality Control Laboratories in Refineries *Proc. of Workshop on Quality in the Palm Oil Industry*

Effect of fermentation time of mixture of solid and liquid wastes from tapioca industry to percentage reduction of TSS (Total Suspended Solids)

S Pandia[1*], S Tanata[1], M Rachel[1], C Octiva[1] and N Sialagan[1]

[1]Department of Chemical Engineering, Faculty of Engineering, Universitas Sumatera Utara, Padang Bulan, Medan 20155, Indonesia.

*E-mail: setiaty_pandia@yahoo.com

Abstract. The waste from tapioca industry is as an organic waste that contains many important compounds such as carbohydrate, protein, and glucose. This research as aimed to know the effect of fermentation time from solid waste combined with waste-water from the tapioca industry to percentage reduction of TSS. The study was started by mixing the solid and liquid wastes from tapioca industry at a ratio of 70:30, 60:40, 50:50, 40:60, and 30:70 (w/w) with a starter from solid waste of cattle in a batch anaerobic digester. The percentage reduction of TSS was 72.2289 at a ratio by weight of the composition of solid and liquid wastes from tapioca industry was 70:30 after 30 days of fermentation time.

1. Introduction

The production of biogas from biomass is important due to the benefit from the environmental point of view and resulting renewable energy resources [1]. The raw material for biogas can derive from waste such as the wastes of tapioca industry where these wastes make environmental impacts, methane gas emissions, soil and ground water contamination and odor problem [2]. Therefore tapioca is one of the industries that generate enormous wastes of solid and liquid waste [3]. Both wastes (liquid and solid waste) from tapioca industry still contain a lot of carbohydrates, protein, and other sugar forms. Furthermore, the solid waste of tapioca industries itself still contains a lot of sugar compounds such as sucrose, glucose, fructose, dextran, galactose and nitric acid which is convertible into other products [4]. The purpose of this study was to determine the percentage reduction of TSS (Total Suspended Solids) due the time of fermentation. The raw material used was the mixture of solid waste combined with liquid waste from tapioca industry at several compositions. Then the mixture was fermented anaerobically.

2. Material and Methods

2.1 Preparation of Raw Materials

The raw materials which were solid waste and liquid waste were collect from tapioca starch factory in Sei Rampah – Sumatera Utara Province. The solid waste was mix with liquid waste before experiment at the ratio of 70:30, 60:40, 50:50, 40:60, and 30:70 by weight to create slurry. The initial pH of the slurry was 5 and then adjusted to 7 using buffer solution (Calcium Hydroxide) and initial TSS (Total Suspended Solid) was analyzed gravimetrically.

2.2 Inoculum, Batch Digester, and Operation

Fresh cattle manure was collect from a farm in Sunggal, Sumatera Utara Province. Sufficient water was adding to cattle manure at a ratio of 1:1 by weight to produce the slurry which was used to inoculate test digester with five days acclimation. The experiments were conduct in batch mode and an aerobic after inoculation. Schematic of anaerobic digester experimentation design layout is show in figure 1.

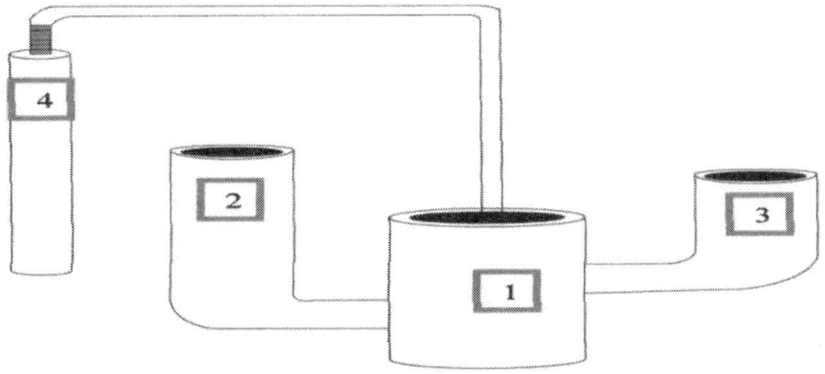

Figure 1. Schematic of anaerobic digester experimentation design layout

After acclimation, the digesters were fed on a batch basis. Daily mixing was done to prevent a crust. The sampling of effluent was analyzed every 3 days until it reached the steady-state condition.

2.3 Analytical Methods

pH was measured using pH indicator, and TSS (Total Suspended Solids) were analyzed gravimetrically in Research Laboratory of Chemical Engineering Department of USU.

3. Results and Discussion

TSS is one of the important parameters in industrial waste water treatment. On the figure below, the change of TSS value during the fermentation process in an aerobic digester.

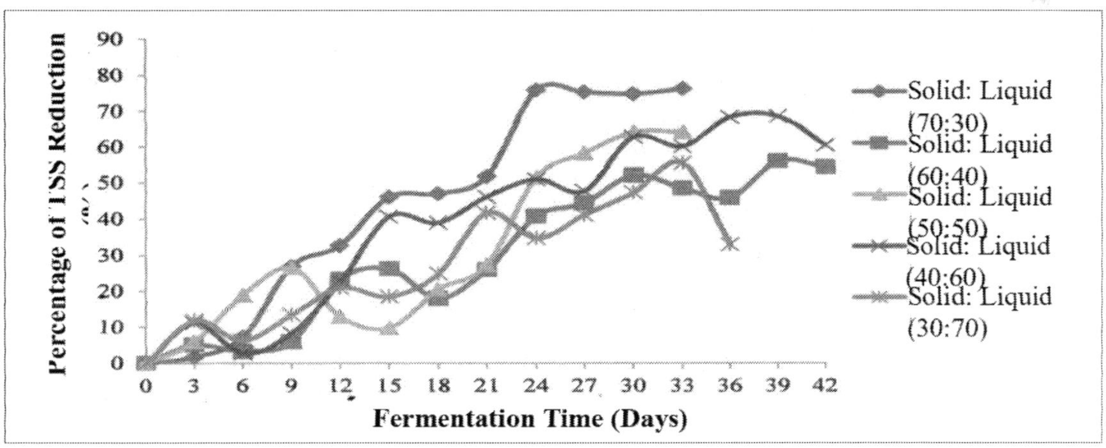

Figure 2. The effect of time fermentation to the reduction of TSS in several compositions

Like the figure, that the longer time of fermentation resulting in the reduction of TSS because higher, although tended, fluctuated. The *highest* percentage of TSS reduction was achieved by the

mixture of solid waste and liquid waste at a ratio of 70:30 at the day of 33 which was 76.2289 %. At a ratio of 60:40, the highest percentage of TSS reduction was 56.1495% at the day of 39, while it was 64.2857 % at the day of 33 at a ratio of 50:50. For the ratio of 40:60, the highest off TSS reduction was 68.6444 % at the day of 39 it was 55.4646 % for the ratio of 30:70 at the day 33. On the other hand, the volume of biogas resulted was optimum and constant at the day of 33.
The experimental result like the table below:

Table 1. Experimental result

Ratio of Solid Waste to Liquid Waste (w/w)	Day	TSS (mg/L)
70:30	0	5208
	33	1239
60:40	0	4976
	33	2266
50:50	0	5180
	33	1850
40:60	0	5326
	33	2100
30:70	0	4832
	33	3240

Theoretically, TSS is total solid which is resisted by a filter with a maximal particle size of 2 μm of bigger from colloidal particle size. The categories of TSS are sludge, clay, oxide metal, sulfide, algae, and bacteria. TSS is treated by flocculation and filtration in general. The negative impact of TSS due to its contribution to turbidity which can be affecting light to penetrate for photosynthesis process and visibility in the water body. Moreover, when the TSS in inlet flow is relatively high, they will need a longer time of hydraulic in a reactor for hydrolysis process by an anaerobic microorganism to become simpler compounds. Hydraulic time of 12-24 hours is not enough to hydrolysis and degradation process biologically of suspended solid in inlet flow. In this a case which was a batch process, the microorganism had enough time to degrade solids in waste. Therefore the efficiency of TSS reduction can be higher [5]. From the figure 2, it can be seen that there were fluctuate reduction.

It was because of the presence of intermediate products which could be limited the degradation process, the inhibition process in methanogens due to the higher unhydrolised inhibitor like sulfide and inhibition due to the rise of free ammonia compounds [6]. According to the regulation by the Minister of State for Environment, the maximum of TSS for a liquid waste of tapioca industry is 50 mg/L [7]. Therefore the liquid waste of this industry cannot be thrown away to the environment. Otherwise, it should be used as liquid fertilizer or using adsorbent to diminish the TSS content.

4. Conclusion

TSS is treated by flocculation and filtration in general. The negative impact of TSS due to its contribution to turbidity which can be affecting light to penetrate for photosynthesis process and visibility in the water body. The highest percentage of TSS reduction was achieved by the mixture of solid waste and liquid waste at a ratio of 70:30 at the day of 33 which was 76.2289 %.

References

[1] Irvan, Trisakti B, Husaini T, Sitio A, and Sitorus T B 2017 Performance Evaluation on Otto Engine Generator Using Gasoline and Biogas from Palm Oil Mill Effluent, *IOP Conf. Series: Mat. Sci. En.* **206** 012028

IOP Conf. Series: Materials Science and Engineering **309** (2018) 012086 doi:10.1088/1757-899X/309/1/012086

[2] Pompan P and Annop N 2008 Anaerobic co-digestion of cassava pulp and pig manure: effects of waste ratio and inoculum substrate ratio *IGSEE* 3
[3] Obalua, AO 2007 Cassava waste: treatment option and value addition alternatives *African J. of Biotechnology* **6** 18
[4] Oratai C and Maneerat O 2008 Clean technology for the tapioca starch industry in Thailand *J. of Cleaner Production* 216
[5] Vindis P 2008 Biogas production with the uses of mini digester *J. of Achievement Mat. and Manufacturing Eng.* **28** 1
[6] Mc Carty PL and Smith DP 1991 *Anaerobic wastewater treatment* (NY: John Wiley and Sons)
[7] Kepmen LH: KEP-51/MENLH/10/1995 Tentang baku mutu limbah cair bagi kegiatan industri

IOP Conf. Series: Materials Science and Engineering **309** (2018) 012087 doi:10.1088/1757-899X/309/1/012087

Utilization of oil palm fronds in producing activated carbon using Na_2CO_3 as an activator

S Maulina[1,2*], FN Anwari[1]
[1] Department of Chemical Engineering, Faculty of Engineering, Universitas Sumatera Utara
[2] Sustainable Energy and Biomaterial Center of Excellence, Faculty of Engineering, Universitas Sumatera Utara

*E-mail: maulina_harahap@yahoo.com

Abstract. Oil Palm Frond is a waste in palm oil plantations that have the potential to be processed into more valuable products. This possibility is because of the presence of cellulose, hemicellulose, and lignin in oil palm fronds. Therefore, this study aimed to utilize oil palm fronds in manufacturing of activated carbon through pyrolysis and impregnation that meets the requirements of the Industrial National Standard 06-3730-1995. The palm-fringed oil palm fronds were pyrolyzed in reactors at 150^{0}C, 200^{0}C, and 250^{0}C for 60 minutes. Subsequently, the charcoal produced from the pyrolysis was smoothed with a ball mill, sieved with a size of 140 meshes, and impregnated using a Sodium Carbonate (Na_2CO_3) for 24 hours at a concentration of 0 %, 2.5%, 5%, and 7.5 % (w/v). The activated carbon has 35.13% of charcoal yield, 8.6% of water content, 14.25% of ash content, 24.75% of volatile matter, 72.75% of fixed carbon, and 492.29 of iodine number. Moreover, SEM analysis indicated that activated carbon porous are coarse and distributed.

1. Introduction

This research aimed to utilize oil palm fronds as one of the waste resulting from oil palm plantations into more valuable products, such as charcoal or activated carbon that follows the Industrial National Standard 06-3730-1995. The main idea of this research was based on the fact that oil palm plantations in Indonesia is keep expanding in almost all provinces. Currently, Indonesia is classified as a first world producer of palm oil followed by Malaysia [1]. The untapped potential of oil palm plantations in Indonesia can be reflected from the fact that there are 10.14 million tons of oil palm fronds waste that was produced by an area of oil palm plantations. Unfortunately, the utilization of oil palm fronds in Indonesia was mainly for animal food, nitrocellulose, and compost. One of the solid wastes produced from oil palm plantation is oil palm frond, which contains cellulose, hemicellulose, and lignin. This content shows great potential to be used as raw material in the manufacturing of charcoal or activated carbon using pyrolysis and impregnation process. In other words, oil palm fronds could be processed further into useful and economically valuable products.

Consumption of activated carbon in Indonesia is still relatively high. It is due to the widespread use of activated carbon in the industrial sector, such as fisheries and health sector [2]. The domestic activated carbon manufacturing industry has been progressing rapidly due to the increasing market demand for activated carbon both inside and outside the country [3]. This is supported by the fact that demand for activated carbon increases by 9% annually throughout 2014, while global activated carbon consumption in 2014 is estimated at 1.7 million tonnes per year. Furthermore, the number of activated carbon exports in Indonesia amounted to 21.652.271 kg per year in 2011 [2].

Hence, the purpose of this research is to utilize oil palm fronds as waste into products that have more value, for instance charcoal or activated carbon-based on Industrial National Standard 06-3730-1995.

2. Theory

Charcoal or activated carbon can be made through two stages, namely pyrolysis and activation or impregnation. Pyrolysis is an indoor combustion process in the absence of oxygen and other chemicals, whereas impregnation is required to convert combustion products into adsorbents that have larger surface area. Impregnation is the treatment of charcoal which aims to enlarge the pore by breaking the hydrocarbon bond or oxidizing the surface molecule so that the char is changing in nature, either physics or chemistry, for example, the surface area is larger and has an effect on the adsorption process [7].

In the activation of the charcoal chemically, the pyrolyzed raw material is then impregnated with chemical activator such as acids, bases or salts and heated in an inert space [4]. Although the use of alkali hydroxides and acids such as KOH and HCl shows progress on the high surface area of activated carbon, but the nature is corrosive, harmful, and environmentally unfriendly. The use of alkali salts, such as carbonates can be another alternative to acid or base replacement. Examples of carbonate salt activating agents are Li_2CO_3, Na_2CO_3, and K_2CO_3, with activating mechanisms as the following [5]:

$$Na_2CO_3 + 2\ C \longrightarrow 3\ CO + 2\ Na$$

A sodium carbonate solution is used because it is readily available, completely soluble in water and will not produce metal oxide if it decomposes and can decrease metal content, and has high alkalinity [6].

Charcoal or activated carbon may be used as good adsorbent agent for gas purification processes and removal of organic pollutants from water, such as in liquid waste. The texture and properties of the activated carbon depend on the initial treatment, the used of activation process, the number of process provided, the kind of activator agent, and the time or the process in the application of activated carbon itself [7], for example in adsorption of organic matter is using silica gel and active alumina, zeolite, molecular filters, and synthetic resins. Among of these examples, activated carbon is more efficient in eliminating many pollutants, either organic, inorganic, or biological [8].

3. Method

The process of manufacturing activated carbon from oil palm fronds is done through several stages, namely: oil palm fronds preparation, charcoal production, activated carbon production, and product analysis (charcoal and activated carbon).

First, oil palm fronds are cut into small pieces and dried in the oven at 105^0C until the weight of the oil palm fronds are constant. Analysis of raw materials such as water content analysis using the oven. After that, the palm-fringed oil palm fronds was inserted into pyrolysis reactor. Pyrolysis was conducted at 150^0C, 200^0C, and 250^0C for 60 minutes to produce charcoal.

The charcoal produced from the pyrolysis was dried using the oven and smoothed with a ball mill, sieved to obtain a particle size of 140 mesh. Chemical activation was done by using Sodium Carbonate activator (Na_2CO_3) with a concentration of 0 %, 2.5%, 5%, and 7.5% in the oven at the temperature of 80^0C for 2 hours and then impregnated for 24 hours at room temperature. Activated carbon was filtered to separate from the filtrate. Then, activated carbon was washed using distilled water until the pH of 7. After that, activated carbon was dried using the oven.

The product analysis of charcoal and activated carbon includes charcoal yield, the proximate analysis in the form of water content, ash content, volatile matter, fixed carbon and qualitative analysis, such as scanning electron microscopy and iodine number value.

4. Result

4.1. Yield of Charcoal

The relationship between pyrolysis temperature and yield of charcoal as shown in Fig. 1.

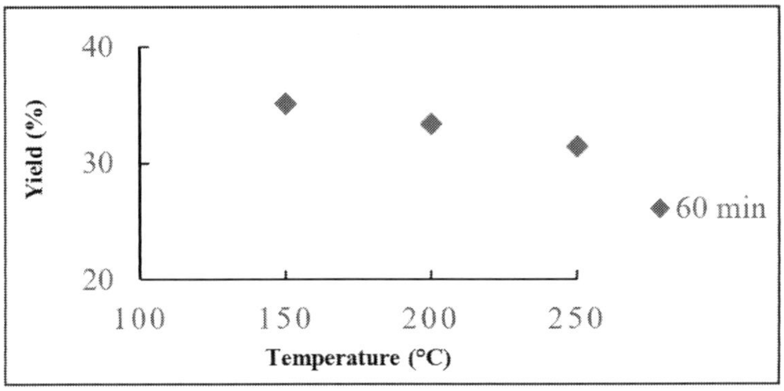

Figure 1. Yield of Activated Carbon

Fig. 1 revealed that there is an decrease in the yield of charcoal along with the increase of pyrolysis temperature. The decrease in yield is due to the increase of pyrolysis temperature. The higher the pyrolysis temperature, the less charcoal produced because the components that contained in raw materials such as water and volatile substances will be decomposed [10] [11], as well as the formation of non-condensable gases such as CO_2, CO, H_2, CH_4 and some hydrocarbons [12].

The yield of charcoal is ranged from 31.40 - 35.13 %.The result of this study is same as the previous researchers, where the yield value of oil palm fronds yielded ranged from 32 to 33% [9].

4.2. Water Content

Water content as a function of temperature and concentration of activator can be seen in Fig. 2.

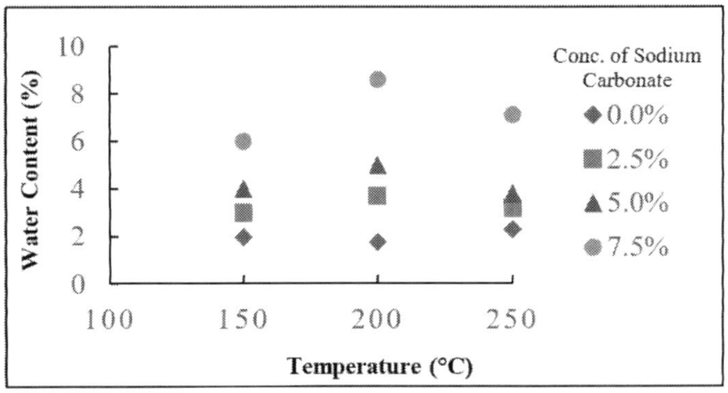

Figure 2. Water Content of Charcoal and Activated Carbon

The value of water content increases with the higher concentration of the activator. It is due to the higher concentration of the activator, the dissolution process of the remaining pyrolysis, such as tar and organic mineral on the surface of the activated carbon will be better. As a result, the resulting pores and the surface area will increase, causing an increase in performance of activated carbon to adsorb water from the air [3] [12].

The water content of the charcoal and activated carbon is lower along with the increasing of pyrolysis temperature. It causes the formation of the pores will increase as well. The bonds of C and H on charcoal are completely released resulting in shifting the crystallite of carbon plate forming new pores and developing the pores that have been formed [13]. It causes the higher performance of the charcoal or activated carbon to adsorb water from the air as well [14].

The value of water content of the charcoal is lower than that of the activated carbon. It is because the presence of the activator will dissolve the remaining pyrolysis and organic mineral tar on the charcoal surface. The resulting the pores will be more numerous and the surface area will increase as well, leading to improve the properties to adsorb water from the air and the hygroscopic ability of activated carbon [3].

Both charcoal and activated carbon have fulfilled the requirement of Industrial National Standard 06-3730-1995 for water content of activated carbon.

4.3. Ash Content
Ash content in charcoal and activated carbon as depicted in Fig. 3.

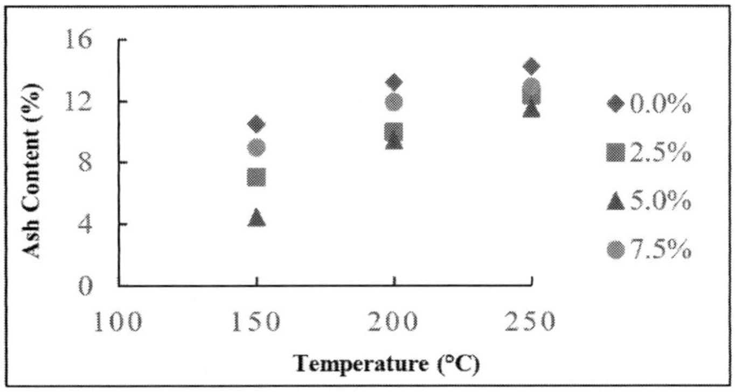

Figure 3. Ash Content of Activated Carbon and Charcoal

Ash content increases along with pyrolysis temperature and causes the deposit of inorganic elements to stick more to the charcoal or activated carbon [14].

Ash content increases with the increasing of activator concentration causing the organic compound of activated carbon is to decrease, but the content of the inorganic compound is relatively fixed. The content of ash content will increase as organic compounds are lower [14].

The ash content of charcoal is higher when compared to the activated carbon. It is because a charcoal still contains high tar and organic minerals due to pyrolysis process [14]. The interaction in the reactor during the pyrolysis process causes the formation of inorganic compounds in the charcoal. These compounds trigger the formation of metal oxides so that ash content on activated carbon is higher [14]. The increasing of ash content occurs because of the formation of mineral salts during the pyrolysis. Ash content is influenced by the amount of silica content, the higher the silica content, the higher ash content will be [15].

IOP Conf. Series: Materials Science and Engineering **309** (2018) 012087 doi:10.1088/1757-899X/309/1/012087

4.4. Volatile Matter

Volatile Matter in activated carbon and charcoal in this research as shown in Fig. 4.

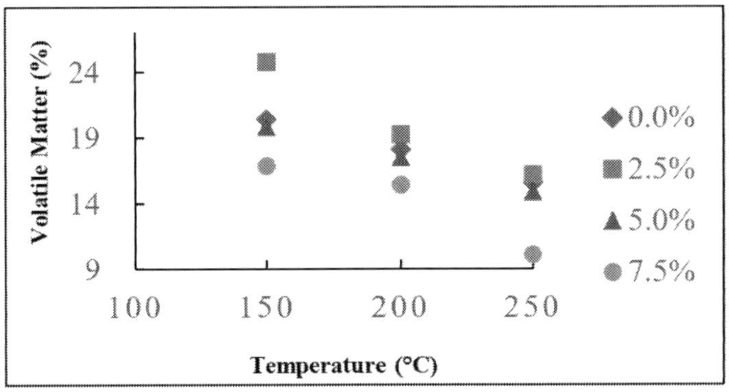

Figure 4. Volatile Matter of Activated Carbon and Charcoal

Volatile matter decreases as the increasing of activator concentration. It is because the addition of activator causes changes in the structure and properties of activated carbon. Sodium carbonate can cause the process of organic material degradation that weakens the surface structure of the activated carbon then releases volatile matter and develops the microporous structure of activated carbon [16].

The pyrolysis temperature increases, the decomposition of non-carbon compounds will be better, so the volatile matter will be lower [17]. At the higher temperatures, the volatile matter will released numerous substance and cause less carbon to form [18]. It is due to the higher pyrolysis temperature, the decomposition of the components contained in raw materials such as water, tar, and volatile matter is also higher [11]. Devolatilization signifies the release of volatile substances from charcoal matrix or activated carbon [19].

The value of the volatile matter on the charcoal is higher than the activated carbon. The impregnation by using the activator is capable of reducing the non-carbon compound that attached to the activated carbon surface and entering the base surface of the charcoal through the pores on the charcoal, so that the presence of the activator will cleanse and enlarge the pore surface [14].

4.5. Fixed Carbon

Fixed carbon of activated carbon and charcoal from oil palm fronds can be shown in Fig. 5.

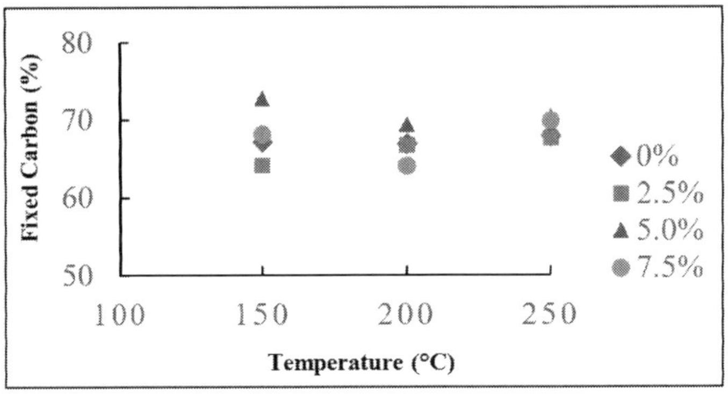

Figure 5. Fixed Carbon of Activated Carbon and Charcoal

The value of fixed carbon increase along with the increasing of activator concentration. The value of fixed carbon fluctuates along with increasing pyrolysis temperature. It is because the value of bound carbon content is influenced by the water content, ash content, and volatile matter of activated carbon

[20].The fixed carbon content is also affected by cellulose and lignin content that can be converted to carbon atoms [14]

4.6. Iodine Number

Iodine number of activated carbon and charcoal from oil palm fronds can be seen in Fig. 6.

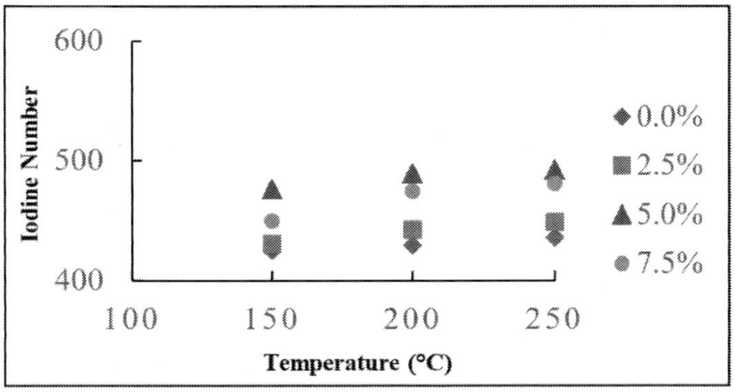

Figure 6. Iodine Number of Activated Carbon and Charcoal

The value of iodine number of charcoal and activated carbon increases as the increasing of pyrolysis temperature. The higher pyrolysis temperature will result in higher pressure in the pyrolysis reactor, thus facilitating disconnection of carbon chain [21] and cause the increase in number of shifting carbon plates and will induce hydrocarbon compounds and other organic compounds to break out during pyrolysis process [12]. This result increases surface area of the charcoal and activated carbon pores [21].

The longer pyrolysis processes can form pores and carbon plates better so as to increase the adsorption capacity of the solution [14]. This is related to the reaction kinetics that cause the higher the activation temperature so the reaction speed will be faster so that the formation of the pores will be better. The bonds of C and H atoms in the charcoal or the activated carbon are completely released resulting in a shift of the crystallite carbon plate forming new pores and developing the pores already formed [14].

The iodine number value of activated carbon increases along with increasing concentration of activator. The higher the concentration of the activator will be easier to bind with tar and volatile matter from pyrolysis. As a result, tar and volatile substances of activated carbon will decrease because of the washing process, so the activated carbon will have an larger surface area [21].

The iodine number on charcoal is lower when compared with activated carbon. The impregnation by using the activator is capable of reducing the non-carbon compound attached to the activated carbon and it will cleanse and enlarge the pore surface [14].

IOP Conf. Series: Materials Science and Engineering **309** (2018) 012087 doi:10.1088/1757-899X/309/1/012087

4.7. SEM Analysis
Figure 7 shows a comparison of surface characteristic between charcoal and activated carbon.

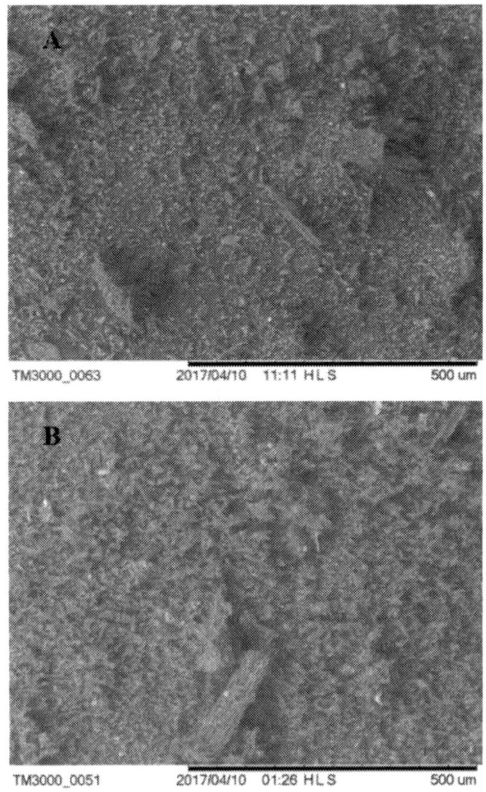

Figure 7. Scanning Electron Microscope (200x) of Charcoal (a) and Activated Carbon (b)

Fig. 7 shows a charcoal has a relatively small pore surface and activated carbon has more pore distribution and cavity on its surface. The structure of the pores formed may result from the evaporation and dissolution of the non-carbon compounds in the raw material. Activated carbon also has a rough and irregular surface that is wider and diffuse compared to charcoal. The presence of the activator can enlarge the pores of activated carbon and expand the surface of the adsorption.

5. Conclusion
The highest yield of charcoal from oil palm fronds is obtained on 35.13% at the temperature of 200 ^{0}C. The highest value of proximate analysis is water content of 8.6%, ash content of 14.25%, volatile matter of 24.75%, and fixed carbon of 72.75%. These results indicated that the charcoal or activated carbon from palm oil fronds had fulfilled the requirements of Industrial National Standard 06-3730-1995.

The porous of has a performance in the adsorption process. It proved by the iodine number value of 492.29. Moreover, SEM analysis indicated a porous activated carbon that is coarse and distributed.

6. Acknowledgments
Many thanks to Chemical Engineering Department of Universitas Sumatera Utara, Medan, Indonesia

References

[1] Irvan, Trisakti B, Sosanty F, and Tomiuchi Y 2016 Effect of discontinuing sodium bicarbonate on fermentation process of palm oil mill effluent *Asian J. Chem.* **28** 377-380
[2] Marina O E and Netti H 2015 *J. Teknik Kimia USU* **4** 1
[3] Gilar S P, Remigius Y.E, Yulianto, Rachimoellah and Endah M M 2013 *J. T. POMITS* **2** 1 2337-3539
[4] Yusufu M I, Ariahu C C and Igbabul B D 2012 *African Journal of Pure and Applied Chemistry* **6 (9)** 123 - 131 1996-0840
[5] Viswanathan, Indra N and Varadarajan 2009 *Methods of Activation and Specific Applications of Carbon Materials* (India: Indian Institute of Technology)
[6] Kirk R E and Othmer D F 1998 **23** (New York: Wiley - Interscience)
[7] Abdessalem O, Mourad B, and Najwa A 2013 *J. of Industrial and Eng. Chem.* **19** 2092 - 9
[8] A M Khah and R Ansari 2009 *J. of ChemTech Research* 0974-4290 **1** 4 859-864
[9] Amalia A N, Awitdrus, and Usman M 2014 *Jurnal Online Mahasiswa FMIPA Universitas Riau* **1** 2
[10] Satriyani S, Melvha H, and Rosdanelli H 2013 *J. Teknik Kimia USU* **2** 1
[11] Rio L and Diah S 2012 *J. Teknik Material dan Metalurgi ITS* **1** 1 1 - 6
[12] Ade I M 2012 *Pembuatan Karbon Aktif dari Jerami Padi Menggunakan Activating Agent H_3PO_4* (Bogor: Institut Pertanian Bogor)
[13] Hendra D and Pari 1999 *Buletin Penelitian Hasil Hutan* **17(2)** 113 - 122 (Bogor: Institut Pertanian Bogor)
[14] Abi G 2012 *Pembuatan Arang Aktif Tempurung Kelapa Sawit sebagai Adsorben dalam Pemurnian Biodiesel* (Bogor: Institut Pertanian Bogor)
[15] Nailul F 2012 *Pembuatan Arang Aktif secara Langsung dari Kulit Acacia mangium wild dengan Aktivasi Fisika dan Aplikasinya sebagai Adsorben* (Bogor: Institut Pertanian Bogor)
[16] R Malik, D S Ramteke, and S R Wate 2006 *Indian J. of Chemical Technology* **13** 319-328.
[17] Indah S, Bambang S, and Iqmal T 2005 *Prosiding Seminar Nasional DIES-50 FMIPA UGM* (Yogyakarta: Universitas Gadjah Mada)
[18] M. Kumar, R.C. Gupta, and T. Sharma 1992 *J. of Biomass and Bioenergy* **3** 6 411 - 7
[19] Ramesh C B, Pallab G and Paruchuri G R 2011 *Int. J. Energy Res.* **35** 929 - 963
[20] Erwin J, Julham P P, and Netti H 2015 *J. Teknik Kimia USU* **4** 2
[21] Abu A B, Diah S, and Hariyati P 2015 *Pengaruh Temperatur Karbonisasi Dan Konsentrasi Zink Klorida ($ZnCl_2$) Terhadap Luas Permukaan Karbon Aktif Eceng Gondok* (Surabaya: Institut Teknologi Surabaya)

IOP Conf. Series: Materials Science and Engineering **309** (2018) 012088 doi:10.1088/1757-899X/309/1/012088

Performance and emissions of an engine fuelled by biogas of palm oil mill effluent

J Arjuna[1*], T B Sitorus[1,2], H Ambarita[1,2] and S Abda[1]

[1]Mechanical Engineering, Universitas Sumatera Utara
[2]Sustainable Energy and Biomaterial Center of Excellence, Universitas Sumatera Utara, Padang Bulan, Medan 20155, Indonesia.

*Email: jayaarjuna2003@yahoo.com

Abstract. This research investigates the performance and emissions of an engine by biogas and gasoline. The experiments use biogas of palm oil mill effluent (POME) with turbocharger at engine loading conditions (100, 200, 300, 400, and 500 Watt). Specific fuel consumption and thermal efficiency are used to compare engine performance, and emission analysis is based on parameters such as carbon monoxide (CO), hydrocarbon (HC), carbon dioxide (CO_2) and oxide (O_2). The experimental data show that the maximum thermal efficiency when engine use biogas and gasoline is 20.44% and 22.22% respectively. However, there was CO emission reduction significantly when the engine using POME biogas.

1. Introduction

One of the innovative solutions in the utilization of renewable energy is the use of alternative fuels such as biogas from palm oil mill effluent (POME) [1,2]. As noted, Indonesia has a lot of raw materials to produce biogas in large quantities. The data of ESDM ministry said that the area of Indonesian palm oil plantation about 13.5 million hectares and for every process of palm fruit production, producing one ton of CPO, it will produce about five tons of POME which still contain many organic substances [3]. The equipment that use a lot of hydrocarbon fuel are internal combustion engines. Internal combustion engines applied in many sectors such as transportation, agriculture and power generation. The fossil fuels such as gasoline and diesel oil use in internal combustion engines produce carbon dioxide (CO), carbon dioxide (CO_2), nitrogen oxide (NOx), sulphur oxide (SOx) and particulates that are polluting the environment. But gas produce an exhaust gas that is environment friendly when used by Otto or Diesel engines [4]. A note that the main components of the biogas constituent are methane (CH_4), carbon dioxide (CO_2) and some other elements such as hydrogen (H_2), nitrogen (N_2) and water vapor (H_2O) [5,6]. This study aims to determine the performance of the engine equipped with a turbocharger when using biogas of POME.

2. Literature Study

2.1. Palm Oil Mill Effluent (POME)

Palm oil mill effluent (POME) is a liquid condensate of sterilization processing which derived from

IOP Conf. Series: Materials Science and Engineering **309** (2018) 012088 doi:10.1088/1757-899X/309/1/012088

palm oil processing to crude palm oil. The properties of POME contain organic compounds and nontoxic. [7]. In general, the main characteristic of POME can be shown in table 1. To be used as gas fuel, POME must be treated specifically by applying technologies such as anaerobic digestion, gasification processes, pyrolysis and carbonization [8,9,10,11].

Table 1. The main characteristic of POME [12]

No.	Parameter	Unit	Range
1	Biological oxygen demand	mg/l	20,000-30,000
2	Chemical oxygen demand	mg/l	41,000-99,000
3	Volatile solid	mg/l	23,200-53,900
4	Total solid	mg/l	28,000-64,700
5	Fat	mg/l	1,700-9,200
6	Kj-N	mg/l	560-1,200
7	pH		3.9 – 4.6

2.2. Parameters of Internal Combustion Engine

The performance of the internal combustion engine is specifically indicated by the value of the parameters of the engine. Some of these parameters can be described as follows [13]. The power is influenced by engine speed and torque that produced by the engine. In most cases, in the internal combustion engine, it is known two type of power that consists of shaft power and power indicator. As a practice, it uses only the shaft power. The shaft power, or effective power is the power that produced by an engine on the output shaft or it is commonly known as brake horse power which is calculated by the equation:

$$\dot{W} = \frac{2\pi \cdot N \cdot \tau}{60000} \quad \text{kW} \tag{1}$$

Where N is engine speed (rpm), and τ is engine torque (Nm).

The specific fuel consumption is the amount of fuel that should consume per unit of power which produced per hour of operation. Indirectly, the specific fuel consumption is an indication of the engine efficiency for generating power from fuel combustion. The value of specific fuel consumption can be defined as follows:

$$Sfc = \frac{\dot{m}_f}{\dot{W}} \cdot 3600000 \quad \text{g/kWh} \tag{2}$$

Where $\dot{m}_f$ is the mass flow rate of the fuel (kg/s).

The thermal efficiency of an internal combustion engine constituted the ratio between the output energy and the chemical energy of fuel, that can be stated as:

$$\eta_t = \frac{\dot{W}}{\dot{m}_f \cdot Q_{HV} \cdot \eta_c} \tag{3}$$

Where Q_{HV} is the calorific value (kJ/kg), and η_c is combustion efficiency 0,97.

Mean effective pressure is a good parameter to compare engines for design or output because it is independent of engine size and speed. The value of mean effective pressure can be defined by using an equation:

$$\text{mep} = \left(\frac{4\ \pi\ \tau}{V_d}\right).1000 \qquad \text{kPa} \qquad (4)$$

Where V_d is displacement volume (cm^3).

3. Method

3.1. Material

The experiments were using biogas of POME and gasoline. Loads of the engine are five lamps with 100 watts respectively.

(a) (b)

Figure 1. a) calorimeter bomb b) emission analyzer Sukyong SY-GA 401

The main equipment used in this study consisted of:

a. Single cylinder stationary Otto engine
b. Calorimeter bomb to determine the calorific value of the fuel
c. Emission analyzer Sukyong SY-GA 401 to determine the composition of exhaust emissions CO, CO_2, HC, and O_2.

3.2. Experimental Scheme

The GFH1900LX engine that used in this research is the type of a four stroke one cylinder stationary engine which specifications are summarized in table 2. The scheme of experimental is shown Figure 3. All experimental data were taken after the engine was properly warmed up 15 minutes after starting. An emission analyzer Sukyong SY-GA 401 with an accuracy of ±5% were used to measure the exhaust gases which placed on the exhaust gas manifold. For each fuel, different emission data were taken over three repetitions. The testing was under the variety engine loads of 100W, 200W, 300W, 400W and 500W. The accuracy of torque measurement is about ±0.2 Nm, and the accuracy of rotation controlling ±10 rpm. The engine was started by using gasoline fuel, whereas it was warmed up, should switch to the biogas. After switching fuels from one type to another, the engine was running about 10 minutes to get stable condition with the new fuel before the measurements were taken. Engine torque and fuel consumption were measured to calculate the power, specific fuel consumption and thermal efficiency of the engine. An 8 ml burette and a timer are used to measure the fuel consumption. The accuracy of the burette about ±0.1 ml and the timer about ±0.01 s.

Table 2. Engine specifications [14]

Type	GFH1900LX / Otto engine
Capacity	900 watts / 220V / 50Hz
Peak power	1,3 kW
Power factor	1
Cooling system	3.0 Hp *air cooled* OHV/ 3600 rpm
Bore	55 mm
Stroke	40 mm
Compression ratio	10,5 : 1
Number of cylinders	1

Figure 2. Modification on carburettor components

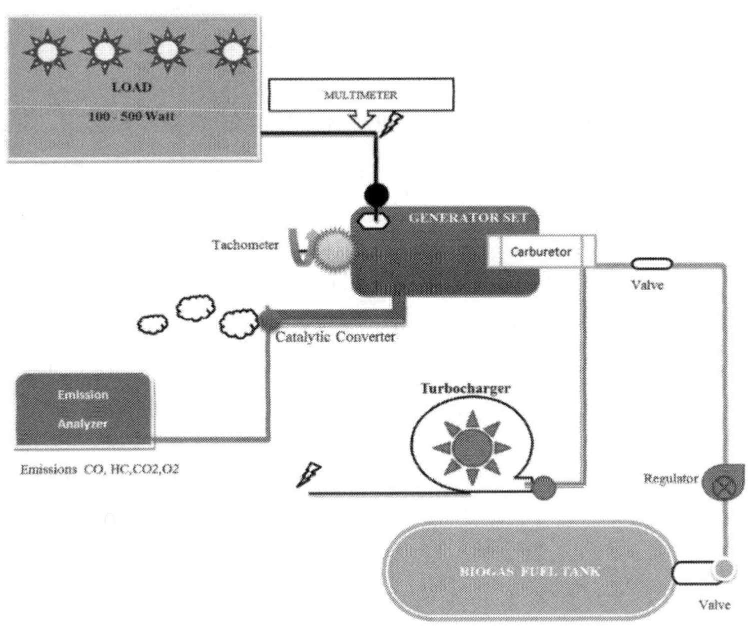

Figure 3. Experimental scheme

4. Results and Discussions

4.1. Calorific Value of Fuels

The calorific value of fuel was examined by using a bomb calorimeter apparatus with the accuracy of ±5%. To get the calorific value of gasoline were the experiments carried out which each fuel as many as five samples. Then the average results of the five samples were to obtain the calorific value of the fuel. The observation results obtained that the calorific value of gasoline is 43964 kJ/kg. The calorific value of biogas of POME is obtained 35900 kJ/kg [3]. The calorific value of fuel shows the energy that generated during the combustion process of fuel per unit mass which is influenced by the composition of the fuel constituent.

4.2. Torque

Figure 4 shows the experimental data that the maximum torque is obtained 2.75 Nm when the engine uses gasoline with a turbocharger for a load of 500 watts. The minimum torque is obtained at 1.04 Nm when engine using biogas of POME with a turbocharger for a load of 100 watts. The torque generated when engine using biogas of POME is lower because the torque is influenced by the energy of fuel combustion. The energy of fuel combustion is influenced by the calorific value of the fuel used.

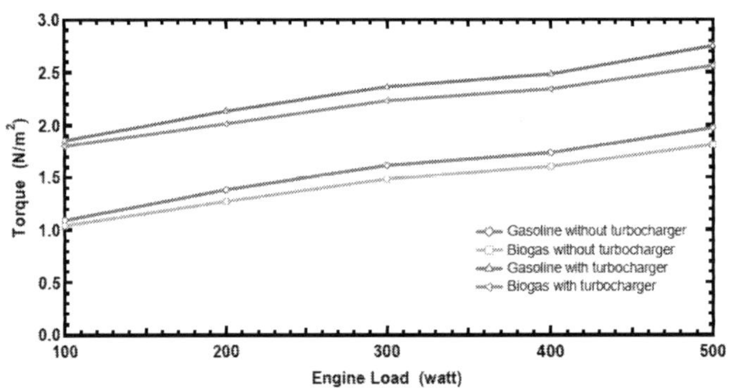

Figure 4. Engine torque versus engine load.

4.3. Power

Figure 5 shows the test results for shaft power when the engine using gasoline and biogas of POME. Based on the test data and calculations has obtained the maximum power of 1.62 kW when engine using gasoline with a turbocharger for a load of 500 watts.

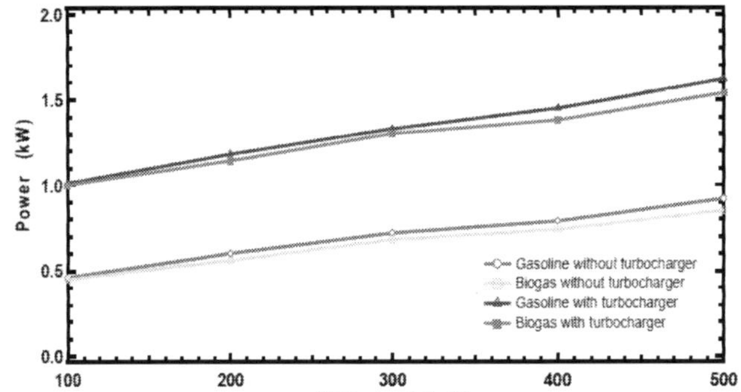

Figure 5. Engine power versus engine load

The minimum power is obtained 0.45 kW when engine using biogas of POME without a turbocharger for a load of 100 watts. The average increase of engine power is about 67.43% after using a turbocharger both when using gasoline or biogas. The addition of a turbocharger increases the amount of air entering the combustion chamber, so this makes the air-fuel ratio (AFR) increased. As noted, AFR parameter also affects the shaft power generated by the engine.

4.4. Specific Fuel Consumption (Sfc)

Based on the test results shown in figure 6, the maximum sfc is obtained 2336.70 g/kWh when engine using biogas of POME without turbocharger for a load of 100 watts. The minimum sfc is obtained 386.85 g/kWh when engine using gasoline with turbocharger for a load of 500 watts. The average decrease of specific fuel consumption generated when using turbocharger is about 55%. As noted, when using a turbocharger that the rate of air entering into the combustion chamber increases while the amount of fuel used is reduced. It can be said that the specific fuel consumption when the engine using biogas of POME has increased. One of them is caused by calorific value of biogas is lower than gasoline. This makes the fuel needed more than when engine using gasoline for all conditions of testing.

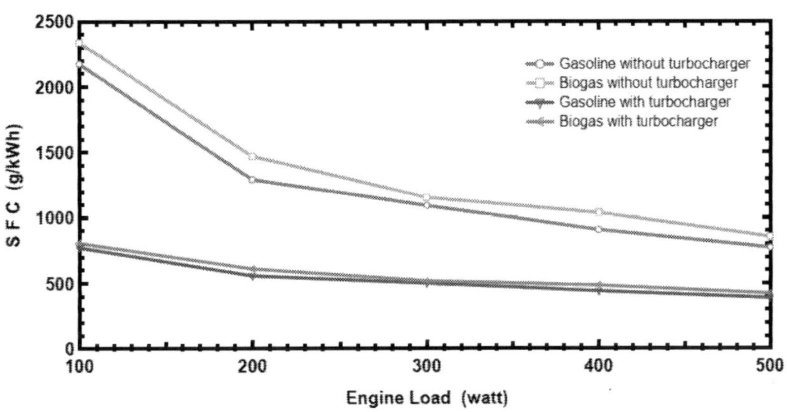

Figure 6. Specific fuel consumption versus engine load

4.5. Thermal Efficiency

Figure 7 shows the thermal efficiency obtained during experiments. The experiment's data show that the maximum thermal efficiency 22.22% when the engine using gasoline with a turbocharger for a load of 500 watts.

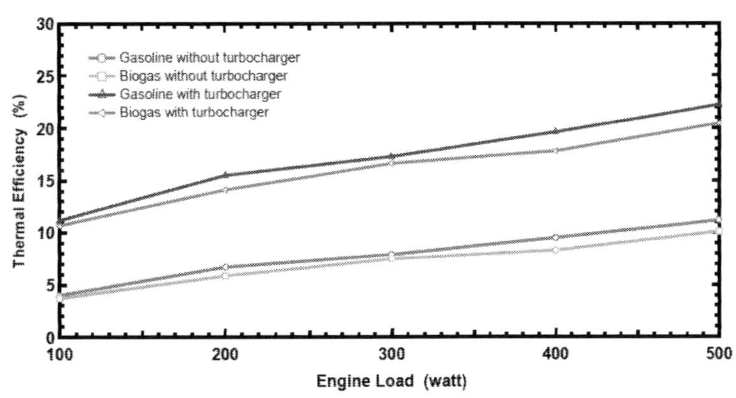

Figure 7. Thermal efficiency versus engine load

The minimum thermal efficiency obtains 3.68% when the engine using biogas of POME without a turbocharger for a load of 100 watts. The increasing of thermal efficiency produced by using a turbocharger both when using gasoline and biogas is about 68.73%. It should be noted that the thermal efficiency of the internal combustion engine is influenced by several parameters such as engine shaft power, the fuel flow rate to the combustion chamber and the calorific value of the fuel used. All three parameters simultaneously affect the achievement of thermal efficiency of an internal combustion engine.

4.6. Mean Effective Pressure

The experimental data shown the maximum mean effective pressure (mep) is 363.64 kPa which occur when the load 500 watt using gasoline with a turbocharger in figure 8. The minimum mean effective pressure is obtained 138.13 kPa by using biogas without a turbocharger on the load of 100 watts. As noted, that the maximum mean effective pressure of good engine designs is well established, and is essentially constant over a wide range of engine sizes. The actual mean effective pressure is a particular engine develops that can be compared with this norm and the effectiveness with which the engine's displaced volume can be assessed. For design calculations, the engine displacement required to provide a given torque or power at a specified speed that can be estimated by assuming the appropriate values for mean effective pressure for that particular application [15].

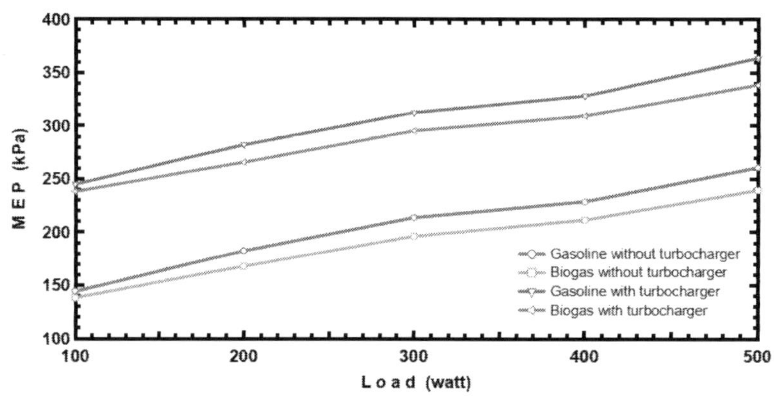

Figure 8. Mean effective pressure versus engine load

4.7. Combustion and Gas Exhaust

In this testing, also examined the effect of combustion conditions on the spark plug and exhaust gas emission composition produced by the engine. To know the combustion process in the combustion chamber, one of them can be reviewed of the spark plug conditions after use. The color of the electrode of the spark plug can indicate the combustion process takes place inside the combustion chamber. It appears in figure 8 that the spark plug electrodes during testing when engine using gasoline tend to be darker than when using biogas of POME.

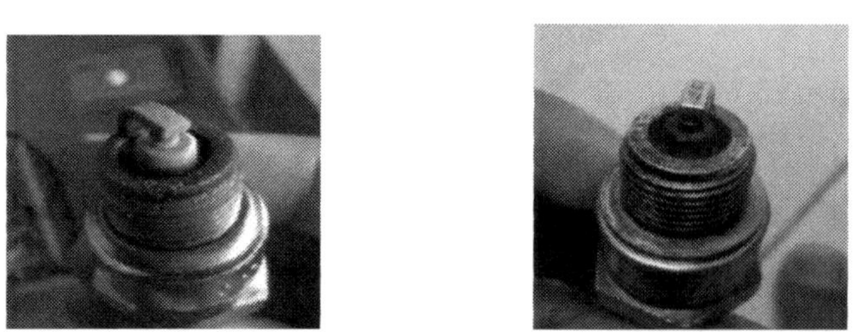

Figure 9. Conditions of spark plug when using (a) gasoline (b) biogas POME

Table 3. Measurement data of exhaust gas with gasoline

Load	Without turbocharger				With turbocharger			
(watt)	CO	HC	CO_2	O_2	CO	HC	CO_2	O_2
	(%)	(ppm)	(%)	(%)	(%)	(ppm)	(%)	(%)
100	3.67	839	1.48	14.62	2.65	726	1.69	14.93
200	4.20	441	1.60	13.35	3.15	322	1.83	13.77
300	4.42	324	1.68	13.48	3.27	218	1.90	13.79
400	4.45	281	1.76	12.91	3.47	164	1.89	13.27
500	4.82	206	1.83	12.73	3.68	121	1.97	13.21

Table 4. Measurement data of exhaust gas with biogas POME

Load	Without turbocharger				With turbocharger			
(watt)	CO	HC	CO_2	O_2	CO	HC	CO_2	O_2
	(%)	(ppm)	(%)	(%)	(%)	(ppm)	(%)	(%)
100	2.98	778	5.68	12.63	1.92	703.00	6.75	13.75
200	3.22	644	5.68	11.87	2.34	369.00	6.73	12.99
300	2.78	598	7.38	9.43	1.56	323.00	8.45	10.55
400	2.88	353	8.18	8.92	1.45	235.00	9.22	10.12
500	2.15	322	7.78	7.88	1.21	211.00	9.87	9.33

Exhaust emissions, which investigated include carbon dioxide (CO_2), carbon monoxide (CO), hydrocarbon (HC) and oxygen (O_2) levels. Tables 3 and 4 show the measurement data of exhaust gas composition when using gasoline and biogas of POME. A note that CO emissions arise because at the time of burning process occurs lack of oxygen. The lack of oxygen supply causes incomplete combustion where the C atom lacks O_2 to form CO_2. Table 4 shows that the use of gasoline fuel produces the most CO for engine load. This is due to the gasoline consisting of C_7 (heptane) to C_{11} hydrocarbon groups having more C atoms than biogas consisting of methane or C_1 group hydrocarbons. Reduction of CO emission when using biogas of POME is significant thereabouts 55-75% when compared using gasoline. The emissions of HC also occur due to lack of oxygen, so that the combustion process takes place imperfectly because many carbon atoms that do not get enough oxygen to form HC gas. The result of the measurement emission with engine load variations is obtained minimum HC level of 121 ppm occurs when engine using gasoline with a turbocharger for a load of 500 watts. The maximum HC has obtained 839 ppm when the engine using gasoline without a turbocharger for a load of 100 watts.

5. Conclusions

The performance and emission of a single cylinder engine fuelled with biogas of POME have been observed and analyzed compared to the base line of gasoline. The results showed an increase in engine power and thermal efficiency with decreased in specific fuel consumption ranged from 45% - 68% when the engine using a turbocharger. Measurement data on exhaust emissions indicate that there are reduced of CO and HC emissions, as well as increases in CO_2 and O_2 when engine using biogas of POME with turbocharger ranges of 40% -75% when compared with using gasoline.

IOP Conf. Series: Materials Science and Engineering **309** (2018) 012088 doi:10.1088/1757-899X/309/1/012088

Acknowledgements
The authors acknowledge Universitas Sumatera Utara through TALENTA project 2017 for their financial support to this research.

References

[1] Barelli L, Bidini G, Campanari S, Discepoli G, Spinelli M 2016 Performance assessment of natural gas and biogas fueled molten carbonate fuel cells in carbon capture configuration *J. of Power Sources* **320** 332-342

[2] Sitorus T B 2002 *Tinjauan pengembangan bahan bakar gas sebagai bahan bakar alternatif* Jurusan Teknik Mesin Fakultas Teknik Universitas Sumatera Utara, digitized by USU digital library

[3] Kementerian ESDM Republik Indonesia 2015 *Buku Informasi Energi-ESDM untuk Kesejahterahan Rakyat* Direktorat Jenderal Energi Baru Terbarukan dan Konservasi Energi.

[4] Porpatham, Ramesh A, Nagalingam B 2012 Effect of compression ratio on the performance and combustion of a biogas fuelled spark ignition engine *Fuel* **95** pp. 247-256

[5] A.-Virgen Q, Taboada-González P, Ojeda-Benítez S, Cruz-Sotelo S 2014 Power generation with biogas from municipal solid waste:Prediction of gas generation with in situ parameters *Renewable and Sustainable Energy Reviews* **30** pp.412-419

[6] Karapidakis, Tsave, Soupios P M, Katsigiannis Y A 2010 Energy efficiency and environmental impact of biogas utilization in landfills *Int. J. Environ. Sci. Technol.* **7** pp.599-608

[7] Siregar P 2009 *Produksi Biogas Melalui Pemanfaatan Limbah Cair Pabrik Minyak Kelapa Sawit dengan Digester Anaerob* Tesis Universitas Sumatera Utara

[8] Li K, Liu R, Sun C 2016 A review of methane production from agricultural residues in China *Renewable and Sustainable Energy Reviews* **54** pp. 857-865

[9] Arroyo J, Moreno F, Muñoz M, Monné C 2015 Experimental study of ignition timing and supercharging effects on a gasoline engine fueled with synthetic gases extracted from biogas *Energy Conversion and Management* **97** pp.196-211

[10] Balmant W, Oliveira B H, Mitchell D A, Vargas J V C, Ordonez J C 2014 Optimal operating condition for maximum biogas production in anaerobic bioreactors *Appl. Therm. Eng.* **62** pp.197-206

[11] dos Santos I F S, Barros R M, Filho G L T 2016 Electricity generation from biogas of anaerobic wastewater treatment plants in Brazil: an assessment of feasibility and potential *J. of Cleaner Production* **126** pp. 504-514

[12] Irvan, Trisakti B, Tomiuchi Y, Harahap U, Daimon H 2017 Effect of Recycle Sludge on Anaerobic Digestion of Palm Oil Mill Effluent in A Thermophilic Continuous Digester *IOP Conf. Series: Mater. Sci. Eng.* **206** 012094

[13] Pulkrabek W W 2004 *Engineering Fundamentals of The Internal Combustion Engine* (New Jersey: Prentice Hall)

[14] Manual Book of Generator Set Type GFH1900LX, 2015

[15] Heywood J B 1988 *Internal Combustion Engines Fundamentals*. Massachusetts Institute of Technology, USA

IOP Conf. Series: Materials Science and Engineering **309** (2018) 012089 doi:10.1088/1757-899X/309/1/012089

Performance of the natural cooler to keep the freshness of vegetables and fruits in Medan City

T B Sitorus[1,2*], H Ambarita[1,2], F Ariani[1] and T Sitepu[1]

[1]Mechanical Engineering, Universitas Sumatera Utara
[2]Sustainable Energy and Biomaterial Centre of Excellent, University of Sumatera Utara, Jl. Almamater Kampus USU, Medan 20155, Indonesia

*E-mail: tburhanudin@yahoo.com

Abstract. One application in a direct evaporative cooling system was a natural cooler. The advantages of this system were not using the electrical energy and so far also environmentally. This research aims to obtain a performance analysis of the natural cooler as a store for vegetables and fruits in Medan city. The materials for natural cooler consists of teak wood and gunny. This study makes experiments during seven days in the open air. The parameter measurement on the weather was using HOBO devices and to record the temperature changes for vegetables or even fruits is using its acquisition data. The results showed that the maximum efficiency of the natural cooler could be obtained for 43.79% in the average air temperature of 30.51°C, the air humidity average is 85.12% with average solar radiation of 183.98 W/m^2. Experimental data were showing that the condition of freshness on vegetables or even on fruits was heavily influenced by weather conditions.

1. Introduction

According to Indonesian national data energy management blueprint book [1], the uses in fuel oil is over 60% of overall energy consumption in the year of 2005. It continues increase every year with population growth and economic development. Of course, at this moment the fuel consumption and energy quite an impact on national economic condition. It is necessary now to think about the environmentally and friendly technology that can reduce the energy consumption, especially on fuel and electrical. One type of technology that now is still useful for humans is coolers. The cooler is a device that works almost on the same principle with a heat engine which use the fluid as its refrigerant and heat transfer tool from indoors to outdoors to keep temperature of the object should be lower than the ambient temperature. In the equatorial area like Indonesia, refrigerant system is still required because it is so cheap and effective.

According [2, 3], Indonesia always exposed to the sun radiation for 10-12 hours daily and it is expected that the average intensity of solar radiation should be 4.8 kWh/m^2. In this study the natural cooler system which was observed without using electrical or mechanical energy, therefore its called as a natural cooler. This research goal was to gain the analysis performance in a natural cooler that using water for its cooling medium. Although research on natural cooler for vegetables and fruits has been applied commonly, it still more interesting time to time because it is friendly, environmentally, simple technology, and rarely to publication.

1.1. The working principle in natural cooler

The working principle in natural cooler is the event of cooling by using the evaporation (evaporative cooling). Cooling by the evaporation, is a physical phenomenon in which the evaporation liquid came from surrounding air to make cool the object. This will make the surface object being colder when the water evaporates from him, cause the water should require heat to change the liquid phase to vapor [4]. It also can be said that heat and mass transfer process is done by the water evaporation on cooling the air system, where the amount of heat is transferred from the air to water that enough to lowering the air temperature [5, 6]. In generally, the evaporation cooling system is classified into three types, which renowned as direct evaporative cooling, indirect evaporative cooling and the combined system of direct and indirect evaporative [7]. The application using the indirect evaporative cooling system can be seen as the heating process, the air ventilation, the air-conditioning in buildings or also in shopping centers. In others, the application of the direct evaporative cooling system is a natural cooler that used to store vegetables and fruits [8]. Based on the preservation research report on vegetables and fruits, requirement of low air temperatures and high humidity will delay the pathological activities that located on the high ground [9]. It can be said also, that the natural cooling system is greatly affected by weather conditions, including temperature and humidity which is suitable to be applied in subtropical and tropical regions [10, 11]. Generally, the natural cooler uses gunny as their container and the water for its cooling medium.

1.2. Efficiency of Natural Cooler

In determining the efficiency of the natural cooler, it is necessary to review the process of heat transfer that occurs around the cooler. The amount of heat transfer conduction (watt) which occur in the natural cooler walls can be calculated from equation [12]:

$$Q_{cond} = k\,A\,\frac{\Delta T}{L} \tag{1}$$

where k is the material thermal conductivity (W/m.K), A is the cross sectional area of heat transfer conduction (m^2). The parameter ΔT is the temperature difference occurs between the surface temperature of the walls of the natural cooler with the ambient air temperature (°C) and L is the thickness of the wall cooler (m).

The heat transfer convection process (watt) which will occur naturally in the natural cooler wall is

$$Q_{conv} = h\,A\,(T_s - T_\infty) \tag{2}$$

where h is the coefficient convection heat transfer (W/m^2 K), A is the cross sectional area (m^2), and ΔT is a temperature difference that occurs between the surface temperature of the walls of the natural cooler with the ambient air temperature (°C). The amount of radiation heat transfer processes (watt) which occur naturally in the walls of the natural cooler is

$$Q_{rad} = \varepsilon\,\sigma\,A\,(T_s^4 - T_\infty^4) \tag{3}$$

where ε is the emissivity of the material used (wood 0.88 and gunny 0.72), A is the cross sectional area of radiation heat transfer (m^2), and σ is the Stefan Boltzmann constant (5.67×10^{-8} W/m^2 K^4). The parameter T_s is the temperature of the walls of the natural cooler (°C) and T_∞ is the air temperature around the natural cooler (°C). The amount of total heat transfer from the natural cooler that obtained from the evaporation which was occurs in the walls of the natural cooler is a combination of the natural convection heat transfer, conduction, and radiation, namely:

$$Q_{total} = Q_{cond} + Q_{conv} + Q_{rad} \quad (4)$$

So that the value of the efficiency of the natural cooler can be determined by the equation:

$$\eta_{nc} = \frac{Q_S + Q_L}{Q_{total}} \quad (5)$$

where Q_S is the sensible heat of air (watt) and Q_L is the latent heat of air (watt).

2. Method

2.1. Material

The material for the natural cooler that used in this study was made the teak wood material with a thickness of 3.5 cm as the framework with consideration of material strength was good enough, resistant to water and termites as well as relatively easy to obtain. Dimensions for them are 490 mm (width) x 530 mm (length) x 1550 mm (high).

Table 1. The main features of a natural cooler

No	Data	Description
1	Frame	Teak , k = 0,12 W/m. K at 27 °C
2	Wall coverings	Gunny, k = 0,06 W/m. K at 27 °C
3	Cooling media	Water
4	Pipe	Copper (∅ = 8 mm)
5	Water Tank	2 Tank (volume 8 liters)
6	Objects cooled	Vegetables and fruits

Figure 1. Photograph of the natural cooler.

The white gunny was used as the wall coverings with a thickness of 1 mm that used to applied to the right, left, back and front. For the roof and the base under the rack, it ia used a plywood with a thickness of 1 cm in order to make it water-resistant to prevent rapid deterioration. The water as a cooling medium should have a temperature not more than 25°C. The object that placed in the natural cooler should be vegetables and fruits. The main features of the natural cooler used is shown in table 1. Photograph of the natural cooler is shown in figure 1.

2.2. Experimental Scheme
The natural cooler is connected to a data acquisition system, Agilent 3497A through the thermocouples. Six thermocouples were used in the experiments, three for the objects cooled and three is located in the natural cooler. Temperatures were measured using J type thermocouples with an accuracy of ±0.4%. A HOBO micro station data logger was used to record weather conditions such as solar radiation intensity (pyranometer with accuracy ±5%), ambient temperature (accuracy ±0,2°C), relative humidity (accuracy ±2.5%) and wind velocity (accuracy ±3%). As a note, western Indonesian time or WIB (Waktu Indonesia Barat) is used in Medan city for a local time.

The experiments were carried out during seven consecutive days in May 2017, at the Solar Energy Laboratory, Universitas Sumatera Utara in Medan city, Indonesia, with geographical coordinates 3°35' North latitude, 98°40' East longitude. Figure 2 shows a schematic of the natural cooler and data measurement systems.

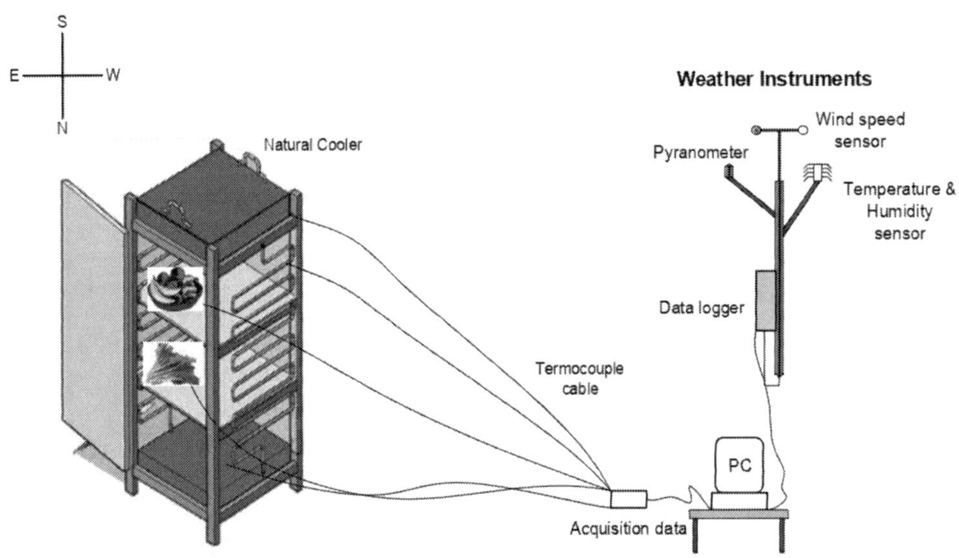

Figure 2. Schematic of the natural cooler and measurement systems.

3. Results and Discussions
The research conducted divided into several studies, namely the weather conditions, the performance of the natural cooler tested and the correlation between the performance of the natural cooler with the weather conditions during experiments.

3.1. Weather Conditions
It is necessary to review the weather conditions, therefore, the experiments were carried out in the open air conditions which exposure to solar radiation. The process of measuring weather conditions for 24 hours during seven days where the weather conditions and the measurement process are done

IOP Conf. Series: Materials Science and Engineering **309** (2018) 012089 doi:10.1088/1757-899X/309/1/012089

with every minute. Table 2 shows the parameters of the weather conditions during the experiment process is done.

Table 2 shows that the weather conditions that include temperature and humidity, wind speed and solar radiation vary every day. Figure 3 shows the typical of solar radiation and ambient air temperature during the experiment on the first day.

Table 2. Weather conditions during experiments

Date	Average Ambient Temperature (°C)	Average Air Humidity (%)	Average Speed Wind (m/s)	Average Radiation Solar (W/m²)
1/05/2017	30.81	84.89	0.73	186.22
2/05/2017	30.51	85.12	1.17	183.98
3/05/2017	30.20	85.81	1.63	182.34
4/05/2017	30.23	85.23	1.29	183.22
5/05/2017	30.66	85.01	1.09	184.71
6/05/2017	31.01	84.78	0.67	189.32
7/05/2017	31.21	84.65	0.48	190.12

Figure 4 shown a significant correlation between changes in air temperature with the air humidity conditions, that correlated with solar radiation which occurs at the experiment location. The experimental data showed that solar radiation began to be detected from 06.20 WIB until 18.25 WIB. During the experiments takes place within seven days were obtained the average solar radiation of 186.22 W/m^2, the average air temperature of 30.81°C and humidity average of 84.89%.

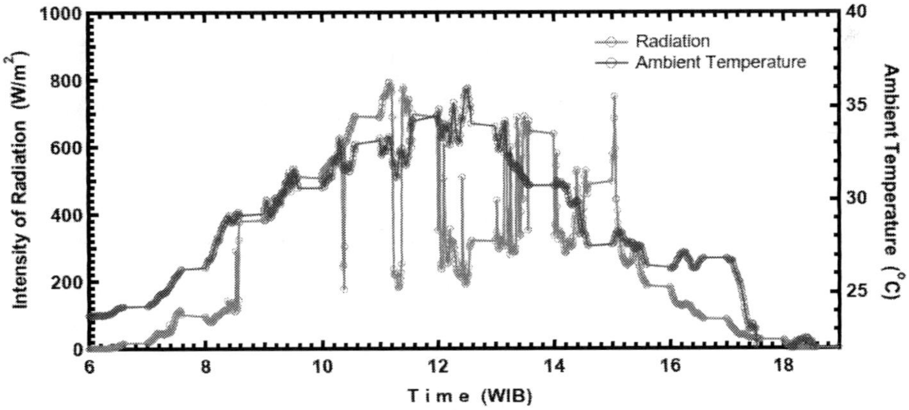

Figure 3. The typical of solar radiation and ambient temperature at first experiment.

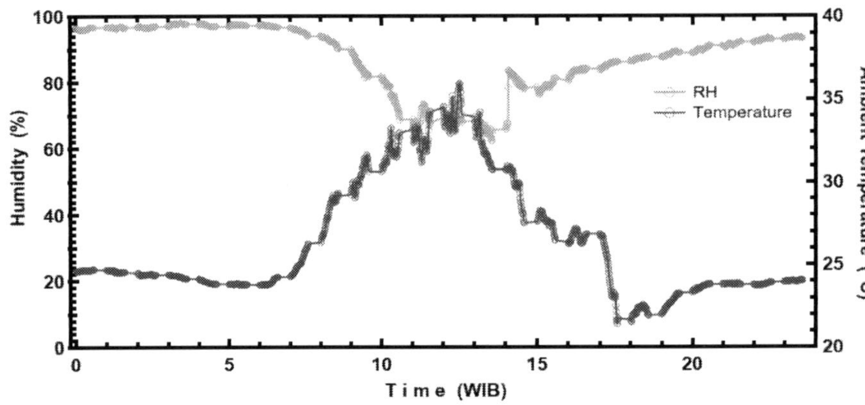

Figure 4. The typical of ambient temperature and air humidity at first experiment.

3.2. Performance of The Natural Cooler

The measuring of the temperature in the natural cooler was done by placing a thermocouple sensor on the top side, rear, right, left, vegetables and fruit. The natural cooler is tested with the condition of the backside facing east.

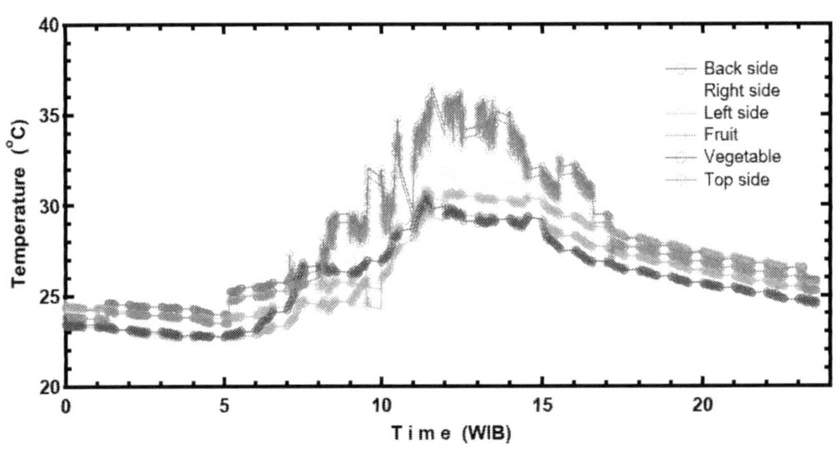

Figure 5. The typical of temperature of components at first experiment.

Figure 5 show the fluctuations in the temperature of each component was measured for the first day of the experiment. This correlates with changes in weather conditions that occur every time. The value of the temperature of each component is increasing from 07.00 WIB morning and reached a maximum temperature at 12.00 WIB - 14.00 WIB which is in line with the increase of solar radiation and ambient temperature. During experiments, the maximum temperature of vegetables and fruit ranges from 34°C - 35°C. The minimum temperature of vegetables and fruit is 22°C - 23°C, which occurred at 04.00 WIB - 05.00 WIB.

The first day The third day

Figure 6. Conditions of vegetables and fruits on the first day and third.

The fifth day The seventh day

Figure 7. Conditions of vegetables and fruits on the fifth day and seventh.

In general, the temperature on the back side of gunny is higher than the other because the back side is directly facing toward the east or the sunrise to absorb more solar radiation. To keep the vegetables and fruits as placed in so that do not quickly wither then used water in the pipe around the walls as a natural cooler. From the calculation taking into account the mass of vegetables and fruits in the cooler and the heat transfer process that occurs, then the volume of water required for the cooling process is 8 liters in the 2 tank. The average air temperature 30.66°C, air humidity average of 85.07% and an average solar radiation is 185.70 W/m^2. The water are used to prevent heat from the outside air into the cooler and absorb heat emitted fruits and vegetables.

Table 3. Efficiency of natural cooler during experiments

Day	Q_S (W)	Q_L (W)	Q_{total} (W)	η_{nc}
1	0.344	31.510	79.615	40.01%
2	0.306	30.679	70.755	43.79%
3	0.344	30.143	79.120	38.53%
4	0.348	29.853	79.937	37.78%
5	0.335	31.096	77.206	40.71%
6	0.364	32.100	84.772	38.30%
7	0.373	32.682	87.121	37.94%

Therefore, the water content in the gunny must be maintained. When the undergoes drying, the heat from the surrounding environment will go into the cooler and emitted heat the fruits and vegetables which can not be absorbed by the gunny. This can result in vegetables and fruits in the cooler will naturally quickly wither. The physical condition of vegetables and fruits during the experiments can be seen in figure 6 and 7. It appears that the condition of vegetables and fruits are still good from the first day until the fifth and started having conditions wither on the seventh day.

Table 3 shows the efficiency of the natural cooler during experiments. The value of the maximum efficiency of the natural cooler obtained during the experiment was 43.79% in the second day of the experiment. The maximum efficiency of the natural cooler obtained when average air temperature 30.51°C, air humidity average of 85.12% and an average solar radiation 183.98 W/m^2. The value of the efficiency of the natural cooler is fluctuated every day.

Based on data from the measurement and the calculation that the efficiency of the natural cooler is affected by heat transfer process that occurs in the natural cooler by radiation, conduction, convection as well as heat from vegetables and fruits and varying weather conditions while experiments.

3.3. Correlation of weather conditions with Natural Coller performance

Correlation of the weather parameters which include solar radiation, ambient temperature, humidity and wind speed against the efficiency of the natural cooler was also being examined. The data used are the weather conditions and the efficiency of the natural cooler for seven days of experiments. Table 4 shown that the parameters of solar radiation, temperature, humidity and wind speed correlated significantly to the efficiency of natural cooler.

By using the multiple regression functions obtained the coefficient of determination of 0.86, it stated that the effect of weather parameters against the efficiency of the natural cooler which tested 86%.

Table 4. Correlation of weather parameters against the efficiency of natural cooler

	Ambient temperature	Humidity air	Solar radiation	Wind speed	Efficiency of natural cooler
Ambient temperature	1				
Humidity air	-0.95	1			
Solar radiation	0.66	-0.69	1		
Wind speed	-0.90	0.83	-0.68	1	
Efficiency of natural cooler	0.68	-0.58	-0.73	-0.55	1

4. Conclusions

The study investigates the performance of a natural cooler without using electricity and environmental friendly in an open air with a varying weather conditions with success. The observation of weather conditions is in the ambient temperature, humidity, and solar radiation. This research obtains an average temperature of vegetables and fruit should be in the ranges of 24.29°C during experiments. According to the experimental data and calculations, the maximum efficiency of the natural cooler is 43.79%. Based on the statistical calculation, the obtained value of the weather conditions that effected on the performance of the natural cooler tested is ranged between of 86% during the experiments. In the next study should focus on improving the efficiency of natural cooler so that the time of the freshness of fruits and vegetables can last longer and untimely.

Acknowledgments

The authors wish to acknowledge the financial support that provided by Universitas Sumatera Utara through TALENTA project in 2017.

References

[1] Buku Putih Penelitian 2006 *Pengembangan dan Penerapan Ilmu Pengetahuan dan Teknologi Bidang Sumber Energi Baru dan Terbarukan untuk Mendukung Keamanan Ketersediaan Energi Tahun 2025* (Jakarta: Kemenristek)

[2] Sitorus T B, Napitupulu F H and Ambarita H 2016 Experimental Study of Solar Refrigerator System Using Activated Alumina and Methanol Adsorption Pair *Int. J. Technol.* **7(5)** pp. 910-922

[3] Rumbayan M and Nagasaka K 2012 Solar Irradiation Estimation With Neural Network Method Using Meteorological Data In Indonesia *Int. J. Technol.* **2** pp.110-120

[4] Ndukwu C N and Manuwa S I 2014 Review of research and application of evaporative cooling in preservation of fresh agricultural produce *Int. J. Agric. & Biol. Eng.* **7(5)**

[5] Boukhanouf A R and Ibrahim H 2015 A Review of Evaporative Cooling Technologies *Int. J. Environ. Sci. Dev.* **6(2)**

[6] Xuan M Y, Xiao F, Niu X F, Huang X and Wang S W 2012 Research and application of evaporative cooling in China: A review (I) - research *Renewable and Sustainable Energy Reviews* **16** pp. 3535-3546.

[7] Duan et al 2012 Indirect evaporative cooling: Past, present and future potentials *Renewable and Sustainable Energy Reviews* **16** pp. 6823 - 6850.

[8] Porumba B, Ungureşana P, Tutunarua L F, Şerban M and Bălana A 2016 A review of indirect evaporative cooling technology *Energy Procedia* **85** pp. 461 - 471.

[9] Adebisi O W, Igbeka J C and Olurin T O 2009 Performance evaluation of absorbent materials in evaporative cooling system for the storage of fruits and vegetables *Int. J. Food Eng.* **5(3)** pp. 1-15.

[10] Pistochini T and Modera M 2011 Water-use efficiency for alternative cooling technologies in arid climates *Energy and Buildings* 43 pp. 631-638.

[11] Odesola I F, Onyebuchi O 2009 A Review of Porous Evaporative Cooling for the Preservation of Fruits and Vegetables *The Pacific J. Sci. Technol.* **10(2)**

[12] Çengel AY 2003 *Fundamentals of Thermal Fluids Science : A Practical Approach* (McGraw-Hill)

IOP Conf. Series: Materials Science and Engineering **309** (2018) 012090 doi:10.1088/1757-899X/309/1/012090

Numerical simulation on thermal plasma temperature field in the torch for different conditions

Y Siregar[1, 2*], N Kodama[1], Y Tanaka[1], T Ishijima[1], Y Uesugi[1]

[1]Division of Electrical Engineering and Computer Science, Kanazawa University, Kakuma, Kanazawa 920-1192, Japan
[2]Department of Electrical Engineering, Universitas Sumatera Utara, Medan 20155, Indonesia

*E-mail: julianta@stu.kanazawa-u.ac.jp

Abstract. Numerical simulation was conducted to obtain the temperature and gas flow fields in inductively coupled thermal plasma (ICTP). Influence of gas composition, power supply, quenching gas, carrier gas, pressure and frequencies were investigated on thermal plasma temperature fields. Results indicated that higher oxygen fraction in gas composition makes the shrinkage of the thermal plasmas, which increases power density in local volume. Further, an increase in the quenching gas, the carrier gas and the pressure did not affect the temperature of thermal plasma. On the other hand, the temperature in thermal plasma decreased at axial position along with degradation of frequency.

1. Introduction

Inductively coupled thermal plasma (ICTP) with high power and high pressure has been widely used in different materials processing like synthesis of diamond films [1], syntheses of fullerene [2] and thermal barrier coatings [3], a chemical source and effective heat source. In the materials processing, raw material powder is injected into the thermal plasma. The injected powder is rapidly heated and then evaporated. These evaporated materials are used for materials processing such as coating, nanoparticle synthesis and so on. One feature of the ICTP is that it needs no electrode. This gives a benefit of no contaminated process compared to the arc plasma processing. From the above benefit, the ICTP is adopted to material processing like synthesis of nanopowder [4]. The ICTP processing is affected markedly by the temperature field of the ICTP.

In this work, we studied the effect of gas composition, the input power, the quenching gas, the carrier gas flow rate, the pressure and the frequency on the ICTP temperature field. Firstly, we desribe the modelling of the ICTP and the assumptions. Secondly, the calculation was made for the temperature field for Ar-O_2 gas mixture thermal plasma at 25 kW. We studied the effect of each of physical parameters: O_2 percentage to Ar, the input power, the quenching gas, the carrier gas flow rate, the pressure and the frequency on the ICTP temperature field. From these results, we obtain the sensitivity from each physical parameter on the ICTP temperature fields.

IOP Conf. Series: Materials Science and Engineering **309** (2018) 012090 doi:10.1088/1757-899X/309/1/012090

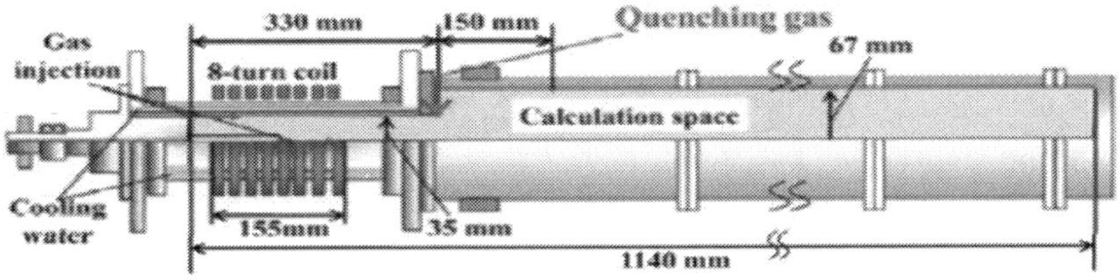

Figure 1. Cross section of inductively coupled thermal plasma torch used in the numerical simulation.

2. Structure of inductively coupled thermal plasma torch

Figure 1 illustrate the structure inductively coupled thermal plasma torch and the reaction chamber used in this numerical simulation. The plasma torch comprises two coaxial quartz tubes with 330 mm in length. The inner quartz tube has a internal radius of 35 mm and an 8 turn-coil is located around the quartz tube to generate electromagnetic field. This electromagnetic field forms a inductively coupled thermal plasma inside the plasma torch. The quartz tube wall of the plasma torch is cooled with cooling water at room temperature.

From the head of the plasma torch, argon and O_2 gas mixture is supplied in the inner quartz tube wall with a swirl to prevent the plasma from attaching the tube wall. This gas is called the sheath gas. From the plasma torch head, a water-cooled tube is inserted for feedstock injection in case of materials processing. The feedstock is usually injected with a carrier gas. In addition, quenching gas can be injected downstream of the plasma torch in radial direction to cool down the thermal plasma.

3. Modeling

3.1 Assumptions

The thermal plasma was modelled based on the following assumptions: (i) The plasma is in local thermodynamic equilibrium; consequently, all the temperatures such as the electron temperature, heavy particle temperature, and excitation temperature are identical. In addition, the model assumes chemical equilibrium conditions for all reactions. (ii) The plasma is optically thin for wavelengths greater than 200 nm. For wavelengths of less than 200 nm, 20% of the total emission coefficient is accounted for radiation loss to consider the effective light absorption. (iii) The flow is steady, laminar, and axis-symmetric, with negligible viscous dissipation.

3.2 Governing equation for thermal plasma region

On the basis of the assumptions described above, the thermal plasma is governed by the following equations.

Mass conservation.

$$\frac{\partial(\rho u)}{\partial z}+\frac{1}{r}\frac{\partial(r\rho v)}{\partial r}=0 \tag{1}$$

Momentum conservation.
Axial:

$$\rho u\frac{\partial u}{\partial z}+\rho v\frac{\partial u}{\partial r}=-\frac{\partial p}{\partial z}+2\frac{\partial}{\partial z}\left(\eta\frac{\partial u}{\partial z}\right)+\frac{1}{r}\frac{\partial}{\partial r}\left[\eta r\left(\frac{\partial u}{\partial r}+\frac{\partial v}{\partial z}\right)\right]+\mu_0\sigma\Re\left[\dot{E}_\theta\dot{H}_z\right] \tag{2}$$

Radial:

$$\rho u\frac{\partial v}{\partial z}+\rho v\frac{\partial v}{\partial r}=-\frac{\partial p}{\partial r}+\frac{\partial}{\partial z}\left[\eta\left(\frac{\partial v}{\partial z}+\frac{\partial u}{\partial r}\right)\right]+\frac{2}{r}\frac{\partial}{\partial r}\left(\eta r\frac{\partial u}{\partial r}\right)-2\eta\frac{v}{r^2}+\mu_0\sigma\Re\left[\dot{E}_\theta\dot{H}_z\right] \tag{3}$$

Swirl:

$$\rho u\frac{\partial w}{\partial z}+\rho v\frac{\partial w}{\partial r}=-\frac{\partial}{\partial z}\left(\eta\frac{\partial w}{\partial z}\right)+\frac{1}{r}\frac{\partial}{\partial r}\left(r\eta\frac{\partial w}{\partial r}\right)-\frac{\rho vw}{r}-\frac{w}{r}\frac{\partial rn}{\partial r} \tag{4}$$

Energy conservation.

$$\rho u\frac{\partial h}{\partial z}+\rho v\frac{\partial h}{\partial r}=\frac{\partial}{\partial z}\left(\frac{\lambda}{C_p}\frac{\partial h}{\partial z}\right)+\frac{1}{r}\frac{\partial}{\partial r}\left(r\frac{\lambda}{C_p}\frac{\partial h}{\partial r}\right)+\sigma\left|\dot{E}_\theta\right|^2-P_{rad} \tag{5}$$

Maxwell equation.

$$\frac{\partial^2}{\partial z^2}+\frac{1}{r}\frac{\partial}{\partial r}\left(r\frac{\partial \dot{A}_\theta}{\partial r}\right)-\frac{\dot{A}_\theta}{r^2}=j\mu_0\sigma\omega\dot{A}_\theta \tag{6}$$

$$\dot{H}_z=\frac{1}{\mu_0}\frac{1}{r}\frac{1}{\partial r}(r\dot{A}_\theta),\ \dot{H}_r=-\frac{1}{\mu_0}\frac{\partial \dot{A}_\theta}{\partial z} \tag{7}$$

$$\dot{E}_\theta=-j\omega\dot{A}_\theta \tag{8}$$

In those equations, the following pertain. *r:* radial position, *z:* axial position, *u:* axial flow velocity, *v:* radial flow velocity, *ρ:* mass density, *p:* pressure, *η:* viscosity, *h:* enthalpy, *λ:* thermal conductivity, C_p: specific heat at constant pressure, *σ:* electrical conductivity, P_{rad}: radiative loss, μ_0: permeability of vacuum, $\dot{A}_\theta$: phasor of the vector potential, ω: frequency of the coil current, $\dot{E}_\theta$: phasor of the electric field strength, $\dot{H}_z$, $\dot{H}_r$: phasors of axial and radial components, of the magnetic field strength, respectively, j: complex factor ($j^2=-1$). The magnitudes of the phasors including $\dot{A}_\theta$, $\dot{E}_\theta$, H_z, and H_r are defined as the root mean square values. The asterisks in equations 2 and 3 indicate the conjugate, and a symbol $\Re$ is the real part of the phasor.

Table 1. Calculation condition-1

Condition No.	3 (ref)	1	2	4
Input power	25 kW	-	20 kW	-
Frequency	450 kHz	-	-	-
Pressure	300 torr	-	-	-
Total sheath gas flow rate	120 L/min	-	-	-
O_2 percentage to Ar	10%	15%	-	-
Carrier Ar gas flow rate	4 L/min	-	-	-
Quenching flow rate	50 L/min	-	-	75 L/min

Table 2. Calculation condition-2

Condition No.	3 (ref)	5	6	7
Input power	25 kW	-	-	-
Frequency	450 kHz	-	-	400 kHz
Pressure	300 torr	-	500 torr	-
Total sheath gas flow rate	120 L/min	-	-	-
O_2 percentage to Ar	10%	-	-	-
Carrier Ar gas flow rate	4 L/min	6 L/min	-	-
Quenching flow rate	50 L/min	-	-	-

The thermodynamic such as the mass density, the enthalpy and the specific heat, and transport properties such as the thermal conductivity, the viscosity and the electrical conductivity are necessary for simulation. These were calculated in advance using the calculated equilibrium composition and the first-order approximation of Chapman-Enskog method [5].

4. Numerical calculation condition

The calculation was made for Ar-O_2 thermal plasmas. The thermal plasma temperature fields is affected by many physical parameters. Here, we studied the physical parameter effects on the thermal plasma temperature field. Table 1 summarizes the numerical calculation conditions in the present work. There are seven conditions. Condition 3 is the reference condition having the input power of 25 kW, the frequency of the coil current of 450 kHz, the pressure of 300 torr, the total gas flow rate of 120 L/min, percentage to Ar of 10%, the Ar carrier gas flow rate of 4 L/min, quenching gas flow rate of 50 L/min.

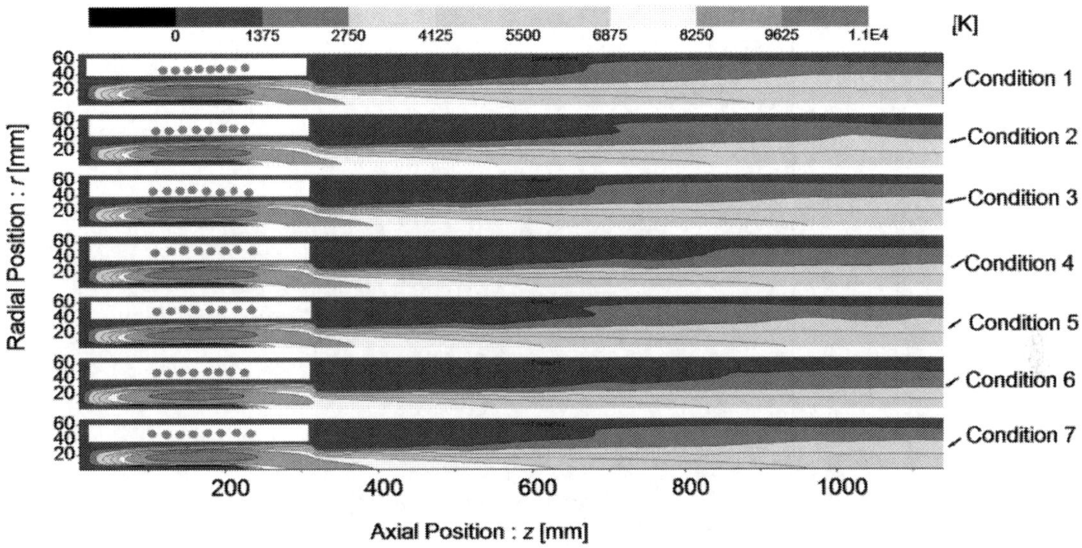

Figure 2. Temperature distribution result with quenching gas injection

From the reference condition, one condition was changed to study the specified parameter on the thermal plasma temperature field. Condition 1 has different O_2 concentration of 15% to condition 3 to see the effect of O_2 concentration. Generally, high percentage of O_2 requires the excitation and dissociation energy of O_2, which leads to the higher specific heat and thermal conductivity. Condition 4 has a higher quenching gas flow of 75 L/min to study the effect of quenching gas flow. Quenching gas frequently cools the thermal plasma temperature. Effect of Ar carrier gas flow rate can be seen in condition 5 with 6 L/min by comparing with the result of condition 3. The carrier gas has low temperature, which injection to the thermal plasma affects the thermal plasma temperature markedly. Pressure is one of the important parameter to determine the thermal plasma temperature. Comparison between the results for conditions 3 and 6 may provide the partial knowledge of the pressure effect between 300 torr and 500 torr. Condition 7 has a different frequency of 400 kHz to study the effect of frequency of the coil current in comparison with condition 3. The frequency of the coil current affects the skin depth of electromagnetic field, and then influences the temperature profile.

5. Numerical result

To get the best result in thermal plasma, we compare the calculation results of gas composition, input power, quenching gas, carrier gas, pressure and frequency as shown in Figure 2. The first compare is about gas composition with condition 1 and condition 3. As we can see in Figure 2, condition 3 is better result in thermal plasma torch than condition 1.

Condition 1 Ar 85% and O_2 15%, the temperature inside the plasma torch is 8032 K at axial position is z = 300 mm and radial position is r = 20 mm. Furthermore, the temperature in the reaction chamber is 6484 K for axial position amount is 600 mm and radial position as much as 5 mm. Condition 3 Ar 90% and O_2 10%, the temperature inside the plasma torch is 8602 K at axial position is z = 300 mm and radial position is r = 20 mm. Meanwhile, the temperature in the reaction chamber is 6932K for axial position amount is z = 600 mm and radial position as much as r = 5 mm.

Frequency 450 KHz with condition 3, the temperature inside the plasma torch is 7788 K at axial position is z = 300 mm and radial position is r = 25 mm. Frequency 400 KHz with condition 7, the temperature inside the plasma torch is 7788 K at axial position is z = 300 mm and radial position is r = 23 mm. Meanwhile, the temperature in the reaction chamber in condition 3 and 7 are same results.

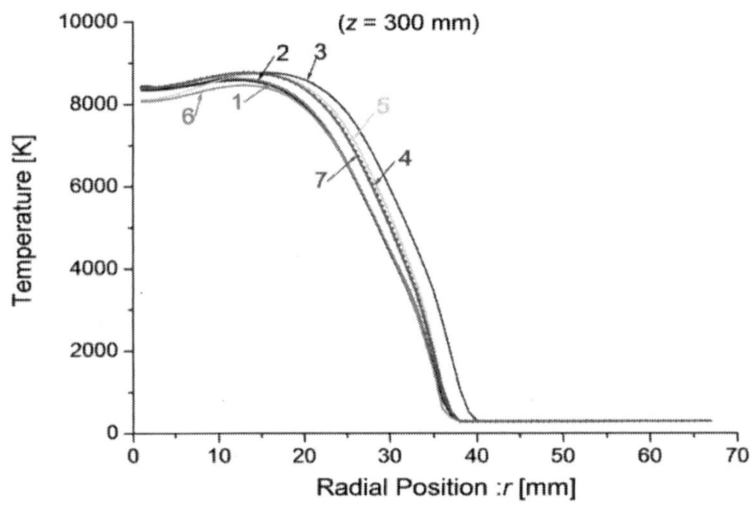

Figure 3. Radial temperature distribution at axial position of z = 300 mm with quenching gas.

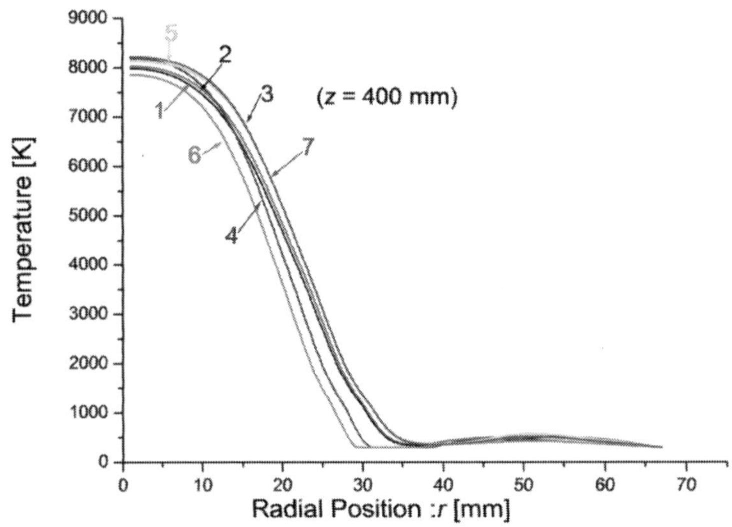

Figure 4. Radial temperature distribution at axial position of z = 400 mm with quenching gas.

Finally, we get best of the best result in condition 3 with compare addition frequency. To compare the temperature distribution clearly in gas composition, input power, quenching gas, carrier gas, pressure and frequency, the radial temperature distributions at axial positions of z = 300 mm and

z = 400 mm are plotted in figure 3 and figure 4, respectively. This figure includes the curves for all condition. As indicated in this figure, the condition 3 has the temperatures about 8602 K and 5203 K (z = 300 and z = 400) at radial positions of r = 20 mm. This implies that the temperature in all condition is influenced easily to decrease by quenching gas. On the other hand, condition 6 is smaller result in thermal plasma torch in other condition with temperature about 7923K and 3621 K (z = 300 and z = 400) at r = 20 mm.

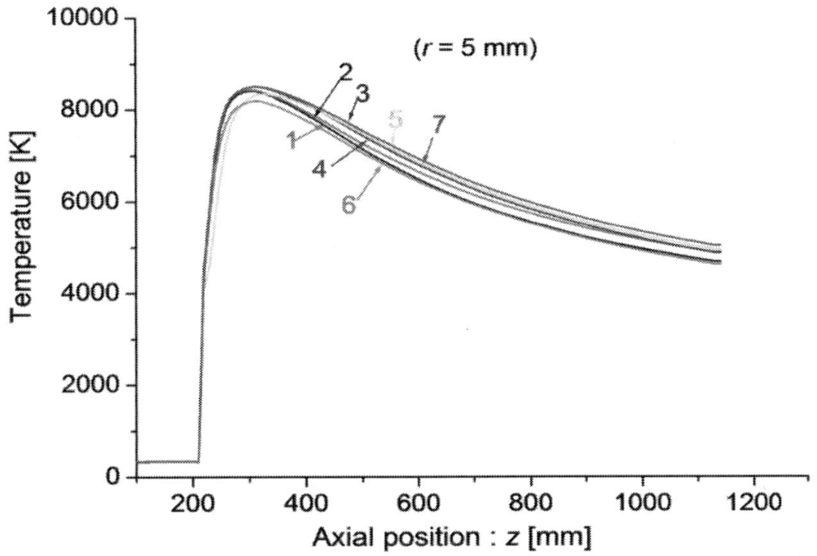

Figure 5. Axial temperature distribution at radial of 5 mm with quenching gas.

To indicate the temperature recovery characteristics in all condition, the axial temperature distributions are plotted in Figure 5. As seen in this figure, the temperature at a radial position of r = 5 mm is 300 K from z = 0 mm to z = 210 mm in all condition. z = 210 to z = 220 mm, the temperature is still 300 K due to cool argon carrier gas flow in all condition. This axial temperature recovers from z = 220 mm to z = 300 mm only to around 8500 K in all condition. On the other hand, in all condition, the temperature decrease to 5000 K up to z = 1100 mm.

6. Conclusions

In this paper, the influences of gas composition, input power, quenching gas, carrier gas, pressure and frequency on the temperature distribution in argon (Ar) induction thermal plasma with quenching gas were investigated numerically. Results indicated that addition frequency at 450 kHz significantly increase temperature at plasma torch in axial position. It means condition 3 is better than the other conditions.

References

[1] Gottlieb S, Wohrl N, Schulz S and Buck V 2016 Simultaneous synthesis of nano diamonds and graphene via plasma enhanced chemical vapor deposition (MW PE-CVD) on copper *SpringerPlus* **5** 568

[2] Kim K S, Moradian A, Mostaghimi J and Soucy G 2010 Modeling of induction plasma process for fullerene synthesis: Effect of plasma gas composition and operating pressure *Plasma Chem. Plasma Process* **30** pp. 91-110

[3] Mao J, Liu M, Deng C G, Deng C M, Zhou K S and Deng Z Q 2017 Preparation and distribution analysis of thermal barrier coatings deposited on multiple vanes by plasma spray-

physical vapor deposition technology *J. Eng. Mater. Technol.* **139** 041003
[4] Wang C, Tanaka Y and Sakuta T 2004 Modeling of Ar Induction Thermal Plasma with an Injection of PTFE Powder *Trans. IEEJ* **124-PE** pp. 440-46
[5] Yos J M 1963 Transport properties of nitrogen, hydrogen, oxygen and air to 30000 K *AVCO Technical Memorandum RAD-TM-63-7*

IOP Conf. Series: Materials Science and Engineering **309** (2018) 012091 doi:10.1088/1757-899X/309/1/012091

Red palm oil production by microwave irradiation

M Sarah[1,2*], S Widyastuti[1] and D Ningsih[1]

[1]Departement of Chemical Engineering, Universitas Sumatera Utara
[2]Sustainable Energy and Biomaterial Centre of Excellent, Universitas Sumatera Utara

*Email: Maya3@usu.ac.id

Abstract. Preliminary study of red palm oil (RPO) production from palm fruitlets by microwave irradiation carried out in domestic microwave oven equipped with thermocouple. The various mass of fruitlets (800, 900 and 1000 g) were heated for 10-18 minutes with 2 minutes interval and microwave power of 400, 560 and 800 Watt respectively. Heated fruitlets were pressed by hydraulic presser to obtain RPO. This study observed heating time parameter was more crucial to RPO quality rather than temperature. Prolonged heating degraded carotenoids in the fruitlets during heating process yielded less carotenoids content in the palm oil. The best time and microwave power combination to produce RPO in this study was 14 minutes and 800 Watt respectively which yielded 11.67% RPO with 1.27% FFA content and carotenoids concentration of 1219.37 ppm. Overall, RPO production by microwave irradiation proceeded faster as compared to conventional process.

1. Introduction

Sterilization is a way to obtain sterile products [1]. It is a process that eliminates viruses, bacteria, fungi, spores [2] and enzyme [3,4]. It can be reached through heating, chemicals, irradiation, high pressure and filtration [3]. The popular method of heating process is heating by microwave irradiation [2]. Microwave offers many advantages such as faster heating and less nutritive loss [4].

In palm oil mill, the aim of sterilization is to inactivate oil splitting enzymewhich accelerates FFA formation, softens the palm fruitlets and facilitates fruits detachment from the bunch [5-7]. In conventional way, sterilization carries out by pressurized steam of 15-45 psi for 90 minutes (temperature more than 100 °C [6] and [8]). In fact, lipase inactivation only requires heating at low temperature (approximately 55 °C [9]). Since the process envolves steam, high FFA content is inevitable. The presence of water will promote the hydrolysis reaction of oil to FFA [5].

Palm oil is well-known as the most abundant resource of carotenoids especially β- and α- carotene (provitamin A) [10-14]. The high content of carotenoids can be utilized to prevent deficiency of vitamin A [10,12,14]. Besides that, palm oil also has good benefit to human body that is antioxidant. It can modulate cardiovascular diseases, cancer, cataract, aging, diabetes, etc [12,14]. However, carotenes are supceptible to degradation by oxidation and thermal process due their highly unsaturated nature [15]. Karer and Jucker (1950) in Santoso *et. al* (2006) reported carotene deterioration at temperature above 60°C while Rutishauser (1992) in Santoso *et. al* (2006)found carotene is not stable at temperature heating more than 100°C [16].

To protect carotenoids content in palm oil, sterilization should carries out at low temperature (below 60°C). Palm oil obtained from extraction of this sterilized fruit is Red Palm Oil (RPO). RPO consist of carotenoids that mainly found between 700-800 ppm as β-carotene (Sambanthamurthi et al., 2000). Among others RPO production is heating or sterilization of oil palm fruit by using microwave irradiation. Some researchers reported utilization of microwave energy to heat and sterilize oil palm fruits [3,5,7,9,17]. As compared to the conventional process, this method can reduce heating time and prevent loss of nutritive components in palm oil. This study evaluates performance of RPO production by microwave irradiation.

2. Materials and Methods

2.1. Materials

Materials used in this study were fresh palm fruit *Tenera* variety taken from Galang-Deli Serdang, chemicals (acetone, NaOH, n-hexane, ethanol, phenolpthalein, and Na_2HPO_4 supplied by Merck, microwave oven (Sharp Model: R-249IN(S)/(W)), thermocouple type K (Krupp and Closs size diameter of 3 x 300 mm (Mineral Insulated) C/w Cable 2 m) connected with thermo controller (Shimaden)) and hydraulic presser (fabricated).

2.2. Methods

2.2.1. Sample Preparation. Fruitlets removed manually from the bunch, cleaned to remove any dirt, weighed and placed at cavity inside microwave oven.

2.2.2. Microwave Sterilization Process. Prior sterilization/heating, we adjusted heating time and microwave power. Fruitlets irradiated by microwave energy at various power levels (medium, medium high and high power level). The temperatures recorded every 2 minutes interval. After the sterilization, the fruitlets were pressed using hydraulic presser to extract the oil.

2.2.3. Lipase Assayed. The palm oil (0.1 g) dissolved in water and stirred at temperature of 37°C.We added acetone and ethanol mixture (50:50 v/v) of approximately 10 ml to inactivate the lipase after 30 minutes and titrated with 0.1 N sodium hydroxide solution using phenolpthalein as indicator until the first permanent pink colour appears[18,19,20].

2.2.4. FFA Test. It conducted according to AOCS Official Method Ca 5a-40. Palm oil sample of 7.05 ±0.05 g dissolved in 75 ml ethanol 95% and was titrated with standard solution of sodium hydroxide of 0.25 N with phenolphthalein as indicator until the first permanent pink colour appears [21].

2.2.5. Water Content. Water content determination was conducted according to AOAC (1995). The empty porcelain dish were dried in the oven at 105°C for 3 hours, transferred to desiccator to cool down the temperature, and weighed. Palm oil sample of 5 g was put in the dish, weighed and placed inside the oven. We dried the sample for 3 hours at temperature of 105 °C. Porcelain dish with dried sample was transferred into desiccator to cool down temperature. The dish and its dried sample were reweighed [22] to determine water content.

2.2.6. Carotenoids test. This study determined carotenoids by using spectrometer UV-VIS according to MPOB Test Method [24]. We diluted 0.1 g oil palm sample using 25 ml n-hexane and shakes vigorously. The absorbance was measured at 446 nm [23].

3. Results and Discussions

Figure 1 shows profile of RPO yield during heating period. The yield increased at the increment of heating period during the first 16 minutes exclude treatment of oil palm fruitlets using power level of medium for 12 minutes. This study obtained maximum RPO approximately 17.6% after oil palm exposure for 12 minutes using microwave power of medium high level. Prolonged heating decreased the yield of RPO, especially those from treatment using power of medium level.

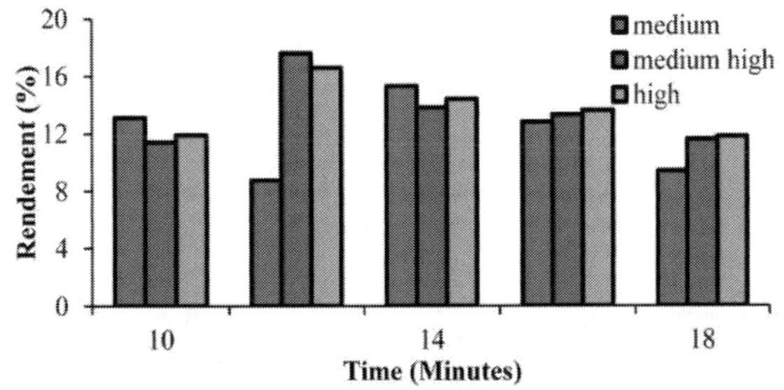

Figure 1. Yield of RPO in this study

Prolonged heating caused evaporation of water inside palm fruits occurred very fast, leaving dried and hard fruitlets. The removal of water from oil palm fruitlets was part of the heating process [17]. Water are being used to generate heat. Higher amount of water removed will produce higher temperature and finally caused temperature increment in the mesocarp and kernel. This mechanism led to the drying effect and hardened mesocarp.Thus heated/sterilized fruitlets were difficult to be pressed resulting low yield of RPO. Increment of microwave power had not affected the rendement significantly at various heating time. The best condition to obtain greater rendement were combination of heating time ranged between 12-14 minutes and microwave power of medium high and high power level.

Figure 2 shows activity of enzyme lipase after the heating complete. The lowest activity was 1.83 U/ml. Increment of heating time decreased activity of lipase indicates microwave irradiation inactivated enzyme successfully during RPO production.

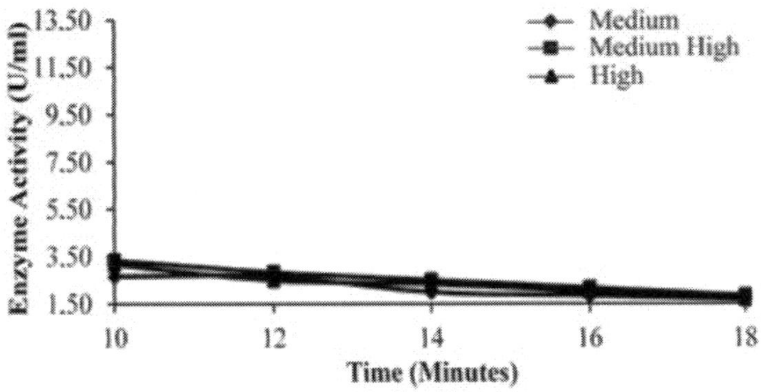

Figure 2. Enzyme activity profile in RPO

Carotene content usually measured as β-carotene in typical CPO. It usually presence in the CPO approximately ranged between 500-700 ppm [5]. The highest carotene content obtained from this study was 1291.96 ppm. During heating process, carotene was broken into another compound. Prolonged heating affected to the increment of temperature which destroy carotene.

FFA content is the most common criterion for determining the quality of palm oil, includes RPO. In this work, FFA content was below maximum standard (below 5%) as the Codex Allimentarius (FAO/WHO) norms [24]. The highest FFA content was 4.09%. Figure 3b (FFA content) shows inconsistent result. This expressed the destruction process of lipase enzyme during microwave heating. The longer period of heating and high temperature usually decreased the activity of enzyme that led to low level formation of FFA.

Typical crude palm oil (CPO) for moisture content is 0.15 – 3%. It obtained from conventional palm oil milling process [5]. In the present study, moisture content was range between 0.07 – 3.46%. Figure 3c shows water content decreased as heating time was extended. Low level of water content in product associated with water utilization to generate heat during microwave irradiation.

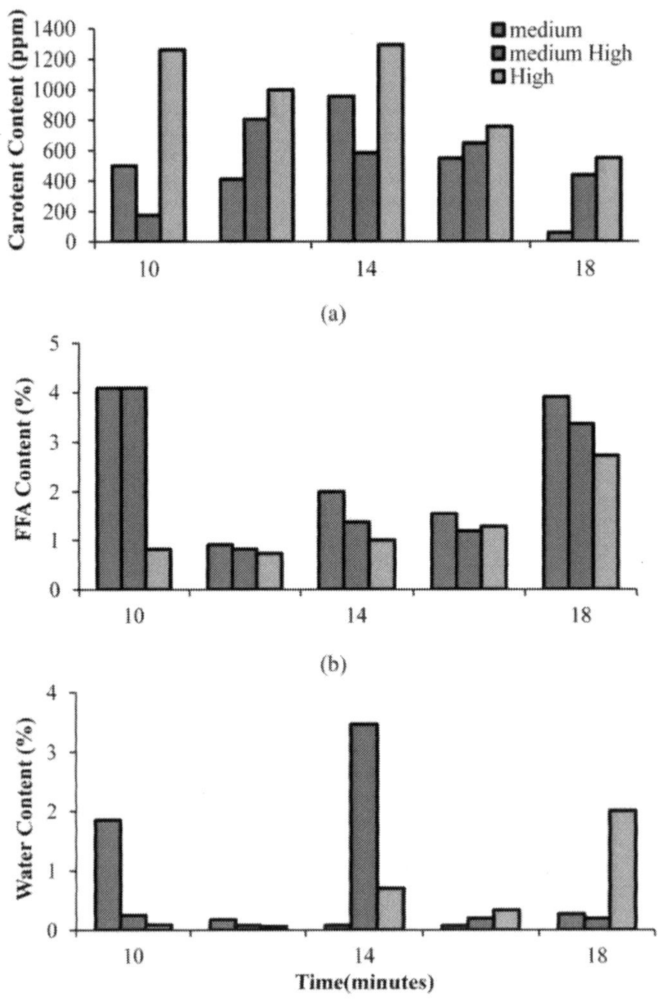

(c)

Figure 3. Effect of heating time to RPO quality: (a) carotene content (b) FFA content (c) water content

4. Conclussion

RPO production by microwave heating required less time as compared to conventional milling.FFA conncentration was below the maximum standard and carotene concentration was higher than typically CPO. Low level FFA content showed that sterilization was also completely done. Due to shorter time heating, carotene in RPO was significant.

References

[1] Van D J P C M and Kopinga K 2009 *J. Appl. Microbiol.* **107** 1054-1060
[2] Sood P, Sood N and Gokhale T 2015 *GSTF J. BioSci.* **3** 1-5
[2] Sarah M and Taib M R 2013 *Int. J. Chem. Eng. Appl.* **4** 129-133
[3] Ozkoc S O, Sumnu G and Sahin S 2011 *Emerging technologies for food processing: recent develpoments in microwave heating* (Amsterdam : Elsevier Academic Press) 361-383
[4] Cheng S F, Nor L M and Chuah C H 2011 *Industrial Crops and Products* **34** 967-971
[5] Choto A, Thongurai C, Kladkaew N and Kiatweerasakul M 2014 *Int. J. of Advance in Chem. Eng. & Biol. Sci.* **1** 123-126
[6] Sukaribin N and Khalid K 2009 *Industrial Crops and Products* **30** 179-185
[7] Umudee I, Chongcheawchamnan M, Kiatweerasakul M and Tongurai C 2013 *Int. J. Chem. Eng. Appl.* **4** 111-113
[8] Noerhidajat, Yunus R, Zurina Z A, Syafiie S, Ramanaidu V and Rashid U 2016 *Int. Food and Res. J.* **23** 129-134
[9] Rice A L and Burns J B 2010 *J. American College of Nutrition* **29** 302S-313S
[10] Nagendran B, Unnithan U R, Choo Y M and Sundram K 2000 *Food and Nutrition Bulletin* **21** 189-194
[11] Mayamol P N, Balachandran C, Samuel T, Sundaresan A and Arumughan C 2007 *J. American Oil Chemists' Society* **84** 587-596
[12] Mancini A, Imperlini E, Nigro E, Montagnese C, Daniele A, Orru S and Buono P 2015 *Molecules* **20** 17339-17361
[13] Mukherjee S and Mitra A 2009 *J. Hum. Ecol.* **26** 197-203
[14] Fauzi N A M and Sarmidi M R 2011 *J. Sci. Technol.* **46** 45-54
[15] Chow M C and Ma A G 2007 *J. of Microwave Power & Electromagnetic Energy* **40** 165-173
[16] Linfield W M, O'brien D J, Serota S and Barauskas R A 1984 *J. American Oil Chemists' Society* **61** 1067-71
[17] Khor H T, Tan N H and Chua C L 1986 *J. American Oil Chemists' Society* **63** 538-540
[18] AOCS 2009 AOCS Official Method Ca 5a-40 Free Fatty Acid *Sampling and Analysis of Commercial Fats and Oils* 1-2
[19] AOAC 1995 Official Method of Analysis
[20] Malaysia Palm Oil Board 2005 *MPOB test method (A compendium of test on palm oil products, palm kernel products, fatty acids, food related products and others)* (Kuala Lumpur: Malaysia Palm Oil Board)
[21] Frank N E G, Albert M M E, Laverdure D E E and Paul K 2011 *J. Stored Products and Postharvest Res.* **2** 52-58

IOP Conf. Series: Materials Science and Engineering **309** (2018) 012092 doi:10.1088/1757-899X/309/1/012092

Improved lifetime of chitosan film in converting water vapor to electrical power by adding carboxymethyl cellulose

T I Nasution[1,2,5]*, M Balyan[1,2,5], I Nainggolan[2,3,4]
[1]Physics Department, Faculty of Mathematics and Natural Science, Universitas Sumatera Utara, Medan 20155, Sumatera Utara, Indonesia
[2]Centre of Excellence for Sensor Systems and Technology, Universitas Sumatera Utara, Medan 20155, Sumatera Utara, Indonesia
[3]Centre of Green Chitin and Chitosan, Universitas Sumatera Utara, Medan 20155, Sumatera Utara, Indonesia
[4]Chemistry Department, Faculty of Mathematics and Natural Science, Universitas Sumatera Utara, Medan 20155, Sumatera Utara, Indonesia
[5]Integrated Laboratory of Physics, Universitas Sumatera Utara, Medan 20155, Sumatera Utara, Indonesia

ikhsan_05@yahoo.com

Abstract. A Water vapor cell based on chitosan film has been successfully fabricated in film form to convert water vapor to electrical power. In order to improve the lifetime of water vapor cell, Carboxymethyl Cellulose (CMC) was added into 1% chitosan solution within concentration variations of 0.01, 0.05, 0.1 and 0.5%. The result showed that the lifetime of water vapor cell increased higher by adding the higher concentration of Carboxymethyl cellulose. The highest lifetime was evidenced by adding 0.5%CMC which maintained for 48 weeks. However, the average electrical power became lower to 4.621 μW. This electrical power lower than the addition of 0.1%CMC which maintained for 5.167 μW. While, the lifetime of chitosan-0.1%CMC film of 44 weeks is shorter compared to chitosan-0.5%CMC film. Based on FTIR characterization, it was founded that the chitosan structure did not change until the addition of 0.1%CMC. This caused the electrical power of water vapor cell degenerated. Therefore, chitosan-0.5%CMC film has excellent lifetime in converting water vapor to electrical power.

1. Introduction

Chitosan, a material derived from the chitin in crustacean cells, insect exoskeletons, and fungus cell walls, is the second most abundant and renewable biopolymer after cellulose [1–3]. Chitosan has aroused considerable interest because of its unique properties such as biocompatibility, biodegradability and multiple functional groups [4–7]. Then, the utilization of chitosan has been used in many applications such as energy storage [8–10], photovoltaic cells [11,12], supercapacitors [13–16], lithium ion batteries [17–19], and fuel cell [20–23].

In our previous work, the use of chitosan film to convert water vapor to electrical power was successfully done through a chemical interaction mechanism [24]. However, the performance of chitosan film is still restricted by low mechanical properties which affected the lifetime water vapor cell [6,25]. Therefore, it is suitable to improve the chitosan film properties using biopolymer which has good mechanical properties.

Carboxymethyl cellulose (CMC) has been chosen as an additive to improve performance of chitosan film. CMC is an ionic polysaccharide derived from cellulose [26,27] which was used in many fields [28–30]. CMC was chosen due to its similar polymeric structure with chitosan and properties such as biocompatibility, nontoxicity, viscosity modifier, emulsion stabilizer and good film forming ability [31–33]. Moreover, CMC has good film forming ability, thickener and plasticizer which reduce swelling properties of chitosan [34–37]. Therefore, CMC has ability to improve performance of chitosan.

2. Experimental

Chitosan (deacylytation degree 85%) was purchased from sigma aldrich. Carboxymethyl cellulose (CMC) was bought from Aqualon. Acetic acid (Glacial) was funded from EMD Milipore. Chitosan solution was prepared by dissolving chitosan powder in 1% acetic acid using Scilogex magnetic stirrer with the rotation speed of 500 rpm for 12 h at room temperature (23 ± 20C). Then, the 1% chitosan solution was added with CMC. The concentration variations of CMC were 0.01, 0.05, 0.1 and 0.5%, respectively. Hereafter, the solution was poured into several Normax Petri Dish and put in oven (memmert UN 55) at 700 °C for 24 h. The obtained films were peeled off from the Petri Dish base after the excess of water was evaporated.

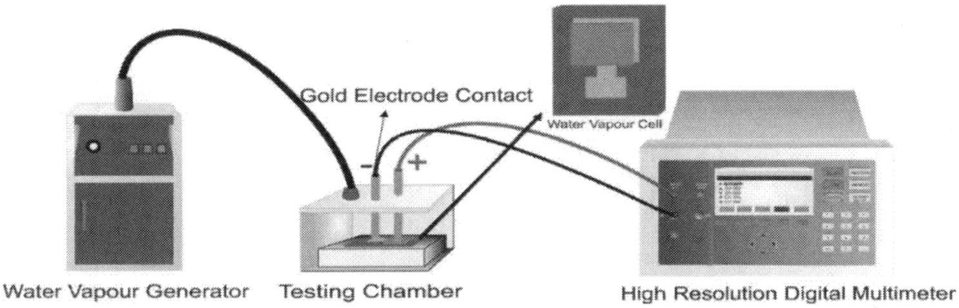

Figure 1. The experiment setup for electrical power measurement.

In this work, the water vapor cell was fabricated of isolator substrate in the bottom layer, aluminum in the middle layer and chitosan or chitosan-CMC film on the top layer. The conversion of water vapor to electrical power of was performed using the experiment setup in Figure 1. The water vapor cell was put in a testing chamber equipped with positive and negative contacts. Inlet valve of the testing chamber was connected to outlet valve of water vapor generator. By pumping the water vapor to testing chamber, the electrical power of chitosan film increased and the data was recorded in High resolution digital multimeter hydra series III fluke. The electrical power measurement as a function of time under water vapor exposure was carried out for chitosan film and chitosan-CMC films lifetime with the concentration variation of 0.01, 0.05, 0.1 and 0.5%, respectively.

3. Results and Discussion

The lifetime of chitosan and chitosan-CMC films was carried out using electrical power measurement as a function of time under water vapor exposure with the concentration variation of 0.01, 0.05, 0.1 and 0.5%, respectively. Lifetimes of water vapor cell based on chitosan and chitosan-CMC films are shown in Figure 2. The result showed that the lifetime of chitosan film maintained for 34 weeks and average electrical power was 4.036 μW. While by adding 0.5%CMC, the longest lifetime of 48 weeks was obtained with average electrical power of 4.621 μW. The electrical power became higher by adding 0.1%CMC, namely 5.167 μW. However, the lifetime of chitosan-0.1%CMC film of 44 weeks is shorter compared to chitosan-0.5% CMC film. Therefore, the best lifetime of water vapor cell is chitosan-0.5% CMC film. It has been believed that plasticizer property of CMC restricted electron movements which caused water vapor cell became less conductive. It is also clearly seen from the Figure 2 that during exposure time, ambient temperature and humidity were kept in constant.

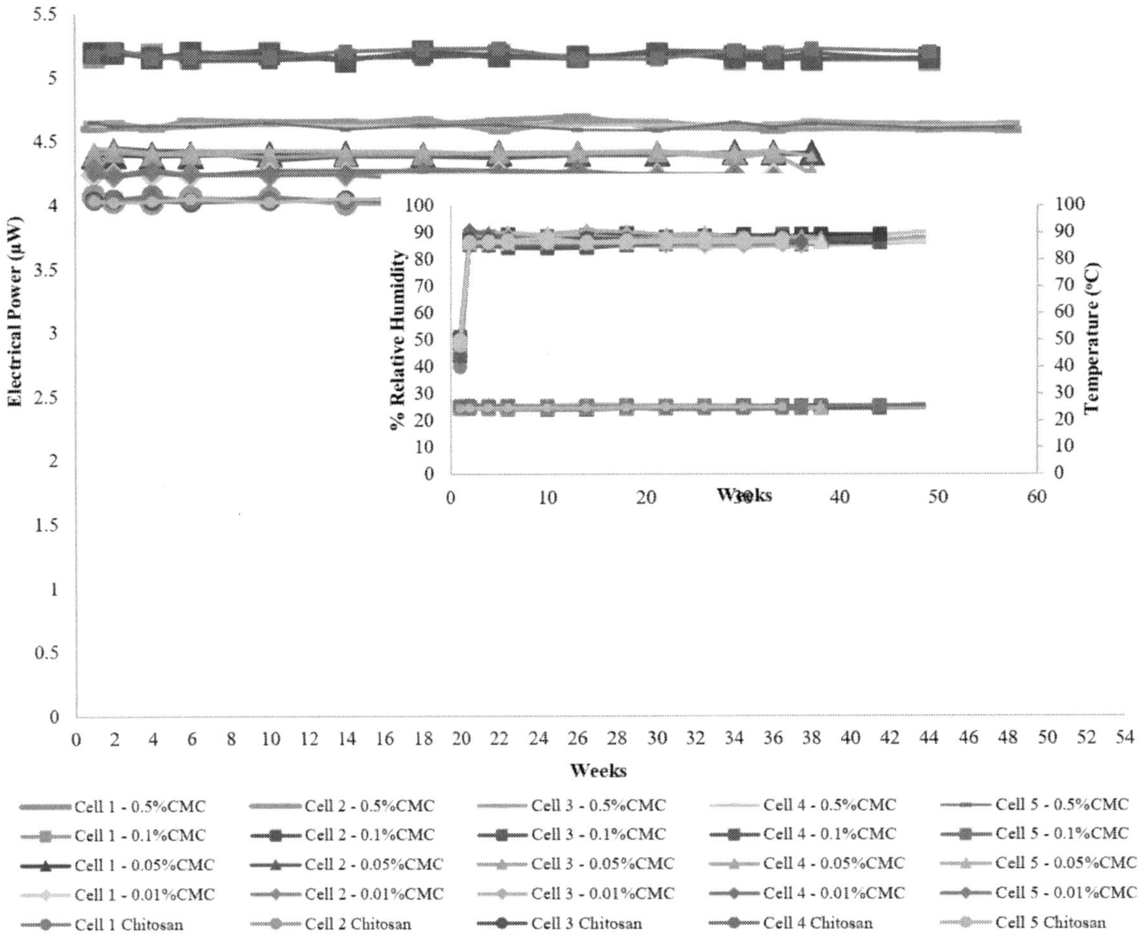

Figure 2. Lifetime of water vapor cell chitosan and chitosan-CMC film.

The interaction between chitosan and chitosan-CMC films were carried out using Fourier Transform Infrared Spectroscopy (FTIR) (Model: Perkin Elmer). The frequency ranges are displayed in Figure 3.

In the featured bands chitosan film at 3367 cm^{-1} is characteristics of –OH. This peak is quite broad and has covered characteristic –NH band at 3722 cm^{-1}. The other peaks shown in spectrum are at 2883 cm-1, 1709 cm^{-1}, 1467 cm^{-1}, 1344 $cm^{-1,}$ and 1061 cm^{-1}, which are characteristic band of C-H and C=O stretching, CH_2 bending, NH deformation and C-O-C stretching, respectively. While, the addition of CMC until 0.1 % into chitosan did not changed certain frequency ranges of chitosan. Moreover, the addition of CMC 0.5 % into chitosan changed broad peaks in the region of 3000-3500 cm-1 become weaker, meaning that the existence of NH as site active was changed to OH^{-}. That's why the addition of 0.5% CMC in chitosan resulted in the decrease electrical power of the film.

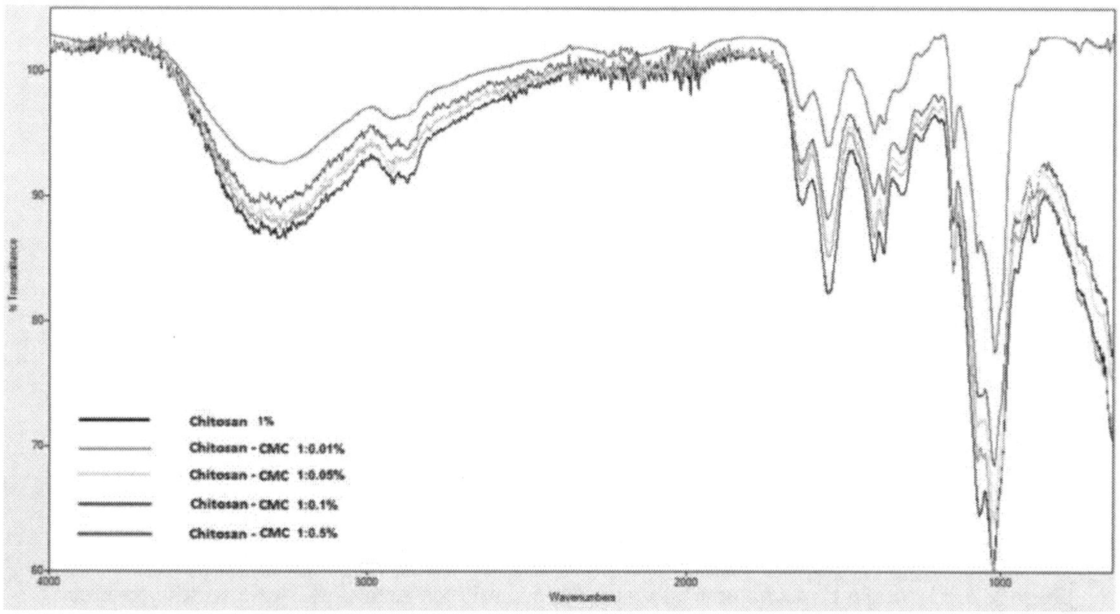

Figure 3. FTIR spectra of chitosan and chitosan-CMC film

In our previous report [24,38], it has been clearly explained about the mechanism of water vapor to electrical power conversion. The chitosan film trapped electrons which contribute to the formation of chemisorb oxygens. The oxygens sticking on chitosan film surface contribute to the formation of chemisorb oxygens which trapped electrons in chitosan film. This generated a formation of oxygen species such as O_2^- or O^- in accordance to Equation (1), (2) and (3).

$$O_2\ (gas) \leftrightarrow O_2\ (adsorbed) \quad (1)$$
$$O_2\ (adsorbed) \leftrightarrow O_2^-\ (adsorbed) \quad (2)$$
$$O_2^-\ (adsorbed) \leftrightarrow 2O^-\ (adsorbed) \quad (3)$$

However, the presence of gaps between the particles make the trapped electrons cannot be transferred from one particle to another. Thus, when water vapor is exposed to chitosan film, water vapor molecules (H_2O) on the surface of chitosan particles donate a pair of electron known as nucleophile to NH_2 known as electrophile, this reaction is shown in Equation (4).

$$Chit\text{-}NH_2 + 2O^- + H_2O \rightarrow Chit\text{-}NH_2 + O_2 + H_2O + e^- \quad (4)$$

These Hydrogen of the amino groups in Equation 5 were bonded with the oxygen of water molecules forming hydrogen bonding and creating surface tension as illustrated in Figure 4. The trapped electrons are forced to escape by the presence of the surface tension. The hydrogen bonds make these electrons move easier. These phenomena establish electrical bridge or pathways which make chitosan film generates electrical power. These phenomena support our assumptions of the possibility of a pathway or electrical bridge.

IOP Conf. Series: Materials Science and Engineering **309** (2018) 012092 doi:10.1088/1757-899X/309/1/012092

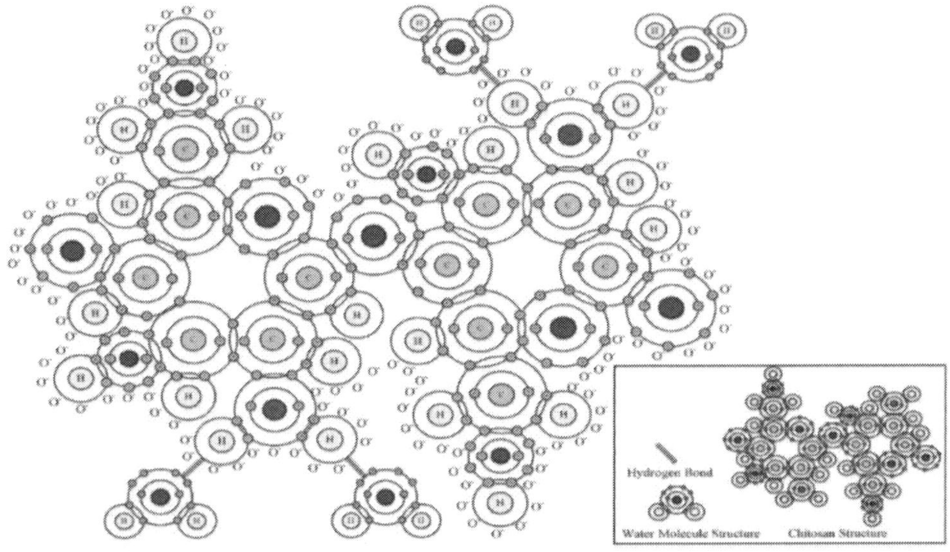

Figure 4. Conversion mechanism of water vapor into electrical power of a water vapor cell.

4. Conclusion

It has proven that chitosan films have an ability to convert water vapor to electrical power which is a new application in energy converter material field. The water vapor to electrical power conversion by direct interaction between chitosan film and water vapor is a novelty. Amine ($-NH_2$) groups acted as an active sites that played an important role in the conversion process. Moreover, chitosan film performance can be improved by addition of CMC. Therefore, the presence of CMC in chitosan film enhances the lifetime and electrical power of chitosan film.

Acknowlegments

We would like to acknowledge the funding and support from Penilitian Unggulan Perguruan Tinggi 2015-2017, Ministry of Research, Technology and Higher Education of the Republic of Indonesia and University of Sumatera Utara.

References

[1] Guo Z, Liu H, Chen X, Ji X and Li P 2006 *Bioorg. Med. Chem. Lett.* **16** 6348–50
[2] Pandiselvi K and Thambidurai S 2014 *Ionics (Kiel).* **20** 551–61
[3] Corazzari I, Nisticò R, Turci F, Faga M G, Franzoso F, Tabasso S and Magnacca G 2015 *Polym. Degrad. Stab.* **112** 1–9
[4] Kurita K 2001 Controlled functionalization of the polysaccharide chitin *Prog. Polym. Sci.* **26** 1921–71
[5] Liu F, Qin B, He L and Song R 2009 Novel starch/chitosan blending membrane: Antibacterial, permeable and mechanical properties *Carbohydr. Polym.* **78** 146–50
[6] Pan Y, Wu T, Bao H and Li L 2011 Green fabrication of chitosan films reinforced with parallel aligned graphene oxide *Carbohydr. Polym.* **83** 1908–15
[7] Sahu P K, Sahu P K, Gupta S K and Agarwal D D 2014 Chitosan: An Efficient, Reusable, and Biodegradable Catalyst for Green Synthesis of Heterocycles *Ind. Eng. Chem. Res.* **53** 2085–91
[8] Hassan S, Suzuki M and El-Moneim A A 2014 *J. Power Sources* **246** 68–73
[9] Chupp J, Shellikeri A, Palui G and Chatterjee J 2015 Chitosan-based gel film electrolytes containing ionic liquid and lithium salt for energy storage applications *J. Appl. Polym. Sci.*

132n/a-n/a

[10] Ramkumar R and Minakshi M 2015 Fabrication of ultrathin CoMoO4 nanosheets modified with chitosan and their improved performance in energy storage device *Dalt. Trans.* **44** 6158–68

[11] Mohamad S A, Ali M H, Yahya R, Ibrahim Z A and Arof A K 2007 Photovoltaic activity in a ZnSe/PEO–chitosan blend electrolyte junction *Ionics (Kiel).* **13** 235–40

[12] Buraidah M H, Teo L P, Majid S R, Yahya R, Taha R M and Arof A K 2010 Characterizations of Chitosan-Based Polymer Electrolyte Photovoltaic Cells *Int. J. Photoenergy* **2010**

[13] Sudhakar Y N and Selvakumar M 2012 Lithium perchlorate doped plasticized chitosan and starch blend as biodegradable polymer electrolyte for supercapacitors *Electrochim. Acta* **78** 398–405

[14] Yamagata M, Soeda K, Ikebe S, Yamazaki S and Ishikawa M 2013 Chitosan-based gel electrolyte containing an ionic liquid for high-performance nonaqueous supercapacitors *Electrochim. Acta* **100** 275–80

[15] Hao P, Zhao Z, Leng Y, Tian J, Sang Y, Boughton R I, Wong C P, Liu H and Yang B 2015 Graphene-based nitrogen self-doped hierarchical porous carbon aerogels derived from chitosan for high performance supercapacitors *Nano Energy* **15** 9–23

[16] Sun G, Li B, Ran J, Shen X and Tong H 2015 Three-dimensional hierarchical porous carbon/graphene composites derived from graphene oxide-chitosan hydrogels for high performance supercapacitors *Electrochim. Acta* **171** 13–22

[17] Ali A M M, Yahya M Z A, Mustaffa M, Ahmad A H, Subban R H Y, Harun M K and Mohamad A A 2005 Electrical properties of plasticized chitosan-lithium imide with oleic acid-based polymer electrolytes for lithium rechargeable batteries *Ionics (Kiel).* **11** 460–3

[18] Kim K, Kim J S, Lee S and Lee J K 2015 Employment of Chitosan–linked Iron Oxides as Mesoporous Anode Materials for Improved Lithium–ion Batteries *Electrochim. Acta* **170** 146–53

[19] Ma L, Zhou X, Xu L, Xu X, Zhang L and Chen W 2015 Chitosan-assisted fabrication of ultrathin MoS2/graphene heterostructures for Li-ion battery with excellent electrochemical performance *Electrochim. Acta* **167** 39–47

[20] Xiang Y, Yang M, Guo Z and Cui Z 2009 Alternatively chitosan sulfate blending membrane as methanol-blocking polymer electrolyte membrane for direct methanol fuel cell *J. Memb. Sci.* **337** 318–23

[21] Ma J and Sahai Y 2013 Chitosan biopolymer for fuel cell applications *Carbohydr. Polym.* **92** 955–75

[22] Noroozifar M, Khorasani-Motlagh M, Ekrami-Kakhki M-S and Khaleghian-Moghadam R 2014 Enhanced electrocatalytic properties of Pt–chitosan nanocomposite for direct methanol fuel cell by LaFeO3 and carbon nanotube *J. Power Sources* **248** 130–9

[23] Shaari N and Kamarudin S K 2015 Chitosan and alginate types of bio-membrane in fuel cell application: An overview *J. Power Sources* **289** 71–80

[24] Nasution T I, Balyan M and Nainggolan I 2017 New Application of Chitosan Film as a Water Vapor Cell *Key Eng. Mater.* **744** 339–45

[25] He L, Wang H, Xia G, Sun J and Song R 2014 Chitosan/graphene oxide nanocomposite films with enhanced interfacial interaction and their electrochemical applications *Appl. Surf. Sci.* **314** 510–5

[26] Hiroki A, Tran H T, Nagasawa N, Yagi T and Tamada M 2009 Metal adsorption of carboxymethyl cellulose/carboxymethyl chitosan blend hydrogels prepared by Gamma irradiation *Radiat. Phys. Chem.* **78** 1076–80

[27] Thomas S, Soloman P A and Rejini V O 2016 Preparation of Chitosan- CMC Blends and Studies on Thermal Properties *Procedia Technol.* **24** 721–6

[28] He F and Zhao D 2007 Manipulating the Size and Dispersibility of Zerovalent Iron Nanoparticles by Use of Carboxymethyl Cellulose Stabilizers *Environ. Sci. Technol.* **41** 6216–21

[29] He F, Zhao D, Liu J and Roberts C B 2007 Stabilization of Fe–Pd Nanoparticles with Sodium Carboxymethyl Cellulose for Enhanced Transport and Dechlorination of Trichloroethylene in

Soil and Groundwater *Ind. Eng. Chem. Res.* **46** 29–34
[30] Kono H 2014 Characterization and properties of carboxymethyl cellulose hydrogels crosslinked by polyethylene glycol *Carbohydr. Polym.* **106** 84–93
[31] Cheng Y, Feng B, Yang X, Yang P, Ding Y, Chen Y and Fei J 2013 Electrochemical biosensing platform based on carboxymethyl cellulose functionalized reduced graphene oxide and hemoglobin hybrid nanocomposite film *Sensors Actuators B Chem.* **182** 288–93
[32] Arinaitwe E and Pawlik M 2014 Dilute solution properties of carboxymethyl celluloses of various molecular weights and degrees of substitution *Carbohydr. Polym.* **99** 423–31
[33] Tran T H, Okabe H, Hidaka Y and Hara K 2017 Removal of metal ions from aqueous solutions using carboxymethyl cellulose/sodium styrene sulfonate gels prepared by radiation grafting *Carbohydr. Polym.* **157** 335–43
[34] Tongdeesoontorn W, Mauer L J, Wongruong S, Sriburi P and Rachtanapun P 2011 Effect of carboxymethyl cellulose concentration on physical properties of biodegradable cassava starch-based films *Chem. Cent. J.* **5** 6
[35] Barkhordari S and Yadollahi M 2016 Carboxymethyl cellulose capsulated layered double hydroxides/drug nanohybrids for Cephalexin oral delivery *Appl. Clay Sci.* **121–122** 77–85
[36] Ebrahimzadeh S, Ghanbarzadeh B and Hamishehkar H 2016 Physical properties of carboxymethyl cellulose based nano-biocomposites with Graphene nano-platelets *Int. J. Biol. Macromol.* **84** 16–23
[37] Prajapat A L, Das P and Gogate P R 2016 A novel approach for intensification of enzymatic depolymerization of carboxymethyl cellulose using ultrasonic and ultraviolet irradiations *Chem. Eng. J.* **290** 391–9
[38] Nasution T I, Nainggolan I, Hutagalung S D, Ahmad K R and Ahmad Z A 2013 The sensing mechanism and detection of low concentration acetone using chitosan-based sensors *Sensors Actuators B Chem.* **177** 522–8

IOP Conf. Series: Materials Science and Engineering **309** (2018) 012093 doi:10.1088/1757-899X/309/1/012093

Effect of pieces size of Empty Fruit Bunches (EFB) on composting of EFB mixed with activated liquid organic fertilizer

B Trisakti[*1,2], P Mhardela[1], T Husaini[1], Irvan[1,2], and H Daimon[3]
[1]Chemical Engineering Department, Faculty of Engineering, Universitas Sumatera Utara. Jalan Almamater Komplek USU Medan, 20155, Indonesia
[2]Sustainable Energy and Biomaterial Center of Excellence, Universitas Sumatera Utara, Medan, 20155, Indonesia
[3]Department of Environmental and Life Sciences, Toyohashi University of Technology, Toyohashi, 441-8580, Japan

[*]Email: b_trisakti@yahoo.com

Abstract. This research was to determine the effect of pieces sizes of oil palm empty fruit bunch (EFB) on the composting of EFB mixed with activated liquid organic fertilizer (ALOF) in a basket composter in order to obtain high quality compost. The composting process was started by cutting the EFB into pieces with varies sizes, inserting the EFB pieces into basket composter (33 cm W × 28 cm L × 40 cm H), and adding ALOF until moisture content (MC) in the range of 55-65%. During composting, the compost pile was turned every 3 days and the MC was maintained at 55-65% range by adding the ALOF. The sizes of the EFB pieces were varied into <1, 1-3, 4-7, 8-11, and 12-15 cm. The parameters analysed during the composting were temperature, pH, MC, compost weight, water holding capacity (WHC), CN ratio, and the quality of the final compost. Composting was carried out for 40 days and the best result obtained at EFB pieces size was 1-3 cm with compost characteristic were pH 9.0; MC 52.59%; WHC 76%; CN ratio 12.15; N 1.96%; P 0.58%; and K 0. 95%.

1. Introduction

Oil palm plantations have become the most important economic contributor in Indonesia. However, in the processing of fresh fruit bunches (FFB) at palm oil mill (POM), besides producing palm oil and palm kernel, also produced by-product of liquid and solid waste. For every ton of treated FFB will be produced about 0.6 to 0.87 m^3 of liquid waste (or 2.4 to 3.7 tons of liquid waste per tonne of palm oil produced) [1] with biological oxygen demand (BOD) 20,000-25,000 mg/L, chemical oxygen demand (COD) of 40,000-50,000 mg/L, and pH 3.8 to 4.5 [2,3]. In addition, for every ton of TBS processed, POM also generates large amounts of solids wastes such as empty fruit bunch (EFB) (23%), shell (5%), and mesocarp fiber (12%) [4,5].

Before 1996, EFB was burned in an incinerator to ashes. The ash of EFB can be used as fertilizer, because its potassium content is relatively high ± 30%. However, to prevent air pollution, the EFB combustion process is prohibited through Minister of Environment Decree number 15 of 1996 on the blue-sky program [6].

Today, most EFB is used as mulch on oil palm lands by placing EFB around the oil palm tree. Although the existence of mulch is useful to control weeds, maintain moisture and prevent soil erosion but its utilization is costly for labor and transportation [7]. Moreover, the processes of composting of mulch take a long time depending on environmental factors. Therefore, EFB should be composted before being spread to the field, or distributed to the farmers. Compost will also become potential products when used for the cultivation of ornamental plants [6].

Composting is a means to change various organic wastes into products that can be used safely and profitably as a biological fertilizer. The main function of the compost is to help improve the physical, chemical and biological of soil. Physically, compost can loosen the soil, compost application into the soil increase the number of cavities so that the soil becomes loose. While chemical properties capable addressed by the application of compost is increasing the cation exchange capacity (CEC) of the soil and can increase the ability of soil to water retain (water holding capacity). As for the improvement of biological characteristics, the compost can improve soil microorganism populations [8]. The composting process are influenced by number of factors such as nutrients, CN ratio, size of the compost material, temperature, pH, water content, and turning frequency [6,9].

Several studies have been conducted to process EFB into compost by mixing it with additional materials such as manure [10] or palm oil mill effluent (POME) [11]. This research used activated liquid organic fertilizer (ALOF) as alternative additional materials as it is also used by Trisakti et al [6]. ALOF is product of advance fermentation of treated biogas effluent from methanogenic anaerobic digester, which still has high nutritional and microbial sources. Hence, mixing EFB with ALOF in the basket composter might enrich the compost material with high nutrient and microbial sources [6]. Therefore, the aim of this research is to determine the effect of EFB pieces sizes on the rate of composting of EFB mixed with ALOF in a basket composter.

2. Materials and Methods

2.1. Materials

Main materials in this research were EFB and ALOF. EFBs were obtained from Rambutan POM, Sumatera Utara, Indonesia. The EFB was shredded and cut to varied size of <1, 1-3, 4-7, 8-11, and 12-15 cm. While, ALOF were liquid product from methanogenic anaerobic digestion obtained from USU - Biogas Pilot Plant, Pusdiklat LPPM, USU, Medan, Indonesia. Characteristics of EFB and ALOF are presented in Table 1.

Table 1. Characteristic of EFB and ALOF

Parameters	**Unit**	**EFB**	**ALOF**
Moisture	(%)	43.83	-
pH	-	9.00	8.09
C	(%)	33.15	0.58
N	(%)	0.99	0.10
C/N	-	33.48	5.80
WHC	(%)	62.00	-
P_2O_5	(%)	-	0.016
K_2O	(%)	-	0.167
COD	mg/l	-	1,580

2.2. Equipment

Main equipment of this research was basket composter same as used by Trisakti et al [6]. Composter is a holes plastic basket that is usually used as a laundry basket. Inside of the basket is covered with perforated carpet in order to control the influx of oxygen and to release the gases as composting product. The front side of composter is equipped with three holes, which are used for sampling and temperature measurements. The detailed sketch of the composter has been presented in Trisakti et al [6].

2.3. Composting Process

Composting process was started by shredding and cutting EFB to size of The EFB was shredded and cut to varied size of <1, 1-3, 4-7, 8-11 and 12-15 cm, measuring the weight of total EFB, and putting into the composter. Next, ALOF was poured and mixed until the moisture content of compost reached 55-65% and maintained at that value range during the composting proceed. During the process, compost pile was turned every three days in order to reduce and equalize the moisture content.

2.4. Sampling and Data Collection

Sampling was carried out by collecting approximately 20 g of composting materials through sampling holes. Meanwhile, temperature measurement was performed by inserting digital thermometer through thermometer holes. Data collected during the experiment were temperature, moisture content (MC), pH, CN ratio, and compost quality. Temperature measurement was performed twice a day, in the morning and afternoon. Moisture content was determined everyday by drying oven method. pH was measured every two days. Content of C and N were determined on days 0th, 40th, and 60th, by Walkley & Black and Kjeldahl methods.

3. Results and Discussions

The composting process of EFB from palm oil mill was carried out by shredding and cutting EFB to varied size of <1, 1-3, 4-7, 8-11, and 12-15 cm and then mixing with ALOF. Composting process was conducted in the basket composter with turning frequency three days. Composting process was evaluated for 60 days by measuring and determining the temperature, moisture content (MC), pH, CN ratio, and others compost quality. The results of the research are described in this following section.

3.1. Effect of EFB Pieces Size on Temperature and Moisture Content

Temperature and moisture content are the most important parameter to be measured in the composting process [12]. The average of temperature and moisture content during the composting process that EFB pieces sizes of <1, 1-3, 4-7, 8-11, and 12-15 cm was presented in Figure 1.

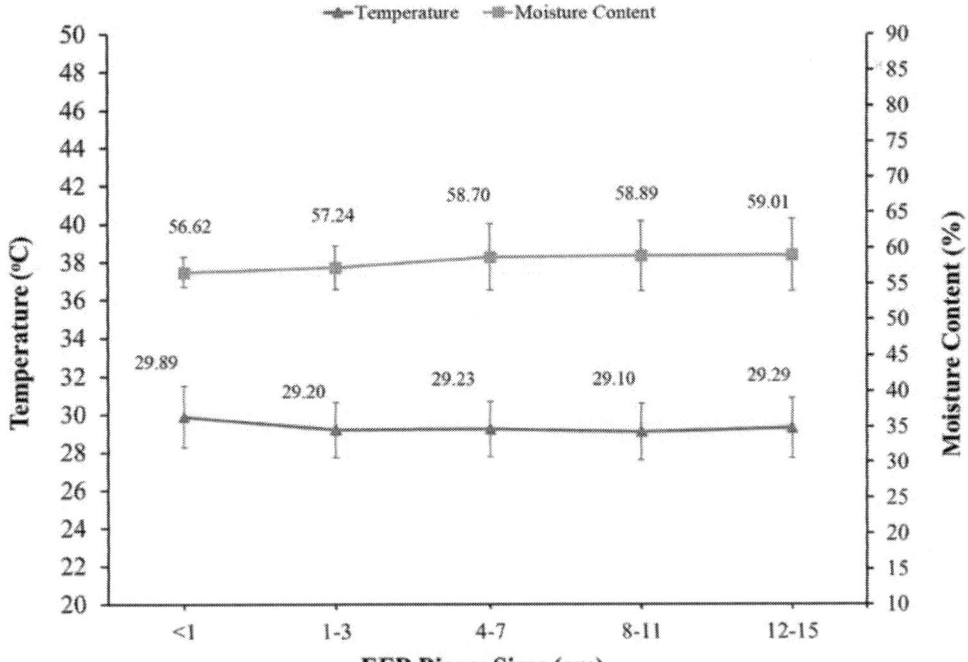

Figure 1. Effect of EFB pieces size on temperature and moisture content

Figure 1 shows that the average temperature in composting of EFB pieces sizes of <1, 1-3, 4-7, 8-11, and 12-15 cm were 29.89 ± 1.63, 29.20 ± 1.46, 29.23 ± 1.47, 29.10 ± 1.49, and 29.29 ± 1.59°C, respectively. This means, an increase in the size of EFB pieces (compost material) cause to lower composting temperatures and the highest average temperature achieved at composting EFB with size <1 cm. This is consistent with the statement of Bueno et al., that the highest temperature is achieved at the smallest compost size [13]. This is due to the small particle size making porosity between small materials and not providing air space between particles [14], so heat will increase in the stack [15]. The error bar at temperature curve shows the temperature difference at each part of compost material pile, because the activity of microorganisms in each section height of compost material pile is different [16].

Figure 1 also shows that the average MC in composting of EFB pieces sizes of <1, 1-3, 4-7, 8-11, and 12-15 cm were 56.62 ± 3.05, 57.24 ± 3.05, 58.70 ± 4.68, 58.89 ± 4.91, and 59.01 ± 5.10%, respectively. The value of average MC increased with the increasing of EFB pieces size but the values are in the range of 56.62 - 59.01% or still in the range of best MC value for composting [4,17]. Baharuddin et al. [4] reported that the best of MC value for composting is in the range of 55-65%. If the MC value is too high, it can inhibit the growth of microorganisms because water molecules are going to fill the air cavity resulting in anaerobic conditions, which will cause odor and can kill aerobic microorganisms due to oxygen deficiency [16,18].

3.2. Effect of EFB Pieces Size on ALOF Addition

ALOF addition in each composter depends on the MC value of compost pile. It is maintained in the range of 55-65%. Therefore, at the end of composting, the total number of ALOF added to each composter is different. The number of ALOF added to each composter containing the sized EFB pieces <1, 1-3, 4-7, 8-11, and 12-15 cm is presented in Figure 2.

Figure 2 shows that the total addition of ALOF in composting of EFB pieces sizes of <1, 1-3, 4-7, 8-11, and 12-15 cm were 12.66, 11.91, 11.3, 11.65 and 10.14 liters, respectively. The biggest ALOF addition is compost with EFB pieces size <1 cm and the smallest is compost with EFB pieces size 12-15 cm. However, the difference is relatively small due to the average of ALOF addition was 11.53 ± 0.92 liters.

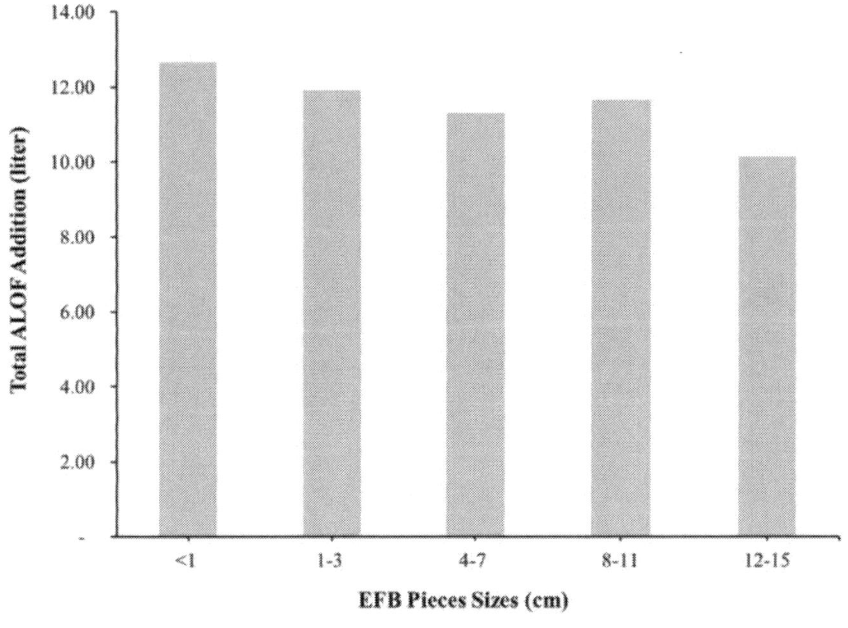

Figure 2. Effect of EFB pieces size on ALOF addition

3.3. Effect of EFB Pieces Size on Compost pH

The average value of compost pH during the composting process of EFB pieces sizes of <1, 1-3, 4-7, 8-11, and 12-15 cm is presented in Figure 3.

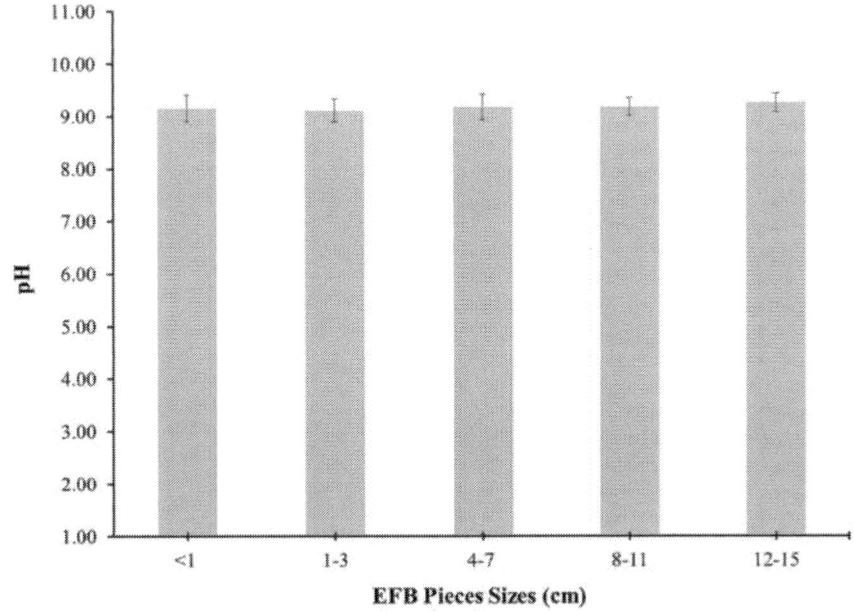

Figure 3. Effect of EFB pieces size on compost pH

Figure 3 shows that the average value of compost pH in composting of EFB pieces sizes of <1, 1-3, 4-7, 8-11, and 12-15 cm were 9.15 ± 0.25, 9.11 ± 0.22, 9.17 ± 0.25, 9.18 ± 0.17, and 9.25 ± 0.18%, respectively. The highest average value of compost pH is compost with EFB pieces size 12-15 cm and tends to fall with decreasing of EFB pieces size, but the lowest pH value is compost with EFB pieces size 1-3 cm.

The pH value of the compost tend to base condition and this is good because it can inhibit the growth of pathogens such as fungi that can survive in acidic conditions [19]. The base condition occur because, if the size of the compost material increases than more N elements will decrease [17]. Reducing N in composting occurs because N turn to NH_3 or NH_4^+ by the ammonification process, so the pH increases [20].

3.4. Effect of EFB Pieces Size on CN Ratio

Maturity and quality of the compost produced is determined by analyzing the compost based on the CN ratio. The CN ratio of the compost material at 0th, 40th, and 60th day of compost with EFB pieces sizes of <1, 1-3, 4-7, 8-11, and 12-15 cm is presented in Figure 4.

Figure 4 shows that the initial CN ratio of composting material at each EFB pieces sizes variation is the same, i.e. 33.48. On the 40th day, the CN ratio of the compost drastically decreased, where the CN ratio of compost with EFB pieces sizes of <1, 1-3, 4-7, 8-11, and 12-15 cm were 18.79, 12.15, 26.50, 24.65, and 17.46, respectively. Finally, on the 60th day, the CN ratio of compost with EFB pieces sizes of <1, 1-3, 4-7, 8-11, and 12-15 cm were 10.00, 11.74, 13.48, 16.07, and 12.11, respectively. The CN ratio decreases is occurs because of the organic matter decomposition as the results of microbial activity [17,21]. On 60th day, C-N ratio for each EFB pieces size variation is below 20, it means the compost is already mature and appropriate with the quality of the standard of CN ratio of compost mature i.e. in the range of 10-20 [11,22]. While on the 40th day, only compost with pieces size <1, 1-3, and 12-15 cm the value of the CN ratio in the range of 10-20, while compost with pieces size 4-7 and 8-11 cm the CN ratio is still above 20. Results of complete analysis of compost quality and standard of compost are presented in Table 2.

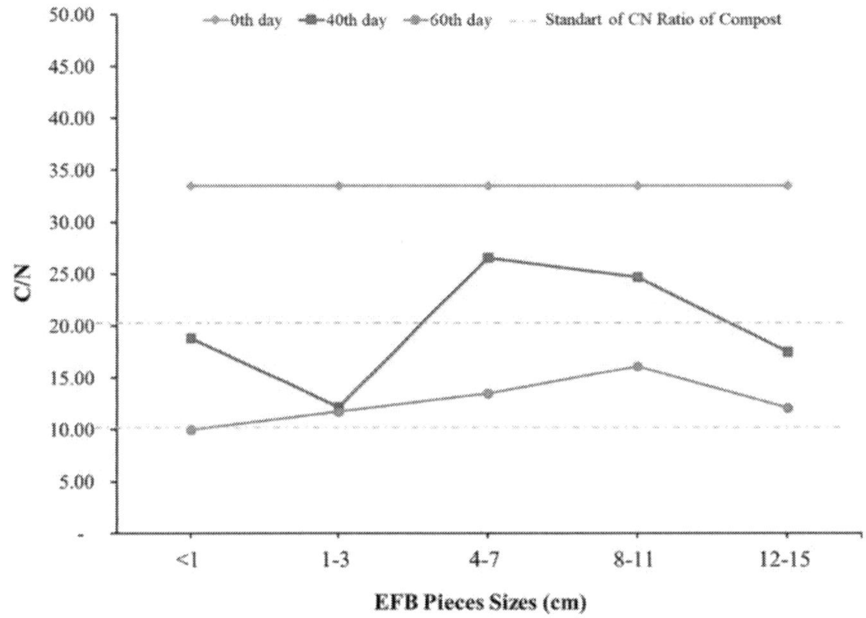

Figure 4. Effect of EFB pieces size on CN ratio

Table 2. Results of complete analysis of compost quality and standard of compost

Parameter	EFB Pieces Size (cm)					Standard of compost*	
	<1	1-3	4-7	8-11	12-15	Minimum	Maximum
MC (%)	55.61	52.69	61.27	58.89	58.26	-	50.00
pH	9.40	9.00	9.30	9.10	9.20	6.80	7.49
WHC (%)	74.00	76.00	68.00	74.00	80.00	58.00	-
C (%)	21,05	23.82	29.68	27.61	29.34	9.80	32.00
N (%)	1.12	1.96	1.12	1.12	1.68	0.40	-
C/N	18.79	12.15	26.50	24.65	17.46	10.00	20.00
P (%)	0.68	0.58	1.30	1.35	1.48	0.10	-
K (%)	0.74	0.95	0.56	0.83	0.62	0.20	-
Na (ppm)	0.51	0.51	0.51	0.51	0.51		
Ca (ppm)	1.65	1.65	1.65	1.65	1.65	-	25.50
Mg (ppm)	0.53	0.53	0.53	0.53	0.53	-	0.60
Cd (ppm)	1.52	1.52	1.52	1.52	1.52	-	3.00
Cu (ppm)	2.09	2.09	2.09	2.09	2.09	-	100.00
Fe (ppm)	0.98	0.98	0.98	0.98	0.98	-	2.00
Pb (ppm)	2.34	2.34	2.34	2.34	2.34	-	150.00
Zn (ppm)	3.19	3.19	3.19	3.19	3.19	-	500.00

* Standards of compost according to the Indonesian National Standard

It can be said that the value of the CN ratio decreases with the smallness of the EFB pieces size. This is because the smaller particle size will cause the surface area of the particle to be large, so the decomposition process becomes faster [14]. However, there is an aberration that the smallest CN ratio value is generated by the EFB pieces size of a 1-3 cm. This is because the EFB pieces size of a 1-3 cm is in accordance with the basket composter used.

Conclusions

Composting of oil palm empty fruit bunches (EFB) mixed with activated liquid organic fertilizer (ALOF) was carried out for 40 days and the best result obtained at EFB pieces size was 1-3 cm with compost characteristic were pH 9.0; MC 52.59%; WHC 76%; CN ratio 12.15; N 1.96%; P 0.58%; and K 0. 95%.

Acknowledgments

This research was financially supported by Lembaga Penelitian Universitas Sumatera Utara in accordance with the research contract TALENTA Universitas Sumatera Utara fiscal year 2017 number: 5338/UN5.1.R/PPM/2017, date May 22, 2017.

References

[1] Trisakti B, Irvan, Mahdalena, Taslim and Turmuzi M 2017 Effect of temperature on methanogenesis stage of two-stage anaerobic digestion of palm oil mill effluent (POME) into biogas *IOP Conf. Ser. Mater. Sci. Eng.* **206** 8

[2] Irvan, Trisakti B, Sosanty F and Tomiuchi Y 2016 Effect of Discontinuing Sodium Bicarbonate on Fermentation Process of Palm Oil Mill Effluent *Asian J. Chem.* **28** 377–80

[3] Trisakti B, Irvan, Zahara I, Taslim and Turmuzi M 2017 Effect of agitation on methanogenesis stage of two-stage anaerobic digestion of palm oil mill effluent (POME) into biogas *AIP Conf. Proc.* **1840** 7

[4] Baharuddin A S, Hock L S, Yusof M Z, Abdul N A, Shah U, Hassan M A, Wakisaka M, Sakai K and Shirai Y 2010 The effect of palm oil mill effluent (POME) anaerobic sludge from 500 m^3 of closed anaerobic methane digested tank on pressed-shredded empty fruit bunch (EFB) composting process *African J. ...* **9** 2427–36

[5] Trisakti B, Manalu V, Irvan, Taslim and Turmuzi M 2015 Acidogenesis of Palm Oil Mill Effluent to Produce Biogas: Effect of Hydraulic Retention Time and pH *Procedia - Soc. Behav. Sci.* **195** 2466–74

[6] Trisakti B, Lubis J, Husaini T and Irvan 2017 Effect of Turning Frequency on Composting of Empty Fruit Bunches Mixed with Activated Liquid Organic Fertilizer *IOP Conf. Ser. Mater. Sci. Eng.* **180** 8

[7] Kavitha B, Jothimani P and Rajannan G 2013 Empty Fruit Bunch-a Potential Organic Manure for Agriculture *Int. J. Sci. Environ. Technol.* **2** 930–7

[8] Oviasogie P O, Aisueni N O and Brown G E 2010 Oil palm composted biomass: A review of the preparation, utilization, handling and storage *African J. Agric. Res.* **5** 1553–71

[9] Tchobanoglous G, Theisen H and Vigil S A 1993 *Integrated Solid Waste Management: Engineering Principles and Management Issues* (McGraw-Hill Education)

[10] Suhaimi M and Ong H K 2001 Composting Empty Fruit Bunches of Oil Palm *Ext. Bull. - Food Fertil. Technol. Cent.* **505** 1–8

[11] Baharuddin A S, Wakisaka M, Shirai Y, Abd-Aziz S, Abdul Rahman N A and Hassan M A 2009 Co composting of Empty Fruit Bunches and Partially Treated Palm Oil Mill Effluents in Pilot Scale *Int. J. Agric. Res.* **4** 69–78

[12] Saiidi S, Hasani M, Hashemi J and Amini-rad M 2012 Investigation of Optimum Conditions of Co-composting Process by Using of Sewage Sludge and Municipally Waste *The 1th International*

and The 4th National Congress on Recycling of Organic Waste in Agricultural (Isfahan, Iran) pp 1–9

[13] Bueno P, Tapias R, López F and Díaz M J 2008 Optimizing composting parameters for nitrogen conservation in composting *Bioresour. Technol.* **99** 5069–77

[14] Rynk R, Van de Kamp M, Willson G B, Singley M E, Richard T L, Kolega J J, Gouin F R, Laliberty, Jr. L, Kay D, Murphy D W, Hoitink H A J and Brinton W F 1992 *Optimizing Composting Parameter for Nitrogen Conservation in Composting* ed R Rynk (New York: Northeast Regional Agricultural Engineering Service)

[15] Mortazavi M, Ghehsareh A M and Kalbasi M 2012 The Effect of Particle Size and Composting Period on (C/N) Ratio of Date- Palm Waste *The 1st International and the 4th National Congress on Recycling of Organic Waste in Agriculture* (Isfahan, Iran) pp 1–5

[16] Tiquia S M, Tam N F Y and Hodgkiss I J 1997 Effects of turning frequency on composting of spent pig-manure sawdust litter *Bioresour. Technol.* **62** 37–42

[17] Wan Razali W A, Baharuddin A S, TarmezeeTalib A, Sulaiman A, Naim M N, Hassan M A and Shirai Y 2012 Degradation of oil palm empty fruit bunches (OPEFB) fibre during composting process using in-vessel composter *BioResources* **7** 4786–805

[18] Shen Y, Ren L, Li G, Chen T and Guo R 2011 Influence of aeration on CH_4, N_2O and NH_3 emissions during aerobic composting of a chicken manure and high C/N waste mixture *Waste Manag.* **31** 33–8

[19] Saidi N, Chérif M, Jedidi N, Mahrouk M, Fumio M, Boudabous A, Hassen A, P T R B and Lif H 2008 Evolution of Biochemical Parameters during Composting of Various Wastes Compost *Am. J. Environ. Sci.* **4** 332–41

[20] Kananam W, Suksaroj T T and Suksaroj C 2011 Biochemical changes during oil palm (Elaeis guineensis) empty fruit bunches composting with decanter sludge and chicken manure *ScienceAsia* **37** 17–23

[21] Hock L, Baharuddin A S, Ahmad M N, Shah U K M, Rahman N a. a., Abd-Aziz S, Hassan M a. and Shirai Y 2009 Physicochemical changes in windrow co-composting process of oil palm mesocarp fiber and palm oil mill effluent anaerobic sludge. *Aust. J. Basic Appl. Sci.* **3** 2809–16

[22] Zahrim A Y and Asis Tahang 2010 Production of Non Shredded Empty Fruit Bunch Semi-Compost *Inst. Eng. Malaysia* **71** 11–7.

IOP Conf. Series: Materials Science and Engineering **309** (2018) 012094 doi:10.1088/1757-899X/309/1/012094

Production of oil palm empty fruit bunch compost for ornamental plant cultivation

B Trisakti[*1,2], P Mhardela[1], T Husaini[1], Irvan[1,2], and H Daimon[3]
[1]Chemical Engineering Department, Faculty of Engineering, Universitas Sumatera Utara. Jalan Almamater Komplek USU Medan, 20155, Indonesia
[2]Sustainable Energy and Biomaterial Center of Excellence, Universitas Sumatera Utara, Medan, 20155, Indonesia
[3]Department of Environmental and Life Sciences, Toyohashi University of Technology, Toyohashi, 441-8580, Japan

[*]Email: b_trisakti@yahoo.com

Abstract. The aim of this research was to produce the oil palm empty fruit bunch (EFB) compost for ornamental plant cultivation. EFB compost was produced by chopping fresh EFB into 1-3 cm pieces, inserting the pieces into basket composter (33 cm W × 28 cm L × 40 cm H), and adding activated liquid organic fertilizer (ALOF) until moisture content (MC) in the range of 55-65%. During composting, the compost pile was turned every 3 days and the MC was maintained at 55-65% range by adding the ALOF. The compost processed was then mixed with sand and rice husk with a ratio of 1:1:1; 1:3:1; 1:0:1 and was used as a potting medium for planting some valuable ornamental plants i.e. cactus (*cactaceae*), sansevieria, and anthurium. Composting was carried out for 40 days and the compost characteristic were pH 9.0; MC 52.59%; WHC 76%; CN ratio 12.15; N 1.96%; P 0.58%; and K 0.95%. The compost-sand-husk rice mixture can be used as a growing medium where the best ratio for cactus, sansevieria, and anthurium was 1:3:1; 1:1:1; and 1:0:1, respectively.

1. Introduction

The palm oil industry is the industry that makes the most important contribution to the Indonesian economy. However, in the processing of fresh fruit bunches (FFB) in palm oil mill (POM), besides producing palm oil and palm kernel oil also produce by-products that are liquid and solid waste. For every ton of treated FFB will be produced about 0.6 to 0.87 m^3 of liquid waste or 2.4 to 3.7 tons of liquid waste per tonne of palm oil produced [1], with biological oxygen demand (BOD) 20,000-25,000 mg/L, chemical oxygen demand (COD) of 40,000-50,000 mg/L, and pH 3.8 to 4.5 [2,3]. In addition, for every ton of TBS processed, POM also generates large amounts of solids wastes such as empty fruit bunch (EFB) (23%), shell (5%), and mesocarp fiber (12%) [4,5].

Before 1996, EFB was burned in an incinerator to ashes. The ash of EFB can be used as fertilizer, because its potassium content is relatively high ± 30%. However, to prevent air pollution, the EFB combustion process is prohibited through Minister of Environment Decree number 15 of 1996 on the blue-sky program [6].

Today, most EFB is used as mulch on oil palm lands by placing EFB around the palm tree. Although the existence of mulch is useful to control weeds, maintain moisture and prevent soil erosion but its utilization is costly for labor and transportation [7]. Moreover, the processes of composting of mulch take a long time depending on environmental factors. Therefore, EFB should be composted before being spread to the field, or distributed to the farmers. Compost will also become potential products when used for the cultivation of ornamental plants [6].

Composting is a means to change various organic wastes into products that can be used safely and profitably as a biological fertilizer. The main function of the compost is to help improve the physical, chemical and biological of soil. Physically, compost can loosen the soil, compost application into the soil increase the number of cavities so that the soil becomes loose. While chemical properties capable addressed by the application of compost is increasing the cation exchange capacity (CEC) of the soil and can increase the ability of soil to water retain (water holding capacity). As for the improvement of biological characteristics, the compost can improve soil microorganism populations [8].

Several studies have been converted EFB into compost by mixing it with additional materials such as manure [9] or palm oil mill effluent (POME) [10]. While in this study, active liquid organic fertilizer (ALOF) is used as an alternative additional materials such as that used by Trisakti et al [6]. ALOF is product of advance fermentation of treated biogas effluent from methanogenic anaerobic digester, which still has high nutritional and microbial sources. Hence, mixing EFB with ALOF in the basket composter might enrich the compost material with high nutrient and microbial sources [6].

Therefore, the aim of this research is to produce the oil palm empty fruit bunch (EFB) compost for ornamental plant cultivation. These include to obtain of the EFB composting degradation data, the quality of the compost produced, and the best of sand-rice husk-compost ratios for planting media of ornamental plants.

2. Materials and Methods

2.1. Materials

Main materials for the composting were EFB and ALOF. EFBs were obtained from Rambutan POM, Sumatera Utara, Indonesia. Before being used, the EFB was shredded and cut to size of 1-3 cm. While, ALOF were liquid product from methanogenic anaerobic digestion obtained from USU - Biogas Pilot Plant, Pusdiklat LPPM, USU, Medan, Indonesia. Characteristics of EFB and ALOF have been presented by Trisakti et al [6]. Meanwhile for planting media, the materials were sand, rice husk, and the EFB compost. The sand is black sand that forms from the lava or magma of the volcano that cools rapidly. Usually referred to as Malang sands because it is produced from highland area or volcano around Malang. Rice husks used are derived from local rice mills which are fried without oil before use.

2.2. Composting Equipment

Main equipment of this research was basket composter same as used by Trisakti et al [6]. Composter is a holes plastic basket that is usually used as a laundry basket. Inside of the basket is covered with perforated carpet in order to control the influx of oxygen and to release the gases as composting product. The front side of composter is equipped with three holes, which are used for sampling and temperature measurements. Detailed sketch of the composter has been presented in Trisakti et al [6].

2.3. Composting Process

Composting process was conducted by shredding and cutting EFB to size of 1-3 cm, measuring the weight of EFB total, and putting into the composter. Next, ALOF was poured and mixed until moisture content of compost reach 55 - 65% and maintained at that value range during the composting proceed. During the process, compost pile was turned every three days in order to reduce and equalize the moisture content.

2.4. Preparing The Planting Media

The planting media was mixture of EFB compost, sand, and rice husk. The ratios of the mixture were 1:1:1, 3:1:1, and 1:0:1. EFB compost is the result of this research, the sand is black sand, and the rice husk is fried without oil before use.

2.5. Sampling and Data Collection

Sampling was carried out by collecting approximately 20 g of composting materials through sampling holes. Meanwhile, temperature measurement was performed by inserting digital thermometer through thermometer holes. Data collected during the experiment were temperature, moisture content (MC), pH, CN ratio, and compost quality. Temperature measurement was performed twice a day, in the morning and afternoon. Moisture content was determined everyday by drying oven method. pH was measured every two days. Content of C and N were determined on days 0th, 40th, and 60th, by Walkley & Black and Kjeldahl methods.

3. Results and Discussions

The composting process of EFB from palm oil mill was carried out by shredding and cutting EFB to size of 1-3 cm and then mixing with ALOF. Composting process was conducted in the basket composter with turning frequency three days. The compost produced then mixed with sand and rice and was used as a potting medium for planting of cactus (*cactaceae*), sansevieria, and anthurium. Composting process was evaluated for 60 days by measuring and determining the temperature, moisture content (MC), pH, CN ratio, and others compost quality. The results of the research are described in this following section.

3.1. Profiles of Temperature and Moisture Content

Temperature and moisture content is the most important parameter to be measured in the composting process [11]. Profiles of temperature and moisture content during the composting process which evaluated every three days is presented in Figure 1.

Previously, temperature of composting material was 27°C, and after being mixed with ALOF, the composting process started, then the temperature of the composting material increased to 35°C after 3 days. This is in accordance with research reported by Hock et al. [12], if material containing cellulose, such as EFB, mixed with highly organic matter material, heat would be generated in the piles due to the metabolic heat generation from biodegradation process. The high error bars on the temperature curve showed a relatively large temperature differences both in the morning and evening as well as on each compost pile height in the composter.

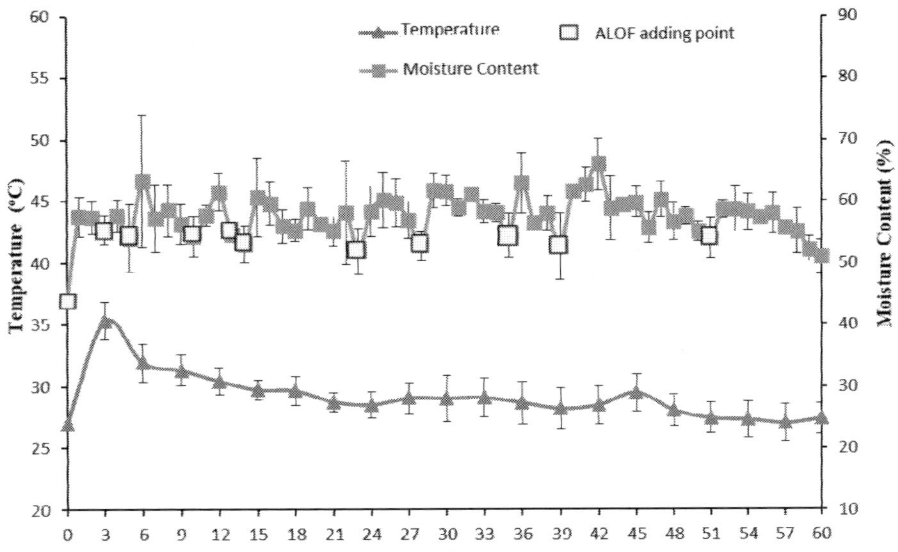

Figure 1. Profiles of temperature and moisture content during the composting

Temperature profile tends to decline during the composting process. The average maximum temperature of 35°C was reached on the 3rd day. Then, the temperature slowly began to decline until the 50th day and the next day until the 60th day became constant at 27°C. This indicates that the degradation process has completed which is similar case as reported by Shen et al. [13] and Hock et al. [12]. They reported in their researches that after a rapid increase in temperature, it will decrease slowly which indicates that the composting process slows down as the depletion of nutrient availability.

Moisture content is an important parameter to optimize the composting process. According to Hock et al. [12], the dependence on water to support microbial growth can affect its ability to biodegrade of organic materials. In this research, the addition of ALOF to EFB has two purposes, first as a source of microbes and nutrients; and second to maintain the moisture content in the range of 55-65%. Initial moisture content of EFB was 43.83% and then after being added 2.97 liters of ALOF, it changed the value of its moisture content to 57%.

Turning of the compost pile during the process decreased the value of moisture content, except on the 3rd, 18th, 21st, 24th, 39th, 48th, and 51st days. The disadvantage of turning process due to the moisture content of the composting material was reported by Tiquia et al. [14], the turning process of compost pile can cause continuous loss of water due to the evaporation. High error bar on moisture content curve (Figure 2) shows the difference of moisture content in each section of compost pile; at the top, middle, or bottom of the composter. This is because the data of moisture content was collected before the turning process carried out, so that the value of moisture content in each section of compost pile was different.

Final moisture content of compost after 60 days was 50.99%. This moisture content value is similar to the value obtained by the previous studies that also performing the turning process frequency once in three days. As reported by Hock et al [12] in their research, they obtained final moisture content about 50% after 39 days, while Baharuddin et al [5] obtained final moisture content was 52%. Meanwhile, Tiquia et al [15] stated that in order for the microbes to survive, moisture content of composting materials should be around 40-60%, if they exceed to 80%, the aerobic microbes will die because of lack of oxygen.

3.2. pH Profile During The Composting

The pH profile during the composting process is presented in Figure 3. In Figure 2 shows the pH range during sixty days of composting process which is in the range of 8.7 to 9.5. This indicates that the operational condition tend to be alkaline. The pH increased at the beginning of 10 days of composting, then dropped at the 35th day, and then tends to be constant until the 60th day.

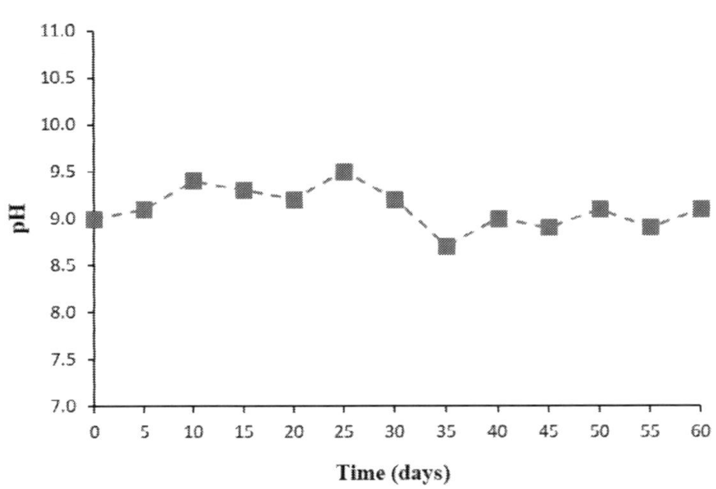

Figure 2. The pH profile during the composting

The changes of pH during the composting process were caused by microbial activity [16,17]. In the first ten days, the pH increased from 9.0 to 9.4. It occurred because the N transformed into NH_3 or NH_4^+ via ammonification process [16]. However, from the next day until the 35th day, pH tends to decrease to 8.7, although at the 25th day it increased to 9.5. The decrease of pH is caused by the process of volatilization of ammonium and the release of hydrogen ions as a result of nitrification process [5]. Finally, pH of mature compost was 9.1.

Overall condition that occurred during the composting process tends to base condition which pH ranges from 8.7 to 9.5. This happens because of the turning process was performed once in 3 days. By performing the turning process, it can emit CO_2 gas which is trapped in the empty space among particles of the composting material, thus preventing the acidic conditions in the compost pile or significant decrease in pH [18,19]. Increasing the pH to alkaline conditions is good for the composting process as this may inhibit the growth of pathogens such as fungi that can survive in acidic conditions [20].

3.3. C-N Ratio and Compost Quality Analysis

Maturity of the compost produced is determined by analyzing the compost based on the CN ratio. The CN ratio of the compost material at 0th, 40th, and 60th day is presented in Figure 3. Meanwhile, the quality of the compost was analysis and the results of complete analysis of compost quality are presented in Table 1.

The decreasing value of CN ratio is due to the decreasing of C content during composting process, otherwise N content is increasing during composting. This occurs because of the decomposition of organic materials from the microbial activity [12,21]. CN ratio is one of the important indicators that express the maturity of compost [10,22]. The value of CN ratio of the EFB initially was 33.48, then on the 40th day was 12.15, and finally on the 60th was 11.74. This means that the EFB have been composted less than 40 days because compost mature standard on CN ratio has a value between 10-20 [10,23].

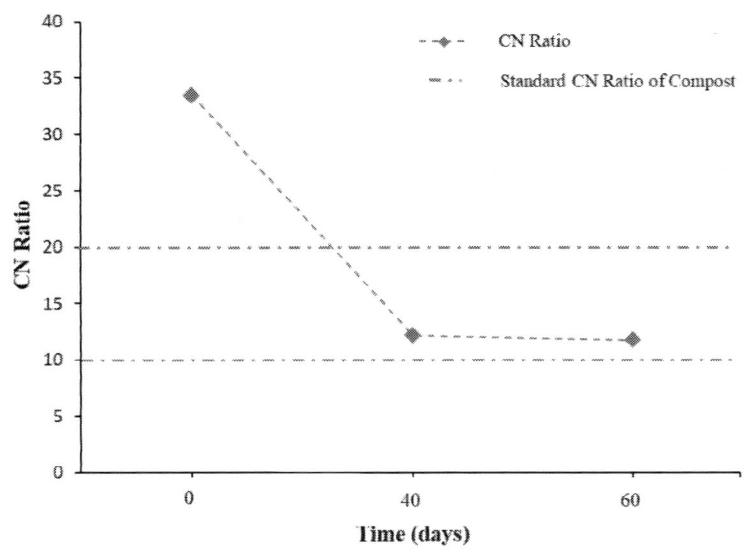

Figure 3. CN ratio changes during the composting

IOP Conf. Series: Materials Science and Engineering **309** (2018) 012094 doi:10.1088/1757-899X/309/1/012094

Table 1. The EFB compost quality and standard of compost

Parameters	Composting material			Standard of compost*	
	Start up	40th days	60th days	Minimum	Maximum
MC (%)	43.83	61.57	50.99	-	50.00
pH	9.10	9.00	9.10	6.80	7.49
WHC (%)	62.00	76.00	80.00	58.00	-
C (%)	33.15	23.81	28.41	9.80	32.00
N (%)	0.99	1.96	2.42	0.40	-
C/N	33.48	12.15	11.74	10.00	20.00
P (%)	-	-	0.58	0.10	-
K (%)	-	-	0.95	0.20	-
Na (ppm)	-	-	0.51		
Ca (ppm)	-	-	1.65	-	25.50
Mg (ppm)	-	-	0.53	-	0.60
Cd (ppm)	-	-	1.52	-	3.00
Cu (ppm)	-	-	2.09	-	100.00
Fe (ppm)	-	-	0.98	-	2.00
Pb (ppm)	-	-	2.34	-	150.00
Zn (ppm)	-	-	3.19	-	500.00

* Standards of compost according to the Indonesian National Standard

3.4. The EFB Compost as Potting Media

The EFB compost produced by this research is further mixed with sand and rice husk. The mixture is further used as a growing medium of some high value ornamental plants such as cactus, sansieviera, and anthurium. The growth of the three ornamental plants on a variety of mixed ratio of growing medium is presented in Table 2.

Table 2. Growth of plants on several of mixed ratio of growing media

Plants	Ratio of Compost – Sand – Rice husk	Plant growth
Cactus	1 : 3 : 1	good
Cactus	1 : 1 : 1	medium
Cactus	1 : 0 : 1	bad
Sansevieria	1 : 3 : 1	good
Sansevieria	1 : 1 : 1	good
Sansevieria	1 : 0 : 1	bad
Anthurium	1 : 3 : 1	bad
Anthurium	1 : 1 : 1	medium
Anthurium	1 : 0 : 1	good

Cactus and sansevieria like the planting medium is porous, rough-textured, and containing little organic material. Porous planting medium will ensure the availability of oxygen for plant roots so that plant growth becomes good. In addition, high porosity of planting media causes good drainage so that planting medium does not store too much water. Cactus and sanseviera dislikes the overly moist planting media conditions that can cause root rot. While anthurium, although also like porous medium but fine textured and contains high organic material. This is so that the roots of this plant is easy to grow and spread.

Conclusions

Composting of oil palm empty fruit bunches (EFB) mixed with activated liquid organic fertilizer (ALOF) was carried out for 40 days and the compost characteristic were pH 9.0; MC 52.59%; WHC 76%; CN ratio 12.15; N 1.96%; P 0.58%; and K 0.95%. The compost-sand-husk rice mixture can be used as a growing medium where the best ratio for cactus, sansevieria, and anthurium was 1:3:1; 1:1:1; and 1:0:1, respectively.

Acknowledgments

This research was financially supported by "Direktorat Riset dan Pengabdian Masyarakat Direktorat Jenderal Penguatan Riset dan Pengembangan Kementerian Riset, Teknologi, dan Pendidikan Tinggi" in accordance with research contract number: 003/SP2H/LT/DRPM/IV/2017, date April 03, 2017.

References

[1] Trisakti B, Irvan, Mahdalena, Taslim and Turmuzi M 2017 Effect of temperature on methanogenesis stage of two-stage anaerobic digestion of palm oil mill effluent (POME) into biogas *IOP Conf. Ser. Mater. Sci. Eng.* **206** 8

[2] Trisakti B, Irvan, Zahara I, Taslim and Turmuzi M 2017 Effect of agitation on methanogenesis stage of two-stage anaerobic digestion of palm oil mill effluent (POME) into biogas *AIP Conf. Proc.* **1840** 7

[3] Irvan, Trisakti B, Sosanty F and Tomiuchi Y 2016 Effect of Discontinuing Sodium Bicarbonate on Fermentation Process of Palm Oil Mill Effluent *Asian J. Chem.* **28** 377–80

[4] Trisakti B, Manalu V, Irvan, Taslim and Turmuzi M 2015 Acidogenesis of Palm Oil Mill Effluent to Produce Biogas : Effect of Hydraulic Retention Time and pH *Procedia - Soc. Behav. Sci.* **195** 2466–74

[5] Baharuddin A S, Hock L S, Yusof M Z, Abdul N A, Shah U, Hassan M A, Wakisaka M, Sakai K and Shirai Y 2010 The effect of palm oil mill effluent (POME) anaerobic sludge from 500 m3 of closed anaerobic methane digested tank on pressed-shredded empty fruit bunch (EFB) composting process *African J. ...* **9** 2427–36

[6] Trisakti B, Lubis J, Husaini T and Irvan 2017 Effect of Turning Frequency on Composting of Empty Fruit Bunches Mixed with Activated Liquid Organic Fertilizer *IOP Conf. Ser. Mater. Sci. Eng.* **180** 8

[7] Kavitha B, Jothimani P and Rajannan G 2013 Empty Fruit Bunch-a Potential Organic Manure for Agriculture *Int. J. Sci. Environ. Technol.* **2** 930–7

[8] Oviasogie P O, Aisueni N O and Brown G E 2010 Oil palm composted biomass: A review of the preparation, utilization, handling and storage *African J. Agric. Res.* **5** 1553–71

[9] Suhaimi M and Ong H K 2001 Composting Empty Fruit Bunches of Oil Palm *Ext. Bull. - Food Fertil. Technol. Cent.* **505** 1–8

[10] Baharuddin A S, Wakisaka M, Shirai Y, Abd-Aziz S, Abdul Rahman N A and Hassan M A 2009 Co composting of Empty Fruit Bunches and Partially Treated Palm Oil Mill Effluents in Pilot Scale *Int. J. Agric. Res.* **4** 69–78

[11] Saiidi S, Hasani M, Hashemi J and Amini-rad M 2012 Investigation of Optimum Conditions of Co-composting Process by Using of Sewage Sludge and Municipally Waste *The 1th International and The 4th National Congress on Recycling of Organic Waste in Agricultural* (Isfahan, Iran) pp 1–9

[12] Hock L, Baharuddin A S, Ahmad M N, Shah U K M, Rahman N a. a., Abd-Aziz S, Hassan M a. and Shirai Y 2009 Physicochemical changes in windrow co-composting process of oil palm mesocarp fiber and palm oil mill effluent anaerobic sludge. *Aust. J. Basic Appl. Sci.* **3** 2809–16

[13] Shen Y, Ren L, Li G, Chen T and Guo R 2011 Influence of aeration on CH4, N2O and NH3 emissions during aerobic composting of a chicken manure and high C/N waste mixture *Waste Manag.* **31** 33–8

[14] Tiquia S M, Wan J H C and Tam N F Y 2002 Dynamics of yard trimmings composting as determined by dehydrogenase activity, ATP content, arginine ammonification, and nitrification potential *Process Biochem.* **37** 1057–65

[15] Tiquia S M, Tam N F Y and Hodgkiss I J 1997 Effects of turning frequency on composting of spent pig-manure sawdust litter *Bioresour. Technol.* **62** 37–42

[16] Kananam W, Suksaroj T T and Suksaroj C 2011 Biochemical changes during oil palm (Elaeis guineensis) empty fruit bunches composting with decanter sludge and chicken manure *ScienceAsia* **37** 17–23

[17] Sundberg C, Smårs S and Jönsson H 2004 Low pH as an inhibiting factor in the transition from mesophilic to thermophilic phase in composting *Bioresour. Technol.* **95** 145–50

[18] Liu H and Cai L 2014 Effect of Sewage Sludge Addition on the Completion of Aerobic Composting of Thermally Hydrolyzed Kitchen Biogas Residue *Bioresources* **9** 4862–72

[19] Sekman E, Top S, Varank G and Bilgili M S 2011 Pilot-scale investigation of aeration rate effect on leachate characteristics in landfills *Fresenius Environ. Bull.* **20** 1841–52

[20] Saidi N, Chérif M, Jedidi N, Mahrouk M, Fumio M, Boudabous A, Hassen A, P T R B and Lif H 2008 Evolution of Biochemical Parameters during Composting of Various Wastes Compost *Am. J. Environ. Sci.* **4** 332–41

[21] Wan Razali W A, Baharuddin A S, TarmezeeTalib A, Sulaiman A, Naim M N, Hassan M A and Shirai Y 2012 Degradation of oil palm empty fruit bunches (OPEFB) fibre during composting process using in-vessel composter *BioResources* **7** 4786–805

[22] Nutongkaew T, Duangsuwan W, Prasertsan S and Prasertsan P 2011 Production of Compost from Palm Oil Mill Biogas Sludge Mixed with Palm Oil Mill Wastes and Biogas Effluent *TIChE International Conference* (Hatyai, Songkhla, Thailand) pp 1–5

[23] Zahrim A Y and Asis Tahang 2010 Production of Non Shredded Empty Fruit Bunch Semi-Compost *Inst. Eng. Malaysia* **71** 11–7.

IOP Conf. Series: Materials Science and Engineering **309** (2018) 012095 doi:10.1088/1757-899X/309/1/012095

Microencapsulation of vitamin e from palm fatty acid distillate with galactomannan and gum acacia using spray drying method

J. Br. Tarigan[1*], J. Kaban[1] and R. Zulmi[2]
[1]Departement of Chemistry, University of Sumatera Utara
[2] Program Studi Magister Ilmu Kimia

*E-mail: juliati@usu.ac.id

Abstract. Vitamin E from palm fatty acid distillate (PFAD) has been encapsulated using spray drying method with gum acacia (GA) and mixed of galactomannan from *Arenga pinnata* (GAP) with GA as encapsulating agent. Composite films with thickness vary from 0.542 – 0.779 mm were prepared by incorporating vitamin E onto matrix of GA (7 g) with various concentration of GAP (0.1; 0.2; 0.3 and 0.4 g). The film obtained from 0.2 g GAP and 1.3 g vitamin E showed better compatibility and have viscosity similar with standard (ISO 9001:2008 and ISO 22000:2005). That composition was used for spray drying method rendering micro-particle size 11 μm and the particle had spherical shape. Although the increment of GAP decreasing moisture content and the particle size from 16 μm to 11 μm, the yield of microcapsule, encapsulation efficiency, the amount of vitamin E absorbed and oxidation stability of vitamin E were increased.

1. Introduction

Vitamin E with low solubility is important for human health and usually occurred from food or supplement vitamin. Tocopherol and tocotrienol are the main compound of vitamin E which have high antioxidant properties that can block free radical, protecting cell membrane from lipid oxidation [1, 2]. Vitamin E also have been used as antioxidant in emulsion or milk product to prolonged storage time [3]. The concentration of tocopherol and tocotrienol in crude palm oil is around 600 – 1000 ppm. Palm fatty acid distillate (PFAD) is a by-product of purification process of crude palm oil which contained vitamin E. Most of vitamin E compound contained in PFAD is tocotrienol (70%) and tocopherol (30%) [4].

Vitamin E is very sensitive that it[5] activity and stability are affected by light, humidity, oxygen and high temperature particularly when in storage [6]. In addition, vitamin E also has low solubility in water which therefore decreases the availability in body. Incorporation into polymer matrix is one of the possible ways to protect their availability and to improve the solubility (Goud et al., 2005). The best and cheapest method to encapsulating vitamin E is using spray drying method [7]. Spray drying is common process for microencapsulation lipid and flavour [5]. This method will produce powder with good quality, low water activity, easy to handle and could protect the active material from unnecessary reaction. The choice of protecting material and emulsion properties affect the efficiency process and stability of microencapsulation product [8].

IOP Conf. Series: Materials Science and Engineering **309** (2018) 012095 doi:10.1088/1757-899X/309/1/012095

There are some requirements to use as protecting material such as must have good characteristic (high solubility, good emulsion and low viscosity), easy to obtained and mixed with other protecting material [9]. The choice of protecting material affects the emulsion and characteristic of the particle after spray drying and in storage time [10]. Gum Arabic or gum acacia which contained galactomannan was the common protecting material using spray drying method. However the price of gum acacia is very expensive. Another galactomannan source is *Arenga pinnata* seed (or 'kolang-kaling') which found abundantly in Indonesia and the utilization remain low [11]. Commonly *Arenga pinnata* seed only used for cocktail and soup food in Indonesia [12]. The main polysaccharide component contained in *Arenga pinnata* is galactomannan which is water soluble polysaccharide [13, 14] with ratio of galactose : mannose is 1 : 1.33 and the value of IC_{50} is 22.109 mg/ml [14].

Other researchers have studied the microencapsulation of vitamin E using surfactant chitosan oligosaccharide [15], liposome [16], whey protein isolate [17]. Utilization microencapsulation technology to protect active compound could increase physical stability, protecting from chemical degradation and other food ingredient. Microencapsulation of vitamin E from palm oil produces a powder which has high antioxidant and good stability. As a powder, this product has advantages such as easy to use as food additive for food product. Gum acacia also already used as protecting material in one of oleo-chemicals companies in Medan. To the best of our knowledge no studied was found related to microencapsulation of vitamin E using galactomannan from *Arenga pinnata* or combination with other polysaccharide.

Based on that, the aim of this study is to explore the potency of galactomannan from *Arenga pinnata* (GAP) in combination with gum acacia as encapsulation material. The properties of encapsulated material such as emulsion stability, compatibility in form of film and viscosity were studied while the microcapsule product was characterized based on their surface morphology, lipid content, oxidation stability, yield, encapsulation efficiency, water content and the mixed compound interaction using Fourier Transform Infrared.

2. Materials and Methods

2.1. Materials

The materials such as *Arenga pinnata* seed was obtained from local market in Medan, North Sumatera – Indonesia while PFAD was from local oleo-chemicals company in Medan. All the chemicals were obtained from local distributor and was used without any purification.

2.2. Procedures

2.2.1. Preparation of Emulsion

Gas chromatography (GC) was used to analyse the component contained in vitamin E PFAD using AOCS Ce5-86 method. The ratio of α, β, γ and δ of tocopherol and tocotrienol were analysed using high performance liquid chromatography (HPLC) using a mixture of n-heptane and ethyl acetic in the ratio of 97:3 (v/v) as stationary phase (AOCS Ce8-89). Extraction of galactomannan from *Arenga pinnata* seed was conducted following our previous method [14].

Preparation of emulsion for microencapsulation vitamin E was carried out based on the procedure from one of oleo-chemicals industry following regulation of ISO 9001:2008 and ISO 22000:2005 with slightly modification. GAP was dissolved in deionized water at temperature of 70^0C under stirrer at 1500 rpm. After that, gum acacia (GA) was added to the GAP solution followed by addition of vitamin E and stirred for 30 min. The mixed solution was examined for viscosity and stability. Next, a 95 ml of mixed solution was poured to petri dish (diameter 9 cm) and dried in oven blower at 35^0C for 10 hours and was stored in desiccator before used. The physicochemical properties of the film formed from this procedure were determined by thickness using micrometre and the surface morphology using scanning electron microscope (SEM).

IOP Conf. Series: Materials Science and Engineering **309** (2018) 012095 doi:10.1088/1757-899X/309/1/012095

2.2.2. Viscosity Analysis of Emulsion
Sample was added to viscometer Brookfield tube and raised until spindle was covered by the solution. The spindle was rotated at speed of 60 and 100 rpm at temperature of 70°C and the viscosity value shown was recorded in %CP and torque.

2.2.3. Stability Analysis of Emulsion
Stability of emulsion was determined by observing the mixed solution of GAP, GA and vitamin E. Observation was conducted after the mixed solution was heated at 70°C and stirred at 1500 rpm for 30 min. The time that vitamin E was separated from the solution was recorded..

2.2.4. Microencapsulation using Spray Drying Method
Vitamin E, water, GAP and GA was mixed based on the formulation that given uniform compatibility and viscosity which related to industry standard. Those formulation was used for spray drying method with initial temperature was 70°C for 15 min. Next, the mixed solution was filtered using filter bag 1000 micron and had spry dryer using hot air (180 – 200°C).

2.2.5. Yield Analysis of Microcapsule Powder
The yield analysis of the powder product was conducted following equation below.

$$Y = \frac{Wp-(Wp\ x\ MCp)}{Wf-(Wf\ x\ MCf)}\ x\ 100\% \quad (1)$$

Where:
Wp = weight of microcapsule product
Wf = weight of mixed solution before spray drying
MCp = weight of water contained in microcapsule product
MCf = weight of water contained in mixed solution

2.2.6. Water and Lipid Content Analysis in Microcapsule Product
Water content contained in microcapsule was determined using moisture analyser AND MX-50. 5 gram of sample was putted in glass plate and the drying process was begun. The moisture value was automatically appeared in the screen. Lipid content analysis was conducted using soxhlet extraction method with petroleum benzene as the solvent.

2.2.7. Oxidation Stability Analysis of Microcapsule
The microcapsule was putted in closed glass bottle and was stored at 60°C to increase the oxidation process. The oxidation stability was measured using 2 different methods such as p-aniline value for peroxide and the vitamin E content using HPLC started at zero time (after drying) and after 2 day storage [8].

2.2.8. Efficiency Encapsulation
The efficiency encapsulation was determined based on Bae and Lee (2008) method with modification. A 15 ml of hexane was added to 1.5 gram of microcapsule in closed glass bottle and be shaken with hand for 2 min at room temperature. The mixed of solvent was separated using Whatman No. 1 filter paper and the residue was washed three times with 20 ml of hexane. Next, the solvent was evaporated at 60°C and the residue was weighed. The efficiency encapsulation was calculated based on the equation below.

$$Efficiency\ Encapsulation = \left[\frac{TO-SO}{TO}\right]\ x\ 100\% \quad (2)$$

Where:
EE = efficiency encapsulation
TO = Weight of total lipid content
SO = Weight of lipid extracted

IOP Conf. Series: Materials Science and Engineering **309** (2018) 012095 doi:10.1088/1757-899X/309/1/012095

3. Results and Discussion

3.1. Composition of Vitamin E PFAD

Crude palm oil contained vitamin E in the range of 600 – 1000 ppm which tocopherol (21 – 31%) and tocotrienol (66 – 79%) are the main compound. However, most of vitamin E was lost during refinery process (Schwartz *et al.*, 2008). Vitamin E could be extracted from PFAD as refinery by product (Fizet, 1993). Based on the GC analysis, the main components of PFAD was vitamin E as showed in figure 1 while figure 2 presented the components of vitamin E.

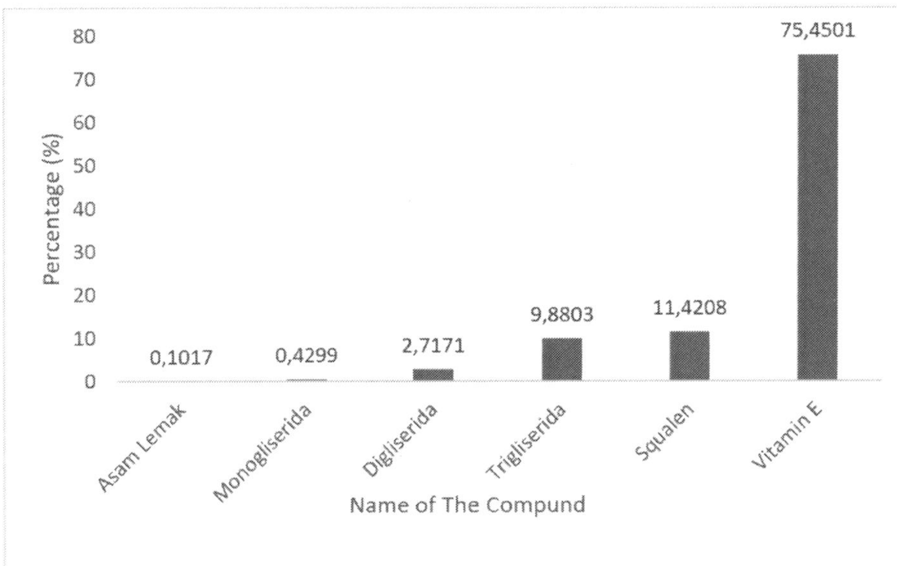

Figure 1 Components of vitamin E from PFAD

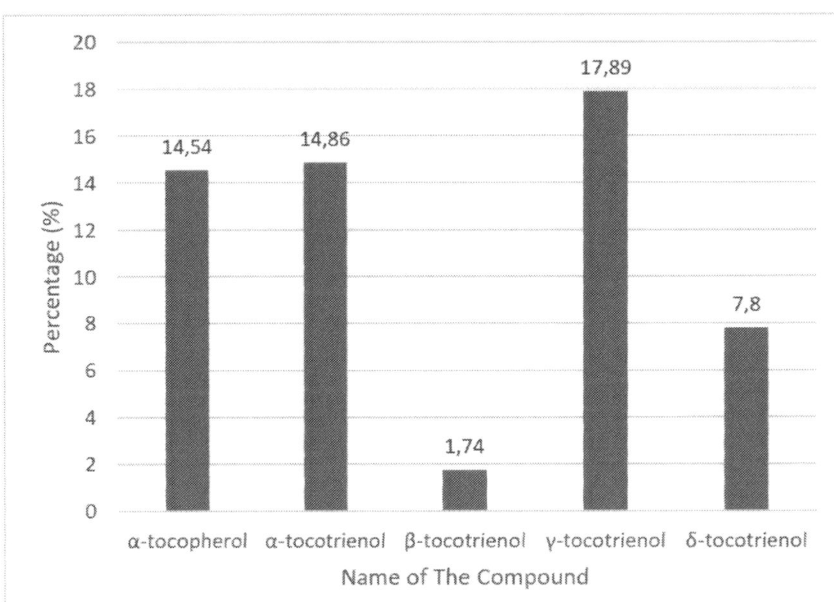

Figure 2 Components of vitamin E

3.2. Characterization of Emulsion

The emulsion was formed from mix of GAP varied from 0.1 gram to 0.4 gram, 7 gram of GA and 1.3 gram of vitamin E as shown in table 1. Those emulsion composition was also used for film preparation.

Table 1. Formulation Emulsion

Materials	Formulation (gram)						
	1	2	3	4	5	6	7
GAP	-	0.1	0.2	0.3	0.4	0.4	0.3
GA	7.0	7.0	7.0	7.0	7.0	-	7
Vitamin E	1.3	1.3	1.3	1.3	1.3	1.3	-
Deionized Water	91.7	91.6	91.5	91.4	91.3	98.3	92.7
Total	100.0	100.0	100.0	100.0	100.0	100.0	100.0

Based on table 2, the viscosity of GAP was higher than GA which 0.4 gram of GAP has viscosity of 3.95 cP while 7 gram of GA is only 2.92 cP. Therefore it can be concluded that in the increment of GAP the viscosity of emulsion increased rendering increasing of the thickness of the film. The viscosity value that similar with the standard (cP 13.2 and torque 27.0%) was obtained from formula 3 and 4 using GAP 0.2 gram and 0.3 gram. The stability analysis was conducted by observing the solubility of GAP, GA and vitamin E after heated at 70°C and stirred at 1500 rpm for 30 min. The time for vitamin E separated from the emulsion was used to determine the emulsion stability. As shown in table 2, the stability of GAP was lower than GA which probably due to the emulsifier properties of GA was 4 times higher than GAP.

Table 2 the viscosity and stability of emulsion and the thickness of the film

Formulasi	Viscosity at 70°C				Stability (second)	Thickness of Film (mm)
	60 rpm		100 rpm			
	cP	%Torque	cP	%Torque		
1	2.92	5.8	2.74	9.0	126	0.542
2	4.45	8.9	4.08	13.6	378	0.601
3	5.73	11.4	5.49	18.43	395	0.635
4	7.75	15.5	7.08	23.36	412	0.646
5	9.22	18.4	9.93	33.10	507	0.779
6	3.95	7.9	3.57	10.20	30	0.101
7	7.43	14.9	7.86	26.20	-	0.684

The formulation 3 showed better compatibility based on the film morphology as shown in figure 3. Therefore formulation 3 was used for spray drying. This result is similar with other researchers founded that efficiency encapsulation was affected retention property and ability to form film of the material protecting [8].

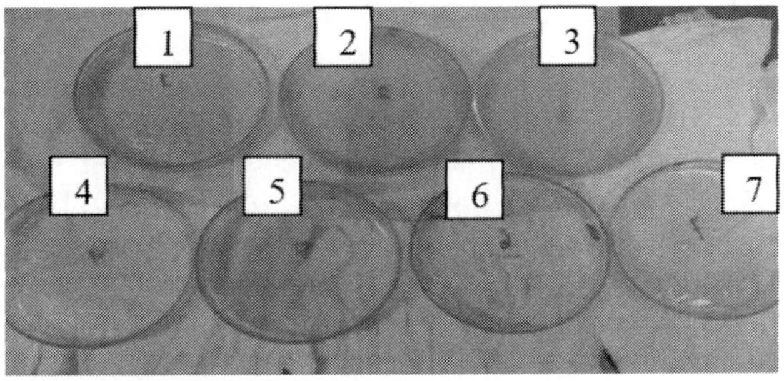

Figure 3 the films of mixed GAP, GA and vitamin E

3.3. Characterization of Microcapsule

Based on the compatibility film formed from formulation 1 to 7, the formula 3 was chose for spray drying method with minimum capacity was 1000 gram as shown in table 3. The microencapsulation process was started with heated the water until 70^{0}C followed by addition of GA under stirred at 3000 rpm for 30 min which was continued for 15 min after the addition of vitamin E. The sample was used for spry dry using heat air with temperature input ranging from 180 – 200^{0}C and output temperature was 90^{0}C. The sample was pumped with using peristaltic pump with flow rate was 20 rpm. The microcapsule was accommodated in closed bottle followed by determined the weight. Similar procedure also conducted with or without the addition of GAP which already dissolved in water.

Table 3 Formulation of Microencapsulation

Materials	Formulation Microencapsulation (%)		Weight Feed (gram)	
	A (GA)	B (GA:GAP)	A (GA)	B (GA:GAP)
Air	60	60	600.00	600.00
GAP	-	0.4	-	4.04
GA	14	13.6	140.10	136.15
Vit. E	26	26	260.16	260.21
Total	100	100	1000.26	1000,40

Table 4 Some parameters of microcapsule

Parameters	Microcapsule	
	A	B
Yield (%)	53.1524	64.0918
Water Content (%)	3.40	3.08
Particle Size (μm)	16	11
Lipid Content (%)	61.8253	62.4414
Unencapsulated Lipid (%)	24.3097	18.7224
Efficiency Encapsulation (EE) (%)	60.6800	70.0160

The yield occurred from microencapsulation process was below 100% as showed in table 4 which because some of microcapsule stick to the wall of spray drying column. In addition, some of the particles have small size which brought by hot air flow and could not separate from the cyclone. Furthermore, small size particle was stuck in the filter and hence could not collect. However with addition of GAP 4 gram in 1000 gram solution could increase the yield of microcapsule around 17.07%. The water content was decreased to 9.41%. Water content is one of important parameters which affect the quality of dry microcapsule powder. Low water content could inhibit fungi or bacteria growth which could degrade the product (Master, 1979). Buckle et al. showed that high water content could degrade the product therefore the maximum of water content is less than 8%. Therefore based on our study the addition of GAP could increase the storage time of microencapsulation due to low water content.

The efficiency encapsulation was increase at 13.33 with increasing of GAP which due to increasing of emulsion stability. This was similar with other researcher findings that application of more than 2 protecting material could increase efficiency encapsulation. Furthermore, Carneiro et al. concluded that efficiency encapsulation was affected by the type of protecting material used. Poor stability could produce low efficiency encapsulation [8]. The quantity of encapsulated lipid raised in the high efficiency encapsulation (Barbosa et al., 2005). Figure 4 G and I showed that there are some black dot which probably is unencapsulated vitamin E. In comparison no black dot appeared in figure 4 H and J which mean all the vitamin E has been encapsulated. Therefore it can be concluded that addition of GAP could increase the encapsulation process. Based on the figure 4 I and J the size of microcapsule was decrease in addition of GAP. Bigger size of microcapsule shown the mixing is not homogenized which have low emulsion stability. In addition using different protecting material also affects the morphology of microcapsule (Carneiro at al., 2013). This was shown in figure 4 that in addition of GAP produce oval shape which have smooth surface.

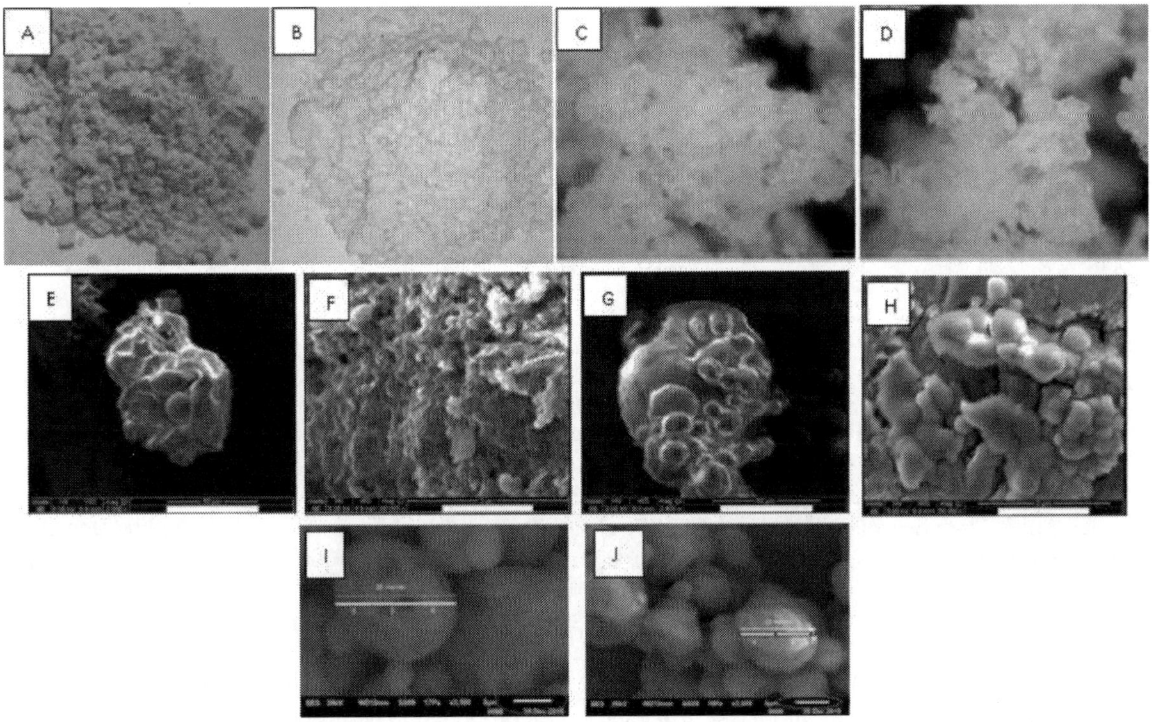

Figure 4 the photo of (A) microcapsule A; (B) microcapsule B and the SEM images of (C) microcapsule A; (D) microcapsule B; (E) gum acacia powder; (F) GAP powder; (G) microcapsule A; (H) microcapsule B; (I) the size of microcapsule A (16 μm); (J) the size of microcapsule B (11 μm).

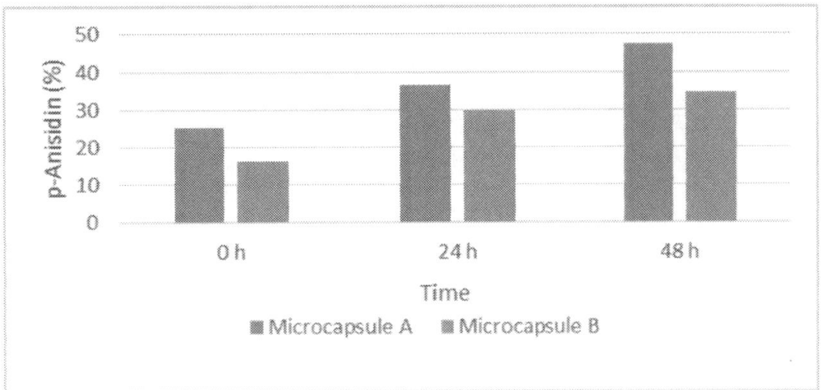

Figure 5 Oxidative Stability of Encapsulated Vitamin E PFAD by p-Anisidin Value Method (Microcapsule A and B).

Oxidation stability of microcapsule A and B showed by p-anisidine value as presented in figure 5. No oxidation was detected at 0 hour. As shown in figure 5 the total vitamin E was only slightly decreased in the increasing storage time. The addition of GAP as protecting material have lowering p-anisidine value which mean encapsulated vitamin E have good stability.

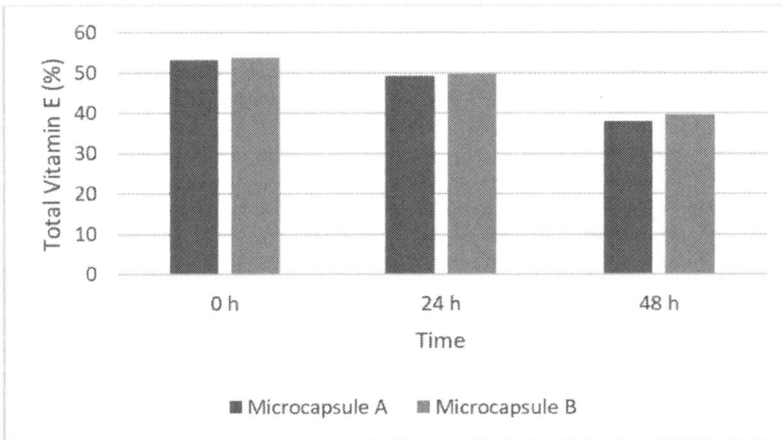

Figure 6 the total vitamin E contained in microcapsule A and B in several time

FT-IR spectroscopy was used to identify interaction between vitamin E and protecting material in microcapsule. The absorbance spectra of GAP, GA, vitamin E, microcapsule A and B were recorded in the region of 4000 – 400 cm^{-1} as shown in figure 7. A peak at 3500 cm^{-1} appeared in all spectra which were a stretching vibration of OH group. Two strong peaks at region of 2800 – 3000 cm^{-1} represent stretching of C-H group while peak at 1732 cm^{-1} which only appeared in spectra of vitamin E represent stretching of C=O group. This was because vitamin E also contained triglyceride, diglyceride, monoglyceride, fatty acid and squalene. Figure 7 D and E showed the mixed of vitamin E, GAP and GA which have similar peaks with each pure compound. Therefore it can be concluded that there was interaction between all the components without any changes in their structure. This result was similar with other researcher findings [18, 19].

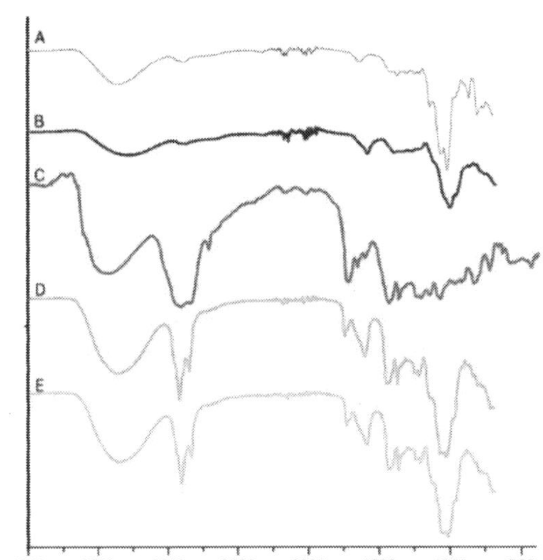

Figure 7 the FT-IR spectrum of (A) GAP; (B) gum acacia; (C) vitamin E PFAD, D) microcapsule A and (E) microcapsule B

4. Conclusion

The encapsulation vitamin E PFAD has been successfully conducted using spray drying method with GAP and GA as protecting material. Composite films with thickness vary from 0.542 – 0.779 mm were prepared by incorporating vitamin E onto matrix of GA (7 g) with various concentration of GAP (0.1; 0.2; 0.3 and 0.4 g). The formulation 3 and 4 which used GAP 0.2 gram and 0.3 gram, respectively, showed the viscosity value which similar with the standard. That composition was used for spray drying method rendering micro-particle size 11 μm and the particle had spherical shape. Although the increment of GAP decreasing moisture content and the particle size from 16 μm to 11 μm, the yield of microcapsule, encapsulation efficiency, the amount of vitamin E absorbed and oxidation stability of vitamin E were increased.

5. References

[1] Gonnet M M, Lethuaut L and Boury F 2010 New trends in encapsulation of liposoluble vitamins *Journal of Controlled Release* **146(3)** 276-290
[2] Loughrill E, Govinden P and Zand N N 2016 Vitamins a and e content of commercial infant foods in the uk: A cause for concern? *Food chemistry* **210** 56-62
[3] Zhuang Y, Sterr J, Schulte A, Kulozik U and Gebhardt R U 2016 Casein microparticles from blend films forming casein/α-tocopherol emulsion droplets in solution *Food biophysics* **11(4)** 332-338
[4] Musalmah M, Nizrana M, Fairuz A, Aini A N, Azian A, Gapor M and Ngah W W 2005 Comparative effects of palm vitamin e and α-tocopherol on healing and wound tissue antioxidant enzyme levels in diabetic rats *Lipids* **40(6)** 575-580
[5] Bae E and Lee S S 2008 Microencapsulation of avocado oil by spray drying using whey protein and maltodextrin *Journal of Microencapsulation* **25(8)** 549-560
[6] Prosapio V, Reverchon E and Marco I D 2015 Coprecipitation of polyvinylpyrrolidone/β-carotene by supercritical antisolvent processing *Industrial & Engineering Chemistry Research* **54(46)** 11568-11575
[7] Carolina B C, Carolina S, Zamora M C and Jorge C 2007 Glass transition temperatures and some physical and sensory changes in stored spray-dried encapsulated flavors *LWT-Food Science and Technology* **40(10)** 1792-1797
[8] Carneiro H C F, Tonon R V, Grosso C R F and Hubinger M D 2013 Encapsulation efficiency and oxidative stability of flaxseed oil microencapsulated by spray drying using different combinations of wall materials *Journal of Food Engineering* **115(4)** 443-451
[9] Charve J P and Reineccius G A 2009 Encapsulation performance of proteins and traditional materials for spray dried flavors *Journal of agricultural and food chemistry* **57(6)** 2486-2492
[10] Jafari S M, Assadpoor E, He Y and Bhandari B 2008 Encapsulation efficiency of food flavours and oils during spray drying *Drying Technology* **26(7)** 816-835
[11] Mogea J, Seibert B and Smits W 1991 Multipurpose palms: The sugar palm (arenga pinnata (wurmb) merr.) *Agroforestry Systems* **13(2)** 111-129
[12] Orwa C, Mutua A, Kindt R, Amnadass J R and Simons A 2009 Agroforestree database: A tree species reference and selection guide version 4.0 *World Agroforestry Centre ICRAF, Nairobi, KE*
[13] Rao K V 1961 Development and life history of a nudibranchiate gastropod cuthona adyarensis rao *Journal of the Marine Biological Association of India* **3(1&2)** 186-197
[14] Tarigan J B 2014 *Karakterisasi edible film yang bersifat antioksidan dan antimikroba dari galaktomanan biji aren (arenga pinnata) yang diinkorporasi dengan minyak atsiri daun kemangi (ocimum basilicum l.)*
[15] Sharipova A, Aidarova S, Grigoriev D, Mutalieva B, Madibekova G, Tleuova A and Miller R 2016 Polymer–surfactant complexes for microencapsulation of vitamin e and its release *Colloids and Surfaces B: Biointerfaces* **137** 152-157

[16] Tsai W C and Rizvi S S 2017 Simultaneous microencapsulation of hydrophilic and lipophilic bioactives in liposomes produced by an ecofriendly supercritical fluid process *Food Research International*
[17] Parthasarathi S and Anandharamakrishnan C 2016 Enhancement of oral bioavailability of vitamin e by spray-freeze drying of whey protein microcapsules *Food and Bioproducts Processing* **100** 469-476
[18] Renò F, Aina V V, Gatti S and Cannas M 2005 Effect of vitamin e addition to poly (d, l)-lactic acid on surface properties and osteoblast behaviour *Biomaterials* **26(28)** 5594-5599
[19] Prosapio V, Reverchon E and Marco I D 2017 Incorporation of liposoluble vitamins within pvp microparticles using supercritical antisolvent precipitation *Journal of CO2 Utilization* **19** 230-237

An application of the AHP in water resources management: a case study on urban drainage rehabilitation in Medan City

A P M Tarigan[1*], D Rahmad[2], R A Sembiring[3], and R Iskandar[4].
[1,4]Teaching Staff, Civil Engineering Postgraduate Study Programs, Universitas Sumatera Utara
[2]Master's Student, Civil Engineering Postgraduate Study Programs, Universitas Sumatera Utara
[3]Teaching Staff, Civil Engineering Department, Universitas Sumatera Utara

*E-mail: a.perwira.mulia@gmail.com

Abstract. This paper illustrates an application of Analytical Hierarchy Process (AHP) as a potential decision-making method in water resource management related to drainage rehabilitation. The prioritization problem of urban drainage rehabilitation in Medan City due to limited budget is used as a study case. A hierarchical structure is formed for the prioritization criteria and the alternative drainages to be rehabilitated. Based on the AHP, the prioritization criteria are ranked and a descending-order list of drainage is made in order to select the most favorable drainages to have rehabilitation. A sensitivity analysis is then conducted to check the consistency of the final decisions in case of minor changes in judgements. The results of AHP computed manually are compared with that using the software Expert Choice. It is observed that the top three ranked drainages are consistent, and both results of the AHP methods, calculated manually and performed using Expert Choice, are in agreement. It is hoped that the application of the AHP will help the decision-making process by the city government in the problem of urban drainage rehabilitation.

1. Introduction

1.1 Background and objective

With the growth of population and urbanization, a large and densely populated city in a developing country like Medan City will experience lots of land changes that cause a significant increase in runoff coefficients. As the run off coefficients become higher, the ability of the land surfaces to recharge water to the ground water aquifer become less effective, increasing the stream flows during and after significant rainfall events. In other words, because of the urbanized areas with more paved streets and parking lots, buildings and roof tops which prevent seepage and increase runoff, more water from the rainfall enters the city drainage.

On the other hand, the city drainages suffer from substantial sedimentation resulting in the reduction of the effective cross-section of the drainages. Consequently, the capacity of the drainages to flow water becomes reduced. As a result of the condition in which the run-off increases while the drainage capacity decreases, flood is inevitable especially after heavy rain.

Stagnant water cannot be addressed because the drainage is not working properly drain water from the road. Roads drainage only designed and built without regard to good or bad performance. Many found that the roads drainage has been constructed does not work properly (not able to drain the stagnant water in the road because of the condition of the existing drainage damaged, clogged, not well maintained, high sediment and others) [1].

In order to mitigate the flood hazards, the city drainages need to be improved or rehabilitated regularly. However, the city budget is not sufficient to do regular maintenance works for such extensive drainages within the whole city area. In this case prioritization is required, especially over the 'hot spots' where flood events frequently occur.

In order to reach these goals, we need to design a system with systematic processes [2]. And The purpose of this study is to illustrate the application of Analytical Hierarchy Process (AHP) as a potential decision-making method in water rèsource management related to drainage rehabilitation. The prioritization problem of urban drainage rehabilitation over selected 'hot spots' in a district in Medan City is used as a study case. It is hoped that the application of the AHP will help the decision-making process by the city government in the problem of urban drainage rehabilitation.

1.2 Analytical Hierarchy Process (AHP)

The AHP, developed by Saaty, is one of the methods based on the concept of the multi criteria decision analysis [3-6]. The concept is used to help decision makers learn about their problems in a decision problem situation in which discretely defined alternatives are to be evaluated. More than one criterion (or attribute) are involved in the decision process to reach the preferred alternatives in a ranked fashion. Based on the judgement of the decision maker, the AHP aims at quantifying relative priorities for a given set of alternatives on a ratio scale and stresses the importance of the consistency of the comparisons of alternatives [7].

The procedures of the AHP can be outlined as follows. Step1: Define the problem and its objective. Step 2: Set up the decision hierarchy. Step 3: Make pairwise comparisons of attributes and alternatives using the relative scale measurement shown in Table 1. Step 4: Transform the comparison into weights and check the consistency of the decision maker's comparison using Table 2. Step 5: Use the weights to obtain scores for the different options and make a provisional decision. And Step 6: Perform sensitivity analysis.

Table 1. Pairwise comparison scale for AHP preferences [3-6]

Numerical rating	Verbal judgments of preferences
9	Extremely preferred
8	Very strongly to extremely
7	Very strongly preferred
6	Strongly to very strongly
5	Strongly preferred
4	Moderately to strongly
3	Moderately preferred
2	Equally to moderately
1	Equally preferred

Table 2. Average random consistency (Ri) [3-6]

Ordo Matrix	1	2	3	4	5	6	7	8	9	10
RI	0	0	0,58	0,9	1,12	1,24	1,32	1,41	1,45	1,49

2. Method

The locations of the study, consisting of 7 sewer drainages, are located within the district of Medan Baru which is one of the total 21 districts in Medan City. University of Sumatera Utara (USU) is located in Medan Baru with elevation approximately 30 m above sea level and distance 30 km from the coastline [8]. Note that we observe frequent flooding at USU and its vicinities during heavy rain. The locations are selected due to the frequent flooding observed in their vicinities.

The main data in this study are obtained from interview and questionnaire which is designed for the AHP method. The AHP allows group decision making, where group members can use their experience, values and knowledge to break down a problem into a hierarchy and solve it by the AHP steps [9]. There were nine respondents who were interviewed and did fill in the questionnaire. All of them are considered knowledgeable about the flood problems occurring in the study locations.

There are two stages of interview conducted in this study. Stage 1 is the cut off method which aims at defining the criteria and sub-criteria contained within the AHP. Stage 2 is the full AHP interview with questionnaire in which the whole pairwise comparisons are given. The values taken into consideration for the cut off method as well as the AHP are the arithmetic averaged values given by all respondents.

3. Results and discussions

3.1 The cut off method

As mentioned above, the purpose of the cut off method is to define the criteria and sub-criteria of the AHP based on the defined objective [10, 11], i.e. to rank the prioritization of the drainage rehabilitation. The questionnaire used was given in 5 values of classification from 0 to 4 as follows: 0 for not important, 1 for less important, 2 for important, 3 for more important, and 4 very important. Based on the questionnaire answered by the respondents, from 19 sub-criteria included in 4 criteria (or aspects), we selected 8 sub-criteria to be involved in the next stage of interview. The selection process was done based on a threshold value of the cut off and the principle of the absence of redundancy. Note that if two criteria duplicate each other because they may represent the same thing, then one of these criteria can be said redundant and be neglected. Figure 1 shows the final hierarchy of the drainage rehabilitation comprising of the criteria and sub-criteria used in the AHP method.

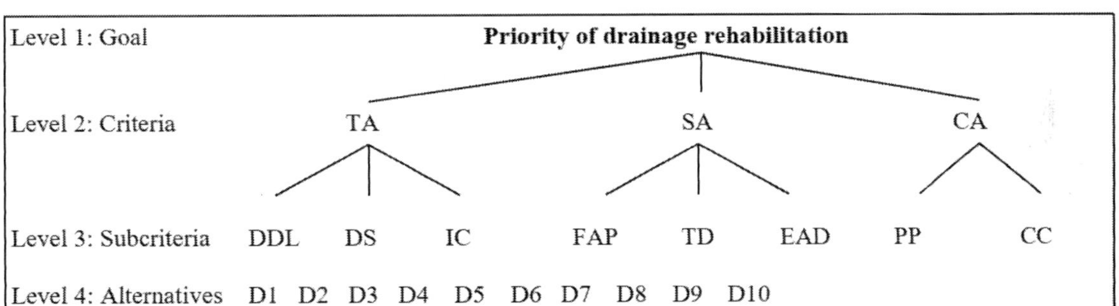

TA = Technical Aspects
CI = Inundation Condition
CC = Community Character
SA = Social Economic Aspects
FAP = Flood Affected Population
CA = Cultural Aspects
TD = Traffic Disturbance
DDL = Drainage Damage Level
EAD = Economic Activities Disturbance
DS = Drainage Status
PP = Public Participation
D1, D2, D3, D4, D5, D6, D7, D8, D9 and D10 are the alternatives.

Figure 1. The final hierarchy of the drainage rehabilitation

3.2 The AHP process

First, we process the pairwise comparison for the criteria (level 2 in the hierarchy), i.e. technical aspects, social aspects, and cultural aspects. The result is shown in Table 3.

Table 3. Pairwise comparison matrix for the criteria (level 2)

	Technical aspects	Social aspects	Culture aspects
Technical aspects	**1**	0.42	2.4
Social aspects	2.4	**1**	2.8
Culture aspects	0.42	0.36	**1**
$\sum =$	**3.82**	**1.77**	**6.20**

Dividing each the element of the above matrix by its respective column total results in the following.

$$\begin{bmatrix} 0.262 & 0.235 & 0.387 \\ 0.629 & 0.564 & 0.452 \\ 0.109 & 0.201 & 0.161 \end{bmatrix}$$

The priority vector which indicates the weight for each aspect is the row average as follows:

$$\begin{bmatrix} 0.295 \\ 0.548 \\ 0.157 \end{bmatrix}$$

The priority vector is then multiplied with the elements of the initial matrix in Table 1 to get the following.

$$0.295 \begin{bmatrix} 1 \\ 2.4 \\ 0.42 \end{bmatrix} + 0.548 \begin{bmatrix} 0.42 \\ 1 \\ 0.36 \end{bmatrix} + 0.157 \begin{bmatrix} 2.4 \\ 2.8 \\ 1 \end{bmatrix} = \begin{bmatrix} 0.900 \\ 1.696 \\ 0.476 \end{bmatrix}$$

The above vector is divided by the respective priority vector to yield:

$$\frac{0.900}{0.295} = 3.056\,, \frac{1.696}{0.548} = 3.094\,, \frac{0.476}{0.157} = 3.025$$

Then the maximum principal eigen value λ_{max} is obtained by averaging the above values:

$$\lambda_{max} = \frac{(3.056 + 3.094 + 3.025)}{3} = 3.058$$

The consistency index Ci is computed as follows

$$\mathrm{Ci} = \frac{\lambda max - n}{n-1} = \frac{3.058-3}{3-1} = 0.029$$

With the random index Ri = 0.58 (Table 2) for size of matrix n = 3, the consistency ratio Cr can be calculated as follows:

$$\mathrm{Cr} = \frac{Ci}{Ri} = \frac{0.0292}{0{,}58} = 0.050$$

$$\mathrm{Cr} = 0.050 < 0.1, \text{OK}$$

Since Cr less than 0.1, we can conclude that the comparisons are consistent and thus acceptable.

The same procedures are conducted for the comparisons of the sub-criteria (level 3 in the hierarchy) as well as for the alternative drainages (level 4 in the hierarchy). Tables 4, 5, and 6 show the results of the comparisons of the sub-criteria, i.e. technical aspects, social aspects, and culture aspects, consecutively.

Table 4. Pairwise comparison matrix for Technical aspects

	Drainage status	Level of drainage damage	Condition of inundation	Priority vector
Drainage status	**1**	0.769	0.625	**0.252**
Level of drainage damage	1.3	**1**	0.5	**0.278**
Condition of inundation	1.6	2	**1**	**0.470**

$\lambda_{maks} = 3.026$, Ci = 0.013, Ri = 0.58, Cr = 0.002 < 0.1, OK

Table 5. Pairwise comparison matrix for social aspects

	Flood affected population	Traffic disturbance	Economic activities disturbance	Priority vector
Flood affected population	**1**	2.4	2.4	**0.529**
Traffic disturbance	0.42	**1**	2.4	**0.301**
Economic activities disturbance	0.42	0.42	**1**	**0.170**

$\lambda_{max} = 3.018$, Ci = 0.004, Ri = 0.58, Cr = 0.0158 < 0.1, OK

Table 6. Pairwise comparison matrix for culture aspects

	Public participation	Community characters	Priority vector
Public participation	**1**	2.4	**0.706**
Community characters	0.417	**1**	**0.294**

$\lambda_{max} = 2$, Ci = 0, Ri =0.00, Cr = 0 < 0.1, OK

There are 10 alternative drainages to be compared. They are denoted as follows: D1 = Sei Putih drainage, D2 = Sei Asahan 1 drainage, D3 = dr. Mansyur drainage, D4= Mandolin&Rebab drainage, D5 = Sei Selayang drainage, D6 = Sei Mencirim drainage, D7 = Darussalam drainage, D8 = Djamin Ginting drainage, D9= Sei Padang drainage, and D10 = Sei Asahan 2 drainage.

The pairwise comparison matrices of all alternative drainages for all sub-criteria are given in Tables 7 to 14. Since there are 8 sub-criteria, then we have the following 8 sub-criteria matrices.

Table 7. Pairwise comparison matrix for level of drainage damage

DDL	D1	D2	D3	D4	D5	D6	D7	D8	D9	D10	Priority vector
D1	**1.00**	0.33	0.42	0.42	0.42	0.42	0.29	2.40	2.40	0.25	0.055
D2	3.00	**1.00**	0.42	0.42	2.40	2.40	0.42	2.40	2.40	0.42	0.097
D3	2.40	2.40	**1.00**	3.50	3.50	2.40	2.40	2.40	4.00	2.40	0.213
D4	2.40	2.40	0.29	**1.00**	2.50	2.40	0.33	2.40	2.40	0.42	0.106
D5	2.40	0.42	0.29	0.40	**1.00**	2.40	0.29	2.40	0.42	0.42	0.067
D6	2.40	0.42	0.42	0.42	0.42	**1.00**	0.42	2.40	2.40	1.00	0.077
D7	3.40	2.40	0.42	2.40	3.50	2.40	**1.00**	3.50	3.00	1.00	0.156
D8	0.42	0.42	0.42	0.42	0.42	0.42	0.29	**1.00**	0.42	0.50	0.040
D9	0.42	0.42	0.25	0.42	2.40	0.42	0.33	2.40	**1.00**	0.50	0.056
D10	4.00	2.40	0.33	2.40	2.40	1.00	1.00	2.00	2.00	**1.00**	0.131
											$\sum = 1.00$

$\lambda_{max} = 11.113$, Ci = 0.0126, Ri =1.49, Cr = 0.0845 < 0.1, OK

Table 8. Pairwise comparison matrix for drainage status

DS	D1	D2	D3	D4	D5	D6	D7	D8	D9	D10	Priority vector
D1	**1.00**	0.42	0.42	1.00	0.42	1.00	1.00	0.42	0.42	1.00	0.059
D2	2.40	**1.00**	1.00	2.40	1.00	2.40	2.40	1.00	1.00	2.40	0.141
D3	2.40	1.00	**1.00**	2.40	1.00	2.40	2.40	1.00	1.00	2.40	0.141
D4	1.00	0.42	0.42	**1.00**	0.42	1.00	1.00	0.42	0.42	1.00	0.059
D5	2.40	1.00	1.00	2.40	**1.00**	2.40	2.40	1.00	1.00	2.40	0.141
D6	1.00	0.42	0.42	1.00	0.42	**1.00**	1.00	1.00	0.42	1.00	0.067
D7	1.00	0.42	0.42	1.00	0.42	1.00	**1.00**	0.42	1.00	1.00	0.065
D8	2.40	1.00	1.00	2.40	1.00	1.00	2.40	**1.00**	2.40	2.40	0.147
D9	2.40	1.00	1.00	2.40	1.00	2.40	1.00	0.42	**1.00**	2.40	0.123
D10	1.00	0.42	0.42	1.00	0.42	1.00	1.00	0.42	0.42	**1.00**	0.059
											Σ =1.00

λ_{max} = 10.195, Ci = 0.0216, Ri =1.49, Cr = 0.0145 < 0.1, OK

Table 9. Pairwise comparison matrix for condition of inundation

IC	D1	D2	D3	D4	D5	D6	D7	D8	D9	D10	Priority vector
D1	**1.00**	0.42	0.22	2.40	2.40	2.40	0.33	2.40	2.40	0.33	0.089
D2	2.40	**1.00**	0.25	2.40	2.40	1.40	0.33	2.40	2.00	0.42	0.096
D3	4.50	4.00	**1.00**	3.00	4.50	2.40	1.00	2.40	2.40	3.00	0.212
D4	0.42	0.42	0.33	**1.00**	2.40	2.40	0.42	2.40	2.40	0.42	0.081
D5	0.42	0.42	0.22	0.42	**1.00**	2.40	0.33	3.00	2.40	0.50	0.069
D6	0.42	0.71	0.42	0.42	0.42	**1.00**	0.42	2.40	2.00	0.50	0.061
D7	3.00	3.00	1.00	2.40	3.00	2.40	**1.00**	3.00	2.40	2.00	0.175
D8	0.42	0.42	0.42	0.42	0.33	0.42	0.42	**1.00**	0.50	0.33	0.040
D9	0.42	0.50	0.42	0.42	0.42	0.50	0.42	2.00	**1.00**	0.42	0.049
D10	3.00	2.40	0.33	2.40	2.00	2.00	0.50	3.00	2.40	1.00	0.127
											Σ =1.00

λ_{max} = 11.079, Ci = 0.120, Ri = 1.49, Cr = 0.0805 < 0.1, OK

Table 10. Pairwise comparison matrix for flood affected population

FAP	D1	D2	D3	D4	D5	D6	D7	D8	D9	D10	Priority vector
D1	**1.00**	2.40	0.42	2.40	2.00	1.00	2.40	3.00	2.40	0.33	0.128
D2	0.42	**1.00**	0.42	0.50	0.50	0.33	0.42	1.40	0.42	0.33	0.045
D3	2.40	2.40	**1.00**	2.40	2.40	2.00	2.40	2.40	2.40	2.40	0.189
D4	0.42	2.00	0.42	**1.00**	1.40	0.42	0.42	2.40	1.00	1.00	0.076
D5	0.50	2.00	0.42	0.71	**1.00**	0.42	0.42	2.40	1.00	1.00	0.072
D6	1.00	3.00	0.50	2.40	2.40	**1.00**	2.40	2.40	3.00	3.00	0.162
D7	0.42	2.40	0.42	2.40	2.40	0.42	**1.00**	2.40	0.42	0.42	0.089
D8	0.33	0.71	0.42	0.42	0.42	0.42	0.42	**1.00**	0.42	0.42	0.042
D9	0.42	2.40	0.42	1.00	1.00	0.33	2.40	2.40	**1.00**	1.00	0.088
D10	2.40	2.40	0.42	1.00	1.00	0.33	2.40	2.40	1.00	**1.00**	0.109
											Σ = 1.00

λ_{max} = 10.743, Ci = 0.826, Ri = 1.49, Cr = 0.0554 < 0.1, OK

Table 11. Pairwise comparison matrix for traffic disturbance

TD	D1	D2	D3	D4	D5	D6	D7	D8	D9	D10	Priority vector
D1	**1.00**	1.40	0.33	1.40	1.40	0.42	0.33	2.40	2.40	2.00	0.093
D2	0.71	**1.00**	0.29	2.00	2.00	1.00	0.42	2.40	2.00	2.00	0.099
D3	3.00	3.50	**1.00**	3.00	2.40	3.00	2.40	2.40	2.40	2.40	0.214
D4	0.71	0.50	0.33	**1.00**	1.00	0.42	0.33	1.40	1.70	2.40	0.071
D5	0.71	0.50	0.42	1.00	**1.00**	1.00	0.50	2.40	2.00	2.40	0.087
D6	2.40	1.00	0.33	2.40	1.00	**1.00**	0.42	2.40	2.40	3.00	0.116
D7	3.00	2.40	0.42	3.00	2.00	2.40	**1.00**	2.00	2.00	0.71	0.150
D8	0.42	0.42	0.42	0.71	0.42	0.42	0.50	**1.00**	0.42	0.59	0.045
D9	0.42	0.50	0.42	0.59	0.50	0.42	0.50	2.40	**1.00**	0.50	0.055
D10	0.50	0.50	0.42	0.42	0.42	0.33	1.40	1.70	2.00	**1.00**	0.070
											Σ =1.00

λ_{max} = 10.886, Ci = 0.098, Ri = 1.49, Cr = 0.0661 < 0.1, OK

Table 12. Pairwise comparison matrix for economic activities disturbance

EAD	D1	D2	D3	D4	D5	D6	D7	D8	D9	D10	Priority Vector
D1	**1.00**	0.71	0.33	1.00	0.42	0.33	0.42	1.00	1.00	1.00	0.057
D2	1.40	**1.00**	0.42	1.00	0.50	0.42	0.42	1.00	0.42	1.00	0.060
D3	3.00	2.40	**1.00**	2.40	2.00	0.42	2.40	3.40	2.40	2.40	0.170
D4	1.00	1.00	0.42	**1.00**	0.50	0.42	0.42	1.00	0.50	1.00	0.058
D5	2.40	2.00	0.50	2.00	**1.00**	0.42	0.42	2.40	2.00	2.40	0.113
D6	3.00	2.40	2.40	2.40	2.40	**1.00**	2.40	2.40	2.00	3.00	0.202
D7	2.40	2.40	0.42	2.40	2.40	0.42	**1.00**	2.00	2.00	2.40	0.134
D8	1.00	1.00	0.29	1.00	0.42	0.42	0.50	**1.00**	0.42	1.00	0.056
D9	1.00	2.40	0.42	2.00	0.50	0.50	0.50	2.40	**1.00**	2.40	0.096
D10	1.00	1.00	0.42	1.00	0.42	0.33	0.42	1.00	0.42	**1.00**	0.055
											Σ =1.00

λ_{max} = 10.392, Ci = 0.0436, Ri = 1.49, Cr = 0.0292 < 0.1, OK

Table 13. Pairwise comparison matrix for public participation

PP	D1	D2	D3	D4	D5	D6	D7	D8	D9	D10	Priority vector
D1	**1.00**	2.40	0.42	1.00	0.50	1.00	1.00	3.00	2.40	0.42	0.092
D2	0.42	**1.00**	0.42	0.42	0.33	0.42	0.42	2.00	2.40	0.42	0.058
D3	2.40	2.40	**1.00**	2.80	0.42	2.00	2.40	3.00	3.00	1.00	0.155
D4	1.00	2.40	0.36	**1.00**	0.50	1.00	1.00	2.00	2.40	0.42	0.086
D5	2.00	3.00	2.40	2.00	**1.00**	2.00	2.40	3.00	2.40	2.40	0.196
D6	1.00	2.40	0.50	1.00	0.50	**1.00**	1.00	2.40	2.40	0.42	0.090
D7	1.00	2.40	0.42	1.00	0.42	1.00	**1.00**	1.00	2.40	0.42	0.081
D8	0.33	0.50	0.33	0.50	0.33	0.42	1.00	**1.00**	1.20	0.33	0.047
D9	0.42	0.42	0.33	0.42	0.42	0.42	0.42	0.83	**1.00**	0.42	0.043
D10	2.40	2.40	1.00	2.40	0.42	2.40	2.40	3.00	2.40	**1.00**	0.152
											Σ =1.00

λ_{max} = 11.463, Ci = 0.0515, Ri = 1.49, Cr = 0.0345 < 0.1, OK

Table 14. Pairwise comparison matrix for community characters

CC	D1	D2	D3	D4	D5	D6	D7	D8	D9	D10	Priority vector
D1	**1.00**	2.00	1.40	1.00	0.42	1.00	1.00	3.00	2.40	3.00	0.118
D2	0.50	**1.00**	1.00	0.50	0.33	0.42	0.42	3.00	2.40	2.40	0.080
D3	0.71	1.00	**1.00**	1.40	0.42	0.42	0.42	3.00	3.40	2.40	0.099
D4	1.00	2.00	0.71	**1.00**	0.42	1.00	1.00	2.00	2.00	2.40	0.104
D5	2.40	3.00	2.40	2.40	**1.00**	2.40	2.40	3.00	2.00	4.00	0.212
D6	1.00	2.40	2.40	1.00	0.42	**1.00**	1.00	2.00	2.40	3.50	0.127
D7	1.00	2.40	2.40	1.00	0.42	1.00	**1.00**	2.00	2.00	2.00	0.118
D8	0.33	0.33	0.33	0.50	0.33	0.50	0.50	**1.00**	0.50	2.00	0.049
D9	0.33	0.33	0.25	0.50	0.50	0.33	0.50	2.00	**1.00**	2.00	0.057
D10	0.33	0.42	0.42	0.42	0.25	0.29	0.50	0.50	0.50	**1.00**	0.038
											Σ =1.00

λ_{max} = 10.446, Ci = 0.0519, Ri = 1.49, Cr = 0.034 < 0.1, OK

Having made all the matrices from Table 5 to 14, we can calculate the score of priority for each drainage. The overall priority score of drainage D1 can be computed and is given below as an example, whereas Table 15 shows the summary for computation of the scores for all alternative drainages.

The overall priority score of drainage D1 =

(0.295 X 0.252 X 0.55) + (0.295 X 0.278 X 0.059) + (0.295 X 0.470 X 0.089) +
(0.548 X 0.529 X 0.128) + (0.548 X 0.301 X 0.093) + (0.548 X 0.301 X 0.177) +
(0.157 X 0.706 X 0.092) + (0.157 X 0.262 X 0.118) = 0.094

Table 15. Summary of score computation

	Technical aspects (0.295)			Social aspects (0.548)			Culture aspects (0.157)		Overall priority vector
	DDL	DS	CI	FAP	TD	EAD	PP	CC	
	(0.252)	(0,278)	(0.470)	(0.529)	(0.301)	(0.177)	(0.706)	(0.262)	
D1	0.055	0.059	0.089	0.128	0.093	0.057	0.092	0.118	0.094
D2	0.097	0.141	0.096	0.045	0.099	0.060	0.058	0.080	0.076
D3	0.213	0.141	0.212	0.189	0.214	0.170	0.155	0.099	0.184
D4	0.106	0.059	0.081	0.076	0.071	0.058	0.086	0.104	0.077
D5	0.067	0.141	0.069	0.072	0.087	0.113	0.196	0.212	0.102
D6	0.077	0.067	0.061	0.162	0.116	0.202	0.090	0.127	0.120
D7	0.156	0.065	0.175	0.089	0.150	0.134	0.081	0.118	0.119
D8	0.040	0.147	0.040	0.042	0.045	0.056	0.047	0.049	0.052
D9	0.056	0.123	0.049	0.088	0.055	0.096	0.043	0.057	0.071
D10	0.131	0.059	0.127	0.109	0.070	0.055	0.152	0.038	0.099
									Σ =1.00

Sensitivity analysis is made by neglecting the culture aspects. This is done with the consideration that the judgement for the culture aspects are quite difficult if compared for the other aspects. Also the culture aspects have the least weight (15.7%). Hence, we have only 2 criteria, i.e. technical and social aspects, in the second level of the hierarchy. With this change, we find that ranks 1, 2 and 3 are still consistently unchangeable, i.e. D3 = Dr. Mansur, D6 = Sei Mencirim, D7 = Darussalam, consecutively. In other words, the top three are among the most favorable to have rehabilitation. Also we do computation using Expert Choice. We find that the same results are obtained.

4. Concluding remarks

We have shown the applicability the AHP as a potential decision-making method in water resource management related to the drainage rehabilitation. As a study case, we are able to rank 10 urban drainages in the Medan Baru district in Medan City for rehabilitation.

Based on the sensitivity analysis made, it is observed that the top three ranked drainages are consistent. Also both results of the AHP methods, calculated manually and performed using Expert Choice, are in agreement. It is hoped that the application of the AHP will help the decision-making process by the city government in the problem of urban drainage rehabilitation.

References

[1] Kamil I, et al. 2013 *Design of performance evaluation tools for drainage of roads system in developing country (case study: drainage system for city roads in padang indonesia). Proc. of the Int. Sym. on the AHP* (Kuala lumpur, Malaysia)

[2] Huang H.F, Chen H C and Liu S 2015 *Management strategies for Taiwan shimen reservoir catchment area: perspectives of collaborative planning. Int. J. of the AHP.* vol.**7(1)** 121-137

[3] Saaty T L 1980 The Analytic Hierarchy Process (New York: Mc Graw Hill)

[4] Saaty T L 1985 Decision making for leaders (Belmont, California)

[5] Saaty T L 1990 *How to make a decision: the AHP. European j. of oper. research* (North-holland) **48** 9-26

[6] Saaty T L and Kearns K P 1991 *Analytical planning: the organization of systems*. The analytic hierarchy process series 1991 **4RWS** (Pittsburgh,USA)

[7] Triantaphyllou E and Mann S H 1955 *Using the Analytic Hierarchy Process for decision making in Engineering Applications: Some Challenge. Int. J. of Industrial Engineering: Applications and Practice*, vol.**2**(1) pp. 35-44

IOP Conf. Series: Materials Science and Engineering **309** (2018) 012096 doi:10.1088/1757-899X/309/1/012096

[8] BPS Medan City 2016 District of Medan Baru Statistics (Medan, Indonesia)
[9] Al harby K M A S 2001 *Application of the AHP in project management. Int. J. of Project Managemen* Vol.**19(1)** 19-27
[10] Tam MCY and Tummala V M R T 2001 *An application of the AHP in vendor selection of a telecommunications systems. Omega Int. jour. of management science* **29** 171-182
[11] Tamin O Z, Najid, Sjafruddin A and Santoso I 2005 *Determination priority of road improvement alternatives based on region optimization. Proc. of the Eastern Asia Society for Transportation Studies*, vol.**5** pp 1040-1049 (Bandung, Indonesia)

IOP Conf. Series: Materials Science and Engineering **309** (2018) 012097 doi:10.1088/1757-899X/309/1/012097

Multistage switching hardware and software implementations for student experiment purpose

A Sani, Suherman*
Electrical Engineering Department, Universitas Sumatera Utara, Indonesia

*E-mail: suherman@usu.ac.id

Abstract. Current communication and internet networks are underpinned by the switching technologies that interconnect one network to the others. Students' understanding on networks rely on how they conver the theories. However, understanding theories without touching the reality may exert spots in the overall knowledge. This paper reports the progress of the multistage switching design and implementation for student laboratory activities. The hardware and software designs are based on three stages clos switching architecture with modular 2x2 switches, controlled by an arduino microcontroller. The designed modules can also be extended for batcher and bayan switch, and working on circuit and packet switching systems. The circuit analysis and simulation show that the blocking probability for each switch combinations can be obtained by generating random or patterned traffics. The mathematic model and simulation analysis shows 16.4% blocking probability differences as the traffic generation is uniform. The circuits design components and interfacing solution have been identified to allow next step implementation.

1. Introduction

Switching system enables terminal equipment such as telephone, facsimile, and computer to be connected to the same equipment in other places [1, 2]. Switching system is the basic technology for both telephone system and internetworking.

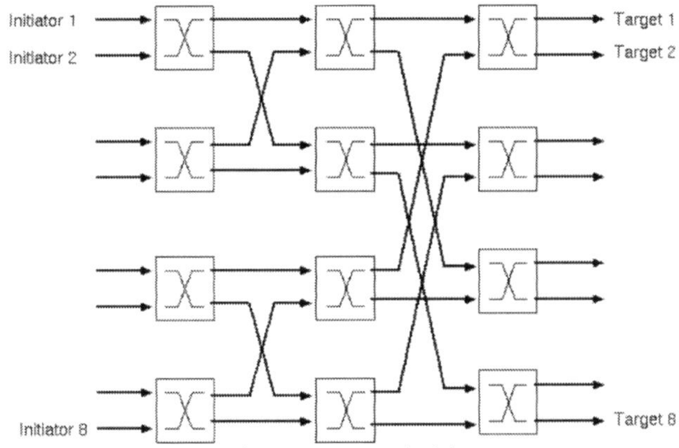

Figure 1. Threestage Switching System

One of the most famous switching architecture is threestage Clos switching system [3]. The sample of the switching architecture is shown in Figure 1. Number of input of the first stage is N,

IOP Conf. Series: Materials Science and Engineering **309** (2018) 012097 doi:10.1088/1757-899X/309/1/012097

number of input the second stage is K, then the switching matrix in the first, second and the third stages are NxK, KxK and KxN. The probability of blocking is approximated by Lee [4]:

$$P_B = \frac{(u!)^2(2-p)^{2u-m}p^m}{m!(2u-m)!} \quad [1]$$

where u=n-1, n is number of input in the first stage module, m is number of switching modules in second stage, p is the probability of an input link busy.

Since then, the switching research and development are improved rapidly [5]. However, the academics approach for laboratory activities are limited to a sample of mini switching system that is hardly representing the way those switches working, or analytic approach containing mathematic and simulation, that are also hardly showing how the switch works.

This paper reports the initial work on designing a multistage switching system that is aimed for student's laboratory. The switch is designed based on the threestage Clos switching system and made of digital circuits and analog switch IC's that are controlled by the Arduino. Initial designed presents the circuit analysis using Lee approximation, circuits' performance evaluation using discrete event simulation, circuits interfacing problem identifications and interfacing solutions.

2. Research Methods

2.1 Hardware Design

In order to realize the multistage switching system, the following devices are required:

- Switch to connect input and output ports. In order to operate in both analogue and digital technologies, the analogue switch integrated circuit that is controlled by digital signal is needed.
- Latch is employed to buffer digital output, programmed by Arduino and drive the switches.
- Decoder is employed to allow Arduino choose which switch is programmed by sending 8 bits control signal.
- Arduino is employed as the switching controller

The designed circuit blog is shown in Figure 2. The Arduino programs the switch modules by sending 8 bits to latch. As there are multiple switch modules, the programmed latch is chosen by using a decoder.

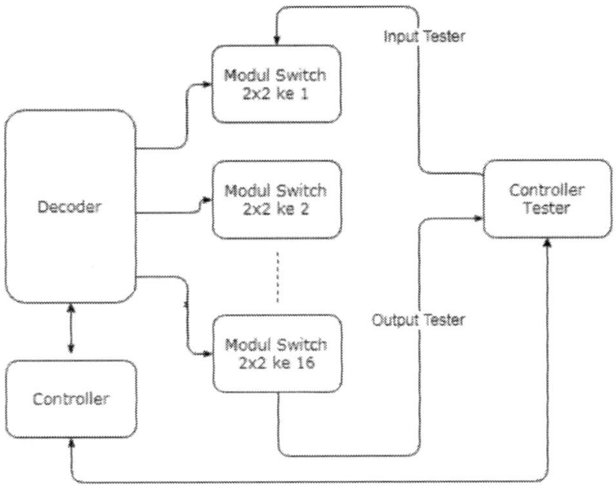

Figure 2. Circuit Block Diagram

2.2 Software Design

Software design for Arduino is shown in Figure 3. Arduino as the processor and the controller receives connection instructions from the controller or test circuit which is separately reported. The Arduino program should perform the following function:

- Configure the selected switch depending NxK size.
- Choose the correlated or the alternative switch route to connect the assessed input and output.
- Response to next connection without loosing the existing connection

2.3 System Simulation

Simulation is conducted by using Java programming language by:

- Identifying switches for any input and output connections.
- Generating traffic based on offered traffic in Erlang.
- Evaluating the successful and the blocking rates.

3. Current Results

The results reported by the time this paper written includes the components that have been selected, potential interfacing problems that have been resolved, and simulations that have been performed.

3.1 Circuit Components

HCF4066B [6] is selected as the switch IC in DIP package. HCF4066B is a quad bilateral switch as shown in Figure 3a.

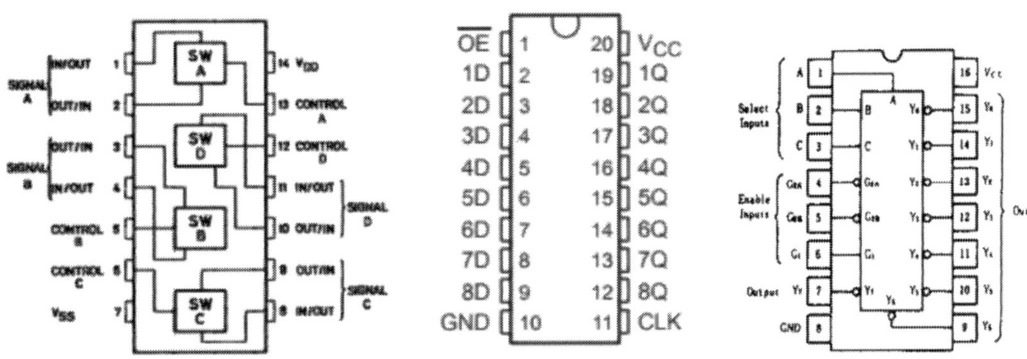

(a) (b) (c)

Figure 3. HCF4066B, SN74HC574 and HD74LS138

IC SN74HC574 [7], a D-type flip-flop octal edge-triggered with 3-state output is employed for driving IC HCF4066B. SN74HC574 receives data when clock transition is from low to high. Pin output-enable (OE) is used to activate output. Figure 3b shows the schematic diagrams for IC SN74HC574.

HD74LS138 [8] is chosen as the decoder to allow Arduino selecting switch modules to be programmed by changing latch states. This IC decides which pin is active by using 3 digit binary numbers in select pins. The two active-low and one active-high enable-pins is used to cascade more than one decoder. Finally, the Arduino UNO is selected as the processor and the controller.

3.2 Potential Problems

The main problem when working with digital ICs is the interfacing between ICs with different technologies such as CMOS (complementary metal oxide semiconductor) and TTL (transistor-transistor logic).

As HCF4066B is a CMOS IC and SN74HC574 is a TTL IC, the input and output voltage differences require a specific pin connection. Table 1 shows the voltage differences.

Table 1. Interfacing 74574 and 4066

SN74HC574 Output	HCF4066B Input
High=4,3-5V	High= V_{cc}
Low= 0-1,8V	Low= V_{DD}

Table 2. Interacing 74138 and 74574

HD74LS138 Output	SN74HC574 Input
High=2,7-5V	High= 4,2-5V
Low= 0-0,5V	Low= 0-1,8V

Even dough, SN74HC574 specification with voltage supply from –0.5 V to 7 V, output clamp current reaches 20 mA, continuous output current ±35 mA, and continuous current through VCC achieves ±70 mA. By considering these values and control current for switch HCF4066B less than 10uS, the SN74HC574 can be connected directly to HCF4066B. It is supported the fact that the high-current 3-State Non inverting Outputs of SN74HC574 is able to drive bus line up to 15 LSTTL ICs. So the interfacing solution is depicted in Figure 4a.

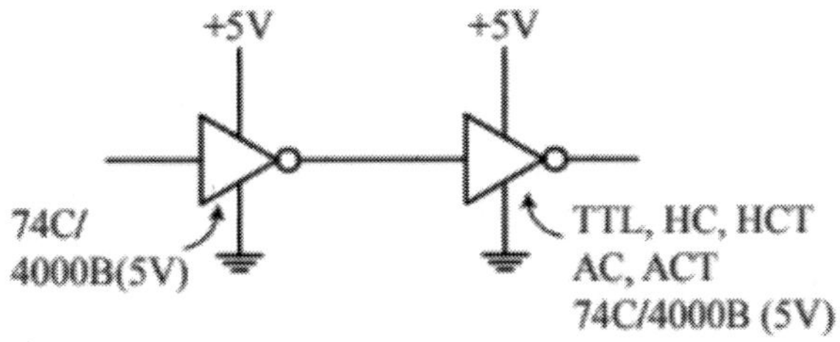

(a). Interfacing 74 HC ke TTL [9]

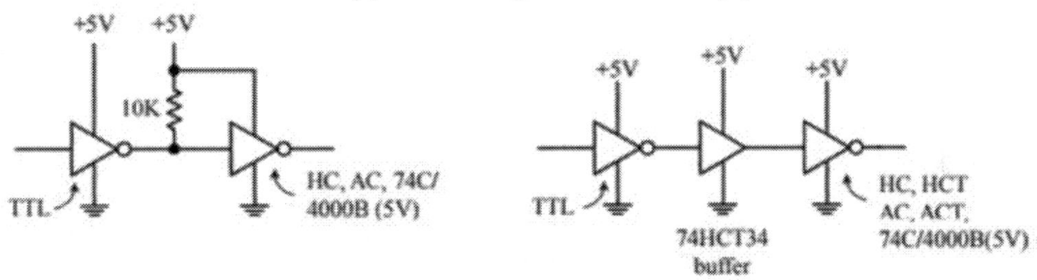

(b). Interfacing LS TTL to HC CMOS

Figure 4. IC Interfacing [9]

The same analysis falls to HD74LS138 and SN74HC574 interfacing with voltage levels as shown in Table 2 and a solution as depicted in Figure 4b. Meanwhile, the interfacing between the Arduino UNO and SN74HC574 as well as the Arduino and IC HD 74LS138 can be connected directly.

3.3 Analytical and Simulation Evaluations

By using Lee approximation and a discrete event simulation to a sample of three stage clos switching system as depicted in Figure 5, which have the plotted switch alternatives for each connection.

IOP Conf. Series: Materials Science and Engineering **309** (2018) 012097 doi:10.1088/1757-899X/309/1/012097

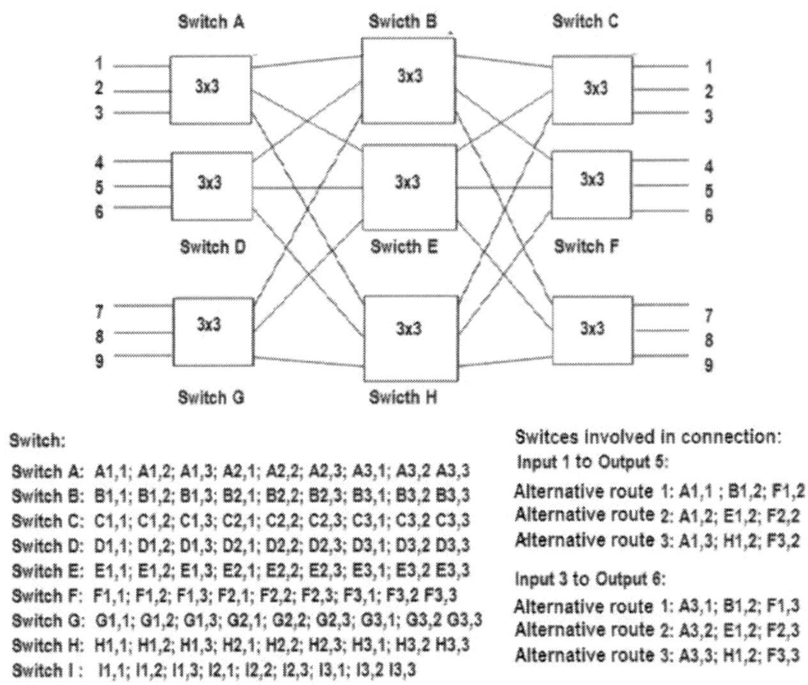

Figure 5. 9x9 Switching System

An offered traffic in each link generates the probability of a link busy, p. Every time a traffic comes, system should decides which route is taken by considering free switches status. If one of the switches in the first route is occupied, system checks the next alternatives. The generated traffic should not request connection to the busy output as the blocking assessment is only to switching blocking rate. By doing so, adjusting offered traffic from 0.1 to 1 erlang, the results are plotted in Figure 6. Despite 16.4% deviation, the simulation results pattern is matched to Lee approximation.

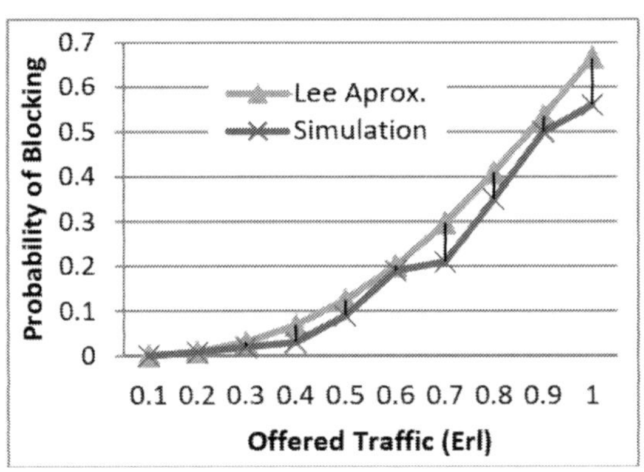

Figure 6. Simulation Results

4. Conclusions

This paper has presented the initial result in designing multistage clos switching system for student's laboratory activities. Potential components and problems have been identified and initial analysis as well as simulation has been carried out. Although the last generated 16.4% deviation, the model and simulation patterns are inline.

Future work is to implement the selected components to electronic board. Changes may apply depending on the circuit performances.

Acknowledgement
This research has been supported by the Penelitian Produk Terapan, funded by DIKTI through the grant No 38/UN5.2.3.1/PPM/KP-DRPM/2017.

References

[1] Schwartz M 1977 Computer-communication network design and analysis. Vol. 25. Englewood Cliffs, NJ: Prentice-Hall
[2] Suherman R F 2006 *Jaringan Telekomunikasi.* Jurnal Telekomunikasi, 2-20
[3] Charles C 1953 *A Study of Non-Blocking Switching Network*, The Bell System Technical Journal, March
[4] Hui J Y 1990 Switching and Traffic Theory for Integrated Broadband Networks, Kluwer Academic Publishers, 1990
[5] Venghaus H, Grote N 2017 Fibre Optic Communication: Key Devices. Vol. 161. Springer
[6] HCF4066B datasheet, STMicroelectronics, 2004.
[7] SN74HC574 datasheet, Texas Instruments, 2017.
[8] HD74LS138 datasheet, Renesas Technology Corp, 2005.
[9] Scherz P, Monk S 2013 Practical Electronics for Inventors, third edit., McGraw Hill,

IOP Conf. Series: Materials Science and Engineering **309** (2018) 012098 doi:10.1088/1757-899X/309/1/012098

Supply of avocado starch (*Persea americana mill*) as bioplastic material

M H S Ginting*, R Hasibuan, M Lubis, F Alanjani, F A Winoto and R C Siregar
Department of Chemical Engineering, Universitas Sumatera Utara, Jalan Almamater, Medan, 20155, Indonesia

*E-mail: hendra.ginting@usu.ac.id

Abstract. The purpose of this study was to determine the effect of time precipitation of avocado slurry seed to yield of starch. Starch analysis included starch content, moisture content, amylose content, amylopectin content, ash content, protein content, fat content, Fourier transform infra red analysis and rapid visco analyzer. Supply of starch from avocado seeds was used by extraction method. Every one hundred grams of avocado slurry was precipitated by gravity with variations for 4 hours, 8 hours, 12 hours, 16 hours, 20 hours and 24 hours. The Starch yield was washed, and dried using oven at 70°C for 30 minutes. Starch yield was the highest as 24.20 gram at 24 hours. The result of starch characterization was 73.62%, water content 16.6%, amylose 0.07%, amylopectin 73.55%, ash content 0.23%, protein content 2.16%, fat content 1.09%. Rapid visco analyzer obtained at 91.33°C of gelatinization temperature. Scanning electron microscopy analyzes obtained 20 μm oval-shaped starch granules. Fourier Transform Infra Red analysis of starch obtained the peak spectrum of O-H group of alcohols, C-H alkanes and C-O ether.

1. Introduction

The use of plastic packaging is increasing because it has advantages: lighter, not easily broken, and durable when compared to metal or glass materials [1]. Plastics have weakness: difficult to biodegradable, can not be renewed, so it need environmentaly plastic materials (bioplastic) [2]. Bioplastics are made from renewable resources (starch) and biodegradable [3]. Starch is energy stored by green plants. Bioplastics material are made from potato starch, jackfruit seeds, sago, maize, avocado seeds and durian seeds [4-10]. Avocado seeds are easy to obtain (waste) and have a high starch content. Avocado seeds are extracted using water. The purpose of this research was to determine the effect of time precipitation of avocado slurry to yield of starch.

2. Material and Method

2.1 Material

Avocado seeds are obtained from the trader of juiceavocado in Binjai, Province Sumatera Utara, Glasial acetic acid, Chitosan, Ethylene glycol and Formic acid was obtained from UD. Rudang Jaya Sumatera Utara.

2.2 Extraction an avocado starch.

The an avocado seeds are cut thin [8, 11] with a thickness of ± 2 mm [12], Repeatedly rinsed with water until clean and dried under the sun for 6 hours. The dried avocado seeds are blended by adding

1:5 (w/v) water to form the slurry. The avocado seeds were taken the filtrate. Starch suspension was cooled and precipitated by time variation. Wet starch dried in an oven at a temperature of 50°C [9] for ± 24 hours to dry [13]. Starch dried ground and sieved to be a size of 100 mesh [10, 14].

2.3 Characterization of starch avocado seed
2.3.1 Starch, amylose and amylopectin
Analyzing of starch, amylose and amylopectin content was observed at Laboratorium Baristan Medan.

2.3.2 Ash content
Procedure of analyzing water content from avocado seed starch applied standard of AOAC: Sampel was scaled as much as 5 grams to be entered into porcelain dish. Dish filled with sample was fired on flame of bunsen burner untill no haze appeared anymore. Then entered into furnace with temperature 550°C for ± 12 hours. Dish filled with sample was cooled in the dessicator then scaled until the weight was constant.

2.3.3 Protein and Fat content
Analyzing of protein and fat content was observed at Laboratorium Jasa Uji Ilmu Fakultas Teknologi Industri Pertanian Universitas Padjajaran. Method used for this analyzing of protein content by method of Kjeldhal semimicro. Method used for this analyzing of fat content by direct extraction method utilizing soxhlet equipment

2.3.4 Fourier Transform Infra Red Analysis (FTIR)
FTIR Analysis at Research Laboratory Faculty of Pharmacy University of Sumatera Utara, Medan.

2.3.5 Gelatinization profile analysis
Analysis of gelatinization profile in Laboratory of test service Faculty of Agricultural Industrial Technology Padjadjaran, Bandung.

2.3.6 Scanning Elecron Microscopy (SEM)
Scanning Elecron Microscopywas conducted at Laboratory of Faculty of Mathematics and Natural Sciences, Universitas Negeri Medan.

3. Result and Discussion
3.1 Rendement of avocado starch
The effect of variation precipitation time toavocado starch yield is presented by figure 1.

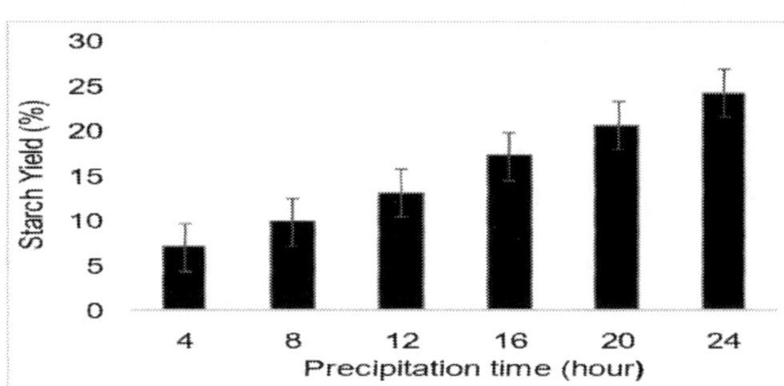

Figure 1. Precipitation time of starch yield

Figure 1 showed the longer precipitation time of starch slurry, the yield of starch is greater. Avocado slurry seed takes a long time to precipitat. The highest yield of avocado seed starch at 24 hours was 24.20%. One hundred grams of avocado seeds produce 24.20 grams of starch with precipitation time of 24 hours. Cinelli, et al.(2006) reported time of starch precipitationfor 24-48 hoursuntil the starch precipitated perfectly.

3.2 Characteristics of avocado starch.
Characteristics of starch avocado seed aims to determine the percentage of components contained in starch produced, Characterization include starch content (amylum), water content, ash content, fat content, protein content, amylose content and amylopectin. Components of starch is presented by table 1.

Table 1. Result of avocado starch characteristics

Avocado starch component	Level (%)	Indonesia Industrial Standards (%)
Starch (amylum)	73.62	min 75
Amylose	0.07	-
Amylopectin	73.55	-
Water	16.6	max 14
Ash	0.23	max 1.5
Fat	1.09	-
Protein	2.16	-

3.2.1 Starch content
The starch content is the quantity of starch contained in the dry matter expressed in percent. The purpose analysis of starch content to determine the percentage of starch/amylum (per unit of starch powder mass). The result of avocado starch content was 73.62%. Based on Indonesian Industry Standard (SII), the starch content was at least 75%, the starch content has approach SII. Difference of starch content is effected by the level of purity during the starch extraction process, in which the more the mixture in the starch powder obtained, such as the fibers, the sand/impurities then lower of starch content per mass unit [7].

3.2.2 Amylose and Amylopectin content
The purpose of this analysis to determine amylose and amylopectin content at avocado seeds. Starch consists of two types of polysaccharides: amylose and amylopectin. Amylose hasseveral branch linear molecule chain, amylopectin is branch molecule chain. One hundred milligrams avocado starch was obtained amylose 0.07% and 73.55% amylopectin content. The ratio between amylose and amylopectin determines the physico-chemical properties of starch, such as gelatinization and retrodegradation [14]. Low amylose levels cause high viscosity and low pasting temperature, while high amylose levels cause low viscosity and high pasting temperatures [15].

3.2.3 Water content
The water content shows the quantity of water in a material, water bound or free water, compared to the mass of the material [16]. The purpose of this analysis know the water content in starch that affects the bioplastic characteristics. The moisture content of a material will determine the acceptability, freshness and durability of the material. The result of the water content is 16.6%. Based on Indonesian Industrial Standard (SII), water content is at least 14%. Water content not yet approaching SII.

3.2.4 Ash content
Ash content is the content in the form of chemical compounds and has an important role in determining the quality and resistance of a material [17]. Ash is an inorganic residue of burning organic matter. Ash content can be calculated based on the reduction of sample weight during the combustion process at high temperature (500-600°C) through the vaporation process of organic material. The higher ash content then the higher the mineral content of the food [15]. Ash content was 0.23%, Based on Indonesian Industrial Standard (SII), ash content was at least 1.5%. Ash content was already meets SII.

3.2.5 Fat content

The purpose of this analysis know the fat content of starches. Fat content obtained 1.09%. The desired fat content is as small as possible because it interferes with the gelatinization process, inhibiting the release of amylose from starch [18]. Increased fat content will significantly decrease the degree of starch gelatinization. The effect of oil/lubricant from fat, which resulted in decreasing product temperature and decreasing the degree of gelatinization.

3.2.6 Fourier Transform Infrared (FTIR)

Characteristics Fourier Transform Infraredof avocado starch is presented by figure 2

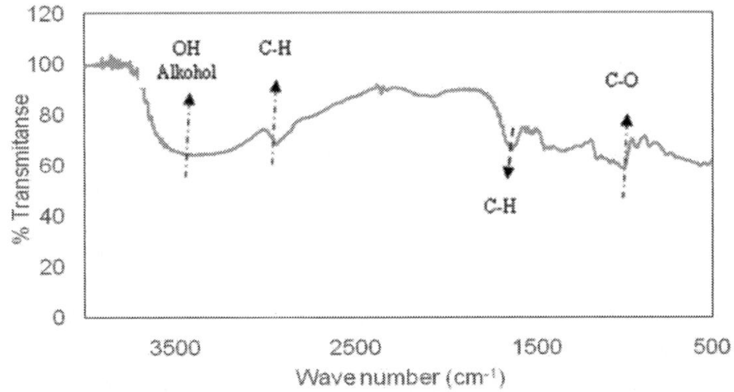

Figure 2. Fourier Transform Infra Red (FTIR)

The FTIR of avocado starchseed obtained by C-O eter, O-H, C-H alkane groups. The resulting functional groups have represented avocado starch such as amylose and amylopectin [19,20]. FTIR analysis wave number 1246.02 cm^{-1}, 3394.72 cm^{-1} and 2927.94 cm^{-1} showing C-O ether, O-H, C-H alkane groups.

3.2.7 Rapid Visco Analyzer (RVA)

Rapid Visco Analyzer (RVA) to determine the starch gelatinization profile avocado seed. Characterization related to the measurement of starch viscosity with a certain concentration during heating and stirring. The results of Rapid Visco Analyzerand measurement parameters are presented by figure 3 and table 2.

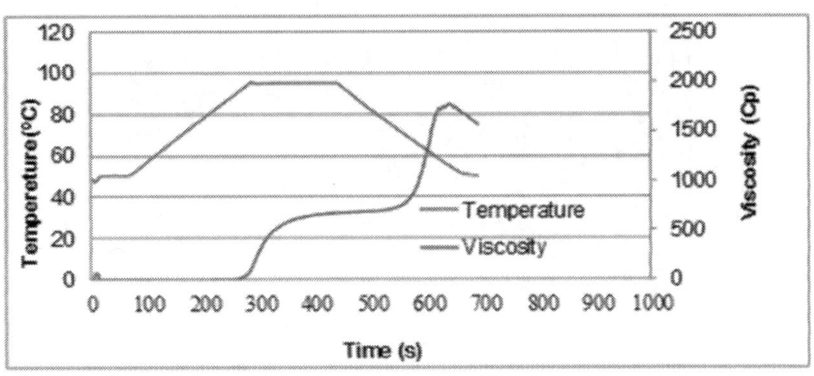

Figure 3. Rapid visco analyzer starch avocado seed

Table 2. Parameters measured RVA

Parameter	Avocado starch
Peak Viscosity	674 cP
Hold Viscosity	669 cP
Final Viscosity	1273 cP
Breakdown	5 cP
Setback 1	604 cP

Table 2 showed parameters measured pasting temperature, peak viscocity, hold viscocity, final viscocity, breakdown and setback viscocity. Gelatinization process heat the starch with excess water at a certain temperature so that there is an increase in swelling power and viscocity starch. Temperatures at viscosity increase and at start gelatinization is called pasting temperature. The optimum viscosity or Peak Viscosity (PV)is a parameter of the starch granule's ability to bind water and retain swelling during heating [21]. During the holding period, the sample is affected by the mechanical stress-strain properties that cause starch granules and the amylose leaching process. Viscosity during the holding period is called Hold Viscosity (HV). When samples are periodically cooled and increased viscosity is called Final Viscosity (FV) which is related to the process of retrogradation of amylose molecules [22]. Breakdown viscosity is the difference between PV and HV, stating the paste stability against heating. Breakdown causes to differences in viscosity when swollen, starch granules have been gelatinized and viscosity when the starch-grained granules are partially or partially disrupted [23]. Setback 1 is the difference between HV and FV, indicating the ability of the starch paste to be retrograde, the process of reforming the gelatinized starch matrix. The Setback viscosity implies degrees of retrodegradation [24]. The gelatinization temperature profile comprises several phases. The first phase began at a temperature below the starch gelatinization temperature, low viscosity and starch measurable starts accruing thermal treatment at a temperature of 48°C. The second phase, the temperature is gradually increased until it reaches a temperature of gelatinization, which is the temperature at which the starch granules begin to swell and the viscosity increases. Table 2 shows the temperature of avocado starch gelatinization of 91.33°C.

3.2.8 Scanning Electron Microscope (SEM)

Characteristics of Scanning Electron Microscope (SEM) avocado starch is presented by figure 4.

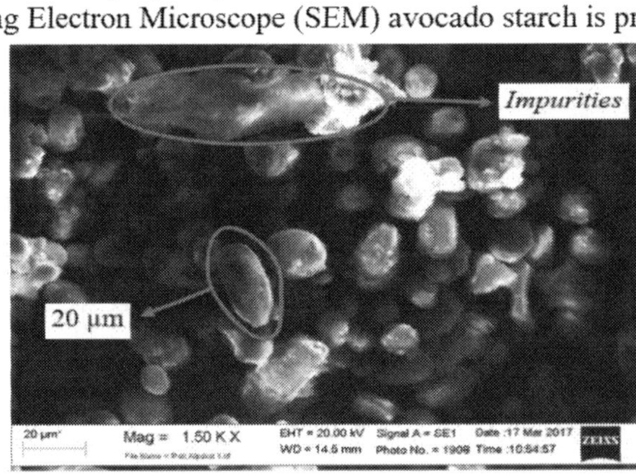

Figure 4.Scanning Electron Microscopeof Starch Avocado Seeds

Scanning Electron Microscope 1500x enlargement with starch size of 100 mesh. The analysis is showed the shape of rounded or elliptical starch granules and has a granule size of 20 µm. Form of oval starch granules, average granular size of 5-35 μm, Starch has a gap and is free of pores.Starch granules not yet modificated process have a smooth and intact surface[24].

4. Conclusion
The best yield of avocado seed starch is 24.20 gram at 24 hours deposition. Characterization of starch obtained 16.6% moisture content, ash content 0.23%, 73.62% starch content, Kadar 0.07% amylose, amylopectin content of 73.55%, 2.16% protein content, fat content of 1.09%, gelatinization temperature of 91.33°C peak viscosity of 674.5 cP.

Acknowledgments
The authors gratefully acknowledge that the present research is supported by Ministry of Research and Technology and the Higher Education Republic of Indonesia. This research grant under of Talenta USU Implementation of year 2017 Contract Number: 5338/UN5.1.R/ PPM/2017 date 22 May 2017.

References
[1] Terer E K and Magut H 2013 *International Journal of Green and Herbal Chemistry*. **2** 15
[2] Lu D R, Xiao C M and Xu S J 2009 *Express Polimer Letter*. **3** 366
[3] Kolusheva T and Marinova A 2007 *J. of the University of Chemical Tech and Metallurgy*. **42** 93
[4] Borghei, Mehdi, Karbassi A, Oromiehie A, and Javid A 2010 *African Journal of Biotechnology*. **9** 4075
[5] Zuraida A, Yusliza Y, Anuar H and Muhaimin H R 2012 *International Food Research Journal*. **19** 715
[6] Ginting M H S, Tarigan M F R, and Singgih A M 2015 *The Ijes*. **4** 36-43
[7] Ginting M H S, Hasibuan R, Lubis M, Tanjung D S and Iqbal N 2017 *IOP Conf. Ser : Mater. Sci. Eng*. **180** 012126
[8] Cinelli P E, Chiellini, Lawton J W and Imam S H 2006 *Polymer Degradation and Stability*. **11** 47
[9] Lubis M, Harahap M B, Manulang A, Alfarodo, Ginting MH S and Sartika 2017 *J. Physi:Conf*.Ser. **80** 012014
[10] Muaz A Z, Faiz M, Suffian M Y and Hamidi A A 2014 *Advances in Environmental Biology*. **8** 129
[11] Srianta, Hendrawan B, Kesumawati N and Blanc P J 2012 *International Food Research Journal*. **19** 941
[12] Mirhosseini H, Nur Farhana, R A, Bahareh T A, Cheong K W, Kazemi M Zulkurnain M 2015 *LWT - Food Science and Technology*. **63** 184
[13] Lopez, V M G *J. Agric. Food Chem.* **46** 3643
[14] Silva G P, Mack M and Contiero J 2009 *Biotechnology advances*. **27** 30
[15] Eduardo P C, Moisés M V, José M F and Socorro V R 2013 *The Scientific World Journal* **2013** 4
[16] Malangngi L P, Sangi M S, and Paendong J J E 2012 *J. MIPA UNSRAT*. **1** 5
[17] Kaur M, Kaushal P, and Sandhu K S 2013 *J. Food Sci. Technol*. **50** 94
[18] Han, X and Hamaker B 2001 *J. Cereal Sci.* **34** 279
[19] Sasaki T, Yasui T, and Matsuki J 2000 *Cereal Chem.* **77** 5863
[20] Agustin S 2011 *Teknol. Pertan*. **7** 28
[21] Beta T and Corke H 2001 *J. Cereal Sci*. **34** 261
[22] Pongsawatmanit R,Thanasukarn P, and Ikeda S 2002 *Science Asia*. **28** 129
[23] Sumnu G, Ndife M K, and Bayindirh L 1999 *Eur. Food Res Techonology*. **209** 68
[24] Wittaya T 2012 *Struct. Funct. Food Eng*. **12** 103

IOP Conf. Series: Materials Science and Engineering **309** (2018) 012099 doi:10.1088/1757-899X/309/1/012099

Palm ethyl ester purification by using *Choline Chloride* – 1,2 propanediol as deep eutectic solvent

R Manurung*, M A Alhamdi and A Syahputra
Department of Chemical Engineering, Faculty of Engineering, University of Sumatera Utara

*E-mail: renitachem@yahoo.com

Abstract. Deep eutectic solvent (DES) has gained more attention for using in biodiesel production because of environmental benefits and process improvements. This study was aimed to test the potency and effectiveness of Deep Eutectic Solvent (DES) based choline chloride: 1.2-propanediol as co-solvent in biodiesel purification. The method used in preparing DES synthesis process was conducted by mixing choline chloride: 1.2-propanediol with mole ratio variation such as: 1:2 ; 1:2.5 ; 1:3 ; and 1:3.5 (mole/mole). The temperature of DES synthesis was at 80 °C with 300 rpm stirring speed for 60 minutes. Variation of DES concentration base on percentage palm oil used: 1, 3, and 5 %. DES possible to increase the ethyl ester yield of biodiesel in the purification process. The best result of yield was 89.95% with the 9:1 molar ratio ethanol: oil and 5% of DES. The operation condition was at 70 °C of temperature reaction, 400 rpm of stirring speed, and 90 minutes of reaction time.

1. Introduction

Indonesia has the highest level of fuel consumption as a source of energy. The data show that oil consumption in Indonesia reached 363.52 million BOE (Barrel of Oil Equivalent), this figure is close to 36.41% of total energy consumption of 998.53 million BOE (Barrel of Oil Equivalent) [1]. Because energy is one of the important elements in human life, there is an increase in the consumption of fuel oil as a source of energy. Human population and industrial growth will also increase energy consumption. Fossil fuels are a non-renewable fuel and are limited in nature. Continuous use of fossil fuels will lead to a fuel crisis. Currently, Indonesia has developed various alternative energy sources as a substitute for fuel oil to solve the energy crisis problem that occurred. One of them is biodiesel, is mostly produced by a transesterification reaction, from oil with an alcohol and the presence of a catalyst to generate mono-alkyl esters and glycerol, which are then separated and purified [2]. Biodiesel is an alternative fuel can use as a substitute for fossil fuels by reacting oil and alcohol. When ethanol use for transesterification, the product is FAEE (Fatty Acid Ethyl Ester).

However, the utilization of ethanol also presents inconveniences. Effectively, it indicated in the literature, the base-catalyzed formation of the stable emulsion during *ethanolysis* is a problem [3].The resulting ethyl ester has advantages over the methyl esters, i.e., higher oxidation stability, lower iodine number, and better lubricity properties, ethyl ester also has a lower cloud point and pour point, which can increase engine life at temperature low, and a carbon atom in the ethanol molecule slightly increases combustion heat and cetane number, the studies performed show the emissions [4].

IOP Conf. Series: Materials Science and Engineering **309** (2018) 012099 doi:10.1088/1757-899X/309/1/012099

Development of low-cost and environmentally friendly solvents, known as Deep Eutectic Solvent (DES) is attractive because DES have several advantages over conventional ionic liquid and the organic solvents such easy preparation, low cost, low toxicity, and high biodegradability [5]. Recent studies had demonstrated the use of DES as co-solvent and catalyst in *methanolysis* can increase yield and make the purification easier [3]. Few studies reported the use of *ChCl* based DES with palm oil examples are the removal of an alkaline transesterification catalyst such as potassium hydroxide (KOH) and the separation of glycerol from biodiesel [6].

2. Material and Methods

2.1 Synthesis of DES

This study used a DES-based CHCl as ammonium salt and 1,2-propanediol as HBD. The synthesis of deep eutectic solvents (DES) performed with a fixed variable of 400 rpm stirring rate with a reaction temperature of 80 ° C for 2 hours [7] and the free variable was the molar ratio of ChCl: 1,2-propanediol 2: 1; 1.5: 1; 1: 1; 1: 1.5; 1: 2; 1: 2.5; 1: 3 and 1: 3.5.

2.2 Preparation of Biodiesel

The raw material used is crude palm oil and carried out pretreatment such as degumming with phosphoric acid. The result of this process is the degummed palm oil used for biodiesel production process. The transesterification reaction for biodiesel synthesis was carried out with a fixed variable of 400 rpm stirring at 70 ° C for 1 hour with a molar ratio of ethanol: DPO of 9: 1 and a catalyst concentration of 1.2% [2] and an independent variable of DES (1 , 3, 5%) and molar ratio of ChCl: 1.2-propanediol (1: 2; 1: 2.5; 1: 3 and 1: 3.5). In the biodiesel purification process, DES with a concentration is mixed into raw biodiesel and stirred for 5 minutes at 400 rpm.

2.3 Products Analysis

Analysis of the resulting biodiesel, Acid Numbers (AOCS Cd 3d-63 or SNI 7182-2015), Soap Numbers (AOCS Cd 3-25 or SNI 7182-2015), Total Glycerol (AOCS Ca 14-56 or SNI 7182 -2015) and analysis of ethyl ester content by using Gas Chromatography instrument (Type 122-5711, Durabond-5HT).

3. Result and Discussions

DES was use in this study were made from a mixture of choline chloride (ChCl) as quarternary ammonium salt and 1.2-propanediol ($C_3H_8O_2$) as Hydrogen Bond Donor (HBD) at different molar ratios. DES produced can be described in table 1 below.

Table 1. Encoding and Form of Chcl Based DES: 1,2-Propanediol

Molar Ratio ChCl: 1.2-Propanediol	Code	DES form (Room Temperature)
2.0:1	DES 1	Turbid White Solid
1.5:1	DES 2	Turbid White Solid
1:1.0	DES 3	Turbid White Solid with some Liquid
1:1.5	DES 4	Turbid White Solid with some Liquid
1:2.0	DES 5	Colorless Liquid
1:2.5	DES 6	Colorless Liquid
1:3.0	DES 7	Colorless Liquid
1:3.5	DES 8	Colorless Liquid

From table 1 above it can be seen that DES at molar ratio of 2: 1; 1.5: 1; 1: 1 and 1.5: 1 (DES 1, DES 2, DES 3 and DES 4) are not produced as good DES as solvents because of the ratio, the resulting product forms crystals or freezes throughout the process and after cooling to room temperature.

3.1 Effect of DES on Biodiesel Acid Numbers

Analysis of the effect of DES in the purification of biodiesel on the acid number of biodiesel is done by applying different amounts of DES with different ratios and comparing it with the acid number of biodiesel without DES. The value of the the acid number of biodiesel produced without DES is 0.00165 mg KOH / g oil. By using DES, the value of acid number of biodiesel can be see in the following figure.

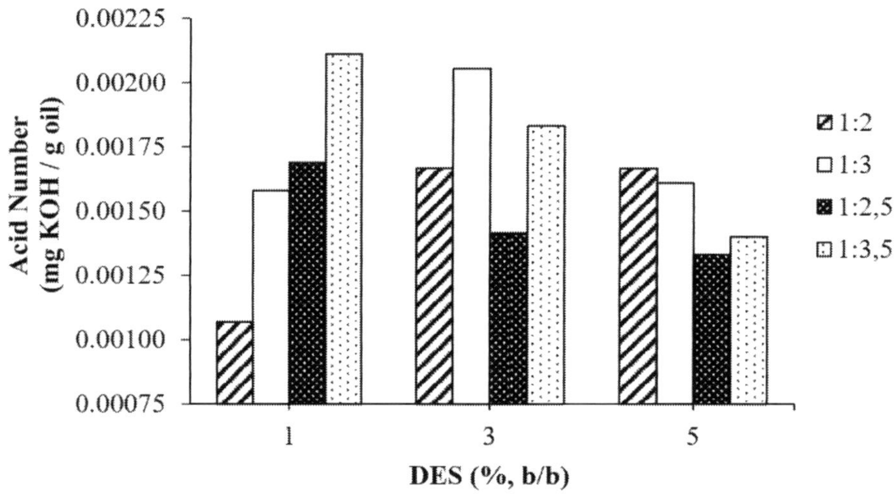

Figure 1. Effect of Total DES (%) on Acid Number of Ethyl Ester Biodiesel

From Figure1 it can be seen that at different molar ratios with the same amount of DES, acid numbers are fluctuating, except of 1% DES tends to increase as the molar increase of 1.2-propanediol HBD. In the 1: 2 molar ratio and 1: 3 the acid value increases when the addition of DES 3% and then decreases in the addition of 5% DES. In the 1: 2.5 molar ratio and 1: 3.5 the value of the acid number low with the addition of DES percent. The lowest acid number is at a 1: 2 molar ratio with 1% DES amount, while the highest acid number obtained at 1: 3.5 molar ratio with 1% DES amount.

Increasing acid numbers in some DES variations may be due to the pH of the DES it self (ChCl or 1.2-propanediol), such as ChCl having an acidic (pH 5). Then testing on some of the resulting DES, has an acidic pH as well, so it is likely to increase the acid number of biodiesel.

The acid number is the amount of KOH (mg) required to neutralize the free acid in the sample weight used. High acid numbers indicate the presence of oxidation reactions that occur. Because the peroxide compounds from the oxidation of oxidized unsaturated fatty acids form aldehyde compounds. The aldehydes will oxidize further to form carboxylic acids. The acid number is an important parameter as one of the indices of biodiesel quality. Because acidic numbers are potentially corrosive to vehicle engines, this is due to the high oxidation process [8].

3.2 Effect of DES on "Total Glycerol" Biodiesel

Analysis of the effect of DES in the purification of biodiesel to *total glycerol* biodiesel is done by applying different amounts of DES with different ratios and comparing it with *total glycerol* biodiesel without DES. *Total glycerol* of biodiesel produced without using DES is 0.2898%. Total glycerol of biodiesel can be see in the following figure 2.

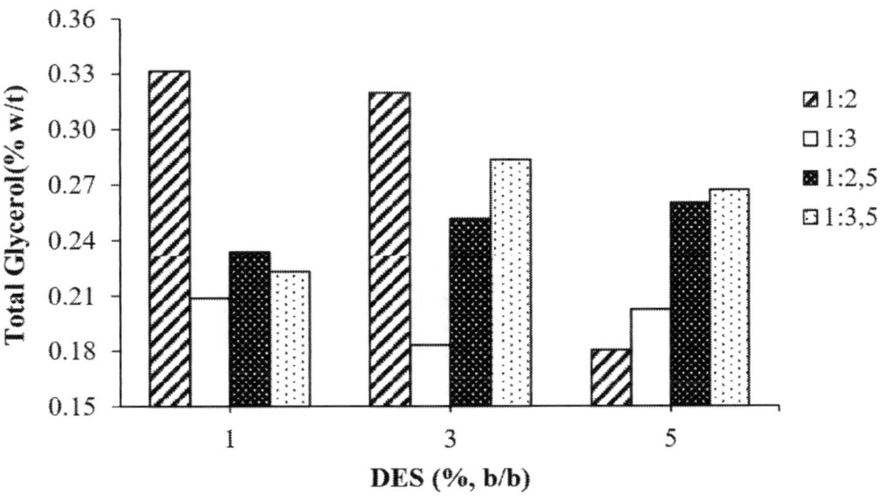

Figure 2. Effect of Total DES (%) on *Total Glycerol* Ethyl Ester Biodiesel

From Figure 2 it can be seen that between the amount of DES with *total glycerol* of biodiesel with some molar ratio has fluctuated, except in DES 5% which increases as the molar ratio of HBD increases. At a 1: 2 molar ratio the *total glycerol* value decreases with DES, whereas in the molar ratio of 1: 2.5 the decrease often increases the number of DES. The lowest *total glycerol* was at a 1: 3 molar ratio with 3% DES, while the highest total glycerol obtained at a 1: 2 molar ratio with 1% DES amount.

At the molar ratio of ChCl: 1.2 propanediol of 1: 3 are the physical properties of chemical DES produced is good. It means DES characteristics affecting the performance of solvent used in the separation of glycerol, which can be see from the *total glycerol* can be separated. The best *total glycerol* content is the lowest of 0.18%. This value obtained by the use of ChCl-based DES: 1.2 propanediol at a 1: 3 molar ratio. It shows the type of DES CHCl: 1.2 propanediol with a molar ratio of 1: 3 is the most appropriate solvent for the separation of ethyl esters of glycerol from biodiesel palm.

3.3 Effect of DES on Purity of Biodiesel

Determination of purity of biodiesel is done by standard quality analysis method for calculation of purity of biodiesel (level of ethyl ester). The method was done manually by method for titration and calculation by finding the acid number, saponification number and total glycerol of biodiesel, after which tested some result of biodiesel with Gas Chromatography as comparison. The purity data of ethyl ester biodiesel can be seen in Figure 3.

From Figure 3 can be seen that the purity of biodiesel produced using without using DES, obtained purity of 95.982% biodiesel. In the mole ratio of 1: 2 with the addition of DES of 1, 3 and 5% yielded purity of biodiesel respectively of 96.853; 96.875; And 98.880%. At a mole ratio of 1: 3 the resulting purity of 97.945 each; 98.001; And 97.790%. At a ratio of 1: 2.5 the resulting purity is 97.574; 97.414; And 96.804%. And at a mole ratio of 1: 3.5 the resulting purity is 85.88; 88.35; And 87.43%.

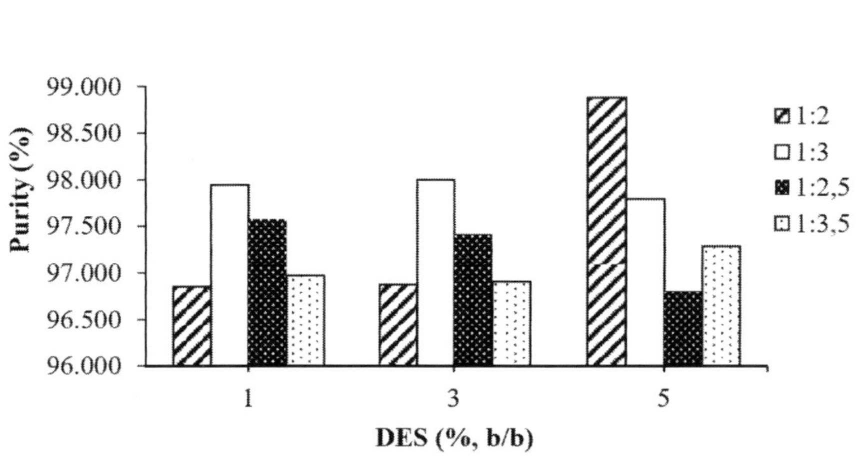

Figure 3. Effect of Total DES (%) on Ethyl Ester Biodiesel Purity

Testing by using Gas Chromatography instrument there is one sample, that is DES with molar ratio 1: 3.5 as much as 3% got ethyl ester level equal to 97.18%. While the method applied in this study obtained purity of 96.905%. So that the percentage of errata is 0.287%. With such a small percentage of errata, then the method used in this study can be done.

3.4 Effect of DES on Yield of Biodiesel
Analysis of the effect of DES on biodiesel yield is done by applying the various amount of DES and comparing it with biodiesel yield without DES. Can be seen in the following picture

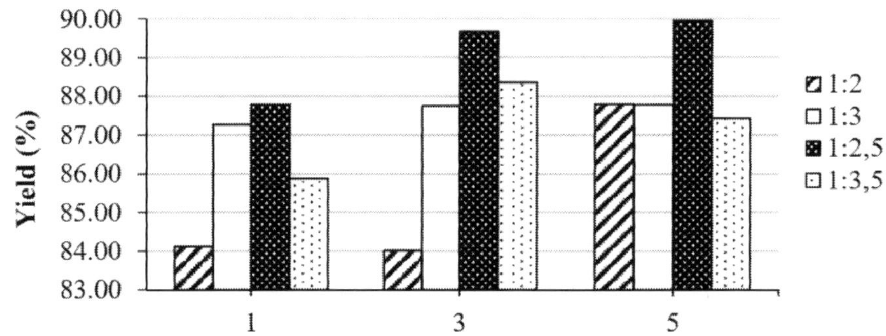

Figure 4. Effect of Number of DES (%) on Ethyl Ester Biodiesel Yield

From Figure 4 it can be seen that the yield of biodiesel produced using DES a fluctuating graph. Without using DES, obtained biodiesel yield of 84.46%. At a 1: 2-mole ratio with DES addition of 1, 3 and 5% biodiesel yields of 84.12; 84.01; And 87.78%. At a mole ratio of 1: 3 the resulting yield increases with the addition of DES. The yield of biodiesel produced was 87.27; 87.75; And 87.78%. At a ratio of 1: 2.5, yield increases along with the addition of DES. The resulting biodiesel yields were 87.78; 89.67; And 89.95%. And at a mole ratio of 1: 3.5 the yield is also increase along with the addition of DES up to 3%. Then on the addition of DES 5% biodiesel yield decreased. The yield of biodiesel produced is 85.88; 88.35; And 87.43%. The study showed that the addition of DES in amounts less than 5% (w / w) was able to increase the yield of biodiesel produced.

As shown in Figures 3 and 4, the purity of the ethyl ester biodiesel product is not directly proportional to the yield obtained. It is possible, due to the same behavior in each run of the experiment, because it should be if the resulting high purity, the yield obtained is also high.

3.5 Effect of DES on Biodiesel Washing Process
In this research, the method used in biodiesel washing process is wet washing method. After the biodiesel separation process with glycerol, biodiesel is washed with hot water to remove impurities such as alcohol residues, catalysts, glycerol which can reduce the purity and yield produced to clear water washing water.

Then also looks more turbid compared to using DES; this explains that in raw biodiesel there are still many impurities that do not separate with glycerol. Table 2. shows how much washing performed or water used for washing biodiesel From the table; it can be see that some washing of biodiesel with DES reaction with a molar ratio and a certain amount requires less water than without DES.

Table 2. Volume of Washing Biodiesel Water Conducted

Molar Ratio DES	% DES (b/b)	Volume (ml)
-	Without DES	360
	1	240
1:2.0	3	230
	5	200
	1	200
1:2.5	3	250
	5	265
	1	240
1:3.0	3	250
	5	230
	1	310
1:3.5	3	300
	5	300

Many refining or washing methods have been reported to purify biodiesel. The wet washing method is the most widely used method of purifying biodiesel because only by this can reduce a number of alcohol residues and glycerol in biodiesel [9].

The table above shows the amount of washing performed or the volume of washing water used for biodiesel washing. It can be see that biodiesel washing using DES requires little washing water compared with no DES. From the above research can be concluded that by using DES can accelerate the process of separation of alkyl ester with glycerol to minimize the use of washing water and facilitate the process of purification of biodiesel.

4. Conclusion

- Choline chloride-based DES: 1.2 propanediol has the potential for biodiesel purification process to increase the purity and yield of biodiesel by 6.5%.
- The highest ethyl ester purity of 98.880% obtained at 5% DES in the DES 1: 2 mol ratio. While the yield of ethyl ester was 89.95 obtained at 5% DES amount at a mol ratio of DES 1: 2.5.
- Choline chloride-based DES: 1.2 propanediol is potentially used as co-solvent in biodiesel purification, as it can speed up the process and save up to 45% of the washing water usage.

Acknowledgment
Many thanks to Chemical Engineering Department of Universitas Sumatera Utara, Medan, Indonesia

References
[1] Widayat, Wibowo A D K and Hadiyanto 2013 Study On Production Process Of Biodiesel From Rubber Seed (Hevea Brasiliensis) By In Situ (Trans)Esterification Method With Acid Catalyst *Journal Energy Procedia* **32** 64–73
[2] Manurung R, Bestari N G and Syabri D K 2017 Palm Biodiesel by Transesterification Process Using Choline Hydroxide as Catalyst *Malaysian Journal of Analytical Sciences 21* **1** 240-247
[3] Manurung R, Winarta A, Taslim, and Indra L 2017 Biodiesel Production from Ethanolysis of Palm Oil Using Deep Eutectic Solvent (DES) as co-solvent *IOP Conf. Series: Materials Science and Engineering* **206**
[4] Sánchez B S, Benitez B, Querini C A and Meadow G 2015 Transesterification of sunflower oil with ethanol using sodium ethoxide as a catalyst, Effect of reaction conditions, Fuel Processing Technology **131** 29-3
[5] Taslim, Indra L, Manurung R, Winarta A and Ramadhani DA 2017 Biodiesel Production from Ethanolysis of DPO Using Deep Eutectic Solvent (DES) based Choline Chloride-Ethylene Glycol as Co-Solvent *International Conference on Chemistry, Chemical Process and Engineering (IC3PE)*
[6] Shahbaz K, Mjalli F S, Hashim M A and AlNashef I M 2011 Eutectic Solvents for The Removal of Residual Palm Oil-Based Biodiesel Catalyst *J Sep. Purif. Technol.* **81** 216-222
[7] Hayyan A, Mjalli F S, AlNashef I M, Al-Wahaibi T, Al-Wahaibi Y M and Hashim M A 2012 Fruit Sugar-Based Deep Eutectic Solvents and their Physical Properties *Thermochimica Acta* **541** 70-75
[8] Gonzaga, Barbieri F and Sobral S P 2012 A New Method for Determining the Acid Number of Biodiesel Based on Coulometric Titratio *Chemical Metrology Division, National Institute of Metrology, Brazil Talanta* **97** 199-203
[9] Abbaszadeh A, Ghobadian B, Najafi G and Yusuf T 2014 An Experimental Investigation of the Effective Parameters on Wet Washing of Biodiesel Purification *International Journal of Automotive and Mechanical Engineering* **9** 1525-1537

IOP Conf. Series: Materials Science and Engineering **309** (2018) 012100 doi:10.1088/1757-899X/309/1/012100

Production of bioplastic from jackfruit seed starch (*Artocarpus heterophyllus*) reinforced with microcrystalline cellulose from cocoa pod husk (*Theobroma cacao L.)* using glycerol as plasticizer

M Lubis[1*], A Gana[1], S Maysarah[1], M H S Ginting[1], and M B Harahap[2]

[1]Department of Chemical Engineering, Faculty of Engineering, University of Sumatera Utara, Medan, 20155, Indonesia
[2]Department of Physics, Faculty of Mathematics and Natural Sciences, State University of Medan, Medan, 20221, Indonesia

*Email: maulida@usu.ac.id

Abstract. The production of bioplastic from jackfruit seed starch reinforced with microcrystalline cellulose (MCC) cocoa pod husk using glycerol as plasticizer was investigated to determine the most optimum mass and volume of MCC and glycerol in producing bioplastics. To produce MCC, Cocoa pod husk was subjected to alkali treatment, bleaching, and hydrochloric acid hydrolysis. The degree of crystallinity of MCC, were determined by XRD, functional group by FT-IR and morphologycal analysis by SEM. Analysis of bioplastic mechanical properties includes tensile strength and elongation at break based on ASTM D882 standard. Bioplastics were produced by casting method from jackfruit seed starch and reinforced with MCC from cocoa pod husk at starch mass to MCC ratio of 6:4, 7:3, 8:2, and 9:1, using glycerol as plasticizer at 20%, 25%, 30% (wt/v of glycerol to starch). From the result, the isolated MCC from cocoa pod husk were in a form of rod-like shape of length 5-10 μm with diameter 11.635 nm and 74% crystallinity. The highest tensile strength of bioplastics was obtained at starch to MCC mass ratio of 8:2, addition of 20% glycerol with measured tensile strength of 0.637 MPa and elongation at break of 7.04%. Transform infrared spectroscopy showed the functional groups of bioplastics, which the majority of O-H groups were found at the bioplastics with reinforcing filler MCC that represented substantial hydrogen bonds.

1. Introduction

In daily life, plastics are used almost everywhere in the world for various purposes because plastics are inexpensive, readily available, durable and versatile [1]. However, the main raw material in producing plastics derived from petroleum is declining and non-renewable. In addition, the plastics can not be destroyed quickly and naturally by destructive microbes in the soil. This causes the accumulation of waste, pollution and environmental damage [2]. The environmental, economic and safety challenges have prompted many scientists to replace part of a petrochemical-based polymer with another biodegradable type, called bioplastics [3]. Bioplastics are a renewable type of plastics because its constituents come from plants such as starch, cellulose, lignin, and animals such as casein, protein and lipids [4]. But there is still lack in the development of bioplastics as its high cost of production leads to certain limitations. An alternative for a low cost and a renewable substrate has been proposed by using agriculture waste (AW) [5].

One of the major components of bioplastics is starch. Starch is often used in the form of biodegradable films in a variety of applications as they are renewable, abundant and inexpensive materials [6]. In jackfruit contains about 100-500 jackfruit seeds or about 8-15% by weight of the jackfruit itself with a high starch content [7]. In producing of bioplastics, the addition of fillers is essential to increase the strength and toughness of the bioplastic products. The addition of fillers such as cellulose has been shown to be a very promising material [8]. One of the materials that have cellulose potential is cocoa pod husk which is an agro-industrial waste produced from the cocoa plant (*Theobroma cacao L.*). Cocoa pod contains 74% fruit peel, 2% placenta and 24% seeds [9]. Cocoa pod husk contains 11% hemicelluloses, 35% cellulose, 15% lignin, 6% pectin, and other mineral elements such as K (3.18%), Ca (0.32%) and P (0.15%) [10]. Also, the addition of plasticizer serves to increase the flexibility of bioplastic. Glycerol is a good plasticizer to reduce the internal hydrogen bonds that will improve the intermolecular distance [11].

2. Materials and Methods

2.1. Materials

Starch from jackfruit seeds obtained from jackfruit merchant in Langsa, Aceh, Indonesia. Cocoa pod husk as raw material for microcrystalline cellulose production is taken from Kampung Padang area, Sei Rampah, North Sumatera, Indonesia. Aquadest (H_2O), Sodium Hydroxide (NaOH), Sodium Hypochlorite (NaOCl), from Rudang Jaya Chemicals Store. Acid Chloride (HCl) from Laboratory of Microbiology Engineering, Chemical Engineering, University of Sumatera Utara, Indonesia.

2.2. Producing of Microcrystalline Cellulose from Cocoa Pod Husk

Cocoa pod husk powder of 100 g was inserted into beaker glass then added 1.5 L NaOH 4% and heated for 2 hours at 100°C. The residue is filtered and washed with aquadest to neutral pH. The residue is bleached with 2.5% sodium hypochlorite soaked 1 L for 24 hours at room temperature then filtered, and the residue washed with aquadest to neutral pH. The obtained residue added 17.5% NaOH as much as 650 ml then heated at 80 °C for 1 hour. The residue is filtered and washed with aqua dest to neutral pH. Bleaching with 500 ml of 2.5% sodium hypochlorite and heated at 100 °C for 5 minutes. The residue is filtered and washed with aqua dest until the pH is neutral. Dried in an oven at 60 °C for 12 hours. The obtained powder of microcrystalline cellulose was hydrolyzed using 2.5 N HCl by boiling for 10-15 minutes and filtered. The obtained residue is washed to neutral with aqua dest then dried and mashed.

2.3. Producing of Jackfruit Seed Starch

Jackfruit seeds as much as 100 g peeled parts of the outer shell, and its skin then cleaned with clean water. Seeds cut to the size of approximately one cm^2 then crushed using a blender with added water. The mixture is filtered until the resulting dregs and liquid filtrate (starch suspension). The obtained suspension is then deposited for 24-48 hours. The starch-rich liquid then filtered using Whatman filter paper no.1 to obtain wet starch. The precipitate obtained was dried in an oven at 70 °C for 30 minutes. Dried starch powder then sieved with a 100 mesh sieve.

2.4. Producing of Bioplastic

The amount of starch and microcrystalline cellulose mass is weighed by the variation of the 6: 4, 7: 3, 8: 2 and 9:1 ratio of 10 g of the total dry weight of starch to microcrystalline cellulose. Then the starch solution made with a ratio of starch: aqua dest 1:20 and microcrystalline cellulose solution with 5% NaOH solution by 40 ml. The starch solution was heated while stirrer was mixing for 10 minutes and then added glycerol with 20%, 25%, 30% variation of the total dry weight of starch to microcrystalline cellulose. Microcrystalline cellulose solution was then added to the mixture after 20 minutes. The mixture was then heated while stirred to a temperature 88,15 °C. The mixture is cooled and printed on a 25 x 25 x 3 mm acrylic mold. Bioplastic is dried in an oven at 60 °C for 24 hours. Bioplastic is removed from the oven and cooled to be removed from the mold.

2.5. Microcrystalline Cellulose Characterization

2.5.1. X-Ray Diffraction (XRD).
This analysis is performed to measure the crystallinity and the resulting microcrystalline cellulose diameter. This analysis was conducted at Physical Chemistry Laboratory, State University of Medan.

2.5.2. Scanning Electron Microscope (SEM).
This analysis aims to observe the morphology of microcrystalline cellulose from cocoa pod husk. This analysis was conducted at Integrated Laboratory, Diponegoro University, Semarang.

2.6. Bioplastic Mechanical Test

2.6.1. Tensile Strength.
Tensile strength was measured with GoTech Universal Testing Machine (UTM) using the standard of ASTM D882. Tensile strength was calculated as follow :

$$\text{Tensile Strength} = \text{Max Load} \times \text{Gravity} \quad (1)$$

2.6.2. Elongation at Break.
Elongation at break is an indication of bioplastics flexibility, and it's expressed as a percentage. Elongation at break was calculated as follows:

$$\text{Percent elongation (\%)} = \frac{\text{(Elongation at rupture)}}{\text{(Initial gage length)}} \times 100\% \quad (2)$$

2.6.3. Fourier Transform Infrared Spectroscopy (FT-IR)
Functional groups of bioplastics were analyzed by using IR Prestige-21 Shimadzu. This analysis was conducted at the Pharmaceutical Research Laboratory, University of Sumatera Utara.

3. Results and Discussions

3.1. Microcrystalline Cellulose Characterization

3.1.1. X-Ray Diffraction (XRD)

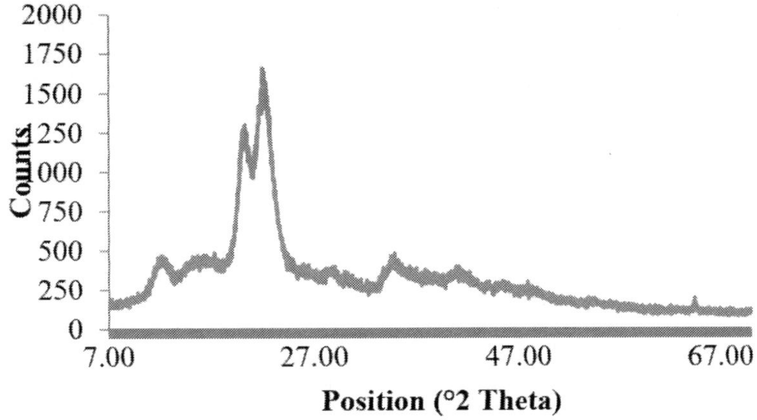

Figure 1. X-Ray Diffraction Graph of Microcrystalline Cellulose from Cocoa Pod Husk

From figure 1. It shows that the absorption peaks of the spectra generated by microcrystalline cellulose are at 2θ = 12.14°; 20.20° and 22.08°. The value of crystallinity was obtained by calculating the intensity of the XRD analysis using Segal method with the equation below [12] :

$$CrI = \frac{I_{002} - I_{AM}}{I_{002}} \times 100\% \quad (3)$$

The percentage crystallinity of microcrystalline cellulose from cocoa pod husk obtained is 74%, this value is much higher when compared with other non-wood materials (52-53%) [13]. The diameter of the crystal can also be calculated on the results of crystallinity analysis by X-Ray Diffraction using Scherrer's equation with the equation below [14] :

$$B(2\theta) = \frac{K\lambda}{L.Cos\ \theta} \quad (4)$$

The absorption peak of the spectra generated by microcrystalline cellulose is at 2θ = 12.14°; 20.20° dan 22.08°. By calculating the crystal diameter of microcrystalline cellulose at the apex of 2θ=22.08° using the Scherrer's equation, the diameter of the crystal was 11.635 nm.

3.1.2. Scanning Electron Microscope (SEM)

Figure 2. Microcrystalline cellulose with magnification 1000x

Figure 2. shows the SEM result of microcrystalline cellulose from cocoa pod husk with 1000x magnification. From the SEM analysis, it can be seen that most of the microcrystalline cellulose from cocoa pod husk morphology of rod-shaped is interconnected with a size of about 5-10 μm.

Microcrystalline cellulose morphology derived from sisal shows the same result, i.e., rod-shaped [15] [16]. Several pieces of micro fibers are found in the observed microcrystalline cellulose. Particle shape is an important determinant in determining the density of material. The particle shape of material also describes the porosity of the material. Particles of larger size have a lower porosity [17].

3.2. Bioplastic Mechanical Test Analysis

3.2.1. Tensile Strength

Figure 3. shows the highest tensile strength value is 0.637 MPa for composition ratio of starch: MCC 8:2 with the addition of glycerol 20 %. While the lowest tensile strength value is 0.147 MPa for composition ratio of starch: MCC 6:4 with the addition of glycerol 30%. The addition of microcrystalline cellulose of gelatinization starch film involved grouping of intermolecular hydrogen bonding that causes a molecular bond of amylose in starch more compact [18]. In this case homogeneity of bioplastic is the other important factor of bioplastic with a good characteristic. This homogeneity is related with the solubility of microcrystalline cellulose in NaOH as a solvent. Kontturi (2015) report that a part of cellulose crystalline could partially soluble in a solution of NaOH 5-20% after the accurate pretreatment than part of amorphous could soluble in NaOH 4% [19]. In this research, the solution of NaOH 5% (w/v) used as solvent.

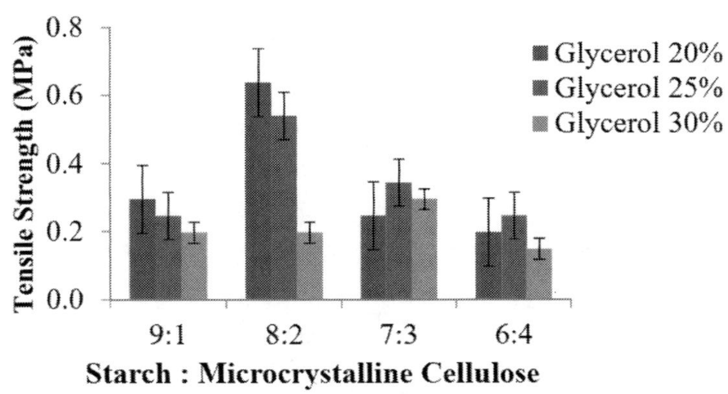

Figure 3. The effect of microcrystalline cellulose and glycerol addition on tensile strength of bioplastic from jackfruit seed starch

The amount of cellulose which is dissolved in a solution of NaOH-water depends on polymerization degree and also its cellulose crystallinity [20]. Based on the graphic, tensile strength at high filler loading (7:3 and 6:4) has declined because the presence of high content of MCC fillers might contribute in retarding the intermolecular interaction of the starch films. This induces the development of a heterogeneous film structure, featuring discontinuities, resulting in the decrease in tensile strength of films [8]. From figure 3. could be seen that the tensile strength value tends to drop with increasing of glycerol volume addition. Because more variation of glycerol added then the tensile strength get lower. This case causes by molecules of plasticizer will disturb compact of the starch, decreasing interaction of hydrogen bonding and increasing the polymer mobility [21]. Molecules of glycerol will disturb compactness of constituents molecules. This condition effected the increasing of edible film flexibility. After that, the elongation and tensile strength of edible film become decrease [22]. But, there is tensile strength in a composition ratio of starch: MCC 8:2 and glycerol 20% increased than it dropped again. Abdorreza et al (2011) stated that its deviation caused by the ability of plasticizer sustain crystallinity of bioplastic film. The flexibility could increase by the addition of plasticizer, but it also formed plasticizer crystal in film [23].

IOP Conf. Series: Materials Science and Engineering **309** (2018) 012100 doi:10.1088/1757-899X/309/1/012100

3.2.2. Elongation at Break

Figure 4. shows that by the increasing of microcrystalline cellulose mass then elongation at break value of bioplastic will be decreased. While by the increasing of glycerol volume addition, elongation at break value of bioplastic will be increased. The highest of elongation at break value is 15.76% for bioplastic with MCC content 4 g with glycerol 30%. The lowest of elongation at break value is 1.82% for bioplastic with MCC content 3 g with glycerol 20%.

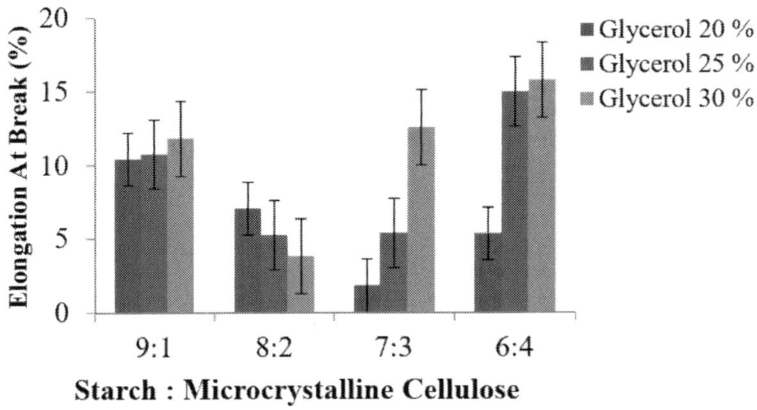

Figure 4. The effect of microcrystalline cellulose and glycerol addition on elongation at break of bioplastic from jackfruit seed starch

Based on the results, it shows that with increasing glycerol addition cause the percentage of elongation also increasing, because glycerol can increase the distance between molecules that make bioplastic become more elastic [24]. The percentage of elongation is inversely proportional to tensile strength. With more fillers added to the plastic film, then the elongation will be dropped, but the tensile strength is increased. A drop in elongation supposed happened because there is strong interaction between composition of bioplastic substance, that is starch molecules with fillers. This bonds happened between starch molecules with fillers is more tight and compact so that bioplastic is more strength that effected the film becomes difficult to stretchable and elongated, it will reduce the percentage of film elongation [25]. But, there is a deviation for composition ratio of starch: MCC 8:2 and the addition of glycerol 20 % and 30% where there is decreasing elongation at break value by glycerol addition. The addition of plasticizer in low concentration to medium concentration (1% - 25%) facilitated the formation of a crystal in a starch film, which is direct to antiplasticizer behavior. This is due to movement from polymer chain which is cause water molecules, and plasticizer disappear slowly, it impacted to the amylose and amylopectin form strong hydrogen bonding, so it happens recrystallization or retrogradation [26].

3.2.3. Fourier Transform Infrared Spectroscopy (FT-IR)

Figure 5 represented the characteristic of absorption peak FT-IR analyze of microcrystalline cellulose of cocoa pod husk, bioplastic from jackfruit seeds starch and glycerol without MCC, bioplastic from jackfruit seed starch with MCC and glycerol, and also jackfruit seeds starch. The dark blue line presented FT-IR spectra of jackfruit seeds starch with spectrum peak that indicate O-H group, C-H alkanes, C=C alkenes, C=O aldehyde and C-O eter in jackfruit seeds starch. This FT-IR result are consistent with Maulida research (2016) which is reported same function group with jackfruit seeds starch [27].

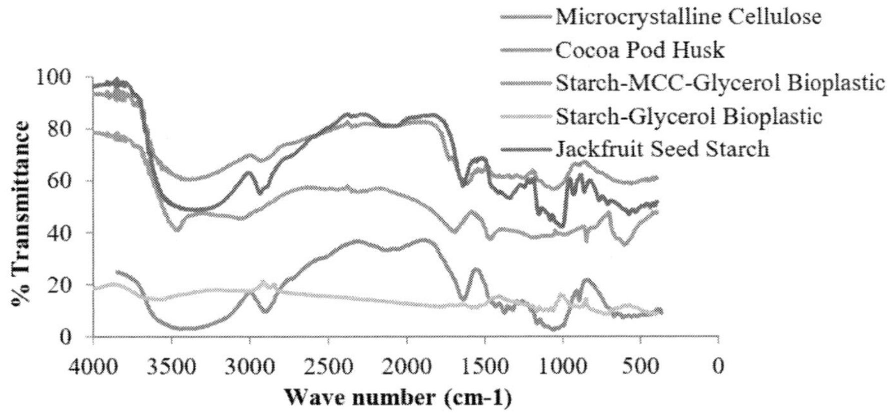

Figure 5. Characteristics of FT-IR spectra analysis

The presence of functional group has represented the contents of jackfruit seeds starch which is consisted of amylose and amylopectin and also reducing sugar $(C_6H_{10}O_5)n$. The blue line represented the characteristic of absorption peak of microcrystalline cellulose at wave number 3445 cm^{-1}, it indicates the existence of -OH groups and at absorption peak with wave number 2903 cm^{-1} indicate the existence of C-H groups which is proved there is C-H bonding at the end of cellulose structures and C-O carbonyl at wave number 1374.1 cm^{-1} is also typical cellulose. Orange line represents the characteristic of the absorption peak of cocoa pod husk which has many similarities with microcrystalline cellulose sample, because of microcrystalline cellulose result is chemical process from cocoa husk, so the groups indicated the existence of cellulose are in similar wave number. The appearance of absorption peak at cocoa husk is 3383 cm^{-1}, 2927 cm^{-1}, 1612 cm^{-1}, 1249 cm^{-1}, 1056 cm^{-1}, and 898 cm^{-1}. On starch-glycerol bioplastic with yellow line could be seen the absorption peak shows that OH groups, C-H alkanes, O-H carboxylates, C=O aldehyde, and C-O carboxylates, is same with bioplastic from jackfruit seed starch with MCC and glycerol with grey line, the appearance of absorption peak is not showing the existence of new groups formation. But, there is an enhancement of O-H bonds wave number of starch and microcrystalline cellulose 2931.6 cm^{-1} of starch, become 3001.04 cm^{-1} at bioplastic, and from 3445.71 cm^{-1} at microcrystalline cellulose become 3464.15 cm^{-1} at bioplastic.

4. Conclusion

The isolated MCC from cocoa pod husk is result in the form of rod-like shape 5-10 μm with diameter 11.635 nm and crystallinity 74%. The best condition of bioplastics obtained at the comparison of mass starch: MCC 8:2 and the addition of glycerol 20% for tensile strength 0.637 MPa and elongation at break 7.04%.

References

[1] Narissara K and Shabbir H G 2013 Greenhouse Gas Evaluation and Market Opportunity of Bioplastic Bags from Cassava in Thailand *Journal of Sustainable Energy & Environment.* **4** 15-19
[2] M. P. Cereda 2000 Characterization of Edible Films of Cassava Starch by electron Microscopy *Journal of Food Technology.* 91-95
[3] R. Laxmana Reddy, V. Sanjeevani Reddy and G. Anusha Gupta 2013 Study of Bio-plastics As Green & Sustainable Alternative to Plastics *International Journal of Emerging Technology and Advanced Engineering.* **3** (5) 82-89
[4] Luc A 2004 Biodegradable Multiphase System Based on Plasticized Starch : A Review *Journal of Macromolecular Science.* **44** (3) 231-274
[5] Jain R and Tiwari A 2015 Biosynthesis of Planet Friendly Bioplastics Using Renewable Carbon Source *J environ Health.* Sci Eng **13**
[6] J. S Alves, K. C. Reis, E.G. T. Menezes, F. V Pereira and J. Pereira 2014 Effect of Cellulose Nanocrystals and Gelatin in Corn Starch Plasticized Films *Carbohydr. Poly.*
[7] M. A. and K. Sajjaanantakul 2004 Physico-chemical Properties of Flour and Starch from Jackfruit Seeds (*Artocarpus heterophyllus Lam.*) Compared with Modified Starches *Int. J. Food Sci. Technol.* **39** 271-276
[8] Maulida, M. Siagian and P. Tarigan 2016 Production of Starch Based Bioplastic from Cassava Peel Reinforced with Microcrystalline Cellulose Avicel PH101 Using Sorbitol as Plasticizer *Journal of Physics: Conference Series* **710** (2016) 012012
[9] Aji, Dhimas Prasetyo, Sri Utami and Suparwi 2013 Cocoa Pod Husk (*Theobroma cacao L.*) Fermentation using *Aspergillus niger* and its Effect on VFA and N-NH_3 levels by In-Vitro *Jurnal Ilmiah Peternakan.* **1** (3) 774-780
[10] Hutomo, Gatot S., Djagal W. Marseno, Sri Anggrahini and Supriyanto 2012 Synthesis and Characterization of Sodium Carboxymethyl Cellulose from Pod Husk of Cacao (*Theobroma cacao L.*) *African Journal of Food Science.* **6** (6) 180-185
[11] T. H. McHugh and K. J. M 1994 Sorbitol vs Glycerol Plasticed Whey Protein Edible Film : Integrated Oxygen Permeability and Tensite Property Evaluatio *J. Agic Food Chem.* **2**, no. 4, 841-845
[12] Segal, L., J.J. Creely., A.E. Martin and C.M. Conrad 1959 An Empirical Method for Estimating the Degree of Crystallinity of Native Cellulose Using the X-Ray Diffractometer *Textile Research Journal*
[13] M. S. Jahan, A. Saeed, Z. He and Y. Ni 2011 Jute as Raw Material for The Preparation of Microcrystalline Cellulose *Cellulose*, **18**, no. 2, 451-459
[14] Leroy Alexander and Harold P. Klug 1950 Determination of Crystallite Size with the X-Ray Spectrometer *Journal of Applied Physics* **21**, 137
[15] Y. Habibi, L. A. Lucia and O. J. Rojas 2010 Cellulose nanocrystals: Chemistry, Self-Assembly, and Applications *Chem. Rev.* **110**, no.6, 3479-3500
[16] Sumaiyah, Basuki, Wirjosentono, Karsono, M. P. Nasution and Saharman G 2014 Preparation and Characterization of Nanocrystallone Cellulose from Sugar Palm Bunch (Arenga Pinnata (Wurmb) Merr.) *International Journal of Pharm Tech Research.* **6**, No. 2, pp 814-820
[17] N. A. Bhimte and P. T. Tayade 2007 Evaluation of Microcrystalline Cellulose Prepared from Sisal Fibers as a Tablet Excipient : A Technical Note *AAPS PharmSciTech,* **8**, no.1, pp. 1-7
[18] S. M. Lii, C.Y and Chang 1981 Characterization of Red Bean (Phaseoulus Radiatus Var. Aurea) Starch and Its Noodle Quality *J. Food Sci.* **46**
[19] E. Kontturi 2015 Cellulose : Dissolution," *Chem. Adv. Biomater.*
[20] C. Olsson and G. Westman 2013 Direct Dissolution of Cellulose : Background, Means and Applications
[21] M. Rodriguez, O. Javier, Z. Khalid and I. M. Juan 2006 Combined Effect of Plastizers and Surfactants on the Physical Properties of Starch Based Edible Film *J. Food Research International.* **39**, pp. 840-646

[22] R. Malaka, M. Taufik and U. Hasanuddin 2014 Effect of Percentage Variation of Glycerol as Plasticizer on the Mechanical Properties of Edible Film from Combination of Whey Dangke and Agar pp. 214–219

[23] M. N. Abdorreza, L. H. Cheng and A. A. Karim 2011 Food Hydrocolloids Effects of Plasticizers on Thermal Properties and Heat Sealability of Sago Starch Films *Food Hydrocoll.* **25**, no. 1, pp. 56–60

[24] Nathalie Gontard, Stephane Guilbert and Jean Louis Cuq 1993 Water and Glycerol as Plasticizer effect mechanical and Water Vapor Barrier Properties of an Edible Wheat Gluten Film *Journal of Food Science*. **58** (1) 1993 : 206-211.

[25] Setiani, W., T. Sudiarti and L. Rahmindar 2013 Preparation and Characterization of Edible Film of Polybags Bread Starch : Chitosan. *Journal of Valence Chemistry* **3**(2) : 100-109. Chemistry Department, Faculty of Science and Technology UIN Sunan Gunung Djati. Bandung.

[26] Y. Zhang and C. Rempel 2012 Retrogradation and Antiplasticization of Thermoplastic Starch in *Thermoplastic Elastomers Edited*, 7th ed., P. A. El-Sonbati and ISBN, Eds. Canada : InTech, pp. 117–134.

[27] M. Lubis, M. B. Harahap, A. Manullang, Alfarodo, M. H. S. Ginting, and M. Sartika 2017 Utilization Starch of Jackfruit Seed (*Artocarpus heterophyllus*) as Raw Material for Bioplastics Manufacturing Using Sorbitol as Plasticizer and Chitosan as Filler *IOP Conf. Series: Journal of Physics: Conf. Series* **801** (2017) 012014

IOP Conf. Series: Materials Science and Engineering **309** (2018) 012101 doi:10.1088/1757-899X/309/1/012101

Effect of cellulose nanocrystals from corn cob with dispersion agent polyvinyl pyrrolidone in natural rubber latex film after aging treatment

H Harahap*, M Ridha, Halimatuddahliana, Taslim and Iriany
Department of Chemical Engineering, Faculty of Engineering, University of Sumatera Utara, Medan, Indonesia

*E-mail: hamidah.usu@gmail.com

Abstract. This study about the resistance of natural rubber latex films using nanocrystals cellulose filler from corn cob waste by aging treatment. Corn cob used as organic filler composed of cellulose, hemicellulose, and lignin. Each component has a potential for reuse, such as cellulose. Cellulose from corn cob has potential application as a filler prepared by hydrolysis process using a strong acid. The producing of natural rubber latex films through coagulant dowsing process. This research started with the pre-vulcanization process of natural rubber latex at 70 °C and followed by process of vulcanization at 110 °C for 20 minutes. Natural rubber latex films that have been produced continued with the aging treatment at 70 °C for 168 hours. The mechanical properties of natural rubber latex films after aging treatment are the tensile strength, elongation at break, M_{100} and M_{300} have performed.

1. Introduction

Natural rubber latex (NRL) obtained from *Hevea Brasiliensis,* and it consists of rubber particle (cis-1,4 polyisoprene) and nonrubber in water. Natural rubber latex products have disadvantages such as other polymers susceptible to oxidation or degradation due to heat (high temperatures) and low mechanical properties in thin films [1]. To improve the quality of NRL products, they are commonly incorporated with inorganic and organic fillers. However, the addition of fillers has its limitation since some of them are polar while the rubber matrix is nonpolar. Therefore, it is necessary to modify the filler. Several studies reported that modification of fillers succeeds in enhancing the quality of filled rubber [2-4].

Polyvinyl pyrrolidone (PVP) is an amorphous thermoplastic that has been used for various purposes, such as an additive in ultrafiltration membranes, as a stabilizer in suspension polymerization, as a thickener in the cosmetic or food industries, as a coating aid in the photographic industry, and more recently as a sizing material for fiber reinforced composites. [5].

So many factors can significantly affect the durability of a polymeric material, such as temperature, irradiation, moisture, chemicals [6]. The objectives of testing the quality of a product with thermal aging treatment are to see and estimate the rate of change that will take place from the rubber product after being treated with high-temperature effects over a period [7].

Therefore, it will test the mechanical properties of natural rubber latex film using cellulose nanocrystals (CNC)s as a filler and polyvinyl pyrrolidone (PVP) as a dispersion agent after aging treatment.

IOP Conf. Series: Materials Science and Engineering **309** (2018) 012101 doi:10.1088/1757-899X/309/1/012101

2. Materials and Methods

2.1. Materials
Raw materials (Corn Cob Waste) were obtained from corn ice trader around Jl. Dr. Mansyur, University of Sumatera Utara (USU), Medan, North Sumatera, Indonesia. Aquadest (H_2O), nitric acid (HNO_3), sulfuric acid (H_2SO_4), hydrogen peroxide (H_2O_2) brand MERCK, sodium hydroxide (NaOH), sodium hypochlorite (NaOCl), sodium nitric ($NaNO_2$), sodium sulfite (Na_2SO_3) brand Duran.

2.2 Isolation of α-Cellulose from Corn Cob
75 grams of corn cob fiber was inserted into beaker glass, then added by mixture 1 L of 3,5% HNO_3 and 10 mg $NaNO_2$, after that it was heated on a hot plate at 90 °C for 2 hours. Then it was filtered, and the fibers are washed until the filtrate is neutral. Next, it was digested with 750 ml of a solution containing 2% NaOH and 2% Na_2SO_3 at 50 °C for 1 hour. Then it was filtered, and the fibers are washed until the filtrate is neutral. Bleaching was conducted with 250 ml of 1.75% NaOCl solution at boiling temperature for 30 minutes. Then it was filtered, and the fibers are washed until the filtrate is neutral. After that, α-cellulose purification of the sample with 500 ml of 17.5% NaOH solution was obtained at 80 ° C for 30 minutes. Then it was filtered, and the fibers are washed until the filtrate is neutral. Bleaching was taken with 10% H_2O_2 at 60 °C in the oven for 1 hour. Then it was filtered, and the fibers are washed until the filtrate is neutral.

2.3 Isolation of Cellulose Nanocrystals from α-Cellulose
1 grams of α-cellulose dissolved in 25 ml of 45% H_2SO_4 at 45 °C for 45 min. Then cooled and added with 25 ml aquadest, then left it for one night until the suspension is formed. The suspension was centrifuged at 10000 rpm for 25 min until pH was neutral. Then it was ultrasonicated for 10 minutes, put into the dialysis membrane and soaked in 100 ml aquabidest, left it for four days while stirring. Then the aquadest was evaporated at 70 °C to obtain cellulose nanocrystals.

2.4 Filler
Table 1 shows the Percentage of cellulose nanocrystals (CNC)s with polyvinyl pyrrolidone.

Table 1. The Formulation of Filler

Material	Percentage (%)
Cellulose Nanocrystals	10
Polyvinyl Pyrrolidone	1
Water	89

2.5 Natural Rubber Latex Compound
Table 2 shows the formulation of NRL compound for the pre-vulcanization process at 70 °C.

Table 2. The Formulation of NRL compounds

Compounds	Amount (gram)
60 % *High Ammonia* Latex	166.7
50 % Sulfur Dispersion	3
50 % ZDEC Dispersion	3
30 % ZnO Dispersion	0.83
50 % Antioxidant	2
10 % KOH	3
10 % Filler Dispersion System	0; 2; 4; 6; 8

2.6 Mechanical Properties of Natural Rubber Latex Film

Natural rubber latex film was cut into dumbbell-shaped specimens according to ASTM D 412. The mechanical properties test were performed with Universal Testing Machine Instron 5565 at room temperature.

3. Results and Discussions

The obtained cellulose nanocrystals were spherical shaped with an average particle size of 9-29 nm and the crystallinity is 70.05% from previous research.

3.1. Mechanical Properties of Natural Rubber Latex Film

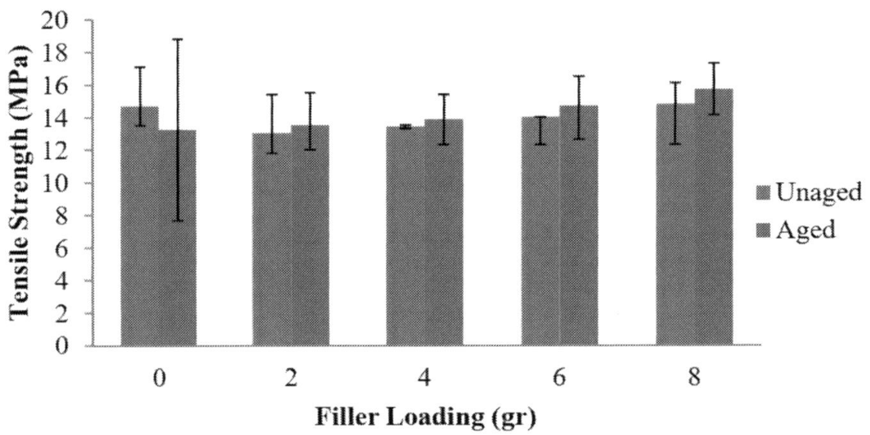

Figure 1. Effect of Aging Treatment in Natural Rubber Latex Film Filled Cellulose Nanocrystals From Corn Cob and Polyvinyl Pyrrolidone to Tensile Strength

Figure 1 shows the tensile strength of the natural rubber latex film filled cellulose nanocrystals with polyvinyl pyrrolidone. The tensile strength of the natural rubber latex film before aging is decreased after adding the CNCs and PVP but increased along filler loading. And after aging treatment obtained that the tensile strength increased along with the addition of the filler loading indicate that cellulose nanocrystals have an excellent ability as a filler in natural rubber latex film. This is due to the small size of the cellulose nanocrystals particles that can improve the interaction of film interface. The good interaction of CNCs causes the filler is more easily dispersed in the natural rubber latex evenly throughout the surface, thereby it can reduce the aging effect due to the heat-sensitive rubber, with the addition of a cellulose nanocrystals filler leading to a reduced portion of the degradable rubber [8].

With an addition of PVP, the results showed that has higher tensile strength compared with the products which only added cellulose nanocrystals to the film. Higher tensile strength after aging treatment means that PVP has an ability to increase the strength of a product because with an addition of PVP it can increase the stress on natural rubber latex film [10].

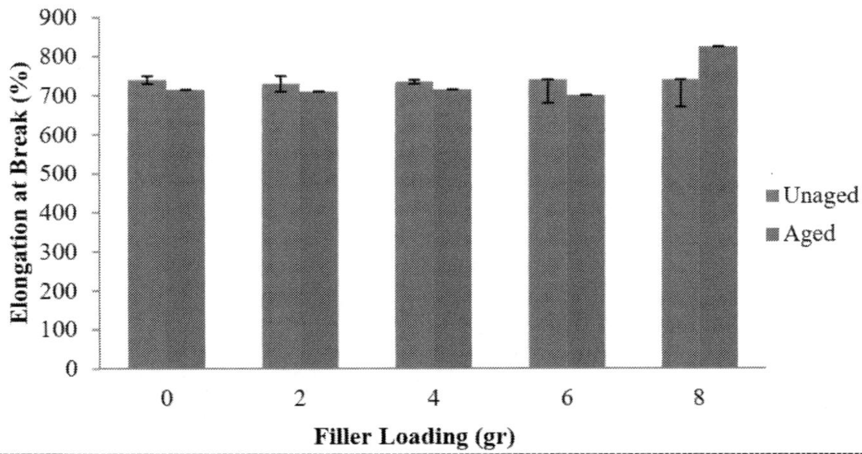

Figure 2. Effect of Aging Treatment in Natural Rubber Latex Film Filled Cellulose Nanocrystals From Corn Cob and Polyvinyl Pyrrolidone to Elongation at Break

Figure 2 shows the elongation at break of natural rubber latex film filled cellulose nanocrystals as a filler and PVP as a dispersion agent. At a load of filler 2, 4 and 6 grams the elongation at breaks of natural rubber latex films before and after aging relatively same, even an addition of (CNC)s it can generate an extension value. At a load of filler 8 grams, the elongation at break of natural rubber latex film was higher after and before aging treatment. Along with an addition of (CNC)s and PVP, it will produce a high shear stress. High elongation values caused by increased of filler loading improve the film ability on holding the decrease of the film quality due to aging process [11,1]. The chemical degradation of natural rubber through oxidation with thermal-oxidation has superior properties in the product that increases tensile strength and breaking elongation [9]. Overall, the elongation at break values was not significantly different between before and after the aging treatment.

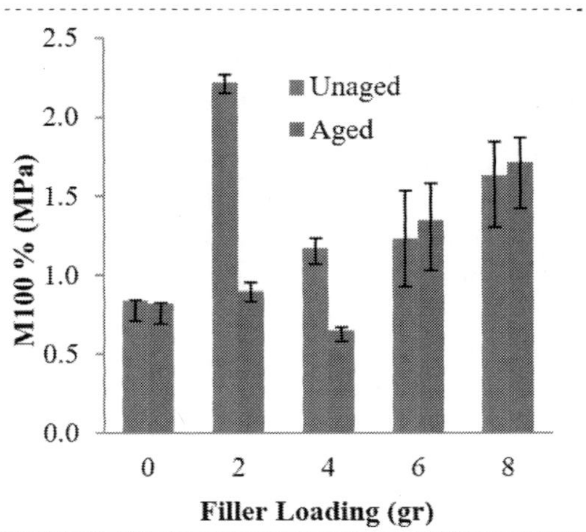

Figure 3a. Effect of Aging Treatment in Natural Rubber Latex Film Filled Cellulose Nanocrystals From Corn Cob and Polyvinyl Pyrrolidone to M_{100}

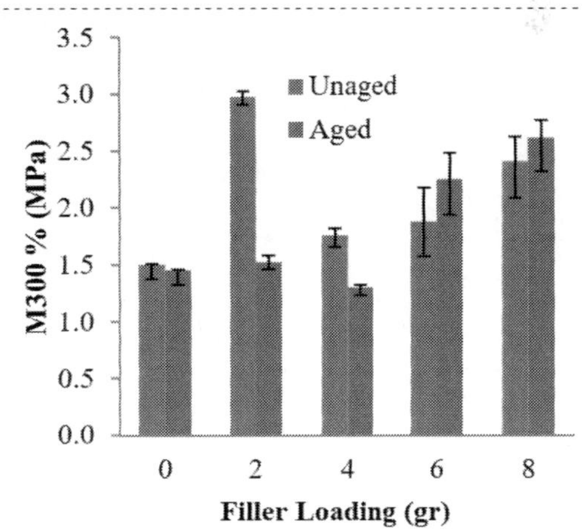

Figure 3b. Effect of Aging Treatment in Natural Rubber Latex Film Filled Cellulose Nanocrystals From Corn Cob and Polyvinyl Pyrrolidone to M_{300}

Figure 3a shows the tensile modulus at 100% elongation (M_{100}) is the number of forces that given when the sample has 100% elongation and Figure 3b shows the tensile modulus at 300% elongation (M_{300}) represents the amount of force that given when the sample has 300% elongation. Modulus shows the stiffness relativity of the material. The small value shows the elastic nature of the material while the large value shows the stiff properties of the material. During aging, the tensile modulus increases as the main-chain scission take place and effect of cross-linking predominates [12]. This led to the increasing value of the M_{100} and M_{300}.

4. Conclusion

Cellulose nanocrystals have a good role as a filler in natural rubber latex film products before and after aging treatment with adding PVP which also produced the products that have high mechanical properties. As proved by the high value of tensile strength, elongation at break and tensile modulus of NRL films produced.

References

[1] Hamidah H, Marfuah L, Nuim H and Sukardi 2016 *International Journal of ChemTech Research* **9** 201-205
[2] Siby V and Jozsef K-K 2004 *Journal of Applied Polymer Science* **91** 813-819
[3] Kim W-K, Jang S H, Kang Y G, Han M H, Kyu H and Kim W 2012 *Journal of Applied Polymer Science* **128** 2344-2349
[4] Saritha A, Joseph K, Thomas S, Muraleekrishnan R 2012 *Journal of Applied Polymer Science* **124** 4590-4597
[5] Christelle M L 1997 *Spectroscopic Characterization of Molecular Interdiffusion At a Poly(Vinyl Pyrrolidone)/Vinyl Ester Interface* (Virginia Polytechnic Institute and State University)
[6] Zenasni R, Sahnoune M, Olay J V, Amado A A and Hebbar A 2015 *Journal of Material Science & Engineering* **4** 161
[7] Foo Y T 2013 *A Study Of Thermal Ageing Of Carboxylated Nitrile Rubber Latex Thin Films* (Universiti Tunku Abdul Rahman)
[8] Khalil A, Shaikh S N, Nudrat Z R and Khalid M 2012 *International Journal of Industrial Chemistry* **3** 21
[9] Tri S and Chasri N 2017 *Journal Dinamika Penelitian Industri* **28** 32-41
[10] Haque M A, Ahmad M U, Khan M A, Raihan S A and Dafader N C 2012 *Polymer-Plastic Technology and Engineering* **51** 669-674
[11] Huang S, Zhou L, Li M-C, Wu Q, Kojima Y and Zhou D 2016 *Materials* **9** 523
[12] Vinod V S, Varghese S and Kuriakose B 2002 *Polymer Degradation and Stability* **75** 405-412

IOP Conf. Series: Materials Science and Engineering **309** (2018) 012102 doi:10.1088/1757-899X/309/1/012102

Manufacture and impact analysis of bmx helmet made from polymeric foam composite strengthened by oil palm empty fruit bunch fiber

Mahadi*
Department of Mechanical Engineering, Faculty of Engineering, Universitas Sumatera Utara Padang Bulan Medan 20155 Indonesia

*Email: mahadi_usu@yahoo.co.id

Abstract. *Helmets are protective head gears wear by bicycle riders for protection against injury in case of the accident. Helmet standards require helmets to be tested with a simple drop test onto an anvil. The purpose of research is to know toughness of bicycle helmet made from polymeric foam composite strengthened by oil palm empty fruit bunch fiber. This research contains report result manufacture and impacts analysis of bicycle helmet made from polymeric foam composite materials strengthened by oil palm empty fruit bunch fiber (EFB). The geometric helmet structure consists of shell and liner; both layers have sandwich structure. The shell uses matrix unsaturated Polyester BQTN-157EX material, chopped strand mat 300 glass fiber reinforce and methyl ethyl ketone peroxide (MEKPO) catalyst with the weight composition of 100 gr, 15 gr, and 5 gr. The liner uses matrix unsaturated Polyester BQTN-157 EX material, EFB fiber reinforces , Polyurethane blowing agent, and MEKPO catalyst with the composition of 275 gr (50%), 27.5 gr (5%), 247 gr (45%), and 27.5 gr (5%). Layers of the helmet made by using hand lay-up method and gravity casting method. Mechanical properties of polymeric foam were the tensile strength (σ_t) 1.17 Mpa, compressive strength (σ_c) 0.51 MPa, bending strength (σ_b) 3.94 MPa, elasticity modulus (E) 37.97 Mpa, density (ρ) 193 (kg/m^3). M4A model helmet is the most ergonomic with the thickness 10 mm and the amount of air channel 11. Free fall impact test was done in 9 samples with the thickness of 10 mm with the height of 1.5 m. The result of the impact test was impacted force (Fi) 241.55 N, Impulse (I) 6.28 Ns, impact Strength (σi) 2.02 Mpa and impact Energy (Ei) 283.77 Joule. The properties of bicycle helmet model BMX-M4A type was 264 mm length, 184 mm width, 154 mm height, 10 mm thick, 580 mm head circle, 331 g mass and 11 wind channels.*

1. Introduction

The appliance of laminated composites is improve in all sorts of engineering applications especially in aerospace, sports, transportation and in marine also. Helmets are protective head gears wear by bicycle riders for protection against injury in case of the accident. The shell is usually hard and helps resist objects from piercing during impact thereby preventing direct injury on the skull and the liner helps to absorb most of the impact energy. Most of the helmets was prepared from resins or plastics reinforced with fibers. The helmet protects bicycle rider head by absorbing mechanical energy and also protects against penetration. Boufous, Senserrick, Stevenson, and Ivers (2011) stated that head injuries due to bicycle crashes are a major factor in accidents.The structure and protective capacity was changed in high-energy impact. Besides energy-absorption capacity, their volume and weight are major problems that must be considere so as to reduce injury risk. In this research was designed and manufacture of the ergonomic bicycle helmet that has strong, comfortable and aesthetic characteristics. The material used is Polymer Unsaturated Polyester BQTN-157 EX as the matrix, oil palm empty fruit bunch fiber as reinforces and *polyurethane as Blowing Agent* (BA). Unsaturated polyester resins are the condensation products of unsaturated acids or anhydrides and diols with/without diacids. The unsaturation present in this type of polyesters provides a site for subsequent cross-linking (Reinhold, 1956). The properties of matrix was Modulus Young (E) 2 s/d 4,5 Gpa, Density (ρ) 1,2 s/d 1,5 mg.mm^3 and Tensile Strength (σ_T) 40 s/d 90 (MPa) (Hashim, J.,2003). Oil palm empty fruit bunch fiber (EFB) is a composite reinforcer of nature that is widely obtained in the tropics. Each fiber contains *lignin* (16,19%), *cellulose* (44,14%) dan *hemicellulose* (19,28%).

2. Method

A bicycle helmet is designed to attenuate impacts to the head of a cyclist in falls while minimizing side effects such as interference with peripheral vision (*CPSC, 1998*). There is scientific research into the degree of protection offered by bicycle helmets in the event of an accident, and the effects of helmet wearing on cyclist and motor vehicle driver behaviour. The modern bicycle helmets normally have a shell, usually between 0.3 and 0.8 mm thick, that is frequently attached to the liner material during the manufacturing process. The micro-shell liner offers little rigidity or load distribution, but may help to sustain helmet reliability in an impact, which may be most important if a second impact occurs in the same accident. Asiminei, A. G, Van der Perre, Verpoest, and Goffin (2009) stated that by changing the material used as the liner, it is ideally possible to tune head protection properties for a particular impact condition of interest. Current helmets have more ventilation holes, are thicker at the rear, and sometimes have a non-smooth external profile. The helmets subjected to oblique impact tests in 2002 had less than ten, large ventilation holes. The number of ventilation holes has increased, and their size decreased (N.J.Mills 2007). A helmet should manage as much energy as possible in a very hard crash, certainly below 300 g for a two-meter drop. A helmet should have a strong strap that keeps it on your head after the impact, easy to adjust properly or be self-adjusting, comfortable to wear, smooth and round as possible on the outside to prevent snagging in a crash. The impact test drop height increased from 1.0 to 1.5 m, when British Standard BS 6863 was replaced by EN 1078 in 1997 (BS EN 1078, 1997). The helmet model was be tested as shown in Figure 1.

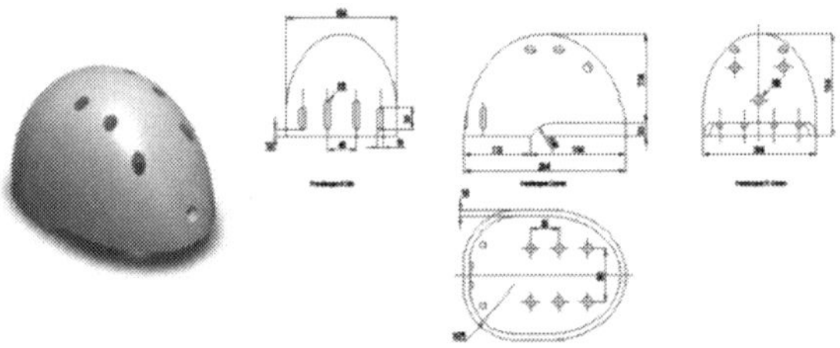

Figure 1. BMX-M4A Model

The research material is Unsaturated Polyester BQTN-157 EX resin as the matrix, oil palm empty fruit bunch fiber as reinforcement, *polyurethane as blowing agent* and MEKPO as catalyst. Tests of mechanical properties was performed in enclosed spaces with the room temperature of 27^{0}C with Servopulser static test machine. Polymeric foam helmet material is made using composition of fiber 5% wt (27,5gr), unsaturated Polyester 157 BQTN-EX 50% wt (275gr), Polyurethane blowing agent 45% wt (247 gr) and Methyl ketone peroxide (MEKPO) 5% catalyst % wt (27.5 gr). After that, the manufacturing of helmet was done by using hand lay-up and castings process.

Impact Test Procedure

Bicycling may lead to the dangers of death or severe harm because of head injuries. The suitable utilization of protective helmets can minimalize the danger of death or interminable harm. A free-fall drop test as in Figure 2 was employed to drop helmets onto a flat, firm, and non-yielding steel base which replicates a road surface. The drop height based of the helmet was 1.5 m, which coincides with the standard of Consumer Product Safety Commission (CPSC) drop height.

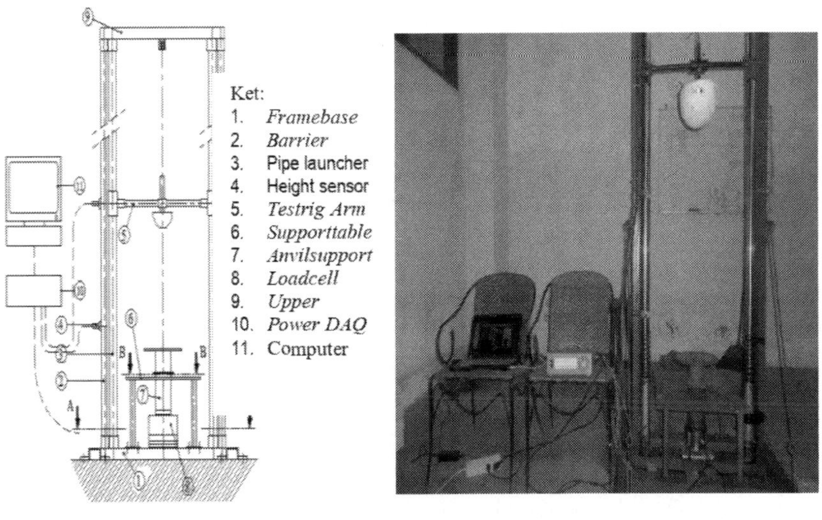

Figure 2: Impact Free Fall Test Rig.

According to rules governed by Consumer Product Safety Commision (CPSC), helmets are released with a certain mass of head forms in lab test and the g-forces are recorded by instruments placed inside the head form with measurement recorded below than 300 g. A standard for helmet tests exists in every country. Each has its own particular principles, but the ASTM F1446-95a head protector standard is the most-utilized standard. The three most frequently used helmet testing standards are EN1078, CPSC, ASTM F1447 and Snell B-95. Table 1 demonstrates the correlations among each standard. The application of conservation of energy to a falling object allows us to predict its impact velocity and kinetic energy, but we cannot predict its impact force without knowing how far it travels after impact. As an object falls from rest, its gravitational potential energy is converted to kinetic energy. Conservation of energy as a tool permits the calculation of velocity just before it hits the surface. The dynamic energy in a moving object can be expressed as follows:

$$\text{Potential Energy, PE} = mgh \tag{1}$$

$$\text{Kinetic Energy, KE} = \frac{1}{2}mv^2 \tag{2}$$

$$\text{The impact velocity,} \quad v = \sqrt{2gh} \tag{3}$$

Table 1: Helmet Test Criteria

	CPSC	**ASTM F1447**	**Snell B-95**	**EN 1078**
Drop height on flat anvil (m)	2.0	2.0	2.2	1.5
Drop height on hemispherical anvil (m)	1.2	1.2	1.5	N/A
Head form weight (kg)	5	5	5	4
Failure threshold (g)	300	300	300	250

3. Results and Discussion

The forces required for causing a particular injury are variable and very little quantitative information which exists about the magnitude of force and time. The serious brain injuries willoccur while hitting of a helmet on the road because of a contact between the helmet, head, and the road surface. The results of impact test on nine helmets thick 10 mm and drop height on flat anvil 1.5m and impact area 119 mm^2 is shown in Figure 3 Up to 5.

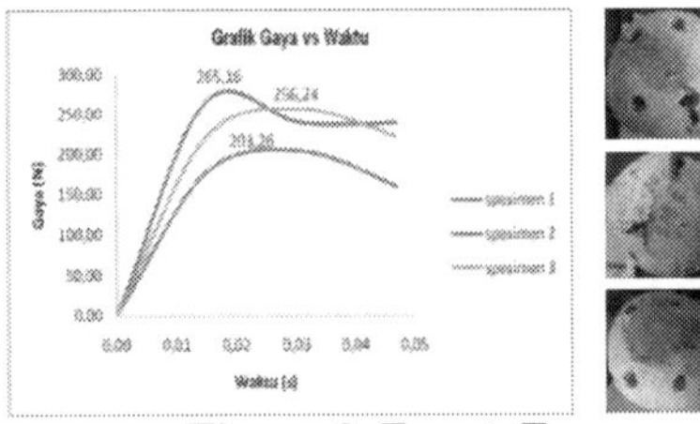

Figure 3. F vs. t Top

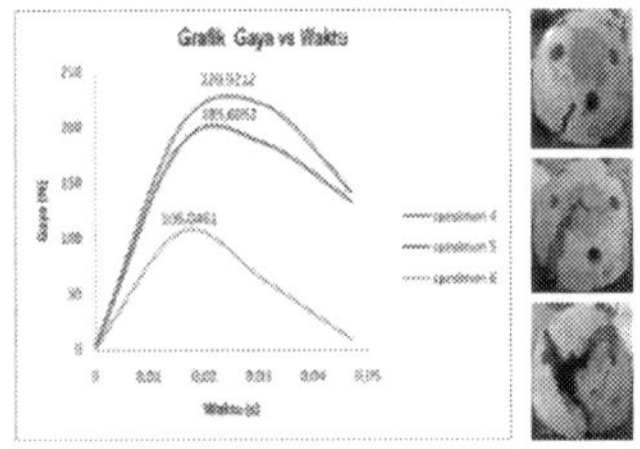

Figure 4. F vs. t Front

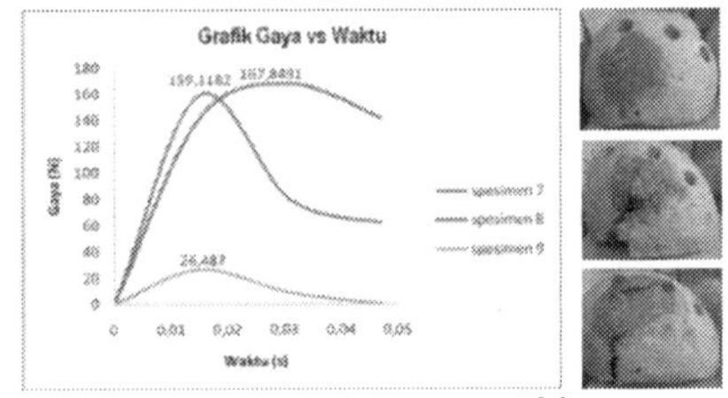

Figure 5. F vs. t Side

Table 2. Result of Helmet Impact Test

Helmet	Impact Point	Mass (kg)	Impact Force (F_i) (N)	Impact Time (T_i)(det)	Impact Stress (σ_i) (MPa)	Info
1	Top	0. 328	203.26	0.031	1.70	Rupture
2	Top	0. 334	265.16	0.016	2.22	Rupture
3	Top	0. 354	256.24	0.031	2.15	Rupture
4	Front	0. 350	220.92	0.016	1.85	Rupture
5	Front	0. 272	185.60	0.031	1.56	Rupture
6	Front	0. 345	106.04	0.047	0.89	Rupture
7	Side	0. 296	159.12	0.015	1.33	Rupture
8	Side	0. 360	167.85	0.031	1.41	Rupture
9	Side	0. 341	26.48	0.047	0.22	Rupture

Table 3. Impact Energy Absorption

Helmet	Weight Helmet + Test Rig (kg)	Theoretical Impact Energy Et = m.g.h (Joule)	Impact Force (F_i)(N)	Experimental Impact energy Ee =F_i.h (Joule)	Impact Energy Absorption E_i=Ee-Et (Joule)
1	5.328	78.40	203.26	304.89	226.49
2	5.334	78.49	265.16	397.74	319.25
3	5.354	78.78	256.24	384.36	305.58
4	5.350	78.73	220.92	331.38	252.65
5	5.272	77.58	185.6	278.4	200.82
6	5.345	78.65	106.04	159.06	80.41
7	5.296	77.93	159.12	238.68	160.75
8	5.360	78.87	167.85	251.77	172.90
9	5.341	78.59	26.48	39.72	-38.87

Table 4. Impuls and Impact Time

Thick (mm)	Average Time (t) and Impuls (I)					
	Top		Front		Side	
	Time (sec)	I (Ns)	Time (sec)	I (Ns)	Time (sec)	I (Ns)
10	0.026	6.28	0.031	5.29	0.031	3.65

Table 5. Force (F) and Impact Energy (Ei)

Thick **(mm)**	**Average Force (F) and Impact Energy (E_i)**					
	Top		Front		Side	
	Force (N)	**Impact Energy (J)**	**Force (N)**	**Impact Energy (J)**	**Force (N)**	**Impact Energy (J)**
10	241.55	283.77	170.85	177.96	117.81	98.26

Table 6. Result of Test

Parameter	**BMX- M4A**	**COMMERCIAL**	**CPSC**
Impact Force (F_i) (max)	241.55 N	441.84 N	-
Impact Energy (E_i) (max)	283.77 Joule	555.64 Joule.	110 Joule
Impuls (I) (max)	6.28 Ndet	55.23 Ns	-
Impact stress (σ_i) (max)	2.02 MPa	3.78 Mpa	-
Mass (m)	331 g	300 g	300 g

4. Conclusion

Base on data of research can be concluded as follow:

1. Specification helmet of bicycle produced is type half face- BMX helmet, model M4A. Dimension: 264 mm long, 184 mm large, 154 mm height, 580 mm head circumference, the number of ventilation air 11, 10 mm thick and mass 331 kg.
2. The technique of making helmet is shell layer with hand lay up method, liner with cast method and the both layers are sandwich structured.
3. The shell helmet uses matrix unsaturated Polyester BQTN-157EX material, chopped strand mat 300 glass fiber reinforced, and Methyl Ethyl Ketone Peroxide (MEKPO) catalyst. The liner uses matrix unsaturated Polyester BQTN-157 EX material, EFB fiber reinforces, Polyurethane blowing agent, and MEKPO catalyst with the composition of 275 gr (50%), 27.5 gr (5%), 247 gr (45%), 27.5 gr (5%).
4. Result A free fall drop test inform that there are different parameters of the M4A bicycle helmet with a commercial helmet. Impact energy the bicycle commercial helmet (Ei) 555.64 joule, Helmet M4A (Ei) 283.77 Joule. Impact energy helmet M4A close to the standard threshold valve of CPSC 110 Joule.

Reference

[1] *Consumer Product Safety Commission (CPSC)* 1998 16 CFR Part 1203 *Federal Register. Rules and Regulations* **63** 46
[2] N J Mills and A Gilchrist 2007 Oblique impact testing of bicycle helmets *Metallurgy and Materials* University of Birmingham B15 2TT UK
[3] Hashim J 2003 Pemrosesan Bahan Edisi pertama Johor Bahru Ratu Sdn. Bhd

[4] Reinhold 1956 Polyesters and Their Applications Bjorksten Research Laboratories New York
[5] Asiminei A G, Van der Perre G, Verpoest I and Goffin J 2009 A transient finite element study reveals the importance of the bicycle helmet material properties on head protection during an impact *Proceedings of the International Research Council on Biomechanics of Injury* (IRCOBI) *conference* New York UK 357-360
[6] BS EN 1078 1997 Helmets for pedal cyclist and for users of skateboards and roller skates *British Standards Institution* London
[7] BS EN 960 1995 Headforms for use in the testing of protective helmets *British Standards Institution* London
[8] Newton R E 1988 Theory of Shock Isolation Shock & Vibration Handbook Chapter 31 New York McGraw-Hill Book Company

IOP Conf. Series: Materials Science and Engineering **309** (2018) 012103 doi:10.1088/1757-899X/309/1/012103

Determining the need for improvement of infant incubator design with quality function deployment

U Tarigan*, R Ginting and I Siregar
Department of Industrial Engineering, Faculty of Engineering, University of Sumatera Utara, Jalan Almamater Komplek USU Medan, 20155, Indonesia

*E-mail: ukurta.tarigan@yahoo.co.id

Abstract. A Newborns need particular attention because it takes time to adapt to the outside world. In accordance with the standard post-neonatal procedures, newborns should be incorporated into the incubator within a specified period of time according to the infant's health level. Infant incubator as a medical device used to care for newborns, often get complaints from doctors and child nurses. The complaint indicates consumer dissatisfaction with incubator products in the hospital. Broadly speaking, objection against infant incubators lie in inappropriate designs. To overcome these complaints the researchers apply the method of Quality Function Deployment to determine the characteristics of priority techniques in accordance with the wishes of consumers with it. The primary focus of QFD is to engage customers in the product development process as early as possible, which their needs and desires serve as the starting point of the QFD process. Therefore, QFD is called the *voice of customer*. The underlying philosophy is that customers are not always satisfied with a product even though the product has been perfectly produced. The results show that the category used as a priority improvement is the additional function of the oxygen cylinder and the size of the door hole. QFD phase one produces technical characteristics of Ergonomy, Features and Cost as critical part determinants.

1. Introduction

Based on World Health Organization (WHO) data, the number of births in Indonesia in 2012 reached 4.7 million per year while neonatal mortality (newborn) reached 72 thousand babies per year. Prematurity, low birth weight, infection, asphyxia, and birth trauma cause many of newborn death (neonatal) [1]. Data obtained from the Medical Record Section Kabanjahe Hospital mentioned that the number of premature babies in April - September 2014 as we would see in Table 1.

Physicians and nurses of the children's section have complaints regarding the use of infant incubators in the hospital. By introducing a couple of preliminary questionnaires to doctors and child nurses, researcher collected the complaints from them which include price, design, and maintenancing the infant incubator. The grievances obtained indicate the dissatisfaction of the doctor and child nurse to the available incubator. Such dissatisfaction will definitely affect the decrease in the performance of doctors and nurses as well as the quality of services offered by General Hospital of Kabanjahe.

Table 1. Number of Infants Births
(April – September 2014)

Bulan	Number of Premature Infants
April	9
May	8
June	8
July	6
August	10
September	10

Departing from these problems, Quality Function Deployment (QFD) is used to identify the most essential needs in an effort to improve infant incubator. QFD provides benefits in providing a better understanding of the interactions customers request and design. Early engagement of the production section during the design process will reduce iteration and focus the design while also fostering teamwork [2]. The use of the Quality Function Deployment method is expected to identify the technical characteristics of the infant incubator to produce an appropriate improvement of the target incubator design in improving customer satisfaction.

1.1 Literature

QFD is a way to improve the quality of goods or services by understanding the needs of consumers and then connect it with technical provisions to produce a good or service at every stage of manufacturing goods or services produced. Quality Function Deployment is a design tool used to help businesses stakeholder to focus on the needs of their customers when designing and its specifications [3].

The main benefits of QFD are as follows:

a. Focuses the design of new products and services on customer needs.
b. Prioritize design activities.
c. Analyze the company's key product performance to meet the needs of key customers.
d. Recent estimates show a saving of between one-third and a half compared to before the application of QFD.
e. Reduce the number of design changes after issuing it by ensuring focused efforts at the planning stage.
f. Encourage implementation of work teams and eliminate barriers between sections by involving marketing, engineering, and fabrication from the beginning of the project.
g. Provide a way to make process documentation and provide a firm foundation for making design decisions

QFD consists of four phases that we can solve by using a series of matrices, i.e.:

1. Phase I Product Planning
2. Phase II Product Designing
3. Phase III Process Planning
4. Phase IV Process Control

Four phase of QFD as we would see in Figure 1.

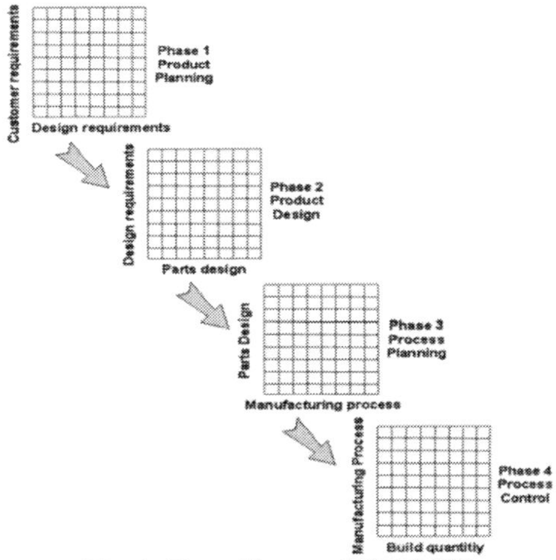

Fig 1. Four Phase of QFD [2]

1.2 Infant Incubator

The skin temperature and body of premature infants tend to decrease significantly due to conduction heat loss, convection, radiation and liquid evaporation. Cold temperatures experienced by babies for a long enough time will cause lack of oxygen, hypoglycemia and metabolic disorders. The baby incubator is a closed device with equipment to control the temperature that warms the baby by circulating warm air. Fatalities and injuries to infants in incubators have been associated with thermostat failure that causes overheating incubators and malfunctions that pose a fire and electric hazard [4].

2. Methods

2.1 Data Collection

We collected primary data by direct observation or measurement, including preliminary questionnaire data, open questionnaire data, closed questionnaire data, questionnaire data of technical characteristics, questionnaire data relationship between fellow technical characteristics and questionnaires data about the critical part and questionnaire data relationships among fellow critical part. Secondary data is obtained based on the one that obtained from interviews with suppliers of products, including product and cost components.

In this study data collection methods are as follows:

1. Observation technique, namely by made a direct observation of the process of using infant incubator.
2. Survey Technique, that is a technique to get primary data needed relating to consumer requirement to infant product incubator by spreading questionnaire.
3. Documentation techniques, namely by collecting data about the characteristics of infant incubator products in supplier companies.
4. The study of literature, namely as the initial foundation of the problem as well as the compilation of the questionnaire variables.

The variables that we questioned to the consumer are made based on the results of the preliminary questionnaire which asks about the complaints of doctors and nurses against infant incubator [5]. Then we adjusted the infant incubator variable for the complaint and the available literature [6].

The population taken is all doctors and nurses of the children who served in the hospital in Karo District. Research is conducted on all elements of the population because the subject is relatively little or called the population study. This choice applies to data collection using a preliminary questionnaire, an open questionnaire and a closed questionnaire. As for data collection using

IOP Conf. Series: Materials Science and Engineering **309** (2018) 012103 doi:10.1088/1757-899X/309/1/012103

Questionnaire Technical Characteristics, Questionnaires Level Relationships Between Technical Characteristics, Critical Part Questionnaire, and Level Critical Part is taken from the population of suppliers infant incubator. The sample population is the vendor who understands the product well [7]. The sampling method used is Purposive Sampling.

2.2 Data Processing

Stages of data processing begins by collecting data from the spread of open questionnaires to obtain the mode to be used for closed questionnaires. The result of the closed questionnaire is then poured into an assessment of the importance level that is tested for its validity and reliability. House of Quality is built on product attributes that have been obtained [7].

Development of HoQ matrix is done with the following stages:

1. Identify consumer desires into product attributes
2. Determine the relative importance of the attributes
3. Creating a resistance matrix between product attributes and characteristics.
4. Establish the relationship between technical characteristics and product attributes.
5. Identify the relevant interactions between technical characteristics
6. Determine the target image to be achieved for technical characteristics [7].

There should be an understanding that data processing in this paper is limited to only stage 1 QFD.

3. Results and Discussions

The following is the Customer's Requirement to infant incubator product as we would see in Table 2

Table 2. *Customer Requirement* (CR) to *Infant Incubator*

Atribut
Additional Functions of Infant Incubator is the Place of Oxygen Tube
The size of the door hole is 15 cm
Infant Incubator Price Rp 10.000.000
The Infant Incubator dimension is 70 cm x 50 cm x 40 cm
The floor distance from the base of the compartment is 100 cm
Compartment infant incubator materials are Acrylic
The controller's position on the infant incubator is on the top right side.

Table 3 contain the level of importance of the costumer requirement variable.

Table 3. Customer Importance (CI)

Costumer Requirement	**The Result of Quest. Measurement Scale**					**CI**
	1	**2**	**3**	**4**	**5**	
Additional Functions of Infant Incubator is the Place of Oxygen Tube	0	2	6	16	11	4
The size of the door hole is 15 cm	3	9	11	10	2	3
Infant Incubator Price Rp 10.000.000	0	2	14	10	9	3
The Infant Incubator dimension is 70 cm x 50 cm x 40 cm	0	4	12	11	8	3
The floor distance from the base of the compartment is 100 cm	2	11	10	8	4	2
Compartment infant incubator materials are Acrylic	1	2	7	12	13	5
The controller's position is on the infant incubator is on the top right side.	0	2	11	9	13	5

The data has been obtained in the previous steps, then made into the HOQ matrix. The House of Quality of Infant incubator product as we would see in Figure 2.

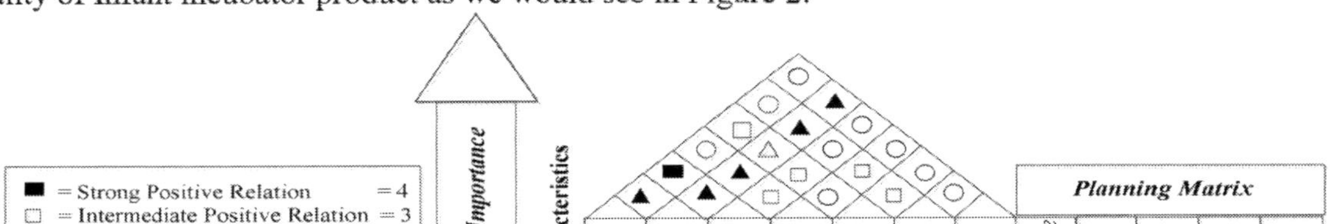

Customer Requirement	Customer Importance	Ergonomy	Cost	User Friendly	Features	Safe Operation	Durability	Aesthetics	Satisfaction Performance	Improvement Ratio	Sales Point	Absolute Weight	Relative Weight
Additional Functions of Infant Incubator is the Place of Oxygen Tube	4	0	3	0	9	1	0	1	3.690	1.084	1.5	6.504	22.63
The size of the door hole is 15 cm	3	3	1	3	3	3	0	3	3.899	0.769	1.5	3.463	12.05
Infant Incubator Price Rp 10.000.000	3	0	9	0	3	0	9	9	3.993	0.751	1.0	2.254	7.84
The Infant Incubator dimension is 70 cm x 50 cm x 40 cm	3	9	1	1	0	1	0	1	4.008	0.748	1.0	2.245	7.81
The floor distance from the base of the compartment is 100 cm	2	9	1	9	0	1	0	1	4.015	0.498	1.2	1.195	4.16
Compartment infant incubator materials are Acrylic	5	3	9	1	1	3	9	1	4.333	1.154	1.2	6.924	24.09
The controller's position on the infant incubator is on the top right side.	5	3	1	9	3	3	0	3	4.060	1.232	1.0	6.158	21.42
Difficulty Level		3	3	4	3	4	3	1					
Degree of Importance		19	18	16	13	8	13	13					
Cost Estimate		14	14	19	14	19	14	5					

Fig 2. QFD Phase I of Infant Incubator

The Results of QFD Phase I of Infant Incubator products shows that the most important characteristics to be fixed immediately are Ergonomy, Cost and User-Friendly with the highest degree of importance, difficulty level, and cost estimate. From the planning matrix obtained the product attributes that become the priority of improvement because it has the greatest weight and high sales points, namely the compartment material and the additional function of the oxygen cylinder.

Technical characteristics with highest to the lowest degree of importance are ergonomy, cost, user-friendly, features, durability, aesthetics and safe operations. We can use the attributes as input in QFD Phase II for further research (8). By using QFD, we succesfully analyzed product performance to meet the needs of customers. We also reduced the number of design changes by ensuring focused efforts at the planning stage.

4. Conclusion

Newborns need particular attention because it takes time to adapt to the outside world. Postnatal neonatal procedures suggest that newborns should be incorporated into the incubator within a certain period according to the infant's health level. Infant incubator is one of the medical equipment used to help babies adapt to the outside world especially regarding temperature differences [9].

The technical characteristics that are output in phase I are Ergonomy, Cost, User-Friendly, Features, Safe Operation, Durability and Aesthetics. The development of Quality Function Deployment Phase I produces performance measures of technical characteristics ie difficulty level, degree of importance and cost estimates. Technical characteristics Ergonomy, Cost and User-Friendly has the level of difficulty, degree of interest and the largest cost estimates so that a priority improvement. While product attributes that become priority improvement based on planning matrix from QFD Phase I is Additional Function on Infant Incubator that is Place of Oxygen Tubes and Compartment infant incubator is Acrylic.

Acknowledgement
This study was funded by Directorate of Research and Community Service, Directorate General for Research and Development at the Ministry of Research, Technology and Higher Education of Indonesia, in accordance with Research Contract No. 003/SP2H/LT/DRPM/IV/2017 dated 20 April 2017.

References

[1] WHO Statistics [Cited 2015] World Health Statistics *2014* [Internet] available from http://who.int/gho/publications/world_health_statistics/2014/en/
[2] Jaiswal E 2012 A Case Study on Quality Function Deployment (India: Mewar University) pp 2-4
[3] Ginting R 2010 Perancangan Produk (Yogyakarta: Graha Ilmu) pp 135-137
[4] World Health Organization [cited 2014] Medical Devices [Internet] available from http://www.who.int/medical_devices /innovation/en/
[5] Otto K and Wood K 2001 Product Design (London: Prentice Hall)
[6] Gea Medical Operator's Manual YP-100 Infant Incubator
[7] Cohen L 1995 Quality Function Deployment: How to Make QFd Work for You (USA: Addison-Wesley Publishing Company) pp 123
[8] Day Ronald G 1993 Quality Function Deployment, Linking a Company with Its Costumers (Milwaukee: ASQC Quality Press)
[9] Nurlandi F 2010 Desain Inkubator Bayi dengan Kontrol Otomatis yang Ekonomis untuk Klinik Persalinan *(Ecobator)* (Surabaya: ITS)

IOP Conf. Series: Materials Science and Engineering **309** (2018) 012104 doi:10.1088/1757-899X/309/1/012104

Effects of alkanolamide loading on swelling, rheometric and tensile properties of chloroprene rubber compounds

I Surya*[1,2], H Ismail[3]
[1]Department of Chemical Engineering, Universitas Sumatera Utara, Medan, 20155, Indonesia
[2]Sustainable Energy and Biomaterial Center of Excellence, Faculty of Engineering, Universitas Sumatera Utara, Medan, 20155, Indonesia
[3]School of Materials and Mineral Resources Engineering, Engineering Campus, Universiti Sains Malaysia, 14300, Nibong Tebal, Penang, Malaysia

*E-mail: indradanas@yahoo.com

Abstract. The effects of Alkanolamide (ALK) addition on swelling, rheometric and tensile properties of unfilled chloroprene rubber (CR) compounds were investigated. The ALK was prepared from Refined Bleached Deodorized Palm Stearin and diethanolamine and -together with magnesium and zinc oxides- incorporated into the CR compounds. The ALK loadings were 0.5, 1.0, 1.5 and 2.0 phr. It was found that ALK enhanced the cure rate and torque difference of the CR compounds. ALK also enhanced the tensile modulus and tensile strength; especially up to a 1.5 phr loading. The swelling test proved that the 1.5 phr of ALK exhibited the highest degree of crosslink density which caused the highest in tensile modulus and tensile strength.

1. Introduction

Polychloroprene or chloroprene rubber (CR) is a popular specialty elastomer. Its areas of application within the rubber field are widespread such as transmission belts, conveyor belts, moulded goods, cable jackets, seals, coated fabrics, etc. It is excellent in mechanical properties, ozone and oil resistances, flame retardancy, weatherability, special cohesiveness, moderate resistance to most chemicals and ease of processability [1-4]. The molecular structure of CR is similar to that of natural rubber (ref); except that chlorine has replaced the methyl groups [5-6]. The presence of chlorine causes the cure system of CR to be generally different from that of NR [7]. The chlorine atoms decrease the reactivity of double bonds on the CR backbone and hence, the reactivity with sulphur becomes less.

Magnesium oxide (MgO), zinc oxide (ZnO) and ethylene thiourea (ETU) are always used as the cure system for CR. ETU is used as the curing accelerator for CR. It is a toxic material and suspected to be carcinogenic [8]. Therefore, the appearance of an alternative curing accelerator for CR, which is capable of providing CR vulcanisates equivalent or superior to those provided by the ETU and both metal oxides, has been in demand.

This study reports the using of Alkanolamide (ALK) as a rubber additive for CR. The effects of ALK addition on crosslink density, rheometric and tensile properties of CR compounds were investigated.

2. Materials and Methods

2.1. Materials

Polychloroprene rubber (CR) [Skyprene B-30] was purchased from TOSOH Co. (Japan). Other compounding ingredients such as MgO, ZnO, stearic acid and sulphur were all obtained from Bayer Co. (M). Sdn. Bhd., Petaling Jaya, Selangor, Malaysia. All materials were used as supplied. The ALK was prepared using RBDPS and diethanolamine. The procedure of the ALK preparation was given in our previous report [9].

2.2. Compounding

A sulphur-accelerated curing system was used for compounding. The recipe for the preparation of the CR compounds is given in Table 1. The compounding procedure was done in accordance with the American Society for Testing and Material (ASTM) – Designation D 3184 – 80. Compounding was done on a two-roll mill. Table 1 also shows the designation and composition of the CR-based recipes used in this study.

Table 1. The composition and designation of the CR compounds

Ingredients	Content (phr)*	Designation				
		CR-0.0 (Control)	CR-0.5	CR-1.0	CR-1.5	CR-2.0
CR	100.0	100.0	100.0	100.0	100.0	100.0
Magnesium oxide	4.0	4.0	4.0	4.0	4.0	4.0
Zinc oxide	5.0	5.0	5.0	5.0	5.0	5.0
Stearic acid	1.0	1.0	1.0	1.0	1.0	1.0
Sulphur	2.0	2.0	2.0	2.0	2.0	2.0
ALK	0.0 – 2.0	0.0	0.5	1.0	1.5	2.0

*parts per hundred parts of rubber

2.3. Rheometric properties

The rheometric properties of the CR compounds were determined at 150 °C using a Monsanto Moving Die Rheometer (MDR 2000). The rheometric properties such as scorch time (ts_2), cure time (t_{90}), minimum torque (M_L) and maximum torque (M_H) were measured according to ASTM D2084. The CR compounds were subsequently compression moulded using a stainless steel mould at 150 °C with a pressure of 10 MPa using a laboratory hot-press based on the respective curing times.

2.4. Tensile properties

Dumbbell-shaped samples were cut from the molded sheets according to ASTM D 412-93. Tensile tests were performed at a crosshead speed of 500 mm/min. Tensile tests were carried out with a universal tensile machine Instron 3366 to determine the tensile properties in terms of tensile strength (TS), stresses at 100% and 300% elongations (M100 and M300) and elongation at break (EB).

2.5. Swelling properties

The swelling test was performed in toluene according to ISO 1817. Cured test pieces of the compounds of dimension (30 mm x 5 mm x 2 mm) were weighed using an electrical balance and swollen in toluene until equilibrium, which took 72 h at room temperature. The samples were removed from the liquid, the toluene was removed from the samples surfaces, and the weight was determined. Calculation of the change in mass is as follows;

$$\text{Swelling (\%)} = \frac{W_2 - W_1}{W_1} \times 100\% \qquad (1)$$

IOP Conf. Series: Materials Science and Engineering **309** (2018) 012104 doi:10.1088/1757-899X/309/1/012104

Where W_1 is the initial mass of specimen (gram) and W_2 is the mass of specimen (gram) after immersion.

3. Results and Discussion

3.1. The rheometric properties

Table 2 shows the rheometric, swelling and tensile properties of CR compounds with and without ALK. The addition of up to 2.0 phr of ALK into the unfilled CR compound produced CR with ALK compounds with shorter scorch (ts_2) and cure (t_{90}) times, compared to the control compound (CR-0.0). The ALK acted as a curative additive since it affected the rheometric properties of unfilled CR compounds. It increased the cure rate and it was considered as an accelerator. Any rubber additive that improves the action of a curing agent to speed up the resultant cure, even though it constitutes a very small part of a rubber compound, is classified as an accelerator [10]. Amine is an accelerator ingredient of rubber compounds [10] and hence, the amine constituent of ALK caused in the cure enhancement [11-13].

The higher the ALK loading, the longer were the scorch and cure times. It was due to ALK functioning as an additional curing agent (discussed later).

The torque difference of the CR-A0.5 compound was higher than that of the control compound. The torque difference is used as an indication of the crosslink density of a rubber compound [12, 14-15]. The higher the torque difference value, the higher is the degree of crosslink density. The addition of 0.5 phr of ALK improved the torque difference of the unfilled compound significantly. The improvement in torque difference has the same meaning as improvement in crosslink density and hence, ALK was considered as an additional curing agent for the unfilled CR compounds.

As discussed earlier, increasing the ALK loading up to 2.0 phr increased the scorch and cure times. Simultaneously, the torque difference value increased with up to 1.5 phr, but decreased beyond this loading. These phenomena was due to the function of ALK as an additional curing agent. The higher the loading, the more was the amount of ALK in the CR compounds as a reactant, and the more pronounced was its action as a curing agent over that of as an accelerator and hence, a longer time was needed to finish the curing process.

The decrease in torque difference value after the 1.5 phr of ALK was most probably due to the excessive amount of ALK which decreased the crosslink density.

3.2. The tensile properties

As presented in Table 2, the addition of up to 1.5 phr of ALK into the unfilled CR compound increased the M100, M300 and TS. However, further increase the ALK loading decreased those properties. The tensile properties of a rubber vulcanisate are dependent mainly on the degree of crosslink density [12, 15-16]. Again, the improvements of M100, M300 and TS up to 1.5 phr were due to the action of ALK as an additional curing agent, which increased the degree of crosslink density of the unfilled CR compounds. This explanation was agree with the trend of swelling-test result as shown in Fig. 1. It is well-known that the swelling-test result is related to the crosslink density of a rubber vulcanisate. A less solvent uptake or penetration into the CR compounds indicating a higher crosslink density [17]. The addition of a 1.5 phr of ALK exhibited the lowest swelling percentage of CR vulcanisate. This indicates the highest degree of crosslink density, which altered the interactions between CR segments into the strongest ones and hence, caused an increase in tensile strength.

The decreases in properties after a 1.5 phr was due to the excessive amount of ALK, which decreased the degree of crosslink density. Supposedly, the excessive amount of ALK formed some boundary layers, which dissolved a part of the curative system and hence, decreased the degree of crosslink density. This explanation is again agree with the result of swelling-test as shown in Fig. 1.

Table 2. The rheometric properties, swelling and tensile properties of unfilled CR compounds

Compounds	Rheometric properties			Tensile properties			
	ts_2, min.	t_{90}, min.	M_H–M_L, dN.m	M100, MPa	M300, MPa	TS, MPa	EB, %
CR-0.0	11.29	28.28	7.91	0.72	1.30	13.7	1612
CR-0.5	3.74	25.31	14.90	0.78	1.31	15.5	1590
CR-1.0	6.68	27.09	15.56	0.82	1.37	17.2	1537
CR-1.5	7.16	27.43	15.59	0.83	1.38	18.9	1487
CR-2.0	7.58	27.96	12.63	0.78	1.29	17.8	1516

The ALK decreased the EB by up to 1.5 phr, and then increased it slightly as the ALK loading further increases. EB depends mainly on the degree of crosslink density [14]. The decreasing of EB by up to 1.5 phr was simply due to the higher degree of crosslink density; which immobilised the CR segments freely. However, after the 1.5 phr of ALK loading, the EB increased. An explanation for this is again given by the excessive amount of ALK, which decreased the crosslink density and caused in a more free movement of the CR chains.

3.3. The swelling properties

Effect of ALK loading on the swelling percentage of unfilled CR compounds is shown in Fig. 1. The swelling percentage was measured by toluene uptake until equilibrium swelling was reached at room temperature. It is widely used that the swelling is related to the crosslink density of a network chain [17], with a less toluene penetration into the blends indicating a higher degree of crosslink density. As shown in Fig. 1, the swelling percentage of unfilled CR compounds decreased with the addition of 0.5 phr of ALK and decreased with further increases in the ALK loading up to 1.5 phr. After the 1.5 phr of ALK loading, the swelling percentage increased. The increases in swelling percentage was due to the excessive amount of ALK which decreased the degree of crosslink density.

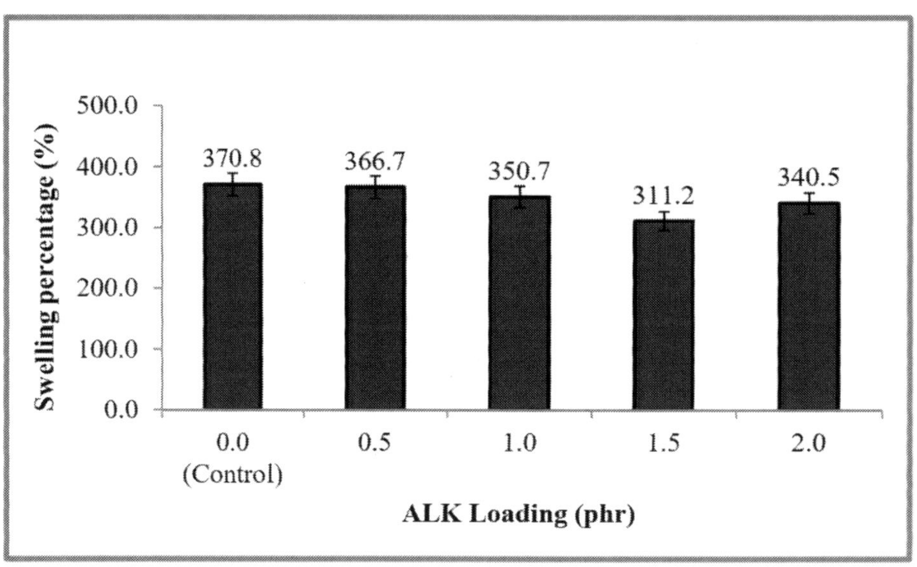

Figure 1. The effect of ALK loading on swelling percentage of unfilled CR compounds

4. Conclusions

From this study, the following conclusions were drawn:

1. Alkanolamide enhanced the cure rate of magnesium-zinc oxides vulcanised chloroprene rubber compounds.
2. Alkanolamide also enhanced the tensile properties and crosslink density of unfilled chloroprene rubber compounds especially up to 1.5 phr loading.
3. The enhancements of the cure rate, crosslink density and tensile properties were due to the function of ALK as a curative additive in unfilled chloroprene rubber compounds.

Acknowledgements

The authors would like to thank Universiti Sains Malaysia for providing the research facilities for carrying out the experiment and for making this research work possible. Authors - Indra Surya is grateful to the DRPM, Ministry of Research, Technology and Higher Education (Kemristekdikti) of the Republic of Indonesia, for research grant under Penelitian Hibah Bersaing scheme with contract No. 017/SP2H/LT/DRPM/II/2016, 17 February 2016.

References

[1] Rodgers B 2004 *Rubber Compounding: Chemistry and Applications* CRC

[2] Schatzel R, Cassell G 1939 Synthetic Elastic Polymers in the Cable Industry *Industrial & Engineering Chemistry* **31(8)** 945-949

[3] De S K, White J R 2001 *Rubber Technologists Handbook* Vol 1 Rapra Technology

[4] Ciesielski A 1999 *An introduction to rubber technology* William Andrew Pub

[5] Dick J S 2001 *Rubber technology : compounding and testing for performance* Hamser, Munich

[6] Surya I, Ismail H 2016 Alkanolamide as a novel accelerator and vulcanising agent in carbon black-filled polychloroprene rubber compounds *Plastics, Rubber and Composites* **45(7)** 287-293

[7] Akiba M, Hashim H 1997 Vulcanization and crosslinking in elastomers *Progress in polymer science* **22(3)** 475-521

[8] Smith D M 1984 Ethylene thiourea: thyroid function in two groups of exposed workers *British journal of industrial medicine* **41(3)** 362-366

[9] Surya I, Ismail H, Azura A 2013 Alkanolamide as an accelerator, filler-dispersant and a plasticizer in silica-filled natural rubber compounds *Polymer testing* **32(8)** 1313-1321

[10] Long H 1985 *Basic Compounding and Processing of Rubber, Rubber Division* American Chemical Society Inc The University of Akron, Ohio, USA

[11] Surya I, Ismail H, Azura A 2014 The comparison of alkanolamide and silane coupling agent on the properties of silica-filled natural rubber (SMR-L) compounds *Polymer testing*

[12] Surya I, Ismail H, Azura A 2015 The effect of alkanolamide loading on properties of carbon black-filled natural rubber (SMR-L), epoxidised natural rubber (ENR), and styrene-butadiene rubber (SBR) compounds *Polymer testing* **42** 208-214

[13] Surya I, Ismail H 2016 The effect of the addition of alkanolamide on properties of carbon black-filled natural rubber (SMR-L) compounds cured using various curing systems *Polymer testing* **50** 276-282

[14] Ismail H, Ng C 1998 Palm oil fatty acid additives (POFA's): Preparation and application *Journal of elastomers and plastics* **30(4)** 308-327

[15] Hayeemasae N, Surya I, Ismail H 2016 Compatibilized natural rubber/recycled ethylene-propylene-diene rubber blends by biocompatibilizer *International Journal of Polymer Analysis and Characterization* 1-12

[16] Ismail H, Chia H 1998 The effects of multifunctional additive and epoxidation in silica filled natural rubber compounds *Polymer testing* **17(3)** 199-210

[17] Nabil H, Ismail H, Azura A 2013 Compounding, mechanical and morphological properties of carbon-black-filled natural rubber/recycled ethylene-propylene-diene-monomer (NR/R-EPDM) blends *Polymer testing* **32(2)** 385-393

IOP Conf. Series: Materials Science and Engineering **309** (2018) 012105 doi:10.1088/1757-899X/309/1/012105

Passenger car equivalents of becak bermotor at road segment in Medan

M S Surbakti[1,2], I Sembiring[1]

[1]Civil Engineering Department, University of Sumatera Utara, Jl. dr. Mansur Kampus USU Medan 20155.
[2]Sustainable Transportation Research Center, Faculty of Engineering University of Sumatera Utara, Jl. Almamater Kampus USU Medan 20155

*Email: medissurbakti@yahoo.com

Abstract. The road traffic systems, travel patterns and other traffic characteristics are different for each country due to differences in the geometric patterns, available transport facilities for commuters, proportional and type of the vehicle itself and so on. In Indonesia, the standard of pce (Passenger Car Equivalent) value found on IHCM (Indonesian Highway Capacity Manual) published in 1997. IHCM stated that the value of pce for heavy vehicles and motorcycles are 1.3 and 0.5 respectively. On these day, regarding Medan as a third biggest city in Indonesia, there have been lot of changes with regarding to the composition of the vehicle, as well as variations of the type of the vehicle itself. Becak bermotor (motorized tricycles) is a vehicle which is widely available in the city of Medan. Data from Medan City Transportation Department stated that there are more than 20,000 motorized motorized tricycles vehicles operating in the city at these day. Pce value of these rickshaws will be calculated based on observations at road and intersections in Medan. The calculation result shows that the pce value of motorized rickshaw is more than 1. This value will make the calculations regarding the performance of the traffic can be performed more accurately.

1. Introduction

For Planning of road transport infrastructure in Indonesia, the Highway Capacity Manual (MKJI) was used as one of its references. The equivalence of passenger cars (pce) is a factor that indicates the effect of various types of vehicles compared to light vehicles on speed, ease of use, dimensions of light vehicles in traffic flow (for passenger cars and light vehicles of similar chassis pce = 1.0).
The value of the determination of the equivalence of passenger cars in MKJI is taken from the results of research in certain areas. This causes the value of the equivalent of passenger cars is not necessarily able to represent the characteristics of traffic that exists throughout Indonesia. Based on these conditions it is necessary to review the provisions of the existing passenger car equivalence to be adjusted to the conditions and characteristics of current in each region in Indonesia.

In a road plan, the types of vehicles expected to cross the road are heavy vehicles, light vehicles and motorcycles. Motor vehicles are an important means of land transportation for the community to facilitate the mobility of people and goods. Along with the increasing need for movement of people and goods, the number and volume of motor vehicles increases rapidly, this will result in among others; Congestion, frequent accidents and air pollution is detrimental to health. Road segment will be congested if the capacity of the road is not sufficient to accommodate the volume or current through the road, in other words the traffic volume exceeds the existing road capacity.

One of the important problems in the city of Medan is a motorized rickshaw. This vehicle in the city of Medan is a vehicle used to transport people from one place to another. The existence of motor rickshaws is currently experiencing a very significant growth number, this will certainly affect the speed of other vehicles in a road, because this motor vehicle pedicab has a low acceleration and deceleration.

For that reason, it is necessary to carefully influence the effect of pedicab motor vehicles on the traffic performance and need to study the equivalence value of passenger cars for motor tricycles vehicles, related to their role as part of the traffic on the road segment. The study was conducted on Gajah Mada Street in Medan city which is a 4 lane road 2 direction without median. The selection of the location is based on the volume of pedicab motor vehicles in the location.

2. Passenger Car Equivalent

In the Indonesian Road Capacity Manual (MKJI) 1997 [1], the value of traffic flows reflects the composition of traffic, stating the flow of traffic in units of passenger cars (pcu). All traffic values (per direction and total) are converted to units of passenger cars (pcu) using pce irically derived passenger cars (pce) derived for the following types of vehicles:

- *Light vehicles (LV) including passenger cars, minibuses, pik-ups, small trucks and jeeps.*
- *Heavy vehicles (HVs) including trucks and buses*
- *Motorcycle (MC).*

The equivalence of passenger cars (pce) for each type of vehicle depends on the type of road and the total traffic flow expressed in the vehicle / hour. All pce values for different vehicles are shown in Table 1 bellow.

Table 1. Passenger car equivalent (pce) for undivided road (UD)

Road Type	Total Traffic for all direction (veh/hour)	pce: HV	pce: MC Width of lane, Wc(m) ≤6	pce: MC Width of lane, Wc(m) >6
Two Lane Undivided (2/2UD)	0 to 1800 >1800	1,3 1,2	0,5 0,35	0,4 0,25
4 lane undivided (4/2UD)	0 to 3700 >3700	1,3 1,2	0,4 0,25	

Source: Manual Kapasitas Jalan Indonesia, 1997 p. 5-38

Passenger Car Equivalence (pce) is a factor that indicates the effect of different types of vehicles compared to light vehicles on speed, ease of use, dimensions of light vehicles in traffic flow (for passenger cars and light vehicles of similar chassis pce = 1.0) [1].

There are several ways or methods that can be used to estimate the equivalence value of passenger cars, depending on the characteristics and conditions of traffic. The way or method that can be used to find or estimate the equivalence of passenger car (pce) are:

2.1. Capacity base

The pce counting method based on the capacity according to Chang Chen (1978) in Murtiono ES [2], is a method of modal mode comparison, modified and adapted to the research conditions of survey results obtained from various modes of transport. With these various combinations can be searched the

IOP Conf. Series: Materials Science and Engineering **309** (2018) 012105 doi:10.1088/1757-899X/309/1/012105

equivalence value of passenger cars from motorized tricycles with multiple linear regression which formulates as follows

$$Q = b_1\, qlv + b_2\, qhv + b_3\, qmc + b_4\, qbck \qquad (1)$$

Where:
Q = Traffic Volume (pcu/hour)
b = Coefficient
qlv = Number of Light Vehicle
qhv = number weight of vehicle
qmc = number of motorcycle
qbck = number of becak motor/motorized tricycles
which b_1 = pce for lv = 1 ;

$b_1\, qlv = Q - b_2 qhv - b_3 qmc - b_4 qbck$

From the equation, the coefficient generated on each vehicle type is the pce value of the vehicle type.

2.2. Speed Basis
Pce calculation method with base speed according to Van Aerde and Yagar (1984), that is by knowing relation of velocity (v) and traffic volume (q) by using multi linear regression.
The linear model of velocity and volume relationships is chosen because in practice the relationship between volume and velocity is close to linear. Multiple regression models of the relationship of speed and volume are:

$$V = a - b_1\, qlv - b_2\, qhv - b_3\, qmc - b_4\, qbck \qquad (2)$$

Where:
V = average speed (km/h)
a = free current velocity (km/h)
b = coefficient
qlv = number of light vehicles
qhv = number of heavy vehicles
qmc = number of motorcycles
qbck = the number of motorized tricycles

To determine pce vehicles other than passenger cars the coefficient of each vehicle is divided by the coefficient of passenger car (lv) and can be formulated:

$$pce_i = \frac{b_i}{b_1} \qquad (3)$$

Where:
bi = vehicle type coefficient i
b1 = passenger car coefficient (lv)

3. Analysis and Discussion
The result of regression analysis of motorbike proportion relation to traffic flow velocity, ploted in graph form as shown in Figures 1 and 2:
From the analysis survey result in front of gas station (location 1), at interval of vehicle volume 1448-1872 pcu/hour, got relationship between proportion of motorized tricycles with speed of traffic (r) - 0.143. This shows that there is a very low relationship. While the direction of the relationship is negative, it indicates that the larger the motorized tricycles proportion the smaller the speed of traffic at that location.

From the analysis results in front of LP3i (location 2), simple correlation (r) obtained, at the vehicle volume interval 1475-1930 pcu / hour obtained between the proportion of motorized tricycles with the traffic speed is (r) -0.304. This shows that there is a low relationship between the two parameter. While the direction of the negative relationship, indicating that the greater the proportion of the pedicab, the speed of traffic in the location will be smaller.

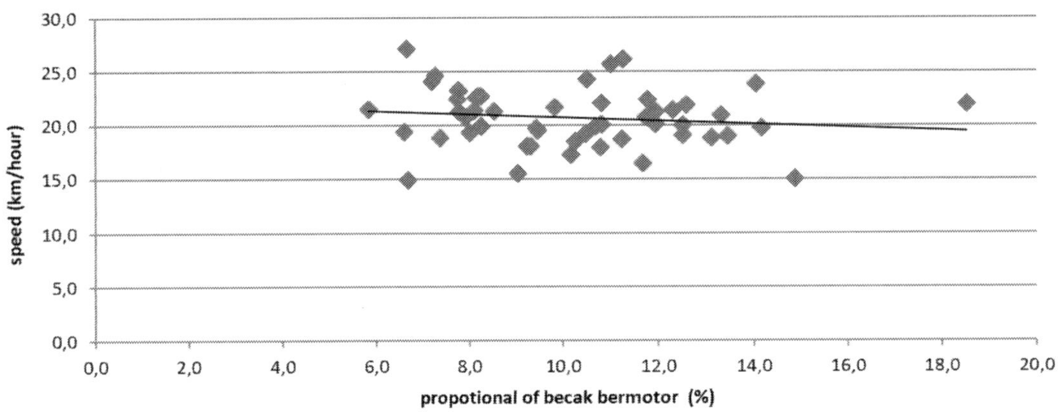

Figure 1. Graph of influence of motorized tricycles vehicle proportion to speed of traffic at location 1 (Jalan Gajah Mada).

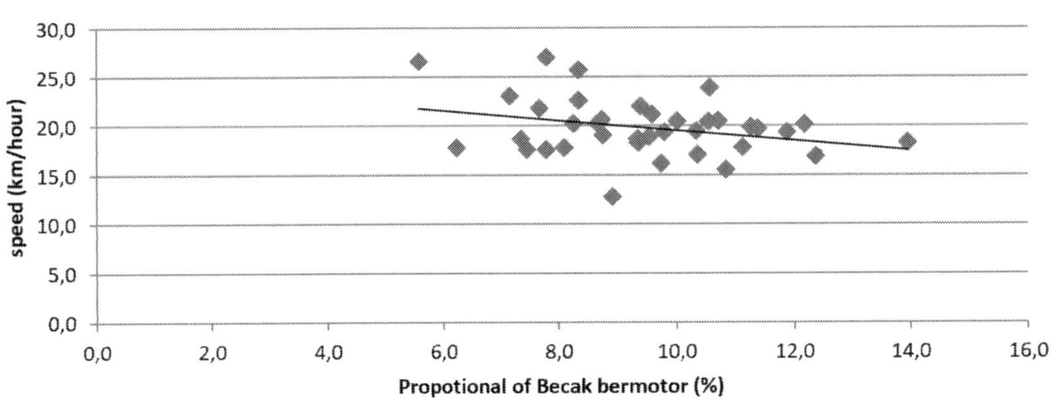

Figure 2. Graph of influence of motorized tricycles vehicle proportion to speed of traffic at location 2 (Jalan Gajah Mada).

In order for the value of pce to be a reference, it is necessary to have an pce value that can represent each type of vehicle. To obtain a representative pce value, the calculated average pce values for both sites were surveyed (Table 2).

Table 2. Recapitulation result of pce

Vehicle Type	pce		Number of Vehicle	
	Location 1	Location 2	Location 1	Location 2
LV	1	1	5470	5551
HV	1.99	1.89	110	83
MC	0.54	0.55	13021	11912
BCK	0.84	1.13	1915	1765

From the data available in Table 2, we can find the average pce per type of vehicle and compare it to the MKJI as described at Table 3.

Table 3. The differences of pce between MKJI and Analysis

	Total Volume, 2 Way (veh/hour)	LV	HV	MC	BCK
Pce MKJI 1997	$\geq$ 3.700	1	1.2	0.25	-
pce survey 2016	$\geq$ 3.700	1	1.95	0.54	0.98

There is a difference between the three pce values, which occur because:
a. pce at MKJI in 1997 is a summary of typical traffic in some cities in Indonesia while pce in survey 2016 is an actual condition in Medan.
b. There is a difference of pce value in the field with the pce value on MKJI because of the difference of initial assumption that "becak bermotor" or BCK value is same as motorcycle (MC).

4. Conclusion
The results obtained from the analysis can be concluded as follows:
1. The proportion of motorized tricycles vehicles has a low effect on the average speed of traffic on Jln. Gajah Mada.
2. Pce values obtained in Jln. Gajah Mada stated that the traffic density had a significant effect on the average speed.
3. From the analysis results obtained pce values for traffic on the road Gajah Mada is LV = 1, HV = 1.95, MC = 0.54, BCK = 0.98.
4. There is a difference of average pce value from Gajah Mada street with pce value on MKJI because of differences in the initial assumption that motorized tricycles value is same as motorcycle, this is done because the proportion of motorized tricycles is still small.

References
[1] Direktorat Jenderal Bina Marga 1997 Manual Kapasitas Jalan Indonesia (MKJI), Jakarta
[2] Morlok E K 1988 *Introduction to Transport Engineering and Planning*, Erlangga, Jakarta
[3] Koeswandono W 2007 The Influence Of Non Motorized Motorists On The 2 road lane Without a Median. Yogyakarta.Murtiono, ES. 2002. Effect of Motorcycles At Intersection With Traffic Control In Kendal, Simposium V FSTPT

IOP Conf. Series: Materials Science and Engineering **309** (2018) 012105 doi:10.1088/1757-899X/309/1/012105

[4] Tamin O Z 2008 *Transportation Modelling and Planning*: problems example and applications, ITB, Bandung

[5] Lubis HA et. al. 2005 Urban Toll Road User Value of Time Saving. Journal of The Eastern Asia Society for Transportation Studies, Vol. 4, Issue No. 6, pp 137-148

[6] McShane WR 190 *Traffic Engineering*, Prentice Hall, New Jersey

IOP Conf. Series: Materials Science and Engineering **309** (2018) 012106 doi:10.1088/1757-899X/309/1/012106

Effect of pH and complementary ion concentration on nitrate removal using puroliteA400 Resin impregnated Cu in batch system

M Turmuzi[1*], Z N Tarigan[2], L Nadapdap[2], F Batubara[1]
[1] Department of Chemical Engineering, Universitas Sumatera Utara, Jalan Alumni 3 Kampus USU, Medan 20155
[2] Department of Environmental Engineering, Universitas Sumatera Utara, Jalan Almamater Kampus USU, Medan 20155

*Email: turmuzi@usu.ac.id

Abstract: The total nitrogen content in water bodies should be below 50 mg NO_3^-/L (11.3 mgN/l) World Health Organization (WHO) 2006. The content of nitrogen exceeding the quality standard threshold will cause damage to the aquatic ecosystem and be carcinogenic to humans. The Purolite A-400 resin will be modified with Cu metal by batch method to see the adsorption allowance of nitrate in synthetic liquid waste with nitrate concentration of 50 mg/l. This study will evaluate the effect of pH and complementary ions on the adsorption process. From the result of the research, the second order pseudo model is the most suitable adsorption kinetics model. For the adsorption isotherms the most suitable model is the Freundlich adsorption isotherm model. The optimum pH conditions were at the range of 8.5. The addition of complementary ions sulfate and phosphate did not show any significant change, but sulfate is the most effective complementary ion with a content of 20 mg/l.

1. Introduction

Nitrate is an essential nutrient for plant growth and microorganisms, but the excess content in water bodies causes disruption of ecosystems in water (eutrophication) and is harmful to humans [1]. The excessive use of chemical fertilizers in the agricultural sector and poor waste water treatment systems are sources of nitrate pollutants [2].Under World Health Organization (WHO) regulations 2006, the permitted nitrate content in water bodies is 50 mg NO_3^-/L (11.3 mg N/l). Reclamation and reuse of wastewater is one of the solutions in water supply. In recent decades, the wastewater treatment industry has identified that there are inorganic anionic materials such as nitrate, phosphate, and fluoride in water bodies [3].

There are many technologies available for nitrate removal such as Reverse Osmosis, Electrodialysis, Biological Denitrification, and Ion Exchange [2]. The previous case study states that ion exchange resins can adsorb nitrate optimally [2]. Ion exchange resins are a method capable of removing various types of pollutants in wastewater, with operational, easy, economical and capable of handling large waste loads [3],[4] [5]. There are several types of anion exchange resins used in the previous study: Purolite A520E [1]; [6]), IONAC SR-7 [7], Amberlite IRN-78 [8], IND NSSR [9], Indion NSSR [10], Amberlite IRA 400 [11]. The results of the study showed that ion exchange technology using resin as a cation or anion exchanger from the solution phase proved very effective in removing the nitrate content. The resin adsorption capacity by anion exchange can be increased by metal impregnation method. Metal impregnation can increase the positive charge on the resin surface. The ion exchange resin method by impregnation of metals using various types of resins can eliminate the content of Cd, Hg, and Pb [12].

Resin impregnated with metal causes the surface of the resin to be positive which can increase the ability to adsorb nitrate [8]. The ion exchange resin method with metal impregnation has several advantages such as soluble selectivity selectivity characteristics at simple operation (fixed bed column method and fluidized bed) [12]. In this study, a Purolite A-400 resin was impregnated with Cu metal to remove nitrate ions.

2. Research Method

2.1. Resin Purolite A400 Impregnation
The Purolite A400 resin is obtained from Purolite Pte Ltd, Singapore which is a strong base anion exchange resin, faded yellow spherical beads with a particle size of 300-1200 μm. The Purolite A400 resin is impregnated with copper (Cu) metal. 10 g of Purolite resin is introduced into 1 L $CuCl_2.2H_2O$ concentration, stirred for 1 hour at a stirring rate of 120 rpm. The pH was increased to 8.0 by adding a 1 M NaOH solution and stirred for 1 hour at a stirring rate of 30 rpm. The resin is filtered and washed with distilled water and dried in the oven for 24 hours at ± 45 ° C [1].

2.2. Influent Solution
Synthetic waste is prepared by dissolving the crystalline powder of $NaNO_3$ into distilled water at concentrations of 50 mg/l.

2.3. Nitrate Analysis
The final concentration (Ct) of nitrate was analyzed by Pastel UV Analyzer Secomam RS232 and nitrite content analyzed by Cary 50 Conc UV-Visible Spectrophotometry-Variant.

2.4. Batch Study
The Batch test was performed using Water Bath where the results of the research will be incorporated into kinetics model and adsorption isotherm on pure resin and Cu impregnation resin. The efficiency determination of the nitrate removal uses the following equation.

$$E = \frac{C_{1-C_2}}{C_1} \times 100\% \quad (1)$$

which:

E = Efficiency
C_1 = Pre-treatment Concentration of Nitrate or Nitrite
C_2 = After treatment Concentration of Nitrate or Nitrite

2.5. Effect of pH on adsorption test
1g of adsorbent is placed on an erlenmeyer containing 50 mg/l of nitric synthetic waste. pH was varied (7, 7.5 and 8.5) by adding a NaOH 1 M or HCL 1 M solution then stirred at a rate of 120 RPM.

2.6. Kinetics Adsorption
The adsorption kinetics test is carried out by inserting 1g of Purolite A-400 pure adsorbent resin and Purolite A-400 resin Cu impregnation at optimum pH (procedure 2.5) into 50 mg/l of nitrate synthetic waste. Erlenmeyer stirred at Water Bath with a speed of 120 rpm of effluent from nitrate will be taken and tested at 40, 80, 120, 160, and 200 minutes.

The amount of nitrate (qt, mg N/g) adsorbed over time is calculated using the following equation.

$$q_t = \frac{(C_o - C_t)\,V}{M} \tag{2}$$

C_o is the initial concentration of nitrate (mg N / l),
C_t is the concentration of nitrate at time t (mg N / l),
V is the volume of the solution, and
M is the adsorbent mass (g).

The adsorption kinetics model used is the first order Pseudo kinetics model and the second order Pseudo with the following equation.

First order pseudo model:

$$\frac{dq_t}{t} = K_1\,(q_1 - q_t) \rightarrow log q_1 - \frac{k_1 t}{2.303} \tag{3}$$

Second order pseudo model:

$$\frac{dq_t}{t} = K_2 (q_2 - q_t)^2 \rightarrow \frac{t}{qt} = \frac{1}{k_2 q_{2}{}^{2}} + \frac{t}{q_2} \tag{4}$$

Which:
q_t =Absorption of nitrate at time t (mg/g)
$q_1 \& q_2$ =Maximum adsorption capacity for first and second order pseudo (mg/g)
$(k_2 \& k_t)$ =Constants for nitrate and nitrite in the sorption process for pseudo first order $(minute^{-1})$ and second order.

2.7. Isotherm Adsorption
To determine the adsorption equilibrium of Purolite A-400 pure resin and Purolite A-400 Cu impregnation, this research used Langmuir and Freundlich model equations. Langmuir models that apply to surface adsorption monolayers are:

$$\frac{1}{q} = \frac{1}{q_{max} \times K_L} \times \frac{1}{C_e} + \frac{1}{q_{max}} \tag{5}$$

q_{max} =Maximum capacity of adsorbed nitrate per unit of sorbent weight for monolayer formation in the surface area of the adsorbent (mg/g)
K_L =Constants associated with the affinity of the region bound by the sorbent (Langmuir isotherm constants) (mg^{-1})

Langmuir Q_{max} dan K_L parameter values are calculated by slash and intercept of linear plot 1/q vs 1/Ce, to predict the efficiency of the adsorption process, calculated using the equation:

$$r = \frac{1}{1 + K_L\,C_o} \tag{6}$$

r = Dimension of adsorption quantity

The Freundlich isotherm model corresponds to the adsorption capacity and intensity of each adsorption, applied to non-ideal adsorption with surface monolayer in repeated experiments with the following equation.

$$q_e = Kf\,Ce^{1/n} \tag{7}$$

Ce = nitrate concentration at the time of equilibrium (m N/L)
Qe = amount of nitrate absorbed per unit of adsorbent mass at the time of equilibrium (mgN/g)
Kf = Freundlich adsorption isotherm constant corresponding to the adsorption capacity

level

1 / n = the intensity of adsorption, which varies according to the heterogeneity of the material (the range of values n = 1-10)

2.8.Influence of complementary ions on adsorption test

Effect of sulfate and phosphate complementary ions on nitrate removal with concentrations (10, 20, and 30 mg / l). The test was performed at a concentration of 50 mg / l nitrate synthetic waste and 1 g of adsorbent. Sulfate and phosphate ions are chosen because they are often found at high concentrations in surface water and groundwater just like nitrates.

3. Result and Discussion

3.1. Effect of pH on Nitrate Adsorption

pH is a factor that greatly affect the adsorption rate. The pH may affect the charge on the surface of the adsorbent during the ion exchange process [3]. The effect of pH on nitrate removal can be seen in Figure 1. The maximum adsorption capacity on Purolite A-400 resin without modification is at pH condition 7.5. When the pH>7.5 adsorption capacity continues to decline, this may be due to competition between Cl^- and NO_3^- on the resin surface due to the addition of HCL and NaOH to the pH setting [13].

In the Purolite A-400 modified Cu resin obtained the optimum pH range at pH 8.5, it can be seen in Figure that the pH base also increases the adsorption rate [12, 13, 14].

3.2. Effect of Complement Ion on Nitrate Adsorption

The effect of sulfate and phosphate complementary ions on nitrate adsorption capacity on Purolite A-400 pure resin and Purolite A-400 resin Cu impregnation at optimum pH is shown in Figure 2. From the figure shows that the efficiency of nitrate removal increases with the increase of complementary ion concentration. Optimum occurs at a complementary ion concentration of 20 mg/l. After that, the nitrate removal efficiency tends to decrease with the increase in the complementary ion concentration, this is due to the competition between complementary ions [13]. From the figure also shows, the removal efficiency is higher with the presence of sulfate ions than the phosphate ions. The results of this test are consistent with previous research which states that the removal capacity with the addition of sulfate complementary ions is greater than phosphate [3].

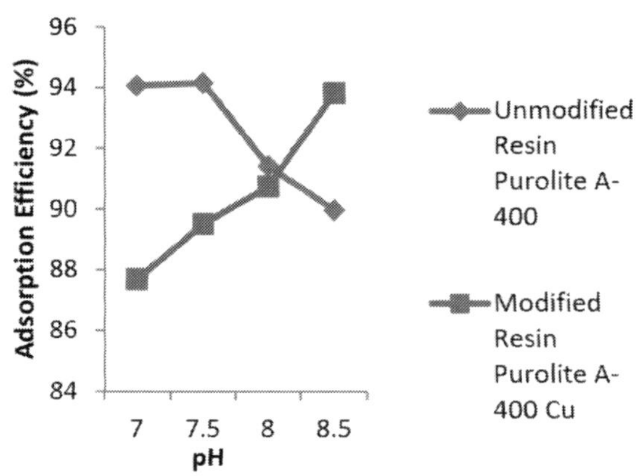

Figure 1. Effect of pH on Adsorption Efficiency

3.3. Contact Time Against Nitrate Adsorption Rate

The contact time test is conducted to determine the time required by the adsorbent to achieve the adsorption equilibrium and to know the maximum adsorbent ability in absorbing nitrate. Figure 3 shows the relationship between contact time and the absorbed nitrate concentration. The picture shows the increasing of contact time so that the adsorption efficiency will increase until equilibrium is reached. The optimum nitrate ion absorption process at Purolite A-400 pure resin at 160 min contact time and Purolite A-400 impregnation Cu resin at 120 min. Modified Cu resin Purolite higher efficiency of removal than unmodified resin this proves that the surface of the modified resin has greater adsorption efficiency [1], [3], [13].

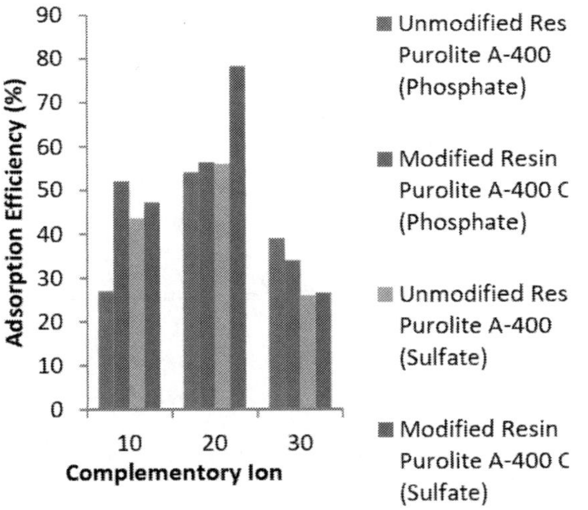

Figure 2. Effect of Complementory Ion on Adsorption Efficiency

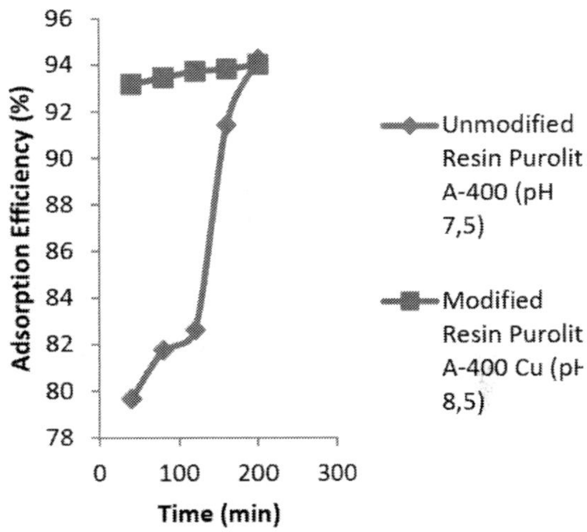

Figure 3. Effect of Time on Adsorption Efficiency

3.4 Kinetics Adsorption

The determination of the adsorption kinetics serves as a determinant of the rate of adsorption adsorption rate. The adsorption rate greatly affects the residence time of a treatment [15]. To review the adsorption kinetics in this study, we use first order and second order pseudo pseudo model. From equation 3 in the plot of the vs. t log curve (qe-qt) for the first-order equation presented in Figure 4 and 5, and equation 4 is plotted curve t vs $\frac{t}{q_t}$with slope $\frac{1}{q_e}$ and intercept $\frac{1}{k_2 {q_e}^2}$ for the second order Pseudo model presented in Figures 6 and 7.

Based on the analysis in Figures 4, 5, 6, and 7 it can be concluded that the most suitable kinetic model is a first-order model. This is indicated by the value of the correlation coefficient closest to the number one. The selection of a second-order model indicates that the adsorption process occurs chemically [6]. In a previous study using an Amberlite IRA 400 kinetics model after the nitrate removal was a second order pseudo model [11].

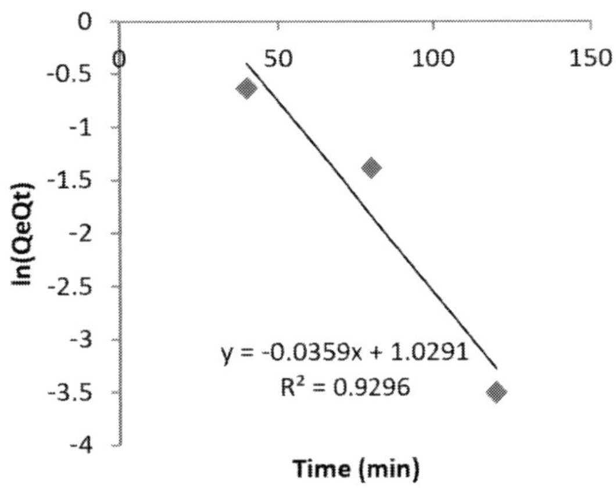

Figure 4. Pseudo First-Order Unmodified Resin Purolite A-400 Kinetic Model

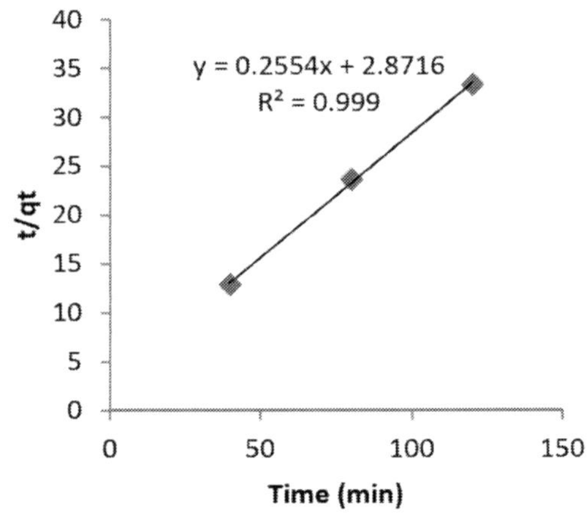

Figure 5. Pseudo Second-Order Unmodified Resin Purolite A-400 Kinetic Model

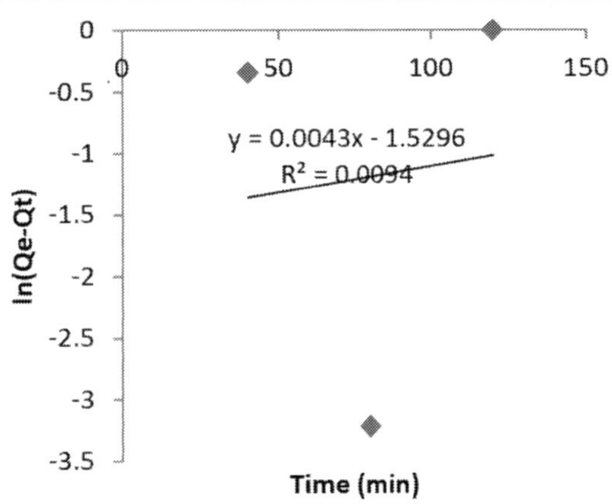

Figure 6. Pseudo First-Order modified Resin Purolite A-400 Cu Kinetic Model

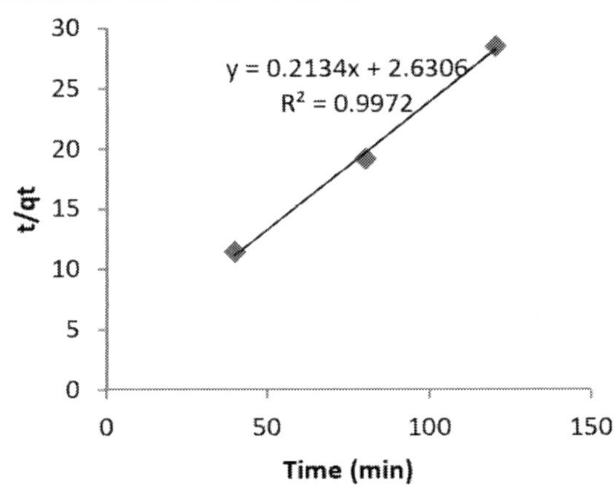

Figure 7. Pseudo Second-Order modified Resin Purolite A-400 Cu Kinetic Model

3.5. Isotherm Adsorption

The adsorption isotherm model aims to find out the characterization of adsorption of nitrate on the adsorbent so that the adsorption process is carried out chemically or physically [12].In this study, adsorption isotherm data was fitted on Langmuir and Freundlich models. The Langmuir model (equation 5) is shown in Fig. 8, while the Freundlich model (equation 7) is shown in Fig. 9. Based on Figures 8 and 9 it can be seen that the value of the correlation coefficient r^2 is greatest following the Freundlich model than the Langmuir model. Freundlich isotherm model signifies that the adsorption process takes place physically [11].

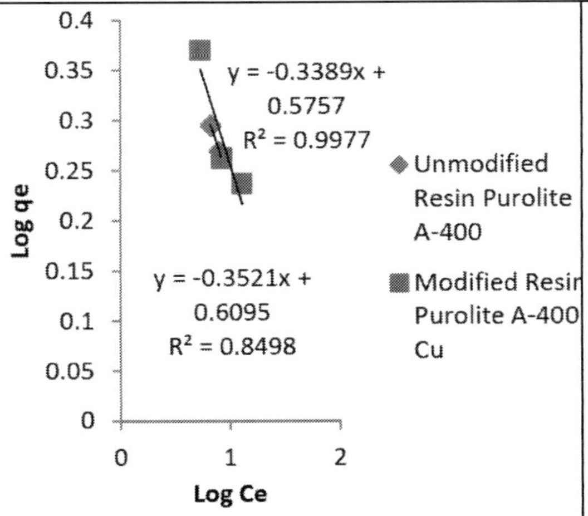

Figure 8. Isotherm Adsorption Langmuir Model

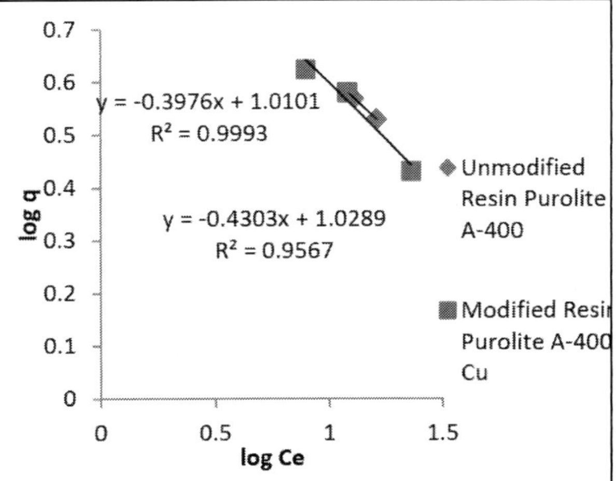

Figure 9. Isotherm Adsorption Freundlich Model

Table 1 shows the value of the adsorption intensity and Freundlich constant value obtained from the equation in the figure above, can be seen in the table of adsorption intensity value at nitrate removal more effectively by using Purolite A-400 impregnation Cu resin because the value of adsorption intensity is greater than resin without Modification with value 2,325 (mg N/g). The greater the value of Kf, the greater the intensity of the adsorption, it can be concluded that the Purolite A-400 resin impregnation of Cu is greater in the adsorption capacity compared with the purolite A-400 pure resin.

Table 1. Isotherm Parameters for the Adsorption Nitrate by Resin

Freundlich Model	Nitrate	
	Unmodified Resin Purolite A-400	Modified Resin Purolite A-400 Cu
Intensity of Adsorption (mgN/g)	2,518	2,325
K_F (l/mg)	4,321 x 10^{-3}	0,011
R^2	0,999	0,956

4. Conclusion

Purolite A-400 impregnation Cu resin has a higher efficiency of nitrate adsorption, the optimum pH found in this study for Purolite A-400 pure resin under pH 7.5 conditions and for Purolite A-400 impregnation Cu resin at pH 8, 5. The most suitable adsorption isotherm model for nitrate discharge is Freundlich isotherm model. This is indicated by R2 value of 0.999 in unmodified resin and 0.956 in Cu impregnation resin. The most appropriate kinetic model equation in nitrate removal is a second order model for purolite A-400 pure resin and Purolite A-400 impregnation Cu resin. The addition of sulfate complementary ions at a dose of 20 mg/l can improve the efficiency of the nitrate removal.

5. References

[1] Nur T, Shim W G, Loganathan P, Vigneswaran S and Kandasamy J 2014 *International Journal of Environmental Science and Technology* **12(4)** 1311
[2] Kapoor A and Viraraghavan1997 *Journal of Environmental Engineering* **123(4)** 371
[3] Nur T 2014 Thesis School of Civil and Environmental Engineering Faculty of Engineering & Information Technology: University of Technology Sydney
[4] Daud M, Khan Z, Aisha A, Danish I and Qazy I 2015 *Journal of Nanotechnology* **2015** 7
[5] Hekmatzadeh A A, Karimi-Jashani A, Talebbeydokhti N and Kløve B 2012 *Desalination* **284 (4)** 22
[6] Bulgariu L, Ceica A, Lazar L, Cretescu I and Balasanian I 2010 *Revista de Chimie* **61** 1136
[7] Cyplik P, Czaczyk K, Cyplik P, Marecik R and Grajek W 2008 *Nauka Przyroda Technologie* **2(1)** 1-8
[8] de Heredia J B, Dominguez J R, Cano Y and Jimenez I 2006 *Appl. Surf. Sci.* **252** 6031-5
[9] Farajpourlar M, Ram Mohan R S and Basava R V V 2013 *IOSR Journal of Environmental Science, Toxicology and Food Technology* **6(1)** 2319-99
[10] Chabani M, Amrane A and Bensmaili A 2006 *Chemical Engineering Journal* **125** 111-7
[11] Hanafi H A and Azeema S M A 2016 *Journal of Environmental & Analytical* **6** 346
[12] Khan AM, Ahn Y, Kumar M, Lee W, Min B, Kim G, Cho D, Park W B and Jeon B 2012 *Sep. Sci, Technol.* **46**, 2575-84
[13] Xu X, Gao B, Zhao Y, Chen S, Tan X, Yue Q, Lin J and Wang Y 2012 *Journal of Hazardous Materials* **203-204** 86-92
[14] Gefin Y, 2012 Thesis, Universitas Indonesia. Depok
[15] Milmile S N, Pande J V, Kamakar S, Bansiwal A, Chakrabarti A and Biniwale R B 2011 *Desalination* **276** 38-44

IOP Conf. Series: Materials Science and Engineering **309** (2018) 012107 doi:10.1088/1757-899X/309/1/012107

Biodiesel production from rice bran oil by transesterification using heterogeneous catalyst natural zeolite modified with K_2CO_3

Taslim*, Iriany, O Bani, S Z D M Parinduri, and P R W Ningsih
Department of Chemical Engineering, Faculty of Engineering,
Universitas Sumatera Utara, Medan 20155, Indonesia

*E-mail: taslim_hr@yahoo.co.id

Abstract. In the present study, an effort had been made to use natural zeolite from Tapanuli Utara, North Sumatera as a potential catalyst for biodiesel production. Biodiesel production is usuallythrough transesterification, and a catalyst is employed to improve reaction rate and yield. In this research rice bran oil (RBO) was used as feedstock. The objective of this work was to discover the effectiveness of natural zeolite modified by K_2CO_3 as catalysts in biodiesel production from RBO. K_2CO_3/natural zeolite catalyst modification was by impregnation method at various K_2CO_3 concentrations followed by drying and calcination. Transesterification was conducted at 65°C and 500 rpm. Effect of process variables such as the amount of catalyst, reaction time, and the molar ratio of methanol to RBO was investigated.The maximum yield of 98.18% biodiesel was obtained by using 10:1 molar ratio of methanol to RBO at a reaction time of 3 hours in the presence of 4 w% catalyst. The obtained biodiesel was then characterized by its density, viscosity and ester content. The biodiesel properties met the Indonesia standard (SNI).The results showed that natural zeolite modified by K_2CO_3 was suitable as a catalyst in the synthesis of biodiesel through transesterification from RBO.

1. Introduction

Biodiesel is usually synthesized by transesterification of oil or fat with the aid of a catalyst. While the catalyst can either be homogeneous or heterogeneous catalysts, the later have some advantages over the other, such as less corrosive, easierto handle and separate, reusable and generating less amount of toxic wastes [1,2].

Natural zeolite is a potential catalyst for biodiesel production due to its abundance and low cost. Zeolite is a crystalline aluminosilicate with 3-dimensional structures that form uniform pore size [3]. Because of its physical and chemical properties, it has been utilized as absorbent, ion-exchange resin, and catalyst with high activity [4,5,6]. It can be modified to further increase its activity to enable higher biodiesel yield. Zeolite modification can be by acid or base impregnation. Based on previous research, impregnation with base is better than acid [7].

The use of KOH for natural zeolite modification had been reported in literature [8], however there is no study on K_2CO_3 usage for natural zeolite modification used as a catalyst in biodiesel production through transesterification. Therefore, this research aimed to assess the effect of K_2CO_3 in biodiesel synthesis from methanol and RBOas well as evaluating the properties of biodiesel produced and comparing it with existing standards.

2. Materials and Methods

2.1. Materials
Natural zeolite was obtained from Tapanuli Utara, North Sumatera, Indonesia. Before usage, natural zeolite was pulverized using ball mill, then treated with 30%(v/v) hydrogen peroxide solution under manual stirring for several minutes to remove impurities. Afterward, the solution was separated by heating in a water bath till dried. The zeolite was then washed with aqua dest and oven dried for 24 h at 110°C [8]. Dried zeolite was milled with mortar (micro hammer mill) to produce natural zeolite powder at 140 meshes. Refined rice bran oil (RBO) was purchased from local market, while analytical grade methanol and potassium carbonate were obtained from Rudang Jaya, Medan. The standard fatty acid methyl ester for FAME analysis was purchased from Sigma Aldrich.

2.2. Characterization of RBO
In this work, determination of density and kinematic viscosity was according to SNI-04-7182-2006.Analysis of free fatty acid content andfatty acid composition was according to SNI-01-0018-2006.

2.3. Catalyst preparation
Prepared natural zeolites were impregnated in K_2CO_3 solution. Solution concentration was varied from 15/60 to 55/60 (g K_2CO_3/ml aqua dest). The process was carried out in a three-neck flask equipped with condenser reflux, thermometer, and magnetic stirrer at fixed mass ratio of natural zeolite to the K_2CO_3 solution of 1:4 at 60°C for two h. Afterward, the mixture was oven driedat 60°C for 24 h. The modified catalyst was then separated fromthe K_2CO_3 solution by vacuum filtration. Next, the catalyst was dried inan oven at 110°C for 24 h to remove the water, then calcined in a furnace at 450°C for four h [8]. After calcination, itwas refined with a mortar to 140 meshes then stored forlater use. The catalyst was characterized for its potassium content by using Atomic Absorption Spectroscopy (AAS).

2.4. Transesterification
Transesterification adopted procedures reported by Kusuma et al. [8]. Methanol, at a molar ratio of RBO to methanol of 1:8; 1:10; and 1:12, and K_2CO_3/natural zeolite catalyst, at 2 – 4 %w, were prepared. The methanol andcatalyst were mixed inathree-neck flask on a hot plate and heated to 65°C at 500 rpm. Afterward, 125 ml RBO was addedslowly into the flask and reaction carried on for2 – 4 h. At the end of the reaction, the catalyst was separated from the reaction mixture by vacuum filtration, and thefiltratewas placed in a separating funnel for 24 hours to form 2 layers. The lower layer (glycerol)was removed, and the upper(methyl ester/biodiesel) was washed with aquadest at 60°C till the washing water was clear to remove impurities and remaining catalysts. The resulting biodiesel was heated to 105°C to remove residual water. Obtained biodieselwas then weighed and analyzed.

3. Results and Discussion

3.1. Characteristic of RBO
The results of RBO characterization were as follow: the density at 25°C was 890 kg/m^3, the kinematic viscosity was 43 mm^2/s, and free fatty acid content was 0.22%. The chemical composition of RBO was analyzed using gas chromatography (GC), and the results are presented in Table 1.

3.2. Catalyst characterization
Preparation of K_2CO_3/natural zeolite catalyst was performed with nine variations of K_2CO_3 solution concentration to obtain highest potassium content in the catalyst. Figure 1 shows the potassium content in natural zeolite analyzed by AAS. The binding of potassium to the main skeleton of natural zeolite is illustrated in Figure 2.

Table 1. Fatty acid composition of RBO.

Fatty Acid	Composition (%)
Lauric acid (C12:0)	0.0114
Myristic acid (C14:0)	0.3912
Palmitic (C16:0)	20.8620
Palmitoleic acid (C16:1)	0.2638
Stearic Acid (C18:0)	2.0197
Oleic acid (C18:1)	42.4643
Linoleic acid (C18:2)	32.2081
Linolenic acid (C18:3)	1.2199
Arachidic acid (C20:0)	0.0522
Eicosenoicacid (C20:1)	0.5074

From Figure 1, potassium content in modified natural zeolite increased with the K_2CO_3 concentration of impregnating solution till K_2CO_3 concentration of 45 g per 60 ml aquadest, which reached 11.24%. The potassium content in this K_2CO_3/natural zeolite catalyst is less than that of KOH/natural zeolite by Kusuma et al. [8] which peaked at 45.34%. The lower potassium content is because K_2CO_3 is not a strong base which can fully ionize, and its alkaline properties are below KOH base properties. At concentration beyond 45/60 (w/v), potassium content decreased to 5.54%. This decrease is possibly caused by close to the saturation condition of the K_2CO_3 solution at higher concentration.

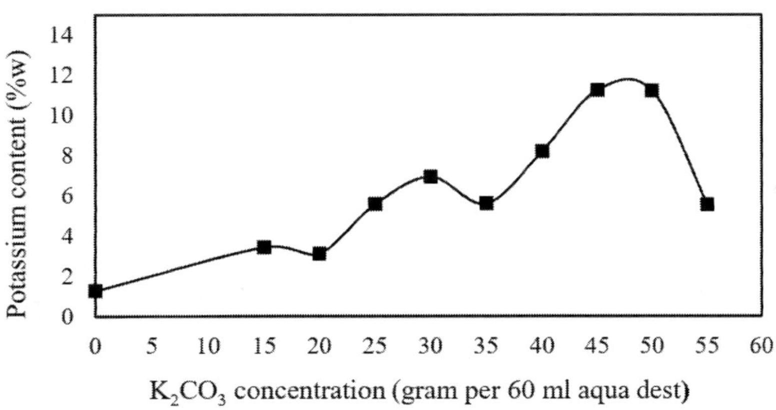

Figure 1. Effect of K_2CO_3concentration on potassium content (%w) in catalyst.

K_2CO_3 solubility in water at 30°C is 113.7 g in 100 ml water [9]. In water, K_2CO_3 will ionize, and K^+ will be absorbed in the natural zeolite. As in Figure 2, K^+ions bind to the negative side of the natural zeolite structure $(SiO_4)^{4-}$ or $(AlO_4)^{5-}$. The negatively charged oxygen (O) binds to positively charged potassium (K) resulting in an equilibrated chemical structure of natural zeolites [10]. The chemical structure of natural zeolite modified with K_2CO_3 will possess more potassium (K) than the chemical structure of natural zeolite without modification. Also, a new group of Si-O-K or Al-O-K is formed.

Impregnation will lead to ion exchange. In this ion exchange, a counter-reaction may occur because of equal cations [9]. This phenomenon is due to competition between K_2CO_3 ions in high concentration K_2CO_3 solution which renders the ions unable to compete with free ions or to reform K_2CO_3 and cause potassium in natural zeolite todecrease.

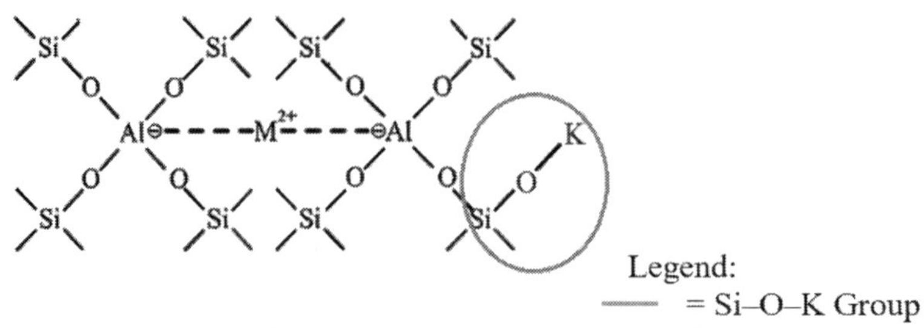

Figure 2. Main skeletal structure of natural zeolite [10].

3.3. Transesterification

3.3.1. Effect of amount of catalyst on biodiesel yield

Based on the AAS analysis of K_2CO_3/natural zeolite catalyst, the catalyst with highest potassium concentration was employed for the rest of this experiment. Figure 3 presents the effect of the amount of catalyst on biodiesel yield.Within the observed range, increasing the amount of catalyst increased in biodiesel yield. This trend is also observed by other researchers [11,12]. In this research, biodiesel yield increased from 79.05% to 98.18% as the amount of catalyst from 2% to 4%. A greater amount of catalyst led to increasing of catalyst active side, resulting in higher yield of biodiesel [13]. Kusuma et al. [8] reported that the yield of biodiesel could reach 95% at 3% catalyst usage. However, their study did not varythe amount of catalyst. Also, they used a natural zeolite derived from Pacitan East Java and impregnated with KOH solution.

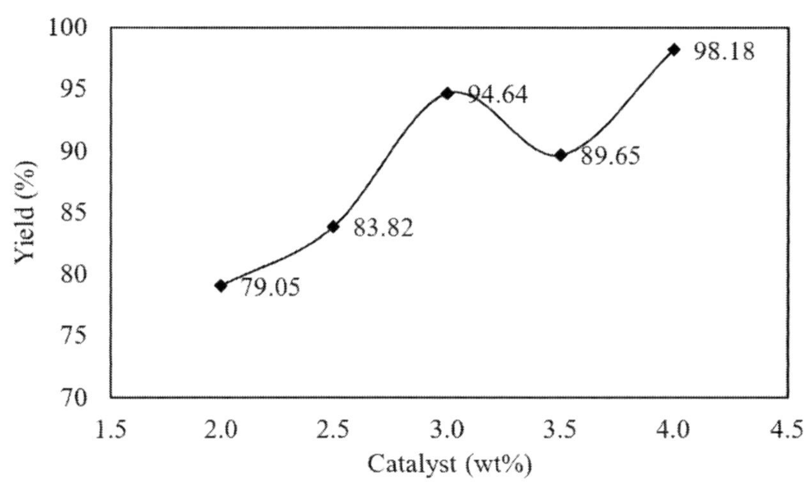

Figure 3. Effect of catalyst amount on biodiesel yield at reaction temperature of 65°C, RBO to methanolmolar ratio of 10:1, and reaction time of 3 h.

In this study, biodiesel was also produced by transesterification of RBO using natural zeolite catalyst without K_2CO_3 impregnation, which yielded 3.94% biodiesel. This yield is well below those obtained using a modified natural zeolite catalyst

3.3.2. Effect of reaction time on biodiesel yield

To observe the effect of reaction time on biodiesel yield, the experiment was conducted by varying the reaction time at fixed amount of catalyst, reaction temperature, and reactant molar ratio. Figure 4 depicts the effect of reaction time on biodiesel yield. The biodiesel yield increased with reaction time in the time range of 2-3 h. However, after 3 hours, biodiesel yield decreased. Initially, the reaction proceeded slowly due to mixing and dispersion of methanol in oil. After that, the reaction continued until it reached maximum conversion. Transesterification is a reversible reaction.Once the optimum yield is achieved, additional reaction time will not affect the yield and even result in reverse reaction to form fatty acid,decreasing methyl ester yield [14].

In this study, for a reaction time of 2 hours, the yield of biodiesel produced was 85.86%. After 2.5 hours biodiesel yield increased to 91.56%. At a reaction time of 3 hours, biodiesel yield was highest at 98.18%. The same phenomenon was also reported by Noiroj et al. [7], but they used palm oil as raw material with KOH/NaY modified catalysts.

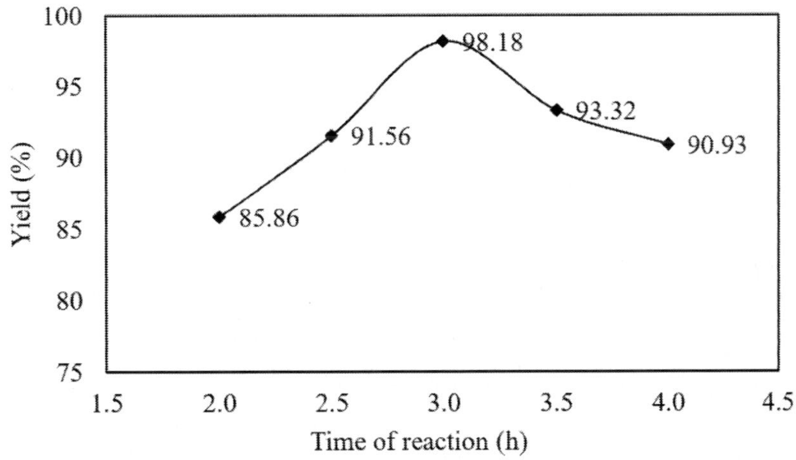

Figure 4. Effect of reaction time on biodiesel yield at 65°C, 4% catalyst, and methanol to RBO molar ratio of 10:1.

3.3.3. Effect of methanol to RBO molar ratio on biodieselyield

Figure 5 illustrates the effect of methanol to RBO molar ratio on biodiesel yield at various amount of K_2CO_3/ natural zeolite catalyst.

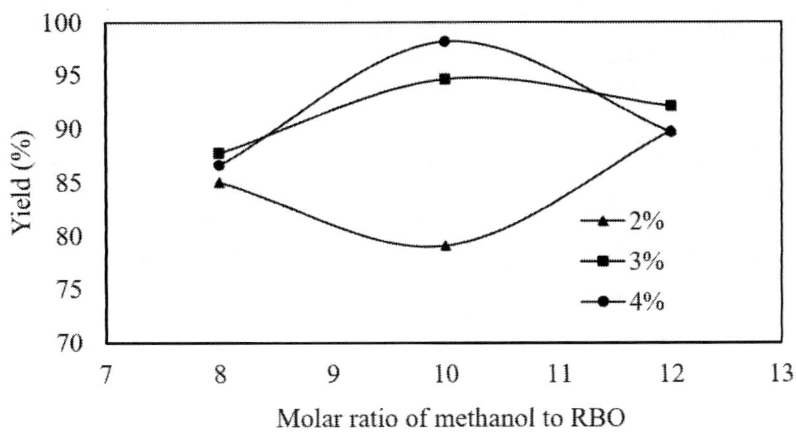

Figure 5. Effect of methanol to RBOmolar ratio on biodiesel yield at reaction temperature of 65°C and 3 hreaction time.

As in Figure 5, biodiesel production at 2% catalyst followed different trend than those at 3% and 4% catalyst. For 3% and 4% catalyst, biodiesel yield increased from methanol to oil molar ratio of 8:1 to 10:1, then decreased at 12:1. This decrease is because adding methanol will cause the oil concentration to drop, resulting in a low reaction rate and shifting the reaction equilibrium [14]. Also, excess glycerol as a byproduct blocks the reaction between methanol with oil and catalyst. Decreasing biodiesel yield is also due to the polar hydroxyl group in methanol which acts as an emulsifier and complicates product separation [13].

At a molar ratio of methanol to oil of 10: 1, the reaction formed a lot of methyl ester with little glycerol. This result indicates that the ratio is particularly suitable for biodiesel production. In other words, the best reaction condition is at methanol to oil molar ratio of 10:1 with a yield of 98% biodiesel. The results obtained are greater than those reported by Kusuma et al. [8] with the highest yield of 95.09% at 7:1 molar ratio using palm oil and KOH/ natural zeolite catalyst from Pacitan, East Java, Indonesia.

3.4.Biodiesel Properties

Table 2 lists several biodiesel properties resulting from this study and their comparison with the SNI-04-7182-2006 standard. From the test results of some biodiesel properties, the synthesized biodiesel met the established standards. Theresultssuggestthat modified K_2CO_3/natural zeolite catalyst can be used as a heterogeneous catalyst in the manufacture of biodiesel from RBO.

Table 2. Comparison of the properties of biodiesel produced from RBO with K_2CO_3/zeolite catalyst with Indonesia Standard (SNI-04-7182-2006).

Properties	Biodiesel produced	SNI
Methyl Esther content, %	98.82	>96,5
Density at 15 °C, g/ml	0.87	0.86-0.90
Kinematic viscosity at 40 °C, mm²/s	4.22	1.9-6.0
Flash point, °C	150.00	>100

4. Conclusion

The natural zeolite modified with K_2CO_3 can be used as a heterogeneous catalyst in the manufacture of biodiesel from RBO and gives a much higher biodiesel yield compared to natural zeolite without modification. At a methanol to RBO molar ratio of 10:1, 3 hourreaction time, and 4% catalyst, the biodiesel yield was highest at 98.18%. From the test results, the biodiesel properties (purity, density, viscosity, and flash point)met the SNI Standard.

References

[1] Helwani Z, Othman M R, Aziz N, Kim J and Fernando W J N 2009 *Applied Catalysts A: General* **363** 1
[2] Rathore V, Tyagi S, Newalkar B and Badoni R P 2014 *Fuel* **140** 597
[3] Cejka J, Bekkum H V and Corma A 2007 *Introduction to Zeolite Science and Practice* (Oxford: Elsevier)
[4] Xia H S, Zhou C H, Tong D S, Yu W H and Yu W H 2011 *Applied Clay Science* **53** 279
[5] Zhang D, Zhou C H, Lin C X, Tong DS and Yu W H 2010 *Applied Clay Science* **50** 1
[6] Zhao C H 2011 *Applied Clay Science* **53** 97
[7] Noiroj K, Intarapong P, Luengnaruemitchai A and Jai-In S 2009 *RenewableEnergy,* **34** 1145
[8] Kusuma R I, Hadinoto, J P, Ayucitra A, Soetaredjo F E and Ismadji S 2013 *Applied Clay Science* **74** 121
[9] Perry R H and Green D W 1997 *Chemical Engineering HandBook* (New York: McGraw-Hill)
[10] Elaiopoulos K, Perraki T H, and Grigoropoulou E 2010 *Microporous and Mesoporous Materials* **134** 29
[11] Xie W and Li H 2006 *Journal of Molecular Catalysis A: Chemical* **255** 1
[12] Baskar G and AishwaryaR 2016 *Renewable and Sustainable Energy Reviews* **57** 4996

[13] Roschat W, Siritanon T, Yoosuk B and Promarak V 2016 *Energy Conversion and Management* **119** 453
[14] Chen G Y, Rui S, Yan B B, Shi J F, Li S Y and Liu C Y 2016 *Fuel Processing Technology* **143** 110

IOP Conf. Series: Materials Science and Engineering **309** (2018) 012108 doi:10.1088/1757-899X/309/1/012108

Analysis of carbon monoxide (CO) with Delhi Finite Line Source (DFLS) in MT Haryono Street, Medan City

M Turmuzi[1*], I Suryati[2], E T Mashaly[2] and F Batubara[1]
[1]Department of Chemical Engineering, Faculty of Engineering, Universitas Sumatera Utara, Medan 20155, Indonesia
[2] Department of Environmental Engineering, Faculty of Engineering, Universitas Sumatera Utara, Medan 20155, Indonesia

*Email: turmuzi@usu.ac.id

Abstract. One source to decrease urban air ambient quality is transportation sector. Important pollutants are released by gas emissions from vehicles are carbon monoxide (CO), hydrocarbons (HC), nitrogen dioxide (NO2), particulate matter and others. The presence of CO pollutants in the ambient air can be predicted by modeling air quality. This study aims to estimate CO concentration resulting from transportation activities using Delhi Finite Line Source (DFLS) model, comparing CO prediction using a DFLS model with CO observation in the field, and determine the suitability of the DFLS model application on the MT Haryono street in Medan City. Research was conducted for 3 days at two sample points with frequency twice daily. Based on research results, the range of CO concentration from observation between 22.903 μg/m^3 – 27.484 μg/m^3. CO observation is still below the ambient air quality standard. According to the DFLS calculations, the range of CO concentration between 1.499 μg/m^3– 2.051 μg/m^3. The calculation index of agreement (IOA) validation test obtained value of d = 0.22. The DFLS model is not suitable to be applied on MT Haryono street because many factors affected such as wind direction and wind velocity, ambient air temperature and humidity

1. Introduction

One of the sectors in declining air quality is the road transportation sector (motor vehicles). Mortality rates increased by 5% and 12% in China and India respectively because air pollution from transportation sector [1;2]. The transport sector contributes the highest pollutant, which is about 80.22% - 90% of CO [3;4]. Several cities in China each year experience increased concentrations of pollutants from three cities (Haikou, Lhasa, and Zhousan) in 2013 to eight cities (Haikou, Lhasa, Zhousan, Shenzen, Zhuhai, Fuzhou, Huizhou, dan Kunming) in 2014 [4].

Roadside air quality in Medan City is monitored on three streets only with CO concentrations ranging from 3,041 μg/m^3 – 9.276 μg/Nm^3 [5]. Based on the results of air quality index mapping for CO parameters in Medan City, there are some districts that are categorized as very unhealthy and unhealthy condition: Belawan, Medan Helvetia, Medan Barat and Medan Amplas [6]. In addition, MT. Haryono Street has heterogeneous traffic and a V / C ratio of 1.08 [7], which mean heavy traffic. To predict the concentration of pollutants in the ambient air can be used various models of FLLS, GFLS, and DFLS. According to [8], using the DFLS (Delhi Finite Line Source) model in New Delhi is suitable for predicting CO concentrations. According to [9;10], the use of the DFLS model in Surabaya is not suitable for predicting CO concentrations because of many factor influence the air pollutant dispersion.

Because of the condition of atmospheric stability in India, especially the city of New Delhi is unstable (B), with an average wind speed of 3-5 m/s and the intensity of the sun is 1.367 W/m^2 [11;12]. The condition of the stability of the Medan City atmosphere is also unstable (B) with negative atmospheric stability index [13]. In addition, the appropriate DFLS model is applied to heterogeneous traffic conditions [14], so in this research DFLS model is selected.

The purpose of this research is to estimate CO concentration in MT Haryono Street generated from transport activities using the DFLS Model, comparing CO concentration data using DFLS model with concentration of CO sampling data (CO observation), and knowing the suitability of DFLS model implementation in MT Haryono Street, Medan City.

2. Methodology

This study used a quantitative approach by observing vehicles (traffic counting), which calculates the type and number of vehicles to obtain the rate of emission [15]. Furthermore, the emission rate is incorporated into the DFLS model equation to obtain the calculated CO concentration. Then at the same time with traffic counting, will be done sampling of CO concentration in field (CO observation) and taking meteorological field data such as temperature, solar intensity, direction and wind speed. The concentration of CO DFLS model and CO observation will be compared with the validation test of Index of Agreement (IOA).

2.1. Data Collection

The data collected in this study consists of primary and secondary data. Primary data consists of traffic counting data, i.e. number and type of vehicle (motorcycle, passenger car, bus, and truck); sampling CO concentration data in field (CO observation) using portable CO Monitor (NDIR Analyzer analysis method) and meteorological data (temperature, speed, and wind direction).

Primary datas were collected simultaneously at 2 (two) points along MT Haryono Street. Point 1 is located at Cirebon Intersection and Point 2 is located at Thamrin Intersection. Measurements are made in the morning and afternoon for 1 hour each.

The secondary data obtained from BPS Medan (such as population, area, and number of motor vehicles), BMKG Region I Medan (such as data direction and wind speed of 2011-2015), BMKG Sampali (such as solar intensity data), and DISHUB Medan City (such as data ratio V/C road segment in Medan City). This secondary data is required for determination of the sampling location and the determination of the atmospheric stability class.

2.2. Delhi Finite Line Source (DFLS) Modelling

The first step in this research is to calculate the emission rate obtained from traffic counting of each type of vehicle during sampling. The emission rate of each vehicle is calculated using the following formula [16;17]:

$$Q_L = n \times FE \qquad (1)$$

Q_L = emission rate per unit length (g/m. s); N = number of vehicles per hour (vehicle/hour) and FE = emission factor (g/km.vehicle)

Next, the pollutant concentration on the highway is calculated using DFLS modeling. The DFLS equation is expressed as follows [17]:

$$C = \frac{Q_L}{2\sqrt{2\pi}\sigma_z \bar{u}_e}\left[\exp\left\{-\tfrac{1}{2}\left(\frac{z-h_0}{\sigma_z}\right)^2\right\} + \exp\left\{-\tfrac{1}{2}\left(\frac{z+h_0}{\sigma_z}\right)^2\right\}\right] \qquad (2)$$

C = concentration of pollutant in ambient air (g/m^3); Q_L = emission rate per unit length (g/ m. s); σ_z= vertical dispersion coefficient (m); u_e = effective wind speed (m / sec); z = receiver height or surface monitoring device (m) and ho = high effective source (m)

The assumptions used in this model are as follows [17]: line source (path) for the source of the pollutant is straight; meteorological data is valid; CO pollutants only come from motor vehicles along the line source and the chemical physics transformation of CO pollutants is ignored or not included in the modeling. The rate of change of shape and its removal is not included in the modelling. Movement of pollutant is depended on the direction of wind.

The third step is to test data validation. In this validation, the results of measured CO concentration (observation data/O) is compared to calculated CO concentration (predicted data/P). Then both data (O and P) are validated using the IOA (Index of Agreement) equation. The formula used is as follows [8]:

$$d = 1 - \frac{\sum_{i=1}^{N} (P-O)^2}{\sum_{i=1}^{N} \left(|P - O_{mean}| + |O - O_{mean}| \right)^2} \quad (3)$$

P = calculated CO concentration / CO DFLS model; O = measured CO concentration / CO observation, O_{mean} = average of CO concentration from sampling result (average of CO observation)

If the validation test value of IOA (d) obtained with a range of values of 0.8-1, then the DFLS model can be used to predict CO concentrations in MT Haryono Street. If the validation test value of IOA (d) obtained with a value < 0.8, then the DFLS model cannot be used to predict CO concentrations in MT Haryono Street.

3. Result and Discussions

3.1. Number of Vehicles, Type of Vehicles, and CO Emission Rate at MT Haryono Street

The number and types of motor vehicles as well as the rate of CO emissions obtained on MT Haryono Street are presented in Figure 1.

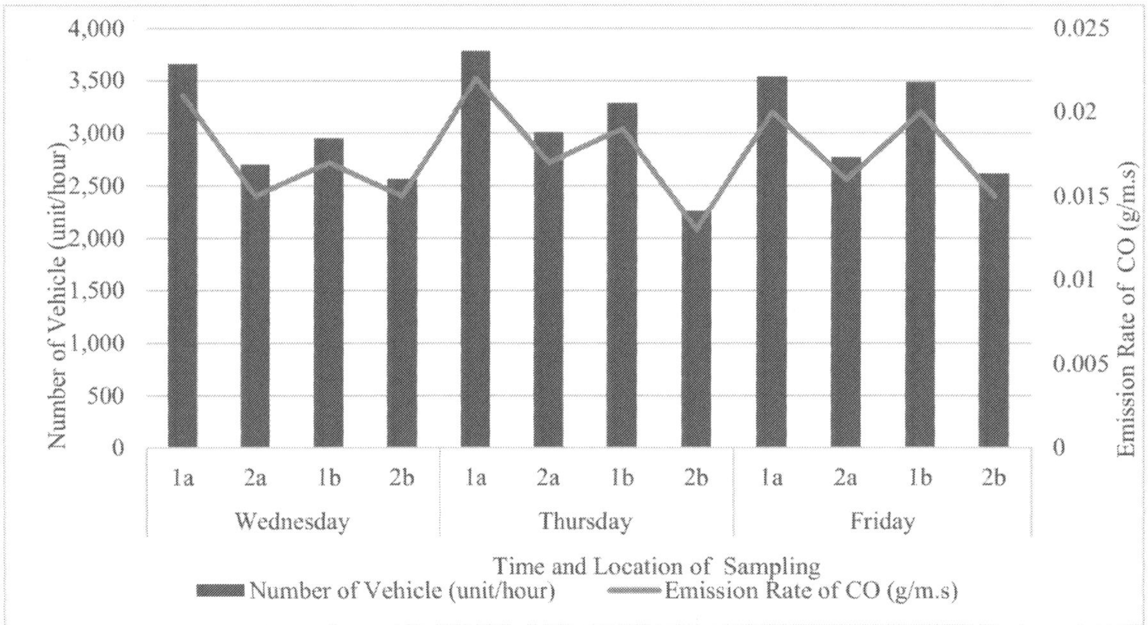

Figure 1. Number of Vehicle and CO Emission Rate at MT Haryono Street

Average of CO emission rate in the morning is 0.018 g/m.s and at noon is 0.017 g/m. s. The total emission rate is the total average of all emissions sourced from vehicles passing through MT Haryono Street. The types of vehicles that contribute most of CO emissions are passenger cars ranging from 57% - 62%.

Vehicle emissions are influenced by the number of vehicles and the value of emission factors. Based on [18], the value of CO emission factor of passenger cars is greater than motorcycles, buses and trucks, so that the emission rate of passenger cars is greater than motorcycles although the number of motorcycles is higher than passenger cars. According to [19], emission factors are influenced by vehicle type, fuel type, driving activity, and others. Vehicle emissions are also highly dependent on congestion and driving patterns.

3.2. Measured CO Concentration on MT Haryono Street

Measurements of CO concentration at MT Haryono Street were performed using portable CO Monitor placed on the road shoulder ± 1 m from the roadside and at an altitude of approximately 1.5 m above the road surface. The monitoring result of measured CO concentration in μg/m^3 can be seen on Figure 2.

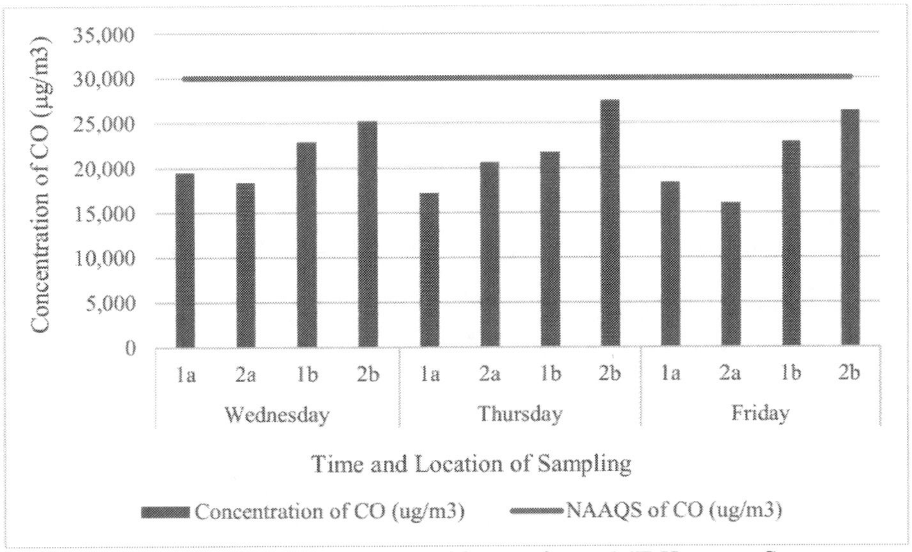

Figure 2. Concentration of CO Observation at MT Haryono Street

The highest CO concentration for monitoring is in the morning 20.613 μg/m^3 and in the afternoon 27.484 μg/m^3 located in point 2 on Thursday. No monitoring results have passed the standard [20] of 30,000 μg/m^3, but on Thursday afternoon at point 2 the value is almost close to the standard. This needs to be aware as if exposed to CO for ≥ 8 hours, it will affect the function of body organs such as brain, liver, and central nervous [21;22]. CO is a silent killer because it has properties that are not smelly and visible. CO gas can bind hemoglobin more rapidly than O_2, thereby reducing the oxygen capacity in the blood that can cause shortness of breath, fainting until death [23;24;25].

3.3. CO Concentration with Delhi Finite Line Source (DFLS)

The inputs in the modeling using emission rate data and meteorological factors that have been obtained from the field are included in equation (2). The result of calculated CO concentration can be seen in Figure 3.

Based on Figure 3, the range of CO concentration using DFLS Model is 400 μg/m^3 – 2.051 μg/m^3. The highest CO concentration using DFLS Model is located at point 1 (Cirebon Intersection), on morning monitoring. The high concentration at this point because it is influenced by the source of emission in the form of the number of vehicles, meteorological factors such as wind direction and speed, solar radiation intensity, ambient air temperature and humidity.

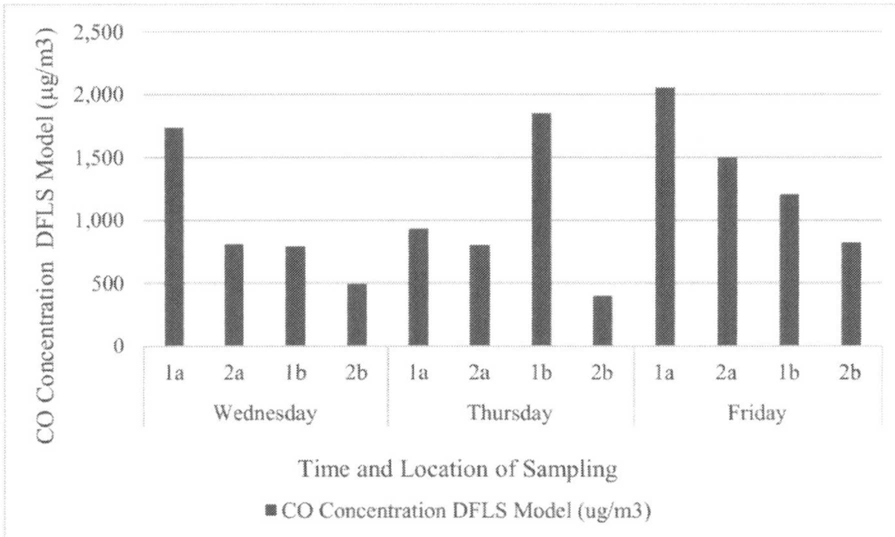

Figure 3. CO Concentration with DFLS at MT Haryono Street

3.4. Comparison of CO Observation with CO DFLS Model

Comparison of CO observation with CO DFLS Model can be seen in Figure 4.

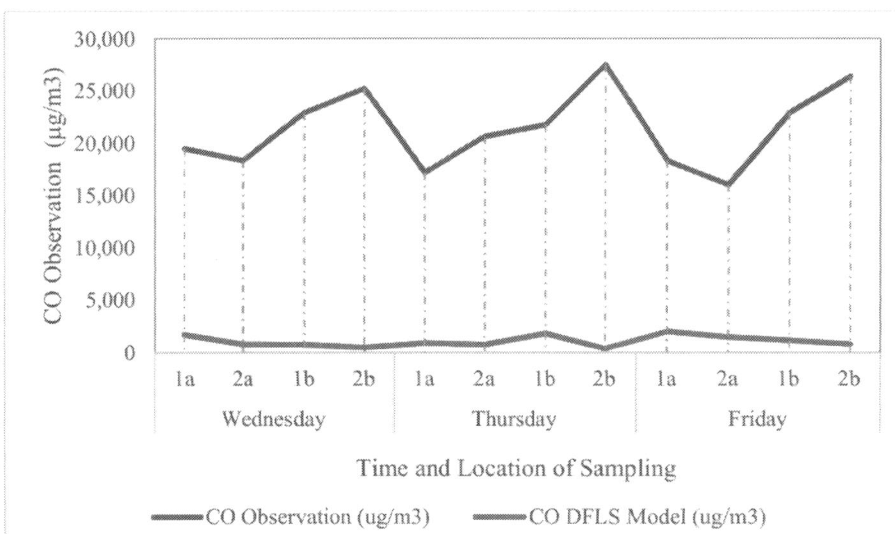

Figure 4. CO Observation vs CO DFLS Model at MT Haryono Street

The concentrations of CO at points 1 and 2 in morning measurements have similar trend but when daytime measurements have the opposite trend. The calculated CO concentration during daytime measurements at point 1 is higher than point 2, due to the number of vehicles passing in point 1 more than the point 2. Conversely, the CO concentration is measured when the daytime measurement at point 1 is lower than the point 2 due to congestion conditions in Point 2. Based on the results of the study [4], traffic congestion conditions greatly affect the increase of CO concentration in ambient air.

The highest measured CO concentration is 27.484 µg/m^3. Value of this concentration can be included in high category although not yet reached the standard quality of 30.000 µg/m^3. If human beings exposed every day, it can endanger human health. Precautions need to be taken to reduce the concentration. The way of control can be done with the application of Bus Rapid Transit (BRT) and if possible increase the green area around the location.

According to [26], the application of BRT can reduce CO pollutants in areas on the BRT line by 19.4%, the area at 2.5-10 km from the BRT line by 17.2%, and the area at a distance of 10 - 30 km by 16.6%. In addition, according to [27], the use of BRT can reduce CO emissions in Medan City by 25.02% - 29.28% (CNG) and 25.17% - 29.44% (diesel). The results also show that the type and number of vehicles are strongly correlated with CO emissions.

3.5. Test Validity with Index of Agreement (IOA)

The data used for this validation calculation is the measured CO concentration data and the calculated CO concentration. The formula used in the validation test with IOA can be seen in (3). Modeling can be used if the validation values obtained in the range of values 0.8 - 1. Based on the above calculation, we can see the value d = 0.223, far from the range of values 0.8 - 1. This means that DFLS model cannot used to predict CO concentrations in MT Haryono Street.

The ratio of calculated CO concentration tends to be smaller than the measured CO concentration. This can be caused by several factors, namely: road conditions at point 2 (two) have congestion, thus affecting the measured CO concentration. According to [4], traffic congestion conditions result in increased CO concentrations.

Based on the modeling assumption, that CO pollutants only come from motor vehicles and ignore the effects of other emission sources. While on field research, CO pollutant measurements are strongly influenced by other emission sources such as rail transport activities.

The spread of pollutants around the site is affected by the local topographic condition of MT Haryono Street is surrounded by a building with an altitude of ± 12-20 m. According to [28], the effect of the height of the building affects the spread of pollutants that ultimately the pollutants are not spread evenly while in the model ignores the influence of maximum mixing height (MMH).

4. Conclusion

The result of calculated CO concentration with DFLS model is smaller than result of CO observation. Based on the sampling result, the maximum measured CO concentration was 27.484 μg/m^3 on Thursday afternoon at point 2, while the minimum measured CO concentration was 16.032 μg/m^3 on Friday morning at point 2.

Maximum CO concentration obtained using Delhi Finite Line Source (DFLS) modeling is 2.051 μg/m^3 on Friday morning at point 1 whereas the minimum calculated CO concentration is 400 μg m^3 on Thursday afternoon at point 2.

The result of data validation test with Index of Agreement (IOA) is d = 0,223 which means DFLS model is not suitable applied in MT Haryono Street.

Reference

[1] OECD 2014 *The Cost of Air Pollution Health Impact of Road Transport* (Prancis)
[2] Saikawa E, Kurokawa J, Takigawa M, Borken-Kleefeld J, Mauzerall1 D L, Horowitz L W, and Ohara T 2011 *Atmos Chem Phys* **11** 9465-9484
[3] Lu *et al* 2015 *Air Pollution Prevention and Control Progress in Chinese Cities* CAA
[4] Sengkey L S, Jansen F, Wallah S 2011 *Jurnal Ilmiah Media Engineering* **1(2)** 119-126
[5] BLH Provinsi Sumatera Utara 2015 *Laporan EKUP Provinsi Sumatera Utara Tahun* 2015 Medan
[6] Suryati I, Khair H 2016 *IOP Conf. Series: Materials Science and Engineering* **180** 12114
[7] Dinas Perhubungan (DISHUB) Kota Medan 2016 Medan
[8] Khare M, Sharma P 1999 *Transportation Research Journal* **Part D** 65-70
[9] Hidayatullah E R 2012 *Estimation Cost of Emission and SO_2 Concentration From Transportation Sector with DFLS Model Case Study South of Surabaya (Gayungsari Barat Street)*(Surabaya: ITS Library)
[10] Rahayu A 2012 *Estimation Concentration of Carbon Monoxide by Transport Activities with DFLS Model* (Surabaya: ITS Library)
[11] AWS Truepower 2012 *Wind Resource of India* (India)

[12] Solar Energy Centre 2008 *Solar Radiation Handbook: Typical Climatic Data for Selected Radiation Stations 1986-2000* (India)
[13] Yonatan A, Cahyanti R P, Putri C C, and Sutilah 2015 *Practical Reports of RAOB Applications in Data Analysis Rawinsonde* (Bogor: Library IPB)
[14] Nagendra dan Khare 2002 *Line Source Emission Modelling* (India: Indian Institute of Technology)
[15] Khare M and Sharma P 2012 *Performance Evaluation of Air Quality Models St Urban Intersection of an Indian City: a Case Study of Delhi City* (WIT Press)
[16] Cooper C D and Alley F C 1994 *Air Pollution Control: A Design Approach. Second Edition* (USA: Waveland Press, Inc)
[17] Khare dan Nagendra 2007 *Introduction, Artificial Neural Network in Vehicular Pollution Modelling* (SCI) 41 (Berlin: Springer-Verlag Berlin Heidelberg) 1-6
[18] KLH 2013 *The Pollution Emission Inventory of Preparation Technic Guidance in Urban Area* (Jakarta)
[19] Zhang *et al* 2013 *Factor Affecting Vehicular Emissions and Emission Models* (China: American Society of Civil Engineers)
[20] USEPA 1996 *Emission Inventory Program: Preferred and Alternative Methods for Gathering and Locating Specific Emission Inventory Data* (Washington, DC: United States EPA)
[21] Haloulakou *et al* 1999 *Occupational Exposure to CO Concentration in Enclosed garages Estimation of Flood COHb Levels* (Athena)
[22] WHO 2000 *Air Quality Guidelines – Second Edition Carbon Monoxide* (Denmark)
[23] Fechter *et al* 2012 *Toxicological Profile for Carbon Monoxide* (Atlanta: U. S. Department of Health and Human Services)
[24] Nevers N de 2000 *Air Pollution Control Engineering Second Edition* (Singapore: McGraw-Hill)
[25] Wardhana W A 2004 *The Environmental Pollution Impact* (Yogyakarta: ANDI)
[26] Bel G 2015 *Evaluation of The Impact of Bus Rapid Transit Air Pollution. Research Institute of Apllied Economics*
[27] Suryati dan Khair 2016 *Prosiding Seminar Nasional Sains dan Teknologi Lingkungan* **II** 157-161
[28] Supriyadi E 2009 *Application of Finite Length Line Source Model for Predicting Concentration of Pollutants from the Line Source (Case Study: M.H. Thamrin Street, DKI Jakarta)* (Bogor: Institut Pertanian Bogor)

IOP Conf. Series: Materials Science and Engineering **309** (2018) 012109 doi:10.1088/1757-899X/309/1/012109

The lean ergonomics in green design of crude palm oil plant

L N Huda* and R Matondang

Department of Industrial Engineering, Universitas Sumatera Utara,
Jl. Almamater, Kampus USU, Medan 20155

*Email: listiani@usu.ac.id

Abstract. Ergonomics can help manufacturing and industrial engineersmaximum work output without physical harm to workers.Physiology, biomechanics, anthropometrics, and allocation are the areas of ergonomics most useful to manufacturing in apllying the concept of Lean and Green in manufacturing. These systems require efficient production and low use of resources such as energy and material. Its philosophy encourage worker to look at waste. This concept is applied in one of national plant of Crude Palm Oil (CPO) that located in North Sumatera. The problem found in the company are the working posture and excessive workload of the workers. These conditions are affects to the their job performance. The study was carried out by evaluated the worker body position using the Work Posture Assesment (WPA) and Biomechanics method. The WPA results shows the operator's working position was mostly bent more than 30 degrees. This indicates that almost all workers are felt musculoskeletal disorders during work hours. While the biomechanics analysis found the significant relation between the values of Recommended Weight Limit (RWL) and Lifting Index (LI) which the increase of RWL value will decrease the LI value. This indicated that the recommended load for a worker under certain circumstances affects the appointment made so as not to contain the risk of spinal injury. In fact these condition are due to in-efficiency in production which can be maintaned the green design of CPO plant by improving the existing work.

1. Introduction

Workers are an important asset for the company, but sometimes, company is less attentive to the needs and interests of workers. There are still many companies don't have standard methods and ergonomic facilities for supporting the production process,it caused workers often have complaints on the part of their body. Complaints that arise due to the lack of ergonomic facilities and posture of the workers, causing workers feel less comfortable [1].

Manual handling and lifting are major causes of work-related low back pains (LBP) and impairments. Therefore, it is important to design MMH tasks ergonomically and safely from an economic and ethical point of view so that industry-related accidents including LBP can be prevented [2]. In another study, some of the ergonomic factors that could cause increased worker injury were identified [3]. Some of the factors identified were: lifting, bending, stooping, squatting and carrying loads.As for biomechanics, the RWL and the LI are based on the concept that the risk of work-related lower back pain (LBP) increases as the demands of the lifting task increase. In other words, as the magnitude of the LI increases, so does the risk of LBP.

IOP Conf. Series: Materials Science and Engineering **309** (2018) 012109 doi:10.1088/1757-899X/309/1/012109

PT. X is one of the largest plantation companies in Indonesia, where there are operators who work in positions that lack of ergonomic in several stations. This palm oil factory has 8 stations consisting of a loading ramp, trippler and track, pressing, kernel plant, clarification plant, boiler, workshops, and sorting. Operators' complaint found on the loading ramp station, trippler and track, a boiler, workshops and the kernel. Operator complained of pain in the back, arms, neck, shoulders, legs, and thighs.

The improvement for posture and working methods of the operator who works on palm oil plant, is carried out by take the picture of operator working posture, then do the measurement and calculation of the angle based on REBA (Rapid Entire Body Assessment) and Biomechanics because this method can be used to assess the interference factor of the body on the operator. Similar research also has been done in various studies before such as [1, 2, 3].

2. Methodology

This research uses the science of ergonomics to finish problem on Palm oil factory which deals with complaints of fatigue, workload, and work posture.

2.1. Work Posture Analysis (WPA)

REBA (Rapid Entire Body Assessment) is a method developed by Sue Hignett and Lynn McAtamney, this method is effectively used to assess workers posture, effort used, and the type of movement of workers. REBA (Rapid Entire Body Assessment) judge the load that handled within a work system, the couplingand activities undertaken. This method is used because it is relatively easy to find out the value of a member of the body, becauseit is not necessary to find a specific angle of body, but just a range of angles. In the end the final value of REBA (Rapid Entire Body Assessment) gives an indication of the level of risk of a work and the action to be carried out/taken [5]. Questionnaire of REBA (Rapid Entire Body Assessment) shown in Figure 1.

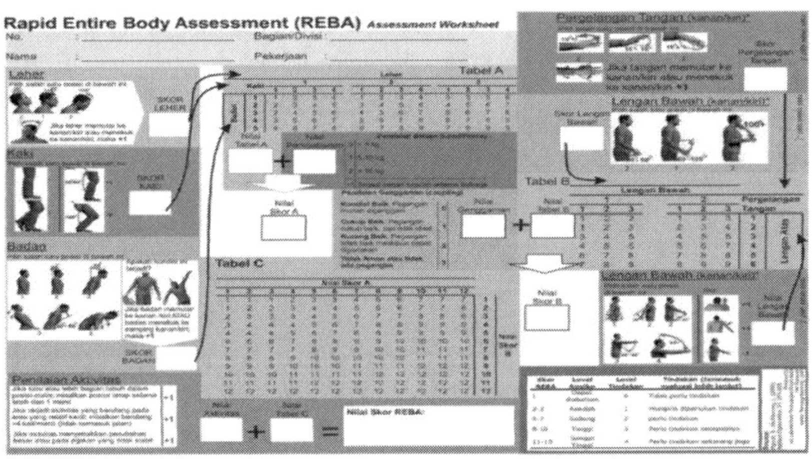

Figure 1. REBAQuestionnaire

2.2. Biomechanics

Work biomechanics is the study of interactions of workers with equipment,machinery and materials, so that workers can improve their performance and on the other hand can minimize the risk of injury in workplace (musculoskeletal). According to Frankel and Nordin, biomechanics uses the concept of physics and engineering to describe movement in a variety of body parts and the forces on the body in daily activities. According to [6] occupational biomechanics is the study of the relationships between workers and equipment, the working environment and others to improve performance and minimize the possibility of injuries.

IOP Conf. Series: Materials Science and Engineering **309** (2018) 012109 doi:10.1088/1757-899X/309/1/012109

Recommended weight limit is the limit load recommendations that can be raised by humans without causing injury even though the work is repetitive and is done in a long time period. RWL is defined by NIOSH in 1991 in the United States [7].The equation to determine the recommended load to a worker under certain conditions according to NIOSH as follows:

$$RWL = LC \times HM \times VM \times DM \times AM \times FM \times CM \quad (1)$$

where:
RWL : The recommended load limit
LC : Lifting Constant= 23 kg
HM : Horizontal Multiplier = 25/H, which H in centimeter.
DM : Distance Multiplier = 0,82 + 4,5/D which D in centimeter.
AM : Asymmetric Multiplier = 1 – (0,0032 A)which A in degree.
FM : Frequency Multiplier
CM : Coupling Multiplier
VM : VerticalMultiplier = (1-(0,003 |V-75|))which V in centimeter.

Lifting index is an estimate of the relative physical stress associated with manual lifting jobs. When the value of LI increases, the level of risk that is given to workers increased, and the percentage of the workforce who may be at risk will increase as risk associated with lower back pain (LBP).

$$LI = Workload/RWL \quad (2)$$

Where:
LI<1 = the work isn't too risky
LI>1 = the work have risky potential (specially musculoskeletal disorder)

Then, Biomechanic problems were analyzed by the formulation of RWL and LI of the operators who are experiencing complaints using ErgoFellow Software.

3. Result and Discussion

3.1. Work Posture Analysis (WPA)

Work posture using REBA (Rapid Entire Body Assessment)and the improvement for operator that has a score of REBA 'need treatment' and 'need treatment as soon as possible' is shown in Table shown in Table 1.

Table1. Reba Recapitulation

No	Activity		REBA Score	Improvement	Description
	Real	CAD			
1			10 (Need treatment as soon as possible)		The operator should use tojok to withdraw palm oil from the top of the mast delimiter
2			6 (Need treatment)		Operator can roll up the cord with the position more upright

Table1. Reba Recapitulation (Continued)

No	Activity		REBA Score	Improvement	Description
	Real	CAD			
3			10 (Need treatment as soon as possible)		Operator should push using both hands to reduce the load
4			3 (May need treatment)	-	-
5			6 (Need treatment)		Operator can roll up the cord with the position more upright
6	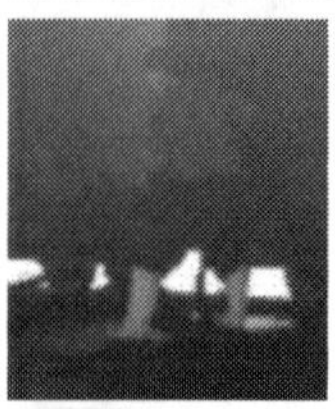		3 (May need treatment)	-	-
7			6 (Need treatment)		Operator can pull the sticks without having to bend down
8			1 (No treatment needed)	-	-
9			6 (Need treatment)		Operator can pull the sticks with the position of the arm is lowered and the position of the upper arm straight down

Table1. Reba Recapitulation (Continued)

No	Activity		REBA Score	Improvement	Description
	Real	CAD			
10			8 (Need treatment as soon as possible)		Operator can pull the sticks without having to bend down
11			7 (Need treatment)		Operator can steer sticks without having to bend down
12		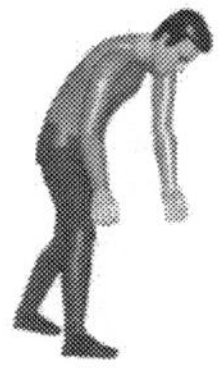	4 (Need treatment)	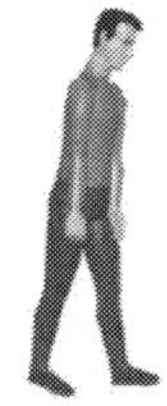	Operator lifted the wheelbarrow with an upright position
13			11 (Need treatment as soon as possible)		Operator can do the welding with the sitting position rather than squat
14			8 (Need treatment as soon as possible)		Operator can sculpt with sitting position rather than squats

From Table 1, it can be seen that almost 80% operators needed a treatment to improve their performance. Some of REBA score even reach 8-11 score which means the work postures are in dire need of immediate corrective action. Through the assessment of REBA, the results show that bending posture associated with squatting, pushing, and pulling task has high exposure to health problems. Similarfinding is also found in various studies such as [8] and [9].

3.2. Biomechanics

Biomechanics problems were analyzed by the formulation of RWL and LI of the operators who are experiencing complaints using ErgoFellow Software. The result of the calculation of operators' RWL and LI shown in Table 2.

Table2. RWL and LI Recapitulation

Operator	RWL	LI	Operator	RWL	LI
1	6,93	2,309	8	9,602	2,43
2	8,182	2,331	9	9,638	2,442
3	8,823	2,333	10	9,728	2,448
4	8,874	2,343	11	9,818	2,592
5	9,395	2,364	12	9,858	2,607
6	9,419	2,386	13	9,869	2,811
7	9,466	2,395	14	9,961	3,319

Based on the Table 2, the average of RWL is 9,254, but the standard of RWL is 6,93. It means RWL in the work place is already exceeds the limit. While the average of LI is 2,507, it means a great possibility for operators affected by musculoskeletal disorder.The relationship between the RWL values and LI values was illustrated at the Figure 2.

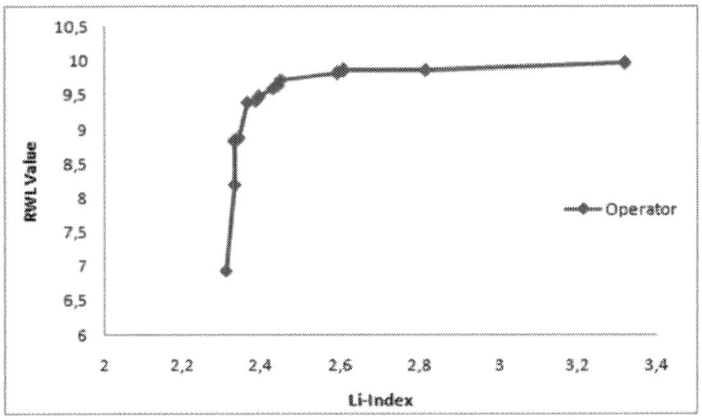

Figure 2. The Relationship Between RWL and LI

Figure 2 shows that there have a significant relation between the values of Recommended Weight Limit (RWL) and Lifting Index (LI) which the increase of RWL value will increase the LI value. The higher the RWL value the higher is also the value of LI. This indicates that the more weight that is lifted then the risk of experiencing musculoskeletal disorders will be higher. For the recommended improvement, as suggested in [10]; an ergonomic intervention with the aim of reducing external back loading should therefore primarily focus on major factors such as the height and horizontal distance of the load and not just on workers' technique. Chung &Kee [2] also suggested that the job should be evenly distributed within 2 hours or less work cycles followed by a sufficient recovery period, to keep the lifting frequency less than 0.5 lifts/min.

4. Conclusion

Working postures of operators on the oil palm factory is requires treatment, especially for operator 2, 5, 7, 9, 11, and 12. This result prove that working posture on this oil palm factory have mistake and needs improvement in order operators are not distracted and experience fatigue during the work.

The result of biomechanics of the operators is the RWL was between 6-9.5 kg for all operators. As for LI the entire operator exceeds the limit of LI > 1 = the work has risky potential (especiallymusculoskeletal disorder).Both of these methods indicate that the necessary improvements of work posture, working methods, or working facilities from oil palm factory.

Acknowledgments

The authors gratefully acknowledge that the present research is supported by Ministry of Research and Technology and Higher Education Republic of Indonesia. The support is under the research grant BP-PTN USU of Year 2016 Contract Number 6049/UN5.1.R/PPM/2016. Special thanks to my students Savudan, Muhammad Akbar, Tuah Affandi for their efforts to collect data for the puposes of this research.

References

[1] Nazlina 2008 Usulan perancangan postur kerja dengan menggunakan pendekatan biomekanika dan fisiologi pada aktivitas pencetakan batu bata *Paper in Industrial Engineering National Seminar and Congress of BKSTI*
[2] Chung M K and Kee D 1999 Evaluation of lifting tasks frequently performed during fire brick manufacturing processes using NIOSH lifting equations Univ. of Keimyung, South Korea
[3] Swanson N, Piotrkowski C, Curbow B, Graville S, Kushnir T and Owen B 1994 Occupational health and safety issues in child-care work *Pediatrics* **94 (6 pt.2)** 1079-80
[4] Lucia M and Remor E Proposed procedures for measuring the lifting task variables required by the Revised NIOSH lifting equation – A case study Federal University of Parana, Brazil
[5] Stanton N A 2005Handbook of human factors and ergonomics methods USA: CRC Press
[6] Caffin D B and Anderson G B J 1984 Occupational biomechanics USA: Wiley
[7] NIOSH 1997 Musculosceletal disorders and workplace factors: A critical review of epidemiologic evidence for work related musculosceletal disorders
[8] Widodo L 2016 Workload analysis of the container unloading process worker *Proceeding of 9th International Seminaron Industrial Engineering and Management*
[9] Saad NH 2006 Measuring and analysing work-related musculoskeletal disorders in Malaysian small medium industries University of Teknologi MARA: Selangor
[10] Plamondon A, Delisle A and Bellefeuille S Lifting strategies of expert and novice workers during a repetitive palletizing task Univ. of Sherbrooke, Canada

IOP Conf. Series: Materials Science and Engineering **309** (2018) 012110 doi:10.1088/1757-899X/309/1/012110

Morphology and thermal stability of nano titanium dioxide filled natural rubber prepared by latex mixing method

N Hayeemasae[1], I Surya*[2,3] and H Ismail[4]
[1]Department of Rubber Technology and Polymer Science, Faculty of Science and Technology, Prince of Songkla University, Pattani Campus, 94000, Pattani, Thailand
[2]Department of Chemical Engineering, Faculty of Engineering, Universitas Sumatera Utara, Medan 20155, Sumatera Utara, Indonesia
[3]Sustainable Energy and Biomaterial Center of Excellence, Faculty of Engineering, Universitas Sumatera Utara, Medan, 20155, Indonesia
[4]School of Materials and Mineral Resources Engineering, Engineering Campus, Universiti Sains Malaysia, 14300, Nibong Tebal, Penang, Malaysia

*E-mail: indradanas@yahoo.com

Abstract. This paper deals with the morphology and thermal stability of nano Titanium Dioxide (TiO_2) filled natural rubber composites. This study also suggests a new method of incorporating TiO_2. Aqueous dispersions of nano TiO_2at the loadings of 0, 2, 4, 6 and 8 phr were dispersed in natural rubber latex, the resulting compounds were then dried prior to mixing it with other ingredients on a two-roll mill. By applying this technique, the homogeneity of the compound is significantly improved. This can be clearly seen from the morphology observed. Adding TiO_2 results in shifting the decomposition temperature and char residue irrespective of the loadings of nano TiO_2.

1. Introduction

It is a well-known fact that the addition of rigid filler particles, even in small amounts, to a rubber, strongly influences its mechanical behaviour [1]. Currently, researchers have faced complications with the use of white fillers in rubber composites. As for silica, it is mostly incompatible with rubber especially natural rubber. It requires an addition of compatibilizers or coupling agents, which complicates the processing method and subsequently increases production cost. Meanwhile, calcium carbonate and clay play major roles as non-reinforcing fillers, which result in deterioration of the mechanical properties [2-4]. Therefore, new filler materials need to be determined to maintain acceptable properties.

Titanium Dioxide (TiO_2) is widely used as filler in many polymers because of the improved physical and mechanical properties it yields [5-7]. The nature of TiO_2 has great influence on the reinforcement and thermal stability of polymer composites. As reported by Meera et al. [8], they elucidated that TiO_2 do not possess any active hydroxyl groups on their surface unlike silica particles that have available surface hydroxyl groups. This could provide a less tendency to form aggregate which leads to a better dispersion of the TiO_2 accompanied with strong interfacial adhesion between matrix and fillerin the polymer composites.

In this regard, nano TiO_2 has received special interest as filler in the rubber matrix because of their specific properties. Incorporation of a small amount of nano TiO_2 into the rubber matrix has been largely unexplored.The aim of this work is to incorporate nano TiO_2 in natural rubber composites together with the suggestion of a new route of dispersing nano TiO_2. By applying this method, the homogeneity and respective thermal stability of the compound is expected to improve.

2. Materials and Methods

2.1. Materials

The formulation used in this study is listed Table 1. Low ammonia type Centrifuged Natural rubber latex (LATZ type), bentonite clay and Disodium 2,2'-methylene-bis-naphthalene sulphonate (Tamol) were supplied by Zarm Scientific (M) Sdn. Bhd., Penang, Malaysia. Nano $TiO_2$was purchased from Sigma-Aldrich Co., LLC., USA., with the size ranging from 15 – 40 nm(TEM), 4.26 g/ml in density, surface area of 35 – 65 m^2/g (BET), and more than 99.5% purity based on an analysis of trace materials [9]. Other compounding ingredients, such as zinc oxide, stearic acid, N-cyclohexyl-benzothiazyl-sulphenamide(CBS), N-2-Propyl-N'-phenylenediamine(IPPD), and sulfur were purchased from Bayer (M) Ltd.

Table 1. The formulation used in this study.

Ingredients	Amount (phr)
Natural Rubber	100
ZnO	5
Stearic acid	2
CBS	0.5
IPPD	2
Sulfur	2.5
Nano TiO_2	0, 2, 4, 6 and 8

2.2. Preparation of the composites

Figure 1 illustrates the diagram for the compounding preparation of natural rubber composites. The aqueous dispersion of nano TiO_2 was prepared by using a mechanical ball mill. The compositions of 40 parts (w/w) of nano-sized TiO_2, 2 parts (w/w) bentonite clay as a colloidal stabilizer, 1 parts (w/w) of Disodium 2,2'-methylene-bis-naphthalene sulphonate(Tamol) as a dispersion agent and 57 parts (w/w) of distilled water as the medium, were added to the jar of the ball mill. Then, the whole mixture was kept on the ball milling machine for 24 hrs at the rotation speed of 10 rpm. Later, the dispersion form of nano TiO_2 was stored in an amber glass bottle to prevent from self photo-oxidation process.

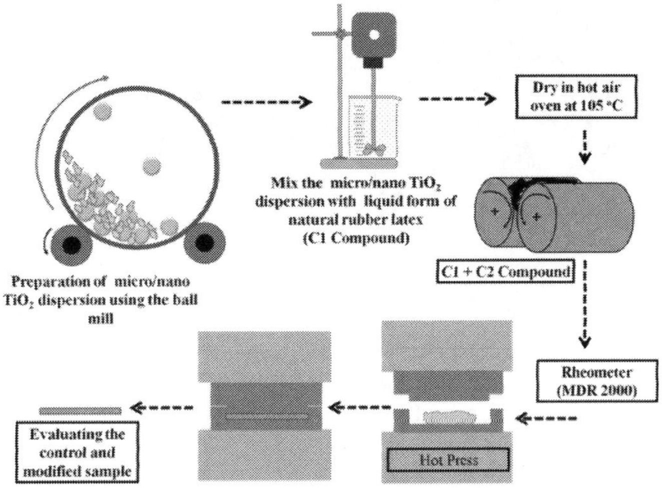

Figure 1. Schematic illustration representing the preparation of the composites.

The compounding preparation was done by mixing the centrifuged natural rubber latex with the aqueous dispersion of nano TiO_2. The mixture was agitated mechanically for 8 hrs in order to ensure the homogeneity. Subsequently, this colloidal mixture was poured into a glass tray and dried in hot air oven at 100°C for 8 hrs to remove all the water (C1 compound). Ultimately, the dried sheet of C1 compound was compounded with other additives on the laboratory size two-roll mill (Model XK-160). The final compound was later tested for its curing characteristics by using a Monsanto Moving Die Rheometer (MDR 2000). The compounds were subsequently compression-moulded using a stainless steel mould at 150°C with a pressure of 10 MPa using a laboratory hot-press based on the respective curing times.

2.3. Morphological studies
The examination of nano TiO_2 particles was carried out using Transmission Electron Microscope (TEM) model Philips CM12 and the images were analysed using Docu version 3.2 image analysis. The dispersion of nano TiO_2 in natural rubber matrix was captured through tensile fractured samples using a scanning electron microscope (SEM) model Zeiss Supra-35VP. The fractured pieces were coated with a layer of gold palladium to eliminate electrostatic charge build-up during examination.

2.4. Thermogravimetric analysis)TGA(
Thermogravimetric analysis of nano TiO_2filled natural rubber composites was carried out with a Perkin-Elmer Pyris 6 TGA analyser. The sample was scanned from 30°C to 600°C at the nitrogen air flow of 30 ml/min with heating rate of 20°C /min.

3. Results and Discussion

3.1. Morphology
Figure2 shows the TEM image of nano TiO_2 particles. It was visibly seen that nano TiO_2 particles are in rod- or elongated shapes and are far smaller than other commercial fillers such as silica, calcium carbonate and etc., indicating that high reinforcement and thermal stability can be attained by incorporating nano TiO_2.

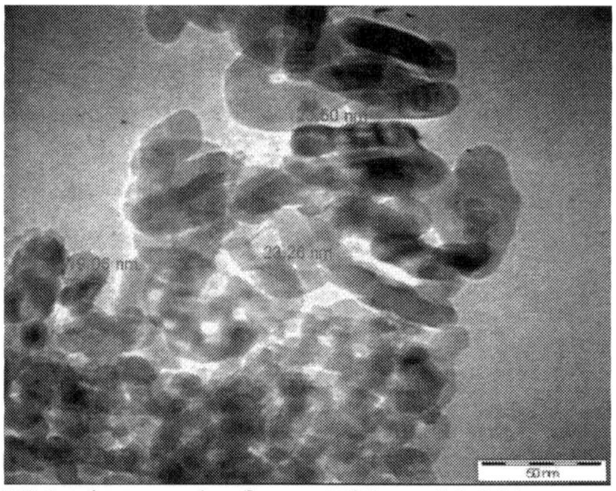

Figure 2. TEM micrograph of nano TiO_2 at 35,000× magnifications.

The dispersion of nano TiO_2inside the natural rubber matrix can be visibly seen in Figure 3. Nano TiO_2 was dispersed homogenously in the rubber matrix, indicating that using the new route for preparing the composites results in better dispersion. The homogeneity of the composites is significantly improved; especially with loading up to 6 phr. Well-dispersed nano TiO_2 is responsible for the improved thermal stability which will be discussed in later session.

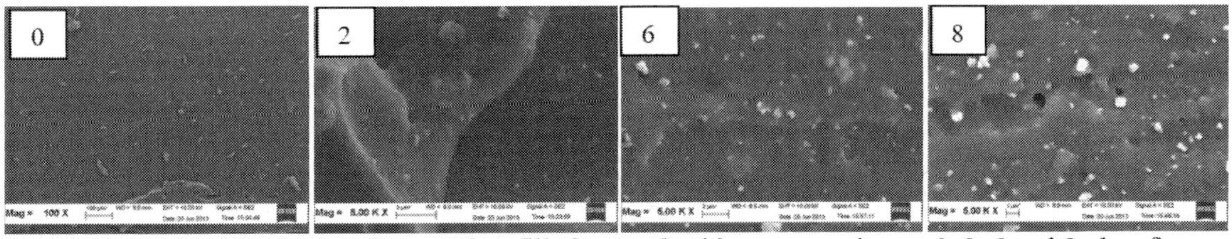

Figure 3. SEM micrographs of nano TiO_2 filled natural rubber composites at 0, 2, 6 and 8 phr of nano TiO_2 (35,000× magnifications).

3.2. Thermo gravimetric analysis

Thermo gravimetric analysis measures the amount and rate of the mass transition of a sample as a function of temperature in a certain atmosphere. Such analysis is used primarily to determine the thermal stability and compositional properties of materials. The thermal decomposition behaviour and derivative weight thermo grams of nano TiO_2 filled natural rubber composites are shown in Figure 4. The decomposition temperatures at various weight loss stages and char residue are also listed in Table 2. Two regions of degradation were subsequently observed. The initial minor weight loss at around 180–200°C was associated to the presence of volatile matter such as stearic acid and absorbed water at around 300°C [10]. The major stage in the degradation of the composites started at about 330°C and was completed at around 450°C. This is attributed to the degradation of natural rubber segment. The degradation of the natural rubber segment is sensitive to the presence of the oxidized structure as well as the depletion of sulphidic cross-link in natural rubber [11].

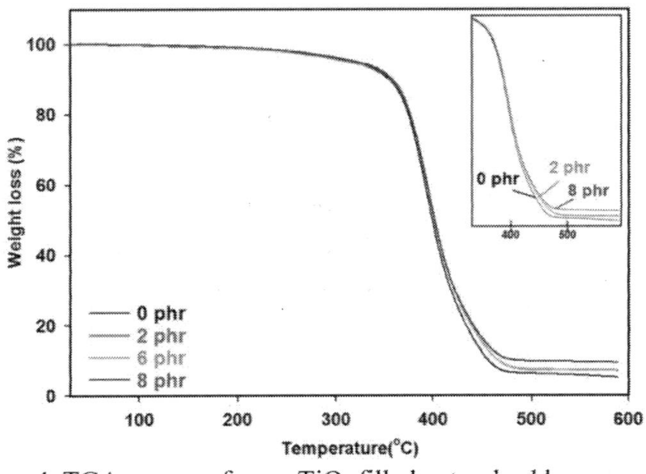

Figure 4. TGA curves of nano TiO_2 filled natural rubber composites.

Table 2.Thermal stability parameters of nano TiO_2 filled natural rubber composites.

Nano TiO_2 loading (phr)	Temperature at 10% weight loss (°C)	Temperature at 50% weight loss (°C)	Temperature at 70% weight loss (°C)	Weight res. (%)
0 phr	348.21	401.14	415.51	2.53
2 phr	350.15	402.81	418.21	3.16
6 phr	354.91	403.74	421.22	7.45
8 phr	359.28	405.24	424.57	9.22

It is interesting to note that the decomposition temperature at 10%, 50%, and 70% weight loss of natural rubber composites was greatly influenced by the presence of nano TiO_2. It was found that the decomposition temperature clearly differed from the earlier stages of decomposition. Introducing nano TiO_2 into the natural rubber composites greatly affected their thermal stability. The decomposition temperature of the composites showed further improvement at the decomposition temperatures above 50% of weight loss. Nano TiO_2 do not possess any active hydroxyl groups on their surface unlike silica particles that have available surface hydroxyl groups. This could provide a less tendency to form aggregate which leads to a better dispersion of the TiO_2 accompanied with strong interfacial adhesion between matrix and fillerin the rubber.

From this point of view, highly compatibilized nano TiO_2 filled natural rubber composites and better thermal resistance are formed. As a result, more pronounced thermal stability is observed in the presence of nano TiO_2. Furthermore, the char residue of all composites was influenced by introducing of nano TiO_2. This is simply due to the fact that the amount of filler was varied; thus keeping the char residue mostly changed. According to Chakraborty et al. [12], the amount of char is very much dependent on the type and amount of filler. The difference in filler amount mainly concerns the char formation.

4. Conclusions

The following conclusions can be presented:

1. Rod or elongated shape of nano TiO_2 is clearly seen in TEM image, indicating that high contact area and thermal stability can be attained by incorporating nano TiO_2.
2. Applying new technique of incorporating nano TiO_2in the natural rubber matrix has greatly influenced on the dispersion of nano TiO_2. This is visibly seen in the SEM images.
3. The thermal decomposition temperature of the rubber has shifted to a higher temperature when nano $TiO_2$is used, indicating that more heat is needed to decompose the sample.

References

[1] Zhang J, Wang L and Zhao Y 2013*Mater Design* **50** 322-31
[2] Nabil H, Ismail H and Azura A R 2011*J Elas Plas* **43** 429-49
[3] Sheril R V, Mariatti M and Samayamutthirian P 2014 *J Vinl Addv Technol* **20** 160-7
[4] Navarchian AH, Joulazadeh M and Mousazadeh S 2013 *J Vinl Addv Technol* **19** 276-84
[5] Buzarovska A 2013 *Polym-Plas Technol Engineer* **52** 280-6
[6] Sreekumar P, Al-Harthi M A and De S K 2012 *J Compos Mater* **46** 3181-7
[7] Ahmad J, Deshmukh K and Hägg M B 2013 *Int J Poly Anal Charac* **18** 287-96
[8] Meera A P, Said S, Grohens Y, Luyt AS and Thomas S 2009 *Ind Eng Chem Res* **48** 3410–6
[9] Technical Report of Sigma-Aldrich (M) Sdn. Bhd. Available at: http://www.sigmaaldrich.com/Graphics/COfAInfo/SigmaSAPQM/SPEC/71/718467/718467-BULK_______ALDRICH__.pdf (Accessed 10 June 2017).
[10] Tomer NS, Delor-Jestin F, Singh R and Lacoste J 2007 *Polym Degrad Stab* **92** 457-63
[11] Deuri A S, Bhowmick A K, Ghosh R, John B, Sriram T and De S 1988 *Polym Degrad Stab* **21** 21-8
[12] Chakraborty S, Roy P, Pathak A, Debnath M, Dasgupta S, Mukhopadhyay R and Bandyopadhyay S 2011 *J Elas Plas* **43** 499-508

IOP Conf. Series: Materials Science and Engineering **309** (2018) 012111 doi:10.1088/1757-899X/309/1/012111

Drying kinetics and characteristics of dried *gambir* leaves using solar heating and silica gel dessicant

R Hasibuan[1,*], J Hidayati[2], R Sundari[3], and A S Wicaksono[1]

[1]*Department of Chemical Engineering, University of Sumatera Utara, Indonesia*
[2]*Department of Industrial Engineering, University of Sumatera Utara, Indonesia*
[3]*Department of Mechanical Engineering, Mercu Buana University, Indonesia*

*E-mail: rosdanelli@yahoo.com

Abstract. A drying combination of solar heating and silica gel dessicant has been applied to dry *gambir* leaves. The solar energy is captured by a collector to heat the air and the hot air is used to dry *gambir* leaves in a drying chamber. An exhaust fan in drying chamber assists to draw water molecules from *gambir* leaves accelerated by silica gel dessicant. This study has investigated the drying kinetics and drying characteristics of *gambir* leaves drying. In drying operation the air velocity is tuned by a PWM (pulse width modulation) controller to adjust minimum and maximum level, which is based on the rotation speed of the exhaust fan. The results show that the air velocity influenced the drying kinetics and drying characteristics of *gambir* leaves using solar-dessicant drying at 40 cm distance between exhaust fan and silica gel dessicant.

1. Introduction

A *gambir* leaf (*Uncaria gambir Roxb.*) is a plant that lives in Indonesia and have small comodity with high economic, and the origin of *gambir* plants is known with certaintly, but allegedly from Southeast Asia. *Uncaria gambir* is known as safed katha belongs to family Rubiceae, a native Southeast Asian herbal plant, which contains tannins namely, catechutannic acid, acacia catechin, catechu red, catechin, epicatechin & flavonoids–quercetin and quercitrin. In Indonesia, *gambir* is found growing wild in the forest of Sumatera, Borneo and The Malay Peninsula [1-3].

The shape of *gambir* plants height is 1,5 to 2 m, the branch colour is light brown to dark brown, branching many angles of 30°-50° from the main trunk. For the *gambir* leaves, the leaves shape is ovalis, and have young green colour, green brown, and green, with a length 5 cm to 8 cm [1].

Drying is commonly described as the operation thermally removing water content to yield of a product [4]. Drying with sun light method isa conventional method that is still used in many developing countries. Due to low cost, this method still used. However, this method has shortage like contamination of aves, flea, microorganism, dust particle and so on. This methodtakes long drying time and yields uniform dried material [5].

Generally, the drying process of *gambir* leaves is still using conventional drying, and thenthe resulting of dried material is not uniform and not hygienic. Therefore, this study applies a drying combination of solar heating and silica gel dessicant. The air velocity in terms of PWM, drying rate, and moisture content related to drying time to study the drying kinetics and drying characteristics have been investigated in this work. The drying method follows the previous work used for drying rosella using solar-dessicant dryer (SDD).

IOP Conf. Series: Materials Science and Engineering **309** (2018) 012111 doi:10.1088/1757-899X/309/1/012111

2. Materials and Method

2.1 Materials

Gambir leaves were obtained from Sidikalang farm, silica gel from Rudang jaya and PWM controller was made by Electrical Engineering students from Universitas Sumatera Utara, Medan, Indonesia.

2.2 Preparation of solar-dessicant dryer (SDD)

Solar energy is captured by a collector to heat the air in the collector and the hot air is used to dry the *gambir* leaves put on the surface of silica gel dessicant. The air velocity can be adjusted by tuning the exhaust fan in drying chamber to minimum or maximum level.

(a) (b)

Figure 1. (a) Solar-Dessicant Dryer Front View (b) Solar-Dessicant Dryer Side View

(a) (b)

Figure 2. (a) Gambir Leaves before dried and (b) Gambir Leaves after dried

2.3 Drying
150g of *gambir* leaves was dried using solar-dessicant dryer (SDD) with a distance of 40 cm between the silica gel dessicant and the exhaust fan. The weight ratio of *gambir* leaves and silica gel dessicant was made 1:1. The exhaust fan is tuned with a PWM controller (minimum or maximum speed of exhaust fan). Drying started from 10.00 AM to sunset.

2.3.1 Moisture Content.
The moisture content of *gambir* leaves has been investigated by the standard of SNI 01-3391-1994 with 10g of *gambir* leaves. *Gambir* leaves were put in an oven for 24h. The moisture content is calculated as follows [6]:

$$Mc=\frac{w-w_d}{w_d}\times 100\% \quad (1)$$

where:
w = wet weight (gram)
wd = dry weight (gram)
Using this formula, the initial moisture content and moisture content after dried of *gambir* leaves with SSD method can be determined.

3. Results and discussion

3.1 Initial Moisture content
The average initial moisture content of three replications of 10g *gambir* leaves put in an oven for 24h is found to be 69-70%.

3.2 Drying kinetics in terms of moisture content
Figure 3 shows the reduction of water content in the drying process of *gambir* leaves as the drying time getting longer with respect to different speed of exhaust fan (minimum and maximum).

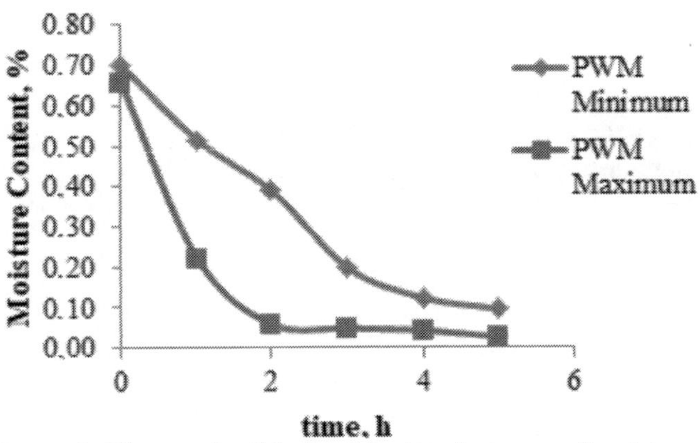

Figure 3. The graph of drying kinetics in terms of moisture content with respect to minimum and maximum speed of exhaust fan.

In Figure 3, it is shown that the graph of drying kinetics in terms of moisture content of *gambir* leaves dried. At t = 0, the moisture contentsof PWM minimum and PWM maximum were found to be 70% and 66%, respectively. In the drying process, the moisture content of PWM minimum decreased much slower than that of PWM maximum [5]. At the end of this investigation (t = 300 min), the moisture contents of PWM minimum and PWM maximum were found to be 10% and 3%, respectively.

3.3 Drying kinetics in terms of drying rate

Figure 4 shows the decreasing of drying rate in the drying process of *gambir* leaves as the drying time getting longer with respect to different speed of exhaust fan (minimum and maximum), which is referred to PWM minimum and PWM maximum.

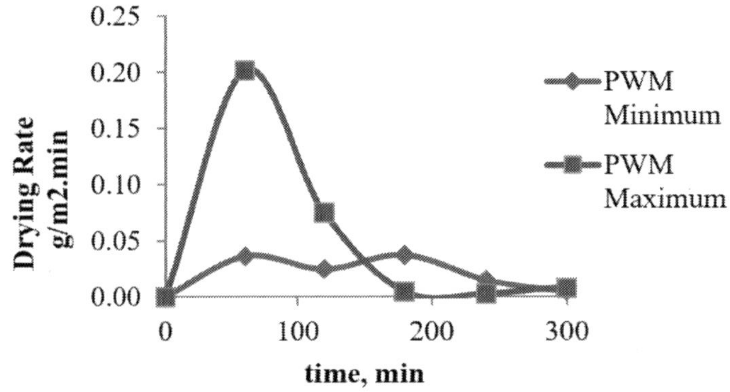

Figure 4. Drying kinetics in terms of drying rate of *gambir* leaves drying at 40 cm distance between exhaust fan and silica gel dessicant.

At Figure 4 when t = 0, based on both graphs of drying kinetics, both PWM minimum and PWM maximum show zero drying rate. Moreover, when t = 60 min, the drying rate of PWM minimum is found to be 0,04g/m^2.min, however, the drying rate of PWM maximum is 0.20 g/m^{2}min,and after t = 120 min, both drying rates decreased with increasing drying time. With respect to the drying rate of PWM minimum, it is fluctuated from t = 60 min to t = 240 min. This phenomenon is related to slow heat transfer giving time of phase change from free water liquid to free water vapor and therefore, it shows unstable drying rate. Furthermore, at t = 180 min, the graph of drying rate of PWM minimum increasedagain from 0,04g/m^2.min(t = 120 min) to 0,02 g/m^2.min(t = 180 min) and after that it continuously decreased.

3.4 Drying characteristics of gambir leaves

Figure 5 shows the drying characteristics of *gambir*leaves drying. This investigation found two drying periods, i.e. the increasing drying period and the decreasing drying period.

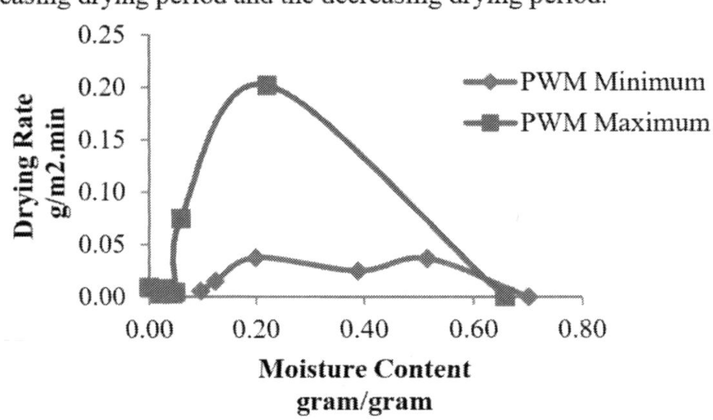

Figure 5. Drying characteristics of *gambir* leaves at 40 cm distance between exhaust fan and silica gel dessicant.

Figure 5 reveals that the drying characteristic of PWM maximum shows two periods of drying rate, i.e. the increasing and decreasing drying rates, however, the drying characteristics of PWM minimum shows three periods of drying rate, i.e. the increasing, constant, and decreasing drying rates. With respect to the drying rate of PWM maximum, the faster speed of exhaust fan caused faster heat transfer and therefore, all free water molecules on the surface of *gambir* leaves released and directly changed from liquid to vapor. On the other hand, with respect to the drying rate of PWM minimum, the slower speed of exhaust fan caused slower heat transfer and therefore, only part of free water molecules on the surface of *gambir* leaves changed to water vapor, thus, it shows a fluctuated drying rate referred to a period of constant drying rate. The period of increasing drying rate is related to the release of free water molecules on the surface of a material due to heat transfer, while the period of constant drying rate is related to phase change of water liquid to water vapor indicated that the heat transfer is used for phase change in this period, and the period of decreasing drying rate is associated to releasing of bound water in the capillary system of material subsurface. With respect to the PWM maximum, it can be elucidated that the faster heat transfer causing faster drying process in relation to flammable organic properties of *gambir* leaves.

4. Conclusion

The study shows that air velocity due to rotation of exhaust fan in terms of PWM level affected the drying kinetics and drying characteristics of *gambir* leaves drying.This study found that the PWM minimum yielded three periods of drying rate related to part of free water released from material surface changed to water vapor due to slower heat transfer in the first period of drying rate, however, the PWM maximum showed two periods of drying rate related to all free water changed to water vapor due to faster heat transfer in the first period of drying rate.

5. Acknowledgment

The authors gratefully acknowledge that the present research is supported by the University of Sumatera Utara, Medan, Indonesia through Non-PNBP fund under talent research contract 2017.

References

[1] Fauza H 2014 *International Journal on Advanced Science Engineering Information Technology* **4** No 6
[2] Patil S H *et al* 2012 *International Journal of Drug Development & Research* **4** 234-238
[3] Rauf A Rahmawaty and A Z Siregar 2015 *Procedia Chemistry* **14** pp 3-10
[4] Tiwari A 2016 *Journal of Food Processing & Technology* **7** 1000623
[5] Hasibuan R and Zam-Zami A 2017 Effect of drying medium velocity on quality of ginger (*Zinger officinale Roscoe*) using combination of solar energy-molecular sieve drying system **180** 012153
[6] Hartley J and J Marchant 1995 *Methods of determining the moisture content of wood* (Sydney: State Forest of New South Wales)

IOP Conf. Series: Materials Science and Engineering **309** (2018) 012112 doi:10.1088/1757-899X/309/1/012112

System simulation application for determining the size of daily raw material purchases at PT XY

H L Napitupulu*
Department of Industrial Engineering, Faculty of Engineering,
Universitas Sumatera Utara, Jl Almamater, Kampus USU, Medan 20155, Indonesia

*Email: humala_n@yahoo.com

Abstract. Every manufacturing company needs to implement green production, including PT XY as a marine catchment processing industry in Sumatera Utara Province. The company is engaged in the processing of squid for export purposes. The company's problem relates to the absence of a decision on the daily purchase amount of the squid. The purchase of daily raw materials in varying quantities has caused companies to face the problem of excess raw materials or otherwise the lack of raw materials. The low purchase of raw materials will result in reduced productivity, while large purchases will lead to increased cooling costs for storage of excess raw materials, as well as possible loss of damage raw material. Therefore it is necessary to determine the optimal amount of raw material purchases every day. This can be determined by applying simulation. Application of system simulations can provide the expected optimal amount of raw material purchases.

1. Introduction

Raw material procurement is very important for every manufacturing company. The procurement of perishable raw materials should be done day by day, as in PT XY which is engaged in the processing of sea catch. One kinds of raw material processed by this company is squid to produce frozen squid. In this case the company does not specify the amount of daily raw material purchases but receives and pays for the amount of raw materials delivered by the supplier.

The amount of daily raw squid purchases by companies varies from day to day. These variations are related to daily catches that vary according to season and weather. Historically the minimum amount of raw material purchased is less than 2 tons per day and the maximum amount is almost 10 tons per day. While the production capacity needs to be determined and adjusted to the amount of purchase.

The amount of raw squid purchases can cause problems in the procurement of raw materials. This can lead to shortages of raw materials or otherwise excess raw materials to be processed on a daily basis. The low amount of raw material purchases below the level of production capacity will lead to low production. While the high amount of raw material purchases exceeds the production capacity will result in additional cooling costs for storage of excess raw materials. Excessive storage of raw materials over long periods of time will cause worsening quality of squid. Although raw squid can be stored in a freezer, the duration should not last long to avoid damage. Squid that began to suffer damage will not be processed but disposed of as waste. Therefore, the maximum amount of daily purchases of raw materials and the optimal levels of production capacity should be determined to maximize profits, minimize additional costs and prevent losses. This can be determined by simulation.

System simulation is the process of imitating real events in real systems. Computer simulations with artificial system operations can overcome certain research obstacles and experimental constraints, such as the relatively high cost and longer periods required for real research and experimentation. By applying simulations, the results of research and real experiments that take place in long time intervals can be obtained in seconds in virtual form [1]. Therefore, to overcome real experimental problems, system simulations can be applied to find out the maximum amount of daily raw material purchases and optimal production capacity levels. Simulation system in the form of computer simulation can be done by doing step-by-step process [2].

2. Method

The simulation steps are:

1. Problem formulation. The lack of decision on daily purchasing volume of squid atPT XY Medan, Sumatera Utara.
2. Setting of objectives and overall project plan. The objectives of the simulation is to determine the maximum amount of daily squid purchase which is in accordance with the optimal production capacity.
3. Model conceptualization. Conceptual model of real world system (mathematical and logical relationship concerning the component and the structure of the system). The model was build in the form of a diagram illustrating the relationship between the components used in dynamic simulation.

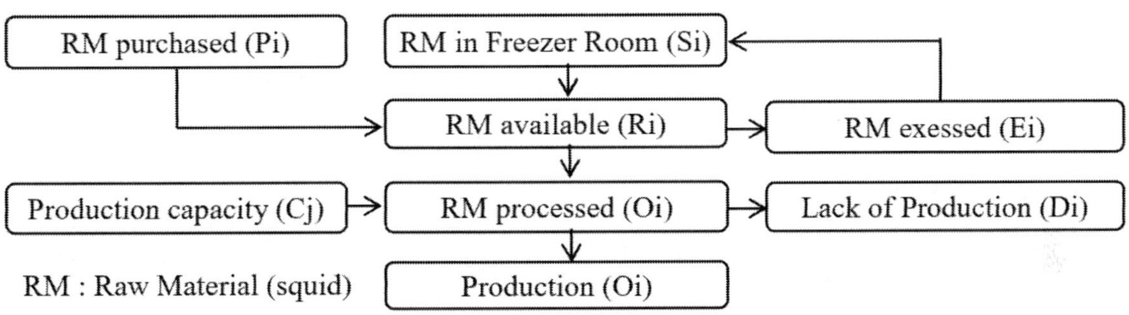

Figure 1. The model of the real world system components.

The mathematical models used in system simulation are as follows :

At day i and production capacity level Cj

Theamount of raw materials in the freezer room : $Fi = F(i\text{-}1) + Ei$

Daily amount of raw materialspurchased : Pi

The amount of raw materials available for processing : $Ri = Fi + Pi$

The amount of excess raw material (Ei) : $[Ri > Cj] \rightarrow Ei = Ri - Cj$

The amount of shortage of raw materials : $[Ri < Cj] \rightarrow Di = Cj - Ri$ and $Ei = 0$

The amount of production (Oi) : $[Cj \leq Ri] \rightarrow Oi = Cj$ * average rendement

$[Cj \geq Ri] \rightarrow Oi = Ri$ * average rendement

The amount of gross profit : $Si = Oi$ x profit per unit

The total additional cost : $Di = Ei$ * Freezer cost per unit of RM

The total profit : $Zi = Si - Di$

The model of system simulation was build as a discrete dynamic model.

4. Data collection. The data required should be collecting in the format required. Data collection cocerning the daily supplied material, production capacity, the selling price of products, production costs, and rendement of production.
   The amount of daily supplied raw material was supposed as the quantity of daily raw material purchases. The amount of daily supply was collected from historical data in 2 months ago. The data then analized statistically. The results of statistical descriptive analysis on daily amount of squid supply within sample size N = 46 are as follows:
   Average : $\mu = 4.87$, Standard Deviation : $\sigma = 2.41$, Maximum : 9.67, Minimum : 1.55
   Pattern of data distribution of raw material supply was tested by using chi square test.
   The result of chi square test (χ^2 count = 4.70 and χ^2 tables = 9.488) approve that the pattern of amount of daily supply of squid is normal pattern. The value of μ and σ are then used as parameter in generating of artificial data of daily amount of squid supply.
   The artificial data was generated by use the Excel'sequation of NORMINVERSE() and random number represented the probability. Articial data of daily amount of supply of squid was prepared per month as shown in Table 1.

Table 1.The artificial data of daily supply of squid (ie. in June)

No.	Date	(Ton/day)	No.	Date	(Ton/Day)
1	04/06	3.00	13	18/06	6.20
2	05/06	4.80	14	19/06	2.90
3	06/06	3.20	15	20/06	7.88
4	07/06	3.70	16	21/06	4.35
5	08/06	5.50	17	22/06	5.40
6	09/06	3.20	18	23/06	9.11
7	11/06	4.40	19	25/06	5.20
8	12/06	5.80	20	26/06	1.98
9	13/06	6.00	21	27/06	4.25
10	14/06	9.10	22	28/06	9.67
11	15/06	4.60	23	29/06	6.80
12	16/06	8.80	24	30/06	2.00

5. Model translation. The model of system simulation was build as a discrete dynamic model. The conceptual model was codified into computer recognizable form, as an operational model in worksheet Excel.
6. Verification concerns the operational model. The model which translated and build in form of Excel worksheet was executed by use the artificial data. The results of model execution then compare to the output of real production. The result obtained was vary but the variance was nearly equal to the daily output.
7. Validation that the conceptual model is accurate representation of the real system. Validation is done by compare the result of simulation to total profit on real system. The validation based on statistical analysis of total profit in monthly period.
8. Experimental design. The experiments in the simulation are designed by determining the length of the system operation period, the number of repetitions, and initialization values in the initial stage. The length of the operation period in the simulation is limited to 24 or 25 working days per month. The simulation was run in 10 replicates for each level amount of squid purchase.The maximum amount of daily purchase of squid determined in 7 levels to be compared is set at a rate of 3.5 tons, 4.0 tons, 4.5 tons. 5.0 tons. 5.5 tons 6.0 tons, and as much as supplied by the suppliers per day. The dynamic simulation result of squid purchase is determined and presented in form of total profit per month.

9. Production runs and analysis. The process of system was run and the results then analized. The statistical analysis of the acceptance of the simulation results shows that the number of replication n = 10 is greater than the minimum number of the required theoretical replication n '= 6. Thus the average total earnings on each level of daily squid purchase which is obtained by system simulation is acceptable as representing the total earnings of real system

3. Results And Discussions

The simulation results in the form of monthly total earnings at each level of squid purchases are presented in Table 2 below. Total profit per month is greater at a higher level up to 5 tons per day, and then decreases at the next level. Thus the average expectation of the highest total profit per month is at the maximum level of 5 tons of squid per day, as shown in the graph in Figure 2. According to the maximum amount of this purchase, the optimal production capacity can be determined at 5 tons per day .

Table 2. Expected total profit per month

Level	(Tons/day)	(Rp .000/month)
I	3.50	414.143
II	4.00	459.619
III	4.50	499.970
IV	5.00	570.046
V	5.50	401.356
VI	6.00	369.540
Var	8.86	295,418

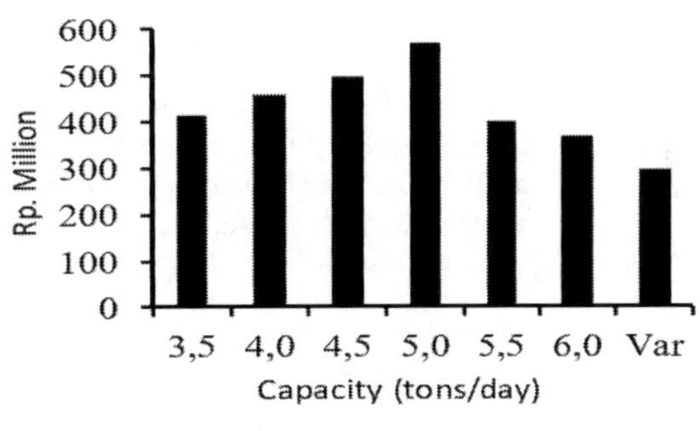

Figure2. Expected total profit per month.

4. Conclusion

Discrete system simulations are run by stochastic processes using probabilistic models. The stochastic dynamic process is used in determining the quantity of squid supply and the stock of raw materials available on an ongoing basis from day to day. System simulation application can provide optimal solution in form of maximum amount of squid purchase every day and optimal production capacity in PT XY. The system simulation result is the average monthly total earnings based onsystem verification and output validation.

References

[1] Napitupulu H 2009 Simulasi Sistem, Permodelan dan Analisis Medan USU Press

[2] Banks J 1998 *Handbook of Simulation* John Wiley and Sons

IOP Conf. Series: Materials Science and Engineering **309** (2018) 012113 doi:10.1088/1757-899X/309/1/012113

The study of stiffness modulus values for AC-WC pavement

AS Lubis[1,2,*], Z A Muis and T D Iskandar

[1]Civil Engineering Department, Faculty of Engineering, Universitas Sumatera Utara, Padang Bulan, Medan 20155, Indonesia
[2]Sustainable Energy and Biomaterial Center of Excellence, Universitas Sumatera Utara, Padang Bulan, Medan 20155, Indonesia.

*Email: adinasarilubis@gmail.com

Abstract. One of the parameters of the asphalt mixture in order for the strength and durability to be achieved as required is the stress-and-strain showing the stiffness of a material. Stiffness modulus is a very necessary factor that will affect the performance of asphalt pavements. If the stiffness modulus value decreases there will be a cause of aging asphalt pavement crack easily when receiving a heavy load. The high stiffness modulus asphalt concrete causes more stiff and resistant to bending. The stiffness modulus value of an asphalt mixture material can be obtained from the theoretical (indirect methods) and laboratory test results (direct methods). For the indirect methods used Brown & Brunton method, and Shell Bitumen method; while for the direct methods used the UMATTA tool. This study aims to determine stiffness modulus values for AC-WC pavement. The tests were conducted in laboratory that used 3 methods, i.e. Brown & Brunton Method, Shell Bitumen Method and Marshall Test as a substitute tool for the UMATTA tool. Hotmix asphalt made from type AC-WC with pen 60/70 using a mixture of optimum bitumen content was 5.84% with a standard temperature variation was 60°C and several variations of temperature that were 30, 40, 50, 70 and 80°C. The stiffness modulus value results obtained from Brown & Brunton Method, Shell Bitumen Method and Marshall Test which were 1374,93 Mpa, 235,45 Mpa dan 254,96 Mpa. The stiffness modulus value decreases with increasing temperature of the concrete asphalt. The stiffness modulus value from the Bitumen Shell method and the Marshall Test has a relatively similar value.The stiffness modulus value from the Brown & Brunton method is greater than the Bitumen Shell method and the Marshall Test, but can not measure the stiffness modulus value at temperature above 80°C.

1. Introduction

The asphalt mixture is a combination of aggregate mixed evenly and coated with hard asphalt. In a planning of the asphalt mixture use several parameters to achieve strength and durability. One of the parameters of the asphalt mixture in order for the strength and durability to be achieved as required is the stress-and-strain showing the stiffness of a material [1].The stiffness value of ordinary material is assessed from the elasticmodulus of the material. The elastic modulus is a measure of the stiffness of a material; so that the higher of the elastic modulus of the material is the less form changes that occur when given the force. Thus, the greater of the stiffness modulus is the smaller of the elastic strain that occurs or the more rigid [2].

The stiffness modulus value of an asphalt mixture material can be obtained from the theoretical (indirect methods) and laboratory test results (direct methods). The empirical approach of the stiffness modulus is, among others, the Van der Poel method which later developed into the Shell Bitumen method [3], and the other one is Brown & Brunton method [4]. While one of the laboratory tool to test the stiffness is the UMATTA tool, but in this study used the Marshall tool [5].

Al-Jassar et al. [6] and Widodo [5] tested the elastic modulus of concrete asphalt with Marshall tool. The test result states that the elastic modulus of asphalt concrete is strongly influenced by the temperature factor. Wardoyo J. [7] tested the effect of gilsonite as an additive material on the stiffness modulus value of the concrete asphalt at low temperature using Brown and Brunton method. The test results show that the addition of gilsonite can decrease the penetration value and increase the softening point of asphalt. Sjahdanulirwan [8, 9] conducted a literature study on the effect of the time loading on the stiffness of concrete asphalt according to several methods.This research is a study about the stiffness modulus value of concrete asphalt using indirect methods and direct methods. For the indirect methods used Brown & Brunton method, and Shell Bitumen method; while for the direct methods used Marshall tool.

2. Literature Review

The elastic modulus is the number that used to measure the object or resistance of the material to undergo elastic deformation when force is applied to the object. One type in the elastic modulus is the stiffness modulus. The stiffness modulus is the elastic modulus used in the elastic theory. As it is known that almost all of the pavement materials are not elastic but have permanent deformation after receiving load repetition. But if the load is relatively small against the strength of the material, then the permanent deformation occurring at each high load repetition can almost be perfectly and proportional to the load. In this condition the material can be considered as an elastic material [10].

Stiffness is the resistance of a material to elastic deformation, because the rheological nature of the asphalt stiffness is the relation between stress and strain as a function of the duration of loading and temperature. Van der Poel [3] termed stiffness of bitumen (Sb) as a comparison between stress-strain on asphalt, which is a function of the duration of loading (frequency) applied, temperature difference with T800 and Penetration Index. T800 is the temperature at the time of penetration reached 800.Based on Bina Marga 2013[11] the modulus characteristics of materials for Indonesia's climate and loading conditions on the AC-WC type is 1100Mpa. AASHTO [12] says the need for attention to asphalt concrete surface coarse has a stiffness modulus value above 450,000 psi. The high stiffness modulus value causes the asphalt concrete to be more rigid and resistant to deflection, but susceptible to temperature and fatigue cracking.

There are two approaches to measure the stiffness modulus value of an asphalt mixture material, i.e. by indirect methods and direct methods. In the indirect approach the stiffness modulus is estimated through the asphalt routine data using formulas and nomograms that have been developed by Brown& Brunton, and Shell Bitumen.

According to Brown & Brunton [4] method, the stiffness modulus of the concrete asphalt mixture (Sme) is influenced by the asphalt stiffness modulus (Sb) and the VMA (Voids in Mineral Agregate). The asphalt stiffness modulus (Sb) is influenced by temperature, recovered penetration, recovered softening point, penetration index and duration of loading. The equations of Brown & Brunton methods are:

$$Sme = Sb\left[1 + \frac{257{,}5 - 2{,}5\,VMA}{n\,x\,(VMA - 3)}\right]^{n} \tag{1}$$

$$n = 0{,}83 \text{ x } \log\left(4.10^{10}/Sb\right)$$

with:Sme = Stiffness modulus of concrete asphalt mixture/Mix Stiffness (MPa) and Sb = asphalt stiffness modulus/Bitumen Stiffness (MPa).

IOP Conf. Series: Materials Science and Engineering **309** (2018) 012113 doi:10.1088/1757-899X/309/1/012113

According to Shell Bitumen method [3] the stiffness modulus of concrete asphalt mix (Sm) is influenced by asphalt stiffness modulus (Sb) and air pore content (Cv). The asphalt stiffness modulus (Sb) is affected by T800, penetration index, pavement temperature and duration of loading. The equation of Shell Bitumen method is:

$$Sm = Sb\left[1 + \frac{2{,}5\ X\ Cv}{n\ x\ (1\text{-}Cv)}\right]^{n} \tag{2}$$

where: Sm = stiffness mixture of asphalt concrete (N/m2); Sb = stiffness of bitumen/asphalt (N/m2); N = 0.83 x log (4.1010/Sb) and Cv = 0.70-0.90

The stiffness modulus sought through a direct approach is obtained from the results of laboratory testing. The direct approach is the best method of determining the value of the Rigid Modulus. Laboratory tests for the stiffness modulus has several provisions such as AASHTO T294-92, 1992 and SNI 03-6836-2002 using the UMATTA (Universal Materials Testing Apparatus) triaxial repeater.

To be able to determine the value of modulus by laboratory testing can be determined by testing the dynamic modulus and resilient modulus [13]. Dynamic modulus is the ratio between the working voltage compared to the measured vertical strain. The dynamic modulus is determined by the loading of the sinusoidal to the cylindrical test object having a high ratio and a 2:1 diameter. The resilient modulus is the unity of the required volume strain energy so that the material gets a voltage from the unloading state to the point of yield. resilient modulus is determined by Indirect Tensile/RLIT Repeated Load testing [13]. In this RLIT test, the specimen is not burdened to collapse, but it is only burdened about 2-20% of its indirect tensile strength.

To test the stiffness modulus in the laboratory, the following tools can be used:

1. Universal Testing Machine (UTM); is a testing machine to test the tensile stress and compressive strength of material. The usage of UTM is by giving force or tensile force to the material tested.
2. Universal Material Testing Apparatus (UMATTA); this test is performed under repetitive load at low voltage so that the response of the specimen tested remains elastic. This test is considered a nondestructive test carried out using a machine of UMATTA according to ASTM D4123, Computer-controlled equipment and use a pneumatic actuators to apply the loads.
3. Indirect Tensile Stiffnes Modulus; the test is performed in accordance with BS DD 213:1993. The test method using either cylindrical specimen is made from field or laboratory cores formed, with a diameter of 100-150mm and thickness between 30-80mm.
4. Bending Point Test, bending test carried out by giving the load on the specimen either with the principle of 3-point bending or 4-point bending. At 3-point bending the maximum moment is only formed at one point, i.e. at the point of giving the test load. However, at 4-point bending, the maximum moment is formed along the distance of one loading point with another point of loading.

If the four types of tools above are not available, the stiffness modulus test of concrete asphalt can be carried out by Marshall Test tool, by using the Voltage and Strain value approach [5].

From Marshall Test results obtained stability and flow value. The amount of stress is known from the stability value divided by the area of the press field, while the amount of strain is known from the flow value divided by the diameter of the specimen. The voltage and tension equations are:

$$\sigma = \frac{p}{d.t} \tag{3}$$

$$\varepsilon = \frac{f}{d} \tag{4}$$

where: σ= the voltage occurring on the specimen; p = work load; d = the diameter of the specimen; t = thick/height of specimen; ε= strain occurring on the specimen and f = melting of specimens.

Based on equation (3) and equation (4), a curve of the relationship between stress and strain can be made, where the stiffness modulus value is the slope of the straight line curve.

3. Research Methods

The research was conducted in Asphalt Mixing Plan Laboratory (AMP) Rapi Arjasa Binjai City and Laboratory of Pavement Highway, Department of Civil Engineering, Faculty of Engineering, Universitas Sumatera Utara. The material used consists of asphalt, aggregate and filler for mixed AC-WC type. The aggregate material sourced from the Wampu river Binjai, the asphalt was from Iran and the filler was Portland Cement.

For aggregate materials, the tests were consists of specific gravity, fine grain analysis, and absorption of coarse and fine aggregates. While, for asphalt conducted penetration test of asphalt, softening point and Thin Film Oven Test (TFOT).

To get the value of Optimum Asphalt Content (OAC) made by specimen with variation of asphalt content 4,5%, 5%, 5,5%, 6% and 6,5%, each 3 pieces. For the PRD (Percentage Refusal Density) test the specimens were made using only 5%, 5.5% and 6% asphalt, one each. The total samples were 18 pieces.

After the OAC value obtained, then made 30 samples to be tested at 60°C. For the tests at several temperature variations i.e. 30°C, 40°C, 50°C, 70°C and 80°C, each 5 samples. All of the 55 samples were tested with Marshall Test. The method of observation was done by the addition of melting mechanism of the sample on 0.5mm.

The stability and flow values obtained from the test then used to calculate the values of stress and strain. Then we got the modulus of stiffness value from the correlation curve between stress and strain that happened at the beginning of loading.

Next, calculated of stiffness modulus value with Brown & Brunton method and Bitumen Shell method. The values of these two methods then compared with the stiffness modulus from Marshall Test, including the influence of temperature as well as the correlation of Marshall values against the modulus values and comparing the value of the analysis results with Bina Marga 2013.

4. Results and Analysis

4.1. Before OAC

The test results before OAC presented in Table 1. From the Barchat for Presentation of Mix Design Data and Selection of Design Bitument AC-WC,the value of OAC obtained 5,84%.

Table 1. Test Results Before OAC.

No.	Type ofTest	Asphalt Content (%)					Spec.
		4,5	5,0	5,5	6,0	6,5	
1	Stability (Kg/cm2)	1235	1221	1197	1660	1211	≥ 800
2	Flow (mm)	3,03	3,20	3,17	3,60	3,80	2 - 4
3	Density (%)	2,27	2,30	2,31	2,27	2,29	-
4	VIM (%)	5,77	5,68	4,59	3,64	4,03	3 - 5
5	VIM PRD (%)	-	3,89	3,02	2,14	-	2
6	VMA (%)	15,72	16,92	17,02	16,93	18,55	≥ 15
7	VFB (%)	63,34	66,46	73,01	78,52	78,27	≥ 65
8	Marshall Quotient (Kg/mm)	399	373	370	452	312	≥ 250

Source : Laboratory of AMP PT.Rapi Arjasa, 2016

IOP Conf. Series: Materials Science and Engineering **309** (2018) 012113 doi:10.1088/1757-899X/309/1/012113

4.2. After OAC

Table 2. Stiffness Modulus in several temperature variations

Temperature (°C)	Stiffness Modulus		
	Marshall Test (MPa)	Brown &Brunton Sme (MPa)	Shell Bitumen Smix (MPa)
30	425,54	4.076,82	419,05
40	382,68	3.099,46	377,28
50	309,73	2.264,58	301,34
60	254,96	1.374,93	235,45
70	178,37	403,36	170,76
80	136,99	-	126,34

The results obtained were analyzed to see the effect of temperature, the correlation of marshall stability with modulus of stiffness and comparison of Modulus Stiffness value.

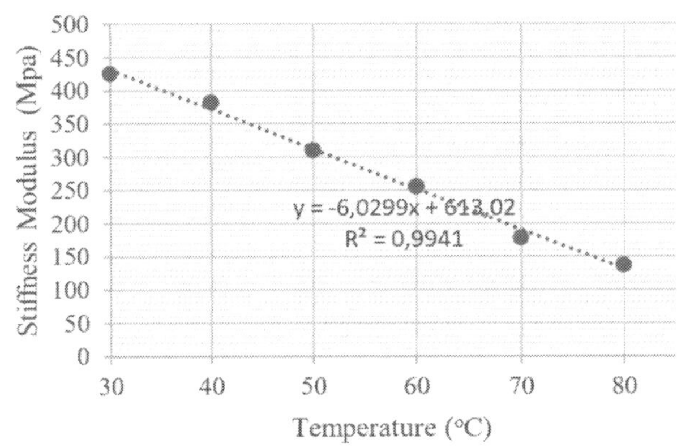

Figure 1. Correlation betweenTemperature and Stiffness Modulus according to Marshall Test

From Figure 1 it can be seen that the temperature and modulus stiffness have a inverse correlation. The higher the temperature the stiffness modulus value is lower. So at high temperature conditions the material is not elastic. The matter of temperature has also been proven by Widodo [5].

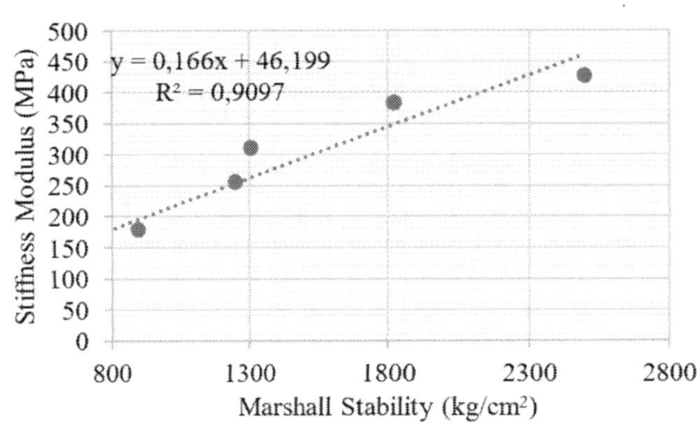

Figure 2. Correlation between Marshall Stability with Stiffness Modulus

From Figure 2 it can be seen that the marshall stability value and the stiffness modulus have a proportional relationship. The higher the marshall stability value the higher the modulus of stiffness. The value of R2 obtained is also good enough that is 0,9097.

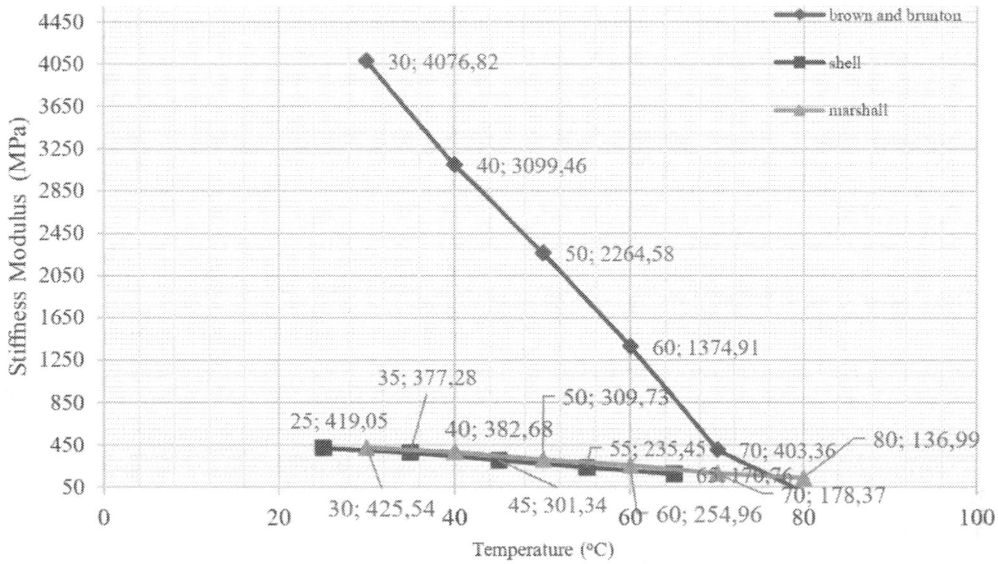

Figure 3. Comparison of Stiffness Modulus Value
Brown & Brunton Methods, Bitumen Shell and Marshall Test

From Figure 3 it can be seen that Brown & Brunton and Shell Bitumen methods are not much different, but in Brown & Brunton method there are additional parameters: recovered penetration, recoverd softening point, pentration index and VMA. As a result, there are additional parameters of Sbit value to rise and higher power factor to produce high Smix value.

Table 3. Comparison Stiffness Modulus Values with Bina Marga 2013.

No.	Methods	Stiffness Modulus Results (MPa)	Bina Marga 2013 (MPa)
1	Brown & Brunton	1.374,93	
2	Metode Shell Bitumen	235,45	1100
3	Marshall	254,96	

From Table 3 stiffness modulus value resulting from Brown & Brunton method has fulfilled the requirement determined by Bina Marga 2013. While the value produced by Shell and Marshall has not reached the required value. The results of the analysis should still be validated again with a special tool that tests the Universal Material Testing Apparatus Modulus (UMATTA).

5. Conclusion

The stiffness modulus value decreases with increasing temperature of the concrete asphalt.The stiffness modulus value from the Bitumen Shell method and the Marshall Test has a relatively similar value.The stiffness modulus value from the Brown & Brunton method is greater than the Bitumen

IOP Conf. Series: Materials Science and Engineering **309** (2018) 012113 doi:10.1088/1757-899X/309/1/012113

Shell method and the Marshall Test,but can not measure the stiffness modulus value at temperature above 80°C.

Acknowledgements
The authors also would like to thank to Sustainable Energy and Biomaterial Center of Excellence (SEBCOE), Universitas Sumatera Utara for the financial aids.

References
[1] Yoder E and Witczak M W 1975 Principles of pavement design 2nd Edition John Wiley & Sons Inc Canada
[2] Widodo S et al 2012 Analisis modulus elastisitas beton aspal yang diperkuat dengan geotekstil dari hasil uji tekuk batang **12** 130-4
[3] Shell Bitumen 2003 The Shell Bitumen Handbook *Fifth edition* Thomas Telford Ltd London
[4] Brown S F and Brunton1984 An Introduction to the Analytical Design of Bituminous Pavement 2nd EditionUniversity of Nottingham
[5] Widodo S and Setyaningsih I 2011 Penggunaan Alat Marshall untuk menguji modulus elastisitas beton aspal Simposium RAPI X FT UMS ISSN No.1412-9612 Surakarta
[6] Al-Jassar A M et al 2003Using Marshall test to assess asphalt-concrete modulus for mixes used in Kuwait Kuwait University 169-182
[7] Wardoyo J2003 Pengaruh bahan tambah pada Asphalt Concrete Wearing Course (AC-WC) terhadap nilai properties marshall dan modulus kekakuan Tesis Teknik Sipil Universitas Diponegoro Semarang
[8] Sjahdanulirwan M N 2009 Nilai mekanistik beton aspal lapis permukaan terhadap pengaruh temperatur dan pembebanan Direktorat Jenderal Bina Marga Puslitbang Jalan dan Jembatan Bandung
[9] Sjahdanulirwan M N 2011 Pengaruh waktu pembebanan terhadap kekauan beton aspal lapis permukaan Direktorat Jenderal Bina Marga Puslitbang Jalan dan Jembatan Bandung
[10] Huang Y H 1993 Pavement analysis and design University of Kentucky Prentice Hall Englewood Cliffs New Jersey
[11] Direktorat Jenderal Bina Marga 2013 Manual desain perkerasan jalan Pusat Litbang Jalan dan Jembatan Badan Penelitian dan Pengembangan Bandung
[12] AASHTO1993 Guide for design of pavement structure Washington DC
[13] Roberts F L et al 1991 Hot mix asphalt materials mixture designand construction NAPA Education Foundation Lanham Maryland

Priority of road maintenance management based on halda reading range on NAASRA method

M Surbakti[1,2*] and A Doan[3]

[1]Civil Engineering Department, University of Sumatera Utara, Jl. dr. Mansur Kampus USU Medan 20155.
[2]Sustainable Transportation Research Center, Faculty of Engineering University of Sumatera Utara, Jl. Almamater Kampus USU Medan 20155
[3]Civil Engineering Graduate Program, Faculty of Engineering, University of Sumatera Utara, Jl. Almamater Kampus USU Medan 20155

*Email: medis@usu.ac.id

Abstract. The road pavement, constantly experiencing stress-strain due to traffic load through it, can cause damage to the pavement. Therefore, early detection and repair of the damage will be able to prevent more severe damage that can develop into pavement failure. A road condition survey is one of the earliest attempts to detect the initial damage of a pavement. In this case the driving comfort is the most important part for the driver in assessing road conditions that are affected by the level of road surface roughness. To determine the level of roughness of the road, one of the methods developed is the measurement using the NAASRA method. In this method the roughness of the road is an accumulation of the average unevenness of the road, with the general setting on halda of 100 m. However, with this 100-meter setting, in some places the final value of the roughness value is too large or too small so that it will result in the priority of the road maintenance. This is what underlies roughness research by comparing halda settings at 50 m and 200 m different from the general settings above.This study uses the International Roughness Index (IRI) method in determining the level of road stability concerning driving discomfort. IRI score obtained from direct survey in field by using Roughometer-NAASRA.The final result shows that there is a significant difference between the reading of halda which is set at 100 m reading with halda set with 50 and 200 meter readings. This may lead to differences in handling priorities, which may impact on the sustainability of road network maintenance management (Sustainaible Road Management)

1. Introduction

Flexible pavement is a pavement consisting of several layers which become a unified layer which is responsible for carrying the load of vehicles passing over it, and can channel the load from the vehicle properly from the top layer to the underlying layer. As pavement layers, these structures are expected to serve traffic well, safely, and comfortably. But in fact some of the existing roads that can not meet those expectations. Currently, road pavement construction is not only required to serve high traffic and vehicle load, but also to pay attention to convenience (Sukirman, 1999)

Road users generally prefer the flatness or comfort of the road, so it is necessary to check the condition periodically. The inspection is intended to measure the unevenness of the road that can be used in the maintenance planning program or improvement so that service for road users can be improved. To know the level of road flatness, the method that now developed is the measurement of the flatness of the road using the method of NAASRA with Halda 100 m setting.

Based on the above, the background of technical analysis in analyzing the comparison of the value of road surface unevenness based on the NAASRA reading range which generally use the range of 100 m to other different ranges, ie in the range of 50 m and 200 m.

IOP Conf. Series: Materials Science and Engineering **309** (2018) 012114 doi:10.1088/1757-899X/309/1/012114

Assessment of pavement conditions is the most important aspect in terms of determining road maintenance and repairs. In order for roads to accommodate the needs of movement with a certain level of service it is necessary to make an effort to maintain the quality of road services, where one of its efforts is to revise the road surface condition. The value of these road conditions will be used as a reference to determine the type of revaluation program that must be done, whether it is an improvement program, regular maintenance, or routine maintenance.

1.1. Flexible Pavement Damage

In general, types of road damage can be divided into two categories:

1.1.1. Structural Damage

Structural damage is damage to the road structure, partly or wholly, causing pavement no longer able to support the traffic load. Therefore, it is necessary to strengthen the structure of the pavement by overlaying or repairing existing pavement layers.

1.1.2. Functional Damage

Functional damage is damage to the road surface which may cause disruption of the road function. This damage can be related to or not with structural damage. In functional damage, the pavement is still able to withstand the working load but does not provide the desired level of comfort and safety. For that the pavement surface layer should be treated for a good back surface.

In principle, the type of functional damage will reduce the level of comfort and safety of road users such as:

- Increases noise due to friction wheels and road surfaces
- Increase the risk of water splashing (water splashing) at the time of wet surface
- Raise the risk of slipping while crossing the bend, when the surface is wet
- Increases the risk of slipping when braking on wet or dry surfaces

Indications that point towards road damage, both functional damage and structural damage can vary that can be seen from the form and process of occurrence. Indications that arise on pavement surfaces may affect the roughness value of the pavement. Broadly speaking, damage to asphalt pavement can be grouped into four incident modes, namely: cracks, surface defects, deformations, and defects at the edge of the pavement

1.2. International Roughness Index (IRI)

International Roughness Index is a parameter used to determine the level of unevenness of the road surface. The Roughness parameter is presented in a scale depicting the unevenness of the pavement surface felt by the rider. The unevenness of the pavement surface is a function of the longitudinal and cross-sectional surface of the road.

The degree of unevenness of the road is one of the functional factors of a highly riding-quality pavement, where the technical indicator for assessing road surface performance is the IUD (International Roughness Index) value, The value of surface roughness is indicated as the cumulative length of the rising surface per unit length. The unevenness of the road surface is considered as a resultant condition of the pavement as a whole. If it is quite flat, then the road is considered good from the lower layer to the top layer of pavement and vice versa, (Hikmat Iskandar 2005).

IRI value is expressed in meters, from cumulative rise and fall of road surface per kilometer road length (m / km). If the value of IRI = 10 m / km, meaning the amount of amplitude (up and down) of the road surface of 10 m in each km of road length. The larger the IRI value, the worse the pavement surface.

1.3. Condition and Stability of the Road

Road conditions are a matter of great concern in determining road maintenance programs. According to the Department of Public Works of Director General of Highways (1992), road conditions can be classified as follows:

- The road with good condition is the road with a completely flat pavement surface, no waves and no surface damage.
- Medium-sized roads are roads with a moderate pavement surface, there are waves but no surface damage.
- Roads with lightly damaged conditions are roads with pavement surfaces beginning to wavy, there is surface damage and patching (less than 20% of the road area under review).
- Roads with heavily damaged conditions are roads with heavily damaged surfaces such as wavy, crocodile cracks, and large chips (20-60% of the roads under consideration) accompanied by damage to layers of foundation such as ambulances, sungkur and etc.

The definition of each term stability path is as follows:

- Steady Road Construction is a road with construction conditions inside a steady corridor which for handling only requires maintenance activities. The steady construction path set by Minimum Service Standards is a road in good and medium condition, where in this study used a limit of IRI <8 m / km.
- Unstable roads are roads with conditions outside the steady corridor which for minimum handling is the periodic maintenance and maximum upgrading of the road with the aim of adding value to the construction structure.

The Directorate General of Highways uses the International Roughness Index (IRI) parameter in determining the condition of road construction, which is divided into four groups. Below is shown Table 2 determining road condition conditions and handling requirements:

Table 1. Determination of roads condition and handling needs.

Road Condition	IRI (m/km)	Maintenance Program	Stability Term
Good	IRI avg. $\leq 4,0$	Routine	Steady/Stable
Fair	$4,1 \leq$ IRI avg. $\leq 8,0$	Periodic	Steady/Stable
Slightly damaged	$8,1 \leq$ IRI avg ≤ 12	Upgrading	Unsteady
Heavily damaged	IRI avg. > 12	Upgrading	Unsteady

2. Methodology

This study took a study on three national road segments in North Sumatra province with direct observation in the field (field survey), namely: Parapat - Tapanuli Utara regency borderline (10,000 km), Simalungun regency borderline - Silimbat Regency (34,000 km) and Silimbat –Tapanuli Utara regency borderline (11,000 km) so the total length of the roads under consideration is 55,000 km. Car and survey equipment used in this study can be seen in the picture below.

IOP Conf. Series: Materials Science and Engineering **309** (2018) 012114 doi:10.1088/1757-899X/309/1/012114

Picture 1. Car and survey equipment for roughness survey

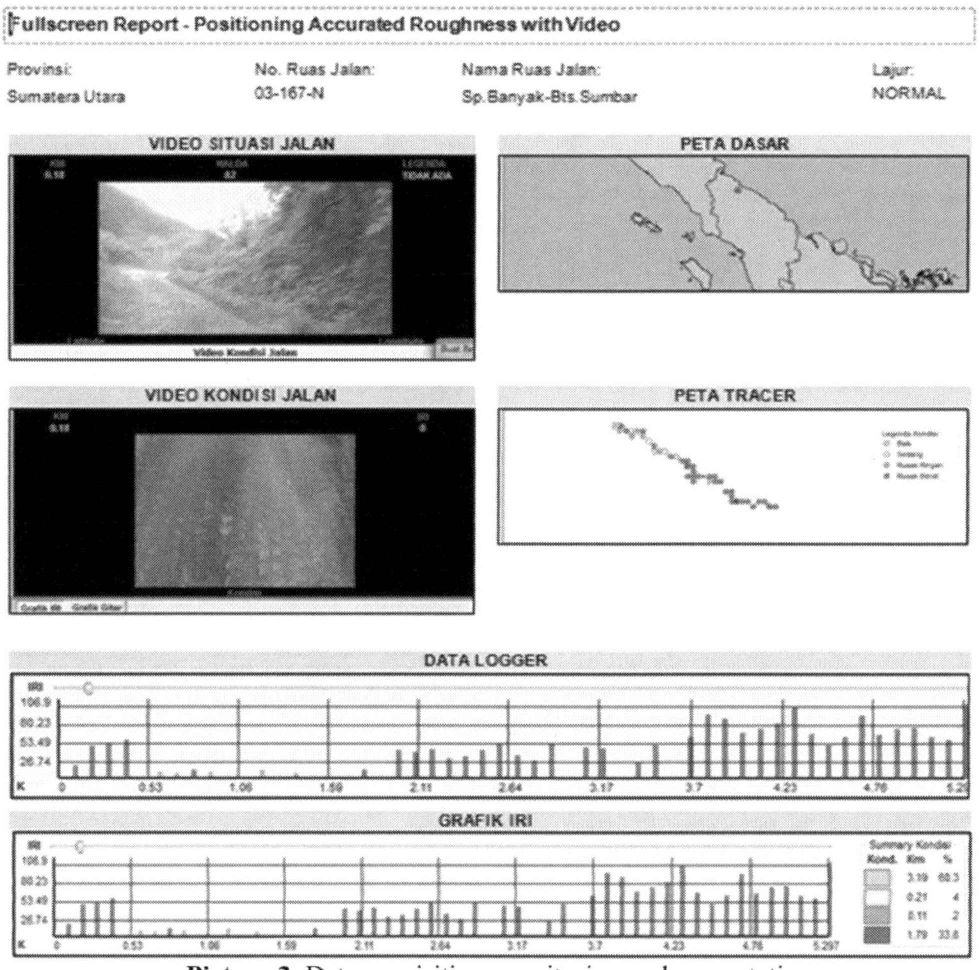

Picture 2. Data acquisition, monitoring and presentation

3. Analysis and Discussion

From the data obtained, it can be seen in Table 2 that in the Halda 50 meter setting for the three national road segments, the length of the road segment with good condition is greater than the Halda setting at 100 and 200 meters, this is indicated by the length of the survey road Along 1200 m, 1600 m and 350 m.

However, in moderate, lightly damaged, and heavily damaged, there are variations that are not dominated in the halda settings, such as for road number 066. The length of the road with medium road conditions will be produced by Halda 200 along 7800 m, followed by Halda 50 Along 6500 m, and last Halda 100 along 5300 m. This variation is also available for national roads 067 and 068.

The magnitude of the value of unevenness of the road under consideration is influenced by the value of NAASRA obtained at the time of the survey, where the greater the value of NAASRA produced, the greater the unevenness (IRI) of the road which will result in the condition of the road being damaged lightly and heavily damaged longer.

Table 2. Length of road based on physical condition of pavement

No Segment	Halda Setting	Road Length	Road Condition							
			Good (meter)	(%)	Fair (meter)	(%)	Slightly damaged (meter)	(%)	Heavily Damaged (meter)	(%)
066	Halda 50	10 Km	1200	12%	6500	65%	1650	16.50%	650	6.50%
	Halda 100		300	3%	5300	53%	3100	31%	1300	13%
	Halda 200		-	0%	7800	78%	2000	20%	200	2%
067	Halda 50	34 Km	1600	4.76%	24900	73.24%	5700	17%	1800	5%
	Halda 100		200	0.60%	25500	75%	7000	20.60%	1300	3.80%
	Halda 200		-	0%	28000	82.35%	4800	14.12%	1200	3.53%
068	Halda 50	11 Km	350	3.18%	5900	53.64%	4100	37.27%	650	5.91%
	Halda 100		200	1.20%	6800	61.80%	2900	27%	1100	10%
	Halda 200		-	0%	6000	54.55%	4400	40%	600	5.45%

Factors that cause NAASRA value to increase is; Quantity and location of the increasingly widespread type of asphalt damage, where in the absence of serious handling the level of quality of the road will decrease drastically, caused by daily traffic that weighs on national roads, which tend to be bypassed by heavy vehicles. This happens because the national road is the connecting road between the provincial capital that serves in fulfilling the need for goods and services for each region.

The basic thing that distinguishes the reading of this halda is the distance setting, which with a smaller distance then the quantity of asphalt damage that is covered will decrease, causing the acquisition of IRI value tends to decrease.

To achieve road conditions with smaller IRI values, the Halda settings that should be used are smaller because they will result in a more dominant stability (as seen on the 066 & 067 segments). This also can also be obtained with Halda 200 settings, but in providing good road conditions this Halda setting does not give results. Stability of road obtained solely only from the results of "fair" road conditions (Table 3).

Table 3. Percentage of road with stable/unstable condition

No Segment	Halda Setting	Road Length	Value of Road Stability/Stadyness			
			Stable		Unstable	
066	Halda 50	10 Km	77%	7700 m	23%	2300 m
	Halda 100		56%	5600 m	44%	4400 m
	Halda 200		78%	7800 m	22%	2200 m
067	Halda 50	34 Km	78%	26500 m	22%	7500 m
	Halda 100		75.6%	25700 m	24.4%	8300 m
	Halda 200		82.35%	28000 m	17.65%	6000 m
068	Halda 50	11 Km	56.82%	6250 m	43.18%	4750 m
	Halda 100		63%	7000 m	37%	4000 m
	Halda 200		54.55%	6000 m	45.45%	5000 m

IOP Conf. Series: Materials Science and Engineering **309** (2018) 012114 doi:10.1088/1757-899X/309/1/012114

From the above exposure was obtained an initial conclusion that the 50-meter Halda setting tends to be better than the Halda 200 meter setting. With the results of the steadiness obtained from both types of Halda settings, then the handling provided will be lower because the results obtained. This is in contrast to the Halda 100 setting process, which results in greater road inconsistency conditions, when viewed from the results then this setting will require extra handling compared to both 50 and 200 meter settings.

4. Result

To get better result that is relatively small value of IRI, hence setting halda 50 will be better use. But with a better level then the sensitivity of road handling will tend to be reduced. This is in contrast to the 100 meter halda setting which will result in IRI values that tend to be larger, but with the results obtained then the priority level of road handling will be much more to do.

With respect to the above conclusions, it can be concluded that the level of precision of road conditions is obtained from halda readings per 50 meters, but in terms of maintaining a sustainble road network maintenance program, maintenance programs based on road stability require early maintenance in order to keep more roads Which has steady conditions

References

[1] Hardiyatmo HT 2009 *Pavement Maintenance* Yogyakarta: Gadjah Mada University Press
[2] NAASRA training brief report Dipstick Z-250 ATC-M420 and BB di Riau Province
[3] Mulyono AT and Riyanto B 2005 *Technical Review on the Quality Performance of National and Provincial Roads* Media Komunikasi Teknik Sipil
[4] Mulyono AT 2007 *Monitoring and Evaluation Model for the Implementation of Standards for Quality Based Road Pavement Systemic Approach* Semarang
[5] Sukirman S 2010 *Road Flexible Pavement* Nova Bandung
[6] Suswandi A, Sartono W and Chritady H H 2008 *Evaluation of Road Damage Level With Pavement Condition Index Method (PCI) To Support Decision Making* Forum Teknik Sipil
[7] Procedure for Surveying Road Pavement Surface with NAASRA Measurement Tool SNI 03-3426-1994
[8] Public Works Department 2000 Road Mainntenance Engineering. Agency for Research and Development Research Center for Transportation Infrastructure Development

IOP Conf. Series: Materials Science and Engineering **309** (2018) 012115 doi:10.1088/1757-899X/309/1/012115

Roundabout performance analysis in the city of Medan

M Surbakti[1,2*] and F Iswahyudi[1]
[1]Civil Engineering Department, University of Sumatera Utara, Jl. dr. Mansur Kampus USU Medan 20155
[2]Sustainable Transportation Research Center, Faculty of Engineering University of Sumatera Utara, Jl. Almamater Kampus USU Medan 20155

[*]Email: medis@usu.ac.id

Abstract. Medan city as one of the largest cities in Indonesia, currently has problems in the field of transport is quite worrying. Like the other cities in Indonesia, the growth of traffic volume can not be followed by improvement of existing infrastructure. It is characterized by frequent traffic jams, long queues and delays contained in the segment or intersection. One type of arrangement is a roundabout intersection. Roundabout intersection canalization is composed of a central circle surrounded by a one-way street. Roundabout can act as a controller, divider and guidance for the traffic system which rotates in the direction. This study aims to determine the performance of the roundabout on the road Ir. Haji Juanda, Medan. Analyses were performed to obtain the performance of these intersections. Analyses were performed by using a calculation method MKJI 1997 and the Australian Road Research Board (ARRB) manually. The performance measures analyzed in the form of the degree of saturation, delay and queue opportunities using the method MKJI 1997. whereas the ARRB method, the calculated performance of the roundabout is the degree of saturation and delay. Good infrastructure handling can create a reduction of delay, and congestion, so that it can reduce pollution and support a more green environment.

1. Introduction

Intersection as part of a road network is a critical area in serving the flow of traffic. Therefore improvement of the intersection will reduce the barriers and will increase capacity and is expected to be able to reduce the occurrence of accidents. One type of intersection setting is with roundabout.

Intersection with roundabouts have a very important role in ensuring the smooth flow of traffic. In addition to being a meeting point between different roads, the roundabout part can act as a controller, a divider and a driver for a one-way rotating traffic system.Capacity and measure of performance in the degree of saturation, delay and queue opportunities for the roundabout part of the roundabout can be calculated which represents a geometric condition, environment and traffic flow. The traffic flow data required for calculation is the traffic flow data for each movement. The required traffic movement data is the volume and direction of the traffic movement during peak hours.The obstacle in this calculation is the fundamental difference between the MKJI 1997 method and the ARRB in terms of the roundabout performance assessment. In the ARRB not counted motorcycles in its performance assessment while in Indonesia traffic conversion in motorcycle dominance.

The purpose of writing this paper is to compare the performance measures of both methods and provide solutions to improve roundabout performance.

1.1. Basic theory

Roundabout is one type of intersection control commonly used in urban and out-of-town areas as a meeting point between several road segments with medium traffic levels because it has a relatively lower level of traffic accidents than the type of signaled intersection or non-signaled intersections. The roundabout can be considered a special case of canalization that the island in its center can act as a controller, divider and guide for a one-way rotating traffic system. Continuous movement and a large turn on the entire leg of the road meeting will reduce the source of the accident and provide more comfort to the driver's condition (Hobbs, 1995).

For larger intersections, closure of the braided area is easy and the safety of the roundabout decreases. Although the impact of the roundabout traffic in the form of delay is always better than other types of intersections such as a signal intersection, signal installation is still preferred to ensure a certain capacity can be maintained even in peak hour currents. The roundabout is preferred because it can reduce the delay and allow many vehicles to cut the intersection without having to stop completely (MKJI, 1997).

1.2. Roundabout Concept

The roundabout is a canalization junction consisting of a central circle surrounded by a one-way street. The roundabout is a special form of canalization intersection where the incoming vehicle rotates clockwise around the central traffic island. The intersection entrance is governed by the Give Way marker and priority is given to the vehicle circulating at the roundabout.

Channelization is the process of separating or regulating conflicting vehicle flows into clear road routes by placing paved concrete or pavement rails to create a safe and orderly movement for vehicles and pedestrians. Proper canalization can increase capacity, improve security, provide full comfort, and also boost driver confidence. Channelization is often used in conjunction with stop signs or speed control signs or at traffic light intersections.

The principle of the roundabout is to create a continuous stream of traffic without stopping and the purpose of the roundabout plan is to ensure the safety of traffic between intersecting traffic and minimum delay. The efficient operation of the roundabout depends on the gap acceptable to the driver in the circulating traffic flow. Because traffic flows together and separates at small angles at relatively low speeds, traffic accidents in roundabouts rarely have fatal consequences.

2. Methodology

2.1. Research location

The research location is in Medan city where the roundabout to be studied is Ir. H Juanda roundabout which is a road intersection between Jl. Ir. H Juanda direction to Polonia, Jl. Samanhudi, and Jl. Ir. H Juanda direction to Taman Makam Pahlawan. Locations can be seen on the map below:

Figure 1. Reseach Location

IOP Conf. Series: Materials Science and Engineering **309** (2018) 012115 doi:10.1088/1757-899X/309/1/012115

2.2. Research Analysis

There are two analysis that carry out on this reseach, to obtain the performance of roundabout; MKJI 1977 and ARRB. The performance measures analyzed in the form of the degree of saturation, delay and queue opportunities using the method MKJI 1997. whereas the ARRB method, the calculated performance of the roundabout is the degree of saturation and delay

3. Data Acquisition Analysis and Discussion

3.1. MKJI 1997 data

The data obtained from the field is the volume during peak hour morning, afternoon and evening for each approach of the intersection. The volume data can be seen in Table 1 below.

Table 1. Volume at all approach, on Ir. H. Juanda roundabout

Time	Road			Total Volume
	Jln Samanhudi	Jln Ir H Juanda	Jln Ir H Juanda	(pcu/hour)
07.00 - 08.00	878	2955	1838	5671
07.15 - 08.15	951	2955	1777	5683
07.30 - 08.30	997	2869	1609	5474
07.45 - 08.45	926	2653	1462	5041
08.00 - 09.00	823	2427	2450	5699
11.30 - 12.30	1329	1736	1844	4909
11.45 - 12.45	1438	1672	1885	4995
12.00 - 13.00	1494	1778	1908	5181
12.15 - 13.15	1438	1926	1894	5257
12.30 - 13.30	1302	1975	1670	4947
16.30 - 17.30	1168	1975	1985	5127
16.45 - 17.45	1197	2079	2033	5309
17.00 - 18.00	1200	2189	2108	5497
17.15 - 18.15	1139	2182	2299	5620
17.30 - 18.30	1024	2196	2512	5732

Note: Peak Hour on 17.30 to 18.30

3.2. ARRB data

In contrast to the way MKJI is calculated, ARRB takes into account the cycles of each of the existing cross-linking arms. The volume of each approach for each cycle calculated can be seen in Table 2 and Figure 2 below.

Table 2. Ir Handa's traffic jam traffic volume at 1 hour interval for each approach

Time	Road			Total volume each cycle (veh/hour)
	Jln Samanhudi (A)	Jln Ir H Juanda	Jln Ir H Juanda	
07.00 – 08.00	194	1090	968	2252
07.15 – 08.15	220	1092	969	2281
07.30 – 08.30	258	1100	924	2282
07.45 – 08.45	244	1049	877	2170
08.00 – 09.00	220	947	828	1995
11.30 – 12.30	406	936	1107	2449
11.45 – 12.45	444	896	1123	2463
12.00 – 13.00	484	952	1140	2576
12.15 – 13.15	491	989	1143	2623
12.30 – 13.30	446	1019	1007	2472
16.30 – 17.30	308	1048	1155	2511
16.45 – 17.45	333	1075	1147	2555
17.00 – 18.00	355	1104	1153	2612
17.15 – 18.15	333	1081	1205	2619
17.30 – 18.30	292	1029	1277	2598

Note : peak hour on 12.15 – 13-15 cycle

In addition to volume, the ARRB also requires other parameters calculated, Intra Bunch Headway, Headway Follow Up and Mean Critical Gap, as shown in Figure 2.3 and 4 below.

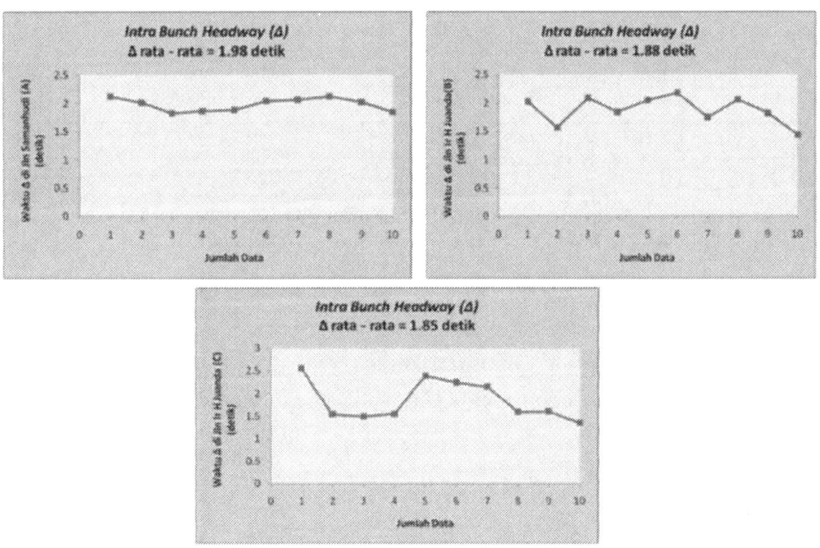

Figure 2. *intra bucnh headway* value fluctuation

IOP Conf. Series: Materials Science and Engineering **309** (2018) 012115 doi:10.1088/1757-899X/309/1/012115

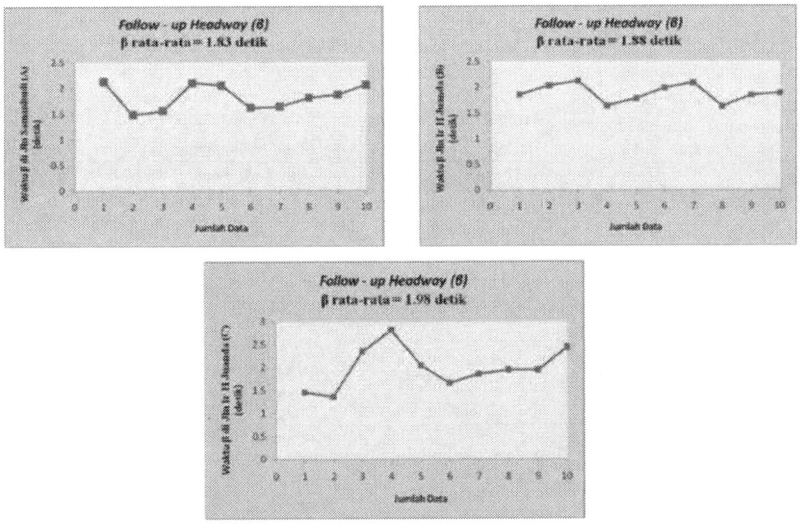

Figure 3. *follow – up headway* value fluctuation

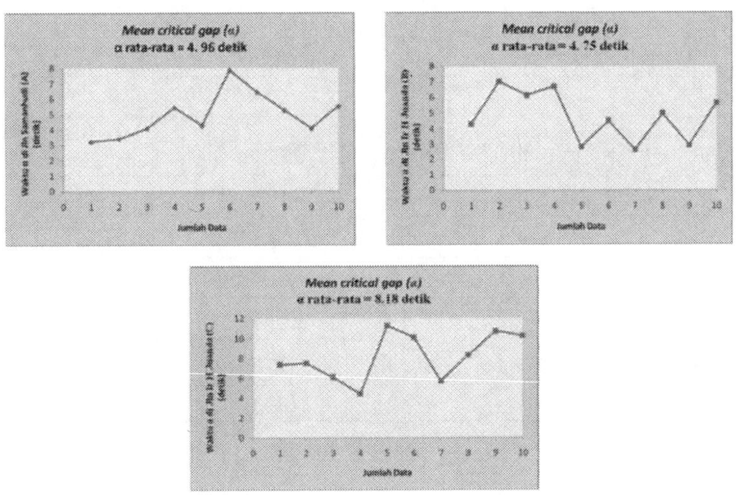

Figure 4. *mean critical gap* value fluctuation

4. Analysis and Discussion

4.1. Analysis by MKJI method 1997

For analysis with the MKJI 1997 method, the RWEAV-II form is used, to obtain roundabout capacity and traffic characteristics such as degree of saturation, traffic delays and queuing opportunities, as shown in Table 3.

IOP Conf. Series: Materials Science and Engineering **309** (2018) 012115 doi:10.1088/1757-899X/309/1/012115

Table 3. RWEAV – II for MKJI 1997 analysis

RWEAV-II Form Roundabout analysis	Date : 25 Nopember 2016		City Size (mil)	: 2.1
	City	: Medan	Road enviroment	: Komersial
	Road A	: Jln Samanhudi	Side Friction	: Rendah
	Road B	: Jln Ir H Juanda	Period	: 17.30 – 18.30
	Road C	: Jln Ir H Juanda		

1. Geometric Parameters, weaving area

	Weaving	Entry Width Appr. 1 (m)	Entry Width Appr. 2 (m)	Avrg Entry Width W_E (m)	Weaving Width W_W (m)	W_E/W_W	Length of Weaving L_W (m)	W_W/L_W
1	A-B	8	13	10.5	12	0.875	30	0.40
2	B-C	10	10	10	12	0.833	30	0.40
3	C-A	13	10	11.5	16	0.719	22	0.73

2. Roaundabout Capacity

	Weaving Section	Factor W_W	Factor W_E/W_W	Factor P_W	Factor W_W/L_W	Base Capacity (Co) (pcu/hour)	Faktor penyesuaian City Size (F_{CS})	Faktor penyesuaian Road Envi. (F_{RSU})	Capacity C (pcu/hour)
1	A-B	3414.046	2.567	0.848	0.546	4074	1	0.95	3854
2	B-C	3414.046	2.482	0.816	0.546	3776	1	0.95	3587
3	C-A	4563.020	2.253	0.835	0.374	3493	1	0.95	3318

3. Traffic Behaviour

	Weaving area	Flow on Weaving Q (pcu/hour)	Degree of Saturation DS	Traffic Delay DT (sec/pcu)	Total Delay DT tot = QxDT (sec/hour)	Q PRob. QP% Lower limit	Q PRob. QP% Upper limit
1	A-B	3341	0.867	7.20	24055.2	24	53
2	B-C	2705	0.754	4.80	12984	15	36
3	C-A	3310	0.994	14.90	49319	38	78
DS at Weaving D_{SR}			0.994	Total	86358.2		
Avrg. Traffic Delay DTr (sec/pcu)					14.543		
Aveerage Roundabout Delay Dr (DTr+4) (sec/pcu)					18.543		
Prob. Roundabout Delay QP%						38	78

From the table above it can be concluded that the degree of saturation for each current of the intersection of the arm has approached the saturation point. The most critical link is the C-A braid, where the value of degree of saturation has reached 99.4%, as well as the queue probability reaching 78%, which means during peak hours, there will almost certainly be queues at the roundabout.

4.2. Analysis by ARRB method

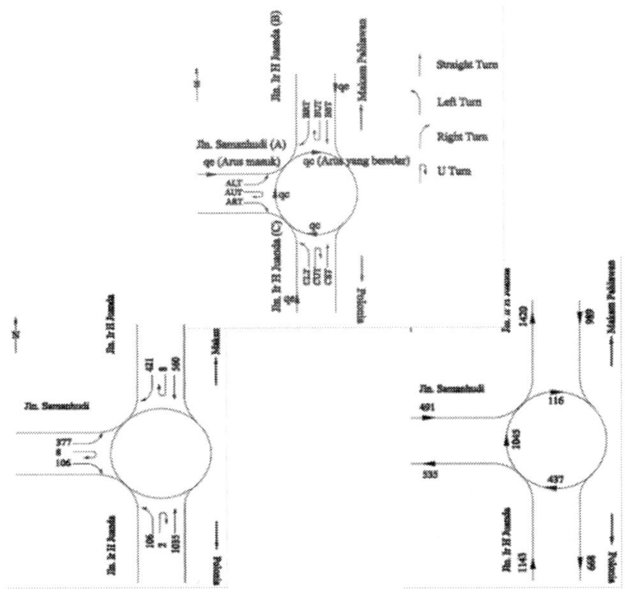

Figure 5. Calculation of Traffic Volume on the movement of vehicle inflows at the roundabout for the three approaches includes U rotation for qc and qe (peak hour at 12.15-13.15).

Table 4. Capacity Analysis

Pendekat	qc (kend/jam)	qc (kend/dtk)	β (dtk)	α (dtk)	Δc (dtk)	φ (kend/dtk)	λ	Qe (kend/jam)
Jln Samanhudi (A)	1045	0.29028	1.83	4.96	1.98	0.3196	0.2177	531
Jln Ir H Juanda (B)	116	0.03167	1.88	4.75	1.88	0.7053	0.0232	1731
Jln Ir H Juanda (C)	437	0.12139	1.98	8.18	1.85	0.5802	0.0908	867

Table 5. Degree of Saturation Analysis

Pendekat	qc (kend/jam)	Qe (kend/jam)	qe (kend/jam)	Derajat kejenuhan (x)
Jln Samanhudi (A)	1045	531	491	0.9249
Jln Ir H Juanda (B)	116	1731	989	0.5713
Jln Ir H Juanda (C)	437	867	1143	1.3183

Table 6. Delay Analysis

App.	qe	Qe	x	Z	T	dm	k	d
	(kend/jam)	(kend/jam)			(jam)	(dtk)	(kend/dtk)	(dtk)
Jln Samanhudi (A)	491	531	0.9249	- 0.0751	0.25	6.8	1.003	45.7
Jln Ir H Juanda (B)	989	1731	0.5714	- 0.4286	0.25	48.1	23.1	99
Jln Ir H Juanda (C)	1143	867	1.3183	0.4159	0.25	11.38	2.73	191.9

From Table 4-8 above it can be seen that the largest volume comes from the C approach, but the capacity C has is the smallest capacity of the three approaches. As a result of the degree of congestion, the delay value of the C approach is the greatest value of the three approaches. The delay value of the C approach is 2 times the value of B approach and more than 4 times from A.

5. Resume

In the calculation using the ARRB method, the characteristics and performance values of each approach can be known, whereas with the method of MKJI the characteristic and performance of the roundabout becomes the final result of its evaluation. An improvement is needed at the intersection so that there is an increase in capacity, crossing, reducing the degree of saturation and the amount of delay and the probability of queuing. The most important improvement is the C approach, because in addition to having a small braid length, it also has the largest volume of the three arms.

References

[1] Anonim 1997 *Indonesia Road Capacity Manual (MKJI 1997)* Directorate General of Highways Ministry of Public Works
[2] Soedirdjo T L 2002 *Traffic Engineering* Jakarta: Higher Education Research Improvement Project
[3] Alamsyah A A 2005 *Traffic Engineering* Malang: UMM Pres
[4] Hobbs F D 1995 *Traffic Planning and Engineering* Yogyakarta: Gajah Mada University Press
[5] Akcelik R 1998 *Roundabouts: Capacity And Performance Analysis* Australia: ARRB Transport Research
[6] Anonim 1986 *Roundabouts a design guide* Australia: National Association of Australian State Road Authorities
[7] Chung E, Young W and Akcelik R 1992 Comparison of Roundabout Capacity and Delay Estimates from Analytical and Simulation Models *Proch. 16th ARRB Conf.* **16(5)** 369-385

IOP Conf. Series: Materials Science and Engineering **309** (2018) 012116 doi:10.1088/1757-899X/309/1/012116

Production layout improvement by using line balancing and Systematic Layout Planning (SLP) at PT. XYZ

Buchari*, **U Tarigan**, **M B Ambarita**
Industrial Engineering Department, Faculty of Engineering, Universitas Sumatera Utara, Jl. Almamater Kampus USU, Medan 20155

*E-mail: ibossanti@yahoo.com

Abstract. PT. XYZ is a wood processing company which produce semi-finished wood with production system is make to order. In the production process, it can be seen that the production line is not balanced. The imbalance of the production line is caused by the difference in cycle time between work stations. In addition, there are other issues, namely the existence of material flow pattern is irregular so it resulted in the backtracking and displacement distance away. This study aimed to obtain the allocation of work elements to specific work stations and propose an improvement of the production layout based on the result of improvements in the line balancing. The method used in the balancing is Ranked Positional Weight (RPW) or also known as Helgeson Birnie method. While the methods used in the improvement of the layout is the method of Systematic Layout Planning (SLP). By using Ranked Positional Weight (RPW) obtained increase in line efficiency becomes 84,86% and decreased balance delay becomes 15,14%. Repairing the layout using the method of Systematic Layout Planning (SLP) also give good results with a reduction in path length becomes 133,82 meters from 213,09 meters previously or a decrease of 37.2%.

1. Introduction

Due to the continuing needs of new industries around the world the planning and regulation of the production process has always been a concern to avoid accidental releases (spills) that can cause public and environmental damage [1]. The design of the plant layout has become the basis of current industrial plants that can affect the efficiency of work. It is necessary to plant and position employees, materials, machinery, equipment, and other manufacturing support and facilities to create the most effective plant layout [2]. However, none of the methodologies can be repeatedly applied to all plant layouts so they turn into the most efficient plant layout. In this sense, making plant layouts is more an art than a science [3].

Line balancing problems are found in various industries. It is categorized into various versions depending on the product timing information in each station and the number of products. Pathway balancing is a powerful managerial tool used to improve productivity of the production system [4]. Line Balancing consists of assigning tasks to each work station, while optimizing one or more destinations without breaking the boundaries. Line balancing is used to achieve the same level of production at each station [5].

Systematic layout planning (SLP) is used to improve spatial distances between facilities (machines and work stations) and also improve the flow of material at a plant. With the SLP is expected material handling costs are reduced significantly, workers move faster and overall productivity increases [6]. In addition the systematic layout planning method (SLP) has also been done in a steel company in Mumbai, India. Research is done for improved productivity and better layout planning. The new factory layout is designed and then compared to the current layout. The SLP method shows that the new plant layout significantly decreases the material flow distance from the preservation process to the packing [7].

The layout of the plant was carried out in accordance with the systematic steps of plant layout design Yujie et al. Studying a common field using SLP's that best layout shows a good workflow and practical significance [8]. The industry layout planning is applied to systematic layout planning (SLP) which shows step-by-step planting of the input data and evaluation activities of plant layout. This method provides a new plant layout that improves the process of flowing through the plants, and helps increase the space in the industry [9]. Manufacturing systems that produce many components and functions in highly unstable environments are increasingly challenged to meet consistently high operating levels, efficiency and flexibility [10]. The benefits of a manufacturing system depend on the main factors of designing the engine cell and its parts, setup time and method of operation. The manufacturing system becomes a failure when both of these factors are not used correctly [11]. Various techniques developed for assembly line balance problems have proven useful in various assembly industries such as manufacturing, automobiles, consumer electronics, and others. In traditional straight-line ALBPs, a limited set of tasks for the assembly line are assigned including the associated processing time and the prior relation that determines the permissible sequence of tasks. The goal is to assign a series of tasks to successive workstations to meet specific production requirements, in which the number of workstations required is minimized [12]. The use of tools, materials, and workings is considered to meet the quality requirements if all the requirements specified in the criteria and specifications are met. Methods are crucial in implementing a construction project because the right implementation method can provide project quality from a financial or time perspective. One construction technology that offers several advantages for easy implementation and economics is precast technology [13].

The layout of PL facilities is not only a matter of cost reduction, but also contributes greatly to the efficiency of the PL process. However, the optimization of the PL facility layout has always been neglected in recent decades [14]. The layout is the procedure of setting up the plant facilities to support the smoothness of the production process. Setting by utilizing the existing area to place the engine and supporting production facilities that impact on the smooth movement of material. This is set and machine department. In addition to this arrangement, the selection of one type of facility layout is also influenced by production volume and product variation [15]. The factory layout involves setting and selecting machines, material handling devices, material handling lines, resulting in reduced time and costs. Handling layout issues aims to find spatially related units such as departments and machines [16].

PT. XYZ is one of the wood stem manufacturing companies processing into semi-finished wood material. The problem in the company is that there is a considerable difference in cycle time, resulting in different capacity at each work center shown in Table 1.

Table 1. Capacity of each Work Center at PT. XYZ

Work Center	Process	Capacity (/days)
I	Reaping Process	566 unit
II	Cutting Process	677 unit
III	Gluing Process	314 unit
IV	Assembling Process	518 unit
V	Finishing Process	251 unit
VI	Preperation Process	297 unit
VII	Packaging	316 batch

In Table 1 above, there is a difference in capacity at each work center. The average demand per month is 264 units of wood / day. As a result, there is a build up of material at the next work center which has a smaller capacity. In the table above shows that there is accumulation in work center V as much as 12 units / day.

Besides the buildup problem, there are also problems in the production flow. Layout of factory production section of PT. XYZ can be seen in Figure 1.

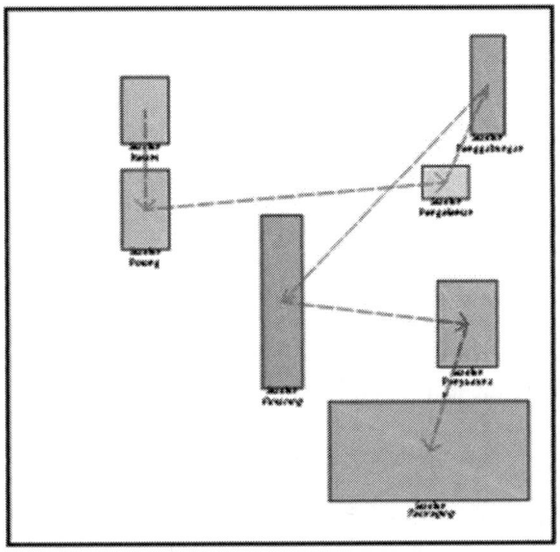

Figure 1. PT. XYZ Layout

In Figure 1 it can be seen that there is an unfavorable flow of material seen in the presence of irregular material flow patterns resulting in the intersection of the trajectory between the cutting station to the dive station and the merging station to the finishing station. This intersection causes the distance of the production line to be longer that is 213.09 meters.

2. Research Methods

This research was conducted at PT. XYZ in May 2016 until July 2016. The type of research conducted is descriptive research in the form of job and activity analysis [17]. The object of research is the trajectory of production in the process of wood bar production as raw material for making furniture. The general objective of the research is to obtain a design of the production floor layout by considering improvements in track balance.

The dependent variable in this research is the production floor layout with the optimum balance of production trajectory. Independent variables used in this study, namely:

a. Cycle Time is the time it takes the operator to complete his work.
b. Rating factor is the adjustment factor given by the researcher on the operational reasonableness of the operator.
c. Allowance is a factor of adjustment given by the researcher to the existence of a number of non-work needs that occur during the work.
d. The area of production is the data area of each operation is used to know the area used by each operation to be used as a block design in facilitating the preparation of the layout of the production floor.

Data collection is done by measuring cycle time at each work station by using stopwatch time study method [18]. In addition to cycle time collection, production floor data collection is also conducted.

Data processing is done in three stages. Namely processing time cycle data, processing track balance and processing of production floor production.

The analysis is done by comparing the balance delay, smoothness index, line efficiency and production line distance between actual trajectory with the proposed trajectory.

2.1. *Processing time cycle data*

The result of time cycle measurement will be uniform test where each data obtained must be below Upper Control Limit (BKA) and above Lower Boundary Control (BKB). Having obtained that the data is uniform, then tested sufficiency with the formula:

$$N' = \left(\frac{\frac{k}{s}\sqrt{N.\sum X^2-(\sum X)^2}}{\sum X}\right)^2 \tag{1}$$

Where:
N' : Number of observations
: Measurement data
s : Level of thoroughness
k : Level of convidence
If the data obtained is uniform and sufficient, then the normal time calculated and then the standard time. The normal time formula follows:

$$\text{Normal time} = \text{Cycle time x } \textit{Rating factor} \tag{2}$$

While the standard time formula is as follows:

$$\text{Standard time} = \text{W.Normalx}(1+ \textit{Allowance}) \tag{3}$$

2.2. *Line Balancing Processing*

The positioning weight method is the earliest developed heuristic. This method was developed by W. B. Helgeson and D.P. Birnie (Nasution, 2008) [19].

2.3. *Creation Layout Processing*

After obtaining the allocation of work elements at the work station, then arranged the production floor using systematic layout planning method (SLP).

2.4. *Problem Solving Analysis Method*

Having obtained the allocation of proposed layout with a more balanced path, then compared the parameters balance delay, smoothness index, line efficiency and long distance of production floor. The formula for finding balance delay (BD) and smoothness index (SI) are:

$$BD = \frac{\text{CTxN-}\sum \text{ti}}{\text{CTxN}} \times 100\% \tag{5}$$

$$SI = \sqrt{\sum_{i=1}^{K} (\text{WSK}_{\text{max}} - \text{WSK}i)^2} \tag{6}$$

The calculation of the length of the production path is done by using the rectilinear distance method, by the formula:

$$d_{ij} = |x\text{-}a|+|y\text{-}b| \tag{7}$$

3. Results And Discussion

3.1. *Results Weighted with Line Balancing*

The use of the Positional Weight (RPW) method of allocation of work elements into the work center can be seen in Table 2.

Table 2. Preparation of Alternative Work Station

Work Station	Work Element	Work Time	Total Time
1	1	6.55	81.63
	2	37.9	
	3	32.01	
	4	5.17	
2	5	22.6	80.04
	6	57.44	
3	7	7.74	48.62
	8	29.72	
	9	11.16	
4	10	93.29	93.29
5	11	7.05	91.67
	12	67.39	
	13	17.23	
6	14	47.02	79.71
	15	32.69	

In table 2 it can be seen that the number of work center proposals is as much as 6 work centers with each work element already allocated.

3.2. Line Balancing Performance Comparison

The purpose of line balancing is to maximize line efficiency and minimize work center, balance delay and smoothness index.
Comparison of track criteria of actual conditions and proposals can be seen in Table 3.

Table 3. Performance Comparison

Reference Parameters	Condition	
	Actual	Revision
Number of work station (unit)	7	6
Balance Delay (%)	27.26	15.14
Smoothness Index	89.52	49.93
Line Efficiency (%)	72.73	84.86

In Table 3 it can be seen that the decrease in the number of work centers from the previous 7 work centers to 6 indicates if this revision is applied in the factory then the path with 6 work stations is more efficient and the delay time or idle is smaller. The value of the line efficiency at the proposed condition is 84.86% from the previous condition that is 72.73% which means the decrease of material accumulation between the work stations. Balance delay from the previous 27.26% to 15.14%, which means reduced idle time caused by poor allocation between the work stations and the smoothness index to 49.93 from the previous 89.52 which means the relative waiting time decrease from a production line. So from the results of the comparison, it can be said the balance of the proposed trajectory better than the actual trajectory.

3.3. Changes in Production Flow Patterns

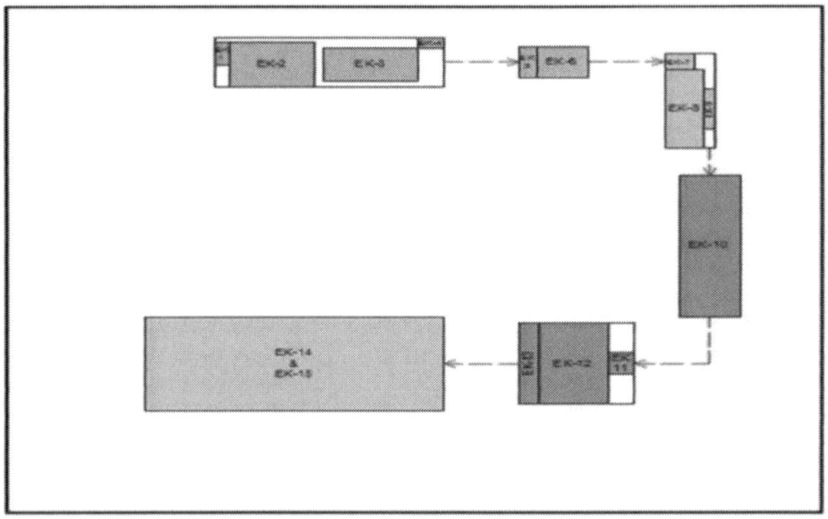

Figure 2. Revision Layout

After the redesigned layout, then no longer found back tracking and also obtained material flow changes. Prior to the redesign, the flow of material used is irregular, as can be seen from the irregularity of material displacement on the production floor. After the design, we get a proposed layout that can be seen in Figure 2. In the figure, it can be seen that after the redesign, the material flow with U shape is formed where the raw material and finished product are in adjacent position on the production floor.

3.4. Comparison of Initial Layout and Proposed Layout

The distance between machines in the production section is not well organized where there is a great distance between machines, the absence of standard flow patterns and the occurrence of back tracking, so this causes the movement of materials must travel a long distance and result in longer production time. The initial track length is 213.09 meters. After the redesigned layout by using systematic layout planning (SLP) method obtained the length of the production path to 133.82 meters or decreased by 79.27 meters. The calculation of the total correction of material displacement moment is as follows.

$$\text{Correction} = \frac{213.09 - 133.82}{213.09} \text{x } 100\%$$

$$\text{Correction} = 37.2\%.$$

4. Conclusions

The conclusions obtained from the research results are as follows:

a. The use of Positional Weight (RPW) Methods to improve path balance, new path with reduction of work station to 6 stations with balance delay of 15.14%, line efficiency of 84.86% and smoothness index of 49.93.
b. The use of Systematic Layout Planning (SLP) method in layout improvement, got new layout with previously irregular flow pattern into U-shape flow pattern.
c. The initial production trajectory has 7 work stations with balance delay of 27.26%, 72.73% line efficiency and 89.52 smoothness index. After the track balancing, the number of stations obtained 6 stations with balance delay of 15.14%, line efficiency of 84.86% and the smoothness index of 49.93.

d. The result of layout improvement using systematic layout planning (SLP) method was found to change the flow pattern of the material and also the reduction of track production length from 213.09 meter to 133.82 meter, or decrease the trajectory length 79.27 meter or 37.2%

References

[1] Douglas T A S, de Medeiros, Jose L, Araujo, Ofelia de Q 2016 *Loss Prevention in the Process Industries.* **41** 202-214
[2] Anucha W, Phichit K, Patcharee D, Wisitsree W 2011 *Proceedings of The International Multi Conference of Engineering and Computer Science.* **II** 1174-1176
[3] Merwan M 2008 *Annual Conference & Exposition.* 13.368.1-12
[4] Prasad M M, Ganesan K, Suresh R K 2013 *International Journal of Lean Thinking.* **4** 22-32
[5] Santosh T G and Suresh M S 2013 *International Journal of Advanced Engineering Research and Studies.* **II** 01-03
[6] Ashish M, Sandip K M, Sachin M 2012 *International Journal of Computer Applications.* **40** 27- 32
[7] Wiyaratn W and Watanapa A 2010 *International Journal of Mechanical, Aerospace, Industrial, Mechatronic and Manufacturing Engineering.* **4** 1382-6
[8] Pramod P S, Manmath S S, Sane S M 2012 *International Journal of Advanced Engineering Research and Studies.* **I** 259-261
[9] Rachmad H 2014 *Prosiding IENACO.* **I** 289-295
[10] Farhad S, Fantahun M D, Moussa, Soha E M 2015 *Journal of Industrial Engineering International.* **11** 283-295
[11] Satheeshkumarà V, Karthikeyan K, Thomas C J R, Silambarasan R., Bhagyanathan C 2014 *International Journal of Current Engineering and Technology.* **2** 183-190
[12] Qing-Lian L, Hu-Chen L, Dou-Jin W, Long L 2015 *Journal of Intellegent Manufacturing.* **26** 87-95
[13] Hsiu-Hsuh K, Din-Horng Y, Yi-Hsien W 2011 *Review of Economic & Finance.* **1** 71-80
[14] Vinza F F and Trijoko W A 2016 *International Seminar on Science and Technology.* **2** 37-42
[15] Vinod A and Singh S C 2013 *International Journal of Innovatitions in Engineering and Technology.* **2** 297-304
[16] Lorenzo T 2013 *International Journal of Production Economics.* **162** 201-215
[17] Rapeepan P and Kanchana S 2015 *Engineering Optimization.* **48** 253-271
[18] Amir J K and D J Tidke 2013 *International Journal of Engineering Research and General Science.* **2** 55-62
[19] Bobby J, Jubin J, R Mahesh R. 2013. *International Journal of Emerging Technology and Advanced Engineering.* **3** 514-520
[20] Lei W, Lu B 2015 *International Journal Applied Decision Sciences.* **8** 359-375
[21] Sukaria S 2011 *Metode Penelitian 1st Edition* Medan: USU Press
[22] Ifktikar S 2006 *Teknik Tata Cara Kerja* Bandung : ITB.
[23] Arman H N and Yudha P 2008 *Perencanaan dan Pengendalian Produksi* Yogyakarta: Graha Ilmu.
[24] Rosnani G 2007 *Sistem Produksi* Yogyakarta: Graha Ilmu

IOP Conf. Series: Materials Science and Engineering **309** (2018) 012117 doi:10.1088/1757-899X/309/1/012117

The effect of material productivity on scrap reduction on aluminum reduction pot process

L N Huda*

Industrial Engineering Department, Faculty of Engineering, Universitas Sumatera Utara, Jl. Almamater Kampus USU Medan 20155, Indonesia

*Email: lnurulh@gmail.com

Abstract. Scrap in the production process is defined as part of production that can not be accepted as the final product due to production failure. The amount of scarp occurring in the reduction pot production process of one of the aluminum smelting furnaces exceeds the acceptance standard of 5 to 10%. This affects the amount of production produced. The purpose of this study is to reduce the amount of scrap produced to increase the productivity of the materials used. The solution is carried-out by making improvements using Lean Six Sigma method. In the initial calculation, material productivity ranged from 0.88. Based on the study, the factors that cause the high amount of scrap produced is the use of Na2O which is inserted into the alumina is different in size and the influence of noise that occurs in pots that are difficult to control. The analysis base on Lean Six Sigma obtained process cycle efficiency (PCE) equal to 86,92% for actual condition. Process improvement is carried-out through define, measure, analyze, improve and control procedure on Na2O feeding process and PCE value decrease 6,05%. This indicates that increased process capability will reduce scrap. A decrease in scrap amount of about 6% will be able to increase material productivity by 0.05%. This indicates that continuous process control is needed to decrease the number of scraps and increase productivity.

1. Introduction

Aluminum is one of the metals whose usefulness can almost be found in every human activity. Aluminum metal is widely used because of some advantages compared with other metals. Non ferrous metals, aluminum have scrap that can be recycled [1].

Scrap in the production process is defined as part of production that can not be accepted as the final product due to production failure. The process of smelting aluminum reduction pot produces at 11.88% of scrap. The amount of scrap has exceeded the standard that is 5%-10%. The cause of the large amount of scrap generated is Na2O levels in alumina and noise. The amount of scrap that exceeds the standard limit will reduce the amount of aluminum production so the productivity of material is declining.

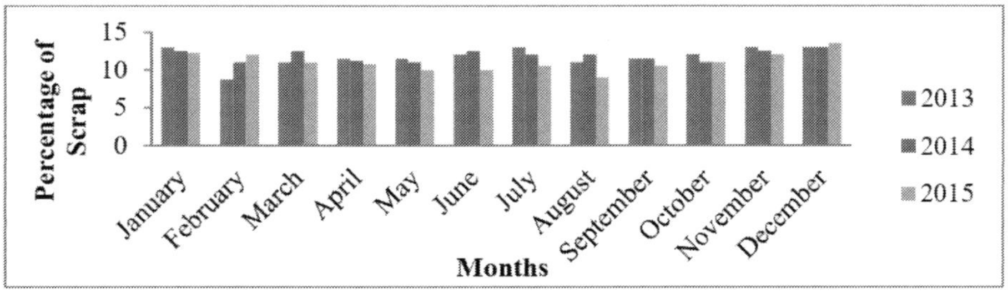

Figure 1. Percentage of *Scrap* in 2013-2015 Years

The average reduction in aluminum smelting pot produces the amount of scrap exceeded the standards that lead to material productivity becomes lower and reduced the number of production (Figure 1). It is therefore necessary to reduce the amount of scrap produced, smoothen the flow of the production process and improving the productivity of materials using lean six sigma.

Lean Six Sigma (LSS) method promotes, in organizations, the continuous improvement of products (and/or services) and processes that aligns with the business strategy to maximize the value of products [2]. Six Sigma is a business strategy that seeks a statistical approach to identify and eliminate causes of errors or defects, defined as anything which could lead to customer dissatisfaction or failures in business processes by focusing on outputs that are critical to customers; it uses various distributions and a strong relationship between product NonConformities, defects and product yield, inventory, schedule, etc [3]. Six Sigma methodology has two approaches: 1. DMAIC (D-Define, M-Measure, A-Analyze, I-Improve, C-Control). 2. DMADV (D-Define, M-Measure, A-Analyze, D-Design, V-verify) [4]. Six Sigma DMAIC methodology has been used in the manufacturing industries like automotive part manufacturing, metal processing, gloves manufacturing, file manufacturing, laser mouse manufacturing, semiconductor manufacturing, grinding operations, rolling mills [5]. DMAIC methodology is applicable to existing product or Phases six sigma implemented include the define, measure, analyze, improve, control [6]. Productivity is a measure of the ability of one unit of input in producing the output [7]. The purpose of this study is to reduce the amount of scrap produced to increase the productivity of the materials used. Similar Research has also been done in [8], [9], & [10].

2. Method and equipment

Research used is the comparative, descriptive study [11]. Step of research includes collecting data allowance and rating factor, NA2O levels in alumina data, noise data (interference in pot), and the amount of scrap produced, result and discussion, conclusion. The DMAIC methodology can be used for improving the product quality during the injection molding as follows:

a. Define

Define (D) is the first step of the Six Sigma methodology. In this phase selection of projects, initial goals and targets are set and develop a project charter or statement of work is carried out. Cost of quality related to existing process is being calculated. Analysis of SIPOC (Supplier, Input, Process, Output, Customer) and CTQ (Critical To Quality) are listed. Improvements target and goals are set in terms of sigma level and cost associated. This phase includes identification of the key problem areas and defining quality characteristics.[4]

b. Measure

Measure (M) phase is the second step of Six Sigma methodology. In this phase collection and observation of data is carried out and based on that DPMO (Defects Per Million Opportunity) or Process Capability Analysis is estimated based on the features of data (Attribute or Variable) [4]. In the processing measure data calculations will be performed, the time cycle, normal time, standard time, calculation metric lean, determination of Critical to Quality (CTQ), sigma level calculation and DPMO (Defects per Million Opportunities).

Calculation metric lean: [13]

1) The calculation of the process cycle efficiency should do the separation process of value-added and employments are not value-added.

$$\text{Process cycle efficiency (PCE)} = \frac{\text{Value Added Time}}{\text{Total Lead Time Manufacturing}} \quad (1)$$

2) Process lead time and Process Velocity

$$\text{Process lead time (PLT)} = \frac{\text{Amount of products processed}}{\text{Average speed of completion}} \quad (2)$$

$$\text{Process velocity (PV)} = \frac{\text{Amount of activity on process}}{\text{Process lead time}} \quad (3)$$

c. Analyze
The third step is Analyze (A). The identification of possible causes due to which variation or defects are occurring and which are affecting the output of the process is identified. Most frequently step in Analyze phase is cause and effect diagram[4]. At this stage of the analysis will be done by making a causal diagram (fishbone diagram) and a diagram of the five why used as tools to further analyze the results that have been obtained at this stage of the Measure.

d. Improve
The forth step is Improve (I). After collecting and analyzing the data suggestion is recommended to reduce the defect or nonconformity, which is critical for customer. In this phase counter measures for root causes are listed out which can be implemented to reduce defects [4].

e. Control
Fifth and last stage of DMAIC methodology is Control (C). In this stage control of improved implementation is done. If process is heading towards out of control, indication of that should be carried out by early signals [4].

3. Result and Discussion

Lean six sigma approach to deliver improvements on the speed of the production process and the quality of the resulting product so that companies are increasingly approaching the ideal circumstances as follows:

a. Define
The problems that plagued the industry aluminum smelting currently is about the amount of production that are generated. This is due to the amount of scrap that occur during the production process takes place that is affected by several factors such as levels of Na2O in alumina, time cycles, noise, and the skill of the operator when the metal tapping. This research was conducted to improve the speed of process and product quality by reducing the activities that do not add value (non value added) and manage the amount of scrap produced during the production process takes place. SIPOC diagram aims to describe information about suppliers, inputs, process, outputs and the customer is involved in the production of molten aluminum as describe in Figure 2 [15].
There are 8 inputs used in smelting aluminum cathode blocks, namely alumina, carbon anode, kriolit, soda ash, alumina, coke and liquid fluoride bath. The supply is input from the raw materials warehouse and warehouse equipment. The aluminum smelting process there are 6 i.e. reconstruction, baking, start-up, transition, normal operating, cut off. The output resulting from the process of smelting aluminum is liquid alumina which will be distributed to the customer. Value

Stream Mapping is one method of mapping the flow of production and the flow of information on the overall production. In value stream mapping, there are two mapping that should be described that is the making of the current state and future state folder. The making of the current state of the folder is done to map the floor of the actual production conditions. After the identification of waste is carried out, then it can be described a future state map. Future state of the folder is the mapping of the condition of the company in the future as the draft proposal of the improvement of the current state of an existing folder. The number of items contained in the process of production is dictated by the amount of production undertaken by the company. In the production process there are 5 work in process (WIP), namely Coke base waiting to be included, a liquid bath waiting to be put into a pot of molten metal reduction, waiting to be put into the pot, the reduction of material waiting to be entered into upon the molten metal anode, and waiting to be brought to the casting shop.

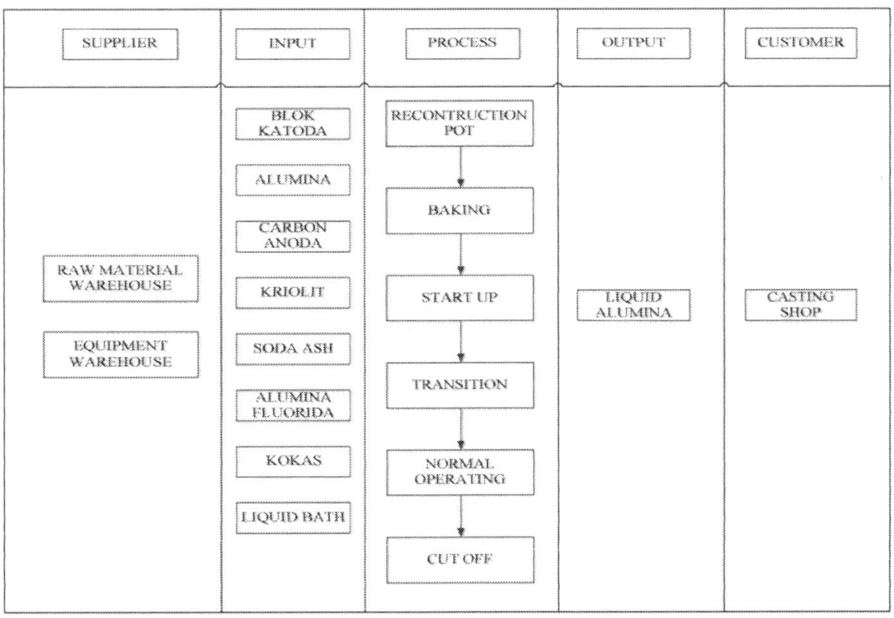

Figure 2. SIPOC Diagram of Liquid Aluminum's Process

b. Measure

Manufacturing lead time calculation is done by summing the entire time the labor process consisting of 40 work process. Description of the work process and the raw time can be seen in Table 1.

Table 1. Manufacturing Lead Time Calculation Based On Raw Time

Process	Average Cycle Time (hours)	Average Cycle Time (minutes)	*Rating Factor*	Normal Time (minutes)	Allowance	Standard Time (minutes)
1	2.28	136.5	1	136.5	5%	143.68
2	72.00	4320	1	4320	21.50%	5503.18
3	0.14	8.22	1	8.22	17.50%	9.96
4	0.13	7.8	1	7.8	21.50%	9.94
5	0.26	15.3	1	15.3	5%	16.11
6	0.15	8.76	1	8.76	21.50%	11.16

Table 1. Manufacturing Lead Time Calculation Based On Raw Time (Continued)

Process	Average Cycle Time (hours)	Average Cycle Time (minutes)	*Rating Factor*	Normal Time (minutes)	Allowance	Standard Time (minutes)
7	0.20	11.94	1	11.94	21.50%	15.21
8	0.74	44.52	1	44.52	5%	46.86
10	0.34	20.34	1	20.34	21.50%	25.91
11	0.15	9.18	1	9.18	21.50%	11.69
12	0.39	23.4	1	23.4	5%	24.63
14	0.11	6.72	1	6.72	21.50%	8.56
15	0.21	12.72	1	12.72	21.50%	16.20
16	0.10	6	1	6	21.50%	7.64
17	0.22	13.02	1	13.02	21.50%	16.59
18	0.13	7.68	1	7.68	21.50%	9.78
19	0.38	22.92	1	22.92	5%	24.13
20	0.14	8.46	1	8.46	21.50%	10.78
21	0.11	6.54	1	6.54	21.50%	8.33
22	0.24	14.16	1	14.16	5%	14.91
23	0.53	31.86	1	31.86	5%	33.54
25	0.20	12.18	1	12.18	21.50%	15.52
26	0.27	16.02	1	16.02	5%	16.86
27	0.23	13.98	1	13.98	21.50%	17.81
28	0.57	34.14	1	34.14	5%	35.94
30	0.33	19.92	1	19.92	21.50%	25.38
31	0.12	7.2	1	7.2	21.50%	9.17
32	0.10	6	1	6	21.50%	7.64
33	0.05	3	1	3	21.50%	3.82
34	0.43	26.04	1	26.04	19.00%	32.15
35	0.12	7.26	1	7.26	21.50%	9.25
36	0.35	20.76	1	20.76	5%	21.85
37	0.24	14.64	1	14.64	21.50%	18.65
39	0.43	25.5	1	25.5	5%	26.84

A firm is said to have executed the program Lean if has a value of process cycle efficiency by 30% which means that the processing time for a work process or activity which is value-added reach 30% of the time the process or activity as a whole.
The results of the calculation process cycle efficiency (Equation 1), process lead time (equation 2), the process velocity (Equation 3) before and after the improvement can be seen in Table 2.

Table 2. The Recap of PCE, PLT, and PV before and after Improvement

Process	Process Cycle Efficiency (PCE)	Process Lead Time (PLT) (Days)	Process Velocity (PE) (Process Per Day)
Before	86,92%	30	1,33
After	92,97%	30	1,23

Sigma level is a measure used to indicate how often the defects which may occur [9]. The calculation of the level of the sigma value obtained of 2.790. This shows the value of the sigma is still much to be achieved, namely 6 sigma. DPMO value earned 98,500 meaning for each 1 million of times the production possibility of disability is 98,500. To increase the value of the sigma, it needs to be done process improvement in order to reduce the number of defective products are produced so that it can increase the productivity of the material.

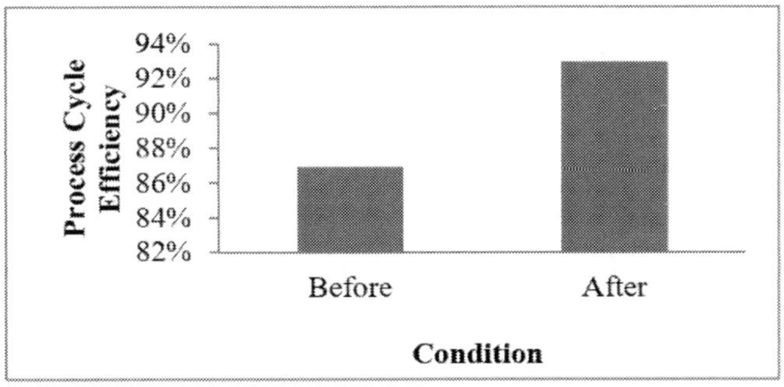

Figure 3. Improvement of Process Cycle Efficiency

Sigma level on the inspection of the condition in production process that are still considerably 2.79 worst with sigma level to be achieved 6 sigma. process cycle efficiency are increased by 6.05% (Figure 4) indicates the capability of a process to reduce the amount of scrap.

c. Analyze

Causal diagram (fishbone diagram) is used to help organize information about the causes of a potential problem. The analysis that will be performed include human analysis, methods of work, and the machinery and equipment against the amount of scrap produced (Figure 4).

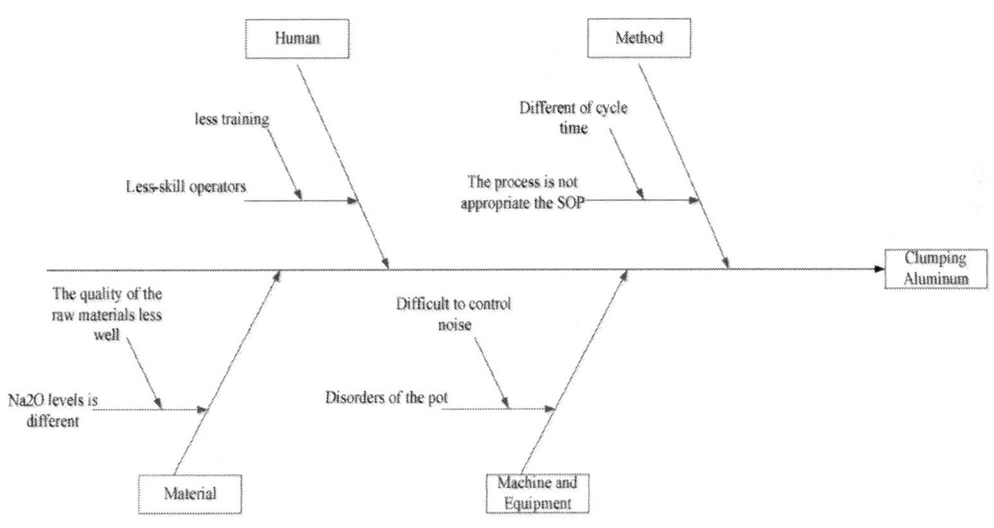

Figure 4. Fishbone Diagram

Diagram of Five Why is a diagram that is used to reveal the root of the problem of the cause of the discrepancy, which is obtained from the causal diagram so that it can be fixed with the right to ask constantly why something mismatch happens to found the root of the problem, see Table 3.

Table 3. Five Why Diagram for Disability Aluminum Lumpy

Problem	Why	Why	Why	Why	Why
Aluminum lumpy	The quality of the raw materials less well	Specification of Sodium oxide (Na2O) on different raw materials	Suppliers of raw materials originating from different countries	-	-
	Less-skilled operators	Carriers do not get enough training	The company does not run a program of continuous training activities	The company does not have a setting to carry out the training program schedule	The company focus is to produce aluminium in order to achieve the target of production
	Disorders of the pot	Difficult to control noise	Levels of Si and Fe in the pot already high	Cathode blocks cracked and eroding bar collector armor	The abundance of scrap generated
	The sequence of the workmanship not in accordance with SOP	Machining cycle time is different	Less supervision towards the operator	-	-

d. Improve

Based on the analysis that has been performed using diagrams Five Why, found problems that cause the amount of high quality i.e. scrap levels in different alumina Na2O, noise is difficult to control, and the order of the work are not appropriate the SOP.

The resulting kriolit is influenced by kriolit where also influential Na2O in terms of input and output resulting from the reduction furnace. When the concentration of Na2O in raw materials is too small, then the kriolit generated from the reaction of Na2O with AlF3 will too little so that operation of the pot furnace reduction will experience a shortage of kriolit which resulted in disruption of stability in pot furnaces and reduction in concentrations in Na2O raw material is too large, then kriolit is generated from the reaction of Na2O with AlF3 will too much pot furnace operation so that the reduction will have an excess of kriolit that can exceed the capacity of the pot furnace reduction and destabilize the pot reduction furnace. An excess or deficiency of kriolit will produce too much scrap giving rise to difficulties in the management of scrap. To keep the kriolit generated on the pot furnace reduction remains stable, then the need for a good control on raw materials of alumina. In raw materials of alumina concentration required range Na2O detail to know input and output that is more appropriate, so there is no problem about the shortage and excess kriolit on the operation of the pot furnace reduction.

Improvements are being made to lower the amount of scrap is Na2O levels on alumina type KR1116 GOVE subtracted to 0.33% and noise are used of 102mV with the content of Fe and Si respectively 0.034% and 0.085%. The fewer levels of Na2O and noise are used, and then the amount of scrap produced will be less and less. This indicates a decline in scrap from 11.88% to 6.126% (Figure 5).

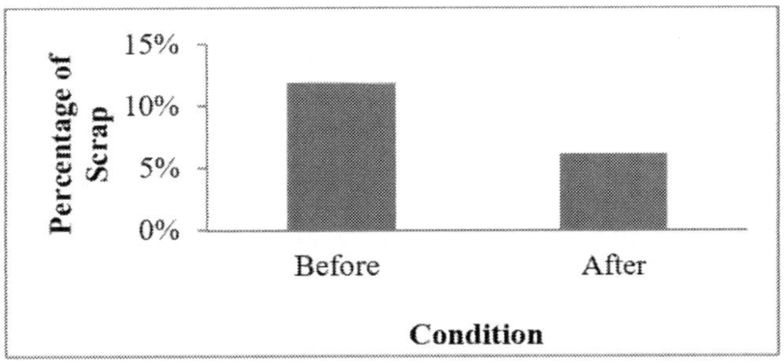

Figure 5. A comparison of Actual and Improvement Scrap

e. Control

The control method to reduce amount of scrap is making standard operating procedures (SOP) to control the levels of Na2O in alumina and tackling noise. Standard operating process is the manual that contains the standard operational procedures that exist in an organization that is used to ensure any decision, action and the use of facilities [14].

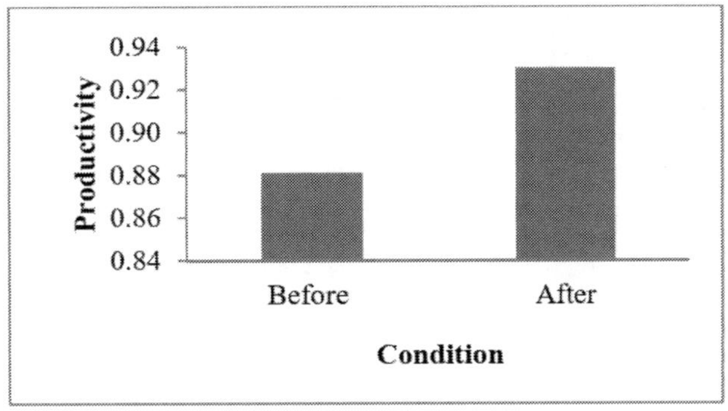

Figure 6. Increased Productivity

The decline in the number of scrap will increase the productivity of materials from 0.88 to 0.93 (Figure 6). Improved productivity gained amounting to 0.05. With increasing productivity and decreasing the amount of scrap material generated then factor in material and Na2O levels of noise (interference in pot) must continue to watch out for so that the material continues to increase productivity.

4. Conclusions

The increase in scrap is caused by irregular controlling of Na2O levels and noise. After the improvement is done, this indicates the material productivity experienced an increase of 0.05 and process cycle efficiency experienced an increase of 6.05%. The amount of scrap was declined by 5.75% after doing improvement.

Acknowledgments

The authors gratefully thanks to my student Marina for her asistance for the purposes of this research.

References

[1] Mubarak A Z and Akhyar 2013 Perancangan dan Pembuatan Dapur Peleburan Logam dengan Menggunakan Bahan Bakar Gas (LPG) ISSN 2301-8224 **1(3)**
[2] George M L 2002 *Lean Six Sigma - Combining Six Sigma Quality with Lean Speed* USA: McGraw-Hill Professional
[3] Krishnappa A and Ravishankar D 2016 Lean Management and Six Sigma Practices within Small and Medium Manufacturing Industries ISSN 2348-1439 **3(9)**
[4] Patel N 2015 A Review on Implementation of Six Sigma in Manufacturing Industries ISSN 2349-5162 **2(2)**
[5] Jadhav G dkk 2015 Six Sigma DMAIC Literature Review ISSN 2229-5518 **6(12)**
[6] Coskun A *Six Sigma Projects and Personal Experience* Croatia: Janeza Trdine
[7] Sinulingga S 2014 *Rekayasa Productivitas* Indonesia: USU Press
[8] Valles A et al 2009 Implementation of Six Sigma in Manufacturing Process: A Case Study *International Journal of Industrial Engineering* **16(3)** 171-181
[9] Hassan M K 2013 Applying Lean Six Sigma for Waste Reduction in a Manufacturing Environment *American Journal of Industrial Engineering* **1(2)** 29-35
[10] Ferreira J E and Lopes I S 2010 Improvement of Scrap Request Process with Six Sigma Methodology *Proceedings of the World Congress on Engineering* **3**
[11] Pande P and Holpp L *What is Six Sigma* Unites States of America: McGraw-Hill
[12] Michael L G dkk 2005 *The Lean Six Sigma Pocket* Toolbook New York: McGraw-Hill
[13] Tambunan R 2008 *Standard Operating Procedures* Indonesia: Malestas Publishing
[14] Praveen G 2005 *The Six Sigma Performance Handbook* New York: McGraw-Hill

IOP Conf. Series: Materials Science and Engineering **309** (2018) 012118 doi:10.1088/1757-899X/309/1/012118

The estimation of parameter compaction values for pavement subgrade stabilized with lime

A S Lubis[*], Z A Muis and C A Simbolon

Civil Engineering Department, Faculty of Engineering. Universitas Sumatera Utara, Medan, Indonesia

[*]Email: adinasarilubis@gmail.com

Abstract. The type of soil material, field control, maintenance and availability of funds are several factors that must be considered in compaction of the pavement subgrade. In determining the compaction parameters in laboratory desperately requires considerable materials, time and funds, and reliable laboratory operators. If the result of soil classification values can be used to estimate the compaction parameters of a subgrade material, so it would save time, energy, materials and cost on the execution of this work. This is also a clarification (cross check) of the work that has been done by technicians in the laboratory. The study aims to estimate the compaction parameter values ie. maximum dry unit weight (γd_{max}) and optimum water content (W_{opt}) of the soil subgrade that stabilized with lime. The tests that conducted in the laboratory of soil mechanics were to determine the index properties (Fines and Liquid Limit/LL) and Standard Compaction Test. Soil samples that have Plasticity Index (PI) > 10% were made with additional 3% lime for 30 samples. By using the Goswami equation, the compaction parameter values can be estimated by equation $\gamma_{d\ max}{}^{\#}$ = -0,1686 Log G + 1,8434 and $W_{opt}{}^{\#}$ = 2,9178 log G + 17,086. From the validation calculation, there was a significant positive correlation between the compaction parameter values laboratory and the compaction parameter values estimated, with a 95% confidence interval as a strong relationship.

1. Introduction

The strength of pavement construction depends on the characteristic and carrying capacity of the subgrade. The effort which is required to increase the carrying capacity of basic soil is compaction. Laboratory compaction test is determined by conducting Proctor Compaction Test on several soil samples with varying moisture content. The result obtained is the Parameter Value of Compaction which is maximum dry unit weight (γd_{max}) at the time of optimum water content (w_{opt}). Whereas the field compaction test is obtained by Sand Cone Test or Dynamic Cone Penetrometer Test which yielded the maximum dry unit weight (γd_{max}) value in field [1]. In determining the compaction parameters in laboratory desperately requires considerable materials, time and funds, and reliable laboratory operators. If the result of soil classification values can be used to estimate the compaction parameters of a subgrade material, so it would save time, energy, materials and cost on the execution of this work. This is also a clarification (cross check) of the work that has been done by technicians in the laboratory [2]. This study was used the Goswami equation to estimate the value of laboratory compaction parameters (maximum dry unit weight and optimum water content) of soil mixed with 3% lime.

2. Literature Review

Subgrade is the basic soil at the bottom of the pavement layer. Soil compaction is a mechanical process whereby the amount of soil composed of solid particles, water and air that reduced by their volume by using a load. For field compaction, generally heavy equipment is used such as Three Wheel Roller, Tandem Roller, Pneumatic Tired Roller (PTR) and others. Generally vibratory roller is used for compaction of subgrade, while in laboratory, the method which is used by hitting it using a hammer. Compaction testing in the laboratory has two methods: Standard Proctor Test and Modified Proctor Test [1].

The physical properties of the soil (index properties) show the soil conditions and type of the soil. In general there are five laboratory tests of subgrade material before the compaction test [1] i.e. Moisture Content Test, Specific Gravity Test, Atterberg Limit Test, Sieve Analysis Test and Soil Classification (USCS and AASHTO).

Several studies in estimating soil compaction values (maximum dry unit weight and optimum water content) have been widely developed. The studies used several geotechnical parameters, such as : plastic limit, liquid limit, specific gravity, compaction energy, particle size distribution analysis and soil classification. The study determining the relationship between compaction parameters with one or several value of the geotechnical parameters for variety of soil types [3, 4, 5, 6, 7, 8, 9].

The estimation for the value of maximum dry unit weight (γd_{max}) and optimum water content (w_{opt}) can also be calculated from the model suggested by Goswami with the following equation [2]:

$$Y = m \operatorname{Log} G + k \tag{1}$$

Y = maximum dry unit weight (γd_{max}) and optimum water content (w_{opt})
m = the slope of the curve
k = constant
G = coefficient of gradation (1 + F) (AX1 + BX2 + CX3)
X1 = % weight retained shieve 4.75 mm
X2 = % weight retained between shieve 4.75 mm and 0.075 mm
X3 = % weight through shieve 0.075 mm
A, B, C = constants for shieve number
F = % fine grain

The m and k constants are obtained from the graph between Log G with the value of the maximum dry unit weight and the optimum water content from the experimental results in the laboratory. Whereas, F is fine grain percent that is determined based on percent through shieve 0.075 mm and Plasticity Index (IP) value (Table 1).

Table 1. Determination of F value.

% weight through shieve 0.075 mm	Value of F	
	IP < 10%	IP > 10%
0 – 25	0,0	0,0
26 – 40	0,2	0,2
41 – 60	1,0	1,0
61 – 85	1,0	0,0
86 – 100	1,0	1,0

3. Research Method

The samples were 30 soil samples with plasticity index (PI) >10%. The lime was lime powder (CaO) which is generally sold in the material store and pass through the sieve No. 200. Laboratory testing for the original soil samples consists of the the index properties test and compaction test. The index properties test were moisture content test, specific gravity test, sieve analysis test, atterberg limit test, while the compaction test was standard proctor test. After 3% of lime was added [10], the samples had the same tests. Using the Goswami model, the graph of correlation between the Log G with the compaction parameter values laboratory was obtained, and then the m and k values were determined. Using the m and k values in Goswami equation, the compaction parameter values estimated were obtained. These values then were compared with the compaction parameter values laboratory. Compaction parameter values estimated were also analyzed base on the soil classifications (USCS and AASHTO). Afterward, the level of accountability was observed by the validation method, to get a positive correlation level of accountability.

4. Results and Analysis

4.1. Original soil test results in laboratory

The test results of original soil in laboratory ie. index properties values and compaction parameter values can be seen in Table 2.

Table 2. Test results of original soil in laboratory.

Parameter	Average
Water content (%)	33,87
Specific Gravity (SG)	2,65
Liquid Limit (LL) (%)	40,29
Plastic Limit (PL) (%)	22,88
Plasticity Index (PI) (%)	17,41
Fines (Passing No.200) (%)	50,39
AASHTO	A-6 (6)
USCS	CL
Maximum dry unit weight (γd_{max}) (gr/cm^3)	1,529
Optimum water content (W_{opt}) (%)	21,57

4.2. Soil + lime test result in laboratory

After the soil was mixed with lime, the laboratory tests were repeated ie. index properties tests (sieve analysis and Atterberg limit test) and standard compaction test. The test results of soil samples added with lime can be seen in Table 3. The addition of lime as much as 3% was resulted in changes of the index properties value, were decreased the Plasticity Index (IP) from 17.41% to 11.27%, Liquid Limit (LL) from 40.29% to 34.85%, Maximum Dry Unit Weight from 1.529 gr/cm3 to 1,396 gr/cm3. Conversely, were increased Plastic Limit (PL) from 22.88% to 23.56%, Fines percentage from 50.39% to 53.36% and Optimum Water Content from 21.57% to 24.87%. This results was supported the previously research [10].

4.3 Compaction parameter values estimated by Goswami model

Using the Goswami model, from the graph of correlation between the Log G with the compaction parameter values laboratory, then the m and k values were determined as shown as Figure 1 and Figure 2. From Figure 1, there were m = -0,1686 and k = 1,8434, while from Figure 2, m = 2,9178 and k = 17,086.

Table 3. Test result of soil samples added with lime.

No	Fines (%)	LL (%)	PL (%)	PI (%)	$\gamma_{d\ max}$ (gr/cm³)	Wopt (%)
1	49,52	37,19	25,33	11,86	1,401	24,02
2	52,29	35,52	24,69	10,83	1,403	24,21
3	54,23	37,17	24,12	13,05	1,362	25,16
4	53,44	35,55	24,71	10,84	1,399	25,51
5	53,10	34,27	24,37	9,90	1,396	25,53
6	52,25	35,86	24,77	11,09	1,393	25,32
7	47,99	34,30	24,23	10,07	1,392	24,92
8	46,65	30,44	24,39	6,05	1,402	23,97
9	48,38	34,23	23,25	10,98	1,384	24,71
10	53,13	36,06	24,05	12,01	1,389	25,94
11	56,06	34,72	24,41	10,31	1,408	23,43
12	55,07	36,72	23,50	13,22	1,391	24,53
13	54,92	34,27	24,04	10,23	1,392	24,38
14	53,27	32,83	22,28	10,55	1,383	25,16
15	55,06	32,24	22,41	9,83	1,388	25,15
16	52,24	33,10	22,86	10,24	1,396	25,69
17	50,61	40,90	26,52	14,38	1,406	24,66
18	56,45	34,02	22,56	11,46	1,410	24,66
19	56,65	34,50	21,68	12,82	1,407	24,61
20	56,09	39,25	23,27	15,98	1,412	24,67
21	52,65	33,40	24,30	9,10	1,386	25,69
22	55,67	33,95	22,43	11,52	1,406	24,63
23	51,37	33,21	23,18	10,03	1,393	25,09
24	50,88	35,16	22,59	12,57	1,394	25,04
25	56,36	32,89	22,40	10,49	1,398	24,53
26	55,61	33,16	22,27	10,89	1,394	15,46
27	53,51	34,05	23,74	10,31	1,392	25,20
28	58,82	33,33	22,42	10,91	1,405	23,69
29	52,52	38,78	22,42	16,36	1,387	25,34
30	56,12	34,32	24,18	10,14	1,397	25,19
Avr	**53.36**	**34.85**	**23.56**	**11.27**	**1,396**	**24.87**

where : γ_{dmax} = Maximum dry unit weight (w_{opt}) ; W_{opt} = Optimum water content; Fines=Percentage of fines; LL=Liquid Limit; PL=Plastic Limit; PI=Plasticity Index

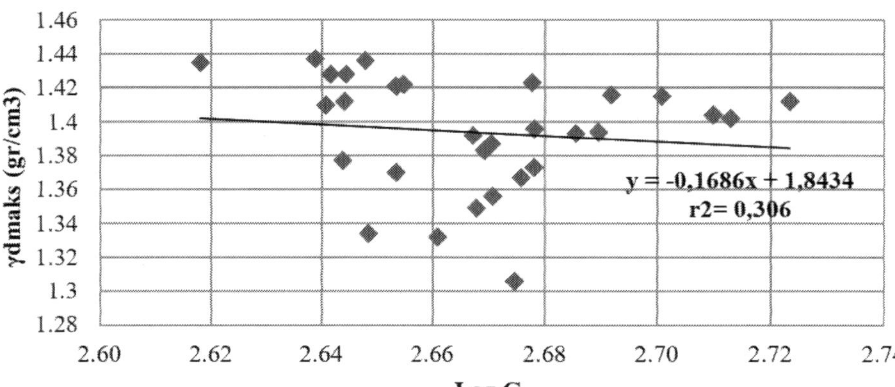

Figure 1. Correlation of maximum dry unit weight (γd_{max}) with Log G

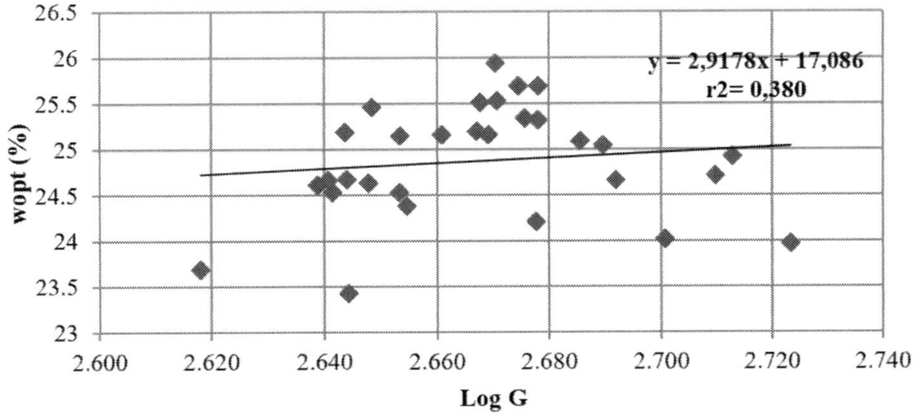

Figure 2. Correlation of Optimum Water Content (w_{opt}) with Log G

Using the m and k values in Goswami equation, the compaction parameter values estimated ie. maximum dry unit weight estimated ($\gamma_{d\ max}^{\#}$) and optimum water content estimated ($W_{opt}^{\#}$) were obtained as seen as Table 4. Subsequently, the soil type of the sample can be determined, which according to AASTHO classification of soil belongs to A6 and A7 class and according to USCS including class of SL, ML and CL.

Table 4. Compaction Parameter Values Estimated by Goswami Model.

No	AASHTO	USCS	Fines (%)	LL (%)	PL (%)	PI (%)	$\gamma_{d\ max}$ (gr/cm^3)	$\gamma_{d\ max}^{\#}$ (gr/cm^3)	W_{opt} (%)	$W_{opt}^{\#}$ (%)
1	A-6 (3)	SM	49,52	37,19	25,33	11,86	1,401	1,388	24,02	25,84
2	A-6 (3)	ML	52,29	35,52	24,69	10,83	1,403	1,392	24,21	25,78
3	A-6 (5)	CL	54,23	37,17	24,12	13,05	1,362	1,395	25,16	25,73
4	A-6 (4)	ML	53,44	35,55	24,71	10,84	1,399	1,394	25,51	25,75
5	A-4 (3)	ML	53,10	34,27	24,37	9,90	1,396	1,393	25,53	25,76
6	A-6 (3)	ML	52,25	35,86	24,77	11,09	1,393	1,392	25,32	25,78
7	A-4 (2)	SM	47,99	34,30	24,23	10,07	1,392	1,386	24,92	25,88
8	A-4 (1)	SM	46,65	30,44	24,39	6,05	1,402	1,384	23,97	25,91
9	A-6 (3)	SC	48,38	34,23	23,25	10,98	1,384	1,387	24,71	25,87
10	A-6 (4)	CL	53,13	36,06	24,05	12,01	1,389	1,393	25,94	25,76
11	A-4 (4)	ML	56,06	34,72	24,41	10,31	1,408	1,398	23,43	25,68
12	A-6 (5)	CL	55,07	36,72	23,50	13,22	1,391	1,396	24,53	25,71
13	A-4 (4)	ML	54,92	34,27	24,04	10,23	1,392	1,396	24,38	25,71
14	A-6 (3)	CL	53,27	32,83	22,28	10,55	1,383	1,393	25,16	25,75
15	A-4 (3)	CL	55,06	32,24	22,41	9,83	1,388	1,396	25,15	25,71
16	A-4 (3)	CL	52,24	33,10	22,86	10,24	1,396	1,388	25,69	25,78
17	A7-6 (5)	ML	50,61	40,90	26,52	14,38	1,406	1,392	24,66	25,82
18	A-6 (4)	CL	56,45	34,02	22,56	11,46	1,410	1,395	24,66	25,67
19	A-6 (5)	CL	56,65	34,50	21,68	12,82	1,407	1,394	24,61	25,66
20	A-6 (7)	CL	56,09	39,25	23,27	15,98	1,412	1,393	24,67	25,68

Table 4. Compaction Parameter Values Estimated by Goswami Model (continued).

No	AASHTO	USCS	Fines (%)	LL (%)	PL (%)	PI (%)	$\gamma_{d\,max}$ (gr/cm^3)	$\gamma_{d\,max}^{\#}$ (gr/cm^3)	W_{opt} (%)	$W_{opt}^{\#}$ (%)
21	A-4 (3)	ML	52,65	33,40	24,30	9,10	1,386	1,392	25,69	25,77
22	A-6 (4)	CL	55,67	33,95	22,43	11,52	1,406	1,386	24,63	25,69
23	A-4 (3)	CL	51,37	33,21	23,18	10,03	1,393	1,384	25,09	25,80
24	A-6 (4)	CL	50,88	35,16	22,59	12,57	1,394	1,387	25,04	25,81
25	A-4 (4)	CL	56,36	32,89	22,40	10,49	1,398	1,393	24,53	25,67
26	A-6 (4)	CL	55,61	33,16	22,27	10,89	1,394	1,398	15,46	25,69
27	A-4 (3)	CL	53,51	34,05	23,74	10,31	1,392	1,396	25,20	25,75
28	A-6 (4)	CL	58,82	33,33	22,42	10,91	1,405	1,396	23,69	25,60
29	A-6 (6)	CL	52,52	38,78	22,42	16,36	1,387	1,393	25,34	25,77
30	A-4 (4)	ML	56,12	34,32	24,18	10,14	1,397	1,396	25,19	25,68

where : $\gamma_{d\,max}$=Maximum Dry Unit Weight; Wopt=Optimum Water Content; Fines=Percentage of Fines; LL=Liquid Limit; PL=Plastic Limit; PI=Plasticity Index

According to the testing for range of validity for the compaction parameter values estimated were known that the maximum dry unit weight estimated ($\gamma d_{max}^{\#}$) obtained 95% validity range with a significant positive correlation with the maximum dry unit weight laboratory (γd_{max}). Likewise, the optimum water content estimated ($w_{opt}^{\#}$) obtained 95% validity range also with a significant positive correlation with the optimum water content laboratory (W_{opt}). But, this model only used on soil types A-4, A-6, and A-7 (AASHTO) or SL, ML and CL (USCS) with 3% lime added.

5. Conclusion

From the results of laboratory testing and data analysis obtained as follows: (1) The addition of 3% lime was decreased the index properties, Plasticity Index (IP) Liquid Limit (LL) and Maximum Dry Unit Weight value and was increased Plastic Limit (PL), Fines percentage and Optimum Water Content value; (2) The compaction parameter values can be estimated by Goswami equation : $\gamma_{d\,max}^{\#} = -0{,}1686 \text{ Log } G + 1{,}8434$ and $W_{opt}^{\#} = 2{,}9178 \log G + 17{,}086$; (3) Estimation with Goswami Model is easier and more effective because it only requires the value of Fine; (4) Based on the soil classification, the soil samples were A4, A6 and A7; (5) The correlation between the compaction parameter values estimated with the compaction parameter values laboratory provides a significant positive correlation with a 95% confidence range.

6. References

[1] Bowles J E 1993 *Sifat-sifat Fisis dan Geoteknis Edisi Kedua* Erlangga Jakarta.
[2] Muis Z A 1998 *Penentuan Berat Isi Kering Maksimum Bahan Aggregat Base Berdasarkan Data Klasifikasi Tanah Pada Proyek Jalan Raya* Seminar Highway Engineering Medan Academic Commitee (MAC) Teknik Sipil USU
[3] Muis Z A, Siagian D 2013 *Estimasi Nilai Parameter Kompaksi Bahan Subgrade Berdasarkan Nilai Index Properties Pada Proyek Jalan Raya* Skripsi Sarjana Departemen Teknik Sipil USU
[4] Metacalf J B, Romanoschi S A 2008 *Prediction Od Maximum Dry Density And Optimum Moisture Content From Simple Material Properties*
[5] Nendi A M 2010 *Korelasi Antara Hasil Ujian Mampatan Dengan Had Atterberg* Skripsi Sarjana Fakultas Teknik Sipil Universitas Teknologi Malaysia.
[6] Ugbe F C 2011 Estimating Compaction Characteristics From Fines in A-2 Type Lateritic Soils *Research Journal Of Environmental And Earth Sciences* **3(4)** 433-7

[7] Ugbe FC 2012 Predicting Compaction Characteristics Of Lateritic Soil Of Western Niger Delta, Nigeria *Research Journal Of Environmental And Earth Sciences* **4(5)** 553-9
[8] Lubis A S 2017 The Estimation of Compaction Parameter Values Based on Soil Properties Values Stabilized with Portland Cement *IOP Conf. Ser.: Mater. Sci. Eng* **180** 012139
[9] Lubis A S 2017 *Estimation of Compaction Parameters Based on Soil Classification*
[10] Australia Stabilisation Pavement Recycling And Stabilisation Association 2009 *Laboratory Investigation report for the insitu stabilisation of unsealed road trials using lime, cementious, and poyimers binders*

IOP Conf. Series: Materials Science and Engineering **309** (2018) 012119 doi:10.1088/1757-899X/309/1/012119

Effect of sintering temperature on the microstructure, electrical and magnetic properties of Zn0.98 Mn0.02O material

K Sebayang[2], D Aryanto[1,3], S Simbolon[1,3], C Kurniawan[1], S F. Hulu[2], T Sudiro[1], M Ginting[1], and P Sebayang[1,3,*]

[1]Research Center for Physics, Indonesian Institute of Sciences (LIPI), Puspiptek Bld. 440-442, Serpong 15314, Indonesia

[2]Department of Physics, Universitas Sumatera Utara, Medan 20155, Indonesia

[3]Department of Mechanical Engineering, Universitas Pamulang, Banten, Indonesia.

*E-mail: sebayang.perdamean@gmail.com

Abstract. $Zn_{0.98}Mn_{0.02}O$ material was synthesized from ZnO and MnO_2 powders using solid state reaction method. The microstructure, electrical and magnetic properties of $Zn_{0.98}Mn_{0.02}O$ were studied as a function of sintering temperature. The X-ray diffraction analysis indicates that the main phase of synthesized sample is composed of hexagonal wurtzite ZnO phase. While the secondary phase of $ZnMnO_3$ were found at the sintering temperature of 700°C and 900°C. The electrical properties measurement of $Zn_{0.98}Mn_{0.02}O$ sample revealed that the resistivity and the dielectric constant of samples increase with the increase of sintering temperature. The ferromagnetic properties at room temperature were observed in the $Zn_{0.98}Mn_{0.02}O$ samples sintered at 500°C and 700°C. It also found that the increase in sintering temperature leads to a tendency toward the changes in the magnetic properties into paramagnetic. The presence of $ZnMnO_3$ secondary phases in $Zn_{0.98}Mn_{0.02}O$ system is believed to be a factor that affects the decrease of the electrical and magnetic properties of the sample.

1. Introduction

Recently, the information technology devices have combined a magnetic (spin) and semiconductors (charge), such as the spintronics. The substitution of some transition metal elements in the compound of semiconductor crystal host can enhance the local magnetic moment at a low energy level system [1]. The electrons charge and spin in the spintronic can improve the device performance, such as high speed, better efficiency, and stability [2]. According to the theory, ZnO and GaN with Mn doping exhibit ferromagnetic properties above the room temperature [3]. These properties encourage many researchers doing an intensive work on the doping variation experiment of dilute magnetic semiconductor (DMS). The study on the DMS continues to grow because of the stable magnetic properties, especially in ZnO material after doped with Fe, Co, Ni, or Mn to achieve a high Curie temperature [4].

DMS based on ZnO material has the hexagonal wurtzite structure with a wide band gap (~3.4 to 3.7 eV), the excitation binding energy (~60 meV) at room temperature, and a high optical gain (300 cm^{-1}). Moreover, ZnO has a short luminescence lifetime which is needed in optoelectronic applications and magneto-optical devices [5,6]. The ZnO material also has magnetic properties when it was doped with the transition metal elements, such as Cr, Co Fe, Ni, and Mn [7–9]. Previous studies by Choudhury [10] showed that Zn^{2+} ions in the ZnO lattice could be replaced by a transition metal ion. This produces a magnetic ion interaction with a carrier in the semiconductor band gap of ZnO and generates the exchange of spins magnetic interaction.

The Mn-doped ZnO is interesting to study since its Curie temperature (T_c) is above the room temperature, so that it possess a ferromagnetic properties at room temperature [11]. The similarity of ionic radii of Mn and Zn atoms produces a high solubility of Mn into ZnO crystal lattice [12], which makes more carrier atoms available so that it can increase the ferromagnetic properties of ZnO at room temperature [13]. The Mn-doped ZnO were studied intensively using various methods, such as solid state reaction [14,15], ion implantation [16], solvothermal process [17,18], sol-gel [19] and coprecipitation method [11,20]. The results of previous studies show that the doping concentration, fabrication method and treatment temperature of the samples play an important role in generating Mn-doped ZnO sample characteristic. The good magnetic properties of the sample are obtained when the Mn-doped ZnO material does not contain a secondary phase [21]. Several studies have reported that the solubility limit of Mn doping on ZnO to avoid the formation of secondary phase is around 4 %at [15,22]. According to the report of Riyadi et al. [23], Banerjee et al. [17] and Chattopadhyay et al. [5] who doped ZnO with Mn up to 3 %at. obtained a single phase of Mn-doped ZnO with ferromagnetic properties. However, only a few studies reported the effect of sintering temperature on the fabrication of Mn-doped ZnO using solid state reaction method. The advantage of solid state reaction compared to the other methods is due to relatively simple, low cost and synthesized sample can be used further as a target in the ion implantation and thin film deposition [24].

In this study, the Mn-doped ZnO samples were synthesized using a solid state reaction method. The effect of sintering temperature on the microstructure, electrical and magnetic properties of Mn-doped ZnO was studied using X-ray diffractometer, I-V meter, C-V meter and VSM.

2. Experiment Method

The $Zn_{0.98}Mn_{0.02}O$ bulk was prepared from ZnO and MnO_2 powders by the solid state reaction method using a high-speed shaker mill. The sample preparation process was done by mixing ZnO and MnO_2 powders which contain 2 %at. of Mn dopant using stainless steel balls that were inserted into the cylindrical stainless steels vial with ball and powder ratio of 10:1. The powder mixing was carried out at the oscillation frequency of about 700/min. This process was done by wet milling for 3 hours using toluene as solvent. Afterward, the milled powders were dried in an oven at 100°C for 3 hours. The powders were then compacted by an automatic axial hydraulic press into a bulk with 15 mm in diameter and 2 mm in thickness using a load of 1500 kgf/cm^2. Then the bulk samples were sintered at elevated temperatures of 500°C, 700°C, and 900°C for 4 hours in air with a heating rate of 10°C/min. The X-ray diffraction data was collected by using a Smartlab Rigaku diffractometer with CuK_α radiation (λ = 1.5406 Å). The scanning range is 20°-60° (2θ) with a step of 0.02°. The lattice parameters such as lattice constant, *d*-spacing, crystalline size, strain and dislocation density were calculated based on the XRD data. The resistance measurement was carried out at room temperature using dual-probes of 8842A High Impedance Fluxe Multimeter. A 590 CV Keithley analyzer was used to obtain the capacitance at room temperature, which is then used to determine the dielectric constant of the sample. The magnetic properties were measured using Vibrating Sample Magnetometer 250, Dexing Magnet, ltd.

3. Results and Discussion

Figure 1 shows the XRD patterns of pure ZnO and $Zn_{0.98}Mn_{0.02}O$ samples which prepared using a solid state reaction method at varies sintering temperatures. Based on the results of XRD characterization, all samples have a major phase of ZnO with hexagonal wurtzite structure. These results are indicated by the presence of peaks that are identical to ZnO crystal planes of (100), (002), (101), (102) and (110) (JCPDS No. 00-005-0664). However, the $Zn_{0.98}Mn_{0.02}O$ samples sintered at a temperature of 700°C and 900°C show a secondary phase of $ZnMnO_3$. The diffraction peaks of $ZnMnO_3$ are confirmed at the diffraction angle 2θ of around 30.09° and 35.44°, which show the crystal planes of (220) and (311). In general, the intensity of the main peak does not change significantly. But, the $ZnMnO_3$ peak intensity in the present study increases with the increase of the sintering temperature. This is the indication of the occurrence of $ZnMnO_3$ crystal growth [25]. The formation of $ZnMnO_3$ phase probably due to Mn solubility in the ZnO crystal lattice is higher under non-equilibrium conditions [26]. The result of this work is also supported by previous studies using different methods [17,26,27], where in the secondary phases as MnO_2, Mn_2O_3 or $ZnMn_2O_4$ appear at the sintering temperature above 500°C. Additionally, Fabbiyola et al. [11] with the co-precipitation method showed that the secondary phase of Mn_3O_4 appears when ZnO powder with 2 %at. Mn-dopant is dried at 200°C.

Table 1. Crystal parameters of $Zn_{0.98}Mn_{0.02}O$ sample at (100) plane sintered at the elevated temperature of 500°C, 700°C and 900°C 4 hours.

Temperature of Sintering (°C)	2θ (deg.)	FWHM (deg.)	*a* (Å)	*c* (Å)	*c/a*	*d-spacing* (Å)	*D* (Å)	ξ (%)	Δ (line/Å^2) × 10^{-6}
500	31.773	0.179	3.2494	5.2052	1.602	2.603	462	0.274	7.9
700	31.769	0.173	3.2498	5.2058	1.602	2.603	478	0.265	7.4
900	31.784	0.159	3.2483	5.2033	1.602	2.602	520	0.243	6.2

The crystal size of the sample increases with the increase of the sintering temperature, where the crystal size is in the range of 46 to 52 nm. The increase of the sintering temperature from 500°C to 900°C accelerates the diffusion of grain Mn-doped ZnO at the grain boundaries, resulting from an agglomeration of a small grain into a larger grain. It affects the decrease of the lattice strain and dislocation density of the Mn-doped ZnO crystal system.

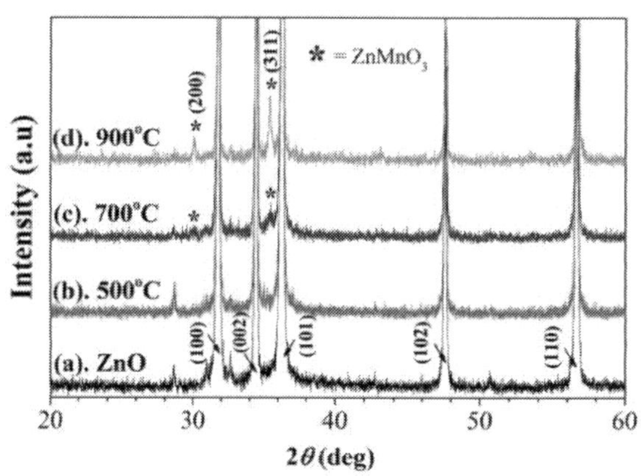

Figure 1. XRD patterns of (a) ZnO and $Zn_{0.98}Mn_{0.02}O$ samples sintered at varying temperature of (b) 500°C, (c) 700°C, and (d) 900°C for 4 hours.

The previous studies also reported that the separation of oxide phase occurs at a doping concentration of high than 4 %at. [23]. Therefore, the presence of secondary phase in this work confirms that the sintering temperature also affects the formation of a secondary phase in the synthesis of ZnO-doped Mn. The detail observation of diffraction peaks of hexagonal wurtzite ZnO in the (100) crystal planes is shown in Figure 2 [28]. The diffraction peak intensity and the position are changes with the addition of 2 %at. Mn-dopant and the sintering temperature. This indicates that the presence of Mn-dopant and the increase of sintering temperature is likely to affect the quality of ZnO crystals. These results confirmed from the calculation of crystal parameters as presented in Table 1.

In Figure 2, it can be seen that the diffraction peak of the sample was shifted toward a smaller 2θ angle at the sintering temperature of 500°C and 700°C. This indicates the increase in the crystal lattice parameters (*a* and *c*), as shown by Neogiet al. [22] and Omri et al. [29]. The increase in lattice parameter is attributable to the substitution of Mn^{2+} ions replacing the position of Zn^{2+} ions in the ZnO crystal lattice. The radius of Mn^{2+} ions (0.66 Å) is larger than Zn^{2+} ion (0.60 Å) which leads to an increase in the lattice constants (*a* and*c*) of the ZnO crystal. While the diffraction peak of samples which is sintered at 900°C is shifted towards to a greater 2θ angle. This result is believed to be due to the presence of Mn^{3+} ions replacing the position of Zn^{2+} ions in the ZnO crystal lattice. The Mn^{3+} ion has a smaller radius (0.58 Å) compared to that of Zn^{2+} ions (0.60 Å) [30,31], which decrease the lattice constant of ZnO crystal [32].

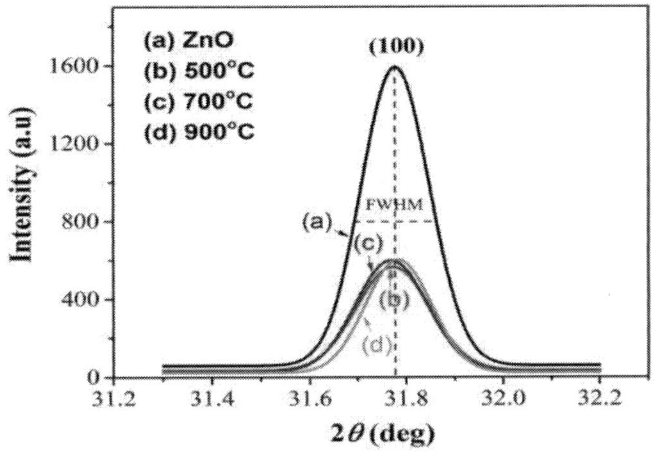

Figure 2. The intensity patterns of ZnO (a) and $Zn_{0.98}Mn_{0.02}O$ samples on (100) plane sintered at the elevated temperature of (b) 500°C, (c) 700°C and (d) 900°C for 4 hours.

Figure 3 shows the resistivity and dielectric constant of $Zn_{0.98}Mn_{0.02}O$ sample at varying sintering temperature. From Figure 3 can be seen that the resistivity of Mn-doped ZnO increases with increasing sintering temperature, where the values are in the range of 54.8-62.6 MΩcm. The resistivity obtained in this work is smaller than the previous studies which were using a mechanical milling technique (resistivity of 84 MΩcm) [33]. However, the aforesaid resistivity is greater than the resistivity of the Mn-doped ZnO which was prepared by sol-gel method (29.42 MΩcm) [34] and sputtering [35]. The increase of the resistivity value in this work is related to the presence of $ZnMnO_3$ phase, which leads to the defects in the grain boundary area. Also, Mn doping builds a potential barrier at the grain boundaries [36,37] that causes the dopping of Mn^{2+} in ZnO is likely to decrease [38]. This case strongly affects the decrease of the sample conductivity. The dielectric constant of Mn-doped ZnO samples, sintered at the elevated temperature of 500°C up to 900°C is in the range of 9.8 to 26.2 (see also Figure 3). The increase of the sintering temperature leads to an increase of the dielectric constant. This evidence is in accordance with the results of Das et al. [39], in which the impurity phase of $ZnMnO_3$ at grain boundaries of the $Zn_{0.98}Mn_{0.02}O$ acting as semi-blocking layers. In addition, the presence of external electric field causes the carrier charge gather at the grain boundaries and leads to the polarization and an increase of the dielectric constant [40].

IOP Conf. Series: Materials Science and Engineering **309** (2018) 012119 doi:10.1088/1757-899X/309/1/012119

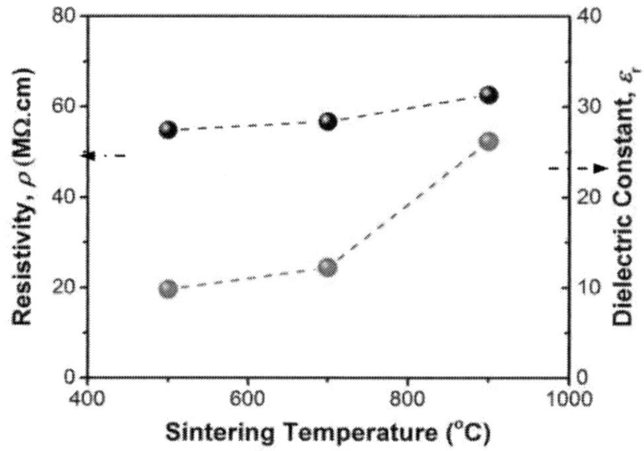

Figure 3. Resistivity and dielectric constant of $Zn_{0.98}Mn_{0.02}O$ sample at varying sintering temperature.

Figure 4 shows the results of Vibrating Sample Magnetometer (VSM)-250 measurement of the pure ZnO and $Zn_{0.98}Mn_{0.02}O$ samples sintered at elevated temperatures of 500°C, 700°C, and 900°C. The magnetic hysteresis (M-H) curve indicates that the pure ZnO sample behaves as the diamagnetic. The presence of Mn dopant changes ZnO into ferromagnetic sample. In Figure 4, it is seen that the increase in sintering temperature on the synthesis of Mn-doped ZnO causes the decrease of the ferromagnetic properties. It is indicated by the decrease in the saturation (M_s) and the magnetic remanence (M_r) of the samples. At 500°C, the magnetic parameters as magnetic saturation (M_s), remanence (M_r) and coercivity (H_c) of the sample are 0.16 emu/g, 0.03 emu/g, and 277 Oe, respectively. Whereas at 700°C, the M_s, M_r, and H_c of the sample are 0.04 emu/g, 0.01 emu/g, and 481 Oe, respectively. For Mn-doped ZnO sample sintered at 900°C, the sample exhibits paramagnetic properties indicated by the M-H curve which almost looks like a straight line. Compared to the results of previous studies, the present magnetic saturation (M_s) is relatively larger. Cahttopadhyay et al. [41] reported that the M_s of $Zn_{0.98}Mn_{0.02}O$ sample prepared by a solid state reaction method was 0.11 emu/g. While Hammad et al. [42], obtained the magnetic saturation of 0.194 emu/g for $Zn_{0.98}Mn_{0.02}O$ sample fabricated by sol-gel method.

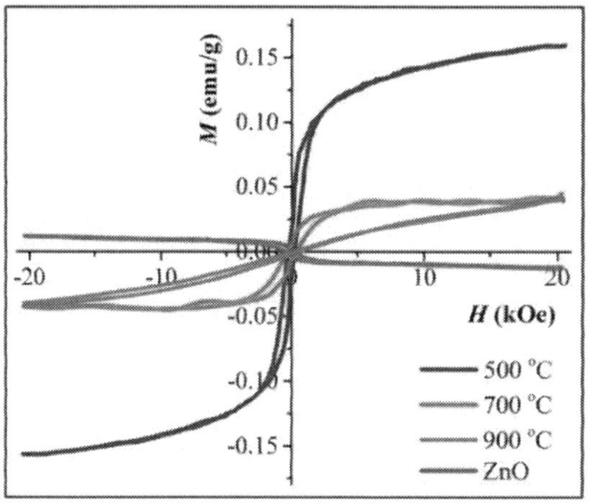

Figure 4. Magnetic hysteresis curves (M-H) of pure ZnO (red line) and $Zn_{0.98}Mn_{0.02}O$ samples sintered at the elevated temperatures of 500°C (blue line), 700°C (magenta line) and 900°C (brown lines) for 4 hours.

The ferromagnetic properties of DMS are due to the exchange interaction between the carriers (holes or electrons from the valence band) and magnetic ions derived from a transition metal [10]. The substitution process causes the exchange interaction between the Mn^{2+} ion and Zn^{2+}ion for a relatively long time, so that almost all the Mn^{2+} ions aim towards the same spin. The polarized electrons in the local spin and the conduction electrons will be easier to occur with the addition of Mn^{2+} concentration [21]. Moreover, the ferromagnetic induction of ZnO-doped metal oxides sample is also possible to occur because of the formation of some intrinsic defect, such as: zinc vacancy (V_{Zn}), zinc interstitial (Zn_i), oxygen antisite (O_{Zn}), oxygen interstitial (O_i) [11]. While, the decrease of the ferromagnetic properties may be related to the formation of $ZnMnO_3$ impurity phase. This is because that $ZnMnO_3$ has paramagnetic properties at room temperature [43,44]. The similar results were also reported by Blascoet al. [45], where the presence of $ZnMnO_3$ in the $Zn_{0.96}Mn_{0.04}O$ sample resulting in the appearance of paramagnetic properties. The presence of $ZnMnO_3$ phase in the $Zn_{1-x}Mn_xO$ (x = 6, 8, 10 %at.) samples synthesized by the sol-gel method cause the sample becomes paramagnetic [22]. Based on the previous studies, the present results confirm that the change of the ferromagnetic into paramagnetic properties is due to the presence of a $ZnMnO_3$ phase as the effect of the increase of the sintering temperature.

4. Conclusion

The $Zn_{0.98}Mn_{0.02}O$ was successfully synthesized by a solid state reaction method. The XRD data show that all samples have a predominant crystal phase of hexagonal wurtzite. At high sintering temperatures, the secondary phase of $ZnMnO_3$ was formed in the $Zn_{0.98}Mn_{0.02}O$ system. The resistivity and dielectric constant of $Zn_{0.98}Mn_{0.02}O$ increase with the increase of the sintering temperature. The $Zn_{0.98}Mn_{0.02}O$ sample synthesized at low temperature exhibits ferromagnetic properties at room temperature and close to paramagnetic properties at high sintering temperatures. The decrease of the magnetic properties of $Zn_{0.98}Mn_{0.02}O$ synthesized at high temperatures is believed due to the presence of a $ZnMnO_3$ secondary phase.

5. Acknowledgements

This work was financially supported by Research Center for Physics, Indonesian Institute of Sciences.

6. References

[1] Pearton S J, Norton D P, Heo Y W, Tien L C, Ivill M P, Li Y, Kang B S, Ren F, Kelly J and Hebard A F 2006 *Journal of Electronic Materials* **35** 863868
[2] Sharma P, Gupta A, Owens F J, Inoue A and Rao K V 2004 *Journal of Magnetism and Magnetic Materials* **282** 115-121
[3] Dietl T, Ohno H, Matsukura F, Cibert J and Ferrand D 2000 *Science* **287** 1019-1022
[4] Sato K, Katayama-Yoshida H 2001 *Jpn. J. Appl. Phys.* **40** L334–L336
[5] Chattopadhyay S, Dutta S, Banerjee A, Jana D, Bandyopadhyay S, Chattopadhyay S and Sarkar A 2009 *Physica B* **404** 15091514
[6] Chaari M, Matoussi A and Fakhfakh Z 2011 *Materials Sciences and Application* **2** 765-770
[7] Liu Y, Yang Y, Yang J, Guan Q, Liu H, Yang L, Zhang Y, Wang Y, Wei M, Liu X, Fei L, and Cheng X 2011 *Journal of Solid State Chemistry* **184** 1273-1278
[8] Saleh R, Djaja N F and Prakoso S P 2013 *Journal of Alloys and Compounds* **546** 48-56
[9] Ahmed F, Kumar S, Arshi N, Anwar M S, Koo B H and Lee C G 2012 *Microelectronic Engineering* **89** 129-132
[10] Choudhury S, Sain S, Mandal M K, Pradhan S K and Meikap A K 2016 *Materials Research Bulletin* **77** 138-146

[11] Fabbiyola S, Kennedy L J, Dakhel A A, Bououdina M, Vijaya J J and Ratnaji T 2016 *Journal of Molecular Structure* **1109** 89-96
[12] Moontragoon P, Pinitsoontorn S and Thongbai P 2013 *Microelectronic Engineering* **108** 158-162
[13] Yang J H, Zhao L Y, Zhang Y J, Wang Y X, Liu H L and Wei M B 2007 *Solid State Communications* **143** 566-569
[14] Jayakumar O D, Gopalakrishnan I K and Kulshrestha S K 2006 *Physica B* **381** 194-198
[15] Sanz R, Jensen J, Gonzalez-Diaz G, Martinez O, Vazquez M and Hernandez-Velez M 2009 *Nanoscale Res. Lett.* **4** 878-887
[16] Banerjee S, Rajendran K, Gayathri N, Sardar M, Senthilkumar S and Sengodan V 2008 *Journal of Applied Physics* **104** 043913(1-3)
[17] Ghoshal T, Kar S, Biswas S, De S K, Nambissan P M G 2009 *J. Phys. Chem. C.* **113** 3419-3425
[18] Karmakar R, Neogi SK, Midya N, Banerjee A and Bandyopadhyay S 2016 *J. Mater Sci: Mater Electron* **27** 6371-6381
[19] Bououdina M, Omri K, El-Hilo M, Amiri A E, Lemine O M, Alyamani A, Hlil E K, Lassri H and Mir L E 2014 *Physica E* **56** 107-112
[20] Milivojević D, Blanuŝa J, Spasojević V, Kusigerski V and Babić-Stojić B 2007 *Solid State Communications* **141** 641-644
[21] Cong C J, Liao L, Fan L X and Zhang K L 2005 *Nanotechnology* **16** 981-984
[22] Neogi S K, Karmakar R, Misra A K, Banerjee A, Das D and Bandyopadhyay S 2013 *Journal of Magnetism and Magnetic Materials* **346** 130-137
[23] Riyadi S, Muafif A, Nugroho A, Rusydi A and Tjia MO 2007 *J. Phys.: Condens. Matter.* **19** 476214(1-8)
[24] Owens F J 2009 *Journal of Magnetism and Magnetic Materials* **321** 3734-3737
[25] Wang X L, Lai K H and Ruotolo A 2012 *Journal of Alloys and Compounds* **542** 147-150
[26] Zhang J, Skomski R and Sellmyer D J 2005 *Journal of Applied Physics* **97** 10D303(1-3)
[27] Kolesnik S and Dabrowski B 2004 *Journal of Applied Physics* **96** 5379-5381
[28] Sebayang P, Hulu2. F, Nasruddin, Aryanto D, Kurniawan C, Subhan A, Sudiro T and Ginting M 2017 *American Institute of Physics* **1862** 030050(1-5)
[29] Omri K, Ghoul J E, Lemine O M, Bououdina M, Zhang B and Mir L E 2013 *Superlattices and Microstructures* **60** 139-147
[30] Abrishami M E, Hosseini S M and Kompany A 2011 *Journal of Applied Sciences* **11** 1411-1415
[31] Menon A S, Kalarikkal N and Thomas S 2013 *Indian Journal of NanoScience* **1** 16-24
[32] Lu Y, Lin Y, Xie T, Shi S, Fana H and Wang D 2012 *Nanoscale* **4** 6393-6400
[33] Dutta S, Chattopadhyay S, Sarkar A, Chakrabarti M, Sanyal D and Jana D 2009 *Progress in Materials Science* **54** 89-136
[34] Hao Y M, Lou S Y, Zhou S M, Yuan R J, Zhu G Y and Ning-Li 2012 *Nanoscale Res. Lett.* **7** 100
[35] Barker A, Crowther S and Rees D 1997 *Sensors and Actuators A.* **58** 229-235
[36] Han J, Senos A M R and Mantas P Q 2002 *Materials Chemistry and Physics* **75** 117-120
[37] Zhou Z, Kato K, Komaki T, Yoshino M, Yukawa H, Morinaga M and Morita K 2004 *Journal of the European Ceramic Society* **24** 139-146
[38] Mote V D, Purushotham Y and Dole B N 2016 *Materials and Design* **96** 99-105
[39] Das J, Mishra D K, Srinivasu V V, Sahu D R and Roul B K 2015 *Indian Journal of Physics* **89** 1143-1151
[40] Arshad M, Ahmed A S, Azam A and Naqvi A H 2013 *J. of Alloys and Compounds* **77** 469-474
[41] Chattopadhyay S, Neogi S K, Sarkar A, Mukadam M D, Yusuf S M, Banerjee A and Bandyopadhyay S 2011 *Journal of Magnetism and Magnetic Materials* **323** 363-368

[42] Hammad T M, Griesing S, Wotocek M, Kuhn S, Hempelmann R, Hartmann U and Salem JK 2013 *Appl. Nanosci.* **3** 153159
[43] Rall J D, Thota S, Kumar J and Seehra M S 2012 *Applied Physics Letters*. **100** 252407(1-4)
[44] Hernándes E and Sagredo V 2015 *International Journal of Physical Sciences* **10** 413-419
[45] Blasco J and Garcıa J 2006 *Journal of Solid State Chemistry* **179** 2199-2205

IOP Conf. Series: Materials Science and Engineering **309** (2018) 012120 doi:10.1088/1757-899X/309/1/012120

Isothermal approach to predict the removal efficiency of β-carotene adsorption from CPO using activated carbon produced from tea waste

S A A Harahap[1*], A Nazar[1], M Yunita[1], RA Pasaribu[1], F Panjaitan[1], F Yanuar[1], E Misran[1,2]

[1]Department of Chemical Engineering, Universitas Sumatera Utara, Padang Bulan Medan 20155, Indonesia
[2]Sustainable Energy and Biomaterial Centre of Excellent, Universitas Sumatera Utara, Jl. Almamater Kampus USU, Medan 20155, Indonesia

*E-mail: asianpudi@yahoo.com

Abstract. Adsorption of β-carotene in crude palm oil (CPO) was studied using activated carbon produced from tea waste (ACTW) an adsorbent. Isothermal studies were carried out at 60 °C with the ratio of activated carbon to CPO were 1:3, 1:4, 1:5, and 1:6, respectively. The ACTW showed excellent performance as the percentage of adsorption of β-carotene from CPO was > 99%. The best percentage removal (R) was achieved at ACTW to CPO ratio equal to 1:3, which was 99.61%. The appropriate isotherm model for this study was Freundlich isotherm model. The combination of Freundlich isotherm equation and mass balance equation showed a good agreement when validated to the experimental data. The equation subsequently executed to predict the removal efficiency under given sets of operating conditions. At a targetted R, CPO volume can be estimated for a certain initial concentration β-carotene in CPO C_0 and mass of ACTW adsorbent M used.

1. Introduction

Indonesia is the largest producer of crude palm oil (CPO). In 2012, the production of CPO in Indonesia reached 23.5 million ton [1]. Domestic CPO is used for food industry especially cooking oil industry and non food industries such as cosmetics and pharmaceuticals. CPO consists of major ingredient, namely triglyceride (94%) and minor ingredients are tocopherol, sterol, phosphatide and carotenoid [2]. CPO is raw material in cooking oil production. In Indonesia, consumers prefer yellow and clear cooking oil color than reddish one. Carotenoid forms reddish color of CPO thus in cooking oil. In order to meet consumer preference, a bleaching process on CPO is applied to remove carotenoid compounds. The bleaching process is conducted by adsorbing carotenoid compounds [3].

CPO contains carotenoid compounds in 500-700 ppm (0.5-0.7 kg per ton) [4] and is potential to be the huge source of provitamin A and antioxidants as carotenoid compounds consists of 80% β-carotene. Also, many by-products of CPO processing can be used as a base in the chemical industry while β-carotene is needed in food, pharmaceuticals and cosmetics industries [5]. Therefore, carotenoid compounds in CPO are important to be recovered. Unfortunately, normal CPO processing into cooking oil resulted in 80% broken carotenoid. This is due to bleaching processes of carotenoid takes place at the temperature > 240 °C [6]. Broken carotenoid is discarded as waste.

Various methods to get carotenoid compounds in CPO were performed such as extraction using supercritical fluid [7] and adsorption using adsorbents [6]. Adsorption is a bleaching method that is widely used as it is fast and easy. Carotenoids in CPO are bound to the adsorbent. Since there is no chemical reaction, then CPO does not change. Also, adsorption process can be performed at room temperature. Many adsorbents were used to adsorb β-carotene in CPO such as rice husk ash, silica gel, alumina, clay, oil palm shells and activated carbon [3].

Several studies have reported the manufacture and utilization of activated carbon for β-carotene adsorption. Tea waste is the potential raw material for activated carbon production. Tea waste consists of cellulose (29.42%), lignin (36.94%), ash (4.53%) and extractive (15.22%) [8]. One of the considerations for manufacturing activated carbon from tea waste is its high level of holo cellulose (60.81%). The previous study showed that the activated and carbonized tea waste had high fixed carbon content of 57.1%. The maximum capacity of the adsorbent to adsorb nitrophenol was 142.85 mg/g [9]. Also, tea wastes adsorption capacity for heavy metal i.e. Ni (II) varied in a range of 50 - 100%. This condition was affected by contact time, and the amount of adsorbent used [10].

The research on isothermal adsorption of β-carotene from CPO using bentonite and activated carbon was also reported. The study showed that the value of activation energy were 74.28 kcal/mole and 30.04 kcal/mole for bentonite and activated carbon, respectively and was able to adsorb β-carotene as high as 79.05% [11]. Furthermore, a study on β-carotene and phosphorus adsorption from CPO using modified natural clays was also reported. The adsorbents used in the study were able to adsorb β-carotene and phosphorus as high as 90% [6]. The previous studies mentioned above showed that activated carbon of tea waste had excellent capacity in heavy metals adsorption. Here, we investigated the utilization of activated carbon from tea waste to study its potential as the adsorbent of β-carotene in CPO.

2. Materials and Method

2.1 Equipment and materials

The equipments used in this study were tray dryer, ball mill, sieve of 50 mesh, pyrolysis reactor equipped with N_2, hot plate, electric motor and stirrer. Whereas the main materials used were CPO, tea waste, and 85% phosphoric acid (H_3PO_4).

2.2 Activated carbon production

Tea waste used in this study was collected from Acehnese restaurants in Medan. The activated carbon was prepared in the same procedure as published previously in [12]. Tea waste was dried using tray dryer at 40 °C. Later, tea waste was milled using a ball mill and was sieved using a 50 mesh sieve. Subsequently, it was activated with a solution of 85% H_3PO_4 for 24 hours with a weight ratio of it and H_3PO_4 was 1: 2. The carbonization process was conducted in a pyrolysis reactor at 500 °C for 15 min using N_2 stream. The activated carbon acquired was rinsed using hot water (85 °C) and was dried in the oven at 110 °C. Activated carbon is then pulverized using a mortar.

2.3 Adsorption experiments

CPO and activated carbon were prepared in ratios of 1:3, 1:4, 1:5 and 1:6 in beaker glasses. The mixture was heated using a hot plate at 60 °C and was homogenized using an electric motor at 120 rpm for 2 hours. The mixture was filtered using Whatman filter paper #1. The filtrate was put into a plastic bottle and analyzed using UV-VIS spectrophotometer to determine the content of β-carotene.

2.4 Adsorption isotherms model and mass balanced

Adsorption isotherm models were used to determine the distribution between liquid and solid phase adsorbents as a measure of equilibrium position in the adsorption process. The models were represented by Langmuir and Freundlich isotherm models [6].

Langmuir isotherm model defines that the maximum adsorption capacity is the result of a single layer (monolayer) on the surface of the adsorbent adsorbate [13]. The linear form of the Langmuir isotherm equation is shown in the Equation (1) [6].

$$\frac{C_e}{q_e} = \frac{1}{K_L q_m} + \left(\frac{1}{q_m}\right) C_e \tag{1}$$

where q_e is the amount of adsorbate per mass of adsorbent, C_e is the equilibrium concentration adsorbate, K_L and q_m are the Langmuir constants related to adsorption capacity and rate of adsorption.

Freundlich isotherm model is the model used for the non-ideal adsorption and multilayer adsorption [13]. The Relationship between amount of adsorbed substances and concentrations is shown in the Equation (2) [6].

$$\log q_e = \log K_f + \frac{1}{n} \log C_e \tag{2}$$

where q_e is the amount of adsorbate per mass of adsorbent, C_e is the equilibrium concentration adsorbate, n is the intensity of adsorption and K_f is the constant of Freundlich isotherm.

The isotherm model combined with a mass balance can be used to predict the adsorbate efficiency removal and the required mass of the adsorbent for the desired removal efficiency based on the previous modeling for adsorption study [14]. Firstly, at a given initial concentration and amount of adsorbent, the concentration at equilibrium is calculated using Equation (3) and (4) for Langmuir isotherm and Freundlich isotherm, respectively.

$$K_L C_e^2 + \left(1 + \frac{K_L q_m M}{V} - K_L C_0\right) C_e - C_0 = 0 \tag{3}$$

$$\frac{K_f M}{V} C_e^{1/n} + C_e - C_0 = 0 \tag{4}$$

where M is the mass of adsorbent and V is the volume of adsorbate.

The percent removal is determined by the following equation [13]:

$$R = \frac{C_0 - C_e}{C_0} \tag{5}$$

where R is the percentage of removal, C_e is the equilibrium concentration of adsorbate and C_0 is the initial concentration of adsorbate.

For a given initial solute concentration and target removal efficiency, the equation to predict the required mass of adsorbent can be calculated using Equation (6) and (7) for Langmuir isotherm and Freundlich isotherm, respectively.

$$M = \frac{VR}{K_L q_m (1-R)} + \frac{VR}{q_m} C_0 \tag{6}$$

$$M = \frac{V}{K_f} \frac{R}{(1-R)^{1/n}} C_0^{\left(1-\frac{1}{n}\right)} \tag{7}$$

Equation (6) and (7) can also be rearranged to get the volume of CPO that could be processed at a given mass of adsorbent.

2.5 Parameter determination and model calculation

Isotherms were fitted to experimental data for activated carbon with the Equation (1) and (2) to determine the isotherm parameters. Equation (3) and (4) were solved using numerical analysis based on the Newton-Raphson method.

3. Results and Discussion

3.1 The ability of ACTW to adsorb β-carotene

The concentration of β-carotene initially contained in CPO was 509 ppm. The concentration of β-carotene decreased significantly after contacting with 14 g of activated carbon produced from tea waste (ACTW) at 60 °C for 120 min. Table 1 showed the concentration of β-carotene in various ratios activated carbon to CPO after the adsorption process. The best percentage removal of β-carotene of 99.61% was obtained at ratio 1: 3 with the final concentration of β-carotene of 2 ppm. At the same amount of adsorbent, the increment of CPO volume resulted in a decrease of adsorption capacity although not significant.

Table 1. Removal efficiency of β-carotene using activated carbon produced from tea waste

Ratio of ACTW to CPO (w/w)	Mass of CPO (g)	Volume of CPO (mL)	Final Concentration of β-Carotene (ppm)	Percentage removal, *R* (%)
1:3	42	46.94	2	99.61
1:4	56	62.58	3	99.41
1:5	70	78.23	4	99.21
1:6	84	93.88	5	99.02

The ability of ACTW to adsorb β-carotene from CPO in this study is better compared to the ability of modified natural clay. Modified natural clay was able to adsorb β-carotene and phosphorus by 90% [6]. This study also showed the higher performance of ACTW compared to bentonite and activated charcoal to adsorb β-carotene from CPO which has 79% adsorption [11]. The ACTW showed excellent performance since the percentage removal of β-carotene from CPO were > 99% for all of the variables under this study.

3.2 Determination of adsorption isotherms model

Isothermal models fitness determination was conducted by plotting the related variables as shown in Equation (1) and (2) for Langmuir and Freundlich isotherm models, respectively. The experimental data form a linear graph with excellent correlation (Figures 1 and 2). From the linear equation obtained, can be determined the constants associated with each isothermal models (Table 2).

Table 2. The adsorption isotherm model parameter of β-carotene from CPO

Model isotherm	Parameter	
Langmuir	q_m (mg/gr)	10
	K_L (L/mg)	0.1
Freundlich	K_f (L/mg)	1.0020
	n	1.3368

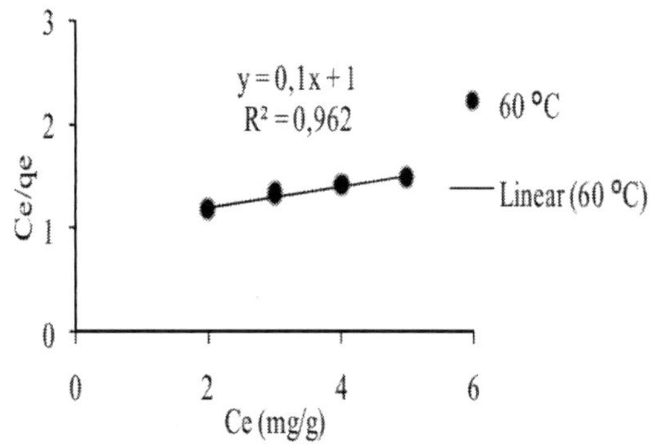

Figure 1. Langmuir isotherm curve at 60 °C

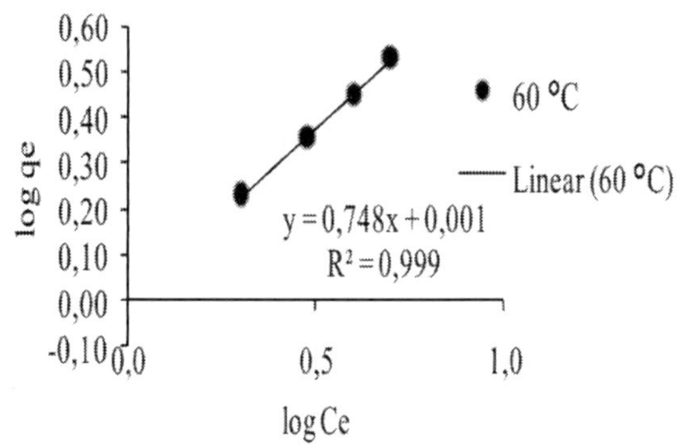

Figure 2. Freundlich isotherm curve at 60 °C

Therefore, the isotherm model corresponds to this study is the Freundlich isotherm model. The correlation coefficient in the Freundlich isotherm model is better than the Langmuir isotherm models, which is of $R^2 = 0.999$. Results of previous research on the study of the adsorption isotherm of β-carotene from crude palm olein using bentonite also showed that the appropriate model was Freundlich isotherm corresponding to the value of the correlation coefficient (R^2) of 0.9899 [11].

3.3 Validation of calculation of R and V prediction with R and V experimental

Since the appropriate isotherm model for this study was Freundlich isotherm, then Equation (4) and (5) would be executed to predict the adsorbate efficiency removal (*R*) or the percentage of adsorption. Table (3a) shows a comparison between the experimental and the predicted value of *R*. In the adsorption process, the same initial concentration of β-carotene in CPO (C_0) of 509 mg/L and the mass of ACTW adsorbent (*M*) of 14 g were used for the entire experiments. When the ratio of ACTW to CPO = 1:3, thus the mass of CPO was 42 g or the volume of CPO became 0.04694 L. The removal efficiency was predicted to be 99.60%, and the experimentally determined value was 99.61%, and thus the prediction error was 0.0038%. For other ratios of ACTW to CPO, the errors were in the range of 0.0049 to 0.0111%. The predictions for CPO volume to obtain the desired *R* value were shown in the Table (3b) based on the calculation using Equation (7). With the same initial concentration of β-carotene as shown in Table (3a) and desired *R* was 99.61%, the predicted volume of CPO was 0.0466 L, and thus the error was 0.72%.

The results suggested that the constant in the equation of the Freundlich isotherm associated in mass balance equation can be applied to predict the removal efficiency at various initial concentrations of solutes and the volume of CPO. Moreover, for given *R* and mass of adsorbent, the volume of CPO can be predicted without significant error.

Table 3. Calculation validation of *R* and *V* predictions with experimental *R* and *V*

Initial condition		Experimental *R* (%)	Predicted *R* (%)	Error (%)
C_0 (mg/L)	*M* (g)			
509	14 (in 0.04694 L)	99.61	99.60	0.0038
509	14 (in 0.06258 L)	99.41	99.42	0.0084
509	14 (in 0.07823 L)	99.21	99.22	0.0049
509	14 (in 0.09388 L)	99.02	99.01	0.0111

Table 3. Calculation validation of *R* and *V* predictions with experimental *R* and *V* (Cont.)

Initial condition		Experimental *V* (L)*	Predicted *V* (L)	Error %
C_0 (mg/L)	*R* (%)			
509	99.61	0.0469	0.0466	0.72
509	99.41	0.0626	0.0632	0.98
509	99.21	0.0782	0.0786	0.47
509	99.02	0.0939	0.0931	0.83

* mass of ACTW adsorbent was 14 g

3.4 Prediction of R value at given initial concentration and mass of adsorbent
Furthermore, the prediction of ACTW ability to adsorb β-carotene in processing more volume of CPO can be conducted. The initial concentration (C_0) of 509 ppm and mass of ACTW (*M*) of 14 g was applied to Equation (4) and (5), with the various value of CPO volume (*V*) based on the various ratios ACTW to CPO (greater than 1:6). This step aimed to determine the maximum volume of CPO that could be processed for the target value of *R* was $\geq$ 90%. From the simulation results, the mass of CPO that can be processed is 532 grams or equivalent to 0.59 L of CPO. In another word, ACTW has the ability to remove 90% of β-carotene until the ratio of ACTW to CPO is 1:37.

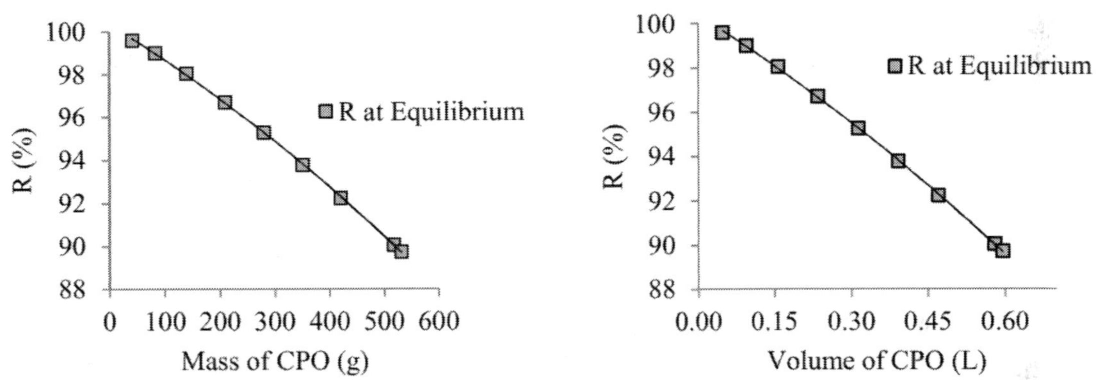

Figure 3. Correlation of *R* with mass of CPO (left) and with CPO volume (right) at initial concentration of 509 ppm and mass of ACTW of 14 g

4. Conclusions

Activated carbon produced from tea waste (ACTW) showed excellent performance to absorb β-carotene from CPO. The ability of ACTW was > 99% for various mass ratios of ACTW to CPO under study. At the same initial concentration, the greater ratios of ACTW to CPO will decrease the percentage of removal. Freundlich isotherm model is by the process of adsorption β-carotene from CPO. The combination of Freundlich isotherm equation and mass balance equation can be used to determine the removal efficiency (*R*) for a given set of operating conditions. The equation has been validated for accuracy since it can predict the *R* value well without a significant error in comparison with the experimental result. These results enabled prediction of CPO volume at greater ratio of ACTW to CPO using the combined equation. For C_0 = 509 ppm, M = 14 gr, and *R* was targetted $\geq$ 90%, the maximum ACTW to CPO ratio is 1:37. At that condition, the maximum mass of CPO that can be processed is 532 grams or equivalent to 0.59 L of CPO.

Acknowledgements
The authors gratefully acknowledge The Ministry of Research, Technology and Higher Education for supporting this research through Program Kreativitas Mahasiswa (PKM) 2017.

References

[1] Irvan, Trisakti B, Tomiuchi Y, Harahap U, Daimon H 2017 Effect of Recycle Sludge on Anaerobic Digestion of Palm Oil Mill Effluent in A Thermophilic Continuous Digester, *IOP Conf. Series: Materials Science and Engineering* **206** 012094
[2] Elmariza J, Zaharah T A, Arreneuz S 2015 Optimasi ukuran partikel, massa, dan waktu kontak karbon aktif berdasarkan efektivitas adsorpsi β-carotene pada CPO *JKK* **4** 2 21-25
[3] Serlahwaty D 2007 Kajian isolasi karotenoid dari minyak sawit kasar dengan metode adsorbsi menggunakan penjerap bahan pemucat *Skripsi* Fakultas Teknologi Pertanian IPB Bogor
[4] Nwankwere E T, Nwadiogbu J O, Yilleng M T, Eze K A 2012 Kinetic investigation of the adsorptive removal of β-carotene pigments from palm oil using unmodified natural clay *Advances in Applied Science Research* **3** 2 1122-1125
[5] Hudiyono S, Septian A 2012 Optimization carotenoids isolation of the waste crude palm oil using α-amylase, β-amylase, and cellulase *IOSR Journal of Applied Chemistry* **2** 2, 7-12
[6] Silva S M, Sampaio K A, Ceriani R, Verhé R, Stevens C, De Greyt W, Meirelles A J A 2013 Adsorption of carotenes and phosphorus from palm oil onto acid activated bleaching earth: Equilibrium, kinetics and thermodynamic *Journal of Food Engineering* **118** 4 341-349
[7] Ragaguci 2011 Sistem CO_2-etanol dalam bentuk gas-expended liquid (GXL) sebagai pelarut untuk ekstraksi senyawa xanthone dari kulit manggis *Skripsi* Program Sarjana Fakultas Teknologi Pertanian IPB
[8] Tutus A, Kazaskeroglu Y, Cicekler M 2015 Evaluation of tea wastes in usage pulp and paper production *Bio Resources* **10** 3 5407-5416
[9] Ahmaruzzaman M, Gayatri S L 2010 Activated tea waste as a potential low-cost adsorbent for the removal of p-nitrophenol from wastewater *Journal of Chemical Engineering* **55** 11 4614-4623
[10] Shah J, JanM R, ul Haq A, Zeeshan M 2015 Equilibrium, kinetic and thermodynamic studies for sorption of Ni (II) from aqueous solution using formaldehyde treated tea waste leaves *Journal of Saudi Chemical Society* **19** 3 301–310
[11] Muslich, Suryadarma P, Hayuningtyas I 2010 Kinetics of isothermal adsorption of β-carotene from crude palm olein using bentonite *J. Tek. Ind. Pert.* **19** 2 93-100
[12] Misran E, Panjaitan F, Yanuar F M 2016 Pemanfaatan karbon aktif dari ampas teh sebagai adsorben pada proses adsorpsi β-karoten yang terkandung dalam minyak kelapa sawit mentah *Jurnal Rekayasa Kimia dan Lingkungan* **11** 2 92-98
[13] Yan L G, Qin L L, Yu H Q, Li S, Shan R R, Du B 2015 Adsorption of acid dyes from aqueous solution by CTMAB modified bentonite: Kinetic and isotherm modeling *Journal of Molecular Liquids* **211** 1074-1081
[14] Chung H, Y, Kim W H, Park J, Cho J, Jeong T Y, Park P K 2015 Application of Langmuir and Freundlich isotherms to predict adsorbate removal efficiency or required amount of adsorbent *Journal of Industrial and Engineering* **28** 241–246

IOP Conf. Series: Materials Science and Engineering **309** (2018) 012121 doi:10.1088/1757-899X/309/1/012121

Establishing Value of Ceramic Solid Waste Into Light Weight Concrete

U Tarigan*, H R Prasetya and U P P Tarigan

Industrial Engineering Department, University of Sumatera Utara, Jl. Almamater USU Campus Medan 20155, Indonesia

*E-mail: ukurta.tarigan@yahoo.com

Abstract. Ceramic solid waste is a waste in the form of the ceramic or ceramic powder that has a defect and cannot be resold where the amount will continue to increase as the ceramic industry continues to produce. Handling waste so far is done by pilling it on vacant land so that if the waste continues to grow the more areas are also needed to stockpile. In addition, waste handling by boards can be a potential hazard to the surrounding environment such as chemical content in ceramics can be carried to the waters and the dust can be blown by the wind and disrupt breathing. This study aims to convert ceramics solid wastes into bricks that have more added value. Data collection is done with primary and secondary data. The method used is Taguchi experiment design to determine the optimum brick composition. The experiment consisted of 4 factors and 3 levels of ceramic with 4 kg, 5 kg and 6 kg, cement with level 3 kg, 4 kg and 5 kg, silica with level 3 kg, 4 kg and 5 kg, water level 500 ml, 750 ml, and 1000 ml. After that proceed with the financial analysis that is determining the selling price, Break Event Point (BEP, Internal Rate of Return (IRR), Pay Back Period (PBP), and Profitability Index. The results of this research are the optimum composition of the concrete blocks, 6 kg of ceramics, 5 kg of cement, 4 kg of silica sand and 1000 ml of water with the compressive strength of 125,677 kg/cm2 and signal to noise is 41,964 dB. In the financial analysis, the selling price of brick is Rp 7,751.75/unit and BEP 318,612 units of product, IRR level 43.174% and PBP for 1 year and 10 months

1. Introduction

Ceramic solid waste is a waste in the form of ceramic or the ceramic powder that has a defect and can not be resold where the number will continue to increase as the ceramic industry continues to produce. Handling of waste during this done by hoarding empty land so that if the wastes continue to grow increasingly also the area needed to hoard it [1]. In addition, confectionary waste in a hoard can potentially harm the surrounding environment such as chemical content in ceramics can be brought to the waters and the dust can be blown by the wind and disrupt breathing [2].

The current emerging technology is the management of industrial waste to be used as raw material for building materials. With the discovery of materials, innovations are expected to replace building materials and reduce production costs and reduce industrial waste[3]. One such innovation is to use ceramic wastes instead of sand aggregates in brick [4].

Brick is one of the building material with sand and aggregate forming material (the mixture of sand, gravel, and water) [5]. Brick is molded through a compacting process into blocks with certain sizes and requirements and the hardening process is placed in moist areas or not exposed to direct sunlight or rain [6].

Brick includes non-structural wall building materials. Although only a nonstructural part of the building does not mean that the brick has no standard of strength and tolerance to be met because in its use a certain quality brick can be used in a load-bearing construction [7]. There are certain restrictions as a requirement on the brick so that in its use, the brick has the resilience of various influences either directly or indirectly influences such as the provisions in Indonesian National Standard (SNI 03-0349-1989) [8]. The price of the brick in the market today is Rp 7500, much different than the price of Rp 1200, but the brick has advantages over the bricks that have stronger strength, easier installation, more waterproof and can isolate the air [9].

Based on the study that has been done, the processing of solid waste (ceramic defects) in XYZ company does not have good processing yet. In research at this company before, it is known that the factory has a production capacity of 40,000 m2 and has a target disability rate of 5% that is 2000 m2 per day and actual ceramic disability of 12.43% ie 4,972 m2 per day, from the defective product there is no solution to add The sale value of the defective product. So far, the defect product has been dumped to cause the accumulation of defective products and does not generate profits and damage the environment.

Given these conditions, research needs to be done to increase the value of solid waste into more value. Therefore, researchers have thought to treat the solid waste into brick considering the basic nature of the ceramics that can be used into the product. In the research used Taguchi experimental design method to find a good composition in order to meet the existing standards and can be marketed to the market.

The purpose of this research is to get the composition of brick from ceramic solid waste so it is feasible for sale to the market, besides also to analyze the response of test result and signal to noise brick to each given treatment, know the optimum strength and optimum signal to noise by the brick from ceramic solid waste materials, knowing the influencing factors in the press test and signal to noise brick.

1.1. Taguchi Method

Taguchi method is a most powerful and popular statistical method used for the design of experiments (DOE) can be effectively employed in optimizing the process/product by using number of steps such as planning, conducting and evaluating results of orthogonal array (OA) experiments to determine the optimum levels of control parameters under very noisy environment [10]. The prime objective goal is to maintain the variance in the results very minimal even in the presence of noise inputs to make robust design process against all variations. Generally, its focus is to optimize the quality characteristic of a process economically and for determining the optimal parameter settings of a process and thereby achieving improved process performance with reduced process variability [11]. Taguchi's method involves the use of specially constructed tables called "orthogonal array" (OA) required for very less number of experimental runs in designing which are consistent and easy to apply [10]. It is successfully used in the various fields of Engineering especially in manufacturing industries. The paper deals with implementation of Taguchi's DOE methodology and technique in respect of shrinkage and optimization of process parameters [12].

2. Methodology/Experimental

The determination of variables is determined by:

- Brainstroming
  In this research, brainstorming is done by writer, colleague writer, expert staff of brick and writer.
- Flowchart Diagram
  Flow diagram can be seen in Figure 1.

- Cause-and-effect Diagram
  A cause-and-effect diagram is used to identify potential causes. Cause and effect diagram can be seen in Figure 2.

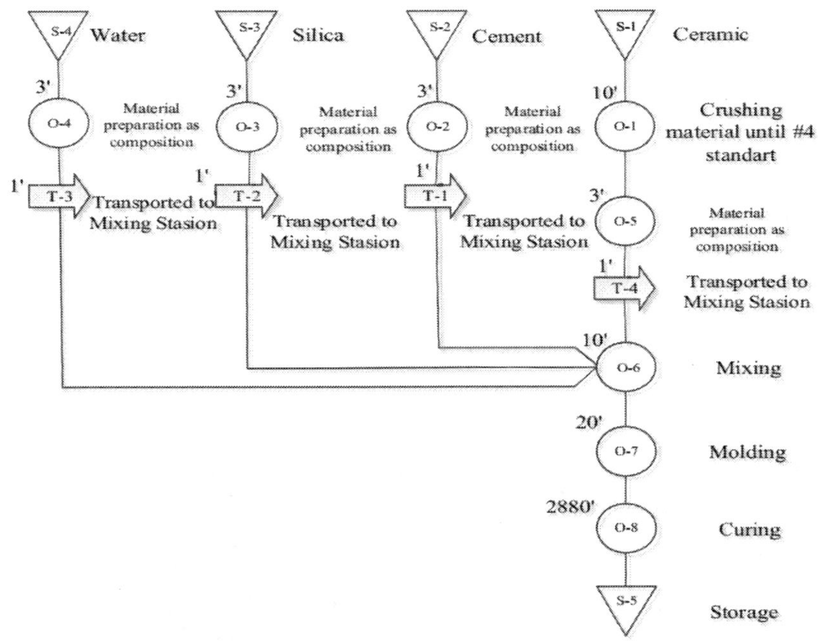

Figure 1. Flowchart Diagram

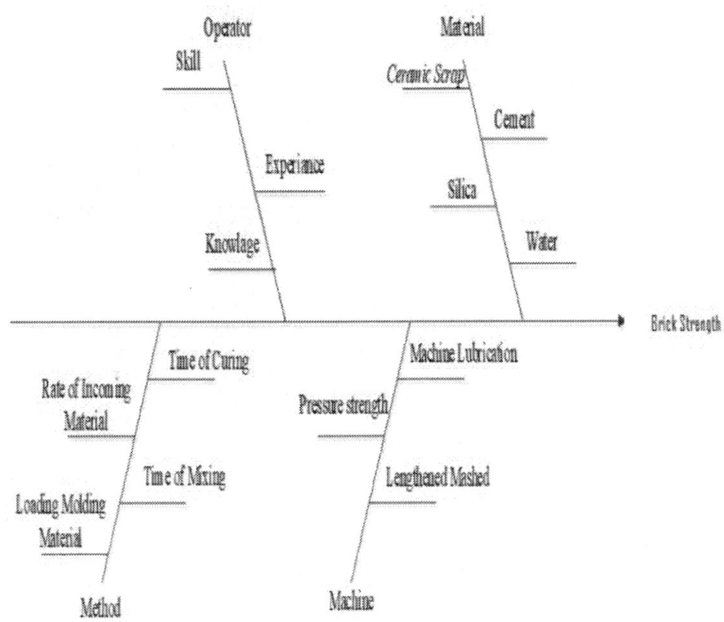

Figure 2. Cause and Effect Diagram

IOP Conf. Series: Materials Science and Engineering **309** (2018) 012121 doi:10.1088/1757-899X/309/1/012121

2.1. Pre-Experiment

In the experiments to determine the level of each factor by experimenting directly with the expert staff of the brickwork by looking at a critical point (the point where the brick is destroyed/ unreachable) from the brick and from that critical point, the range is divided into certain levels. Based on the method of identifying the problem, the researcher chose the material factors as the independent variable. These factors are as follows:

- Water composition
  In this study, there are three levels of water composition that is 500 ml, 750 ml, and 1000 ml.
- Silica sand composition
  In this research, there are three levels of silica sand composition that is 3 kg, 4 kg and 5 kg.
- The composition of cement
  In this study, there are three levels of the cement composition that is 3 kg, 4 kg, and 5 kg.
- Ceramic scarp composition
  In this study, there are three levels of ceramic scrap composition of 4 kg, 5 kg, and 6 kg.

2.2. Data collection

Material and material specifications can be seen in Table 1.

Table 1. Material Specification

No	Name	Specification	Figure
1	Ceramic Waste	Ceramic of with mesh by 4	
2	Cement	Portland Cement type I	
3	Silica Sand	Sludge level max 20% Mesh 50	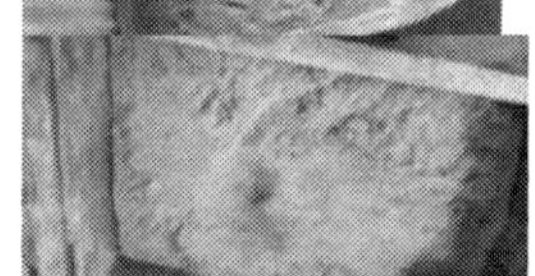
4	Water	Ground water, Ph = 7.4 Hardness $CaCO_3$ = 322 mg/ L	

Total Inventory effort can be seen in Table 2 and total fixed cost can be seen in Table 3.

Table 2. Total Investation

No.	Investation	Price (Rp)
1	Machine	195,000,000
2	Tools	34,249,000
3	Truck	221,000,000
4	Building	500,000,000
Total		950,249,000

Table 3. Total Fixed Cost

Information	Cost Estimation (Rp)
1. Maintenance	22,512,450
2. Salary	318,600,000
3. Depression	48,449,900
4. Operational	2,396,900
Total	391,959,250

Variable costs can be seen in Table 4.

Table 4. Total Variable Cost

Information	Cost Estimation (Rp)
1. Material Cost	6,327
2. Electricity	194.58
Total	6521.58
Total Cost 2520 products /day	16,434,370.08
Total Cost/Year	4,864,573,544

3. Data Processing

3.1. Taguchi Method

In this study there are 4 factors and 3 levels observed, it can be in Table 5.

Table 5. Taguchi Treatment

Treatment	Information	Level		
		1	2	3
A	Ceramic Waste	4 kg	5 kg	6 kg
B	Cement	3 kg	4 kg	5 kg
C	Silica Sand	3 kg	4 kg	5 kg
D	Water	500 ml	750 ml	1000 ml

The response of each factor and level treatment can be seen in Figure 3.

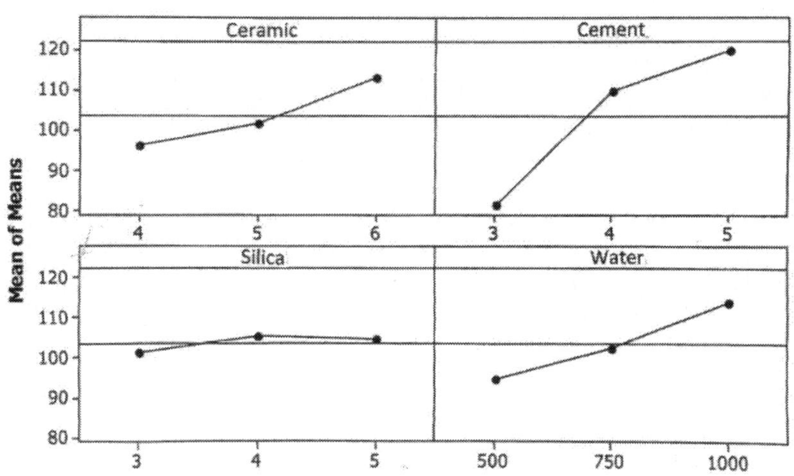

Figure 3. Treatment Response

The average equation model of compressive strength of the resulting brick is as follows:

$$\text{Mean Prediction} = Y + (A_3 - Y) + (B_3 - Y)) = 129,401 \quad (1)$$

IOP Conf. Series: Materials Science and Engineering **309** (2018) 012121 doi:10.1088/1757-899X/309/1/012121

$$\text{Cl} \quad = \pm\sqrt{\text{F}(0{,}10;(1{:}\,18)x\ Ve\ \text{x}\ \frac{1}{neff}} \qquad (2)$$
$$= \pm\ 7{,}796$$

Mean Prediction=121,694 ≤ 129,401 ≤ 137,107

The S / N response to each factor and level treatment can be seen in Figure 4.

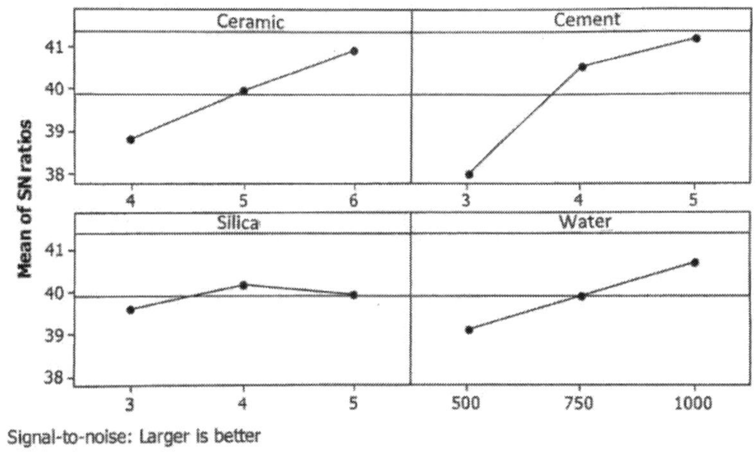

Figure 4. Treatment Response S/N

The average equation model of the signal noise of the resulting brick is as follows:

$$\text{Mean Prediction} = Y + (A_3 - Y) + (B_3 - Y) \qquad (3)$$
$$= 42.343$$

$$\text{Cl} \quad = \pm\sqrt{\text{F}(0{,}10;(1{:}\,18)x\ Ve\ \text{x}\ \frac{1}{neff}} \qquad (4)$$
$$= \pm\ 0.629$$

Mean Prediction= 41.714 ≤ 42.343 ≤ 42.973

Confirmation Experiment
The results of the confirmation experiment can be seen in Table 6.

Table 6. Confirmation Experiment Result

No	Result (kg/cm2)
1	114.1511
2	125.8763
3	128.4928
4	116.5586
5	119.9401
6	123.4148
7	136.6102
8	137.2174
9	139.3076
10	115.2077
Mean	125.6777

The average confidence interval for the confirmation experiment is as follows:

$$Cl = \pm\sqrt{F(0{,}10;1{:}18)\ x\ MSe\ x\ \left[\frac{1}{neff}+\frac{1}{r}\right]} \quad (5)$$
$$= \pm\ 12.474$$

The average Trust Interval is: $113.203 \leq 125.677 \leq 138.152$

S/N confidence interval for confirmation experiment is as follows:

$$Cl = \pm\sqrt{F(0{,}10;1{:}18)\ x\ MSe\ x\ \left[\frac{1}{neff}+\frac{1}{r}\right]} \quad (6)$$
$$= \pm\ 1.018$$

The average trust interval is:$40.944 \leq 41.963 \leq 42.982$
Comparison between experiment result and confirmation result can be seen in Table 7.

Table 7. Comparison of Experiment Results and Confirmation

		Prediction	Optimization
Taguchi Experiment	Mean	129.4011	129.4011 ± 7.706
	Variability S/N	42.343	42.343 ± 0.629
Confirmation Experiment	Mean	125.677	125.677± 12.474
	Variability S/N	41.963	41.963 ± 1.018

3.2. Financial Aspects

In the financial aspect will be calculated the feasibility of the brick business that will be made as follows:

Production per day = $40.000m^2$
Height of ceramic = 0.35 cm = 0.0035 m
Production volume/day = 40,000 x 0.0035m =140 m^3
Defect = 140 m^3x 5 % = 7 m^3
Defect volume = 0.4 m x 0.25 m x 0.0035 m = 0.00035 m^3
Scrap/day = $7m^3$ x 2257.761 kg/m^3 = 15,804.33 kg

Lightweight concrete should produce = 15804.33 kg / 6 kg = 2634.055unit

Production time without curing process = 56 minutes
Total cycle/day = 7 hour x 60 minutes / 56 minutes = 7.5 = 7 cycle

The capacity of each cyle production is 270 kg raw material, so total mold brick each cyle is :

Total mold brick each cycle	= 270 / 15	= 18 mold
Total brick each cyle	=18 mold x 5 unit/ mold	= 90 unit
Total brick 1 shift 1 machine	= 7 cycle x 90 unit	= 630 unit
Total brick 2 shift 2 machine	= 2 machine x 2 shift x 630 unit	= 2520 unit

The steps taken to determine the Break Even Point (BEP) are:

- Define product price
  If the profit define as 10 %, so the calculation be :
  Profit = TR-TC
  10%.TC = TR -TC
  TC(10% + 1) = TR
  P (Price) = Rp 7751,750

- Define Break Even Point
  From the above data. Then BEP per year can be calculated that is:

$$N_{BEP} = \frac{TFC}{P - VC} \tag{7}$$

$$= 318{,}620.59 \approx 318{,}621 \text{ unit/year}$$

The figure of Break Even Point can be seen in Figure 5.

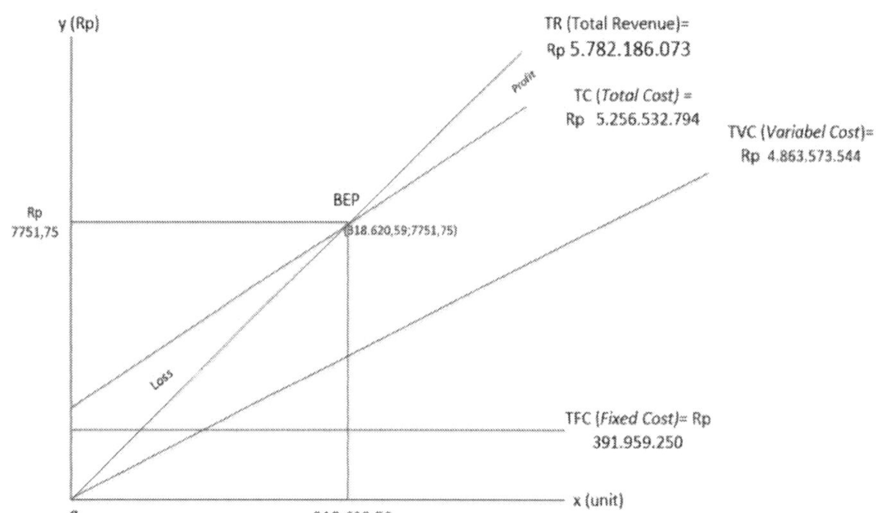

Figure 5. BEP Point

IRR calculation
For I = 40 %
ΣNPV =Rp420,869,696 (P/A.40%.10) – Rp950,249,000=Rp65,730,445
For i = 45%
Σ NPV = Rp420,869,696 (P/A.45%.10) – Rp 950,249,000= - Rp 37,803,500

IRR is the value of discount rate i which makes the NPV of the project equal to zero. From the calculation can be seen IRR is between i = 40% and i = 45%. Then with interpolation obtained:

IRR :

$$= 40\% + \frac{-37.803{,}500}{-37.803{,}500- \quad .730.445}(45\% - 40\%) = 43.1743\%$$

Payback period calculation can be seen in Table 8.

Table 8. Payback Period (PBP) Calculation

Year	Net Cash Flow	Outcash Flow	Cash
0		-950,249,000	-950,249,000
1	367,957,296		-582,291,704
2	735,914,591		153,622,887
3	1,103,871,887		1,257,494,773
4	1,471,829,182		2,729,323,956
5	1,839,786,478		4,569,110,433
6	2,207,743,773		6,776,854,207
7	2,575,701,069		9,352,555,276
8	2,943,658,364		12,296,213,640
9	3,311,615,660		15,607,829,300
10	3,724,197,956		19,332,027,256

PBP = 1,7912year or 1 year 10 month.
Profitability Index calculation for brick business plan can be seen as:

$$\textit{Profitability Index (PI)} = \frac{\text{Rp}4.208.696.956}{\text{Rp } 950.249.000} = 4{,}4290 > 1$$

so the plan is feasible.

Value added of ceramic scrap
Brick from ceramic waste = Rp 7751.75
Brick conventional = Rp 7500
Stone Price / m^3 = Rp 200,000
Weight of 1 m^3 split stopne = 180 kg
Split stone price per kg = Rp 1111.11
Stone need for brick conventional = 3 kg
%profit = 10%

Value added ceramic waste $= \frac{(7751{,}75-75 \quad)+(1111{,}11 \times 3)}{6} \times 90\% \quad = \text{Rp } 537.79$

4. Troubleshooting Analysis

The response of each treatment to the compression means with the greater the better of each level is shown in Figure 6.

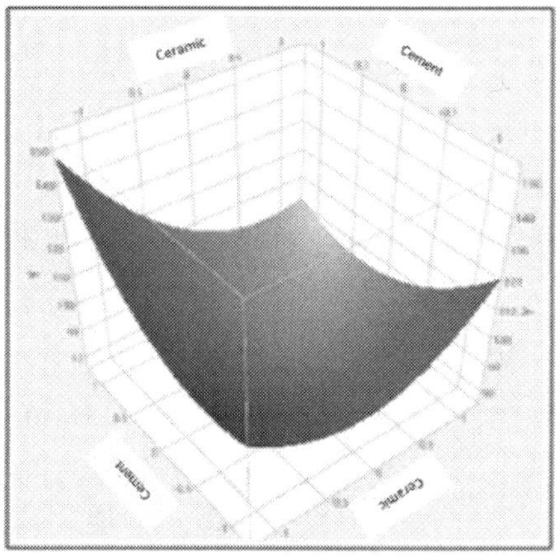

Figure 6. Surface Test Chart Press Light Weight Concrete

In the graph, the best response is ceramic at level 3 (6kg), Cement level (5kg), Silica level 2 (4) and water level 3 (1000 ml) with power 129,401. Based on this it can be concluded that 6 kg of ceramics, 5 kg of cement, 4 kg of silica and 1000 ml of water is the optimum composition to obtain the best compression test.

In Anava analysis, mean factor that has the significant difference is treatment B (cement), with F arithmetic 11,22> 2,42 F table, so known cement is a factor which has significant influence to the compressive strength of brick. Contribution to the compressive strength can be stated as 40.845%.
The response of each treatment to the Signal to Noise test press with the goal the greater the better of each level is shown Figure 7.

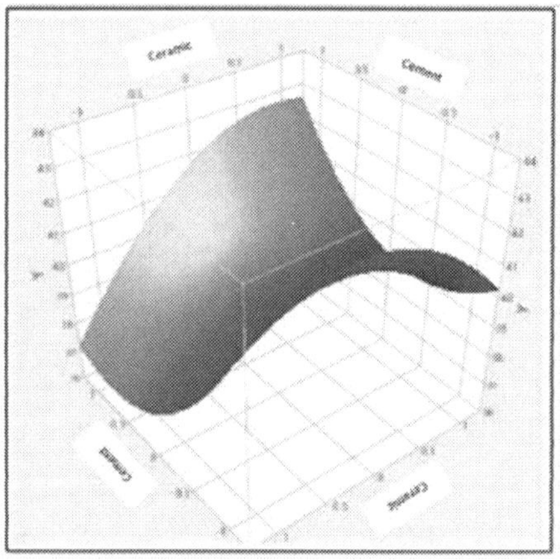

Figure 7. Surface S / N Chart of Light Weight Concrete

On the surface graph, the best response is ceramic at level 3 (6kg), cement level (5kg), silica level 2 (4) and water level 3 (1000 ml). Based on this it can be concluded that 6 kg of ceramics, 5 kg of cement, 4 kg of silica and 1000 ml of water is a solid composition in the removal of variation.

In the analysis of Anava S/N factors that have significant difference is treatment A (ceramic scrap), with F count 3.467> 2.62 with contribution of 8.427% and treatment B (cement), with F arithmetic 13.469> 2.62 F table, So that cement is known to have a significant effect on the elimination of the variation of the response. The contribution of cement to the removal of variations from the brick can be expressed by 42.590%.

IRR analysis is to compare the level of bank MARR with interest generated by doing this business. In this effort, the level of IRR is 43.174%> 6.25%, so it can be said this business is feasible to do. In this business, payback time occurs in 1 year 10 months. The rapid return of this capital occurs because of the number of products produced, on the other hand, the need for brick in the development and market is broadly given the lack of production of bricks produced by other business actors. The PI value of this brick business is 4.429> 1 so it can be said that this brick business is feasible. The addition of ceramic value after made brick is Rp 537,79/kg. The results obtained from the removal of the value of brick from gravel with ceramic scrap and the difference in selling price of each product.

5. Conclusions

Based on experiments that have been done and financial analysis that has been done in research at XYZ company, it is concluded that the optimum composition in the brick making experiment with the ceramic solid waste material is 6 kg ceramic, 5 kg cement, 4 kg silica, and 1000 ml water, the optimum compressive strength of the ceramic wastewater bricks in this research is 125,677 kg/cm2. The optimum signal to noise of ceramic wastewater bricks in this research is 41,963 dB, the most influential factor in average of compressive test and signal to noise making of brick is ceramic scrap, cement, silica sand, and water. The price of ceramic scrap bricks can be sold at Rp 7751.75 per unit, the breakeven point is 318,621 unit of products, Internal Rate Of Return (IRR) 43.174% and Pay Back Period (PBP) for 1 year 10 months.

Reference

[1] Subramani T 2015 *Experimental Investigation Of Using Ceramic Waste As A Coarse Aggregate Making A Light Weight Concrete* **4**
[2] Singla and Rakesh 2015 *Utilization Of Waste Ceramic Tiles As Coarse Aggregate In Concrete* **2**
[3] Nursyamsi 2016 *Pemanfaatan Serbuk Kaca Sebagai Bahan Tambah Dalam Pembuatan Batako* **14**
[4] Mallisa H 2011 *Studi Kelayakan Kualitas Batako Hasil Produksi Industri Kecil Di Kota Palu. Media Litbang Sulte* **2**
[5] Ristinah dkk 2012 *Pengaruh Penggunaan Bottom Ash Sebagai Pengganti Semen Pada Campuran Batako Terhadap Kuat Tekan Batako* **6**
[6] Mallisa H 2011 *Studi Kelayakan Kualitas Batako Hasil Produksi Industri Kecil Di Kota Palu.Media Litbang Sulteng* **2**
[7] Yacob I 2003 *Studi Kelayakan Bisnis* **1**
[8] Suad Husnan and Suwarsono 1991 *Studi Kelayakan Proyek*, Yogyakarta: UPP-AMP YKPN **1**
[9] Vishnu A 2015 *Application Of Taguchi Methods And ANOVA In Optimization Of Process Parameters For Metal Removal Rate In Electrochemical Machining Of Al/5%Sic Composites* **2**
[10] Moslemi A 2011 *Using Experimental Designs in Order to Analyze the Affect of Effective Factors on Rice Seeds Weight* **5**
[11] Oji J 2013 *Taguchi Optimization of Process Parameters on the Hardness and Impact Energy of Aluminium Alloy Sand Castings* **2**
[12] Pachal A 2013 *Taguchi Optimization of Process Parameters in Friction Welding of 6061 Aluminum Alloy and 304 steel: A Review* **3**

IOP Conf. Series: Materials Science and Engineering **309** (2018) 012122 doi:10.1088/1757-899X/309/1/012122

Annealing optimization in the process of making membrane PSF19%DMFEVA2 for wastewater treatment of palm oil mill effluent

A A Said* and Mustafa
Department of Chemical Engineering, Faculty of Industrial Technology, Institute of Technology Medan. Jl. Gedung Arca No. 52 Medan, 061-7363771

*E-mail: aja.avriana@yahoo.com

Abstract. A small proportion of the Palm Oil Mill Effluent (POME) treatment has used its wastewater to converted to methane gas which will then be converted again into electrical energy. However, for Palm Oil Mill whose has a value of Chemical Oxygen Demand in its wastewater is less than 60.000 mg / L this can't so that the purpose wastewater treatment only to reach the standard that can be safe to dispose into the environment. Wastewater treatment systems that are general applied by Palm Oil Mill especially in North Sumatera are aerobic and anaerobic, this method takes a relatively long time due to very dependent on microbial activity. An alternative method for wastewater treatment offered is membrane technology because the process is much more effective, the time is relatively short, and expected to give more optimal result. The optimum membrane obtained is PSF19%DMFEVA2T75 membrane,while the parameter condition of the permeate analysis produced in the treatment of POME wastewater with membrane PSF19%DMFEVA2T75 obtained at pH = 7.0; TSS = 148 mg / L; BOD = 149 mg / L; And COD = 252 mg / L. The results obtained is accordance with the standard of the quality of POME.

1. Introduction

The liquid waste of the Palm Oil Mill (POM) which contain Chemical Oxygen Demand (COD) above 60.000 mg/L currently by some Palm Oil industry has been processed into methane biogas which then will be converted into electrical energy [1]. A large number of industries especially in the Zone North Sumatera, COD content in the liquid waste has not reached 60.000 mg/L, so it is not possible to convert into electrical energy, but the waste must still be processed to get the standard quality of waste to be disposed safely into the environment [2].

Handling of wastewater carried out during this time in the POM generally by accommodating the waste in aerobic and anaerobic pond. However, this method is considered less effective because, in addition to requiring relatively long time and highly dependent on microbial activity [3]. This method also requires a large area of land for the pond. Alternative waste treatment offered as a new solution that is estimated to be much effective in waste water treatment systems especially for the POM wastewater is a purification method using membranes because the process is much effective, the time is relatively short, independent of microbial activity, broad, and the results of processing is much optimal.

Based on some previous research, membrane technology has been used to treat/purify the POM wastewater. The optimum membrane obtained for purification of the POM wastewater in the previous study was PSF19%DMFEVA2 [4] membrane type, but the result of the test parameters for the effluent treatment was considered to be optimum when compared to the standard of quality of liquid waste. It must be fulfilled yet is still less than targeted because the processing results of each test parameter is still above 100 mg/L. For that purpose, membrane optimization is needed, especially in annealing

IOP Conf. Series: Materials Science and Engineering **309** (2018) 012122 doi:10.1088/1757-899X/309/1/012122

process in making PSF19%DMFEVA2 membrane for wastewater treatment of the POME. The resulting PSF (Polysulfone) membrane is expected to purify the POM wastewater more optimally and able to get the liquid waste quality standards set at the same time by the final value of the targeted test parameters.

The objectives include optimizing the optimum annealing temperature in PSF19%DMFEVA2 membrane making the process for the wastewater POM treatment and simultaneously optimizing membrane performance of PSF19% DMFEVA2 in purification operation of the wastewater of POME.

2. Methodology

The stages of this research consist preparation of membrane polymer with phase inversion technique which used PSF as polymer and DMF (Dimethyl formamide) as solvent. After than was done the preliminary analysis of feedstock of the wastewater POME derived from aerobic ponds, tested of membrane performance with membrane module,and for the last was done permeate analysis for some parameter as pH, TSS, COD, and BOD and membrane characterization with SEM (Scanning Electron Microscopy).

The materials used in this study were PSF as a mixed polymer solution dope, DMF as the solvent, Aquadest as nonsolvent, Solution dope PSF as the casting solution, and the wastewater POME derived from Aerobic ponds effluent as feedstock sample. While the membrane operating equipment consists of a set of membrane module. The membrane operating apparatus was a set of membrane modules to testing the membrane performance as follows.

Figure 1. Membrane module

1. Mixer
2. Membrane Cell
3. Pressure gauge
4. Pump

Process variables in membrane making include temperature Annealing 75°C and without Annealing whereas the variables on the important membrane performance test operation are operating pressure with variation 5; 10; 15; 20 and 25 psia.

3. Result and Discussion

3.1 Structure of Membrane Morphology

Based on SEM analysis was gotten the pore distribution as follow on the Table 1. PSF19%DMFEVA2 membrane has the range average pore diameter wider, 0.0004-0.0011 μm (0.0007 μm) than PSF19%DMFEVA2T75, 0.00075-0.00125 μm (0.0005μm), while on the average space pore of PSF19%DMFEVA2 membrane has space pore more dense, 0.000632μm than PSF19%DMFEVA2T75, 0.00147μm.

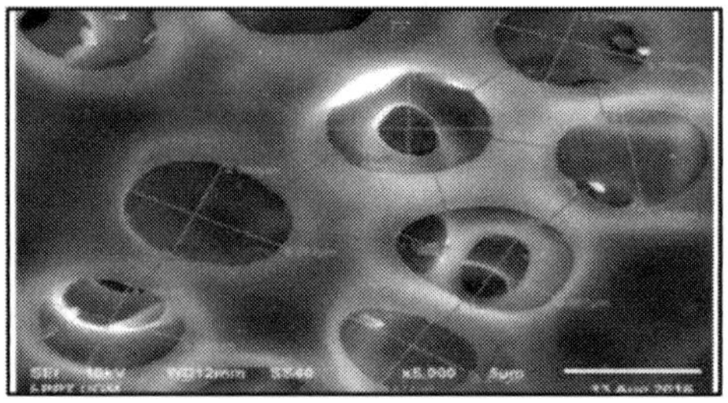

Figure 2. SEM photo of membrane PSF19% DMFEVA2 (without Annealing)

Figure 3. SEM photo of membrane PSF19%DMFEVA2T75 (with annealing)

Table 1. Analysis of SEM

No	Types of Membrane	Parameter of Analysis	
		Range of Pore Diameter (µm)	Average Space Pore (µm)
1	PSFDMF19%EVA2	0.0004 – 0.0011	0.000632
2	PSFDMF19%EVA2T75	0.00075 – 0.00125	0.00147

3.2 Tested for Membrane Performance

Based on analized for the membrane performance was gotten the permeability as follow on the Table 2. PSF19%DMFEVA2 membrane has the permeability lower, 77.19 x cm^3/cm^2.det.psia 10^{-5} with the maximum flux was 471.80 cm^3/cm^2.det on 20 psia than PSF19%DMFEVA2T75, 144.3 x 10^{-5} cm^3/cm^2.det.psia with the maximum flux was 522.35 cm^3/cm^2.det x 10^{-5}

Table 2. Flux and Permeabilty Coefficient Membranes for Permeate

Types of Membrane	Permeability cm^3/cm^2.det.psia). 10^{-5}	Flux (cm^3/cm^2.det) x 10^{-5}				
		5	10	15	20	25
PSF19%DMFEVA2	77.19	274.83	274.83	353.86	471.80	425.67
PSFDMF19%EVA2T75	144.3	324.36	350.43	434.71	471.81	522.35

Table 3. Average Rejection of permeate for each membrane

Types of Membrane	**Average Rejection**				
	5 psia	10 psia	15 psia	20 psia	25 psia
PSF19%DMFEVA2	968.590	968.383	968.122	967.607	967.165
PSF19%DMFEVA2T75	975.160	974.616	974.567	974.205	974.193

Table 4. pH, TSS, BOD, and COD of The Wastewater POME Feedstock

Types of Membrane	Wastewater POME Feedstock			
	pH	TSS	BOD	COD
PSF19%DMFEVA2 PSF19%DMFEVA2T75	3.6	13.67	4,214	8,427

Table 5. pH, TSS, BOD, and COD of Permeate

Types of Membrane	Permeate			
	pH	TSS	BOD	COD
PSF19%DMFEVA2	7.7	177	172	360
PSF19%DMFEVA2T75	7	148	149	252

Table 6. Limitation of Wastewater POME before Flue to River

Types of Membrane	Permeate			
	pH	TSS	BOD	COD
PSF19%DMFEVA2 PSF19%DMFEVA2T75	6-9	Max.250	Max.250	Max.500

3.3 Effect of The Annealing Temperature on Morphological Structure

Annealing temperatures have important role in the formation of membrane morphological structures. This can be seen in Figures 2 and 3 which were the result of the SEM analysis. If both membranes are compared the PSF19%DMFEVA2 membrane (without annealing) in Fig. 2 has a less stable pore diameter range and more varies between 0.0004 - 0.0011 μm of membrane PSF19%DMFEVA2T75 (with annealing) in Fig. 3 (see Table 1) is between 0.00075 - 0.00125 μm. It was caused the annealing process will be affected the pore stability of the obtained membrane became more stable and more uniform the pore size of the membrane was formed than membrane without Annealing like PSF19%DMFEVA2 [5]. The annealing process on the casted membrane will minimize the swelling at the edges of the membrane pores, due to frequent use as a filter medium in the separation operation [6].

3.4 Effects of The Operating Pressure on Flux and Permeability

In Table 2 above can be seen the value of permeability on each membrane by using pure water feed. The permeability value will be directly proportional to the flux value and the increase in operating pressure will usually also directly affect the increase in flux value. It can be explain that pressure is the driving force for the nanofiltration process (NF) so that with the increase in operating pressure causes an increase in total flux [7]. The high total flux was due to increased operating pressure and flow rate and caused the solute to carry as well.

3.5 Effects of The Operation Pressure of Rejection on the Membranes

The flux is directly proportional to the operating pressure and the rejection is inversely proportional to the operating pressure on each membrane. Rejection ratio values on both membrane can be seen in Table 3. The PSF19% DMFEVA2T75 membrane has a higher rejection value with rejection optimum at a PSF membrane is 97.42% when compared with not annealed membrane is PSF19% DMFEVA2 membrane. This is because the pore of the transgene membrane is much more stable and the degree of swelling on the pores of the membrane is much smaller so has a much less soluble solubility rate than not annealed membrane, for which the membrane considered more effective for purification of the wastewater POME is a PSF19% DMFEVA2T75 membrane, where the optimum state is obtained at 20 psia with permeability of 144 x 10-5 cm3 /cm2.det.Psia, the flux value is 472 x 10-5 cm3 / cm2.det and the rejection value is 97,42%. While the parameter condition of the analysis of the waste processing results obtained pH = 7.0; TSS = 148 mg/L; BOD = 1149 mg/L; And COD = 252 mg/L.

4. Conclusions

1. The most optimization membrane obtained is a PSF19%DMFEVA2T75 membrane having a pore diameter range 0.00075 - 0.00125 μm and an average pore spacing of 0.00147 μm.
2. Optimization operating conditions obtained from the treatment of POM liquid waste with membrane PSF19% DMFEVA2T75 at P = 20 psia; Value of flux (Jv) = 472 x10-5 cm3/cm2.det; Permeability (Lp) = 144 x10-5 cm3/cm2.det.psia; and rejection (R) = 97.42%.
3. The permeate condition produced in the treatment of POM liquid waste with membrane PSF19%DMF EVA2T75 as follows: P = 20 psia; PH = 7.0; TSS = 148 mg/L; BOD = 149 mg/L; and COD = 252 mg/L
4. The processing of liquid waste POM by using Polysulfone membrane on this research was quite effective, because the result was accordance with the standard quality set.

References

[1] Irvan, Trisakti B, Husaini T, Sitio A, and Sitorus T B 2017 Performance Evaluation on Otto Engine Generator Using Gasoline and Biogas from Palm Oil Mill Effluent, *IOP Conf. Series: Materials Science and Engineering* **206** 012028

[2] Irvan, Trisakti B, Sosanty F, and Tomiuchi Y 2016 Effect of discontinuing sodium bicarbonate on fermentation process of palm oil mill effluent, *Asian J. Chem.* **28** 377-380

[3] Irvan, Trisakti B, Tomiuchi Y, Harahap U and Daimon H 2017 Effect of Recycle Sludge on Anaerobic Digestion of Palm Oil Mill Effluent in A Thermophilic Continuous Digester, *IOP Conf. Series: Materials Science and Engineering* **206** 012094

[4] Said A A and Mustafa 2016 Optimization The Annealing Time Of Membrane PSFDMF19% for Palm Oil Mill Effluent Treatment, *Journal of Engineering Research and Education* **8** University Malaysia Perlis, Malaysia.

[5] Dahnia Y 2006 *Optimasi Waktu Annealing Terhadap Kinerja Membrane CA Pada Proses Pemurnian Senyawa Organik dengan Mikrofiltrasi* Hibah Penelitian Mahasiswa PHK A1 JTK-ITM, Medan

[6] Mulder 1996 *Basic Principles of Membranes Technology*, Kluwer Academic Publisher, London

[7] Rautenbach 1989 *Membrane Processes*, Jhon Willey & Sons, Singapore

IOP Conf. Series: Materials Science and Engineering **309** (2018) 012123 doi:10.1088/1757-899X/309/1/012123

Effect of solvent volume ratio and time extraction of glycerol purification

M S Sinaga*, G Rico, A N Nababan and T A Manullang
[1]Department of Chemical Engineering, Faculty of Engineering, Universitas Sumatera Utara, Padang Bulan, Medan 20155, Indonesia.

*Email: loisdanjoel@yahoo.com

Abstract. Glycerol as a byproduct of biodiesel production about 10% of the biodiesel weight. Impurities which contained in the glycerol such as catalyst, soap, methanol, water, salt, and matter organic nonglycerol (MONG) on have a significant effect on the glycerol concentration. So, it is necessary to treat the impurities. The purpose of this study is to know the effect of ethylene glycol to glycerol purification process with acidification method using phosphoric acid aspretreatment process. This research was begun with an acid addition to the glycerol to neutralize the base content and to split the soap content into free fatty acid and salt, which easier separated from glycerol. Then the process was continued with extraction by the solvent ethylene glycol using the variable of test volume ratio (v/v) (1:0,5, 1:1, 1:1,5) and the extraction time (20, 40, and 60 minutes). The results showed that the more volume of solvent used, gave less extraction time to produce high purity of glycerol. The highest purity produced in this study amounted to 90.646% is obtained at the ratio of the volume solvent (v/v) 1:1 with extraction time 60 minutes.

1. Introduction

Biodiesel has emerged as an alternative fuel to replace fossil fuels is continually dwindling. Biodiesel is non-toxic, renewable and environmentally friendly. Biodiesel also has low emissions renewable and also an excellent lubricant [1]. Therefore, biodiesel has become one of the most common biofuels in the world. Biodiesel is the most widely produced by transesterification of vegetable oils or animal fats (triglycerides) with methanol using alkaline or acidic catalysts [2]. During the process of transesterification in biodiesel plant produces crude glycerol as a by-product accounting for about 10% by weight of biodiesel products [3].Crude glycerol has a purity of 15-80% and contains amounts of contaminants such as water, methanol, soap, free fatty acid (FFA), salt, and reactant [4]. Average - Average crude glycerol composition oscillates between 40-70% glycerin, 10% water, 4% salt, 0.5% less than about 0.5% methanol and free fatty acids. Many applications for crude glycerol are used in the pharmaceutical, food, industrial and cosmetic products [5].Production of crude glycerol from biodiesel conversion increases every year. From 2008 – 2011 theconversion is 2.06 to 2.88 million tonnes [6]. It is estimated that in 2020 global production of glycerol would reach 41.9 million liters [7]. Current market value is $ .0850 to .0975 per pound of pure glycerol [8].However, glycerol is produced as a product of biodieseltransesterification process is not pure enough to be used directly in the form of food, medicine, and cosmetics. To solve this problem, thereare solution to remove impurities. Also, manufacturing and pharmaceutical and food industries are increasingly demanding high quality and superior physical properties and low contamination and odorless [9].Crude glycerol produced from various raw materials have differentcompositions, the process of purification also is different [10]. In general, the methods used to improve the purity glycerol such as by extraction of ions, the combined chemical and physical processes, chemical processes and vacuum distillation, bipolar electrodialysis chemical and physical processes [11]. According Singhabandhu and Tetsuo purification process begin with chemical pre-treatments that neutralization with acid, to remove water and alcohol in this process to produce 80-88% pure glycerin that is ready to be sold as crude glycerin. In a further process, refined glycerin is produced back to the purity to 99% or higher. Glycerin with a purity of 99% can be sel to cosmetic and pharmaceutical markets.

2. Method

2.1 Raw Material

The materials used in this study:Crude glycerol byproduct of biodiesel production with a purity of 74.7161% of PT. Wilmar, ethylene glycol ($C_2H_6O_2$) from PT. Rudang Jaya, sodium hydroxide (NaOH) 12.5 M, Hydrochloric Acid (HCl) acid, Aquadest (H_2O), Activated Carbon, blue bromthymol, H_2SO_4 0.2 N NaOH 0.05 N, $NaIO_4$ obtained from the Laboratory of Chemical process Industry Department of Chemical Engineering Faculty of Engineering, University of North Sumatra. The equipment used in this study is Erlenmeyer, magnetic stirrer, hot plate, separating funnel, glass beakers, measuring cups, digital balance, rod stirrer, thermometer, funnel glass, pipette, stopwatch, pH meter, Whatman no.41. It was obtained from the Laboratory of Chemical Process Industry Department of Chemical Engineering Faculty of Engineering, University of North Sumatra.

Table.1 The Structure of variable research

Run	StirringSpeed (Rpm)	Temperature (°C)	Time (minutes)	Etylene Glicol:Gliseryn (v/v)	Mol H_3PO_4 : Gliceryn (n/n)
1			20	0.5:1.0	0.1:1.0
2	200	50	40	1.0:1.0	0.3:1.0
3			60	1.5:2.0	0.5:1.0

2.2 Pretreatment Raw Materials

As a pretreatment step, put glycerol weighing 30 grams into Erlenmeyer, then added phosphoric acid with the ratio determined. The mixture was heated up to 70°C and stirring with a speed of 250 rpm for 1 hour, then fed into a separator funnel and allowed to stand to form three layers. A layer of glycerol taken and neutralization with NaOH, and after that glycerol is evaporated to remove the water contents.

2.3 Extraction with Ethylene Glycol

Entered ethylene glycol is glycerol with the volume ratio (v / v). The mixture was heated on a hot plate until the temperature reaches 50 ° C with stirring 200 rpm during a specified time interval, then allowed to stand to separate into two phases. Glycerol is evaporated at a temperature of 95 °C.

2.4 Characterization

Analysis of the purified glycerol produced include the analysis of the composition of raw materials and products using GC 2010, moisture content, ash content, levels MONG (Matter Organic Non-Glycerol) with Test Method Standard Method ISO 2098-1972, pH by using a pH meter, and glycerol levels SNI Titration method.

3. Results and Discussion

3.1 Characterization of Gas Chromatography

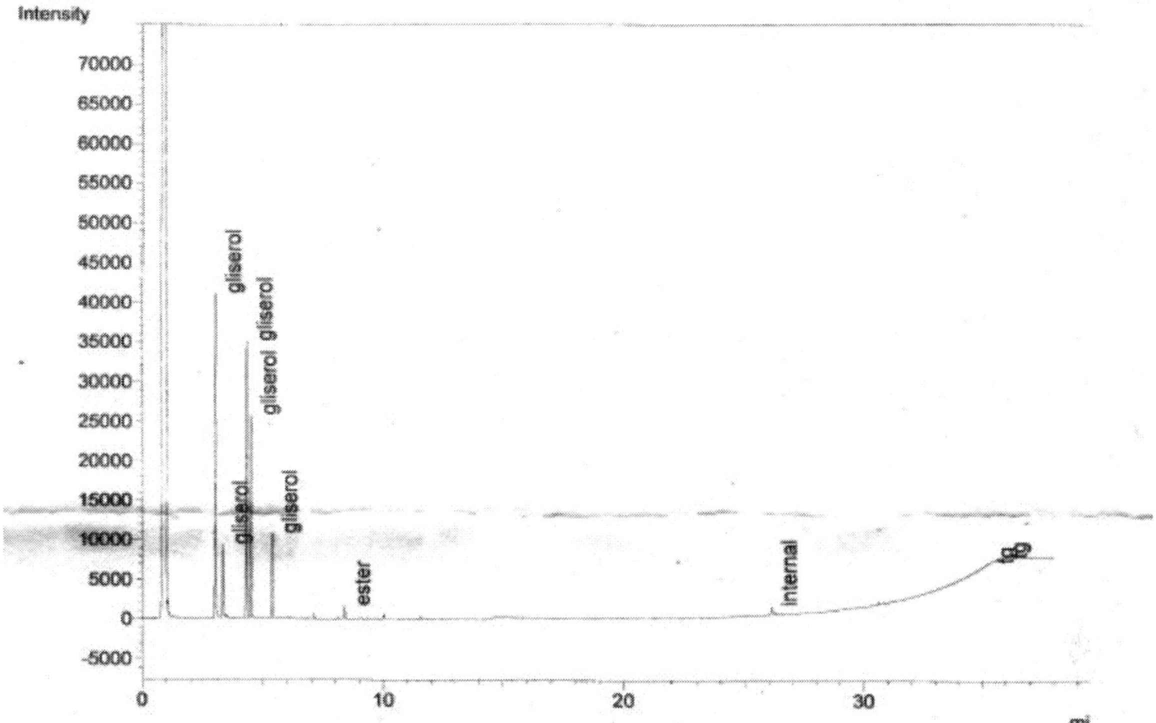

Hasil Analisis - Channel 1

Peak#	Ret.Time	Height	Area	Area%	Cmpd Name
1	3.051	40726	104011	34.6574	gliserol
2	3.341	9218	20659	6.8839	gliserol
3	4.319	34753	72434	24.1355	gliserol
4	4.491	25352	53394	17.7912	gliserol
5	5.373	10622	21542	7.1780	gliserol
6	8.378	1590	3446	1.1483	ester
7	26.098	1110	4795	1.5976	internal
8	35.257	363	9212	3.0696	tg
9	35.803	197	10619	3.5384	tg
Total		123930	300113	100.0000	

Figure 1. The Analysis of Glycerol Using GC 2010

3.2 Acidification Phosphoric Acid

Acidification performed as a pretreatment step before extraction of crude glycerol. In this study, the acid used is phosphoric acid. When acid is add to glycerol, is formed of three layers, wherein the top layer is a layer that contains free fatty acids, the next layer is a layer rich in glycerol, and the bottom layer is the inorganic salts precipitated [4].

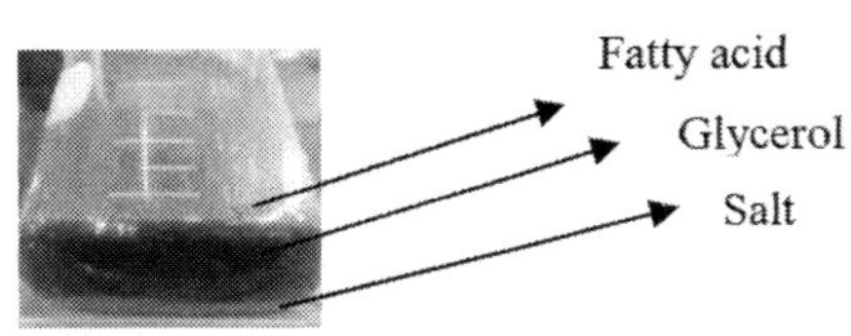

Figure 2. Glycerol layer is formed after the acidification process

The experiment begins with the addition of acid to glycerol to neutralize the alkaline catalyst used and broke the soap formed, into free fatty acids and salts are more easily separated from the glycerol. Decomposition reaction of thesoap with the aid of hydrochloric acid into fatty acids can be seen as follows:

$$RCOOK + H_3PO_4 \rightarrow RCOOH + KH_2PO_4 \qquad [1]$$

In addition to the acid decomposition, occurs neutralizing the base catalyst derived from the reaction of biodiesel contained in crude glycerol. The catalyst reacts with hydrochloric acid to form chloride salts, can be see as follows:

$$NaOH + H_3PO_4 \rightarrow Na_3(PO_4)_2 + H_2O \qquad [2]$$

After acidification, further filtration to filter out the salt buildup. pH glycerol acidification is currently under acidic conditions, namely 2. It is necessary for neutralization with NaOH 12.5 M. After the neutralization, heating is done to remove the water formed from the neutralization reaction, and also to accelerate the formation of salt deposits.

3.3Effect of Solvent Volume Ratio (v / v) Extraction And Time Of Purity Glycerol

Relationships amount of solvent volume ratio (v / v) the length of time the extraction of the purity of the resulting glycerol can be see in the graph below. The extraction is finish at 70 ° C and stirring conditions of 200 rpm with variations in time and also the ratio of solvent volume (v / v) were use.

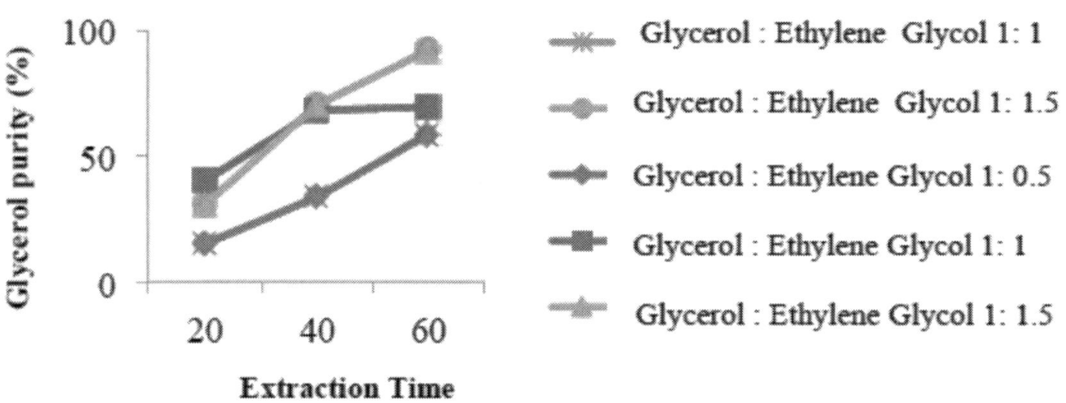

Figure 3. Effect of Solvent Volume Ratio (v / v) Extraction and Time of Glycerol Purification

From Figure 3 can be see that, volume ratio of 1: 0.5, the glycerol concentration when the extraction time 20 minutes is 15.3214%, then increased at 40 minutes to 33.7381% and increased again at 60 minutes to 58.6044 %. For solvent volume ratio of 1: 1, the glycerol concentration when the extraction time 20 minutes is 40.2715% and then increased at 40 minutes to 68.1732% and at 60 minutes increased very little to 69.6040%. For solvent volume ratio of 1: 1.5, the glycerol concentration when the extraction time 20 minutes is 30.7321% and then increased at 40 minutes into 70.2515% and increased again on 60 minutes to 90.646%.To increase levels of glycerol, it is necessary to solvent extraction after acidification stage. From the chart above, we can conclude that the use of ethylene glycol as a solvent in the extraction of glycerol, with a low solvent ratio, it takes a longer time to get the best levels of glycerol. And for ratio higher, it takes only a short extraction. However, for the ratio of solvent volume (v / v) 1: 1.5, the extraction solvent glycerol with ethylene is effective.

On the use of a polar solvent, i.e., methanol, found that the best conditions for the extraction solvent is to volume ratio (v / v) 1: 2 with the same extraction time [13]. Hunsom et al. (2013) in his study found that using the solvent hexane and diethyl ether, increased levels of glycerol with increasing the ratio of solvent, even though only a small increase. The best ratio obtained difference due to treatment conditions, and the type of solvent that is different from that used by researchers and ratio of solvents used are very influential in glycerol purification to increase levels of glycerol.

3.4 Characteristics of Pure Glycerol

From table 2 below can be see a comparison between the characteristics of the crude glycerol and glycerol is refine, as well as glycerol standard set by British Standard 2621: 1979.

Table 2. Physical properties of Crude Glycerol, Glycerol Purification results, and Glycerol Standard

Physical Properties	***Crude*Glycerol**	**Glycerol Standard BS 2621 : 1979 [3]**	**GlycerolPurificationResults**
Color	tawny	limpid	limpid
Glycerol content	74.7161 %	>80%	90.646 %
Density	1.2 gr/cm^3	1.2671gr/ cm^3	1.2710gr/ cm^3
Ash content	12%	<10%	8%
Water content	1.98%	<10%	0.2183 %
MONG	11.3099 %	<2.5%	1.1357 %

It can be shown from the table above, that glycerol purification results in this study have met the standards set for commercial glycerol. Then the method has been effective enough to purify crude glycerol rest of the manufacture of biodiesel, where the crude glycerol is a product with a purity of 74.7161% increase to 90.646%.

4. Conclusion

The highest purity results generated in this study is 90.646% were obtained at a ratio of solvent volume (v / v) 1: 1.5 with the old extraction 60 minutes. The results of the analysis that has been purified either glycerol density, moisture content, ash content, thecontent of glycerol and also MONG (Matter Organic Non-Glycerol) meets the standards of commercial glycerol by British Standard 2621: 1979.

References

[1] Omar, Wand Amin N.Optimization of heterogeneous biodiesel production from waste cooking palm oil via response surface methodology.*J. Chem. Chem. Eng.* **8** (2011) 736-743

[2] Lin L, Zhou C, VittayapadungS, XiangqianSandMingdongD2013. Opportunities and challenges for biodiesel fuel. School of Food And Biological Engineering Jiangsu University Zhenjiang

[3] Kosmider, A, Katarzyna L, andKatarzyna C. Poznan University of Life Science. Poland

[4] Nanda M.R, Yuan Z, Qin W, Poirier MA andChunbao X 2014. Purification of crude glycerol using acidification: effects of acid types and product characterization. Department of Chemical and Biochemical Engineering. Institute for Chemicals and Fuels from Alternative Resources (ICFAR). Western University. London.

[5] Quispe Cesar A.G, Christian J.R, Coronado and Joao A. CarvalhoJr 2013. Glycerol: production, consumption, prices, characterization and new trends in combustion. São Paulo State University, Guaratinguetá Campus, Department of Energy Av. Ariberto Pereira da Cunha.

[6] Oleoline 2012 Glycerin Market Report. Oleoline.com

[7] Nanda M.R, Yuan Z, Qin W, Poirier M.A andChunbao X2014*Purification of Crude Glycerol using Acidification: Effects of Acid Types and Product Characterization*. Department of Chemical and Biochemical Engineering. Institute for Chemicals and Fuels from Alternative Resources (ICFAR). Western University. London

[8] OPIS Ethanol & Biodiesel Information service 2015. *Ethanol Market Overview*. **12**(4)

[9] Ishak N.R.W, RamliZ.A.C, IsmailM, Jamaliah M.J and YarmoM.A2014Recovery and purification of crude glycerol from vegetable oil transesterification: a review.*Separation and Purification Reviews*

[10] Ardis, M.S, ArouaM.K andHashimN.A. Progress, prospect, and challenges in glycerol purification process: a review.*Renewable and Sustainable Energy Reviews***42** (2015) 1164-1173

[11] Kongjao S, Damronglerd S andHunsom M. Purification of crude glycerol derived from waste used-oil methyl ester plant.*Korean JChemEng* 2010; **27**: 944–949
[12] Singhabhandhu A and Tezuka T 2010. Aperspective on incorporation of glycerin purification process in biodiesel plants using waste cooking oil as feedstock.*Energy*, **35**: 2493-2504
[13] Hunsom M, Payia S, Penpisuth C andWinataKositnan. Comparison and combination of solvent extraction and adsorption for crude glycerol enrichment. *International Journal Of Renewable Energy Research* (2013) **Vol.3**, No.2
[14] Sadhukhan S andUjjaini S. Production of purified glycerol using sequential desalination and extraction of crude glycerol obtained during trans-esterification of Crotalaria Juncea oil. *Energy Conversion and Management***118** (2016) 450-458

Developing strategic planning of green supply chain in refinery CPO company

J Hidayati[1*], G Mumtaz[1] and S Hasibuan[2]
[1]Industrial Engineering Department, Engineering Faculty, Universitas Sumatera Utara
Jl. Almamater Kampus USU, Medan 20155
[3]Industrial Engineering Department, Marcubuana University
Jl. Meruya Selatan No. 1, Kembangan, Jakarta

*E-mail: rivaijuliza@gmail.com

Abstract. We are conducted a research at the company of the manufacturing CPO into cooking oil, margarine and materials of oleochemical industries. Today palm oil based industries are facing global challenges related to environmental issues. To against these challenges, it is necessary to have an environmentally friendly supply chain. However, the limited resource owned by the company requires the integrated environmental strategy with the company's business strategy. The model is developed based on management orientation towards external pressure, internal key resources and competitive advantage that can be obtained as the decision factor. The decision-making method used is Analytical Network Process (ANP). The results obtained institutional pressure becomes the criterion with the greatest influence on green supply chain initiatives and sub criteria of customer desires and stakeholder integration having the most significant influence on green supply chain initiatives. There are five green alternative initiatives that can be done: green product design, greening upstream, greening production, greening downstream and greening post use. For green supply chain initiative, greening upstream is the best priority.

1. Introduction

Green Supply Chain Management (GSCM) is a concept of the traditional supply chain management with the environmental aspect integrated, include product design, material procurement, manufacturing activity, packaging activity, distribution activity and the post use management. The GSCM objective is for eliminating or minimizes waste (energy, gas emission, dangerous chemical, and another waste) along the supply chain network.

This company is a one of palm oil refinery industry who manufacturing Crude Palm Oil (CPO). The company has two core processes, the refinery and the fractionation. In refined process, palm oil will be processed into RBDPO and PFAD as the side product of RBDPO, after that RBDPO will go to fractionation process into RBD-OLEIN and RBD-STREARIN as end final product. Supply chain activities in this company not yet fully focused with the environmental based supply chain.

The research about green supply chain has been doing in some country. S. Maryam Masoumik (2015) in research journal "A Strategic Approach to Develop Green Supply Chains." Those study explain the external pressure from regulator, consumer, supplier, competitor, social group and media more push organization to include eco practice into their business process. In this case, supply chain manager found many list of green initiative which can be take in supply chain process. However, the problem is limited resources are not possible do the initiative effectively. Therefore, the manager must choose an initiative strategic supply chain to get more competitive.

In another research by Hendra Saputra (2012) in research journal "Planning the measurement model of performance green supply chain pulp and paper." In that research explain the system of performance measurement of the green supply chain is necessary for managing the environment risk, improving the competitiveness with green supply chain practice and to get right strategic for the company. Mensuration performance system of the green supply chain in PT RAPP only focuses on midstream and downstream operation related manufacturing process and supplier. The objective of that research is to design conceptual model and computer program for measurement green supply chain pulp production when information of all supply chain can be controlled as well. Supply Chain Operations Reference (SCOR) matrix applied to identify key performance index (KPI). The weight of KPI defines with AHP method. Computer application designed for implemented conceptual model with basic data combine, mathematic formulates, and user interface. The result has two level and fifteen objectives and twenty-seven KPI.

With the outgrowth today and the challenges for industries to applied green supply chain, so to make research about green supply chain strategic in refinery CPO Company is necessary.

2. Research Methods

We conducted a research at Refinery Palm Oil Company in Medan. The study was conducted in July 2016 - May 2017.

2.1. Research Variables

The variables observed in this study are:

2.1.1. Independent Variables

The independent variable is a variable whose value influences the dependent variable either positively or negatively (Sinulingga, 2011). Based on researched by S. Maryam Masoumik (2015) in research journal "Strategy Approach to Develop Green Supply Chains" hence can compile independent variable for this study, that is as follows:

a. Institutional pressures

Pressures and issues faced such as regulatory rules, changes in consumer desires, community demands, and pressure from competitors.

b. Key resources

Important aspects such as, how to make sustainable improvement and integrate stakeholders become concern areas.

c. Competitive values

To be able to survive in competition, competitive values should be the company's concern.

d. Green supply chain initiative

Alternative practical initiatives were undertaken to improve the company's green supply chain.

2.1.2. Dependent Variables

The dependent variable is the dependent variable whose value is influenced by other variables. The dependent variable in this research is Green supply Chain strategies (eco-friendly supply chain strategy).

2.2. Data Processing Method

Method ANP used to processing Data. ANP calculated the weight of each variable for each green supply chain by using network linkage result from ANP method.

The steps in the ANP method:

a. Structure of green supply chain criteria and sub criteria

b. Green network supply chain design

c. Validation of network green supply chain
d. Pairwise comparison questionnaires
e. The calculation of the weighted average criteria
f. Calculation of partial weight and matrix consistency
g. Preparation of supermatrix with superdecitions
h. Calculate the unweighted supermatrix, weighted supermatrix and limit matrix with superdecitions software.
i. Priority criteria and sub criteria of the green supply chain.

3. Result and Discussion

3.1 Determinate Criteria and Sub criteria of Green Supply Chain

The used criteria are based on references to previous research. There are three performance appraisal criteria used: Pressures and issues faced, important sources, and competitive values. Determination of sub criteria also used previous research, while criterion and sub criteria used can be seen in Table 1.

Table 1. Criteria and Sub criteria

Criteria	Sub criteria
Institutional Pressure (T)	1. Regulatory Pressures 2. Costumer Pressures 3. Market Pressures 4. Society Pressures
Key Resources (S)	1. Continuous Improvement 2. Stakeholder Integration 3. Disruptive Change
Competitive Advantages (K)	1. Cost Reduction 2. Reputation and Legitimacy 3. Future Positioning

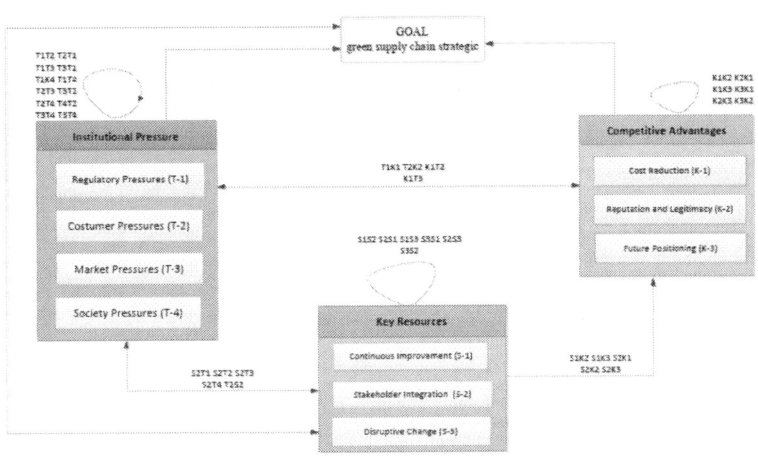

Figure 1. Network Structure (Network)

In the next stage, we made the network structure (Network). At this stage each criterion and sub criteria will be determined whether affecting each other.

Determination of influence relationship between Sub-Criteria is done by an interview with the manager. The instrument used is a closed questionnaire.

Based on the relationship between the sub criteria that have been compiled, then compiled into the network structure (network) that can be seen in Figure 1.

3.2. Paired Comparison Questionnaire

A set pairwise comparison questionnaire is used to assign weights to each of the Criteria and Sub-Criteria so that it can be determined from which perspective has the highest Criteria weight. This questionnaire has two parts namely pairwise comparison between Criteria, pairwise comparison between Sub-Criteria.

3.3. ANP Data Processing

3.3.1. An ANP calculation

The calculation of Consistency Ratio for paired matrix The Institutional Pressure Criteria is shown as follows:

a. Calculate the weighted average by calculated the geometric averages. Geometric averages are calculated by the formula:

$$GM = \sqrt[n]{X_1 . X_2 \dots . X_n}$$

Example for the calculation of geometric mean for pairwise comparison Criteria for institutional pressure:

$$GM = \sqrt[5]{3x3x3x3x5} = 3,32370$$

b. Each column element is divided by the number of columns respectively. Calculation of the normalization matrix and partial weight of the institutional pressure cluster:

Normalize Matrix = column value / total

c. Calculating Consistency ratio

(Matrix of Average Geometric Calculations) x (Voted Weight of Each Line)

d. Calculating Vector Consistency

(Consistency Ratio: Vector weight of each row)

e. Calculate the average of entries (λ_{max})

$$\lambda_{max} = \frac{\sum_{i=1}^{n} ConsistencyVector}{n}$$

f. Calculating Consistency Index (CI)

$$CI = \frac{\lambda_{max} - n}{n-1}$$

g. Calculating Consistency Ratio (CR)

$$CR = \frac{CI}{Random\ Consistency\ Index}$$

Obtained CR <0,1, then respondent answer consistent. The same calculation is performed on all elements between Criteria and Sub-Criteria.

3.3.2. An ANP calculation

Supermatrix are the result of priority vectors of comparison between clusters and sub criteria. Supermatrix consist of three stages: Unweighted Supermatrix, Weighted Supermatrix, and Supermatriks Limit (Limiting Supermatrix). To obtain these three supermatrix used Super Decision software assistance to facilitate the processing.

Unweighted Supermatrix is a translation of the weight of each cluster and sub criteria that is inputted into a complex matrix table. Weighted Supermatrix is obtained by multiplying all elements in the unweighted supermatrix with the values contained in the cluster matrix so that the value of each column is one. The Weighted Supermatrix that has been obtained is then multiplied by itself in several iterations. This iteration aims to keep the weight of supermatrix stable, meaning that the row value of each element becomes the same. When the weight of each column has the same value, then Limiting Supermatrix has been obtained.

Based on the results of processing with Analytic Network Process method above we get the weight for each sub criteria that can be seen in Table 2.

Table 2. Global Weights Sub criteria

Criteria		Sub criteria	Weight
Institutional Pressure (T)	T-1	Regulatory Pressures	0.140383
	T-2	Costumer Pressures	0.237714
	T-3	Market Pressures	0.085612
	T-4	Society Pressures	0.10236
Key Resources (S)	S-1	Continuous Improvement	0.066989
	S-2	Stakeholder Integration	0.222708
	S-3	Disruptive Change	0.03889
Competitive Advantages (K)	K-1	Cost Reduction	0.06749
	K-2	Reputation and Legitimacy	0.023337
	K-3	Future Positioning	0.014517

3.4. Prioritization of Green Supply Chain Initiative
The priority setting of green supply chain initiatives is done by using ANP method. The required data at this stage is the priority value of the sub criteria and the global weighting value of the previous ANP stages as the weighted value.

3.4.1. Data Criteria and Alternative Green Supply Chain Initiative
Alternatives used based on previous research references. There are 5 (five) alternative initiatives that are used are green product design, greening upstream, greening production, greening downstream and greening post use.
The criteria used to assess alternative green initiatives are institutional pressure, core resources, competitive advantage and green strategy capabilities. The sub criteria of green capability strategic used are pollution prevention, product stewardship and clean technology.

3.4.2. Determination of Green Supply Chain Initiative Network Structure
Determination of influence relationship between Sub-Criteria is done by the interview with the previous manager in the determination of the criteria weighting. At this stage, the influence relationship used based on previous research and validating by the company. After that, we determined the relationship between sub criteria. So obtained network structure (Network) as follows.

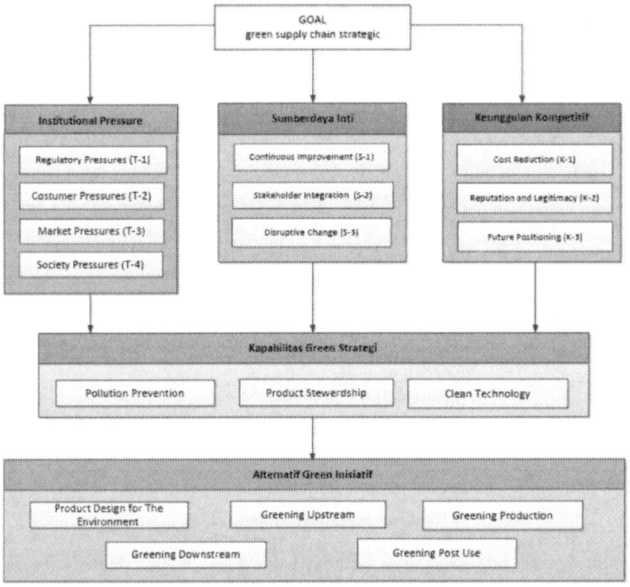

Figure 2. Network Structure (Network) Alternative Green Initiative.

3.4.3. *Calculation of ANP Green Initiative*

At this stage, we weigh the alternatives that have been prepared. Weighting is done in pairs according to the relationship.

The recapitulation obtained is then calculated by the ANP method as in the previous step. The software Superdecitions used to calculations of supermatrix. So the green priority is obtained as follow.

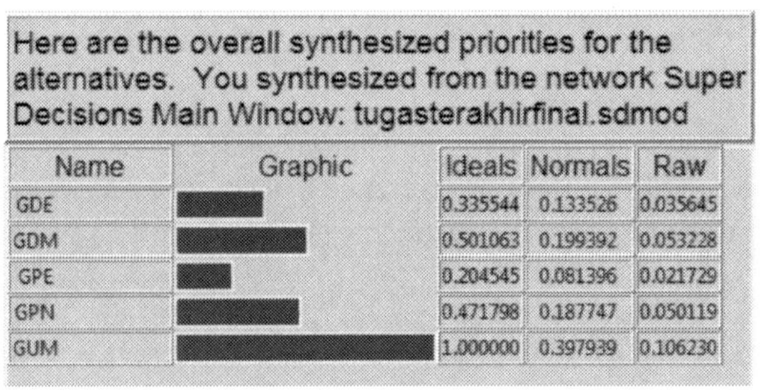

Name	Graphic	Ideals	Normals	Raw
GDE		0.335544	0.133526	0.035645
GDM		0.501063	0.199392	0.053228
GPE		0.204545	0.081396	0.021729
GPN		0.471798	0.187747	0.050119
GUM		1.000000	0.397939	0.106230

Figure 3. Priority Green Supply Chain.

3.5. *ANP Weighting Analysis*

The highest weight for criteria is the institutional pressure with the weight of 0.5661, and the lowest weight of criterion is Competitive Advantages with the weight of 0.1053.

Table 3. Weight of Institutional Pressure Sub-Criteria

Priority	Weight
Regulatory Pressures	0.1404
Costumer Pressures	0.2377
Market Pressures	0.1024
Society Pressures	0.0856
Total	0.5661

At criterion of institutional pressure, sub criteria costumer pressures has the highest weight that is costumer pressures equal to 0.2377, followed by subcriteria of regulation pressures and social pressures. This is influenced by the influence and relationship of customer pressures with the core resource is stakeholder integration, in this case, is the integration between customer and company. The desire of the customers also gives effect to the criteria of competitive advantage in this case that is the reputation and legitimacy of the corporation.

Table 4. Weight of Sub-Criteria of Key Resources

Priority	Weight
Disruptive Change	0.0145
Stakeholder Integration	0.2227
Continuous Improvement	0.067
Total	0.3286

From the above table obtained the highest weight of stakeholder integration is 0.2227. The weight of the sub criteria of stakeholder integration is influenced by the integration of stakeholder relations with all other sub criteria. By conducting stakeholder integration, external pressure on the company can be managed and can provide a positive influence on corporate excellence.

Table 5. Weight of Sub-Criteria of Competitive Advantage

Priority	Weight
Reputation and Legitimacy	0.0233
Cost Reduction	0.0675
Future Positioning	0.0145
Total	0.1053

In the competitive advantage criteria, sub criteria cost reduction obtains the highest weight among other competitive advantage sub criteria with 0.0675. This occurs because of the effect of cost reduction on the institutional pressure that occurs.

3.6. Pareto Analysis

In the Pareto diagram can be seen subcriteria very weak influence on the network structure (network), this can be seen from the global weight of each subcriteria that has been obtained. Here is a Pareto chart of sub criteria.

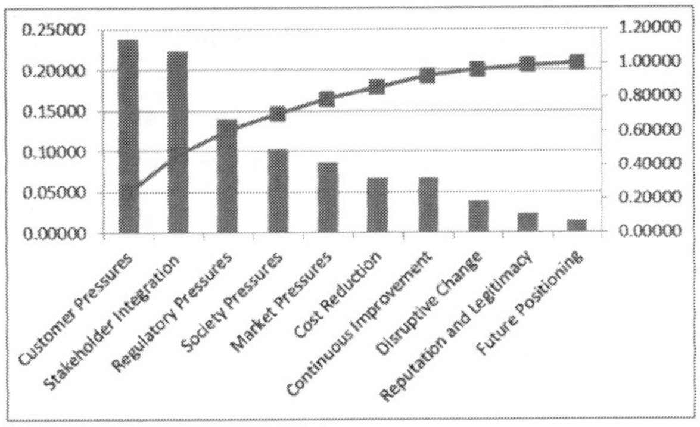

Figure 4. Pareto Diagram of Sub criteria.

Based on the above Pareto diagram it can be seen that the sub criteria that make up the cumulative total weight of more than 20% of the total weight of the whole sub criteria are the Customer pressures and Stakeholder Integration.

3.7. Results Analyze of Alternative Priorities Green Supply Chain

Based on the above Pareto diagram it can be seen that the subcriteria that make up the cumulative total weight of more than 20% of the total weight of the whole sub criteria are the Customer pressures and Stakeholder Integration.

Greening Upstream (GUM) obtained the highest score of 0.397939, following greening downstream (GDM) get score 0.199392, greening production (GPN) obtained score 0.181396, green product design (GDE) get score 0.133526, and greening post use (GPU) obtained score 0.081396. Then to improve supply chain performances become green supply chain performance, corporation should make optimization directed at Greening Upstream practice.

4. Conclusion

From this research can be concluded that There are three perspectives in designing the company's green supply chain strategy, namely: institutional pressure from outside to the firm, the core resources owned by the company and the competitive advantage expected by the corporation.

The most dominant criteria affecting the company's effort in designing the green supply chain strategy of ANP method are institutional pressure criteria (0.5661), followed by key resource criterion (0.3286), and criteria of competitive advantage (0.1053).

The most dominant sub-criteria based on the calculation method of the ANP is a sub-criteria customer pressures (0.2377), followed by sub-criteria Integration of stakeholders (0.2227), Regulatory pressures (0.1404), Social pressures (0.1024), competition pressures (0.0856), cost reduction(0.0675), Continuous improvement (0.067), Disruption change (0.0389), reputation and legitimacy (0.0233), and Position in the future (0.145).

There are five (5) alternative initiatives in strategic efforts to increase green supply chain is product design for the environment (PDE), Greening Upstream (GUM), Greening Production (GPN), Greening Downstream (GDM), and Greening post-Use (GPU). Of the five alternatives are obtained Greening Upstream (GUM) as the best priority to increase the green supply chain.

References

[1] Hadiguna R A 2016 *Manajemen Rantai Pasok Agroindustri* Andalas University Press

[2] Harsono M 2003 *Manajemen Lingkungan Natural Dalam Perspektif Resource-Based View: Tuntutan Stakeholders Ataukah Kebutuhan?* Perspektif FE UNS **8** 125-135

[4] Hart S L and Glen D 2011 *A Natural-Resource-Based View of the Firm: Fifteen Years After* Journal of Management **37** 1464-1479

[5] Hidayati J, Sukardi S, Ani S, Anas M F and Sugiharto 2015 *Optimization of Palm Oil Plantation Revitalization in North Sumatera Indonesia* International Journal on Advanced Science, Engineering and Information Technology **5** 6 460-468

[6] Pujawan N 2005 *Supply Chain Management* (2nd edition) Guna Widya Press

[7] Isik Z, Dikmen I and Birgonul M T 2007 *Using ANP for Performance Measurement in Construction* (RICS)

[8] Masoumik S M, Salwa H A R and Ezutah U O, & Raja A R G 2015 *A Strategic Approach to Develop Green Supply Chains* Procedia CIRP **26**: 670-676

[9] Masoumik, S M, Salwa H A R and Ezutah U O 2015 *The Development of a Strategic Prioritisation Method for Green Supply Chain Initiatives* PLoS ONE **1-33**

[10] Sundarakani B, De Souza R, Goh M, Van Over D, Manikandan S, Koh and S CL 2010 *A sustainable green supply chain for globally integrated networks(Enterprise Networks and Logistics for Agile Manufacturing)* **191-206**

IOP Conf. Series: Materials Science and Engineering **309** (2018) 012125 doi:10.1088/1757-899X/309/1/012125

Pekong, the symbol identity of Chinese at Pekan Labuhan Medan Indonesia

M Siagian*

Department of Architecture, Faculty of Engineering, Universitas Sumatera Utara, Medan, Indonesia

*Email : pohontepiair@gmail.com

Abstract. Urban architecture has always been combined by physical and non-physical components. The physical is formed by spatial pattern, space and history. The non physical is formed by social and cultural life of the community. The combination of that space would give the meaning as a place for the people that use it. Pekong has existed since 1890 in the area of Pekan Labuhan. The presence of two Pekongs in this area gives the symbol for distribution patterns of Chinese residential and community in Pekan Labuhan district. Symbolizing the Pekongs is sourced from being able to shape and influence the components such as the market, shophouses, and houses. Beside that the Pekongs also identify of districts and social influence of the community. The component activities enliven Pekongs make spaces for Pekongs grow to become a magnificent buildings in the comunity settlement. The aim of this research is to examine and describe the Pekong that has became the identity and attraction to the area. By using the qualitative method this research found that the Pekongs are the important buildings in identifying of the district.

1. Introduction

The architecture can describe the values, cultural and community activities that occurred at that time. It is then known as the symbol which is a sign of man-made are used not only to introduce an object but at the same time host it (Langer, 1942). The symbol is also a doctrine to guide human thought towards an understanding of something in depth. The symbol is also often used as a conducting medium of communication between each other, so that despite the absence of firsthand information about a building, but through the symbols we got information about the history, identity, function and meaning of the building.

Responses of human vision or against symbols that are present in space depends greatly on its ability to recognize the symbol with a capital of memory in the brain and the shape of the symbol in the description on the man who saw it. The spaces in the area in the form of location, buildings and landmarks can be understood by human beings when there is one symbol that is able to explain the whole meaning of the area.

Furthermore, the spaces around the symbols are clearly have established human activity or the communities that live in it. The occupants of the space make use of the venue for the struggling defending life through a wide range of interests, such as economic, social, cultural or religious. The human activity will slowly form a social relationships because their lives are different background but coexist in the same space. Intercultural community life together in it cannot be separated from the attractiveness of the region's symbol, so that the symbol is not only becoming a marker of the region, but also has become a strength of communities that live in the spaces around the symbols. These

IOP Conf. Series: Materials Science and Engineering **309** (2018) 012125 doi:10.1088/1757-899X/309/1/012125

spaces have become the identity of the area both for the people that live in it as well as people from outside the area.

The district of Pekan Labuhan is one of the important areas in Medan (Fig.1). The area of the cross which is very near from the port of Belawan is not only rich in values and meanings of the historical development of the city. This area is rich in values and the meaning of symbols, the phenomenon of space and activity of the community that lives in it. The symbol of this area is characterized by the presence of two Pekong as places of worship of the Chinese community, the spaces of old building with Chinese architecture and the Dutch Colonials influence.

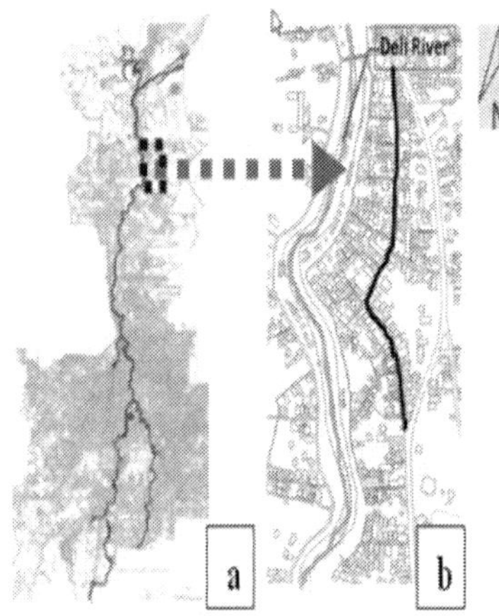

Fig. 1. (a) Medan, (b) Pekan Labuhan

Based on current population, more than 60 percent of the community in Pekan Labuhan is Malays, while the remaining less than 40 percent is ethnic Batak, Javanese and Chinese. This means that today Chinese people does not dominate this place. However a row of shophouses of Chinese style buildings, Pekongs dominate the main spaces. Physically the building describe the identity of the Chinese villages. Based on the phenomenon of space and human activity in this area, this article describe how the space can identify the existence of Chinese people eventhough the area dominated by Malay communitiy nowdays.

2. Methodology

This research venue is located in Pekan Labuhan, Medan. Since the type of this research is qualitative, hence the most effective method to collect data was by interview and direct observation. With interview, the researcher can get the thoughts and ideas that are stored in the minds of respondents who are the people from the community who live in the area. The process of interview was conducted in a relaxed way with the snowball system. Furthermore, researcher directed observing the space and interpreted as the qualitative method analysis. In this method, the social phenomenon and space phenomenon that occurred is concluded as a result of in-depth interpretation and clearly described.

3. Discussion

There are two kinds of architectural components, namely physical and non-physical component. Physical components focused to look buildings and physical objects that exist in a district formed by the interlacing of masses and spaces (the history and development of the city) and the spatial scale (the character and appearance of the space). Non-physical component is the social and cultural life of society which aspects influenced the formation of physical, environmental and utilising human relationships and physical urban. According to Rapoport (1969), culture, values, social interaction, and the relationship of humans with nature, are factor which influence the physical form of a building.

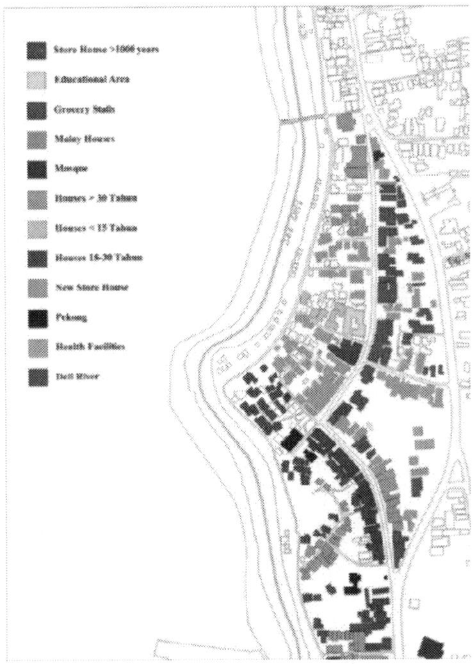

Fig. 2. Map Houses of Pekan Labuhan

The district of Pekan Labuhan is one of the important areas in the city of Medan (Fig.2). The place is originated in the old town area of early development of the Sultanate of Deli (1669). The initial location of the conveniences of the growing city of Medan as a plantation town on the Dutch colonial period (1887). The arrival of Chinese community from Singapore and Penang to Indonesia as workers started in 1870 when the Dutch colonial goverment built a tobacco farm at Tanjung Sepassi, Tanah Deli, North Sumatera. At first, they occupied an area around the Deli River and then in 1890, along with the expansion of tobacco plantations in Tanah Deli, this area became more be advanced and crowded with houses, shops, business centre and a growing market due to the strategic river port of Deli. Since then, this district had been named as Pekan Labuhan.

The physical components of the area of Pekan Labuhan, Medan is the heritage of the old city, the early development the city of Medan on Dutch Colonial Goverment. Facilities such as shop houses, pekong, former port and market centre in colonial era can still be found in this area. The non physical components in this area also has a wealth of relationships between communities that have long been intertwined. The Malays community as early as the Chinese arrivals established social relationships.

Fig. 3. Architecture of Chinese at The Area

The physical architecture component which gives the looks of a building in general can be seen directly by public coming to the district. However basically the physical architecture components are formed by non-physical components (culture, values, social interaction, and the relationship of human beings with nature), so by the building we can recognize what ethnicity live on. The physical architecture components that describe the characteristics of Chinese in the area of Pekan Labuhan, directly pass on the information to public that Chinese ever inhabited in this area.

The Chinese occupied along Syahbuddin Yatim Street, they settled in a shop houses with the concept of residential and commercial uses. Architectural style is influenced by Chinese and Dutch Colonial. In terms of the roof top of the is influenced by Chinese and in terms of the façade is influenced by the Dutch Colonial.

The existence of Pekongs with Chinese style of architecture in the district, became a marker of the presence of the Chinese community in this area. The construction of Pekong in the area is to fulfill the need of houses of worship because of the increasing Chinese community in the region. Therefore in the year 1872 built a pekong who was given the title of Liat Sim Kong, Pekong but considering the large number of Chinese society who serve in this place, then in 1890 in the wake of a much bigger pekong and location development pekong is also positioned on the main area Pekan Labuhan overlooking the main road i.e. JL. Yos Sudarso, the named pekong Pekong Siu San Keng. Nowdays both Pekongs are the oldest temple for Chinese house of worship in Medan. based on fenomenon, Chinese architecture influence is found in two Pekongs which are used as places of worship of Chinese society. Both pekong have become the landmark of this area so that later became the symbol for the area of Pekan Labuhan.

Fig. 4. (a). Pekong Liat Sim Kong. (b). Pekong Siu San Keng.

Pekongs very lively visited by the Chinese community from various places in Medan. The activity community perpetrated by Chinese is then attract the attention of public and interpret the Pekongs as the object that represents the entire Chinese community in the district. Then public associate the buildings as location for the market where Chinese society once runned the business. Pekongs as the symbol of the area of Pekan Labuhan have provided an based information that the area is the location where the Chinese community live and survive. Symbol pekong in Labuhan also used as marker of market that have existed since the Kingdom of Deli in Dutch Colonial era.

The existence of the symbol in the Pekong Pekan Labuhan be succefull to reinforce the identity of the Chinese community in the area, but political incident in May 1998, a reform movement undertaken by the youth gangs and political parties. The movement is anarchist and the destruction of public facilities, the dispossession of non-indigenous property rights, the destruction of the offices of feudal society. These inchident took place almost in the whole area of Indonesia, including in the area of Pekan Labuhan, Medan. At the time the espropriation of property rights, the destruction of the home store, and selling physical violence against the Chinese community in the Pekan Labuhan were happend. Therefore, many of the Chinese community left from the area.

The Reformation in 1998 led the change of the owner of the shop houses in this area, since then the process of buying and selling the shop houses occured and many indigenious house were sold in very chip price. The phenomenon shows why the Malays and other indigenous communities such as ethnic Batak, Javanese, and other ethnic group remaind inhabiting the area. The new owner still preserved the authenticity of the structure, the pattern of buildings and facades until nowdays.

At the beginning of the process of substitution in Pekan Labuhan residents trading activities had ceased in the area. This led to the site being very quiet until some time. Since left the Chinese community, this area as the "City of Death", look no further the crowd selling process at home stores, look no further the hustle in the stalls, in other open spaces. Malay people as native Pekan Labuhan feel deep sadness over move Chinese community who they consider as their companions for a long time.

Since the migration of Chinese communities outside rollicking the region of Pekan Labuhan in 1998, the Malay people and indigenous who still live in the region are struggling to survive. As a space of the ordeal, they sell the spaces on the street side of Syahbuddin Yatim Street who the main street in the area. This street so much shop houses owned by Chinese before nowdays the shop houses are occupied and used become a morning market by the new malays and indigenous owner. Then society also add commercial spaces up to the patio and sidewalk of Syahbuddin Yatim Street. So the house on the first floor, terrace and sidewalks have been met to trade places daily needs. This market is open from 6.00. am until 12.00. am.

Fig. 5. Pekong Market at The Area

Commercial spaces formed by the society became more developed and vibrant. So in addition to traders from the region, many traders from outside region choose to sell here with the rental kiosk system. In addition, the number of buyers is also much more, buyers from within and outside the region daily needs love to shop at the market this morning, the reason they're prefer shopping in this market because of the garbage clean and safe from theft. Communities in the area not only succeed economically, but also socially, as between buyer and seller have known each other and feel comfortable doing economic transactions in this region.

But basically, the reason for the growing market on the main road cannot escape from the influence of the existence of a Pekong Siu San Keng which located at the junction of the road entrance to the area. That is the answer why the location of the market is nicknamed "pekong market" Pekan Labuhan. It is understood that Pekong as the signposts area Pekan Labuhan has become a magnet for people to still pay a visit to the region. Identity markers into Pekong that market area clean and comfortable is located at "*Simpang Pekong Pekan Labuhan*".

The symbolism of pekong has also been attached to the memory of the community and still describes the area as a place Pekan Labuhan of historic for Chinese community. To the Chinese community Pekong have strong inner bond, they often visit and worship at both Pekong. A large part of their save the ashes of a dead body of family at the Pekong in Pekan Labuhan. When comes to the area, the Chinese community feel like go to hometown, due regard had lived in this area and missed the atmosphere while still living here. Pekongs are markers for ethnic Chinese that they are worshipping in the historic district and has its own impression for them. This makes the presence of the two Pekong grow and survive, even more magnificent and thriving since the renovation in both building the pekong.

Although Chinese society is no longer lives in the Pekan Labuhan, but the spaces created on Pekong Liat Sim Kong and Pekong Siu San Keng and commercial space created at the point of the main street can be a magnet that pulls back the city community Medan. Not only Chinese, but also the wider community from various points in the city simply shop at "*Market Pekong*" as well as the roads around the area. So that results from the creation of the space pekongs are able to attract the attention of tourists who want to see the layout of the building and the city's oldest pekongs and continue to be objects representing the Chinese housing in the area as a symbol of the history of the Chinese civilization there.

4. Conclusion

Pekong Liat Sim Kong that was build in 1872 and Pekong Siu San Keng in 1890 have been able become new magnet and markers of identity and attraction, specifically for the community of Pekan Labuhan and general for the people of Medan. Besides Pekongs illustrate the old town, the Chinese civilization and its historical relics in the past, they also illustrates the identity the district at this time. The Pekong have managed to become a symbol of the naming of the place or location that explain the existence of the spaces. That Pekongs also form commercial space in the main street which create the name of "*Pekong Market*".

Important incident in the year 1998 made many Chinese community move out from the area. Through the two Pekongs, the existence of the Chinese community can be maintained. The presence of two Pekongs are like hometown for the Chinese community and their family so that it become a historic place for them. Many of those who have lived outside the district paid a visit to Pekong to do the religious activity and culture, so the day after day Pekong more growing. Then the existence of religious activity and granting blessing of the Pekong has become a fortune for the community around the two Pekong.

IOP Conf. Series: Materials Science and Engineering **309** (2018) 012125 doi:10.1088/1757-899X/309/1/012125

Reference

[1] Buiskool D A 2004 *A Plantation City on the Coast of Sumatera 1870-1942 (Planters, the Sultan, Chinese and the Indian)* The 1st International Urban Conference, Surabaya

[2] Buiskool D A 2005 Medan, A Plantation City on The East Coast of Sumatera !870-1942. Dalam Freek Colombijn, dkk. (Ed). *Kota Lama Kota Baru: Sejarah Kota-Kota di Indonesia.* Yogyakart: Ombak. 273—300

[3] Ekosiwi EK 1989 *Pemikiran Susane K. Lenger tentng Seni Sebagai Simbol Presentasional*, Universitas Indonesia

[4] Gih DS 1964 *Chinese Architecture Post and Contemporary*

[5] Fernando, David B 1992 *Chinese Economic Activity in Netherland India.* Selected Translation From The Dutch. Singapore:ISEAS

[6] Jencks C 1980 Late –Modrern Architecture, Rizzoli, Academy, London

[7] Jufrida and Soedewo E 2004 *Jejak Kejayaan Kerajaan Deli di Pekan Labuhan, Kecamatan Medan Labuhan,Kota Medan.* Brkala Arkeologi: Sangkhakala Nomor 13. Medan: Kementrian Pendidikan dan Kebudayaan, Balai Arkeologi Medan. 30-38

[8] Pratiwo 2010 *Arsitektur Tradisional Tionghoa dan Perkembangan Kota.* Yogyakarta:Ombak

[9] Rapoport A 1977 *Human Aspect Of Urban Form, Toward A Man Enivironment Approach to Urban Form and Design,* Oxford, USA

[10] Ruppert ES 2003, *Henri Lefebvre. The Urban Revolution*, Canadian Journal of Sociology Online, http://www.arts.ualberta.ca/cjscopy/reviews/urbanrev.html

[11] Sandhu KS and Mani A 1993 *Indian Communities in Southeast Asia*, Times Academic Press, Singapore

[12] Soja EW 1980 *The Socio-Spatial Dialectic*, Annals of The Association of American Geographers, Vol. 70, No.2, http://www.jstor.org/stable/2562950

[13] Soja EW 1989 *Postmodern Geographies: The Reassertion of Space in Critical Social Theory*, Verso, London

[14] Soja EW 1996 *Thirdspace*, Blackwell Publishers, Inc., Malden

[15] Stake RE 2010 *Qualitative Research*, The Guilford Press, New York

[16] Zukin S 1995 *The Cultures of Cities*, Blackwell Publishing, Malde

IOP Conf. Series: Materials Science and Engineering **309** (2018) 012126 doi:10.1088/1757-899X/309/1/012126

Fermentative utilization of glycerol residue for the production of acetic acid

Irvan[1,2], B Trisakti[1,2], R Hasibuan[1] and M Joli[1]
[1]Chemical Engineering Department, Universitas Sumatera Utara, Medan, 20155, Indonesia
[2]Sustainable Energy and Biomaterial Center of Excellence, Universitas Sumatera Utara, Medan, 20155, Indonesia

[*]Email: i_v_a_n_mz@usu.ac.id; irvan@usu.ac.

Abstract. Glycerol residue, frequently known as pitch, is a waste produced from the down-stream product of crude glycerine distillation. With the increasing need of pure glycerine in the world, the glycerol residue produced is also increasing. Glycerol residue is a solid waste at room temperature, highly alkaline (pH > 13), corrosive, and categorized as hazardous and poisonous waste. In this research, acetic acid was produced from glycerol residue through the anaerobic fermentation process by using purple non-sulphur photosynthetic bacteria. The purpose of this study was to find out the influence of concentration change of glycerol residue on time and to find out the possibility of glycerol residue to be utilized as acetic acid. In this research, at first 400 g of glycerol residue was diluted with 200 ml of distilled water to change the glycerine phase, from solid to liquid at room temperature, acidified by using hydrochloric acid until pH 2. The top layer formed was fatty acid and triglycerides that should be removed. Meanwhile, the bottom layer was diluted glycerol residue which was then neutralized with caustic soda. To produce acetic acid, glycerol residue with various concentrations, salt, and purple non-sulphur photosynthetic bacteria were put together into a 100 ml bottle which had been previously sterilized, then incubated for four weeks under the light of 40-watt bulb. The result showed that on the 28th day of fermentation, the produced acetic acid were 0.28, 1.85, and 0.2% (w/w) by using glycerine with the concentration of 0.5, 1.0, and 1.5% (w/w), respectively.

1. Introduction

Oleochemical industries show fast development nowadays. It can be seen from the rise of global palm oil production from 15.2 million tons in 1995 to 62.6 tons in 2015 [1]. By that, glycerol as by product of palm oil also increased from 660.000 tons in 1995 to 1.6 million tons in 2009 [2].

Glycerol also known as glycerin is a neutral, clear, odorless, naturally sweet taste, and viscous liquid. It is derived from both natural and petrochemical feedstock. Glycerin occurs in combined form (triglycerides) in animal fats and vegetable oils and is obtained from these fats and oils during transesterification, such as in biodiesel production. Glycerol is widely used as an additive source in various industries such as textile, food and beverage, pharmaceutical, cosmetic, etc. [3]. Glycerol also can be applied as carbon sources for the fermentation process to produce other various products; one of them is acetic acid [4].

Many researchers have tried to transform this waste into a valuable product. Yusuf M. (2007) studied the utilization of glycerol residue as elastic material properties of polyvinyl chloride (PVC), which previously reacted with sodium and potassium hydroxide at high temperature. This reaction will form polyglycerol which then acetylated by using acetic acid to produce polyglycerol acetate. This final product is used as additive material in making PVC [5].

Other research which already been conducted to handle glycerol from glycerol residue is by the recovery process. This process has several steps such as dilution, acidification, mixing, separation by filtration (giving a crude glycerine with lower dissolved salt and higher glycerol), then neutralization and concentration step. The composition of recovered glycerol was: glycerol 51.4%, ash 13.8%, water 8.9% and matter organic non-glycerol 25.9% [6].

Glycerol as an organic material can be applied as carbon sources together with DL malate. In this research *Rhodopseudomonas sphaeorides* as the purple non-sulphur photosynthetic bacteria was able to perform glycerol dissimilation process and grew with a doubling time of 6 hours [7]. Purple non-sulphur photosynthetic bacteria (*Rhodopseudomonas capsulata*) were also able to perform glycerol assimilation process at 30 °C by using a lamp as light source continuously. This assimilation process could increase cell number, dry weight, protein, and bacteriochlorophyll [8]. In this research, glycerol residue was fermented through the anaerobic process to produce acetic acid by using purple non-sulphur photosynthetic bacteria. This fermentation process used a 40-Watt lamp as the energy source, running for 24 hours.

2. Experiment, materials and method

2.1. Experiments

A preliminary experiment was conducted to observe the growth of anoxygenic photosynthetic bacteria in salt and glycerol residue medium at various concentrations: 0.05; 0.1; 0.5% (w/w). The preliminary result showed that higher amounts of purple non-sulphur bacteria grew in medium with concentration 0.5% compared to 0.05 and 0.1%.

Experiments were conducted in Microbiology Laboratory of FMIPA-USU, PT. Wilmar Bioenergi Indonesia Laboratory and BATAN Laboratory.

2.2. Materials and equipment

Main equipment used in this research were Anaerobic Microbiology Mart Culture Jar (series #AJ9023) 2.75 liter of volume, Autoclave Yamato SN 210, Isotherm incubator, 40-watt bulb lamp (Philips). While, reagents and chemicals used were glycerol residue, bromothymol blue indicator, NaOH, $NaIO_4$, H_2SO_4, ethylene glycol, H_3PO_4, K_2HPO_4, KH_2PO_4, $MgSO_4.7H_2O$, $FeSO_4.7H_2O$, $ZnSO_4.7H_2O$, $MnCl_2.2H_2O$, methanol, and hydrochloric acid.

2.3. Method

The raw glycerol residue was obtained PT. Nubika, Permata Hijau Group, Labuhan Batu. All the reagents and chemicals used were analytical grade. All the experiment was performed in triplicate. Experiment procedures consist of six procedures, they are:

1. Separation of fatty acids and triglycerides
   400 g of glycerol residue was added to 200 ml distilled water and stirred until uniformly distributed. Adding hydrochloric acid until the pH of the mixture was between 2 - 3 and then leave it for several minutes until they split into two layers. The top layer was separated using filter paper. The filtrate was neutralized with caustic soda to neutral pH, and then the filtrate was analysed to know the glycerine concentration.

2. Preparation of salt media
   Into a 500 ml volumetric put 12.5 g of $(NH_4)_2SO_4$, 12.5 g, 8 g of KH_2PO_4, 9 g of K_2HPO_4, 5 g of yeast extract, 2 g of $MgSO_4.7H_2O$, 0.7 g of $CaCl_2$, 0.03 g of $FeSO4.3H_2O$, and 0.02 g of EDTA. All reagents were mixed then put some water until the volume became 500 ml [9].

IOP Conf. Series: Materials Science and Engineering **309** (2018) 012126 doi:10.1088/1757-899X/309/1/012126

3. Preparation of Glycerol Residue 0.5%
Glycerol residue 0.5% was prepared by pipetting 4.0 ml of raw glycerol residue into 125 ml bottle, then adding salt media until the volume of the bottle became 115 ml. The bottle was sealed and then sterilized in an autoclave at a temperature of 120 °C with a pressure of 2 bar. After passing through the process of sterilization, the bottle was cooled at room temperature.

4. Fermentation Procedure
Five ml of purple non-sulphur photosynthetic anoxygenic bacteria (PAB) was inoculated in a bottle closed tightly. The bottle was placed in a room equipped with 40-watt incandescent light bulbs. For the preparation, 8.0 and 12.0 ml of glycerol residue of 1.0% and 1.5% were pipetted, respectively and also for 125 ml of 15% glycerol residue into 125 ml volume bottles. Samples were prepared in 12 bottles for each concentration. In the meantime, preparation for control sample did not use glycerol as carbon source. The experiment used only 115 ml of salt medium. The control sample was prepared in 12 bottles.

5. Determination of Acetic Acid Concentration.
Fifteen-μl acetic acid standard with concentration of 0.02, 0.04, 0.06 and 0.08 % were injected into HPLC columns (Shimadzu SPD-6A). The injection was performed by duplicate. The retention time and area of each acetic acid concentration were recorded, in the same way, the samples were injected. The retention time of outgoing sample was recorded. The total area produced by the sample was used to calculate the concentration of acetic acid formed by using linear regression equation analysis.

6. Determination of Microbial Growth, Glycerol Concentration, and pH
Determination of microbial growth was performed by using UV Visible Spectrophotometer (Thermo Spectronis Genesys 5). Glycerol concentration was determined by using titration method. While, measurement of pH was performed by using pH meter (Mettler Toledo S220)

3. Results and Discussion

3.1. Change of glycerine concentration on fermentation time

Glycerine as carbon source is utilized well by the purple non-sulphur bacteria. It can be seen from the increase of sample absorbance values on Figures 1, 2 and 3. This pattern shows that anoxygenic photosynthetic bacteria (APB) can live and utilize glycerine at various concentrations by absorbing light as an energy source. This change was started by the assimilation process where glycerine residues were oxidized by bacteria using enzymes. This oxidation process caused glycerine to lose hydrogen atoms in the presence of energy derived from adenosine triphosphate (ATP). Enzymes produced by bacteria will come into contact with the surface of glycerine which will then form a temporary bond between the enzyme and glycerine. The glycerine molecules will be broken and converted into new compounds and then the enzyme leaves the glycerine and ready to react with other glycerine molecules [5].

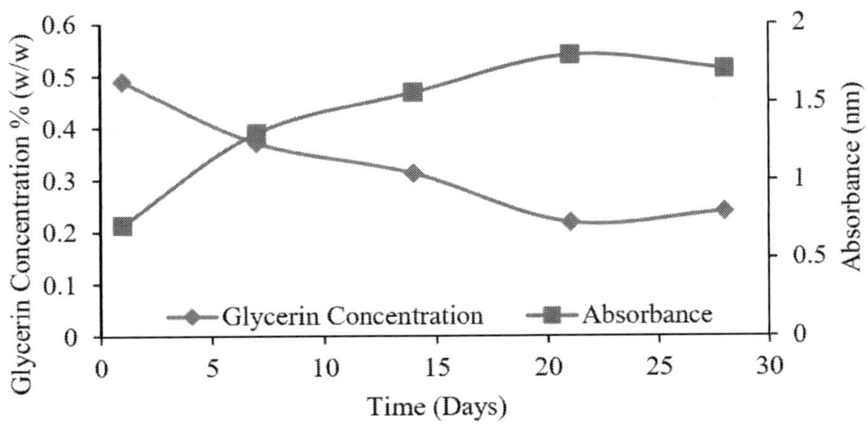

Figure 1. Change of 0.5 % glycerine against time

Figure 1 shows the change of glycerine concentration on fermentation time. From Figure 1 it can be seen that visible glycerine concentration decreased from 0.5 to 0.22%. The percentage decrease of glycerine until day 22nd reaches 56.0% from the initial value. The decrease in absorbance values indicates that bacteria lead to the reduction of growth phase due to the reduction of glycerine as the carbon source and the presence of competition during the growth process so that the microbes that are not able to survive will become toxic to other microbes. According to Jawetz, et al., (1996), the accumulation of toxic metabolic outcomes will cause cell growth to stop [10].

The decrease in concentration was also accompanied by an increase in the absorbance value of the sample from 0.614 to 1.814 abs on day 1st until day 21st day with a growth value reaching 2.95 times of the initial absorbance. Glycerine as a carbon source can be utilized by bacteria to grow, marked by an increase in turbidity value from 20 to 150 for 52 hours. This increase in turbidity indicates a larger bacterial population [7].

Growth curves directly formed logarithmic growth without going through adaptation phases. The adaptation phase is likely to start after a few hours through the 1st day. Rhodopseudomonas palustris bacteria grows after a 72-hour incubation period [11]. Meanwhile, bacteria Rhodopsedomonas sphaeraides known as APB bacteria will enter the initial phase or adaptation in the process of glycerol dissemination after going through 6 - 10 hours since the beginning of growth [7].

This study utilized 40-watt lamplight as a source of APB bacterial energy. The 40-watt lamp has a very high light intensity that greatly affects the growth of photosynthetic bacteria. According to Garcia-Contreras et al. (2004), the higher the intensity of the lights, the faster the growth of photosynthetic bacteria [12].

On the 28th day, glycerine concentration should decrease or remain, but in this study, the concentration of glycerine has increased. The possibility of this happening due to the errors conducted during the experiments. This is confirmed by the decrease in the absorbance value of control sample which supposedly increased. This error is likely due to the presence of oxygen penetrating into the sample bottles. The presence of oxygen directly affects the activity of non-sulphur purple bacteria (Yurkov and Beatty, 1998). Meanwhile, APB bacteria can only grow and develop if there is no oxygen [13].

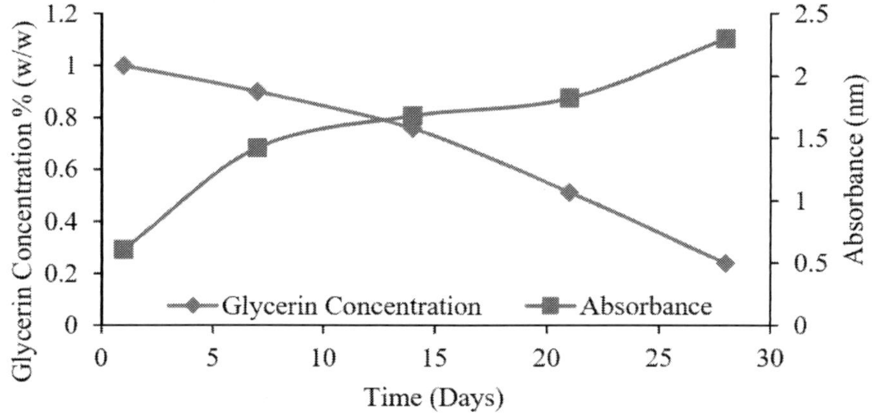

Figure 2. Change of 1.0% glycerine solution against time

In Figure 2, 1.0% glycerine solution decreased from 1.01% on day 1 to 0.22% on day 28 with a conversion rate of 78.2%. The decrease in glycerine concentration was also followed by an increase in bacterial growth value from 0.612 abs on day 1 to 2.344 abs on day 28 with a growth rate of 3.83 times. The decrease in glycerine concentration of 10.0% and the increase in absorbance value of microbial growth showed the highest value compared to glycerine fermentation at 0.5 and 1.5%. Glycerine with a concentration of 1.0% can be metabolized up to 97%. The growth pattern of bacteria 1.0% is similar to the bacterial growth pattern of 0.5% glycerine. Bacterial growth goes directly to the logarithmic growth phase without passing through the adaptation phase [14].

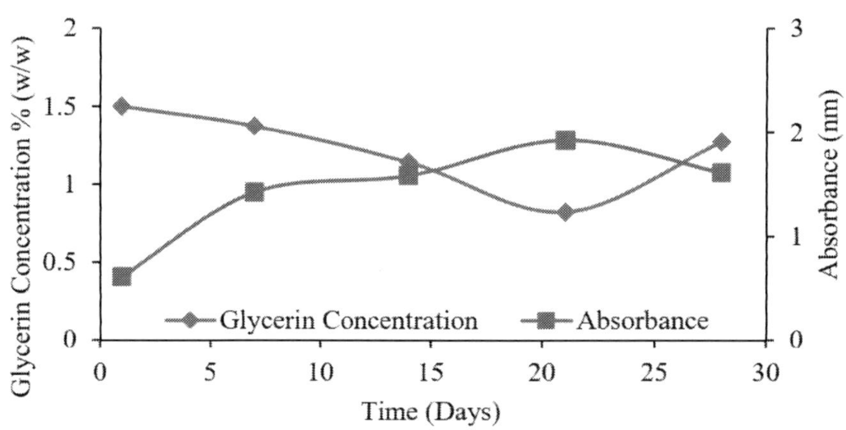

Figure 3. Change of glycerine concentration 1.5% against time

Figure 3 shows the change of 1.5% glycerine solution on time. As shown from Figure 3, the pattern of glycerine concentration change of 1.5% is almost equal to 0.5% of glycerine concentration. From day 1 to day 21 the concentration of glycerine decreased from 1.49 to 0.83% with the conversion of glycerine change reaching 44.3%. Increasing of glycerine concentration was also marked by the increasing absorbance value of microbial growth from 0.673 abs on day 1 to 1.911 abs on day 2, with a growth value of 2.84 times.

On the 28th day, glycerine concentration should decrease, but in this study, the concentration of glycerol has increased. This is also indicated by the decrease in the growth absorbance value of the sample that should rise. This problem was possible as well as fermentation of glycerine at 0.5% concentration that is the entry of external air due to less tightening of the bottle lid insulation binder.

3.2 Formation of acetic acid

The formation of acetic acid is the result of glycerine fermentation process initiated by phosphorylation glycerine to glycerol 3-phosphate by ATP derived from light. Glycerine undergoes oxidation or releases a hydrogen atom become dihydroxyacetone then converted to glyceraldehyde which is subsequently oxidized by releasing a hydrogen atom to form pyruvic acid. The pyruvic acid is then reduced to acetic acid [15].

Figure 4 shows the formation of acetic acid at various concentrations. As shown from Figure 4, the highest increase of acetic acid was found at 1.0% glycerine concentration with 1.85% acetic acid concentration on day 28 compared with 0.1% and 1.0%. This indicates that glycerine with a concentration of l0.0% is well utilized by APB bacteria. Glycerine with a concentration of 1% exploited by bacteria as carbon element [14].

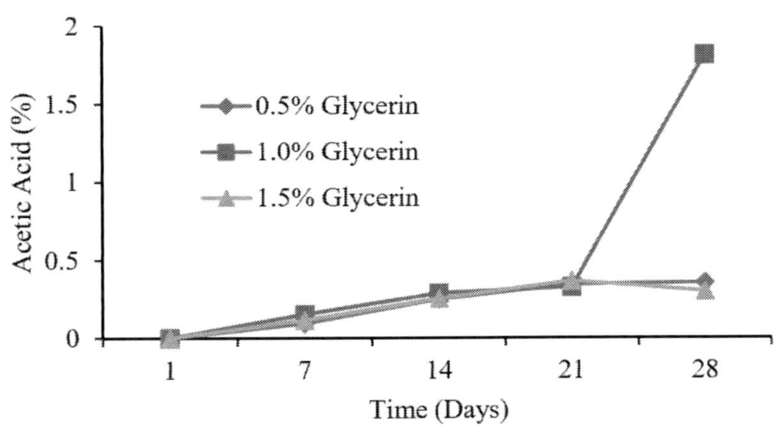

Figure 4. Effect of time on formation of acetic acid

3.3 Degree of acidity observation

The setting of neutral pH in the early stages provides good growth on non-sulphur purple bacteria. In Figure 4 shows the degree of acidity of 6.71 and decreased until day 28 for glycerine with a concentration of 1.0%. The decrease in pH was due to the formation of acetic acid and the possibility of other acid acids produced during the fermentation process. Acid compounds are formed from the process of glycerine dissimilation in the form of pyruvic acid, acetic acid, and formic acid, anaerobically [15]. The concentration of glycerine at 0.5 and 1.5% also decreased pH to day 21 with pH values of 3.7 and 3.3. This decrease indicates that the fermentation process is going well, but on the 28th day the pH has increased to 4.4 for both 0.5 and 1.5% concentrations. APB bacteria should still be able to perform fermentation until the pH becomes 2.5 and stop when the pH becomes 2.2 [16]. The increasing in pH is likely due to the disruption of bacterial activity due to the oxygen existence. Other then influences the pH, oxygen also affects changes in glycerine concentration as well as the formation of acetic acid for 0.5% and 1.5% glycerine concentrations.

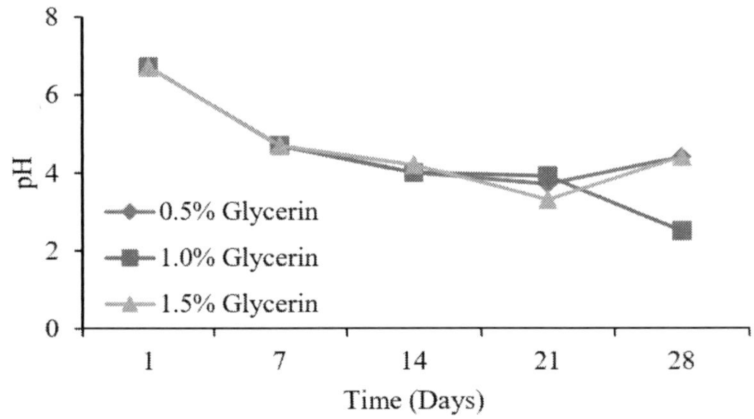

Figure 5. Effect of time on degree of acidity

3.4 Observations on control samples

Observations on control samples aimed to see whether purple non-sulphur bacteria could ferment in the absence of glycerine as carbon elements. Fig. 5 shows that there is no change in pH of acetic acid and absorbance from control sample from day 1 to day 28. This revealed that no growth or formation of acetic acid occurred in sample.

Table 1. Change on control samples

Time (days)	pH	Acetic acid (% w/w)	Absorbance (abs)
1	6.71	0.02	0.614
7	6.74	0.01	0.608
14	6.78	0.01	0.612
21	6.72	0.05	0.613
28	6.79	0	0.609

4. Conclusion

Glycerine residues of 0.5, 1.0, and 1.5 % concentrations can be utilized by purple non-sulphur photosynthetic bacteria. Glycerine with concentrations of 0.5, 1.0, and 1.5% can be converted to acetic acid. The highest conversion of glycerine to acetic acid was obtained from glycerine residue with a concentration of 1.0% with an acetic acid concentration value of 1.85%.

Acknowledgment

This research was supported by Hibah Penelitian Talenta Universitas Sumatera Utara No: 5338/UN5.1.R/PPM/2017, May 22nd, 2017.

References

[1] Scientific Advisory Panel 2016 *The Palm Oil Story: Facts and Figures*, European Palm Oil Alliance, Netherlands
[2] Wendisch V F, Lindner S N, and Meiswinkel T M 2011 *Use of Glycerol in Biotechnological Applications*, in Gisela Montero G, Stoytcheva M, *Biodiesel- Quality, Emissions and By-Products,* InTechOpen
[3] Miner C S, Dalton N N 1952 *Glycerol*, Reinhold Publishing Corporation
[4] Hidawati E N, Sakinah A M M 2011 *Int. J. Chem. Environ. Eng,* **2** 309-313
[5] Yusuf M 2007 *Penyediaan poligliserol asetat dari residu gliserol pabrik biodiesel sebagai bahan pemlastis pada polivinil klorida*, Master Thesis, Chemistry Department - University of Sumatera Utara
[6] Ooi T L, Yong K C, Dzulkefly K, Yunus W M Z W, Hazimah A H 2001 *J. Oil Palm Res.* 13 16–22
[7] Pike L, and Sojka A G 1975 *J. Bacteriol.* **124** 1101 – 1105
[8] Lueking D, Tokuhisa D, Sojka G 1973 *J. Bacteriol.* **115** 897 – 903
[9] Suryanto D, Suwanto A, Meryandini A 2001 *Biotropia* **17** 9 – 17
[10] Jawetz E, Melnick J L, Adelberg E A, Brooks G F, Butel J S, Ornston L N 1995 *Mikrobiologi Kedokteran* 20^{ed}. San Francisco-University of California
[11] Getha K, Vikineswary S, Chong V C 1998 World J. Micro. Biotech. **14** 505 – 511
[12] Garcia-Contreras R, Celis H, Romero I 2004 *J. Bacteriology 186* 6651 – 6655
[13] Yurkov V V, Beatty J T 1998, *Microbiol. Mol. Biol. Rev.* **62** 695–724
[14] Forsberg C W 1987 *Appl. Environ. Microbiol.* **53** 639-643
[15] Magasanik B, Brooke M S, Karibian D 1953 *J. Bacteriol.* **66** 611-619
[16] Chiruvolu V, Eskridge K, Cregg J, Meagher M 1998 *Appl. Biochem. Biotech.* **75** 163-173

IOP Conf. Series: Materials Science and Engineering **309** (2018) 012127 doi:10.1088/1757-899X/309/1/012127

The engine maintenance scheduling by using reliability centered maintenance method and the identification of 5S application in PT. XYZ

N Sembiring*, N Panjaitan and A F Saragih
Department of Industrial Engineering, Faculty of Engineering, Universitas Sumatera Utara, Jl. Almamater Kampus USU, Medan 20155

*Email: nurhayatipandia68@yahoo.com

Abstract. PT. XYZ is a manufacturing company that produces fresh fruit bunches (FFB) to Crude Palm Oil (CPO) and Palm Kernel Oil (PKO). PT. XYZ consists of six work stations: receipt station, sterilizing station, thressing station, pressing station, clarification station, and kernelery station. So far, the company is still implementing corrective maintenance maintenance system for production machines where the machine repair is done after damage occurs. Problems at PT. XYZ is the absence of scheduling engine maintenance in a planned manner resulting in the engine often damaged which can disrupt the smooth production. Another factor that is the problem in this research is the kernel station environment that becomes less convenient for operators such as there are machines and equipment not used in the production area, slippery, muddy, scattered fibers, incomplete use of PPE, and lack of employee discipline. The most commonly damaged machine is in the seed processing station (kernel station) which is cake breaker conveyor machine. The solution of this problem is to propose a schedule plan for maintenance of the machine by using the method of reliability centered maintenance and also the application of 5S. The result of the application of Reliability Centered maintenance method is obtained four components that must be treated scheduled (time directed), namely: for bearing component is 37 days, gearbox component is 97 days, CBC pen component is 35 days and conveyor pedal component is 32 days While after identification the application of 5S obtained the proposed corporate environmental improvement measures in accordance with the principles of 5S where unused goods will be moved from the production area, grouping goods based on their use, determining the procedure of cleaning the production area, conducting inspection in the use of PPE, and making 5S slogans.

1. Introduction

PT. XYZ is one of the companies engaged in processing oil palm. Based on preliminary observations, companies often experience machine breakdown issues. Where the production machine is most often damaged is the cake breaker conveyor machine. This can hamper the production process which has an impact on the decrease in production capacity. PT. XYZ applies two maintenance systems namely corrective maintenance and planned maintenance [1]. Maintenance can be divided in two types: corrective and preventive maintenance. Corrective maintenance means that machine or component need a repair work, is done after a breakdown or when obvious failure has been found. Corrective maintenance as its best should be applied in non-critical machine where consequences of failure are slight ,rapid failure repair are possible, consequences of failure are slight, capital costs are small, and quick failure identification. Preventive maintenance is intended to minimize the probability of failure or degradation of functioning of a component, to carried out predetermined intervals.

In corrective maintenance repair is done when there is damage to engine and planned maintenance, maintenance is done by scheduling every one week. Other problems faced by the company are the environment conditions that are less convenient for operators such as slippery, muddy, and scattered (fiber) strewn conditions like this can hamper production activities where if the fiber fell into the engine area, the engine can be damaged like a broken CBB pen . Therefore the application of 5S is required in this study in addition to using RCM method. Problems of PT. XYZ is basically can be overcome by trying to propose a machine maintenance system using Reliability Cantered Maintenance (RCM) method. Defining the maintenance schedule and knowing the appropriate maintenance tasks that must be performed on every damaged component of the machine are done by the RCM method. And to overcome the problem of working environment needed an improvement method [2].

As a procedure to applied preventive maintenance in a complex system, Reability Centered Maintenance has been known and applied in many industrial field. Such as railway network, steel plant, ship maintenance or aviation. The countries applying RCM include Japan, Britain, The United States, etc [3]. The RCM methodology performs a structured and applied approach for achieving a satisfactory maintenance strategy for each part of a RCM system is an methodology that all of the management level concern, care, and preventive an maintenance strategies and practices in an integrated procedure to increase the probability that a machine or component will function as its design with minimum maintenance [4]. Any companyhave benefit from RCM if its breakdowns frequencies for more than 20-25% of the total system maintenance [5].Research on maintenance that apply Reliability Centered Maintenance ever done and poured in a journal entitled "Proposed Planning Machine Care with Reliability Centered Maintenance (RCM) Method at PT. Perkebunan Nusantara VII (Persero) Business Unit River NiruKab. Muara Enim. 2012, Asisco Hendro, et al [6]. And also research Evi Febrianti, et al with title "Proposed Planning Care Machine Roughing Stand with Approach Reliability Centered Maintenance. 2016" [7].Srinivasan et al. (2016) have analysed that 5S implementation can influence effectively by improving management and employee commitment, communication, work environment , priority of safety, management appreciation of risk , towards achieving improvement in productivity by optimizing space utilization, inventory ratio and minimizing cycle time [8]. The manufacturing management should follow up certain procedures which are comfortable to implement in order to achieve high level of improvement in performance indicators. So managers, consultants, and HR executives are conducted and build a strong team to utilize the effectiveness of management principles [9]. P.3SEM5S has emerged as a simplified and smooth tool for eliminating the waste and non-value added activities in the LM process to bring continuous improvement in an organization's performance [9]

2. Research Methods

Research begins with the review and data collection at PT. XYZ. Data collected from company documents are damaged machine components, damage interval data, damage frequency, and breakdown time of cake breaker conveyor machine. While observation of environmental condition of company conducted by observation directly in field. Reliability Centered Maintenance method is used to determine maintenance schedule of critical machine components. This method begins with System selection and information collection, Defines system boundaries, System description and function block, Descriptions of system function and malfunction, Failure Mode and Effect Analysis (FMEA), Logic Tree Analysis (LTA), Action Selection. After doing the next step RCM is testing the data distribution of each component, so the parameters obtained for the calculation of Reliability and MTTF. The above steps are performed to get the appropriate maintenance schedule for each broken machine component. The final step of this research is to provide suggestions for improvement of the company's environmental conditions.For the environment in PT XYZ would applied the 5S method, which is the method of structuring and maintaining the work area will improve workers' discipline and create a safe and comfortable working environment for all workers with support of all management level.

3. Results and Discussion

3.1. Steps of RCM
The first step in solving the problem of research results is by performing RCM steps. This method begins with System selection and information collection, Defines system boundary, System description and function block, Descriptions of system function and malfunction, Failure Mode and Effect Analysis (FMEA), Logic Tree Analysis (LTA), Action Selection. The flowchart of RCM steps can be seen in Figure1.

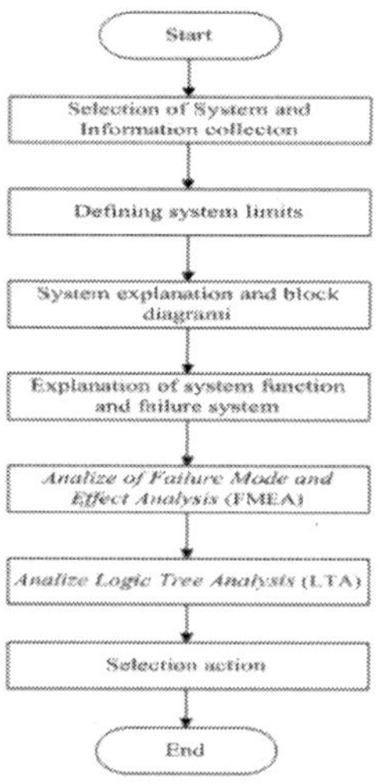

Figure 1. Steps of RCM

3.2. Distribution Test
The distribution test is performed on the components included in the Time Directed (TD) action selection. These four components are also critical components of the conveyor cake breaker machine. The results of the distribution test can be seen in Table 1.

Table 1. Distributions Test

NO	Name of Components	Distribution
1	*Pen* CBC	Normal
2	*Pedal conveyor*	Normal
3	*Gearbox*	Logormal
4	*Bearing*	Weibull

Based on Table 1, the pattern of damage distribution depends on the interval of component damage over a certain period.

3.3. Recapitulation Reliability and Mean Time to Failure (MTTF)
After knowing the distribution's type of critical component, the next step is recapitulation, reliability and MTTF critical component as shown on Table 2.

Table 2. Recapitulation Reliability and MTTF Critical Component of Cake Breaker Conveyor Machine

Name of Components	*Reliability*	MTTF
Bearing	0,4698	37,6776
Gearbox	0,4150	97,401
Pen CBC	0,4821	35,6429
Pedal Conveyor	0,4834	32,7647

Based on Table 2 can be known reliability and MTTF each component of cake breaker conveyor machine that is damaged. So that companies can apply preventive maintenance schedule based on research that has been done with the aim to reduce the engine breakdown time.

3.4. Maintenance Schedule of Critical Components

The maintenance schedule of cake breaker conveyor machine components is obtained based on Mean Time to Failure (MTTF) and based on the parameters of each selected distribution. Maintenance of critical machine components that often experience failure mode on cake breaker conveyor machine can be seen in Table 3.

Table 3. Maintenance Schedule of Components Machine

NO	Name of component	Maintenance Schedule (Days)
1	*Pen* CBC	35
2	*Pedal conveyor*	32
3	*Gearbox*	97
4	*Bearing*	37

Based on Table 3the maintenance schedule of the critical machine component becomes the basis of the company's consideration in making the calendar of repair schedule or the replacement of machine components in one year. Substitution of components is done before the damage occurs so as to reduce the downtime of the conveyor cake breaker machine. Actual care system is still corrective and planned maintenance, with the maintenance schedule of the machine is obtained a preventive treatment.

3.5. Proposed Improvement of the Work Environment with the 5S Implementation

Improving the working environment by applying the 5S principle is expected to provide comfort and security for operators at work so as to increase work productivity and avoid work accidents.

For implementation 5S in PT XYZ there are some of the procedure:

1.Top management involvement initiatives such as provide resources for supporting 5S as a continuous improvement initiatives, communicate and educate its employees about mission statement and strategy, demonstrate the business vision, strategy and implementation plan to the employees, procedural hassles in the administrative work identified and consistently reviewed to improve various support functions.

2. Employee involvement initiatives such as training to regard 5S and other quality control tools for understanding/solving problems, consistently evaluated and efforts made to improve the skills from time to time, motivate to participate in continuous improvement initiatives.

Implementation of preventive maintenance will not be expected to prevent the occurrence of damage to the machine but only reduce the number of machines damaged. Here is an example calendar maintenance schedule cake breaker conveyor machine in July 2017

Jul-17						
Sun	Mon	Tue	Wed	Thu	Fri	Sat
						1
2	3	4	5	6	7	8
9	10	11	12	13	14	15
16	17	18	19	20	21	22
23	24	25	26	27	28	29
30	31					

Figure 2. The Schedule of Maintenance

The preparation of the engine maintenance calendar in figure 2 is based on the MTTF value of each machine component. On the maintenance calendar specified the date of machine components that must be repaired or replaced. The color giving of the treatment calendar aims to distinguish the components to be treated. The purple color symbolizes the gearbox component, the green color symbolizes the conveyor pedal components, while the yellow color symbolizes the CBC pen component.

4. Conclusion

The critical component maintenance schedule based on the research result is bearing component is 37 days, gearbox components 97 days, 35 day CBC pen component and 32 day conveyor pedal components. Improvement of the company's work environment is done by proposing improvements with the implementation of 5S. After identifying the application of 5S, the proposed corporate environmental improvement actions are in accordance with 5S principles where unused goods will be transferred from the production area, grouping goods based on their use, determining the production area cleaning procedure, conducting ongoing inspection on the use of PPE, and creating a slogan.Slogan 5Sshould be keep in mind that the implementation of preventive maintenance will not be expected to prevent damage to the machine but only decrease the number of defective machines

References

[1] Niu G, et. al. 2010 *Development of an optimized condition-based maintenance system by data fusion and reliability-centered maintenance.*
[2] Cheng Z, et. al. 2008 *A framework for intelligent reliability centered maintenance analysis.*
[3] Fonseca D J, Knapp G M 2000 *An expert system for reliability centered maintenance in the chemical industry.*
[4] Igba J, et.al. 2013 *A Systems Approach towards Reliability-Centred Maintenance (RCM) of Wind Turbines.*
[5] Yssaad et. al. 2014 *Reliability centered maintenance optimization for power distribution Systems.*
[6] Hendro A, *et.al.* 2012 *UsulanPerencanaanPerawatanMesindenganMetode Reliability Centered Maintenance (RCM)* di PT. Perkebunan Nusantara VII (Persero) Unit Usaha Sungai Niru Kab. Muara Enim.
[7] Febrianti E *et.al.* 2016 *Usulan Perencanaan Perawatan Mesin Roughing Stand dengan Pendekatan Reliability Centered Maintenance*
[8] Randhawa J S 2016. *Evaluating Impact of 5S Implementation on Business Performance*

IOP Conf. Series: Materials Science and Engineering **309** (2018) 012128 doi:10.1088/1757-899X/309/1/012128

Design of preventive maintenance system using the reliability engineering and maintenance value stream mapping methods in PT. XYZ

N Sembiring*, N Panjaitan and S Angelita
Department of Industrial Engineering, Faculty of Engineering, Universitas Sumatera Utara, Medan 20155, Indonesia

*Email: nurhayatipandia68@yahoo.com

Abstract. PT. XYZ is a company owned by non-governmental organizations engaged in the field of production of rubber processing becoming crumb rubber. Part of the production is supported by some of machines and interacting equipment to achieve optimal productivity. Types of the machine that are used in the production process are Conveyor Breaker, Breaker, Rolling Pin, Hammer Mill, Mill Roll, Conveyor, Shredder Crumb, and Dryer. Maintenance system in PT. XYZ is corrective maintenance i.e. repairing or replacing the engine components after the crash on the machine. Replacement of engine components on corrective maintenance causes the machine to stop operating during the production process is in progress. The result is in the loss of production time due to the operator must replace the damaged engine components. The loss of production time can impact on the production targets which were not reached and lead to high loss costs. The cost for all components is Rp. 4.088.514.505. This cost is really high just for maintaining a Mill Roll Machine. Therefore PT. XYZ is needed to do preventive maintenance i.e. scheduling engine components and improving maintenance efficiency. The used methods are Reliability Engineering and Maintenance Value Stream Mapping (MVSM). The needed data in this research are the interval of time damage to engine components, opportunity cost, labor cost, component cost, corrective repair time, preventive repair time, Mean Time To Opportunity (MTTO), Mean Time To Repair (MTTR), and Mean Time To Yield (MTTY). In this research, the critical components of Mill Roll machine are Spier, Bushing, Bearing, Coupling and Roll. Determination of damage distribution, reliability, MTTF, cost of failure, cost of preventive, current state map, and future state map are done so that the replacement time for each critical component with the lowest maintenance cost and preparation of Standard Operation Procedure (SOP) are developed. For the critical component that has been determined, the Spier component replacement time interval is 228 days with a reliability value of 0,503171, Bushing component is 240 days with reliability value of 0.36861, Bearing component is 202 days with reliability value of 0,503058, Coupling component is 247 days with reliability value of 0,50108 and Roll component is 301 days with reliability value of 0,373525. The results show that the cost decreases from Rp 300,688,114 to Rp 244,384,371 obtained from corrective maintenance to preventive maintenance. While maintenance efficiency increases with the application of preventive maintenance i.e. for Spier component from 54,0540541% to 74,07407%, Bushing component from 52,3809524% to 68,75%, Bearing component from 40% to 52,63158%, Coupling component from 60.9756098% to 71.42857%, and Roll components from 64.516129% to 74.7663551%.

1. Introduction

The increasing competition in manufacturing, the company must make continuous improvement to keep the company's stability in achieving its vision.According to remain competitiveness, manufacturing companies are forced to optimize their production in terms of cost efficiency, lead time and quality [1].

One of the factors that need to be considered is the machine maintenance system. PT. XYZ is a private company engaged in the production of rubber processing into crumb rubber. The production section is supported by some interacting machines and equipment to achieve optimal productivity. The types of machines used in the production process are Conveyor Breaker, Breaker, Twin Roll, Hammer Mill, Conveyor Claw, Roll Mill, Crumb Shredder, and Dryer. To take care of the whole machine in the production, PT. XYZ implement maintenance/maintenance system. System maintenance at PT. XYZ is still corrective maintenance that is doing repair or replacement of machine components after damage to the machine. Production machines in PT. XYZ have many damaged where the frequency of breakdown and downtime of production machine exceeds the standard downtime set by the company.

The high downtime is due to the company applying corrective maintenance. If this maintenance strategy is used as the primary maintenance strategy will result in the high impact of unplanned maintenance activities from replacement parts. Replacement of components on the machine cause the engine to stop operating at the time of production process is in progress. The result is in the loss of production process time due to having to replace the damaged machine components. This is a problem at PT. XYZ due to production activities that do not run smoothly resulting in loss of production time that causes the company to experience loss production. The loss of production process time has an impact on not reaching the target on the set schedule so that causing an order delay. Loss of production time also affects the cost of loss of production during component replacement.

Based on the problem, the company suffered a lot of losses because the maintenance system used is still corrective maintenance. Therefore, it is important to apply preventive maintenance scheduled maintenance planning to perform component replacement so that the machine can operate at standard working hours without any damage during production process so that the production process can run smoothly, the target is reached on the specified schedule, and reduce the loss production time [3]. The goal of this research is to design preventive maintenance machine system with Reliability Engineering and Maintenance Value Stream Mapping (MVSM). The novelty of this research is there will be a calendar that has been marked in accordance with the replacement schedule of machine components.

Preventive maintenance is not a new method, but this method is very effective to reduce downtime and breakdown machine in a company.Previous research had been done in PT. Chaeron Phokphand Indonesia. Maintenance systems that are still breakdown maintenance and corrective maintenance cause the loss is the cost of loss of production, higher repair costs and overtime costs due to loss of production. Therefore, the calculation of machine maintenance costs on machine components, namely preventive cost (cost prevention) and failure cost (damage cost) [2]. Another research has been done in PT. XXX. The observed factory maintenance system is still corrective maintenance. This research was conducted to plan the maintenance of production machines using Reliability Engineering and MVSM (Maintenance Value Stream Mapping) method [3].

One of definition of system reliability is the ability of the system performing its intended function by using probabilistic expressions, like as the probability that the system will fail in a specified time, the expected period of time to failure and the expected number of failures for a certain period [3].

It must be known that reliability analysis is conducted and then the results are presented and also communicated to the management [4].

System Reliability analysis has objectives that all the estimating process of the system's probability will be able to indicate its function. Reliability of system's part influenced by the main of system reliability [5]

2. Research Methods

The type of this research is descriptive research, that is research which aim to describe systematically, factually and accurately about facts and properties of a certain object or population. The result of this research is to give the proposal of replacement schedule of the machine component and cost calculation to minimize company expenditure especially maintenance part.

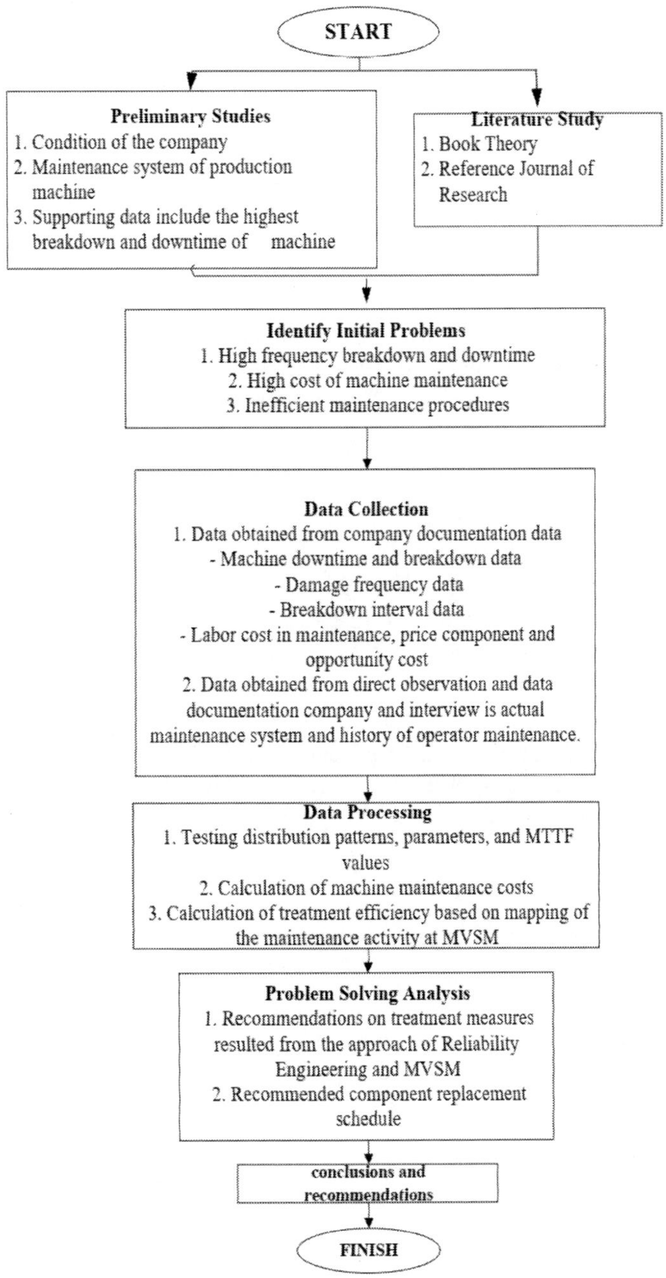

Figure 1. Reliability Engineering and Maintenance Value Stream Mapping Steps

The observed object is Mill Roll machine because this machine has the highest frequency of damage. The method begins with data collection, i.e. data collected from company documentation (machine downtime and breakdown, the frequency of damage, interval damage, maintenance personnel cost, component price, and opportunity cost), and data collected from direct observation and interview with the head of maintenance.All of that data is important to applicative the method. Further testing of distribution, parameters, and MTTF values, calculation of machine maintenance cost, and calculation of maintenance efficiency at MVSM.

VSM contains all activates that non value added as well as value added. The main objective of VSM is to select different types of wastes and then to eliminate them. The step begin with selecting a specific product as the target for improvement. The second step is developing a current state map that is mainly figure how processes are currently being done. The third step is drawing the future state map that is display of how the production process should be done after the wastes and inefficiencies have been removed. The future state map done by creating the answerof a collection of questions on topics relevant to efficiency.The last step, the suggested map is applied as a process for making essential changes to the system [6].Value Stream Mapping (VSM) method is very simple and effective to draw and redesign of value streams. So the current value stream is captured, and then value stream designed, in which sources of waste within the production process are reducedand uncovered [1].As a tool VSM help us to understand material's flow and product's information through the value stream. By calculating the difference between the current and the future state , the performance of the system could be improved [7]. The flowchart of Reliability Engineering and Maintenance Value Stream Mapping steps can be seen in Figure 1.

3. Results and Discussion

3.1. Testing Distribution

The first step in the Reliability Engineering Method is testing the distribution of each critical engine component. The results of the selected distribution pattern test for each machine component can be seen in Table 1. The preferred distribution pattern is the distribution that gets the largest Index of Fit (Correlation Coefficient) value.

Table 1. Recapitulation of Distribution Pattern of Damage Components

No.	Component	Distribution
1	*Bearing*	Normal
2	*Bushing*	*Weibull*
3	*Coupling*	Normal
4	*Spier*	Normal
5	*Roll*	*Weibull*

3.2. Calculation of Parameters and MTTF Machine Components

Calculating MTTF (Mean Time To Failure) is used as a parameter for determining the replacement of machine components. The result of recapitulation of MTTF calculation, hence got the interval of change of each component can be seen in Table 2.

Table 2. Recapitulation of MTTF Value

No.	Component	MTTF
1	*Bearing*	202
2	*Bushing*	240
3	*Coupling*	247
4	*Spier*	228
5	*Roll*	301

3.3. Reliability Calculation on Component Replacement Interval Schedule
The calculation of the reliability value of the critical machine component on the proposed replacement schedule is used to determine the reliability value of the machine components when the proposed component replacement schedule is performed. Reliability calculation can be seen in Table 3.

Table 3. Reliability Calculation on Component Replacement Interval Schedule

No.	Component	Reliability
1	*Bearing*	0,503058
2	*Bushing*	0,36861
3	*Coupling*	0,50108
4	*Spier*	0,503171
5	*Roll*	0,373525

3.4. Maintenance Cost Calculation
Maintenance cost calculation is divided into two, namely cost of failure and cost of preventive. Comparison of maintenance cost of machine component can be seen in Table 4.

Table 4. Total Comparison of Maintenance Costs

No.	Component	**Cf**	**Cp**
1	*Bearing*	53.541.257	39.641.114
2	*Bushing*	59.889.400	45.988.543
3	*Coupling*	38.565.714	30.225.229
4	*Spier*	58.299.771	49.957.000
5	*Roll*	90.391.971	78.572.486
	Total	**300.688.114**	**244.384.371**

From the result of correlation ratio computation cost and preventive maintenance cost, it can be seen that there is cost saving from IDR 300,688,114 to IDR 244.384.371.

3.5. Calculation of treatment efficiency on MVSM
The first step taken on MVSM is the establishment of current state map and future state map. The comparison between current state map and future state map can be seen in Table 5.

Table 5. Comparison of Current and Future State

Component	Category	Cf	Cp
Spier	MTTO	1,5	0,5
	MTTR	2	2
	MTTY	0,2	0,2
	% Maintenance Efficiency	54,054	74,074
Bushing	MTTO	1,8	0,8
	MTTR	2,2	2,2
	MTTY	0,2	0,2
	% Maintenance Efficiency	52,380	68,75
Bearing	MTTO	1,3	0,7
	MTTR	1	1
	MTTY	0,2	0,2
	% Maintenance Efficiency	40	52,6315

Coupling	MTTO	1,5	0,9
	MTTR	2,5	2,5
	MTTY	0,1	0,1
	% Maintenance Efficiency	60,9756	71,4285
Roll	MTTO	1,9	1,05
	MTTR	4	4
	MTTY	0,3	0,3
	% Maintenance Efficiency	64,5161	74,7663

Mill Roll Machine has the greatest frequency of damage and is chosen to be a critical machine. Mill Roll is used to grind the rubber from the form of small chunks into sheets. This machine is often damaged due to several things including the incorrect installation of machine components by the operator and wear caused by two surfaces that rub against the machine at the time of operation. Machine maintenance schedule with the critical component replacement for Spier component is 228 days with reliability 0,503171, Bushing component is 240 days with reliability 0,36861, Bearing component is 202 days with reliability 0,503058, Coupling component is 247 days with reliability 0, 50108, and the Roll component is 301 days with reliability 0.373525. The machine maintenance schedule can be seen in Figures 2 and 3.

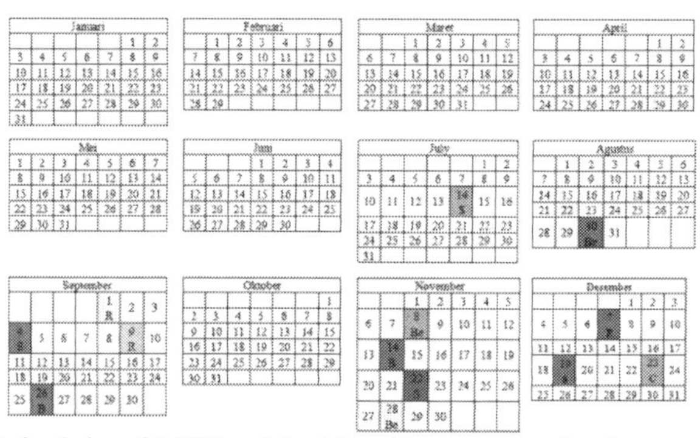

Figure 2. Schedule of Milling Machine Component Assistance Year 2016

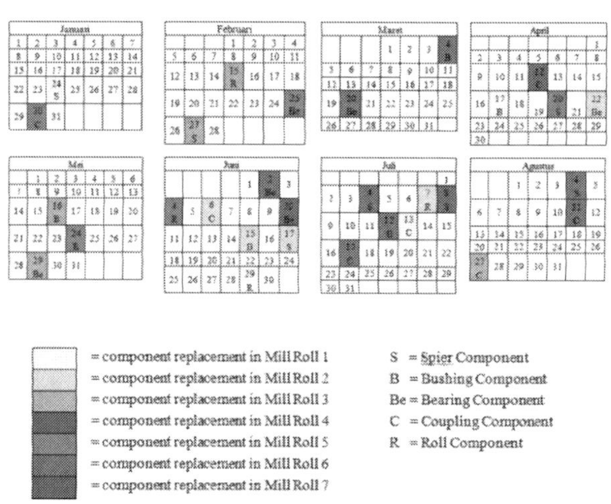

Figure 3. Schedule of Milling Machine Component Assembly Year 2017 and Description

The comparison of current machine maintenance costs and estimated cost proposals resulted in cost savings of Rp. 300,688,114 (cost of failure) to Rp. 244.384.371 (cost of preventive). While maintenance efficiency increased with preventive maintenance application that is for spier component is 74,07407% from 54,0540541%, bushing component is 68,75% from 52,380,9524%, bearing component is 52,63158% from 40%, coupling component is 71.42857% from 60.9756098%, and the Roll component is 74.7663551% of 64.516129%. Implementation of preventive maintenance is more profitable and cost-effective compared to the application of corrective maintenanceprocedure of the production area, conducting ongoing inspection in the use of PPE, and creating a slogan -Slogan 5 Keep in mind that the implementation of preventive maintenance will not be expected to prevent damage to the machine but only decrease the number of defective machines.

4. Conclusion

The critical engine that is the priority of the research is Mill Roll machine is the machine with the greatest damage frequency with critical engine components namely spier, bushing, bearing, coupling, and roll.The machine maintenance schedule with replacement of critical components for Spier component is 228 days, Bushing component is 240 days, Bearing component is 202 days, Coupling component is 247 days, and Roll component is 301 days. The reliability value of machine component on spier component replacement schedule is 0,503171, Bushing component equal to 0,36861, Bearing component equal to 0,503058, Coupling component equal to 0, 50108, and Roll component is 0.373525. The comparison of current machine maintenance costs and estimated cost proposals resulted in cost savings of Rp. 300,688,114 (cost of failure) to Rp. 244.384.371 (cost of preventive). Implementation of preventive maintenance is more profitable and cost-effective compared to the application of corrective maintenance. The results of the Stansard Operation Procedure (SOP) development with Maintenance Value Stream Mapping (MVSM) resulted in the percentage of maintenance efficiency for each component increased. This increase is derived from the comparison of current state map application value to development with future state map. The percentage of maintenance efficiency in the future state map for the spier component is 74.07407%, the bushing component is 68.75%, the bearing component is 52.63158%, the coupling component is 71.42857%, and the Roll component is 74.7663551%.

References

[1] Benjamin H et.al. 2014 Quality Value Stream Mapping p1-7
[2] Soesetyo I, Bendatu L Y 2014 Penjadwalan Predictive Maintenance dan Biaya Perawatan Mesin Pellet di PT. Chaeron Pokphand Indonesia – Sepanjang. Surabaya. Universitas Kristen Petra
[3] Oktalisa P 2013 Perancangan Sistem Perawatan Mesin dengan Pendekatan Reliability Engineering dan Maintenance Value Stream Mapping (MVSM) pada PT. XXX. Medan: FT USU
[4] Aven T 2017 Improving the Foundation and Practice of Reliability Engineering p.1-2
[5] Compare M et.al. 2017 Aggregation of Importance Measures For Decision Making In Reliability Engineering p.1
[6] Rohani, J M et.al. 2015 Production Line Analysis Via Value Stream Mapping: A Lean Manufacturing Process Of Color Industry p.7
[7] De Steur H et. al. 2016 Applying Value Stream Mapping to Reduce Food Losses and Wastes In Supply Chains: A Systematic Review p.2-4
[8] Faulkner W, et. al. 2014 Suistainable Value Stream Mapping (Sus-VSM): Methodology to Visualize and Assess Manufacturing Suistainability Performance
[9] Meudt T et.al. 2016 Value Stream Mapping 4.0: Holistic Examination Of Value Stream And Information Logistic In Production
[10] Verma N et.al. 2016 Energy Value Stream Mapping a Tool to develop Green Manufacturing

IOP Conf. Series: Materials Science and Engineering **309** (2018) 012129 doi:10.1088/1757-899X/309/1/012129

The utilization of Sambung Nyawa (*Gynura procumbent* [Lour]. Merr) leaves extracts as an antioxidant for coconut oil by using ethanol solvent

M S Sinaga*, R Ariska and P Defriska
Department of Chemical Engineering, Faculty of Engineering, University of Sumatera Utara, Medan, Indonesia

*E-mail: loisdanjoel@yahoo.com

Abstract. Sambung Nyawa (*Gynura procumbent* [Lour]. Merr) is commonly well-known as traditional medicine. It contains several chemical constituents such as flavonoids, saponins, tannins and steroids or triterpenoids, which are potential as an antioxidant. The aim of this study was to analyze the effect of Sambung Nyawa leaves to extract as an antioxidant and its ability to preserve the quality of coconut oil. First, flavonoids were extracted from Sambung Nyawa leaves at various leaf to the solvent ratio (w/v) and extraction temperature. The extracts which gave the highest total flavonoids content was added into the coconut oil and stored for 3 days, 6 days, 9 days, 12 days, and 15 days. Total flavonoids content of Sambung Nyawa leaves extracts were evaluated by UV-Vis spectrophotometry. Coconut oil was analyzed for its value of acid, iodine, and peroxide number. The results showed that the highest total flavonoids content of 5.18% was obtained for the leaf to solvent ratio of 1:15 (w/v) and an extraction temperature of 65 °C. The lowest acid number of 0.35%, the highest iodine number of 8.09 g I_2/100 g, and the lowest peroxide number of 5.20 mg O_2/100 g was obtained for the storage time of 3 days for coconut oil mixed with the Sambung Nyawa leaves extracts.

1. Introduction

The plant of Sambung nyawa is a traditional herb comes from South East Asia that contains several active chemical constituents such as flavonoids, saponins, tannins, and steroids or triterpenoids [1]. The plant has anti-Hyperglycemic, anti-Hyperlipidemic, analgesic [2]. The classification of Sambung Nyawa is shown in Table 1.

Table 1. The Classification of Sambung Nyawa

Division	Spermatophyta
Subdivision	Angiospermae
Class	Dicotyledoneae
Order	Asterales
Family	Asteraceae
Genus	Gynura

Coconut oil is a vegetable oil that produced from coconut meat. The fatty acids found in coconut oil consist of 90% saturated fatty acids and 10% unsaturated fatty acids [3]. The quality standards of coconut oil are 5% of maximum acid number, 8-10 g I_2/100g of iodine number and 5 mg O_2/100 g of maximum peroxide number [4]. Lots of treatments have been done to improve the quality of coconut oil as the development of technology, such as an addition of antioxidants into coconut oil to prevent oxidation [5]. Coconut oil oxidation may occur due to contact between the amount of oxygen and the coconut oil and the heating process [6].

Antioxidants are synthetic or natural. The function of the substances that may delay, retard or prevent the process of oxidation. Based on the research, synthetic antioxidants turned out to be carcinogenic, so that the usage of synthetic antioxidant began to be restricted [7, 8]. Natural antioxidants are commonly known as phenolic or polyphenolic compounds, such as flavonoids, cinnamic acid derivatives, coumarin, tocopherol, and polyfunctional acids [9]. Flavonoids are the largest phenolic group found in nature [10]. Flavonoids are widely distributed in all parts of the plant including fruit, root, leaf and stem epidermis. Flavonoids have been reported to have antioxidant activity which can inhibit the cell damage caused by oxidation [11, 12]. Flavonoids have the ability to change or reduce free radicals and also as anti-free radicals [8]. This extraction method is the most common method and easy to use to obtain the active substances [13].

In this study, flavonoids were extracted from Sambung Nyawa leaves by using ethanol solvent and reflux extraction method by varying temperature time and the ratio of raw material and solvent. The effectiveness of flavonoids contained in Sambung Nyawa leaves extracts as an antioxidant for coconut oil wiil be evaluated.

2. Methods

2.1. Materials
Sambung Nyawa leaves were collected from Medan, Indonesia. Ethanol 96% was produced by Rovin Chemical Indonesia.

2.2. Preparation of Sambung Nyawa Leaves Powder
Sambung Nyawa leaves were washed using clean water and left for 3-4 days (no solar heat) until completely dry. The dried leaves were blended into powder using a blender and then they were sieved using a strainer of 40 mesh/inch to equalize the size.

2.3. Flavonoids Extraction from Sambung Nyawa Leaves Procedure of Powder
Extraction of flavonoids was done by using reflux extraction method and ethanol as solvent. Sambung Nyawa leaves powder (25 grams) was mixed with (1:5; 1:10; 1:15) (w/v) of ethanol 96% and heated using a hot plate (35 °C, 45 °C, 55 °C, 65 °C) for 3 hours with 300 rpm stirring. The extracts were filtered using Whatman no.1 filter paper and then heated in the water bath (50 °C) for the viscous extracts. Total flavonoids content was analyzed by using UV-Vis Spectrophotometry in BTKLPP Medan.

2.4. Procedure of The Utilization of Sambung Nyawa Leaves Extract as an Antioxidant
Coconut oil (15 grams) was mixed with Sambung Nyawa leaves extract (5% of coconut oil mass) and put into the black bottle and stored for (3 days, 6 days, 9 days, 12 days, 15 days). The effectiveness of flavonoids contain in Sambung Nyawa leaves extracts was evaluated based on acid, iodine and peroxide number of coconut oil. The analysis were done for the coconut oil mixed with Sambung Nyawa leaves extracts, and the coconut oil with no addition of Sambung Nyawa leaves extract. The standard that used for the coconut oil characteristics based on SNI 01-2902-1992.

3. Results and Discussion

3.1 Total Flavonoids Content in Sambung Nyawa Leaves Extracts

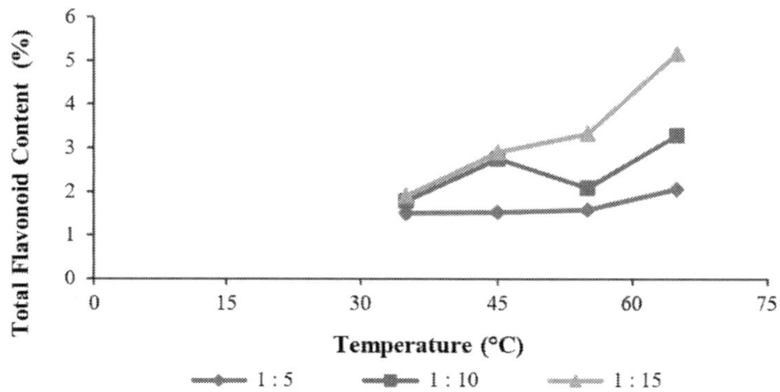

Figure 1. The effect of material to solvent ratio (w/v) and Temperature extraction to total flavonoids content (%)

Figure 1 showed the effect of material to the solvent ratio (w/v) and extraction temperature to total flavonoids content (%). For the ratio of 1:5 (w/v), total flavonoids content increased with increasing the extraction temperature. Total flavonoids content of 1.50%; 1.53%; 1.59%; 2.07% were obtained at 35 °C, 45 °C, 55 °C, 65 °C, respectively. The ratio of 1:10 (w/v) gave the fluctuation total flavonoids content. Total flavonoids content of 1.79%; 2.76%; 2.09%; 3.30% were obtained at 35 °C, 45 °C, 55 °C, 65 °C, respectively. For the ratio of 1:15 (w/v), total flavonoids content increased with increasing the extraction temperature. Total flavonoids content of 1.90%; 2.90%; 3.33%; and 5.18% were obtained at 35 °C, 45 °C, 55 °C, 65 °C, respectively. The extraction temperature and ratio of material to solvent (w/v) are factors that can influence the extraction process. The increasing of extraction temperature caused the movement of molecules of ethanol as solvent random and faster. The pores of the solid of Sambung Nyawa leaves powder swelled. Hence, solvent easier to diffuse into the pores of the solid of Sambung Nyawa leaves powder and dissolved flavonoids compound. So, the rate of extraction was higher and gave higher results. Similarly to the ratio of material and solvent (w/v) that the increase of material to the solvent ratio (w/v) will give the higher results, because the extracted material will contact more often with the solvent which more numerous than the solvent of small amounts [14, 15].

Overall, the results of this study showed that the increase of material to the solvent ratio (w/v) and extraction temperature gave the increase of total flavonoids content (%). But, in this study, there is the deviation for the ratio of 1:10 (w/v) at an extraction temperature of 55 °C where total flavonoids content decreased. It was caused by the presence of impurities in Sambung Nyawa leaves extracts which caused the change of flavonoids characteristic. Hence, the number of total flavonoids content became smaller. The highest total flavonoids content of 5.18% was obtained at temperature extraction of 65 °C and material to solvent ratio of 1:15 (w/v).

3.2 Acid Number of Coconut Oil Analysis

Figure 2 showed the analysis results of an acid number of coconut oil with and without extracts addition. From figure 2, we can see that an acid number of coconut oil increased with the duration of storage time. But, an acid number for coconut oil with extracts addition was lower than coconut oil without extracts addition. The acid number of 0.50%; 0.70%; 1.45%; 1.50% and 1.70% were obtained for coconut oil without extracts addition with storage time of 3 days, 6 days, 9 days, 12 days, and 15 days, respectively. The acid number of 0.35%; 0.55%; 0.90%; 1.25% and 1.45% were obtained for coconut oil with extracts addition with storage time of 3 days, 6 days, 9 days, 12 days, and 15 days, respectively. The lower value of the acid number for coconut oil with extracts addition showed the presence of flavonoids compound in Sambung Nyawa leaves extracts that acted as an antioxidant for coconut oil. Free fatty acid was found in oil or fat since the material was harvested and the amount of free fatty acid will increase during the processing due to the oxidation process that caused the amount of free fatty acid increased and the value of acid number became higher [3]. According to SNI 01-2902-1992, the maximum value of acid number for coconut oil is 5%. The results of this study gave the value of acid number that fulfill the standard of coconut oil for storage time of 3 days, 6 days, 9 days, 12 days, and 15 days. The lowest acid number of 0.35% was obtained for coconut oil with extracts addition and storage time of 3 days.

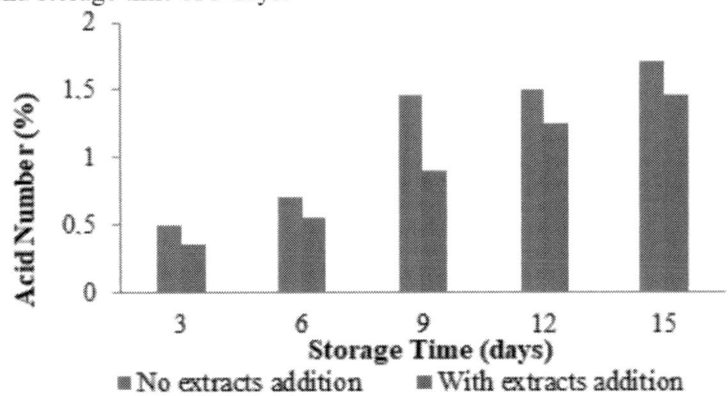

Figure 2. The effect of extract addition and storage time to acid number of coconut oil

3.3 Iodine Number of Coconut Oil Analysis

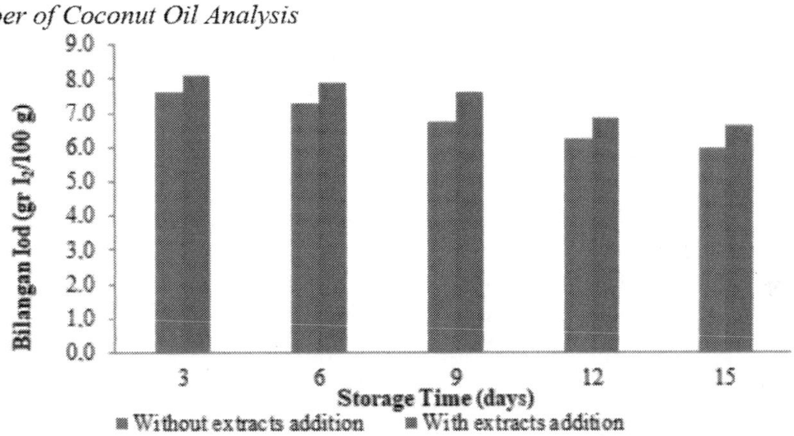

Figure 3. The effect of extract addition and storage time to iodine number of coconut oil

Figure 3 showed the analysis results of iodine number of coconut oil with and without extracts addition. From figure 3 we can see that the iodine number decreased with the longer of storage time. But, the iodine number for coconut oil with extracts addition was higher than coconut oil without extracts addition. The iodine number of 7.61 g I_2/100 g; 7.30 g I_2/100 g; 6.73 g I_2/100 g; 6.25 g I_2/100 g and 5.96 g I_2/100 g were obtained for coconut oil without extracts addition with storage time of 3 days, 6 days, 9 days, 12 days, and 15 days, respectively. The iodine number were obtained for coconut oil with extracts addition with storage time of 3 days, 6 days, 9 days, 12 days, and 15 days, respectively. Iodine number indicates the amount of unsaturation fatty acid in the coconut oil. The higher value of the iodine number indicates the better quality of coconut oil. The iodine number can decrease due to rupture of unsaturated bond become saturated caused by the upsurge of oxidation period hence the degree of unsaturation fatty acid in coconut oil becomes lower [16, 17]. Addition of antioxidant into coconut oil inhibits the oxygen bind to the double bonds of fatty acid and delay the oxidation process due to the lower of the amount of oxygen that bind to the double bonds, hence the

IOP Conf. Series: Materials Science and Engineering **309** (2018) 012129 doi:10.1088/1757-899X/309/1/012129

iodine number is increasing [18]. Addition of Sambung Nyawa leaves extracts into coconut oil gave the lower iodine number decreasing than coconut oil without Sambung Nyawa leaves extracts addition. According to SNI 01-2902-1992, the value of iodine number was 8-10 g I_2/100 g. Coconut oil with extracts addition with storage time of 3 days fulfill the standards of coconut oil, but for the storage time of 6 days, 9 days, 12 days, and 15 days, the iodine number decreased so that the value of iodine number of coconut oil lower than the minimum limit of the standard and for the coconut oil without extracts addition, the value of iodine number didn't fulfill the range of the standard with storage time of 3 days, 6 days, 9 days, 12 days and 15 days. The highest iodine number of 8.09 g I_2/100 g was obtained for coconut oil with extracts addition.

3.4 Peroxide Number of Coconut Oil Analysis

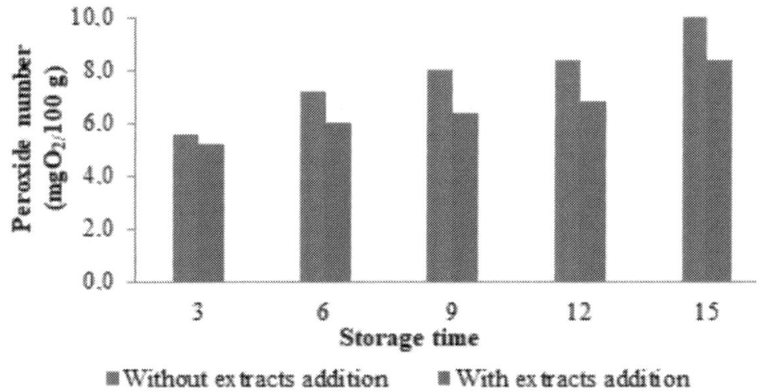

Figure 4. The effect of extract addition and storage time to peroxide number of coconut oil

Figure 4 showed the analysis results of peroxide number of coconut oil with and without extracts addition. From figure 4 we can see that the peroxide number will increase days by days. But, the peroxide number of coconut oil with extracts addition was lower than coconut oil without extracts addition. The peroxide number of 5.60 mg O_2/100g; 7.20 mg O_2/100g; 8.00 mg O_2/100g; 8.40 mg O_2/100g; and 10.00 mg O_2/100g were obtained for coconut oil without extracts addition. The peroxide number of 5.20 mg O_2/100g; 6.00 mg O_2/100g; 6.40 mg O_2/100g; 6.80 mg O_2/100g; and 8.40 mg O_2/100g were obtained for coconut oil with extracts addition. The peroxide number showed the rancidity of coconut oil due to oxidation and hydrolysis. The high peroxide number indicates the fat or oil has oxidized. The peroxide number of coconut oil will increase along with the duration of the storage time [6, 18]. Coconut oil with extracts addition gave the lower peroxide number than coconut oil without extract addition. According to SNI 01-2902-1992, the maximum value of peroxide number was 5.0 mg O_2/100 g. The results of peroxide number analysis showed the value that higher than the maximum limit of standard with storage time of 3 days, 6 days, 9 days, 12 days, and 15 days. The lowest peroxide number of 5.20 mg O_2/100g was obtained for the coconut oil with extracts addition dan storage time of 3 days.

4. Conclusion

In this study, the highest total flavonoids content of 5.18% that was obtained at the extraction temperature of 65 °C and material to solvent ratio of 1:15 (w/v) was added into coconut oil. To evaluate the flavonoid activity in Sambung Nyawa leaves extracts as an antioxidant, the coconut oil with extracts addition was compared to coconut oil without extract addition based on some characteristics of coconut oil which included the acid number, iodine number, and peroxide number. The results showed that the lowest acid, the highest iodine number, and the lowest peroxide number of 0.35%; 8.09 g I_2/100 g and 5.20 mg O_2/100, respectively were obtained for coconut oil with extracts addition and storage time of 3 days. Sambung Nyawa leaves extracts that was added into coconut oil can retard the oxidation process, hence gave the better characteristics than coconut oil without extracts addition.

5. References

[1] Riadini R K, Sidharta B B R and Pranata F S 2015 Uji Aktivitas Antioksidan Ekstrak Daun Sambung Nyawa (Gynura procumbens (Lour.)Merr.) Berdasarkan Perbedaan Metode Ekstraksi dan Umur Panen *Jurnal Biota UAJY.* 1-6

[2] Rosidah, Mun F Y, Amirin S and Asmawi M Z 2015 Antioxidant Potential of Gynura procumbens. *Informa.* 616-25

[3] Syah A N A 2005 Mengenal Lebih Dekat Virgin coconut Oil: Minyak Penakluk Aneka Penyakit Agromedia Pustaka, Jakarta 17-26

[4] Standar Nasional Indonesia (SNI) 01-2902-1992. *Minyak Kelapa.* Badan Standarisasi Nasional.

[5] Susilowati 2009 Pembuatan Virgin Coconut Oil dengan Metode Penggaraman *Jurnal Teknik Kimia* **3** 246-47

[6] Muslimah U and Guntarti A 2014 Ekstrak Etanol Kulit Buah Manggis (Garcinia mangostana L.) Sebagai Antioksidan Alami Pada Minyak Krengseng Prosiding *Seminar dan Workshop Perkembangan Terkini Sains Farmasi dan Klinik IV* Yogyakarta, 22-8

[7] Tamat S, Wikanta T and Maulina, L.S 2007 Aktivitas Antioksidan dan Toksisitas Senyawa Bioaktif dari Ekstrak Rumput Laut Hijau Ulva reticulata Forsskal *Jurnal Ilmu Kefarmasian Indonesia* **5** 31-6

[8] Zuhra C F, Juliati and Sihotang H 2008 Aktivitas Antioksidan Senyawa Flavonoid Dari Daun Katuk *Jurnal Biologi Sumatera* **3** 7-10

[9] Utami R D, Yuliwati K M and Syafnir L 2015 Pengaruh Metode Ekstraksi terhadap Aktivitas Antioksidan Daun Sukun (Artocarpus altilis (Parkinson) Fosberg) Prosiding *Penelitian SPESIA Unisba*, Bandung. 280-82

[10] Kumar S and Pandey A K 2013 Chemistry and Biologycal Activities of Flavonoids: An Overview: *The Scientific World Journal,* **13** 1-3

[11] Widyawati W 2007 Ekstrak Daun Sambung Nyawa (Gynura procumbens (Lour.) Merr) Terhadap Kadar Metil Merkuri Darah dan karakteristik Eritrosit Tikus Putih (Rattus norvegicus L.) Paska Pemaparan Metil Merkuri Klorida, Skripsi, Universitas Sebelas Maret, Surakarta

[12] Worotikan D E 2011 Efek Buah Lemon Cui (Citrus microcarpo) Terhadap Kerusakan Lipida Pada Ikan Mas (Cyprinus carpio L) Dan Ikan Cakalang (Katsuwonus pelamis) Mentah, Skripsi, Universitas Sam Ratulangi, Manado

[13] Umar F 2008 Optimisasi Ekstraksi Flavonoid Total Daun Jati Belanda, Skripsi, Institut Pertanian Bogor, Bogor

[14] Mastuti E, Kim E V and Christanti M E 2012 Ekstraksi Senyawa Brazilin dari Kayu Secang (Caesalpinia sappan linn) Sebagai Bahan Baku Alternatif untuk Zat Warna Alami *Ekuilibrium,* **11** 1- 5

[15] Ramadhan A E and Phaza H A 2010 Pengaruh Konsentrasi Etanol, Suhu dan Jumlah Stage pada Ekstraksi Oleoresin jahe (Zingiber officinale Rosc) Secara Batch, Skripsi, Universitas Diponegoro, Semarang

[16] Nugraheni D T 2010 Analisis Penurunan Bilangan Iod Terhadap Pengulangan Penggorengan Minyak Kelapa dengan Metode Titrasi Iodometri, Skripsi, Universitas Sebelas Maret, Surakarta

[17] Triyanto A 2013 Peningkatan Kualitas Minyak Goreng Bekas Menggunakan Arang Ampas Tebu Teraktivasi dan Penetralan dengan $NaHSO_3$, Skripsi, Universitas Negeri Semarang, Semarang

[18] Laitupa D and Susane H 2010 Pemanfaatan Eugenol dari Minyak Cengkeh Untuk Mengatasi Ranciditas pada Minyak Kelapa *Jurnal Teknik Kimia Undip* 1-10

IOP Conf. Series: Materials Science and Engineering **309** (2018) 012130 doi:10.1088/1757-899X/309/1/012130

Research mapping in North Sumatra based on Scopus

M K M Nasution[1]*, R Sitepu[2], Rosmayati[3], D Bakti[3], S M Hardi[1]

[1]Information Technology Department, Universitas Sumatera Utara, Padang Bulan USU Medan 20155, Indonesia.
[2]Program Studi Ilmu Hukum, Fakultas Hukum, Universitas Sumatera Utara, Padang Bulan USU Medan 20155, Indonesia
[3]Fakultas Pertanian, Universitas Sumatera Utara, Padang Bulan USU Medan 20155, Indonesia.

*Email: mahyuddin@usu.ac.id

Abstract. Research is needed to improve the capacity of human resources to manage natural resources for human well-being. Research is done by institutions such as universities or research institutions, but the research picture related to human welfare interests is not easy to obtain. If research can be proven through scientific publications, scientific research publication databases can be used to view research behaviour. Research mapping in North Sumatra needs to be done to see the suitability of research conducted with development needs in North Sumatra, and as a presentation is the Universitas Sumatera Utara which shows that research conducted has 60% strength, especially in the exact sciences.

1. Introduction

There has been a gap in the competence of human resources between North Sumatra and other provinces in Indonesia [1]. This is also in line with Indonesia's lower human development index compared to the overseas state, and it certainly affects the level of economic growth and the level of people's welfare. Dissent with it, in other words, the rich natural resources have no effect on improving the social welfare of the community, with which social engineering plays a role as object of unusual crime [2].

Social engineering aims to improve the quality of human resources, one of which is by increasing the knowledge of community members [3]. North Sumatra as one of the provinces in Indonesia has a large population, which is in fourth place in Indonesia, or 13,527,937 people (census of 2014), which is psychologically significant to the improvement of the Indonesian economy [4]. The low human development index of North Sumatra is caused by the inadequate knowledge of the community and its ability to process natural resources [5]. While the ability to process natural resources is obtained through research and development that runs in college. University of North Sumatra is one of the state universities in North Sumatra Indonesia [6]. Therefore, research abilities in universities should synergize with natural resources to benefit the community. This paper aims to map the research potential of the North Sumatra based on Scopus indexed scientific publication as a proof that the research has been done.

IOP Conf. Series: Materials Science and Engineering **309** (2018) 012130 doi:10.1088/1757-899X/309/1/012130

2. Problem Definition

In the Scopus database [7] there are 6 (six) institutions in North Sumatra that have produced Scopus index documents (July 2017), i.e.

a. Universitas Sumatera Utara (University of North Sumatra) or abbreviated to USU with 867 documents.
b. Indonesian Oil and Palm Research Institute with a total of 23 documents.
c. University of North Sumatra Medical School (this is as an affiliation name mistake made by the author so it is not included as a scientific paper from Universitas Sumatera Utara) with a total of 10 documents.
d. Islamic University of North Sumatra as many as 9 (nine) documents.
e. Universitas Sari Mutiara Indonesia as many as 9 (nine) documents.
f. Nias District Health Department - Indonesia has only 1 (one) documents.

The University of North Sumatra therefore is reasonable to represent the existing research in North Sumatra. Moreover, the research power of Universitas Sumatera Utara lies in 15 faculties plus one graduate school: Medicine, Law, Agriculture, Engineering, Economics and Business, Dentistry, Cultural Sciences, Mathematics and Natural Sciences, Social Science and Political, Pharmacy, Public Health, Psychology, Nursing, Computer Science and Information Technology (Fasilkom-TI), and Forestry, whereby the spearhead of strength is in 156 study programs that supported by the research institution [6].

For example, at Medicine Faculty of USU, there are courses and departments: clinical, tropical, nutritional, biochemical, histology, physiology, pharmacology, microbiology, parasitology, anatomy, community, eye diseases, internal diseases, neurological diseases, skin and venereal diseases, tht (ear-nose-throat) diseases, anesthesiology and reanimation, surgery, child health, orthopaedic surgery and traumatology, heart and blood vessel disease, anatomical pathology, clinical pathology, psychiatry, and neurosurgery (http://www.usu.ac.id/program-studi.html). Thus the description of the power distribution of the study program reflects the strength of medicine faculty. In this case, accreditation of the study program of neurosurgery department is A, it is to demonstrate that potential of its alumni in an international scale. However, it is necessary to consider the strength or superiority of research related to the intended study program, and among all of same study programs is quite competitive, it should be observed portrait of the related research.

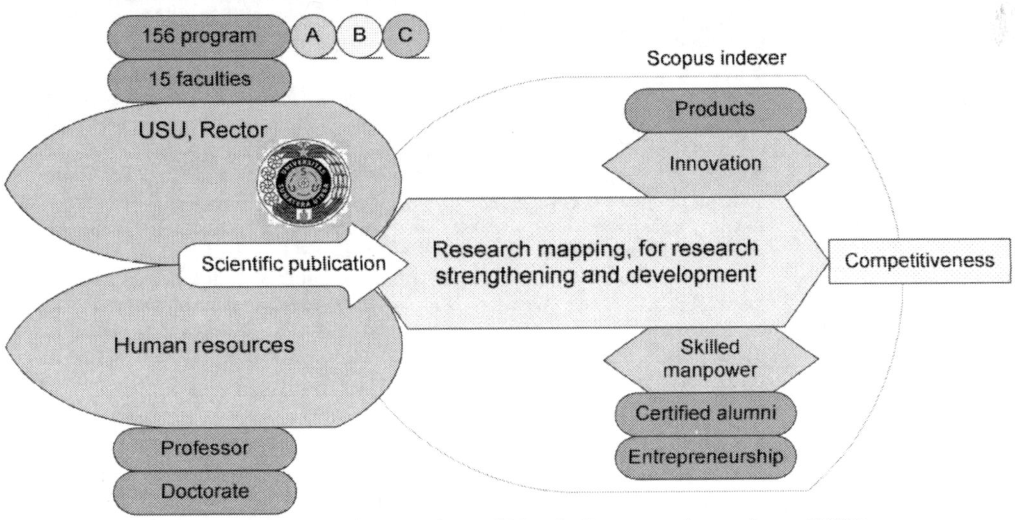

Figure 1. Research mapping of North Sumatra depends on USU.

Research mapping based on research outcomes such as Scopus index documents provides an overview of the growing research portrait at a university. Scopus-indexed document is one of the parameters of the existing potential in USU as representation of North Sumatra to field of science and technology, or the knowledge and its development [8, 9]. As discussed in order to spur the presence of appropriate technology or research products in the community, or products that can improve the welfare of the community, the excellent centre is developed which ultimately leads to produce a number of scientific papers: scientific publications, patents and intellectual property rights. Furthermore, in implementation of downstream policy of research will appear related small businesses or industries associated with it [10].

Table 1. Description of faculties in USU

No	Faculties	Subject area based on study program - department	Totality
1.	Medicine	clinical, tropical, nutritional, biochemical, histology, physiology, pharmacology, microbiology, parasitology, anatomy, community, eye diseases, internal diseases, neurological diseases, skin and venereal diseases, tht (ear-nose-throat) diseases, anesthesiology and reanimation, surgery, child health, orthopaedic surgery and traumatology, heart and blood vessel disease, anatomical pathology, clinical pathology, psychiatry, and neurosurgery	25
2.	Law	law science, notariat, economic law, international law, state administration, civil, state, and criminal.	8
3.	Agriculture	agricultural science, agro-technology, soil science, aquatic resources management, agribusiness, food science, food technology, livestock science, agricultural engineering.	9
4.	Engineering	mechanical, electrical, civil, chemical, industrial, factory management techniques, industrial mechanical technology, civil engineering sciences, chemical analysis, architecture, urban, and environmental.	12
5.	Economic and Business	economics, development economics, development studies, management, management sciences, secretarial, financial, accounting.	7
6.	Dentistry	dental education, dental professions, periodonsia, orthodontics, and prosthodontics.	5
7.	Cultural Sciences	linguistic sciences, Indonesian literature, regional literature: Batak, Malay, Arabic literature, English and English literature, history, ethnomusicology, creation and study of art, literature and Japanese language, library science, Chinese literature, tourism.	13
8.	Mathematics and Natural Sciences	physics, chemistry, factory instrumentation technology, metrology and instrumentation, mathematics, statistics, informatics engineering, and biology	8
9.	Social Sciences and Political	sociology, social welfare science, taxation, state administration, communication science, social anthropology, political science, commercial administration.	8
10.	Public Health	Public health sciences, environmental health management, administration and health policy, public health, population and biostatistics, epidemiology, public health nutrition, occupational safety and health, environmental health, health education and behavioural sciences.	10
11.	Pharmacy	pharmaceutical science, pharmaceutical and food analyst, pharmacist profession and education, pharmaceutical pharmacology, pharmaceutical chemistry, pharmaceutical biology, and pharmaceutical technology.	7
12.	Nursing	nursing science, nursing, ners, profession, educator midwife, soul and community nursing, nursing maternity an child, nursing basic medical surgery.	7
13.	Psychology	science and professions, industry and organization, clinical, educational, general and experimental, developmental, social.	7
14.	Fasilkom-TI	computer science, informatics engineering, and information technology	3
15.	Postgraduate	management, property management, science management of natural and environmental sub lay, science of regional and rural development planning, development studies, natural resource and environmental management, regional and rural development planning.	7

3. An Approach

To map the strength of research in an institution there are several parameters to consider: human resources, availability and budgetary adequacy, dissemination and overlapping of activities or cooperation, and partisanship. These parameters have constraints, i.e. resource productivity, regulation, research performance, research strength, natural or local resources, and the strength of graduate students [11, 12]. Institutionally, qualified university is an institution that has the ability to conduct research. Research is carried out by qualified human resource, such as qualified physician lecturers supported by funding, otherwise funds that spent without any outcomes in the associated resources not having sufficient, then the human resource is not competence as a quality resource [13]. Quality resources lead to the emergence of skilled workers (alumni) who are able to produce products and innovation. The determinant factor of human resources are qualified if they produce scientific publications, intellectual property, patents, and books as research outputs, see Fig. 1.

Research outcomes through information technology can be easily accessed, but difficult to understand well immediately [14]. Information about scientific publications or scientific papers can be obtained either by involving search engines directly to researchers, or accessing database online [15,16]. For example, through the Google search engine the name of the author of the scientific papers "Opim Salim Sitompul" resulted in hit count 29,000 with various recording activities, while through Google Scholar there are 209 hits relating to the document, whereby the citation recorded as 61. While the database data publication field of computer science or dblp computer science bibliography only presents 3 scientific papers [17]. If we collect several authors from an institution and the appropriate information about them, it can be representative of the institution as a whole. The field that develop will be visible from the continuity of the writings of the related authors. Therefore, the development of research can be seen directly through the database of scientific publishing indexers, such as DBLP, Scopus or others. Thus, to map the research in North Sumatra, we collect scientific papers from Scopus database based on affiliation (Medan for North Sumatra). As research mapping, the strength of the research is tailored to the description of the faculty at USU by involving the measurement of similarity [18].

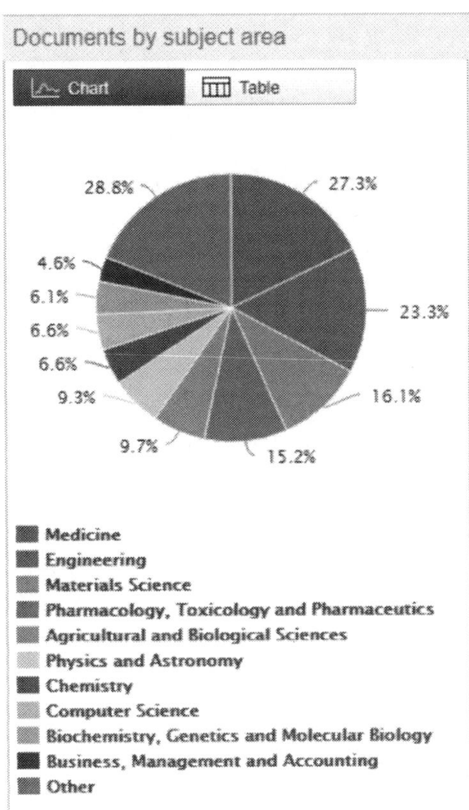

(a)

1	Subject area	2016	2015	2014	2013	2012	2011	2010	2009	2008	2007
2	Total	187	103	62	39	28	22	16	12	7	6
3	Medicine	31	22	7	8	5	5	4	2	2	1
4	Biochemistry, Genetics and Molecular Biology	11	4	3	1	3	3	3	2		1
5	Immunology and Microbiology	1	2	1		1					
6	Agricultural and Biological Sciences	17	12	5	8	9	3	3	3	2	
7	Energy	3	2		5	1		1			
8	Materials Science	16	4	2	3	3	1	1			2
9	Economics, Econometrics and Finance	20	2	1	1	1			1		
10	Social Sciences	11	4	1	1			1	1		
11	Pharmacology, Toxicology and Pharmaceutics	46	33	22	4	1			1		
12	Decision Sciences	2									
13	Mathematics	11	1	3	1	1		3	2		
14	Physics and Astronomy	11	3	2	9	3					
15	Engineering	23	23	11	7	5	5	5		2	2
16	Chemistry	20	9	7	4	2		1			
17	Chemical Engineering	13	8	6		2	1		1		
18	Dentistry	1			1	1					
19	Arts and Humanities	6									
20	Nursing		1		1						
21	Computer Science	16	7	3	2	1	1	1		1	2
22	Environmental Science	11	3	3	2		1				
23	Business, Management and Accounting	24	4	1	1	5			1		
24	Health Professions			1				1			
25	Multidisciplinary		1	2		1	2				
26	Earth and Planetary Sciences	4		1		1					
27	Psychology				1						
28	Veterinary		1								
29	Neuroscience	1									

(b)

Figure 2. Subject area based on scientific papers of USU in Scopus.

4. Discussion

The branches of knowledge developed at each faculty of Universitas Sumatera Utara [6] at most are in the medical faculty that is 25 subject areas, see Table 1. This is in line with the number of scientific papers indexed by Scopus still dominated by the medical fields, or most of 867 documents, see Fig. 2 (a) – July 2017. A lecturer in USU first has been produced document Scopus indexed records in 1970 in the Faculty of Medicine, since then the following years have sprung up though not many (1971, 1973, 1974, 1976, 1978, 1979, 1980, 1981, 1982, to present) USU produces documents gradually. Number of documents more than 10 papers first time in 1990, i.e. 17 documents, while previous years under 10 papers only that were dominated by medicine, but although not many some other subjects area also present their Scopus documents. Thus, there is a change of behaviour about research: ideas, concepts, and innovations in USU [10], which shows the change of research map a goal to meet the welfare society [19, 20].

By the time line, before 2012, not all subject areas (fields) that have emerged are able to produce the scientific papers, indicating that research in the field is not conducted or nothing is done, but on the medicine area continues to exist. The consistency of research in the various fields of science emerged one by one, see Figure 2 (b). This condition continues to be take place from 2012 until now (July 2017). Thus, almost all fields of sciences, especially the exact field is represented by scientific publications every year. But the weakness but seen in the field of social sciences, although the subject of new areas continue to emerge but only in the exact field not on social areas.

Although similarity [18] between the subject areas in Table 1 and Fig. 2 (b) are only about 60% roughly, it indicates that research services that may impact on the welfare of society come only from the presence of technology and exact sciences, but from the point of social policy change only few can be proven to exist. Therefore, social engineering is given only by technology factors or development of exact science in medicine and engineering.

5. Conclusion

Welfare comes from every human ability, the ability emerges from the knowledge they possesses. The development of knowledge gained from the implementation of research that is proven and disseminated through scientific papers. By comparing the strength of USU in faculty description with the growth of scientific publications from year to year in the Scopus databases according to the subject areas obtained the mapping research that developed in North Sumatra with a strength of 60% in the exact sciences mainly. Research mapping needs to be exposed so as to see the researcher's behavior as the next research.

References

[1] M K M Nasution 2016 Karya ilmiah dosen & mahasiswa *Harian Waspada*.
[2] I Lubis, M K M Nasution, and M Maulina 2017 Basic framework of urban design based on natural resources *Friendly City* **4**.
[3] T R Peltier 2006 Social engineering: Concepts and solutions *Information Systems Security* **15(5)**.
[4] M K M Nasution, T E Nuradi, and R Syah 2017 SumutSiana: A framework for applying ICT to preserve the cultural heritage of Sumatera Utara Indonesia *Journal of Telecommunication, Electronic and Computer Engineering* **9(2-4)**.
[5] R Syah, T E Nuradi, and M K M Nasution 2017 A framework to apply ICT for bequeathing the cultural heritage to next generation *Journal of Physics: Conference Series* **801(1)**.
[6] Universitas Sumatera Utara 2015 Rencana strategis USU 2015-2019.
[7] J Bar-Ilan 2008 Which h-index? – A comparison of WoS, Scopus and Google Scholar *Scientometrics* **74(2)**.
[8] L I Meho and K Yang 2007 Impact of data sources on citation counts and rankings of LIS faculty: Web of science versus Scopus and google scholar *Journal of the Association for Information Science and Technology* **58(13)**.
[9] M E Falagas, E I Pitsouni, G A Malietzis, and G Pappas 2008 Comparison of PubMed, Scopus, Web of Science and Google Scholar: strengths and weaknesses *The FASEB Journal* **22(2)**.

[10] M K M Nasution 2016 Hilirisasi penelitian berbasis teknologi pada perguruan tinggi *Harian Analisa*.
[11] M K M Nasution 2016 Fenomena riset *Harian Analisa*.
[12] M K M Nasution 2016 Carut marut menulis karya ilmiah *Harian Waspada*.
[13] S Bell, A Montarzino, and P Travlou 2007 Mapping research priorities for green and public urban space in the UK *Urban Forestry & Urban Greening* **6(2)**.
[14] M K M Nasution 2017 Semantic interpretation of search engine resultant *Interior*.
[15] M K M Nasution 2017 Modelling and simulation of search engine *Journal of Physics: Conference Series* **801(1)**.
[16] M K M Nasution 2016 Social network mining (SNM): A definition of relation between the resources and SNA *International Journal on Advanced Science, Engineering and Information Technology* **6(6)**.
[17] M K M Nasution, and S A Noah 2011 Extraction of academic social network from online database *2011 International Conference on Semantic Technology and Information Retrieval* (STAIR).
[18] K M N Mahyuddin, O S Sitompul, S Nasution, and H Ambarita 2017 New similarity *IOP Conference Series Materials Science and Engineering* **180(1)**.
[19] M K M Nasution 2016 Menggali potensi ujian nasional *Harian Waspada*
[20] M. K. M. Nasution and M Maulina 2017 Calligraphy design for coconut garbage use *Friendly City* **4**.

IOP Conf. Series: Materials Science and Engineering **309** (2018) 012131 doi:10.1088/1757-899X/309/1/012131

SumutSiana

M K M Nasution*

Information Technology Department, Universitas Sumatera Utara, Padang Bulan USU Medan 20155, Indonesia

*Email: mahyuddin@usu.ac.id

Abstract. Technically, a definition is at the heart of all forms of understanding. However, in the engineering aspect, the introduction becomes a first step that expands the meaning and function of definitions used for something, especially a particular study. SumutSiana is a term specifically constructed to understand not only North Sumatra culture, but the natural wealth required in social engineering, and this paper becomes infrastructure for it. SumutSiana is disclosed as a scope of discussion on natural resources, culture, human resources, and all related to North Sumatra.

1. Introduction

To give meaning or semantic technology, defining is the basic technique for achieving an understanding of a term [1,2]. In the world of engineering, all the technologies and tools as part of the system or not is to increase social prosperity or for better social engineering must be expressed more clearly in order to be adapted and they have a bridge of implementation to different systems [3]. In the world of education, giving definition to something is the scientific culture. The goal is that the conversation about something is not misdirected or widespread uncontrollably [4]. Moreover, to study things scientifically, it is necessary that something be well understood and true.

SumutSiana is an unknown term either it used by the public or it contained in any dictionary [5]. The term that is used to understand something, the like SumutSiana must be clearly recognizable its meaning. Usually everything that is not recognized should be introduced first [6,7]. This paper intends to introduce a scientific discourse related to the wealth of Indonesian nation located in the region of North Sumatra. Therefore, the first part of this paper is to introduce this term as the motivation for further research on this, and then to propose reasons as an approach, an end with a discussion of the scope of the intended wealth.

2. Definition and Motivation

The term written as "sumutsiana" and introduced also as the logo in uppercase S, namely "SumutSiana" has not been the vocabulary of any dictionary either of Indonesia or other language [5]. So have no clear meaning. The term "SumutSiana" consists of a combination of the words "sumut" and siana". Sumut stands for "Sumatera Utara" (in English it is North Sumatra). The name of the states in Indonesia, i.e. Sumatera Utara Province, in geographically according to layout located at coordinates 1° - 4° N 98° - 100° E, see Fig. 1. While the term "siana" comes from the article "si" (the) and the word "ana". "ana" has a meaning to exist and also refers to the female or not male gender. The word "ana" present with the article "si", i.e. "siana" means treasury or scientific wealth owned by the original owner from the beginning or has been a long time ago, and is not owned by another.

Therefore, the "SumutSiana" can be expressed as or refers to the treasures that are specifically owned by the North Sumatra region or natural wealth which is the kinds of nature or culture in the region of North Sumatra [6,7].

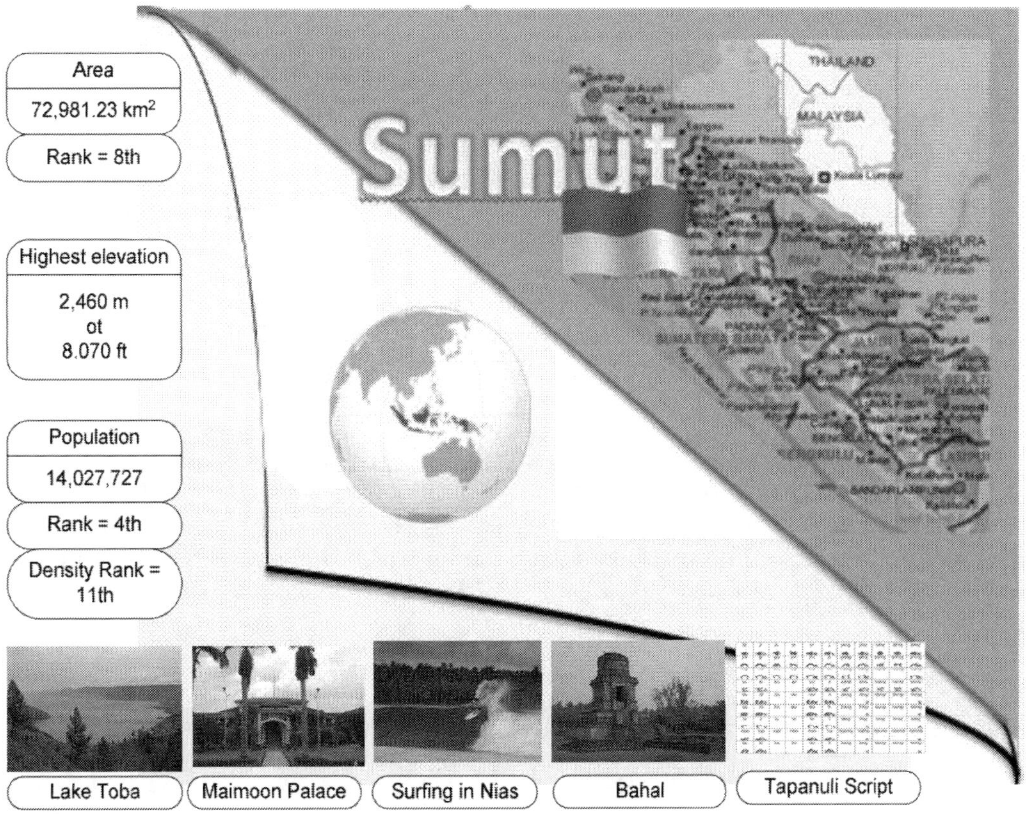

Fig.1. Sumatera Utara and some potentials

The area of North Sumatra has geographically rich contours, and this area is inhabited by 8 (eight) ethnics (the oldest social community in Indonesia) plus some immigrants [8]. Culturally there are differences and similarities between one another, both language and in social. This wealth not only includes natural resources but also local policies that have long been at the core of the social community activities of these tribes, like "Dalihan na tolu" [9]. Therefore, the definition of "SumutSiana" also involves a specific scope of regionalism in particularly, generally to Indonesian, covering the whole of what is contained in geographical space uniquely exist in North Sumatra, including ideas and concept, planning, design, strategy, implementation, naming, order, socialization, adaptation, policy, etc. Therefore, North Sumatra has potential as described briefly in Fig. 1. For example, the strategic plan of Universitas Sumatera Utara (USU) is based on and named with "TALENTA" (it means talent in English), but it based on mapping for the research and developing in USU, i.e. **T**ropical science and medicine, **A**groindustry, **L**ocal wisdom, **E**nergy (sustainable), **N**atural resources (biodiversity, forest, marine, tourism), **T**echnology (appropriate), **A**rts (ethnic) [10]. All of these go into the concept and idea about the study of SumutSiana.

3. An Approach

A definition is an approach to the complete understanding of something. Next, we describe the definition based on content and assigns a measurement value to each content, including the existence of the description and scope. SumutSiana becomes the basic concept for developing framework to embrace the scope of all the wealth of the North Sumatra region. This approach as a scientific culture that developed from the past until now, and developed specifically to explore the hidden potential of the science and the technology (the hidden knowledge) in the area of North Sumatra and to enrich the knowledge of human life generally. Therefore, SumutSiana will have an impact on the continuous effort of conducting studies related to the scope of the definition, as seen in the tropical science phrase in "TALENTA", although in general tropical science cannot be expressed in terms of its definition by ontology of knowledge [11,12], except only the naming about the kinds of publications and laboratories for example (use term "Tropical Science" as content of queries for submitting to Google and that search engine returns such information.[13]), so the term "Tropical Science" should be stated first by involving the ontology of knowledge.

The definition of SumutSiana also reveals the scope associated with North Sumatra technically, but socially engineering it include the related culture [5]. As culture comes from custom, and habit is caused by the compulsion either by nature or by rules. Thus, the scope of SumutSiana involves the implementation of an understanding related to the areas in definition. Therefore, SumutSiana which includes scientific wealth, it refers to the special thing of the natural resources, that is everything that comes naturally and effects to the human life in general, ranging from non-scatter creatures to moving creatures or termed as biotic and abiotic. The biotic category is such as animals, plants, micro-organisms, etc. While abiotic category is such as petroleum, natural gas, various metals, water, soil, etc. However, not all existing natural wealth will be expressed as a study in SumutSiana. In other words, the intended natural wealth is anything that is seen as a special North Sumatran treasury, such as Lake Toba exists only in North Sumatra [6,7].

Conceptually, every scientific wealth can be categorized and measured into a set of T, whereas the immigrant culture is also so measured into a set of S, by which the size of the influence between cultures that is prevail among T and S or $T \cap S$ [2,14]. The formula that the existence of a culture is expressed by

$$e(C) = T \cap S/T, \quad (1)$$

while the influence of a culture to another culture is

$$i(C) = T \cap S/S. \quad (2)$$

The same formula (Eq. (1) and Eq. (2)) can be used to see the effect of external food on local food, where local food is a region's wealth.

Table 1. Populations, categories and ethnic group assumption of districts/cities in Sumatera Utara

No	Districts/Cities	Population	*cats*	*eg*	No.	Districts/Cities	Population	*cats*	*eg*
1.	Asahan	796,283	P/D/F		17.	Padang Lawas Utara	252,589	P	M
2.	Batubara	400,803	D	D	18.	Pakpak Bharat	45,516	P	P
3.	Dairi	279,020	P	P	19.	Samosir	123,789	P	T
4.	Deli Serdang	2,029,308	P/D/F		20.	Serdang Bedagai	608,691	D	D
5.	Humbang Hasundutan	182,991	P	T	21.	Simalungun	849,405	P	S
6.	Karo	389,591	P	K	22.	Tapanuli Selatan	275,098	P	M
7.	Labuhanbatu	462,191	P/D/F		23.	Tapanuli Tengah	350,017	P/D	
8.	Labuhanbatu Selatan	313,884	P/D/F		24.	Tapanulis Utara	293,39	P	T
9.	Labutanbatu Utara	351,097	P/D/F		25.	Toba Samosir	179,704	P	T
10.	Langkat	1.013,385	D	D	26.	Binjai	264,687	D/F	
11.	Mandailing Natal	430,894	P	M	27.	Gunungsitoli	135,995	P/D	
12.	Nias	136,115	P	N	28.	Medan	2,210,624	P/D/F	
13.	Nias Barat	84,917	P	N	29.	Padangsidempuan	209,796	P	M
14.	Nias Selatan	308,281	P	N	30.	Pematangsiantar	247,411	P/D/F	
15.	Nias Utara	133,897	P	N	31.	Sibolga	86,519	P/D	
16.	Padang Lawas	258,003	P	M	32.	Tangjungbalai	167,012	P/D/F	
33.	Tebing Tinggi	156,815	P/D/F			Sumatera Utara	14,027,727	P/D/F	

4. Discussion

Each region is technically inhabited by its own ethnic group, and the tribe has culturally own customs. All of this is the richness of the region, the wealth uniquely owned by the tribe through the ability to constantly explore the potential that exists in order to face the challenges of life, the changing world and to reach a certain level of life [15]. Therefore, in the tribe it will grow and the form of different cultures that become treasures of each tribe. In addition to such wealth, there is wealth naturally lost in the tribal areas. This wealth is sometimes not in line with the growing needs of people or cultures. Thus, much of the potential in the culture become dysfunctional and the tribes become lazy or unable to prosper with it [6].

Based on the time line, North Sumatra has its own history about the inhabited by humans. Apart from the conflict between cultures, eight tribes become a distinct wealth from North Sumatra, which from an anthropological point of view was recognized as proto or deutero. Therefore, the scope of the study on SumutSiana revolve around the culture that has emerged from a long time until now in the North Sumatra region. Viewed from related human resources are settled ethnic groups (*eg*) consisting of ethnics: Malay (D), Tapanuli (Mandailing-Angkola (M), Toba (T), Karo (K), Simalungun (S), and Pakpak-Dairi (P)), and Nias (N) [6]. This paper does not use the term "Batak" for Tapanuli because this term means "*nations are not cultured*" [16], but this tribes have the oldest writings recognized by the world [17]. The province of North Sumatra has 33 districts/cities (Table 1), of which there are the origin of eight ethnic groups. Any evaluation of the existence of a culture can be done independently by involving the culture of the immigrant tribes such as Aceh, Minangkabau, Java and other immigrants from outside Indonesia such as Tionghoa (Chinese), India, and others, which generally involve three categories (*ca*): proto (X), deutero (Y), and foreign-settlers (Z) [18], for example.

Generally the spread of eight ethnic groups as follows: Karo is in Karo district and Deli Serdang regency; Mandailing is in east coast and west coast region; Malay around east coast region; Nias is mostly in Nias Island and small population around west coast; Pakpak, majority in Dairi regency, and Pakpak Bharat District; Simalungun, majority in Simalungun District, a part of Pematangsiantar, Asahan, Tebing Tinggi, other districts; and Toba is throughout regencies in north part. Therefore, Mandailing ethnic is in some areas Labuhanbatu, Labuhanbatu Selatan, Labuhanbatu Utara, Mandailing Natal, Padang Lawas, Padang Lawas Utara, Tapanuli Selatan, and Padangsidempuan with population (T) is 2,553,552. For social engineering, based on Eq. (1), existence of Mandailing in Medan $2{,}210{,}624/T = 0{,}87$, while based on Eq. (2) the other influence on Mandailing ethnic in Medan is $2{,}553{,}552/9{,}263{,}551 = 0{,}28$, whereby $S = 9{,}263{,}551$ is the population in Sumatera Utara – the population in Medan – 2,210,624. This technique can be used to view cultural mixing in areas where cultural assimilation applies between proto, deutero, and foreign-settlers. One function of this technique is to reduce the rate of cultural loss of ethnicity.

The same technique can also be used to measure the effects of local ingredients used in foods, medicines, transportation and others, so as to obtain the influence of local materials on economically sustainable production [19,20], for example, the influence of the wasabi Japanese spice on the existence of the andaliman from Sumatera Utara: T and S or ($T \cap S$) has value based on Web (andaliman,wasabi,Indonesia) = 2,500, T has value from internet (andaliman,Indonesia) = 140,000, while S(wasabi,Indonesia) = 572,000 [21], hence existence of andaliman can be expressed as 0.018 while the influence of wasabi on food andaliman is 0.004.

North Sumatra has a natural wealth: North Sumatra is flanked by two beaches and oceans, east and west; North Sumatra has two types of waters, the salt water (sea) and the fresh water (lake), so it has a diversity of marine wealth and freshwater wealth such as fish, shrimp, and shellfish (other sea fruit) [22]. In addition to culture, natural resources can also be positioned based on the region of existence [23]. Like some natural plant-specific plant resources per area compared to the extent of the plant coming from outside, such as oil palm ("kelapa sawit") against coconut plants ("kelapa") [24]. Regional potential naturally grows in virgin forests, whereas additional potential when new plants emerge comes from outside. The number of high and low plateau form the contour of the region, and this becomes its own potential, including activities on the coast and the surrounding ocean, such as the presence of wind as a potential for new, is due to the sea and sloping land.

Freshwater waters have an area of 30 km with a depth of 505 meters named the Lake Toba. This lake is the largest in Indonesia and Southeast Asia. Being at a considerable altitude and being on a hillside and among the towering mountains in the centre of North Sumatra [25]. Therefore, North Sumatra has a distinct biological richness between lowlands and highlands. The hot lowland plains have special peculiarities with the plants and animals that live there. Some plants and animals that live there. Some plants and animals are in cold places. If it continues to be sustainable, the lake is not only a source of freshwater, but source of energy that can improve the welfare of North Sumatra. From the point of tourism, the existence of Lake Toba is 0.991, while the influence of Bali towards Lake Toba tourism is 0.007, in this case $T \cap S$ = 45,000, T = 459,000 and S = 66,900,000 based on search engine Google [21].

Nature geographically influences culture, and with it technically every ethnic has local wisdom, engineered the life of related society, forms social groups, and determined social structure. Conversely, limited natural resources lead to the change of society social, especially changes in social structure [26]. In this case, the natural, the resources and the community groups become the study of SumutSiana.

5. Conclusion

SumutSiana term is presented from a special area named North Sumatra and all that is in it. SumutSiana is a discussion discourse or study that aims to explore all potentials related to North Sumatra, from human resources to natural resources. This definition becomes the framework of reference for all learning, research, and development activities within the coverage of North Sumatra.

References

[1] Suthers D D 2009 Technology affordances for intersubjective meaning making: A research agenda for CSCL *Computer-Supported Collaborative Learning* **1**

[2] Nasution M K M 2017 Semantic interpretation of search engine resultant *Interior*

[3] Nasution M K M 2016 Social network mining (SNM): A definition of relation between the resources and SNA *International Journal on Advanced Science, Engineering and Information Technology* **6(6)**

[4] Ramim M M 2005 Towards an understanding and definition of academic misconduct in online learning environments *IEEE Proceedings SoutheasthCon*

[5] Nasution M K M 2007 SumutSiana *Renungan, Kajang-Selangor Darul-Ehsan, Malaysia*

[6] Nasution M K M, Nuradi T E, Syah R 2017 SumutSiana: A framework for applying ICT to preserve the cultural heritage of Sumatera Utara Indonesia *Journal of Telecommunication, Electronic and Computer Engineering* **9(2-4)**

[7] Syah R, Nuradi T E, Nasution M K M 2017 A framework to apply ICT for bequeathing the cultural heritage to next generation *Journal of Physics: Conference Series* **801(1)**

[8] Bruner E M 1961 Urbanization and ethnic identity in North Sumatra *American Anthropologist* **63(3)**

[9] Nasution M K M 2016 Budaya lokal dan keterbukaan informasi *Sosialisasi Undang-Undang – WR3 USU*

[10] Universitas Sumatera Utara 2015 Rencana strategis USU 2015-2019

[11] Nasution M K M 2017 Ontologi *Bahan Kuliah S3 Fasilkom-TI USU*

[12] Nasution M K M 2012 The ontology of knowledge based optimization *Cornell University Library* (arXiv:1207.5130)

[13] Nasution M K M 2017 Modelling and simulation of search engine *Journal of Physics: Conference Series* **801(1)**

[14] Nasution M K M, Syah R, Elfida M 2017 Information retrieval based on the extracted social network *Advances in Intelligent Systems and Computing (AISC)*

[15] Nasution M K M, Noah S A 2010 Superficial method for extracting social network for academics using web snippets *Lecture Notes in Computer Science* **6401**

[16] Iskandar T 1970 Kamus Dewan *Dewan Bahasa dan Pustaka Kementerian Pelajaran Kuala Lumpur*

[17] Kratz E U 2010 Jawi spelling and orthography: A brief review *Indonesia and the Malay World* **30(86)**

[18] Cristiany C, Budiyanti A E, Hidayat A, Koesoemahardja H D 2013 Differences of lateral cephalometry values between Australo-Melanesian and Deutero-Malay races *Journal of Dentistry Indonesia* **20(1)**
[19] Lubis I, Nasution M K M 2017 Probability model for designing environment condition *Journal of Physics: Conference Series* **801(1)**
[20] Lubis I, Nasution M K M, Maulina M 2017 Basic framework of urban design based on natural resources *Friendly City* **4**
[21] Nasution M K M, Syah R, Elvey M 2017 Studies on behaviour of information to extract the meaning behind the behaviour *Journal of Physic: Conference Series* **801(1)**
[22] Basyuni M, Wati R 2017 Bioinformatics analysis of the oxidosqualene cyclase gene and the animo acid sequence in mangrove plants *Journal of Physics: Conference Series* **801(1)**
[23] Beard V A, Kunharibowo Y 2001 Living arrangements and support relationships among elderly Indonesians: case studies from Java and Sumatra *Population, Space and Place* **7(1)**
[24] Ambarita H, Sinulingga E P, Nasution M K M, Kawai H 2017 Performance and emissions of a small compression ignition engine run on dual-fuel mode (diesel-raw biogas) *IOP Conference Series: Materials Science and Engineering* **180(1)**
[25] Nishimura S, Abe E, Nishida J, Yokoyama T, Dharma A, Hehanussa P, Hehuwat F 1984 A gravity and volcanostratigraphic interpretation of the Lake Toba region, North Sumtra, Indonesia *Tectonophysics* **109(3-4)**
[26] Nasution M K M, Maulina M 2017 Calligraphy design for coconut garbage use *Friendly City* **4**

IOP Conf. Series: Materials Science and Engineering **309** (2018) 012132 doi:10.1088/1757-899X/309/1/012132

Post occupancy evaluation of energy-efficient behavior in informal housing of high density area

D N Aulia[1,2]*, B O Y Marpaung[1]
[1]Department of Architecture, Faculty of Engineering, University of Sumatera Utara, Jl. Perpustakaan Kampus USU Medan 20155
[2]Sustainable Energy & Biomaterial Research Centre, Faculty of Engineering University of Sumatera Utara, Jl. Almamater Kampus USU Medan 20155

*Email: dwira.nirfalini@usu.ac.id

Abstract. The concept of energy-efficient building emphasizes the critical of efficiency in the use of water, electrical energy, and building materials, beginning with design, construction, to the maintenance of the building in the future. This study was conducted to observe the behavior of Energy Saving of the residents in performing everyday activities in the building. The observed variables are the consumption of natural resources (energy, material, water, and land) and the emissions of air, water, and land related to the environment and health. This research is a descriptive qualitative research with the method of data collection is the distribution of questionnaires and observation. The method of analyzing data is posted occupancy evaluation undertaken to obtain patterns of community-based behavior in urban areas. The informal high-density housing area is a typology of population settlements that found in many big cities in Indonesia. This community represents various community groups regarding occupation, education, income, and race. The results of the study concluded that there are five components of energy-saving behavioral formers in housing namely: residential building components, environmental components in occupancy, external occupancy components, components of social activities and elements of business

1. Introduction

The concept of energy saving house emphasize in the increase in the efficiency in water, electricity and material usage started from the design to building maintenance in the future. Energy conservation behavior significantly affects the plan, the construction and the maintenance of the energy saving building itself. [1]. The condition of the nearby environment also affects the energy consumption of a house such as a tree position, building position and orientation and also building density.[2]. The highly populated urban area is the typological landscape of resident environment that can found frequently. Another aspect that affects energy saving is living lifestyle. Research by [3] suggests that living lifestyle can be evaluated to obtain energy consumption and can be used to predict future energy use. Energy saving lifestyle is important to the researcher to realize the concept of energy saving house. The aim of the study is to observe the energy conservation behavior in a highly populated residential area.

Energy-efficient house

Energy-efficient houses utilize many opening in its design to maximize air circulation and natural light and also minimize the usage of the lamp and air conditioning during the daytime. The plan also limits

IOP Conf. Series: Materials Science and Engineering **309** (2018) 012132 doi:10.1088/1757-899X/309/1/012132

building area, utilizes simple design, utilizes room flow, uses high-quality material efficiently, uses eco-friendly material and utilizes energy saving lifestyles such as alternative energy utilization such as the wind and natural light. The roof of the building can develop into roof garden and solar panel roof which in return lowers the ambient air temperature, lower pollution and increase green open spaces [4].

Some aspects needed to considered in the concept of energy saving house are [5]:

1. Efficient and economical consumption of resources (energy, material, water and land space).
2. Good emission into the air, water, and land associated with its impact towards the environment and health.

Environmental component around the building consists of laying of the constructing and planning the activity space. Laying of the building will be affected by the climates such as sun light, rainfall, and the wind. Planning activity area will be affected by activity types, furniture layout, room requirements and residents' lifestyle.

Sun light can function as natural light during daytime inside the building. It can also be utilized to lower the humidity in the building. Rainfall, especially heavy rain with high winds, can cause water leaks on the openings of the building. Using big tilted roof will help prevents water leak in the building.

Building materials susceptible to rot and mold should equip with open rain channels located below the ceiling to drain the rainfall into infiltration wells and ditches on the side of the road. Proper ventilation should be in the same direction as the wind direction. Wind will helps cool the air and circulates the air to the environment.

Energy-efficient behavior in residential area

Living lifestyle should follow energy saving behavior. Theories of energy conservation behaviors in living have researched by many researchers. As a basic theory, energy conservation behavior in a residential area was said to be affected by three aspects such as (1) external influence (2) community impact and (3) personal influence. [6]. External influence is the influence of the condition of the environment nearby the housing area. Community influence is the influence of the lifestyle of the community nearby the housing area. Individual lifestyle is the influence of the person residing the housing area.

Some activities of the residents can categorize into general activities (rest, working, and hygienic activities) and specific activities (gardening, businesses, and hobby).

Each activity occupies different spaces inside the house each with different specification and requirements that support those activities. [7] categorized areas inside the home in regards to energy usage into three clusters: Cluster 1 which is kitchen area, Cluster 2 which is entertainment area and Cluster 3 which is another aspect. The results obtained through analysis suggest that highest energy usage occurs in group 1 followed by group 2 and group 3. The research concluded that understanding energy consumption in residential housing would be more appropriate using socio practical theory approach which is accurate theory in behavior and habit, relationship between humans and materials, competence and activities significance.

Based on activity types in the family, each space provided according to the number of household member, furniture or tools used and circulation or activities traffic. Two activities can function in the same space is used in different times, and space has the similar characteristic. Time management in space and tool usage can also use as a strategy to save energy.[8]

Using environmental friendly kitchen equipment as opposed to regular kitchen equipment can help reduce energy usage. Replacing conventional oven with electric oven helps minimize the emission of greenhouse gases from the house to the environment. A good refrigerator helps balance energy usage with temperature control. These examples help sustainable living by reducing energy consumption. In designing the interior of the kitchen, using energy saving appliances can contribute to reducing the energy bill.

Kitchen ventilation settings. Kitchen with good ventilation will use less energy and lighting during daytime. Fresh air from windows and ventilation helps reduce harmful gases and mold.

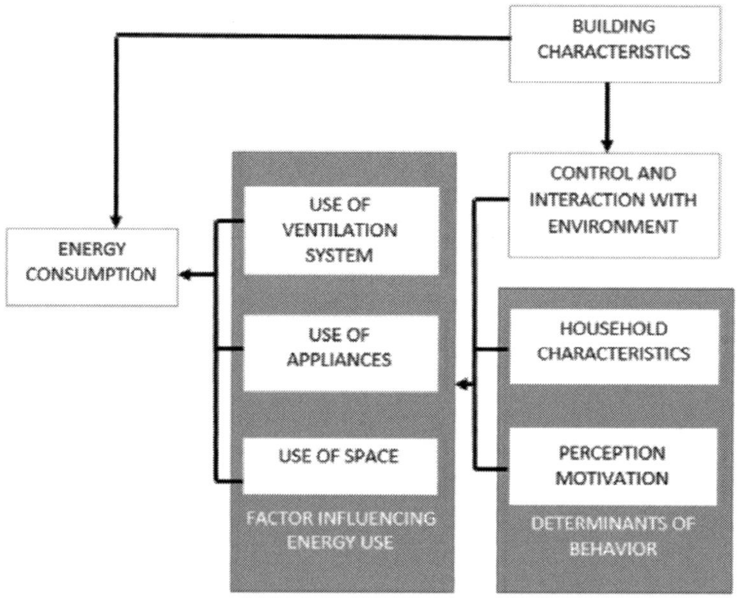

Figure 1. Research Schematic [9]

Energy-efficient concept

In Indonesia, primary energy (kerosene) is consumed mainly for cooking. However, consumption of electrical power increased lately (around 7% increase) with the assumption that 50% of the total power used in residential sector. Residential area significantly affects electricity consumption specifically for lighting, household appliances, and air conditioning. As a developing country, Indonesia will increase the quality of life of its citizens. As such, electricity consumption in the home used for life convenience purposes such as air conditioning system and lighting systems [10]. For cooking purposes, LPG or coal briquettes will replace oil usage. In regards to energy saving concepts, there are parameters to be considered such as:

1. Thermal convenience is used to control the absorption of solar energy needed for a household. Buildings in the hot climate in Indonesia need to be able to block the sun heat for cooling purposes.
2. Visual convenience is used to control sunlight absorption for natural lighting purposes.
3. Passive environment control is the utilization of natural energy optimally without the usage of high technology tools. Thermal and visual convenience can control by utilizing the potential of climate monitored by the elements of the building such as roofs, wall, floor, etc.
4. Active environment control is used to monitor the environment using tools or appliance to achieve thermal and visual convenience.
5. Hybrid environment control is the combination of passive and active environmental control which means some parts uses passive environment control while other parts used active environment control.

2. Method

This research is a descriptive qualitative research. Primary data collection acquired through interviews and observation. This study examined the behavior and perception of housing residents related to energy saving lifestyle. The location of the research chosen was in Kecamatan Medan, which was one of most populated districts in Medan City. Twenty-five houses used as research sample however only ten households were willing to be respondents to research on post-residency evaluation as a method to analyze data of residents' behavior pattern after five years minimum of occupancy.

Figure 2. Research Location at Jl. Wakaf Medan Area District

3. Results and Discussion

Use of Appliances

The power consumption of each respondent can consider as fair, ranging between 300-600VA per month with the total power supply of 900 VA for each house except Respondent 7 (R7). The power consumption of Respondent 7 considered high which is almost equal to the power provided to the house which is 900VA.

This high power consumption was caused by factors such as:

- A large number of residents, which was ten persons. (two family households living in the same house).
- Water supplied from underground wells as opposed to water companies. As such, to provide water for the house water pumps were used which then increase power consumption.
- Wasteful energy consumption. For examples, the usage of electronic appliances such as TV, laptop/computer, and fan used simultaneously for extended periods of time (until early morning).

Family characteristic from economic aspects will affect the energy consumption in a household. However, the knowledge on energy saving concept in a house is related to the education level of the household. [11]

The result of the interview from all ten respondents suggested that there was no significant effort in reducing energy consumption. Furthermore, no respondents were using alternative energy as power sources. All respondents still relied on conventional energy supplied by the government.

Control and interaction with environment

All respondents' houses were north-south oriented with slanted roofs. However, some of the roofs were not full enough. Natural ventilation was adequately applied but not fully optimized because of houses attached to one another. Building orientation was North-South oriented. (Figure 3)

Figure 3. North-South oriented house

Data on the health and convenience condition listed on (Table 1). Eight aspects reviewed with ratings of Good (G), Average (A) and Bad (B)

Table 1. Data on health and convenience of respondents' house

RESPONDENT	RATING		
	GOOD	AVERAGE	POOR
R1	–	Air conditioning, Garden condition, Water puddle, Vegetation, Roof condition, House cleanliness	Lighting, Temperature and Humidity
R2	–	Lighting, Air conditioning, Temperature and humidity, Garden condition, Water puddles, Vegetation, Roof condition, House cleanliness	–
R3	Lighting, Garden condition, Water puddle, Vegetation, Roof condition, House cleanliness	Air conditioning, Temperature and humidity	–
R4	–	Lighting, Air conditioning, Temperature and humidity, Vegetation, House cleanliness	Garden condition, Water puddles, Roof condition
R5	House cleanliness	Lighting, Garden condition, Water puddles, Roof condition	Air conditioning, Temperature and humidity, Vegetation
R6	Lighting, Air conditioning, Temperature and humidity, Water puddles, Roof condition, House cleanliness	Garden condition, Vegetation	–
R7	–	Lighting, Air conditioning,	Garden condition, Roof

		Temperature and humidity, Water puddle, Vegetation, House cleanliness	condition
R8	–	Air conditioning, Temperature and humidity, Garden condition, Water puddle, Roof condition, House cleanliness	Lighting, Vegetation
R9	–	Lighting, Temperature and humidity, Garden condition, Water puddle, Roof condition, House cleanliness	Air conditioning, Vegetation
R10		Lighting, Air conditioning, Temperature and humidity, Water puddle, Vegetation	Garden condition, Roof condition, House cleanliness

According to the table above, no house has all good rating; all of them were mostly average or bad. Common problems happen in lighting, air conditioning, temperature and humidity, and cleanliness (inside and outside of the house). The cause of these challenges was:

The position of the houses was attached to one another. As such, openings such as doors, windows, and ventilations located in front of the houses only. The possibility of cross ventilation was small, and the size of the openings was little as well (in comparison with room size).

Household characteristic

There were no problems in regards to income or outcome of respondents. All respondents were capable of providing daily needs. In fact, Respondent 4 (R4) has no job (livelihood supported by the children), and R4 still feels convenient. To sustain economic activities, each respondent cannot depend on the primary source of income but also needs another source of income. There were three out of ten respondents that have alternative source of income. Respondent 1 has beauty salon with the turnover of Rp 2.000.000/month. Respondent 2 has home contracting business with the turnover of Rp7.500.000/month. Respondent 8 has gunny craft business with the turnover of Rp1.000.000/month.

Perception and Motivation

The only one that has the problem was Respondent 7 (R7). Energy saving concept (openings, space arrangement, alternative energy, etc.) will have little impact because the root of the problem is in the behavior pattern. The primary cause of high energy consumption is the simultaneous usage of 2-3 laptops/computers over the extended period (often until early morning). During that time fans and light were turned on as well and never turned off again because of oversleeping. This behavior occurred due to the job demand that cannot avoid. Hence the concept of energy saving by reducing the usage of the electric appliance cannot be applied. The respondent's perception in energy conservation will affect the respondent behavior toward energy efficiency. [1] Information on energy saving was still inadequate that motivation toward energy conservation was small as well.

Apart from the sanitation, garbage also became a problem for many respondents such as R1, R4, R7, and R10. The problem occurred because there was no garbage dump in front of the house. Inadequate waste management will likely create floods in the future.

4. Conclusion

Perception of energy efficiency was still inadequate. Information on energy saving behavior was less understood because of lack of information and socialization in energy efficiency. Control and interaction toward the environment were still not prioritized because of other livelihood needs.

Managing energy usage has been done by several households although the motivation was not for reducing energy consumption but for reducing costs. Usage of environmentally friendly material was still scarce because of the expense of those materials were still high.

5. References

[1] Kao T F and Tu Y C 2015 Effect of Green Consumption Values on behavior : The Influence of Consumption Attitude *International Journal of Arts and Sciences* **8(8)** 119-130

[2] Nelson C, McHale M R and Peterson M 2012 Influences of lanscape and lifestyle on home energy consumption *Urban Ecosystem* **15** 773-793

[3] Gajowniczek K and Zabkowski T 2017 Electricity forecasting on the individual household level enhanced based on activity patterns *PLoS ONE* **12(4)** e0174098

[4] Nikoei M, Falamaki M M and Masoudi A 2014 Saving Energy in Building through Local Architecture Principles *European Journal of Natural and Social Sciences* **3(4)** 301-308

[5] Yudelson J 2009 *Green Building through integrated design* 1st edition New York: Mc Graw Hill Company

[6] Mayne R, Darby S and Jo H 2012 A literature Review for the EVALOC Project. Individual and Social Influences on Energy Use *Evaluating Low Carbon Communities* 1-51

[7] Thomas E G, Walker I, Verplanken B and Shaddick G 2016 Householders' Mental Models of domestic energy consumption using a Sort-End-Cluster Method to identify Shared Concept of Appliance Similarity *PLoS ONE* **11(7)** e0158949

[8] Buys L 2015 A Framework for understanding and generating Integrated Solutions for Residential Peak Energy Demand *PLoS ONE* **10(3)** e0121195.

[9] Santin O G 2013 Occupant behavior in energy efficient dwellings : evidents of a rebound effect. *Journal Housing and the Build Environment* **28** 311-327

[10] Nuryanti and Herdinie S S 2007 *Analisis karakteristik konsumsi energi pada sektor rumah tangga di Indonesia* Yogyakarta Seminar Nasional III SDM Teknologi Nuklir

[11] Jovanovic R, Bousselham A and Bayram I S 2016 Residential Demand Response Scheduling with Consideration of consumer Preferences *Applied Sciences* **6(16)** 1-14

IOP Conf. Series: Materials Science and Engineering **309** (2018) 012133 doi:10.1088/1757-899X/309/1/012133

The analysis of mechanical properties of non autoclaved aerated concrete with the substitution of fly ash and bottom ash

R Karolina* and F Muhammad
Civil Engineering Department, Universitas Sumatera Utara (USU)
Jl. Perpustakaan, Kampus USU Medan 20155 Indonesia

*E-mail: rachmie_caroline@yahoo.co.id

Abstract. Based on PP. No.85 of 1999 on the management of hazardous and toxic (B3), fly ash and bottom ash wastes are categorized into B3 waste because there are heavy metal oxide contents that can pollute the environment. One form of environmental rescue that can be applied is to utilize waste fly ash and bottom ash in the manufacture of concrete. In this research, fly ash and bottom ash waste are used as substitution of cement and fine aggregate to make lightweight concrete. The purpose of this research is to know the mechanical properties of non-autoclaved aerated lightweight concrete (NAAC) with FA and BA substitution to cement and fine aggregate which is expected to improve the quality of concrete. The NAAC lightweight concrete in this study is divided into 4 categories: normal NAAC lightweight concrete, NAAC lightweight NAAC substituted concrete with FA, NAAC lightweight concrete substituted with BA, and NAAC combined light weight from FA and BA with variations of 10%, 20% And 30%. The test specimen used in cylindrical shape, which was tested at the age of 28 days, amounted to 90 pieces and consisted of 10 variations. Each variation amounted to 9 samples. Based on the test results with FA and BA substitutions of 10%, 20%, and 30%, the highest compressive strength was achieved in samples with FA 30% of 12.687 MPa, maximum tensile strength achieved in samples with FA 30% of 1,540 MPa, The highest absorption was achieved in normal NAAC of 5.66%. Based on the weight of the contents of all samples, samples can be categorized in lightweight concrete, since the weight of the contents is less than 1900 kg / m3.

1. Introduction

With the rapid growth of knowledge and technology in the field of construction, then we are encouraged to pay more attention to quality standards and work productivity to be able to participate in improving construction for better quality. Therefore, we need a building material that has a better advantage than existing building materials. Concrete is one of the most widely used building materials and also its development of innovation and technology continues to run rapidly. One of the innovative developments is (Lightweight Concrete).

According to (SNI 3402-2008) Lightweight concrete has lighter density than concrete in general. Ordinary concrete has a high specific gravity (about 2400 kg / m3), where as lightweight concrete has a unit weight of not more than 1900 kg / m3. Lightweight concrete manufacture in principle is by providing a light mixture of stuffing, the mixture can be Styrofoam, pumice stone, foam agent, and fly ash. In this study, it is planned a non-autoclaved aerated concrete (NAAC) lightweight concrete using foaming agent, gravel as coarse aggregate, and fly ash and bottom ash as substitution of cement and sand. Fly ash and bottom ash are waste from coal obtained from industry.

Fly ash is the result obtained from coal combustion and is a very fine granular material whose component is 72.2% Silica based on test result from Balai Penelitian dan Standarisasi Industri Medan. The nature of fly ash is similar to that of cement that serves as a good adhesive, so it is considered able to reduce the use of cement in order to reduce costs incurred. While bottom ash is a material that does not burn completely from coal combustion. Bottom ash has a larger and heavier particle size than fly ash with dark gray pore-shaped gray characteristics that are considered capable of reducing the use of fine aggregates.

Fly ash and bottom ash used in this research comes from PT. SOCI MAS. Therefore, the use of fly ash and bottom ash as substitution in the manufacture of non-autoclaved aerated concrete (NAAC) lightweight concrete is expected to improve the quality of concrete, as well as support the environmentally friendly movement, because we can utilize waste from unused coal in order not to pollute the environment

Figure1. Fly ash and bottom ashfrom PT. SOCI MAS

Fly Ash

Class F fly ash is produced from anthracite or bituminous coal combustion, has a pozzolanic nature and to obtain cementitious properties, quick lime , hydrated lime, or cement should be added. Fly ash grade F has a low lime content (CaO <10%). Class C Fly ash class is produced from lignite or sub-bituminous coal combustion in addition to having pozolanic properties also has self-cementing properties (the ability to harden and add strength when reacted with water) and these properties arise without the addition of lime. Fly ash class C contains lime greater than fly ash class F (CaO> 20%).So fly ash from PT. SOCI MAS used in this research is classified into fly ash class F, because lime content in the fly ash is 4.79% (CaO <10%).

Bottom Ash

Bottom ash is a material that does not burn completely from coal combustion. Bottom ash has a larger particle size and heavier than fly ash with a dark gray colored in the shape of pivot grains characteristics that are considered capable of reducing the use of sand.

IOP Conf. Series: Materials Science and Engineering **309** (2018) 012133 doi:10.1088/1757-899X/309/1/012133

2. Method

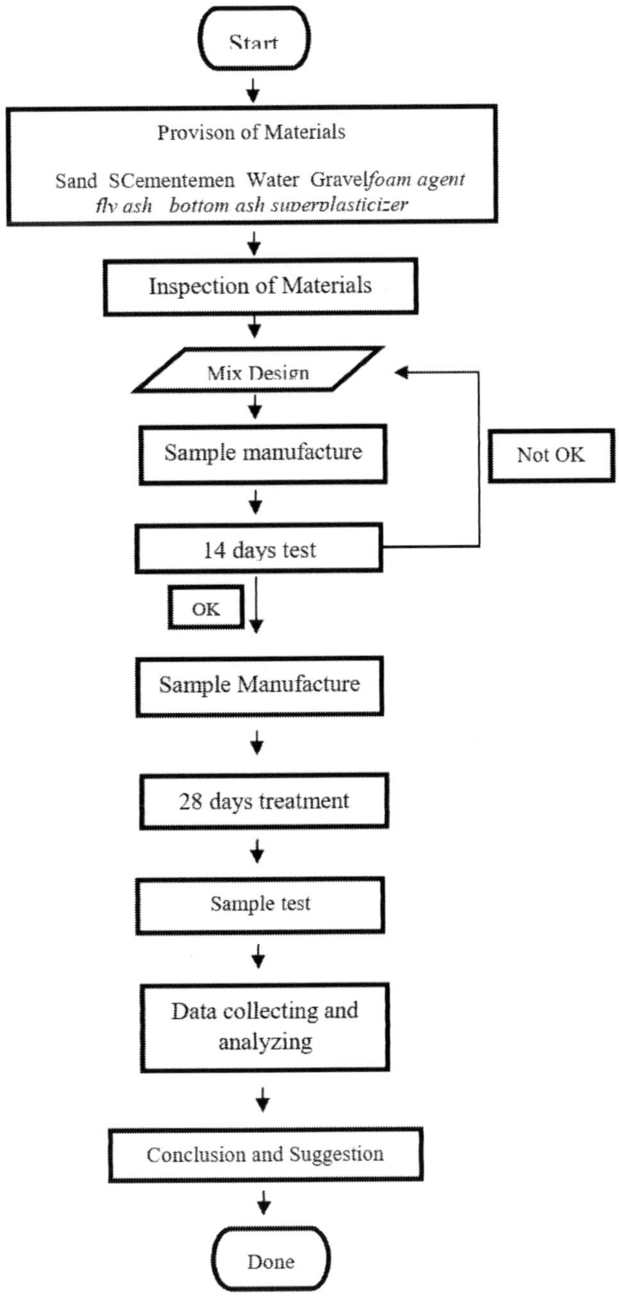

Figure2. Fly ash and bottom ash from PT. SOCI MAS

IOP Conf. Series: Materials Science and Engineering **309** (2018) 012133 doi:10.1088/1757-899X/309/1/012133

3. Result and discussion

Absorption of Concrete Cylinder

Absorption is the ability of a material to absorb water. The absorption value is closely related to the specific gravity or porosity of a material, since large absorption values indicate the number of cavities present in the material. The amount of absorption can also cause a decrease in the strength of concrete, because the pores that exist cause the bonding between particles in a material is reduce.

Table 1. The test results of the sample test absorption using fly ash andbottom ash substitution

No.	NAAC Lightweight Concrete *Fly Ash* and*Bottom Ash*Substitution			Wet Weight (kg)	Dry Weight (kg)	Absorption (%)	Average Absorption (%)
1.	FABA	10%	Sample I	8,769	8,397	4,43	**4,60**
			Sample II	8,646	8,257	4,71	
			Sample III	8,653	8,267	4,67	
	FABA	20%	Sample I	8,779	8,450	3,89	**3,92**
			Sample II	8,714	8,392	3,84	
			Sample III	8,693	8,356	4,03	
	FABA	30%	Sample I	8,863	8,389	5,65	**5,17**
			Sample II	8,969	8,550	4,90	
			Sample III	8,887	8,467	4,96	

From all absorption test results, it showed that the use of fly ash and bottom ash as substitution of fine aggregate and cement can decrease the absorption value of concrete, this is caused by finer grains of fly ash and bottom ash are able to fill the smaller pore, thus the concrete produced is more dense and solid, the largest absorption value at every test is in normal NAAC lightweight concrete with absorption value of 5.66% and the smallest absorption value is in 30% of fly ash substitution with absorption value of 2.76%.

Compressive Strength of Concrete Cylinder

The tests areperformed at age 28 days. The result of compressive strength test for concrete variation can be seen in the table below

Table 2. The test result of weight and compressive strength of test sample by using fly ash and bottom ash substitution

No.	NAAC Lightweight Concrete with *Fly Ash* and*Bottom Ash*Substitution			Concrete Weight (kg)	Weight of Content (kg/m³)	Average Weight of Content (kg/m³)	Age (days)	Compressive Strength		Average Compressive Strength (MPa)
								kN	MPa	
1.	FABA	10%	Sample I	8,197	1546,604	**1548,239**	28	141	9,618	**9,095**
			Sample II	8,257	1557,925			122	8,322	
			Sample III	8,163	1540,189			137	9,345	
	FABA	20%	Sample I	8,350	1575,472	**1574,654**		140	9,549	**9,594**
			Sample II	8,392	1583,396			144	9,822	
			Sample III	8,295	1565,094			138	9,413	
	FABA	30%	Sample I	8,489	1601,698	**1597,925**		135	9,208	**9,003**
			Sample II	8,450	1594,340			128	8,731	
			Sample III	8,468	1597,736			133	9,072	

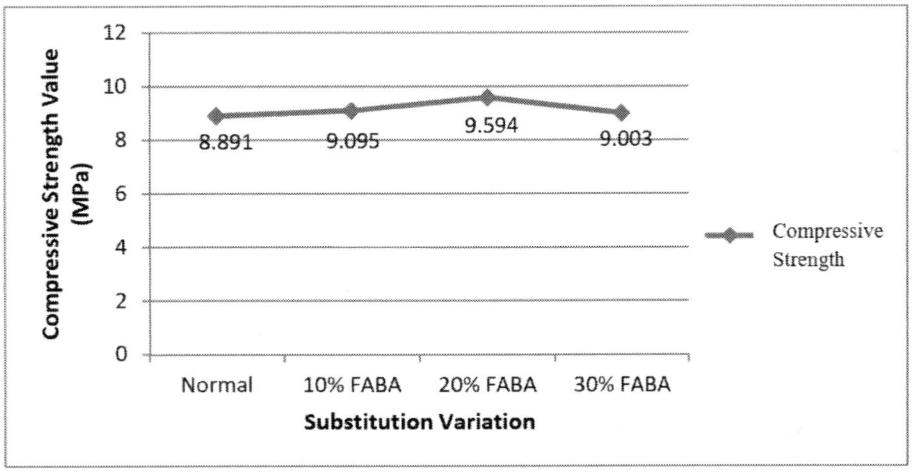

Figure 3.Compressive Strength Value

Table 3. Classification Control of Lightweight Concrete Quality

No.	**Variation**	**Average Compressive Strength (MPa)**	**Average Weight of Content (kg/m³)**	**Weight of Content according to non structural lightweight concrete quality** (Young, J. Francis. 1972)	**Compressive Strength according to non structural lightweight concrete quality** (Young, J. Francis. 1972)	**Non structural lightweight concrete quality** (Young, J. Francis. 1972)
1	Normal	8,891	1522,704	1100-1600 kg/m^3	7-14 MPa	Qualified
2	10 % FA	11,505	1590,252	1100-1600 kg/m^3	7-14 MPa	Qualified
3	20 % FA	12,232	1614,340	1100-1600 kg/m^3	7-14 MPa	Unqualified
4	30 % FA	12,687	1630,440	1100-1600 kg/m^3	7-14 MPa	Unqualified
5	10 % BA	9,686	1583,396	1100-1600 kg/m^3	7-14 MPa	Qualified
6	20 % BA	9,776	1600,252	1100-1600 kg/m^3	7-14 MPa	Unqualified
7	30 % BA	9,026	1640,252	1100-1600 kg/m^3	7-14 MPa	Unqualified
8	10 % FABA	9,095	1548,239	1100-1600 kg/m^3	7-14 MPa	Qualified
9	20 % FABA	9,594	1574,654	1100-1600 kg/m^3	7-14 MPa	Qualified
10	30 % FABA	9,003	1597,925	1100-1600 kg/m^3	7-14 MPa	Qualified

From the data on every figure, it can be seen that the compressive strength of the sample obtained that can meet the requirement of non-structural lightweight concrete compressive strength is 7-14 MPa with the weight of concrete content of 1100-1600 kg / m¯3 (Young, J. Francis.1972). From the compressive strength test of the sample, it is obtained that the highest compressive strength on NAAC lightweight concrete with 30% of fly ash substitution with a compressive strength is reaching 12,687 MPa with the weight of 1630,440 kg / m3. The increase in compressive strength is due to the high silica content found in fly ash and bottom ash, the finer grain of fly ash and bottom ash is able to fill the smaller pores, thus the concrete produced is denser and solid. Similarly, the addition of superplasticizer also improves the workability of the concrete, the addition of this admixture resulting a workability.

Tensile Strength of Concrete Cylinder

The tests are performed at age 28 days. The tensile strength test results for concrete variation can be seen in the table below

Table 4.Test result of tensile strength of test samples by using fly ash and bottom ash substitution

No.	**NAAC Lightweight Concrete with *Fly Ash* and *Bottom Ash* Substitution**			**Age (days)**	**Tensile Strength**		**Average Tensile Strength (MPa)**
					kN	**MPa**	
1.	FABA	10%	Sample I	28	60	1,023	**1,125**
			Sample II		73	1,244	
			Sample III		65	1,108	
	FABA	20%	Sample I		72	1,227	**1,313**
			Sample II		79	1,347	
			Sample III		80	1,364	
	FABA	30%	Sample I		53	0,903	**1,006**
			Sample II		61	1,040	
			Sample III		63	1,074	

From the data on all figures it can be seen that the tensile strength of the sample has increased with the substitution of fly ash and bottom ash. The increase in tensile strength occurs due to the high silica content contained in fly ash and bottom ash, the finer ash grains and the finer bottom ash are able to fill smaller pores, thus the concrete produced is denser and solid. Similarly, the addition of superplasticizer also improves the workability of the concrete, the addition of this admixture resulting a workability. In certain proportions the superplasticizer will disperse the cement more evenly, resulting in a more perfect hydration reaction. This reaction will make the mixture more compact and solid so that the binding strength of the mixture becomes stronger and increases the strength of the resulting concrete. From the sample of tensile strength test, the highest tensile strength in NAAC lightweight concrete with 30% of fly ash substitution with tensile strength reaches 1,540 MPa and the smallest tensile strength is found in normal NAAC light concrete with the value of 0,801 MPa.

4. Conclusion

From the results of the research that has been done, there are some conclusions that can be taken as follows:

1. Fly ash and Bottom Ash can be substituted for fine aggregate ie cement and sand in the manufacture of lightweight concrete Non Autoclaved Aerated Concrete.
2. The largest absorbtion in this study is found in normal NAAC lightweight concrete that is 5.66% while the smallest absorption of water in this research is the substitution of fly ash 30% that is equal to 2.76%.
3. The largest compressive strength of this research is found in the fly ash substitution of 30% that is 12,687 MPa while the smallest compressive strength in this research is on normal NAAC lightweight concrete that is equal to 8,891 MPa.
4. The largest tensile strength in this research is found in 30% of fly ash mixture that is 1,540 MPa while the smallest tensile strength in this research is on normal NAAC lightweight concrete that is equal to 0,801 MPa.

IOP Conf. Series: Materials Science and Engineering **309** (2018) 012133 doi:10.1088/1757-899X/309/1/012133

5. From the maximum compressive strength and the resulted weight of contents, the concrete can be categorized as non-structural lightweight concrete with a weight of 1100-1600 kg / m¬3 with a compressive strength in the range of 7-14 MPa (Young, J. Francis 1972).

References

[1] Departemen Pekerjaan Umum 1990 *Spesifikasi Bahan Tambahan Untuk Beton* SK SNI S-18-1990-03 Bandung Yayasan LPMB

[2] Istimawan D 1993 *Struktur Beton Bertulang* Departemen Pekerjaan Umum : Jakarta

[3] Murdock LJ, Brook KM, Stephanus H 1991 *Bahan dan Praktek Beton* Jakarta: Erlangga.

[4] Paul N dan Antoni 2007 *Teknologi Beton dan Material, Pembuatan Beton Kinerja Tinggi* Yogyakarta Andi Offset

[5] Mehta P K dan Monteiro P J M 1993 *Concrete* United States Prentice-Hall

[6] Husin A, Setiadji R 2008 *Pengaruh Penambahan Foam Agent Terhadap Kualitas Bata Beton* Pusat Litbang Permukiman Bandung

[7] Mardiono 2010 *Pengaruh Pemanfaatan Abu Terbang (Fly Ash) Dalam Beton Mutu Tinggi* Skripsi Tugas Akhir Jurusan Teknik Sipil Universitas Gunadarma Jakarta

[8] Modul Praktikum Laboratorium Bahan Rekayasa Departemen Teknik Sipil Fakultas Teknik Universitas Sumatera Utara

[9] PP No 85 tahun 1999 *Pengolahan Limbah Bahan Berbahaya dan Beracun* SNI 15-2049-2004 *Semen Portland*

[10] BASF *the chemical company, aplikasi light weight concrete untuk bata ringan*

[11] Standar Nasional Indonesia (SNI) 1990 Metode Pengujian Kuat Tekan Beton Pusjatan-Balitbang PU Bandung

[12] Standar Nasional Indonesia (SNI) 2002 Metode Pengujian Kuat Tarik Belah Beton Pusjatan – Balitbang PU Bandung

[13] Departemen Pekerjaan Umum 2002 *Tata cara perancangan campuran beton ringan dengan agregat ringan* SNI 03-3449-2002 Badan Standarisasi Nasional Bandung

[14] Pasaribu R P 2010 *Analisis kemampuan beton ringan abu sekam padi* Universitas Tarumanegara Jakarta

IOP Conf. Series: Materials Science and Engineering **309** (2018) 012134 doi:10.1088/1757-899X/309/1/012134

Optimization of fly ash and bottom ash substitution against paving block manufacture according to SNI 03-0691-1996

R Karolina, Syahrizal and N Bahri*
Departemen Teknik Sipil, Universitas Sumatera Utara (USU)
Jl. Perpustakaan, Kampus USU Medan 20155 INDONESIA

*E-mail: nirwanbahri@gmail.com

Abstract. The waste of coal burning has a very negative impact on the environment if the waste is not managed as well as possible. The remaining waste of coal combustion consists of fly ash and bottom ash. FA and BA can be developed into substitution materials in the process of making paving blocks. The purpose of this study was to determine the quality of paving block in accordance with SK SNI 03-0691-1996 with optimization in the use of FA and BA. This study uses a 351 paving block sample size of 20x10x6 cm. Paving blocks are divided by 4 categories, namely normal paving block, paving block with FA substitution, BA substitution and combination of FA-BA with each variation 0%, 25%, 50%, 75% and 100%. Each variation amounted to 27 samples. Paving block quality measurement is done through 4 tests: absorption, compressive strength, sodium sulphate resistance and Los Angeles tests. The result of the test shows the absorption of normal paving block is 3,229%, paving block with 25% FA is 3,889%, paving block with 50% BA is 5,560% and paving block with 25% FA-BA combination is 5,794%. Compressive strength in normal paving block is 25,50 MPa, paving block with 25% FA is 25,28 MPa, paving block with 25% BA is equal to 27,61 MPa and paving block with 25% FA-BA is 26 , 00 MPa. In testing of sodium sulfate resistance, almost all test specimens are eligible except for paving block with 50% FA and 75% FA. In the test of wear resistance, no specimen is eligible according to SK SNI 03-0691-1996. The comparison of the strength of the test specimen can be seen in substitution with 25% BA which reaches maximum strength.

1. Introduction

As times progressed, innovation in the construction world continues to increase, such as the development of construction on concrete. Concrete is one of the most commonly used construction materials. Many other usage of concrete in everyday life. One of them is Paving Block.

The development of Paving Block usage is considered more economical, able to withstand the compressive force given from the outside and is considered to have a beauty value. Paving Block usage is widely used in certain areas such as roads in residential complexes, sidewalks, walkways / alleys, office and home pages, roads in tourist areas and much more.

Paving Block manufacture has a mixture of compositions such as cement, fine aggregate, coarse aggregate, and water. In this study cement and fine aggregate are substituted by Fly Ash and Bottom Ash.Fly Ash is obtained from coal combustion which has silica content of 72,2%. Because it resembles cement so it can serve as an adhesive and can reduce the use of cement. While Bottom Ash is the material of the remaining of imperfect coal combustion which has a particle resembling sand with dark gray, shaped pelleted grains physical characteristics that are considered able to reduce sand usage.

Based on PP. No. 85 of 1999 on the management of hazardous and toxic waste (B3), Fly Ash and Bottom Ash are categorized as B3 waste because there are heavy metal oxide contents that will naturally encroach and pollute the environment. Hazardous and toxic substances (B3) are the remains of an effort and or activity containing toxic hazardous substances due to their nature and or concentration and or amount, either directly or indirectly, may pollute and damage the environment, or may harm life sustainability, health, human survival and other living things.

According to Government Regulation Number 18 of 1999 article 9 paragraph 1 concerning any person conducting an effort and or activity using hazardous and toxic materials and / or producing B3 waste shall perform B3 waste reduction, manage B3 waste and or accumulate B3 waste. So with this research is considered able to reduce the negative impact of Fly Ash and Bottom Ash. The following table contains the content of Fly Ash and Bottom Ash. According to SNI 03-0691-1996 "Paving Block is a building material made of a mixture of portland cement or similar hydrolytic adhesive, water and aggregate with or without other additives which does not reduce the quality of the concrete.

2.1 Fly Ash

Fly Ash is obtained from coal combustion with the temperature of 1600°C which has silica component content of 72,2%. Because it resembles cement it can serve as an adhesive and can reduce cement usage. Class F fly ash is produced from anthracite or bituminous coal combustion, has a pozzolanic nature and to obtain cementitious properties, quick lime , hydrated lime, or cement should be added. Fly ash grade F has a low lime content (CaO <10%). Class C Fly ash class is produced from lignite or sub-bituminous coal combustion in addition to having pozolanic properties also has self-cementing properties (the ability to harden and add strength when reacted with water) and these properties arise without the addition of lime. Fly ash class C contains lime greater than fly ash class F (CaO> 20%). So fly ash from PT. SOCI MAS used in this research is classified into fly ash class F, because lime content in the fly ash is 4.79% (CaO <10%).

Figure 1. Fly Ash

Figure 2. Bottom Ash

Table 1. Elements contained in fly ash

No	Parameter	Unit	Result	Method
1	Silica as SiO_2	%	72.2	Gravimetry
2	Aluminium as Al_2O_3	%	18.8	Calculation
3	Iron as Fe_2O_3	%	0.79	A A S
4	Calcium as CaO	%	4.79	Tritimetry
5	Magnesium as MgO	%	3.50	Gravimetry
6	Sodium as Na_2O	%	0.03	A A S
7	Potassium as K_2O	%	0.04	A A S
8	Fosfor as P_2O_5	%	0.19	Spektrofotometry
9	Sulfur (S)	%	2.12	Gravimetry
10	Mangan	mg/Kg	81.8	A A S

Source : Laboratoy of Research Assessment and Standarization, Medan

2.2 Bottom Ash

Bottom Ash is an imperfect combustion coal waste material that has a particle resembling sand with a dark gray physical characteristic, in the shape of a pivot grains that is considered capable of reducing sand usage.

Table 2. Elements contained in fly ash

No	Parameter	Unit	Result	Method
1	Silica as SiO_2	%	53.4	Gravimetri
2	Aluminium as Al_2O_3	%	6.77	Perhitungan
3	Iron as Fe_2O_3	%	1.27	A A S
4	Calcium as CaO	%	8.74	Tritimetri
5	Magnesium as MgO	%	4.12	Gravimetri
6	Sodium as Na_2O	%	0.06	A A S

Source : Laboratoy of Research Assessment and Standarization, Medan

2. Research and Method

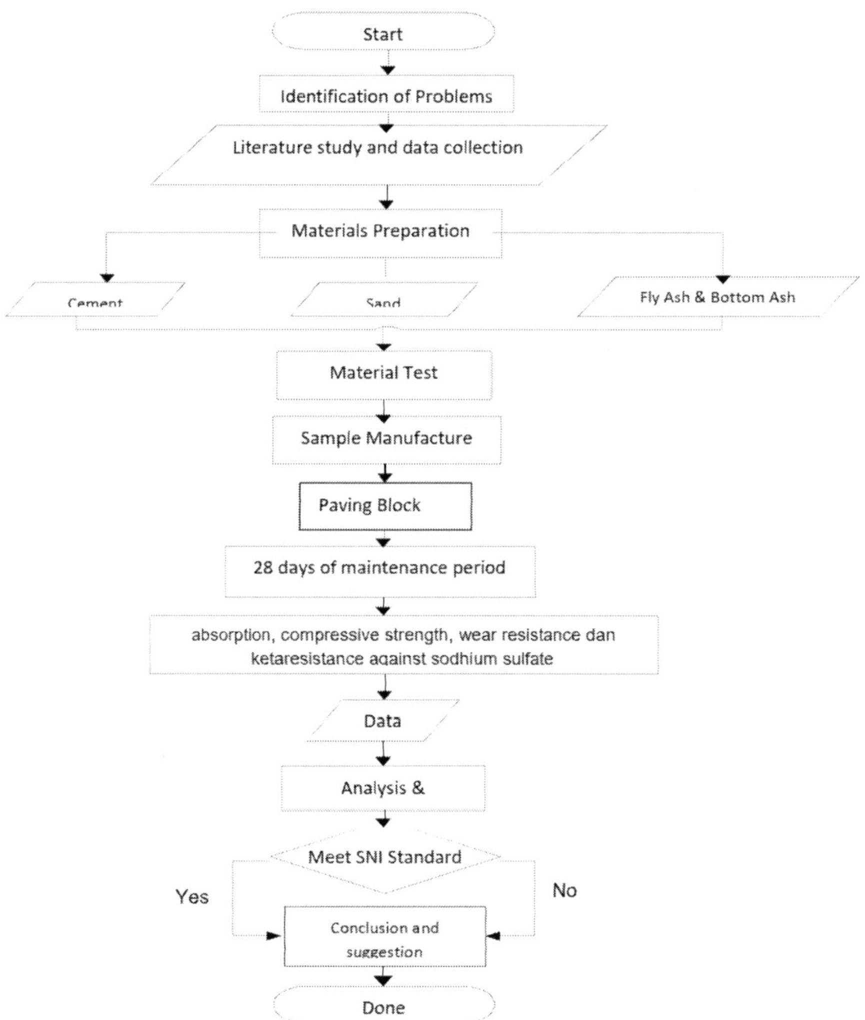

Figure 3 . Research Flow Diagram

3. Result And Discussion

3.1 Absorption Test

The results of water absorption test on Paving Block are as follows:

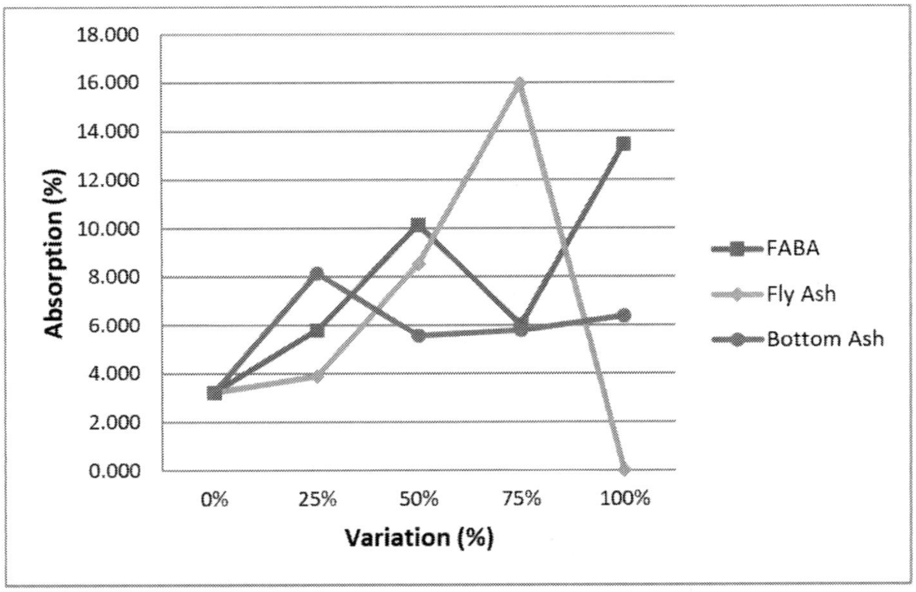

Figure 4. Graph of absorption value with fly ash and bottom ash substitution

From the graph of the tested composition of Paving Block, the smallest water absorption value occurs in normal Paving Block with water absorption value of 3.229%. The value of water absorption in Paving Block with the smallest Fly Ash substitution is on Paving Block of 25% Fly Ash of 3.998% and the largest in Paving Block of 75% Fly Ash of 15.963%. But in Paving Block of 100% Fly Ash water absorption value can not be calculated because the Paving Block is crushed, caused the Fly Ash only has a low lime content so Fly Ash can not bind perfectly. The value of water absorption in Paving Block with Bottom Ash substitution is the smallest in Paving Block of 50% Bottom Ash of 5.560% and the largest is Paving Block of 25% Bottom Ash of 8.139%. The value of water absorption on Paving Block with the smallest Fly Ash Bottom Ash substitution is on Paving Block of 25% Fly Ash Bottom Ash of 5.794% and the largest is Paving Block of 100% Fly Ash Bottom Ash of 13,461%.

The use of substitution variation of Fly Ash greater than 25% can cause a large absorption value because the physical properties of Fly Ash can not bind perfectly as cement and absorb more water, therefore Paving Block is not solid and causes many pores causing the absorption value to rise as well as the substitution of Bottom Ash.

3.2 Compreessive Strength Test

The test results of compressive strength on Paving Block in this study is as follows. From the graph above it can be seen that the highest compressive strength is found in Paving Block with Bottom Ash substitution of 27.61 MPa. And the lowest compressive strength is found in Paving Block with Fly Ash substitution of 7.22 MPa. But on Paving Block with 100% Fly Ash substitution, compressive test can't be done because Paving Block is very fragile and easily crushed.

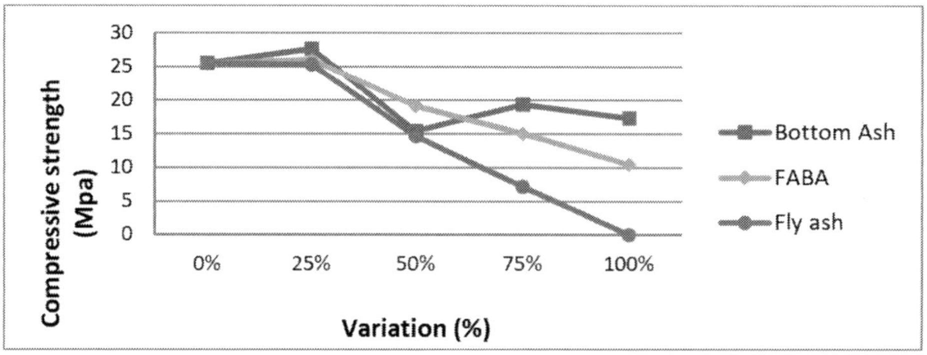

Figure 5. Graph of relation betweeni *Fly Ash* and *Bottom Ash* substitution against compressive strength

According to ASTM, ACI, and SNI, the value of Fly Ash content used in concrete mixture is 15% - 25% by weight of cement. Several studies show the optimum level of Fly Ash is at 10% -20% for improving the quality of concrete. Variations of Fly Ash sits greater than 25% can not achieve a large compressive strength value because Fly Ash can't bind perfectly causing the Paving Block not as solid as cement. This is because Fly Ash can not replace the nature of cement that serves as the main binder material on the concrete. The occurrence of compressive strength in Paving Block substitution of Bottom Ash with variation greater than 25% is due to Bottom Ash having a finer gradation than sand gradation so that Paving Block substitution of Bottom Ash with variation greater than 25% has a compressive strength value under normal Paving Block.

Paving Block that can be included in category B of quality is normal Paving Block, Paving Block of 25% Fly Ash, Paving Block of 25% Bottom Ash, Paving Block of 75% Bottom Ash, Paving Block of 100% Bottom Ash, Paving Block of 25% Fly Ash Bottom Ash , And Paving Block of 50% Fly Ash Bottom Ash. Paving Block with B quality can be used for parking lot.

Paving Block that can be included into the category C of quality is Paving Block of 50% Fly Ash, Paving of Block 50% Bottom Ash, and Paving Block of 75% Fly Ash Bottom Ash. Paving Block category C of quality can be used for pedestrians. Paving Block that can be included into quality D is Paving Block with 100% of Fly Ash and Bottom Ash substitution and can be used for garden and other usage.

Paving Block with 75% substitution of Fly Ash can't be included into category D because the value of Paving Block compressive strength is below the minimum requirement of quality D.

3.3 Wear Resistance Test

The results of wear resistance test on Paving Block in this study is as follows:

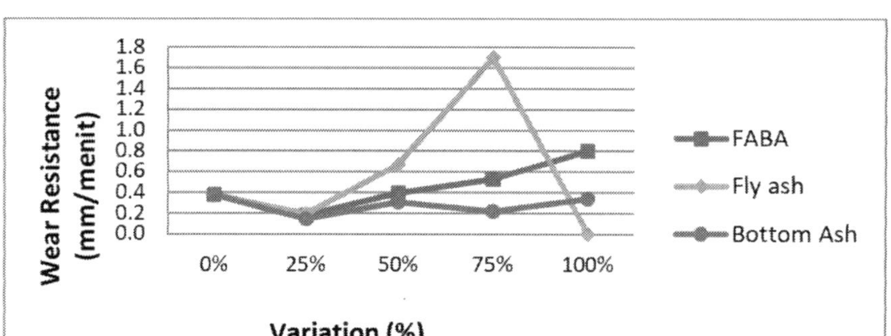

Figure 6. Graph of relation between *Fly Ash* and *Bottom Ash* substitution against wear resistance

From the graph of the wear resistance test results, it can be seen that Paving Block is normal, Paving Block of 25% Fly Ash, Paving Block of 25% Bottom Ash, Paving Block of 75% Bottom Ash, Paving Block of 100% Bottom Ash, Paving Block of 25% Fly Ash Bottom Ash, And Paving Block of 50% Fly Ash Bottom Ash have wear resistance value which do not fulfill the standard of Paving Block of B quality according to SNI 03-0691-1996 that the minimum wear resistance is 0,149 mm / minute and average 0,130 mm / minute.

Paving Block of 50% Fly Ash, Paving Block of 50% Bottom Ash, and Paving Block of 75% Fly Ash Bottom Ash have wear resistance value that does not meet the standard of Paving Block of C quality according to SNI 03-0691-1996 that the minimum wear resistance is 0,184 mm / Minutes and an average of 0.160 mm / min. For Paving Block of 100% Fly Ash Bottom Ash has wear resistance value that does not meet the standard of Paving Block of D quality according to SNI 03-0691-1996 which is the minimum wear resistance is 0.251 mm / min and an average of 0.219 mm / min. But on Paving Block with 100% of Fly Ash substitution, wear resistance can not be tested because Paving Block is very fragile and easily crushed. From Paving Block test result against wear resistance, it is found wear resistance value that does not meet SNI 03-0691-1996 because the tool used in this test does not meet the standard of SNI.

4. Conclusion

From the results of research obtained and from the results of the discussion that has been done it can be concluded things as follows:

1. The largest water absorption from all experiment variation is in Paving Block with 75% Fly Ash composition of 15,963%.
2. The largest compressive strength is in Paving Block with 25% of Bottom Ash substitution.
3. From the table of resistance test to sodium sulfate result , it is seen that almost every Paving Block is qualified, except the Paving Block with Fly Ash substitution of 50% and Fly Ash of 75%.
4. From the wear resistance test, the wear resistance value is too big so that it is not in accordance with the standard SNI 03-0691-1996.
5. The comparison of all normal Paving Block test with Paving Block substitution of Fly Ash and Bottom Ash is visible.

References

[1] Berca L H B 2012 *Pengaruh penggunaan Fly Ash Pada Beton Mutu Normal Dan Mutu Tinggi Ditinjau Dari Kuat Tekan dan Absorbsi* Bandung
[2] Departemen Pekerjaan Umum 1990 *Spesifikasi Bahan Tambahan Untuk Beton* SK SNI S-18-1990-03 Bandung Yayasan LPMB
[3] Dewan Standardisasi Nasional 1996 Bata Beton (*Paving Block*) SNI 03-0691-1996 Jakarta BSN
[4] Gunaedi and Hidayat I 2012 *Pengaruh Fly Ash Pada Kuat Tekan Campuran Beton Menggunakan Expanded Polystyrene Sebagai Substitusi Parsial Pasir* Jakarta
[5] Mulyono T 2003 *Teknologi Beton* Yogyakarta CV Andi Offset
[6] Murdock L J, Brook K M, Hendarko and Stephanus 1991 *Bahan dan Praktek Beton* Jakarta: Erlangga
[7] Nugraha P and Antoni 2007 *Teknologi Beton dan Material Pembuatan Beton Kinerja Tinggi* Yogyakarta: Andi Offset
[8] Segel R, Kole P and Kusuma G 1993 *Pedoman Pengerjaan Beton*. Jakarta: Erlangga
[9] Suarnita I W 2012 *Pemanfaatan Abu Dasar (Bottom Ash) Sebagai Pengganti Sebagian Agregat Halus Pada Campuran Beton* Palu
[10] Tjokrodimuljo K 1992 *Teknologi Beton* Yogyakarta: Gramedia
[11] Zongjin L I 2011 Advance Concrete Technology Canada : John Wiley & Sons
[12] Mehta, P Kumar and Monteiro P J M 1993 *Concrete* United States: Prentice-Hall

IOP Conf. Series: Materials Science and Engineering **309** (2018) 012135 doi:10.1088/1757-899X/309/1/012135

The effect of various parameters of large scale radio propagation models on improving performance mobile communications

*M Pinem and R Fauzi
Electrical Engineering Department, Universitas Sumatera Utara, Indonesia

*E-mail: maksum.pinem@gmail.com

Abstract. One technique for ensuring continuity of wireless communication services and keeping a smooth transition on mobile communication networks is the soft handover technique. In the Soft Handover (SHO) technique the inclusion and reduction of Base Station from the set of active sets is determined by initiation triggers. One of the initiation triggers is based on the strong reception signal. In this paper we observed the influence of parameters of large-scale radio propagation models to improve the performance of mobile communications. The observation parameters for characterizing the performance of the specified mobile system are Drop Call, Radio Link Degradation Rate and Average Size of Active Set (AS). The simulated results show that the increase in altitude of Base Station (BS) Antenna and Mobile Station (MS) Antenna contributes to the improvement of signal power reception level so as to improve Radio Link quality and increase the average size of Active Set and reduce the average Drop Call rate. It was also found that Hata's propagation model contributed significantly to improvements in system performance parameters compared to Okumura's propagation model and Lee's propagation model.

1. Introduction

Code Division Multiple Access (CDMA) is one of the communication technologies for 3G standards in mobile communications. 3G is an international technology standard that aims to improve efficiency and improve mobile network performance. 3G offers improved service quality (QoS) [1]. One facility in a cellular system to ensure continuity of communication when a customer moves from one cell to another is a handover. CDMA-based cellular networks are capable of supporting soft handover, which makes smoother transitions and improves communication quality. With Soft handover, some radio links can operate in parallel, so the MS (mobile station) connection with several base station transceivers can be done simultaneously [2].

Soft handover is often associated with active set and its size. Active set is a set of all stations that communicate with MS. The input / discharge of a base station to / from the active set is determined by the initiation trigger that has been determined. Initiation can be done with strong signal measurements from the existing BTS set. One of the initiating initiatives set up on signal strength measurement is the Hysteresis-threshold method [3].

Due to the high mobility of MS moving from one cell to another, it is difficult to predict the signal propagation and affect the signal strength of the reception. The strong signal levels received by MS are

IOP Conf. Series: Materials Science and Engineering **309** (2018) 012135 doi:10.1088/1757-899X/309/1/012135

influenced by path loss, Shadow fading and Fast fading, as a result of propagation attenuation and irregular environmental conditions [4][5]. The accumulation of MS mobility, propagation damping and environmental conditions will affect soft handover performance on CDMA systems.

In this paper, the propagation channel used as a signal transmission channel connecting base stations with MS is a large-scale propagation model, i.e. Lee, Okumura and Hata models. These propagation models predict signal strength at any distance on the MS receiver against the base station. Furthermore, the signal strength of the prediction results of these propagation models is used as the signal strength for the initiation trigger to activate the soft handover. So we get the relationship between the height of BS antenna and MS antenna on the improvement of system performance, namely Drop Call, Radio Link Decrease Rate and the average size of Active Set.

.

2. Model System

2.1. Network configuration

For simplification, the system configuration consists of two BSs. Each BS has the same and separate transmit power at distance D and MS moves on a straight path with a regulated speed as shown in Figureure1. During movement, MS receives signals with descending power. This decrease is due to the increase of distance and obstacle from the surrounding nature that spread between transmitter base station with MS. In this paper the decrease in power due to surrounding natural obstacles is determined by large-scale propagation model. MS samples the received signal strength (RSS) at a fixed interval distance (d = kds), where ds is the sampling distance. The ds value used is 1meter. The variable k is an integer with the value k ∈ [0, D / ds].

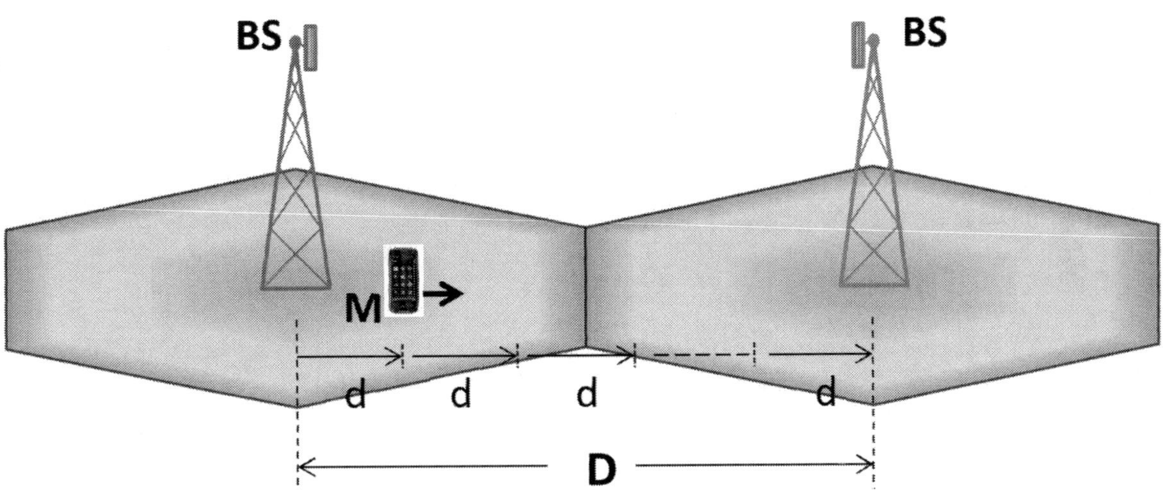

Figure. 1 Network model

2.2. The propagation Channel Model

The propagation model describes the average propagation of signals in an area. The magnitude of these propagation losses varies according to the spectrum and the nature and the environment. Estimating the losses to be signaled is very important. One of them is the losses generated by signal propagation. Loss of propagation are fairly difficult to predict. This loss is directly affected by the state of the environment surrounding the signal. Experts have produced some mathematical models that can provide good enough value to approach real-world circumstances. Among the radio signal propagation channel models are large-scale propagation models. In this research, the propagation

IOP Conf. Series: Materials Science and Engineering **309** (2018) 012135 doi:10.1088/1757-899X/309/1/012135

channel model is determined based on large-scale propagation models for urban areas ie Model Lee, Okumura and Hata.

2.2.1. Okumura model

To determine the damping of the path with the Okumura model, we must first calculate the free path loss, then the Amu (f, d) value of the Okumura curve that is added to the correction factor to determine the type of area. The path loss of Okumura model (PL (dB)) can be written in Equation (1) [6] [7].

$$PL\ (dB) = L_F + A_{mu}(f,d) - G(h_{te}) - G(h_{re}) - G_{AREA} \quad (1)$$

Where; L_F, A_{mu}, $G\ (h_{te})$, $G\ (h_{re})$ and G_{AREA} are respectively free path losses (dB), average attenuation relative to free path loss (dB), gain of antenna BS (dB), gain of MS antenna (dB) and gain area (dB).

The gain of antenna BS ($G\ (h_{te})$) is calculate by Equation (2)

$$G\ (h_{te}) = 20\log\ (h_{te} / 200),\ \text{for;}\ \ 100\ m > h_{te} > 10\ m \quad (2)$$

The gain of antenna MS ($G\ (h_{re})$) is calculate by Equation (3) and (4)

$$G\ (h_{re}) = 20\log\ (h_{re} / 3),\ \text{for;}\ \ 10\ m > h_{re} > 3\ m \quad (3)$$

$$G\ (h_{re}) = 10\ \log\ (h_{re} / 3),\ \text{for;}\ h_{re} \le 3\ m \quad (4)$$

with h_{te} and h_{re} are high antenna BS (m) and high antenna MS (m) respectively. The Okumura model is a simple model but provides good accuracy for predicting track attenuation on mobile radio communication systems for irregular areas.

2.2.2. Hata Model

Hata model is a form of empirical equation of curved track curve made by Okumura, therefore this model is more commonly referred to as Okumura-Hata model. Hata creates a standard equation to calculate the path attenuation in the urban area, whereas to calculate the path attenuation in other types of regions (suburbs, open areas, etc.), Hata gives the correction equation. To determine the loss of Hata model for urban (L(*urban*)(dB)) area using Equation (5), [6] [8].

$$L(urban)(dB) = 69{,}55 + 26{,}16\log(f_c) - 13{,}82\log(h_{te}) - a(h_{re}) + (44{,}9 - 6{,}55\log(h_{re}))\ \log(d) \quad (5)$$

where; f_c, h_{te}, h_{re} and d are working frequency (150-1500 MHz), high effective antenna transmitter BS (30-200 m), high effective antenna receiver MS (1-10 m) and distance between Tx-Rx (km). $a(h_{re})$ is correction factor for the effective height of the MS antenna as a function of the area served.

For small towns to medium cities, the correction factors a (h_{re}) are given by Equation (6).

$$a(h_{re}) = (1.1\log\ (f_c) - 0.7)\ h_{re} - (1.56\log(f_c) - 0.8)dB \quad (6)$$

while for big cities, the correction factors a (h_{re}) are given by Equation (7) and (8).

$$a(h_{re}) = 8.29\ (\log 1.54 h_{re})^2 - 1{,}1\ dB,\ \ \text{for;}\ \ f_c < 300\ MHz \quad (7)$$

$$a(h_{re}) = 3.2\ (\log(11{,}75hre))^2 - 4{,}97\ dB\ \text{for}\ fc > 300\ MHz \quad (8)$$

Although the Hata model has no specific track correction as the Okumura model provides, but the above equations are very practical to use and have excellent accuracy

IOP Conf. Series: Materials Science and Engineering **309** (2018) 012135 doi:10.1088/1757-899X/309/1/012135

2.2.3. Lee Model

Lee's propagation model is derived from experimental data in several major cities in the world. The reference parameter is 900 MHz, at antenna height 30.5 m, with power transmission 10 W. Lee's model propagation loss PL_{so} is shown at Equation (9), [6] [8].

$$PL_{so} = L_0 + \gamma \log d - F_0 \tag{9}$$

where; L_0, γ, d and F_0 are transmission losses at a distance of 1 km (dB), slope of path loss (dB / decade), distance from base station (m) and adjustment factor. The values of L0 and γ are obtained from the experimental data [5].

While the value of F_0 is given by equation (10),

$$F0 = F1\ F2\ F3\ F4\ F5 \tag{10}$$

with;

$$F_1 = \frac{[actual\ base\ station\ antenna\ height\ (m)]^2}{(30.5\ m)^2} \tag{11}$$

$$F_2 = \frac{[actual\ transmitter\ power\ (W)]}{10\ W} \tag{12}$$

$$F_3 = \frac{[actual\ gain\ of\ bus\ station\ antenna]}{4} \tag{13}$$

$$F_4 = \frac{[actual\ mobile\ antenna\ height\ (m)]^2}{(3\ m)^2} \tag{14}$$

$$F_5 = \frac{[fc]^2}{[f_0]^2} \quad \text{where } f_0 = 900\text{ MHz} \tag{15}$$

MS measures the strength of the received signal (RSS) of each BS. The measured RSS (dBm) value represents the difference between the transmitted power of the BS and the attenuation of the PL radio wave propagation model (dB) [3].

Suppose d_i indicating the distance between MS with BSi. If the power transmitted by the BTS is P_t, then the signal strength of the BSi is denoted by Xi (d), and is written with Equation (16)

$$X_{i(r)}(d) = P_t - P_L \tag{16}$$

where; $X_{i(r)}$ (d), P_t and P_L respectively are the signal strength of the i-th BS received by the MS at a distance d (dBm), power transmitted BTS (dBm) and path loss of the radio wave propagation model (dB). Then, the result of signal strength measurement, is averaged to eliminate the effect of rapid variation of the received signal. The method used in averaging is the window method. The Equation model of the window method can be written with Equation 17.

$$\hat{X}_{i(r)}(k) = \frac{1}{N_W} \sum_{n=0}^{N-1} X_{i(r)}(k-n) W_n \tag{17}$$

Where; $\hat{X}_{i(r)}$ is an averaged signal power gain and $X_{i(r)}$ is the signal strength before the smoothing process. N is the number of samples and Wn is the large window used for sampling. The value of Wn is unity for all n if the method used is a rectangular method. The averaged signal strength $\hat{X}_{i(r)}$ of the whole BS is the quantity to calculate the probability of radio link decrease and as the initiator trigger for soft handoff. The link degradation rate (λ_{LD}) is calculated by Equation (18)

$$\lambda_{LD}(\Phi, S) = E\left[\frac{1}{N} \sum_{k=1}^{N} \mathbb{I}_{\{Degradasi\ Link\ at\ time\ k\}}\right] \tag{18}$$

λ_{LD} measures the signal quality at time k on link that is in a descending state. $\mathbb{I}$ is an indicator function, worth 1 or 0 depending on whether the argument is true or false. The state of link degradation (LD) occurs when RSS ($\hat{X}_{k,i(r)}$) is below the threshold Δ, as shown in (19).

$$\max_{i \in A_k}\left(\hat{X}_{k,i(r)}\right) < \Delta \qquad (19)$$

3. Initiation of SHO Algorithm

The reference parameter used in initiating handoff is the strong signal received from the pilot signal. The soft handover algorithm used is a model of threshold with hysteresis. Threshold is a strong value of the minimum receive signal that must be achieved to maintain a call. If the signal strength falls below the threshold value, then the call will be dropped. As for hysteresis, it is an additional value to the algorithm. This value is useful to improve the performance of the soft handover algorithm.

The type of initiation used is MACHO / MAHO, with algorithm parameters are Threshold, HYST_ADD, and HYST_DROP. The way the soft handover algorithm is described as follows [3]

1. If the active set contains BS1 and $\hat{X}_1(d)$ dan > Δ and the absolute difference of $\hat{X}_1(d)$ and $\hat{X}_2(d)$ is greater than HYST_ADD then active set still contains BS1
2. If $\hat{X}_1(d)$ and $\hat{X}_2(d)$> Δ and the absolute difference of $\hat{X}_1(d)$ and $\hat{X}_2(d)$ is smaller than HYST_ADD then active set contains BS1 and BS2.
3. If $\hat{X}_1(d)$ and $\hat{X}_2(d)$> Δ and the absolute difference of $\hat{X}_1(d)$ and $\hat{X}_2(d)$ is greater than HYST_DROP then active set contains BS2 (Soft handoff occurs).
4. If $\hat{X}_1(d)$ and $\hat{X}_2(d)$ < Δ then active set does not contain BS1 or BS2. MS will not have connection with BS1 and BS2. This condition is called a drop call.

The size of the active set for two BS can be modeled with the state diagram as illustrated in Figure 2.

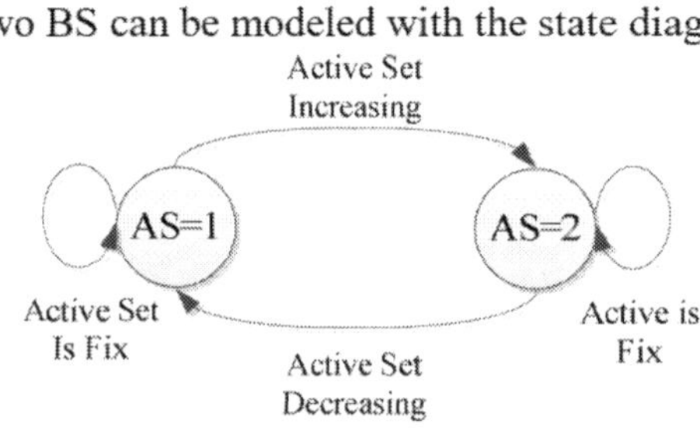

Figure 2. State diagram of Active Set for two BSs

The system parameters used in the simulation are shown in Table 1

Table 1 The system parameter

Symbol	Description	Value
d	Distance between BS	2000 m
Pi	Power transmission	23 dBm
Δ	Minimum Signal Level	-90 dBm
f	Frequency	900 MHz
σ	Standard Deviation	8 dB
M	Large Window	20
h_{re}	High Antenna MS	1,5- 11 m
h_{te}	High Antenna BS	20-120 m
Hist_Add	Hysteresis ADD	10 dBm

Hist_Drop	Hysteresis DROP	10 dBm

Based on the parameter values in Table 1, the discussion emphasizes the observation of the high value changes of BS Antenna and the height of the MS Antenna. For the changing of the high antenna BS, the height of the BS Antenna is varied from a height of 20 meters up to 120 meters with a space of each addition being 5 meters. Minimum signal reception power level is -90 dBm and Hysteresis Add and drop value is 10 dBm. For the changing of the high antenna MS, the height of the BS Antenna is varied from a height of 1.0 meters up to 11.0 meters with a space of each addition being o.5 meters. Minimum signal reception power level is -90 dBm and Hysteresis Add and drop value is 10 dBm. The height variations of antenna BS and MS above is observed its effect on mobile network performance parameters namely; Drop Call Rate, Decreasing of Radio Link, Average Size Active Set.

4. Results And Discussion

The graph of the relationship between each BS Antenna High and MS Antenna with Drop Call Rate is shown in Figure 3 and 4. It can be observed that with the increase of BS Antenna and MS Antenna, the Drop Call rate decrease.

For Lee's model, almost at all BS altitudes, the mobile network has Drop Call, except starting at a height of 120 meters. As for the Okumura model, the mobile network is free from Drop Call when the BS antenna is at a height of 60 meters. The average value of Drop Call Rate against the increase of antenna height BS, ie from 20 meters to 120 meters for each Lee and Okumura model is 0.2304 and 0.0897. Meanwhile for the Hata model, at all altitudes the BS Antenna does not experience Drop Call. This is due to the fulfillment of signal reception level from BS above the threshold value of the network system provision.

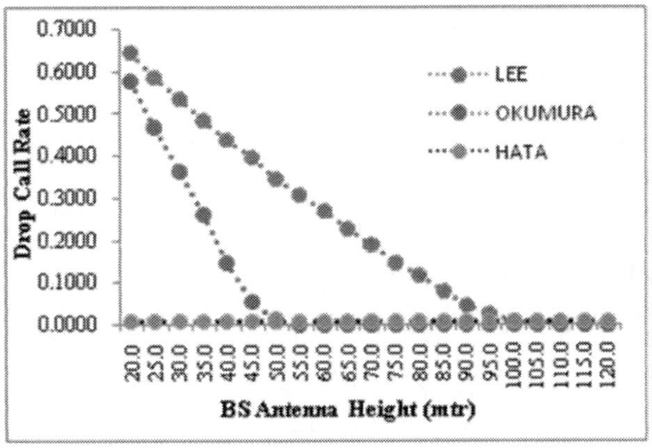

Figure 3. Drop Call Rate Against the increase of BS antenna height

Similarly to the increase in the height of the Antenna MS, the mobile network is free from Drop Call when the height of the MS antenna for each Lee and Okumura model is 8.5 meters and 4.5 meters. The average value of Drop Call Rate against the increase of antenna height MS, ie 1.0 meter to 11.0 meters for each Lee and Hata model is 0.1808 and 0.0667. Meanwhile for the Hata model, as well as the increase of BS antenna height, which is at all altitude Antenna MS mobile network does not experience Drop Call.

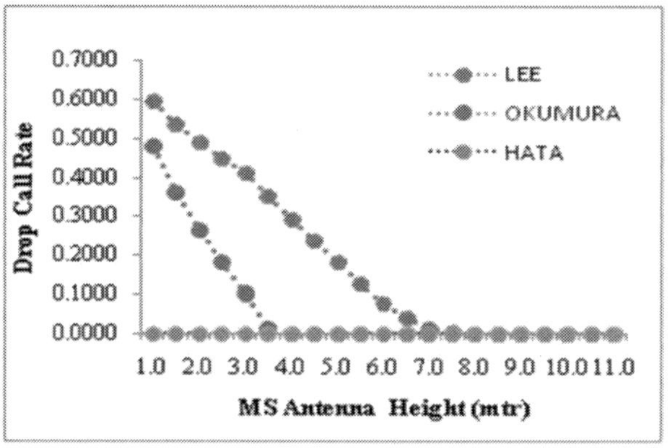

Figure 4. Drop Call Rate Against the increase of MS antenna height

The relation graph of each of the height of BS Antenna and MS Antenna on Radio Link decrease is shown in Figure. 5 and 6. From both Figure it can be seen that the trend pattern of improvement of Radio Link decrease with high increment of BS and MS Antenna is close to equal

In Figure. 5 it is shown that, for the three propagation models, ie Lee, Okumura and Hata models, repair of mobile network radio links increases with the increase of the BS Antenna height. This is because the increase in antenna height of the BS contributes to the decrease of radio signal path losses, so that the receiving power increases along with the decrease of the propagation losses. It is also shown that, the decrease of the average radio link of the Hata model against the increase of BS antenna height (20 meters to 120 meters) is lower than that of the Okumura and Lee models, which for the Hata model is 0.007, Okumura model is 0.0233 and Lee Of 0.1553.

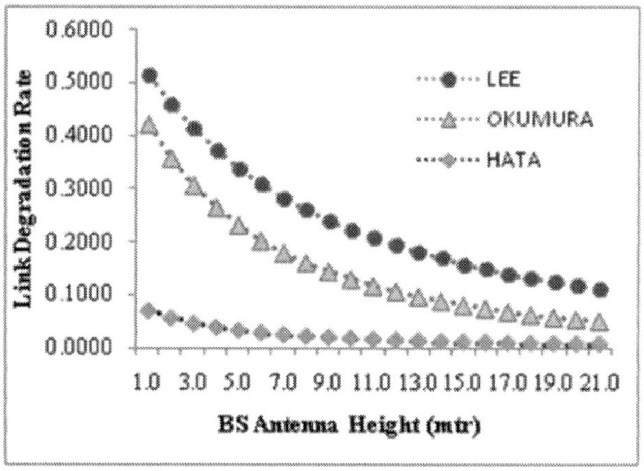

Figure 5. Link Degradation Rate Against the increase of BS antenna height

Similarly to the increase of the height of the Antenna MS (Figure 6), that improvement of mobile network radio links increases with the high increase of MS Antenna for all three propagation models. However, the decrease in the average radio link of the Hata model against the increase in the height of the MS antenna (1.0 meters to 11.0 meters) was lower than the Okumura and Lee models, where for the Hata model is 0.0130, meanwhile for each of Okumura and Lee model's are 0.01185 and 0.1879.

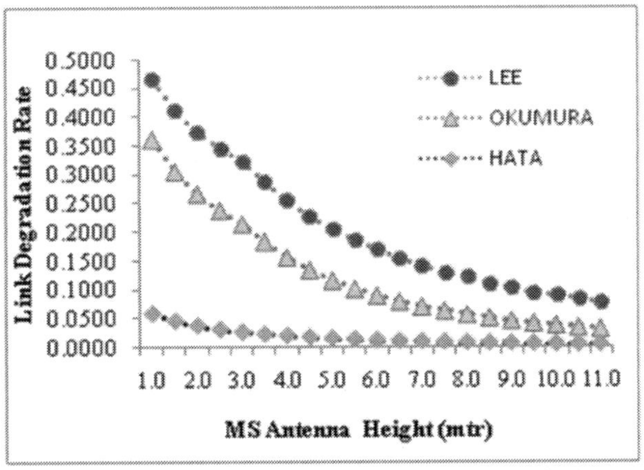

Figure 6. Link Degradation Rata Against the increase of MS antenna height

The average size of Active Set against the high increments of BS and MS Antennas for the three radio propagation models is shown in Figures. 7 and 8. It is shown that the height incremental effects of both BS and MS antennas increase the average number of active sets on mobile networks. Increasing the average number of active sets will improve the service of the continuity of communication, which certainly minimizes the possibility of drop calls.

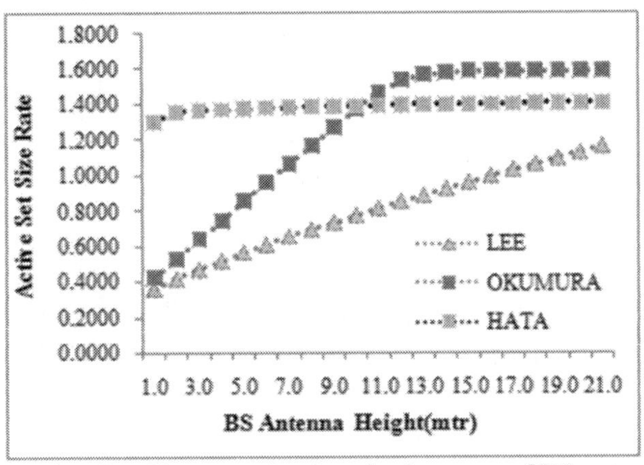

Figure 7. Active Set Size rate Against the increase of BS antenna height

Based on Figure 7 and 8, the performance of the three propagation models for the behavior of the active set size against the increase of the antenna height is slightly different. For Hata propagation model, the active set size remains stable on both the increase of BS and MS antenna height with average size of each is 1.3850 and 1.3656. For the Okumura model, the size of the active set rises above the Hata model, starting at height of 70 meter for BS antenna and at height of 5 meter for MS antenna. The average value of the active set size against increasing of BS and MS antenna are 1.2522 and 1.3202, and the average value of these is still below of the average value of the Hata model. Meanwhile for Lee's model, the size of the active set continues to increase in line with the increase of the BS and MS antennas. The average value of Lee's model for the height increase of BS and MS antennas is 0.7942 and 0.9062. However, the average value of the size of the active set Lee model is still far below the size of the Hata and Okumura models.

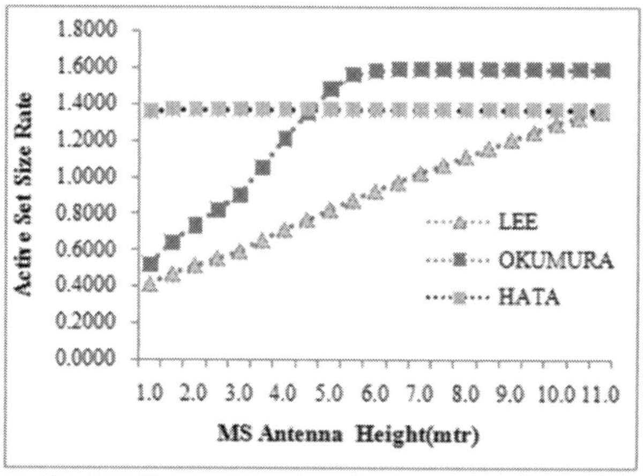

Figure 8. Active Set Size rate Against the increase of MS antenna height

5. Conclusions

In this paper, the effect of changing parameters of large-scale radio propagation models on mobile network performance improvements has been reported, as measured in terms of drop call rates, decreased radio links and active set sizes. An increase in BS and MS antenna altitude, is determined as a model propagation parameter of large scale. The simulation results show that with the increase of BS and MS antenna height improves the performance of the mobile system, where there is an increase of radio link and the size of active size, thereby decreasing the drop call rate. Then from the three large-scale radio propagation models of Lee, Okumura and Hata models, it was found that with the increase of BS and MS antenna altitude, the Hata Model contributed the most in improving the performance of the mobile system compared to both Okumura and Lee models, both in terms of increasing radio links and active set sizes as well as decreasing the rate Drop call

Acknowledgment

This research was funded by Research Institute of North Sumatera University with TALENTA of North Sumatera University with Research Contract of Fiscal Year 2017, Number: 5338 / UN5.1.R / PPM / 2017 dated May 22, 2017.

References

[1]. Smith C and Collins D 2002 3G Wireless Networks. McGraw-Hill Companies, Inc
[2]. Singh N P and Singh B 2008 Performance of Soft Handover Algorithm in Varied Propagation Environments. World Academy of Science. Engineering and Technology. **45** 377-381
[3]. Singh N P and Singh B 2010 Performance Enhancement of Cellular Network Using Adaptive Soft Handover Algorithm. Wireless Personal Communications
[4]. Veeravalli V V and Kelly O E 1997 A locally optimal handoff algorithm for cellular communications. *IEEE Trans. Veh. Technol.* **46** 603–610
[5]. Singh N P and Singh B 2010 Effect of Soft Handover Parameters on CDMA Cellular Networks. Journal of Theoretical and Applied Information Technology **110-115**
[6]. Rappaport T S 1995 Wireless Communications: Principles and Practice. 2nd Ed. New Jersey
[7]. Goldsmith A 2005 Wireless Communication. Cambridge University Press
[8]. Lee C Y and William 2006 Wireless and Cellular Telecommunications. 3rd Edition
[9]. Singh N P and Singh B 2007 Effects of Soft Handover Margin under various Radio Propagation Parameters in CDMA Cellular Networks. IEEE Conference on WCSN. **45-50**
[10]. Corraza G E, Giancristofaso D and Santucci F 1994 Characterization of Handover Initiation in Cellular Mobile Radio Networks. IEEE Technology Conference, **1896-1872**

IOP Conf. Series: Materials Science and Engineering **309** (2018) 012136 doi:10.1088/1757-899X/309/1/012136

Processing of palm oil mill wastes based on zero waste technology

Irvan[1,2,*]
[1]Chemical Engineering Department, Universitas Sumatera Utara, Indonesia
[2]Sustainable Energy and Biomaterial Center of Excellence, Universitas Sumatera Utara, Indonesia

*E-mail: irvan@usu.ac.id

Abstract. Indonesia is currently the main producer of palm oil in the world with a total production reached 33.5 million tons per year. In the processing of fresh fruit bunches (FFB) besides producing palm oil and kernel oil, palm oil mills also produce liquid and solid wastes. The increase of palm oil production will be followed by an increase in the production of waste generated. It will give rise to major environmental issues especially the discharge of liquid waste to the rivers, the emission of methane from digestion pond and the incineration of empty fruit bunches (EFB). This paper describes a zero waste technology in processing palm oil mill waste after the milling process. The technology involves fermentation of palm oil mill effluent (POME) to biogas by using continuous stirred tank reactor (CSTR) in the presence of thermophilic microbes, producing activated liquid organic fertilizer (ALOF) from discharge of treated waste effluent from biogas digester, composting EFB by spraying ALOF on the EFB in the composter, and producing pellet or biochar from EFB by pyrolysis process. This concept can be considered as a promising technology for palm oil mills with the main objective of eliminating the effluent from their mills.

1. Introduction

Palm oil is a necessary ingredient in the manufacture of various products such as edible oil, biodiesel, cosmetics, cleaning products, hair care, soaps and others personal care items. Palm oil is making an increasing contribution to Indonesia's national income, its production of palm oil has become the amazing growth from 752,000 tonnes in 1980, increase to 8.3 million tonnes in 2000 and become 34 tonnes in 2016, claiming as the largest producer of palm oil in the world [1]. That huge production was supported by the increase of palm oil plantations which become 11.3 million hectares in 2015. Until 2015 it was reported that around 11.3 million hectares of palm oil trees were planted and approximately 740 palm oil mills were built in Indonesia. Table 1 shows the area of palm oil plantation and total number of palm oil in Indonesia in 2015.

IOP Conf. Series: Materials Science and Engineering **309** (2018) 012136 doi:10.1088/1757-899X/309/1/012136

Table 1. Total number of Indonesia's palm oil mills in 2015 [2]

Region	Number of Companies	Area (Ha)	Production (Tonnes)	Number of Mills	Capacity (Tonnes/hour)
Sumatera	775	3,116,572.51	7,927,136.02	482	18,797.5
Jawa	5	11,825.44	43,587.96	2	50.0
Borneo	371	2,824,955.43	3,309,309.57	246	3,075.3
Sulawesi	22	238,336.22	316,137.50	5	270.0
Papua	8	99,041.60	93,070.80	7	390.0
Total	1,181	6,290,731.20	11,689,241.85	742	22,582.8

Table 1 shows that most of palm oil mills are located in Sumatera, mainly in Riau and North Sumatera Province. In 2015, Riau and North Sumatera produce 4,237,736.70 and 793,176.28 tonnes palm oil, respectively [2]. During its operation, besides producing crude palm oil (CPO) and palm kernel, palm oil mill (POM) also produces wastes. Every ton of CPO, it discharges approximately 2.5 ton of palm oil mill effluent (POME), 0.9 ton of empty fruit bunches (EFB), 0.6 of mesocarp fibers, and 0.27 ton of shells. Therefore, in 2015 besides producing 31 million ton CPO, Indonesia also produced 77.5 million ton POME, 27.9 ton EFB, 18,6 ton mesocarp fibers and 8.4 million ton shells. It is a huge amount of waste, if it is not treated properly it will become a serious problem for the environment.

POME consists of 0.6–0.7% palm oil, 95–96% water, 4–5% total solid, and 2–4% suspended solids. The existing method that we often found to treat POME in the palm oil mills is the lagoon or pond system. This pond system consists of a series of anaerobic, facultative, and aerobic ponds. The purpose of this treatment is to reduce the quality of wastewater like BOD, COD, pH before being flown to the river. Although this method success to reduce quality of wastewater until 95% however it needs very long time. It needs around 55 until 100 days operation needs a larger area, and other disadvantage is release of gas house gases (GHG) like methane and CO_2 to the atmosphere.

Some mills which care about the environment, try to reduce the GHG by covering the pond to trap the GHG and then burn into the atmosphere. Usually they make joint venture with other company. The biogas combustion process is essentially converting methane to CO_2 as seen in the reaction. Although CO_2 is also a greenhouse gas, but methane is considered 21 times more dangerous than CO2. So the advantage, we can do carbon trading. But the lack of this system is useless, because biogas is not used as fuel.

Approximately 20 – 22.5 tons of EFB are produced from 100 tons of FFB [3]. Indonesia is estimated to produce 90.5 million metric tons of FFB, resulting in EFB waste about 20.8 million metric tons per year [4]. EFB contains high cellulose, which relatively easy to decompose with a combination of biological, physical, and chemical processes [5]. Before 1996, EFB was burned in an incinerator to ashes. The ash of EFB can be used as fertilizer, because its potassium content is relatively high ± 30%. However, to prevent air pollution, the EFB combustion process is prohibited through Minister of Environment Decree number 15 of 1996 on the blue-sky program.

Today, most EFB is used as mulch on oil palm lands by placing EFB around the palm tree. Although the existence of mulch is useful to control weeds, maintain moisture and prevent

IOP Conf. Series: Materials Science and Engineering **309** (2018) 012136 doi:10.1088/1757-899X/309/1/012136

soil erosion but its utilization is costly for labor and transportation. Moreover, the processes of composting of mulch take a long time depending on environmental factors. Composting of EFB is also interesting way to reduce of solid waste from palm oil mills. However using open composting process take a long time and release GHG emission also. Zero waste technology offers a solution to solve the problem. This technology is a method how to manage waste which is designed to reduce the negative impact of palm oil wastes discharged into the environment.

2. Method

Zero waste technology involves fermentation of POME to biogas by using continuous stirred tank reactor (CSTR) in the presence of thermophilic microbes, producing activated liquid organic fertilizer (ALOF) from discharge of treated waste effluent from biogas digester, composting EFB by spraying ALOF on the EFB in the composter, and pyrolysis of EFB to biochar or pellet. The flow chart of this method is shown in figure 1.

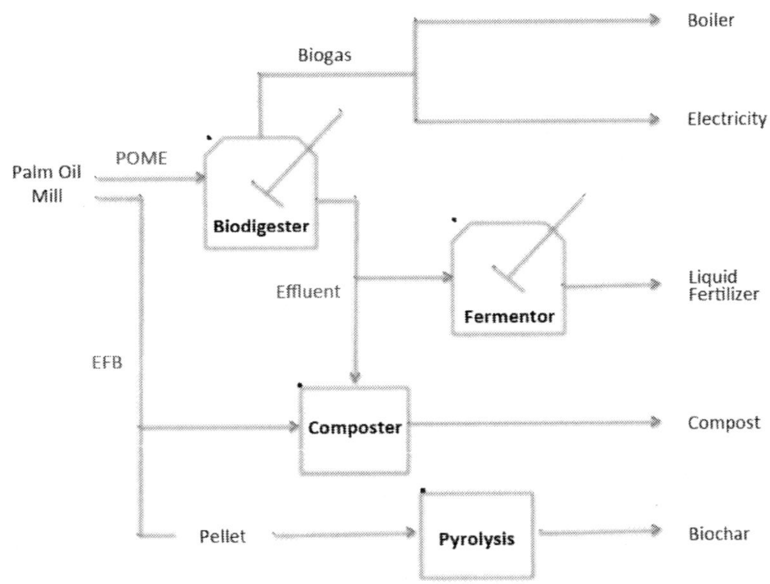

Figure 1. Zero waste technology in palm oil mill

2.1. Fermentation of POME to Biogas

POME contains many saturated fatty acids such as, palmitic acid and stearic acid. The melting points of palmitic and stearic acids are 63 °C and 69.9 °C respectively. When POME is treated using the mesophilic (ambient) fermentation method, it is difficult to decompose these saturated fatty acids because the temperature of the mesophilic digester is usually around 35 °C [6, 7, 8, 9, 10]. Even in the case of thermophilic fermentation at 55 °C, it is difficult for anaerobic fermentation to treat such fatty acids at a high speed. Our research was conducted to treat POME by using an anaerobic digester in a thermophilic state. Fresh POME was mixed with concentrated sludge previously separated by a gravity thickener in a mixing tank at 70 °C; then the mixed slurry was pumped into an anaerobic digester. Schematic diagram of our system is shown on figure 2.

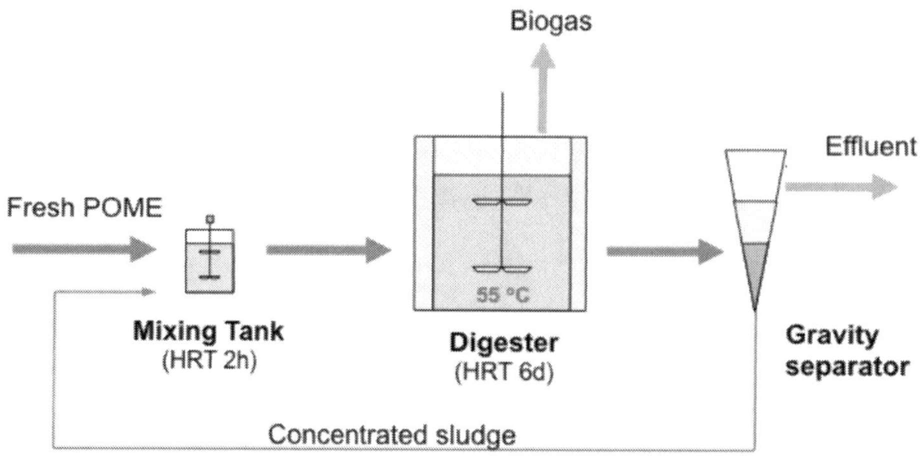

Figure 2. Schematic drawing of POME fermentation to biogas

a. Laboratory experiment (2-5 l scale)

The experiment was conducted in a laboratory with a continuous stirred tank reactor (CSTR) of 2-liter scale at thermophilic temperature, 55 °C [11, 12]. We tried to extend the sludge retention time (SRT) by returning the concentrated sludge by means of a gravity separator. Here concentrated sludge and POME were mixed in a mixing tank at 70°C. Hydraulic retention time (HRT) of CSTR was maintained at 6 days. As a result of the laboratory scale test, methane production of 0.25 kg CH_4/kg COD reduction was obtained and 63% CH_4, 37% CO_2 in biogas composition [13].

b. Pilot plant scale experiment (3.7 m^3 scale)

Figure 3 is a photo of the pilot plant. The pilot plant with volume of the digester is 3.7m^3 in this system was built in 2010 and has been operated at University of Sumatera Utara. POME was transported by trucks from palm oil mill to USU once a week.

Table 2 shows the results of the laboratory test and pilot plant test. The laboratory test and pilot plant test results indicated that high COD and volatile solid (VS) decomposition rate, more than 80%, can be achieved when using pre-heating and by extending SRT in the return sludge process, despite the short HRT of six days. In addition, biogas produced from pilot plant was used as fuel in the Otto engine generator 4 – stroke [14].

Table 2. Results of laboratory test and pilot plant test

Experiment	VS POME average	Digester volume	Return rate	HRT	SRT	VS Digester	VS Discharge	VS Decom position rate	COD Decom position rate	Gas generation	CH_4	H_2S
	(mg/L)	(Liter)	(%)	(days)	(days)	(mg/L)	(mg/L)	(%)	(%)	(Nl/g-VS)	(%)	ppm
Laboratory	30,100	2	20	6	15.8	16,700	6,100	80	84	0.71	63	250
Pilot Plant	22,900	3,700	20	6	10.8	13,500	7,500	67	80	0.68	60	40

Figure 3. Pilot plant

2.2. Activated liquid organic fertilizer from POME

Figure 4 shows flowsheet of liquid fertilizer production unit which extended process from biogas generation unit. Main equipments of this unit are digester tank capacity 5,000 L equipped with mixer and heater, gravity thickener tank capacity 1,000 L equipped with screw pump, mixing tank capacity 500 L equipped with mixer, and storage tank capacity 2,000 L. Effluent which came from biogas generation unit was pumped into fertilizer digester tank of liquid production unit intermittently every 2 hour. Inside the fertilizer digester, the feed (effluent from biogas digester) was degraded by anaerobic microbe, therefore COD concentration decreased from 10,000 mg/L become 1,500 mg/L. After that, the low COD effluent was flowed to gravity thickener by overflow. Inside the gravity thickener, the effluent was let to settle to the bottom of the tank.

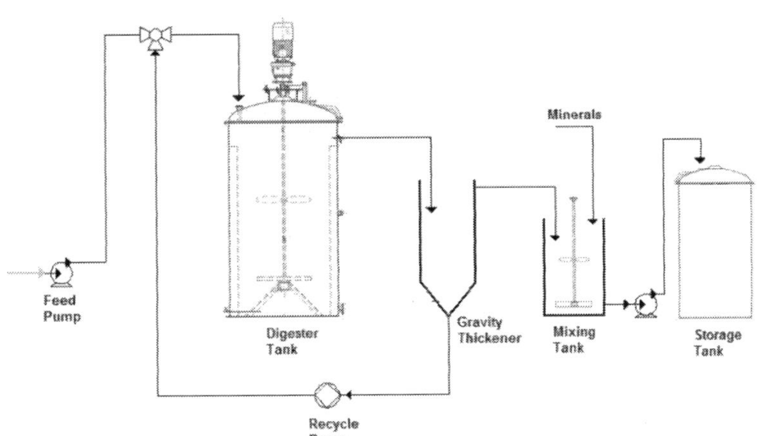

Figure 4. Flowsheet of liquid fertilizer production unit

Effluent of biogas digester was characterized before being fed to the fertilizer digester. COD concentration of biogas digester is high enough, 8,600 mg/L which still unfulfilled value to be thrown to the river. Everyday, 616 L of effluent was fed to the fertilizer digester and produced liquid fertilizer with almost similar amount. Main chemical properties of this

IOP Conf. Series: Materials Science and Engineering **309** (2018) 012136 doi:10.1088/1757-899X/309/1/012136

liquid fertilizer are shown in Table 3. Table 3 summarizes that liquid from fertilizer digester has high content of N, P, and K which are considered as the important components in fertilizer. Another interesting result, COD concentration decreased significantly from 8,600 to 1,500 mg/L, it makes the function of digester is not only for fertilizer production but also for wastewater treatment.

2.3. Composting empty fruit bunches (EFB)

During its operation, besides producing crude palm oil (CPO) and palm kernel, palm oil mill (POM) also produces waste, i.e. solid wastes such as EFB and POME. Several studies have been conducted for the processing of EFB into compost by adding additional materials such as animal wastes, and by mixing with POME [15]. In this research, the addition was carried out with active organic fertilizer (AOF) that serves as an activator and also maintains the moisture content (MC).

Table 3. Properties of liquid fertilizer

Parameter	Unit	Value
CODcr (*Chemical Oxygen Demand*)	mg/l	1,500
Nitrogen	%	0.53
P_2O_5 total	%	0.40
K_2O	%	0.56
MgO	%	0.01
CaO	mg/l	< 0.001
C organic	%	0.42
pH	-	8.09
C/N	-	0.79

a. Basket composter

Composting of empty fruit bunches (EFB) mixed with activated liquid organic fertilizer (ALOF) is an alternative utilization of solid waste of palm oil mill [16]. Composting treatment was conducted by cutting EFB into pieces with size 1 - 3 cm, inserting the EFB into basket composter (33 cm W × 28 cm L × 40 cm H), and by adding ALOF until moisture content (MC) in the range of 55-65%. The parameters analyzed during composting were temperature, pH, MC, compost weight, C/N ratio, and the quality of the final compost. Composting was carried out for 60 days with best results obtained at day 40. The best compost characteristic is pH 9.1; moisture content 61.57%; water holding capacity (WHC) 76%; CN ratio 12.15%; P 0.58%; and K 0, 95%.

b. Tower composter

Composting of empty fruit bunches (EFB) with activated liquid organic fertilizer (ALOF) in a tower composter was carried out. The use of tower composter, which was made of concrete with dimensions of 0.40 m W × 0.40 m L × 3.00 m H, was to accelerate the composting time due to the achievement of thermophilic conditions. Detailed sketch of the composter is presented in Figure 5.

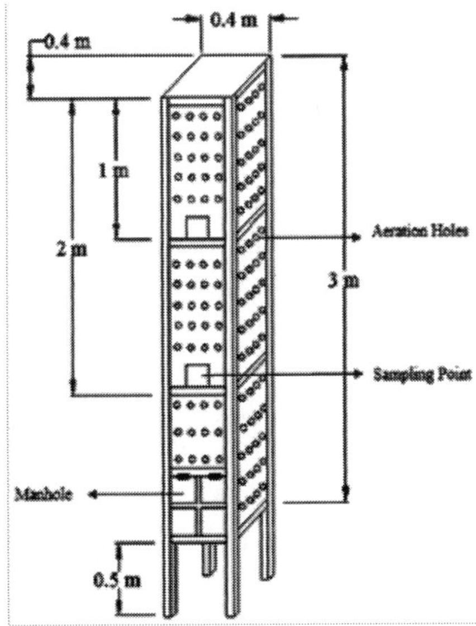

Figure 5. The tower composter

Composting process was conducted by putting non-shredded EFB into the composter tower and by adding ALOF as activator. The composting treatment was completed in a short time in 40 days, even though on day 10, EFB turned to dark brown indicating mature compost. The quality of final compost from day 10 until day 40 were slightly similar i.e. pH scale ranges from 7.6 to 8.4; moisture content 41.16 - 58.44%; 20.27 - 21.34% C, 1.02 - 1.25% N, C/N ratio of 19.94 - 20.24, bacterial count 17 - 41 × 10^6 CFU g^{-1}, and electrical conductivity 3.735 - 4.9 $dS.m^{-1}$. The quality of product compost obtained has met Indonesian National Standard (SNI).

2.4. Producing biochar from EFB by pyrolysis process

Biochar is a solid product of biomass pyrolysis which can be utilized as an alternative fuel. In addition as a fuel, biochar can also be converted to activated carbon then can be used in purification processes. The conversion of empty fruit bunches (EFB) to biochar through pyrolysis has the following stages: shredding, chopping, milling, pelletization, and pyrolysis of EFB into biochar, then analysis of biochar products. The steps of this process can be seen figure 6. Producing biochar from EFB by pyrolysis process is still under proceeding in our laboratory.

Figure 6. Conversion EFB to biochar

3. Conclusion

Zero waste technology can be considered as a promising technology for palm oil mills with the main benefits such as eliminating the effluent from the mills, production of biogas can generate electricity or as boiler fuel, production of compost or liquid fertilizer as chemical fertilizer substitution, production of biochar and pellet as fuel, green house gases reduction, and zero pollution to the environment.

Acknowledgment
This research was supported by Hibah BP-PTN Universitas Sumatera Utara fiscal year 2016.

References

[1] Index Mundi 2017) Indonesia Palm Oil Production by Year http://www.indexmundi.com/agriculture/ ?country=id&commodity=palm-oil&graph=production, Retrieved 7 September 2017
[2] Directorate General Estate Crop, Ministry of Agriculture 2016 Agricultural Statistics Database, Jakarta: Ministry of Agriculture
[3] Kerdsuwan S, Krongkaew L 2011 Renewable Energy – Trends and Applications, Intech
[4] Ishola M M, Isroi, Taherzadeh M J 2014 *Bioresour. Technol.* **165** 9-12
[5] Alkarimiah R, Rahman RA 2014 *Int. J. of Eng. Sci. Innov. Tech.* **3** 132-145
[6] Peyton T O, Cooper I W, Quah S K 1979 *Proc. 34th Indust. Waste Conf.*, Purdue Univ. Lafayette, IN.
[7] Borja R, Banks C J 1994 *Bioresour. Technol.* **48** 209-214
[8] Faisal M, Unno H 2001 *Biochem. Eng. J.* **9** 25-31
[9] Wulfert K, Darnoko D, Tobing P L,Yuliasari R, Guritno 2002 *International Oil Palm Conference*, Nusa Dua, Bali, Indonesia, July 8-12 265-275
[10] Najafpour G D, Zinatizadeh A A L, Mohamed A R, Isa M H, Nasrollahzadeh H 2006 *Proc. Biochem.* **41** 370-379
[11] Irvan, Trisakti B, Sosanty F and Tomiuchi Y 2016, *Asian J. Chem.* **28** 377-380
[12] Irvan, Trisakti B, Wongistani V and Tomiuchi Y 2012, *Int. J. Sci. Eng.* **3** 32-35
[13] Irvan, Trisakti B, Tomiuchi Y, Harahap U, Daimon H 2017 *IOP Conf. Series: Materials Sci. Eng.* **206** 012094
[14] Irvan, Trisakti B, Husaini T, Sitio A, Sitorus T B 2017 *IOP Conf. Series: Materials Sci. Eng.* **206** 012028
[15] Baharuddin A S, Wakisaka M, Shirai Y, Abd-Aziz S, Abdul Rahman A N, Hassan, M A 2009 *Int. J. Agri. Res.* **4** 69-78
[16] Trisakti B, Lubis J, Husaini T, Irvan 2017 *IOP Conf. Series: Materials Sci. Eng.* **180** 012150

AUTHOR INDEX

AUTHOR INDEX

AUTHOR INDEX

Institute of Physics
Dirac House, Temple Back
Bristol BS1 6BE UK

CURRAN ASSOCIATES INC.
proceedings.com

ISSN: 757-8981
ISBN 978-1-5108-5923-4